Europe and Africa Regional Meeting of Polymer Processing Society (PPS2017)

Dresden, Germany
26 - 29 June 2017

Volume 1 of 2

ISBN: 978-1-5108-5487-1

Printed from e-media with permission by:

Curran Associates, Inc.
57 Morehouse Lane
Red Hook, NY 12571

Printed by Curran Associates, Inc. (2018)

Polymer Processing Society
810 E. 10th St.
Lawrence, KS 66044
USA

Phone: +1-800-627-0326
+1-785-865-9403
Fax: +1-785-843-6153

pps@allenpress.com

Additional copies of this publication are available from:

Curran Associates, Inc.
57 Morehouse Lane
Red Hook, NY 12571 USA
Phone: 845-758-0400
Fax: 845-758-2633
Email: curran@proceedings.com
Web: www.proceedings.com

TABLE OF CONTENTS

VOLUME 1

S01 – MIXING AND COMPOUNDING

S02 – POLYMER BLENDS AND ALLOYS

S03 – RHEOLOGY AND PROCESS SIMULATION

S04 – FUNCTIONAL, NANO AND BIO COMPOSITES

S06 – INJECTION MOLDING AND EXTRUSION

VOLUME 2

S09 – POLYMER MODIFICATION WITH IONIZING RADIATION

S10 – NANO- AND MICROSTRUCTURED SURFACES AND FILMS

S11 – IN-LINE ANALYTICS AND PROCESS MONITORING

S12 – WELDING AND JOINING TECHNOLOGY

S13 – POLYMER MATERIALS FOR MEDICAL APPLICATIONS

S14 – ADDITIVE MANUFACTURING

Toughening Improvement of Polyamide 6 with Low Molecular Weight Polyethylene

***Ishigami Akira* [(1)]*, *Konno Takuya* [(1)], *Watanabe Kazuki* [(1)], *Nishitsuji Shotaro* [(1)], *Ishikawa Masaru* [(1)], *Ito Hiroshi* [(1)]**

[(1)] Yamagata University - Yamagata - Japan

In this study, we attempted to toughen Polyamide 6 (PA6) by addition low molecular weight polyethylene (LMWPE) instead of rubber or elastomer. First, in order to investigate the relationship between dispersion state and toughness, a plurality of melt kneading methods such as single screw melt compounding, 8-axis screw melt kneading process and high-shear rotational processing were utilized. As a result, LMWPE was uniformly and finely dispersed using 8-axis melt kneading process. On the other hand, LMWPE aggregates were not observed in high-shear rotational processing. In the result of three-point bending test, PA6 was brittle fractured, but ductility of PA6/PE blend was shown by adding a small amount of LMWPE. The displacement of fracture was also greatly improved. In the comparison by the melt kneading process, it was possible to toughen PA6 with 8-axis screws melt kneaded products at the smallest addition amount (ca. 2 wt%). As observation results of the cross-sectional view during the bending test, it was confirmed that voids were formed from the LMWPE. By adding a LMWPE with low strength, the formation of voids was induced with low stress, and bending toughness could be improved.

Analysis of the Process Behavior of Co-Kneaders

Johannes Rudloff[1, a)], Matthias Wilhelm[1], Marieluise Lang[1], Peter Heidemeyer[1] and Martin Bastian[1]

[1] *SKZ - German Plastics Center - Würzburg - Germany*
[a)] j.rudloff@skz.de

Abstract. The co-kneader is well known for its superior mixing performance and its exact temperature control capabilities. Therefore it is widely used in the polymer industry for the compounding of shear and temperature sensitive materials like PVC or high filled compounds. In contrast to the considerable amount of scientific work that deals with investigation, modeling or simulation of the process behavior of single and twin screw extruders, there are only few publications about the co-kneader. Due to increased quality requirements and the trend to cost reduction by process optimization, this is increasingly becoming a problem for plant construction and processing companies. In order to address this problem, this paper presents experimental investigations of the process behavior of co-kneaders. Based on the experimental results a physical-mathematical model to estimate the process behavior of co-kneaders is in development. The basic approach is to divide the extruder in small sections with constant geometry conditions. For these sections all material parameters can be considered as constant. Then solid conveying, melting, melt transport, energy input und temperature development can be modeled separately and coupled afterwards in a simulation tool.

INTRODUCTION

The co-kneader is a compounding extruder that is based on a patent of Heinz List [1]. The extruder consists of a reciprocating single screw with interrupted flights that oscillates in a barrel containing pins in such a manner as to cause the pins to wipe the flights [2]. The modern Kneader is modular in construction containing a range of different screw element arranged on a shaft. Beyond publications dealing with applications [2-6] there was little study of the basic process behavior of co-kneaders until the 1990s. In the 1990s Elemans and Meijer [7] as well as Lyu and White [8-11] dealt with the fluid mechanics, heat transfer and melting in co-kneaders. Elemans and Meijer developed a melt transporting model for Newtonian fluids including the axial movement and the pin influence. Lyu and White modeled the melt transporting for viscous non-Newtonian polymer melts [8-10] but neglected the pin influence. Furthermore they performed experimental investigation of the residence time and melting [11]. There are several problems using the models for current development processes:

- The investigations were all performed on co-kneaders with 3-flighted mixing elements. Since 2007 there are also co-kneaders with 4-flighted mixing elements available. For these machines investigations are performed with the aim of process modelling.
- No models of the melting process in co-kneaders are known.
- The published models are not implemented in scientific or commercial simulation software.

To address these problems experimental investigations of a co-kneader with 4-flighted mixing elements are presented. Furthermore an approach for a comprehensive process modelling of the co-kneader is described and a melting model as part of this comprehensive model is presented.

EXPERIMENTAL INVESTIGATIONS

In order to investigate the process behavior of co-kneaders dead-stop-trials were performed. A Buss co-kneader type MX-30 was used. The extruder screw consisted of a solid conveying zone, a melting zone and two mixing zones, each separated by a melt conveying zone. As material a high density polyethylene type Sabic Vestolen 6060 N and a polystyrene type INEOS Styrolution 486 N were used. In the experiments shown here a throughput of 10 kg/h, a rotation speed of 400 1/min and a barrel temperature of 200 °C were selected. In order to separate solid particles and melt the pellets were dyed blue and yellow with a pigment in a previous compounding step. Therefore 0.75 % yellow pigment and 0.1 % blue pigment were used.

The results of the dead-stop-trials are shown in Figure 1. The extrusion direction is from left to right. Even if PS sticks more to the barrel than PE the process behavior for both materials is similar. There are solid conveying zones and fully and partially filled sections for both materials. The melting starts for both materials approximately at the first fully filled section. In this melting section no compact solid bed is found and a disperse melting mechanism occurred. PE is melting slower than PS because of its higher melting enthalpy. If one takes a closer look at the melt conveying, there are at first partially filled conveying elements upstream of the mixing sections. Then a short fully filled conveying element section of approximately 10 mm followed by a fully filled mixing element section occurs. Further downstream a partially filled conveying elements section follows. This indicates that the melt transport capacity of the conveying elements is high enough to transport the throughput given by the dosing unit. In contrast, the melt transport capacity of the mixing elements is not high enough to transport the throughput.

(a)

(b)

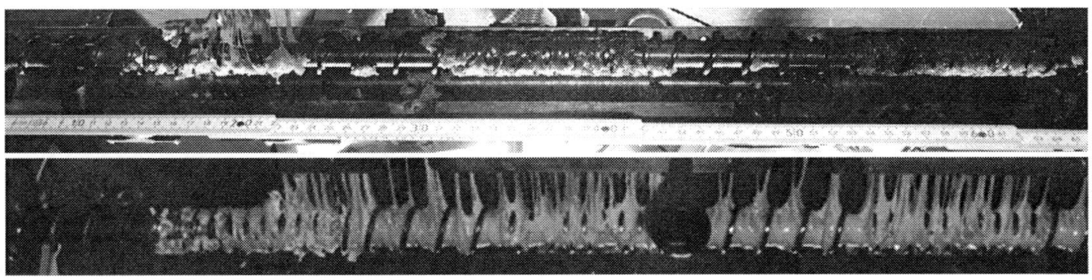

FIGURE 1. Results of the dead-stop-trials for PE (a) and PS (b)

MODELLING

In the first step the co-kneader is divided in small sections with constant geometry and temperature conditions along the extruder length. The length of these sections must be 1 mm or smaller. This allows a separate modelling for all processes like melt transport, temperature development or melting and coupling the single calculation steps in an Excel based simulation tool called "SimKo". Furthermore for coupling the separate process models, an iterative simulation process is performed in the simulation tool. This is a common process in the extrusion modelling that is also used for the simulation of single screw, co-rotating twin screw and planetary roller extruders [12-15].

Geometry Model

The co-kneader is basically a single screw extruder, so the base geometry parameters of this extruder type as published by White and Potente [14] can be used. If one looks at unrolled elements as shown in figure 2 one receives for example the geometry parameters channel width b and number of channels i_K.

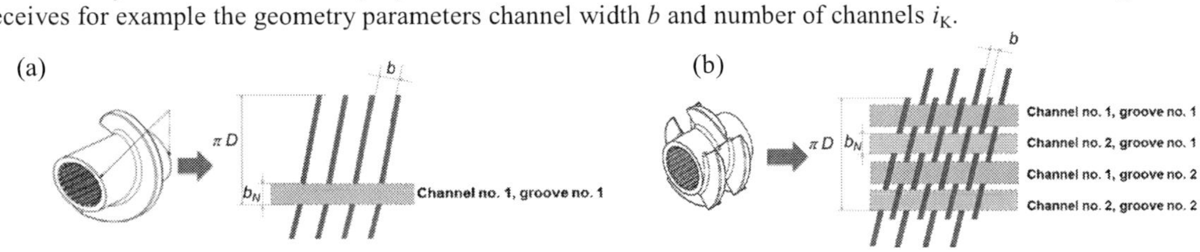

FIGURE 2. Schematically view of the groove geometry model of co-kneaders for conveying (a) and mixing (b) elements

The conveying element shown in figure 2 (a) consists of one channel, the mixing element in figure 2 (b) consist of two channels each with a width of b. These geometry parameters have to be extended by further geometry parameters to describe the grooves and pins of the kneader. Because one wants to neglect the pin influence in the first model only the grooves in the first step need to be described. Therefore groove width b_N and groove number i_N are used. The convening element has one groove in its one channel and the mixing element has two grooves in each of its two channels all with a width of b_N.

Melt Transport Model

In order to model the melt transport in co-kneaders the melt transport in channels and grooves are modelled separately and coupled by a pressure and volume flow balance. The influence of pins and axial stroke are neglected in the first modeling step presented in this paper. It is assumed that the volume flow given by the dosing unit is the sum of the volume flow in each channel minus the volume flow in each groove:

$$\dot{V} = i_K \cdot \dot{V}_K - i_n \cdot \dot{V}_n \tag{1}$$

Furthermore it is assumed that the pressure difference is the same at each position along the channel direction (z-axis):

$$\frac{\Delta p}{\Delta z} = \frac{\Delta p_K}{\Delta z_K} = \frac{\Delta p_N}{\Delta z_N} \tag{2}$$

Then it is assumed that the melt transport can be described as a superposition of drag and pressure induced flow by dimensionless flow numbers established for rectangular channels. Therefore the equations shown in table 1 are used to describe the flow in channels and grooves.

TABLE 1. Dimensionless operating numbers

	Extruder	Channel	Groove
Dimensionless volume flow	$\pi_{\dot{V}} = \frac{\dot{V}}{i_K \cdot 0{,}5 \cdot h \cdot b \cdot v \cdot \cos\beta}$	$\pi_{\dot{V},K} = \frac{\dot{V}_K}{0{,}5 \cdot h \cdot b \cdot v \cdot \cos\beta}$	$\pi_{\dot{V},N} = \frac{\dot{V}_N}{0{,}5 \cdot h_N \cdot b_N \cdot v \cdot \sin\beta}$
Dimensionless pressure gradient	$\pi_p = \frac{\Delta p}{\Delta z} \cdot \frac{h^{n+1}}{6 \cdot K \cdot (v \cdot \cos\beta)^n}$	$\pi_{p,K} = \frac{\Delta p}{\Delta z} \cdot \frac{h^{n+1}}{6 \cdot K \cdot (v \cdot \cos\beta)^n}$	$\pi_{p,N} = \frac{\Delta p}{\Delta z \cdot \tan\beta} \cdot \frac{h_N^{\;n+1}}{6 \cdot K \cdot (v \cdot \sin\beta)^n}$
Linking	Equation 3	$\pi_{\dot{V},K} = A_K - B_K \cdot \pi_{p,K}$	$\pi_{\dot{V},N} = A_N - B_N \cdot \pi_{Vp,N}$

The model parameters A_K, A_N, B_K and B_N can describe further influences on the flow like structure viscosity or screw pitch. By coupling the equations in table 1 with equation 1 and 2 we get equation 3 that describes the dimensionless volume flow of the co-kneader:

$$\pi_{\dot{V}} = A_K - A_N \cdot \tan\beta \cdot \frac{h_N \cdot b_N \cdot i_n}{h \cdot b \cdot i_k} - \left(B_K + B_N \cdot \frac{h_N^{\;n+2} \cdot b_N \cdot i_n}{\tan\beta^n \cdot h^{n+2} \cdot b \cdot i_k} \right) \cdot \pi_{p,K} \tag{3}$$

The velocity is calculated by equation 4 and the shear rate to calculate the viscosity parameters from the power law can be estimated by equation 5:

$$v = \pi \cdot D \cdot N_0 \tag{4}$$

$$\dot{\gamma} = \frac{h}{v \cdot \sin\beta} \tag{5}$$

If one sets all model parameters (A_K, A_N, B_K and B_N) on 1, which means that further influences like structure viscosity and cross flow are neglected, one gets pressure-throughput characteristics as shown in figure 3. A single screw extruder without grooves would be represented by a line from 1 on the y-axis to 1 on the x-axis. Because of the grooves the maximal volume flow and pressure gradient are decreasing. This is the reason why mixing elements in co-kneaders are fully filled in most of the operating points. The fully filled conveying element section upstream of the mixing elements builds up the pressure that is necessary to overcome the mixing elements.

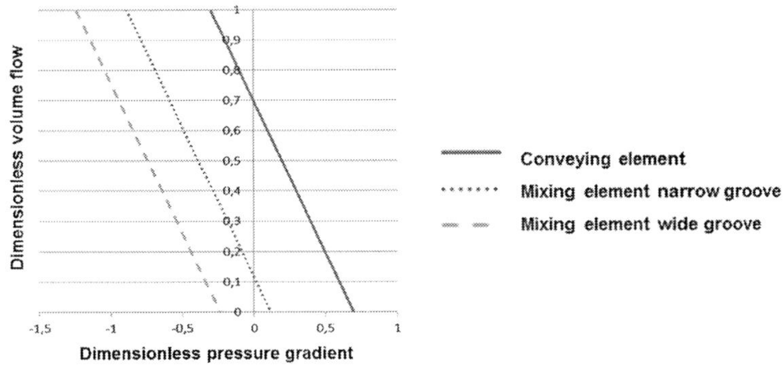

FIGURE 3. Pressure-throughput characteristics of co-kneader elements

For partially filled sections the filling degree f can be calculated with equation 6 and the average residence time in a section is given by equation 7:

$$f = \frac{\pi_{\dot{V}}}{A_K - A_N \cdot \tan\beta \cdot \dfrac{h_N \cdot b_N \cdot i_n}{h \cdot b \cdot i_k}} \tag{6}$$

$$\bar{t} = \frac{V \cdot f}{\dot{V}} \tag{7}$$

Melting Model

For melting it is assumed that the melting starts with the first fully filled section and a dispers melting model is used. These assumptions suit the investigated operating point and are used for twin screw extruders [13] and planetary roller extruders [15] as well. In the disperse melting model, isolated solid particles in surrounding melt are observed. The solid particles are considered as spheres and ongoing melting leads to a reduction of the radius of the spheres. The sphere radius then correlates with the melting degree and is described by equation 8:

$$r_{i+1} = \left\{ r_i^2 - \frac{2 \cdot \lambda_M}{\rho_S \cdot c_{p,M}} \cdot \ln\left(1 + \frac{c_{p,M} \cdot (T_M - T_{Fl})}{\Delta h}\right) \cdot \Delta\bar{t} \right\}^{\frac{1}{2}} \tag{8}$$

CONCLUSIONS

Simulated pressure, filling degree and melting degree profiles for PE (a) and PS (b) are shown in figure 4. In these simulations the melt temperature was set on the value of the barrel temperature. This is necessary because until now no temperature models have been developed for the co-kneader.

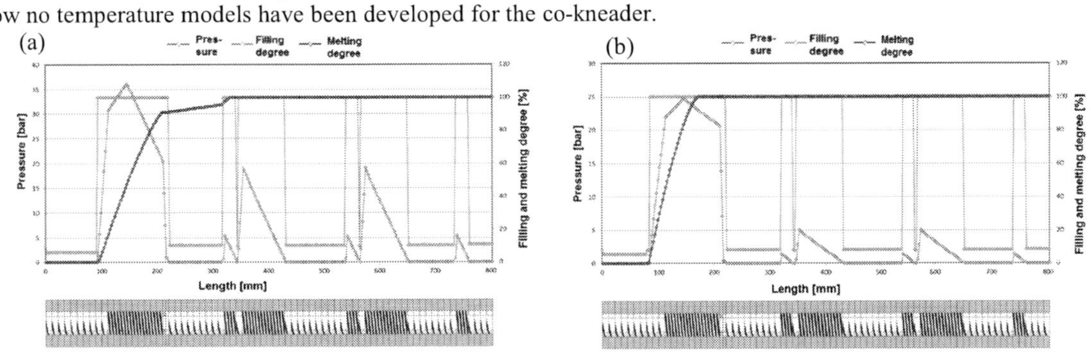

FIGURE 3. Simulated pressure, filling and melting degree profiles for PE (a) and PS (b)

The different melting length for PE and PS can be covered well. The simulated filling degrees show also a good agreement with the experimental results, especially under consideration of the assumptions that were necessary for modelling. In the next step the models will be extended, so that they can cover pin, axial stroke, cross flow and structure viscosity influences. Therefore further experimental investigations are necessary. In this experiments pressure gradients and melting degrees along the melting and mixing sections will be measured for a wider range of operating conditions.

SYMBOLS

A_K	Model parameter	A_N	Model parameter	B_K	Model parameter
A_N	Model parameter	b	Channel width	b_N	Groove width
$c_{p,M}$	Spec. melt heat capacity	D	Screw Diameter	f	Filling degree
h	Channel height	h_N	Groove height	Δh	Melting enthalpy
i_K	Channel number	i_N	Groove number	K	Power law consistency factor
N_0	Rotation speed	n	Power law flow exponent	$\frac{\Delta p}{\Delta z}$	Pressure gradient
r_i	Particle radius	T_{Fl}	Softening Temperature	T_M	Melt Temperature
$\bar{t}$	Residence time	$\dot{V}$	Volume flow	$\dot{V}_K$	Channel volume flow
$\dot{V}_N$	Groove volume flow	v	Velocity	β	Channel pitch angle
λ_M	Thermal conductivity of the melt	ρ_S	solid density	π_p	Dimensionless pressure gradient of the extruder
$\pi_{p,K}$	Dimensionless pressure gradient of the channel	$\pi_{p,N}$	Dimensionless volume flow of the groove	$\pi_{\dot{V}}$	Dimensionless volume flow of the extruder
$\pi_{\dot{V},K}$	Dimensionless volume flow of the channel	$\pi_{\dot{V},N}$	Dimensionless volume flow of the groove	$\dot{\gamma}$	Shear rate

ACKNOWLEDGMENTS

The results presented here were obtained in the research project "BA 1841/30-1", which is supported by the German Research Foundation (DFG). Special thanks for supporting us. Additional acknowledgements go to BUSS, Albis, SABIC Europe and Lifocolor Farben for supplying the extruder and polymer materials.

REFERENCES

1. H. List, Swiss Patent No. 1945 247,704 (20 Aug 1947).
2. K. H. Stade, Polym. Eng. Sci. **17**, 50–57 (1977).
3. E. Schneider and D. H. M. Brooks, British Plastics, **30**, 480 (1957).
4. T. Timm, D. Stoltzenberg and H. Fettback, Kautschuk Gummi Kunst. **18**, 206 (1965).
5. D. B. Todd and J. W. Hunt, SPE ANTEC Tech Papers **19**, 577 (1973).
6. P. Schnottale, Kautschuk Gummi Kunst. **38** 116 (1985).
7. P. H. M. Elemans and H. E. M. Meijer, Polym. Eng. Sci. **30**, 893-904 (1990).
8. M. Y. Lyu and J. L. White, Int. Polym. Process. **10**, 305–313 (1995).
9. M. Y. Lyu and J. L. White, Int. Polym. Process. **11**, 208–221 (1996).
10. M. Y. Lyu and J. L. White, Polym. Eng. Sci. **37**, 623–635 (1997).
11. M. Y. Lyu and J. L. White, Polym. Eng. Sci. **38**, 1366–1377 (1998).
12. B. Vergnes, G. Della Valle and M. Delamare, Polym. Eng. & Sci. **38**, 1781–1792 (1998).
13. A. Thümen, "Disperses Aufschmelzen in Doppelschneckenextrudern", Ph.D. thesis, Uni. Paderborn, 2008
14. J. L. White and H. Potente, *Screw Extrusion* (Carl Hanser Verlag, Munich, 2002).
15. J. Rudloff, M. Lang, K. Kretschmer, P. Heidemeyer, M. Bastian, M. Koch "A mathematical model describing the solid conveying and melting behavior of planetary roller extruders", AIP Conference Proceedings 1593, (American Institute of Physics, Melville, NY, 2014), pp. 592–595.

New Screw Design Induced Extensional Flow for Enhancement of CNT Dispersion in PP Matrix through Twin-Screw Extruder.

Matsumoto Koki (1)*, Tanaka Tatsuya (1)

(1) Doshisha University - Kyoto - Japan

Carbon nanotube (CNT) based nanocomposites have been developed to obtain excellent mechanical properties and functionalities. The twin-screw extrusion (TSE) processing for improving the dispersion of CNT is not effective solution due to the limited residence time. Hence, expected properties have not been achieved because dispersion of CNT is poor. This describes that shear flow which is dominated mainly in TSE cannot disperse the CNT effectively. Extensional flow which is caused in converging section has been attracted as the effective solution for enhancement of nanofiller dispersion. However, the dispersion effect of extensional flow hasn't been revealed in case of nanocomposite system. In this study, we suggest a new screw design induced extensional flow for TSE based on our research achievements. CNT dispersion is enable by extensional flow induced high pressure drop and high-speed flow through the small hole. This technique applied to disc type segment which we called Blister Disk (BD) with many small holes. However, melted polymer leaked out between the gaps of this two discs. To avoid this leakage, separated BDs were modified to fixed type BD (XBD) by using ball bearing. To investigate the dispersion efficiency of XBD, that of neutral type Kneading Disc (KD) and separated BD are compared in case of CNT based polypropylene (PP) nanocomposite. The dispersion state of CNT is investigated through morphological observation, rheological-, electrical- and mechanical measurement. The morphological assessment shows CNT dispersion is improved by using XBD. Therefore, the electrical conductivity measurement also shows low volume resistivity when XBD is used. The tensile strength also improved slightly. These reason could be the effect of extensional flow. Furthermore, the resin temperature at outlet of TSE by using XBD can be reduced 5 °C by compared with KD against the same specific mechanical energy (SME). Therefore, XBD can provide to high quality compounding in respect to enhancement of CNT dispersion and reduction of shear heating.

Reactive Compatibilization Schemes for High Speed Twin Screw Extrusion of Polymer Blends

Sobkowicz Margaret J [(1)]*, Farahanchi Azadeh [(1)], Calderon Barbara [(1)]

[(1)] UMass Lowell - Massachusetts - USA

This work examines the effects of high shear on polymer degradation and bend compatibility. A reactive blending scheme was investigated for biobased blends of poly(propylene carbonate) and poly(butylene succinate) (PPC/PBS) in twin screw extrusion. Additionally, degradation in unfilled polymer resins was studied and modeled using reaction kinetics combined with computational fluid dynamics simulation. The effects of screw profile and process parameters were investigated for homopolymers and blends compounded at speeds up to 4000 rpm using a novel high speed twin screw extruder. It was found that blends are more sensitive to thermomechanical degradation than neat polymers, and extent of chain scission is related to the inherent shear thinning behavior. Morphology investigation showed that the reactively compounded blends had smaller droplet size with a broader shape distribution than batch mixed blends. Effects of interfacial modification on morphology and physical properties were also characterized. The results indicate that in order for extreme shear to provide improved properties, the screw design must be optimized to minimize degradation in polymer blends.

EFFECT OF BLENDING METHODS ON MECHANICAL PROPERTIES OF PS/TPU BLENDS

Mansourieh Mohamadreza [(1)]***, Javadi Azizeh*** [(1)]*

[(1)] Amirkabir University - Tehran - Iran

Polystyrene (PS) is a general-purpose glassy polymer with low cost and good processability, which is used extensively in polymeric products. The drawback of this polymer is its brittleness, which in turn limits its applications. In this research, Thermoplastic Elastomers of Polyurethane (TPU) was used to improve the mechanical properties of PS. Blends with different compositions were prepared using a twin-screw extruder and internal mixer. Test specimens for impact and tensile characterizations were obtained with an injection-molding machine and compression molding respectively. The results showed that increasing the amount of TPU leads to higher impact strengths and elongations at break of the blends. It was found that extruded and injected samples with TPU amounts bellow 20 wt. % leads to a rather smooth increase in the mentioned properties whereas in higher TPU contents (30-50 wt. %), the mechanical properties increase dramatically. On the other hand, tensile curves showed that adding only 10 wt. % of TPU changes the behavior of PS from brittle to ductile and therefore yielding occurred in all blends. Furthermore samples prepared by one-step and two-step method via internal mixer showed different trend of changing of mechanical behavior. In two step method,samples had better mechanical properties rather than extruded ones specially in samples with amount 10-30 % of TPU because different morphology. Scanning Electron Microscopy (SEM) showed a homogeneous dispersion of TPU particles in the PS matrix in all samples and the size of TPU domains increased by its higher amounts.Also samples prepared by two-step method showed sub-inculsion morphology which causes better impact strength.

Preparation by melt compounding of composites reinforced by natural fibers: from breakage mechanisms to process optimization

VERGNES Bruno (1)*

(1) MINES ParisTech - Sophia Antipolis - France

In order to reduce parts weight in automotive applications, composites where carbon or glass fibers are substituted by natural lignocellulosic fibers are more and more considered. These materials are usually compounded using classical twin-screw extruders. However, as vegetal fibers are sensitive to high temperatures and mechanical stresses, the compounding process must be adapted and optimized. In the present talk, we will first show using rheo-optical techniques how lignocellulosic fibers from various origins behave in a shear flow. Then, we will propose a systematic study of the mechanisms of fiber breakage, based on experiments with an internal mixer. We will show that the kinetics of breakage can be described by exponential laws, functions of the strain supported by the material. The use of these models allows us to calculate the evolution of the fibers dimensions along the screws during the compounding process and thus to propose an optimization of process conditions and screw profile in order to limit fibers degradation.

Melt homogenization in co-rotating twin-screw extruders: Design of a novel kneading disk

Hristov Velichko (1)*

(1) Borealis Polyolefine GmbH - Upper Austria - Austria

A novel kneading element for co-rotating twin-screw extruders (TSE) was invented by designing the disk lobal surfaces in a specific way to create converging flow channels. The polymer melt is forced to pass through the converging channels both in axial and circumferential directions experiencing elongational deformation, which is beneficial for mixing polymers having high viscosity ratio. By using the novel screw elements effective mixing of the polymers and fillers was achieved. It is known that blending polymers having large viscosity difference on conventional co-rotating TSE is a challenge due to their limited dispersive mixing capabilities. According to H. P. Grace, shear does not favor good mixing of polymers having viscosity ratio higher than 4. Poor homogenization leads to product inconsistency, fluctuation of extrusion processes and deterioration of the mechanical properties. In order to achieve efficient dispersion of highly viscous components elongational flow passages have to be created in the processing equipment. The objective of this work was to investigate the influence of a novel screw element on melt homogenization in co-rotating TSE. The mixing zone of a co-rotating TSE ZSK-40 consisted of three sequences with two kneading disks in each. It was observed that the novel kneading disks provide better homogenization of the higher molecular weight polymer and CB masterbatch. It was also found that increasing the screw speed worsens the melt homogenization. GPC analyses confirmed that the Mz and Mz-1 molecular weights of the polymer extruded with the novel kneading disks were significantly higher than those achieved with the standard screw elements.

A Novel Experimental Setup for Characterization of Polymer Blends in Single-screw Extruders

O. Celik[a] and C. Bonten

IKT, Institut für Kunststofftechnik, University of Stuttgart, Pfaffenwaldring 32, 70569 Stuttgart, Germany

[a]Corresponding author: Oguz.Celik@ikt.uni-stuttgart.de

Abstract. In plastics engineering, mixing processes are omnipresent. Plastics reach their high performance only through mixing of many individual formulation components. By means of simulation methods, the design of mixing processes can be realized more efficiently. They allow for saving energy, staff and material costs at the same time during process optimization and increase the productivity over the lifetime of a machine. But findings under real operation conditions about the ongoing processes during the mixing are necessary for model development. However, frequently representative fluids are used in experimental investigations of mixing processes instead of real plastic melts. In this work, real plastic melts were used for the investigation of dispersive mixing instead of representative fluids. For model development and validation, a new experimental setup was realized. By means of a linear actuator and a specially designed injection nozzle melt strands were generated, which were inserted into the main flow of a continuous phase shortly before a shearing element. Afterwards, samples were taken from the extrudates for light-microscopic examination. On the basis of these findings, it will be possible to develop a new simulation model to predict the droplet behavior under real processing conditions later on.

INTRODUCTION

In case of plastics compounding, co-rotating twin-screw extruders are often used [1]. Due to their excellent mixing performance, twin-screw extruders are suitable for adding additives, for reactive processing or mixing several polymers. For producing so-called blends, two polymers with different material-properties are mixed to combine favorable properties of both components. A distinction is drawn between miscible and non-miscible blends. In case of non-miscible blends, commonly a bonding agent to achieve a fine dispersion is used. The formation of the morphology of the mixture depends on the compatibility and contents of both components. The following processing influences the morphology significantly and at the same time the properties of the finished product. A further possibility is to prepare and process blends in one step. However, the mixing performance of twin-screw extruders can not be reached by ordinary single-screw processing machines. The advantage is that there are no additional costs for energy, human resources and transport.

STATE OF THE ART AND OBJECTIVE

The behavior of droplets in highly viscous fluids has already been investigated in numerous publications. The capillary number represents the ratio of forces in a flow, which are trying to deform the droplet to those, that maintain the droplet shape. It is defined as follows:

$$Ca = \frac{G \cdot \eta_k \cdot R_0}{\sigma}. \tag{1}$$

Where G, η_k, R_0 and σ are the deformation velocity, the viscosity of the continuous component, the initial radius of the droplet and the interfacial tension coefficient.

Already in 1981, Grace [2] investigated the behavior of droplets in different flow types. Based on the results, a critical capillary number Ca_{crit} in dependence of the viscosity ratio of the disperse to the continuous component $\lambda = \eta_d/\eta_k$ could be derived, as shown in Fig. 1.

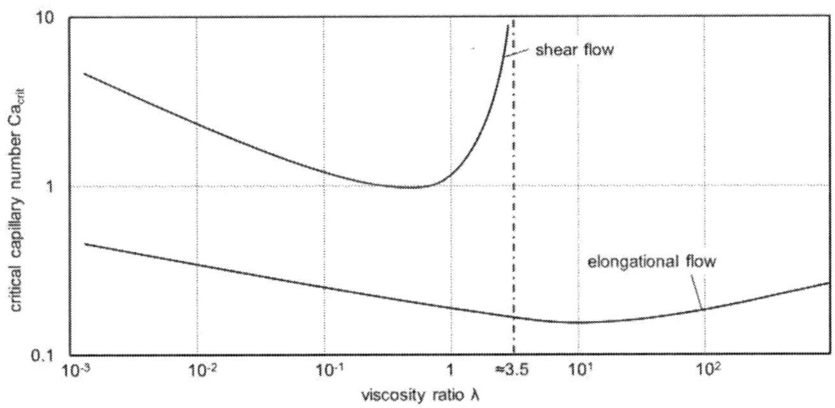

FIGURE 1. Critical capillary number according to Grace [2]

It could be observed that a droplet in a pure shear flow can no longer be broken up above a viscosity ratio of around 3.5. To still achieve the dispersion of the droplet, an elongational flow is needed.

Examples for further experimental investigations of droplet behavior are publications of Kaufmann [3], Heindl [4], Rauwolf [5] and Boonen [6]. Whereas Kaufmann and Boonen used model fluids in their experiments, Heindl and Rauwolf utilized a mixture of polyethylene and polystyrene. Heindl characterized the blends in an especially developed extensional rheometer. Rauwolf used the flow in the gap of a capillary rheometer. In both cases, images have been taken with the help of scanning electron microscopes for characterizing the morphology under variable conditions. Even if the investigations showed valuable findings concerning the establishing morphology related to the flow types, the transfer to real process conditions is not trivial.

In this work, mixing procedures in a single-screw extruder shall be investigated under real process conditions with the help of a new experimental setup. The purpose is to create principles, which lead to a better understanding of the droplet behavior and to develop a quantitative, mathematical model for the droplet dispersion. In this article, the new experimental setup is explained and initial results are described.

EXPERIMENTAL SETUP

The experimental setup consists of the three main components feed extruder, mixing device with independent drive and a droplet supply device. This assembling can be seen schematically in Fig. 2.

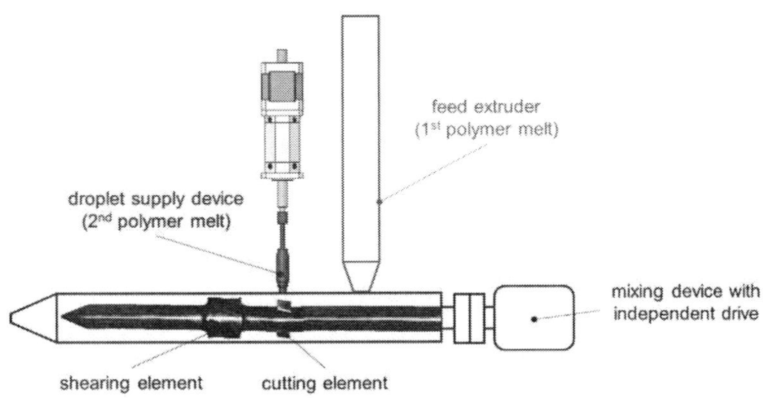

FIGURE 2. Schematic test setup

The mixing device is fed through a single-screw extruder with a low-viscosity polymer. The feed extruder has an inner cylinder diameter of D = 35 mm. The supply of the second component, a highly viscous polymer, is realized using a specially developed droplet supply device. This device will be explained later.

PROPERTIES OF THE MIXING DEVICE

The mixing device has been developed by the company Ematik GmbH, Magdeburg, Germany, in close cooperation with the IKT. The inner cylinder diameter is about 50 mm. It has an effective length of 250 mm and can reach a maximum speed of 510 rpm. The maximum torque is about 161 Nm. Furthermore, the screw can be equipped with different cutting and mixing elements using a feather key connection. Temperature and pressure can be captured over the entire length of the mixing part.

The cutting element, which was used, consists of three blades which are distributed over the circumference, see Fig. 2. Its task is to separate the strand that leaves the injection cylinder and carry it to the shearing part. The dimensions of the shearing element are summarized in the following Tab. 1.

TABLE 1. Dimensions of the shearing element

Geometry and Unit	Value
Length in mm	30
Height of the shearing gap in mm	0.9
Amount of threads	3
Chamfer angle in °	15
Chamfer in mm	4

PROPERTIES OF THE DROPLET SUPPLY DEVICE

The droplet generation and the supply are provided by a linear actuator, as illustrated in Fig. 3.

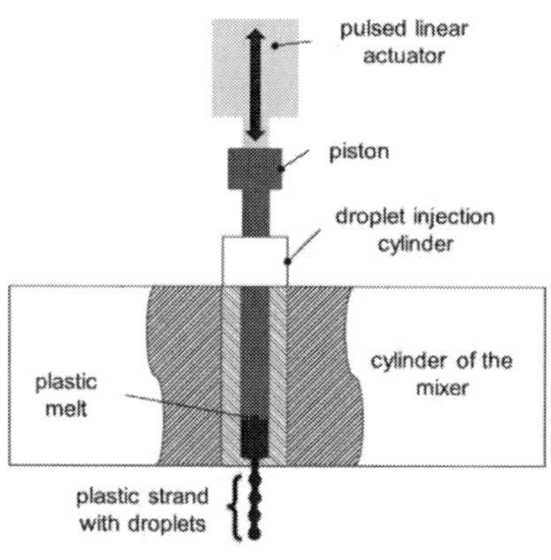

FIGURE 3. Schematic view of the droplet supply

The linear actuator moves a piston, which pushes the disperse component that was melted in the injection cylinder before. Strands with droplets can be generated by using a particular driving profile of the linear actuator. Different driving profiles have been investigated in preliminary tests for the droplet injection device.

The injection cylinder was rheologically optimized considering the force that the linear actuator can generate at maximum. Furthermore, the minimum possible piston stroke per step was taken into account for constructing the injection cylinder. The linear actuator L4118L1804-T6x1-A50 of the company Nanotec which was used can reach a maximum force of 450 N, the step is 5 μm and the maximum stroke is 50 mm.

The inner diameter of the injection cylinder is about 5 mm, whereas the outlet diameter was set to 0,35 mm. However, after manufacturing it could be shown that the outlet diameter is about 0,43 mm. This difference can be attributed to manufacturing tolerances.

For the calculation of the force, a Newtonian fluid with a constant viscosity of 5000 Pas was assumed. This assumption is permissible because the polymer melt is always moved starting from a resting position. The selected viscosity corresponds to the zero shear viscosity plus a safety factor for the disperse component.

POLYMERS AND PROPERTIES

The mixing device is fed through a single-screw extruder with low-viscosity polypropylene RE420MO, produced by Borealis (PP). The disperse component, a highly viscous polystyrene 486M manufactured by BASF SE (PS) is injected through the droplet injection device. To facilitate the later evaluation of the tests, the PS was dyed in blue. Both polymers, PP and PS, were investigated on a rotational rheometer (plate-plate) at a temperature of 220 °C. The resulting viscosity ratio is shown in Fig. 4. It was determined by considering the zero shear viscosity for the droplet component PS. It was assumed that at the time, when the droplet has been injected, no deformation of the droplet occurs. Within the considered process window the viscosity ratio is above the threshold value 3.5. According to Fig. 1 the PS droplets can not be broken up in a pure shear flow.

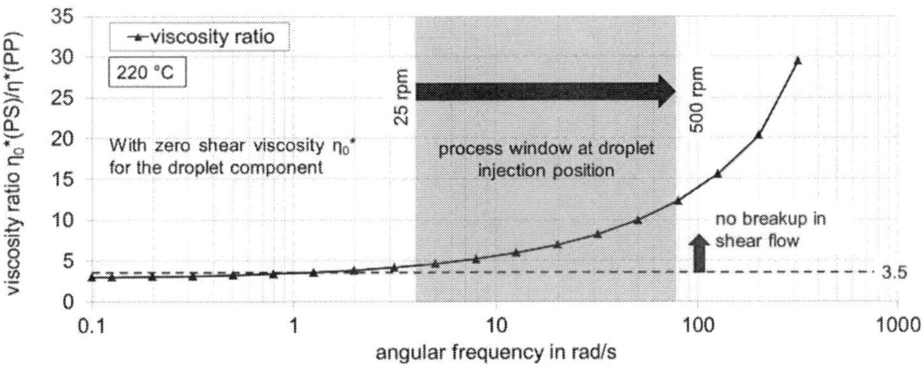

FIGURE 4. Viscosity measurements at a temperature of 220 °C

EXPERIMENTAL PROCEDURE

The interesting part of the experimental setup is shown in Fig. 5. The mixing device is fed with a throughput of 6,27 kg/h with PP at 220 °C. In the first tests, such a low throughput has been chosen, because only the effect of the mixing speed on the droplet behavior should be seen. Afterwards, the mixing speed was varied between 25 and 500 rpm for the experiments. As soon as a stationary state could be seen in the torque curve, the linear actuator was started to inject the second component (PS) through the droplet injection device.

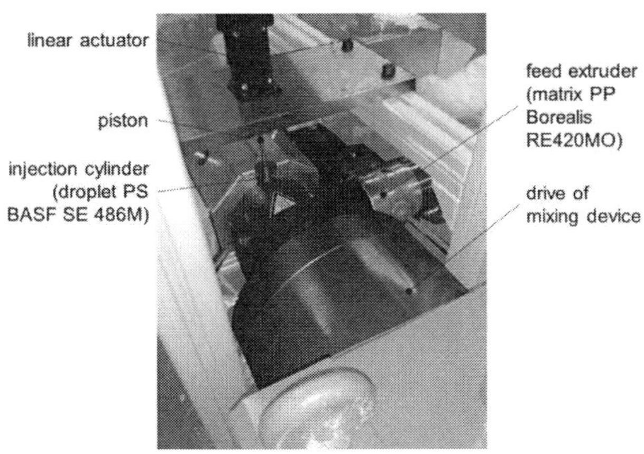

FIGURE 5. Detail of the test setup

Furthermore, these primary experiments should give an answer which microscope is appropriate for a quantitative evaluation. Therefore, the piston was driven with a constant speed of 0.4 mm/s at first to achieve the highest possible concentration of PS in the sample. This ensured that the PS particles could be seen in the whole matrix afterwards.

RESULTS AND DISCUSSION

At the outlet of the mixing device, cylindrical samples were taken. The samples were separated lengthways through the center, so that semi-circular thin sections could be carved out. Those sections are only a few micrometers thick. Figure 6 represents exemplary a thin section for a mixing speed of 25 rpm and 500 rpm by means of polarized light in a light transmission microscope. The very low thickness of the thin section leads to uneven areas during the preparation of the samples. Additionally, the grooves of the knife can be observed as horizontally arranged lines. Nevertheless, the PS droplets can be observed clearly at a speed of 25 rpm. In contrast, at a speed of 500 rpm, PS droplets can not be detected clearly at the same magnification.

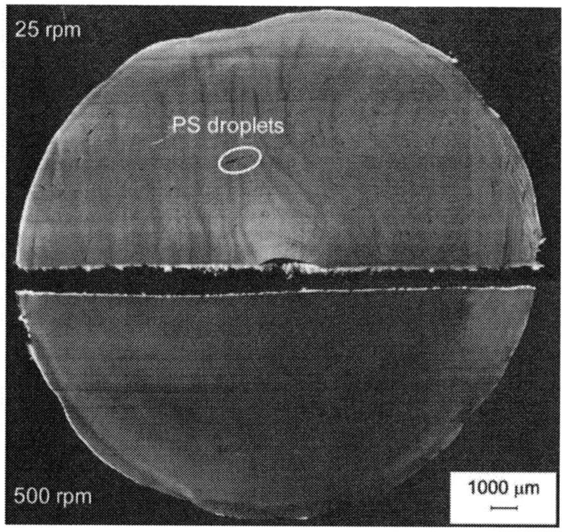

FIGURE 6. Overview of a thin section after a mixing speed of 25 and 500 rpm

Magnification in Fig. 7 shows that the PS droplets are stretched in the peripheral direction. The breakup after stretching can be imagined and understood quite well, based on the images.

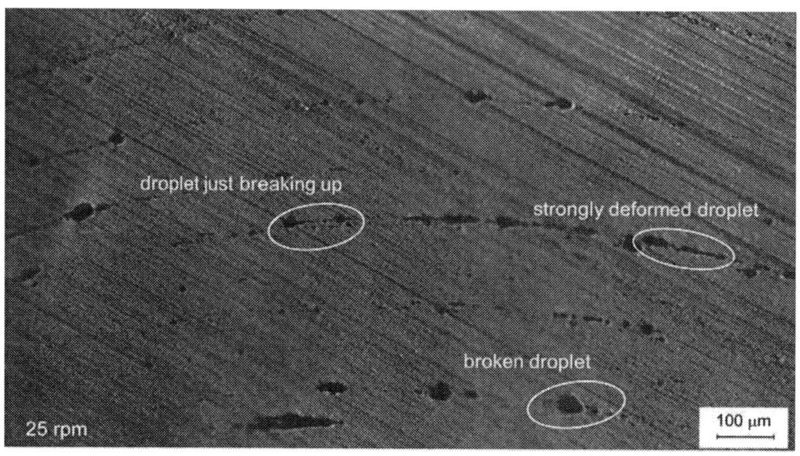

FIGURE 7. PS droplets at 25 rpm

In Fig. 8 the particle sizes at a mixing speed of 25 rpm and 500 rpm are compared at a higher magnification. At a speed of 500 rpm, the PS particles seem to be dispersed more finely and homogenously. Furthermore, it can be observed that in this case the thin section is thicker than the particle diameter itself. Therefore, it cannot be excluded that several particles are overlaid in the image. Hence, another microscopy method, as for example the scanning electron microscopy (SEM) or the atomic force microscopy (AFM), should be preferred to determine the particle size distribution quantitatively. The disadvantage is found, however, in the fact that only a relatively small area of the sample can be evaluated. In case of large particles, light microscopy still should be used.

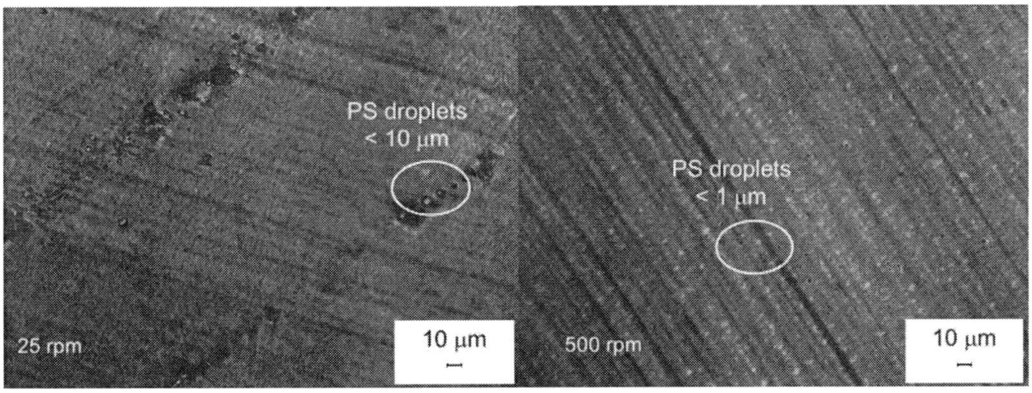

FIGURE 8. Higher magnification at 25 and 500 rpm

CONCLUSION AND OUTLOOK

With the help of the results, it can be recorded that the new experimental setup is suitable for the systematic investigation of the droplet behavior. With the help of a light transmission microscope it could be observed qualitatively that a finer dispersion and a more homogenous distribution of the PS particles occurs with higher speed. The first results show as well that a quantitative evaluation of the droplet separation is only partially possible with light microscopy. In case of smaller particles, SEM or AFM images should be preferred.

In further experiments, the droplet behavior shall be studied quantitatively under different conditions. The next step is to investigate the influence of the driving profile of the linear actuator. Furthermore, besides the mixing speed, the throughput of the feed extruder shall be varied. It is planned to analyze further polymer combinations, with which the influence of the viscosity ratios and the interfacial tension between disperse and continuous component can be identified more precisely. The paramount objective is to investigate a mathematical model that can predict the droplet behavior in single-screw extruders based on the experimental results.

ACKNOWLEDGMENTS

This research and development project is funded by German Research Foundation (DFG) under the funding code BO 1600/23-1.

DFG

REFERENCES

1. Bonten, C.: *Kunststofftechnik – Einführung und Grundlagen*, 2nd edition, München: Hanser, 2016, ISBN 978-3-446-44674-8
2. Grace, H. P.: Dispersion phenomena in high viscosity immiscible fluid systems and application of static mixers as dispersion devices in such systems. In: *Chemical Engineering Communications* (1982), no. 14, pp. 225-277
3. Kaufmann, S. F. M.: Experimentelle und numerische Untersuchungen von Tropfendispergiervorgängen in komplexen laminaren Strömungsfeldern. Eidgenössische Technische Hochschule, Institut für Lebensmittel- und Ernährungswissenschaften, Ph.D. thesis, 2002

4. Heindl, M.: Einfluss von Dehnströmung auf die Morphologieausbildung in Polymerblends. University of Erlangen-Nürnberg, Technical Faculty, Ph.D. thesis, 2005
5. Rauwolf, M.: Morphologieausbildung in strömenden Kunststoffgemischen. University of Stuttgart, Institut für Kunststofftechnik, Ph.D. thesis, 2006
6. Boonen, E.: Morphology Development of Immiscible Polymer Blends in Complex Flow Fields. Catholic University of Leuven, Engineering Faculty, Ph.D. thesis, 2009

Effect of surface modification on thermal properties of paraffin/copper oxide Nanocomposite

Mohammad Mahdi Noori[1, b)], Hossein Ali Khonakdar[2, a)], HamedAzizi[1], Mehdi Ghaffari[3]

[1] Department of Polymer Processing, Iran Polymer and Petrochemical Institute (IPPI).
[2] *Leibniz Institute of Polymer Research Dresden, Hohe Str. 6, D-01069 Dresden, Germany*
[3] *Polymer group, Faculty of Technical and Engineering, Golestan University, Gorgan, Iran.*

[a)]hakhonakdar@gmail.com
[b)]mohamadmahdinoori0541@gmail.com

Abstract. In this work phase, change material Nanocomposite (NPCM) was prepared based on paraffin and copper oxide (CUO). CUO has been used to improve the thermal conductivity of paraffin. Since the final properties and efficiency of NPCM is dependent on the dispersion of Nanoparticles in the matrix, the surface of CUO particles was treated by tri-methoxy(octadecyl)silane. Grafting of Silane on the CUO surface was characterized by Fourier transform infrared spectroscopy (FT-IR) and thermogravimetric analysis (TGA) techniques. The Effects of Nanoparticles on latent heat and thermal conductivity were measured by Differential Scanning Calorimetry (DSC) and KD2. According to the results, increasing the Nano copper oxide in composite caused an increasing the thermal conductivity and decreasing the latent heat. Moreover, the results showed that time of solidification and melting process decreased.

Keywords: Phase change material- Nano copper oxide- Paraffin- Thermal properties.

INTRODUCTION

Nowadays, the use of renewable energy with a lot of speed in progress. the world's researchers believe that should be the base of oil outside of the energy supply and renewable energy through it and be like the wind, solar and water power[1]. There are several methods for storing renewable energy that exists among them heat energy storage capability, due to the creation of the high density of the unseen, save energy and save its energy at constant temperature (temperature of phase change material) of more attention. [2] Substances that energy to this method in their store so-called phase change materials providers say. These materials are organic compounds or mineral uptake and storage of heat energy values hidden within. Thermal energy storage in these materials, during the process of changing the phase (solid to the liquid state of change or vice versa), lifted. These materials, when a change in phase of liquid or solid from liquid to solid, the heat absorbed from the environment or to the environment. Phase change material heat energy can be without any changes, even after thousands of cycles of change phase, keep within.[3,4]

One of the most important and most common materials of paraffin phase change energy storage is a relatively favorable due to this end. But one of the weak points of this article is its low coefficient of thermal conductivity. The use of materials with high thermal conductivity coefficient can be one of the ways to improve the thermal conductivity of paraffin[5]. On the research of Nano-copper oxide that has a high coefficient of thermal conductivity is to increase the thermal conductivity of paraffin. Therefore Nano-copper oxide modified by silane and added to paraffin by 0.5, 1 & 3% with different mixing methods like mechanical mixing and ultrasonic.

We observe that by increasing the Nano-copper oxide, the thermal conductivity of paraffin was increased and a small amount of enthalpy of PCM is decreased. And also with the surface modification of Nano, increase compatibility between the Nano and paraffin.

Experimental

1. Materials

The materials used in this research were Paraffin wax (Rubitherm, Rt44HC), CUO (Aldrich), tri-methoxy silane (Aldrich), triethylamine (Et3N), Ethanol (97.6%), Toluene (99%).

2. Modification of Nano-copper oxide

Nanoparticles were dispersed in toluene; then silane coupling agent and Et3N were added. The mixture was heated (120C°) and stirred for 12 hr under the reflex system; after that The Nanoparticles was separated by centrifuge. The Nanoparticles were dried at 120C° for 24 hr[6].

3. Preparation of PCMs

The paraffin in a water bath at a temperature of 50 C ° was completely melted. The copper oxide nanoparticles at concentrations of 0.5, 1 and 3 were added by an ultrasonic vibrating bath at a temperature of 50 C ° and 37 frequency power 70 and was broadcast for 30 minutes. Then, immediately for 20 minutes with a mechanical stirrer and stirred at 200 rpm for 15 minutes and then was power 70 of the ultrasonic probe.[7]

The molten solution is then poured into molds and cooled to a temperature of -13 C ° in the freezer.

4. Characterization

To check the test of Nano-modified infrared spectroscopy(FTIR) was used. Pure and Modified Nanoparticles to form tablets with KBr was prepared and evaluated by the operator. For this test device models EQUINOX 55 Manufacturing Co. broker (Bruker Co) Germany was used. To investigate the surface modification of Nano, thermal gravimetric analysis (TGA) was performed. TGA test using a model of PL Polymer Laboratories, in the temperature range 30C ° to 600 C ° with a heating rate of 10 C° / min and was conducted under nitrogen gas. To measure the ΔH of paraffin and Effect of Nanoparticles on ΔH, DSC test was used. The device tested by differential scanning calorimetry DSC200F3 model making company Netzsch, Germany, under nitrogen gas in thermal analysis laboratory at 10 ˚C / min temperature of 0 ° C to 90 C ° to go back, go after. device for measuring the thermal conductivity of PCM was used PRO KD2. The device to determine thermal conductivity in the range of 0.02 to 2 watts per meter Kelvin degrees is used. The measurement accuracy is also 0.01.

RESULT AND DISCUSSION

The results FTIR have shown that Nanoparticles modified by Silane (figure 1). According to the results, FTIR spectra of pure nanoparticles in a thin and strong peak observed at 535-540 cm-1 related to the Cu-O bond stretching absorption. A broad peak in the region of 3000-3500 cm-1 is observed to be physically adsorbed water. The peak seen in 1650 cm-1 is related to the absorption of water molecules. In the spectrum of nanoparticle modified index peaks at 800 and 1100 cm-1 can be seen which is related to Si-O-C bond. The peak seen in 1250 cm-1 is related to the Si-CH2. Peaks appear that the reform process has been done well.[6,8]

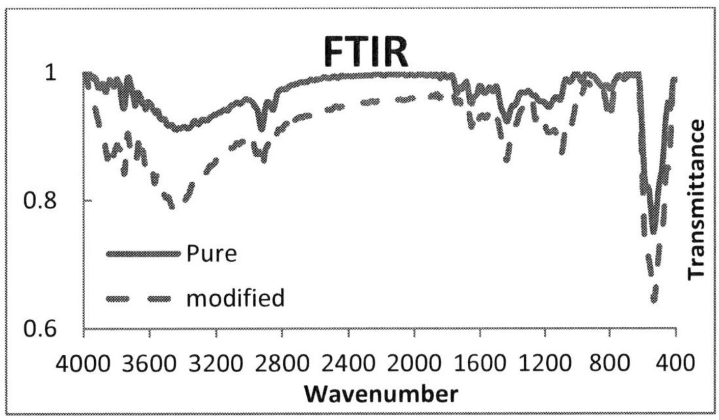

FIGURE 1. FTIR of copper oxide Nanoparticles

TGA result was approved this surface modification (figure 2). The TGA tipping point of modified Nano indicates that this Nano was destroyed by increasing temperature in range 250 to 400 C° and this range is because of Si.[6]

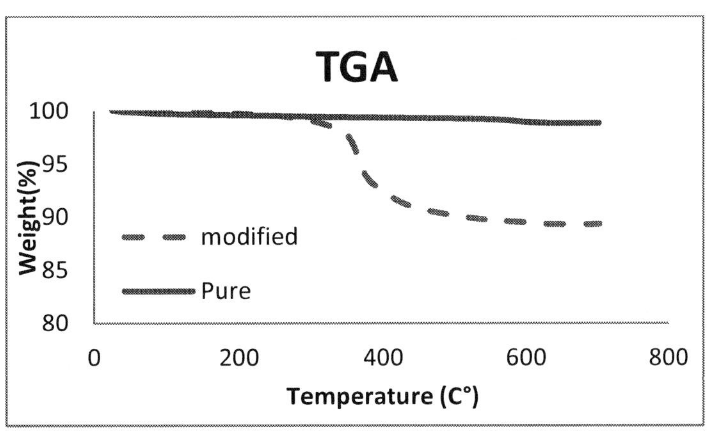

FIGURE 2. TGA of copper oxide Nanoparticles

DSC results showed that increasing the Nano, because that one hundred percent crystalline substance paraffin, ΔH decreases due to the prevention of nanoparticles complete crystallization paraffin, but for PCM cuts have to be too low to provide energy well save.[9,10] Therefore, the reduction ΔH is very important and should be minimized. Two peaks can be seen that the main peak of heterogeneous nucleation transfer fluid (rotator-liquid) and represents a change from solid to liquid is called solid-liquid transferring to say. The small peak on the left represents the major peak of solid-solid transitions that correspond to homogeneous nucleation (crystal-rotator) is paraffin.[7]

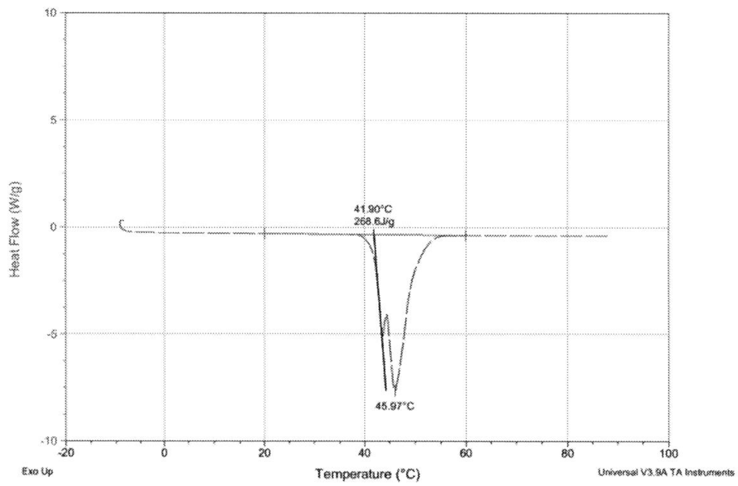

FIGURE 3. DSC of Pure Paraffin

TABLE 1. Latent heat of PCM

Sample code	Nano%	Latent heat(j/g)
P	-	268.6
B(0.5)	0.5	258.8
B(1)	1	252.2
B(3)	3	250
BS(0.5)	0.5	260.6
BS(1)	1	256.4
BS(3)	3	253.3

KD2 results showed that by increasing the copper oxide Nano unmodified and modified in both solid and liquid samples of thermal conductivity of pure paraffin increased.[7] paraffin thermal conductivity in the solid state is more than the liquid state because of the Regular structure. The increase in the higher thermal conductivity of copper oxide Nanoparticles has been attributed. The thermal conductivity of the Nano-copper oxide modified in percentages of less than 3% lower than unmodified, and in 3% Nano-modified more increase the thermal conductivity is, the increase in broadcast better examples of silane modified matrix is attributed.

TABLE 2. Thermal conductivity of PCM

Sample code	Thermal conductivity in solid state	Thermal conductivity in liquid state
P	0.2	0.15
B(0.5)	0.281	0.156
B(1)	0.291	0.164
B(3)	0.383	0.175
BS(0.5)	0.252	0.153
BS(1)	0.270	0.159
BS(3)	0.4745	0.188

Conclusion

Viewed copper oxide Nano-surface modification with silane is compatibility between Nano and paraffin. With increasing Nano-copper oxide to paraffin, the thermal conductivity of PCM-enhanced and enthalpy decreased slightly.

ACKNOWLEDGMENTS

We gratefully acknowledge the financial support provided by

Iran Polymer and Petrochemical Institute(IPPI)

REFERENCES

1. J. Baker, "New technology and possible advances in energy storage," Energy Policy, vol. 36, no. 12, pp. 4368–4373, 2008.
2. A. Sharma, V. V Tyagi, C. R. Chen, and D. Buddhi, "Review on thermal energy storage with phase change materials and applications," Renew. Sustain. energy Rev., vol. 13, no. 2, pp. 318–345, 2009.
3. M. F. Demirbas, "Thermal energy storage and phase change materials: an overview," Energy Sources, Part B Econ. Planning, Policy, vol. 1, no. 1, pp. 85–95, 2006.
4. S. Mondal, "Phase change materials for smart textiles–An overview," Appl. Therm. Eng., vol. 28, no. 11, pp. 1536–1550, 2008.
5. N. Sahan and H. O. Paksoy, "Thermal enhancement of paraffin as a phase change material with nanomagnetite," Sol. Energy Mater. Sol. Cells, vol. 126, pp. 56–61, 2014.
6. J. Lin, J. A. Siddiqui, and R. M. Ottenbrite, "Surface modification of inorganic oxide particles with silane coupling agent and organic dyes," Polym. Adv. Technol., vol. 12, no. 5, pp. 285–292, 2001.
7. M. Nourani, N. Hamdami, J. Keramat, A. Moheb, and M. Shahedi, "Thermal behavior of paraffin-nano-Al 2 O 3 stabilized by sodium stearoyl lactylate as a stable phase change material with high thermal conductivity," Renew. Energy, vol. 88, pp. 474–482, 2016.
8. W. Shen, X. Dong, Y. Zhu, H. Chen, and J. Shi, "Mesoporous CeO 2 and CuO-loaded mesoporous CeO 2: synthesis, characterization, and CO catalytic oxidation property," Microporous Mesoporous Mater., vol. 85, no. 1, pp. 157–162, 2005.
9. S. Kim and L. T. Drzal, "High latent heat storage and high thermal conductive phase change materials using exfoliated graphite nanoplatelets," Sol. Energy Mater. Sol. Cells, vol. 93, no. 1, pp. 136–142, 2009.
10. M. Mehrali, S. T. Latibari, M. Mehrali, H. S. C. Metselaar, and M. Silakhori, "Shape-stabilized phase change materials with high thermal conductivity based on paraffin/graphene oxide composite," Energy Convers. Manag., vol. 67, pp. 275–282, 2013.

Upcycling contaminated post-industrial PP through compounding: a Design from Recycling case study.

Ragaert Kim [1]*, Hubo Sara [1], Delva Laurens [1], Kuzmanovic Maja [1], Veelaert Lore [2], Dubois Els [2]

[1] Ghent University - Gent - Belgium, [2] University of Antwerp - Antwerp - Belgium

Design from Recycling is a novel methodology in which industrially available recycled materials are matched to potential new products. If needs be, the recycled polymer is upgraded to be made fit-for-use for the new product application. This methodology was applied to a PET-contaminated recycled PP, destined for a high-impact, medium-stiffness application. Phase segregation in the blend was found to severely inhibit the impact properties of the material. Different impact modifiers were tested in a research environment, after which an industial trial was conducted with the best result, an addition of 10 wt% SEBS. The industrial trial consisted of compounding of a large batch, followed by extrusion and thermoforming of sheets. Contrary to expectations, these products failed consequent testing. A failure analysis of the product morphology led to the intermediary hypothesis that interparticle distance between the rubber particles was still too large for complete toughening and that the interfacial tension with the immiscible PET was insufficiently mitigated by the potential interaction with the SEBS. In a second research trial, the total compounded rubber amount was increased to 15 wt% and a compatibilizing agent (CA) was added to compatibilize the PET and improve the microstructural dispersion of the rubber phase. Spectacular results were achieved for the formulation including the CA. Based on these results, this formulation was the used for the second industrial trial.

Friction characteristics improvement of ABS for low squeaking noises generating interior automotive parts

Rojsatean Jareenuch (1)*, Larpsuriyakul Patcharee (1), Prakymoramas Natcha (1), Thanomjitr Dumrong (1), Kaewket Sanya (1), Singsom Thawatchai (1), Srinun Duangruthai (1)

(1) National Science and Technology Development Agency, Ministry of Science and Technology - Pathumthani - Thailand

Acrylonitrile butadiene styrene (ABS) widely used as the interior automotive parts tends to generate friction-induced squeaking noises when rubbing with each other. To minimize annoying squeaking noises, friction characteristics of ABS were modified by compounding technique with some promising additives. In this work, ethylene bis stearamide (EBS) and poly(ethylene-co-octene) (PEOc) were selected as internal lubricants and compounded with ABS utilizing a twin screw extruder, obtaining ABS+0.5wt%EBS and ABS+5wt%PEOc. The friction testing was conducted on a universal testing machine equipped with the friction testing fixture. The effects of lubricant types and part surface roughness on the friction characteristics, static and kinetic coefficient of friction (COFs) and double amplitude of stick-slip curve were investigated in this study. The double amplitude was especially used to evaluate the tendency to produce the squeaking noise of ABS and its compounds. It was found that PEOc was more effective than EBS in reducing COFs and double amplitude of stick-slip curves of ABS. Due to the migration of PEOc to the part surface, its surface energy decreased leading to the reduction in adhesion between the moving surfaces of ABS with PEOc added. Moreover, the static and kinetic COFs of ABS+5wt%PEOc showed strong dependence on the surface roughness while this was not the case for ABS and ABS+0.5wt%EBS.

Investigation of the temperature development within the solid conveying zone of twin-screw extruder based on the Discrete Element Method (DEM)

Scharr Kim Jacqueline (1)*, Schöppner Volker (1)

(1) Paderborn University - Paderborn - Germany

In the field of polymer processing, a few established models for the conveying and melting behavior within twin-screw extruders exist. These models picture the majority of relevant industry applications well. The polymer solid temperature development is calculated under consideration of thermal conduction and convection, for example within the solid temperature calculation of Melisch [1]. However, the frictional heat, resulting from relative motion within the partially filled screw channel and the heat transfer between cylinder and screw, is being neglected in the models. In order to analyze the other dissipation processes, numerical simulations based on the Discrete Element Method (DEM) have been carried out. With the help of a DEM-simulation-program, a 32 mm twin-screw test stand has been built up. It is possible to simulate the conveying of different polymer pellets, whose material characteristics have been defined in preliminary investigations and were implemented in the simulation. By evaluating the contacts between the particles, the particles and screws and particles and barrel surfaces, the dissipation within the system is calculable. Therefore, it is now possible to evaluate the temperature at the end of different solid conveying zones, in due consideration of friction and heat transfer because of barrel contacts. In upcoming experimental investigations the simulated temperatures at the end of the solid conveying zone will be verified. After the verification it will be possible to build up a new model to calculate the dissipation in the solid conveying zone for common extruder sizes. [1] U. Melisch: Grundlagen zur Simulation des Förder- oder Plastifizierprozesses dichtkämmender Gleichdrall-Doppelschnecken. Dissertation, University of Paderborn, 1998

Modified Polylactide With Improved Properties For Extrusion Foaming

Svenja Göttermann[1, a)] and Tobias Standau[2, b)], Volker Altstädt[2], Christian Bonten[1]

[1]*Institut für Kunststofftechnik, University of Stuttgart, Pfaffenwaldring 32, 70569 Stuttgart, Germany*
[2]*Lehrstuhl für Polymere Werkstoffe, University of Bayreuth, Universitätsstraße 30, 95447 Bayreuth, Germany*

[a)]svenja.goettermann@ikt.uni-stuttgart.de
[b)]tobias.standau@uni-bayreuth.de

Abstract. Polylactide (PLA) is the most important bioplastic on the market due to its good mechanical properties and a permanent growth of the production capacity. But in terms of extrusion foaming the melt strength and extensibility of commercial polylactide is too low. Therefore such commercial grades need to be modified to induce crosslinking, chain extension or branching by means of reactive extrusion on a twin-screw extruder. In this work different modifiers (chemical as well as physical modifiers) were used. The most effective results were achieved with organic peroxide. With this modifier the molecular weight, the melt viscosity, the melt strength and the crystallization rate were improved leading to foams with a closed-cell structure and improved compression behavior. Furthermore, the modifier was found to be more effective than a commercial modifier.

INTRODUCTION

Foams are two-phase systems (gas-solid) with a cellular structure. Due to the cellular structure, foams have unique properties (high impact resistance, low density, high insulation properties), which are not achievable by any other blend or composite system. Due to the very low density foam products are highly material efficient. The majority of the foamed products are based on polyurethane (46 %) and polystyrene (33 %). Polystyrene (PS) is used as thermal insulation material in civil engineering as well as in the packaging sector to protect products against impact. But due to environmental issues more and more foamed PS-products are banned by legal restrictions and the demand for more sustainable materials is increasing.

A suitable plastic for substituting PS is PLA, a bioplastic that is biobased and at the same time biodegradable. It has similar properties compared to PS such as similar values for the elongation at break, similar impact strength, similar Youngs as well as flexural modulus [1]. The advantage is that the energy input to produce PLA as well as its carbon foot print is much lower. Because of this, but also because of its biobased nature, sustainable foams with higher material efficiency could be realized.

The main drawback is that standard, cost-effective PLA has a low melt strength and absence of the so called strain hardening effect, which is disadvantageous for foaming. To overcome these drawbacks PLA must be modified. According to Nofar et al. [2] this can be done by increasing the molecular weight with chain extenders or other chemical modifiers, the variation of D/L-PLA ratio, controlling the crystallization kinetics or the adding additives. In this work, the aim is to (i) increase the melt strength and improve the crystallization rate by the use of suitable chemical modifiers as well as (ii) the building of stereocomplexes by varying the D/L-PLA ratio.

EXPERIMENTAL

Reactive Extrusion and Blending

The material used in this study was a commercial available PLA (PLA IngeoTM 7001 D) from NatureWorks (Minnetonka, MN / USA). This resin is specifically designed for the use in injection stretch blow molded bottle applications. Different modifiers with different functionalities were used for modification. An overview of the produced materials is given in Table 1. As reference material, extruded unmodified polylactide (PLA ex) was used. The selected modifier concentration is based on preliminary examinations. Therefore, the modifier concentrations as well as the processing conditions were varied. The concentration leading to the highest change in complex viscosity (highest zero shear viscosity) was selected. Then, the reactive extrusion was carried out on a twin screw extruder type ZSK26 MC from Coperion (Stuttgart / Germany). For all reactants the same screw design and the functional zones (plasticizing, distributive and dispersive blending as well as the chemical reaction zone) were used.

The mixing elements lead to a better blending, causing a faster reaction speed. Depending on the used modifier, the temperature profile of the twin screw extruder was varied between 180 °C and 200 °C, but the screw speed (150 min^{-1}) and the throughput (7 $kg \cdot h^{-1}$) were kept constant. The solid materials were added gravimetrically in zone 1. Liquid modifiers were added volumetrically using medical infusion pumps in zone 1. The extruded double-strands were cooled in a water bath, granulated and then dried in a vacuum oven at 60 °C overnight. For the materials PLA BODI and PLA ICDI the reactive extrusion was performed in a two-step process. In the first step isocyanurate or bis-oxazoline was added and in the second step diisocyanate.

TABLE 1. Overview of the modified Polylactides

Material	Modifier	Modifier conc. in wt.-%	Proposed reaction type / chain topology
PLA ex	-	-	linear
PLA OP	organic peroxide	0.2	branched
PLA EPO	multifunctional epoxide (Joncryl®)	1.0	chemically crosslinked
scPLA	Poly-(D)-lactide	5.0	physically crosslinked
PLA BODI	bis-oxazoline / diisocyanate	0.2 / 0.2	chain extension
PLA ICDI	isocyanurate / diisocyanate	0.2 / 0.2	chain extension, branching
PLA SMA	styrene maleic anhydride	0.7	chain extension, branching

Thermal and Rheological Characterization

To investigate the crystallization process, differential scanning calorimetry (DSC) measurements on a DSC 204 Phoenix from the company Netzsch (Selb / Germany) were performed. Samples of about 10 mg were weighed and sealed in aluminum DSC crucibles and placed in the DSC cell. The samples were first heated from 20 to 200 °C at a rate of 10 $K \cdot min^{-1}$ under nitrogen atmosphere. Then the samples were cooled from 200 °C to – 20 °C at 2 $K \cdot min^{-1}$ and were heated again up to 200 °C at 10 $K \cdot min^{-1}$. The crystallization process depends on the cooling rate. With increasing cooling rate the crystallization temperature shifts towards lower values. In addition, the degree of crystallization decreases. The chosen lower cooling rate of 2 $K \cdot min^{-1}$ is a typical analytical parameter for studying the crystallization properties of plastics.

The Rheotens experiment is a fibre spinning experiment, where the polymer strand is stretched under the action of a constant drawdown force until rupture of the filament occurs. Rheotens measurements were carried out to prove the ability of the polymer melt to withstand uniaxial strain as strain appears during foaming. The characteristic values, melt strength and melt extensibility (drawability), can be determined. A Rheotens device from Göttfert (Buchen / Germany) was used. The test samples were prepared using a high pressure capillary rheometer Rheograph 2000 from Göttfert connected to a piston (diameter = 9.5 mm) operated at a piston speed of 0.5 $mm \cdot s^{-1}$. The die is 2 mm in diameter and 40 mm in length. For each test, a molten polymer strand was drawn from the die by the two counter-rotating measurement wheels mounted on a sensitive force transducer connected to the Rheotens 71.97 unit. The tensile force on the polymer melt strand is measured as a function of time or velocity of the measurement wheels. The melt strength is represented by the medium force at which the strand breaks. The measurements were carried out at 175 or 180 °C and a constant acceleration of 6 $mm \cdot s^{-2}$.

Foam Extrusion and Foam Characterization

Foam extrusion was carried out on a tandem extrusion line from Dr. Collin (Ebersberg / Germany), which is a twin-screw extruder (L / D = 42, D = 25 mm; A-extruder), that melts the polymer and homogenizes the injected blowing agent with the polymer melt to achieve an one-phase mixture. The twin-screw extruder is followed by a single-screw extruder (L / D = 30, D = 45 mm), where the polymer gas mixture is cooled subsequently until the die, where foaming occurs. The temperature in the twin-screw extruder was set to 200 °C while in the single-screw extruder the temperature was decreased successively (down to 160 °C and 140 °C at the die, respectively), depending on the modification. Carbon dioxide was used as physical blowing agent at a concentration of 8 wt.-%.

For analyzing the foam morphology the surfaces of cryogenically fractured foam samples were investigated with a scanning electron microscopy (SEM) JSM-6510 from JEOL. Cell sizes were determined by the use of ImageJ software analyzing at least 3 x 50 cells per sample. The density was determined in accordance to the Archimedes principle with an AT 201 of Mettler Toledo (Columbus, OH / USA). The compressive strength was determined according to DIN EN ISO 844. For the test changed samples geometries were used. Cylindrical specimens with d = 7 - 10 mm and h = 10 mm were cutted out in extrusion direction. The compression tests were performed on a Zwick Universal testing machine type 1455 (Ulm / Germany) at 25 °C with a constant compression rate of 10 mm /min and an initial load of 1 N.

RESULTS AND DISCUSSION

Thermal and Rheological Analysis of the Materials

A fast crystallization helps to stabilize the foam structure during foaming. The melt strength is higher with the beginning of the crystallization, so that cell collapse can be avoided. Additionally, changes in crystallinity or in the crystallization kinetic are an indication for the presence of branched chains. The thermal properties like the glass transition temperature (T_g), cold crystallization temperature (T_{cc}) or the melting point (T_m) were described in a previous work [2]. PLA crystallizes slowly, so that crystallization takes only part in the heating curve of the DSC experiment. This effect is known as the so called cold crystallization. To analyze the effect of the modification on the crystallization the focus in this work was on the cooling process and crystallization. Therefore, tests with a cooling rate of 2 $K \cdot min^{-1}$ were performed. The results are shown in Figure 1.

With help of the organic peroxide used in PLA OP it was possible to influence the crystallization process significantly. PLA OP is the only material that crystallizes during the cooling process. The organic peroxide introduces branched chains in the PLA, which have a crystal nucleation effect. The positive effect of branches on crystallization kinetic in polylactide was also shown by Corre et al. and Dorgan et al. [4,1]. Branches enhance the nucleation effect caused by heterogeneities. At the same time the crystallinity is lower compared to the linear counterpart, as described in [3], which is also an indication of branched chains. Detailed investigations in our previous work on the rheological characterization with a plate-plate rheometer in shear modus [3,5,6] showed that the modified chain structure can affect the rheological properties. This study focuses on the investigation of the elongational properties and the determination of the melt strength and extensibility, as high melt strength and / or melt extensibility are crucial to build a stable foam structure.

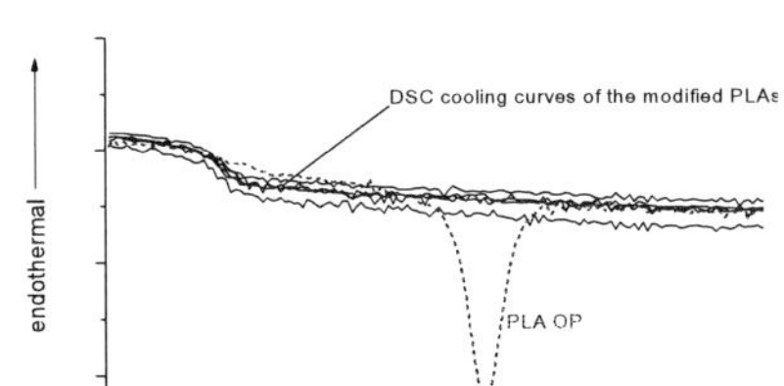

FIGURE 1. DSC cooling thermograms

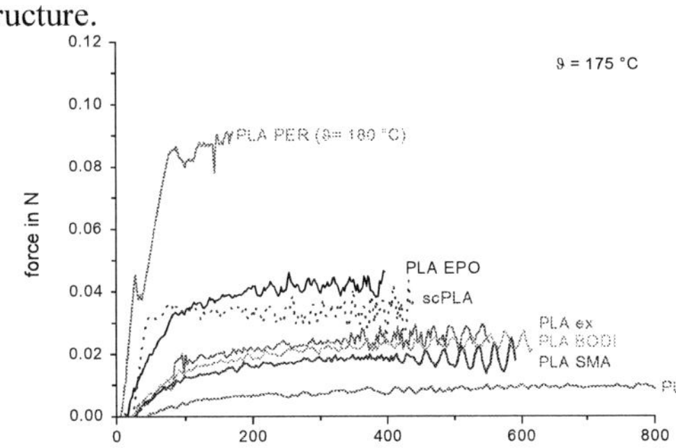

FIGURE 2. Rheotens curves - melt strength and melt extensibility

Figure 2 shows the results of the Rheotens measurement in which the force F has been plotted against the draw velocity v.The maximum force is characteristic for the melt strength. The melt strength values of the modified polylactides PLA SMA (0.02 N) and PLA BODI (0.025 N) are slightly lower than that of the unmodified PLA ex (0.03 N). By contrast, the melt strength of PLA ICDI is with 0.01 N much lower, but has higher melt extensibility. PLA EPO and scPLA are the two crosslinked systems that show a higher melt strength and a lower extensibility than PLA ex. PLA EPO and scPLA show a melt strength about 0.04 N and a strand rupture at a draw velocity of 400 mm · s^{-1}. The crosslinks in these two materials are responsible for the decreased extensibility of the materials. PLA OP occupy a special position. The melt strength is much higher compared to the other materials, which is due to the significantly higher molecular weight, broader molecular weight distribution and the branched topology of this polymer [3]. On the other side this material also shows significant decreased melt extensibility. PLA OP shows a melt strength of 0.09 N, which is three times higher than PLA ex. PLA OP sows, the highest value in melt strength but at the same time a strand rupture at 170 mm · s^{-1} corresponding to the low extensibility. The early strand rupture was first not expected. The molecular weight alone cannot be the reason for this behavior. But in case of branched chains, a better melt extensibility was expected, as already described in [7]. A reason for a reduced extensibility is the accelerated crystallization of PLA OP, which results in a more stable material, in turn leading to higher melt strength but also to lower melt extensibility. In case of extrusion foaming it is expected that PLA OP will show the best foam properties due to the enhanced crystallization behavior and high melt strength of this material. The positive properties can be related to the high molecular weight and polydispersity and also to the expected presence of branched chains.

Foaming and Foam Characterization

The neat and the modified PLA were successfully foamed by foam extrusion. Closed cells with thin cell walls were obtained (see Figure 3).

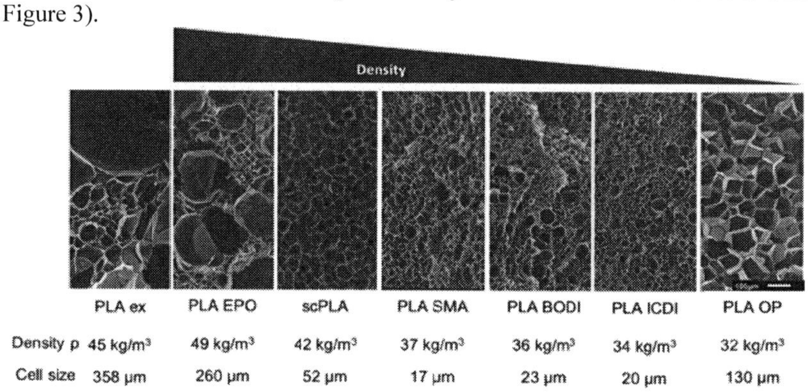

FIGURE 3: SEM images, foam densities and cell sizes of neat and modified PLA

The neat PLA and the PLA EPO modified with the multifunctional epoxide show a bimodal cell size distribution. This was at first not expected as PLA EPO shows relatively high melt strength (Figure 2). But, shear measurements in previous work [3,5,6] showed, that the PLA chains are partially crosslinked. The chemical crosslinks can hinder the flow and inhibits cell growth during foaming. The heterogeneous cell size distribution can also be attributed to a poorer solubility of the blowing agent and a smaller diffusion coefficient caused by the modification, which reduces the free volume due to crosslinking [8]. All the other materials show a lower density, a smaller cell size and a more homogeneous cell size distribution compared to PLA ex and PLA EPO. The lowest density was found in PLA OP ($\rho = 32$ kg/m^3, modified with the organic peroxide), which is in the density range of a standard XPS foam ($\rho = 25\text{-}45$ kg/m^3). Additionally, this foam shows the highest homogeneity without cell rupture. PLA OP tends to form larger cells compared to PLA SMA, PLA BODI, PLA ICDI and scPLA, which is due to its high melt strength. In the next step the compressive strength σ of the modified PLA OP foam was compared to PLA EPO (modified with a commercial epoxide) as a benchmark and the unmodified PLA as a reference. As we can see in Figure 4 most of the modified PLA foams have a higher compression strength at 10 % deformation compared to the unmodified PLA ($\sigma_{10\%} = 0.10$ MPa). Compared to PLA EPO the PLA OP foams shows higher compression strength at 10 % deformation (0.17 MPa vs. 0.23 MPa) and a higher ability to absorb energy till 60 % deformation.

Also, PLA OP revealed the highest expansion during foam extrusion experiments. The improved compression behavior is caused by a monomodal cell size distribution with a higher number of smaller cells and lesser cell rupture. The compression strength is comparable to a standard XPS foam with $\sigma_{10\%}$ in the region of 0.15 – 0.70 MPa [9].

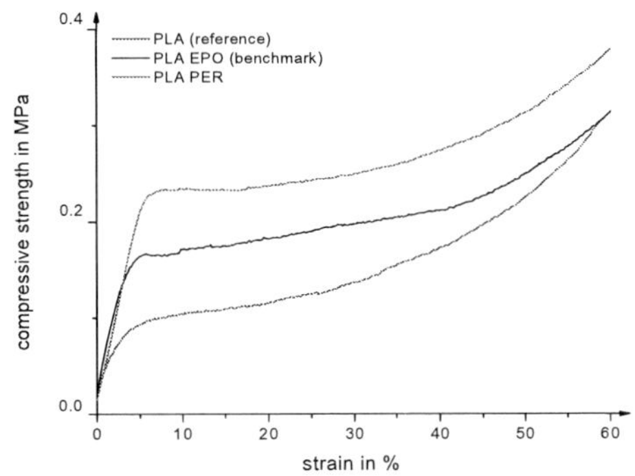

FIGURE 4: Compression test results and cell morphology of the extrusion foams

CONCLUSION

By the use of suitable modifiers, the topology of PLA chains can be changed leading to higher melt strength and improved crystallization properties, whereby the foaming behavior could be improved, leading to lower densities. The branching with the organic Peroxid (DCUP) in PLA OP accelerates the crystallization significantly, whereby the foam structure is additionally stabilized. The foaming of modified PLA leads to closed-cell foams with very thin cell walls, reduced densities and an improved compression behavior comparable with XPS foams. Thereby, the organic peroxide is more effective than the commercially available multifunctional epoxide (Joncryl®). The use of renewable resources (here in form of the biopolymer Polylactide) increases the material efficiency of the produced foams.

AKNOWLEDGEMENT

The findings in this work were publically funded. The authors would like to thank the German Research Foundation (DFG) for funding the project (BO1600/15-1 and AL474/24-1).

REFERENCES

1. J. R. Dorgan, H. Lehermeier, M. Mang, M, Journal of Polymers and the Environment **8**, 1, 1-9 (2000).
2. M. Nofar and C.B. Park, Progress in Polymer Science **39**, 1721-1741 (2014).
3. S. Göttermann, S. Weinmann, C. Bonten, T. Standau, V. Altstädt, 24. Stuttgarter Kunststoffkolloquium, Stuttgart, ISBN: 978-3-00-048562-6 (2015).
4. Y.-M. Corre, A. Maazouz, J. Reignier, et al., Journal of Polymer Engineering and Science **54**, 3, 610-625 (2014).
5. S. Göttermann, S. Weinmann, C. Bonten, T. Standau, V. Altstädt, AIP Conference Proceedings, **1779**, 060001, 1-5 (2016).
6. S. Göttermann, S. Weinmann, C. Bonten, T. Standau, V. Altstädt, Kunststoffe International, **01-02**, 70-73 (2016).
7. J. Stange, „Einfluss rheologischer Eigenschaften auf das Schäumverhalten von Polypropylen unterschiedlicher molekularer Struktur“, Ph.D. thesis, University of Erlangen-Nuremberg, 2006.
8. T. Standau, S. Göttermann, S. Weinmann, C. Bonten, V. Altstädt, Journal of Cellular Plastics, Special Issue **(0)**, doi: 10.1177/0021955X16670588 (2016).
9. H. Winterling and N. Sonntag, Kunststoffe inernational **10**, 18-21 (2011).

Water assisted extrusion of PA/South African Bentonite nanocomposites: a process cost saving perspective

Ojijo Vincent Omondi [1]*, Sinha Ray Suprakas [1], Scriba Manfred [1]

[1] CSIR - GAUTENG - SouthAfrica

As opposed to the in-situ polymerization technique, downstream compounding of polyamides (PA) and nanoclays to make polymer nanocomposites with enhanced properties poses some challenges. The clays need to be modified with surfactants, which may migrate into foodstuff, in the case of food-contact film applications. The costs of such modifications in turn, may result in commercially unfeasible products. Water assisted extrusion has been employed as a mitigation strategy. In this study, water-assisted extrusion of polyamide and unmodified South African Bentonite was carried out in a co-rotating twin screw extruder with L/D of 40. Small angle Xray scattering (SAXS) studies revealed better clay dispersion in the products processed through water-assisted extrusion of PA with unmodified clays (process A) when compared to the normal extrusion of PA with organo-modified clays (process B). Furthermore, material properties (e.g. thermal stability, clarity etc.) and performance (e.g. oxygen barrier) was marginally better in the case of process A. For the first time, cost saving benefits of water-assisted processing of PA/South African Bentonite nanocomposites is illustrated. For a PA/clay masterbatch with 30% inorganic content, process A resulted in a typical material cost reduction of ~32% over process B. In conclusion, with such cost reductions and the elimination of concern for leaching of surfactants, process A offers a more suitable business case and could be employed to fabricate films for barrier applications. Keywords: Water assisted extrusion, Polyamides, South African Bentonite, Process cost savings

Cardanol derivatives as effective plasticizers for polylactic acid

Greco Antonio (1)*, Ferrari Francesca (1)

(1) University of Salento - Italy - Italy

This paper is aimed to study the plasticizing effectiveness towards poly(lactic acid) (PLA), of innovative and bio-based plasticizers obtained by chemical modification of cardanol. The cardanol derivatives used as PLA plasticizers are raw cardanol, cardanol acetate (CA), and epoxidated cardanol acetate (ECA). The plasticizing performances of cardanol derivatives were compared to those of poly(ethylene glycol) (PEG 400). Plasticized PLA samples were obtained at different amounts of plasticizers, ranging between 10 and 30%. DSC analysis results indicate that addition of cardanol, cardanol acetate and PEG400 involves an initial decrease of the glass transition of the polymer, which however is not substantially modified upon further addition of plasticizer; in contrast, the glass transition continuously decreases upon addition of ECA, which indicates its excellent compatibility with PLA. As a consequence, the glass transition of PLA plasticized by ECA can be reduced well below room temperature. In addition, the presence of the plasticizer increases the crystallization kinetics of PLA, leading to a semi-crystalline structure also at very high cooling rates. Mechanical characterization showed that addition of PEG400, cardanol and CA involves a decrease of the tensile modulus, which is in agreement with the concept of plasticization, and reduction of the glass transition temperature. On the other hand, the increased crystallinity of the material is responsible for a reduction of the elongation at break; in every case, and for any amount of plasticizer, the elongation at break of plasticized PLA is lower than that of neat PLA. Therefore, these materials cannot act as efficient plasticizers for PLA. In contrast, ECA plasticized PLA shows a continuous reduction of the modulus (from 1800 to 3 MPa) and increase of the elongation at break (from 3 to 400%) indicating the excellent plasticizing effectiveness of ECA.

Exfoliation of layered zirconium phosphate nanoplatelets by melt compounding

Chen Lei (1)*, Sun Dazhi (1), Li Jin (2)

(1) Southern University of Science and Technology - Guangdong - China, (2) Fudan University - Shanghai - China

A melt compounding method to achieve the exfoliation of layered zirconium phosphate (ZrP) nanoplatelets was presented in this work. ZrP nanoplatelets were first intercalated by diglycolamine (DGA) to increase the interlayer distance and functionality. In the next melt compounding process, the cyclic anhydrides of maleic anhydride grafted polyolefin elastomers (POE-g-MA) reacted with the hydroxyls in DGA-ZrP. Subsequently, the long polyolefin chains of POE-g-MA intercalated into the interlayer of ZrP smoothly and exfoliated the nanoplatelets successfully. ZrP nanoplatelets with various sizes were fully exfoliated in POE-g-MA matrices. The modulus of elasticity of POE-g-MA was improved by the incorporation of ZrP nanoplatelets and further increased if the nanoplatelets are exfoliated. The ductility of ZrP/POE-g-MA compounds also became better, if ZrP nanoplatelets were first intercalated by DGA and then exfoliated, especially for the smaller nanoplatelets. Our work provides a general method to prepare polymer nanocomposites containing exfoliated ZrP nanoplatelets in a large scale.

Interaction between Coolant, Polyamide 6.6, Glass Fibers and Additives

J. Heyn[a)], S. Kliem and C. Bonten

IKT, Institut für Kunststofftechnik, University of Stuttgart, Pfaffenwaldring 32, 70569 Stuttgart, Germany

[a)] Corresponding author: Johannes.Heyn@ikt.uni-stuttgart.de

Abstract. Glass fiber reinforced polyamide 6.6 (PA66) has a high service temperature as well as good chemical resistance. Because of its low raw material price, it is a common engineering material for under the hood applications involving coolants. In order to keep up with the developments in the automotive sector, the material has to be improved as well. In the field of polymer synthesis, fiber sizing and additives, there is a high potential for further developments of new materials with the aim of a high material efficiency. This will only be successful, if the complete system is taken into account. The influencing coefficients, such as the PA66 type, the fiber sizing or the additives, will be discussed in this study. The main influencing factors will be evaluated. Based on this study, the main factor is the fiber sizing, while the different additive masterbatches or PA66 types have minor influences.

INTRODUCTION

The properties of a material are defined by its recipe components. Therefore, the choice of polymer determines the basic properties such as its chemical and thermal resistance. The mechanical properties of the neat polymer can be strongly influenced by the use of fibers or soft phases. Likewise, specific additives are able to affect the coloring, the long-term resistance or the processability. Thus, the more complex the material system is, the more complicated are the interactions. As a result, the goal is to achieve synergistic interactions and to avoid antagonistic interactions in order to develop new materials. [1, 2]

Each component can be obtained from different sources and therefore shows different properties. PA66 can be encountered with different molecular masses or molecular mass distributions, and with a different concentration of amino end groups. Last but not least, there is a basic stabilization of every polymer that every raw material manufacturer must perform on their own.

Looking at glass fibers reveals that E-glass is used in most cases since it has a defined composition. However, the main influence of the glass fibers in the interaction with the polymer matrix the sizing. The exact composition, and thereby the mode of action of the sizing, is the company-specific know-how of the glass fiber manufacturer.

Considering the variety of possible additives, such as antioxidants, light stabilizers, nucleating agents, processing aids or colorants, and keeping in mind that the different additivations can be realized in different ways, the complexity of the task becomes apparent. [2, 3]

This is aggravated by the fact that a material modification cannot be successful without knowing the actual operation conditions, meaning its thermal, mechanical or chemical conditions. Specifically, media absorption in the matrix can cause different fracture mechanics that lead to an early failure. This also depends on the type of the medium [4].

STATE OF THE ART

PA66 types with an increased concentration of amino end groups tend to have an increased hydrolytic resistance than those with a lower concentration [5].

The fiber sizing and the associated fiber-matrix adhesion can also have a strong influence on the solvolytic resistance. In order to ensure a good linkage of the fiber to the PA66 matrix, a high polarity of the sizing is needed. However, this has a negative impact on the coolant interactions, since the coolant is also polar. A sizing of low polarity has a low affinity to the coolant, but also a low fiber-matrix adhesion.

A common method to stabilize polymers against external influences is the use of additives or blending of the matrix. Blending with non-polar polymers decreases the polarity of the material and contributes to a significant increase of the hydrolytic resistance by reducing the coolant absorption. The use of processing agents like montan wax can also slow down the diffusion rate of the coolant into the PA66 matrix and reduce the saturation point, especially for high-molecular PA66 types.

The addition of small amounts of aromatic or sterically hindered carbodiimides or epoxides complicates its ability to be processed, but it improves the coolant resistance by chain extension or branching of the polymer chains.

The exact procedure of the chemical interaction of a coolant mixture with glass fiber reinforced PA66 at elevated temperature and pressure is hard to describe. The coolant mixture consists of various components whose interactions with the PA66, the glass fiber sizing and the additives deliniate a complex system. Additionally, PA66 is prone to solvolysis. When exceeding the equilibrium state in the presence of coolant at appropriate temperatures, the covalent bonds of the PA66 undergo solvolytic reactions. The molecular degradation of the PA66 matrix and possible interactions between the coolant and the sizing influence the materials' properties. However, specific additives and the use of adjusted fiber sizings can delay the material damage. Nevertheless, over the long term, there are limits to the coolant resistance of glass fiber reinforced PA66 under the given conditions.

There are various further ways to increase the coolant resistance of glass fiber reinforced PA66 that are not listed here in detail. The aim of all these measures is to delay the temporal course of the solvolytic degradation as much as possible.

In this project, the effect of single and combined approaches of hydrolysis stabilization on the coolant resistance of glass fiber reinforced PA66 will be investigated. The main influencing factors will be identified such that in the future, the focus will be solely on the promising approaches for developing new materials.

EXPERIMENTAL DESIGN

Materials

As a reference, a commercially available, hydrolysis-stabilized PA66 type with 30 wt.-% glass fibers has been used. The production of the compounds was based on two PA66 types (PA1, PA2) with different amino end group concentrations. Four different types of chopped glass fibers (GF1 to -6) with a length of 4 mm and a diameter of 10 µm and specially designed glass fiber sizings have been used. In addition, commercially available additive mixtures, as well as personally created mixtures, have been tested.

Compounds with different compositions, glass fiber content between 0 wt.-% and 30 wt.-% and varying additive mixtures have been produced on a twin-screw extruder ZSK 26 from Coperion, Stuttgart, Germany. The PA66 had been dried for four hours at 80 °C in a vacuum furnace VD115, Binder GmbH, Tuttlingen, Germany, prior to processing. The compounds were produced with a throughput of 8 kg/h, a screw speed of 135 1/min and a side feeder speed of 300 1/min.

After compounding, the test specimens of type 1, according to the standard DIN EN ISO 179-1, were injection molded on an Allrounder 520S, Arburg, Loßburg, Germany. Mold temperature was 80 °C, while the temperature profile was set from 275 °C to 320 °C.

Storage Conditions

The compounds were stored at 135 °C for 504 h and for 1,000 h in an autoclave. The presented results are mainly focused on the compounds kept for more than 1,000 h in storage. The coolant mixture consisted of the commercially available coolant G13 and water in a 1:1 mixing ratio. All measured values refer to the dry initial state (0 h).

CHARACTERIZATION METHODS

Ash content: The determination of the ash content follows the standard DIN EN ISO 3451-4, such that a part of the injection molded test specimen is taken from the middle area of the sample. The dried specimens are burned at 600 °C in an oven for 6 h. The mass content of the fibers is determined by the mass of the ash content to the initial weight of the sample.

Charpy impact test: The determination of the impact strength after Charpy a_{cU} is handled in standard atmosphere at 23 °C and 50 % relative humidity. The differently stored test specimens are tested on the short side according to the DIN EN ISO 179-1 standard. For the determination of the characteristic values, the impact pendulum 5102 with a 2.0 J and 4.0 J impact pendulum of Zwick, Ulm, Germany, has been used.

Three point flexural test: The test specimens in different storage states are tested according to DIN EN ISO 178 in standard atmosphere at 23 °C and 50 % relative humidity on the universal testing machine UPM 1455 Zwick, Ulm, Germany. The testing speed was uniformly 2 mm/min.

Optical examination: In order to investigate the interaction of different additive mixtures on the fiber-matrix adhesion, the test specimens are split into two parts after storage in liquid nitrogen. Images of the fractured surface are taken with a scanning electron microscope (SEM) JSM 6300 F, Jeol, Eching near München, Germany.

RESULTS AND DISCUSSION

Ash content: On the basis of the results of the ash content of the examined compounds, the glass fiber content of 30 wt.-% ±0.5 wt.-% can be assumed to be uniform. The average glass fiber length distribution shows a broad spectrum with an average fiber length of 0.244 µm. Different fiber sizings or additive mixtures, which also contain processing aids, only have a minor impact on the glass fiber length distribution.

Optical examination: As an example, Fig. 1 shows the influence of different additive masterbatches on the fiber-matrix adhesion at the dry initial state of a selective glass fiber.

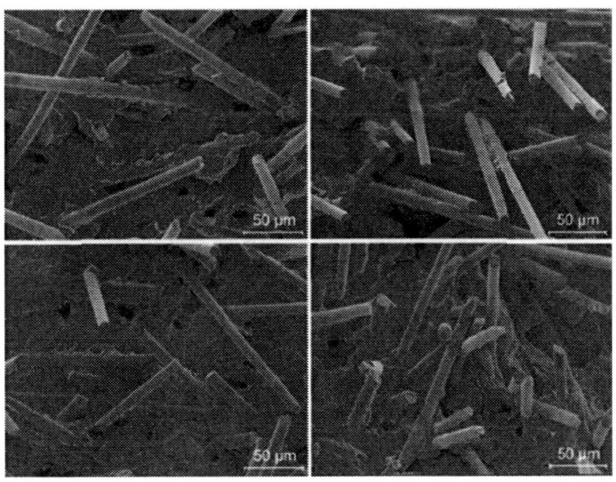

FIGURE 1. Fiber-matrix adhesion, uniform glass fiber, different additive mixtures

All four SEM images show that the glass fibers are wetted by the PA66 matrix. Therefore, uniform starting conditions for each glass fiber reinforced PA66 are assumed based on the results of the ash content and the SEM images regardless of the used additive masterbatch.

Additivation of the pure PA66 matrix without glass fibers: In order to evaluate the effect of different additive mixtures on the coolant resistance of PA66 without the influence of glass fibers, compounds without glass fibers are produced. The stored test specimens showed fatal damages after a storage time of 1,000 h at 135 °C in the coolant mixture regardless of the additivation. Therefore, no mechanical testing was possible. The discoloration of the initially non-colored compounds is due to the applied coolant. Characterization after 504 h of storage was possible but did not show significant differences.

As glass fiber reinforced PA66 is a common material for usage in coolant-contacting parts and is designed for such storage conditions, the influence of the glass fibers on the coolant resistance has to be significant.

Glass fiber reinforced PA66 without additivation: In order to investigate the influence of different glass fibers and the related glass fiber sizings, glass fiber reinforced PA66 compounds without additivation are submitted to coolant storage. The focus of the mechanical characteristic values is not on the initial dry state, but on the characteristic values after coolant storage for 1,000 h.

Figure 2 shows the results of the three point flexural test and the Charpy impact test before and after 1,000 h of coolant storage. I should be noted that the results of the initial state represent the flexural stress at conventional deflection σ_{fc}, whereas the results after 1,000 h of storage show the maximum flexural stress σ_{fm}. The red lines show the reference values (RV) after 0 h and 1,000 h of a commercially available, hydrolysis-stabilized glass fiber reinforced PA66 type.

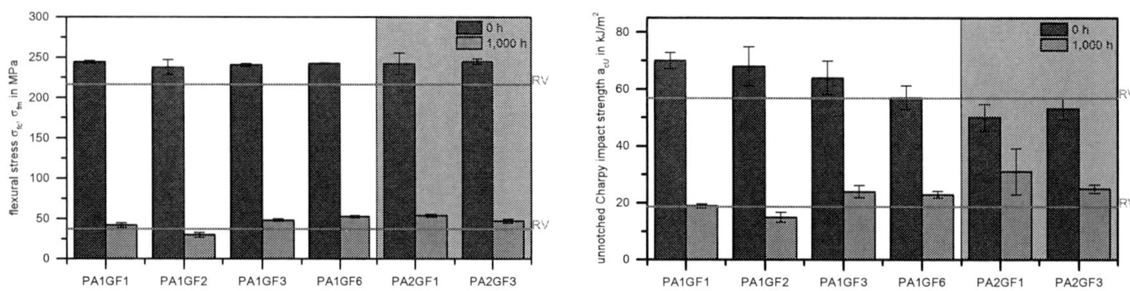

a) b)

FIGURE 2: Comparison of a) the flexural stress and b) the unnotched Charpy impact strength after 0 h and 1,000 h of coolant storage

It is noticeable that the values in the initial state are in part significantly higher than that of the reference material. However, this is not the focus of this investigation. More importantly, the unstabilized PA66 types achieve a comparable or even higher level as the hydrolysis-stabilized reference material after 1,000 h of coolant storage.

Furthermore, the PA66 type with increased amino end group concentration (PA2) reaches higher values than the standard PA66 type (PA1). Additionally, influences of the used glass fibers on the mechanical properties are noticeable. These can vary with the different PA66 types.

These paradoxical results that unstabilized compounds deliver comparable or even better results than the stabilized reference material are guaranteed by the repetition of the measurements and are elaborated in the next paragraph.

Glass fiber reinforced PA66 with additivation: In the following, additivated glass fiber reinforced compounds after 1,000 h of storage are presented in order to obtain a direct comparison to the characteristic values, including additivation.

The blue bars in Fig. 3 present the reference values of the unmodified glass fiber reinforced PA66 compounds. The cross-hatched bars show the results of compounds to which a black colorant (C) has been added (no pigment). The value of the reference material is again depicted by a red line. Two PA66 types (PA1, PA2) are compared with each other.

Regardless of the used glass fiber or the additivation, no significant difference between the two PA66 types is noticeable.

Again, the unstabilized compounds (blue) achieve the highest characteristic values. Adding a black colorant diminishes the initial value significantly. When additive masterbatches are added on top of the colorant, the mechanical properties after 1,000 h of coolant storage are slightly improved. However, the initial value of the unstabilized compounds cannot be achieved again by using a specific additivation.

Thus, when developing new materials, the polymer matrix, glass fiber sizing and the choice of additives have to be taken into account, but in addition, the choice of an appropriate colorant also has to be considered. Even if the dry initial values do not show major modifications differences, the variation in values after storage can be significant.

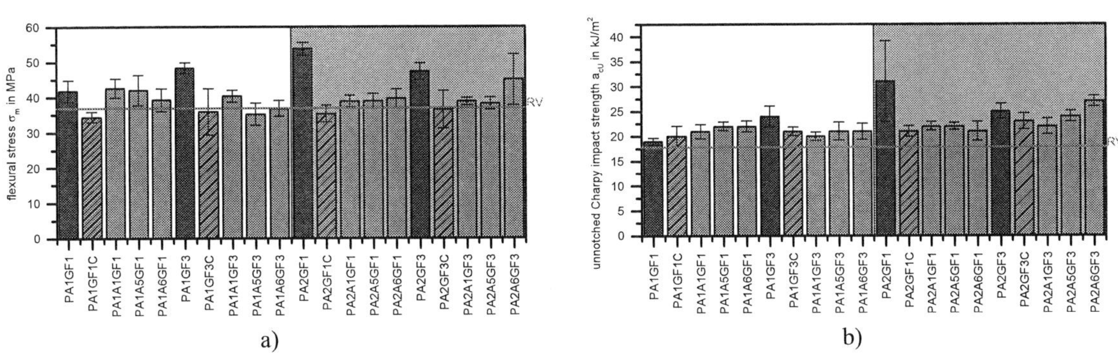

FIGURE 3: Comparison of a) the flexural stress and b) the unnotched Charpy impact strength after 1,000 h of storage for different compounds

CONCLUSION

An additivation of the pure PA66 matrix for an improvement of the coolant resistance was not effective. The influence of a PA66 type with increased amino end group concentration brings a slight improvement to the coolant resistance. However, the choice of the PA66 type plays only a minor role. The incorporation of glass fibers into the PA66 matrix provides important insights into the improvement of the coolant resistance. Based on a promising combination hereby, fine-tuning by means of additives is worth it. Attention should be paid on the used colorants, as those may have a strong negative influence on the mechanical values after coolant storage.

Ideally, the new hydrolysis-stabilized material comes from one source, a manufacturer which can accurately coordinate the polymer synthesis, the glass fiber sizing and the additive mixtures. If the three components have different sources, the coordination is still highly important, but due to various unknown factors, it would be far more problematic. Hereby, the focus should be on the choice of the optimal glass fibers and their sizing. Different additive mixtures only had a small impact on the coolant resistance of an existing PA66-glass fiber system. Therefore, it could be an important last step to consider when trying to reach the maximum optimization. On the other hand, if the PA66-glass fiber combination is inappropriate, even a promising additivation cannot compensate for it.

Further investigations that are targeted especially for more praxis-oriented storage conditions, temperatures and pressures, as well as component geometries for coolant-contacting parts, are already initiated. The aim is to develop a fast and reliable forecasting method, which can compare the different storage conditions and provide reliable information for the choice of material.

ACKNOWLEDGEMENTS

This research and development project has been supported by funds from the German Federal Ministry for Economy and Energy within the funding program AiF-ZIM.

REFERENCES

[1] C. BONTEN, *Kunststofftechnik: Einführung und Grundlagen*, 2nd edition (Hanser, Munich, 2016)

[2] H. ZWEIFEL, R. D. MAIER and M. SCHILLER, *Plastic Additives Handbook*, 6th edition (Hanser, Munich, 2009)

[3] G. W. BECKER, D. BRAUN, L. BOTTENBRUCH and R. BINSACK, *Kunststoffhandbuch: Polyamide – Technische Thermoplaste. Band 3*, (Hanser, Munich 1998)

[4] J. HEYN and C. BONTEN, *Media influence on the fracture mechanical behavior of polyamide.* In: AIP Conference Proceedings 1779, 070004, 2016

[5] P. EIBECK, J. ENGELMANN, R. NEUHAUS and A. BERNNAT, *EP application 2 057 223 B1, 19.01.2011*

[6] G. W. EHRENSTEIN, S. PONGRATZ, *Beständigkeit von Kunststoffen.* 1st edition. München: Hanser, 2007

Glycerol as plasticizer in PVC

Righetti Ronald Assumpção (1), Lazareti Clodoaldo (1), Bonse Baltus Cornelius (2)*

(1) Faculdade SENAI de Tecnologia Ambiental - São Bernardo do Campo SP - Brazil, (2) Centro Universitário FEI - São Bernardo do Campo SP - Brazil

With the increase in biodiesel production, which generates about 10 wt% glycerol as the main byproduct, numerous added-value applications for glycerol have been investigated. One such application is its use as a plasticizer in polymers. In the current investigation, glycerol was used to produce flexible polyvinyl chloride (PVC) which was compared with standard flexible PVC obtained with plasticizer dioctyl phthalate (DOP). Flexible PVC sheets containing 50 phr of each plasticizer, in addition to stearic acid and heat stabilizer, were manufactured by means of a conventional rolling mill. Test specimens were subjected to density, hardness, tear and tensile testing. Exudation tests and migration stability tests in water, ethanol, and gasoline were also carried out. Density, hardness, tensile modulus and tear strength values of the flexible PVC containing glycerol were similar to the standard flexible PVC containing DOP. However, tensile strength and strain at break of the former were about 50% higher than the latter.

Mechanical properties of plasticized poly (lactic acid) with epoxydized soybean oil

Khaghanipour Maryam [1], Badeli Ali [1], Khodabakhshi Khosrow [1]*

[1] Iran Polymer and Petrochemical Institute - Tehran - Iran

Poly (lactic acid) has a poor mechanical properties such as low elongation at break. In this paper, Poly (lactic acid) with epoxydized soybean oil were melted and mixed in Brabender internal mixer for 9 minutes at 190ºC to produce a ductile poly (lactic acid) with more elongation at break than poly (lactic acid). All functional groups of neat poly (lactic acid) and ductile poly (lactic acid) were investigated by FTIR analyzer. Mechanical properties such as tensile stress, elongation at break and young modulus of neat poly (lactic acid) and ductile poly (lactic acid) were investigated by Tensile strength test.

The Influence of Mixing Parameters on the Properties of Micron-sized Blast Furnace Slag Filled PP Compounds

Mostafa Abdelhamid (1)*, Pacher Gernot (2), Lucyshyn Thomas (3), Holzer Clemens (3), Flachberger Helmut (4), Frtiz Bertram (5)

(1) Polymer Competence Center Leoben GmbH - Styria - Austria, (2) Polymer Competence Center Leoben GmbH - Styria - Austria, (3) Chair of Polymer Processing, Montanuniversitaet Leoben - Styria - Austria, (4) Chair of Mineral Processing, Montanuniversitaet Leoben - Styria - Austria, (5) voestalpine Stahl GmbH - Upper Austria - Austria

Blast furnace slag (BFS) is majorly composed of acidic and basic oxides and shows a complex, multiphase structure, which is expected to promote promising functionality for BFS for semi-crystalline thermoplastics such as PP. In the current study, the influence of variation of two internal mixing process (IM) parameters, mixing speed and duration, upon the rheological, thermal and mechanical properties of a PP-BFS compound is investigated. The rotors' torque is observed to be directly proportional to the mixing speed, while the amount of consumed specific energy (E) during the mixing process is witnessed to be a function of the combined mixing speed and duration. Rheological findings showed that the complex shear viscosity and -moduli were modestly affected by the variation of the processing parameters. Thermal analyses revealed a modest influence of the processing parameters upon the degree of crystallinity (DOC) as well as the crystallization and melting temperatures, Tc and Tm. The neat PP recorded the lowest DOC and E levels, whereas the setting with 120 rpm screw speed and 20 min mixing duration ('120,20') recorded the highest ones. For the mechanical properties, the tensile modulus of the filled PP compounds showed a clear dependence on the variation of process parameters. The yield strength decreased notably as the process parameters progressed beyond the '60,10' condition. The ultimate strength levels were comparable for all filled conditions, while the highest ductility was achieved with the '60, 20' and '90,10' conditions.

Melt processing of cellulose nanofibrils and amphiphilic copolymer based bio-composites

Venkatesh Abhijit (1)*, Thunberg Johannes (2), Boldizar Antal (3)

(1) Chalmers Institute of Technology, CTH - Gothenburg - Sweden, (2) Dr. - Gothenburg - Sweden, (3) Professor - Gothenburg - Sweden

The development of sustainable reinforcing material in plastics coupled with the novel developments in cellulose pulp processing have made cellulose nanofibrils (CNF) commercially available. These highly fibrillated cellulose pulps, with fibrillary dimensions in the nanoscale, have high aspect ratio and excellent potential to be used as reinforcements in thermoplastic composites. However, problems in the form of insufficient reinforcement dispersion, fibre-matrix incompatibility and poor fibre-matrix adhesion are commonly encountered during the melt processing of cellulose-thermoplastic composites and this leads to poor mechanical properties. In this work, continuous extrusion of cellulose nanofibrils (CNF) with thermoplastic is made. The composites were prepared through a combination of water-assisted mixing, master-batch approach and extrusion. The CNF and poly(ethylene acrylic acid) (EAA), which formed the master-batch, were compounded with a modified polyethylene in a single-screw extruder to obtain the final composite. The mechanical properties of the obtained composites were compared to dry extruded samples and compression moulded samples of CNF- EAA and pulp-EAA.

Synergistic effect of fire retardant additives on fire behavior of cross-linkable polyolefin compounds toward intumescent compounds

Raboonpour Soraya [1]*, *Ahmadi Shervin* [2], *Morshedian Jalil* [3]

[1] M.Sc. Student, Iran Polymer and Petrochemical Institute - Tehran - Iran, [2] Assistant Professor, Iran Polymer and Petrochemical Institute - Tehran - Iran, [3] Professor, Iran Polymer and Petrochemical Institute - Tehran - Iran

In this study, the effect of different quantities of ammonium polyphosphate (APP) and melamine cyanurate (MC) as flame retardants, on the flammability of polyolefin compounds was studied. The effect of using polystyrene as a carbon source companion with flame retardants was also evaluated. The Flammability parameter results were obtained by cone calorimeter test at a uniform external irradiance of 50 kW/m2, and by combustion tests (limited oxygen index (LOI), UL-94). Test results showed that by adding polystyrene to formulations, the carbon dioxide production is decreased, and flame retardancy properties are improved via the synergistic effect of APP and MC.

Influence of the screw configuration in a co-rotating twin screw extruder on the rheological and mechanical properties of a talc-reinforced polypropylene

Luger Hans-Jürgen (1)*, Unger Thomas (2), Mayrbäurl Erwin (3), Miethlinger Jürgen (4)

(1) Insitute for Polymerextrusion and Compounding, JKU Linz - OÖ - Austria, (2) Leistritz Extrusionstechnik GmbH - Bayern - Germany, (3) Poloplast GmbH & Co KG - OÖ - Austria, (4) Insitute for Polymer Extrusion and Compounding, JKU Linz - OÖ - Austria

The aim of the present work was to study the influence of the screw configuration in the mixing zone of a co-rotating twin screw extruder on the material properties of a polypropylene talc compound. Highly-filled PP-talc compounds were prepared by means of an industrial Leistritz ZSE27MAXX (48 L/D) twin-screw extruder. Mixing in a twin-screw extruder involves a considerable amount of elongational flow, which is known to be favorable for dispersion of agglomerates. Therefore, high pitch conveying elements as well as eccentric elements were compared to standard kneading blocks. In addition, transition elements from two to triple-flighted screw profiles were used. Subsequently profound material testing was conducted to assess the compound quality. This involved determination of the Filter Pressure Value and the mechanical properties, such as tensile modulus, tensile strength and impact strength. Shear viscosity was measured with a plate-plate-rheometer in order to assess material degradation. Moreover, elongational viscosity of the different compounds was obtained by a High Pressure Capillary Rheometer (HPCR) via Cogswell analysis. In addition, standard quality control tools such as MFR and OIT measurements were performed. With the high pitch conveying elements appealing compound properties could be achieved compare to the standard kneading blocks. In order to achieve high performance compound properties it is necessary to keep dispersion and material degradation in balance, which was achieved by different eccentric screw geometries. While the transition elements show a lower degree of dispersion, indicated by the Filter Pressure Value, they led to promising results in terms of material degradation.

PBAT/Thermoplastic Starch Blend:"Concentration of OTPS & comparison with TPS"

CAN BUSE NUR (1)*, OZKOC GURALP (1)

(1) Kocaeli University - Kocaeli - Turkey

The growth of biodegradable plastics, which are safely and efficiently eliminated in solid or in compost, has become gradually important to reduce the effect on the environment. One of the typical biodegradable polymers that show no contrary effects on the environment is poly(butylene adipate-co-terephatalate) (PBAT). However, the usage of PBAT is limited because of its high price [1]. Therefor many researches focus on to reduce its price by blending PBAT with starch, a natural polymer. Thermoplastic starch (TPS) is obtained in presence of plasticizers to break the strong hydrogen bonding between the starches. However, the difference in hydrophilicity between starch and PBAT makes this blend incompatible with poor properties [2]. On the other hand, the oxidized starch obtained from the reaction of starch in presence of oxidizing agents may help the hydrophilicity between two components [3]. The aim of this study is to investigate the comparison of the effect of oxidized starch on the properties of TPS/PBAT blends in the presence of styrene-acrylic multi-epoxidized polymeric compatibilizer. Blends were obtained by laboratory twin-screw Xplore microcompounder and were characterized by tensile, impact and thermal tests. The rheological and morphological properties were also investigated. The results showed that OTPS/PBAT blends have better effects on the impact strength and elongation at break. According to thermal and rheological results, OTPS yielded comparable results with TPS. [1] J. B. Olivato, M. V. E. Grossmann, F. Yamashita, D. Eiras, L. A. Pessan, Carbohydrate Polymer, 87 (2012), 2614-2618. [2] A.R. Rahmat, W.A.W.A. Rahman, L.T. Sin, A.A. Yusuf, Material Science and Engineering C 29, 2370-2377 (2009). [3] P. Forssell, A. Hamunen, K. Autio, T. Suortti, K. Poutanen, Starch/Strake, 47 (1995), 371-377.

Reactive Extrusion of PA6 – Different Ways to Increase the Viscosity

J. Benz [a)] and C. Bonten

IKT, Institut für Kunststofftechnik, University of Stuttgart, Pfaffenwaldring 32, 70569 Stuttgart, Germany
[a)]Corresponding author: johannes.benz@ikt.uni-stuttgart.de

Abstract. Depending on processing, different requirements are placed on plastic melt. While in injection molding polyamide (PA) with low viscosities is generally processed, higher viscosities are required for extrusion or thermoforming. In order to increase the molar mass and hence the viscosity of PA, the solid-state post-condensation is used, which is however time-consuming and thus expensive. An alternative method is reactive extrusion. In doing so, so-called chain extenders are added during compounding and thus simultaneously to the metering of further additives. In this work, PA6 with different chain extenders is compared. The different chain extenders either connect polymer chains linearly or act as a backbone and polymer chains are grafted onto. The compounding is carried out in a twin-screw extruder. The changes in the viscosity are examined by means of plate-plate rheometer and those of molar mass by means of gel permeation chromatography (GPC). It can be shown that a significant increase in the viscosity due to the chain extension occurs. The measurements of the GPC confirm the increase of the molar mass, whereby even a bimodal distribution of the molar mass occurs during grafting. The reactive extrusion is thus a cost-effective alternative to increase the viscosity of PA6 and thereby to increase its range of application.

INTRODUCTION

Depending on the process, different requirements are imposed on the molecular structure of the plastics. While a low viscosity is advantageous in the injection molding process to ensure the mold filling, a higher viscosity is required for other processing methods. For example, increased melt stiffness is necessary in the extrusion process in order to achieve the desired result in shaping. An even higher viscosity and thus molar mass is required for the production of flat films, especially if they are used in a subsequent process for blow molding or for thermoforming [1]. In addition, an increased molar mass leads to improved stiffness and strength, which can otherwise only be achieved by the addition of reinforcing agents [2].

Low molar mass fractions produce increased flow, lower strength and low rigidity. Higher molar mass components, however, provide more strength and toughness. This is due to the increasing number of entanglements of the higher molar mass polymer chains, whereas shorter polymer chains are more likely to crystallize. Strength and stiffness may also increase. However, the influence on the mechanical properties is only slight from a certain molar mass [3].

Reactive extrusion as well as conventional solid-state post-condensation (SSP) are two ways to increase the molar masses of standard plastics. Polymer chains are covalently linked to one another by short or long chain coupling molecules and thereby the molecular mass and specific properties of the subsequent plastic are adjusted in a targeted manner. One speaks of reactive extrusion, since the coupling reaction is carried out in a twin-screw extruder during the compounding process. This process can also be used for the recycling of plastics which exhibit lower molar masses at the end of the product life cycle due to aging and processing effects [4].

Possible reactions are the coupling of molecules to the reactive amino and carboxyl terminal groups of PA in a twin-screw extruder [5, 6, 7]. Two different methods can be distinguished in chain extension. It is possible to bind several polymer chains with one another by means of coupling molecules (oxazolines, bisapaprolactams, epoxides and certain polymer blends) in a linear chain extension [7].

Oxazoline predominantly attacks the carboxyl end groups of the PA, whereby a doubling of the molar mass can be achieved. At the same time, this means an increase in viscosity by a factor of six [8]. Analogously, bis-caprolactams couple to amino end groups [9, 10]. Epoxides are capable of covalently attaching to the amino as well as the carboxyl end groups and thus to cause a drastic polymer chain extension [11]. In addition, PC-PA blends show potential as a chain extender [12].

A further possibility is to carry out chain extension also by grafting onto a macromolecular backbone. Strictly speaking, this is not a chain extension but a graft copolymer. Depending on the backbone used (frequently styrene-maleic anhydride) very long molecular chains can be produced thereby.

EXPERIMENTAL

The materials used for the investigations are Ultramid B3S from BASF, Leverkusen. A chain extender from Brüggemann, Heilbronn, Bruggolen M1251 (linear) and a SMA of type EF-40 from TOTAL Cray Valley, Exton, USA, are used for the coupling.

After drying the materials the compounding took place on a twin-screw extruder of type ZSK 26 from Coperion, Stuttgart. A screw speed of 100 rpm and a mass throughput of 5 kg/h were selected as parameters. The two chain extenders are already added in the feeding zone with a mass-related percentage of one.

From now on PA6 with linear chain extender is referred to as $PA6_{linear}$ and the backbone is referred to as $PA6_{graft}$. In addition PA6 was extruded as a control material.

The rheological analysis of the PA is carried out with a plate-plate rheometer SR200 from Rheometrics, Munich, with a plate diameter of 250 mm and a gap spacing of 1 mm. Both the dynamic frequency sweep and the dynamic time sweep were used. The temperature of the rheometer was set to be $\vartheta = 260$ °C. Nitrogen is added as a protective gas so that the plastic sample is shielded from the outside air during the test and water absorption is avoided.

In the dynamic frequency sweep, a maximum shear stress of 500 Pa was set. The frequency range is 0.1 to 100 rad/s. For the dynamic time weep the total time of the measurement was 3600 s. The oscillation frequency was $\omega = 1$ rad / s and the shear stress value $\tau = 100$ Pa.

For Gel Permeation Chromatography (GPC) a column set of the brand Olexis from Agilent, Santa Clara, USA, with a length of 950 mm is used. The solution concentration is 1.5 mg/ml at a flow rate of 1 ml/min. As the eluent meta-cresol, a methyl-phenol mixture is used.

RESULTS AND DISCUSSION

The influence of the chain extension on the viscosity can be clearly seen in figure 1. The addition of the chain extenders causes an increase in the viscosity which can be attributed to a higher molar mass. As expected the pure PA6 shows the lowest viscosities. For $PA6_{linear}$ the viscosity values are approximately doubled. The highest values, however, are obtained with $PA6_{graft}$ whereby a 40-fold increase in viscosity can be measured in the range of 0.1 rad / s figure 2 shows the storage modulus G' and the loss modulus G" of all samples. Thus the viscous and the elastic behavior can be described. At higher shear rates the polymer chains and thus the structure becomes more and more rigid. Due to the resulting restriction of movement of the macromolecules, more deformation energy can be stored and the elastic behavior dominates, see G' over G" in Figure 2. The $PA6_{graft}$ has a comb-like character. Due to the chain extension and the resulting constrictions it shows elastic properties.

In the lower frequency range G'' is above G' resulting in increased movability between the macromolecules, the structure is more flexible. As a result of friction and associated dissipation deformation energy is lost. The viscous behavior dominates and also the viscosity values η^* rise, as shown in figure 1. This transition from elastic to viscous behavior is most clearly shown for $PA6_{graft}$. This transition point is also referred to as a sol-gel transition or crossover point. This point of intersection shifts from $PA6_{graft}$ over $PA6_{linear}$ to PA6 to higher shear rates.

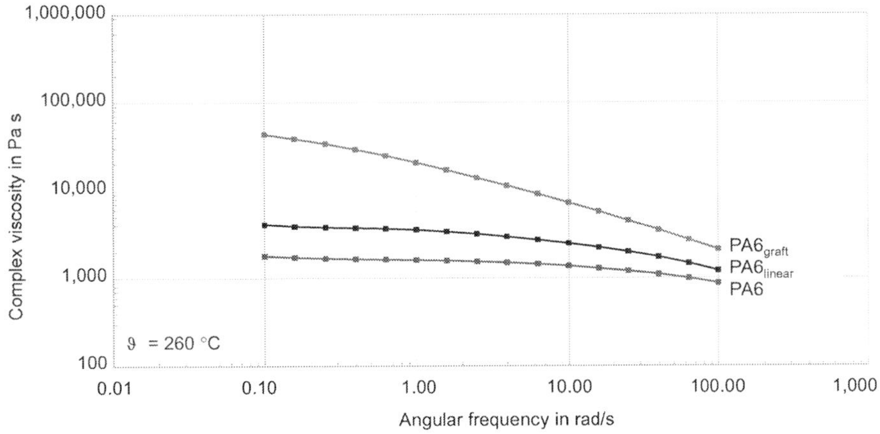

FIGURE 1. Viscosities of the compounded PA

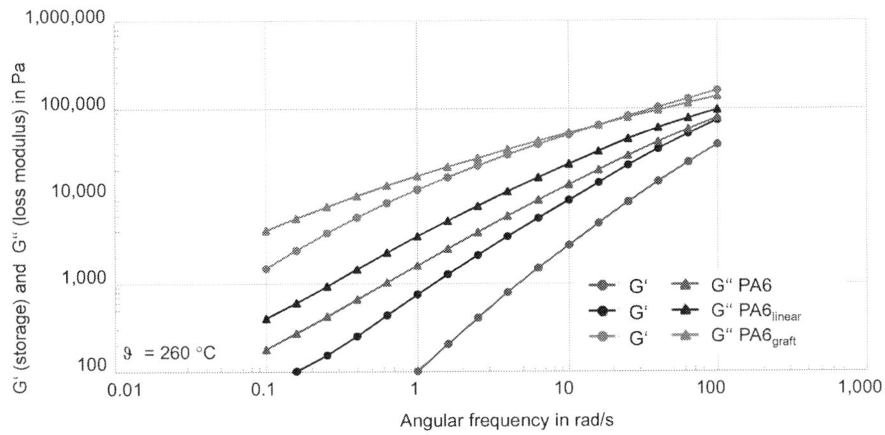

FIGURE 2. Storage and loss modulus with a visible crossover point for $PA6_{graft}$

When viscosity is viewed as a function of time in figure 3 the values for PA6 rise. Since no chain extenders have been added to this material, this increase can be explained by post-condensation during the measurement process. This also leads to an increase in the molar mass and thus also to a higher viscosity. In $PA6_{linear}$ this increase is also observed to the same extent. In the same manner a post-condensation can be concluded here and at the same time an after-reaction of the chain extender can be excluded. This would result in a sudden higher increase in viscosity. In $PA6_{graft}$, such post-condensation cannot be detected. However, it can be shown for both chain extenders that the coupling reaction has been completed.

The changes in molar mass can be demonstrated using the GPC. $PA6_{graft}$ has the highest value for M_w, which has more than doubled compared to the standard material. The mean value is $PA6_{linear}$ while the lowest is for the PA6. In the case of the number-average molar mass M_n, the highest result of $PA6_{linear}$ is obtained. Again the PA6 achieves the lowest result. The polydispersity index represents the ratio of the mass average and the number average molar mass and is a measure of the nonuniformity of a polymer.

Regarding the polydispersity index the sample from the pure material PA6 has reached the highest value while $PA6_{linear}$ has the lowest polydispersity. Since the polydispersity of all samples has a value higher than three and this is also shown in the form of the distribution curve, this is a so-called wide distribution.

The values determined can be read in Table 1. The influence of the viscosity by the molar mass is confirmed. The viscosity has increased in all chain-extended samples, the most pronounced in $PA6_{graft}$.

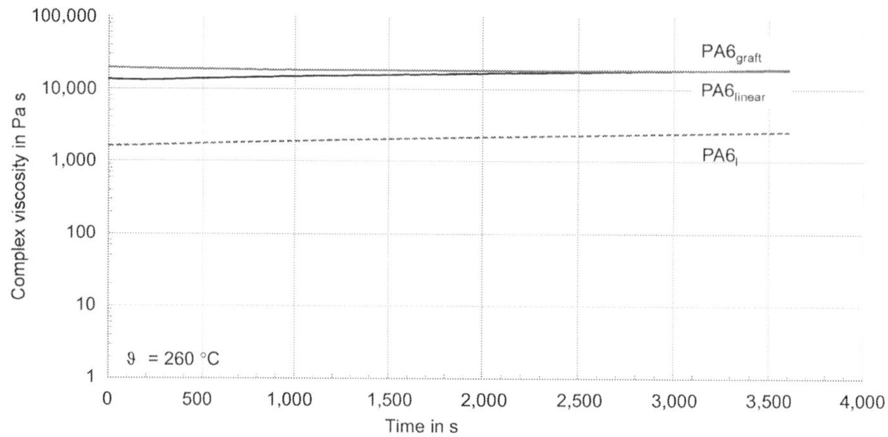

FIGURE 3. Viscosity as function of time

TABLE 1. Molar masses determined by GPC.

	M_w [g / mol]	M_n [g / mol]	polydispersity index
PA6	224.65	33.45	6.7
$PA6_{linear}$	358.55	108.95	3.3
$PA6_{graft}$	470.40	87.85	5.4

According to the obtained data from the GPC analyzes, Fig. 4 shows the differential distribution curves. The differential molar mass distribution is plotted over the molar mass. Particularly striking is the second local maximum which occurs with $PA6_{graft}$. A multimodal distribution, in this case a bimodal distribution, is discussed. This is the case when several polymers with different molar masss are present, so that their peaks overlap and such a curve is formed. The molar mass distribution remains almost the same for all samples in the width.

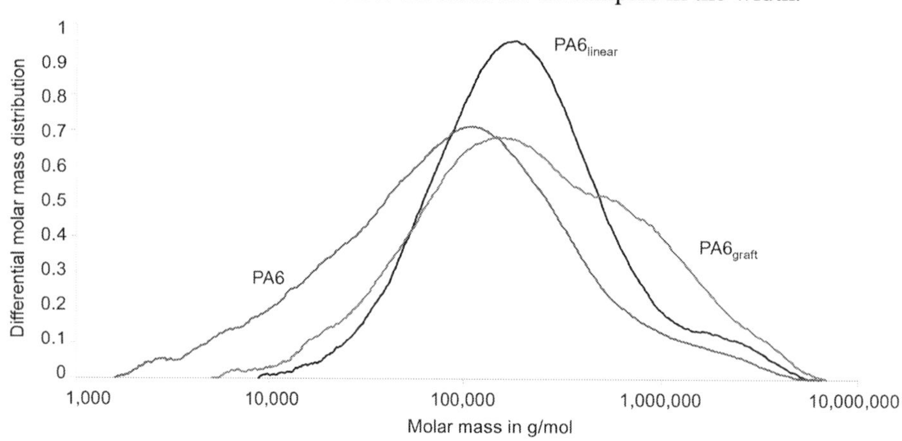

FIGURE 4. Differential molar mass distribution

However, the interpretation of the polydispersity in a multimodal distribution due to the overlapping peaks is not useful. In the case of the direct comparison of the curves in figure 4, a shift of the curves to higher molar mass values in comparison to PA6 can be seen. $PA6_{graft}$ and PA6 both have a broader distribution of the molar masses in the range of the peak at lower molar masses than $PA6_{linear}$.

This differentiation of the curve forms shows a clear distinction between the two types of chain extension. In $PA6_{graft}$, PA6 is grafted onto the backbone, resulting in a very high molar mass. If no coupling occurs, the PA6 chains remain at their molar mass. This results in this bimodal distribution of the molar masses for $PA6_{graft}$.

For the linearly extended $PA6_{linear}$ a sharply marked peak appears. The polymer chains are more homogeneous and the majority is in a very similar molar mass range.

CONCLUSION

Both linearly and by grafting a successful polymer chain extension has taken place. Since the chain lengthening takes place in one step, a fast possibility of the preparation has been verified. In particular, the elimination of a solid-state post-condensation means a considerable increase in both the energy and the production efficiency.

Since melt stability is particularly required during extrusion, this process is extremely suitable. The results of the rheological studies show a clear increase in the viscosity compared to the pure PA6 material.

This is related to the chain extension and the resulting molar mass. The elastic properties due to the increasing entanglement of polymer chains can be seen by a change in the loss and storage modulus in the dynamic frequency sweep.

The change in the molar mass can be confirmed with the GPC analysis. The comparison of the distribution curves shows the difference between the two chain extension modes. The elastic material behavior is thus caused solely by means of the entanglements caused by the chain extension. In the direct comparison of the two chain extensions the highest viscosity increase occurs during the grafting.

One aspect which has not been dealt with in this work is the water absorption capacity of PA6. However, this was not decisive in the first comparison of the two chain extension modes, but should be considered in the future. Furthermore, the effects of chain extension on crystallinity should be investigated. In this case it would be useful to consider the produced PA by means of differential scanning calorimetry. As the next step in the product creation process, the chain-extended polyamides need to be used in various processing processes and thus the potential of chain elongation has to be demonstrated.

REFERENCES

1. Ehrenstein, G.: *Polymer-Werkstoffe : Struktur - Eigenschaften - Anwendung.* 3rd ed., München: Hanser, 2010. ISBN 978-3-446-42967-3
2. Bonten, C.: *Kunststofftechnik - Einführung und Grundlagen*, 2nd ed., München: Hanser, 2016. ISBN 978-3-446-44674-8
3. Kaiser, W.: *Kunststoffchemie für Ingenieure : Von der Synthese bis zur Anwendung.* München: Hanser, 2016. ISBN: 978-3-446-44638-0
4. Förstner, U.: *Umweltschutztechnik*, Springer Berlin Heidelberg 1995, ISBN ISBN 978-3-642-22973-2
5. Musialek, M.; Neubig, B.; Bonten, C.: *Modifiziertes Polyamid 6.6 für das Thermo- und Blasformen*, 24. Stuttgarter Kunststoffkolloquium 25. – 26. Februar 2015, p. 279 – 283
6. Poindl, M.; Bonten, C.: *Morphological Studies on Block Copolymer Modified PA 6 Blends*, Proceedings of PPS-29 (2014) AIP Publishing LLC 978-0-7354-1227-9, p. 654 – 657
7. Schacker, O.; Braun, D.; Hellmann, G.: *Chain Extension of Oligoamide or Polyamide in an Extruder.* In: Macromolecular Materials and Engineering (2001), Vol. 286, No. 7, p. 382-387
8. Zhang, D.; Fritz, H.-G.; Bonten, C.: *Reaktivcompoundierung von PA66 für die Extrusion dickwandiger Hohlprofile.* In: Zeitschrift Kunststofftechnik (2011), Vol. 7, No. 5, p. 202-224
9. Buccella, M.; Dorigato, A.; Pasqualini, E.; Caldara, M.; Fambri, L.: *Chain Extension Behavior and Thermo-Mechanical Properties of Polyamide 6 Chemically Modified With 1,1'-Carbonyl-Bis-Caporlactam.* In: Polymer Engineering and Science (2014) DOI: 10.1002/pen
10. Buccella, M.; Dorigato, A.; Pasqualini, E.; Caldara, M.; Fambri, L.: *Thermo-mechanical behaviour of Polyamide 6 chain extended with 1,1'-Carbonyl-Bis-Caprolactam and 1,3-Phenylene-Bis-2-Oxazoline.* In: Journal of Polymer Research (2013), Vol. 20, No. 225, p. 1-9
11. Qian, Z.; Chen, X; Xu, J.; Guo, B.: *Chain Extension of PA1010 by Reactive Extrusion by Diepoxide 711 and Diepoxide TDE85 as Chain Extenders.* In: Journal of Applied Polymer Science (2004), Vol. 94, p. 2347-2355
12. Buccella, M.; Dorigatio, A.; Pasqualini, E.; Caldrara, M.; Fambri, L.: *Thermo-mechanical properties of Polyamide 6 chemically modified by chain extension with Polyamide/Polycarbonate blend.* In: Journal of Polymer Research (2012), Vol. 19, p. 9935-9944

Compounding Anionic Polymerized Polyamide 6 Using Terephthalic Acid

Benjamino Rocco Formisano[a)] and Christian Bonten

IKT, Institut für Kunststofftechnik, University of Stuttgart, Pfaffenwaldring 32, 70569 Stuttgart, Germany

[a)]Corresponding author: Benjamino.Formisano@ikt.uni-stuttgart.de

Abstract. Polyamide 6 can be produced by anionic ring-opening polymerization of ε-caprolactam using e. g. sodium-caprolactamate as a catalyst. One of the best known anionically polymerized polyamide 6 is so called cast polyamide 6, used for particularly stressed parts. The reason for its good mechanical properties is the comparably high molecular weight. When using cast polyamide 6 for extrusion or injection molding, its high molecular weight can be a disadvantage due to its very high viscosity. To enhance processability, the flow behavior of these materials has to be adjusted. Besides the use of a lubricant, shorter chains would be another way to achieve better processability. In this work, the latter is performed by compounding anionically polymerized cast polyamide 6 using a chain scissorer, e. g. terephthalic acid. Compounding of dry-blends made of cast polyamide 6 and said terephthalic acid is done on a twin-screw extruder type ZSK 26MC of Coperion, Stuttgart. Mechanical properties and flow behavior were investigated afterwards. Surprisingly, it could be shown, that adding very small amounts of terephthalic acid did not reduce but raise viscosity significantly. However, higher amounts of terephthalic acid reduce these viscosities, as expected. Additionally, a reduction of viscosity can be seen over time, because remaining ingredients seem to still be active. Mechanical properties such as Young's modulus or tensile strength are not affected by adding terephthalic acid and are comparable to those of injection molding polyamide 6. Solely impact strength was slightly reduced, when a larger amount of terephthalic acid was added.

INTRODUCTION

Polyamide 6 can be synthesized from ε-caprolactam in two different ways. In the hydrolytical polymerization water is added to the caprolactam inducing a ring-opening process for getting aminocaproic acid which polymerizes. The second way uses anionic caprolactamate as a catalyst for a ring-opening polymerization process. This is described in more detail in [1,2].

As the polymerization temperature is lower for the anionic process, much higher molecular weights can be achieved. If polyamide 6 is anionically polymerized directly in a mold, this material is called cast polyamide 6. The catalyst remains in the material and is still active. Thus, anionically polymerized polyamide decomposes when processed in molten state. Because of said decomposition, compounded anionically polymerized polyamide 6 has similar molecular weight and thus mechanical properties as a hydrolytically polymerized extrusion type polyamide 6. [2]

At the IKT, different works were done in compounding anionically polymerized polyamide 6 by using different additives to influence mechanical and rheological properties of the material. To enhance the processability, terephthalic acid was added in larger amounts of about 0.5 wt.-%. Using this additive, viscosity could be reduced and processability improved [3]. This effect is due to the attack of the acid group on the amino group:

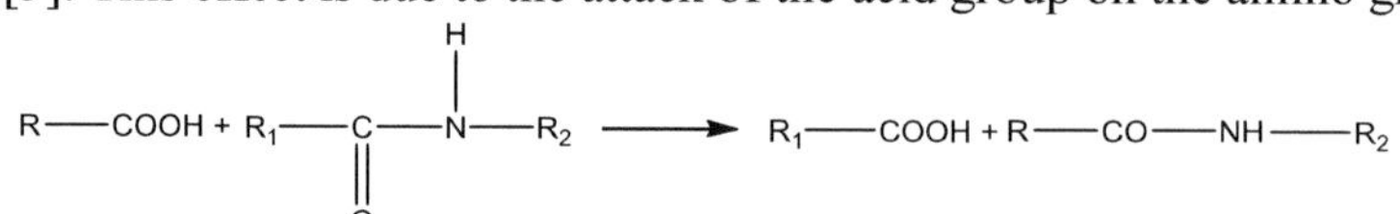

(1)

Disadvantageously, impact strength of the compounded material was reduced [3].

The main objective of the presented work is to raise the impact strength of such a compounded material, resulting in a well processable polyamide having adequate impact strength. As terephthalic acid has two acid groups which can react with polyamide, it is further aimed to check possible influences when terephthalic acid is used in different amounts.

EXPERIMENTAL

Materials

Experiments were conducted using grinded anionically polymerized polyamide 6, as this material is not available as granules. The polyamide was dried in a vacuum oven at 80 °C until the residual moisture was below 0.1 %. As an additive, terephthalic acid, supplied by Merck, Darmstadt, Germany, was used. The Polyamide and the additive are dry blended before compounding, see Tab. 1. For rheological comparisons Ultramid B40, an extrusion type polyamide 6, from BASF SE, Ludwigshafen, was used as reference.

TABLE 1. Produced materials made of anionically polymerized polyamide 6

material	components	compounding steps
M0-1/2	anionically polymerized polyamide 6	one/two
M1-1/2	anionically polymerized polyamide 6 + 0.1 wt.-% additive	one/two
M2-1/2	anionically polymerized polyamide 6 + 0.2 wt.-% additive	one/two
M3-1/2	anionically polymerized polyamide 6 + 0.3 wt.-% additive	one/two
M5-1/2	anionically polymerized polyamide 6 + 0.5 wt.-% additive	one/two

Compounding

Compounding of the dry blends was performed using a ZSK 26MC from Coperion, Stuttgart, Germany. A vacuum degassing was installed, a double-strand die was used and the received strands were pelletized.

The dry-blends were fed into the compounder, setting the throughput to 5 kg per hour. The compounding step itself was carried out two times for each batch. Thus, observing the materials' behavior during further processing and potential changes in the material after a longer dwell time was possible. After each compounding step the materials were dried at 80 °C in a vacuum oven until the residual moisture was below 0.1 %.

Characterization

To get information about the mechanical behavior, test specimens were prepared by injection molding using an Allrounder 220M, 350–390 from Arburg, Loßburg, Germany. The temperature of the melt at the dye was set to 280 °C. A family tool, heated to 80 °C, was used to produce a tension rod and a body for notched impact tests, which was notched afterwards. Before testing, the specimens were dried at 50 °C for 66 h and then cooled down for two hours.

Tensile tests were carried out on a universal tensile testing machine Type 1455 from Zwick and Co. KG, Ulm, Germany. Testing speed was 1 mm/min for measuring Young's moduli and 50 mm/min afterwards. Preload was set to 0.2 MPa. Notched Charpy impact strength measurements were carried out on an impact testing machine Type 5102.202 from Zwick; an impact hammer of 1 J was used. 20 specimens were tested mechanically for each material.

The linear viscoelastic deformation behavior was investigated using a plate-plate rheometer DSR200 from Rheometric Scientific, Munich, Germany, with a plate diameter of 25 mm. All measurements were carried out at 270 °C with a shear stress of 500 Pa under nitrogen atmosphere. Time-sweep measurements were performed at a frequency of 10 rad/s.

Viscometry was performed using a Ubbelohde-viscometer AVS 310 from Schott Geräte, Mainz, Germany. Materials M0-1 till M5-1 were solved in formic acid, 90 % at a concentration of 5 g/l for 3 hours at 23 °C. Measurements were done at 25 °C.

Crystallinity was analyzed by a differential scanning calorimeter DSC 204 Phoenix from Netzsch, Selb, Germany, equipped with a liquid nitrogen cooling system. The samples were heated from room temperature to 270 °C at a heating rate of 10 K/min followed by cooling to room temperature at the same rate. All tests (heating and cooling) were run in two cycles for each sample.

RESULTS AND DISCUSSION

Mechanical properties of the compounded anionically polymerized polyamides are shown in Fig. 1. Young's moduli of the different materials are about 3,100–3,300 MPa, Fig. 1(a), and tensile strengths are about 83–84 MPa, Fig. 1(b). Within accuracy of measurement, no difference can be determined; terephthalic acid has no influence on both properties. Despite the measurement deviation, Fig. 1(c) shows that higher amounts of terephthalic acid reduce Charpy notched impact strength.

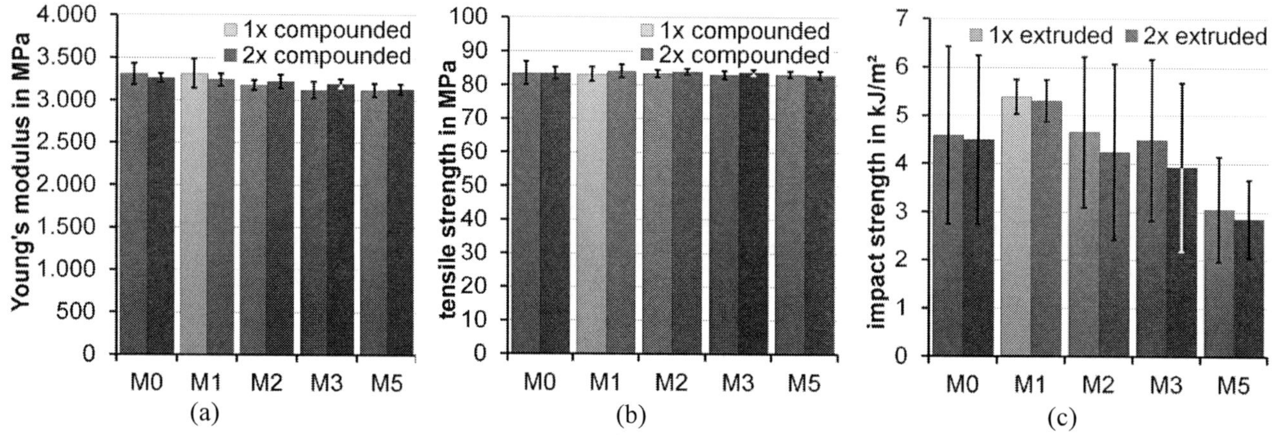

FIGURE 1. Mechanical properties of the different materials

Measurements of the material's viscosities are shown in Fig. 2(a). Anionically polymerized polyamide 6 has, as said above, significantly higher viscosities compared to hydrolytically polymerized PA6. Purely compounded M0 shows viscosities higher than those of hydrolytically polymerized polyamide 6, but they are clearly lower than those of anionically polymerized polyamide 6. M0-2 reaches slightly higher viscosities than M0-1, suggesting a condensation reaction.

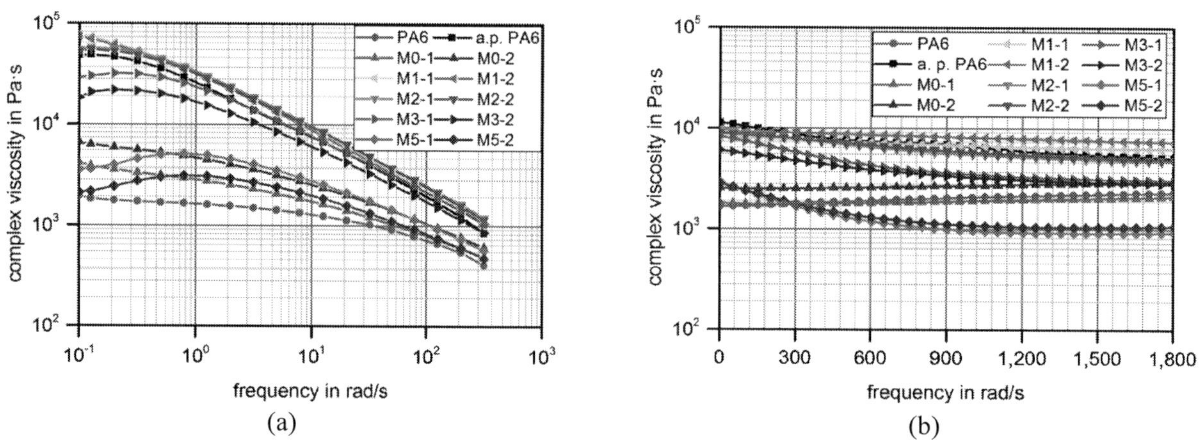

FIGURE 2. Rheological measurements of the produced materials

Adding 0.1 to 0.2 % of terephthalic acid does not reduce viscosity, as it could be expected by former works. Both times, no difference can be seen between first and second compounding step. Complex viscosities of M1 and M2 are similar to those of unprocessed anionically polymerized polyamide 6, showing an easy way to raise viscosity, if needed by a specific process.

Further addition of terephthalic acid results in stepwise reduction of viscosity, seen for M3 and M5. For both materials, a drop of the curves at lower frequencies (< 1 rad/s) is clearly shown. This will be discussed, when we have a look at the time-sweep measurements. It is also shown in Fig. 2(a), that viscosity is reduced for M3 and M5 after the second compounding step, suggesting further decomposition, if processed longer times.

Time-sweep measurements of all materials at 10 rad/s are shown in Fig. 2(b). Anionically polymerized polyamide shows a drop of viscosity by time, clearly demonstrating decomposition as said in literature. M0 shows a slight post-condensation instead, which is because of the high viscosity loss (more than 10,000 Pa·s). The equilibrium reaction of decomposition is already completed after the first compounding step. Furthermore, it is shown that adding additive results in descending curves. The more additive is added, the faster this descending is completed and a plateau is reached. It is suggested that terephthalic acid alters the equilibrium state and a higher amount of it fastens readjustment.

By using rheometry, it is also possible to get some information about the average molecular weight of polymers by looking at the crossover point of storage and loss modulus. The lower the frequency of the cross over point is, the higher the average molecular weight. [4] Fig. 3(a) shows the loss angle tan (δ). It is the quotient of loss G'' and storage modulus G' according to

$$tan\ (\delta) = \frac{G''}{G'} \tag{1}$$

The cross over point is reached when tan (δ) equals one. Comparing crossover points of materials having low amounts of terephthalic acid (M1 and M2) to those of purely processed material (M0), it is clearly shown that an increase in average molecular weight is achieved. Said crossover points are similar to those of non-processed anionically polymerized polyamide 6. Reduction of molecular weight and viscosity does not occur.

Additional viscometric measurements are shown in Fig. 3(b). According to the Staudinger-Mark-Houwink equation [5], the viscosity number is proportional to the molecular weight of a polymer. As viscometry is a relative method [6], it can be demonstrated, that a higher viscosity number and therefore average molecular weight is reached for M1–5 in comparison to purely processed anionically polymerized polyamide.

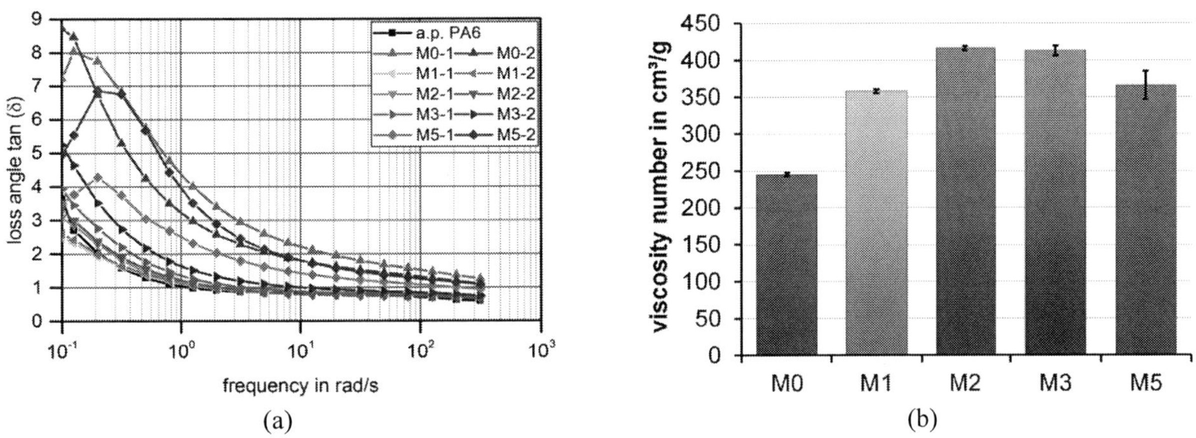

FIGURE 3. Loss angle (a) und viscometry (b) to get information about average molecular weight

In both Figures 2 and 3, an increasing amount of terephthalic acid finally yields a reduction in molecular weight and viscosity. So it can be said that high amounts of terephthalic acid can reduce molecular weight, but small amounts tend to raise it.

Results of DSC measurements are shown in Fig. 4. Solidification temperatures Fig. 4(a), and also melting temperatures (not shown) are not affected by adding terephthalic acid. The shapes of the curves are also similar. Within the accuracy of measurement of 5.5 % for DSC measurements [7], also solidification enthalpies, Fig. 4(b) are not changed. Terephthalic acid has no influence on such thermal behavior, as shown.

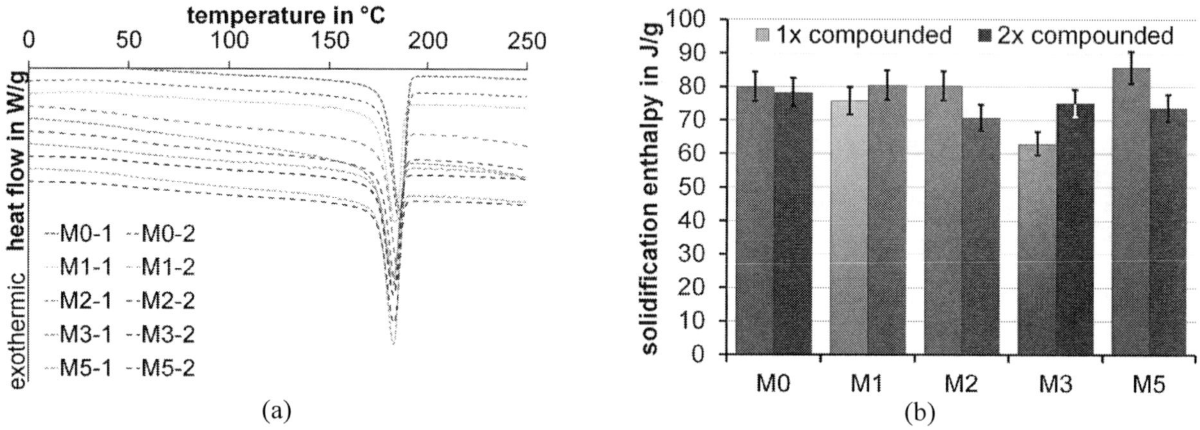

FIGURE 4. Second cooling-down curve and solidification enthalpies and measured by DSC

CONCLUSION

Polyamide 6 can be synthesized from ε-caprolactam by an anionic ring-opening polymerization process. This polymerization process allows for much higher molecular weights than those achieved from hydrolytic polymerized polyamide 6 for injection or extrusion applications. If compounding such a material, it is possible to adjust viscosity and impact strength of anionically polymerized polyamide 6 by using terephthalic acid. Larger amounts of it reduce impact strength and viscosity, while small amounts of less than 0.3 % raise molecular weight and viscosity. Young's modulus and tensile strength are not affected by adding terephthalic acid as well as melting behavior.

In the future, also other acids should be tested for this behavior. A further objective will be to compound commercially available (hydrolytic polymerized) polyamide 6 investigating a possible occurance of this effect also for "standard" polyamide. In such a case, a cheap and easy possibility to enhance viscosities for polyamide will be given.

REFERENCES

1. B. Tieke, *Makromolekulare Chemie*, 2nd edition (Wiley-VCH, Weinheim, 2005), p. 117.
2. A. Fahnler and A.A. van Greenen, "Polyamide aus technischer anionischer Lactampolymerisation", in *Technische Thermoplaste – Polyamide*, edited by L. Bottenbruch and R. Binsack (Hanser, Munich, 1998), pp. 406-431.
3. B. Formisano, C. Bonten, *Zeitschrift Kunststofftechnik* **12**, pp. 259–293 (2016).
4. T.G. Mezger, *Das Rheologie-Handbuch* (Vincentz, Hanover, 2000), pp. 152–174.
5. H. Staudinger and W. Heuer, *Ber. dtsch. Chem. Ges.* **63**, pp. 222–234 (1930).
6. C. Bonten, *Kunststofftechnik*, 2nd edition (Hanser, Munich, 2016)
7. G.W. Ehrenstein, G. Riedel, P. Trawiel, *Praxis der Thermischen Analyse von Kunststoffen*, 2nd edition (Hanser, Munich, 2003) pp. 1–119

Saw dust reinforced toughened polypropylene (PP/EPDM/talc): effect of plasticizer incorporation

Gabriel Matheus Pinto [1] and Baltus Cornelius Bonse [1, a)]

[1] *Centro Universitário da FEI. Av. Humberto de A.C.Branco 3972 São Bernardo do Campo – SP- Brazil*

[a)]Corresponding author: prebbonse@fei.edu.br

Abstract. The effect of a paraffinic and a vegetable plasticizer on the mechanical and thermal properties of sawdust reinforced toughened polypropylene (PP/EPDM/talc) was studied, employing two different methods of plasticizer incorporation. The first method is based on pre-mixing all components in a tumble mixer, extrusion in a co-rotational twin-screw extruder and injection molding. In the second method the plasticizer was added in the injection molding hopper, after extrusion of the other components. The addition of 10wt% paraffinic plasticizer increased impact strength of the plasticizer-free composite by almost 50% for both incorporation methods, while the increase generated by the addition of 10wt% vegetable plasticizer was only 30%. The compounds containing 10wt% paraffinic plasticizer presented tensile and flexural strength, slightly higher than that presented by the compounds containing 10wt% vegetable plasticizer. A significant conclusion of this study is when comparing the responses regarding the plasticizer incorporation methods. There was practically no difference in the investigated properties, either mechanical or thermal, when the plasticizer was added in the extrusion or injection stage, which means one step less in the processing of the composite.

INTRODUCTION

Natural fiber thermoplastic composites have become prevalent not only for environmental issues, but also for economic reasons, recyclability, ease of processing etc. [1] A very promising application is in the automotive industry, which is increasingly pressured to produce lighter, cheaper and less polluting vehicles [2].

One of the most widely used polymers in the automotive industry for making internal and external parts of vehicles is toughened polypropylene (PP/EPDM/talc) [3]. EPDM (ethylene-propylene-diene monomer) elastomer is incorporated in PP to increase its impact strength, but it ends up reducing other mechanical properties, such as Young's modulus and tensile strength. To offset these losses, talc is added at levels ranging from 10 to 40%, which also increases thermal strength, creep, dimensional stability and resistance to moisture [4]

PP/EPDM/talc when reinforced with bamboo fibers showed loss in composite elongation at break and impact strength [5]. To overcome this problem plasticizers may be used, which increase flexibility as well as improve processing. Plasticizers reported for use in PP include paraffinic oil, dioctylsebacate and isooctylphthalate [6]

Within this context the aim of this investigation was to study the effect of a paraffinic and a vegetable plasticizer on the mechanical and thermal properties of sawdust reinforced toughened polypropylene (PP/EPDM/talc) employing two different methods of plasticizer incorporation.

EXPERIMENTAL

Materials

PP/EPDM/talc pellets (EF210AEB) were supplied by Borealis - MFR (230°C/2.16kg) = 24g/10min, density = 1,04g/cm³. Compatibilizer used was maleic anhydride grafted PP (Polybond 3200) - MFR (190°C/2.16kg = 110g/10min; density = 0,92g/cm³), from Crompton-Uniroyal Chemical (São Paulo, Brazil).

Saw dust was supplied by Indústria Madeireira UlianaLtda (Brazil), containing a mix of several timber types: pink cedar (*Cedrela sp*), arana ceder (*Cedrelingacateniformis*), ipe (*Tabebuia* spp), angelim (*Hymenolobiem petroeum*), marupa (*Simarolba amara*), loro vermelho (*Nectanda rubra*), cumaru (*Dipterysa odorata*) and jatobá (*Hymenoeasp*).

The vegetable oil, Biokap EM 3221, was supplied by Biocapital (São Paulo, Brazil) and the paraffinic oil, NML, by Fragon (São Paulo, Brazil).

Methods

PP/EPDM/talc composites containing sawdust were prepared according to two methods. The first involved tumble mixing in a Y-mixer of the polymer, fibers, compatibilizer and plasticizer, followed by extrusion and pelletization, and injection molding of test specimens. In the second method, the plasticizer was added in the injection molding hopper together with the extruded pellets containing the other components.

Sawdust and compatibilizer contents were 40wt% and 4wt%, respectively, whereas plasticizer concentrations tested were 5, 10 and 20 wt% with regard to the total composite mass. Sawdust was dried for 12 h at 100°C in an oven with forced air circulation prior to processing. Prior to injection molding pelletized composites were dried for 2 h at 100°C.

Extrusion was carried out in a Thermo Scientific HAAKE PolyLab OS RheoDrive co-rotating twin-screw extruder (L/D =25 mm; D=16 mm) with temperature profile ranging from 170 to 190°C. Injection molding was performed in an HM60/350 Battenfeld machine with temperature profile from 190 to 198°C, and mold temperature 60°C.

The injected specimens were subjected to tensile and flexure tests in an Instron 5565 Universal Testing Machine at test speeds of 5 mm/min and 1.3 mm/min, in accordance with ASTM D638 and D790, respectively. Charpy notched impact tests were carried out in a Ceast 9050 Instron impact tester in accordance with ASTM D 6110.

Heat deflection tests were carried out in a CEAST HDT-Vicat 6 Stations tester in accordance with ASTM D 648.

Impact fractured surfaces of the composites were sputter-coated with gold and examined in a CamScan CS 3200LV scanning electron microscope (Cambridge) under high vacuum (10^{-5} mbar).

RESULTS AND DISCUSSION

Figure 1 shows some of the mechanical properties, as well as one thermal property obtained of the PP/EPDM/talc matrix, the sawdust reinforced composite with no plasticizer and containing different paraffinic and vegetable plasticizer content. As can be seen in Fig.1, and already discussed in previous works [7], impact strength and tensile strain at break greatly decrease when 40wt% sawdust and 4wt% compatibilizer are added to the polymer matrix, basically due to the restraints imposed on the polymer chains by the stiffer fibres. When adding plasticizer composite impact strength and tensile strain at break increase, likely as a result of increased chain mobility [6]. It can also be seen that with increasing paraffinic oil concentration, there is a clear increase in both aforementioned properties. In relation to the unplasticized composite, 5wt% and 10 wt% paraffinic oil result in an increase in impact strength of about 25% and 50%, respectively. However, when vegetable oil is used as plasticizer, increases are about 20% and 30%, respectively. With regard to tensile strain at break, the same amounts of paraffinic oil result in an increase in 40% and 90%, respectively, whereas with vegetable oil increase is about 20%. Tensile strain at break did not change when increasing vegetable oil from 5wt% to 10 wt%, as confirmed statistically via ANOVA (analysis of variance). The difference in plasticizer effect with regard to the aforementioned properties might be due to the fact that paraffinic oil has very good compatibility with EPDM , whereas ester based oils, such as vegetable oil, have reduced compatibility with EPDM [6].

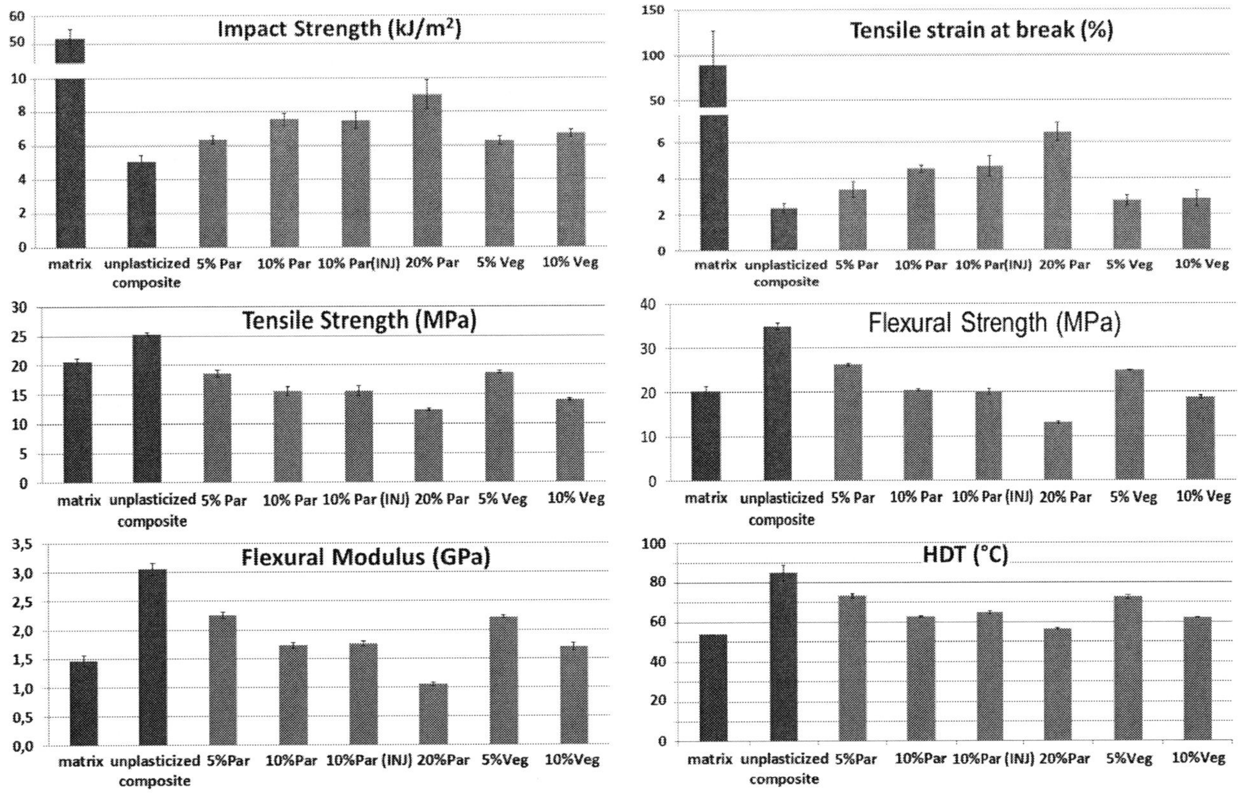

FIGURE 1. Properties of PP/EPDM/talc (matrix), sawdust reinforced composite (unplasticized composite) and composite containing different concentrations of paraffinic and vegetable plasticizer. (INJ): plasticizer added prior to injection molding

Tensile strength of the unplasticized composite clearly shows the reinforcing effect of the sawdust, with an increase of about 20% in relation to the matrix. When plasticizer is added, as expected, tensile strength reduces, and as plasticizer content increases tensile strength is further reduced. The same trend is observed for flexural strength and modulus, as well as HDT. Contrarily to the results observed with impact strength and tensile strain at break, vegetable oil had a similar effect as paraffinic oil on the other investigated properties.

With regard to the plasticizer addition method, statistically no difference has been found in the investigated composite properties between addition prior to extrusion, i.e. during pre-mixing, and addition prior to injection moulding, i.e. in the injection moulding feed hopper. The only difference was a higher standard deviation in some of the properties for the latter method, which is reasonable since better mixing is achieved in extrusion compared to injection moulding.

Figure 2 shows SEM images of impact fractured composites without and with both plasticizers incorporated by the two different aforementioned methods. Fig. 2(a) of the unplasticized composite clearly shows good adhesion between fibre and matrix, i.e. no gaps at the interface. Fig. 2(b), (c) and (c) showed that even after addition of both plasticizers and regardless of incorporation method, adhesion between fibre and matrix was unaltered.

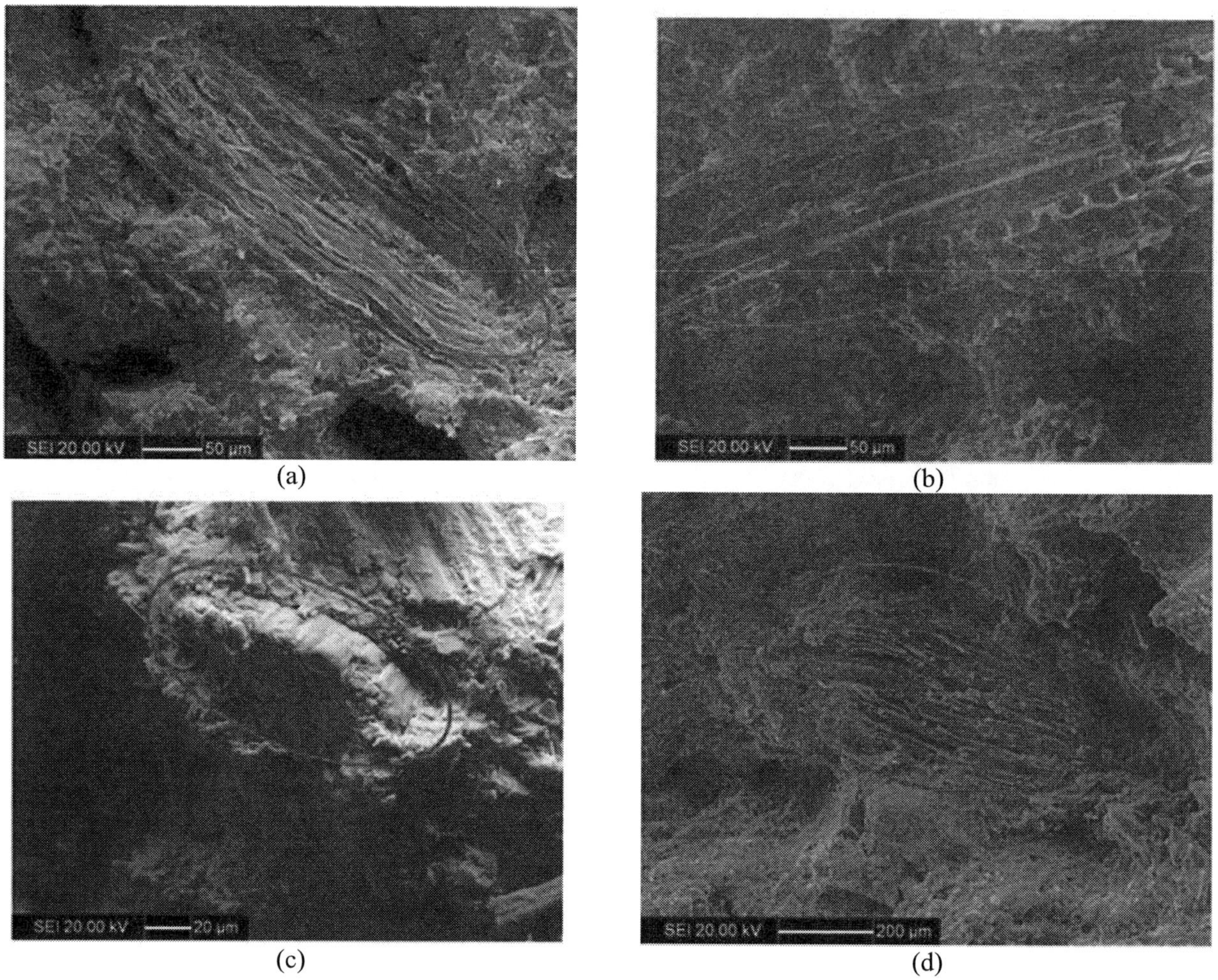

FIGURE 2. Impact fractured surfaces of the composites: (a) unplasticized; (b) containing 10wt% paraffinic oil; (c) containing 10wt% vegetable oil; (d) containing 10wt% paraffinic oil added prior to injection molding

CONCLUSIONS

Paraffinic and vegetable oil were incorporated into PP/EPDM/talc at different concentrations, using two different incorporation methods: prior to extrusion and prior to injection molding. The paraffinic plasticizer presented a slightly higher compatibility with the PP/EPDM/talc matrix than the vegetable, since at 10wt% it increased impact strength by almost 50% in relation to the plasticizer-free composite for the two incorporation methods, whereas the increase generated by the 10wt% vegetable composition was only 30%. The compounds containing 10wt% paraffinic plasticizer also presented tensile and flexural strength, slightly higher than that presented by the compounds containing 10wt% vegetable plasticizer. A significant conclusion of this study is when comparing the responses regarding the plasticizer incorporation methods. There was practically no difference in the investigated properties, either mechanical or thermal, when the plasticizer was added in the extrusion or injection stage, which means one step less in the processing of the composite. It should be mentioned that with the use of the extruder, reproducibility is greater, evidenced by the lower standard deviation in the results. This is due to the better mixing and, consequently, better homogeneity of the material obtained during extrusion in relation to the injection moulding.

ACKNOWLEDGMENTS

Thanks are due to Centro Universitário FEI for the undergraduate research scholarship granted to G.M. Pinto.

REFERENCES

1. Lucas A.A., Ambrosio J.D., Bonse B.C., Bettini S.H.P. (2011) Natural Fibers Poly-mer Composites Technology applied to the Recovery and Protection of Tropical Forests allied to the Recycling of Industrial and Urban Residues. In: P. Tesinova (Ed), Advances in Compo-site Materials - Analysis of Natural and Man-Made Materials (pp. 163-194). InTech.
2. Yang, Y., Boom R., Irion, B., Heerden, D.-J., & Kuiper, P. (2012). Recycling of composite materials. Chemical Engineering and Processing: Process Intensification, 51, 53-68
3. Suddell, B.C., & Evans, W.J. (2005) Natural Fiber Composites in Automotive Applica-tions. In: A.K. Mohanty, M. Misra, & L.T. Drzal (Eds). Natural Fibers, Biopolymers and Bi-ocomposites. Florida: Taylor & Francis Group.
4. Maier C. & Calafut T. (1998) PolypropyleneThe Definitive User's Guide and Databook. Norwich, NY: Plastics Design Library.
5. Bonse B. C., Mamede M. C. S., da Costa R. A., Bettini S. H. P. Proceedings of the Polymer Processing Society 26th Annual Meeting, 2010 Banff (Canada)
6. Wypych, G. Plasticizers Use and Selection for Specific Polymers. Handbook of Plasticizers, 1. ed.: ChemTec Publishing, 2003.
7. Bettini S.H.P, Bonse B.C., Melo E.A., Muñoz, P. A. Polym Eng Sci, v. 50, p. 978-985, 2010.

A Study on the Optimization of Thermal Conductivity of HDPE Using Micro- and Nanoscale Fillers

Stieger Sebastian (1)*, Berger Gerald (1), Battisti Markus (2), Friesenbichler Walter (1)

(1) Montanuniversität Leoben - Styria - Austria, (2) Hirschmann Automotive GmbH - Vorarlberg - Austria

The cooling time in injection and blow molding processes is a crucial factor for economic success. Since unfilled polymers show a low thermal conductivity compared to metals – due to the lack of free electrons - the polymer dominates the cooling efficiency, intensely for thick-walled part geometries. The aim of this study was to increase both the thermal conductivity λ and the thermal diffusivity of high density polyethylene (λ = 0.44 Wm-1K-1) by adding a suitable filler system up to a filler content of 30 vol.%. In a first test series the influence of the filler content on the thermal conductivity was determined. SiO2 filler showed a linear increase of the thermal conductivity with rising filler content, which is in good agreement with calculated values using the Maxwell- and Lewis-Nielsen-Modell. In a second test series, the filler content was set to 30 vol.%. The influence of numerous filler types as well as mixtures of fillers (polymer + filler1 + filler2) was analyzed. For combined-filler compounds, no synergetic effect was found, in contrast to other studies. However, adding hexagonal boron nitride increased λ by 900%. In a third test series three compounds were injection molded, which leads to an inhomogeneous filler distribution if anisotropic filler particles (L/D >> 1, e.g. talcum) are used. While CaCO3 and SiO2 (L/D = 1) yields almost isotropic thermal conductivity, analyzing core and surface sections of talcum compound specimens proved, as expected, the anisotropic filler distribution over the part thickness. Calculating the cooling time for a 4mm thick part, compared to unfilled HDPE, reductions by 45% and 35% were estimated using 30 vol.% of SiO2 and CaCO3 respectively. In conclusion, thermal conductive polymers increase the production efficiency of injection molding. Further studies on improved mechanical performance will be published elsewhere.

Effect of ammonium and aminosilane montmorillonites organo-clays on the curing kinetics of unsaturated polyester (UP) resin nanocomposites

Vargas Maria Angeles (1)*

(1) tecnologico de mexico - edo. mex. - Mexico

The curing kinetics of UP nanocomposites prepared by incorporating different amounts of twokinds of organo-montmorillonite (organo-MMT): trimethyloctadecacylammonium chloride (TMOA) andaminopropyl-triethoxysilane (APTES) were studied by non-isothermal differential scanning calorimetry(DSC) experiments. small angle X-ray scattering (SAXS) was used for measuring the d-spacings in themodified organo-clays, and no intercalation UP into these clays was observed for the nanocomposites.HRTEM images showed dispersed and agglomerated platelets in UP/APTES 2 and 10 wt%. DSC analysisshowed two peaks in UP resins and UP/organo-MMT, and a decrease in the exothermal peaks tempera-ture (Tp1and Tp2) for nanocomposites with the heating rate as compared with those of neat UP system;thus, the higher the heating rate, the higher the curing reaction rate. This effect was more clearly onthe UP/MMT-APTES nanocomposites. The effective activation energies (Ea) were determined with themode-free isoconversional Starik's method. Sesták–Berggren model was chosen to simulate the reactionrate with a good match achieved. Thermal gravimetric analysis showed that the cured UP/APTES at highconcentration were slightly more stable than UP and TMOA.

Variations in rheological properties by melt reactive modification of PET

Dolatshah saeed [(1)]*, Ershad Langroudi Amir [(1)], Ahmadi Shervin [(1)]

[(1)] Iran Polymer And Petrochemical Institute - Tehran - Iran

Poly (ethylene terephthalate) (PET) is prone to the hydrolysis and thermal degradation reactions which occur during melt processing. These reactions and linear structure limited its applications and recycling of PET. In this work, we follow the synergistic effect of using pyromellitic dianhydride (PMDA) and pentaerythritol (PENTA) as chain¬ extender and Dibutyltin dilaurate (DBTDL) for chain extension process during reactive extrusion. Rheological properties of the samples were measured by a parallel plate rheometer. The experimental results shown that by increasing amount of PMDA the complex viscosity of PET increase and formulations which contain also PMDA and PENTA shown a significant increase in this trend. Furthermore, by increasing the amount of DBTDL in the modification process an improvement was observed in rheological properties.

Rheological and Mechanical Behavior of Polyethylene Filled with Estonian Oil Shale Heat Carrier Retorting Waste Ash.

Suld Tiia-Maaja (1)*

(1) Tallinn University Of Technology - Harju - Estonia

Abstract Untreated air-cooled Oil Shale Heat carrier retorting waste ash (SHS was ground, Sieved into different mesh sizes and introduced into LLDPE matrix. Compounding was performed via simple laboratory kneader and gravimetric fed co-rotating twin –screw compounder. In current work rheological and mechanical behaviors of PE/SHC compounds are discussed. Oil Shale heat carrier retorting ash (SHC) is byproduct of oil shale industry and is oxide rich carbon black material 1 It was studied as potential functional filler in thermoplastic (i.e. PE) That is because CHS consist multi oxides that could offer potential property enhancement and could lower consumption of naturally extracted mineral filler. Compounding of PE with SHC was performed at mass temperature 1800C. For SHC average particle sizes of ≥20, 20-45, 45- 70 µm and amounts of 10, 20, 30 w% were tested. Compounded materials were formed into samples for rheological, thermal and mechanical testing. Shear viscosity values for the composites slightly higher than values for neat PE but insignificant influences were wicked when varying filler parameters. SHS increased tensile modulus but decrease tensile strength and elongation at brake. The reason could be filler matrix interfacial debonding which needs compatibilities agents. References 1. Sedman A. Talviste P. Motlep R. Geology 131-132, 37-44, 2012.

Cross-Linked Polyethylene (XLPE) as Filler in High-Density Polyethylene: Effect of Content and Particle Size

Rodrigo Sousa Freitas [1] and Baltus Cornelius Bonse [1, a)]

[1] *Centro Universitário da FEI. Av. Humberto de A.C.Branco 3972 São Bernardo do Campo – SP- Brazil*

[a)] Corresponding author: prebbonse@fei.edu.br

Abstract. Cross-linked polyethylene (XLPE) scrap from electrical wire and cables was micronized and separated in average particle sizes of 100, 500 and 900 micrometer, after which sheets of XLPE concentrates in high-density polyethylene HDPE were produced, using a calendaring process. The sheets were ground and incorporated into HDPE by means of a twin-screw co-rotating extruder and subsequently injection molded into test specimens. A 2 by 2 factorial design with center point was used, where XLPE contents varied between 1 and 9 wt% and average particle size between 100 and 900 micrometers were assessed as to the effect of XLPE content and particle size on HDPE mechanical properties. Increase in XLPE content significantly increased only impact strength, and decreased tensile and flexural strength, and flexural modulus. The effect on tensile modulus and strain at break was not significant for the adopted 95% confidence interval. The increase in XLPE average particle size affected positively only impact strength of the material, and decreased strain at break and flexural modulus. The effect on tensile and flexural strength and on tensile modulus was not significant for the adopted 95% confidence interval. The interaction between the two variables decreased impact strength and increased tensile modulus, and was not significant for the other investigated properties.

INTRODUCTION

Cross-linked polyethylene (XLPE) is used as insulator coating in electrical wires and cables. Incorrect disposal of XLPE waste poses serious environmental problems since this cross-linked material is not recyclable through remelting, as is the case of thermoplastic polymers. Therefore, almost all the waste is burned or disposed of in landfills. An alternative to reuse XLPE waste is by micronizing and incorporating the micronized particles as filler in a thermoplastic matrix.

Several researches have investigated reuse of XLPE [1-3]. Oganda [4] was able to incorporate up to 25% by mass of ground XLPE into virgin PE. The resulting blend could then be used in rotomoulding, injection molding and blow molding. He observed that the heating of the crosslinked PE to approximately 175 ° C causes the surface of the PE to become sticky, allowing it to bind to other materials. Qudaih et al. [5] were able to incorporate 5, 10 and 15wt% XLPE powder with a mean particle size of 1.25 mm in low density PE. Incorporation of 15wt% XLPE in PE resulted in a 17% increase in Young's modulus, 37.2% in yield stress, 22.4% in tensile strength, and a decrease of 3.3% in the melt temperature of the virgin material. Cucchiara and Bonse [6] incorporated 5, 10, 20 and 30wt% XLPE in HDPE and observed an increase in strength of almost 10 times in relation to HDPE with the incorporation of 5wt% XLPE. From this content on, there was a gradual fall, however, values much higher than that of neat HDPE still remained. Therefore, in the current investigation, it was decided to study XLPE contents between 1 and 10wt%.

EXPERIMENTAL

Materials

HDPE pellets (GE7252) from Braskem were used with MFR (230°C/2.16kg) = 2.0g/10min, density = 0,952g/cm³. Ground XLPE scrap from electric wire and cable coating was used as filler in HDPE.

Methods

XLPE was micronized and separated in average particle sizes of 100, 500 and 900 micrometer. Particle size distribution was measured with a Microtrac Bluewave particle size analyser. Sheets of XLPE concentrates, containing different particle size, in high-density polyethylene HDPE were produced, using a calendaring process (front and back rollers at 160 and 166°C, respectively). The sheets were ground and incorporated into HDPE by means of a twin-screw co-rotating extruder and subsequently injection molded into test specimens. A 2 by 2 factorial design with center point was used, where XLPE contents varied between 1 and 9 wt% and average particle size between 100 and 900 micrometers, according to Tab.1. Tensile and three-point flexural tests were carried out in an Instron 2267 universal testing machine, according to ASTM D638 and D790, at speeds of 50 mm/min and 1.3 mm/min, respectively. For tensile strain at break measurements a speed of 500mm/min was used. Charpy notched impact tests were carried out in an Instron Ceast 9050 tester, according to ASTM D6110.

TABLE 1 – Two by two experimental design with central point and control

	Coded variable		Actual variables	
Experiment	XLPE content (wt%)	Av. particle size (µm)	XLPE content (wt%)	Av. particle size (µm)
1	-1	-1	1	100
2	-1	1	1	900
3	1	-1	9	100
4	1	1	9	900
5	0	0	5	500
HDPE				

RESULTS AND DISCUSSION

The results of particle size distribution of the micronized sorted XLPE are shown in Fig.1.

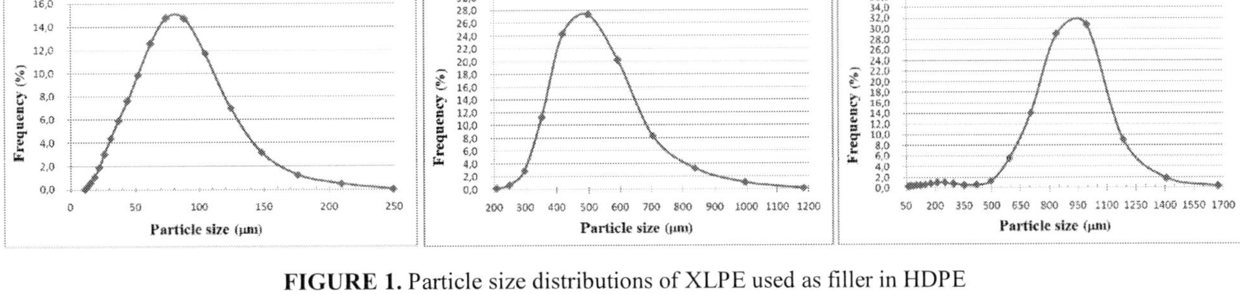

FIGURE 1. Particle size distributions of XLPE used as filler in HDPE

As can be seen average particle sizes were 100, 500 and 900 μm, as required.

The results of the mechanical properties are listed in Tab. 2 and results of the statistical analysis, using software Statistica®, are shown in Tab.3.

Comparison of the center point results (5wt% XLPE with average particle size of 500 μm) in Tab. 4 and the intercepts (a_0) of the model for each investigated property reveals their similarity, and hence the assumption holds that there is a linear relationship between each independent variable and the property, i.e., no need to expand the model.

TABLE 2 – Two by two experimental design with central point and control

wt%XLPE	Av. particle size (μm)	Impact Strength (kJ/m²)	Tensile Strength (MPa)	Tensile strain at break (%)	Flexural Strength (MPa)	Flexural Modulus (MPa)
1	100	12.4±0.2	21.9±0.5	104±16	20,2±0,2	793±16
1	900	12.4±0.1	22.0±0.3	88±6	20,1±0,3	763±17
9	100	16.6±0.2	19.2±0.2	109±6	18,3±0,1	711±13
9	900	16.4±0.2	19.7±0.1	90±4	18,1±0,1	693±10
5	500	14.6±0.2	20.5±0.5	94±5	18,9±0,1	730±11
HDPE control		12,4±0,2	22.7±0.3	110±10	20.2±0.2	791±16

TABLE 3 – Significant coefficients of the multiple regression model $a_0+a_1x_1+a_2x_2+a_3x_1x_2$ for the investigated mechanical properties (x_1: XLPE content; x_2: average XLPE particle size)

	linear terms				adjusted R^2
	a_0	a_1	a_2	a_3	
Impact Strength (kJ/m²)	**14.7384**	**3.462**	**0.498**	**-0.676**	0.97021
Tensile Strength (MPa)	**20.6548**	**-2.533**	0.265	0.199	0.91154
Tensile strain at break (%)	**97.0544**	3.361	**-17.237**	-1.515	0.36035
Flexural Strength (MPa)	**18.9948**	**-1.642**	-0.166	-0.092	0.82684
Flexural Modulus (MPa)	**738.14**	**-76.37**	**-23.78**	5.85	0.78232

TABLE 4 – Comparison of the center point results with the intercepts (a_0) of the model for each investigated property

	a_0	**Result for central point (5%-500 μm)**
Impact Strength (kJ/m²)	14.74	14.6±0.2
Tensile Strength (MPa)	20.65	20.5±0.5
Tensile strain at break (%)	97.05	94±5
Flexural Strength (MPa)	18.99	18.9±0.1
Flexural Modulus (MPa)	738.1	79±16

The model reveals that increasing both XLPE content and particle size increase impact strength, with the effect of the former much stronger than that of the latter. This can be attributed to the rubbery behaviour of cross-linked PE. However, the interaction between the two variables reduces impact strength. For both tensile and flexural strength only XLPE content has a significant and negative effect, whereas for tensile strain at break, only particle size showed to have a significant and negative effect. Increasing both XLPE content and particle size reduce flexural modulus, and like for impact strength, the effect of XLPE content is much stronger than the effect of particle size.

CONCLUSIONS

XLPE scrap, from electrical wire and cables, with average particle size of 100, 500 and 900 micrometer was successfully incorporated in HDPE. A 2 by 2 factorial design with center point was used, where XLPE contents varied between 1 and 9 wt%. Increase in XLPE content significantly increased only impact strength, and decreased tensile and flexural strength, and flexural modulus. The effect on tensile modulus and strain at break was not significant for the adopted 95% confidence interval. The increase in XLPE average particle size affected positively only impact strength of the material, and decreased strain at break and flexural modulus. The effect on tensile and flexural strength and on tensile modulus was not significant for the adopted 95% confidence interval. The interaction between the two variables decreased impact strength and increased tensile modulus, and was not significant for the other investigated properties.

ACKNOWLEDGMENTS

Thanks are due to Centro Universitário FEI and CAPES for the scholarship granted to R.S. Freitas, and to FAEX Soluções Ambientais for donating the XLPE scrap.

REFERENCES

1. Christéen, Jonas. Swedish cable waste for recovery in China or Sweden. 2007. 46f. Master thesis — Linköpings Universitet, Swend 2007.
2. FSRJ — Research Association for Feedstock of Plastics Japan. Mechanochemical recycling and processing of waste crosslinked polymers: waste tire rubber and waste XLPE from cable scraps. Japan, 2016
3. Sirisinha, K; Chuaythong, P., J Mater Sci, 49, 5182-5189, 2014.
4. Oganda, J. Crosslinked PE can be recycled. Plastics Technology, Jun1, 1992
5. Qudaih, R.; Janajreh, I.; Vukusic, S. E. Proc. of the 8th Global Conference on Sustainable Manufacturing, 233-239, 2011.
6. Gallego, M. C.; Bonse, B.C. Proc. of the 9th International Conference on Modification, Degradation and Stabilization of Polymers, Cracow, 2016

Synthesis of MMA/GMA graft and random copolymers by melt and solution copolymerization

Karimi Amirhossein (1), Garmabi Hamid (1)*, Javadi Azizeh (1), Ahmadi Mostafa (1)

(1) Amirkabir University of Technology - Tehran - Iran

In this study Poly(MMA-g-GMA) copolymers was prepared via reactive blending of PMMA and GMA using benzoyl peroxide (BPO) and dicumyl peroxide (DCP), and Poly(MMA-ran-GMA) was synthesized by solution copolymerization of methyl methacrylate (MMA) and glycidyl methacrylate (GMA), using benzoyl peroxide (BPO), as initiator. In order to compare synthesis methods, we evaluate the effect of several items on the chemical composition including the weight percent of MMA and GMA and the initiator concentration. Chemical compositions were determined qualitatively using FT-IR spectroscopy and quantitatively by titration by estimation of the epoxy content of GMA units. It could be concluded that the MMA/GMA copolymers have been successfully synthesized using both melt and solution copolymerization methods. Accordingly the percentage of GMA comonomer units could be directly increased by increasing its fraction in the feed. Keywords: Copolymerization; grafting efficiency; GMA; MMA

Analysis of dynamic oscillatory rheological properties of PP/EVA/organo-modified LDH ternary hybrids based on generalized Newtonian fluid and generalized linear viscoelastic approaches

khonakdar sangdehi hossein ali [(1)*]

[(1)] Leibniz Institute of Polymer Research, 01067 Dresden, Germany - Tehran - Iran

Abstract A comprehensive investigation was performed on single-step melt processed polypropylene (PP)/(ethylene vinyl acetate copolymer (EVA)/organo-modified layered double hydroxide (LDH) counterpart ternary hybrids to explore the effect of LDH loading on small-amplitude oscillatory shear (SAOS) rheological properties and to correlate the properties with microstructure. The rheological results were analyzed in detail from qualitative and quantitative perspectives. Using qualitative interoperation of storage modulus and complex viscosity alteration against LDH loading, and also quantitative analysis by viscosity models based on both the generalized Newtonian fluid (GNF) and generalized linear viscoelastic (GLVE) approaches, detailed predictions were carried out on the microstructure of samples and also partitioning of the organo-modified LDH particles and their intercalation and exfoliation extent within hybrids. By comparing the elasticity and relaxation spectrum of PP-rich/LDH with those of EVA-rich/LDH hybrids, it was predicted that organo-modified LDH platelets, in the case of PP-rich samples, have been located at the interface or within the EVA dispersed particles, while in the case of EVA-rich samples, they mainly localized within the matrix. In addition, the crossover frequency and slope of G0 and G00 curve at terminal region were correlated with the extent of intercalated and exfoliated structures of filler. The validity of these predictions on microstructure of the developed hybrids was confirmed visually using transmission electron microscope (TEM) micrographs.

Thermal and Dynamic Mechanical Properties of PP/EVA Nanocomposites Containing Organo-Modified Layered Double Hydroxides

Khonakdar sangedehi Hossein Ali [1]*

[1] Leibniz institute of polymer research,D-01067Dresden,Germany - Tehran, Tehran - Iran

In this study, organo-modified layered double hydroxides (LDH) as nanoparticle were used to prepare hybrid nanocomposites based on polypropylene (PP), ethylene vinyl acetate copolymer (EVA) blend in presence of a maleated polypropylene (PP-g-MA) as compatibilizer via a one-step melt mixing process. Transmittance electron microscopy (TEM) was employed to determine the place of LDH nanoparticles in polymer blend nanocomposites. Microstructures of hybrid nanocomposites were studied by means of scanning electron microscopy (SEM). Thermomechanical properties of nanocomposites were investigated by dynamic mechanical thermal analysis (DMTA). Differential scanning calorimeter (DSC) and thermo gravimetric analysis (TGA) were also used to investigate crystallization behavior and degradation process of nanocomposites, respectively. The results showed that, LDH nanoparticles in the exfoliated and/or intercalated states preferred to localize in the EVA phase. The average particles size of EVA domains was reduced in presence of LDH and PP-g-MA. The storage modulus of nanocomposites was controlled by combination of LDH and PP-g-MA. The crystallization studies revealed that 5wt.% LDH and PP-g-MA increased crystallization temperature to 133.4oC while addition of LDH in absence of the compatibilizer decreased the crystallization temperature to 122.7oC. Investigation of thermal degradation behavior of samples indicated that presence of LDH and PP-g-MA in PP/EVA blends improved initial degradation temperature of nanocomposites up to 43% compared to the neat PP.

In situ ring-opening polymerization of laurolactam in presence of MWCNT toward Polyamide 12/MWCNT-nanocomposites

ahmadi shervin [(1)], Faridirad Farzaneh [(1)*], Barmar Mohammad [(1)]

[(1)] IPPI - Tehran - Iran

In this article the in situ polymerization of laurolactam in the presence of multiwalled carbon nanotubes (MWCNTs) towards polyamide 12 (PA12)/MWCNT nanocomposite was studied. The processing conditions were optimized with respect to the electrical resistivity and the physical properties. XRD, DSC and ATR analysis were applied in order to investigate the effect of MWCNT of the nanocomposite. Percolation threshold for electrical resistance was occurred at CNT=1.2% which led to 10000 times reduction in the resistivity.

Recycling of Cast Polyamide 6 Using a Lubricant

Benjamino Rocco Formisano[a)] and Christian Bonten

IKT, Institut für Kunststofftechnik, University of Stuttgart, Pfaffenwaldring 32, 70569 Stuttgart, Germany

[a)]Corresponding author: Benjamino.Formisano@ikt.uni-stuttgart.de

Abstract. Cast polyamide 6 is a high-quality material due to its long polymer chains. Because of sprues and post-processing, a comparably large amount of the produced material, up to 30 %, becomes waste. Until today, this high-quality waste is burned or added into standard virgin polyamide 6 raising its molecular weight. Instead of using cast polyamide waste as an additive or combustion material, it would be beneficial to process high-molecular weight recyclates made of this cast polyamide 6. To enhance its processability, using a lubricant as a processing additive is a reasonable approach. Our previous findings within this field revealed an unusual behavior of a lubricant consisting of a polyester modified wax, which is already used for standard polyamide 6 without known problems. In this work, milled cast polyamide 6 chips obtained from machining are washed and dried and then used for producing high-molecular weight recyclates. In literature the deposition of such waste is recommended because of pollution and contaminations. It is hence possible to get clean material for recycling. The recycling step itself is performed on a twin-screw extruder type ZSK 26MC of Coperion, Stuttgart, by processing a dry-blend of cast polyamide and said lubricant. Rheological measurements show increasing complex viscosities of recyclates depending on the lubricant's content. It is possible to achieve viscosities higher than those of the solely processed polyamide recyclate. Even the viscosity of unprocessed cast polyamide 6 can be reached. Compared to virgin extrusion polyamide 6, the processed recyclates achieve higher tensile strengths, while an increasing amount of lubricant does neither affect Young's modulus nor tensile strength. It is also shown that the lubricant raises impact strength significantly.

INTRODUCTION

Cast polyamide 6 (PA6G) is a high-quality polyamide synthesized by an anionic ring opening polymerization of ε-caprolactam. Its high-quality properties, for example wear resistance, are primarily basing on the higher molecular weights compared to typically used standard polyamide 6.

During production of parts made of cast polyamide 6, waste amounts can reach up to 30 %, including sprues, wastes from fluctuations in product quality and post-processing. The annual waste volumes of cast polyamide are estimated to be about 5,000 tons in the European Union. Up to now, cast polyamide waste is burned for energy recovery. Some material is also added into extrusion type polyamide to enhance its molecular weight. But cast polyamide waste also offers a great potential for material recycling.

According to [1], cast polyamide decomposes because of its content of residual catalyst during processing. Because of said decomposition, recycled cast polyamide gets similar mechanical properties as an extrusion type polyamide 6. Literature also recommends mixing recycled PA6G and new extrusion polyamide 6 in a ratio of about 1:1 to reduce product fluctuations; but in this way only an "enhanced" extrusion polyamide is created. Furthermore it is expected that waste from post-processing may lead to difficulties during processing because of contaminations. So it is recommended to dispose this waste separately.

Although this topic is very interesting, only little work is published about recycling of cast polyamide, e. g. [2]. There is some literature about deactivating catalysts in polyamide [3,4], but this work is not used within this field till now. Our previous works within the field of recycling cast polyamide [5] revealed that compounding cast polyamide 6 together with a polyester-modified wax, which is used as a lubricant for polyamide, can result in unexpected raise in viscosity e.g., as lubricants typically reduce viscosity [6].

The objective of this work is therefore, to use this unexpected effect to produce highly viscous polyamide recyclates which will be usable for specific applications in the plastics technology.

EXPERIMENTAL

Materials

Experiments were conducted using milled, washed and dried flakes of cast polyamide 6 waste. Because of the washing step, waste from post-processing can be used without any problems. A polyester-modified wax, Ceralene 694, supplied by EuroCeras Sp., Kędzierzyn-Koźle, Poland, was used as an additive. Before compounding, the additive was mixed with the milled flakes into dry-blends of specific compositions.

Compounding

Processing was done on a ZSK 26 MC from Coperion, Stuttgart, Germany. An atmospheric degassing was installed near the dye to remove volatile oligomers. A double-strand die was used and the received strands were pelletized.

The dry-blends were fed into the compounder, setting the throughput to 5 kg per hour. The compounding step was carried out two times for each batch. Thus, observing the materials' behavior during further processing and potential changes in the material after a longer dwell time was possible. After each compounding step the materials were dried at 80 °C in a vacuum oven until the residual moisture was below 0.1 %. All produced materials and used components are listed in Tab. 1.

TABLE 1. Produced recyclates made of cast polyamide 6

recyclate	components	compounding steps
M0-1/2	pure cast polyamide 6	one/two
M1-1/2	PA6G + 1 wt.-% additive	one/two
M2-1/2	PA6G + 2 wt.-% additive	one/two
M3-1/2	PA6G + 3 wt.-% additive	one/two

Characterization

The linear viscoelastic deformation behavior was investigated using a plate-plate rheometer DSR200 from Rheometric Scientific, Munich, Germany with a plate diameter of 25 mm. All measurements were carried out at 270 °C with a shear stress of 500 Pa under nitrogen atmosphere.

To get further information about the mechanical behavior, test specimens were prepared by injection molding using an Allrounder 220M, 350–390 from Arburg, Loßburg, Germany. The temperature of the melt at the dye was set to 280 °C. A family tool, heated to 80 °C, was used to produce a tension rod and a body for notched impact tests, which was notched afterwards. Before testing, the specimens were dried at 50 °C for 66 h and then cooled down for two hours.

Tensile tests were carried out on a universal tensile testing machine Type 1455 from Zwick and Co. KG, Ulm, Germany, according to DIN EN ISO 527 part 1. Testing speed was 1 mm/min for measuring Young's moduli and 50 mm/min afterwards and preload was set to 0.2 MPa. Notched Charpy impact strength measurements were carried out on an impact testing machine Type 5102.202 from Zwick and Co. KG; an impact hammer of 1 J was used. 20 specimens were tested mechanically for each material.

Crystallinity was analyzed by a differential scanning calorimeter DSC 204 Phoenix from Netzsch, Selb, Germany, equipped with a liquid nitrogen cooling system. The samples were heated from room temperature to 270 °C at a heating rate of 10 K/min followed by cooling to room temperature at the same rate. All tests (heating and cooling) were run in two cycles for each sample.

RESULTS AND DISCUSSION

The flow behaviors of the recyclates are shown in Fig. 1(a). Having a look at the basic cast polyamide 6 first, it can be determined that the complex viscosity of this material is very high. A drop at lower frequencies can also be seen. This drop is due to the decomposition described above. This can be supported by time-sweep measurements – not shown here. Pure recyclates (M0) show complex viscosities reaching 3,100–3,300 Pa·s at a frequency of 0.1 rad/s at 270 °C. The clearly reduced viscosities compared to unprocessed cast polyamide are due to a decomposition during processing. M0 shows no effect of the second compounding step, suggesting that named decomposition is completed within the first step. In the second compounding step, the material is in an equilibrium state, showing no further influences.

Adding 1 % of polyester-modified wax, clearly higher viscosities can be achieved, which are still below those of cast polyamide 6. In this case, the wax, which is used as a lubricant, effects a more gentle processing and therefore less decomposition.

Compounding of cast polyamide using 2 % wax results in a further increase of viscosities. At frequencies higher than about 10 rad/s, they are comparable to those of unprocessed cast polyamide (PA6G). Bellow 10 rad/s, the viscosities of M2-1 and -2 are higher than those of PA6G. It can clearly be seen – supported by time-sweep measurements – that these materials do not suffer from decomposition. So it is possible to produce thermally stable highly viscous recyclates by using 2 % polyester-modified wax.

Raising the wax content to 3 %, no further influence on the flow behavior can be seen during the first compounding step (M3-1). After a second compounding step, the viscosity of M3-2 raises clearly at frequencies bellow 100 rad/s and reaches more than 380,000 Pa·s at 0.1 rad/s, more than twice the value of M2 and M3-1. Especially for this material, and also for M2, a second compounding step raises viscosity. It can therefore be said that higher contents of the used wax need a longer processing time to develop their full effect on the viscosity.

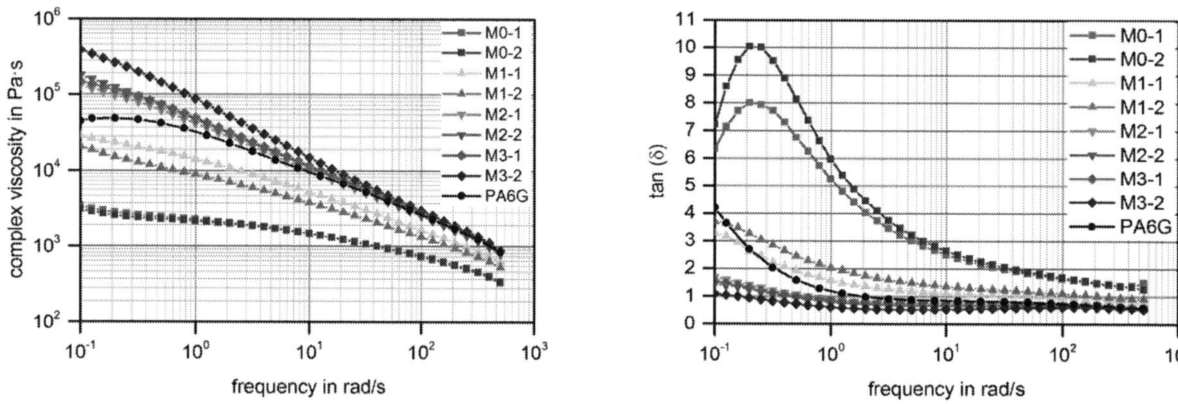

(a) (b)

FIGURE 1. Rheological measurements of the produced recyclates showing frequency-sweep (a) and tan (δ) (b)

Using rheometry, it is also possible to get information about the average molecular weight of polymers by looking at the crossover point of the storage and loss modulus curves: The lower the frequency of the cross over point, the higher the average molecular weight. [7]

For overview purposes, the loss angle tan (*δ*) is shown in Fig. 1(b). It is known that this is the quotient of loss modulus *G"* and storage modulus *G'* according to

$$tan\ (\delta) = \frac{G''}{G'} \quad (1)$$

When tan (*δ*) becomes equal to one, the cross over point is reached. It can be seen in Fig. 1(b) that the crossover point is moving to lower values if the wax content is increased. This means that the average molecular weight raises if raising amounts of polyester-modified wax are added to cast polyamide during compounding. This also explains the shown increase in viscosity.

Having a look at the mechanical properties, shown in Fig. 2, it can be seen that Young's moduli, Fig. 2(a), and tensile strengths, Fig. 2(b), remain unchanged for all recyclates within measurement accuracy. Young's moduli of about 3,300–3,500 MPa and tensile strengths of about 82–84 MPa are achieved, respectively. The Charpy notched impact strength, shown in Fig. 2(c) shows a clear increase for higher amounts of modified wax. While M0 and M1 show similar impact strengths of about 4.5–5 kJ/m², M2 and M3 have much higher impact strengths reaching about 16 kJ/m² for M3-2.

In comparison: Cast polyamide 6 has Young's moduli of about 3,200 MPa, tensile strengths of about 80 MPa and notched impact strengths of about 6 to 7 kJ/m² according to DIN 53445 and ISO 180-1A [1]. Because these measurements were done at specimens produced in a different way (cast polyamide specimens were not injection molded but milled from solid material) comparing is limited. But nevertheless, values remain comparable for M0 and M1.

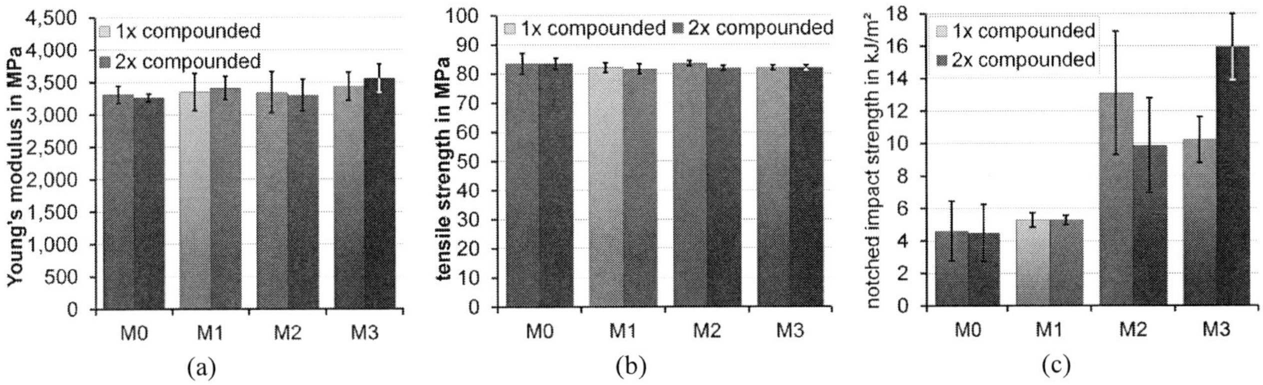

FIGURE 2. Mechanical properties of the different recyclates made of cast polyamide 6

The resulting curve of the second heating phase during DSC is shown in Fig. 3(a). It can be seen that the melting temperatures have similar values between 218 and 221 °C. Thus, the melting temperature is not affected by the used wax.

The melting enthalpies are shown in Fig. 3(b). Within measurement accuracy, no differences between first and second compounding step can be seen. But it can be seen that melting enthalpy is reduced if said wax is added. Pure recyclates reach 81–85 J/g. M3 only shows melting enthalpies of about 71–72 J/g.

According to

$$C = \frac{\Delta H}{\Delta H^0} \tag{2}$$

crystallinity C is proportional to the quotient of melting enthalpy ΔH and melting enthalpy of a 100 % crystalline material ΔH^0 [8]. Thus, a direct connection of melting enthalpy and crystallinity is only an approximation [9]. If we allow this approximation, it can be assumed that crystallinity is reduced by adding polyester-modified wax.

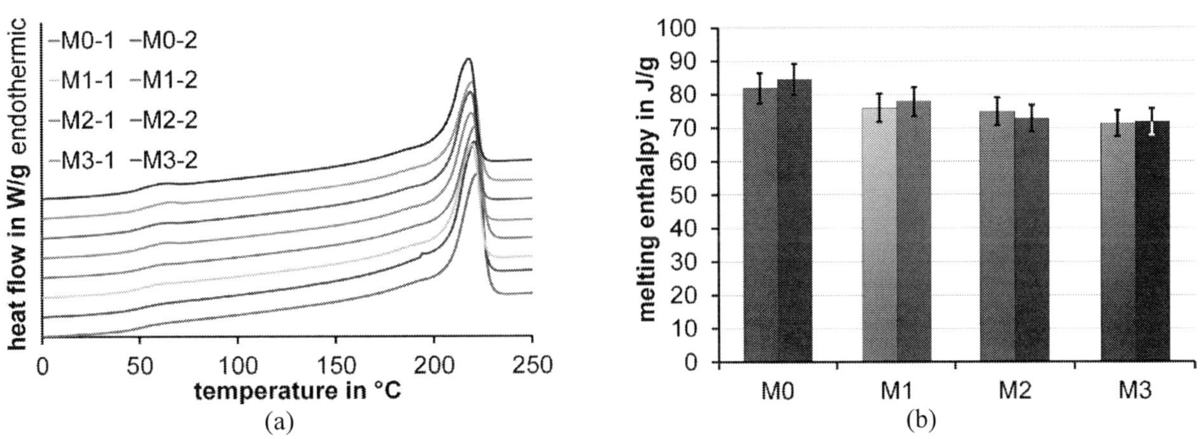

FIGURE 3. Second heating-up curve (a) and melting enthalpies (b) measured by DSC

A reduced crystallinity jointly appearing with significant higher viscosities, raised molecular weights and clearly enhanced impact strength may be an indication for (widely meshed) crosslinking within the polymer. This is confirmed by some solution tests. The used polyester-modified wax induces a crosslinking if added in compounding cast polyamide.

CONCLUSION

Material recycling is a useful way to enhance product efficiency of cast polyamide. It is shown within this work, that a polyester-modified wax, which is currently used as a lubricant for polyamide when used in small amounts, can raise viscosities and molecular weights of cast polyamide recyclates, if used in larger amounts. Furthermore, impact strength can be enhanced significantly.

Raising viscosity, molecular weight and impact strength along with a reduction in melting enthalpy can be an indication for crosslinking within the polymer, which was confirmed by solution tests. Therefore, the used polyester-modified wax induces a (widely meshed) crosslinking if added in compounding cast polyamide. This can be a good starting point to develop a new class of high molecular weight polyamides.

ACKNOWLEDGEMENT

The findings presented in this work are part of the strategic preliminary research of the IKT. We want to thank the Schwartz GmbH, Xanten, and the keim additec surface GmbH, Kirchberg, kindly supporting our work with material.

REFERENCES

1. A. Fahnler and A.A. van Greenen, "Polyamide aus technischer anionischer Lactampolymerisation", in *Technische Thermoplaste – Polyamide*, edited by L. Bottenbruch and R. Binsack (Hanser, Munich, 1998), pp. 406-431.
2. H.G. Seelig and F. Walter, "Entwicklung eines Recycling-Verfahrens für Guß-Polyamide", final report Az: 06502, Willich, 1998.
3. M. Biensan and P. Potin, German patent No. 2241133 (1 March 1973).
4. E. Schmid and I. Laudonia, German patent No. 10118453C1 (12 December 2002).
5. B. Formisano, S. Göttermann and C. Bonten, "Recycling of Cast Polyamide Waste on a Twin-Screw-Extruder", in: *AIP Conference Proceedings* **1779**, 140002; DOI: 10.1063/1.4965582 (2016)
6. C. Bonten, *Kunststofftechnik*, 2nd edition (Hanser, Munich, 2016)
7. T.G. Mezger, *Das Rheologie-Handbuch* (Vincentz, Hanover, 2000), pp. 152–174.
8. G.W. Ehrenstein, G. Riedel, and P. Trawiel, *Praxis der Thermischen Analyse von Kunststoffen*, 2nd edition (Hanser, München, 2003), pp. 1–119
9. K.-H. Illers, H. Haberkorn, „Schmelzverhalten, Struktur und Kristallinität von 6-Polyamid", in: *Die Makromolekulare Chemie* **142**, pp. 31–67 (1971)

Influence of modified and unmodified rice husk on mechanical properties of Poly (lactic acid)

Khaghanipour Maryam* (1)*, Khodabakhshi Khosrow* (1)

(1) Iran Polymer and Petrochemical Institute - Tehran - Iran

In this paper at first,alkaline treatment on rice husk was done to remove waxes, lignin and hemicellulose.That means more hydroxyl groups were existed in compare of unmodified rice husk.Then γ-aminopropyltrimethoxysilane (APS) treatment applied on rice husk to make better interaction between Poly (lactic acid) (PLA) and rice husk.Functional groups of APS on rice husk were investigated by FTIR. At the end, PLA microcomposites with different component of rice husk from 10-30 wt % were prepared and melted in Brabender internal mixer.mechanical properties were investigated by Tensile strength.

Rheological and morphological correlations in PP/EVA blends containing nanoclay in presence of a halogen-free flame retardant

Khonakdar Hossein Ali (1)*, *Wagenknecht Udo* (1), *Heinrich Gert* (1)

(1) Leibniz Institute of Polymer Research Dresden - Dresden - Germany

Abstract In this work attempts were made to establish correlation between rheological and morphological properties of nanoclay-filled polypropylene/ethylene vinyl acetate copolymer (PP/EVA) blends containing a new generation of halogen-free flame retardants commercially known as "NOR116". Scanning electron microscope (SEM), transmission electron microscope (TEM) and wide angle X-ray scattering (WAXS) analyses were implemented in order to investigate the morphological properties. Effect of presence of two different types of compatibilizer, the flame retardant and organoclay on morphology of the nanocomposites was explored. By considering chemical nature of compatibilizers and the SEM and TEM micrographs, organoclay localization was accurately predicted. Using rheological measurements in three different modes of frequency, temperature and time sweeps, linear viscoelastic properties alterations of the samples in terms of temperature variations and consequent morphological evolution were investigated. Using the frequency sweep experiment miscibility level of blends and their interfacial properties were also studied. A good correlation between these properties and morphological observations was established. Also, it was found that the temperature dependence linear viscoelastic properties alterations strictly depend on the blend morphology and chemical structure of components. In case of neat blend a single phase morphology indicating of enhanced miscibility was detected at higher temperatures.

Efficient hydrosilylation reaction in polymer blending: An original approach to structure PA12/PDMS blends at multiscales

Li Jingping [(1)], *Cassagnau Philippe* [(1)*], *Bounor-Legaré Véronique* [(1)]

[(1)] University of Lyon - Lyon - France

An in situ amide hydrosilylation reaction was developed to prepare polyamide 12 (PA12)/polysiloxane by reactive blending. This reaction is focused on the addition of hydrogenosilane groups (SiH) from polysiloxane to the carbonyl group from the PA12 amide function. To evidence this carbonyl hydrosilylation onto an amide based polymer, an approach on model compounds (use of N-methylpropionamide) was carried out. The mechanism and kinetics were investigated with multinuclear NMR (1H, 13C and 29Si). During kinetics studies, the concentration of N-silylated copolymers can reach 70 mol% after 2 hours reaction at 100 °C. Amide hydrosilylation reaction was extended to the reactive blending of polyamide 12 with PDMS under molten processing conditions. Formation of a structured blend was investigated by rheology and electronic microscopy at different scales. The impact of both shearing and reaction on the final morphology was deeply studied and the interfacial enhancement by compatibilization was confirmed. As a result, the dispersion of PDMS domains decreased from 3-4 ìm to around 0.8 ìm in diameter forming submicronic morphology. Furthermore, it is possible to control the dispersion of PDMS at different scales by modifying the physico-chemical parameters (molar mass and functionality) of the two components.

Melt strengthening of poly (lactic acid) and its blends: Shear and elongation rheological investigations for forming process

Maazouz Abderrahim [(1)]*, Khalid Lamnawar [(1)]

[(1)] Université de Lyon, CNRS, UMR 5223, Ingénierie des Matériaux Polymères, INSA Lyon - Lyon - France

The poly (lactic acid) (PLA), through its organic origin and its biodegradation properties, can be a good alternative to petroleum-based polymers. Through this work, we present some promising routes to enhance its processing ability which presents several challenges mainly due to the poor shear and elongation properties of this biopolymer. To our knowledge, there is no paper dedicated to the investigation of foaming and/or blown extrusion of PLA that involves structural, rheological and thermo-mechanical properties. Two main routes were selected: (i) the modification of its structure, rheological and thermo-mechanical properties and (ii) blending it with other ductile thermoplastic biopolymer such as the poly (butylene adipate-co-terephtalate) (PBAT) or polyamide (PA11). Various formulations of PLA with multifunctionalized epoxy, nucleants and plasticizer were prepared and characterized on the basis of their linear viscoelasticity and extensional properties. The balance of chain extension and branching has also been investigated using solution viscosimetry, steric exclusion chromatography (SEC) and rheology (Shear and elongation rheology). On one hand, a batch foaming assisted with supercritical CO_2 was achieved. The influence of the foaming parameters, the extent of chain modification as well as the contribution of crystallization on cell morphology was evaluated. Based on these parameters, structures ranging from micro to macro-cellular-cell were obtained. On the other hand, the stability maps of blown processing for neat and modified PLA were established at different die temperatures. We have achieved a great enhancement of the blown processing windows of PLA with high BUR (Blow Up Ratio) and TUR (Take Up Ratio) attained. We were able to demonstrate that a higher kinetic of crystallization can also be reached for chain-extended and branched PLA formulated with adequate amounts of nucleants and plasticizers.

Spin-correlated charge transfer in low-band-gap copolymer composites

Krinichnyi Victor (1)*, Yudanova Evgenija (1), Bogatyrenko Victor (1)

(1) Institute of Problems of Chemical Physics RAS - Moscow Region - Russia

Light-Induced EPR (LEPR) study of magnetic, relaxation and dynamic parameters of spin charge carriers stabilized in polyaniline doped with toluenesulfonic acid (PANI:TSA), spin pairs of polarons and methanofullerene ion radicals photoinitiated in bulk heterojunctions formed by fullerene derivative, [6,6]-phenyl-C61-butyric acid methyl ester (PCBM), with low-band-gap poly[2,7-(9,9-dioctylfluorene)-alt-4,7-bis(thiophen-2-yl)benzo-2,1,3-thiadiazole] (PFO-DBT) and poly[N-9'-heptadecanyl-2,7-carbazole-alt-5,5-(4',7'-di-2-thienyl-2',1',3'-benzothiadiazole)] (PCDTBT) copolymers as well as in respective PANI:TSA/PFO-DBT:PCBM and PANI:TSA/PCDTBT:PCBM composites are described. Magnetic resonance parameters of all charge carriers photoinitiated in these composites at wide temperature and photon energy regions were determined from deconvolution of LEPR spectra. The magnetic resonance, dynamics and recombination order of spin charge carriers were first shown to be governed by the structure and morphology of their polymer matrices as well as by the concentration and spatial distribution of spin traps reversible formed in composites under illumination. The concentration and energy depth of such traps are determined by the photon energy. These parameters can be additionally handled by the bringing into their bulk of a guest spin ensemble. Spin exchange interaction was shown to originate spin- and photon-correlate effective charge transfer in such multispin composite. Spin-assisted magnetic, relaxation and dynamics parameters of the PANI:TSA/PFO-DBT:PCBM and PANI:TSA/PCDTBT:PCBM multispin composites was interpreted in the framework of exchange of mobile spin pairs differently distributed in parental polymer matrices. An exchange interaction of localized guest and domestic spins with mobile spin pairs opens independent channels for precision controlling/handling their electronic properties for further creation of organic electronic and spintronic devices.

Development of shape memory polymers based on TPU/ABS blends containing MWCNT

Memarian Farzaneh (1)*, Fereidoon Abdolhossein (1), Khonakdar Hossein Ali (2), Ghorbanzadeh Ahangari Morteza (3)

(1) Faculty of Mechanical engineering, Semnan University, Semnan, Iran - Semnan - Iran, (2) Leibniz Institute of Polymer Research, D-01067 Dresden, Germany - Dresden - Germany, (3) Faculty of Engineering and Technology, University of Mazandaran, Babolsar, Iran - Mazandaran - Iran

Development of shape memory polymers based on TPU/ABS blends containing MWCNT Farzaneh Memarian1*, Abdolhossein Fereidoon1, Hossein Ali Khonakdar2 1Faculty of Mechanical engineering, Semnan University, Semnan, Iran 2Leibniz Institute of Polymer Research, D-01067 Dresden, Germany Corresponding author: Farzaneh Memarian (Email: farzane.memarian@gmail.com) Abstract In this work, shape memory polymers based on the TPU/ABS blends containing pristine and organo-modified multi-walled carbon nanotubes (MWCNTs) were prepared by melt blending technique. Mechanical, Izod impact and shape memory properties of the samples were investigated. Microstructures of hybrid nanocomposites were studied by means of scanning electron microscopy (SEM). The morphology results showed that the pristine MWCNTs were dispersed homogenously throughout the TPU matrix. Unexpectedly, the ABS droplets in nanocomposites with organo-modified MWCNTs were greater than ABS droplets in TPU/ABS/CNT nanocomposites. These differences were attributed to different degrees in interfacial interactions between the modified MWCNT and TPU matrix. The presence of modified MWCNT in TPU matrix could have increased the differences in surface energy and polarity of TPU and ABS phases and it has disturbed the compatibility of the blend considerably. Loading 2 wt.% MWCNTs resulted in a significant improvement in mechanical properties such as tensile strength and Young's modulus. Also, addition of this amount of MWCNT results in 32% increment in impact strength of the nanocomposites at -50 °C. In addition it was revealed that, the mechanical properties of nanocomposites with functional MWCNTs are feeble than those containing pristine MWCNTs. The shape memory results showed that the shape recovery and fixity of TPU/ABS blend improved significantly upon increasing MWCNTs loading. Keywords: Thermoplastic Polyurethane (TPU); Acrylonitrile butadiene styrene (ABS); Shape memory; MWCNT; organo-modified MWCNT.

Compatibilization of PET-PE Blends for the Recycling of Multilayer Packaging Foils

Laurens Delva[1, a)], Cédric Deceur[1, b)], Nicolas Van Damme[1, c)] and Kim Ragaert[1, d)]

[1]Centre for Polymer and Materials Technologies (CPMT), Department of Materials, Textiles and Chemical Engineering, Faculty of Engineering and Architecture, Ghent University, Technologiepark 915, 9052 Zwijnaarde, Belgium

[a)]Corresponding author: Laurens.Delva@ugent.be
[b)]Cedric.Deceur@ugent.be
[c)]Ncvdamme.VanDamme@ugent.be
[d)]Kim.Ragaert@ugent.be

Abstract. The compatibilization of a virgin polyethylene terephthalate/polyethylene (PET/PE) blend and a recycled multilayered post-industrial food tray of similar composition are investigated. The compatibilizers of focus are a propylene ethylene rubber (PER) and a styrene ethylene / butylene styrene triblock copolymer grafted with maleic anhydride (SEBS-g-MA).This work examines the effect of these compatibilizers on both the mechanical properties (impact and tensile properties) and morphology of the virgin blends and verifies whether these effects are transferable to an actual waste stream of post-industrial food trays with comparable composition. Regarding the virgin blends, it was concluded that both compatibilizers increase the impact strength and elongation at break of the material, wherein PER increases the impact strength more effectively than SEBS-g-MA. Moreover, PER succeeds in maintaining both tensile strength and stiffness of the blends. Also, the morphology of the PER blends shows clear indication of rubber toughening. The results of the virgin blends are to a certain extent transferable to an actual waste stream consisting of post-industrial food trays with comparable composition. PER shows to be the most promising candidate for increasing toughness while maintaining stiffness and tensile strength.

INTRODUCTION

In multilayered packaging materials, different polymers are stacked in layers in one foil or sheet to maximize the performance of the packaging material. Each polymer contributes in his own way to the overall technical functionality. Polyethylene terephthalate (PET) is often used for water and gas barrier and mechanical strength, while polyethylene (PE) is often used because of its excellent sealing properties, water barrier properties and low-temperature performance. Common combinations in multilayered foils and sheets are PET/PE, polyamide (PA)/PE and PET/PE/ethylene vinyl alcohol (EVOH)/PE[1].

Since the different polymers are physically connected, it is very difficult or economically not feasible to separate them from another. This implies they will have to be mechanically recycled together. Due to the low miscibility of these polymers, the blends have in turn inferior mechanical properties. According to Plastics Recyclers Europe annually 700000 tonnes of PET containing food trays are not sorted out and not effectively recycled in Europe[2]. In order to be able to use this waste stream for new closed-loop or open-loop products, the mechanical properties must be improved. This can be done either by using additives that improve miscibility or specific custom processing techniques[3].

Mainly, we can divide the multilayered packaging waste in post-industrial and post-consumer waste. Post-industrial waste typically is production waste coming from foil producers (e.g. cutting edges, changeovers or non-conform products) and packagers (e.g. unfilled trays, leftovers or cutting edges). This waste is typically quite clean and uncontaminated. Post-consumer waste, however, is mostly still contaminated with organics or other plastics[4].

The purpose of this study is to examine whether the mechanical properties of a recycled post-industrial waste stream consisting of PET trays and films can be improved by adding propylene ethylene rubber (PER) or styrene ethylene / butylene styrene grafted with maleic anhydride (SEBS-g-MA). The influence of these additives on the morphology is also investigated.

Firstly, virgin components are used to simulate the selected waste stream. The virgin blend consists of PET and PE and is processed with and without the compatibilizers. These blends are then mechanically characterized with tensile and impact tests. Furthermore, the morphology is determined using Scanning Electron Microscopy (SEM). Finally, a real waste stream is used consisting of post-industrial food trays. This waste stream is also processed with and without compatibilizers and then mechanically characterized. As with the virgin blends, morphology is also determined using SEM. It is intended to determine the effect of the compatibilizers on an actual waste stream as well as the transferability of the results of the virgin blends to this waste stream.

EXPERIMENTAL

Materials

The selected PET used as matrix is Lighter™ C93 produced by Equipolymers. This PET is made of purified terephthalic acid (PTA) and monoethylene glycol (MEG). It has a melting temperature of 247 °C and a glass transition temperature of 78°C. For PE, Exceed™ 1012 HA mVLDPE produced by Exxonmobil is used. It consists of a metallocene ethylene-hexene copolymer and is designed for the production of sheets. The used PE has an MFI of 1.0 g/10 min and melting temperature of 115°C. The used SEBS-g-MA as compatibilizer is FG1901 G Polymer produced by Kraton. The grafting level is 1.4 to 2.0 wt%. The used PER is Vistamaxx™ 6202 produced by Exxonmobil. It is primarily composed of isotactic propylene repeat units (85 wt%) with random ethylene distribution (15 wt%). Postindustrial food trays were offered by Belgian packaging company Ter Beke and consist predominantly of PET and PE.

Sample Preparation

Samples were prepared by melt blending, followed by injection moulding. The blend components were dried at 60°C during 6h in a vacuum dryer prior to melt processing. The dried pellets were mixed in the weight ratios shown in Table 1. The food trays were shredded before they were fed into the extruder for melt blending. The melt blending of the different mixtures was performed on a co-rotating twin-screw extruder Coperion ZSK 18 MEGAlab. The temperature profile starting from the feeding zone to the die was 190°C – 240°C – 245°C – 245°C – 245°C – 250°C – 250°C – 255°C – 260°C. The screw speed was maintained at 200 rpm. The extrudate was cooled down and stranded in a water bath before pelletization. Before injection moulding, the pellets were dried again at 60°C for 6 hours.

Samples for impact (ISO 179 type 1) and tensile (ISO 527 type 1A) testing were injection moulded with an Engel ES 330 / 80 HL with barrel temperatures spanning the range 250°C to 280°C and the mould temperature at around 15-20°C.

TABLE 1. Composition of the samples.

		PET [wt%]	PE [wt%]	rTray [wt%]	SEBS-g-MA [wt%]	PER [wt%]
Virgin	PET/PE	80	20	-	-	-
	PET/PE/SEBS-g-MA-2.5	78	19.5	-	2.5	-
	PET/PE/SEBS-g-MA-5	76	19	-	5	-
	PET/PE/PER-5	76	19	-	-	5
Recycled	rTray	-	-	100	-	-
	rTray/SEBS-g-MA-5	-	-	95	5	-
	Tray/PER-5	-	-	95	-	5

Characterization

Scanning electron microscopy (SEM) was used to study the surface morphology of the samples using a JSM7600F FEG SEM. The brittle fracture surfaces of the samples were examined. The brittle fracture was achieved by cryogenic breaking after liquid nitrogen cooling. The samples were sputtered with gold by a Baltec SCD005 sputter coater. The accelerating voltage was 15-20 kV.

Charpy impact properties (ISO 179 notched) were measured using a Tinius Olsen IT 503 Pendulum Impact Tester, equipped with a hammer with an energy of 2 J.

Tensile properties (ISO 527) were determined using an Instron 5565 tensile testing machine, with a load cell of 5 kN. The dynamometer was equipped with a Dynamic Extensometer 2620-603 by Instron, which was used to accurately measure the strain. The crosshead speed was set at 1 mm/min to determine the tensile properties until yielding. Then the crosshead speed was increased to 25 mm/min until failure.

All mechanical tests were performed at 23°C and 50% relative air humidity. All the reported results are the mean of at least ten measurements. Statistical analysis of all results was performed by the software package SPSS Statistics 22 (t-independent sample tests, p=0.05).

RESULTS

Mechanical properties

Table 2 gives an overview of the mechanical properties of the different virgin and recycled blends. The impact test shows that the addition of both SEBS-g-MA and PER has a positive influence on the toughness of the virgin blends indicated by an increase in strain at break (ε_b) and impact strength. The addition of 5 wt% SEBS-g-MA and PER respectively doubles and triples the value of the impact strength in comparison with the virgin PET/PE blend. In case of the SEBS-g-MA, both the tensile modulus (E_t) and the tensile stress (σ_t) are significantly reduced upon the addition of the compatibilizer. This reduction indicates a poor effectiveness of the compatibilizer. PER, on the other hand, succeeds in maintaining both tensile strength and stiffness in comparison with the virgin blend. A plot of stiffness and impact strength for the virgin blends is shown in Figure 1a. These results propose PER as the ideal candidate to increase toughness while maintaining stiffness and tensile strength for this specific blend.

The results of the mechanical tests of the virgin blends are to a certain extent transferable to the recycled post-industrial trays. It is seen that PER similarly increases the strain at break and impact resistance but the increase is less pronounced. Adding SEBS-g-MA, however, does not cause a significant increase in impact resistance. The strain at break in the rTray/SEBS-g-MA-5 blend reaches yet the highest value, which indicates there is some degree of compatibility. The results show that the recycled blends perform better than their virgin equivalents despite their comparable composition. A possible explanation is the usage of a different grade of PET and PE which affects compatibility between PET and PE and changes the starting mechanical properties of the materials[5]. A plot of stiffness versus impact strength for the recycled blends is shown in Figure 1b. PER remains the preferred candidate to increase the toughness of the recycled trays despite a slight decrease in stiffness.

TABLE 2. Mechanical properties of the blends with indication of the standard deviation.

		Notched impact [kJ/m²]	**E_t [MPa]**	**σ_t [MPa]**	**ε_b [%]**
Virgin	PE	> 100	95 ± 9	15.22 ± 0.49	571.46 ± 13.10
	PET	2.17 ± 0.19	2479 ± 129	41.01 ± 6.01	2.47 ± 0.72
	PET/PE	2.15 ± 0.07	1409 ± 36	28.07 ± 0.40	4.55 ± 1.19
	PET/PE/SEBS-g-MA-2.5	3.33 ± 0.29	1307 ± 58	26.64 ± 0.54	8.62 ± 2.37
	PET/PE/SEBS-g-MA-5	4.14 ± 0.43	1280 ± 57	25.75 ± 0.48	34.94 ± 18.05
	PET/PE/PER-5	6.24 ± 0.16	1409 ± 37	27.41 ± 0.54	41.54 ± 15.90
Recycled	rTray	3.99 ± 0.32	1677 ± 91	31.40 ± 1.24	14.14 ± 10.20
	rTray/SEBS-g-MA-5	4.63 ± 0.56	1550 ± 50	31.97 ± 0.75	355.57 ± 24.73
	Tray/PER-5	5.78 ± 0.20	1532 ± 80	28.35 ± 0.85	23.58 ± 6.91

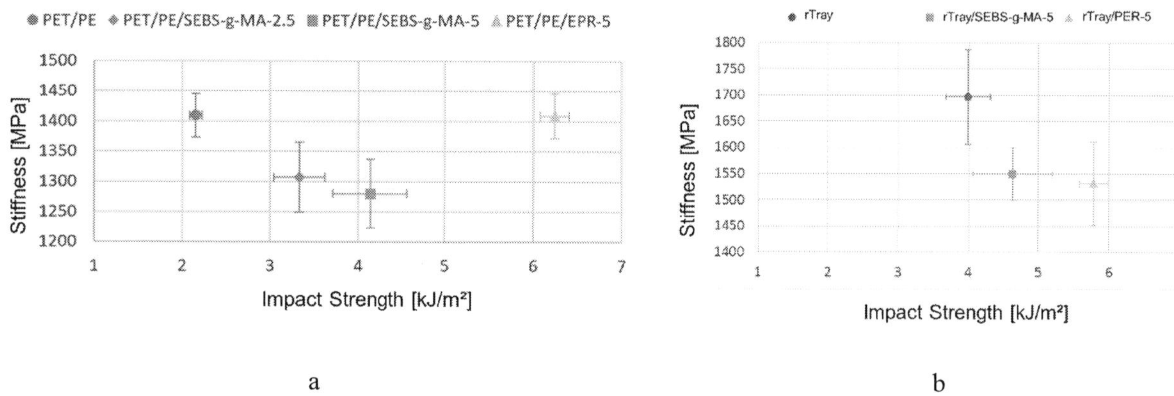

FIGURE 1. Plot of stiffness versus impact strength for (a) virgin and (b) recycled tray blends.

Morphology

Figure 2 shows the SEM-micrographs of the different blends. Figure 1a displays the morphology of a typical immiscible blend in which PE droplets are dispersed in the PET matrix. In the virgin blends (Figures 1a, b and c) there is no clear improvement in morphology by the addition of compatibilizer. Most results from literature focus on the polar component being the minor phase, while in this research it is used as matrix polymer. The average size of the dispersed PE particles remains the same. Both additives show no compatibilization effect at the interface. In case of the PER blend, the presence of a new, fine morphology in the PET matrix can be noticed. This finely divided rubber phase with small interparticle distances can explain why PER performs better in terms of impact strength than SEBS-g-MA[6].

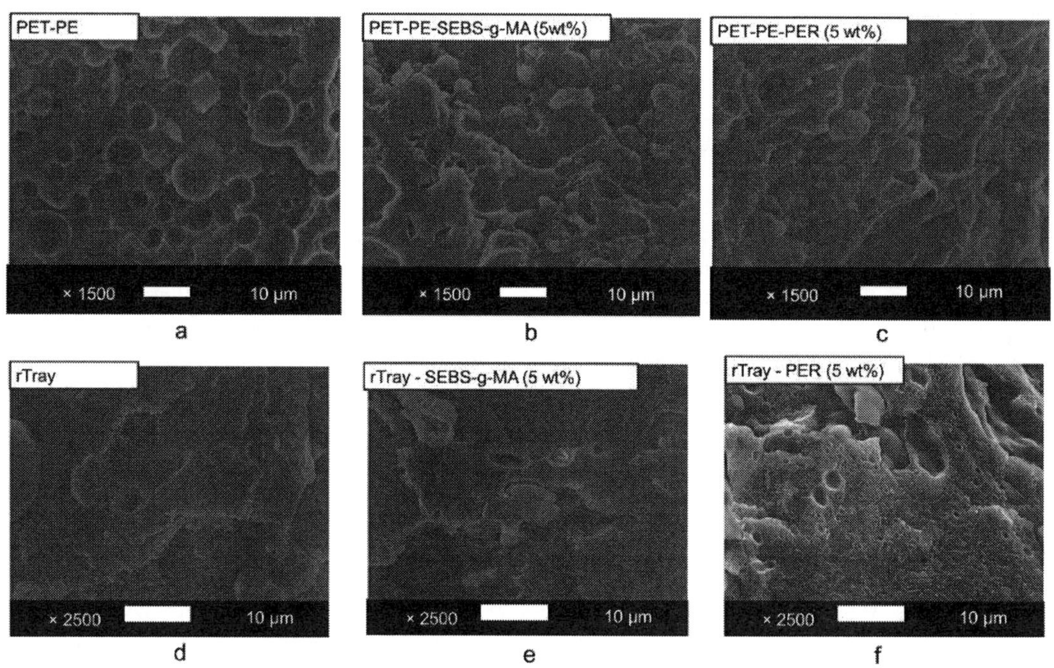

FIGURE 2. SEM-micrographs of the different blends: (a) PET/PE, (b) PET/PE/SEBS-g-MA-5, (c) PET/PE/PER-5, (d) rTray, (e) rTray/SEBS-g-MA-5 and (f) rTRay/PER-5

In case of the compatibilized recycled blends (Figures 1e and f), the size of the dispersion particles appears to be as large as the recycled trays (Figure 1d), confirmed by determining the particle size with imaging software. This indicates that in the recycled blends the compatibilizers similarly fail to reduce the size of the dispersed phase. Once again the presence of a very fine dispersion phase can be seen at the recycled blend to which PER was added. Consequently, it is plausible to assume that both PER and SEBS-g-MA have an impact-modifying rather than a compatibilizing effect, confirmed by the mechanical results.

CONCLUSIONS

This work examines the effect of two compatibilizers SEBS-g-MA and PER on the mechanical properties (impact and tensile properties) and morphology of an 80/20 PET-PE virgin blend and verifies whether these effects are transferable to an actual waste stream of post-industrial food trays with comparable composition. Regarding the virgin blends, it can be concluded that both compatibilizers increase the impact strength and elongation at break of the material, wherein PER increases the impact strength more effectively than SEBS-g-MA. Moreover, PER succeeds in maintaining both tensile strength and stiffness of the blends. Besides the mechanical performance, the morphology of the PER blends shows clear behavior of rubber toughening. The results of the virgin blends are to a certain extent transferable to an actual waste stream consisting of post-industrial food trays with comparable composition. PER also increases the toughness of the recycled food trays. However, the increase is less significant and a small decrease in stiffness can be noticed. Overall, PER shows to be the most promising candidate for increasing toughness while maintaining stiffness and tensile strength. To conclude, these results look promising for the recycling of multilayered plastic waste streams.

REFERENCES

1. J. R. Wagner Jr, *Multilayer flexible packaging*. (William Andrew, 2016).
2. E. Tarlowska, *PET trays existential dilemma*. (Plastics Recyclers Europe, 2014).
3. C. Koning, M. van Duin, C. Pagnoulle and R. Jerome, Prog. Polym. Sci. **23** (4), 707-757 (1998).
4. S. Huysman, J. De Schaepmeester, K. Ragaert, J. Dewulf and S. De Meester, Resources, Conservation and Recycling **120**, 46-54 (2017).
5. Y. Zhang, H. Zhang, W. Guo and C. Wu, Polymers for advanced technologies **22** (12), 1851-1858 (2011).
6. G. Sarkhel, A. B. and A. Banerjee, Polym. Plast. Technol. Eng. **49** (10), 1049-1054 (2010).

Development of highly-filled polymer compounds for fused filament fabrication of ceramics and solvent debinding

Cano Santiago (1)*, Gonzalez-Gutierrez Joamin (2), Sapkota Janak (2), Kukla Christian (3), Holzer Clemens (1)

(1) Montanuniversitaet Leoben,Chair of polymer Processing - Styria - Austria, (2) Montanuniversitaet Leoben, Chair of polymer Processing - Styria - Austria, (3) Montanuniversitaet Leoben, Industrial Liaison Department - Styria - Austria

Fused Filament Fabrication (FFF) is one of the most widely used additive manufacturing processes in the world, due to its simplicity to use and lower cost of the processing equipment. It has been demonstrated that it is possible to produce complete ceramic parts shaped by FFF combined with thermal debinding and sintering [1, 2]. In order to be able to use conventional FFF machines for the shaping of ceramic parts, a highly-filled compound of polymer and ceramic powder needs to be prepared. This compound has to fulfil many requirements: for a constant flow of material a filament must have constant dimensions and shape; the filament should have sufficient stiffness so that it can be pushed without buckling during printing; the filament should be flexible enough to be spooled for easy storage and feeding to the FFF machine; and the polymeric binder should be able to be removed without destroying the printed part. The objective of this investigation was the development of polymer compounds being able to carry out the FFF process and at the same time to be debindable by a solvent. By leaching the major part of the polymers in a first stage with the solvent debinding process, a shorter thermal cycle is required for the degradation of the rest of the polymers, greatly speeding the overall removal of the polymeric binder prior to sintering. Several compounds (feedstocks) were prepared by mixing different polymer combinations with a ceramic powder. The viscosity of the compounds was measured. Filaments were prepared and their tensile properties were analysed. Finally, debinding by immersion in an organic solvent was investigated. References 1. T.F. McNulty, F. Mohammadi, A. Bandyopadhyay, D.J. Shanefield, S.C. Danforth, and A. Safari, Rapid Prototyping Journal, 4, 144 (1998). 2. N. Venkataraman, S. Rangarajan, M.J. Matthewson, B. Harper, A. Safari, S.C. Danforth, G. Wu, N. Langrana, S. Guceri, and A. Yardimci, Rapid Prototyping Journal, 6, 244 (2000).

Influence of Glass flake nanoparticles on thermal and dynamic mechanical properties of PP/EVA blends: Correlation with microstructure

Otady Maryam [(1)]*, *Solymani Maryam* [(1)], *Goodarzi Vahabodin* [(2)]

[(1)] Islamic Azad University - Tehran - Iran, [(2)] Applied Biotechnology Research Center, Baqiyatallah University of Medical Sciences - Tehran - Iran

ABSTRACT In this work attempts were made to establish correlation between thermal, dynamic mechanical properties and morphological properties of polypropylene/ethylene vinyl acetate copolymer (PP/EVA) blends containing glass flake (GF) as an inorganic nano-filler. Compatibilized and non-compatibilized PP/EVA blends loaded with 5 wt.% of GF were prepared by melt mixing process and their thermal and dynamic mechanical properties were determined. TEM investigations revealed that the GF nanoparticles were mainly located within the EVA phase of the blend. Also, microstructure of blends studied by SEM showed that incorporation of GF into the blends led to reduction of dispersed domain size attributed to reduction of interfacial tension due to the filler presence. Melting temperature of PP loaded with 5 wt.% of GF was increased up to 11oC as compared to the neat PP. Addition of GF and PP-g-MAH as compatibilizer to PP/EVA blend increased melting and crystallization temperatures determined by DSC. DMTA studies showed that addition of GF into the blends had no positive effect on the storage modulus of nanocomposites while, their loss modulus were higher than that of the neat PP. Moreover, the β transition temperature of PP phase in the blends loaded with GF and compatibilizer was about 5oC higher than the neat PP. The thermo-oxidative and thermal degradation behavior of developed nanocomposites determined by thermal gravimetric analysis (TGA) showed that thermal stability of nanocomposites was improved and their rate of degradation was reduced which were attributed to network-like structure formation of nano-fillers in the blends. Keywords: Nanocomposites; polypropylene; Ethylene vinyl acetate copolymer; blends; Glass flake; Dynamic mechanical properties

Compatibilization effect of Silica nano particles on Rheology, morphology and mechanical properties of immiscible PA6/LDPE blends

malekmohammad melika [(1)*]

[(1)] iran polymer and petrochemical - Tehran - Iran

A series of Nylon 6/ Low-density polyethylene (PA6/LDPE) (90/10 w/w) and (10/90 w/w) containing 0.25, 0.5, 0.75 and 1% phr modified Silica nanoparticles were melt processed and the effect of nanoparticles on morphological properties was investigated. SEM micrographs revealed a decrease in the average number of domain diameter (Dn) of LDPE phase for Silica nanoparticles amount up to 0.5% phr in the PA6-rich blend, whereas, an increase in Dn of PA6 was observed up to 0.75 phr silica nanoparticles which led us to conclude the preferential location of nanoparticles in PA6. Rheological relaxation time of the blends also suggests that the morphology transition can be related to the changes in dynamic asymmetry between PA6-rich and LDPE-rich phase. Improvement in storage modulus measured by DMTA was relatively more in PA6-rich blend than LDPE-rich blend which is attributed to the reinforcing effect of nanoparticles arising from the interaction with the PA6 chains.

Development of PLA-PBAT and PLA-PBSA bio-blends: Effects of processing type and PLA crystallinity on morphology and mechanical properties

H. Oguz, C. Dogan, D. Kara, Z.T. Ozen, D. Ovali, and M. Nofar*

Metallurgical & Materials Engineering Department, Faculty of Chemical and Metallurgical Engineering, Istanbul Technical University, Maslak, Istanbul, 34469, Turkey
**: Corresponding author: nofar@itu.edu.tr*

Abstract. In this study, immiscible blends of 75 wt% polylactide (PLA) with 25 wt% of poly[(butylene adipate)-co-terephthalate] (PBAT) were prepared through an injection-molding (IM) process as well as a twin-screw extruder (TSE) followed by IM. An amorphous and a semi-crystalline PLA were also used as the matrix to investigate the matrix crystallization effect on morphology and properties development of the blends. Blend of PLA with 25 wt% of poly[(butylene succinate)-co-adipate] (PBSA) was also prepared through IM to compare its mechanical properties with those of PLA-PBAT blends. Eventually, the tensile and flexural properties of the prepared blends were explored and their dependency on the evolution of the blend morphology was investigated. The tensile results showed that when IM is used as the sole processing technique, the ductility and toughness were significantly improved only when the semi-crystalline PLA was used as the matrix. Preprocessing through TSE also resulted in enhancement of the PLA blends' strain at break. According to the flexural experimental results, the morphology variation effect on the PLA blends' flexural properties was not as determinative as it appeared in PLA blends' tensile properties.

Keywords: PLA, PBAT, PBSA, blend, morphology, mechanical properties

INTRODUCTION

Recently, biopolymers have attracted much attention from academia and industries due to environmental concerns and non-composability issues that exist in petroleum-based polymers [1]. Polylactic acid (PLA) is one of the most well-developed commercial bio-based/bio-degradable polymers which is produced from the ring-opening polymerization through the use of renewable resources such as cornstarch or sugarcane [2-3]. As a linear thermoplastic aliphatic polyester polymer, PLA is considered as a copolymer which contains L-lactide and D-lactide units and by varying the D-lactide content, PLA's crystallization kinetics could be controlled [3-4].

PLA could be considered as a promising substitute for the petroleum based polymers in various commodity and engineering applications due to the comparable cost and properties of PLA when compared with those of petroleum-based polymers. Furthermore, due to its non-toxic features, PLA is an appropriate candidate for food packaging and other consumer products. It is also a biocompatible polymer which could also be applied in biomedical applications such as bone fixtures, drug delivery and tissue engineering [2-3, 5].

Despite the noted advantages of PLA, there are several limitations such as brittleness, low toughness, low service temperature, narrow processing window, low melt strength, poor foamability and slow crystallization kinetics that PLA possess [2-3, 5-8]. In order to overcome some of these limitations, specifically the enhancement of ductility and toughness properties, PLA could be blended with other biodegradable polymers such as poly[(butylene adipate)-co-terephthalate] (PBAT) and poly[(butylene succinate)-co-adipate] (PBSA) that have high ductility, higher melt strength, and better processability. It has been shown by Nofar *et al.* [9-11] that the viscoelastic and foaming properties of PLA are improved when blended with PBAT and PBSA and the interfacial interaction between the matrix and dispersed phase has dominant role on complex viscosity and modulus improvement specifically at low frequencies. The localization of clay at the interface of phase also hindered the droplet coalescence under various shear flow [12]. In this study, however, we investigated how the mechanical properties of the PLA blends with PBAT could be dependent on the crystallization behavior of PLA matrix and on using various processing techniques. Moreover, we have compared the properties of PLA-PBAT and PLA-PBSA blends.

EXPERIMENTAL

Materials and Preparations

Commercial PLAs of Ingeo 4060D and 2500HP with respective D-lactide contents of 12% and 0.5% were supplied from NatureWorks LLC, USA. In this work, these PLAs are referred to as PLA4060 and PLA2500 which are amorphous and semi-crystalline grades, respectively. The thermograms of neat PLAs from differential scanning calorimetry (DSC, Perkin Elmer DSC400) were investigated and their thermal properties are disclosed in Figure 1. The glass transition temperatures (T_g) of these PLAs were confirmed as 59°C and 63°C, respectively, and the crystal melting temperature of PLA 2500 was appeared at 175.5°C. As shown, the presence of crystalline regions in PLA2500 caused an increase in its T_g when compared to that in PLA4060. PBAT and PBSA, the dispersed phases, were also supplied from Ecoflex® and Showa Denko K.K. (Japan), respectively. The grade of PBAT is F blend C1200 and of PBSA is Bionolle 3001 MD.

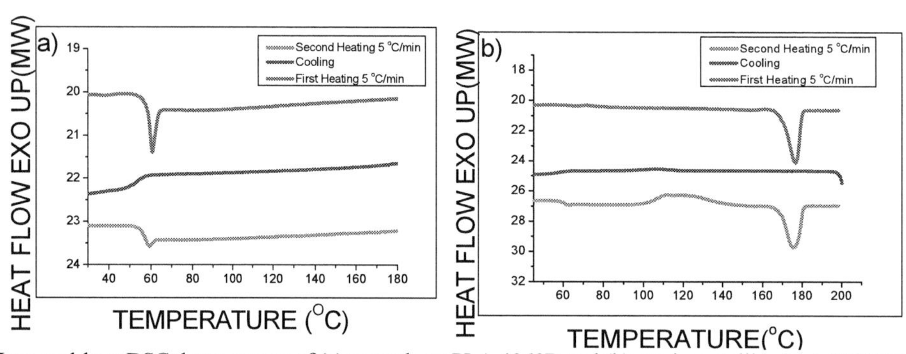

FIGURE 1. Heat-cool-heat DSC thermograms of (a) amorphous PLA 4060D and (b) semi-crystalline PLA 2500HP. The heating and cooling ramp rates was fixed at 5°C/min

All the resins were dried before processing in vacuum oven at 50°C over night to decrease the moisture level of polymers and to prevent their thermal degradation. Blends of 75 wt% PLA as the matrix with 25wt% PBSA or PBAT as the dispersed phases were prepared through injection molding (IM) process by feeding the dry mixed resins of matrix and dispersed phase. The IM process was carried out with a pressure of 1500 bar and with a barrel temperature of 170-190-200-200-200°C in the five zones. Only one mixture with PLA4060-PBAT composition was melt blended using a co-rotating twin-screw extruder (TSE) [Leistritz extruder from Leistritz Advanced Technologies Corp., NJ, USA, with screw diameter (D) of 18 mm and L/D ratio of 40] before running the injection molding (IM). The TSE experiments were conducted at a screw speed of 100 rpm and a set temperature of 160°C for all the heating zones. All of the blend compositions were molded into standard ASTM samples by using injection molding machine [Arburg model, Allrounder 221-K, with L/D ratio of 15]. The samples prepared through only IM process could be summarized as: PLA4060-PBAT, PLA2500-PBAT, PLA4060-PBSA. As noted, PLA4060-PBAT was also premixed in TSE and then processed through IM.

Mechanical Properties

Tensile and flexural properties of the samples were analyzed using a tensile testing apparatus (Schimadzu, ASG-J). Tensile and three-point bending experiments were carried out in order to ASTM638-03 and ASTM-D790 standards with the speed of 5 mm/min, respectively.

RESULTS AND DISCUSSION

Tensile Properties

Figure 2 demonstrates ultimate tensile stress, stiffness and elongation at break values of all samples from tensile testing. Figure 3 also shows the scanning electron microscopy results of injection molded blends samples.

As expected from neat PLA samples, PLA4060 and PLA2500 demonstrated brittle behavior, while they have tensile strength values of 62.8 MPa and72.2 MPa, respectively. The tensile strength of PLA2500 seems higher than that of PLA4060, most likely due to the existence of crystallites in PLA2500. Elongation at break values of PLA 4060 and PLA 2500 are in a similar range of 5%. However, the stiffness of PLA4060 shows higher value than PLA2500 which could be due to higher molecular weight of PLA4060 than PLA2500.

The effect of blending polymer type on mechanical properties of PLA was indicated by using PBAT and PBSA as the dispersed phases in the PLA4060 matrix, individually. Ultimate tensile stress of PLA 4060 was diminished by blending with PBAT to 44.6 MPa (29% decrease) and to 42.5 MPa (32% decrease) by blending with PBSA. In both cases, the strain at break values of PLA blends did not show any improvement with the addition of 25wt% PBAT or PBSA. This might have been due to the non-homogeneous blend morphology when using only IM as the processing route and also when an amorphous PLA is used as the matrix, as also revealed in Figure 3. In addition, a decrease in stiffness values of PLA4060-PBAT and PLA4060-PBSA blends comparing to the neat PLA4060 was 28% and 32%, respectively. The reduction of the tensile strength and stiffness (compared to neat PLA) with the addition of PBAT and PBSA to PLA is rational, however, the strain at break did not increase in presence of ductile droplets. This might be due to the non-homogeneous blend morphology when using only IM as the processing route and also when an amorphous PLA is used as the matrix which during the processing can influence the poor dispersion of the dispersed phases.

The results of crystallinity effect on the mechanical properties of PLA were revealed by comparing the PLA2500-PBAT and PLA4060-PBAT blends. Both PLA2500 and PLA 4060 lost their ultimate tensile stress values at the same ratio after blending with PBAT. However, the blend of PLA2500-PBAT revealed superior ductility enhancement; 3700% increase compared to neat PLA2500 (from 5.5% to 205%). This was while the blend of PLA4060-PBAT demonstrated almost no ductility improvement. After blending with PBAT, there has been a decrease in stiffness values of both PLA4060 and PLA2500 with 28% and 27% decrease, respectively. According to the corresponding SEM results in Figure 3, one can say that the PLA crystallization during the processing could result in obtaining finer dispersed phase and/or more uniform blend morphology and thereby could profoundly improve the mechanical properties of the blend (specifically the ductility and toughness values which are the main purpose of blending PBAT or PBSA with PLA).

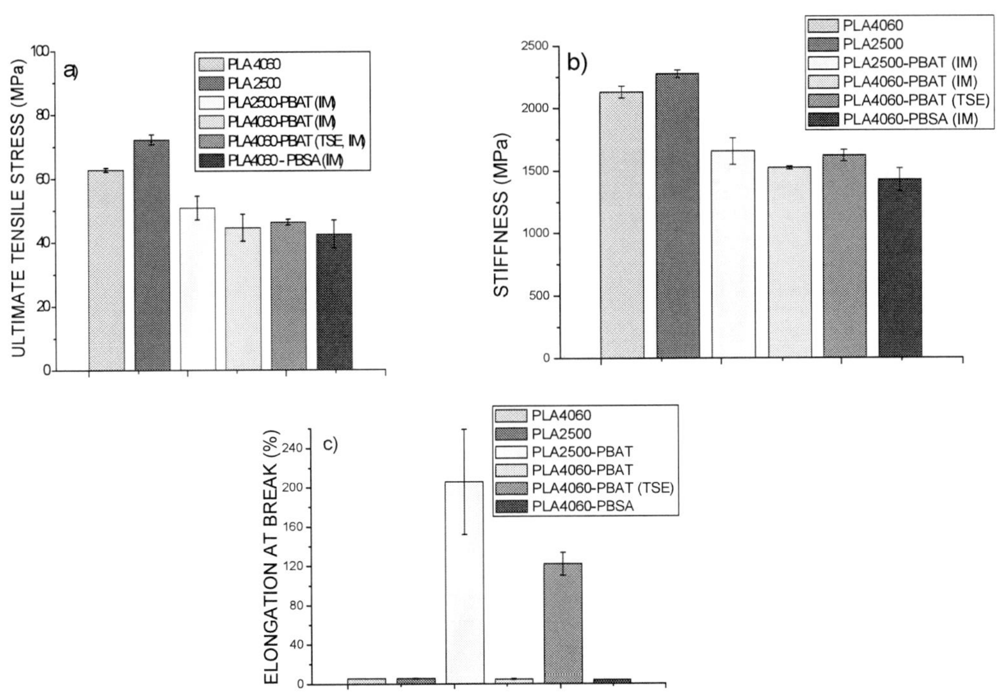

FIGURE 2. (a) Ultimate tensile stress, (b) stiffness and (c) elongation at break values through tensile testing of overall samples

Furthermore, effect of droplet size variation through using different processing methods on mechanical properties of PLA blends were determined by comparing PLA4060-PBAT blends when processed through only IM and when passed through TSE first followed by IM. It can be seen that the blend with TSE preprocessing revealed a much higher elongation at break with the strain value of 121.8% when compared with injection molded case. This was due the much finer and more uniform dispersion of the PBAT droplets when firstly processed through TSE (Figure 3). While injection molded PLA4060-PBAT blend has stiffness of 1522.5 MPa (28% decrease compared with that of neat PLA4060), twin screw extruded PLA4060-PBAT blend has stiffness of 1620.4 MPa (23% decrease from the neat PLA4060). Therefore, PLA4060-PBAT blend preprocessed with TSE also showed less stiffness drop than the injection molded blend.

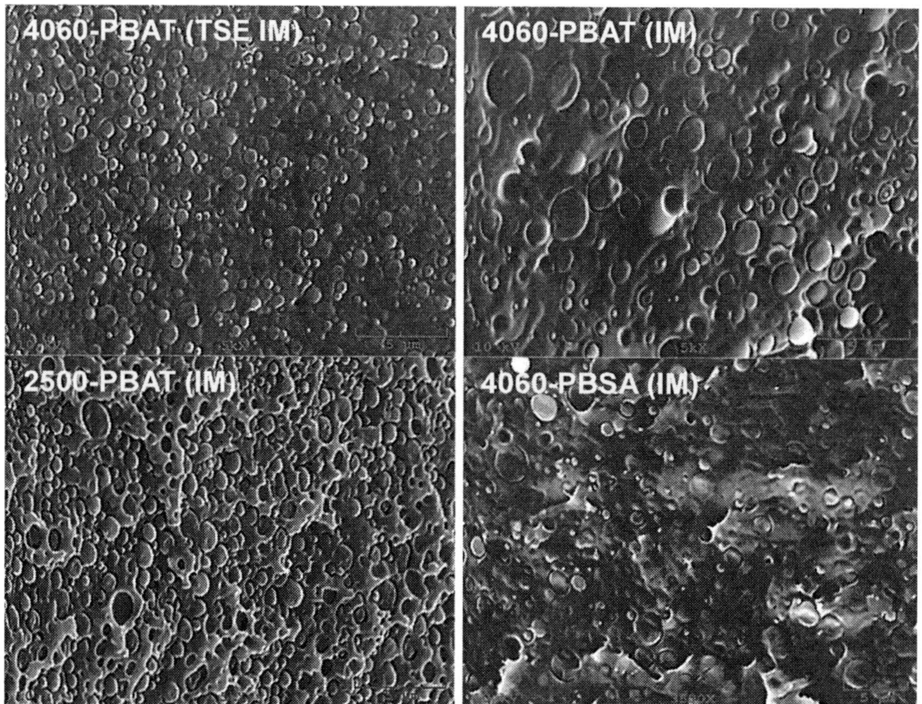

FIGURE 3. Scanning electron microscopy of various blend samples after injection molding

Bending Properties

Figure 4 shows the ultimate flexural stress, flexural modulus and elongation at break values for the neat PLAs and blend samples. Neat PLAs failed just after the elastic region meaning of brittle fracture characteristics. Neat PLA4060 and PLA2500 showed the highest ultimate flexural stress with the values of 100.15 MPa and 112.48 MPa, highest flexural modulus with the values of 2868.7 MPa and 2930 MPa, respectively. Despite the fact that, in tensile testing, the ductility improvement did not appear in all blended samples, the flexural deflection (ductility) of all PLA blend increased. Also, the failure behavior of PLA demonstrated ductile behavior owing to the blending. Elongation at break values obtained from bending test of neat PLA4060 and PLA2500 were 8.27 % and 5.31%, respectively.

Blending PLA with PBAT and PBSA through using different processing or different PLA matrices reduced the flexural strength and stiffness of PLA almost in a similar manner reflecting that the morphology variations in different blends could not vary the flexural properties of PLA however as shown in Figure 2, it significantly influenced the tensile properties. Similar scenario exists in ductility improvement during the flexural testing. In all blend case, the ductility of PLA seems to be improved with the use of any dispersed phase (PBAT or PBSA), any processing technique, or any PLA matrix type. These results reflect that the three-point bending test is more dependent on to the blend composition rather than its morphology.

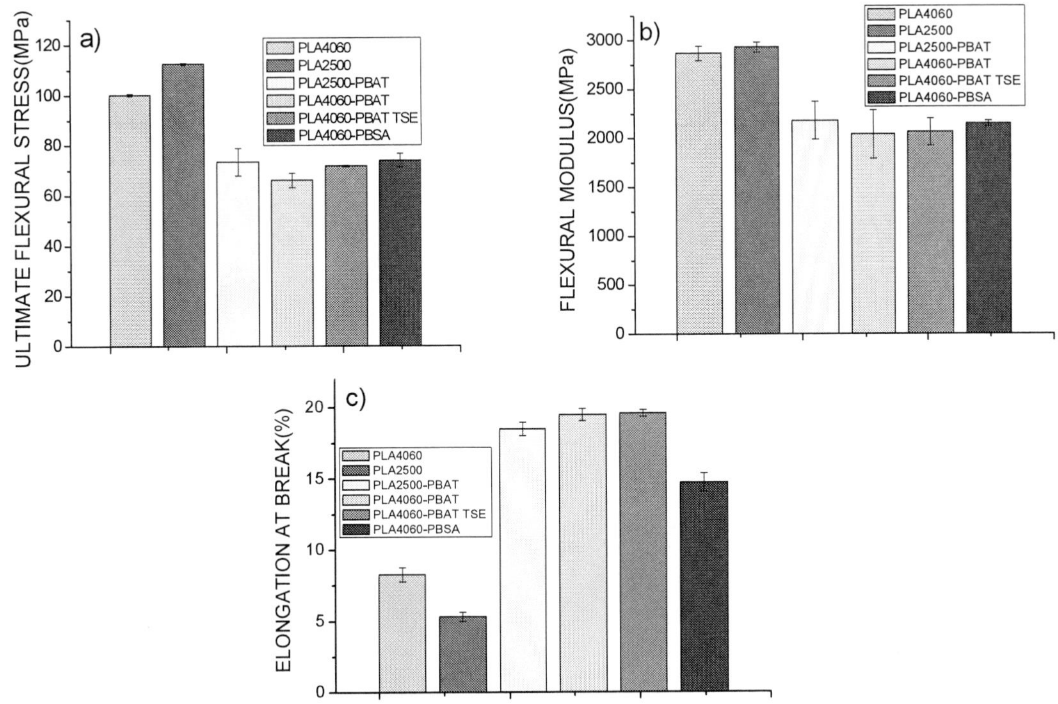

FIGURE 4. (a) Ultimate flexural stress, (b) flexural modulus and (c) elongation at break values through bending test

ACKNOWLEDGMENTS

The authors would like to thank Prof. Huseyin Cimenoglu, Prof. Mustafa Urgen, and Prof. Lutfi Ovecoglu for allowing us using their laboratory facilities. We would also like to thank Research Assistants Dogukan Cetiner, Onur Tazegul, and Çağatay Yelkarası for their help in running the experiments.

REFERENCES

1. Sinclair RG. *J. Macromol. Sci. Pure Appl. Chem.* **33**, 585-597 (1996).
2. Garlotta DJ. *J. Polym. Environ.* **9**, 63-84 (2001).
3. Nofar M, Park C.B, *Polylactide Foams, Fundamentals, Manufacturing, and Applications*, Elsevier, ISBN: 9780128139912
4. Nofar M, Ameli A, Park CB. *Macromol. Mater. Eng.* **299 (10)**, 1232-1239 (2014)
5. Nofar M, Park C.B, *Prog. Polym. Sci*, **39**, 1721-1741 (2014).
6. Saeidlou S, Huneault M, Li H, Park C.B, *Prog. Polym. Sci*, **37**, 1657-1677 (2012).
7. Rasal R.M, Janorkar A.V, Hirt D.E, *Prog. Polym. Sci*, **35**, 338-356 (2010).
8. Dorgan J, Williams J, *J. of Rheol*, **43**, 1141-1155 (1999).
9. Nofar M, Maani A, Sojoudi H, Heuzey M-C, Carreau P.J, *J. of Rheol*, **59 (2)**, 317-333 (2015).
10. Nofar M, Huzey M-C, Carreau P.J, Kamal M.R Randall J, *J. of Rheol*, **60 (4)**, 637-648 (2015).
11. Nofar M, Tabatabaei A, Sojoudiasli H, Park C.B, Huzey M-C, Carreau P.J, Kamal M.R, *Eur. Polym. J.*, **90**, 231-244 (2017).
12. Nofar M, Huzey M-C, Carreau P.J, Kamal M.R, *Polymer*, **98**, 353-364 (2016).

Extruded blends of PLA and PBAT: correlation between com-position and fracture behaviour

Gigante Vito (1)*, Coltelli Maria Beatrice (1), Cinelli Patrizia (2), Aliotta Laura (1), Canesi Ilaria (1), Lazzeri Andrea (1)

(1) Pisa University - Pisa - Italy, (2) National Research Council - Pisa - Italy

Poly(lactic acid) (PLA) is a remarkable biopolymer for packaging applications because it derives from renewable resources and it presents the great advantage to be biodegradable[1]. Nevertheless its high stiffness, low thermal stability and brittleness does not allow the use of PLA in a wide range of applications. Rub-ber toughening of PLA is a promising strategy to improve its mechanical charac-teristics [2]. In this work blends of PLA with poly(butylene-adipate-co-terephthalate) (PBAT) have been studied; PBAT is an oil-based material, but bio-compostable, capable to improve thermal and rheological properties of PLA [3]. Tensile tests, Impact tests and Fracture Characterization of PLA-PBAT blends (prepared by extrusion and injection molding with HAAKE MiniLab MiniExtruder and MiniJet) obtained as a function of composition in terms of PBAT weight percentages were carried out to evaluate blends fracture behavior. The results were discussed by using different testing methods and varying the test speed, the rubber content and temperature. Specific attention at the modifications of brittle/ductile transition [4] changing these parameters has been placed, studying analytical models to predict them. References [1] Coltelli M.B., Bronco S., Chinea C. "The effect of free radical reactions on structure and properties of poly(lactic acid) (PLA) based blends" Polymer Degradation and Stability 95, (2010), 332-341 [2] Lazzeri A., Bucknall C.B., "Dilatational bands in rubber-toughened polymers" Journal of materials science (1993) 28, 6799 [3] Al Itry, R., Lamnawar, K. and Maazouz, A. "Improvement of Thermal Stability, Rheological and Mechanical Properties of PLA, PBAT and Their Blends by Reactive Extrusion with Func-tionalized Epoxy". Polymer Degradation and Stability, 97, (2012), 1898-1914 [4] Mai Y.W., Cotterell B. "On the essential work of ductile fracture in polymers" International Journal of Fracture 32, (1986), 105-125

Development of highly functional butyl –etherified starch/ poly [(butylene succinate)-co-adipate] nanocomposites using reactive extrusion

Maubane Lesego Tabea [(1)]*

(1) Council for Scientific and Industrial Research - Gauteng - SouthAfrica

Increasing environmental awareness has encouraged a lot of researchers to progress to biodegradable materials. Amongst biodegradable polymeric materials, starch is one of the most studied biopolymer; however, thermoplastic starch materials possess poor mechanical properties and suffer from water sensitivity. To overcome these challenges, butyl-etherified starch (BES) types were melt processed with poly (butylene succinate adipate) (PBSA) respectively in the presence of hydrophilic silica particles and maleic anhydride (MA). Results indicated that butyl-etherification of starch improved compatibility while highly branched amylopectin structures of BES increased miscibility with PBSA and properties of BES improved with the increasing PBSA content. Silica particles preferentially localized inside the dispersed BES phase and resulted in increased viscosity which led to limited improvement in droplet breakup. Furthermore, the presence of MA led to hydrolysis of BES phase resulting in reduced molecular weight, viscosity and inadequate improvement in mechanical properties. Therefore, this study is expected to shed some light into the use of butyl-etherified starch blends with PBSA for packaging applications and urge interest for further research in the area of reactive extrusion with different filler materials.

Structure Property Relationships of PA6-EOR-Blends

J. Heyn and C. Bonten

IKT, Institut für Kunststofftechnik, University of Stuttgart, Pfaffenwaldring 32, 70569 Stuttgart, Germany.
Corresponding author: johannes.heyn@ikt.uni-stuttgart.de

Abstract. Impact modifiers in polyamide 6 improve the (un)notched impact properties in many cases. The effect of particle size or the interparticle distance on this improvement is not yet fully understood. For daily use, low testing speeds are very interesting, but in most of the investigations only impact tests at high testing speeds are examined. In this work, it is shown that the criterion of the critical interparticle distance for notched impact tests cannot be considered to be universal. It can also be proven that the criterion of the optimum particle size is only valid for notched impact tests if an adequate phase connection is given. Based on the findings of this work, the two most important factors for a toughness optimization at low testing speeds in PA6-EOR-blends are a phase connection as strong as possible and particle sizes as small as possible.

INTRODUCTION

By modifying polyamide 6 with different additives/fillers it is possible to vary properties, create new properties or to reduce material costs and rather component costs. Blending with a soft phase has been proven for many applications in order to make use of synergistic effects between the blending partners, e.g. to precisely shift temperature limits. Depending on the choice of blending partners as well as processing conditions the properties of the blend are influenced. Taking advantage of the true potential of blends is only possible due to knowledge of the relation between morphology and blend properties [1].

STATE OF THE ART

There are two criteria to be found in literature which describe how to enhance toughness in impact modified blends with a dispersed soft phase. On the one hand Bucknall et al. [2] describe an optimum particle size d_{opt} which leads to a significant increase in notched impact strength. On the other hand Wu [3] found that falling below a critical interparticle distance τ_{krit} of the matrix is responsible for an increase in toughness. Except for a small intersection of particle sizes and interparticle distances both criteria exclude each other.

For low testing speeds, considerably lower than usually used for determining notched impact strength, there is not yet a criterion for enhancing toughness of blends. Still, there is a lack of proof of the two mentioned criteria by Bucknall et al. and Wu. To describe the fracture mechanical behavior at low testing speeds González et al. and Kumar et al. [4-5] assume that a pure dependency of the amount of soft phase, but not a morphology dependency exist. Especially a change of the morphology from a dispersed soft phase to a co-continuous morphology, strongly influences the fracture mechanical properties as own investigations demonstrated [6-7].

This study is examining how phase connection, particle size and interparticle distance are related to the blend properties. Apart from that, the influence of high and low testing speeds is investigated.

EXPERIMENTAL

Materials

PA6-Typ Ultramid® B40 (PA6) provided by BASF SE, Germany, and ethylene-octene-copolymer (EOR) ENGAGE™ 8100 provided by The Dow Chemical Company, Spain, are used. Maleic anhydride (MAH) is used as a coupling agent and dicumylperoxide (DCP) is used as radical former.

Blends with different amount of EOR and varying amounts of MAH and DCP were produced on a twin screw extruder, type ZSK 25 from Coperion, Germany. The blends were produced at a throughput $\dot{m}$ of 10 kg/h and a screw speed n of 180 rpm. After the compounding process specimens type 1 according to DIN EN ISO 179-1 and plates (60 mm x 40 mm x 2 mm) were prepared on an injection molding machine, type Allrounder 520S from Arburg, Germany. The mold temperature was 80 °C and the temperature profile varied from 180 °C to 290 °C.

CHARACTERIZATION

Particle size and distribution

In order to determine particle size and distribution, each blend has been boiled in xylene to remove the soft phase. The measurements were then carried out using a scanning electron microscopy (SEM) type JSM 6300 F from Jeol, Germany. The SEM images were then evaluated with an image analysis software Leica Qwin from Leica, Germany. For determination of the average particle size d the number-average $D_{[10]}$ is used, see equation (1):

$$D_{[10]} = \frac{\sum N_i D_i}{\sum N_i} \tag{1}$$

The interparticle distance τ can be calculated using equation (2) by knowing the particle size and volume ratio of the soft phase as well as a factor k=1 (cubic), which describes the spatial distribution of the particles in the matrix.

$$\tau = d \cdot \left(k \cdot \sqrt[3]{\frac{\pi}{6\Phi_r}} - 1 \right) \tag{2}$$

Rheological Investigation

Rheological measurements were done by using a plate-plate rheometer of type Rheometrics DSR 200-D from TA Instruments, Germany. The measurements were done in an oscillation mode under nitrogen atmosphere. Frequency sweeps were used to determine the storage modulus G', the loss modulus G'' and the complex viscosity η^*. The analyzed angular frequency range was 0.1 rad/s to 100 rad/s. The gap had a fixed width of 2 mm. Through variation of the modulus and the shift of the loss factor tan δ it is possible to detect potential curing and branching reactions of the blends. It is also possible to analyze the effect of the coupling agent.

Fracture Mechanical Testing

Notched Charpy impact strength a_{cA} was determined according to DIN EN ISO 179-1/1eA. To analyze the elastic-plastic fracture mechanics of the blends the method of essential work of fracture (EWF) was used.

Following equation (3) the total work of fracture (W_f) can be divided into two parts, the essential work of fracture (EWF) (W_e) and the non-essential work of fracture (n-EWF) (W_p). The specific EWF w_e is a material constant, at a certain specimen thickness, and the specific n-EWF βw_p depends on the shape (β) of the plastic deformation around the crack and the specimen geometry. If equation (3) is divided by the original cross section (specimen thickness B and ligament length l) the specific energy can be calculated, equation (4).

$$W_f = W_e + W_p = w_e B l + \beta w_p B l^2 \tag{3}$$

$$\frac{W_f}{Bl} = w_f = w_e + \beta w_p l \tag{4}$$

RESULTS AND DISCUSSION

Blend Morphology

Fig. 1 shows a small overview of representative morphologies of the blends with a uniform amount of 10 wt.-% EOR and different blend modifications. Additionally, the average particle size $D_{[10]}$ and the average interparticle distance τ_m are shown. Unmodified, with a radical former, with a coupling agent and with the combination of a radical former as well as coupling agent modified blends are shown. The morphology is not influenced significantly by only applying a radical former. Only the addition of a coupling agent or the combined use of both leads to a significant smaller average particle size $D_{[10]}$ and a smaller average interparticle distance τ_m. This is caused by successful grafting of the EOR.

Phase Connection

Fig. 1 shows the storage modulus G' and the complex viscosity η^* as a function of the angular frequency ω for different modified blends with a consistent amount of 10 wt.-% EOR at 240 °C.

A drastic increase of the storage modulus G' and complex viscosity η^* is only reached by combined adding of coupling agent and radical former to the blend (10EOR100MP). Hence, it can be deduced that the change of rheological properties of the analyzed blends is affected by both, the average particle size $D_{[10]}$ and the average interparticle distance τ_m as well as by the phase connection.

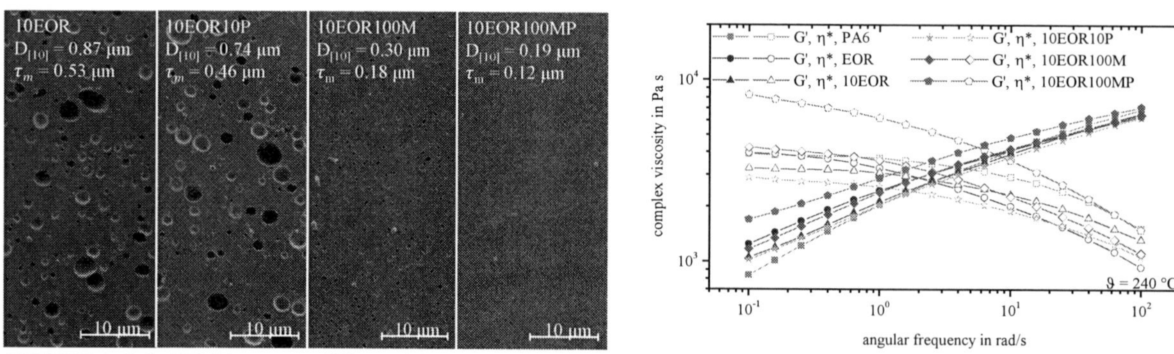

FIGURE 1. left: SEM images – different blend modifications with 10 wt.-% EOR; right: Frequency sweeps of neat materials and differently modified blends with 10 wt.-% EOR

The increase of the storage modulus G' and the complex viscosity η^* for differently modified blends, including the blend with 10 wt.-% EOR, is reasoned by better phase connection. This is possibly caused by chemical or physical bonds formed between both neat materials.

Fracture Mechanical Behavior

An overview of the notched impact strength a_{cA} is given in fig. 2. The notched impact strength is shown as a function of the average particle size $D_{[10]}$ respectively of the average interparticle distance τ_m. An increase of the notched impact strengths a_{cA} is visible by decreasing the average particle size $D_{[10]}$ respectively for decreasing the average interparticle distance τ_m. It rises drastically at an average particle size smaller than 0.4 µm, except for the blend 10EOR100M (10 wt.-% EOR + 1.0 wt.-% MAH). This range of particle size marks the optimum particle size d_{opt} (red shaded zone) as Bucknall et al. found out.

Due to covalent bonds between both blend partners, there is another possible differentiation of notched impact strength a_{cA} subjected to the phase connection. High notched impact strength a_{cA} is achieved by blends with a high phase connection (grey zone) and low notched impact strength a_{cA} is achieved by blends with a low phase connection (white zone). In this context, it can be shown that blends in the range of the optimum particle size d_{opt} also show low values. Especially the highlighted blend 10EOR100M.

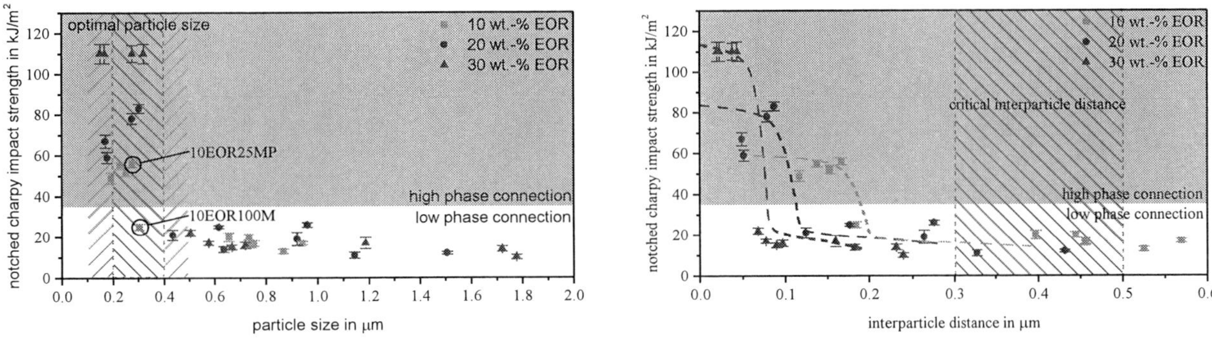

FIGURE 2. Notched impact strength as a function of the average particle size and of the average interparticle distance

An exact separation of both characteristics, on the one hand the particle size d and the interparticle distance τ and on the other hand the phase connection, is barely possible. This is due the effect that a material modification by a coupling agent always affects the morphology and also the phase connection.

Notched impact strength a_{cA} as a function of average interparticle distance τ_m, fig. 2, shows an increase at values lower than 0.2 µm. This increase takes place underneath the critical range of 0.3 µm to 0.5 µm described by Wu. Also by adjusting the critical interparticle distance τ_{krit} to lower values of about 0.2 µm the results show lower values. This is not consistent with the criterion of the critical interparticle distance τ_{krit}, because it should be independent of the amount of soft phase and therefore is disproved definitely.

Fig. 3 shows results of the specific EWF w_e and the specific n-EWF βw_p as a function of the average particle size $D_{[10]}$ for differently modified blends with a uniform amount of 10 wt.-% EOR. Both figures include a red line (-), which schematically represents the value of neat PA6.

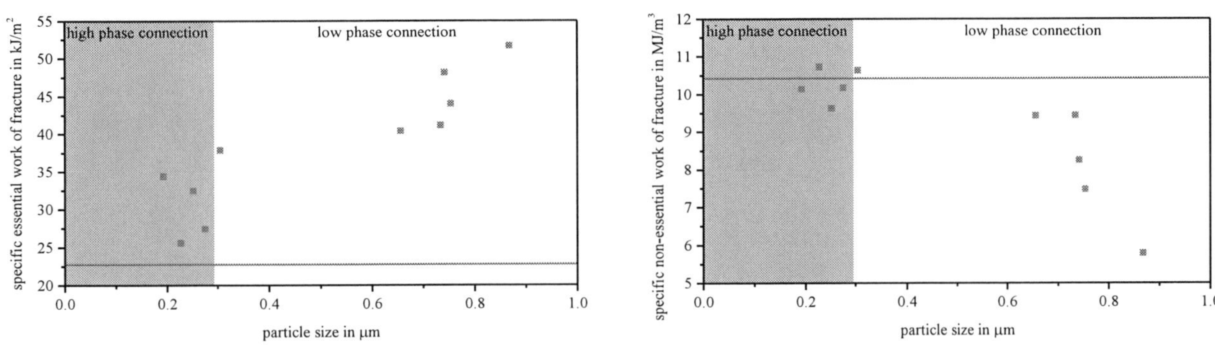

FIGURE 3. left: Specific EWF as a function of the average particle size for PA6-EOR-blends with 10 wt.-% EOR and different modifications; right: Specific n-EWF as a function of the average particle size for PA6-EOR-blends with 10 wt.-% EOR and different modifications

For both specific energies, there is an opposing trend for increasing the average particle size $D_{[10]}$. Along with rising average particle size $D_{[10]}$, the specific EWF w_e increases (left) and specific n-EWF βw_p (right) decreases. This means that high average particle sizes $D_{[10]}$ lead to high specific EWF w_e and low specific n-EWF βw_p, whereas low average particle sizes $D_{[10]}$ lead to low specific EWF w_e and high n-EWF βw_p. The specific EWF w_e and specific n-EWF βw_p as a function of the average interparticle distance τ_m shows similar tendencies and is therefore not shown.

An examination of the phase connection as a function of both specific energies also leads to inversely conclusions. The specific EWF w_e is lower at a high phase connection (grey zone) than at a low phase connection (white zone). It is the other way around with the n-EWF βw_p.

A relation between the criterion of the optimum particle size d_{opt} respectively the criterion of the critical interparticle distance τ_{krit}, deduced from notched impact tests, does not exist. Therefore, a new criterion for low testing speeds is necessary, in order to explain structure-property relationships. Fig. 4 shows the specific EWF w_e and the specific n-EWF βw_p. as a function of the average particle size $D_{[10]}$ for maximum modified PA6-EOR-blends with a high phase connection.

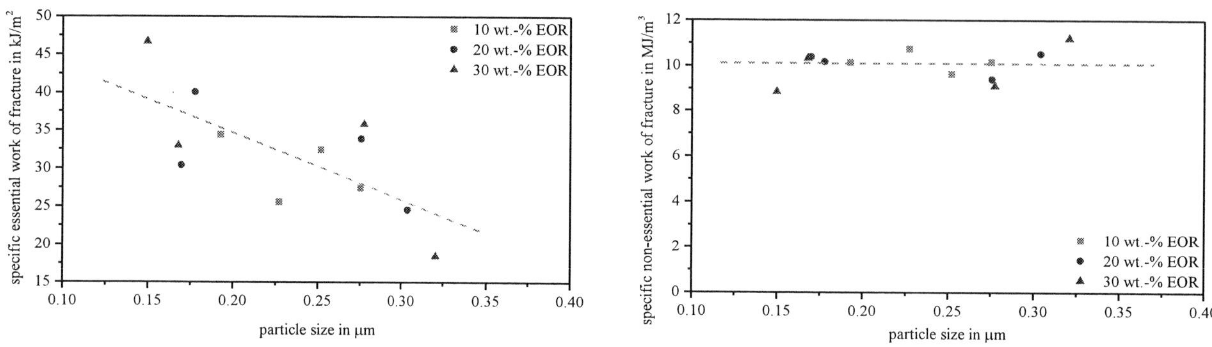

FIGURE 4. Specific EWF (left) and specific n-EWF (left) as a function of the average particle size for PA6-EOR-blends with high phase connection

In contrary to the observations of fig. 3, the specific EWF w_e rises at decreasing the average particle sizes $D_{[10]}$ for different amounts of EOR. This implies that if there is a high phase connection, smaller average particle sizes $D_{[10]}$ lead to higher specific EWF than larger average particle sizes $D_{[10]}$. The specific n-EWF βw_p nearly stays at a constantly high level for small average particle sizes $D_{[10]}$ with a high phase connection.

CONCLUSION

The attempt to explain the optimum particle size d_{opt} by Bucknall et al. was confirmed on condition of a high phase connection for notched impact strength. However, for low testing speeds it was disproved using the method of EWF. The critical interparticle distance τ_{krit} described by Wu was disproved for notched impact strength as well as for low testing speeds by using the method of EWF.

The specific EWF w_e is a material constant at given thickness of specimens and the specific n-EWF βw_p is a function of the specimen geometry. Although the required fracture energy depends on the two specific energies, only the specific EWF w_e is considered. This may lead to wrong conclusions by analyzing the fracture mechanical behavior.

It was possible to show that the specific EWF w_e is highest for large average particle sizes $D_{[10]}$ with a low phase connection. But at the same time the specific n-EWF βw_p is minimal. In order to achieve maximal specific energies at low testing speeds by using the method of EWF a high phase connection is required. It was proven that in order to achieve a large increase of toughness at low testing speeds both specific energies need to be regarded. Only if both of them are as high as possible an optimal increase of toughness is achieved. This only works if a blend morphology with a phase connection as high as possible exists. In combination with lowering particle size this leads to an increase in toughness. These new findings thus form the basis for an even more efficient use of materials.

LITERATURE

1. C. Bonten, Kunststofftechnik - Einführung und Grundlagen. 2. Auflage. München: Hanser, 2016.
2. C. B. Bucknall and D. R. Paul, Notched impact behavior of polymer blends: Part 1: New model for particle size dependence. In: Polymer, 2009, 50(23), pp. 5539–5548
3. S. Wu, A generalized criterion for rubber toughening: The critical matrix ligament thickness. In: Journal of Applied Polymer Science, 1988, 35(2), pp. 549–561
4. I. González, J. I. Eguiazábal and J. Nazábal, Characteristics of the brittle/tough transition of poly(butylebne terephtalate)/maleinized poly(ethylene-octene) blends determined by the essential work of fracture procedure. In: Polymer Testing, 2010, 29(1), pp. 27–32
5. S. Kumar, S. N. Maiti and B. K. Satapathy, Super-toughening polyamide-612 by controlling dispersed phase domain size: Essential work of fracture assessment. In: Materials & Design, 2014, 62, pp. 382–391
6. J. Heyn and C. Bonten, Fracture Behavior of PA6 Rubber Blends Influenced by Water Absorption. In: SPE ANTEC Orlando, Conference Proceedings, 2015, pp. 835-841
7. J. Heyn, Bruchmechanische Betrachtung der Struktur-Eigenschaftsbeziehungen von PA6/TPO-Blends. Dissertation. Stuttgart, Universität, Institut für Kunststofftechnik, 2017

Development of a Block Copolymer for Impact Modification of Polyhydroxybutyrate (PHB)

Linda Goebel[1, a)] and Christian Bonten

[1]IKT, *Institut für Kunststofftechnik, University of Stuttgart, Pfaffenwaldring 32, 70569 Stuttgart*

[a)]Corresponding author: linda.goebel@ikt.uni-stuttgart.de

Abstract. Bioplastics, which will see rising demand in the next years, have to be compounded by using a twin screw extruder like conventional plastics. By using a twin-screw extruder, the molecular weight of the bioplastic polyhydroxybutyrate (PHB) can be degraded under control. In a second step, the short-chain PHB and a soft phase can be compounded to a larger block copolymer by means of reactive extrusion. The processing step takes place in a process- and material-efficient manner without catalysts, solvents and purification processes. The development of the block copolymer improves the ductile properties of the material without changing the thermal or chemical characteristics.

INTRODUCTION

Bioplastics experienced an increasing demand in recent years. Currently bioplastics represent about one percent of the 300 million tonnes of plastics produced annually. Forecasts predict an even larger increase in the forthcoming years. The global production capacity of bioplastics is predicted to grow from around 4.2 million tonnes in 2016 to 6.1 million tonnes in 2021. The demand for biobased plastics is higher than the one for biodegradable plastics. The main drivers are the so-called drop-in solutions like biobased PE, biobased PET and biobased PUR. Drop-in solutions are usually conventional fossil plastics, where the fossil content is substituted by renewable raw materials without changing the properties or the chemical structure of the polymer. [1, 3]

To use bioplastics, biopolymers have to be compounded to plastics, like every other polymer. Figure 1 shows the way from the monomer to the plastic part. The first step is the synthesis of a polymer. The next step is the compounding of polymers to plastics on a twin screw extruder. Compounding means for example adding additives as well as fibres and fillers. Moreover, the reactive extrusion is also part of the compounding. During the reactive extrusion a chemical reaction takes place in the twin screw extruder. The last step is the processing step to form a plastic part by means of injection moulding or extrusion, for instance. [1]

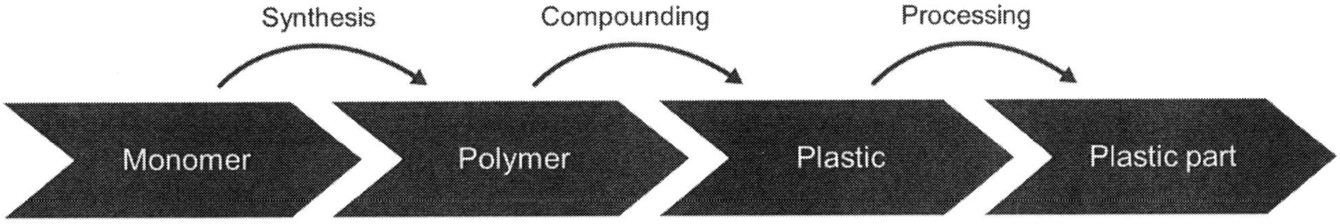

Figure 1: From monomer to plastic part [1]

SATE OF THE ART

Polyhydroxybutyrate (PHB) is a biopolymer, which belongs to the polymer family of polyhydroxyalkanoates (PHA) and was discovered as early as the 1920s. PHB has good water vapor resistance and is resistant to hydrolysis and UV. It is biobased as well as biodegradable. The PHA-family is produced via direct biosynthesis in microorganisms, which accumulate PHB as a carbon and energy storage. After the fermentation the microorganisms consist out of up to 80 % polyester (Fig. 2). By using solvents, the accumulated polyester is extracted and purified for further processing.

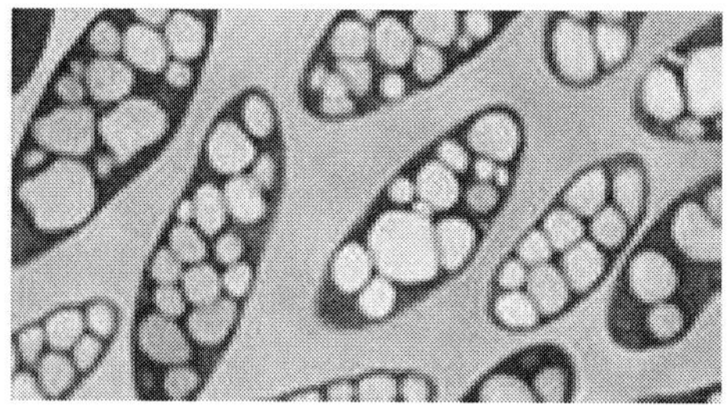

Figure 2: Microorganism with the accumulated polyester [3]

The economic use of PHB is currently restricted by two main factors. Slow crystallization speeds one the one hand and high post crystallization on the other hand. Due to the slow crystallization, PHB cannot compete against other bioplastics or conventional plastics with regard to an economic production. In addition, the high crystallization can cause high warpage in the produced component for a longer time after demolding.

By adding a nucleating agent, these disadvantages of crystallization can be minimized. [4] Another disadvantage of PHB, and most other bioplastics, is the poor ductility. Many industrial applications, such as film blowing for the packaging industry, are restricted by this. The development of a block copolymer with a suitable soft phase could improve the ductile properties of the material. Block copolymers consist of more than one monomer. The different monomer blocks can be arranged differently. Figure 3 shows some examples. They can combine properties of different polymer blocks. One example is thermoplastic vulcanizate (TPV). It is the combination of the functional properties of the elastomers and the processing properties of the thermoplastics. [1]

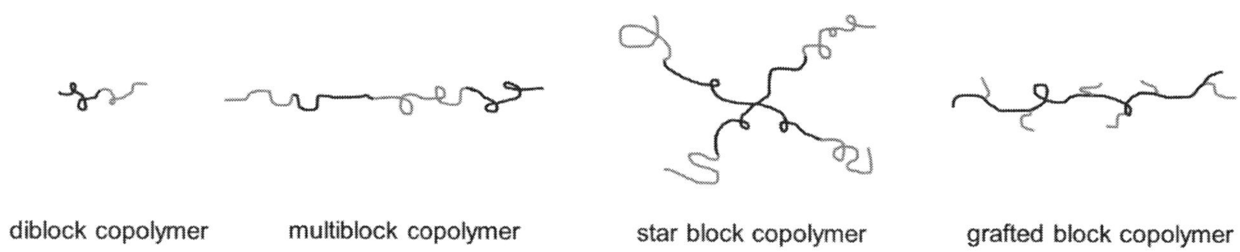

Figure 3: Different block copolymers

The synthesis of a block copolymer generally takes place during the polymer synthesis. As mentioned, PHB polymerizes in microorganisms hence a polymer synthesis with a second polymer block is not possible. However, it was already possible to produce a PHB block copolymer in a two-step process using various chemicals and chemical processes with high investment of time and money. In this two-step process, a PHB diol is produced chemically from pure PHB. In the second step, the synthesis to the block copolymer takes place with a soft phase. Both steps can only be executed in small-scale and in a discontinuous processes. The mechanical properties of the block copolymer produced by this method were not published. [5-7]

EXPERIMENTAL

The new concept introduced in this paper replaces all chemical processes as well as solvents and catalysts by using a twin screw extruder. In the first process step, the molecular weight of PHB is degraded under control. Only mechanical and thermal energy input coming from the twin screw extruder lead to the degradation. By means of reactive extrusion, the short-chain PHB and the soft phase can be compounded to a larger block copolymer in a second process step.

Both steps were executed on a twin screw extruder (ZSK 18) from Coperion, Germany. In the first step the PHB, which is from Biomer, CH ,is compounded by means of two different screw designs. The first screw design consists of different conveying elements whereas the second screw design consists of different conveying elements as well as kneading elements like Fig. 4.

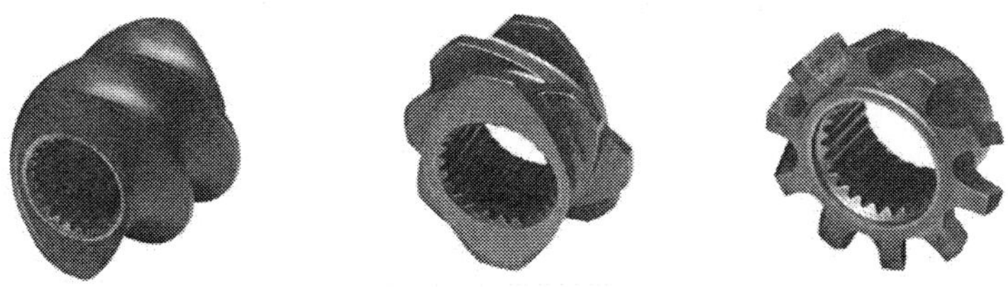

Figure 4: Different screw elements

The second step is the forming of the block copolymer. The reactive extrusion is executed with a special screw design, which guarantees a good dispersive as well as a good distributive mixing. To show that the reactive extrusion to the block copolymer took place in the twin-screw extruder, various analytical methods, such as dynamic differential scanning (DSC) or rheological measurements are applied.

RESULTS AND DISCUSSION

First step

Figure 5 shows the zero shear viscosity of the degraded PHB, depending on the rotational speed of the twin screw extruder and depending on the screw design. The viscosity decreases with increasing rotational speed. It can be seen that the presence of kneading elements reduces viscosity to a larger extent than a strictly conveying screw design. Clearly, kneading elements have a stronger influence on the viscosity than conveying elements.

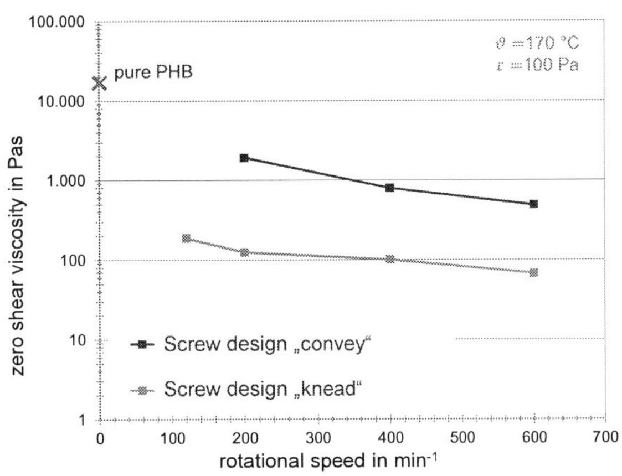

Figure 5: Zero shear viscosity depending on screw design and rotational speed

The zero shear viscosity is proportional to the molecular weight at low molecular weights. Beyond a critical molecular weight, the zero shear viscosity correlates to the power of 3.5 with the molecular weight. Thus, the molecular weight can be described by the following relation:

$$M_W = \frac{\eta_0}{K} \qquad M < M_c \qquad (1)$$

$$M_W = (\frac{\eta_0}{K})^{\frac{1}{3,4}} \qquad M > M_c \qquad (2)$$

The molecular weight decreases with decreasing zero shear viscosity. Thus, it can be concluded that the first step, the degradation of the PHB under control, was successfully executed on a twin screw extruder.

Second step

Figure 6 shows the zero shear viscosity of pure PHB, of the short-chained PHB, which is much lower than the viscosity of the pure PHB, and the viscosity of the PHB-block copolymer. The viscosity of the formed block copolymer is much higher than the viscosity of the short-chained PHB and even higher than the viscosity of the pure PHB. This is a first indicator, that the reactive extrusion to a block copolymer was successfully executed.

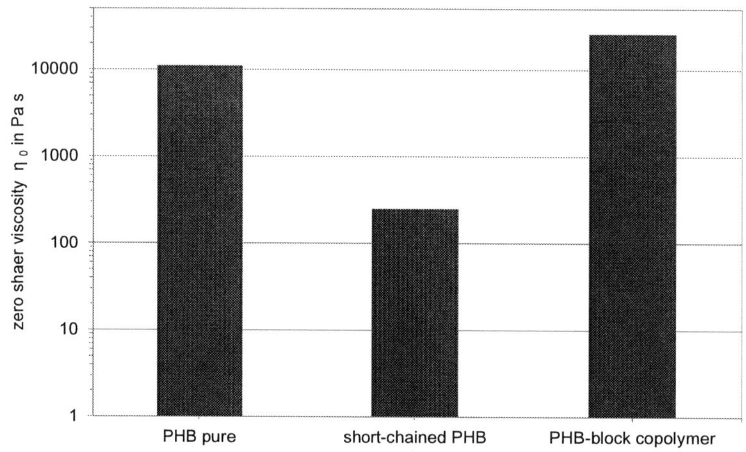

Figure 6: Zero shear viscosity of pure PHB, short-chained PHB and of the block copolymer

The analysis by means of the differential scanning calorimetry also makes a case for the chemical reaction in the twin screw extruder. Figure 8a) shows the thermogram of a PHB-PEG-Blend without any coupler. The heating curve indicates two melting peaks. The first peak around 50 °C belongs to the PEG whereas the twin peak belongs to PHB. In addition to the heating phase, the cooling phase also has two peaks. The first peak (102 °C) represents the crystallization of PHB and the second peak the crystallization of PEG. The thermogram of the block copolymer shows just one melting peak as well as one crystallization peak (Fig. 7b)). Hence, the chemical reaction between PHB, PEG and a coupler during the compounding process was successful. A block copolymer is formed.

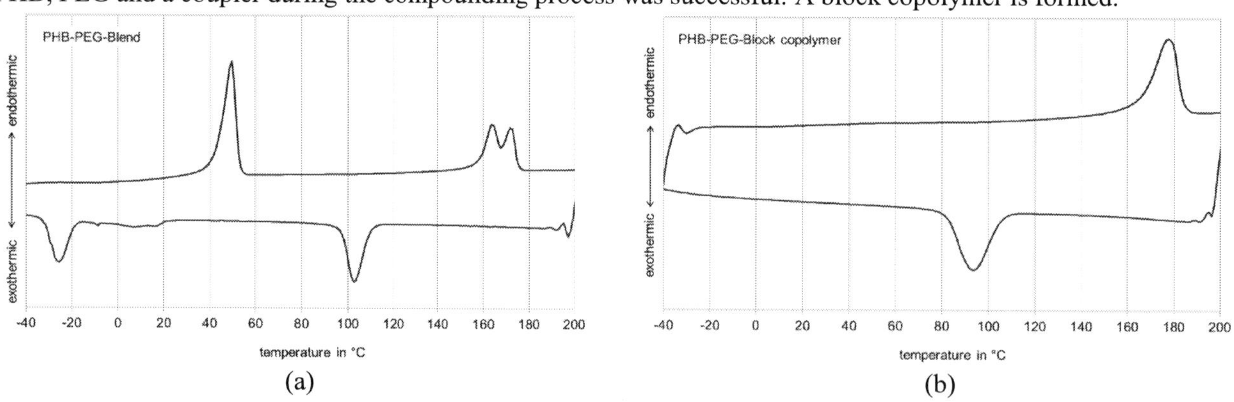

Figure 7: Thermogram by means of dsc a) PHB-PEG-Blend b) PHB-block copolymer

In Fig. 9 the mechanical properties for pure PHB and for the PHB-block copolymer are shown. By forming the block copolymer, the value of elongation at break for pure PHB increases from 1.6 % to 33 % for the block copolymer. The value of impact strength has also been increased greatly.

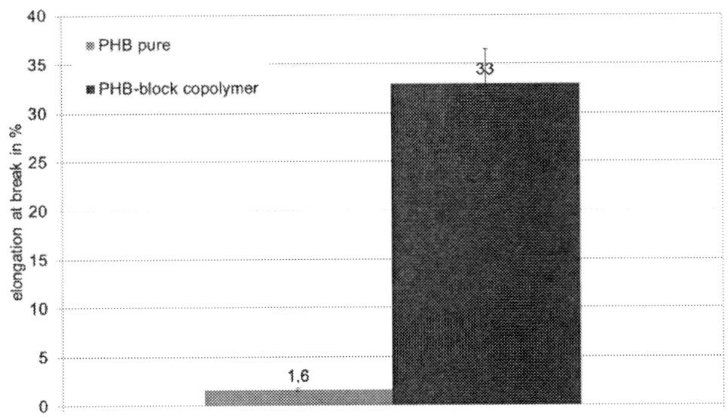

(a)

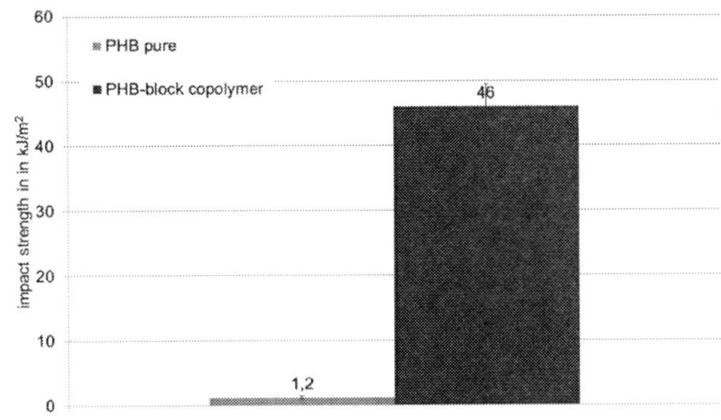

(b)

Figure 8: Mechanical properties of the PHB-block copolymer a) elongation at break b) impact strength

CONLUSION

By using this new method, a PHB-PEG block copolymer can be produced continuously on a twin-screw extruder for the first time. The processing step takes place in a process- and material-efficient manner without catalysts, solvents and purification processes. Depending on the size of the twin screw extruder, any amount of the block copolymer can be produced within a short period of time. The new method enables the production of the PHB-block copolymer in a more economical way and to use its unique properties for new applications.

REFERENCES

1. BONTEN, C. Kunststofftechnik – Grundlagen und Einführung. 2nd Edition München: Hanser, 2016. ISBN 978-3-446-44093-8.
2. IfBB Instiute for Bioplastics and Biocomposites. Biopolymers facts and statistics. 2nd Edition Hannover: Hochschule Hannover, 2015. ISBN 2363-8559.
3. TÜRK, O. Stoffliche Nutzung nachwachsender Rohstoffe. Grundlagen - Werkstoffe - Anwendungen. Wiesbaden: Imprint: Springer Vieweg, 2014. SpringerLink: Books. ISBN 978-3-8348-2199-7.
4. GOEBEL, L.; BONTEN, C.: Crystallization Behavior of Polyhydroxybutyrate (PHB). In: SPE Proceedings ANTEC 2016.
5. HADJICHRISTIDIS, N., S. PISPAS und G. FLOUDAS. Block copolymers. Synthetic strategies, physical properties, and applications. Hoboken, N.J.: Wiley-Interscience, 2003. ISBN 0-471-39436-X.
6. HIRT, THOMAS D., P. NEUENSCHWANDER, U. W. SUTER. Telechelic diols from poly[(R)-3-hydroxybutyric acid] and poly{[(R)-3-hydroxybutyric acid]-co-[(R)-3-hydroxyvaleric acid]}. Macromol. Chem. Phys. 197, 1996, 1609-1614.
7. NGUYEN, S., G.-e. YU und R.H. MARCHESSAULT. Thermal Degradation of Poly(3-hydroxyalkanoates): Preparation of Well-Defined Oligomers Biomacromolecules, 2002, 3(1), 219-224. ISSN 1525-7797. doi:10.1021/bm0156274
8. HIRT, T.; NEUENSCHWANDER P.; SUTER, U. W.. Synthesis of degradable, biocompatible, and tough block-copolyesterurethanes. Macromol. Chem. Phys. 197, 1196(197), 4253-4268.

The influence composition on the properties of polyethylene based blends

Burgstaller Christoph [(1)*], ***Uttendorfer Thomas*** [(1)], ***Stadlbauer Wolfgang*** [(2)]

[(1)] TCKT - Transfercenter für Kunststofftechnik GmbH - - - Austria, [(2)] Fachhochschule Oberösterreich F&E GmbH - - - Austria

In packaging applications, nowadays typically multilayer films are applied to fulfil a variety of requirements like tenacity, printability, barrier properties and food contact approval. One persistent problem with such films is their recycling, even as low contaminated post-industrial waste, which is present as start-up materials and edge-trim, as the different layers are usually incompatible, thus re-melting results in degraded properties, which are insufficient for most applications. Therefore, the aim of this work was to investigate the composition effects of polyethylene based formulations, containing polyamide as the main component, but also barrier layer materials like ethylene vinyl acetate and ethylene vinyl alcohol. The different formulations were melt blended, with or without compatibilisation, in a twin-screw extruder, where for selected formulations processing parameters and screw geometry were varied. The materials were further processed via injection moulding to retrieve universal test specimen for mechanical testing (tensile and impact tests). Further characterisation, via scanning electron microscopy or rheology (represented her by melt flow rate determination), was carried out to investigate the correlations between composition, morphology and properties. We found, that the morphology is formulation dependent, and, that it is possible to compatibilise such formulations by applying maleic anhydride grafted polyethylene, so that the blends properties were sufficient for application again. Also, the mechanisms of the compatibilisation and the resulting morphologies can be, besides some processing influences, related to the chemical compositions of the components and their possibilities of interaction. These findings can help in the mechanical recycling of multilayer film based on polyethylene.

Investigation of PLA-biopolymer Blends to Improve Properties

***Kassos Nikolaos* (1)*, *Kelly Adrian* (1), *Gough Tim* (1), *Gill Andrew* (2)**

(1) University of Bradford - West Yorkshire - UnitedKingdom, (2) Floreon Transforming Packaging - East Yorkshire - UnitedKingdom

Polylactic acid (PLA) is the most widely used biopolymer with applications in fields such as packaging and textiles. Increased usage of biopolymers can help to reduce the accumulation of disposed plastics in the environment and reduce dependency on oil based polymers. PLA can be processed using conventional polymer processing techniques and can provide a diverse range of properties, making it a good candidate for replacing fossil based polymers. PLA has high strength and stiffness below its glass transition temperature (50-70°C), making it suitable for disposable packaging applications at low temperatures; however it is limited by its brittleness and low melt strength. Although developments in polymer chemistry have led to significant improvements in its properties, more immediate and lower cost solutions are required to enhance properties and extend applications. In this study, a commercially available grade of PLA has been blended with two bio-degradable polymers in an attempt to modify its properties and processability. Binary and tertiary blends of a PLA matrix with polybutylene succinate (PBS) and polycaprolactone (PCL) were produced by twin screw extrusion, up to 30wt% loading. Mechanical (tensile, impact and flexural), thermal and rheological characterisation was used to quantify properties of the different blend formulations. Phase separation and miscibility were also investigated using scanning electron microscopy and Raman imaging. Results showed that addition of both PBS and PCL caused a reduction in melt viscosity, elastic modulus and tensile strength, whilst causing an increase in impact strength and extension to break. Analysis of morphology suggested that immiscibility was apparent, particularly at higher PCL and PBS loadings. Results indicated that optimum properties and miscibility was obtained at low loadings in binary and tertiary blends.

Structural and electrochemical properties of PEO/PAN/graphene nano-fibrous blends

Abdollahi Serveh (1)*, Ehsani Morteza (2), Morshedian Jalil (3), Khonakdar Hossein Ali (4)

(1) Department of Polymer Processing, Polymer and Petrochemical Institute - Tehran - Iran, (2) Department of Polymer Processing Polymer and Petrochemical Institute - Tehran - Iran, (3) Department of Polymer Processing Polymer and Petrochemical Institute - Tehran - Iran, (4) Leibniz Institute of Polymer Research Dresden - Dresden - Germany

Lately, incorporation of the nano-size fillers mainly include carbon nanotubes and graphene nanosheets into the polymer electrolytes have received growing attention due to their unique mechanical, electrical, and thermal properties [1, 2]. In the present work, different weight per-cents of graphene Oxide nanosheets (GOs) [x=0.025,0.05,0.1, 0.5 wt.%] were incorporated into polyethylene oxide (PEO)/ polyacrylonitirle (PAN) nanofibrous membranes [PEO/PAN weight ratio : 75/25] via electrospinning process. Incorporation of very small amount of GOs (0.05 wt.%) enhanced ionic conductivity of PEO/PAN membranes remarkably from 3.7 mS/cm to 10.1 mS/cm. It could be due to the improved electronic conductivity of ion conducting poly-mer electrolyte and increased degree of amorphicity of the PEO matrix and thus formation of pathways for the charge carriers. On the other hand, upon adding 0.5 wt% of GO the tensile strength and Young's modulus of the composite membrabes increased sharply by 320.6 % and 128 MPa compared with neat PEO/PAN fibrous membarne respectively. Based on the high ionic conductivity and improved mechanical properties, PEO/PAN/GO membrane could be promisingly used as a polymer electrolyte. Keywords: graphene Oxide nanosheets, polyethylene oxide, polyacrylonitirle, membranes, elec-trospinning, ionic conductivity. References: 1. Seo, S. J., Yun, SH., Woo, J. J. (2011) Preparation and characterization of quasi-solid-state electrolytes using a brominated poly(2,6-dimethyl-1,4-phenylene oxide) electrospun nanofiber mat for dye-sensitized solar cells. Electrochem. commun., 13, 1391–1394. 2. Gao, S., Zhong, J., Xue, G., Wang, B. (2014) Ion conductivity improved polyethylene oxide / lithium perchlorate electrolyte membranes modi fi ed by graphene oxide, J. Memb. Sci., 470, 316–322.

Effects of Epoxy-POSS Nanoparticles on the Compatibility of PLA/PBAT Blends

Tuccar Kilic Nilay (1)*, Can Buse Nur (1), Kodal Mehmet (1), Ozkoc Guralp (1)

(1) Kocaeli University - Kocaeli - Turkey

Poly(lactic acid)(PLA) is the most generally chosen bio-polymer because of its lower price and commercially availability in the market. Although PLA has high strength and modulus, its inherent brittleness and low toughness limits its usage [1]. Poly(buthylene adipate-co-teraphthalate) (PBAT), is an aliphatic-aromatic copolyester that can be used to toughen PLA [2]. Due to incompatibility between PLA and PBAT, in the absence of a compatibilizer they exhibit poor mechanical properties [3]. Polyhedral oligomeric silsesquioxanes (POSSs) are new generation organic/inorganic hybrid materials that have an inorganic Si-O case with functional organic groups. Epoxy functionalized POSS nanoparticles are potential nanoparticles that can be used to compatibilize PLA/PBAT blends since epoxy group can react with -COOH groups in a wide temperature range [4]. In this study, it was aimed to observe the potential of epoxy-POSS nanoparticles to be used as a compatibilizer for PLA/PBAT blends. Three different types of epoxy-POSS types were used. Blends were prepared by melt compounding process. In order to characterize properties of blends, morphological, mechanical, thermo-mechanical, thermal and rheological tests were examined. The results showed that epoxy-POSS types can be used as a compatibilizer for PLA/PBAT immiscible blends. Acknowledgement: This study is financially supported by TUBITAK (Grant Number: 115M576) References [1]A. L. M. Siz, N. A. Ibrahim, W. M. Z. W. Yunus, Iran Polym J. 2013(22) 101-8. [2] N. Zhang, Q. Wang, J. Ren, L. Wang, J. Mater. Sci., 2009(44), 250-6. [3] M. Nofar, A. Maani, H. Sojodi, M. C. Heuzey, P. J. Carreau, J.Rheol. 2015(59-2), 317-33. [4] W. Zhang, B. X. Fu, Y. Seo, E. Schrag, B. Hsiao, P. T. Mather, N. Yang, J. Sokolov, Macromol., 2002(35),8029-38.

Shapememory behavior of nanocomposite based on SBS/LLDPE/CNT

***Kardan Nazanin** [1]*, **Mehranpour Milad** [1], **Ghasemi Ismaeil** [1]*

[1] Islamic azad university - Teheran - Iran

Shape memory polymers are new model of intelligent polymers that can be able to recover their initial shape by external stimuli such PH,humidity,magnetic,heat,etc. SMPs must have two phases, one fixity phase and one recovery phase. Shape memory polymers could be formed and temporarily fixed in a second temporary shape. In the present work, the blend of SBS/LLDPE was used as the shape memory material. The diffrent amount of CNT were added to the blend and the blend were cured by DCP for improving the mechanical, thermal and SMP properties. Fixity and recovery of final samples were measured and improving of recovery were gained because of improvement of thermal conductivity of the blend caused by CNT. The mechanical properties, morphology and the rheological parameters of samples were obtained. The mechanical properties test were measured by tensile device at room temprature(23 C). The rheological properties measured by RMS device and the morphology of samples were obtained by SEM test.

Lamellar assembly of Poly(3-hydroxybutyric acid-co-3-hydroxyvaleric acid) Spherulites Crystallized with Poly(Vinyl Methyl Ether)

Chen Chung-Hao [(1)], Woo Eamor M. [(1)]*

[(1)] National Cheng Kung University - Tainan - Taiwan

Ring-banded spherulites upon crystallization are common in some semicrystalline polymers including poly(3-hydroxybutyri acid-co-hydroxyvaleric acid) (PHBV). In this study, the lamellar assembly of PHBV spherulite in the blend with amorphous poly(vinyl methyl ether) (PVME) has been examined to elaborate the formation mechanisms of banded PHBV spherulites in non-isothermal crystallization process. The behavior of PHBV crystals at different temperatures (Tcs) was recorded and analyzed by using polarized optical microscopy (POM) and scanning electron microscopy (SEM). Under POM, PHBV/PVME (75/25) exhibits double-ring banded morphology with both orange and blue color bands alternatingly arranged along radial direction at a wide temperature range (60 to 100 oC). However, the band spacing and irregularity of the ring patterns on spherulites vary accordingly with Tc from wider-irregular rings at high Tc into slimmer-regular rings at low Tc. To further determine the correlation between crystallization temperature and spherulitic lamellar assembly, SEM was subsequently utilized on fractured samples. Correlations between top-surface banding patterns and interior lamellar assembly under various crystallization temperatures were collected and analyzed for possible formation mechanisms.

Insights on the efficiency of bio-based antioxidants in PLA/ABS blends

Rigoussen Alan (1)*, Verge Pierre (1), Raquez Jean-Marie (2), Habibi Youssef (1), Dubois Philippe (1)

(1) Luxembourg Institute of Science and Technology - Esch-sur-Alzette - Luxembourg, (2) Centre d'Innovation et de Recherche en Matériaux Polymères - Mons - Belgium

Over the last years, interest toward bioplastics has grown and these materials are expected to gradually replace petroleum based plastics. Many efforts to study these bio-materials and optimize their properties were achieved. Nevertheless, their development is still a challenge according to their inherent weaknesses, such as poor processability, brittleness, or poor thermal and physical properties. Among them, polylactide (PLA) is one of the most interesting bio-based material according to its biocompatibility, bio-resorbability, renewable origin and biodegradability, but its weak impact properties and poor thermal/oxidative stability result in a limited applications range. To overcome the brittle nature of PLA, the melt-blending with poly(acrylonitrile–butadiene–styrene) (ABS), a polymer having good impact and thermal properties, is an efficient and easily implementable approach. However, as PLA and ABS are immiscible, the resulting blend has poor mechanical properties and the addition of a third component acting as compatibilizer could tackle this issue. Recently, we have demonstrated that cardanol, a naturally occurring phenolic compound, was acting as an efficient compatibilizer for PLA/ABS immiscible blends. Its antioxidant properties were involved in the grafting of the phenolic moiety onto ABS during reactive extrusion. According to that finding, several natural antioxidant compounds were considered to produce compatibilized blends being stable against oxidation.

Mechanisms of Dual Spherulitic Morphology in Poly(L-lactic acid) Induced by Amorphous Polymer

Lugito Graecia (1)*, Woo Eamor M. (1)

(1) National Cheng Kung University - Tainan - Taiwan

Surface and interior lamellar assembly of dual-morphological spherulite generated from crystallization of poly(L-lactic acid) (PLLA) in the presence of atactic-poly(methyl methacrylate) (PMMA) have been characterized using polarized optical (POM), atomic forced (AFM), and scanning electron (SEM) microscopes. The formation of spherulite with dual morphologies takes place when the blend is crystallized at intermediate temperatures where the mobility of polymer chains is moderate, yet the crystallization rate is high. Besides the crystallization temperature, other factors such as molecular weight of both polymers, composition of the blends, as well as sample thickness have also been considered regarding their influence on the formation of dual-morphological spherulites (differ in their optical birefringence and lamellar arrangement). Direct observations toward the interior lamellar assembly have been conducted by fracturing the crystallized sample and subsequently etching out the amorphous portions from the spherulites. PLLA crystals are found arranging themselves into sheet-like and rod-like lamellae on the glass substrate. The development of these two distinct lamellar arrangements started from the nucleation site has been recorded. Each bundle of sheet-like lamellae at the nucleation site grows as one fibrous region as they cover a greater area; the area in between is covered by the rod-like lamellae growing to/from the substrate at a slower rate. Differ in their long axes and orders; both types of lamellae grow together composing a dual-morphological spherulite. This study demonstrates the microstructure instability induced by rapid crystallization of PLLA in a miscible blend with an amorphous polymer.

Strain sensor with selectively distributed filler network in conductive polymer composites

Cardon Ludwig [1]*, Duan Lingyan [1], Luyckx Geert [1], Degrieck Joris [1], D'hooge Dagmar [1]

[1] Ghent university - Oost-Vlaanderen - Belgium

Piezoresistive strain sensor based on conductive polymer composites (CPCs) becomes more intriguing for great potential applications in recent decades. A trade-off relationship between "high sensitivity" and "high linearity and stretchability" in majority of strain sensors was shown. Large stretchability demands the material remains structurally and morphologically intact under strain, while high sensitivity requires substantial structural changes even under small strain. The conductive network morphology and interfacial interaction play the most important roles in determining the sensitivity of blended CPCs. In this research, incompatible ternary composites based on thermoplastic poly(ether)urethane(TPU) and Ethylene-octane block copolymer (OBC) and carbon black (CB) were produced with twin-screw extruder. Post-annealing was also conducted. The preferential distribution of CB in the blends, the conductive network morphology, the resistance variation of the as-prepared and annealed samples during static and dynamic tensile tests were investigated. The exponential revolution of resistive response to strain was also fitted with a model based on tunneling theory to investigate the change in tunneling distance and the number of conductive pathways. In summary, for TPU/CB or OBC/CB binary composites, similar three regimes of the resistance change ratio-strain relationship was shown. As for the strain range of different regimes and the strain sensitivity, especially in regime I, there is no significant difference. And the sensitivity in regime I is quite low. Lower percolation threshold value and monotonic variation relationship and higher sensitivity could be achieved through the construction of conductive network structure with CB selectively distributed in one polymer phase or the interface in TPU/OBC/CB ternary composites and post annealing treatment.

Improvement of rheological and mechanical properties of PLA by reactive blending with Poly(MMA-g-GMA)

Karimi Amirhossein [1], Garmabi Hamid [1]*, Javadi Azizeh [1], Ahmadi Mostafa [1]

[1] Amirkabir University of Technology - Tehran - Iran

The aim of this study is to improve the melt strength of Polylacticacid (PLA) using methylmethacrylate-g-glycidylmethacrylate copolymers, Poly(MMA-g-GMA), that was synthesized by reactive blending at different weight fractions of GMA from 0 to 8 %. Copolymers were blended with polylacticacid (PLA) in internal mixer at different weight fractions from 0% to 15% as chain extender to modify the rheological behavior. The properties of neat PLA and modified PLA were characterized by Gel Permeation Chromatography (GPC), Differential Scanning Calorimetry (DSC) and Rheometric Mechanical Spectrometer (RMS). The balanced torque was increased by increasing the fraction of copolymer chain extender. PLA was even cross-linked when the content of copolymer reached 15%. The number and weight average molecular weights of PLA increased simultaneously as the fraction chain extender was increased. The results showed that Poly(MMA-g-GMA) is an effective chain extender for PLA. The glass transition temperature of PLA was also changed by blending with the synthesized copolymer. The blending affected the cold crystallization behavior and melting behavior of PLA, as well. Dynamic moduli also showed higher viscosities that can lead to higher melt strength. Keywords: PLA; rheological and mechanical properties; reactive blending; P(MMA-g-GMA)

Elongation Thinning and Morphology Deformation of Nanoparticle-filled Polypropylene/Polystyrene Blends in Elongational Flow

Kong Miqiu (1)*, Huang Yajiang (1), Li Guangxian (1)

(1) Sichuan University - Sichuan - China

Elongational thinning is observed in model polypropylene/polystyrene (PP/PS) blends filled with nanoparticles at nanoparticle loadings ≥3 vol.% and high strain rates, wherein the nanoparticle network that forms at rest is destroyed by the deformation. The elongational thinning is more obvious in blends with hydrophilic silica surfaces than with hydrophobic silica, apparently due to stronger interaction within the hydrophilic silica network and between PS and hydrophilic silica. Moreover, the elongational deformation of droplets is almost unaltered at lower silica loadings but nearly completely inhibited at higher loadings, owing to resisting the deformation by the enough enhanced elastic modulus of the PS matrix filled with high loadings of hydrophilic silica . In contrast, hydrophobic silica significantly boosts the deformation of droplets in elongation presumably because of the reduced interfacial tension due to the preferential distribution of hydrophobic silica at the droplet-matrix interface, relative to the hydrophilic silica which resides mostly in the PS matrix.

Recycled polycarbonate as impact modifier in polypropylene

Jean Baptiste X Michel Hau [1]and Baltus Cornelius Bonse [1, a)]

[1] *Centro Universitário da FEI. Av. Humberto de A.C.Branco 3972 São Bernardo do Campo – SP- Brazil*

[a)]Corresponding author: prebbonse@fei.edu.br

Abstract. Polypropylene (PP) is one of the most widely used commodity thermoplastics, such as in the automotive sector, in part, for being lightweight, flexible and strong, with high flexural fatigue strength. However, it displays low impact strength, which is commonly improved by adding elastomer. Elastomer, however, reduces tensile strength and stiffness. To increase impact strength without reducing tensile strength and stiffness PP was blended with recycled polycarbonate (PC) at 10, 20 and 30 wt%. Since these polymers are incompatible two compatibilizer have been tested in the blends, namely, maleic anhydride grafted PP, PPgMA, and a Lotader copolymer ethylene –co-methyl acrylate co-glycidyl methacrylate, P(E-co-MA-co-GMA). Two methods of blending have been tested: (i) tumble mixing of all components with subsequent extrusion and injection molding; and (ii) extrusion of PC/compatibilizer master-batch, with subsequent injection molding of tumble-mixed master batch plus PP. Specimens have been subjected to impact and tensile testing as well as scanning electron microscopy analysis. In terms of mechanical properties the best performance was found with the 80/20 PP/PC blend containing 10 wt% Lotader, prepared according to the first method of blending.

INTRODUCTION

Polypropylene (PP) is a versatile thermoplastic (TP), which can be processed by various techniques and used in many commercial applications. It is one of the fastest growing commodity TPs, with production volume exceeded only by polyethylene (PE) and polyvinyl chloride (PVC). To increase PP's low impact strength, an elastomeric PP phase, such as EPDM, is often added. However, Young's modulus and tensile strength decrease, and deformation increases. To recover loss in stiffness, brought about by the elastomer, talc is commonly added to the PP/EPDM blend, at levels between 10 and 40%, increasing also thermal and creep strength, dimensional stability and strength to heat and moisture [1].

Another way to increase PP's impact strength without loss in stiffness and tensile strength is to blend it with a strong, stiff and tough thermoplastic such as polycarbonate (PC) [2,3]. However, since PP and PC are immiscible because of polarity differences and solubility parameters a compatibilizer is required to increase physical and/or chemical interactions between the surfaces. This improves adhesion between the phases and reduces interfacial tension. Thus any energy or load exerted on the blends can be more efficiently transferred from one phase to another, increasing the mechanical properties of the blend [4].

Several compatibilizers have been tested for PP and PC [2,5,6]. Some studies showed that compositions with optimal mechanical properties contained 3-10wt% compatibilizer and 10-30wt% PC [4, 7].

The scope of this study was to increase the impact strength of PP without reducing its stiffness, by blending with recycled PC. Therefore, two compatibilizers and two blending methods were investigated: first premixing of all components, extrusion and injection moulding. Second, pre-mixing PC with compatibilizer to prepare a masterbatch by extrusion, and then adding PP directly into the injection moulding machine with the concentrate. The purpose was to establish the best compatibilizer and the best blending method for PP blends with recycled PC, in terms of mechanical properties.

EXPERIMENTAL

Materials

PP pellets (H 503) were supplied by Braskem with MFR (230°C/2.16kg) = 3.5g/10min, density = 0,905g/cm³.

Recycled ground PC was from from injection molding scrap of the materials engineering lab of our institution.

Compatibilizers used were: maleic anhydride grafted PP (Polybond 3200) - MFR (190°C/2.16kg = 110g/10min; density = 0,92g/cm³), from Crompton-Uniroyal Chemical (São Paulo, Brazil), and Lotader AX8900, a random terpolymer of ethylene, acrylic ester and glycidyl methacrylate, P(E-co-MA-co-GMA), produced by Arkema and supplied by Midland.

Methods

The study was divided in two stages. In the first stage 80/20 PP/PC blends were prepared without and with 10% of the two compatibilizers: PPgMA and Lotader. Two preparation methods were used: (i) tumble mixing of all components with subsequent extrusion and injection molding; and (ii) extrusion of 2:1 PC : compatibilizer master-batch with subsequent injection molding of tumble-mixed PP plus master batch. After the definition of the best compatibilizer and processing method, in terms of the investigated mechanical properties, two more blends were prepared: 70/30 and 90/10, maintaining the proportions between the contents of the dispersed phase and the compatibilizers, i.e., 70/30/15 and 90/10/5.

Extrusion was carried out in a Thermo Scientific HAAKE PolyLab OS RheoDrive co-rotating twin-screw extruder (L/D =25 mm; D=16 mm) with temperature profile ranging from 220 to 270°C. Injection molding was performed in an HM60/350 Battenfeld machine with temperature profile from 260 to 280°C, and mold temperature 50°C.

The injected specimens were subjected to tensile and flexure tests in an Instron 5565 Universal Testing Machine at test speeds of 50 mm/min and 1.3 mm/min, in accordance with ASTM D638 and D790, respectively. Charpy notched impact tests were carried out in a Ceast 9050 Instron impact tester, with a 5.4 J hammer, in accordance with ASTM D 6110.

Impact fractured surfaces of the blends were sputter-coated with gold and examined in a CamScan CS 3200LV scanning electron microscope (Cambridge) under high vacuum (10^{-5} mbar).

RESULTS AND DISCUSSION

Part 1: Finding the best compatibilizer and preparation methods

The results of the mechanical test of the PP/PCrec 80/20 blend, with and without compatibilizer and prepared by the two methods, are presented in Fig. 1. The aim is to find out which compatibilizer and which method provides the best results in terms of mechanical properties.

With regard to impact strength and strain at break, the blend without and with compatibilizer PPgMA reduced the properties of neat PP. Impact strength of the blend, without compatibilizer and with 10wt% of PPgMA, is respectively 37% and 94% lower than that of neat PP. That is, neither of them resulted in compatible blends. In contrast, mixtures with Lotader increased by 94% and 123% the impact strength of PP. In addition, the blends obtained by mixing PP with PC/Lotader master in the injection moulding hopper resulted in a 13% higher impact strength than the blend prepared by the other method. Thus, Lotader performed better for impact strength than PPgMA.

As to strain at break, PP showed a much higher strain than those of the blends and PC alone. The blend without compatibilizer or with PPgMA ruptures way before PP or PC: the maximum PP/PC 80/20 elongation at break is 7 times smaller than PP and with PPgMA elongation is 6 times smaller, i.e., in terms of this property these two blends are not compatible. On the other hand, the blends containing Lotader maintain elongation nearer to that of PP. PP/PC/Lotader and PP + (Master PC/Lotader) present a 19% and 10% lower strain at break, respectively in relation to neat PP. This may be due to the much more rubbery nature of Lotader compared to PPgMA. The strain at break of the two blends containing Lotador, but with different processing method, is similar, taking into account the standard deviations.

Thus, impact strength and strain at break result suffice to conclude that the best compatibilizer for this system is Lotador.

Regarding the choice of the best method, although, in relation to the premixing method with subsequent extrusion and injection moulding, the method involving master-batch preparation results in slightly higher impact strength and strain at break, but Young's modulus is lower than that of PP. As our aim is to increase the impact strength of PP without losing stiffness, the best method is that of premixing all blend constituents with subsequent extrusion and injection moulding.

Interestingly, the tensile strength of 80/20 PP/PC blends without compatibilizer and with 10wt% PPgMA are similar to that of PP. On the other hand, the blends with Lotader have a 12% lower tensile strength than that of PP. This drop is probably due to the rubbery behaviour of the Lotader compatibilizer.

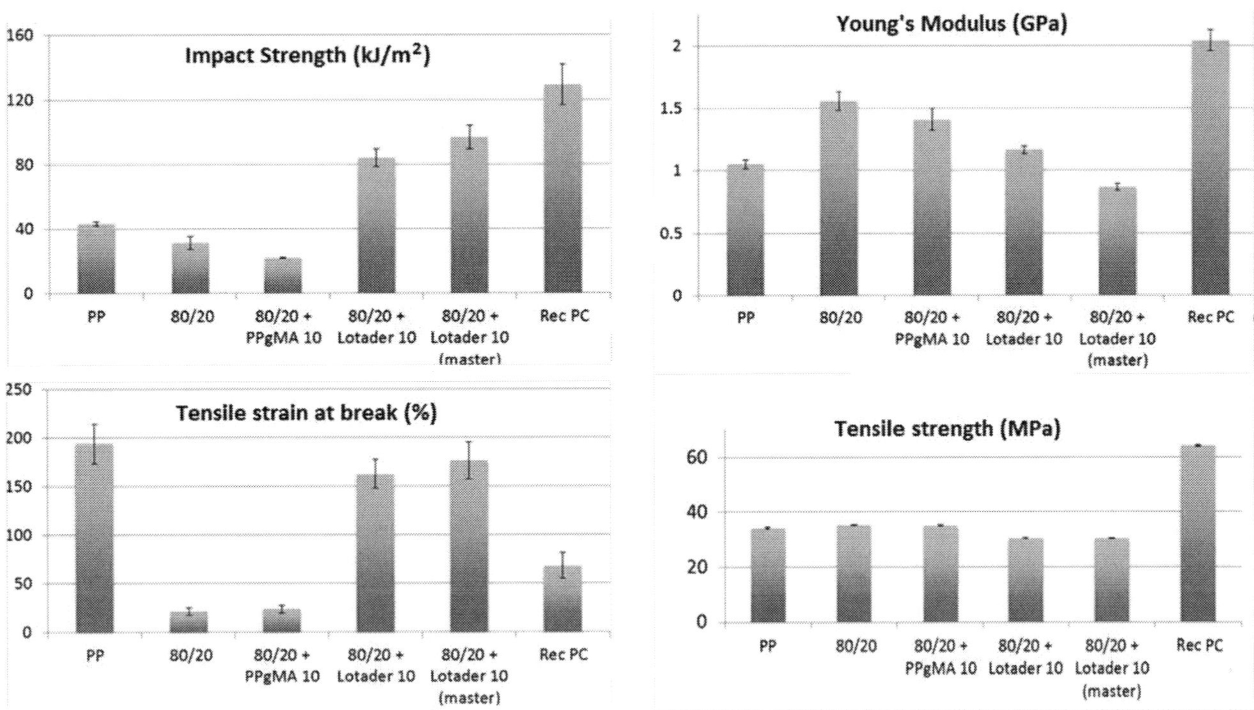

FIGURE 1. Impact and tensile properties of 80/20 PP/recycled PC blends containing 10wt% compatibilizer (PPgMA or Lotader), prepared according to (i) tumble mixing of all components with subsequent extrusion and injection moulding; and (ii) extrusion of PC/compatibilizer master-batch, with subsequent injection moulding of master batch plus PP

Part 2: Comparing different PP/PCrec/Lotader blends

In the second step PP/PC/Lotader 70/30/15 and 90/10/5 blends were prepared by mixing all blend constituents in a twin screw extruder and subsequent injection moulding for comparison with the 80/20/10 blend. Results are shown in Figure 2.

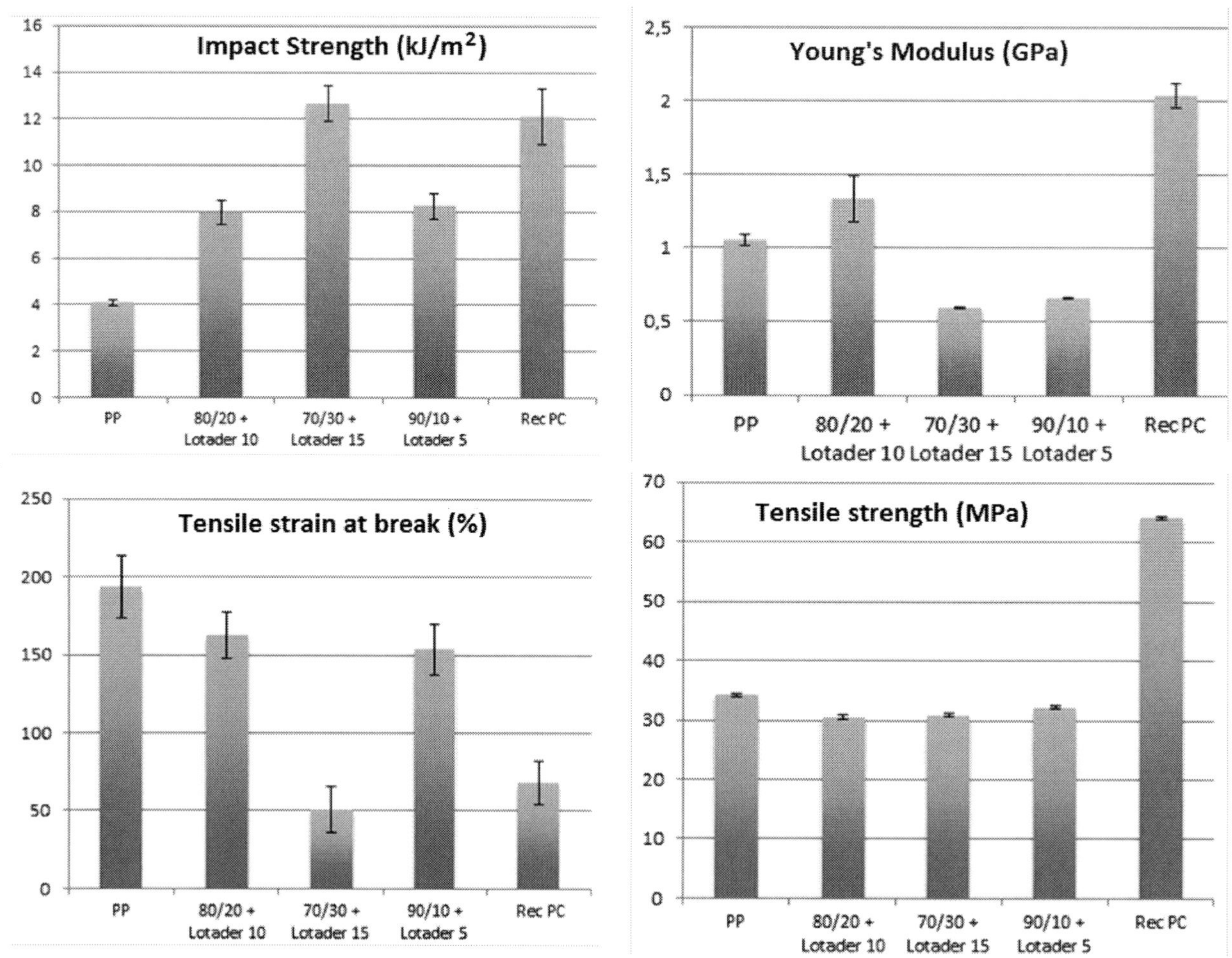

FIGURE 2. Mechanical properties of the 80/20. 70/30 and 90/10 PP/PCrec blends containing Lotader at a 2:1 PC:Lotader ratio.

Figure 2 reveals that the blend with the highest impact strength is 70/30 containing 15wt% Lotader, however this same blend has the smallest Young's modulus. As it has a higher Lotader content, the rubbery character of Lotader likely led to higher impact strength and reduced stiffness. However, this should not be the only reason, since this same blend, while exhibiting the lowest strain at break, its tensile strength did not drop that much (~ 10%) in relation to that of PP. Moreover, the tensile strength of the three blends is practically the same.

With the exception of the elastic modulus, the 80/20/10 and 90/10/5 blends, have similar mechanical properties. An explanation can be provided by morphology. Figure 3 shows very similar dispersed phase morphology between the 80/20/10 and 90/10/5 blends, very different from the morphology of the 70/30/10 blend with much larger dispersed phase size.

All blends presented droplets and elongated particle morphology, probably caused by the shear during injection moulding of the test specimens. They also presented pullout and detachment from the dispersed phase.

As a conclusion of the second step, in terms of a good balance of properties, PP/PC 80/20 containing 10wt% Lotader is the best of the three blends produced by the same process.

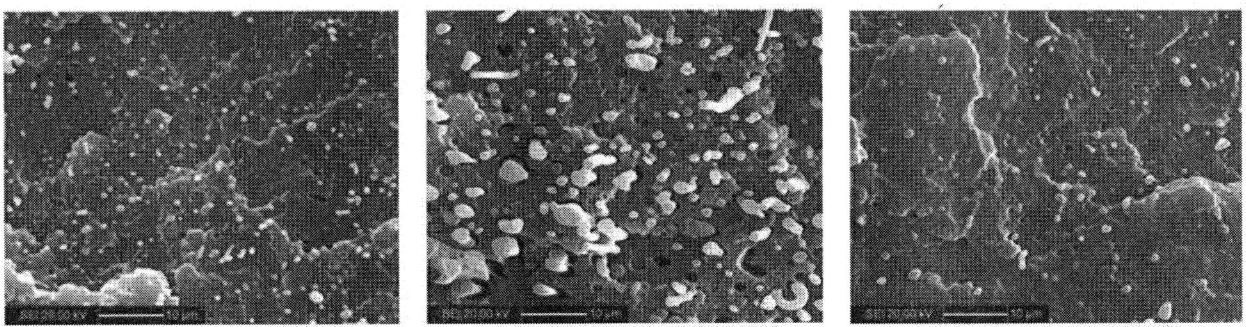

FIGURE 3. Impact fractured surfaces of the PP/PC/Lotador blends obtained by the same method: a) 80/20/10; b) 70/30/15; c) 90/10/5; d) amplified region of b)

CONCLUSIONS

Recycled PC was incorporated in PP to increase PP's impact strength without reducing stiffness and tensile strength. Since PP and PC are immiscible, two compatibilizers were tested: PPgMA and P(E-co-MA-co-GMA) (Lotader). Two blends containing 20wt% recycled PC and 10wt% of each compatibilizer were prepared. Results of the mechanical properties and SEM morphological analyses indicate the best compatibilizer was Lotader. Therefore, two blend preparation methods of PP/PC/Lotader 80/20/10 were tested, i.e., (i) pre-mixing of all components with subsequent extrusion and injection moulding and (ii) preparation in the co-rotating twin-screw extruder of a 2:1 PC: Lotader master-batch, pre-mixing of the master-batch and PP, and injection moulding. Morphological analyses indicated that best mixing was obtained by the first method. So, in the last stage, two blends were prepared in accordance with the best compatibilizer and the best preparation method, i.e., 70/30/15 and 90/10/5 and compared with the 80/20/10 blend. The blend which accomplished the best property balance was 80/20/10, with impact strength and Young's modulus 94% and 27% higher than neat PP. Tensile test obtained was 10% lower.

ACKNOWLEDGMENTS

Thanks are due to Centro Universitário FEI for the undergraduate research scholarship granted to J.B.X.M. Hau.

REFERENCES

1. Maier C., Calafut, T. Polypropylene: The Definitive User's Guide and Databook, Norwich, NY: Plastics Design Library, 1998.
2. Dai, S. Elaboration et caractérisation de mélanges de polymères à base de polypropylène et de polycarbonate à propriétés optimisées (Thesis), 2010
3. Shayuti, M.S.M.; Abdullah, M.Z.; Yusoff P.S.M.M., Compressive Properties and Morphology of Polypropylene/Polycarbonate Blends, 2011 International Conference on Environment and Industrial Innovation, 2011.
4. Yin Z., Zhang X., Zhang Y., Yin J., J Appl Polym Sci, 63, 1857-1863, 1997
5. Ping N., Tze-Man K., Polym Eng Sci 37, pp. 1226-1237,1997
6. Renaut N., Duquesne S., Zanardi S., Bardollet P., Steil C., Delobel R. J Macromol Sci Pure, 42, 977-991, 2005.
7. Dobrzański L.A., Król M., Bilewicz M., Viana J.C. Journal of Achievements of Materials and Manufacturing Engineering, 27, 19-22, 2008

Mechanical properties of virgin ABS/post-consumer ABS blends

Souza Adriana Martinelli Catelli (1)*, Cucchiara Mayara Galego (1)

(1) Centro Universitário FEI - São Paulo - Brazil

Blends of virgin ABS (acrylonitrile butadiene styrene terpolymer) and post-consumer ABS coming from waste of electrical and electronic equipment (WEEE) was studied. Virgin ABS/post-consumed ABS blends were obtained by extrusion in five different proportions: 100 - 75/25 -50/50 - 25/75 -100 (wt%) using a co-rotational twin-screw extruder. Samples for flexural and impact tests were obtained by injection molding. After this procedure, some specimens were select for characterization and others were ground and injected again. Samples obtained by the first and second injection molding cycle were characterized by flexural and impact tests. Significant differences in mechanical properties were assessed by analysis of variance (ANOVA). vABS/pcABS blends, subjected to two injection molding cycles presented an increase in flexural modulus and flexural strength. Blends subjected to one or two injection molding cycles presented a slight increase in flexural modulus and flexural strength with increasing pcABS concentration. Impact strength of vABS/pcABS blends decreased with increasing pcABS concentration and with increasing the number of injection molding cycles

Enhanced thermal conductivity of polymer composites filled with hybrid fillers

Seyed Esfahani Seyed Armin (1), Ghahramani Nikoo (2), Mehranpour Milad (3)*, Nazockdast Hossein (2)

(1) Islamic Azad university,Sciences & Research Branch - Tehran - Iran, (2) Polymer Engineering Department, Amirkabir University Of Technology - Tehran - Iran, (3) Islamic Azad University,Science and Research Branch - Tehran - Iran

Improvement thermal conductivity of immiscible polymer blends filled by hybrid fillers have attracted much attention in many thermal management fields. However, common method for achieving highest thermal conductivity have usually need high amount of conductive filler which result in poor mechanical properties and complex processing. using double percolation method with selective localization can be utilized as a solution for this problem. In the present work, the selective localization of boron nitride (BN) and Reduce graphene oxide (r-GO) in the immiscible polypropylene(PP) and polyamide 6 (PA6) blend is studied and investigation on the synergistic effect of hybrid fillers on thermal conductivity of polymer blend have been made. In this work the different composition of PP/PA6 have been studied in order to produce PP/PA/BN/r-GO thermally conductive composite with minimum volume fraction of fillers. The results of this study showed that hybrid fillers have good synergistic effect on thermal conductivity of PP/PA blend.

Investigation on the thermal stability and morphology of PLA/PCL/Starch ternary blends containing hydroxyapatite nanoparticles and triclosan

Yavarpanah Shakiba (1), *Seyfi Javad* (1), *Khonakdar Hossein Ali* (2)*

(1) Islamic Azad University of Shahrood - Semnan - Iran, (2) Leibniz-Institut für Polymerforschung Dresden - Saxony - Germany

Investigation on the thermal stability and morphology of PLA/PCL/Starch ternary blends containing hydroxyapatite nanoparticles and triclosan Shakiba Yavarpanah1, Javad Seyfi1, Hossein Ali Khonakdar2* 1 Department of Chemical Engineering, Shahrood Branch, Islamic Azad University, Shahrood, Iran 2 Leibniz-Institut für Polymerforschung Dresden e.V., Hohe Strasse 6, D-01069 Dresden, Germany Abstract Nowadays, developing innovative solutions for environmental problems have turned into the major concerns of researchers around the world. Therefore, replacement of petroleum-based plastics with biodegradable polymers is one of the promising solutions. In this study, an entirely biodegradable multicomponent system based on poly(lactic acid) (PLA), starch and poly(caprolactone) (PCL) was prepared via the melt-mixing technique. To achieve a better compatibility among the components, hydroxyapatite nanoparticles (nHA) and triclosan were added to the above-mentioned ternary blend. PLA was the matrix to which starch was added as a filler making the final material more hydrophilic. On the other hand, PCL was added to enhance the blend toughness. Triclosan and nHA were also added to impart some properties such as antibacterial behavior and mechanical strength, respectively. It was found that the combination of nHA and triclosan resulted in a great level of compatibilization. Several samples containing 1, 3 and 5 wt.% of nHA were prepared, and then, characterized by scanning electron microscopy (SEM), and thermogravimetric analysis (TGA). The morphology results showed that addition of nHA enhanced the compatibility among PLA, PCL and starch phases. TGA results also showed that in different temperature zones, the samples exhibited a different thermal stability due to the complexity of the system. All in all, the sample containing 1 wt.% of nanoparticles was found to exhibit an optimum behavior. Keywords: Poly(lactic acid); Poly(caprolactone); Starch; Morphology; Nanocomposite

Morphology, Crystallization Kinetics in Ternary Polymer Blends: Evaluating the Influence of Surface Functionalized Carbon Nanotubes

Kar Goutam Prasanna (1)*, ***Bose Suryasarathi*** (1)

(1) Indian Institute of Science, Bangalore - Karnataka - India

Structural properties, evolution of morphology and crystallization kinetics in immiscible polymer blends filled with multiwall carbon nanotubes (MWNTs) in the droplet phase were systematically investigated in this study. By grafting suitable macromolecules, that can drive the MWNTs to the droplet phase, allowed the understanding of the rate of nucleation and growth of the semicrystalline matrix in droplet-filled blends in contrast to matrix-filled blends. By blending 90 wt% of PVDF with 10 wt% ABS, matrix-droplet morphologies were generated and by grafting (styrene acrylonitrile) SAN onto MWNTs, the localization of the nanotubes was tuned to fill the droplet phase which otherwise prefers the matrix phase (here PVDF); driven by thermodynamics. The evolution of morphology under quiescent annealing conditions was assessed by SEM. The blends with SAN-g-MWNTs also coarsened as a function of time similar to neat blends however, to a lesser extent. Although, droplet laden MWNTs did not suppress coarsening in the blends but it still improved the tensile properties when compared with the neat blends. The fold surface free energy (as evaluated from isothermal crystallization kinetics) was estimated to be less in case of blends with SAN-g-MWNTs in contrast to blends with only MWNTs. This was attributed to the fact that matrix (here PVDF) filled with MWNTs experiences faster crystallization rate due to heterogeneous dispersion of nucleating agents (here MWNTs) in the matrix. However, when the nanotubes were mostly localized in the amorphous dispersed phase (here ABS), PVDF experienced similar crystallization behavior like the neat PVDF/ABS blends. Taken together, our study demonstrates that lower amount of energy is required for the arrangement of PVDF chains into the crystal lattice upon cooling from the melt state and accelerate the crystallization process when the heteronucleating agents are localized in the matrix phase than when they are localized in the amorphous phase.

Nanoparticle Partitioning of MWCNTs Filled PC/PVDF Co-continuous Blends by Rheological Characterization

Ahmadi Hamid [(1)], Hosseini Seyed mohammad [(1)], Javadi Azizeh [(1)], Nazockdast Hossein [(1)], Anvari Mohammad [(2)]*

[(1)] Amirkabir University of Technology - Tehran - Iran, [(2)] Iran University of Science & Technology - Tehran - Iran

The accurate determination of nanoparticle localization in immiscible polymer blends with co-continuous morphology by rheological technique has always been a difficult job. In this study we showed how the relaxation time spectrum ($H(\lambda)$) of polymer blends can be employed to determine the localization of MWCNTs in PC/PVDF Polymer blends with co-continuous morphology. This was carried off by comparing the $H(\lambda)$ of MWCNTs filled blend samples with the filled and unfilled polymer component. It was demonstrated that MWCNTs is preferentially localized in PC phase which was confirm by TEM micrograph.

An Investigation of Linear Viscoelastic Behavior of PP/EVA/LDH nanocomposites by Using Dynamic Mechanical Analysis and Creep test

Mahdavi Reza [(1)]*, Otadi Maryam [(1)], Goodarzi Vahabodin [(2)]

[(1)] Azad University - Tehran - Iran, [(2)] 2Applied Biotechnology Research Center, Baqiyatallah University of Medical Sciences - Tehran - Iran

In this work attempts polypropylene/ethylene vinyl acetate copolymer (PP/EVA) blends loaded with layered double hydroxides LDH nanoparticles were prepared by melt mixing process and their thermal and dynamic mechanical properties were determined. Detailed investigations on LDH dispersion state in the polymeric matrix were conducted by transmission electron microscopy. TEM investigations revealed that the LDH nanoparticles were mainly located within the EVA phase of the blend. Also, microstructure of blends studied by SEM showed that incorporation of LDH into the blends led to reduction of dispersed domain size attributed to reduction of interfacial tension due to the filler presence. Melting temperature of PP loaded with 5 wt% of LDH was increased up to 11 oC as compared to the neat PP. Addition of LDH and PP-g-MAH as compatibilizer to PP/EVA blend increased melting and crystallization temperatures determined by DSC. DMTA studies showed that addition of LDH into the blends had no positive effect on the storage modulus of nanocomposites while, their loss modulus were higher than that of the neat PP. Moreover, the beta transition temperature of PP phase in the blends loaded with LDH and compatibilizer was about 5 oC higher than the neat PP. Time-temperature superposition was applied to obtain viscoelastic master curves at different temperatures. Creep and recovery tests at various temperatures were done. Also, viscoelastic parameters of the nanocomposites at defined temperatures were obtained by applying of Burger and Weibull models. The creep compliance master curves of studied samples were also obtained which showed that compliance of neat polymer increased after 1667 h while that of nanocomposites containing 5wt.% of LDH showed significant reduction. Keywords: polypropylene; Ethylene vinyl acetate copolymer; blends; Nanocomposites; layered double hydroxide (LDH); Dynamic mechanical properties

The Usage of Octaisobutyl-POSS in The Blends of PLA/PBAT: "Effects of Molecular Weight of PBAT"

CAN BUSE NUR [(1)], TUCCAR KILIC NILAY [(1)], KODAL MEHMET [(1)], OZKOC GURALP [(1)]*

[(1)] Kocaeli University - Kocaeli - Turkey

In the last decade, the usage of plastics has become very popular in many areas owing to their several advantages. However the solid waste problem creates morality issues due to environmental problems. In order to prevent the pollution problem, biodegradable polymers have been focused by many researches. Recently, poly(lactic acid) (PLA) becomes most widely desired biodegradable polymer due to its good physical properties and being produced from renewable resources with an economical price [1]. Poly(butylene adipate-co-terephatalate) (PBAT) obtained from petrochemical resources is another biodegradable polymer and that can be easily processed without any converse effects on the environment. It can be also used to toughen PLA via melt blending [2]. Polyhedral oligomeric silsesquioxanes (POSS) organic/inorganic hybrid materials are new generation nano-fillers and could provide molecular reinforcement, increased thermal stability and better flame resistance [3]. The aim of this study is to investigate the effects of octaisobutyl-POSS (OPOSS) on the properties of PLA/PBAT blends. The PBAT molecular weight and OPOSS concentration was taken as the experimental parameters. Blends were prepared in a laboratory twin-screw microcompounder. Mechanical, thermal and rheological properties were characterized. The results showed that the high molecular weight of PBAT improved the mechanical properties of blends such as elongation at break and impact strength. Incorporation of OPOSS enhanced the compatibility between PBAT/PLA regardless of PBAT's molecular weight. In addition, thermal properties were also affected positively. Acknowledgement: This study is financially supported by TUBITAK (Grant Number:115M576) References [1] R. Narayan, K.Schcaf, ASTM Standardization News, (1992), 23-26. [2] J. B. Olivato, M. V. E. Grossmann, F. Yamashita, D. Eiras, L. A. Pessan, Carbohydrate Polymer, 87 (2012), 2614-2618. [3] M. Kodal, H. Sirin, G. Ozkoc, Polym. Comp., 10.1002 , (2015), 264-275.

Effects of different toughens on the properties of poly(lactic acid)

CAN BUSE NUR [(1)]*, TUCCAR KILIC NILAY [(1)], KODAL MEHMET [(1)], OZKOC GURALP [(1)]

[(1)] Kocaeli University - Kocaeli - Turkey

Poly(lactic acid)(PLA) is one of the most extensively studied bio-based aliphatic polyesters that has several properties, such as high strength and stiffness. However the application of PLA is still limited due to its high price, low heat distortion temperature, and brittleness. PBAT is fully biodegradable and flexible polymer makes it a great candidate for the toughness of PLA. Nevertheless PLA and PBAT yields immiscible polymer blends which needs to be compatibilized. The other common polymer used to toughen PLA is thermoplastic polyurethane (TPU). The soft segment of TPU is expected to have a good compatibility with PLA but they make semi-miscible blends. This study focuses on the effect of incorporation of PBAT or TPU in to the PLA to improve the toughness. A styrene-acrylic multi-epoxidized polymeric compatibilizer (Joncryl®) was used in PLA/PBAT blends, whereas a diisocyanate chanin extender was utilized for PLA/TPU system. Blends were prepared in an Xplore Instruments MC15 twin-screw microcompounder. The mechanical, morphological and rheological properties of blends were investigated. The observations showed that Joncryl improved the mechanical and rheological properties of the PLA/PBAT blends, which supported by morphological results. The mechanical properties of PLA/TPU blends were improved with using PDI. PLA/TPU system is found to be tougher in comparison to PLA/PBAT under the experimental conditions. Acknowledgement: This study is financially supported by TUBITAK (Grant Number:115M576)

Immiscible PLA/PBAT blends and Partially Miscible PLA/TPU blends: "Comparison of the Effects of O-POSS on Thermal and Themo-mechanical Properties"

Tuccar Kilic Nilay (1), Can Buse Nur (1), Kodal Mehmet (1), Ozkoc Guralp (1)*

(1) Kocaeli University - Kocaeli - Turkey

Poly(lactic acid) (PLA) is a biodegradable polymer which can be obtained from completely renewable sources such as corn, wheat and rice; hence, it reduces the degree of dependence on petroleum resources. Despite its numerous advantages such as high strength and stiffness its low toughness and brittleness significantly impede its applications; therefore, toughening is required. Poly(butylene adipate-co-terephthalate) (PBATs) and thermoplastic polyurethane (TPU) are good candidates to toughen PLA [1]. PLA/PBAT blends show immiscible behaviour while PLA/TPU blends exhibit partially miscible behaviour [2]. Polyhedral oligomeric silsesquioxanes (POSSs) are nano-particles that create molecular reinforcement and can improve the compatibility between blends. Octaisobutyl-POSSs (O-POSSs) are POSS types that contain eight isobutyl groups [3]. The objective of this study is to compare thermal and thermo-mechanical properties of immiscible PLA/PBAT and partially miscible PLA/TPU blends in the presence of O-POSS. Melt-blending process is used for the preparation of blends. The results indicated that O-POSS positively affected the properties of partially miscible PLA/TPU blends, whereas negatively affected the properties of PLA/PBAT immiscible blends. Acknowledgement: This study is financially supported by TUBITAK (Grant Number: 115M576) References [1] Zhao F., Huang H.X., Zhang S.D., J. Appl. Polym. Sci., 2015 (42511), 1-9. [2] Dogan Kemaloglu S., Reyes E. A., Rastogi S., Ozkoc G., J. Appl. Polym. Sci., 2014(40251), 1-10. [3] Li G., Wang L., Ni H., C. U. Pittman, Journ.of Inorg.and Organomet. Polym., 2001(11-3), 123-54.

Thermal and Thermo-mechanical Properties of PLA/TPU/Epoxy-POSS Ternary Blends

Tuccar Kilic Nilay (1)*, Can Buse Nur (1), Kodal Mehmet (1), Ozkoc Guralp (1)

(1) Kocaeli University - Kocaeli - Turkey

Poly(lactic acid) (PLA) has received great attention as an eco-friendly polymer because of its competitive price and high strength as well as being renewable and biodegradable. However, neat PLA cannot meet the requirements in some special circumstances because of its inherent brittleness; thus, toughening is required [1]. Thermoplastic polyurethanes (TPUs) can be used to toughen PLA by melt blending [2]. Polyhedral oligomeric silsesquioxanes (POSS) organic/inorganic hybrid materials are nanoparticles containing functional groups that are potential sites for polymerization, grafting, surface bonding etc. Epoxy POSS nanoparticles can react with carboxylic acid and isocyanate end-groups of either PLA or TPU in a wide temperature range [3]. Therefore, it can be used as a compatibilizer for PLA/TPU system. The objective of this study is to observe the effects of TPU content on the thermal and thermo-mechanical properties of PLA/TPU/Epoxy-POSS ternary blends. Blends were prepared by melt compounding process using an Xplore Instruments laboratory compounder. Thermal and thermo-mechanical properties were examined. The results showed that increasing TPU content inclined thermal decomposition temperatures of blends. In addition glass transition temperatures and crystalinity behaviours showed differences according to TPU content. Besides, thermo-mechanical properties showed that storage modulus decreased in case of higher TPU content. Acknowledgement: This study is financially supported by TUBITAK (Grant Number: 115M576) References [1] Madhavan Nampoothiri, K.; Nair, N. R.; John, R. P., Research.Bioresour.Technol., (2010), 101, 8493–8501. [2] Han J. J., Huang H. X., Jour.of Appl.Polym. Sci., 2011(120), 3217–3223. [3] Li G., Wang L., Ni H., C. U. Pittman, Journ.of Inorg.and Organomet. Polym., 2001(11-3), 123-54.

Shear-induced crystallization of polymers: discussion on the coupled mechanical and structural effects

HAUDIN JEAN-MARC [(1)]*, BOYER SEVERINE A.E. [(1)]

[(1)] MINES ParisTech CEMEF - Alpes Maritimes - France

Focussed on coupled mechanical (flow), thermal and geometrical conditions, the emphasis of the present work is laid on the influence of shear flow through rheological measurements. The rheological properties measurement has often been used to follow the development of crystallization, both in quiescent conditions and under shear. One can cite: van Puyvelve et al. [1] with the RheoDSC device (combination of the rheological and calorimetric signal), Roozemond et al. [2] with the RheoDSC in the study of flow-induced crystallization through short-term shear flow experiments, Pantani et al. [3] with the rotational Haake Mars rheometer equipped with an optical module. The present investigation used the same rheo-optical RheoScope module (Thermo Scientific)*. The discussions deal with the correlation between the strong increase (crystallization) or decrease (melting) in viscosity and the transformed fraction and the morphology (spherulites and oriented structures) development or vanishing, which were determined simultaneously. The characteristic features of the crystallization process have been quantified (i) number of activated nuclei, (ii) spherulite growth, (iii) overall kinetics. A discrimination of the contributions from spherulitic morphology and oriented structures on the viscoelastic functions of an isotactic polypropylene is done. Acknowledgement The authors kindly acknowledge Florent Jego and Philippe Sierro (Thermo Fisher Scientific) for their technical support with RheoScope module. [1] Janssens V., Block C., van Assche G., van Mele B., van Puyvelde P., Intern. Polym. Proc. 25, 304-310 (2010) [2] Roozemond P. C., van Drongelen M., Verbelen L., van Puyvelde P., Peters G.W.M., Rheol. Acta 54, 1-8 (2015) [3] Pantani R., Speranza V., Titomanlio G., J. Rheol. 59, 377-390 (2015)

Measuring and modelling necking of polymers

Spencer Paul Edward [(1)]*, Coates Phil [(1)], Caton-Rose Fin [(1)], Sweeney John [(1)]

[(1)] University of Bradford - West Yorkshire - UnitedKingdom

The tensile behaviour of semi-crystalline polymers, such as PP and PE, typically involves a necked region, of reduced cross-section, with neck "shoulders" that propagate throughout the length of the specimen. Soon after formation, the neck shoulders assume a constant shape and propagate with constant velocity. We have developed in-house image analysis software to automatically analyse the neck shape profile from image-sequences captured during tensile tests. In addition, we simultaneously measure the strain field, by tracking dots marked on the sample, and the force from the load cell of the tensile testing machine. The exact shape of the neck is governed by the strain rate sensitivity and strain hardening characteristics of the polymer, and so its underlying structure. Here we determine the steady-state neck profiles, formed in flat dog-bone specimens of various PP and PE grades, during tensile tests at elevated temperatures. We study how the neck shape depends on testing conditions (rate, temperature) and material characteristics (molecular weight and thermal history), and investigate the true stress/strain/strain-rate behaviour of the deformation. All these studies are complemented by Finite Element simulations of this necking process, which helps to enhance our understanding by elucidating the triaxial stress states induced. The Finite element predictions, in terms of draw stress and neck shape, agree very well with experiment. Also, we compare experimental and Finite Element results with simple analytical neck propagation models. Necking behaviour is of central importance in solid-phase deformation processes, such as die-drawing, which produces highly oriented polymer with extremely enhanced mechanical properties. Because this system determines both the necking and the stress/strain/strain-rate behaviour, it is very useful in predicting the performance of a polymer in deformation processing operations.

ON THE PULSATING FLOW BEHAVIOR OF A BIOLOGICAL FLUID: HUMAN BLOOD

***Herrera Valencia Edtson Emilio* [1]*, *Sánchez Villavicencio Mayra Luz* [2], *Calderas Fausto* [3], *Medina-Torres Luis* [4]**

[1] FES-ZARAGOZA UNAM - Mexico City - Mexico, [2] Universidad Autónoma Metropolitana-Iztapalapa - Ciudad de México - Mexico, [3] CIATEC, A. C. Omega 201 Industrial Delta. C.P. 37545 León, Guanajuato. - León, Guanajuato. - Mexico, [4] Facultad de Química, Universidad Nacional Autónoma de México - Ciudad de México - Mexico

Abstract: In this work, the rectilinear flow of a complex fluid (human blood) under a pulsating time-dependent pressure gradient is analyzed. A first approximation of the real case of blood flowing in a vein is described. The pressure gradient simulates the pumping work of the heart while the flow geometry (circular tube) is assumed rigid, smooth and cylindrical. The rheological behavior of blood with different cholesterol levels is modelled using the Bautista-Manero-Puig (BMP) constitutive equation. According to the analytical solution, a flow enhancement is predicted to first order which represents the optimum pumping work of the heart which governs the flow of blood in the entire body. This work is a contribution to the understanding of the complex rheology involved in the discontinuous pressure driven flow of blood in the human body.

Simulation of Asymmetrical Multilayer Flat Film Coextrusion Regarding Slip at the Wall and Interfacial Slip at the Polymer-Polymer Interface

C. Kneidinger[1, a)], M. Laengauer[1, b)], E. Mayrhofer[2, c)], G. Zitzenbacher[1, d] and J.Miethlinger[2, e)]

[1]*University of Applied Sciences Upper Austria, School of Engineering, Stelzhamerstraße 23, 4600 Wels, Austria*
[2]*Institute of Polymer Extrusion and Compounding ,Johannes Kepler University, Altenbergerstrasse 69, 4040 Linz, Austria*

[a)]Corresponding author: Christian.Kneidinger@fh-wels.at
[b)] Manuel.Laengauer@fh-wels.at
[c)] elias.mayrhofer@jku.at
[d)] g.zitzenbacher@fh-wels.at
[e)] juergen.miethlinger@jku.at

Abstract. This work deals with the simulation of the viscous flow of polymers in flat film coextrusion regarding asymmetrical films as well as slip at the wall and slip at the polymer-polymer interface. A one dimensional finite volume approach is used to calculate complex flows of generalized Newtonian materials using numerous damped numerical solvers. This iterative approach is neither depending on the numbers of layers used, nor on the mathematical models of the viscosity, the wall slip or the interfacial slip. Furthermore, the work shows how this one-dimensional simulation can be done using standard spreadsheet software. By that means, it contributes largely to the simplification of dimensioning feed blocks for coextrusion. Moreover, the results compare the velocity profiles and the pressure gradient resulting from the application of three different rheological models. It is clearly shown, that the application of the power-law approach can result in much higher pressure compared to the results determined by utilizing the more realistic Carreau approach.

INTRODUCTION

This work deals with the numerical simulation of planar flows of generalized Newtonian materials regarding asymmetrical layer buildup, wall slip and slip at the polymer-polymer interface(s) as well as moving wall conditions (e.g. sheet coating process). This approach extends the possibilities of analytical flow calculations without applying methods of computational fluid dynamics (CFD), so it contributes largely to the simplification of dimensioning feed blocks for coextrusion.

All calculations can be done utilizing standard spreadsheet software. Moreover, realistic rheological models for polymeric melts like that of Carreau [1], Cross [2], Yasuda [3] Prandtl [4] and Eyring [5] as well as other approaches like that of Bingham [6] or Herschel–Bulkley [7] in addition to Newtonian and power-law fluids [8, 9] can be applied. The calculation of the wall slip can be done by applying different static approaches [10]. Additionally, the calculation enables the simulation of symmetric and asymmetric coextrusion flows and the appearance of slip at the polymer-polymer interface(s), which was detected e.g. by Zhao and Macosko [11]. The validation is done by comparing numerical simulations of fundamental Newtonian and power-law flows to analytical calculations according to Michaeli [12].

MODELING, CALCULATION

Figure 1(a) shows a sketch of the flow channel and the coordinate system. The origin of the coordinate system is located at the wall of the channel. The model is converted into a discretized finite volume model consisting of a natural number of elements ($n_{ele,}$) (see fig.1(b)). In case of a coextrusion flow the interfaces between the layers are between two elements. Additional elements are required if wall slip or interfacial slip occurs.

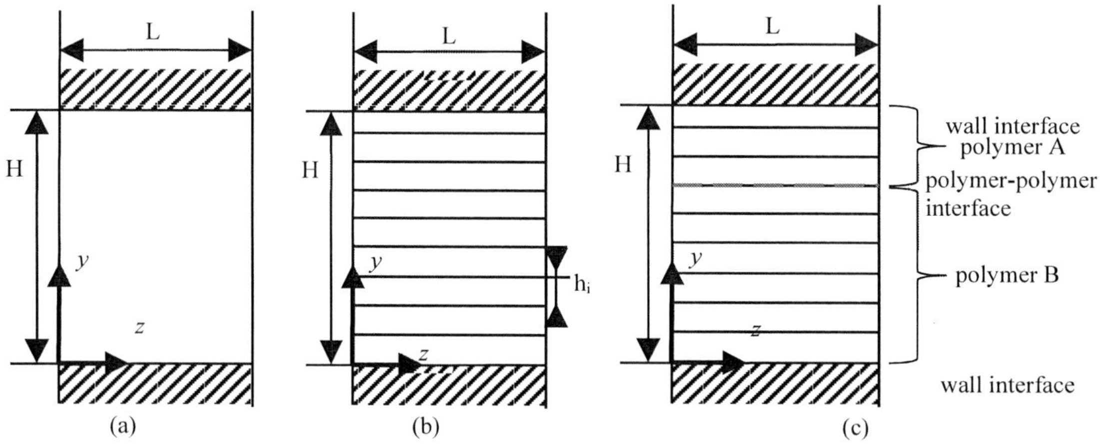

FIGURE 1. (a) sketch of the flow channel and the applied coordinate system, (b) discretized model, (c) discretized coextrusion model in case of coextrusion of two material layers

A fully developed steadystate planar flow in z-direction is considered. Furthermore, gravity and inertia terms are negelcted. The simplyfied equation of momentum is [11]:

$$\frac{\partial p}{\partial z} = \frac{\partial \tau_{yz}}{\partial y} \tag{1}$$

By integrating eq. (1) the shear stress τ_{yz} as a functin of the pressure gradient $\partial p/\partial_z$ is obtained.

$$\tau_{yz} = \frac{\partial p}{\partial z} y + C_\tau \tag{2}$$

The constant (C_τ) can only be determined analytically in case of a symmetrical flow (symmetrical boundary conditions and symmetrical layer buildup)

$$C_{\tau,sym} = -\frac{\partial p}{\partial z}\frac{H}{2} \tag{3}$$

As a matter of fact, this linear equation results in positive and negative values of the shear stress. In the case of asymmetrical layer buildup, moving wall condition (e.g. sheet coating process) or different wall slip behaviors at the two walls, the constant has to be solved numerically.

The velocity $v_z(y)$ is calculated by integrating the shear rate $\partial v_z/\partial_y$ (eq. 4), regarding the boundary condition of the wall ($v_{w,y=0}$) and the wall slip velocity ($v_{s,y=0}$), which are both depending on the wall shear stress $\tau_{yz}(y=0)$.

$$v_z(y) = v_{w,y=0} + v_{s,y=0} + \int_0^y \frac{\partial v_z}{\partial y} dy \tag{4}$$

If the applied rheological model is Newtonian, power-law, Bingham or Herschel–Bulkley, the shear rate $\partial v_z/\partial_y$ can be calculated directly from the shear stress τ_{yz}. If one of the models of Carreau, Cross or Yasuda is applied, a proper shear rate $\partial v_z/\partial_y$ is determined by numerical converging.

The viscosity η is defined as the ratio of the shear stress τ_{yz} to the shear rate $\partial v_z/\partial_y$.

$$\eta = \frac{\tau_{xz}}{\partial v_z / \partial y} \tag{5}$$

To calculate the wall slip velocity or the slip velocity at the polymer-polymer interface v_s considering the possibility of negative shear stresses, the known models have to be adapted. A power-law approach like eq. (6) has to be adapted to eq. (7).

$$v_s = \left(a\tau + b\tau^3\right) \tag{6}$$

$$v_s = \frac{\tau}{|\tau|}\left(a|\tau| + b|\tau|^3\right) \tag{7}$$

Discretized finite volume model

For each element the mean shear stress $\bar{\tau}_i$, which is the shear stress at the mean y position $\bar{y}_i$ is calculated first.

$$\bar{\tau}_i = \frac{\partial p}{\partial z}\bar{y}_i + C_\tau \tag{8}$$

The shear rate is considered to be constant inside of each element, so the "velocity difference" (Δv_i) between the borders of each element can be calculated from the height of the element $\mathrm{h_i}$ and the mean shear rate $\partial \bar{v}_{z,i} / \partial y$, which is calculated from the mean shear stress $\bar{\tau}_i$ according to the applied rheological model.

$$\Delta v_i = \mathrm{h_i} \frac{\partial \bar{v}_{z,i}}{\partial y} , \tag{9}$$

This simplified approach allows a fast and simple calculation and enables analytically correct values at the borders of the elements considering Newtonian flow behavior. In other cases, the height of the element has to be small resulting in a high number of elements. A comparison to results from analytical simulation of power-law fluids showed a deviation of 0.9 % in case of 20 elements and 0.2 % in case of 40 elements.

Solving procedure

The first step of doing this simulation is defining the following terms: the overall geometry of the flow channel, the required volumetric flow of the single layers, models and parameters needed to calculate the shear rate of all layers, models and parameters needed to calculate the sliding velocity at the wall and that at the polymer-polymer interface, starting conditions for the iterative simulation and the precision of the calculation;

The second step is the iterative solution of the following terms:

1. The constant (C_τ) in the shear stress distribution:
   a. When a symmetrical flow profile is calculated, C_τ can be calculated analytically,
   b. In other cases C_τ represents the asymmetry of the shear stress distribution and must converge again when the pressure gradient or the layer buildup are changed.
2. The pressure gradient is numerically adapted to achieve the predefined overall throughput.
3. The thickness of the single layers is adapted to achieve the predefined throughput of the single layer.

RESULTS

The presented results compare the velocity profiles resulting from the application of three different rheological models: the Carreau approach, the power-law approach and Newtonian behavior. To allow a comparison all results were calculated with the same dimensions of the die (height H = 1.6 mm, width = 1 m) and with the same mass flowrate ($\dot{m}$ = 326 kg/h). The density (ρ) of the polypropylen-homopolymer (PP-H) melt is 815 kg/m³, that of the polymethylmethacrylate (PMMA) is 1107 kg/m³. Therefore, the volumetric flow rate ($\dot{V}$ = 0.818 10^{-6} m³/s) and the mean velocity ($\overline{v}_z$ = 0.51 m/s) of PMMA is much lower than that of PP-H ($\dot{V}$ = 1.111 10^{-6} m³/s, $\overline{v}_z$ = 0.694 m/s) at the same mass flowrate.

Figure 2(a) shows the applied rheological models and fig.2(b) shows the resulting velocity profiles. The Newtonian viscosity of PP-H was adjusted to reveal the same pressure gradient as the power law approach. Figure 2(b) clearly shows the Newtonian velocity profile does not match the more realistic one obtained by applying the Carreau approach. The absolute value of the pressure gradient in the flow direction $\partial p/\partial_z$ calculated for PMMA applying the power law approach is more than 30 % higher than that obtained by applying the Carreau approach.

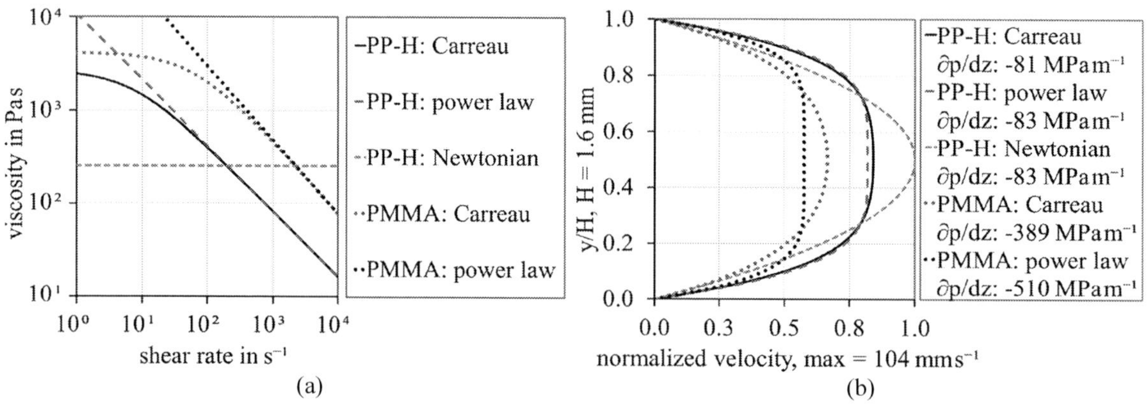

FIGURE 2. (a) Viscosity of the materials/rheological model approaches, mass flowrate: 326 kg/h, density of PP-H: 815 kg/m³, density of PMMA: 1107 kg/m³; (b) Symmetrical velocity profiles resulting from different rheological model approaches;

Figure 3(a) shows velocity profiles of symmetric and asymmetric coextrusion flows. As before, the overall mass flowrate is 326 kg/h, that of each material 163 kg/h resulting in a flow rate of 81.5 kg/h in each of the outer layers of the symmetrical three-layer buildup. The PP-H grade has a much lower viscosity than the PMMA.

The symmetric PP-H/PMMA/PP-H three-layer buildup (red line) leads to a block-like velocity profile and an absolute pressure gradient of 77 MPa·m⁻¹. The reversed PMMA/PP-H/PMMA buildup (black, long-dashed line) leads to a more tip shaped velocity profile and a much higher absolute pressure gradient of 362 MPa m⁻¹. The former PP-H/PMMA/PP-H layer buildup in combination with wall slip (green dashed line) clearly amplifies the block-like flow resulting in a lower maximum velocity and a reduced absolute pressure gradient of only 45 MPa m⁻¹. The power-law model (eq. (7)) was applied to calculate the wall slip. The parameters A and B were both set to 10^{-15} which are exemplary values that were chosen to result in a clearly recognizable change in the velocity profile. The combination of interfacial slip and the PP-H/PMMA/PP-H layer buildup results in a less block-like flow (blue short-dashed line). The inner layer is "pushed" through the outer layers. The power-law model (eq. (7)) was applied to calculate the interfacial slip. The parameters A and B were both set to 10^{-14}, which are again exemplary values. The gray dashed line shows the flow profile of an asymmetrical two-layer buildup. The off-center shear stress distribution with the zero-crossing at the point of the maximum velocity (y/H = 0,754) leads to unequal (absolute) wall shear stresses (-0.083 MPa at y/H = 1 and 0.255 MPa at y/H =0).

Figure 3(b) shows velocity profiles of a flow considering the moving wall condition as it can be found in the sheet coating process. The mass flowrate is 326 kg/h. The solution provided by the Newtonian approach (green dashed line) could be calculated analytically as well, but it does not fit the more realistic value received by applying Carreau's approach (black lines). The power law approach delivers quite good results in case of PP-H, but leads to insufficient results in case of the PMMA where the calculated absolute pressure gradient of 410 MPa m⁻¹ is much higher than the value of 270 MPa m⁻¹ computed for Carreau's approach. Additionally the velocity profile does not fit well.

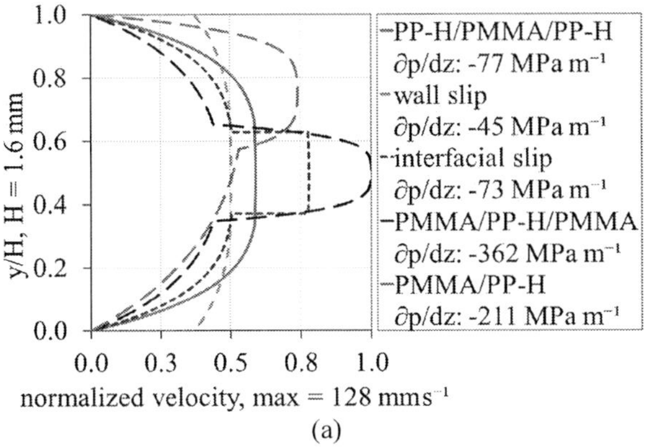

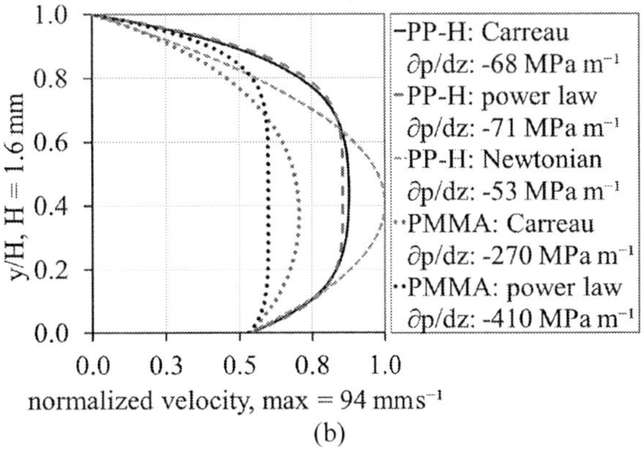

FIGURE 3. (a) Velocity profiles of symmetric and asymmetric coextrusion flows. mass flowrate in all curves 163 kg/h PP-H and 163 kg/h PMMA; (b) velocity profiles for moving wall condition (0.05 m/s at y/H = 0), results of different materials and rheological approaches. $\partial p/\partial_z$ is the pressure gradient in the flow direction, mass flowrate: 326 kg/h.

CONCLUSIONS

This work proposes a method to calculate planar flows which cannot be calculated analytically without doing CFD calculations. Rheological approaches like that of Carreau, Cross, Yasuda, Prandtl and Eyring as well as other approaches like that of Bingham or Herschel–Bulkley can be applied in addition to Newtonian and power-law flow behavior. Furthermore, asymmetrical flows and moving wall conditions can be taken into consideration as well as slip at the wall and slip at the polymer-polymer interface.

Symmetrical and asymmetrical velocity profiles calculated with Newtonian, power-law and Carreau approach are shown. The results show that the calculated pressure gradient, which is proportional to the pressure consumption of the die, calculated with the simplified power law approach (410 MPa m^{-1}) can overestimate the more realistic one calculated with Carreau's approach (270 MPa m^{-1}) by more than 50%.

This approach enables simplified flow simulations without usage of CFD technology and solves the following troubles: Approaches like those of Carreau, Cross, Yasuda, Prandtl and Eyring, Bingham or Herschel–Bulkley provide a realistic flow behavior, but even the simplest flow cannot be calculated analytically. The power law approach enables the analytical calculation of symmetrical flows, overestimating the viscosity at low shear rates resulting in an overestimated calculated pressure consumption. The Newtonian approach enables analytical calculation of symmetrical and asymmetrical flows including coextrusion flows at given film thickness, yet the rheological behavior of polymer melts cannot be described as Newtonian.

REFERENCES

1. P. J. Carreau, "Rheological equations from molecular network theories", Ph.D. thesis, Univ. of Wisconsin, 1968
2. M. M. Cross, J. o. Coll. Sci. 20, 417 – 437 (1965)
3. K. Yasuda, "Investigation of the analogies between viscometric and linear viscoelastic properties of polystyrene fluids", Ph.D. thesis, MIT - Massachusetts Institute of Technology, 1979.
4. L. Prandtl, J. o. Appl. Math. a. Mech. 8, 85 – 106 (1928)
5. H. Eyring, T. J. o. Chem. Phys. 4, 283 - 291 (1936)
6. E. C. Bingham, Bull. O. t. Bureau Stand. 13, 309 – 353 (1916)
7. W. H Herschel and R. Bulkley, Kolloid-Zeitschrift 39, 291 - 300 (1926)
8. W. Ostwald, Kolloid-Zeitschrift 36, 99–117 (1925).
9. A. de Waele J. o. t. Oil & Col. Chem. Ass. 6, 33-69 (1923)
10. S. G. Hatzikiriakos, Prog, I, Poly. Sc., 37, 624–643. (2011)
11. R. Zhao and C. W. Macosko, *Journal of Rheology*, *46*(1), 145-167(2002)
12. W. Michaeli, *Extrusion Dies for Plastics and Rubbers, 3rd Edition* (Hanser Publishers, Munich, 2003)

Modeling of Injection Overmolded Stamp Formed Thermoplastic Composites

Costa Franco S (1)*

(1) Autodesk - VIC - Australia

A method is developed to predict the final warp shaped of a hybrid forming and injection overmolding manufacturing process for thermoplastic composites. The model couples the output (residual stresses, anisotropic properties and thickness changes) of an existing stamp forming simulation software tool (Aniform) into the combined finite element warpage simulation of the injection overmolding process (Autodesk Moldflow). The residual stresses from the forming process together with the in-mold residual stresses and cooling strains of the injection overmolded compound are the drivers of warpage prediction in the combined part. A model of the interface bond strength between the stamp formed composite and the injection molding compound is also developed based on the temperature history at the bonding interface. Outcomes from the warp simulation and bond strength model are compared to experimental results on molded hybrid parts including a complex industrial part.

Strain-Dependent Upscaling Method in Dissipative Particle Dynamics Simulations of Nanoparticles Orientation Under Shear

Holzer Clemens H. (1)*, Gooneie Ali (1)

(1) Montanuniversitaet Leoben - Stmk - Austria

In this study, we develop a theoretical framework for multiscale simulation of the formation of oriented microstructures of anisometric clay nanoparticles in polymers under shear deformations. Dissipative particle dynamics (DPD) was employed to predict the orientation process. The parameters of the DPD models were critically assessed and shown to be in agreement with previous works. We show the capability of DPD to capture the structure with far greater details than traditional orientation models such as the famous Folgar-Tucker model. Here, we propose a strategy in which the trajectories of the orientation process of weakly-interacting nanoclays were parametrized as a function of the applied shear strain instead of the time. The applied strain was used to transfer the flow filed information to the DPD models. Vice versa, it was also incorporated to pass the orientation parameters to the macroscale through a simple combination of affine and nonaffine deformations. This combination was pictured in its simplest form to be a random mixing of DPD unit cells, representing nonaffine deformations, in a larger cell which distributes an affine deformation among the unit cells. It was noted that this strategy could be used to perform multiscale simulations of orientation process provided that the unit cells represent a precise description of the interactions between the components. A comparison of this methodology with Folgar-Tucker and strain-reduction factor models proved the success of the multiscale simulation of the evolution of orientation parameters against the applied shear strain. However, it will also be discussed that the method fails to capture the microstructure evolutions, if the unit cells do not provide an accurate representation of the material, for instance in the case of strongly-interacting polymer nanocomposites

NUMERICAL SIMULATION OF PLASTIC SHEET SOLIDIFICATION BY USING CHILLING ROLLS

Mitsoulis Evan [(1)]*

[(1)] NTUA - Attica - Greece

Often throughput in plastic sheet extrusion of polymer melts is limited by the ability to cool the extrudate. Chilling rolls are used to create effective heat transfer (cooling), and their design is of great importance in the polymer processing industry. The Finite Element Method (FEM) is used to numerically simulate the flow and heat transfer of a polymer sheet melt of a typical high-density polyethylene (HDPE) being rolled between two chilling cylinders. For a given roll temperature and using the melt viscosity, specific heat, heat of fusion and entering melt temperature, the heat to be removed from the polymer is calculated for a given roll size and minimum gap. The results show whether the chilling rolls are undersized or not and can be used for design purposes to improve heat transfer and increase throughput. The numerical simulations also provide a wealth of information regarding contours of isotherms and other variables, and show the extent and shape of the solidification front as a function of the roll speed (or, correspondingly, the dimensionless Peclet number).

Full characterization of multi-phase, multi-morphological kinetics in flow-induced crystallization of iPP at elevated pressure

Peters Gerrit W.M. (1)*

(1) Eindhoven University of Technology - Noord Brabant - Netherlands

Understanding the complex crystallization behavior of isotactic polypropylene (iPP) in conditions comparable to the one found in injection moulding, where the polymer melt experiences a combination of high shear rates and elevated pressures, is key for modeling and, therefore, predicting the final structure and properties of iPP products. Coupling a unique experimental setup, capable to apply wall shear rates similar to the ones experienced during processing and carefully control the pressure before and after flow is imposed, with in-situ X-ray scattering and diffraction techniques (SAXS and WAXD) at fast acquisition rates (up to 30 Hz), a well-defined series of short-term flow experiments are carried on using 16 different combinations of wall shear rates (ranging from 110 to 440 s−1) and pressures (100-400 bar). A complete overview on the kinetics of structure development during and after flow is presented. Information about shish formation and growth of α-phase parents lamellae from the shish backbones is extracted from SAXS; the overall apparent crystallinity evolution, amounts of different phases (α, β and γ), and morphologies developing in the shear layer (parents and daughters lamellae both in α and γ phase) are fully quantified from the analysis of WAXD data. Both, flow rate and pressure were found to have a significant influence on both the nucleation and the growth process of oriented and isotropic structures. Flow affects both shish formation and the growth of α-parents, pressure acts both on relaxation times, enhancing the effect of flow, and (mainly) on the growth rate of γ-phase. The remarkably high amount of γ-lamellae found in the oriented layer strongly suggested the nucleation of γ directly from the shish backbone. All the observations were conceptually in agreement with the flow induced crystallization model framework developed in our group, and represents a unique and valuable dataset that will be used to further validate and implement our numeric

Simulative determination of the optimization potential of additively manufactured static mixing elements for extrusion

Hopmann Christian (1), Schön Malte (1)*, Meiners Wilhelm (2), Shen Liyaowei (2)

(1) Institut für Kunststoffverarbeitung IKV - Nordrhein-Westfalen - Germany, (2) Fraunhofer-Institut für Lasertechnik ILT - Nordrhein-Westfalen - Germany

In plastics extrusion, the homogeneity of the melt is an essential pre-requisite for uniform product quality. Static mixing elements for the extrusion of plastics serve to homogenize the melt by splitting and recombining the flow of polymer melt, thus removing e.g. hot streaks in the melt. However, the geometric complexity of static mixers, which typically consist of crossed bars of metal joined together, makes both the design and the manufacture of specialized mixer geometries difficult. Against this background, both simulation of melt flow in static mixers and additive manufacturing of static mixers by means of Selective Laser Melting are investigated. In order to accurately model the flow through the multitude of small gaps between the bars, the open-source CFD software OpenFOAM is employed. While OpenFOAM already supports very high resolution meshes and parallel computing, the interaction between melt temperature and melt viscosity and the effect of viscous heating are added to the simulation model. In order to systematically categorize the wide variety of static mixer designs, a classification of mixer geometries is conducted, identifying the parameters pertaining to mixing performance. Using this classification, several geometric variations are designed. The variations' influence on the behavior of the mixing process are examined by means of the simulation model mentioned above. Selective Laser Melting not only allows an economic production of small and very small lots, but also offers design freedoms that have not been available before. This allows the production of static mixer geometries that cannot be manufactured with conventional processes. A framework for the exploitation of this advantage is shown. As an example, the potential thermal benefits of tempering channels located within each bar of the static mixer are explored.

Rheology of industrial plastisol formulations

Agassant jean-françois [(1)]*

[(1)] MINES ParisTech - France - France

Plastisol formulations are used for flooring applications.In the knife-over–roll coating process the plastisol flows between a substrate driven by a rotating roll and a fixed knife inducing a wide range of shear and elongational rates. Understanding the process requires a careful rheology analysis. A plastisol is a suspension of PVC and mineral fillers particles in a Newtonian plasticizer. The complexity of the formulations and the interaction between components make plastisol a unique kind of suspension. The rheology of various formulations has been studied using Couette and capillary rheometers. The formulation without filler presents a non-monotonous behaviour: shear thinning at low shear rates, then a Newtonian plateau, then a dilatancy peak followed by a second shear thinning zone at very high shear rates. Substituting a fraction of PVC by filler particles may enhance or decrease the first shear thinning behaviour depending on the size of the filler. The dilatancy peak is significantly reduced. Couette and capillary measurements are consistent for large capillary diameters allowing to cover the wide shear rate range of the process. Important entrance pressure drops have been observed in capillary experiments. They have been analyzed through a modified Cogswell method pointing out high elongational viscosities. For a capillary diameter of the order of the gap between the knife and the substrate an important flow rate increase is observed suggesting slip at the wall or the appearance of a thin lubricated layer. This has been analyzed using a numerical method accounting for the non-monotonous rheological behavior. These rheological measurements allow a careful analysis of the plastisol coating process.

Linear and Non-Linear Visco-elastic Behavior of Fibrillar Morohology Formed via Fiber Spinning of PP/PA6 Blend

Hajiraissi Roozbeh [1]*, Jahani Yousef [2], Hallmann Tobias [3]

[1] Department of Chemical Engineering and Macromolecular Chemistry-University Paderborn - Paderborn - Germany, [2] Faculty of Polymer Processing-Iran Polymer and Petrochemical Institute - Tehran - Iran, [3] Department of Mechanical Engineering-University Paderborn - Paderborn - Germany

The aim of this work is to investigate the influence of fibrillar morphology of deformed droplets of Polyamide 6 (PA6) dispersed in Polypropylene (PP) matrix on the melt viscoelastic behavior of their blends. The blends of PP with various amounts of PA6 (1%, 6%, 10% and 20%) are prepared by melt mixing in a co-rotating twin screw extruder and consequently fibrillated by fiber spinning process. Scanning Electron Microscopy (SEM) showed that the PA6 spherical droplets are deformed to a fibrillar morphology in fiber spinning process. The steady and transient shear rheological responses of samples are evaluated in both linear and nonlinear range of deformation. Storage modulus and complex viscosity of the blends containing PA6 fibrillated structure remarkably enhanced compared to as-extruded samples. The fibrillar-induced elasticity of fibers is a distinguishable behavior which revealed by conducting transient stress and creep-recovery measurements. Upon appearing mature fibrils, elasticity of polymer blend fibers increased significantly.

Effect of thermal history on crystallization behavior of polyethylenes in high-speed DSC measurements

Takarada Wataru (1)*, Kawai Koya (1), Othuki Yasuhiko (2), Kobayashi Yutaka (2), Kikutani Takeshi (1)

(1) Tokyo Institute of Technology - Tokyo - Japan, (2) Prime Polymer Co., Ltd. - Tokyo - Japan

Development of high-speed DSC instruments made it possible to measure exothermic or endothermic behaviors in hi-speed heating or cooling equivalent to the cooling in high-speed melt spinning process or quenching on the mold surface in injection molding process. Tiny specimen was used for high-speed DSC measurement to decrease thermal distribution in the specimen. Only 1 s was enough to melt polymer and measure crystallization behavior in cooling from molten state in tiny specimen of polyethylene. Total value of exothermic heat by crystallization in cooling was nearly constant despite of previous melting time. However, peak temperature of exothermic heat varied depending on melting time. It means that crystallization behavior was influenced by melting history. In this study, crystallization behavior of high-density polyethylene and liner-low-density polyethylene was measured by high-speed DSC repeatedly in different melting time. In high-density polyethylene, crystallization peak temperature was nearly stable after 20 cycles when the melting time was 2 s. In contrast, crystallization peak temperature decreased 95 degree C to 85 degree C when the melting time was 2 min. In linear-low-density polyethylene, decrease of crystallization peak temperature was smaller than high-density polyethylene even at melting time was 2 min.

Rheological behaviour of a high melt strength polypropylene at elevated pressure and gas-loading

Raps Daniel (1)*, Heymann Lutz (2), Altstädt Volker (1)

(1) University of Bayreuth, Department of Polymer Engineering - Bavaria - Germany, (2) University of Bayreuth, Department of Applied Mechanics and Fluid Dynamics - Bavaria - Germany

The rheological properties of a long chain branched polypropylene (LCB-PP) were investigated at the processing conditions of foam extrusion, namely high pressure, gas-loading and high shear-rates, as well as in elongational deformation. For measuring the rheological properties of PP at moderate and high deformation rates an in-line rheometer was used. Comparison of its results to standard rotational rheometry showed good agreement. The effect of the processing parameters was quantified using shift factors for temperature, pressure and gas-concentration. The influence of pressure on shear viscosity was found to be of minor importance due to an unfavorable ratio of relaxation time and residence time in the die. In contrast, shear viscosity was distinctly affected by CO2-concentration, reducing it to one third of its gas-free value at a concentration of 6 wt-% at a specific shear-rate. The change of viscosity by a variation of temperature is similar in magnitude compared to the variation due to dissolved blowing agent. Furthermore, thermo-rheological complex behaviour was observed. In the foaming process, thermo-rheological complexity could contribute to a better morphology control of long-chain branched polymers compared to linear ones. The elongational viscosity was measured using a film stretching-tool (UXF) for comparison. Distinct strain hardening was observed leading to an improved stabilisation of the foam morphology. The elongational visosity is three decades higher than the shear viscosity in the non-linear region, due to the pronounced strain hardening of the melt.

Rheological properties of polyacrylamide and amido amine-based cationic gemini surfactant solutions

***Kamal Muhammad Shahzad* [1]*, *Hussain S.M. Shakil* [1], *Fogang Lionel Talley* [2], *Sultan Abdullah Saad* [1]**

[1] KFUPM - Eastern Province - SaudiArabia, [2] KFUPM - Eastern Province - Russia

Oil is recovered from a reservoir in three different stages namely: the primary, secondary, and tertiary stage. In the primary stage, oil is recovered utilizing the natural reservoir pressure. However, only a small fraction (5-10%) of total oil can be recovered at this stage. In secondary stage, reservoir pressure is increased by injecting water or gas. However, only 20-30% of the total oil can be recovered in the secondary stage. To recover the remaining oil, different techniques such as thermal, gas or chemical enhanced oil recovery (EOR). Chemical EOR is a technique in which different chemicals such as surfactants, polymers, and alkalis are injected to increase the recovery from an oil reservoir. Surfactants decrease the interfacial tension between water and oil that help in displacing the residual trapped oil. Water-soluble polymers are injected to increase the viscosity of displacing fluid (water) that improve the displacement efficiency and decrease the "viscous fingering". Surfactants and polymers are injected together to improve the viscosity and to reduce the interfacial tension simultaneously. Therefore, understandings of surfactant-polymer (SP) interactions are crucial for EOR operation. In this work, interactions between commercial polyacrylamide and in-house synthesized amido amine-based cationic gemini surfactants were assessed using rheological measurements. Increasing the surfactant concentration lowered the viscosity of the polyacrylamide. The SP solution containing longer surfactant tail has lower viscosity compared to the surfactant with a shorter chain. It was observed that spacer nature, rigidity, and length also affects the rheological properties of polyacrylamide. Detail results will be discussed.

Design, structuration and rheological properties of Laponite based polymeric nanocomposites

***ABAKAR ADAM OMAR* (1)*, *Tassin Jean Francois* (2)**

(1) University of Adam Baraka - Abache - Chad , (2) University of Maine - LE MANS - France

The aim of this work is to study the internal structure, thermal properties and rheological properties in molten state for polymer blends nanocomposites based on PMMA-PEO blends and Laponite. Polymer blends and their nanocomposites were prepared by a combination of solution mixing to generate PEO protected Laponite particles and melt dispersion to blend the PEO/Laponite master batch with PMMA and PEO. PMMA/PEO blends appear to be miscible at room temperature only for PEO concentration less than 30wt%. In the molten state, rheological data indicate miscisbility between PMMA and PEO. The introduction of Laponite particles which are, in the master batch, located inside the PEO phase does not lead to their dispersion in the blend. However, diffusion of PEO inside the PMMA phase is observed in the molten state since glass transition temperature and rheological properties of the blends with 20wt% PEO are almost identical to those of the pristine melt with the same composition. On the contrary, at high concentrations of PEO, it has been demonstrated that a continuity of the PEO phase is observed where most of Laponite particles are located. Some particles (Their fraction cannot be estimated(seem to be inside the matrix. These types of nanocomposite blends show a non-vanishing zero frequency storage modulus. Its origin is attributed to the formation of a Laponite PEO network as in PEO melts filled with Laponite1. These results show that Laponite particles, such as those used in this study, i.e. protected by PEO chains, do not act as compatibilizer in these blends in contrast to other layered silicate nanocomposites. On the opposite, TEM micrographs showed that Laponite particles partially hinder the diffusion of PEO in PMMA/PEO blends and thereby limit its miscibility. ____ 1. (a) Jose, J.; Omar, A. A.; Brotons, G.; Tassin, J.-F. In Influence of Preparation Method and Molecular parameters on the Rheology

Distinction of Measurement-related and Material-related Scattering of Young´s Modulus of Thermoplastics during Tensile Testing

***Walter Jan-Uve** [(1)]*, **Schöppner Volker** [(1)]*

[(1)] Paderborn University - Nordrhein-Westfalen - Germany

Young´s modulus is one important value in the field of plastic processing which describes the correlation between stress and strain during deformation of a material at the linear elastic range. The higher the elasticity of a material, the higher is the Young´s modulus. In practice Young´s modulus is measured for example with tensile test. For FEM simulation (finite element method) Young`s modulus is necessary to calculate stress or strain, depending on which of both parameter is given. The measured values scatter around an arithmetic mean since different errors occur. Various errors like systematic or random errors have different influence on the result of the measurement. Systematic errors can`t be excluded, they can only be minimized also they can't be detected. On the other hand, various errors can be detected and described. To characterize these errors parameters like variance or scatter are useful. The result of a measurement according to standard is the total scatter of modulus that can be differentiated into the measurement related and the material related spread. An interesting question is if one of these spread predominate and how these spread can be minimized or eliminated. For this paper, a method was developed to obtain the two different scatters. Based on measurements with the tensile test a nondestructive method was found to receive only the measurement-related scatter. The main results of these experimental investigations are that it is possible to measure different mean variation and to keep apart the different scatter. Three different thermoplastics are used. Polypropylene as a semi-crystalline standard polymer, Polyamide 6.6 as a semi-crystalline technical polymer and Polymethylmethacraylate as an amorphous technical polymer.

Viscosity amplification in dilute suspensions of rods in a Carreau matrix fluid

***Domurath Jan* (1)*, *Saphiannikova Marina* (1), *Ausias Gilles* (2), *Férec Julien* (2), *Heinrich Gert* (1)**

(1) Leibniz-Institut für Polymerforschung Dresden e.V. - Saxony - Germany, (2) Institut de Recherche Dupuy de Lôme, Univ. Bretagne Sud - Brittany - France

The addition of rod-like filler particles to a liquid leads to a noticeable increase in the viscosity of the mixture. The effect can be described by the Dinh and Armstrong model [1]. One important result of this model is the prediction of a strong increase in the viscosity of the suspension with increasing aspect ratio of the rigid rods. Despite the fact that this model was originally proposed for a Newtonian matrix fluid it is also applied to polymer melts filled with rod-like particles. Such an approach completely decouples the influence of the rod aspect ratio from the non-linear properties of the suspending fluid. Yet, since polymer melts often exhibit strong non-Newtonian behaviour, e.g. shear thinning, it is to be expected that such a superposition will give a wrong prediction of the suspension viscosity. To investigate this problem we performed a numerical study of a suspension consisting of a non-Newtonian matrix fluid and rigid rods. In particular, we simulated different flows of a Carreau fluid around rigid rods with different orientations and used numerical homogenization to obtain the intrinsic viscosity of the suspension as function of applied rate of deformation, thinning exponent and aspect ratio of the rod. In the power law regime of the Carreau model we compare our numerical results with analytical models of Souloumiac and Vincent [2] and Gibson and Toll [3]. [1] S. M. Dinh and R. C. Armstrong. Journal of Rheology 28.3 (1984), pp. 207–227. [2] B. Souloumiac and M. Vincent. Rheologica Acta 37.3 (1998), pp. 289–298. [3] A. G. Gibson and S. Toll. Journal of Non-Newtonian Fluid Mechanics 82.1 (1999), pp. 1–24.

Simulation of the Melting Behavior in an Injection Molding Plasticizing Unit as Measured by Pressure and Ultrasound Measurement Technology

Dominik Altmann[1,a)], Bernhard Praher[1] and Georg Steinbichler[1]

[1]*Institute of Polymer Injection Molding and Process Automation, Johannes Kepler University Linz, Altenberger Strasse 69, 4040 Linz, Austria*

[a)]dominik.altmann@jku.at

Abstract. The melting behavior of a Polyamid material in an injection molding plasticizing unit (40/20D) was simulated and experimentally verified. Hence, a simulation program with a focus on standard three-section screws was developed. A clear separation of the melt and solid fraction in the screw channel is a precondition, referring to physical models, which are included. For this reason, the main constituent of the simulation program is the calculation of the melting process in an injection molding plasticizing unit. First of all, a 2D grid generator is used to represent an unwrapped screw channel of a standard three-section screw. Under consideration of the axial screw movement, mechanical, mathematical and physical models based on *Tadmor*, *Gogos*, *White* and *Potente*, it is possible to determine the melting process of the individual functional areas. Validation of simulation and experiment data is indispensable to get significant results. In order to verify the measurement relating to the melting profile, so called pull-out tests are normally executed. This method is very time-consuming and very inaccurate. With this in mind, we combined two different methods for the in-situ measurement of the solid-bed ratio: Analyzing pressure data and ultrasound reflection signals at different axial positions. These methods are more useful to investigate the melting behavior over the screw channel.

INTRODUCTION

In polymer processing, especially in injection molding and extrusion, the plasticizing unit is a major functional component. The plasticizing unit should be able to process a wide range of plastic materials and additives with high output rates. As a consequence, it is essential to design and manufacture screws that are individually adapted to the respective materials. In addition, the melt quality is one of the main factors to guarantee high quality products in injection molding and extrusion processes. In order to predict the optimized geometry and process parameters influencing the melt quality of the respective application, we developed a simulation program.

One key component of this program is the simulation of the melting process. The approximations used in existing simulation are based on the melt only, therefore a clear separation of the melt and solid fraction is essential. Basic research has been done by *White* and *Potente* [1]; *Tadmor* [2]; *Tadmor* and *Klein* [3]; *Tadmor* and *Gogos* [4], which provides a good understanding of the fundamentals of the melting behavior in a single screw plasticizing unit. They also developed a melting model for Newtonian and non-Newtonian materials. Especially the three-section buildup (feed section, compression section and metering section) of a single screw in combination with a standard polymer is quite well investigated. Nevertheless, there is still a potential for development, considering on the one hand the existing melting models with technical polymers and on the other hand the current measuring methods.

For the experimental validation of the melting profile, so called pull-out tests are used. The traditional procedure is to stop the screw immediately and cool down the whole configuration consisting of barrel, screw and processed material. Afterwards some components of the plasticizing unit like the nozzle, the flange and their heating system are disassembled. Finally, one is able to pull out the screw and identify the solid and melt fraction. This procedure is very inaccurate and inefficient, due to the fact that the cooling cycle lasts several hours.

One aim is to find more suitable methods for the determination of the melting profile. Through combining and analyzing pressure and ultrasound reflection signals at different axial positions along the plasticizing unit, the solid-bed ratio can be determined in-situ during the process. It should be mentioned that bores along the barrel are necessary for mounting the pressure sensors. Additionally, some research corresponding to the ultrasound reflection measurement can be found in the literature provided by *Aigner* et al. [5][6].

MATERIAL AND METHODS

The technical polymer PA6 B30S from Lanxess (Köln, Germany) was used to compare the simulation results with the experimental analysis. This material is suitable for injection molding processing. Details are listed in Table 1.

TABLE 1. Material data for simulation - PA6 B30S

Description	Abbrevation	Value	Unit
Melting temperature (peak)	T_m	222.0	°C
Enthalpy of fusion	h_m	255050.0	J/kg
Specific heat capacity (liquid)	c_{p_l}	2250.0	$J/(kg.K)$
Heat conductivity (liquid)	λ_l	0.22	$W/(m.K)$
Bird-Carreau-Yasuda / Arrhenius			
Zero shear viscosity	η_0	139.94	$Pa.s$
Time constant	k	0.0011	s
Power law index	n	0.4258	–
Yasuda parameter	a	2.9750	kg/m^3
Reference temperature	T_r	260	°C
Activation energy	E_A	86110.87	J/mol

At least three different measuring devices were used to characterize this material. The rheological parameters were measured with a plate plate rheometer (MCR 301, 25 x 1 mm) from Anton Paar (Graz, Austria) and a high-pressure capillary rheometer (Rheograph 25) from Göttfert (Buchen, Germany). Combining these two measuring devices it was possible to evaluate a large shear rate region from 0.1 1/s up to 10.000 1/s. Furthermore, the heat conductivity and the specific densities (pvT-graph) could also be measured with the Rheograph 25 in accordance with the standard ISO17744. The thermodynamic parameters like the melting temperature, the enthalpy of fusion and the heat capacity were measured with a Mettler Toledo DSC 822 (Vienna, Austria). In particular the standard ISO 11357-4 procedure (saphire and blank curve) was used measuring the heat capacity.

Simulation Grid and Melting Model

A plasticizing unit including a standard three-section screw was examined in the simulation. Therefore, the three dimensional geometry of the screw channel consisting of feed, compression and metering section was approximated via an unwrapped screw channel. With this reduction to a two dimensional system, it is possible to calculate the melting profile and the corresponding parameters in a much easier way. Under consideration of a constant channel width the unwrapped screw channel was subdivided in numerous small grid cells. This was done in channel direction (z-direction) and in direction of the channel height (y-direction). In order to improve the convergence relating to the real channel cross section, the constant channel width was corrected including the flight radius and the flight height. Afterward using the finite difference method, it was possible to combine the grid cells and the three sections (feed, compression and metering section) to each other.

The melting model by *Tadmor* and *Gogos* [4] is the basis determining the melting behavior in the simulation program. It is a drag induced melting model with a focus on crystalline materials, because of the abrupt change from the solid to the liquid phase. A schematic drawing (Figure 1) is used to give an overview of the main processes in an unwrapped screw channel of an injection molding plasticizing unit. There are at least three main processes, respectively sections. At first, if the barrel temperature exceeds the melting temperature of the polymer a melt film is formed. It is possible to calculate the temperature of this melt film, which was also investigated by *Aigner* et al. [5][6]. Secondly, this melt film is growing due to viscous dissipation and heat conduction of the hot barrel and screw surface. This process occurs in the so-called delay section. Thirdly, after reaching a certain melt film thickness a melt vortex is formed at the pushing flight and the melting section starts. The background to this is that there is a drag flow between the barrel and the solid-bed surface. Afterwards the solid-bed is getting smaller and smaller in direction of the die,

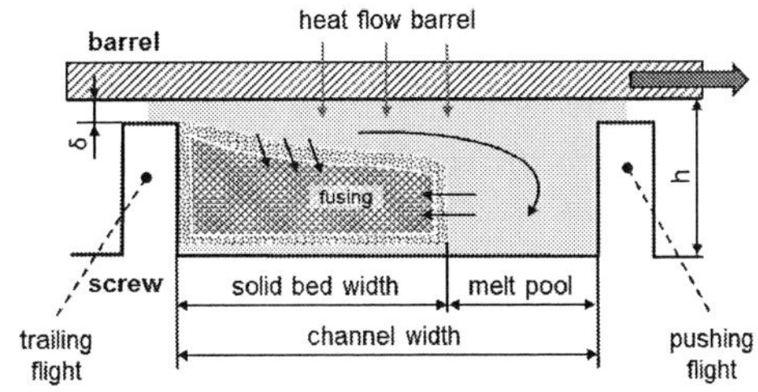

FIGURE 1. Fundamentals of the melting model by *Tadmor*

whereas the melt pool increases. In order to finally calculate the solid-bed ratio over the channel width, the following equations are used (*Tadmor* [2]). Firstly, Equation 1 for a constant channel height in the feed and metering section:

$$\frac{X_1}{b^*} = \frac{X_0}{b^*} \cdot \left[1 - \frac{\phi \cdot (z_1 - z_0)}{2h \cdot v_{sz} \cdot \rho_s \cdot \sqrt{X_0}}\right] \tag{1}$$

and secondly, Equation 2 considering the decrease of the channel height in the compression zone:

$$\frac{X_1}{b^*} = \frac{X_0}{b^*} \cdot \left[\frac{\phi}{\beta \cdot v_{sz} \cdot \rho_s \cdot \sqrt{X_0}} - \left(\frac{\phi}{\beta \cdot v_{sz} \cdot \rho_s \cdot \sqrt{X_0}} - 1\right)\frac{h_0}{h_1}\right] \tag{2}$$

where b^* is the corrected channel width, X the solid-bed width, ϕ the melting rate, h the channel height, v_{sz} the solid bed velocity in z-direction, ρ_s the density of the solid and β is the pitch of the compression zone.

EXPERIMENTAL SETUP AND ANALYSIS

In order to validate the simulation results of the melting behavior we conducted experiments on an injection molding machine (e-motion 310/100T) by the company ENGEL (Schwertberg, Austria). The used plasticizing unit has the following setup and geometries: standard three-section single screw with a diameter of D_s = 40 mm, L/D ratio of 20, pitch t = 40 mm, flight width e = 4.8, channel height of the feed section h_F = 6.125 mm, channel height of the metering section h_M = 2.675 mm, length of the feed section l_F = 9D, length of the compression section l_c = 6D, length of the metering section l_c = 5D and a screw clearance δ = 0.125 mm.

Figure 2a illustrates a schematic picture of the used plasticizing unit. On the bottom side of the barrel there are at least 10 bores for mounting the pressure sensors. Whereas, there is a device on the top of the barrel for placing the ultrasound transducers. In sum we have used seven pressure sensors and three ultrasound transducers (10 MHz). The pressure sensors were mounted at the positions 1 to 7 and the ultrasound transducers at positions 3, 5 and 7 during our measurements. The definition of the scanning line is very important for the later evaluation of the signals and the

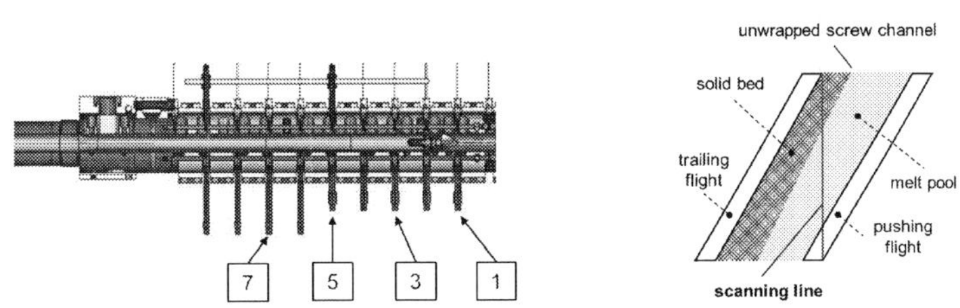

FIGURE 2. (a) Experimental setup and sensor positions, (b) Definition of the scanning line

determination of the solid-bed to melt pool ratio. As already shown in Figure 2b the sensors remain always at the same positions during the measurements. This means that the scanning line stretches from the trailing to the pushing flight relating to a full rotation of the screw. The connecting geometry parameter to the axial cross section is the screw pitch. Nevertheless this evaluation works quite well, because the solid-bed to melt pool ratio across the channel width is almost the same as the solid-bed to melt pool ratio across the scanning line. Furthermore the axial velocity of the screw is very low, so the effect on the solid-bed ratio is negligible.

Figure 3 shows a schematic relationship between the material state in the screw channel and the corresponding pressure and ultrasound reflection data. The measuring time is displayed on the x-axis for both diagrams, which are included in the figure. One can see that there are several ultrasound reflection signals at different travel times, which can be allocated to the inner surface of the barrel, the screw flight and the screw bed. For the later evaluation it is possible to withdraw the detected signals of the barrel from the whole output signal, because they are not interesting for determining the solid bed ratio. In order to measure an ultrasound reflection signal of the screw it is necessary to have a molten polymer across the screw clearance and the channel height. Whereas, the rest of the scanning line without a reflected signal provides an indication of the solid-bed ratio. In practice there is normally no clear separation of the melt and solid fraction. Very often a transition zone, consisting of a polymer melt and some solid granules, is formed. This zone can also be detected via a weakened signal. Moreover, the scanning line is limited by the two signals of the screw flight.

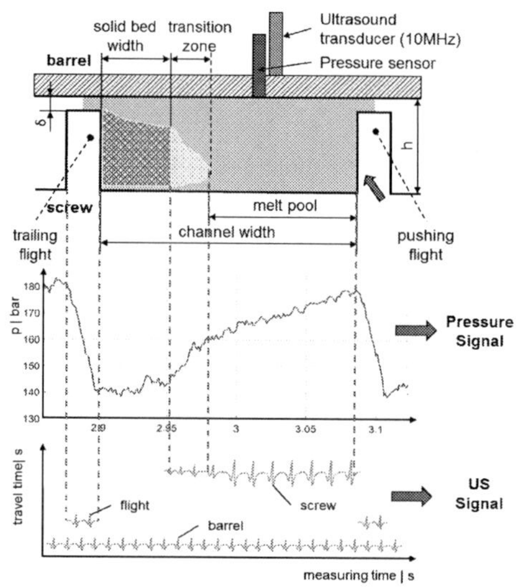

FIGURE 3. Basic evaluation concept for determining the solid-bed ratio

Furthermore, a typical pressure signal is inserted at the centre of Figure 3. The two pressure drops at the trailing and pushing flight are immediately apparent and can be used to determine the scanning line (one full rotation). After the pressure drop of the trailing flight, there is a section, where no pressure increase takes place. Quite the reverse the pressure signals starts fluctuating. In addition, the fluctuation is getting worse and worse from the metering to the feed section (sensor positions 1 to 7). These phenomenons are again an indicator for the solid-bed ratio.

RESULTS AND DISCUSSION

We compared two different process settings with the PA6 material, explained in the chapter Material and Methods.

1. **c1** → screw speed 143.2 U/min, back pressure 150 bar, metering stroke 0.7 D, cooling time 20 s
2. **c2** → screw speed 334.2 U/min, back pressure 150 bar, metering stroke 0.7 D, cooling time 20 s

Applying the ultrasound and pressure analysis on these two experiments, the following results turned up, Figure 4. The melting process is defined by the determination of the solid-bed fraction (y-axis) over the unwrapped screw channel in L/D (x-axis) starting at the entrance of the feed section. The red curve illustrates the simulation results, the black triangle markers the pressure evaluation and the blue dots the ultrasonic evaluation. Furthermore the verification of the gathered data was examined by calculating the mean value of the solid-bed fraction of three full rotations approximately at half metering stroke. One can see that there is a pretty good accordance between the combination of ultrasound and pressure analysis with the simulation results. With an increasing screw circumferential speed the melting process is not so effective and the remaining solid-bed fraction at the end of the metering section is consequently higher. This effect can also be defined with both measurement technologies as shown in Figure 4b.

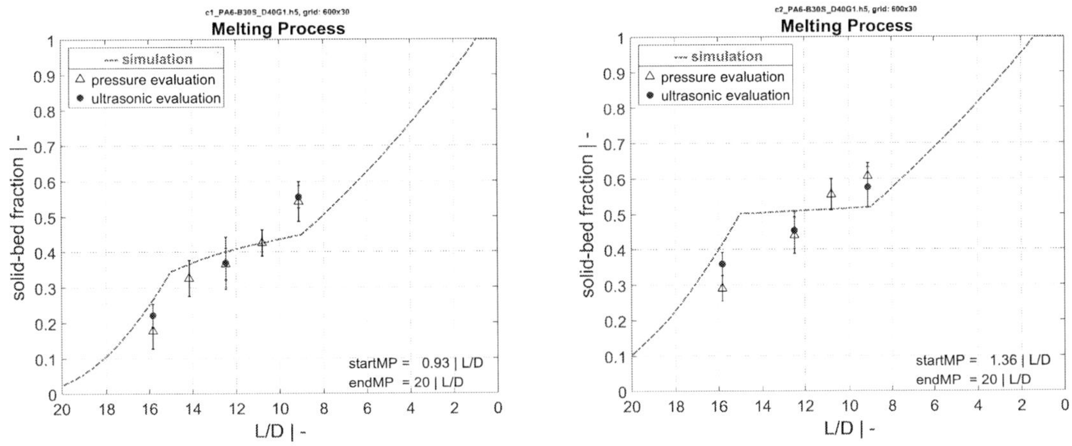

FIGURE 4. (a) Melting process of c1-setting, (b) melting process of c2-setting

CONCLUSION

To sum up, we have applied and developed two measurement technologies and novel evaluation procedures for determining the melting behavior of a PA6 material. Combining ultrasound and pressure analysis have shown good agreement between experimental and simulation results. This provides the opportunity to prove and analyze the simulated melting profile of an injection molding plasticizing unit more efficiently, compared with traditional procedures. In addition, our simulation model contains a special grid, the melting model by *Tadmor* and *Gogos* [4] including the axial screw motion and a numerical temperature calculation in consideration of the idle time. Nevertheless, we have observed that the evaluation of pressure signals has been more complicated at higher L/D ratios, which influences the accuracy of experimental results. Just the reverse can be investigated referring to ultrasound reflection signals. So we want to give a concept proposal. Make use of pressure analysis for determining the solid-bed ratio in the feed section and in the first half of the compression section. Whereas, use ultrasound analysis in the second half of the compression section and in the metering section. Finally, we would like to give an outlook of future prospects. On the one hand, the main focus will be on optimizing our evaluation procedures. With this in mind, verifications relating to different process parameters, plasticizing units, and more unfilled polymer materials will be essential to support our confidence to these methods. On the other hand, the present simulation model will be adapted to simulate high performance screws (e.g. barrier screw geometries). Additional investigations of the pressure and ultrasound reflection signals should be carried out regarding the transition zone (Figure 3).

ACKNOWLEDGMENTS

Financial support by ENGEL AUSTRIA GmbH and WOOD K-PLUS, which play a substantial role by conducting parts of this work, are gratefully acknowledged. WOOD K-PLUS is funded by the Austrian Government and the Provincial Government of Upper Austria within the COMET program.

REFERENCES

[1] J. L. White and H. Potente, *Screw Extrusion* (Carl Hanser Verlag, Munich, 2003).
[2] Z. Tadmor, Polymer Engineering Science **6**, 185–190 (1966).
[3] Z. Tadmor and I. Klein, *Engineering Principles of Plasticating Extrusion* (Van Nostrand Reinhold, New York, 1970).
[4] Z. Tadmor and C. G. Gogos, *Principles of Polymer Processing* (Wiley-Interscience, New Jersey, 2006).
[5] M. Aigner, "Computational and experimental modelling of transport phenomena in single screw plasticating units under consideration of the melt quality," Ph.D. thesis, Johannes Kepler University 2014.
[6] M. Aigner, B. Praher, C. Kneidinger, J. Miethlinger, and G. Steinbichler, Intern. Polymer Processing XXIX **5**, 624–634 (2014).

Molecular Dynamics simulations of isocyanate-based molecules viscosity

Lenzi Veniero (1)*, Driest Piet J. (2), Dijkstra Dirk J. (2), Richter Frank U. (2), Statamatialis Dimitrios (3), Grijpma Dirk W. (3), Ramos Marta Maria Duarte (1), Marques Luís Silvino Alves (1)

(1) Department-Centre of Physics, University of Minho - Braga - Portugal, (2) Covestro, CAS-Global R&D - Leverkusen - Germany, (3) MIRA Institute for Biomedical Technology and Technical Medicine, University of Twente - Enschede - Netherlands

Aliphatic diisocyanates are key ingredients in polyurethanes and are widely used in polyurethane processing. Trimers obtained by reacting three diisocyanates are very stable and exhibit excellent mechanical [1] and optical [2] properties. Despite this, they are mainly used as crosslinkers in coatings and adhesives since a full exploitation of their features is limited by processing issues. While the viscosity of the monomers is low and comparable with water, the trimers have a viscosity several orders of magnitude higher, making difficult their handling. From a fundamental point of view there is no clear understanding of this behavior. In this work we use quantum mechanical and atomistic simulations to understand and predict the properties of diisocyanates molecules. Starting from the GAFF force field, an improved set of parameters was introduced and tested in molecular dynamics simulations to calculate their static and dynamic properties. We show how the customization of force field parameters is fundamental to obtain reliable predictions of the properties of organic liquids. Viscosities of diisocyanate trimers ranging from 3 to 7 aliphatic carbons per monomer were calculated using Green Kubo method and indirect estimations for lower temperatures, and compared with experimental values. The results are in very good agreement with experiments [3], with simulations predicting the correct trend of viscosity with temperature and size of the studied molecules. We attribute the high viscosity of this molecules to the interaction of NCO terminations with the trimer NCO ring, as indicated by quantum mechanical calculations. [1] Moritsugu, M.; Sudo, A. and Endo, T., J. Polym. Sci. Part A: Polym Chem. 2013, 51, 2631–2637 [2] Flipsen, T. A. C; Steendam, R; Pennings, P. J., and Hadziioammou, G, Adv. Mat. 1996, 8(1), 45-48 [3] Driest, P. J., Lenzi, V., et al., Polym. Adv. Technol. 2016, doi: 10.1002/pat.3891

Prediction of final crystal morphology of semi-crystalline polymers using Phase Field Modeling (PFM)

Baghershahi Nasrin [(1)], Pourabbas Behzad [(1)], Salami Hosseini Mahdi [(1)]*

[(1)] Sahand University of Technology - East Azerbaijan - Iran

The properties of semi-crystalline polymers strictly depends on the morphology of their crystals. The crystal morphology depends on many parameters, one of the most important one is thermal condition and history of the material. In the present research, attempts were made to study the effect of thermal conditions on the final crystal morphology of a semi-crystalline polymer. For this purpose, a phase field model proposed by Kobayashi, were employed to simulate the crystal growth and predict the final morphology. A 2D model with certain thickness was taken into with different thermal condition and fixed nucleating sites. The orientation of each spherulite initiated by a single seed was considered to be different so these crystalline phases could not merge together. The simulation were conducted until steady stare was reached and no further crystalline regions were generated. Results depicted that the spherulite growth showed an s-type trend which was observed in experimental studies. Since spherulites growth took place though lamellar formation and thickening, the crystallinity of 100% would not be possible except for single crystals with perfect structure. The model predicted that through spherulite growth, one may reach to the maximum crystallinity of almost 67%. The denderic growth of spherulite from the initial seed would cause some amorphous regions to be trapped between lamellar regions causing the final crystallinity percent not able to reach 100%. Furthermore, the effect of temperature on the spherulite growth was studied. It was observed that decreasing the supercoiling would result in more homogeneous structure.

PEROXIDE CROSSLINKING OF LINEER LOW DENSITY POLYETHYLENE

KURT DEMIR GULSEN [1]*, BICER EZGI [1], KODAL MEHMET [2], OZKOC GURALP [1]

[1] Kocaeli University, Chemical Engineering Department - Kocaeli - Turkey, [2] Kocaeli University, Kosekoy Vocational School - Kocaeli - Turkey

The aim of this study is to determine the physical and rheological properties of cross-linked lineer low density polyethylene (LLDPE) with different concentration of peroxide. Tert-butyl cumyl peroxide (BCUP) was used in crosslinking. The peroxide concentration was changed as 0.5, 1, 2 and 3 phr. LLDPE/Peroxide blends were prepared by melt compounding process using an Xplore Instruments laboratory compounder at 135 deg.C barrel temperature. The crosslinking was carried out using a hot-press operated at 175 deg.C. At the characterization stage; tensile test for mechanical properties, differential scanning calorimetry (DSC) for thermal characterization, rheometer analysis for rheological properties and Fourier transform infrared spectroscopy (FTIR) for structural properties were investigated. It was found that the addition of peroxide to LLDPE increased degree of cross-linking for rheological tests. According to tensile tests, elongation at break of blends increased with increasing peroxide content. The plastic flow behavior during tensile testing for uncrosslinked LLDPE changed into an elastomeric behaviour with increasing strength upon elongation. It was observed that the peroxide concentration was also effective on thermal properties. Keywords: Lineer low density polyethylene, peroxide crosslinking, polyethylene, rheology Acknowledgement: This study has been financially supported by TUBITAK (Grant Number: 116M981).

A New Implementation of the Surface Tension Force to Reduce the Spurious Currents in Two-Phase Flow

Mostafaiyan Mehdi (1)*, Wießner Sven (1), Heinrich Gert (1)

(1) Leibniz Institute for Polymer Research Dresden - Sachsen - Germany

Spurious current, which can be defined as parasite-like velocity vectors and pressure jumps in the vicinity of the interface in two-phase flow system, is an unreal phenomenon and exist due to the flow field calculation errors in different numerical methods. Discontinuity of density and viscosity in the interface area, as well as inaccurate calculation of the interface curvature and normal vectors, are the parameters which are considered to be the source of the calculation errors and the spurious currents, consequently. The spurious currents can affect the numerical prediction of the second phase deformation, especially for the cases the capillary number is small (Ca<1.0). In this research, a coupled computer code for flow field calculation is developed using Galerkin finite element method. Also, some temporary elements are defined in the vicinity of the interface and a moving least square technique is used to obtain a more accurate flow filed parameters in the interface area, without introducing new nodes or new DOFs (degree of freedom). Surface tension force is implemented in the solution using a pressure boundary condition on the temporary elements, which one of their edges lays on the interface. It is shown that, using this technique, the fluctuation of the pressure as well as the magnitude of the parasite-like velocity vectors in the interface area will be reduced satisfactorily.

Evaluation of the Influence of the Tool Surface on Polymer Melt Flow Using a Novel Rheological Extrusion Slit Die

G. Zitzenbacher[1, a)] , E. Brunner[1, b)]

[1]*Department Materials Technology, University of Applied Sciences Upper Austria, School of Engineering, Stelzhamerstr. 23, 4600 Wels, Austria*

[a)]Corresponding author: g.zitzenbacher@fh-wels.at
[b)]elvio.brunner@fh-wels.at

Abstract. In this paper, the influence of the tool surface on polymer melt flow is studied. For this purpose, a novel rheological extrusion slit die in conjunction with a lab-scale single screw extruder in which the die surface can be varied using exchangeable inserts was employed. In the experiments, tool surfaces of different roughness were used to study the influence of the flow channel surface on wall slip of a polymethylmethacrylate (PMMA) melt. Wall slip was evaluated using the concept of reduced volume flow rate and the wall slip velocity was calculated with the Mooney method. Furthermore, surface analysis of the die inserts was performed before and after the extrusion tests to be able to detect possible changes of the die surface caused by the extrusion process. PMMA shows wall slip on polished steel. There is an increase in the wall slip velocity with rising wall shear stress. On a rough steel surface no wall slip was observed. Compared to data obtained with a high pressure capillary rheometer good qualitative correspondence was obtained, but quantitative differences of the wall slip velocity values between the two measuring methods were found. It was confirmed through contact angle measurements that there is an increase in the polar fraction of surface energy after the extrusion tests. Roughness measurements carried out with a confocal microscope revealed only slight smoothing of the rougher die inserts.

INTRODUCTION

Two different surface boundary conditions can be distinguished in polymer melt flow calculation: the wall adherence and the wall slip condition. In the case of wall adherence the polymer melt sticks to the wall and the flow velocity at the boundary to the surface is equal to the surface velocity of the wall. In the case of wall slip a slip velocity is introduced which is dependent on wall shear stress mathematically described by a power law function. Besides the process conditions the flow channel surface also influences wall slip which was reported by several authors [1 - 5]. Most of these studies were carried out with rheometers, where no continuous extrusion conditions can be obtained. In this paper a new rheological extrusion slit die is presented which allows the investigation of the influence of the die surface on wall slip or wall adherence under close-to-process conditions. The wall slip data are compared to data obtained with a high pressure capillary rheometer. Furthermore, the change in the surface properties of the flow channel after the extrusion experiments is studied.

EXPERIMENTAL

Materials used

The experiments were conducted with a polymethylmethacrylate (PMMA) PLEXIGLAS 7M from EVONIK, Darmstadt, Germany, which is suitable for extruded profiles and panels in lighting engineering. PMMA PLEXIGLAS 7M has a melt flow rate (MFR) of 3.45 g (10 min)$^{-1}$ at (230 °C / 3.80 kg). It was predried at a temperature of 80 °C for 24 h before the extrusion experiments. X155CrVMo12-1 steel was used as die material for the inserts.

Methods used

The rheological experiments were performed with a rheological extrusion slit die (see Fig. 1), which was mounted to a laboratory scale single screw extruder Collin E20M, Ebersberg, Germany, with a screw diameter of 20 mm. (3) is the transition component from a circular cross section at the extruder exit to a rectangular cross section in the die. The lower die insert B (7), the upper die insert A (8) and the die sidewalls (2) are exchangeable to enable the investigation of the effect of different die materials and surfaces on the polymer melt flow. These parts form a tapered flow channel with rectangular cross section, which is followed by a region with constant die height. Furthermore, different die heights and flow channel lengths in the last region of constant die height can be realized by exchanging (2), (7) and (8). The pressure (1, 5) and the temperature (4, 9) were measured before the region of constant die height. The experiments were conducted with a polished and a rough steel flow channel at the temperatures 230 and 250 °C. First the tests with polished steel were done at a die height of 0.25 mm and 0.35 mm, both with flow channel lengths in the region of constant die height of 5 and 10 mm. The flow channel width was 20 mm. Tests with different die heights and mass flow rate were necessary for the determination of wall slip. The mass flow rate was varied by screw speed. The different flow channel lengths were needed to evaluate the entrance pressure loss. After these tests, the inserts and side walls were sandblasted to achieve a rough flow channel surface and the influence of rough steel on the polymer melt flow was studied with these surface treated inserts.

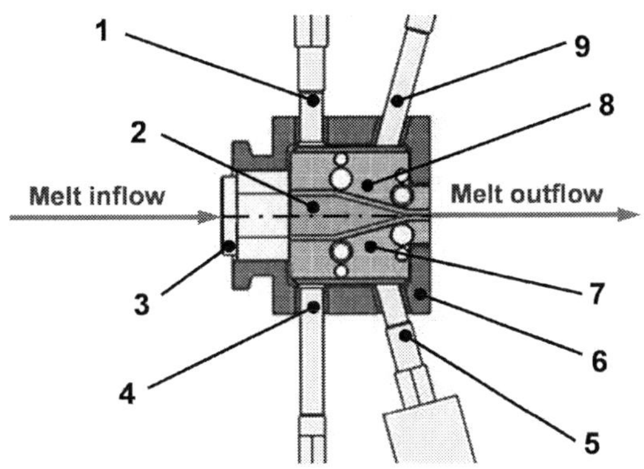

FIGURE 1. Composition of the rheological extrusion slit die including the measuring sensors: pressure transducers (1, 5), thermocouples (4, 9), die sidewall (2), transition component (3), die half body (6), lower die insert B (7), upper die insert A (8)

The concept of reduced volume flow rate and the Mooney-Method for slit dies were employed to evaluate the flow behavior of the polymer melt at the boundary to the die wall [6, 7]. First of all, the mass flow rate dependent on pressure was determined. After changing the screw speed, at least 15 minutes were passed to achieve stable extrusion conditions. The mass flow rate was measured three times over a period of 2 minutes. The pressure and temperature values in the extrusion process were recorded dependent on time using Visual Fecon, a measuring data recording software from Collin, Ebersberg, Germany.

Then the specific volume of the polymer melt dependent on temperature and pressure (pvT-data) was measured utilizing a high pressure capillary rheometer Rheograph 6000 from Goettfert, Buchen, Germany to calculate the volume flow rate from these data. The Bagley-plot was employed to determine the entrance pressure loss using pressure data measured at constant apparent shear rate with die inserts which had two different flow channel lengths but constant die height [7]. The wall shear stress τ_w is

$$\tau_w = \frac{(p - p_E) \cdot H}{2L} \cdot \left(1 + \frac{H}{W}\right)^{-1}, \tag{1}$$

where p is the measured pressure, p_E is the entrance pressure loss, H is the die height, W is the die width and L is the length of the flow channel region with constant die height. In the next step the reduced volume flow rate was evaluated

$$\dot{V}_{red} = \frac{\dot{V}}{WH^2} \tag{2}$$

and plotted dependent on wall shear stress for the two different die heights. When complete wall adhering of the melt occurs, the reduced volume flow rate curves for different flow channel heights are the same. Deviations between these curves indicate wall slip [6, 7]. Wall slip velocity v_s was evaluated quantitatively from the reduced volume flow rate data using the Mooney method for slit die channels

$$\dot{V}_{red} = \frac{v_s}{H} + f(\tau_w), \tag{3}$$

where f corresponds to the contribution of shear flow to the total reduced volume flow rate. The reduced volume flow rate values evaluated from experiments performed with different die heights were plotted dependent on the reciprocal flow channel height at constant wall shear stress and temperature. Then the wall slip velocity v_s at a certain wall shear stress is equal to the slope of the obtained line.

Surface roughness was measured by means of a confocal microscope of the type Leica DCM 3D, Wetzlar, Germany, with an enlargement of 500. An area-weighted roughness value S_a was determined on the surfaces of the die inserts at three positions with the dimensions 0.54 mm x 0.71 mm in the region with constant die height. The averaged line weighted surface roughness in flow direction R_{ay} and transverse to the flow direction R_{ax} was measured along a distance of 1.63 mm.

The surface energy of the flow channel surfaces was determined at room temperature using the contact angle measurement system Kruess DSA30S, Hamburg, Germany. The contact angles of the measuring liquids diluted ethylene glycol (50 %, by volume), diiodomethane and 1,5-pentanediol were detected employing the sessile drop method. The tests were carried out five times for each pairing of measuring liquid and flow channel insert. Then the dispersive $\sigma_{s,dis}$ and the polar fraction $\sigma_{s,pol}$ of surface energy were evaluated according to Owens et al. [8].

RESULTS AND DISCUSSION

The reduced volume flow rate dependent on wall shear stress using the polished steel die inserts is shown in Fig. 2 for two different die heights (0.25 mm and 0.35 mm) and temperatures (230 °C and 250 °C). The obtained reduced volume flow rate values increase with rising wall shear stress. No overlapping shear stress range of the reduced volume flow rate values for the two different die heights was obtained at a temperature of 230 °C. This means that at a flow channel height of 0.35 mm the highest wall shear stress value was 0.109 MPa and at a flow channel height of 0.25 mm the lowest shear stress value was 0.138 MPa. An overlapping shear stress range of the two reduced volume flow rate curves is needed to evaluate wall slip of the polymer melt.

The reduced volume flow rate curves at a temperature of 250 °C for different flow channel heights are not the same. The values obtained for a channel height of 0.25 mm are higher than for a channel height of 0.35 mm, which indicates wall slip. The volume flow rate values were approximated with a power law function to enable the evaluation of the wall slip velocity, where values at two different die heights and constant wall shear stress are needed. The wall slip velocity dependent on wall shear stress is shown in Fig. 3. The data were approximated with the function

$$v_S = A + B \cdot \tau_w^{\ C}, \tag{4}$$

where A = 0.035 (mm s^{-1}), B = 10404.85 (mm s^{-1} MPa^{-C}) and C = 2.36. The coefficient of determination R^2 is 1.00, which indicates the good approximation of the wall slip velocity data by the proposed function.

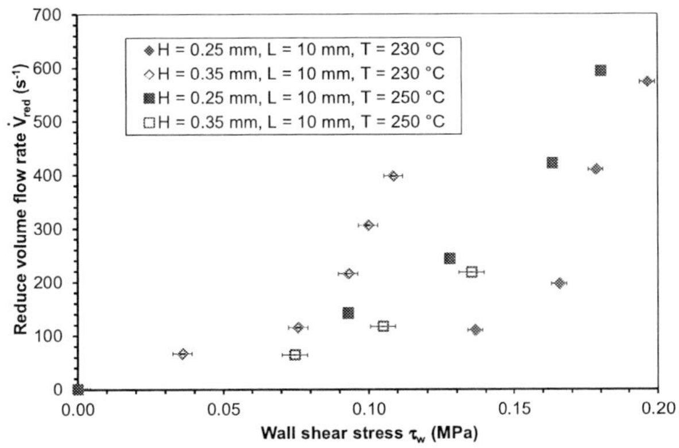

FIGURE 2. Reduced volume flow rate dependent on wall shear stress on polished steel at two different die heights and for the temperatures 230 °C and 250 °C

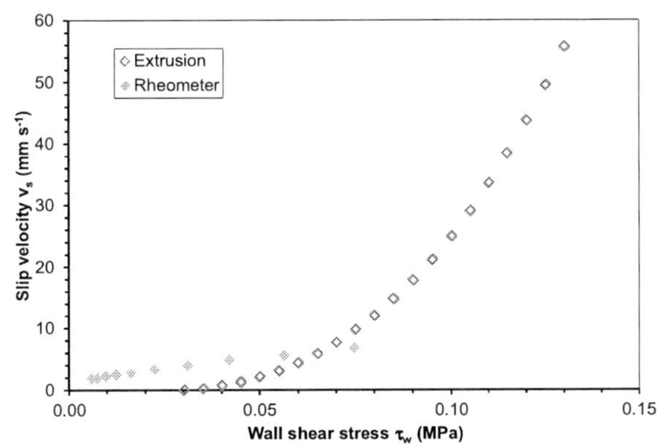

FIGURE 3. Wall slip velocity dependent on wall shear stress on polished steel at a temperature of 250 °C obtained with the rheological extrusion slit die and the high pressure capillary rheometer.

The wall slip velocity data obtained in the extrusion process were compared to data obtained with a high pressure capillary rheometer combined with a slit die. Using the high pressure capillary rheometer the onset of wall slip was already determined at a rather low wall shear stress, whereas with the extrusion slit die the onset of wall slip was observed at a wall shear stress of 0.03 MPa. There are also differences in the shape of the curves. The wall slip velocity data obtained with the high pressure capillary rheometer can be described by a root function, but the data obtained in the extrusion process obey a power law. When comparing these two different methods some differences have to be considered. First of all, it is obvious that the evaluated shear stress range is not completely the same. The wall slip velocity was determined in the extrusion process up to significantly higher values compared to the high pressure capillary rheometer. Second, the flow rate was controlled in the extrusion process by screw speed. An increase in melt temperature occurs with rising screw speed which shifts the reduced volume flow rate curves to higher values and influences the evaluated wall slip velocity. With the rough steel flow channel inserts no difference in the reduced volume flow rate curves measured with two different die heights was observed. This result indicates wall-adhering conditions on the rough steel surfaces and matches well with the data obtained with the high pressure capillary rheometer.

The surface properties of the flow channel inserts were determined before and after the extrusion experiments. The surface roughness values (area-weighted surface roughness S_a, line-weighted surface roughness in flow direction R_{ax} and in transverse direction R_{ay}) of the polished steel inserts are approximately the same; no influence of the extrusion process can be seen. An increase in surface energy, especially in the polar fraction of surface energy, occurs on the polished steel inserts after the extrusion experiments (see fig. 4a).

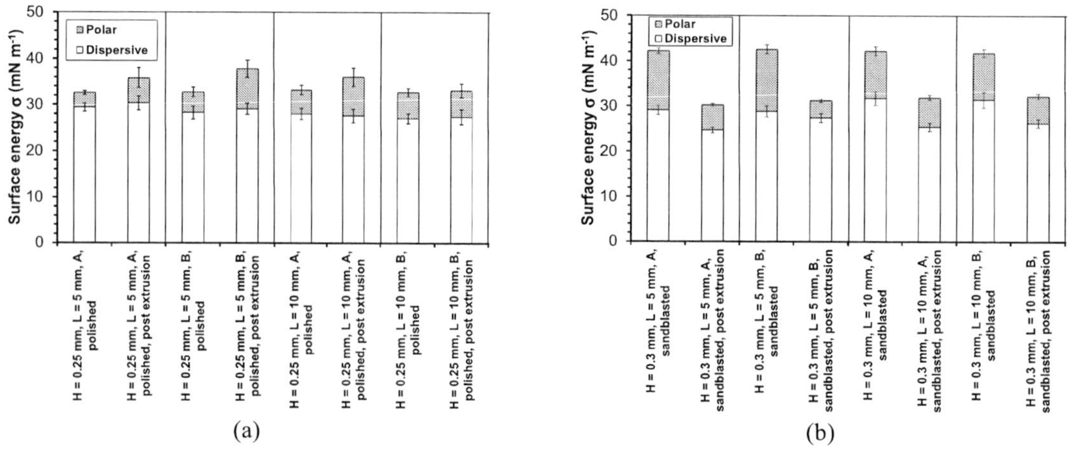

FIGURE 4. Total surface energy, polar and dispersive fraction of surface energy of the polished (a) and sandblasted (b) upper and lower steel inserts for two different flow channel lengths L before and after the extrusion experiments.

The rough steel die inserts exhibit a higher surface energy before the extrusion experiments compared to polished steel. This result can be explained by the influence of surface roughness on the contact angle of a liquid on a solid according to Wenzel [9] and Cassie et al. [10], which further influences the evaluated surface energy. After the extrusion experiments, a decrease in the total surface energy as well as in the polar and dispersive fraction of surface energy was found (see fig. 4b). This observation can be explained by the flattening of the flow channel surface by the polymer which was deposited in the surface cavities.

CONCLUSION

The presented rheological extrusion slit die enables investigation of the influence of the die surface on polymer melt flow under extrusion conditions. Good qualitative accordance with the results obtained with a high pressure capillary rheometer was observed using PMMA as test material. Wall slip was obtained with polished steel, but the rough steel surface caused wall-adhering conditions. The quantitative evaluation of wall slip velocity revealed differences between the two testing methods which can mainly be attributed to the temperature rise in the polymer melt with increasing screw speed. For future work, a temperature correction method has to be introduced in the wall slip evaluation procedure. The extrusion process influences the surface properties of the die. Polished steel exhibits an increase in the total surface energy and in the polar fraction of surface energy. The surface roughness of the rougher die inserts decreases because of the deposition of polymer melt in the surface cavities.

REFERENCES

1. A. V. Ramamurthy, J. Rheol., **25**, 1, (1986)
2. S. G. Hatzikiriakos and J. M. Dealy, Intern. Polymer Processing, **9**, 1 (1993)
3. C. Seidel, A. Merten and H. Münstedt, Kunststoffe, Plast Europe, **92**, 10 (2002)
4. D. R. Arda and M. R. Mackley, J. Non-Newtonian Fluid Mech., **126**, 1 (2005)
5. G. Zitzenbacher, T. Bayer and Z. Huang, Int. J. Materials and Product Technology, **52**, 1-2, (2016)
6. M. Mooney, J. Rheol., **2**, 2 (1931)
7. M. Pahl, W. Geißle, H.-M. Laun, Praktische Rheologie der Kunststoffe und Elastomere, 4th edition, Düsseldorf, Germany, VDI-Verlag (1995)
8. D. K. Owens and R. C. Wendt, Journal of Applied Polymer Science, **13**, 8 (1969)
9. R.N. Wenzel, J. Phys. Chem., **53**, (1949)
10. A.B.D. Cassie; S. Baxter, Trans. Faraday Soc., **40**, (1944)

Extrusion and Injection Molding Nanocellulose - Polypropylene Nanocomposites: Critical Processing Aspects

Girginer Burcu [1]*, Oztoksoy Mert Emre [1], Yildiz Ramazan Bedirhan [2], Guner Fatma Seniha [2], Seydibeyoglu Mehmet Ozgur [3], Taskin Deniz [4]

[1] Farplas Automotive Company R&D Department - Kocaeli - Turkey, [2] Department of Chemical Engineering, Istanbul Technical University - Istanbul - Turkey, [3] Department of Materials Science and Engineering, Izmir Katip Celebi University - Izmir - Turkey, [4] Gama Alfa Plastic Inovation Industry and Trade Inc. - Kocaeli - Turkey

Cellulose is a well known green renewable material with potential usage in many important industrial applications. When forming compounds Cellulose nanofibers formed by cellulose polymeric chains can have big influence on mechanical properties. In our study whilst the growth evidence of nanocellulose is detectable, this important material was used to reinforce polypropylene matrix with a coupling agent. In our previous study, the type (FUSABOND E265) and concentration of coupling agent was optimized as 0.1 wt%. Now, we have further studied the processing parameters of polypropylene nanocellulose nanocomposites with different nanocellulose content including 0.5, 1, 2 and 5 wt% via lab scale twin screw extruder. The compounded nanocomposites were investigated using a rheometer to understand the shear properties and processing optimization. The nanocomposites were further injection molded in order to obtain auto parts jointly with Farplas group in Istanbul. The mechanical properties of the injection molded parts were investigated with tensile testing providing tensile strength, tensile modulus and % elongation. This study gave a full understanding of the processing of nanocellulose composites which is very critical for the future studies and commercial products.

Effects of Interparticle Interactions on the Flow Behaviour of Low Density Polyethylene Filled with Various Fillers

Nico Laufer[1, a)], Harald Hansmann[1, 2, b)], Michael Koch[3, c)], Christian Boss[1, d)], Stefan Ofe[1, e)]

[1]*Institute for Polymer Technologies e. V., 23966 Wismar, Germany*
[2]*Department of Mechanical Engineering, Hochschule Wismar University of Applied Sciences, Technology, Business and Design, 23966 Wismar, Germany*
[3]*Department of Plastics Technologies, Ilmenau University of Technology, 98693 Ilmenau, Germany*

[a)]Corresponding author: laufer@ipt-wismar.de
[b)]hansmann@ipt-wismar.de
[c)]michael.koch@tu-ilmenau.de
[d)]boss@ipt-wismar.de
[e)]ofe@ipt-wismar.de

Abstract. In this study, the flow behaviour of low density polyethylene (LDPE) filled with different volume fractions of wood particles and glass beads have been investigated by means of high pressure capillary rheometry in consideration of different particle sizes. The strongly exhibited shear thinning flow behaviour of the investigated polymer-compounds can be well described by the Ostwald – de Waele power law relationship. The measurements showed that interparticle interactions are differently exhibited depending on the size and aspect ratio of the particles. Interparticle interactions can be taken into account by the interaction exponent, defined as the ratio between flow exponent of the suspension and flow exponent of matrix polymer. The interaction exponent shows a good correlation with the flow consistency index. Consequently, this correlation function can be described as an interaction function. On the basis of a generalized interaction function, that considers the transition from negligibly interactions to the domain of non-negligible interactions, the shift factor and the flow behaviour of the suspensions have been estimated with high accuracy.

INTRODUCTION

Viscosity functions of particle filled polymer melts are shifted to higher values in comparison to those of the unfilled polymer melts. The offset depends on the volume fraction and in the case of non-negligibly interparticle interactions, together with the applied shear stress or shear rate. In the field of rheology of suspensions a variety of models have been published which mostly describe the influence of the particle volume fraction on the relative viscosity $\eta_r(\phi)$. [1-5] W. Gleißle and M. K. Baloch developed a concept in which the vertical offset is caused by an increase of the inner shear rate, the so called *concept of shear-stress-equivalent inner shear rate* [6]. Taking into account the shear thinning flow behaviour of particle filled polymer melts the relative increase of the inner shear rate to the shear rate of the suspension is expressed by the shift factor B, Eq. (1), which is influenced by a number of factors [7-10]. The flow functions of filled polymer melts with varying volume fractions of filler particles can be described on the basis of shift factor B and the power law function of the matrix polymer, Eq. (2).

$$B(\tau_0)^{n0} = \frac{K_c}{K_0^{\chi}} \cdot \tau_0^{(\chi-1)} \tag{1}$$

$$\tau_c = K_0 \cdot \left(B \cdot \dot{\gamma}_0\right)^{n0} \tag{2}$$

Within this study the effects of interparticle interactions on the flow behaviour of low density polyethylene filled with different fillers are investigated. The aim of this study is to find a general approach describing the flow behaviour of particle filled polymer melts, which takes into account interaction effects of different types of fillers.

MATERIALS AND METHODS

The matrix polymer used in this study is a low-density polyethylene (LDPE, Lupolen 2420 K) with a shear thinning flow behaviour. Softwood flour and glass beads were used as filler particles, in two different size fractions for each case (Table 1). The particle size distributions were determined by laser diffraction using Mastersizer 3000 (Malvern Instruments GmbH). The geometry of the filler particles were characterized by microscopy using VHX 2000D (Keyence Microscopes) (Fig. 1).

TABLE 1. Filler particles used in this study

Filler	Filler Type	d10 in µm	d50 in µm	d90 in µm
Lignocel® BK 40/90	softwood floor	262	527	1168
Lignocel® HB 120	softwood floor	20	110	300
soda-lime glass 346	glass beads	229	346	515
soda-lime glass 60	glass beads	44	60	81

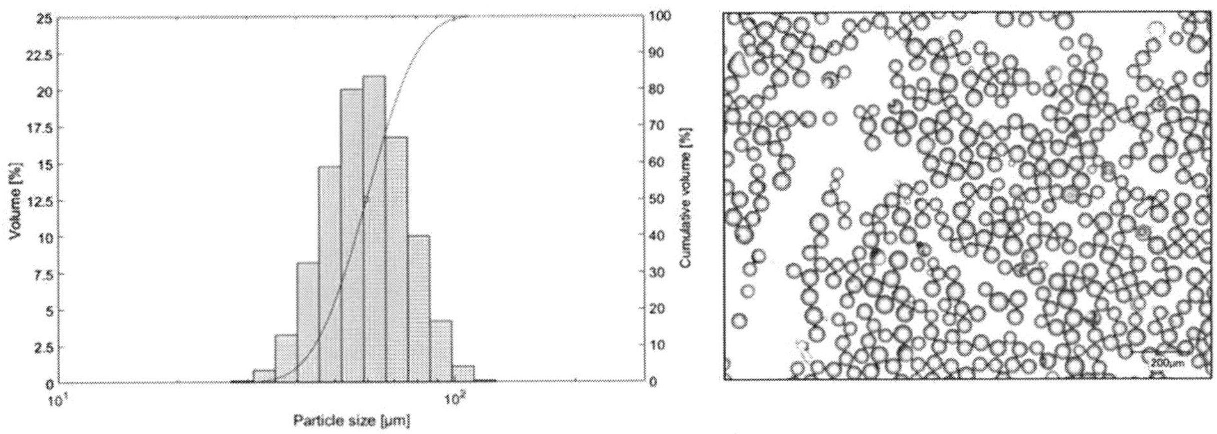

(a) (b)

FIGURE 1. (a) Particle size distribution of soda-lime glass 60, (b) transmitted light microscopy, magnification 200x

The preparations of all formulations were carried out by compounding on the co-rotating twin screw extruder ZSE 27 MAXX (Leistritz Extrusionstechnik GmbH), with a melt temperature of 190 °C and a mass output of 5 kg/h. All rheological analyses were performed on a high pressure capillary rheometer (HKR) Smart RHEO 5000 twin bore (CEAST, Instron®) using a round die with a diameter of 2 mm and a length of 40 mm. The measured data were analysed on the basis of the apparent shear stress and shear rates.

RESULTS AND DISCUSSION

All analysed formulations clearly show shear thinning flow behaviour. The corresponding parameters of the power law model are reported in Table 2.

TABLE 2. Power law model parameters for LDPE filled with soda-lime glass 346

Formulation	filler content in vol.%	K in Pa s^n	n
LDPE	0	5231	0.496
LDPE-GB(346)-5.9	5.9	5806	0.502
LDPE-GB(346)-19.4	19.4	8181	0.502
LDPE-GB(346)-36.0	36.0	13490	0.465
LDPE-GB(346)-45.7	45.7	17712	0.453
LDPE-GB(346)-56.7	56.7	25936	0.443

With increasing volume content of the glass beads the consistency (K) increases as well (Fig. 2 (a)). Whereas the flow exponent and the interaction exponent remains almost constant up to a volume content of 19.4 vol.%. In this region interparticle interactions are negligibly small. From a volume content of 36 vol.%, non-negligible interparticle interactions lead to a decreasing interaction exponent, which decreases further with increasing filler content (Fig. 2 (b)).

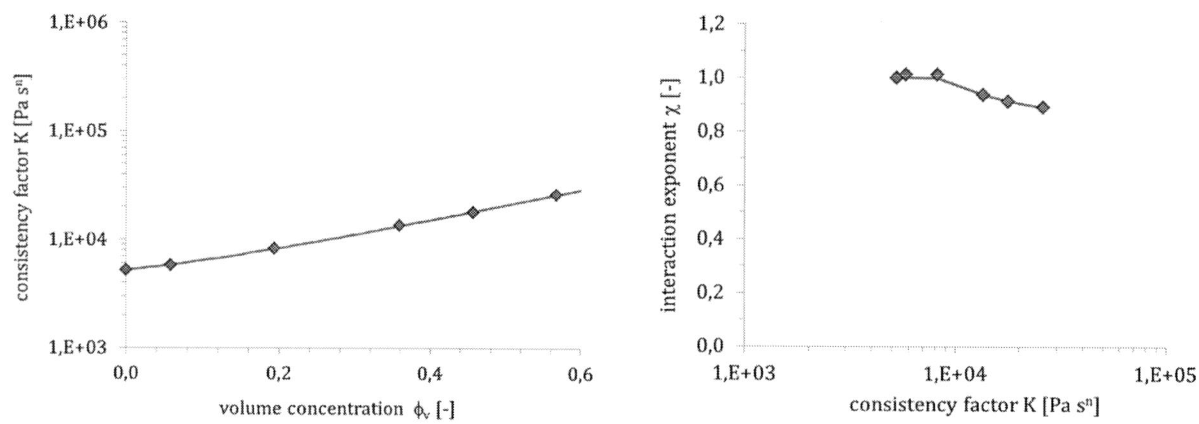

(a) (b)

FIGURE 2. (a) Consistency as a function of volumetric filler content, (b) interaction exponent as a function of consistency factor

In order to describe the quantitative impact of the filler content on the consistency of the compounds, Eq. (1) is used.

$$K_c = f(\phi_v) = K_0 \cdot \left(1 + A \cdot \phi_v^{\alpha} \cdot e^{\beta \cdot \phi_v}\right) \tag{3}$$

The model parameters of this empirical equation have been found by regression analysis with regard to minimize the sum of the relative error squares, $A = 2.179$, $\alpha = 1.079$, $\beta = 2.112$.

A *generalized interaction function* is used, in order to describe the interaction exponent as a function of the consistency, taking into account the transition from negligible interparticle interactions to the domain of non-negligible interactions Eq. (2).

$$\chi = f(K_c) = \frac{1}{\left(1 + \left(\lambda \cdot \ln\left(\frac{K_c}{K_0}\right)\right)^{a}\right)^{\frac{b}{a}}} \tag{4}$$

The model parameters of the generalized interaction function have also been found by regression analysis with regard to minimize the sum of the relative error squares, $\lambda = 2.137$, $a = 241.368$, $b = 0.094$.

By incorporating Eq. (3) and Eq. (4) into Eq. (1), the shift factor B can be calculated by considering the volume filler content and the transition of negligible to non-negligible interparticle interactions. Figure 3 (a) clearly shows that in the region of negligible interparticle interactions the shift factor B depends only on the volume filler content. If interparticle interactions occur the shift factor B becomes dependent on both the volume filler content and the applied shear rate.

Finally, the flow functions for LDPE filled with variable volume fractions of soda-lime glass 346 can be estimated on the basis of the shift factor B and the power law function of the matrix polymer (Fig. 3 (b)).

Based on this approach, a good correlation of estimated flow functions and measured values have been achieved.

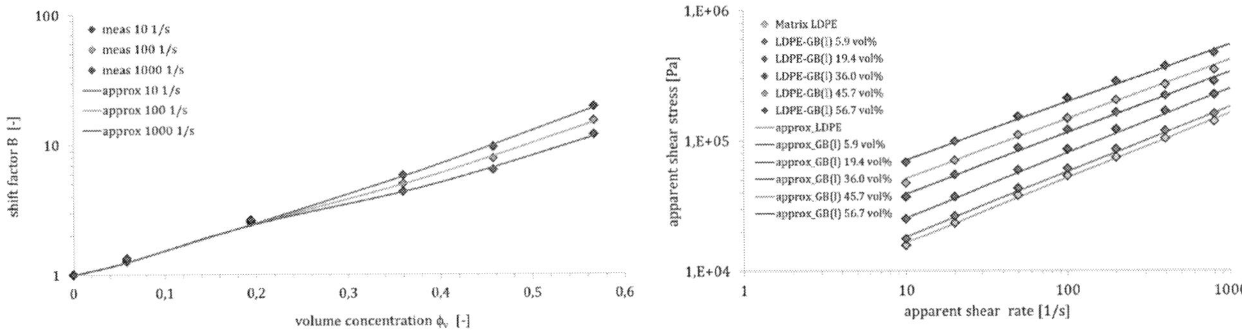

(a) (b)

FIGURE 3. LDPE filled with soda-lime glass 346 (a) shift factor B as a function of filler content and applied shear rate, (b) comparison of approximated and measured flow functions

As the particle diameter of the glass beads decreases, the surface-area-to-volume ratio increases and thus the interface to the matrix polymer increases with constant volume filler content. Figure 4 (a) shows that the particle size has in impact on interparticle interaction effects and thus on the rheological behaviour of these compounds. In order to investigate the applicability of this approach to these suspensions, the generalized interaction function has been used to estimate the flow behaviour LDPE filled with mutable volume fractions of soda-lime glass 60 (Fig. 4 (b)).

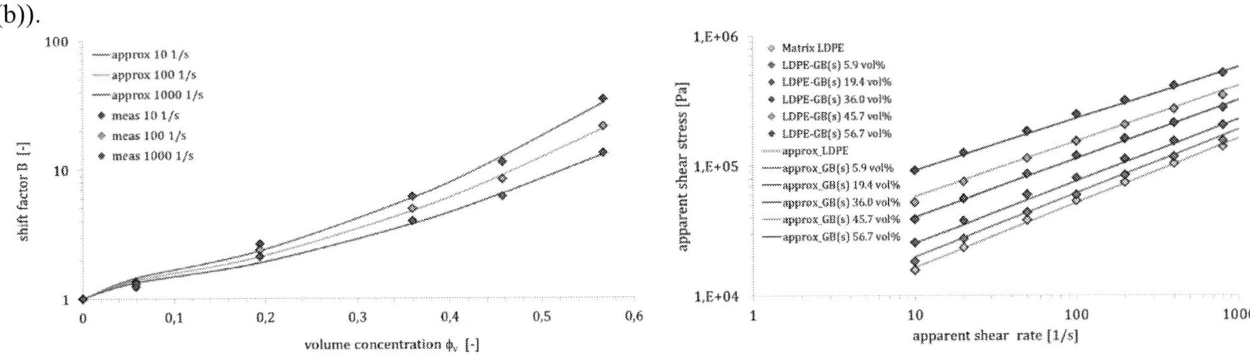

(a) (b)

FIGURE 4. LDPE filled with soda-lime glass 60 (a) shift factor B as a function of filler content and applied shear rate, (b) comparison of approximated and measured flow functions

Lastly the applicability of the generalized interaction function have been investigated to suspensions in which the rheological properties are influenced by interparticle interaction effects even at low volume filler contents such as wood filled polymer melts. On the basis of the shift factor B, the flow functions LDPE filled with different volume and size fractions of softwood floor have been estimated. The estimated flow functions correlate well with the measured values (Fig. 5).

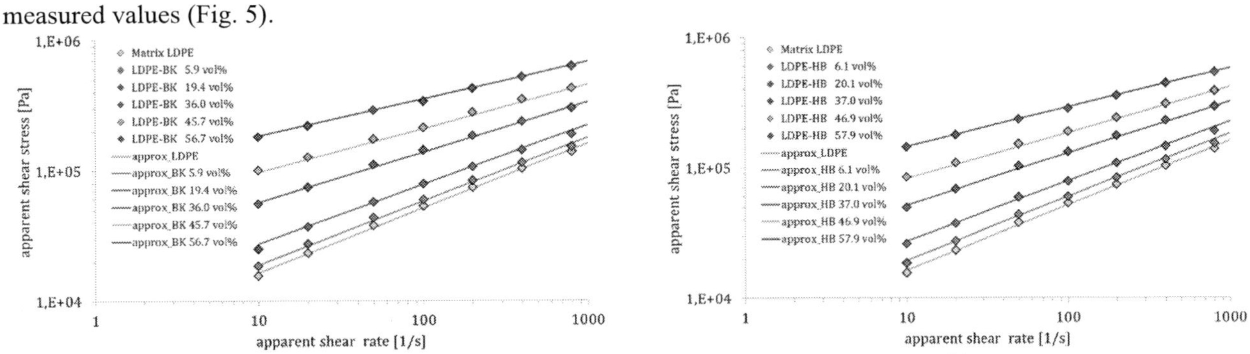

(a) (b)

FIGURE 5. Comparison of estimated and measured flow functions (a) LDPE filled with Lignocel® BK 40/90 (b) LDPE filled with Lignocel® HB 120

CONCLUSIONS

In this study, the influence of interparticle interactions on the flow behaviour of low density polyethylene filled with different volume fractions of wood particles and glass beads have been investigated. It has been shown that the extent of interparticle interactions is influenced by particle size, aspect ratio and volume content of the filler particles. Furthermore it has been shown that on the basis of the generalized interaction function interparticle interaction effects can be mathematically described by considering the transition from negligible interactions to non-negligible interactions. The equation for shift factor B was adapted by using the generalized interaction function. Based on this adaption flow functions particle filled LDPE have been estimated with high accuracy. The application of this concept for predicting the flow behaviour of suspensions is of practical importance in plastics processing. Thus, on the basis of the generalized interaction function, the pressure build-up in flow channels can be estimated as a function of the volume output and the volumetric filler content.

REFERENCES

1. S. Mueller, E.W. Llewellin and H.M. Mader, The rheology of suspensions of solid particles, Proc. R. Soc. A, doi:10.1098/rspa.2009.0445
2. T. Kitano, T. Kataoka, and T. Shirota, An empirieal equation of the relative viscosity of polymer melts filled with various inorganic fillers, Rheol. Acta 20, 207-209 (1981)
3. H. A. Barnes, A Review of the Rheology of Filled Viscoelastic Systems, Rheology Reviews 2003, pp 1 – 36
4. B. Hinkelmann, Zur analytischen Beschreibung des Füllstoffeinflusses auf das Fließverhalten von Kunststoff-Schmelzen, Rheol. Acta 21, 491 - 493 (1982)
5. W. Zhiying, Rheological Behaviors of Concentrated Suspensions in Pressure-Driven Tube Flow, Dissertation Nanyang Technological University, 2008
6. B. Hochstein, Rheologie von Kugel- und Fasersuspensionen mit viskoelastischen Matrixflüssigkeiten, Dissertation Universität Fridericiana Karlsruhe (Technische Hochschule), 1997
7. M. Pahl, W. Geißle und H.-M. Laun, Praktische Rheologie der Kunststoffe und Elastomere, Düsseldorf: VDI-Verlag GmbH, 1995
8. H. Hansmann, N. Laufer und S. Kühn, Investigation on the Flow Behavior of WPC Melts, A. K. Bledzki und V. E. Sperber, Hrsg., München: Hanser, 2012, pp. A6-1 - A6-11
9. N. Laufer, H. Hansmann, M. Koch, High Pressure Capillary Rheometer on Wood Plastic Composites with Variation of Wood Content and Matrix Polymer, 58th Internationale Wissenschaftliche Kolloquium, Ilmenau 2014
10. N. Laufer, H. Hansmann, M. Koch, Rheological Characterisation of the Flow Behaviour of Wood Plastic Composites in Consideration of Different Volume Fractions of Wood, 2nd International Conference on Rheology and Modeling of Materials, Miskolc-Lillafüred, 2015

APPLICATION OF MULTI-SCALE APPROACH TO RUBBERS REINFORCED AT DIFFERENT PROCESSING CONDITIONS

Ivaneiko Ievgeniia **(1)*, *Toshchevikov V.*** **(1), *Svistkov A. L.*** **(2), *Heinrich G.*** **(1), *Saphiannikova M.*** **(1)**

(1) Leibniz-Institut für Polymerforschung Dresden e.V. - Saxony - Germany, (2) Institute of Continuous Media Mechanics - Perm - Russia

Investigation of the dynamic-mechanical properties of reinforced elastomers is of high technological importance for the improvement of performance of rubber compounds in tire treads with low rolling resistance and high wet grip. Such investigation can be successfully executed with a help of a recently developed multi-scale approach [1-3]. This approach allows to fit the master curves constructed for the storage and loss moduli of unfilled and filled rubbers at a chosen reference temperature in a wide range of frequencies stretching over 15 decades. The multi-scale approach is based on the distribution function of relaxation times and presumes four distinct power-law regimes. Starting from the high frequencies, these regimes are determined by: 1) rotational and vibrational motions inside the monomer, 2) the semiflexible chain behaviour on a scale of the Kuhn segment, 3) the relaxation of flexible network strands, and 4) the relaxation of network defects such as entangled dangling chains in randomly cross-linked rubbers. The scaling exponents typical for unfilled rubbers are found to decrease with the addition of filler due to localization of network chains on the surface of filler particles. In the present study the reinforced compounds, comprised of solution-polymerized styrene butadiene rubber and carbon black, have been chosen to investigate the influence of processing conditions by applying different mixing times and different curing packages. The latter contained increasing amount of sulfur and thus resulted in increasing cross-link density. Additionally, we investigate the influence of extra hard carbon fillers, such as multi-walled carbon nanotubes, graphene and nanodiamonds, on the performance of rubber compounds.

Viscosity Measurements for Rubber Compounds using Slit-die Rheometry

Friesenbichler Walter [(1)]*, Neunhaeuserer Andreas [(1)]

[(1)] Montanuniversitaet Leoben, Institute of Injection Molding of Polymers - Styria - Austria

Reliable viscosity data within a wide range of shear rates are of crucial importance for the simulation of rubber injection molding. Due to the numerous user-specific compounds with varying formulations, no viscosity database is available and subsequently, viscosity measurement for rubber compounds has to be carried out very carefully to enable reliable simulation results. Slit die rheometry which is well known since the early 70th of the last century allows for improving pressure detection, widening the shear rate range to be covered and detecting wall slip effects. Adapting slit die systems to rubber injection molding machines offers new possibilities in rheological characterization of compounds with complex rheological behavior, performing measurements at pre-shearing conditions close to processing. Filler-filler interactions (Payne effect) strongly influence viscosity data which can be shown by RPA measurement at suitable pre-shearing conditions preventing material degradation. In this study NBR, SBR and FKM rubber compounds were characterized with a rubber process analyzer (RPA), a high pressure capillary rheometer (HPCR) and a special injection molding machine rheometer (IMMR) which allows shear viscosity measurements up to high shear rates for rubber compounds containing curatives. Viscosity data measured with RPA (conventional, pre-sheared) will be compared with data from HPCR and IMMR. The rheological evaluation of measured values and the approximation of true viscosity curves at higher shear rates take into account the temperature rise due to viscous dissipation. The importance of the temperature correction method is demonstrated for medium and high shear rates showing measured steel and wall temperatures in the rheological slit-die as well as presenting simulation results for the non-isothermal temperature field in the rheological slit. Finally the measurement of pressure dependency of viscosity will be shown for a SBR compound containing curatives.

Viscoelastic Effects on the roll-separating force and power input on the Calendering Process of a Non-Newtonian Fluid

Muñoz Sánchez Jaime Ernesto (1)*

(1) IPN - Ciudad de México - Mexico

Our study deals with the calendering process in viscoelastic sheets of finite initial thickness. The rheological constitutive equation of the fluid under consideration follows the Simplified–Phan–Thien–Tanner (SPTT) fluid model. The corresponding governing equations of mass and momentum balance, based on lubrication theory, are solved for the velocity and pressure fields using numerical techniques. When the above variables were obtained, we followed determining quantities of engineering interest such as the roll-separating force and the power transmitted to the fluid by the rolls. Is our interest to determine the influence of viscoelastic effects (represented by the Weissenberg number Wi) on such variables.

Following Phase Transitions with Rheometry and Simultaneous Raman-Spectroscopy

Jakob Bernd [(1)]*

[(1)] Thermo Fisher Scientific - Karlsruhe - Germany

The use of a coupled rheometer and Raman spectrometer for obtaining comprehensive insight into a materials behavior is presented. Rheology is the analytical method of choice to correlate the absolute flow and deformation characteristics of a given material with its behavior towards a certain processing or application step. However, Rheology as an integral method only yields answers on the bulk of the investigated sample. It does not give any insights into what is actually happening on the molecular level when rheological changes are observed. Raman spectroscopy has shown its ability as a powerful, effective and non-invasive method for chemical analysis. Coupling a rheometer with a Raman spectrometer provides direct information about the molecular reaction kinetics and the mechanical properties. This is extremely useful for studying phase transition behavior like i.e. the crystallization behavior of polymer melts during processing. It can also provide insight for in-situ characterization and monitoring which can be challenging when working with on-line techniques as only relative flow fields are characterized. Finally, the coupling gives the researcher the advantage of maximizing and synchronizing the information gathered from a single measurement as well as enabling transient information on their materials during fast processing conditions. In this contribution we present results obtained with the new combination of a Thermo Scientific HAAKE MARS rheometer with a Thermo Scientific Raman spectrometer. Details of this unique set-up as well as selected results will be presented.

Correlation of steady shear rheology and morphology of Nylon 6 glass fibre composites

Perumal Deverpiran Vishak [1]*, Bhattacharya Sati N [1], Gupta Rahul K [1], Costa Franco S [2]

[1] RMIT University - Victoria - Australia, [2] Autodesk Australia Pty Ltd - Victoria - Australia

In an attempt to understand the effects of fibre orientation on the rheology of engineering thermoplastic composites, a series of steady shear experiments were carried out on Nylon-6 composites containing short glass fibres (length <1mm) of different weight fractions: 0% (PA6), 15% (PA6-15GF) and 30% (PA6-30GF). Steady shear experiments carried out in a parallel plate rheometer showed the presence of a viscosity overshoot after which a steady state viscosity was reached for filled systems. The magnitude of the viscosity overshoot decreased as the applied shear rate was increased. The viscosity overshoot is believed to be due to the reorientation of fibres in the shearing direction as PA6-15GF samples showed a lower magnitude of viscosity overshoot and unfilled PA-6 did not exhibit any overshoot. X-Ray Computed Tomography (X-CT) was employed to measure 3-dimensional fibre orientation and local fibre concentration through the thickness direction in the fibre filled samples at a resolution of 2.5μm. An increase in the principle fibre orientation tensor (A11) value was observed as higher steady shear rates were imposed on the composite samples. Orientation tensor values were compared with predictions from Folgar-Tucker based Reduced Strain Closure (RSC) model to validate the experimental results. A MATLAB code was developed for simulating flow in a rotational rheometer to obtain the predicted values. A comparative analysis between the experimental and predicted results is then presented. By quantifying the effects of fibre alignment on rheological properties, the research work aims to develop a constitutive model that can improve Injection Moulding predictions. Keywords: Computed Tomography, Fibre orientation, Rheology, Nylon, Composites

Effects of morphology and rheology on Nylon 6 glass fibre composites

Gupta Rahul K [(1)]*, Perumal Deverpiran Vishak [(1)], Bhattacharya Sati N [(1)], Costa Franco S [(2)]

[(1)] RMIT University - Victoria - Australia, [(2)] Autodesk Australia Pty Ltd - Victoria - Australia

In an attempt to understand the influence of fillers on the rheology of composite melts, oscillatory shear experiments were carried out on Nylon-6 composites containing short glass fibres (length <1mm) of different weight fractions: 0% (PA6), 15% (PA6-15GF) and 30% (PA6-30GF). Oscillatory time sweep experiments carried out within the Linear Viscoelastic Region (LVR) using a parallel plate rheometer showed an increase in complex viscosity as a function of time for filled systems. PA6-30GF samples exhibited a greater increase in viscosity than PA6-15GF samples while unfilled PA6 samples showed a constant viscosity. While an increase in fibre orientation usually results in the reduction of complex viscosity, these results suggest that other phenomenon such as fibre migration and flow induced crystallisation aided by fillers could be occurring. Differential Scanning Calorimetry (DSC) was used to characterise the crystallisation behaviour of the samples. X-Ray Computed Tomography (X-CT) was employed to measure 3-dimensional fibre orientation and local fibre concentration in the fibre filled samples at a resolution of 2.5µm. The research work aims to gain a better understanding of how specific flow mechanisms impact rheological properties of high fibre content composites which are commonly used in industrial processes. Keywords: Crystallisation, Nylon, Glass fibres, Computed tomography

HYDRODYNAMIC CHARACTERISTICS OF THE POLYMER MELTS FLOWS IN THE CONVERGING PLANE-PARALLEL CHANNELS

Pyshnograi Grigory (1)*, Koshelev Konstantin (2), Kuznetcov Alexander (1), Bocharnikov Vadim (3)

(1) Altai State Pedagogical University - Altai Region - Russia, (2) Institute for Water and Environmental Problems SB RAS - Altai Region - Russia, (3) Altai State Technical University - Altai Region - Russia

At work impact of temperature of polymeric melts on the sizes and intensity of the vortex area arising in case of a flow on an entrance to the slot channel is considered. For mathematical description of a three dimensional flow of melts in the converging channel the modified Vinogradov and Pokrovskii rheological model is used. The model is generalized for accounting of non-monotonic behavior of gradient dependence of elongation viscosity. Also Newtonian law for residual viscosity is used. Temperature dependence of initial shear viscosity of polymeric melts has the Arrenius appearance. Initial relaxation time was estimated as by comparison with experimental data for gradient dependence of stationary viscosity in case of uniaxial elongation, and on the basis of molecular and kinetic approach. Sticking conditions for velocity on a wall have been used. Discrete analogs of system of the equations of dynamics of polymeric fluid were by method of control volume with separation on physical processes. In case of implementation of numerical algorithm the possibility of application of parallel calculations technology based on graphical processors was considered. In numerical experiments returnable courses of polymeric melts before an entrance to narrow part of the canal are found and it is shown that dependence of the vortex sizes and intensity on temperature of melt has non-monotonic character and passes through a maximum. It is also shown that polymeric flow before an entrance to the slot-hole channel has significantly three-dimensional character, the sizes of the vortex area calculated in the sections which are carried out at various distances from an axis of the channel in the beginning increase in case of approach to a firm wall, and then decrease. All features noted in calculations are observed also in real experiments on branched polyethylene of low density melts.

Using the Cone Partitioned Plate geometry for enhanced material characterization of polymer melts

Reim Viktor (1)*, Eickhoff James P (2), Arnold Gunther (1), Läuger Jörg (1)

(1) Anton Paar Germany - Baden-Württemberg - Germany, (2) Anton Paar USA - Virginia - USA

Viktor Reim1, James P Eickhoff Jr.2, Gunther Arnold1, Jörg Läuger1 Anton Paar Germany GmbH, Ostfildern, Germany1, Anton Paar USA, Ashland, Virginia, USA2
Characterizing polymer melts at large deformations and/or high shear rates, is limited due to edge fracturing. This sample instability is characterized by a deformation of the sample surface at the free edges between upper and lower part of the geometry which propagates radially as a function of both time and applied deformation. Hence, edge fracture results in increasing measuring errors within standard cone-plate or plate-plate geometries when presetting large deformations and/or high shear rates. Consequently, the accuracy of start-up shear measurements and flow curves at high shear rates as well as for large-amplitude oscillatory shear (LAOS) measurements can be strongly influenced by edge fracture. In order to reduce the impact of edge fracture effects the use of a cone partitioned plate (CPP) has been recommended on several occasions [e.g.1-4]. Using this setup, the flow instabilities that arise at the edge of the CPP do not have an impact on the sample's properties which are detected in the center of the sample only. Hence, the CPP reduces the effect of trimming and enables rotational and oscillatory measurements even at deformations which would result in an incomplete filling of the active measuring zone of conventional measuring geometries. The aim of the contribution is to highlight the difference in the measuring performance when using a CPP in comparison to conventional cone-plate or plate-plate geometries. [1] Meissner, J., Garbella, R.W., Hostettler, J. (1989). J Rheol 33, 843-864. [2] Schweizer, T. (2002). Rheol Acta 41, 337–344 [3] Schweizer, T. (2003). J Rheol 47, 1071-1085 [4] Snijkers, F., Vlassopoulos, D. (2011). J Rheol 55, 1167-1186.

Modeling the Flow of Non-Newtonian Fluids in Single-Screw Extruders

Marschik Christian (1)*, Roland Wolfgang (1), Miethlinger Juergen (1)

(1) Institute of Polymer Extrusion and Compounding, Johannes Kepler University - Linz - Austria

Numerous analytical and numerical studies have analyzed the metering section of single-screw extruders. Due to substantial mathematical difficulities, all these analyses were based on specific model simplifications (e.g., Newtonian flow, isothermal flow, infinite channel width). We present a comprehensive numerical parameter study analyzing the three-dimensional, fully developed flow of polymer melts in metering sections. The pseudo-plastic nature is approximated by a power law model. Varying the influence parameters h/w (height-to-width ratio), t/D (pitch-to-diameter ratio), n (power law index), and Pp,z (dimensionless pressure gradient), a total number of 44 640 set-ups derived from similitude theory were simulated. For each design point, both the throughput and dissipation rate were evaluated. The calculated data set provides a powerful tool to predicting the conveying behavior of melt-conveying zones of single-screw extruders.

Shear Stress Growth Phenomena of Amorphous Polymer in Low Shear Rate Region

Yamasaki Natsumi [(1)]*, ***Tominaga Aya*** [(1)], ***Sekiguchi Hiroshi*** [(1)], ***Nakano Ryoko*** [(1)], ***Yao Shigeru*** [(1)], ***Takatori Eiichi*** [(2)]

[(1)] Fukuoka University - Fukuoka - Japan, [(2)] TOSOH Analysis and Research Center Co.,Ltd. - Mie - Japan

Polymer is thought to be constructing a random structure in molten state. Especially, longer the longest relaxation time, polymer is in a relaxed state with lost the orientation history. Moreover, the amorphous polymer, cooled and solidified from such molten stage is thought to have no structural morphology and show stable mechanical property. However, we had investigated mechanical properties of virgin and recycled polystyrene and found that the mechanical properties of the recycled sample were much better than the virgin sample. Moreover, we had found that with increasing the holding time (longer than the longest relaxation time) in a compression molding, the mechanical properties of polystyrene increased so much. These results suggest that 1.The recycled polymer retains its molding history in its inner structure even if it is remolded at a high molten temperature and longer than the longest relaxation time. 2. The amorphous polymer has a possibility to construct an ordered structure in molten state longer than the longest relaxation time. In this study, we have measured the time dependence of shear stress at very slow shear rate region in the molten state. And we found that the shear stress grows even after the longest relaxation time.

The Influence of the Morphological Structure of Polymer Foam-Based Adsorbents for the Treatment of Organic Contaminated Wastewater

Cherukupally Pavani (1), Nagarajan Ramani Krishna Prasad (1), Bilton Amy M. (1), Park Chul B. (1)*

(1) University of Toronto - Ontario - Canada

Recently, we have investigated the use of engineered foam adsorbents for the treatment of dissolved and emulsified oily wastewater. The morphological structure of the foam absorbent highly influences the treatment efficacy and energy requirements when implemented as filtration media for continuous separation. Herein, it is hypothesized that the adsorption capacity increases with the surface area-to-volume ratio of the foam. To evaluate this hypothesis, the influence of free volume (porosity of 70%, 80%, and 90%) on pressure drop and droplet adsorption was estimated using multiphysics numerical modeling in the software COMSOL. The pressure drop across the foam was determined based on continuity, Navier-Stokes, and Brinkman equations. Simultaneously, the oil mass transport was described using mass convection coupled with fluid flow, diffusion, and adsorption models. Simulation results indicated that with increasing free volume, pressure drop across the foam would decrease. Similarly, with increasing free volume, the oil mass uptake was found to increase. The lower pressure drop at higher porosity is attributed to the low mass transfer resistance. The oil mass uptake increases due to larger available free volume of the foam. Furthermore, larger pores also allow higher adsorption mass uptake due to multiple layers of oil deposition. These results agree with our hypothesis. In future work, additional morphological properties will be studied and experimental data will be used to validate results.

MATHEMATICAL MODELING OF THE FILM FORMING POLYMER IN A BIAXIAL STRETCHING

Pyshnograi Grigory (1)*, Merzlikina Darina (1), Kuznetcov Alexander (1)

(1) Altai State Pedagogical University - Altai Region - Russia

The work is devoted to the justification of the use of rheological constitutive relation fluid polymer media prepared on the basis of the molecular-kinetic concepts of the dynamics of polymer chains, taking into account the heat transfer. The relevance of this work is the fact that the task of formulation of equations systems suitable for the flows of melts of linear polymers, still does not have a unique solution, and therefore all of the proposed rheological constitutive relations require justification. As the author of the original model was chosen new rheological model (modified Vinogradov and Pokrovskii model). On the model basis the task of forming a melt of the polymer film was examined. This was the starting point for the construction of a mathematical model of the film forming polymer melt. It was observed the appearance of a "neck" when the calculation according to the half-width of the film to the distance from the spinneret exit. The very mathematical model of a melt molding is recorded in the one-dimensional approximation, which makes it suitable for engineering calculations. On the basis of studies of the effect of dimensionless parameters of the model for the type of dependencies obtained by the author concluded that the applicability of this approach to the description of characteristics of the resulting films. This problem is a test in the calculation of two-dimensional and three-dimensional flows of viscoelastic fluids.

Surface treatment of cellulose nanocrystals (CNC) – effects on dispersion rheology

Forsgren Lilian [(1)*], Sahlin Karin [(2)], Moberg Tobias [(2)], Westman Gunnar [(2)], Rigdahl Mikael [(2)]

[(1)] Chalmers University of Technology - Gothenburg - Sweden, [(2)] Chalmers University of Technology and Wallenberg Wood Sience Center - Gothenburg - Sweden

Cellulose nanocrystals (CNC) exhibit potentially good mechanical properties which make them interesting as candidates for reinforcing polymers, i. e. for manufacturing of nanocomposite materials. Incorporation of such elements into a polymer matrix requires however a careful characterization of the particles, their effect on the rheological behaviour of different types of dispersions/suspensions and on the affinity towards matrix material. In case of the last point, a surface treatment of the CNC may be required in order to obtain a sufficiently strong interphase region. In the present work, CNC dispersions were prepared from microcrystalline cellulose by a sulfuric acid treatment. The produced CNC were subsequently surface modified or functionalized with three different azetidinium salts. The surface treatment substantially improved the thermal stability of the CNC (which initially contained thermally instable sulfate groups) as revealed by thermogravimetric analysis. The rheological behaviour of aqueous dispersions containing 0.65 and 1.3 weight-% of the modified CNC was evaluated at room temperature. The surface treatment of the CNC increased significantly the shear viscosity as well as the dynamic (storage and loss) moduli of the dispersions (the extent depending of the modifying agent used). Furthermore, the functionalization reduced the gel point of the CNC dispersion from 2.5 weight-% (unmodified state) down to 02-0.4 weight-%. The implications of the surface treatment with regard to rheological behaviour and use in composite materials will be further commented on.

Influence of Talc Properties on Processability, Rheological Behavior and Crystallinity of PP Based Automotive Compounds

Ramazan Bedirhan Yıldız*[1], Burcu Girginer*[2], F. Seniha Güner[1]

[1]*Department of Chemical Engineering, Istanbul Technical University, Istanbul 34469 Turkey*
[2] *Farplas Automotive Company, Gebze-Turkey*

[a)]Corresponding author: ramazanbedirhanyildiz@gmail.com
[b)]b.girginer@farplas.com
guners@itu.edu.tr

Abstract. Talc reinforced polypropylene (PP) compounds have been widely used in injection molded automotive interior and exterior applications such as bumper, door panels due to their cost/performance and processability. Talc has the critical role of reinforcement and nucleating agent in automotive compounds in order to enhance of stiffness/toughness balance. Owing to recent developments in milling technology, talcs with high aspect ratio and reduced particle size can be generated. This proceeding enables to formulate talc reinforced PP automotive compounds having the increased specific strength and reduced density as a light weighting solution for automotive industry.

The rheological and crystallization behavior of PP based automotive plastics has the critical effects on process parameters such as barrel temperature, injection pressure, injection velocity, injection time, ejection time and overall cycle time during injection molding process. In this study, the influence of talc properties such as particle size, aspect ratio on flowability of 20% talc reinforced homopolymer and copolymer PP was investigated by rheometric measurements. The influence of the particle size, aspect ratio on crystallinity of 20% talc reinforced homo and copolymer PP was also investigated by DSC analysis. The relationship between rheological properties, crystallization behaviors and process parameters was studied.

INTRODUCTION

Polypropylene (PP) is a common semi-crystalline thermoplastic is widely used in industrial production applications because of its low cost, lightweight, balance of strength, chemical resistance and potential to produce complex shapes such as injection molding, blow molding, thermal compression molding *etc.* It is used in industry with manufacturing PP compounds owing to add various inorganic fillers on polymer matrices such as talc, calcium carbonate , wollastonite, glass bead, carbon black, and silica in amounts of 10–40% by weight.. Filling inorganic materials on PP matrices enable polymer to low cost, stiffness, hardness, dimensional stability, processability, light weighting, low coefficient of thermal expansion and shrinkage, scratch resistance and other many advantages [1-5]. Additives such as wollastonite, talc, calcium carbonate, clay, *etc.* seriously affect the degree of crystallinity, crystallization time and crystal nucleation of the semicrystalline polymeric materials. The addition of matrices to such materials increases the crystallization temperature, which allows the cycle time to decrease in the injection molding process. This affects the crystallinity percentage of the material [6]. Besides type of polymer matrices affects the properties above and also affect the interaction with fillers. Talc is common filler used in PP materials as a reinforcer due to its low cost formation and its flake morphological structure. Talc has a magnesium silicate structure and is described by the formulation Mg3Si4O10 (OH) 2. Having various geological formations, talc makes a unique material in morphological and chemical terms [7]. The talc is used in PP matrices as filler material in micro and nano sizes. The addition of talc causes changes in the properties of the PP and is therefore of interest to the industry. Compared to other minerals, talk has been identified to be more effective filler in terms of mechanical properties and macromolecular organization. It has been found that the PP materials together with the talc filler cause an increase in the crystallization temperature and heat of fusion [8].

Filled polymeric materials are subject to complex processing conditions. High temperatures and high shear stresses are required to process and shape these types of materials. In addition to the various advantages mentioned above, the addition of inorganic particles to polymer matrices can cause various drawbacks. The incorporation of inorganic solid particles modifies the viscosity, viscoelasticity and elastic behavior of the molten polymers [9]. In the determination of these behaviors and changes, rheology science and therefore rheological measurement methods are used. When considered from an industrial point of view, the rheological properties of polymers have great importance in terms of plastics processing techniques [10]. Injection molding technology has taken the industry place to produce many inland plastic parts such as automotive parts, electronic parts and packaging products. Injection molding is convenient in series production and is an efficient method in the production of plastic parts. Plastic parts production can be realized in a short time thanks to injection molding technique [11]. Composite materials produced by the injection molding technique exhibit a self-layering structure that varies in core from the surface along with crystal and amorphous properties of the polymer [12].

The rheological and crystallization behavior of PP based automotive plastics has the critical effects on process parameters such as injection pressure, injection velocity, ejection time and overall cycle time during injection molding process. In this study, the influence of talc properties such as particle size, aspect ratio on flowability of 20% talc reinforced homo and copolymer PP was investigated by rheometric measurements. The influence of the particle size, aspect ratio on crystallinity of 20% talc reinforced homo and copolymer PP was also investigated by DSC analysis. The relationship between rheological properties, crystallization behaviors and process parameters was studied.

MATERIALS

Materials

PP homopolymer with MFR 12 g/10min (230°C /2.16kg), density 0.908 g/cm^3 PP HE125MO was provided from Borealis AG.

PP block copolymer with MFR 14 g/10min (230°C /2.16kg), narrow MWD, density was 0.905 g/cm^3 PHC27 was provided by SABIC® Europe Company.

Standart talc d_{50}= 12 mμ, 0.89 aspect ratio, 5XK provided from Mikron's talk

Fine talc d_{50}= 4.6 mμ, 9.63 aspect ratio, Steamic T1 C F provided from IMERYS talc

Nucleating agent, talc with d_{50} = 2.4 mμ HTP2 was provided from IMIFABI.

Sample Preparation

The talc particles were dried in an oven at 60 °C for 12 h before melt extrusion. The PP pellets was melt-compounded in a co-rotating twin screw extruder (Coperion ZSK MC 26) at a temperature in the range 170-200 °C and a screw rotation rate of 450 rpm. The talc particles were side feeded to the homopolymer PP and copolymer PP through the process, named C0 and C3 in the Table 1 by dry mixing before melt extrusion. The talc with d_{50}= 12 mμ and 4.6 mμ contents were 20 wt% with the addition of 0.5wt% nucleating agent. Hereafter, these composites are named C1, C2, C4, and C5 as in the given Table 1. The extrudates were pelletized at the die exit, dried and then injection molded into 1 mm thickness sheets (30 cm -30 cm -1 mm) at 220-235 °C.

TABLE 1. The code of samples

Samples	Sample Codes
Homopolymer PP	C0
Standard Talc Filled Homopolymer PP	C1
Fine Talc Filled Homopolymer PP	C2
Copolymer PP	C3
Standard Talc Filled Copolymer PP	C4
Fine Talc Filled Copolymer PP	C5

Processing

Compounds were molded with a 1700 ton-f injection molding machine (Engel, Duo 7050/1300) and parameters were re-adjusted according to surface quality and dimensional stability expectations where an automotive part was chosen due to better reflection of changes. Injection pressure, injection velocity, ejection time, pack pressure, switch over stroke and overall cycle time have been obtained.

CHARACTERIZATION

Differential Scanning Calorimetry Analysis

To analyze the non-isothermal melt crystallization behaviors of the samples DSC experiments were performed using a Diamond DSC (Perkin Elmer). The samples were heated from the room temperature to 200 °C at a heating rate 10°C /min and equilibrated at 200 °C for 10 min. Then, the samples were cooled to room temperature at a cooling rate 2 °C /min.

Rheological Analysis

All rheological measurements were performed on a controlled strain rheometer (MCR 702 Rheometer, Anton Paar Physica) with using 25 mm diameter parallel plates at 190°C, 210°C and 230°C. Granule PP compounds are used in measurements directly with 1 mm measurement gap. The strain %1 obtained from amplitude sweep tests was used which was in the linear viscoelastic regime for all samples. Frequency sweep tests were conducted in order to find out viscosity and modulus of samples within frequency range of 0.1–600 rad/s. To ensure the reliability of rheological data all measurements were conducted under nitrogen atmosphere to minimize oxidative degradation of the polymer.

RESULTS AND DISCUSSION

Differential scanning calorimetry analysis results

The differential scanning calorimetry analysis results are given in the Table 3. The results shows that addition of talc particles with different particle size increased the degree of crystallinity from 32.1% to 45.3% and 45.2% of homopolymer based samples. And also it increased the degree of crystallinity from 47.4% to 50.8% and 51.7% of copolymer based samples. The addition of talc particles has shown the significant effect on homopolymer crystallization degree than copolymer. There is no significant effect of talc particles size on the degree of crystallinity due to the nucleating agent content. The nucleating agent has the crucial role on crystallization process comparing to talc reinforcement.

TABLE 2. Differential scanning calorimetry analysis results

Sample	Tm, °C	ΔHm, J/g	Tc, °C	ΔHc, J/g	The Degree of Crystallinity, %
C0	161,7	97,4	112,1	-95,3	32,1
C1	163,9	68,3	126,7	-70,1	45,3
C2	163,6	69,5	126,7	-69,9	45,2
C3	165,0	64,6	121,4	-66,8	47,4
C4	165,1	59,3	127,0	-59,9	50,8
C5	163,6	69,5	126,4	-56,4	51,7

According to the isothermal melt crystallization time curves in the given in the Figures 1 and 2; the particle size effects the crystallization time. In the Figure 1 and 2; Talc filled compounds have decreased crystallization time to neat polymer and talc filled compounds has the same crystallization time. The necessary crystallization energy of homopolymer with lower the talc particle size and high aspect ratio (d_{50}= 4.6 mμ, 9.63 aspect ratio content) is lower than homopolymer with higher particle size and lower aspect ratio (d_{50}= 12 mμ, 0.89 aspect ratio). The reduction of crystallization time in the talc filled copolymer compounds is smaller than the homopolymer compounds.

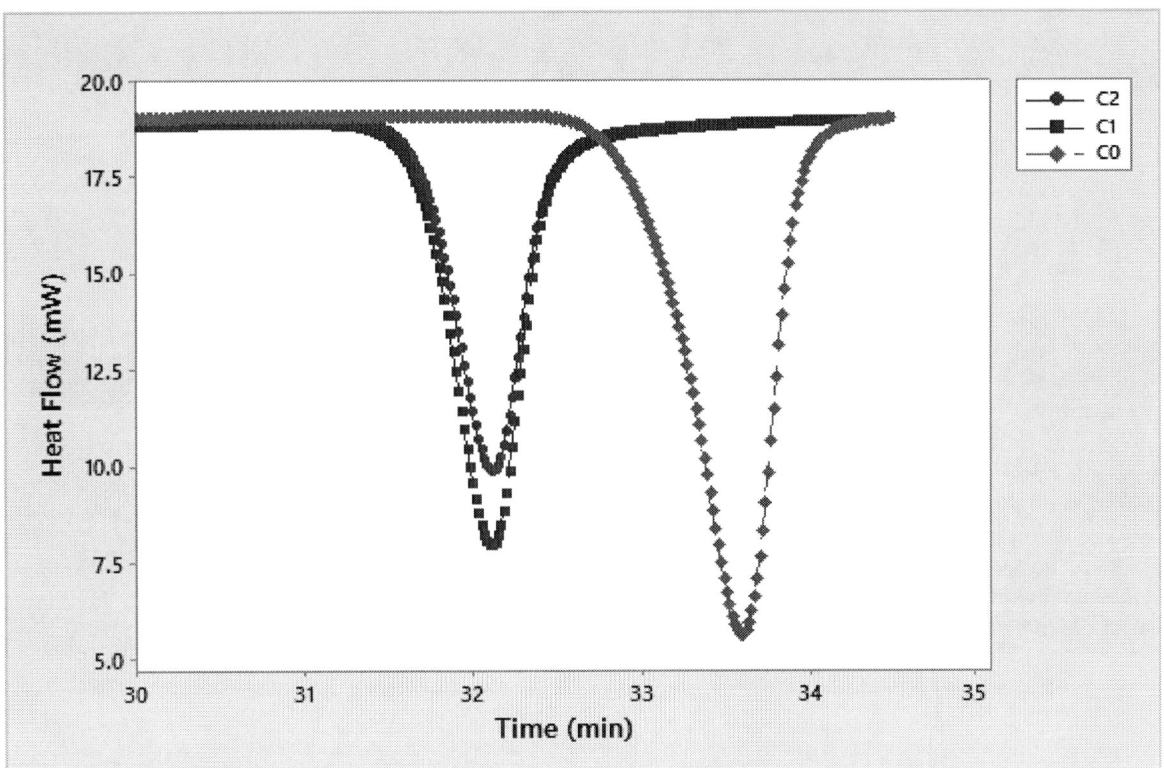

FIGURE 1. Differential scanning calorimetry isothermal melt crystallization time curves of C0-C1-C2 specimens

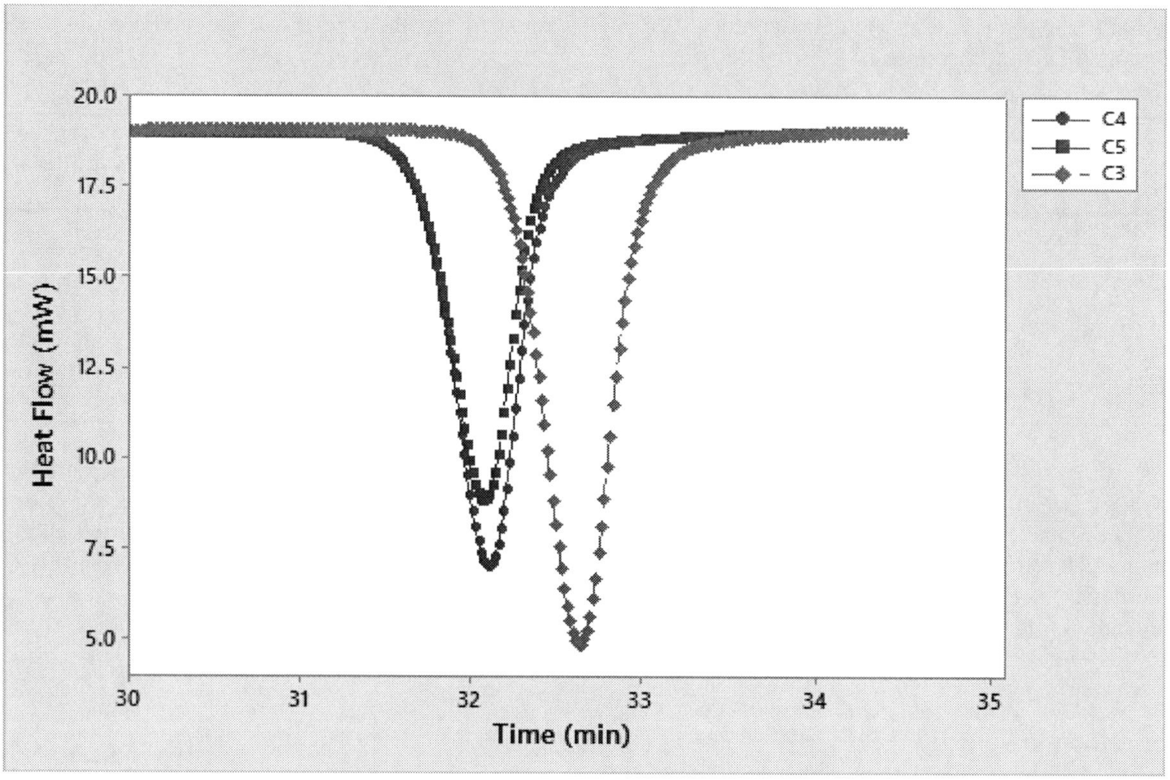

FIGURE 2. Differential scanning calorimetry isothermal melt crystallization time curves of C3-C4-C5 specimens.

Rheological analysis results

The rheological data was collected at elevated temperatures 190, 210 and 230°C. Storage modulus, loss modulus and complex viscosity was measured with regards to frequency. Because of the similarity, only 210°C data are used to generate graphs.

The complex viscosity results of talc filled compounds and neat PP homopolymer and copolymer were shown in Figure 3. The presence of talc with varied particle size affects the viscosity for both homopolymer and copolymer. Addition of talc particles into copolymer PP increased viscosity at lower frequencies contrary to homopolymer. The viscosity reduction of copolymer is observed at higher frequencies evidently due to lower particle size.

In Figure 4, the loss modules of homopolymer and copolymer compounds were presented. Compared to copolymer compounds, talc particles have no significant effect on the loss modules of C1 and C2 homopolymer samples at the lower frequencies.

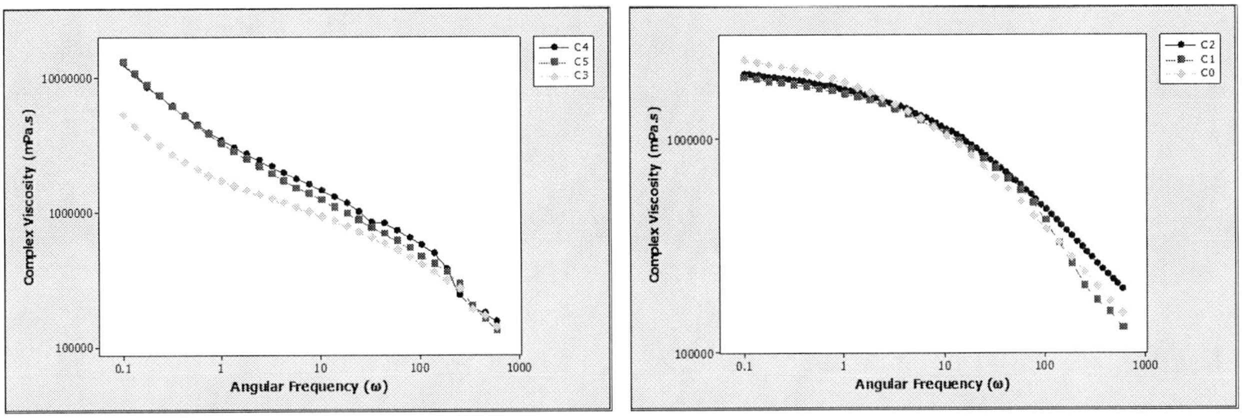

FIGURE 3: The complex Viscosity Measurement Results of C0-C1-C2 and C3-C4-C5 samples.

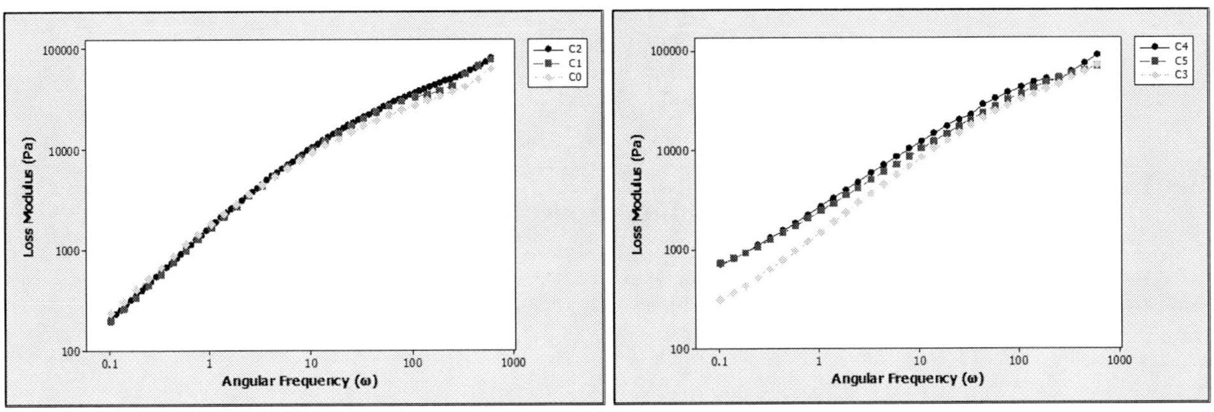

FIGURE 4: The Loss modules (Pa) of C0-C1-C2 and C3-C4-C5 samples.

The storage modules of samples are given in Figure 5. Addition of talc particles into copolymer has shown considerable enhancement of storage modules at the lower frequencies. Talc filled homopolymer samples have shown fewer reduction of storage modules at the same frequency range. In Figure 6, the module values of copolymer samples are given. The effect of talc has delayed the cross over point of module values.

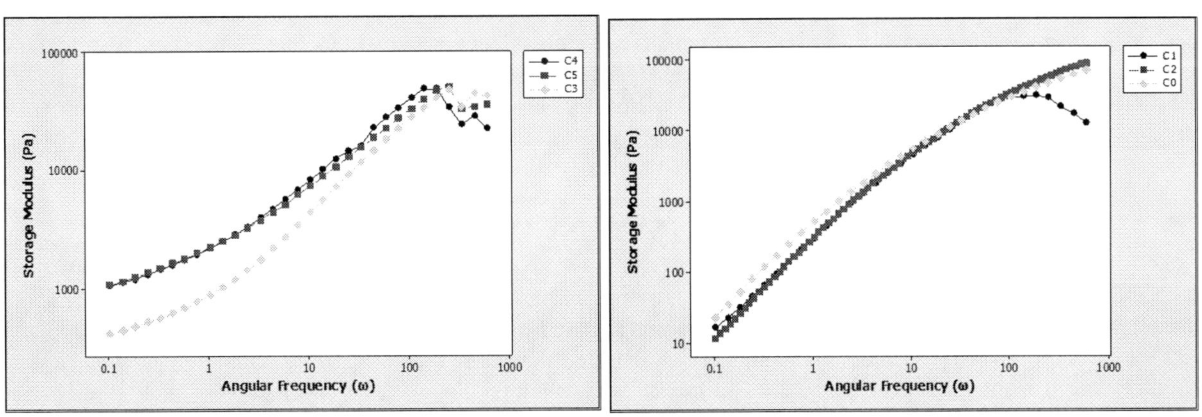

FIGURE 5: The Storage Modules of (Pa) of CO-C1-C2 and C3-C4-C5 samples.

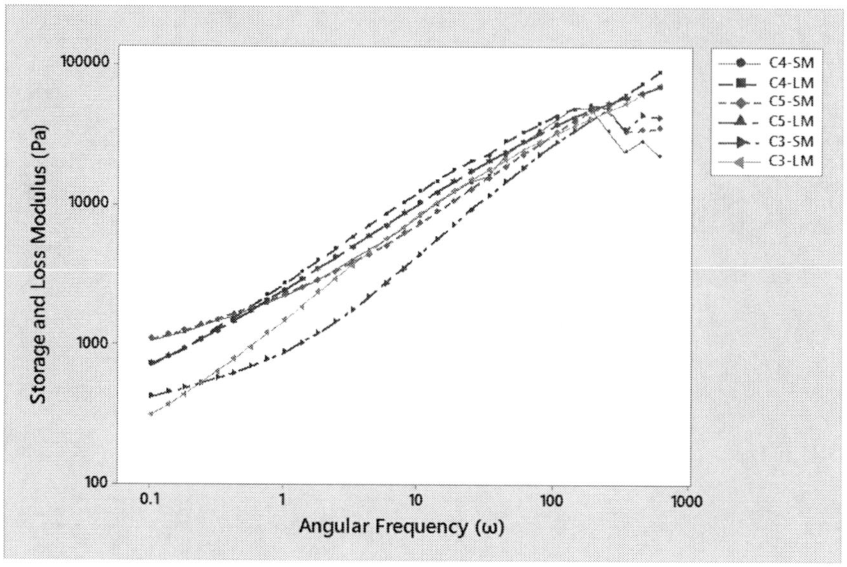

FIGURE 6: The storage and loss Modules of neat and talc filled copolymer compounds

Injection molding experiment results

The main injection molding parameters; barrel temperature, injection pressure, injection velocity and time, ejection time and over all cycle time were obtained for both talc filled copolymer compounds, C4 and C5. Due to the flowability of lower particle size filled compounds, the range of the barrel temperature was set to 220-190 °C instead of 230-200 °C.

TABLE 3. The injection molding parameters of wt 20% talc filled copolymer compounds

Sample	Barrel Temperature, °C	Injection Pressure, bar	Injection velocity, mm/sn	Injection Time, sn	Ejection time, sn	Cycle Time, sn
C4	230-200	120	12-14	13,2	25,8	83
C5	220-190	120	12-18	12,3	18,7	79

The injection pressure remained the same, 120 bar. In the Figure 3, the viscosity of compound termed C4, is higher than compound termed C5. As an outcome of the increasing viscosity, the injection speed was increased contrary to injection time. Therefore the overall cycle time was reduced from 83 s to 79 s.

The ejection time is defined as the total period of tool opening time and duration of robot arm movement for removal of the part from the mold cavity. The ejection time of C4 and C5 were 25.8 s and 18.7 s consecutively. The slower crystallization rate due to the particle size has resulted with the higher ejection time.

CONCLUSION

The effect of particle size and aspect ratio of talc particles on rheological, crystallization behavior and injection moldability of PP homopolymer and copolymer was investigated in this study. It is believed that the talc particle size affects the viscosity of the polymer. The results are shown that the improved flowability with decreased particle size has the significant effect on the process parameters such as injection rate and time. The faster crystallization rate of compounds due to lower particle size and high aspect ratio of talc is the responsible for the observed in dramatic improvement in ejection time. According to rheological measurements, fine talc as a nucleating agent provided lower viscosity to copolymer PP specimens. Fine talc has been proven to increase the flowability of the copolymer compared to standard talc filled copolymer. However it is not seen the homopolymer PP specimens. This may be due to the weak interaction between the polymer matrix and the additives. The amount of deviation observed in the standard talc filled homopolymer may be due to agglomeration.

REFERENCES

1. McNally T., McShane P., Nally, G.M., Murphy, W.R., Cook, M., Miller, A. (2002). Rheology, phase morphology, mechanical, impact and thermal properties of polypropylene/metallocene catalyzed ethylene 1-octene copolymer blends. *Polymer*. 43, 3785-3793.
2. Denac, M., Musil, V., Smit, I., Franjo, R. (2002). Effects of talc and gamma irradiation on mechanical properties and morphology of isotactic polypropylene/talc composites. *Polymer Degradation and Stability*. 83, 263-270.
3. Zhou, Y., Rangari, V., Mahfuz, H., Jeelani, S., Mallick, P.K. (2005). Experimental study on thermal and mechanical behaviour of polypropylene, talc/polypropylene and polypropylene/clay nanocomposites. *Material Science and Engineering*. 402, 109-117.
4. Yousfi, M., Livi, S., Dumas, A., Le Roux, C., Crepin-Leblond, J., Greenhill-Hooper, M., Duchet-Rumeau, J. (2013). Use of synthetic talc as reinforcing nanofillers for polypropylene and polyamide systems: Thermal and mechanical properties. *Journal of Colloid and Interface Science*. 403, 29-42.
5. Maier, C., Calafut, T. (2013). *Polypropylene*. Retrieved from http://0-www.sciencedirect.com.divit.library.itu.edu.tr/science/article/pii/B978188420758750031X
6. Hadal, R.S., Dasari, A., Rohrmann, J., Misra, R.D.K. (2004). Effect of wollastonite and talc on the micromechanisms of tensile deformation in polypropylene composites. *Material Science and Engineering*. 372, 296-315.
7. Eftekhari, M., Fatemi, A. (2016). Creep behavior and modeling of neat, talc-filled, and short glass fiber reinforced thermoplastics. *Composites*. Part B 97, 68-83.
8. Gafur, A., Nasrin, R., Mina, F., Bhuiyan, A., Tamba, Y., Asano, T. (2010). Structures and properties of the compression-molded isotactic-polypropylene/talc composites: Effect of cooling and rolling. *Polymer Degradation and Stability*. 95, 1818-1825.
9. Rueda, M.M., Auscher, M.C., Fulchiron, R., Perie, T., Martin, G., Sonntag, P., Cassagnau, P. (2017). Rheology and applications of highly filled polymers: A review of current understanding. *Progress in Polymer Science*. 66, 22-53.
10. Karamipour, S., Ebadi-Dehaghani, H., Ashhouri, D., Mousavian, S. (2011). Effect of nano-$CaCO_3$ on rheological and dynamic mechanical properties of polypropylene: Experiments and models. *Polymer Testing*. 30, 110-117.
11. Lin, C., Wang, W., Kuo, C., Wu, C. (2014). Experimental and theoretical study of melt viscosity in injection process. *International Journal of Chemical, Molecular; Nuclear, Materials and Metallurgical Engineering*. 8(7), 687-691.
12. Diez-Gutierrez, S., Rodriguez-Perez, M.A., De Saja, J.A., Velasco, J.I. (1999). Dynamic mechanical analysis of injection-molded discs of polypropylene and untreated and silane-treated talc-filled polypropylene composites. *Polymer*. 40, 5345–5353.

Optimization for Starve Fed/Flood Fed Single Screw Extrusion

Nastaj Andrzej (1)*, **_Wilczyński Krzysztof_** (1)

(1) Warsaw University of Technology - Warsaw - Poland

An integrated computer model has been developed to optimize starve fed/flood fed single screw extrusion process. Recently developed starve fed/flood fed single screw extrusion model GSEM has been used to predict extrusion parameters, and evolution strategies (Genetic Algorithms) have been applied for optimization. Extensive experimental studies have been performed to verify optimization procedures.

Structure development in molten conductive polymer composites studied by electrical-rheological measurements

Stary Zdenek (1)*

(1) Institute of Macromolecular Chemistry - Czechia - Czech

Simultaneous electrical and rheological measurements (ERM) on polymer melts were introduced in 2004 to study properties of composites containing carbon nanotubes [1]. Since then these measurements have become a useful and sensitive tool to follow morphological changes in conductive polymer composites induced by a well-defined mechanical deformation. From an evolution of the electrical conductivity during a deformation one can draw conclusions regarding destruction and/or agglomeration of particle structures dispersed in the melt. Simultaneously, the effect of these structural changes on rheological properties of the composite is evaluated. In this contribution the experiments performed on composites with carbon nanotubes are briefly summarized [2]. Electrical and rheological behaviour of composites with carbon black under oscillatory and creep deformation is presented. Novel electrical-rheological measurements on polymer melts in elongation are discussed as well. Finally, opportunities of ERM in studies concerning with morphology development in filled polymer blends are discussed. [1] Kharchenko SB, Douglas JF, Obrzut J, Grulke EA, Migler KB. Nat. Mater. 2004;3(8):564-568. [2] Alig I, Pötschke P, Lellinger D, Skipa T, Pegel S, Kasaliwal GR, Villmow T. Polymer 2012; 53: 4.

LCB formation in PLA by reactive extrusion

***Kaschta Joachim* [1]*, *Lanzl Lydia* [1], *Lemppenau Simon* [1], *Schubert Dirk W* [1]**

[1] FAU Erlangen-Nuremberg - Bavaria - Germany

Poly(lactic acid), frequently referred to as PLA is a polymer which gains increasing interest in both science and industry as it is produced from renewable biological resources. Although is biodegradable and thus compostable it is lacking from some short-comings, e.g. processing operations are limited by the polymers' low melt strength, while its brittle fracture behavior hinders application. The first point is related to the molecules' linear architecture. As well known for PET chain extenders can be used to tailor molar mass distribution (MMD) as well as the degree of long-chain branching and so to increase melt strength. PLA was reactively processed together with a commercial chain extender at various concentrations in a kneader and in a twin screw extruder. All materials were dried under vacuum prior to any processing and/or characterization to a water content below 100 ppm to avoid (hydrolytic)degradation. MMD and(possible)long-chain branching were determined by triple detection SEC in HFIP. All samples for rheological tests were prepared by compression molding. Rheological measurements in shear end elongation were performed in nitrogen atmosphere by means of a rotational rheometer equipped with an elongational test fixture. All rheological test were conducted on time scales assuring thermal stability. Triple detector GPC proves an increase in molar mass, a broadening of the molas mass distribution, as well the formation of LCB. All findings have been confirmed by rheological measurements showing an increase in zero shear viscosity together with improved shear thinning behaviour. LCB formation was further confirmed by a pronounced strain-hardening behaviour in elongational flow. The rheological behaviour suggests a branch-on-branch archictecture of the LCB-PLA. Extruded cast film shows less neck-in with the addition of chain extender and a more uniform film thickness as expected due to strain hardening.

ANALYSIS OF NONLINEAR VISCOELASTIC PROPERTIES OF POLYMERIC MATERIALS AT THEIR LARGE PERIODIC DEFORMATION

Pyshnograi Grigory **(1)*, *Cherpakova Nadezdha*** **(2)**, ***Kondratieva Olga*** **(2)**

(1) Altai State Pedagogical University - Altai Region - Russia, (2) Altai State Technical University - Altai Region - Russia

Analysis of the viscoelastic properties of various materials in the application of large periodic deformations among the most important and most common methods for the characteristics of these substances. Based on multimode Vinogradov Pokrovsky rheological model, were carried out numerical calculations, imposing periodic deformations with large amplitude on the flow of polymeric liquids. In the study of material behavior under large deformations, nonlinearity of the viscoelastic properties was obtained, which can be interpreted by the construction of Lissajous figures. When considering the model was able to confirm the increase in sample response nonlinearity with increasing oscillation amplitude. A convenient method for the visual presentation of the results is the construction of Lissajous figures. For a linear viscoelastic behavior, these figures are ellipses, and transformed into various shapes to form the nonlinear response of the sample case. The work was built Lissajous figures with varying frequency and amplitude of the oscillation based on multimode rheological model Vinogradova-Pokrovskiy and a comparative analysis of the results obtained with previously known experimental and theoretical data, which made it possible to conclude that the model allows us to accurately describe the behavior of polymeric materials upon application periodic large deformations.

Preparation and Thermal Processing of Functional Poly(vinyl alcohol) Based Micro/nanocomposites

***Wang Qi* (1)*, *Li Li* (1), *Chen Ning* (1)**

(1) State Key Lab of Polymer Materials Engineering, Polymer Research Institute, Sichuan University - Sichuan - China

Poly (vinyl alcohol) (PVA), a multi-hydroxyl polymer with high polarity, shows excellent comprehensive properties and has been developed very fast in recent years. However, the narrow thermal processing window between the melting and decomposition temperature results in the difficulty in thermal processing of PVA, which seriously hinders the further preparation of PVA-based micro/nanocomposites with high performance. In this paper, the novel highly filled PVA based composites with high performances were prepared by combing the molecular complexation and the solid state shear milling (S3M) technologies developed in our research group, and their micro-injection molding and 3D printing were studied. Through molecular complexation formed among amido contained compounds, water and PVA, wide thermal processing window (≥ 80oC) of PVA was obtained, and PVA based biomedical micro/nanocomposites, e.g., PVA/hydroxyapatite (HA), PVA/gelatine/HA and PVA/β-tricalcium phosphate(β-TCP), etc., were prepared in a quite large scale via thermal processing by taking advantages of the pretty good compatibility of PVA and organic or inorganic fillers. Adopting S3M technology, the excellent dispersion of bio-micro/nanoparticles, even at high content (40wt%), in PVA matrix was achieved. Accordingly, the micro-injection molding and Fused Deposition Modeling (FDM) of PVA based composites were successfully realized, and the micro-injection samples with higher crystallinity and higher mechanical properties than these obtained by conventional injection molding, and 3D printing parts with complicated shapes as well as good cell compatibility and blood compatibility, were obtained. These functional PVA based Micro/nanocomposites have potential applications in drug delivery, articular cartilage, biological scaffold, etc. This work was supported by the National Natural Science Foundation of China (51433006, 51010004).

Orthopaedic compression screws formed from body-temperature reverting, shape-memory polymers

Thomson Brian (1)*, Thompson Glen (1), Vgenopoulos Dimitrios (1), Nair Karthik (1), Howell Ken (1), Martyn Mike (1), Coates Phil (1)

(1) Bradford University - Yorkshire - UnitedKingdom

Orthopaedic compression screws promote fracture-repair by drawing-together and stabilising adjacent bone fragments. Shape memory polymers (SMP) are smart materials that switch from one macroscopic shape to another following the application of external 'triggering' stimuli (e.g. by shrinking in length when heated above Tg). SMP compression screws that shorten in length following surgical implantation (i.e. display body temperature reversion) could potentially pull bone fragments together and thereby facilitate fracture healing. The biocompatible, bioresorbable polymer polylactic acid (Purasorb 7038; PLDLA) shows shape memory properties but the Tg of unmodified PLDLA is too high for body temperature reversion. Accordingly we have used the plasticiser Tributyl O-acetylcitrate (3.5% v/w; TAC) to reduce PLDLA's Tg and produce a plasticised shape memory material (TAC-PLDLA) that is dimensionally stable at -20°C but which shows shape memory reversion when incubated under physiological conditions (i.e. saline at 37°C). To produce this material, PLDLA and TAC were blended by melt extrusion, pelletized and extruded into 5 mm rods. Shape memory properties were imparted by drawing the rods at 85C, followed by rapid cooling to 4C (draw ratio 4.7, final diameter 2.3 mm). TAC is considered 'Not hazardous' according to regulation (EC) No. 1272/2008 and preliminary cell biological studies have shown that the TAC-PLDLA supports viable cell adhesion. This suggests TAC-PLDLA is likely to be biocompatible. A prototype TAC-PLDLA orthopaedic compression screw was designed based upon knowledge of existing metal compression screws and clinical requirements. Aspects of this design (e.g. thread profile) were optimised in silico. Shape memory polymers can be moulded by selective heating and partial reversion within a mould. Subsections of a candidate mould for the proposed screw were therefore designed in silico, and 3D printed in metal. Ongoing work aims to find the optimal time and temperature requirements for reversion moulding an orthopaedic compression screw from a TAC-PLDLA SMP rod. Future work will aim to evaluate the mechanical properties of the resulting device.

Nanostructured films of natural polymers and graphene derivatives

Paiva Maria Conceição (1)*, Moura Duarte (2), Silva Cláudia (2), Silva Magda (2), Vale Catarina (2), Caridade Sofia (2), Cunha Eunice (1), Sousa Maria (2), Rocha Helena (1), Mano João (2), Alves Natália (2)

(1) University of Minho - Campus of Azurém - Portugal, (2) University of Minho - 3B's - Portugal

Biomaterials play a critical role in implantable devices. They should present suitable structural/mechanical performance and biocompatibility, so that it is not recognized as foreign body by the immune system. Natural polymers such as chitosan (CHI) and alginate (ALG) fulfill the biocompatibility requirements, however present weak mechanical properties. The mechanical reinforcement effect can be enhanced with processing techniques that buildup nanostructure, such as layer-by-layer assembly (LbL). This was achieved with aqueous solutions of polyelectrolytes CHI and ALG. In this work, freestanding multilayered films containing CHI, ALG, and functionalized few-layer graphene (FLG) were produced through LbL assembly, envisaging the preparation of multifunctional films with mechanical and electrical performance. For that purpose, FLG obtained by different approaches was used, namely by production of graphene oxide (GO), stabilizer-assisted liquid phase exfoliation in aqueous solution of a pyrene derivative, and covalent functionalization of exfoliated graphite. Graphene nanoribbons (GNR) were produced by two covalent functionalization approaches, and their reinforcement effect was compared to that of FLG. Freestanding multilayered films were prepared by LbL, with compositions (CHI/ALG/CHI/ALG)100, (CHI/ALG/CHI/FLG)100, and (CHI/ALG/CHI/GNR)100. The films were characterized by thermogravimetric analysis, Raman spectroscopy, scanning electron microscopy, atomic force microscopy, water contact angle measurements, water uptake, degradation tests, uniaxial tensile tests, dynamical mechanical analysis, electrical conductivity measurements, and biological assays using L929 mouse fibroblasts. The high reinforcing effect of graphene, the changes induced on the electrical resistivity and the biocompatibility of the films were demonstrated.

Functionalized nanomaterials based polymer nanocomposites: a way to new generation flame retardant materials

Wang De-Yi ***(1)****

(1) IMDEA Materials Institute - Madrid - Spain

Nanomaterials used as flame-retardant synergists to conventional flame retardants have been regarded as an efficient way to improve the flame-retardant effectiveness. In fact, such synergistic effects work to a lot of polymer systems.[1] Usually, in these flame-retardant polymer systems, nanomaterials need only a low loading, while it can greatly improve the flame retardant properties of polymers via combining conventional flame retardants. Recently, some novel functional nanomaterials emerged and were used in some polymer as independently flame retardant, finding that it is possible to achieve excellent flame retardancy by using only functional nanomaterials. In the group of the author, plenty of investigation on the functional nanomaterials have been carried out in last 3 years that have the potentials to be as independent flame retardants in some polymer system.[2-6] Functionalization and characterization of novel nanomaterials (layered double hydroxide, nanosilica and mesoporous materials), comprehensive assessment on the properties of polymer nanocomposites are presented. Recently, series functional nanomaterials have been developed and used as efficient flame retardant for polymers, such as PVC, Epoxy, etc. The recent advances in nano flame retardants will be discussed in this presentation. Nano layered double hydroxide, nano silica and porous materials are investigated in different polymers. The functionalization of nanomaterials, impact on the properties of polymers, fire behaviours, and possible flame-retardant mechanism will be discussed in details. i) Functionalized hierarchically porous silica The synthesis of hierarchical porous silica adopted a promoted template method. The grafting of tin on hierarchical porous silica utilized in-situ synthesis manner. After adding TEOS 5 min, a certain amount of tin (IV) chloride was added into the solution drop wise. Scheme 1 presents the schematic program of the proposed synthesis procedure of TOS. ii) Functionalized layered double hydroxides (LDH) Following one-step method [7], functionalized LDH was synthesized and presented in Scheme.2. In this novel LDH, several functional modifiers have been developed and used to fabricate specialized LDH. The structure and morphology f novel LDH are characterized by XRD, EA and SEM. The developed functionalized LDHs offered excellent flame retardancy to epoxy, showing that with only 7 wt% of fCD-DBS-Ph-LDH, the resultant epoxy nanocomposites passed UL-94 V0 rating and significant reduction in the pHRR (-72 %) was achieved in the cone calorimeter test compared to the pure epoxy. Importantly, fCD-SDBS-Ph-LDH/EP nanocomposite exhibited excellent anti-UV properties in terms of high preservation on impact, flexural and micro-mechanical properties after 100, 200, 300 400h UV irradiation, respectively, owing to the presence of chalcone specie that served as UV absorber.

NanoMine: Development of Material Data Resource and Analysis Tools for Polymer Nanocomposites

Brinson Catherine [(1)]*, Zhao He [(1)], Wang Yixing [(1)], Li Xiaolin [(1)], Zhang Yichi [(1)], Prasad Aditya [(2)], Chen Wei [(1)], Schadler Linda [(2)]

[(1)] Northwestern University - IL - USA, [(2)] Rensselaer Polytechnic Institute - NY - USA

Research efforts in the past decades on nano-reinforced polymeric materials provide numerous examples of property enhancement across multiple physical property domains including thermo- mechanical, dielectric, optical, and other properties. The project develops an open access data resource (NanoMine) for the polymer nanocomposites community, facilitating faster nanocomposite design and insertion into advanced applications. The capability will support three major types of users: researchers interested in comparing their results in detail to other work; researchers exploring, discovering, and quantifying fundamental scientific principles that govern processing-structure-property (p-s- p) relationships; and materials designers interested in using nanocomposites for advanced applications. NanoMine is created as an open access, easy to use, persistent, flexible data resource for the polymer nanocomposites community that is scalable and will enable improved understanding of p-s- p relationships and design of commercially relevant polymer nanocomposites. We combine a curated database, data-driven analysis tools, and physics-based modeling for polymer nanocomposite analysis and design. To standardize terminology and structure of parameters across p-s-p domains, we created a data template using hierarchical XML schema by curating more than 30 recent papers and experimental data from in house and collaborators. The current template contains parameters that account for metadata of data source, characteristics of polymer and filler particle phases along with surface chemistry and constituent pre-processing, experimental processing and characterization procedures, microstructure dispersion of the nano-phase, as well as reported and measured material properties. This template serves as a comprehensive collection of parameters to account for a nanocomposite sample and a robust ontology is being developed based on the schema and initial data.

Effect of Addition of Filler on the Tribological Properties of CF/PEEK Composites

Sungha Kim[1, b)], Kenichi Sugiyama[1], Kazuhiko Sugiyama[1], Kazuma Mikawa[2], Keisuke Nakamura[2], and Yosuke Nishitani[2, a)]

[1]*Ebara Corporation*
4-2-1 Honfujisawa, Fujisawa, Kanagawa, Japan
[2]*Department of Mechanical Engineering, Faculty of Engineering, Kogakuin University, 2665-1 Nakano, Hachioji, Tokyo, Japan*

[a)]Corresponding author: at13152@ns.kogakuin.ac.jp
[b)]kim.sungha@ebara.com

Abstract. To develop the tribomaterials based on polymeric materials with high performance for sliding bearings in large scale pumps, we investigated the effect of addition of filler on the tribological properties of carbon fiber reinforced polyether ether ketone (CF/PEEK) composites. Two types of filler such as (a) the combination of polytetrafluoroethylene (PTFE) and graphite (Gr), and (b) talc (Tc) were used for filler in this study. These CF/PEEK composites filled with various filler were melt-mixed by a twin screw extruder and compression mold. Tribological properties under two types of testing using a ring-on-plate type sliding wear tester under dry condition were evaluated for these composites. Two types of tribological testing were sliding wear testing at constant normal load and sliding speed, and limiting *pv* value testing by step load method. Frictional coefficient, specific wear rate and limiting *pv* value improved with the addition of filler. In particular, these tribological properties of CF/PEEK composites filled with Tc (CF/PEEK/Tc composites) were superior to those of CF/PEEK ones filled with PTFE and Gr (CF/PEEK/ (PTFE+Gr) ones). These may be attributed to be due to the change in the mode of wear mechanism, which is the change of form of the transfer films on the metal counter-face, the shape of wear debris and the morphology of worn surface, by the addition of filler.

INTRODUCTION

When heavy rain attacks cities, large scale pumps are used to drain out rain water from the low regions to prevent flood. During the operation, inside of the pumps which contain various mechanical parts such as shaft, journal bearings, and sleeves is exposed to air, which indicates that it goes through atmospheric sliding between journal bearings and sleeves. Traditional materials for these bearings were rubber which performs excellently with water lubrication and also has cost merits. The rubber bearings, however, were not able to withstand in atmospheric sliding conditions so that protection pipe was needed all through the main shaft of the pumps including bearing system. Moreover, external water circulating system was installed to maintain the water lubrication film between the rubber bearings and sleeves. However, this protection pipe system has some disadvantages. First, due to additional parts such as pipes and water circulation system, more money, manufacturing time, and maintenance are needed. Secondly, the reliability issues due to complex structure compared to simple structures without protection pipe. From these reasons, ceramic and plastic based composite bearings were introduced for large scale rain water pumps which can endure short and moderate atmospheric sliding conditions. For higher capacity pumps which has larger diameter of shaft and rotating speed, however, temperature of the journal bearings and sleeves during the atmospheric sliding might reaches to limitation point which can causes the problem such as seizure or catastrophic failure in worst case due to thermal expansion or severe wear of bearings.[1]

Composite material which made of super engineering plastics such as PEEK are gathering attentions these days for substitutions of large scale pump journal bearing material. Composite materials made of plastic shows self lubrication which can slide with low fictions without lubrication under atmospheric sliding conditions. One other characteristics of plastic composite material is that adding certain kind of additives can control or enhance the tribological properties such as frictional coefficient, wearing, and anti-slurry wearing in slurry-water. [1-5]

In this study, limiting *pv* value testing by step load method and sliding wear testing at constant normal load and sliding speed were carried out to investigate the tribological characteristic of CF/PEEK composites with talc and PTFE/graphite fillers and analyzed by Scanning electron microscopy and energy-dispersive X-ray spectroscopy (SEM/EDX), Raman spectroscopy, and X-ray photoelectron spectroscopy (XPS) sputtering analysis to find its mechanism.

MATERIAL COMPOSITION AND PROPERTIES

Specimens were made by three steps. First, PEEK powder and fillers are mixed and pelletized by twin screw extruder. After the pelletization, compounds were compress molded by using simple shaped mold which has rough surface and crude details. The temperature and pressure during the compression molding process were 398℃ and 16.2MPa. Lastly, precise features were accomplishes by machining process and final surface polishing.

Two kinds of PEEK compounds which have different molecular grade were used as a base material. Tungsten carbide was used as counter-part for sliding tests. The content of the fillers and material properties are shown in table 1 and table 2.

TABLE 1. Composition of each specimen

mass%	PEEK	Carbon fiber	Talc	PTFE	Graphite
TL	60	23	17	-	-
TH	60	23	17	-	-
PGL	70	10	-	10	10
PGH	70	10	-	10	10

TABLE 2. Material property of each specimen

Material property	Unit	TL	TH	PGL	PGH
Tensile strength	MPa	66±1.8	70.4±1.5	50±3.8	54.2±1.6
Modulus of elongation	GPa	1.1±0.30	1.6±0.35	4.7±0.50	3.2±0.15
Stretch at break	%	2.9±0.25	3.4±0.20	1.7±0.25	3.6±0.25
Complex viscosity	Pa・s	835	1858	1111	1994

EXPERIMENTAL SETUP

Ring-on-plate type sliding wear tester was used for limiting *pv* value testing and sliding wear test at constant normal load and sliding speed. Experimental setups and feature of the counter-part are shown in Fig. 1.

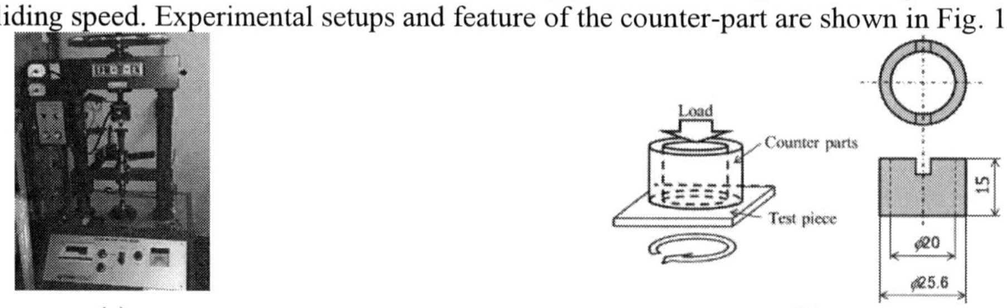

(a) (b)

FIGURE 1. (a)Ring on plate type sliding wear tester (Orientech Co., Japan EFM-Ⅲ-EN) , (b)counter-part

For limiting *pv* value test, constant speed of 2m/s was maintained while increasing the normal force by 50N for every 5minutes from stating normal force of 50N(0.25MPa), until the temperature of the counter-part near sliding surface increases sharply.
For sliding wear testing, constant normal force of 100N (0.5MPa) and constant sliding speed of 2m/s were applied during the test. Total sliding distance was 3000m.

RESULTS

Limiting *pv* Value Test

pv limits of each test pieces are shown in Fig. 2(a). It can be referred that talc fillers area more effective than PTFE/graphite fillers in aspect of the *pv* limit. Higher molecular grade of PEEK is also results in higher *pv* limits, especially with talc fillers.

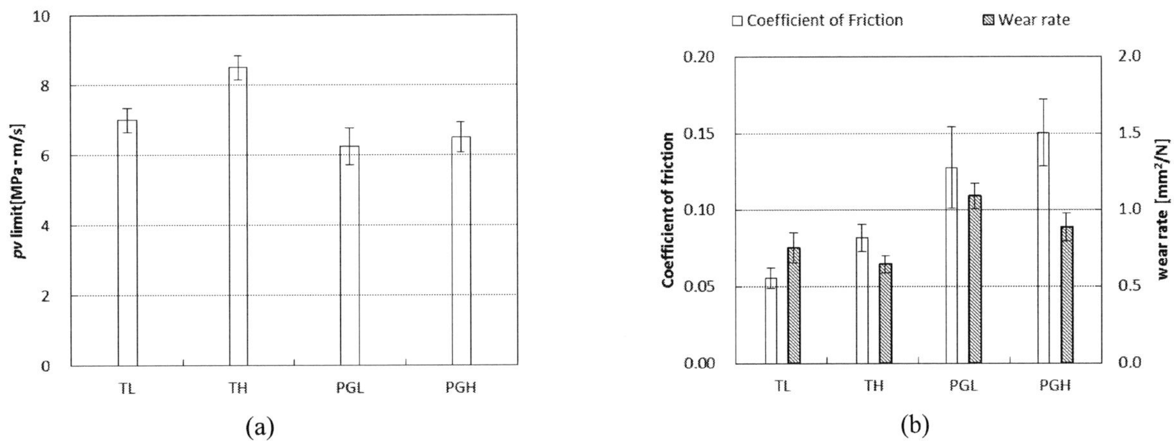

FIGURE 2. (a) *pv* limit values of each test pieces, (b) Coefficient of friction (average of last 1500m) and wear rate of constant normal force and speed test (TL: talc filler with low molecular grade PEEK, TH: talc filler with high molecular grade PEEK, PGL: PTFE and graphite filler with low molecular grade PEEK, PGH: PTFE and graphite filler with high molecular grade PEEK)

Sliding Wear Test at Constant Normal Load and Sliding Speed

From Sliding wear test at constant normal load and sliding speed, coefficient of friction is rather high and unstable in the beginning of the tests. However, it gradually decreases and become stable after approximately 1500m of sliding distance. These phenomena are shown in every other case with both specimens. Average of the coefficient of friction for last 1500m (number of samples=6) is shown in Fig. 2(b). Wear rates were also shown in Fig. 2(b) which measured by comparing the weight of the test pieces before and after the tests. From this result, it can be found that talc additives are more effective to reduce the coefficient of friction than PTFE/graphite at atmospheric sliding condition of constant normal load and sliding speed. Regarding molecular weight, low molecular weight cases shows lower coefficient of friction for both additives, contractively, higher molecular weight shows smaller wear rates.

DISCUSSIONS

Scanning electron microscope and energy-dispersive X-ray spectroscopy was conducted to investigate the sliding surfaces of wear test at constant normal load and sliding speed specimens. To analyze the transfer film on the counter-part sliding surfaces, optical microscope observation, Raman spectroscopy, and sputtering X-ray photoelectron spectroscopy were conducted.

Secondary Electron (SE) and Characteristic X-ray images of the Specimen Sliding Surface

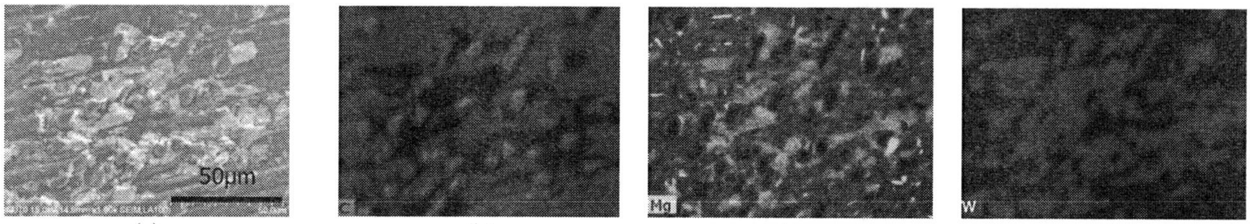

(a) (b) (c) (d)

FIGURE 3. (a) SE image (x1000) and corresponding characteristic X-ray image mapping of (b) Carbon, (c) Magnesium, and (d) Tungsten element of the TL (talc filler with low molecular grade PEEK) sliding surface after constant normal load and sliding speed wear test

It can be referred that clear cross sections of carbon fibers can be observe from SE and characteristic X-ray images(Fig.3 (b)) which indicates that wear occurred perpendicular to carbon fiber direction at the analyzed area. From Fig. 3(c) and (d), it can be seen that particles originated from talc (Magnesium) and counter-part (Tungsten) are accumulated around the carbon fibers and forms the sliding surface.

Optical Microscope Observations of the Counter-part Sliding Surfaces

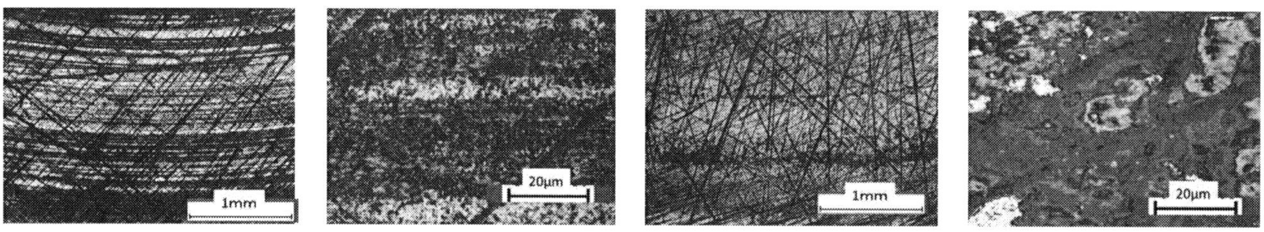

(a) (b) (c) (d)

FIGURE 4. Optical observation of counter-part sliding surface after wear friction test of TL: talc filler with low molecular grade PEEK ((a) x50, (b) x1000), and PGL: PTFE and graphite filler with low molecular grade PEEK((c) x50, (d) x1000)

In order to investigate the transfer film on the counter-part sliding surfaces, optical microscope observation was conducted. As we can confirm from Fig. 4, it can be observed that on the each surface of counter-part sliding surfaces, there are many attached substances of transfer films in different colors. On surface of TL counter-part, broad area of blue and brown substances are shown (Fig. 4(a), (b)). On the other hand, it can be seen that some of grey spots are shown on the surface of PGL counter-part surface (Fig. 4(c), (d)).

Raman Spectroscopy of Counter-part Sliding Surfaces

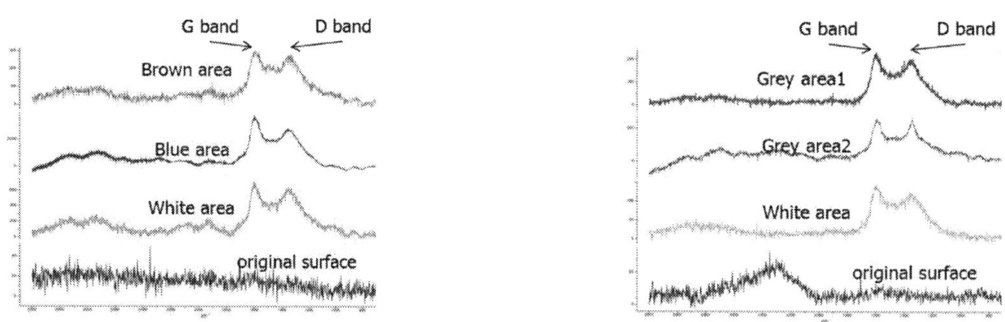

(a) (b)

FIGURE 5. Raman spectroscopy of counter-part sliding surface after wear friction test of TL: talc filler with low molecular grade PEEK (a), and PGL: PTFE and graphite filler with low molecular grade PEEK(b)

Raman spectroscopy was conducted to investigate the substances of transfer films on the surface of counter-part sliding surfaces. Color indications of the spectrums in the graphs are from each colored areas of optical observation results. Comparing with the original surface which shown in the bottom of Fig. 5(a), (b), G and D band appeared in both TL and PGL counter-part surface after the test in every color areas. It can be regarded that the transfer films are consist of amorphous carbon substances which originated from carbon based additives of specimens.

Sputtering X-ray Photoelectron Spectroscopy of Counter-part Sliding Surfaces

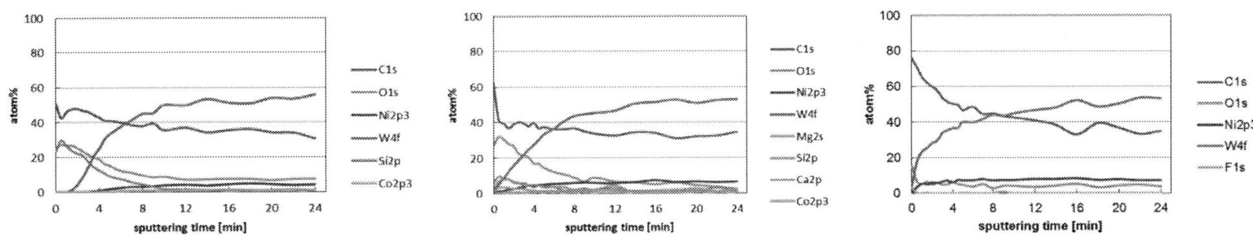

(a) (b) (c)

FIGURE 6. X-ray photoelectron spectroscopy of counter-part along depth direction: (a) original surface (before the test), (b) vs TL, and (c) vs PGL

Compared with original surface of counter-part (Fig.6 (a)), Magnesium from talc was detected on the surface of TL counter-part. On the other hand, PGL counter-part results shows high ratio of carbon which can be regarded as transferred PEEK, graphite, or carbon fiber from PGL specimen (Fig.6 (b), (c)).

CONCLUSIONS

Tribological characteristics in atmospheric condition of CF/PEEK composite material with talc and PTFE/graphite fillers, and different molecular grades were investigated in this study. In *pv* limiting test, talc filler added case can endure higher *pv* than PTFE/graphite filler added case. In case of constant normal force and speed wear test, talc filler was more effective in aspect of reducing friction coefficient and wear rate while higher molecular weight reduces wear rate in both cases of fillers. Surface observation of specimens was carried out by SEM/EDX to investigate the topology of the sliding surface after test. Optical observation, Raman spectroscopy, and X-ray photoelectron spectroscopy were conducted to confirm that transfer film is formed more widely in the counter-part with talc filler test piece which might effective for reducing the coefficient of friction and wear rate in this study cases.

REFERENCES

1. K.Ogawa et al., "Vertical pump bearing for dry start," Ebara engineering review **131**, 30-35(1985)
2. Z.P. Lu and K. Friedrich, Wear **181-183**, 624–631(1995)
3. Z. Zhang, C. Breidt, L. Chang, and K. Friedrich, Tribology International **37**, 271–277(2004)
4. G. Zhanga, Z. Rasheva, and A.K. Schlarb, Wear **268**, 893-899(2010)
5. K.A. Lauxa and C.J. Schwartz, Wear **297**, 919–925 (2013).

Effect of Filler Type and Particle Size Distribution on Thermal Properties of Bimodal and Hybrid - BN/Boehmite-filled EP-Novolac Composites

Christin Pawelski[1,a)], Enpu Kang[1], Gökhan Bakis[1], Volker Altstädt [1,b)*]

[1]*Department of Polymer Engineering, University of Bayreuth, Universitätsstr. 30, D-95447 Bayreuth, Germany*

a) Corresponding Author: christin.pawelski@uni-bayreuth.de

b) altstaedt@uni-bayreuth.de

Abstract. In this study, the combination of *hexagonal* Boron Nitride (BN) and Boehmite are investigated towards their potential to enhance both the thermal and flame retardant properties in Epoxy Novolac matrices for electrical applications. The focus of this scientific work is to correlate the fillers' particle size (distribution), morphology, content, dispersion and orientation, which are investigated with microscopic methods (SEM, TEM), with the resulting thermal properties. Both filler types are studied systematically in monomodal and bimodal filler distribution and finally in a hybrid combination, both with variation in the relative compositions. The thermal properties of the resulting composites are determined via TGA (thermal decomposition), DMA (T_g), and heat flow meter for thermal conductivity. The effect on flame retardancy is examined with Cone Calorimetry. The results show that the thermal conductivity is strongly dependent on filler type, content and particle size (distribution) strongly affecting the resulting microstructure. With increasing filler content and the higher the filler's aspect ratio, the more pronounced is the enhancement of the thermal conductivity and flame retardancy. Furthermore, the variation in the relative composition of the bimodal filler distributions and hybrid combination demonstrates the potential to design and trigger the formation of the micro-structured network in a smart way improving the through-plane (z-direction) heat dissipation.

Keywords: Epoxy Novolac, Thermal Conductivity, Flame Retardancy, Boehmite, Boron Nitride

INTRODUCTION

Recent research trends are strongly driven by the demand for lightweight materials with high reliability during high and long-term operating temperatures and high switching speeds in Printed Circuit Boards (PCBs) or as encapsulation material i.e. in future E-mobility applications. Currently thermosetting polymers, especially epoxy resins of different monomer nature, are widely used because of their excellent chemical resistance, high T_g and their electrical insulation. [1] However, their low intrinsic thermal conductivity (TC) (< 0.2 W/mK) and lack of flame retardancy leads to material failure and fatigue deficiencies, propagating delamination and the reduction of the PCB lifetime. [2] With the incorporation of oxide or nitride-based ceramic fillers with high intrinsic thermal conductivity, the 3-D phonon transfer can be improved and controlled by the interfacial filler contact. In literature, the thermal conductivity of AlN or BN is extensively studied. The effect of BN particles with different sizes is extensively studied, i.e. by Yung *et. al.* [3] and Chung *et.al.* [1]. Both investigated the size effect of BN on thermal conductivity of an epoxy (Bisphenol-A diglycidylether, DGEBA) matrix and found that with an increasing particle size the interaction area with the polymer matrix decreases and therefore the thermal resistance is less. Yu *et.al* studied the effect of thermal and fire resistance on surface-functionalized epoxy (DGEBA) nanocomposites and found an enhancement of these properties.[4] Boehmite, AlO(OH), is a halogen-free flame retardant with a temperature stability up to 340°C and a thermal conductivity in the range of < 30 W/mK. [5] Compared to standard aluminium trihydroxide, ATH, Boehmite is a promising candidate as water is not released below 350°C during PCB soldering or assembling processes. Furthermore, it offers a low CTE, good copper adhesion and reduces the overall production costs of PCBs. Boehmite is studied mostly towards its flame retardant effect and thermal stability properties. Sun *et al.* found that orthorhombic nano-Boehmite increases the LOI by 10% when 10 wt% of filler is added in DGEBA [6]. Other studies only show the effect of Boehmite in thermoplastics on mechanics, crystallinity, degration [7,8] and on mechanics and rheology [9]. Camino *et.al.* showed in Beohmite/EVA composites, that for acceptable LOI and UL classification, at least 50 wt% of Boehmite is necessary [10]. Duwe *et.al.* studied the enhancement of the thermal conductivity. The addition of 15wt% of 14 nm-sized Boehmite the thermal conductvity was increased from 0.20 to 0.24. [5] In the scope of this study, a promising approach for the simultaneous enhancement of both the thermal transport and fire resistance is the combination of Boron Nitride (BN) with Boehmite (BT), tailoring a continuous heat flow path with sufficient thermal stability. Hence, the main goal of this scientific work is to understand and define the contributing effect of the formed micro-structure (filler network) with the focus on filler size, type and relative composition on the thermal properties of these composites. Therefore, Boehmite and BN are studied first in their monomodal formulations, then BN in a bimodal composition. With the partial substitution of bimodal BN with

Boehmite shows the prospect of designing a smart, cost-efficient and multi-functional hybrid composite, which is favourable for both the heat management and thermal stability of the Epoxy matrix.

EXPERIMENTAL SECTION

Materials and Preparation of Cured Resin Plates

A multifunctional Epoxy Novolac resin, D.E.N.™ 438 (purchased from DOW chemicals) with an epoxy equivalent of 176-181 g/eq was used as a high T_g reference matrix resin which was cured with Diethymethylbenzenediamine (XB3473), supplied by Huntsman Advanced Chemicals (Basel, Switzerland), in the stoichiometric ratio. As flame retardant, Boehmite, (platelet shape), supplied by Nabaltec AG with a density of 3.07 g/cm^3 and a D_{50} of 2.0 µm was used. Two *hexagonal* Boron Nitride (BN) types, purchased from Henze BNP with a D_{50} of 2.0 µm and 12 µm were used. All BN types are of platelet shape and have an average filler density between 2.27 and 2.35 g/cm^3.

TABLE 1. Overview of samples investigated in this work with their respective relative composition of Boron Nitride (BN) and/or Boehmite (BT) and labeling. All tested combinations have a constant total of 45 wt% of fillers.

Sample Labels	Relative Composition of Fillers
neat Epoxy Novolac	-
EP + BN 2µm	-
EP + BN 12µm	-
EP + BN 45µm	-
EP + BT 2µm	-
EP + BN 12µm + BN 2µm	1:1 / 3:1/ 1:3
EP + BN 12µm + BT 2µm	1:1 / 3:1/ 1:3

The individual formulation components were prepared in the mass fraction given in Table 1. The solid neat resin was pre-heated in a convection oven at 90°C and was then poured in a jar and kept in an oil bath under mechanical stirring. The fillers were dried and were then slowly added to the resin at 110°C and stirred at 100 rpm for 15 min. The resin-filler mixture was then degassed at room temperature for 5 min at approx. < 20 mbar. The hardener was added in the desired ratio at a temperature of 90 °C and stirred for 5 min at 180 rpm. The mixture was poured into pre-heated aluminium moulds and were step-cured in a convection oven according to the following cycle: 2h at 120°C, 2h at 160°C, 2h at 200°C and 2h at 220°C.

Characterization Methods

Thermal conductivity measurements were carried out in through-plane direction at 25°C using a heat flow meter FOX50 from TA Instruments (New Castle, United States) according to ASTM C518 in specimens with the diameter of 60 mm and a thickness of 3 mm.

The characterization of *degradation behaviour* and the char formation of the cured samples was studied with a thermo-gravimetric analyser (TGA/STDA851e) from Mettler Toledo (Columbus, United States). The samples (15 - 20 mg) were heated from 25°C to 600° under nitrogen atmosphere and from 600°C to 900°C in air at a heating rate of 10 K/min under 50 mL/min nitrogen purge in order to study the thermo-oxidative behaviour. The degradation temperatures were examined at 2wt% loss and 5wt% loss (Td_2 and Td_5). The char yield was evaluated at 800°C.

Dynamic mechanical thermal analysis (DMTA) was performed at a heating rate of 3 K/min from 30 °C- 280 °C in torsional mode at an elastic deformation of 0.1% and an applied frequency of 1 Hz using the ARES RDA III from Rheometrics Scientific (Germany). The specimens had a rectangular geometry in a size of 50 x 10 x 2 mm^3 according to the standard DIN EN ISO 6721-7. The glass temperature was evaluated by the maximum of the tan δ curve.

A *Cone Calorimeter*, iCONE™ FTT (UK) was used to evaluate the flame retardancy and other important fire properties, i.e. TTI (time to ignition), TSP (total smoke production) and HRR (heat release rate). The samples, with a geometry of 100 x 100 x 3mm^3, were measured in a horizontal set-up with an applied heat flux of 35 kW/m^2. The FIGRA value is calculated by dividing PHRR with T_{ig} which indicates the fire growth rate index.

Scanning electron microscopic (SEM) measurements were obtained on a Zeiss Leo 1530 instrument operating at 3 keV. Samples were cryo-fractured, polished and then platin-sputtered.

RESULTS AND DISCUSSION

Effect of filler type and size on thermal conductivity: Study of *monomodal filler distribution*

Formulations with monomodal filler distributions show the intrinsic potential, strongly dependent on each filler type, to enhance the thermal conductivity. At a constant filler geometry (platelet shape), the individual addition of Boehmite (BT) and Boron Nitride (BN) to Epoxy Novolac (0.20 W/mK) shows that the thermal conductivity of the composite was improved with increasing content (Figure 1). Formulations filled with 45wt% (**Table 1**) of 2 μm BN are almost twice as high than these filled with 2 μm Boehmite, 0,53 and 0,37 W/mK, respectively. (= *effect of filler type*). At a critical filler content of < 10 wt%, the thermal conductivity is not affected due to thermal barrier effect of the polymer matrix limiting the phonon transport. [11] Furthermore, increasing the aspect ratio of the platelet-shaped BN at constant 45wt%, from 2 μm, 12 μm to a D_{50} of 45 μm, the thermal conductivity is further enhanced from 0.53 to 0.66 W/mK (orange arrow). The phonon transport can be improved as a) fillers' surface is very smooth, b) the platelets forming a particle chain and c) the heat transport trough bigger platelets is better than through smaller-sized.

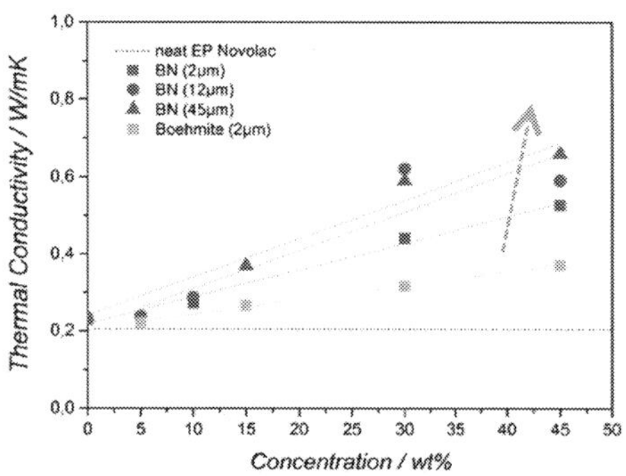

Figure 1: Thermal conductivity of monomodal Boehmite/BN filled composites with increasing content.

Table 1: Thermal conductivity results of formulations at 45wt% filler content.

Composition	λ / W/mK
neat EP Novolak	0.20
EP + BT 2μm	0.37
EP + BN 2μm	0.53
EP + BN 12μm	0.59
EP + BN 45μm	0.66

Combinational effects on Thermal Properties: Study of *binary* and *hybrid* formulations

The combination of two filler, BN + BN or BN+ Boehmite (BT), into the Epoxy Novolac matrix in a bimodal filler distribution shows the influence of both fillers' size and its relative composition on thermal properties. By incorporating 2 μm and 12 μm BN-platets, the heat transport through the composites and the packing can be can be tailored in a smart way. Substituting 2 μm BN-platelets with 2 μm Boehmite-platelets (hybrid combination) shows the potential of establishing simultaneously a low-cost and flame retardant, thermally conductive composite. Figure 2 shows the thermal conductivity results of (a) binary BN-filled and in (b) the hybrid combination and its resulting microstructure in the SEM micrographs. The thermal conductivity of both binary formulations is higher compared to their corresponding monomodal formulations. The thermal conductivity at constant 45wt% filler content was increased from 0.59 (unimodal 12 μm BN) to 0.66 W/mK for the bimodal BN combination. The variation of the relative composition from 1:1 to 1:3, an even higher value is reached (0.71 W/mK). The hybrid combination shows a similar trend. The relative combination of Boehmite (BT) (2 μm) with BN (12 μm) in the 45wt% formulations, reaches a thermal conductivity of 0.77 W/mK (1:3 ratio). Both SEM micrographs show homogeneous filler distribution and dispersion with random orientation of the platelets in regard to the processing. Comparing both micrographs, it can be assumed, that the heat path is established with the inter-filler contact between the 2 and 12 μm platelets when the maximum packing density is reached. The smaller 2 μm particles fill the gaps between the larger platelets.

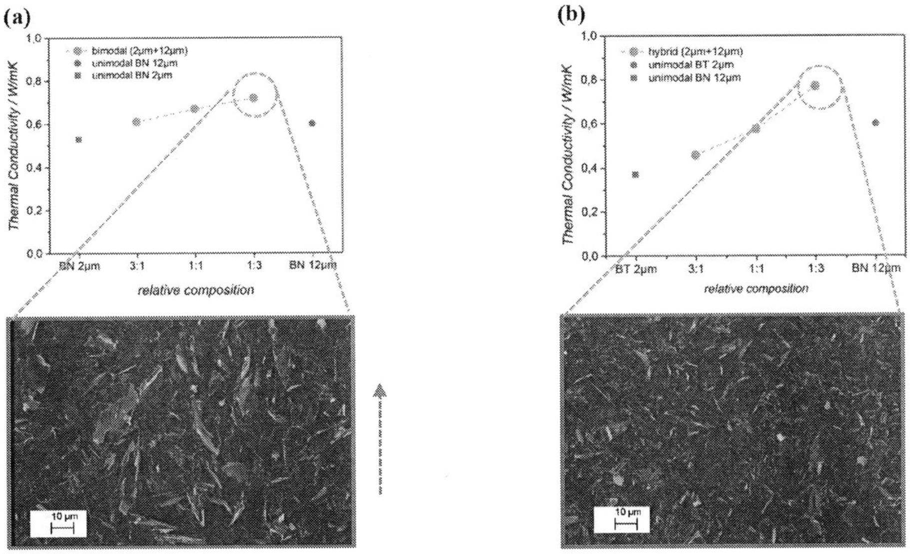

Figure 2: Thermal conductivity and its morphology of a) bimodal BN composite and b) of the hybrid Boehmite/BN composite. The arrows indicate the direction of heat flow during the thermal conductivity measurement.

Table 2: Thermal Conductivity data of bimodal and hybrid formulations at 45wt% filler content.

Composition	λ / W/mK
neat EP Novolak	0.20
EP + BN 2µm+ BN 12µm (1:3)	0.71
EP + BT 2µm + BN 12µm (1:3)	0.81

The thermo-oxidative degradation of selected composites (mono- and bimodal, hybrid) were studied with TGA. The results are shown in Figure 3 a) and Table 3. The incorporation of Boehmite and BN fillers shifts both the decomposition temperatures, Td_2 and Td_5, by up to 10°C compared to the neat Epoxy Novolac resin. The amount of char formed under oxidative condition is higher compared to the unfilled Epoxy Novolac, whereas for the hybrid filled systems slightly lower than the bimodal BN filled formulations. A closer look on the TGA thermogram shows that the unimodal Boehmite filled system has two major mass loss steps; the first step ($\approx$ 360 °C) can be associated with the dehydration of AlO(OH) and phase transition to γ-Al_2O_3, whereas at > 500 °C, Boehmite decomposes entirely (= flame retardant mechanism). [6,7] The glass transition temperature (Tg) is not affected in all systems.

Table 3: Thermal stability and degradation behaviour comparison of selected filled EP Novolac composites.

Composition	$T_{g,DMA}$[a] / °C	T_{d2}[b] / °C	T_{d5}[c] / °C	*Char yield*[d] / wt.%
neat EP Novolak	242	352	363	0.00
EP + BN 2µm	244	352	370	30.45
EP + BT 2µm	242	359	373	29.10
EP + BN 2µm+ BN 12µm (1:3)	236	357	371	30.39
EP + BT 2µm + BN 12µm (1:3)	242	358	370	29.64

[a] T_g values determined by maximum of tan δ.
[b] Degradation temperature at 2% weight loss under N_2.
[c] Degradation temperature at 5% weight loss under N_2.
[d] Residual weight at 800 °C under Air.

The fire retardancy of the systems was studied with Cone Calorimetry. The results in Table 4 and Figure 3 b) show that the time to ignition (t_{ig}) and the time to peak heat release rate (t_{PHRR}) are influenced by the incorporation of the inorganic fillers. Compared to the neat resin, the PHRR was significantly decreased by 38%, from 1184 kW/m^2 to 729 kW/m^2, respectively, by combining simultaneously Boehmite and BN. The smoke release can be suppressed by the Boehmite by the release of water diluting the burning system, reacting with pyrolyzed volatiles and the formation of an Al_2O_3-barrier. Platelet-shaped BN does not severely reduce the smoke release, however has an effect on the total release rate (THR). This can be assumed with the inhibition of heat (and mass) transfer due to exfoliated BN sheets acting as a ceramic barrier. [4] The FIGRA value (ratio of PHRR and t_{ig}) is, compared to the unfilled system, decreased from 10 to a value of 4 indicating a less implication of the material to be fire hazardous.

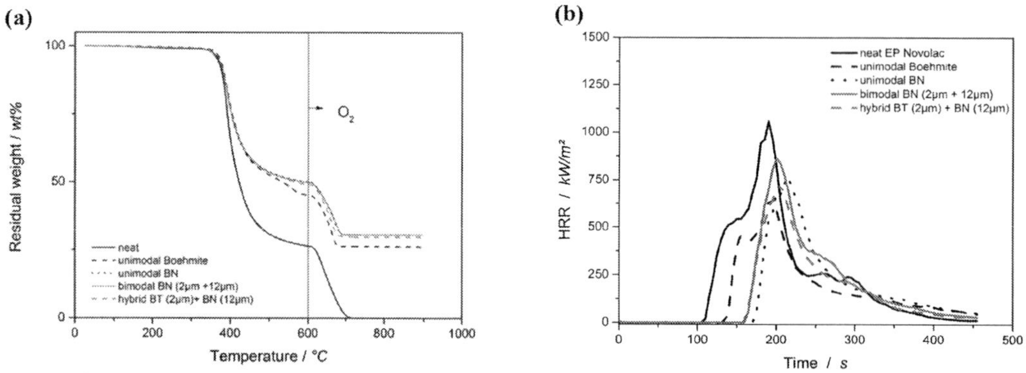

Figure 3: a) TGA thermogram of selected formulations at a constant filler concentration of 45 wt% and b) Heat release rate of selected formulations at a constant concentration of 45wt% (applied heat flux 35 kW/m^2).

Table 4: Combustion behaviour comparison of selected filled EP Novolac composites (heat flux of 35kW/m^2).

Composition	t_{ig} / s	*t (PHRR)* / s	*PHRR / kW/m^2*	*THR / MJ/m^2*	*TSR / m^2/m^2*	FIGRA *kW/m·s*
neat EP Novolak	117	187	1184	95,3	3774	10,12
EP + BN 2µm	175	215	767	71,5	3246	4,38
EP + BT 2µm	140	197	674	72,2	2314	4,82
EP + BN 2µm+ BN 12µm (1:3)	164	200	918	75,7	3019	5,58
EP + BT 2µm + BN 12µm (1:3)	163	207	729	65,1	2591	4,47

CONCLUSIONS

In this study, simultaneously thermal conductive and flame retardant epoxy novolac formulations with a fixed 45wt% by adding BN and Boehmite were successfully established. The addition of these fillers did not influence the T_g negatively. The established hybrid system shows improved thermal conductivity and thermal stability against oxidative degradation and also the values measured with Cone Calorimetry PHRR, SPR, t_{ig} and FIGRA were significantly improved.

ACKNOWLEDGEMENTS

This work was financially supported by Federal Ministry of Economic Affairs and Energy (public funded project "Smart PVI Box"). The authors also thank Nabaltec AG and Isola GmbH for their material supply and fruitful discussions. T. Standau for his support, A. Lang for conducting Scanning Electron Microscopy and U. Kuhn for TGA measurements (all from the Department of Polymer Engineering).

REFERENCES

[1] S. Chung, J. Lin, Thermal Conductivity of Epoxy Resin Composites Filled with Combustion Synthesized h-BN Particles, *Molecules*. 21 **(2016)** 1–11. doi:10.3390/molecules21050670.

[2] X. Huang, P. Jiang, T. Tanaka, A Review of Dielectric Polymer Composites With High Thermal Conductivity, *IEEE Electr. Insul. Mag.* 27 **(2011)**.

[3] K.C. Yung, J. Wang, T.M. Yue, Thermal Management for Boron Nitride Filled Metal Core Printed Circuit Board, *J. Compos. Mater.* 42 **(2008)** 2615–2627. doi:10.1177/0021998308096326.

[4] B. Yu, W. Xing, W. Guo, et al., Thermal Exfoliation of hexagonal boron nitride for effective enhancements on thermal stabilit, flame retardancy and smoke suppresion of epoxy resin nanocompoistes via sol-gel process, *J. Mater. Chem. A Mater. Energy Sustain.* 4 **(2016)** 7330–7340. doi:10.1039/C6TA01565D.

[5] S. Duwe, C. Arlt, S. Aranda, et al., A detailed thermal analysis of nanocomposites filled with SiO 2 , AlN or boehmite at varied contents and a review of selected rules of mixture, *Compos. Sci. Technol.* 72 **(2012)** 1324–1330. doi:10.1016/j.compscitech.2012.04.015.

[6] T. Sun, Q. Zhuo, Y. Chen, et al., Synthesis of boehmite and its effect on flame retardancy of epoxy resin, *High Perform. Polym.* 27 **(2015)** 100–104. doi:10.1177/0954008314540312.

[7] E.S. Ogunniran, R. Sadiku, S. Sinha Ray, et al., Morphology and thermal properties of compatibilized PA12/PP blends with boehmite alumina nanofiller inclusions, *Macromol. Mater. Eng.* 297 **(2012)** 627–638. doi:10.1002/mame.201100254.

[8] C. Özdilek, K. Kazimierczak, D. Van Der Beek, et al., Preparation and properties of polyamide-6-boehmite nanocomposites, *Polymer (Guildf).* 45 **(2004)** 5207–5214. doi:10.1016/j.polymer.2004.05.029.

[9] D. Pedrazzoli, F. Tuba, V. Khumalo, et al., Mechanical and rheological response of polypropylene/boehmite nanocomposites, *J. Reinf. Plast. Compos.* 33 **(2013)** 252–265. doi:10.1177/0731684413505787.

[10] G. Camino, A. Maffezzoli, M. Braglia, et al., Effect of hydroxides and hydroxycarbonate structure on fire retardant effectiveness and mechanical properties in ethylene-vinyl acetate copolymer, *Polym. Degrad. Stab.* 74 **(2001)** 457–464. doi:10.1016/S0141-3910(01)00167-7.

[11] H. Chen, V. V Ginzburg, J. Yang, et al., Thermal conductivity of polymer-based composites : Fundamentals and applications, *Prog. Polym. Sci.* 59 **(2016)** 41–85. doi:10.1016/j.progpolymsci.2016.03.001.

Influence of 1,3,5-Benezenetrisamide Based Additives on the Morphology and Mechanical Properties of Isotactic Polypropylene

Mörl Michaela [(1)]*, Schmidt Hans-Werner [(1)], Altstädt Volker [(1)]

[(1)] University of Bayreuth - Bavaria - Germany

Isotactic polypropylene (iPP) is a widely used commodity plastic due to its good combination of good mechanical properties and low price. To even further expand the possible applications of this material, an interest in toughening iPP has been increasing over the past years. iPP is a polymorphic material and can crystallize in different modifications where the thermodynamically metastable beta modification exhibits a higher toughness than alpha modification that is usually found under normal processing conditions. The easiest and industrial most relevant method to obtain a high amount of beta phase is the use of nucleating agents. Many studies have investigated the influence of different nucleating agents in iPP over the past years. However, only little information is available about difficult-to-toughen iPP grades with high melt flow rates even though these are very relevant for injection molding applications. The influence of two different 1,3,5-benzenetrisamide-based experimental beta-nucleating agents on the morphological and mechanical behavior of isotactic polypropylene with a melt flow rate of 120 g/10 min is investigated and compared to a commercial bisamide-based beta-nucleating agent and a 1,3,5-benzenetrisamide-based alpha nucleating agent. Morphological investigations by wide-angle X-ray scattering (WAXS) show that only 0.04 wt. % of one of the experimental additives is enough to induce high amounts of beta modification in the crystalline part of the polymer. The materials are investigated with quasistatic (tensile testing) as well as short-time dynamic (Charpy impact strength) methods. A much higher toughness of the material containing the experimental beta-nucleating agents can be observed. This behavior is correlated with the different influences of the different additives on the morphology of iPP.

Flame Retardant Properties of Polymer Composites of Urea Complex of Magnesium and Vermiculite

Herminio F. Muiambo[1, a)], Walter W. Focke[2, b)] and Joseph K. O. Asante[3]

[1]*Department of Chemistry, Faculty of Sciences, Eduardo Mondlane University, Maputo, Mozambique.*
[2]*Institute of Applied Materials, Department of Chemical Engineering, University of Pretoria, South Africa.*
[3]*Department of Physics, Tshwane University of Technology, Pretoria,, South Africa.*

[a)] hmuiambo@uem.mz
[b)] walter.focke@up.ac.za

Abstract. Polymeric materials are increasingly used in several applications. However, their relatively high flammability presents a danger to people and property. Their use therefore requires that they are made more resistant to both the initiation and propagation of fires.

In the present work we evaluate, by cone calorimeter tests, the efficiency of urea complex of magnesium and vermiculite (urea- vermiculite) as a stand-alone flame retardant in polyurethane (PU) resin and flexible polyvinyl chloride (PVC).

Flexible PVC cone calorimeter tests proceeded without a visible flame following an initial short-lived ignition which degenerated into a bulk pyrolysis, in combination with a surface glowing-combustion event. Urea-vermiculite formed an exfoliated protective barrier layer which allowed thermal stabilisation of the condensed phase. The simultaneous release of halogen species by the PVC and the action of the exfoliated barrier layer prevented the formation of a flammable air–fuel mixture. The addition of urea-vermiculite lowered the peak heat release rate (pHRR) and the total heat released (tHR) significantly.

In PU composites the urea-vermiculite was unable to form a cohesive protective barrier layer. The poor compatibility between the molten PU and the exfoliated flakes also led to the consumption of the underlying PU. Nevertheless, the addition of urea-vermiculite lowered significantly the pHRR of PU composites.

In general, cone calorimetry results revealed that urea-vermiculite allowed thermal stabilization of the condensed phase at high temperatures but had little influence in the vapour phase behaviour. The amount of released urea's degradation products (non-flammable vapours) was not enough to dilute the flammable vapours' mixture. Thereby when used with PVC, which releases halogen flame poison, it showed great fire performance.

Keywords: Urea-vermiculite, Flame Retardancy, Polyurethane, Polyvinyl Chloride, Cone Calorimeter.

1. INTRODUCTION

Polymer composites are extensively used in various areas due to their relatively low cost and the broad range of properties that can be attained [1]. Fully exfoliated composites have been claimed, but some studies have shown that delamination and dispersion of individual clay layers is still a challenge [1,2].

Flame-barrier compositions have been proposed that rely on vermiculite's expansion properties [3]. Vermiculite can impart fire resistance to polymers, provided the exfoliation temperature can take place at temperatures just below the volatilisation of the polymer.

Current methods for the protection of materials from fire risk are based on engineering approaches, the use of inherently low flammable polymers and flame retardant (FR) additive materials[4]. According to Dufton [5], an effective FR material has to be able to impede or reduce heat supply to below the critical levels required for flame stability. This prerequisite can be accomplished by: i) limiting the heat and mass transfer between the solid and gas phases; ii) scavenging of the free radicals in the gas phase; and iii) creating a heat sink. Existing FR materials in polymers are generally based on halogenated additives and antimony trioxide (as synergist), inorganic fillers, melamine derivatives and phosphate-based materials [6]. To be effective, a FR material has to supply the flame inhibitor in the same temperature range as the decomposing polymer [5]. This criterion complies with the "right place at the right time" theory.

The categorisation of FR materials is somewhat inconsistent due to the multiplicity of polymer processing techniques and, on the other hand, to the fact that the very same FR material can exhibit different FR mechanisms. Nevertheless, all types of FR materials can be classified according to the mode of action (physical or chemical) or the means of incorporation into the polymer matrix (additive, reactive and coatings).

Halogenated chemical FR act essentially in the vapour phase as flame poison stopping the chain reactions and therefore the combustion of the molten polymer [4-7]. Although economically and technically very effective, halogen-based FRs also produce obscuring, toxic and corrosive smoke, and are a source of various environmental concerns [4,5,7,8]. Some phosphorous compounds can be either vapour phase and/or condensed phase FR, depending on their chemical structure and their interaction with the molten polymer matrix [4,6]. The most important organic nitrogen-based FR is melamine (and its derivatives) and ammonium polyphosphate is the most important inorganic additive [9]. Well-dispersed nanoparticles are known to enhance the thermal stability and some studies have shown that nanoparticles have to be used in combination with other FR additives in order to meet some required fire performance levels [6,10]. Intumescent systems provide radiation shielding and thermal insulation by forming a heat-induced swollen multi-layer barrier [5,6]. Mineral fillers can delay the time to ignition (t_{ign}), slow the initial heat release rate (HRR), lower the smoke production rate (SPR) and reduce overall toxic gas emissions. However, in general, they have a limited fire performance window and cannot reduce the total heat release rate (tHR) if enough constant external heat is applied. To be effective, high loadings are needed and such loading levels may impair the processability and the mechanical properties of the polymeric material [4,6,11]. According to Horrocks and Price [12], char-forming agents seem to be the best choice and the way forward in developing effective fire-resistant composites.

Flexible PU foams are highly flammable and are the main and most combustible components of upholstered furniture and mattresses. Nevertheless, rigid PU foams are among the best heat-insulating materials and one of the most used in the building industry [13]. Flame-retarded thermoplastic polyurethane is used in wire and cable applications with aluminium hydroxide and magnesium hydroxide as fire-retardant additives. The use of mica and zinc borate is also reported in the literature. The latter is a synergist for aluminium or magnesium hydroxides[13].

Polyvinyl chloride (PVC) is a very versatile polymer used in diverse applications, including flooring, rigid pipes, flexible hoses, conveyor belting, and wire and cable insulation. Neat PVC features a relatively high chlorine content of 56.7 wt.%. However, the conventional plasticisers used in the manufacture of flexible PVC detract from this outstanding fire resistance. Thus FR and smoke-suppressant (SS) additives must be incorporated in order to meet product test specifications such as the oxygen index, HRR, smoke evolution, or the extent of burning[14]. Plasticised PVC is flexible and offers versatile applications from semi-rigid to very soft materials; it is also cost-effective[15,16].

A key property of urea-vermiculite is its tendency to exfoliate when heated to high temperatures in a worm-like manner to form expanded material with a low density. Urea-vermiculite can be compared to expandable graphite. A disadvantage of graphite-based FRs is the black colour and electrical conductivity they impart to the material. Hence the utility of urea-modified vermiculite was investigated as a FR for PVC and PU composites.

2. EXPERIMENTAL

2. 1. Materials and methods

Detailed preparation of urea-vermiculite followed the procedure reported elsewhere[17].

The ultra-low-viscosity liquid urethanes (Smooth-Cast© 300) were used as received from Smooth-On. TPC Paste Resin Co, Ltd supplied the PVC emulsion grade PG680. Reofos 50, a synthetic isopropylated triaryl phosphate ester plasticiser, was supplied by Chemtura.

TGA experiments were performed on a Mettler Toledo A851 TGA/SDTA instrument. The combustion behaviour was assessed according to the ISO 5660-1 standard procedure in a Dual Cone Calorimeter (Fire Testing Technology (UK) Ltd). The sheets (100 mm × 100 mm × 3 mm) were placed on aluminium foil and exposed to an external heat flux of 35 kW m^{-2}. The time to ignition was manually set from the computers keyboard. The HRR, its pHRR and MLR were obtained directly from the FTT "ConeCalc" software.

2.2. Preparation of the urea-vermiculite PU composites

Part A and Part B (100A:90B, wt.) were mixed thoroughly together with urea-vermiculite and then poured into the mould. The white sheets were fully cured by pressing them at room temperature and 5 MPa for 10 min. A (Universal Mold) release agent was applied to the metallic plates to facilitate demoulding.

2.3. Preparation of the urea-vermiculite PVC composites

The urea-vermiculite/PVC compounds were prepared using the plastisol route. PVC powder (100 g), Reofos 50 (60 g) and an adequate amount of urea-vermiculite were thoroughly mixed together in a high shear mixer for 10 minutes [16]. The paste mixture was immediately poured into a mould and heated for 10 minutes in a convection oven set at 120 °C. Thereafter the sheets were cured at 150 °C and 5 MPa in a hot press for 5 minutes. The mass ratio of PVC and the plasticizer (1000 parts per 60) was kept constant for all samples.

3. RESULTS AND DISCUSSION

All PU compositions exhibited two thermal decomposition steps. The neat PU volatilised completely, before 700 °C, but the composites still retained a residue even at 800 °C (Fig. 1a)).

PVC thermograms (Fig. 1b) show three degradation steps. The first two steps are related to PVC's dehydrochlorination and to the degradation of the plasticiser [16,18]. The last thermal event corresponds to the degradation of the oxidised PVC [16,19]. For the PVC compound and for the urea-vermiculite composites the maximum degradation rate was observed during the first step at $T_{max} = 289 \pm 2$ °C. The typical urea-vermiculite thermal degradation steps were overlapped by those from the PVC and could not be clearly observed. The amount of the residual mass, in the composites, corresponded to more than the urea-vermiculite added and raised with the increment of the quantity of the filler.

a) b)

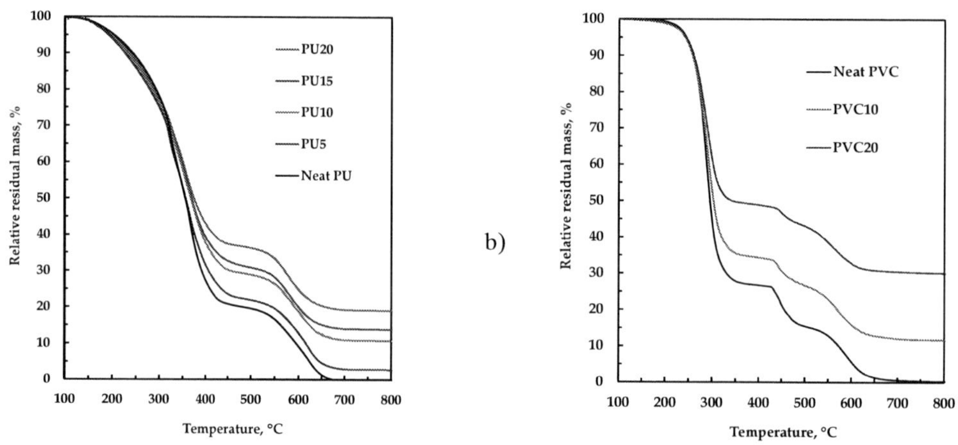

Figure 1. Thermal degradation plots of a) PU and b) PVC with their composites with urea-vermiculite, in air.

Fig. 2a) compares the HRR for PU composites with that of their PU reference polymer. The data in Table 1 summarise the cone calorimeter results. The HRR curves for the neat PU exhibited the shape characteristic of thermally thin samples [20].

a) b)

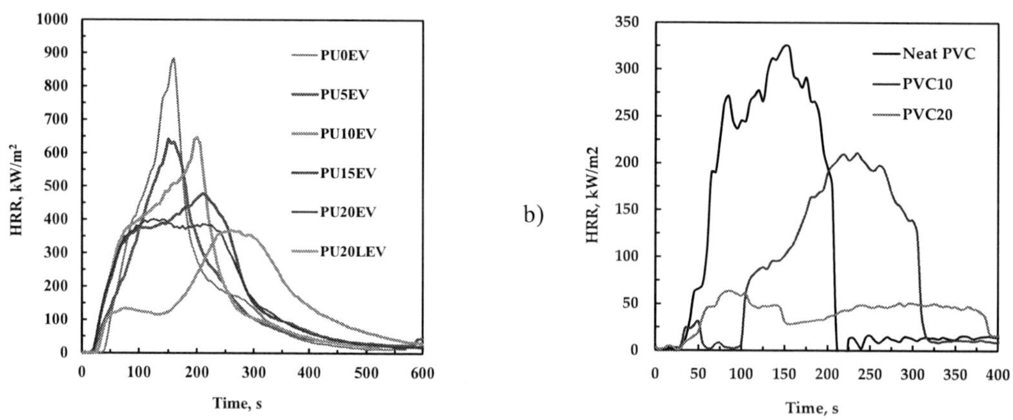

Figure 2. Cone calorimeter average peak HRR for a) PU and b) PVC composites with urea-vermiculite

Compared with the neat PU, the pHRR of the composites was greatly reduced while the tHR was almost not affected. As expected, the t_{ign} was negatively affected, due to the clay-catalysed polymer decomposition [8,10,11]. However, when the filler was concentrated at the top of one of the sides (sample denoted 20T), subjected directly to the cone heat flux, the composite showed better fire performance. The tHR was still high due to fact that the composite had two burning stages: the first one retarded by the filler and second one corresponding to the underlying material.

Table 1. Flammability data summary of PU composites

Property, Units	Urea-vermiculite, wt.%					
	0	5	10	15	20	20T
Time to ignition (t_{ign}), s	40 ± 2	25 ± 2	24 ± 1	26 ± 1	22 ± 4	28 ± 2
Time to flameout, s	521 ± 33	527 ± 98	563 ± 53	517 ± 9	538 ± 8	798 ± 102
Peak heat release rate (pHRR), kW m^{-2}	900 ± 5	779 ± 115	483 ± 20	647 ± 96	433 ± 29	377 ± 13
Total heat release (tHR), MJ m^{-2}	104 ± 1	107 ± 17	112 ± 4	107 ± 1	107 ± 3	98 ± 1
MAHRE (kW m^{-2})	374 ± 9	399 ± 32	371 ± 11	332 ± 7	312 ± 7	203 ± 4
pHRR/t_{ign}, kW $m^{-2}s^{-1}$	23 ± 1	31 ± 6	25 ± 3	20 ± 1	20 ± 2	14 ± 1

Fig. 3 a) shows the appearance of the 10 wt.% composite's residue char after the cone calorimeter test. This image also indicates clearly that the urea-vermiculite flakes could not bond properly to the PU matrix. This lack of bonding between PU char and the exfoliated flakes allowed the consumption of the exposed polymer as well as the underlying material. Some flakes were even swept away from the pyrolysis zone while exfoliating.

Figure 3. a) Image of 10 wt.% PU composite char residue after cone calorimeter test. Images of char residue of 10 wt.% urea-vermiculite/PVC after cone calorimeter tests: b) just after flame flameout and still smoking, and c) after complete extinction of flame and smoke.

The urea-vermiculite/PU composites featured lower CO, CO2 and smoke production rates owing to the reduction in the rate of mass loss. The observed trends mirror those observed for the HRR almost perfectly.

The urea-vermiculite/PVC cone calorimeter results are summarised in Table 2. Representative HRR curves obtained from the cone calorimeter tests are presented in Fig. 2 b). All the neat PVC compound samples ignited and flamed briefly and produced a large amount of smoke. This gave rise to the sharp peak in the HRR at short times. From Fig. 3 b) and c), showing the PVC samples on the cone calorimeter sample holder after flameout, it can be seen that the PVC composites continued releasing smoke even after flameout. A considerably expanded residue was observed due to the action of the urea-vermiculite filler. This latter fact is consistent with the TGA findings. The HRR curves for the neat plasticised PVC compound exhibited the shape characteristic of thermally thin samples [20].

Table 2. Flammability data summary of PVC composites

Property, Units	Urea vermiculite, wt.%		
	0	10	20
Time to ignition (t_{ign}), S	35 ± 3	33 ± 6	59 ± 6
Time to flameout, S	146 ± 59	174 ± 143	122 ± 27
Peak heat release rate (pHRR), kW m^{-2}	325 ± 11	229 ± 16	60 ± 23
Total heat release (tHR), MJ m^{-2}	55 ± 11	42 ± 14	7 ± 2
pHRR/t_{ign}, kW $m^{-2}s^{-1}$	9.3 ± 0.8	7.1 ± 0.9	1.5 ± 0.9

The HRR curves for the samples containing urea-vermiculite were a bit more complex. Most of them ignited briefly, almost flamed out and then reignited. The subsequent abrupt reignition was due to the increase in the amount of flammable volatiles because of the continued pyrolysis and smoking while not burning. On the other hand, the halogen entering the vapour phase also contributed to a "flame poisoning effect", i.e. the reaction and termination of oxygen-based chain branching, free radical reactions occurring in the flame [15,21]. The profuse smoke evolution suggests that the composite samples were pyrolysed while at the same time a glowing combustion occurred at the surface. The urea-vermiculite composites featured lower CO, CO2 and SPRs. Their trends were identical to those observed for the HRR. The lower "fire performance index" (PI = pHRR/t_{ign}) of the composite containing 20 wt.% of urea-vermiculite is a clear indication of better cone calorimetry performance.

The expansion of the urea-vermiculite formed a low-density layer of 'worm-like' structures which provided a protective barrier at the polymer surface. This limited heat transfer to the substrate and thus slowed down the rate of thermal degradation. On the other hand, the release of the HCl by the decomposing PVC apparently also prevented flaming combustion during the latter part of the cone calorimeter tests. This can be attributed to a dilution effect on the air–fuel mixture in the gas phase. However, the halogen entering the gas phase also contributed to a "flame poisoning effect", i.e. the slowing down of the free radical chain reactions occurring in the flame [22]. In combination, these two effects explain why the cone calorimeter test proceeded without a visible flame following an initial short-lived ignition. It degenerated into a bulk pyrolysis experiment in combination with a surface glowing-combustion event.

4. CONCLUSIONS

The cone calorimeter results revealed that the fire performance of urea-vermiculite in PU composites is greatly affected by the distribution of the FR additive along the vertical direction of incidence of the heat flux. The addition of 20 wt.% urea-vermiculite lowered the pHRR. The urea-vermiculite formed an exfoliated but non-cohesive protective barrier layer on top of the PU composites. However, after all the urea-vermiculite had expanded, the loose flakes offered efficient protection only at high loadings.

Urea-vermiculite / PVC cone calorimeter tests proceeded without a visible flame following an initial short-lived ignition which degenerated into a bulk pyrolysis experiment in combination with a surface glowing-combustion event. These results revealed that addition of 20 wt.% urea-vermiculite lowered the pHRR and the tHR The urea-vermiculite forms an exfoliated protective barrier layer. In addition, the simultaneous release of halogen species by the polymer matrix and the exfoliating vermiculite prevents the formation of a flammable air–fuel mixture. The cone calorimetry results also showed that urea-vermiculite allowed thermal stabilisation of the condensed phase at high temperatures, but had little influence on the vapour phase behaviour.

REFERENCES

1 S. Pavlidou and C. D. Papaspyrides, Progress in Polymer Science **33** (12), 1119 (2008).
2 Suprakas Sinha Ray and Masami Okamoto, Progress in Polymer Science **28** (11), 1539 (2003).
3 T. Wada, (Google Patents, 1973).
4 A. B. Morgan and J. W. Gilman, Fire and Materials **37** (4), 259 (2013).
5 P. W. Dufton, (Smithers Rapra Technology, 1995).
6 F. Laoutid, L. Bonnaud, M. Alexandre, J. M. Lopez-Cuesta, and Ph Dubois, Materials Science and Engineering R: Reports **63** (3), 100 (2009).
7 G. Camino and L. Costa, Polymer Degradation and Stability **20** (3–4), 271 (1988).
8 P. Kiliaris and C.D. Papaspyrides, Progress in Polymer Science **35**, 902 (2010).
9 H. Horacek and R. Grabner, Polymer Degradation and Stability **54** (2–3), 205 (1996).
10 D. Tabuani, F. Bellucci, A. Terenzi, and G. Camino, Polymer Degradation and Stability **97** (12), 2594 (2012).
11 M. Berta, C. Lindsay, G. Pans, and G. Camino, Polymer Degradation and Stability **91** (5), 1179 (2006).
12 A.R. Horrocks and D. Price, *Fire Retardant Materials*. (CRC Press, 2001).
13 Sergei V. Levchik and Edward D. Weil, Polymer International **53** (11), 1585 (2004).
14 A. W. Coaker, Journal of Vinyl and Additive Technology **9** (3), 108 (2003).
15 Alister F. Matheson, Robin Charge, and Tor Corneliussen, Fire Safety Journal **19** (1), 55 (1992).
16 A. Jimenez, J. Lopez, A. Iannoni, and J. M. Kenny, Journal of Applied Polymer Science **81** (8), 1881 (2001).
17 Herminio F. Muiambo, Walter W. Focke, Maria Atanasova, and Aida Benhamida, Applied Clay Science **105-106**, 14 (2015).
18 P. Dunn and B. C. Ennis, Journal of Applied Polymer Science **14** (2), 355 (1970).
19 K. S. Annakutty and K. Kishore, European Polymer Journal **29** (10), 1387 (1993).
20 B. Schartel and T. R. Hull, Fire and Materials **31** (5), 327 (2007).
21 Edward D. Weil, Sergei V. Levchik, and P. Moy, in *Flame Retardants*, edited by Edward D. Weil and Sergei V. Levchik (Hanser, 2009), pp. 59.
22 E. D. Weil, S. Levchik, and P. Moy, Journal of Fire Sciences **24** (3), 211 (2006).

Highly filled boron nitride-phthalonitrile nanocomposites for exigent thermally conductive applications

Derradji Mehdi [(1)*]

[(1)] Harbin Engineering University - Heilongjiang - China

Highly filled phthalonitrile nanocomposites, reinforced with both native and silane surface modified nano-BN were investigated for their mechanical and thermal properties. The nanocomposites were prepared by a solution blending technique followed by compression molding. The thermal conductivity was investigated as the core of the study along with a bending test, a morphological behavior, thermal stability investigations, and thermomechanical properties measurements. The highest thermal conductivity value of 4.69 W/m.K was obtained at the maximum nanofillers loading of 30 wt.% treated BN. A high resolution TEM confirmed that the exceptional enhancements in the thermal conductivity are related to the formation of an effective conductive path assured by the silane surface modified nano-BN. The bending test also revealed consequent improvements in both the strength and modulus as the amount of the nanofillers increased. The SEM investigations revealed positive changes in the surface morphology of the neat resin, with a rougher and homogeneous state of surfaces for the hybrids. Moreover, the thermal stability at the maximum fillers loading was found to be one of the best ever reported in the field of high performance polymer-based composites, with a 5 wt.% weight loss temperature of 551 °C.

FIBER SURFACE TREATMENT TO INCREASE THE FIBER CONTENT AND MECHANICAL PROPERTIES OF ROTOMOLDED COIR-LMDPE AND AGAVE-LMDPE COMPOSITES

Cisneros-López Erick Omar (1), González-López Martín Esteban (1), Pérez-Fonseca Aida Alejandra (1), González-Núñez Rubén (1), Rodrigue Denis (2), Robledo-Ortiz Jorge Ramón (1)*

(1) CUCEI, Universidad de Guadalajara - Jalisco - Mexico, (2) CERMA, Université Laval - Québec - Canada

Natural fibers have received the attention of researchers due to their ecological character, biodegradability, low costs, non-abrasive nature, high filling levels, low energy consumption, high specific properties, low density and wide variety of fiber types. However, the polar and hydrophilic nature of lignocellulosic fibers and the non-polar characteristics of most thermoplastics results in different levels of incompatibility and compounding difficulties leading to non-uniform fiber dispersion within polymer matrices, thus limiting their efficiency [1]. Therefore, it is necessary to use physical or chemical treatments to improve the fiber-matrix compatibility. In this study, agave (Agave tequilana Weber var. Azul) and coir (cocus nucifera) fibers were surface treated with maleated polyethylene (MAPE) with two main objectives: 1) to improve the mechanical properties of the composites produced by rotational molding, and 2) to increase the fiber content in the composite. The rotomolded composites were produced with linear medium density polyethylene (LMDPE) and different concentrations (0, 10, 20, 30 and 40% wt.) of fiber (treated or untreated). The samples were then characterized in terms of morphology and mechanical properties (impact, tensile and flexural). The results showed that the MAPE surface treatment was successful leading to a better fiber distribution and a more uniform composite morphology allowing the possibility to use higher fiber contents in rotational molding. At low fiber contents (10 and 20% wt.), the mechanical properties were improved (up to 52%) in treated fiber composites (TFC) compared to the neat polymer and untreated fiber composites (UFC). Although the mechanical properties of TFC decreased at high fiber contents (30 and 40% wt.), they were substantially higher than UFC (about 160%, 400% and 240% for impact, tensile and flexural properties, respectively). [1]. E.O. Cisneros-López et al., Polym. Compos. DOI: 10.1002/pc.23564 (2016) Figure 1. Relative (σcomposite/σLMDPE) tensile and flexural strength.

Stiffening mechanisms in vermiculite-amorphous polyamide bio-nanocomposites

Macheca Afonso Daniel **[1]*, *Focke Walter W*** **[1], *Muiambo Herminio Francisco*** **[1], *Kaci Mustapha*** **[2]**

[1] University of Pretoria - Pretoria - SouthAfrica, [2] Laboratoire des Matériaux Polymères Avancés-University de Bejaia - Bejaia - Algeria

Sub-micron thick flakes were obtained by sonication of vermiculite that was first exfoliated by either thermal shock or chemical treatment with hydrogen peroxide. Dimer fatty acid polyamide nanocomposites with a mixed morphology were prepared via a solution-dispersion technique. The large (in the micrometer range) vermiculite flakes assumed random orientations in the matrix. BET surface area measurements indicated flake thickness below 100 nm but SEM showed that thicker flakes were also present. Filler content was varied up to 30 wt.%. At this loading, the tensile strength doubled, the modulus increased five-fold but the elongation-at-break decreased by a factor of ten. Dynamic mechanical analysis suggests three stiffening mechanisms. The reinforcing effect of the high stiffness inorganic flakes is the primary contributor. Together with the chain confinement effect, that expresses itself in an apparent increase in the glass transition temperature, provides an adequate rationalisation of the stiffness variation below Tg. However, an additional stiffening effect is indicated at temperatures above Tg. The mechanism may involve dynamic network formation based on fluctuating hydrogen bonding interactions between the matrix polymer chains and the filler particles.

PROPERTIES OF CROSSLINKED POLYETHYLENE/Oib-POSS NANOCOMPOSITES:"Effect of peroxide concentration"

BICER EZGI [1]*, KURT DEMIR GULSEN [1], KODAL MEHMET [2], OZKOC GURALP [1]

[1] Kocaeli University, Chemical Engineering Department - Kocaeli - Turkey, [2] Kocaeli University, Kosekoy Vocational School - Kocaeli - Turkey

The aim of this study is to determine the properties of cross-linked low density polyethylene (LDPE)/OctaIsobutyl functionalized Polyhedral oligomeric silsesquioxanes (OibPOSS) nanoparticles. The peroxide used in crosslinking is tert-butyl cumyl peroxide (BCUP). The peroxide content was varied as 0.5, 1, 2 and 3 phr. LDPE/OibPOSS/Peroxide nanocomposites were processed in an Xplore Instruments micro-compounder at a screw speed of 50 rpm and 135 deg.C barrel temperature. The crosslinking was carried out using a hot-press operated at 175 deg.C. The LDPE/OibPOSS/Peroxide nanocomposites were characterized by performing tensile tests, differential scanning calorimetry (DSC), rheometer, scanning electron microscopy (SEM) and Fourier transform infrared spectroscopy (FTIR). Rheological analysis results showed that the degree of cross-linking and storage and loss modules increased with peroxide concentration. According to tensile tests results, tensile strength of composites increased with increasing peroxide content. OibPOSS distribution was investigated by SEM analysis. The incorporation of POSS enhanced ductility of the crosslinked polyethylene. Keywords: Low density polyethylene, Peroxide crosslinked polyethylene, Nanocomposite, Rheology Acknowledgement: This study has been financially supported by TUBITAK (Grant Number: 116M981).

Characterization of Polyamide 6 - Polypropylene Laminates

Judith R. Büttler[1, a)] and Tung Pham[1, b)]

[1] *Research Institute of Textile Chemistry and Textile Physics, University of Innsbruck, Hoechsterstrasse 73, 6850 Dornbirn, Austria*

[a)] *Judith.Buettler@uibk.ac.at*
[b)] *Corresponding author: Tung.Pham@uibk.ac.at*

Abstract. Wover-fiber reinforced composites become more and more important for lightweight construction components, smart textiles and numerous other applications. The interaction between the two components of the composite depends on the chemical, physical and mechanical properties, and on the manufacturing process of the composite. To obtain a strong interaction between different polymers is problematic. Surface modification could be a solution for a more favorable adhesion between the two polymers. In this study the interphase between polyamide 6 (PA6) and polypropylene (PP) was investigated. PA films were laminated with PP or with plasma-treated PP. It was found that laminates containing plasma-treated PP show an increased shear stress compared to laminates with non-modified PP. Fourier transform infrared spectroscopy (FTIR) analysis showed a small amount of carboxylic groups on the surface of modified PP films and contact angle analysis showed a decrease of the water contact angle of PP after plasma treatment. These findings suggest a rise of the polarity of PP, which makes PP more compatible to interact with PA6. Additionally, melt rheological analysis confirmed the increased interfacial strength between PA6 and plasma-treated PP. In conclusion, plasma treatment of PP can improve the adhesive behavior between PA6 and PP.

INTRODUCTION

The development of woven-fiber reinforced composites is of industrial importance. In the last decades, the application of these composites became important in numerous engineering fields. One of the conventionally used composites with woven-fiber reinforcement consists of carbon fibers and epoxy resin. This composite is used for preforms processed in e.g. automotive and aerospace industry. However, composite materials made of thermoset matrices, like epoxy resin and polyurethane, and reinforcements such as carbon, glass, aramids or ultra-high-molecular-weight PEs (UHMWPEs) fibers cannot be recycled without downgrading. Thermoplastic polymers, such as PA, PP and PE on the contrary, can be recycled and reused by melting and reforming [1].

Although the intrinsic weak bonding between PA and PP, PA/PP composites can be used for lightweight constructions, functional applications in the clothing industry, and other technical applications. Previous studies have shown that it is possible to overcome the weak adhesion between PA and PP by interphase modification. Interphase modification by for example addition of a compatibilizer [2, 3], grafting PP with glycidyl methacrylate (GMA) [4] or addition of a diblock copolymer [5] optimize the chemical properties for an enhanced interfacial adhesion.

The aim of this study is to investigate the effect of plasma treatment of PP on the binding to PA6, in order to develop composites made of PP reinforced with woven-fiber PA6 structures. In this study, laminates made of PA6 and PP films are used to investigate the PA6/PP interphase, as a model system, in order to eliminate the influence of the woven structure. The analysis of interfacial adhesion is often investigated by performing mechanical and surface analysis of the separate components. Using rheology, chemical changes can be observed. For example, Valenza and colleagues have observed a variation of rheological behavior depending on the PE modification within PA/PE laminates [6]. This study will provide mechanical and rheological information about the PA6/PP interphase. The obtained understanding may contribute to the development of PA6/PP textile composites.

EXPERIMENTAL

Materials and Modifications

PA6 (Durethan B30S, Lanxess, Germany) and PP (HD204CF, Borealis, Austria) films having a thickness of 150 μm and 100 μm and a melting temperature around 222 °C and 162 °C, respectively, were used. PP films were plasma-treated in a low-pressure plasma system (Diener electronics plasma-surface Technology, Nano Type F) at 0.3 mbar and 200 sccm oxygen for 5 minutes at 900 W. The atmospheric plasma treatment is performed with a plasma pen Piezo brush® PZ2 (Relyon Plasma GmbH, Germany). The treatment is performed at 10mm distance of the sample having a size of 26 x 26 mm.

Analytical Methods

FTIR-ATR analysis was performed with an attenuated total reflection FTIR-spectroscope Vector 22 (Brucker) connected to the OPUS computer program. For all samples 128 scans were performed from wavenumbers 500-4000 cm^{-1}. The contact angles were measured with demineralized water at 20±2°C using a Krüss Easy drop machine with CCD Teli camera. For analysis the DSA4, DSADevMn computer program was used. The lamination for shear testing was performed under 20 bar for 5 min at 190 °C using a Polystat 200T press (ServiTec GmbH, Germany). PP was laminated between two strips of PA6. The width of the sample was 10 mm and the laminated area 100 mm^2. The samples were sandwiched between two 1 mm PTFE sheets with a 300 μm PEEK spacer. The single lap shear tests were performed using a Zwick/Roell Test Control II Z0.5 machine connected to the computer program TestXpert II. The samples were tested at a gauge length of 30 mm and at a speed of 30 mm/min. Rheological characterization was performed with a Modular Compact Rheometer MCR 302 having a Convection Temperature Control Device CTD 450 Twin drive ready connected to the computer program Rheocompas (Anton Paar GmbH, Austria). The frequency sweeps were performed at 240 °C, 10% or 5% strain and a gap of 0.25 mm or 0.15 mm for 2- and 1-interphase laminates, respectively (Fig. 1).

FIGURE 1. Schematic view of the sample arrangement for rheological analysis of laminates having 1 or 2 interphases.

RESULTS AND DISCUSSION

In order to characterize the interphase between PA6 and PP and to investigate the influence of surface modification caused by plasma treatment, firstly mechanical analysis was performed. For this purpose, the shear stress of PA6/PP laminates was determined. As shown in Fig. 2, the adhesion between PA6 and non-treated PP is weak. However, there is a remarkable 50-fold increase of the shear stress if the surface of the PP film is plasma-treated before it is laminated with PA6. Apparently, PP chemically changes due to the plasma treatment and is more favourable to bind to PA6 compared to the non-treated PP.

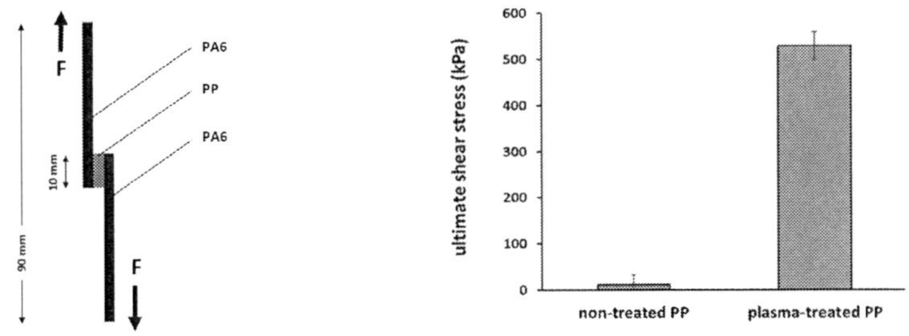

FIGURE 2. Single lap joint shear test with PA6/PP film laminates. PP was modified with low-pressure plasma treatment and as control non-treated PP was used. The films were laminated at 190°C. Measurements were performed in 4-fold.

Surface Analysis of Plasma-Treated PP

The reason for the increased adhesion between PA6 and plasma-treated PP is likely the presence of carboxylic groups at the surface of the plasma-treated PP film, causing a decrease of the contact angle. Figure 3a shows that the water contact angle of non-treated PP is 40° larger compared to the one of PA6. This difference in surface tension indicates a non-favorable interaction between the two materials, because PP is much more apolar than PA6. In contrast, plasma-treated PP has a decline of the water contact angle, the longer the duration of the plasma treatment, the lower the contact angle. After a treatment of 180 seconds PP has a water contact angle similar to the one of PA6. Thus, the polarity of the two materials is similar, indicating an improved adhesive behavior with PA6. Furthermore, the absorption spectrum of non-modified PP shows a typical PP spectrum (Fig. 3b), whereas the spectrum of the plasma-treated PP clearly shows the presence of carboxylic groups in the range of 1650 – 1830 cm^{-1}, which contributes to the decreased water contact angle. The peak at 1630 cm^{-1} is likely moisture, likewise the shoulder at 3020 – 3650 cm^{-1}. Although, this shoulder may also include hydroxyl groups due to degradation of PP. These findings are in accordance with previous studies [7]. The presence of carboxylic groups suggests that there are hydrogen- and dipole-dipole bonds formed between the carboxyls of the plasma-treated PP and the amide groups of PA6 resulting in an increased bond strength. However, these data give only suggestive chemical information about the interphase. Additional investigation of the interphase is still needed.

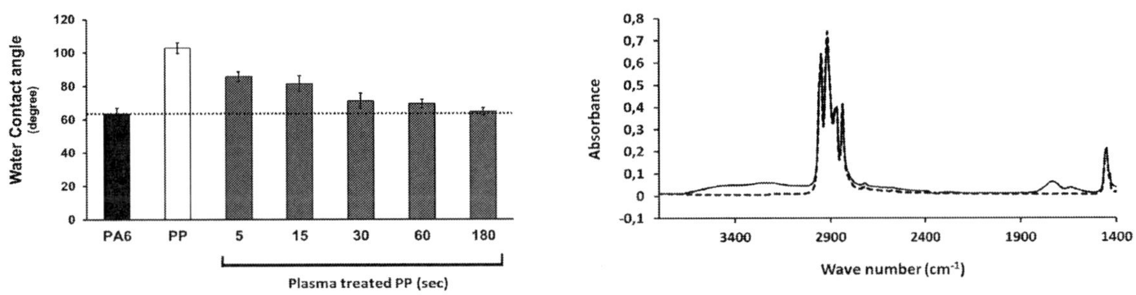

(a) (b)

FIGURE 3. Surface characterization of PP films treated with atmospheric-pressure plasma. Water contact angles of PA6, PP and plasma-treated PP films (a). The duration of the treatment is indicated in seconds. ATR-FTIR absorption spectra (b) of non-treated PP (dashed line) and plasma-treated PP treated for 180 sec. (solid line). The peak at 1733 cm^{-1} shows carboxylic groups.

Rheological Characterization of PA6/PP laminates

In order to obtain more knowledge about the chemical interactions at the interphase rheological analysis was performed, as rheology is a sensitive method to determine chemical changes in polymers. For this purpose, PA6/PP laminates were investigated using the parallel-plate system. Firstly, laminates having two interphases were tested. In Fig. 4, the complex viscosity of the pure polymers and the laminates is depicted. The complex viscosities of the laminates lay between the complex viscosity of PP and PA6, indicating that both polymers contribute to the flow behavior of in the laminate. Interestingly, a significant increase of the complex viscosity (Fig. 4b) and the storage- and loss moduli (data not shown) is observed for the laminates at low frequencies. This means that the polymer melt behaves like a solid structure at rest. The reason for this is not clear, it may be caused by laminar flow mechanisms, as this effect is even more distinct in laminates having 4 or 8 interphases. By comparing laminates containing non-treated PP with those containing plasma-treated PP, it is observed that the complex viscosity decreases (Fig. 4b) and the storage- and loss moduli shift to the right. This is not likely because of a strengthened adhesion between PA6 und PP, as an increased adhesion will result in an increase of complex viscosity and a left-shift of the moduli. Therefore, we attribute these changes to the decrease of molecular weight of PP caused by the plasma treatment. The PP is thought to be degraded as the complex viscosity of plasma-treated PP is lower compared to the complex viscosity of non-treated PP (Fig. 4a). Overall, there are two effects observed in 2-interphases laminates: At low frequencies, there is an effect caused by the laminate structure, and at higher frequencies there is a molecular-weight-effect due to the PP degradation.

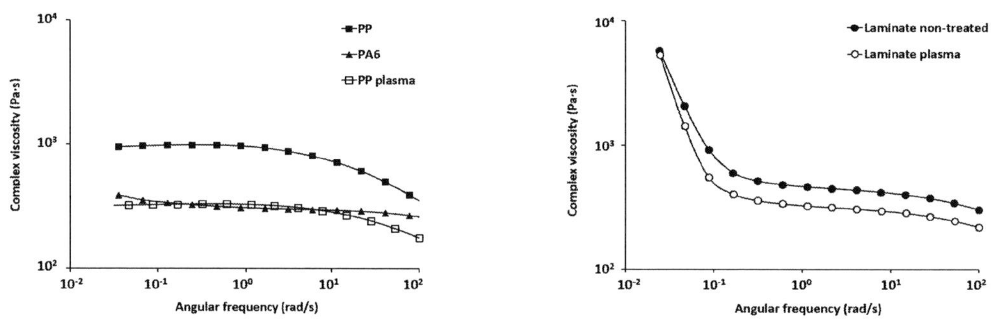

(a) (b)

FIGURE 4. Complex viscosities of pure polymers (a) and 2-interphase laminates (b). The presence of plasma-treated PP is indicated with 'plasma'. The frequency sweeps were performed at 240 °C, 10% strain and a gap of 0.25 mm.

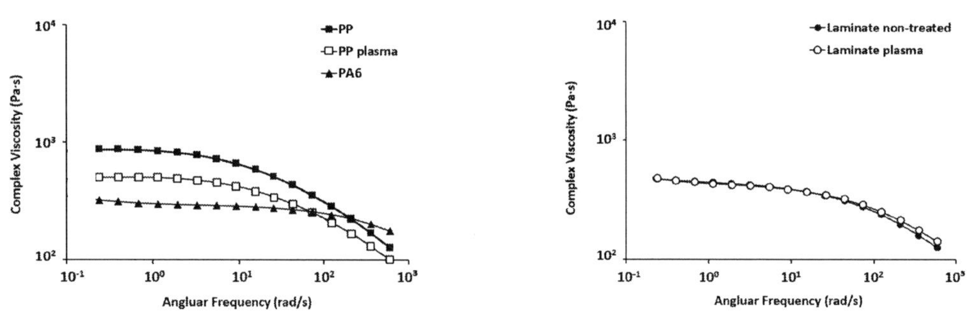

(a) (b)

FIGURE 5. Complex viscosities of pure polymers (a) and 1-interphase laminates (b). The presence of plasma-treated PP is indicated with 'plasma'. The frequency sweeps were performed at 240 °C, 5% strain and a gap of 0.15 mm.

Secondly, laminates containing one interphase were characterized, in order to reduce the influence of the laminate structure on the measurement. Similar to the pure polymers measured with a gap of 0.25 mm, the measurements at 0.15 mm distance show the degradation of PP after plasma treatment. The complex viscosity of plasma-treated PP decreases (Fig. 5a) and the moduli shift to the right (Fig. 6a), compared to the non-treated PP. The right-shift of the gelation points of the plasma-treated PP (Fig. 6a) emphasizes that PP is degraded due to the oxidative effects of the plasma treatment, resulting in chain scissions. However, this degradation is not seen in the measurements of the laminates. The complex viscosity (Fig. 5b) and the moduli (Fig. 6b) of the laminate containing modified PP are equal to the values of the non-treated laminate. These findings indicate that the PP degradation is compensated by the interfacial strengthening between PA6 and plasma-treated PP.

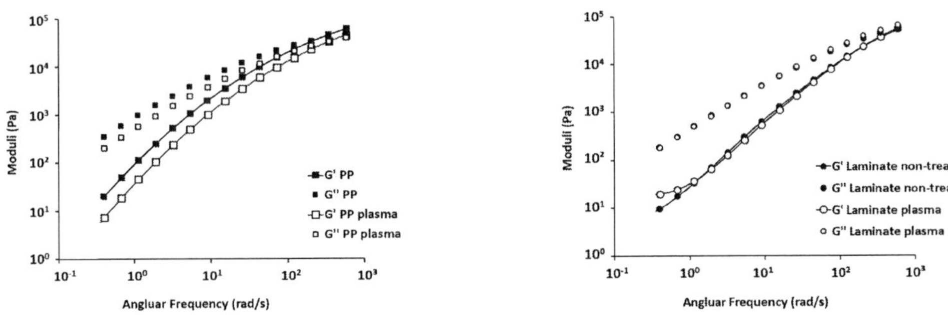

(a) (b)

FIGURE 6. Storage- and loss moduli of pure polymers (a) and 1-interphase laminates (b) at 240 °C, 5% strain and gap 0.15 mm.

CONCLUSION

An increased bond strength was observed between PA6 and plasma-treated PP compared to non-treated PP. This improved adhesion is based on the presence of carboxylic groups on the plasma-treated PP film, resulting in e.g. hydrogen bonding with PA6, and on the reduced polarity. Furthermore, rheological characterization of PA6/PP laminates using the parallel-plate system has shown that PP is degraded due to the plasma treatment and that plasma treatment of PP improves the interfacial adhesion between PA6 and PP. In addition, we observed that the structure of the laminate influences the rheological measurement. Thus, the investigation of the interfacial adhesion between PA6 and PP still remains a challenge. Further investigations using dynamic mechanical thermal analysis (DMTA) and using other PP modifications, such as grafting with maleic anhydride, will provide more basic understanding of the PA/PP interphase at chemical level and contribute to the development of high-performance composite materials.

ACKNOWLEDGMENTS

The authors gratefully acknowledge financial support from the Austrian Federal Ministry of Transport, Innovation and Technology (BMVIT) to the Endowed Professorship Advanced Manufacturing FFG-846932, V-trion textile research GmbH (Austria) for the low-pressure plasma treatment, the Microtechnical department of the Fachhochschule Vorarlberg (Austria) for access to the Zwick machine for shear testing, the Transfercenter für Kunststofftechnik GmbH Wels for PA6 film processing and Borealis Polyolefin GmbH for PP film processing.

REFERENCES

1. Y. Yang, R. Boom, B. Irion, D.J. van Heerden, P. Kuiper and H. de Wit, Chem. Eng. and Process.: Process Intensification **51**, 53-68 (2012).
2. J. Pigłowski, I. Gancarz, M. Wlaźlak and H. W. Kammer, Polymer, **41**(18), 6813-6824 (2000).
3. A. Sacchi, L. Di Landro, M. Pegoraro and F. Severini, Eur. Polym. J. **40**(8), 1705-1713 (2004).
4. X. M. Zhang, Z. Yin, Y. Song and J. Yin, Chin. J. Polym. Sci. **17**(3), 237-244 (1999).
5. C. Laurens, R. Ober, C. Creton and L. Leger, Macromolecules **37**(18), 6806-6813 (2004).
6. A. Valenza, G. Spadaro and D. Acierno, Colloid. Polym. Sci. **278**(10), 986-992 (2000).
7. R. Morent, N. de Geyter, C. Leys, L. Gengembre and E. Payen, Surf. Interface Anal. **40**(3-4), 597-600 (2008).

Improved flame retardancy for poly(butylene terephthalate) with phosphorus-containing polyesters

Pospiech Doris (1)*, Korwitz Andreas (1), Fischer Oliver (1), Ciesielski Michael (2), Döring Manfred (2), Schartel Bernhard (3), Brehme Sven (3), Altstädt Volker (4)

(1) Leibniz-Institut für Polymerforschung Dresden e.V. - Dresden, Saxony - Germany, (2) Fraunhofer-Institut für Betriebsfestigkeit und Systemzuverlässigkeit LBF - Darmstadt - Germany, (3) Bundesanstalt für Materialforschung und -prüfung BAM - Berlin - Germany, (4) Universität Bayreuth - Bayreuth - Germany

Applications of commercial engineering polyesters like poly(butylene terephthalate) (PBT) in transportation, aviation and construction require, among other parameters, a high level of flame retardancy. Flame retardancy is often achieved by adding halogen-containing flame retardants (FR). Recent regulations [1] restrict a number of most efficient compounds. Therefore, the goal of the project was to develop polymeric FRs with polyester backbone for halogen-free flame retardancy with high FR efficiency, preventing blooming of the FR to the surface and not affecting the mechanical property level of the base polyester PBT. In presentation we will outline the influence of the chemical structure of different phosphorus-containing polyesters on their thermal degradation, pyrolysis and burning behavior. Polyesters with 9,10-dihydro-9-oxa-10-phospha-phenantrene-10-oxide (DOPO) DOPO substituents generally showed the best FR potential with respect to high char formation, reduction in total heat release, fire growth rate etc. combined with the occurrence of intumescence [2,3]. After analysis of the neat phosphorus polyesters, they were employed as FR additives in PBT. Their influence on processing, burning behavior and resulting mechanical properties will be discussed [4]. In a next step, phosphorus-containing PBT nanocomposites with either layered silicates or multiwalled carbon nanotubes were prepared and the additional influence of nanoobjects was evaluated. References [1] WEEE (2012/19/EU); RoHS (2011/65/EU). [2] Fischer, O.; Pospiech, D.; Korwitz, A.; Sahre, K.; Häußler, L.; Friedel, P.; Fischer, D.; Harnisch, C.; Bykov, Y.; Döring, M.: Synthesis and properties of phosphorus-polyesters with systematically altered phosphorus environment, Polym. Degrad. Stab. (2011), 96, 2198-2208. [3] Brehme, S.; Köppl, T.; Schartel, B.; Fischer, O.; Altstädt, V.; Pospiech, D.; Döring, M. Macromol. Chem. Phys. 2012, 213, 22, 2386-2397. [4] Köppl, T.; Brehme, S.; Pospiech, D.; Fischer, O.; A

Nanocellulose reinforced thermoplastic composites of poly(ethylene acrylic acid)

Thunberg Johannes [(1)]*, Venkatesh Abhijit [(1)], Boldizar Antal [(1)]

[(1)] Chalmers University of Technology - Goteborg - Sweden

Cellulose pulp fibres are well known reinforcing agents in composites. The benefits of using a natural cellulose reinforcing agent is its sustainable origin, abundance, and low cost. Highly fibrillated cellulose pulps, called cellulose nanofibrils (CNF) or microfibrillated cellulose (MFC), are small cellulose fibrils with nanoscale diameter and lengths on the micrometer scale. The advantage of CNF compared to pulp fibres is the high length to diameter aspect ratio, which is an important factor that influences the reinforcing effect in thermoplastic composites. In this work CNFs are used as reinforcement in a thermoplastic poly(ethylene acrylic acid) (EAA) matrix. A novel approach of mixing water dispersed EAA with CNF was used to avoid fibril agglomeration and cellulose hornification. Using this wet-mixing method, cellulose content of up to 70 vol. % could be achieved in the composites. The mechanical strength of CNF composites exceeded those of plain pulp fibre composites. The stiffness could be improved by a factor of 25 and the strength by a factor of 3. X-ray tomography reveal the microstructure of the fibres and fibrils in the composites and the wet-mixing method produced well dispersed CNF reinforced composites.

Tailored Wood Strand Plastic Composites for High Performance Applications

Hartmann Robert [(1)*], Koch Michael [(1)]

[(1)] Technische Universität Ilmenau - Thuringia - Germany

Processing of thin, oblong wood shavings (strands) is defined by the production of three-layered oriented strand boards (OSB). During the OSB manufacturing process, strands are wetted with a mixture of binder and additives in a large-volume drum, placed quasi-isotropic on a belt and are compacted in an open mold with specific conditions of temperature and pressure. Edge areas of OSB components have to be costly reworked and achievable mechanical properties are limited by the small proportion of strands that are aligned with significant load paths. Utilized binder systems for OSB are based on formaldehyde, which was classified as carcino- and mutagenic by the European Union in 2014. This publication introduces a high performance wood strand plastic composite material (WSPC), which compensates major disadvantages of OSB and enables the production of free-formed, load-conform, zero-waste and non-toxic lightweight components. Options for an automated WSPC production process were developed and requirements for suitable binder systems determined considering viscosity, adhesion and environmental compatibility. The behavior of strands when pressurized was investigated to specify the limit of the compression pressure which allows a maximum in mechanical characteristics while avoiding damage of the wood cell walls. Specimens were produced with an experimental mold by combining deliberately arranged strands with diverse formaldehyde-free thermoplastic and thermosetting matrices using resin injection and hot compression molding techniques. The measured density-specific bending strength parallel to wood fiber direction of a specimen compressed at 10 MPa and consisting of 10 wt-% polypropylene outperformed the reference sample made of OSB3 by 350 % and the comparative metallic sample made of aluminum alloy AlMg3F24 by 70 %. The proven lightweight potential and cost-effectiveness of WSPC are justifying its use for high performance applications in furniture, vehicles and construction.

Use of Recycled Waste Paper as Fiber Reinforcement for Polypropylene –Relationship of Fiber Extraction Process and Mechanical Properties of the Composites

Scholten Andrea[a)], Meiners Dieter[b)]

Institute for Polymer Materials and Plastics Engineering, Clausthal University of Technology Agricolastraße 6, 38678 Clausthal-Zellerfeld, Germany

[a)] Corresponding author: andrea.scholten@tu-clausthal.de
[b)] dieter.meiners@tu-clausthal.de

Abstract. During recycling of waste paper from private households different fractions containing plastics and polluted paper fibers are received. Current research at the Institute of Polymer Materials and Plastics Engineering at Clausthal University of Technology showed that the use of waste paper as filler and reinforcement in polypropylene leads to nearly the same mechanical properties as natural fibers like flax, sisal and hemp. The present investigations focus on the chemical cleaning of the paper recyclate to treat the fiber surface and to solve foreign plastics like polystyrene. For this purpose SOXHLET extraction with different solvents is used. The received extracts are examined via infrared spectroscopy, as well as the treated and the untreated fibers. Afterwards, the cleaned paper is incorporated into polypropylene. The received composites are investigated regarding their mechanical properties by tensile testing and CHARPY impact testing.

INTRODUCTION

Recycling of waste paper from households already started in the 18th century [1]. Since then the paper recycling industry developed further and further and reached recycling shares of up to 99% [2]. Still a high amount of paper fibers which enter the recycling process is rejected with other extraneous material like packaging films, expanded polystyrene (EPS) and hard wood splints. In 2014 ca. 17.000 t of process waste incurred in paper recycling factories in Germany [3]. Since more complex treatments of those mass flows aren't economically reasonable for the paper and cardboard industry at the moment, these rejections are used in incinerating plants to produce energy.

The fraction considered in these investigations exits the process at the marked process step in Fig. 1.

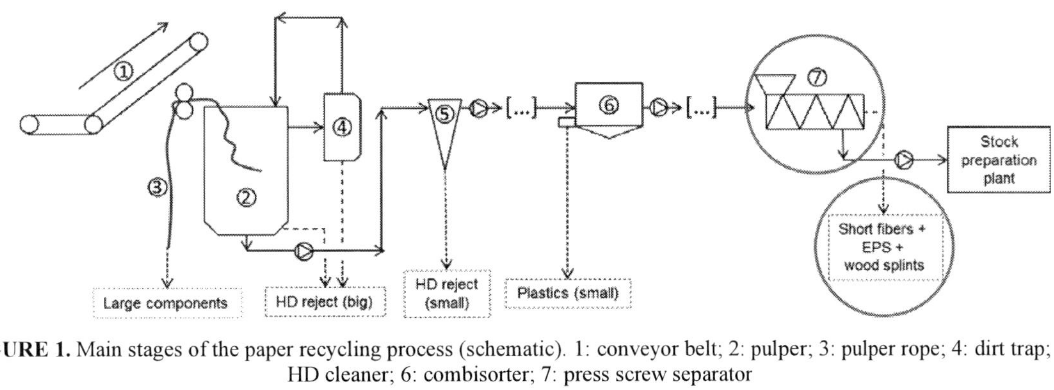

FIGURE 1. Main stages of the paper recycling process (schematic). 1: conveyor belt; 2: pulper; 3: pulper rope; 4: dirt trap; 5: HD cleaner; 6: combisorter; 7: press screw separator

The first step in waste paper recycling is the dissolving of the paper fibers. Therefor the clustered waste paper is transported into a pulper filled with water. Unsolvable, bigger components of the waste are pulled out of the pulper with a pulper rope. Materials with a high density (HD) are separated with different aggregates like dirt traps and high density cleaners. Small plastic particles get rejected by a combisorter. Although the reject from the combisorter should be nearly fiber free the reject from the following reject press and press screw separator possess high fiber content. Other main components in this fraction are expanded EPS and hard wood splints.

The high fiber content of this fraction brought forth the idea to use it as reinforcement in thermoplastic materials. First investigations showed that the properties of polypropylene reinforced with the dried and cut waste paper material lead to mechanical properties similar to natural fiber reinforced polypropylene. The use of maleic anhydride grafted polypropylene (MAPP) as coupling agent was also successful [4]. Challenges in the use of the material are mainly the separation of other components like EPS and wood splints, but also of substances on the surface of the fiber which may influence the fiber matrix adhesion and the effect of coupling agents like MAPP.

EXPERIMENTAL WORK

The waste paper material was received from SMURFIT KAPPA HERZBERG SOLID BOARD GMBH in a moist state. Before further processing, the paper was dried in a drying chamber for at least 24 h at 80 °C and was cut in a cutting mill S15/20 by DREHER to obtain a processable filler material (see Fig. 2).

FIGURE 2. Dried and cut waste paper

Subsequently the mostly fibrous material was extracted by SOXHLET extraction. Five different solvents, namely water, methanol, ethanol, acetone and n-hexane, were used for extraction. The solvents were heated approx. 40 °C above boiling temperature to reassure the progress of the process for 24 h. Additionally one sample was prepared by simply stirring the fibers in acetone at room temperature for 16 h. The cleaned fibers were dried at 80°C for 24 h, whereas the received solution was concentrated with the aid of a rotary evaporator until a brownish solid or very high viscous brownish to orange liquid (color dependent on the solvent) was obtained. IR spectroscopy of the cleaned material and the extracts was performed with an Infrared Vacuum Spectrometer and an ATR-Unit Platinum ATR Diamond F Vacuum from BRUKER and a BRUCKER Alpha T-IR.

Testing samples were manufactured by compounding 30 w-% of the cleaned material and one untreated reference with 70 w-% of a semi-crystalline polypropylene H734-52RNA by DOW CHEMICAL (density = 0.9 g/cm³, MFI = 52 g/10min at 230 °C and 2.16 kg) in a laboratory kneader PolyLab OS with mixing unit Rheomix OS by THERMO HAAKE at 180 °C and 50 rpm. Thereby received compounds were shred in a cutting mill S15/20 by DREHER and injection molded into samples Type 1BB according to DIN EN ISO 527-1 with an injection molding machine ARBURG Allrounder 220S. The temperature profile from feed hopper to nozzle was 180 °C/ 185 °C/ 185 °C/ 190 °C.

For mechanical testing the samples were stored at 23°C and 50 % r.h. for 24 h. At least ten samples per mixture where tested. Tensile testing was performed with a universal testing machine by ZWICK ROELL according to DIN EN ISO 527. The free clamping length was 22 mm and the testing velocity was 2 mm/min. CHARPY impact strength measurements were carried out with a pendulum machine Zwick 5113 according to DIN EN ISO 179 (1 J hammer, effective span 22 mm).

RESULTS

Mechanical Testing

CHARPY Impact Strength

Figure 3 shows the results from the CHARPY impact strength measurements. The mean value and the standard deviation of the untreated material are highlighted and extrapolated to simplify the comparison with the other samples. Nearly all samples of the treated composites have significantly lower CHARPY impact strength than the untreated composite. Only the sample with fibers extracted with methanol has higher impact strength than the other treated samples, which is still slightly lower than the untreated sample.

The extraction with ethanol, acetone and n-hexane removes the EPS from the waste paper. Only the material extracted with water and methanol still contains EPS. It is possible that the PS, which is naturally immiscible with the PP, forms small conglomerates in the matrix during compounding and thereby raises the impact strength of the material compared to a material containing only paper fibers. Since the samples extracted with water don't show similar values than the untreated sample this effect can't be the only one influencing the impact strength. It is also possible that other components of the waste material, which cover the fiber surface and could influence the fiber matrix adhesion, are removed during the extraction. This effect should also lead to a reduction of tensile properties.

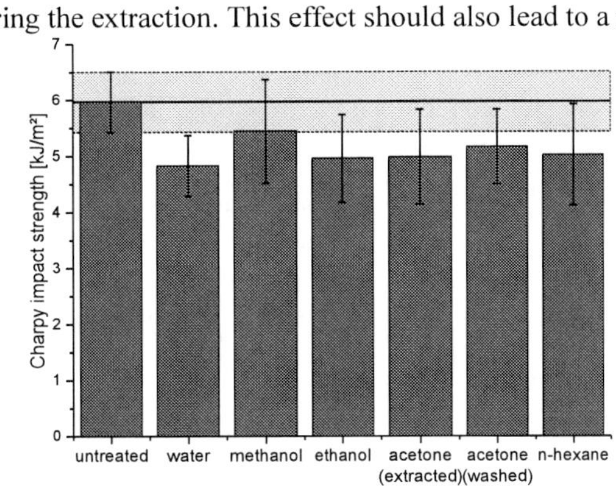

FIGURE 3. CHARPY impact strength of the composites reinforced with 30 w-% of untreated or extracted fibers

Tensile Testing

Figure 4 shows the determined tensile strength and YOUNG's modulus of the prepared composites. Again the black line represents the mean value of the composite containing the untreated paper material. The grey background highlights the standard deviation of this material.

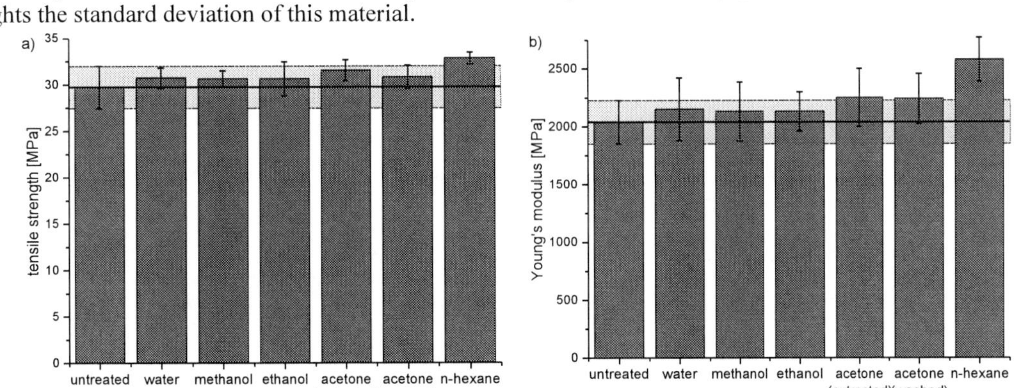

FIGURE 4. Tensile strength (a) and YOUNG's modulus (b) of the composites reinforced with 30 w-% of untreated or extracted fibers

It can be clearly seen, that the extraction of the fibers leads to higher mean values of the tensile properties in all given cases. For the tensile strength (Fig. 3 a) it needs to be questioned if the small increase within the standard deviation of the untreated material can be classified as a real increase in mechanical properties caused by the extraction of the fibers with water, methanol, ethanol or acetone. In case of n-hexane the increase of the tensile strength can't be doubted. Especially the YOUNG's modulus of the samples treated with acetone and again n-hexane is noticeable improved compared to the untreated sample. During the extraction process different components of the waste paper that influence fiber matrix adhesion and the wetting of the fiber with the matrix material seem to be dissolved. In opposition to the impact strength, the extraction of these components leads to an increase of properties. To get more information about the nature and the effect of these components IR spectroscopy was used.

IR Spectroscopy

IR spectra of the analyzed fibers are displayed in Fig. 5a. All spectra show nearly the same absorption peaks with higher or lower intensity. In comparison with a pure cellulose fiber no differences can be seen. This indicates that the components on the fiber surface of the untreated paper waste have a low concentration. The mainly measured material is the cellulosic paper fiber and not the contaminants.

Since the surface of the fibers doesn't give any hints about the extracted components, the extracts itself were analyzed (see Fig. 5b). First thing to note is that the substance extracted with water has a very low absorption rate. The received extract is the only one, which is in a completely solid, mostly crystalline state. This indicates that mostly salts and other water soluble components were extracted which are on the other hand hard to detect via IR spectroscopy. The extracts of the other solvents have kindred spectra. CH_2 groups are responsible for the peaks between 2959 cm^{-1} and 2850 cm^{-1} (stretch vibration), 1480 cm^{-1} and 1420 cm^{-1} (deformation vibration) and 1290 cm^{-1} and 1200 cm^{-1} (twist vibration). The peak around 1730 cm^{-1} indicates a C=O vibration. Other vibrations caused by interactions of C and O like CCO and $-CH_2-O-CH_2-$ vibrations can be found around 1120 cm^{-1} and 1020 cm^{-1}. The intensity of theses peaks and the broad, but weak OH peak between 3700 cm^{-1} and 3200 cm^{-1} decreases from methanol, over ethanol and acetone to n-hexane. In contrast the peak at 1170 cm^{-1}, which likely is linked to branched alkanes increases [5].

These two effects can be explained with the chemical structure of the used solvents. Whereas methanol itself contains both an OH and a CH (or CH_3) group, hexane consists only of non-polar CH_2 (or CH_3) groups. Based on the principle "like dissolves like", methanol, ethanol and also acetone are likely to dissolve hydrophilic and hydrophobic substances. N-hexane as a hydrophobic solvent dissolves primarily hydrophobic substances. All in all the spectra show, that all solvents (except water) dissolve substances containing CH_2 groups and in case of the hydrophilic solvents also OH or ether bonds. These chemical groups are for instance part of fats, waxes, colorants and glues, which contaminated the paper during use.

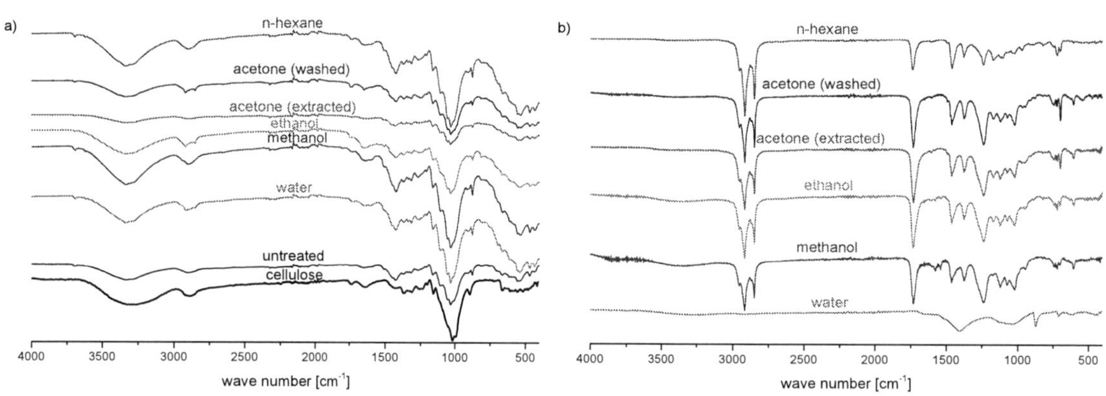

FIGURE 5. IR spectra of cellulose, untreated and treated waste paper fibers (a) and received extracts (b)

A quantification of the polar nature of a solvent can be given by the empirical polarity parameter E_T^N [6]. The higher the polarity of a solvent, the higher is the value of E_T^N, whereat water has a value of 1 and tetramethylsilane as a highly non-polar and chemically nearly inert substance is defined as 0. E_T^N values for the solvents used in this work are given in Table 1 together with the relative weight loss due to SOXHLET extraction. It can be clearly seen that for organic solvents a lower polarity parameter leads to a higher weight loss. Combined with the information from the IR spectroscopy it can be stated that n-hexane dissolves a high amount of non-polar contaminants like fats and waxes. The more polar organic solvents are less able to solve non-polar substances but can also dissolve polar contaminants. Water is unable to dissolve non-polar substances but has a high solubility for polar contaminants, which results in a slightly higher weight loss compared with methanol, ethanol and acetone.

TABLE 1. Empirical polarity parameter E_T^N [6] of the solvents and relative weight loss due to extraction

Solvent	E_T^N	Relative weight loss in %
n-hexane	0,009	0,12
acetone	0,355	0,08
ethanol	0,654	0,08
methanol	0,762	0,05
water	1	0,09

CONCLUSION AND OUTLOOK

The presented results show that the SOXHLET extraction of the waste paper with different solvents influences the mechanical properties of the processed composites. During the extraction process other components like EPS are partly removed and the fibers are cleaned from fats and waxes, which could be detected via IR spectroscopy of the extracts. This cleaning results in a slight increase of tensile strength and YOUNG'S modulus. The amount of increase is depended on the polarity of the solvent. Treatment with n-hexane leads to the highest tensile properties. It is probable that the dissolved, mainly non-polar substances hinder fiber matrix adhesion or act as lubricants. For the CHARPY impact strength no improvement could be proved. The composites with the extracted paper all had lower impact strength than the untreated reference. These results seem to be induced by the removal of fats and waxes on the fiber surface and the removal of EPS which may cause shock absorbing conglomerates in the polypropylene.

Future research in this area needs to focus the effect of EPS in the composites. A test series with higher amounts of EPS has already started. For further understanding of the fiber cleaning and the resulting fiber matrix interaction other analytical methods need to be evaluated, since IR spectroscopy of the fibers shows predominantly peaks linked to cellulose.

ACKNOWLEDGMENTS

Special thanks go to Mr. Jacobi and Mr. Picht (SMURFIT KAPPA HERZBERG SOLID BOARD GMBH) for supplying the waste paper. We would also like to thank Mr. Ehrhardt and Mr. Söftje from the Institute of Organic Chemistry of the Clausthal University of Technology for the help with the IR spectroscopy and the constructive discussions.

REFERENCES

1. A. Zenger, Ph.D. thesis, University of Regensburg, 1995.
2. K.J. Thomé-Kozmiensky, D. Goldmann, *Recycling und Rohstoffe*, Vol. 6 (TK Verlag, Neurupin 2013).
3. Statistisches Bundesamt, *Umwelt – Abfallentsorgung 2014*, Wiesbaden, 2016, available at https://www.destatis.de/DE/Publikationen/Thematisch/UmweltstatistischeErhebungen/Abfallwirtschaft/Abfallentsorgung2190100147004.pdf?__blob=publicationFile .
4. A. Scholten, N. Plogmann, D. Meiners, "Use of recycled waste paper as fiber reinforcement for polypropylene – Examination of mechanical properties in comparison to NFRP" in *21st Symposium on Composites*; Trans Tech Publications Ltd (Bremen, 2017), to be published.
5. G. Socrates, *Infrared and Raman Characteristic Group Frequencies: Tables and Charts, Third Edition* (John Wiley & Sohns LTD, Chicester, 2004).
6. C. Reichardt, Chem. Rev., 2319-2358, 94, (1994)

Properties of Lyocell-Polypropylene Composites

M. Cordin[1, a)], U. J. Griesser[2, b)], T. Bechtold[1, c)] and T. Pham[1, d)]

[1]*Research Institute of Textile Chemistry and Textile Physics, University of Innsbruck, Hoechsterstrasse 73, 6850 Dornbirn, Austria*
[2]*Institute of Pharmacy, Department of Pharmaceutical Technology, University of Innsbruck, Innrain 52c, 6020 Innsbruck, Austria*

[a)]Michael.Cordin@uibk.ac.at
[b)]Ulrich.Griesser@uibk.ac.at
[c)]Thomas.Bechtold@uibk.ac.at
[d)]Corresponding author: Tung.Pham@uibk.ac.at

Abstract. High performance and lightweight composites were prepared by the thermoforming of special source materials, consisting of intermingled polypropylene and Lyocell fibres. The fibres were arranged either parallel in form of fibre slivers or they were aligned in different directions by using embroidered fleeces. Thus the influence of the fibre alignment on the mechanical properties was studied. The mechanical properties were characterized by tensile test experiments and dynamic mechanical thermal analysis (DMTA). Besides the moisture sorption kinetics of composites with different polypropylene proportion was studied by using the Parallel Exponential Kinetics (PEK)-model.

INTRODUCTION

Fibre reinforced plastics play a leading role in the production of lightweight materials. Carbon, glass and aramid fibres are the preferred reinforcing materials in composites. But there is an interest to replace these synthetic fibres by natural ones, because the natural fibres have advantages like environmental sustainability and biodegradability [1, 2]. Especially the usage of cellulose fibres in combination with thermoplastic matrix polymers like polypropylene is advantageous with regard to an efficient mass production of composites. The main challenge in the preparation of this type of composite is the poor adhesion between the hydrophilic cellulose and hydrophobic polypropylene and the distribution of cellulose fibres in the polypropylene matrix. There are various strategies to improve the adhesion between these two components [3-9]. In the present study, we investigated the usage of thin intermingled fibres as source material for the thermoforming to increase the interface interaction between cellulose fibres and polypropylene matrix, and thereby to yield a good profile of the mechanical performance. The high moisture sorption of cellulose lowers the composite performance. Particular attention was paid to this problem by analysing the moisture sorption kinetics by the PEK model [10, 11].

EXPERIMENTAL

Material and Sample Preparation

Polypropylene (PP) and Lyocell type cellulose fibre slivers and embroidered fleeces were the source material for the preparation of composites. In one set fibre slivers with different weight ratios of PP and Lyocell fibres (50:50, 40:60 and 30:70) were converted into composites by thermoforming, and in another set composites were prepared by thermoforming of the embroidered fleeces with different orientations of Lyocell fibres. The weight ratio in the embroidered fleece composite was 62% PP and 38% Lyocell. The source materials were supplied by Schoeller GmbH & CoKG. The fibre fineness in the different source materials was 3.2 dtex for the PP and 3.4 dtex for the Lyocell fibres.

Analytical methods

Tensile properties were determined on a Shimadzu Autograph AG-5000C for samples prepared from the fibre slivers. The tensile properties of the composites prepared from embroidered fleeces were performed on the Zwick-Roell device. Mechanical properties were also characterized by dynamic mechanical thermal analysis (DMTA). These measurements were carried out on an Anton-Paar MCR302 device. The moisture sorption/desorption behaviour of the composites was analysed gravimetrically on a multisample moisture sorption analyser SPS11 (ProUmid, Germany-Ulm). The measurement was started by conditioning the sample at 25°C and 30% relative humidity (RH). After decreasing the relative humidity to 0% and achieving equilibrium, the actual sorption and desorption measurement was started. The relative humidity was changed in 10% steps to obtain a moisture sorption/desorption cycle. The mass of each sample was recorded in intervals of 10 min.

RESULTS AND DISCUSSION

Composites from different source materials were prepared, on the one hand from intermingled fibre slivers with different Lyocell-PP ratio (50/50, 60/40 and 70/30 % w/w) and on the other hand from PP-fleeces with intermingled yarns (62:38 PP:Lyocell). The source material was converted into composites by heating up to 180°C under pressure (1.6 bar for the fibre sliver composites and 50 bar for the composites, prepared from the embroidered fleeces), as illustrated in Fig. 1. Despite the high viscosity of molten PP, the concept of intermingled fibres improves the wetting of the Lyocell fibres by the PP matrix. The Lyocell fibres act as reinforcement in the composites and depending on its orientation it can absorb more or less tensile stress. In the composites prepared from the fibre sliver, the Lyocell fibres are oriented along the direction in which the tensile force occurred in the tensile test experiment. The use of embroidered fleeces as preform for the composite preparation offers the possibility to arrange the reinforcing fibres in special orientations with respect to the tensile force direction. Composites with parallel, 0°/90° and ±45° fibre orientation with respect to the tensile force direction were prepared (Fig. 1b).

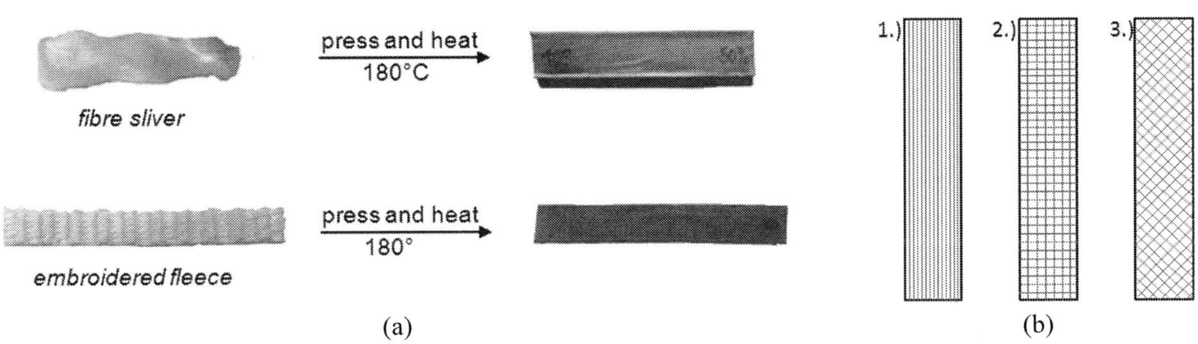

FIGURE 1. (a) Preparation of composites by thermoforming of fibre slivers or embroidered fleeces. (b) Embroidered fleece composite with (1) parallel oriented Lyocell fibres (= composite parallel), with (2) parallel and normal oriented Lyocell fibres (= composite 0°/90°) and with (3) ±45° fibre orientation (= composite 45°).

The mechanical properties of the different composites were characterized by tensile test experiments (Fig. 2 and 3). The ultimate tensile strength values are shown in Fig. 2. The highest ultimate tensile strength was observed for the fibre sliver composites with 50% PP, this type of composite has a tensile strength of 161 MPa, which is approximately three times the value of the PP. So there is a significant improvement of the tensile strength by adding Lyocell fibres (with parallel orientation) to the PP matrix. The influence of the Lyocell fibre orientation can be studied by the different composites, prepared from the embroidered fleeces. As expected the highest strength was observed for the composite with parallel Lyocell fibre orientation. The strength is about 60% smaller than that of the fibre sliver composite. There are two reasons for the difference: firstly, the PP proportion is larger with 62% w/w and secondly in the embroidered fleece the Lyocell fibres have a slightly wavy form, which reduces the reinforcement of the composites. The ultimate tensile strengths of the composites with ±45° and 0°/90° Lyocell fibre orientation are comparable to that of the pure PP sample, in these cases the fibres don't contribute to the ultimate tensile strength.

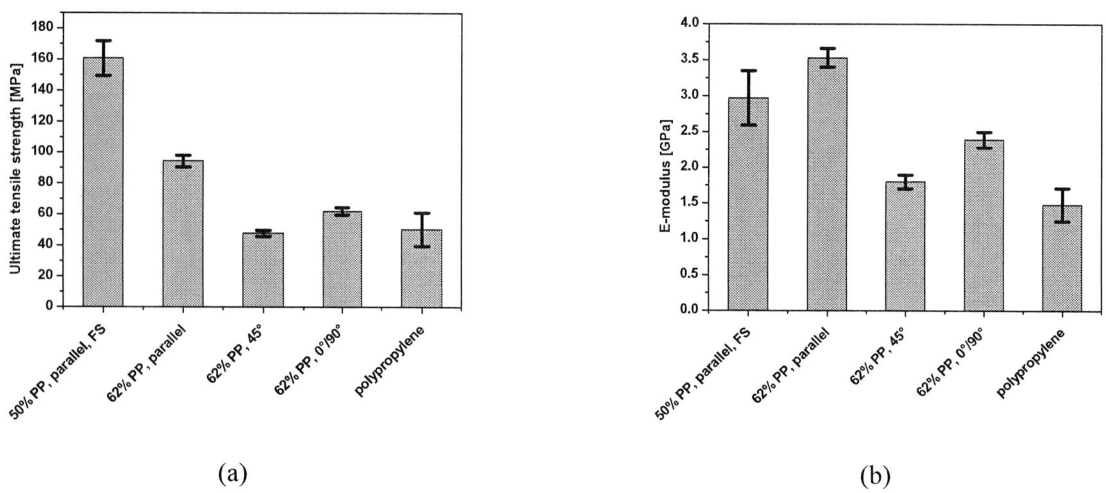

FIGURE 2. (a) Ultimate tensile strength and (b) E-modulus of the different composites and PP

In Fig. 2b the E-modulus of the different composites are shown. The stiffness of all composites are improved compared to the pure PP. The sequence of the E-moduli is the same as for the ultimate tensile strength, with one exception: the highest stiffness was observed for the composites, prepared from the embroidered fleeces with parallel Lyocell fibre orientation. The E-modulus is determined from the linear range in the tensile force versus strain relation, this range is at the beginning of the tensile test experiment.

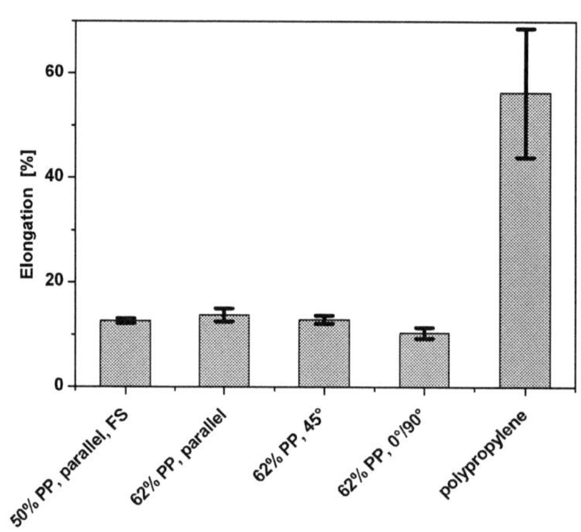

FIGURE 3. Elongation at ultimate tensile strength of the different composites and PP

Obviously the composite prepared from embroidered fleece with parallel Lyocell fibre orientation shows the best performance in this range. In Fig. 3, the elongations at ultimate tensile strength are compared between the different composites. The pure PP is the most extensible material in this comparison, it is possible to stretch the PP to around 150% of the initial length. The fibre reinforcement in the composites prevents the stretching to such high values, the observed length changes are in the range of 10 to 15 %. The reason for the low elongation could be the low adhesion between PP and cellulose fibres as no phase coupling agent is used. The mechanical properties of the composites prepared from the embroidered fleeces were also characterized by dynamic mechanical thermal analysis (DMTA). Note that the bending force in the DMTA measurement is normal to the direction, in which the tensile force was applied in the tensile test experiment. In Fig. 4a the storage modulus and in Fig. 4b the loss modulus for the different composites and a pure PP sample are shown. The highest value for the storage modulus was observed for the composite with parallel fibre orientation. In the case of the loss modulus the composite with 45° fibre orientation shows the highest value. The storage and loss modulus of the pure PP is significant smaller. At around 160°C the PP starts to melt and so the storage and loss modulus are close to zero.

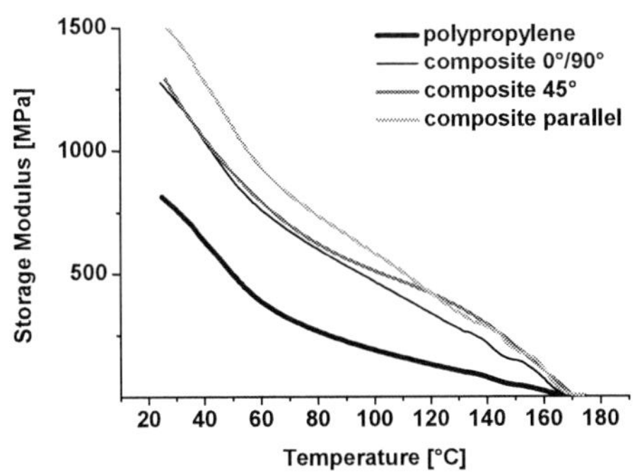

(a)

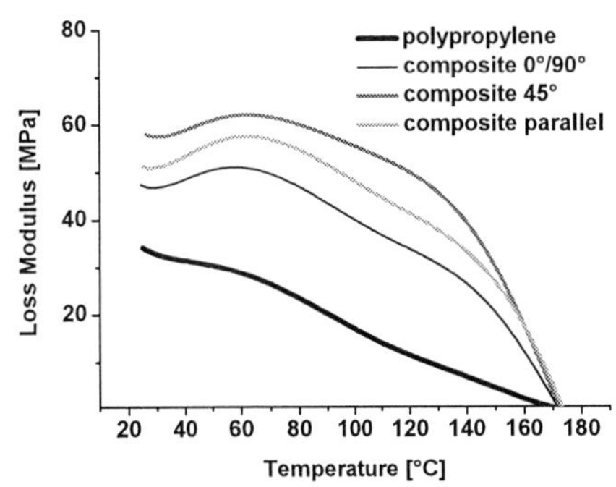

(b)

FIGURE 4. Storage and loss modulus of the different embroidered fleece composites and the pure polypropylene sample

The moisture sorption of fibre sliver composites with a Lyocell fibre proportion of 50, 60 and 70 % w/w was analysed by measurement of the weight change of the different samples as a consequence of climate change. With the so-called Parallel Exponential Kinetics model (PEK) it is possible to describe the sorption and desorption kinetics on each level of relative humidity by two independent first order processes [10, 11].

$$M_t = M_1(1 - e^{-k_1 t}) + M_2(1 - e^{-k_2 t}) \tag{1}$$

The percentage mass increase M_t was fitted by equation (1). M_1 and M_2 are the percentage mass at equilibrium for the fast and slow sorption process and k_1 and k_2 are the rate constants for the sorption of moisture.

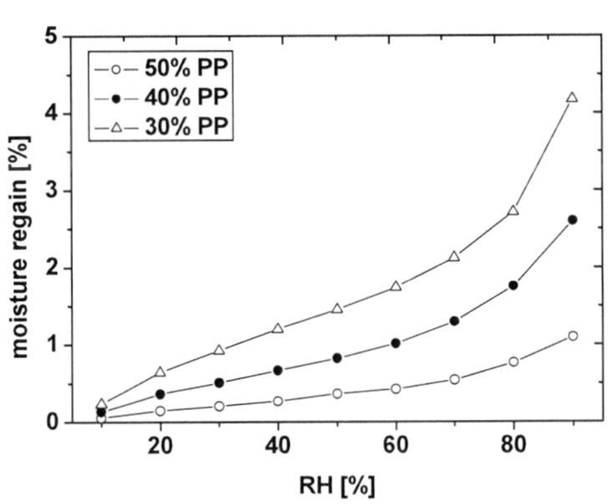

(a)

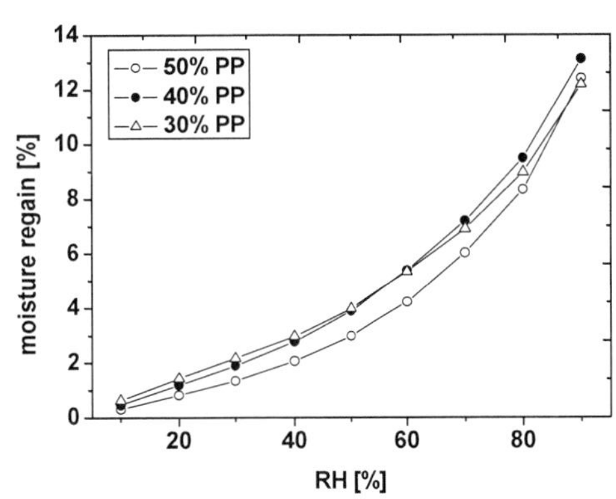

(b)

FIGURE 5. The moisture regains (normalized to 100% Lyocell) for the fast moisture sorption process (left) and the slow moisture sorption process (right) are plotted versus relative humidity (RH) for the different composites.

In Fig. 5 the moisture regain curves, calculated from the PEK parameters M_1 and M_2 (normalized to 100% Lyocell) are plotted versus relative humidity. Increasing PP proportion in the composites decreases the capacity M_1 and the related moisture regain for rapid moisture sorption. So it is assumed that the quickly sorbed water is mainly bounded on the Lyocell fibre surface, because the surface is (partly) covered with PP, which explains the PP dependency of the capacity M_1. The slow moisture sorption capacity M_2 and the related moisture regain is more or less independent from the PP content, so this water is probably sorbed in the bulk of the Lyocell fibres. In Fig. 6 the PEK parameters k_1 and k_2 are plotted versus relative humidity for the different composites. The rate constant k_1 is more or less independent from the PP proportion, because this is the water, which is sorbed by uncovered parts of the Lyocell fibres. k_2 decreases with increasing PP proportion, which means that the sorption velocity into the bulk decreases with increasing PP proportion. This can be explained by the decrease of uncovered entrances into the bulk with increasing PP proportion. The results of the PEK analysis show, that it is difficult to get a continuous layer of PP on the Lyocell fibres, which would protect the fibre from moisture and which would result in a good performance of the composite in the presence of moisture.

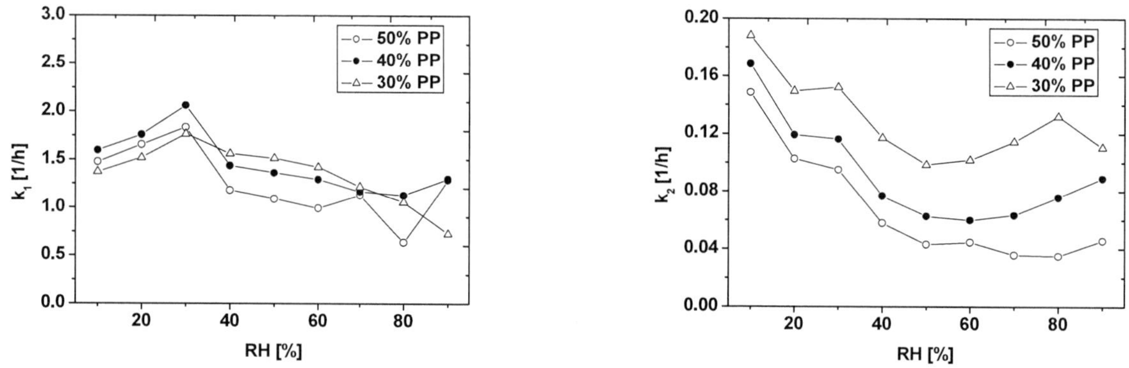

(a) (b)

FIGURE 6. PEK parameters k_1 and k_2 are plotted versus relative humidity for the different composites.

ACKNOWLEDGMENTS

The authors gratefully acknowledge financial support from the Austrian Federal Ministry of Transport, Innovation and Technology (BMVIT) to the Endowed Professorship Advanced Manufacturing FFG-846932.

REFERENCES

1. A. K. Bledzki, and J. Gassan, Progress in Polymer Science **24**, 221 (1999).
2. P. Wambua, J. Ivans, and I. Verpoest, Composites Science and Technology **63**, 1259 (2003).
3. M. N. Belgacem, P. Bataille, and S. Sapieha, Journal of Applied Polymer Science **53**, 379 (2003).
4. D. G. Gray, Cellulose **15**, 297 (2008).
5. C. Joly, R. Gauthier, and B. Chabert, Composite Science and Technology **56**, 761 (1996).
6. W. Qiu, F. Zhang, T. Endo, and T. Hirotsu, Journal of Applied Polymer Science **87**, 337 (2003).
7. H. Quan, Z. Li, M. Yang, and R. Huang, Composites Science and Technology **65**, 999 (2005).
8. D. T. Quillin, D. F. Caulfield, and J. A. Koutsky, Journal of Applied Polymer Science **50**, 1187 (1993).
9. D. T. Quillin, D. F. Caulfield, and J. A. Koutsky, Proceedings of Materials Research Society symposium **266**, 113 (1992).
10. R. Kohler, R. Dück, B. Ausperger, and R. Alex, Composite Interfaces **10**, 255 (2003).
11. R. Kohler, R. Alex, R. Brielmann, and B. Ausperger, Macromolecular Symposia **244**, 89 (2006).

Tailored epoxy system modified with block co-polymer, core shell rubber and hybrid: mechanical properties and fracture mechanisms.

Bajpai Ankur (1)*, ***Wetzel Bernd*** (1)

(1) IVW, TU Kaiserslautern - Rheinland-Pfalz - Germany

The mechanical and fracture performance of a modified epoxy resin cured with an amine hardener containing block copolymer and/or core-shell rubber nanoparticles (CSR) were investigated in the current work. The effect of adding a resin, i.e. fluoren di epoxy resin, to the bisphenol- F based epoxy resin was also investigated. The fracture energy of the neat bis-F (i.e. unmodified) epoxy polymer increased slightly from 0.10 kJ/m2 to 0.14 kJ/m2 due to the addition of 7 wt% of the fluoren di epoxy to the bis-F epoxy. The fracture energy of the modified epoxy polymer increased to 0.530 kJ/m2 and 0.640 kJ/m2 when 12 wt% of the CSR nanoparticles and 8 wt% of diblock copolymer were added to the modified epoxy separately respectively. Furthermore, the addition of D51N block copolymers and MX170 coreshell rubber particles performed to give hybrid epoxy polymer nanocomposites, i.e. which contained an equal amount of BCP's and CSR nanoparticles. No significant improvements in toughness were found for the hybrid epoxy polymer nanocomposites. The measured toughness of the hybrid materials can be related to the degree of dispersion of both nanoparticle phases in the epoxy polymer matrix. The toughening mechanisms were identified using SEM/EDAX fractography studies.

Cellulose Nanomaterials: A Natural Biopolymer as Sustainable Material for Future

Maji Pradip Kumar [(1)*], ***Yadav Chandravati*** [(1)]

[(1)] Indian Institute of Technology Roorkee - Uttar Pradesh - India

Cellulose nanomaterials obtained from the forest materials has proven itself as one of the miracle material that has spread its feet in wide application areas starting as reinforcement phase for preparing high strength composite materials and now in emerging areas of biomedical field. Cellulose, the basic building block of all trees and plants, is a long chain linear polymer composed of repeating units of anhydrous-β-D-glucopyranose linked together with β-(1→4) glycosidic linkage. The hierarchical microstructure possessed by cellulose molecules allows the extraction of cellulosic nanomaterials with at least one dimension in nanometer range, which are basically termed as "Nanocellulose". Nanocellulose has been widely categorized as cellulose nanocrystals, which possess rod or sphere like crystal structure, and the more flexible and longer cellulose nanofibres (CNFs). In the present work waste wood scraps has been used to extract CNFs by employing mild chemo-mechanical treatments. The work exclusively deals with the acid free extraction of CNFs with mechanical fibriallation using ultrasonication at a much lower output power. The diameter of obtained fibres as investigated through FESEM analysis lies in the range of 5 to 80 nm and a diameter range of 2-15 nm was obtained through AFM analysis. The CNFs with high aspect ratio were obtained that can act as one of the promising candidate for preparing green nanocomposites and can open up several opportunities in the field of bio-medical. The reinforcing potential of CNFs within the matrix material of SEBS-g-MA was also determined. Results showed the compatibility of CNFs with modified SEBS due to the presence of polar anhydride groups. Nanocomposite films with only 0.005 phr resulted in a drastic increment in tensile strength and modulus value as compared to neat polymer. As CNFs are supplemented with a flexible entangled network structure, high strength nanocomposites film along with a ductile behavior was fabricated.

Study on effect of organic compounds with OH group on conductivity enhancement of poly(3,4-ethylenedioxythiophene) doped with poly(styrene sulfonic acid) (PEDOT:PSS)

Konagaya Shigeji [(1)]*

(1) Nagoya Industrial Science Research Institute - Aichi - Japan

PEDOT:PSS has not conductivity enough to be replaced for indium tin oxide (ITO). Recently lots of secondary dopants such as ethylene glycol and DMSO have been proposed and applied to PEDOT:PSS for the conductivity enhancement. The authors have been studying the effect of organic compounds with hydroxyl groups such as alcohols, glycols and bisphenol compounds on the conductivity of PEDOT:PSS and its conductivity enhancement mechanism. Alcohols such as methanol, ethanol, iso-propanol etc., glycols such as ethylene glycol, diethylene glycol, poly(ethylene glycol) etc., and bisphenols such as 4,4'-dihydroxydiphenylsulfone (BPSO), 4,4'-dihdroxydiphenylsulfide (BPS), 4,4'-dihydroxydiphenylether (BPO), 2,4'-dihydroxydiphenylsulfone (24-BPSO) etc. were studied as a secondary dopant of PEDOT:PSS. The mixture of organic compound and PEDOT:PSS was dissolved in iso-propanol/water mixture. The solution was coated on the glass plate and dried. The surface resistivity (Rs) of the thin films was evaluated. It was found that there was a relationship between Rs values of the dried PEDOT:PSS/alcohol or glycol mixtures and solubility parameter (SP) of alcohols and glycols which support that the conductivity enhancement was possibly ascribed to the removal of PSS from the surface of PEDOT:PSS nanoparticles clusters by the solvent. On the other hand, there was a large difference in the conductivity enhancement effect among the species of bisphenol compounds. The addition of BPSO or 24-BPSO contributed to the conductivity enhancement of PEDOT:PSS, while BPO and BPS showed increase of its ñ value, although there is no difference in SP value among them. It was found that Rs values of dried PEDOT:PSS/bisphenol mixture has no relationship with SP values of bisphenol, but with electron withdrawing ability of substituents on benzene ring of glycols, that is, acidity of OH groups. The action mechanism of secondary dopant BPSO to PEDOT:PSS is now under study.

Antistatic properties of transparent plastics using a donor-accepter molecular compound antistatic agent

Watanabe Takuro (1)*

(1) Yamagata University - Graduate school of Science and Engineering - Japan

Typical plastic materials have a low electrical conductivity and dust adhesion to a product surface due mainly to static causes various disadvantages.As a countermeasure against static of plastic materials, the use of antistatic agents is gathering attention from viewpoints of convenience and costs, and the effect of surfactant agent to lower surface resistance have been reported. But, surfactant-agents tend to unlimitedly bleed out from plastic base materials. Therefore, it is difficult to keep antistatic property. Donor-accepter molecular compound antistatic agent (BN105) has unique properties that are capable of stably dispersing and remaining in a resin matrix and capable of instantly leaking static not only on a product surface but also through the inside. A little addition of BN105 may allow the prevention of static buildup efficiently without losing the other bulk properties of the matrix polymer resin. In the present study, cyclo olefin polymer (COP) / BN105 composite materials were prepared by using injection molding. Surface resistance of the molded product decreased to 2 ~ 10^11 ƒ¶ /sq along with an increase in BN105 content range up to 2 wt%, and these samples showed excellent dust resistance. In addition, the samples containing BN105 less than 5 wt% hardly had a transmittance decrease. Thus, the results offered hope for the creation of a composite material maintaining the high transparency of the base resin COP.

Deformation behavior of polylactide relating to plastic forming processes

Radusch Hans-Joachim [1]*, Wutzler Andre [2], Asghaier Meflet [1]

[1] University Halle-Wittenberg - Sachsen-Anhalt - Germany, [2] Polymer Service GmbH Merseburg - Sachsen-Anhalt - Germany

Polylactide (PLA) as one of the most important biobased thermoplastic polymers gained progressive importance in the last period for all types of plastics application, among them, film production by uniaxial or biaxial stretching, thermo (vacuum) forming or blow molding. Here, the deformation behavior of the material at elevated temperatures, but below the melting temperature of the crystalline phase, is of importance. So, the deformation behavior of PLA was investigated in dependence on temperature, especially in the glass transition temperature range, and in dependence on the deformation rate. A PLA grade was used with Mw=120,000 g/mol. The specimens for stress-strain investigation, thermal and structural characterization were produced by injection molding or extrusion, respectively. Uniaxial tensile tests were operated at temperatures between 40 and 80 °C as well as deformation rates between 0.1 and 20 s-1. Furthermore, some biaxial stretching tests were performed. On the basis of thermal characterization via DSC and structural investigation by WAXS the mechanisms of structure changes during deformation in dependence on temperature and deformation rate were discussed. It was found that at stretching of PLA molecular alignment is accelerated at higher temperatures, because of the higher segmental mobility that forces the stress-induced crystallization, but it is superimposed by the annealing effect. The optimum temperature of the deformation process is directly above the glass temperature of the PLA. Higher temperatures result in more strain softening, because the relaxation processes dominate the strain hardening. Increasing deformation rate results in constraint molecular recovery and that causes higher molecular orientation. Both effects result in improved mechanical properties of the material. However, an optimum of temperature and deformation rate exists, at which desired shaping is smoothly achieved but with high mechanical ultimate strength of the product.

Advanced short fiber composites with hybrid reinforcement based on glass and cellulose fibers – Effect of coupling agent on mechanical properties

Kahl Christian [(1)]*, Feldmann Maik [(1)], Heim Hans-Peter [(1)]

[(1)] University of Kassel - Kassel - Germany

Reinforcing thermoplastics with short fibers is a common way to improve the mechanical properties in different ways and was investigated for many years. Adding short glass fibers with a high strength and high modulus to a thermoplastic material leads to a raise of the strength in the compound but also decreases the strain at break. Cellulose fiber composites show a higher strain at break and a higher Charpy impact strength due to long fiber pull outs. Mixing these two fibers with their different characteristics in one compound opens a large field of opportunities. To make use of both fiber properties, the interface between the fiber and the matrix is important and has to be tailored for each fiber. The study will show the effect on mechanical properties of a compound reinforced with two different types of fibers. Polypropylene PP 575P was used as matrix material and glass fibers (GF) and cellulose fibers (Cell) were used as reinforcement. Three different ratios (2:1 GF/Cell, 1:2 GF/Cell and 1:1 GF/Cell) of each type of fiber mixed by dry blending were investigated with the same fiber volume content in the composite. Injection molded specimen were manufactured to evaluate tensile and Charpy impact properties. SEM pictures show the fiber matrix adhesion especially with regard to the fracture mechanism of each fiber. It was shown that the hybrids with cellulose fibers show an increase in elongation at break and Charpy impact properties compared to glass fiber composites. Furthermore, 33% of glass fibers in a cellulose reinforced polypropylene leads to a higher tensile strength and Young's Modulus. The break surface of the hybrid composites show fiber pull outs for cellulose fibers as well as fiber breaks for glass fibers. That indicates the intended tailored fiber matrix adhesion.

Polymeric CNT composites: atomistic simulations of interfacial properties

Golebiowski Jacek [1]*, Mostofi Arash A [1], Haynes Peter D [1]

[1] Imperial College London - London - UnitedKingdom

Functionalized carbon nanotube (CNT)/polymer composites have received significant interest as promising structural materials with applications in the most demanding areas of industry such as aerospace and ballistic protection. In order to optimise the properties of this class of materials, it is imperative to understand how load is transferred through the interface, between the functionalized CNT and polymer matrix with the aim to identify the key factors determining the interfacial shear strength and dominant failure mechanisms. Computational investigation of the interface requires simulations of tens of thousands of atoms in order to accurately describe the movement of polymer chains; however, critical interfacial failure involves changes in local chemistry such as bond-forming and breaking effects, necessitating a quantum-mechanical (QM) treatment. These issues are addressed by employing a quantum/classical hybrid simulation technique; in this approach, molecular dynamics based on a classical forcefield is used to simulate the majority of the system under strain, while regions of particular interest, such as the functional groups where changes in electronic structure are likely to occur, are investigated using QM methods resulting in an accurate description of bond-breaking processes.

Immobilization of TiO_2 Nanoparticles in PMMA Nanofiber Mat for Photocatalytic Removal of Dyes from Water

Ozcan Koysuren[1, a)] and H. Nagehan Koysuren[2, b)]

[a)]*Department of Energy and Materials Engineering, Ankara University, Ankara, 06830, TURKEY.*
[b)]*Department of Environmental Engineering, Ahi Evran University, Kirsehir, 40100, TURKEY.*

[a)]Corresponding author: koysuren@ankara.edu.tr
[b)]nagehan.koysuren@ahievran.edu.tr

Abstract. Regarding the photocatalytic ability of titanium dioxide (TiO_2) nanoparticles, poly(methyl methacrylate) (PMMA)/TiO_2 nanofiber mat was studied and the photocatalytic activity of the composite system was determined from degradation of a model dye, methylene blue. For this purpose, TiO_2 nanoparticles in crystal phase was synthesized and immobilized in PMMA nanofiber mat using via electrospinning process. FTIR and UV-VIS spectroscopy, SEM and water contact angle measurements were utilized to characterize PMMA/TiO_2 nanofiber mat and to compare the composite nanofiber mat with that of pure PMMA nanofiber mat. The immobilization of TiO_2 nanoparticles in PMMA nanofiber mat yielded thicker nanofibers with less beaded fiber structures. In addition, the introduction of TiO_2 nanoparticles enhanced the photocatalytic decomposition of methylene blue in the ambient air under ultraviolet light irradiation.

INTRODUCTION

Titanium dioxide (TiO_2) is a special type of semiconductor, which has been the focus of numerous studies due to its excellent photocatalytic properties (1). When TiO2 is illuminated with UV light, the energy promotes electrons from the valence band to the conduction band. These electrons transfer quickly to the surface of the particle and they react with oxygen and surface absorbed molecules to form superoxide and hydroxyl radicals. These active radicals are able to degrade the organic molecules by oxidizing the C-H bonds (2, 3). Regarding the photocatalytic ability of TiO_2 nanoparticles, polymer based photocatalysis UV-TiO_2 systems have been investigated to solve a variety of environmental problems like water purification (1). The immobilization of TiO_2 nanoparticles in polymer matrix eliminates the necessity of a separation process following the water purification and prevents the problems concerning the ecotoxicity to certain extent (4). Polymers in nanofiber form have an exceptionally high specific surface area, high surface energy, surface reactivity and biodegradability (5). Within this scope, researchers have focused on the photocatalytic activity performance of polymer nanofiber mats functionalized with TiO_2 nanoparticles. Photodegradation experiments of methylene blue indicated that polymer nanofiber mat containing TiO_2 nanoparticles resulted enhanced photocatalytic activity (6-10).

In this study, TiO_2 nanoparticles in crystal phase was synthesized and immobilized in PMMA nanofiber mat using via electrospinning process. The photocatalytic activity of PMMA/TiO_2 nanofiber mat was evaluated by investigating the photocatalytic decomposition of methylene blue in the ambient air under ultraviolet light irradiation.

EXPERIMENTAL

TiO_2 nanoparticles were prepared by the sol-gel process through the hydrolysis and condensation reaction of titanium isopropoxide ($Ti(OC_3H_7)_4$) with ethyl alcohol and acetic acid. With this method, amorphous titanium dioxide was produced. The calcination process was performed at 550 °C to turn the produced oxide powder into the crystal phase (11). Laboratory scale electrospinning unit (NE-100, Inovenso) was used to prepare pure PMMA and PMMA/TiO_2 nanofiber mats. For this purpose, PMMA was dissolved in acetone to obtain 2 wt. % solution and calcinated TiO_2 nanoparticles was mixed with 2 wt. % PMMA-acetone solution to obtain 3 wt. % PMMA/TiO_2 composite solution. Prepared solutions were filled up in a syringe and the feed rate of the solutions was maintained as 0.1 ml/hr. An electric potential difference of 40 kV was applied between the collector and the syringe tip. The distance between the collector and the tip was set as 10 cm. Pure PMMA nanofiber mat and the composite nanofiber mat were collected on the collector covered with aluminum foil.

Fourier transform infrared (FTIR) spectra of calcinated TiO_2 nanoparticles, PMMA nanofiber mat and PMMA/TiO_2 nanofiber mat were recorded with a Nicolet 380 (Thermo Scientific) spectrometer in the frequency range from 4000 to 400 cm^{-1} with a resolution of 2.0 cm^{-1}. The surface morphology of TiO_2 nanoparticles, pure PMMA nanofiber mat and the composite nanofiber mat were examined using a scanning electron microscope (SEM: EVO LS10 ZEISS). The thermogravimetric analyses (TGA) of pure PMMA and PMMA/TiO_2 nanofiber mats were performed on a thermogravimetric analyzer (Setaram Labsys TGA/DTA) operating at a heating rate of 5 °C/min. up to 400° C under nitrogen atmosphere. Photocatalytic activity of PMMA nanofiber mat and PMMA/TiO_2 nanofiber mat was evaluated by monitoring the degradation of the model dye, methylene blue. Pure PMMA and PMMA/TiO_2 nanofiber mat samples were immersed into methylene blue solution and were held in the dark for 1 hour for initialization. Then, they were exposed to UV light irradiation supplied by 8 W, 254 nm UV lamp. At specific time intervals, samples of methylene blue solution were taken out and were subjected to UV–visible spectroscopy (UV-Mini 1240 Spectrophotometer) at 665 nm to determine the dye concentration.

RESULTS AND DISCUSSION

Figure 1a illustrate FTIR spectrum of TiO_2 nanoparticles in crystal phase. The broad absorption peak belonging to Ti–O–Ti appears in the Fig. 2a in the range of 1000-750 cm^{-1} (12, 13). Figure 1b shows the characteristic absorptions of pure PMMA nanofiber mat at 2951, 1731, 1446, 1191 and 1149 cm^{-1}, corresponding to the stretching vibrations of C-H, C=O, CH_3, OCH_3 and C-O, respectively (14-16). There are almost no differences in the FTIR spectra of both nanofiber mat samples with its wavenumbers range from 4000 to 1000 cm^{-1}. In the FTIR spectrum of the composite nanofiber mat (Fig. 1c), the absorption peaks in the range range of 1000-750 cm^{-1} were broadened slightly due to TiO_2 nanoparticles.

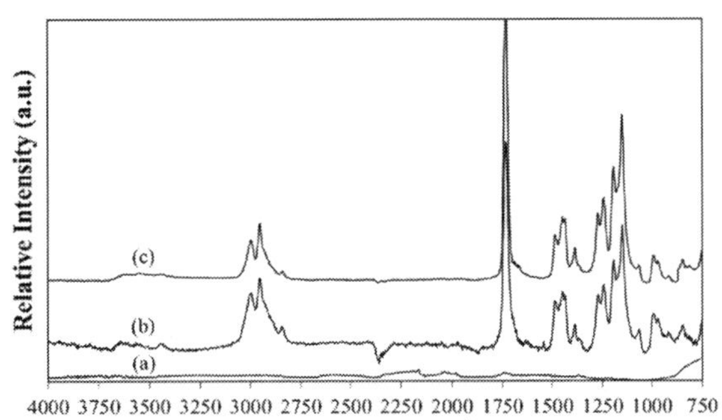

FIGURE 1. FTIR spectra of electrospun nanofiber mats from (a) PMMA and (b) PMMA/TiO_2.

Addition of TiO_2 nanoparticles into PMMA solution might result in an increase in the solution viscosity (17). With increased solution viscosity, the diameter of the electrospun nanofiber also increases (18). As expected, the average fiber of the composite nanofiber mat is thicker than the average fiber diameter of pure PMMA nanofiber mat (Fig. 2). When the viscosity increases, there is a gradual change in the shape of the beads from spherical to spindle-like until a smooth fiber is obtained (18). As shown in Fig. 2, less beaded fiber structures were observed on the surface of the composite nanofiber mat.

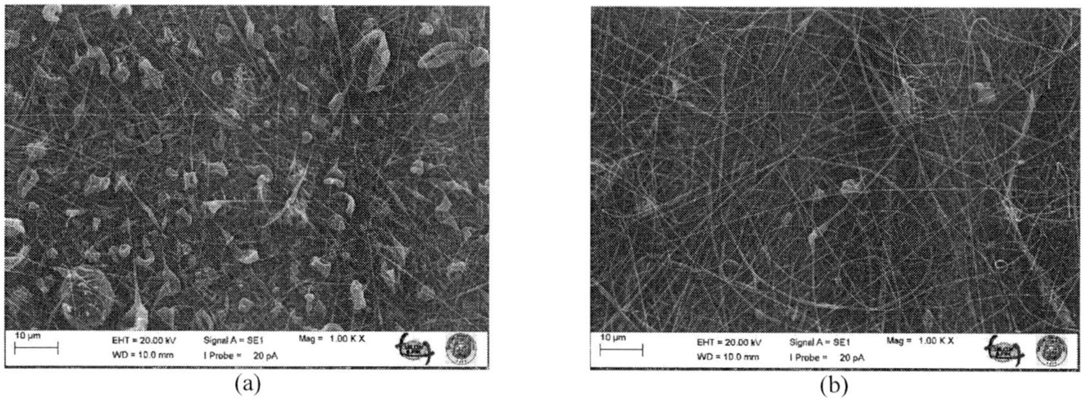

FIGURE 2. SEM images of electrospun nanofiber mats from (a) PMMA and (b) PMMA/TiO_2.

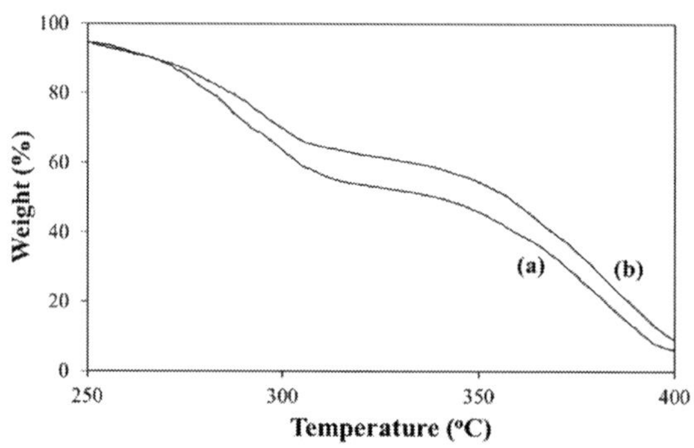

FIGURE 3. TGA curves of electrospun nanofiber mats from (a) PMMA and (b) PMMA/TiO_2.

As shown on the TGA curves (Fig. 3), both of the nanofiber mats exhibit two-step weight loss with lower residual production. The first weight loss is due to the degradation of the polymer, occuring between 270-300 °C, and the second weight loss is due to the cleavage of the PMMA backbone, observed after 350 °C (19). TiO_2 nanoparticles slightly enhanced the thermal stability of PMMA nanofiber mat.

The degradation of the model dye, methylene blue, as a function of irradiation time is illustrated in Fig. 4. The degradation is presented with the variation of the ratio (C/C_0) with irradiation time, where C_0 and C indicate the concentration of methylene blue solution before UV irradiation and after UV irradiation, respectively. When pure PMMA nanofiber mat was irradiated with UV, insignificant degradation of the model dye was obtained. After 180 minutes of irradiation, only 4% of methylene blue was degraded. On the other hand, a much better photocatalytic degradation of methylene blue occurred when TiO_2 nanoparticles were incorporated in PMMA nanofiber mat. 20% of the model dye was degraded after 180 minutes of UV irradiation (Fig. 4).

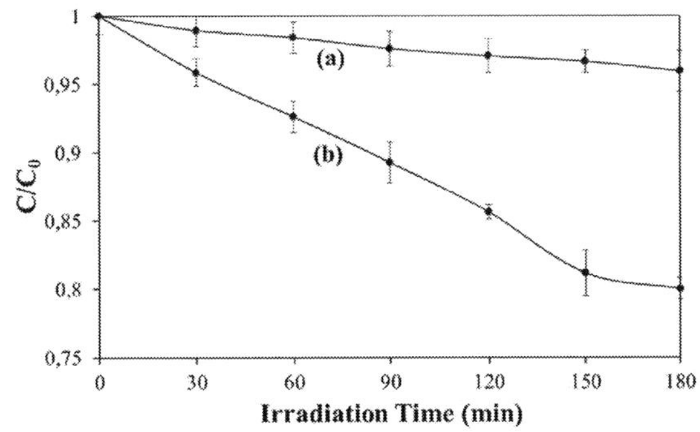

FIGURE 4. Photocatalytic degradation of methylene blue utilizing electrospun nanofiber mats from (a) PMMA and (b) PMMA/TiO_2.

REFERENCES

1. S. S. Madaeni, N. Ghaemi and H. Rajabi, "Advances in Polymeric membranes for water treatment," in *Advances in Membrane Technologies for Water Treatment: Materials, Processes and Applications*, edited by A. Basile, A. Cassano and N. K. Rastogi (Elsevier, Cambridge, UK, 2015).
2. W. A. Lee and M. Raifailovich, "Inorganic Particles against Reactive Oxygen Species for Sun Protective Products," in *Inorganic Nanoparticles: Synthesis, Applications, and Perspectives*, edited by C. Altavilla, E. Ciliberto (CRC Press, Boca Raton, FL, 2011), Chapter 13.
3. L. Zan, W. Fa and S. Wang, Environ. Sci. Technol. **40**, 1681-1685 (2006).
4. M. Cantarella, R. Sanz, M. A. Buccheri, F. Ruffinoa, G. Rappazzoc, S. Scalesed, G. Impellizzeria, L. Romano and V. Privitera, J. Photoch. Photobio. A **321**, 1–11 (2016).
5. J. H. He, Y. Liu, L. F. Mo, Y. Q. Wan and L. Xu, Electrospun Nanofibres and Their Applications (iSmithers, Shropshire, UK, 2008).
6. N. Daels, M. Radoicic, M. Radetic, K. De Clerck and S. W. H. Van Hulle, Sep. Purif. Technol. **149**, 488-494 (2015).
7. A. Mohamed, R. El-Sayed, T. A. Osman, M. S. Toprak, M. Muhammed and A. Uheida, Environ. Res. **145**, 18-25 (2016).
8. A. E. Deniz, A. Celebioglu, F. Kayaci and T. Uyar, Mater. Chem. Phys. **129**, 701– 704 (2011).
9. C. Prahsarn, W. Klinsukhon and N. Roungpaisan, Mater. Lett. **65**, 2498-2501 (2011).
10. O. Koysuren and H. N. Koysuren, J. Macromol. Sci. A **54**, 80-84 (2017).
11. N. Agoudjil and T. Benkacem, Desalination **206**, 531-537 (2007).
12. A. Amlouk, L. El Mir, S. Kraiem and S. Alaya, J. Phys. Chem. Solids **67**, 1464–1468 (2006).
13. M. Karaman, F. Sarıipek, Ö. Köysüren and H. B. Yıldız, Appl. Surf. Sci. **283**, 993-998 (2013).
14. O. Koysuren, M. Karaman, H. B. Yildiz, H. N. Koysuren and H. Dinc, Int. J. Polym. Mater. **63**, 337-341 (2014).
15. Y. Yang and Y. Dan, Colloid. Polym. Sci. **281**, 794-799 (2003).
16. J. B. Yadav, R. K. Puri and V. Puri, Appl. Surf. Sci. **254**, 1382-1388 (2007).
17. S. J. Kang, L. D. Tijing, B. S. Hwang, Z. Jiang, H. Y. Kim and C. S. Kim, Ceram. Int. **39**, 7143–7148 (2013).
18. A. Ramakrishna, K. Fujihara, W. E. Teo, T. C. Lim and Z. Ma, An Introduction to Electrospinning and Nanofibers (World Scientific Publishing, London, 2005).
19. E. M. Abdelrazek, A. M. Hezma, A. El-khodary and A. M. Elzayat, Egypt. J. Basic Appl. Sci. **3**, 10-15 (2016).

Preparation, Mechanical Properties and Moisture Absorption of Urethane Acrylate Resin/Clay Nanocomposites

***Esfandeh Masoud* (1)*, *Khonakdar Hossein Ali* (1), *Mirabedini Seyed Mojtaba* (1), *Gashtil Reza* (1)**

(1) Iran polymer and petrochemical Institute - Tehran - Iran

In this research, nanocomposites based on urethane acrylate resin and nanoclay were prepared at various clay contents including 0.5, 1.5, 3, 5, 7 and 10 wt.%. %. The nanocomposites were characterized using viscometry, XRD and TEM techniques. Also, mechanical and physical properties of nanocomposites including tensile, flexural, impact, water absorption, hardness and burning rate were investigated. The results for XRD and TEM showed an exfoliated structure at clay content up to 3 wt% while at higher loadings intercalation of silicate layers happened. Viscosity measurements showed an increase of 84% in viscosity with the addition of 3 wt% nanoclay. An increase of 25% was observed in tensile and flexural modulus. This was 75%, 90% and 35% for tensile, flexural and impact strength, respectively. A decrease of 32% in water absorption of nanocomposites containing 3 wt% of nanoclay was observed. A self-extinguishing behaviour was observed in nanocomposites at nanoclay loadings between 0.5-5 wt%. Keywords: XRD, Organoclay, Urethane Acrylate, Nanocomposite, Mechanical Properties

Hierarchical Functionalization of Halloysite Nanotube and Its Enhancement in Thermal Stability, Fire Retardancy and mechanical property of Epoxy Resin

Wang De-Yi [1], Li Zhi [1]*, Liu Lejing [1], González Alejandro Jiménez [1]

[1] IMDEA Materials Institute - Madrid - Spain

Aiming to impart epoxy resin (EP) with improved thermal stability, fire retardancy and mechanical property, inspired by core-sheath corncob structure, halloysite nanotubes were successively functionalized with bio-mimic polydopamine nano-coating (PDA) and ultrafine Fe(OH)3 nanoparticles to prepare hierarchically structural halloysite@PDA@Fe(OH)3 (Fig 1). The targeted product was verified via FTIR, XRD, XPS and TEM techniques. Resultantly, in the thermal degradation study, EP composite with halloysite@PDA@Fe(OH)3 possessed notably higher char yield than those of neat EP, EP/5halloysite, EP/5halloysite@PDA. Fire retardancy investigation revealed that 5wt% halloysite@PDA@Fe(OH)3 imparted EP with enhanced LOI value of 33.9% and UL-94 vertical burning rating of V-1, which was a significant improvement compared to the counterparts. In cone calorimeter test, EP/5halloysite@PDA@Fe(OH)3 gave rise to 41% reduction of peak heat release rate (pHRR) with respect to those of EP/5halloysite and EP/5halloysite@PDA. TG-FTIR test manifested that halloysite@PDA@Fe(OH)3 notably decreased volatile evolutions (CO, aliphatic compounds, aromatic compounds, and carbonyl compounds). In-situ FTIR, Raman spectra and SEM observations revealed the char with more compact and continuous structure and higher graphitized degree was acquired with halloysite@PDA@Fe(OH)3. In parallel, the tensile strength and modulus were enhanced with halloysite@PDA@Fe(OH)3, accompanied by the increase of dynamic storage modulus (G'). Finally, the possible mechanism to account for the improved fire retardancy was proposed, which was involved in the interfacial catalytic charring.

Modelling and Optimization of the Mechanical and other Material Properties of a Polymer Nanocomposite using Statistical Design of Experiments

Reinhard Fechter[1], Ines Kühnert[2], Carl Sandrock[1] and Johan Labuschagné[1, a)]

[1]*Department of Chemical Engineering, University of Pretoria, cnr Lynnwood Road and Roper Street, Hatfield, South Africa.*
[2]*Department of Processing, Leibniz-Institut für Polymerforschung Dresden e.V., Hohe Str. 6, Germany*
[a)]*email: johan.labuschagne@up.ac.za*

Abstract. Polymer compounds are complex systems that typically involve many additives that tend to interact with each other. The system is further complicated by the fact that the additives tend to have an effect on multiple material properties. Hence, the effect of a particular ingredient on a certain material property should not be quantified in isolation. For instance, an important consideration in evaluating the effectiveness of an ingredient is not only how it effects the property it was designed to effect but how it effects other properties, such as the mechanical properties of the compound, in the context of the proportions of the other ingredients. This can be achieved by using the principles of statistical design of experiments. In this investigation the mechanical properties of a polymer nanocomposite, a PVC compound including a Layered Double Hydroxide (LDH) nano-additive, are modelled using 2nd degree Scheffe polynomials. The proportions of all the ingredients (7 in total) are varied in a space filling experimental design. The mechanical properties of each formulation are tested using a tensile test on samples manufactured using injection molding. Injection molding is crucial because it produces homogenous test samples that give an accurate representation of the inherent mechanical properties of the material. The models are determined using k-fold cross validation. The mechanical property models, in conjunction with models of other important material properties, allow for an analysis of the effects and interactions of all of the ingredients. For instance, the analysis shows the negative effect that the LDH has on the elongation at break which needs to be taken into account when considering the positive effects it has on the thermal stability of the compound. Importantly the models can also be used to optimize the system.

INTRODUCTION

Polymer compounds typically require a number of additives to be practically and commercially viable. This results in a mixture system which is complex for three specific reasons. The first is that the additives tend to interact with each other and the polymer; there are many examples of synergisms and antagonisms between polymer additives. Secondly, the additives tend to affect multiple material properties of the compound. Finally due to the fact that a polymer compound is a mixture system it is impossible to make an independent change in the relative proportion of a single ingredient.

This means that it is important to evaluate the effectiveness of each ingredient in the context of the relative proportions of the other ingredients. It is also important when evaluating a certain ingredient to not only consider the property that a particular additive was designed to effect but also to consider how it effects other material properties such as, for instance, the mechanical properties of the compound.

This can be achieved using the principles of statistical design of experiments. The purpose of the investigation was to empirically model the mechanical and other important material properties of a polymer nanocomposite as a function of the relative proportions (mass fractions) of the ingredients using statistical experimental design. A secondary aim of the investigation was to enable the optimization of the polymer compound.

In the investigation a flexible PVC compound including a Layered Double Hydroxide (LDH) nano-additive was used due to the significant number of additives typically required to make the compound practically and commercially viable and its consequent complexity. The basic mechanical properties were determined using a tensile test on injection molded tensile bars. In addition the thermal stability and fire retardancy properties were determined. The experimental design was done using JMP Statistical Software, the experiments were conducted in a laboratory environment and the data analysis was done using programs written in Python programming language using Jupyter (Kluyver 2016).

EXPERIMENTAL DESIGN

The aim of the statistical experimental design is to determine empirical models that can be used to describe the basic mechanical properties as well as the thermal stability and flame retardancy of the polymer nanocomposite as a function of the relative proportions of the ingredients which can be defined as the mass fractions of the ingredients.

Selection of Response Variables

The experimental methods used to determine the response variables are the tensile test for the mechanical properties, the torque rheometer for the thermal stability and the cone calorimeter for the fire retardancy. The tensile test is used because it covers the basic mechanical properties of the compound (Grellmann 2013). The torque rheometer is selected because it is the most realistic indicator of thermal stability during processing for the system (Wilkes 2005). Finally the cone calorimeter test is selected because it gives the most complete information concerning fire retardancy with the closest simulation of a real fire (Joseph 2016).

Choice of Factors and Ranges

The factors for the experiment (i.e. the variables) are the mass fractions of the 7 ingredients used in the polymer compound. The actual ingredients that are used (which are the PVC resin, thermal stabilizer, plasticizer, two types of fillers, fire retardant and LDH) are not varied. The ranges of the factors, or more generally the experimental space that will be used is a result of practical considerations of the testing methods which will be used to determine the response variables.

For the tensile test a sample preparation step is required where the tensile bars are injection molded. This is used so that a test sample that is as homogenous as possible is manufactured which represents the intrinsic mechanical properties of the material. This means that the compound must be injection moldable which limits the experimental space. Considering this a set of limits for the experimental space is proposed. Ideally only compounds which are well within the processing limits should be used; however, in this case the experimental space needs to broad enough to make the models meaningful by minimizing the effect of variance.

Choice of Experimental Design

Since the system is a mixture and the experimental space is constrained there are a number of options for the experimental design such as the extreme vertices, D-optimal and space filling designs (Smith 2005, Montgomery 2013). For this investigation the space filling design is used so that the experimental points are distributed evenly throughout the 6 dimensional experimental space. The number of experimental points that are used needs to be greater than the number of parameters of the model, but is limited by the time and resources available to execute the experiment. In this investigation enough experimental points will be used to fit a 2nd order Scheffe polynomial which has the form

$$y = \sum_{1 \le i \le q} \beta_i x_i + \sum_{1 \le i < j \le q} \beta_{ij} x_i x_j \quad (1)$$

where q is the number of factors in $\boldsymbol{x}$, y is the response and β are the model parameters. A Scheffe polynomial is essentially an ordinary polynomial which has been derived to take into account the mixture constraint (i.e. that all proportions must sum to 1). A special feature of the Scheffe polynomials is that the second order term β_{ij} is interpretable in that it quantifies interactions between ingredients i.e. synergism and antagonism (Scheffe 1958).

RESULTS AND DISCUSSION

Model Selection

For every response variable a Scheffe polynomial is required that can be used to predict it as a function of the design variables. To determine the 'best' Scheffe polynomial for each response variable requires that all possible variants of the Scheffe polynomial are tested using the data. The maximum number of terms in a 2nd degree Scheffe polynomial where $q = 7$ is 28. This means that for a model with a complexity of 1 (or in other words including only one term) there are 7 possible models taking into account the hierarchy principle (the hierarchy principle states that a higher order model term can only be included if the lower order terms it contains are also included in the model (Montgomery 2013)). If the model complexity is 2 the number of possible models increases to 21 and then to 56 for 3 terms etc., increasing exponentially with increasing model complexity. In total there are 2.35 million different possible models for the 2nd degree Scheffe polynomial with $q = 7$.

To validate all the possible models the data from the experiments is collected into a single data set where each row represents a sample, identified by its sample number, and each column represents a response variable. The data for each column is normalised to be between -1 and 1. The particular range of -1 to 1 for the normalisation is selected so that in the hypothetical situation where a response variable has no relation to the design variables, the true mean will be 0 and the resultant model parameters for the true Scheffe model will all be 0 as well. All of the possible models are then scored using k-fold cross validation (a validation technique that is used to evaluate the predictive ability of a model (Hastie 2009)) with $k = 3$. The scoring results can be illustrated by plotting the highest score for each level of model complexity as is shown in **FIGURE 1**.

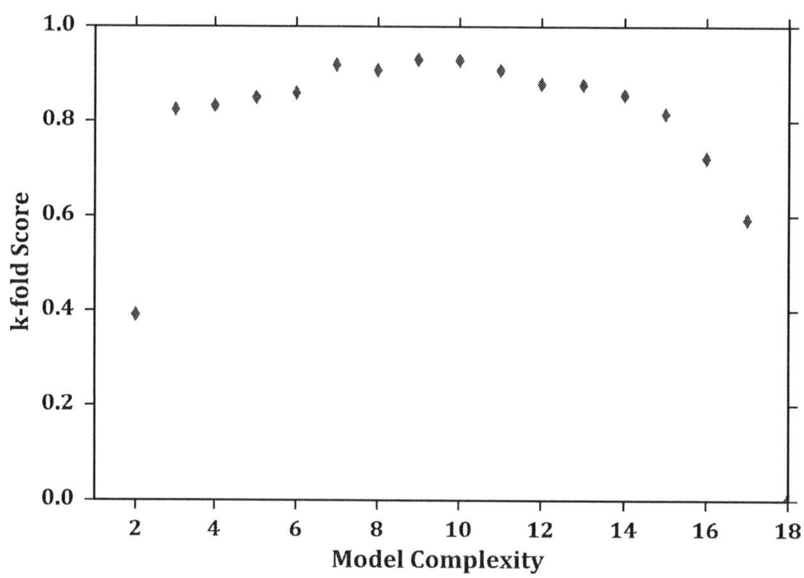

FIGURE 1. k-fold scores for the models fitted to the % elongation at break measured using a tensile test

The k-fold score figures for the response variables that are predicted well tend to have a similar shape to the example shown in **FIGURE 1**. The score increases, plateaus and then eventually decreases due to overfitting. The model with the highest score is selected.

Model Analysis

Statistical Significance of Model Terms

Each of the specific models that are selected contain information about the relationship between the design variables and the response variable. The statistical significance of each term can be analyzed using the *t*-statistic. This is demonstrated again using the % elongation at break measured using a tensile test. The *t*-values for the terms of the selected model are shown in **FIGURE 2**.

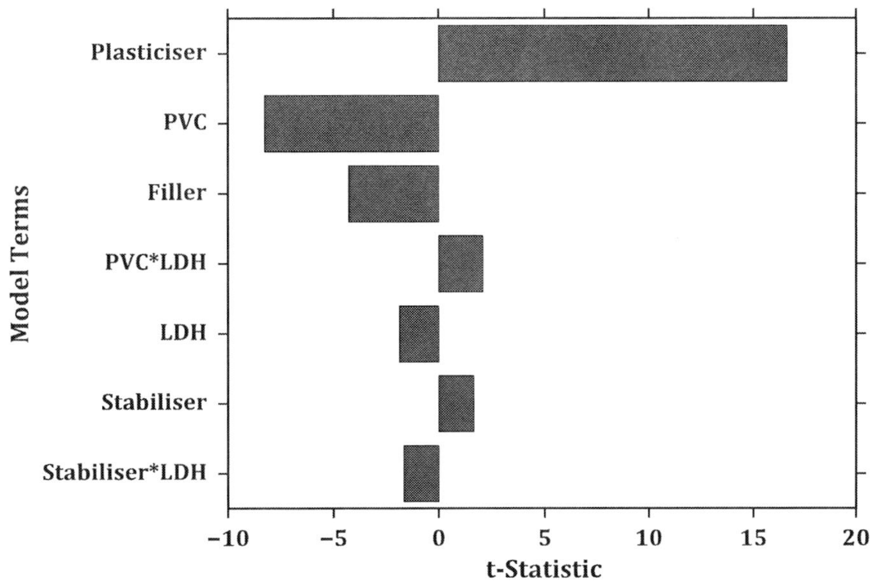

FIGURE 2. *t*-statistic for the model terms for % elongation at break measured using a tensile test

From **FIGURE 2** it is clear that the plasticizer and the polymer are very important when it comes to the elongation at break of the compound. It is also clear that the polymer and the LDH nano-additive interact synergistically whereas the thermal stabilizer and LDH have an antagonistic relationship. The analysis demonstrated above shows how statistical experimental design can be used to draw conclusions about the behavior of all the included design variables and their interaction with each other over the entire experimental space. A similar analysis can be conducted for every response variable. It is important however that the *k*-fold score of the model is considered. If the predictive ability of the model is very poor it is unlikely that conclusions made using the model parameters will be meaningful. For the example of the elongation at break the score for the selected model is 0.92 which is very high (the highest possible score being 1).

Sensitivity Analysis

It is clear from the analysis above that the overall effect of a particular ingredient on a particular response variable can be dependent on the mass fractions of the other ingredients, due to the interaction terms. To determine the overall effect of each ingredient a sensitivity analysis of the response variable at a certain formulation can be used which can be determined by calculating the partial derivatives. The values of the partial derivatives at a certain formulation show the effect of making a very small independant increase in the relevant ingredient. This is interpreted as the sensitivity of the model and is a very useful method to analyze the effects of the individual ingredients on the different response variables. An example of this is shown in **FIGURE 3**.

FIGURE 3 shows the negative effect that the LDH has on the elongation at break. It also shows the strong positive effect that the LDH has on the final thermal degradation time (measured using a torque rheometer) and the smoke reducing effect it has (measured using a cone calorimeter). The empirical models quantify the effects that the ingredients have on the material properties of the compound taking into account the relative proportions of the other ingredients. This means that, as is illustrated in **FIGURE 3**, the negative effect that LDH has on the elongation can be taken into account when evaluating the positive effects it has on other material properties.

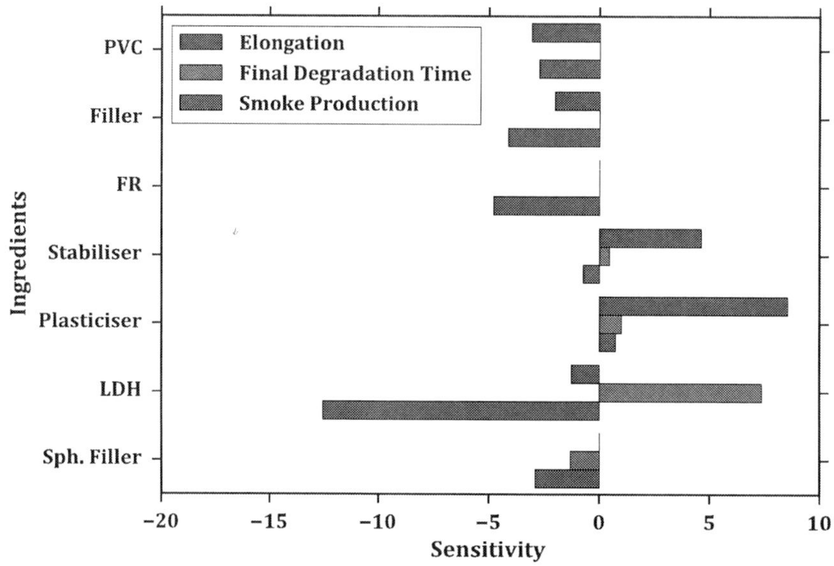

FIGURE 3. Model sensitivities at an example formulation for selected material properties

Finally it is important to mention that the mechanical and other material properties can be used to optimize the system. For instance the overall cost of the formulation can be minimized using a simple cost function as the objective function, where the cost of each ingredient is multiplied by its mass fraction, while using the material models as constraints. The material model constraints ensure that the desired material properties are achieved.

CONCLUSIONS

It can be concluded that the response variables describing the mechanical and other material properties as a function of the mass fractions of the polymer nanocomposite ingredients can be modelled effectively using 2nd order Scheffe polynomials. The particular empirical models for each response variable, selected and evaluated for predictive ability using k-fold cross validation, can be interpreted using statistics to make deductions about the relationship between the response variables and the formulation. The interaction terms in the 2nd order Scheffe polynomial are particularly useful in identifying synergistic and antagonistic relationships between ingredients. Sensitivity analysis can be used to determine the overall effect of an ingredient on a response variable at a given formulation. Finally it can be concluded that the empirical models can be used to optimize the system.

REFERENCES

1. W. Grellmann and S. Seidler, *Polymer Testing*, (Hanser, Munich, 2013)
2. T. Hastie, R. Tibshirani and J. Friedman, *The Elements of Statistical Learning*, (Springer, New York, 2009)
3. P. Joseph, S. Tretsiakova-McNally and R. Zhang, "Techniques for Assessing the Combustion Behaviour of Polymeric Materials: Some Current Perspectives and Future Directions", Macromolecular Symposia **362**, 105–118 (2016)
4. T. Kluyver et al. "Jupyter Notebooks—a publishing format for reproducible computational workflows" in *Positioning and Power in Academic Publishing: Players, Agents and Agendas*, edited by F. Liozides and B. Schmidt (IOS Press, 2016)
5. D. C. Montgomery, *Design and Analysis of Experiments*, (John Wiley & Sons, Hoboken, 2013)
6. H. Scheffe, "Experiments with Mixtures", Journal of the Royal Statistical Society **20**, 344–360 (1958)
7. W. F. Smith, *Experimental Design for Formulation*, (Society for Industrial and Applied Mathematics, 2005)
8. C. E. Wilkes, J. W. Summers and C. A. Daniels, *PVC Handbook*, (Hanser, Munich, 2005)

Processing of electromagnetic interference shielding water-borne polyurethane composites filled with silver nanoparticles deposited on functionalized graphene

Ma Chen Chi M. (1)*, **Lin Sheng Chi** (1), **Shen Ming-Yuan** (1)

(1) National Tsing Hua University - Hsinchu - Taiwan

The objective of this study was to prepare an electromagnetic interference shielding material, water-borne polyurethane (WPU) filled with silver-nanoparticle-decorated functional reduced graphene oxide (Ag@FRGO). Functional reduced graphene oxide (FRGO) was functionalized through free radical polymerization before chemical reduction to prevent restacking and aggregation during the reduction. After the functionalization, the homogeneous dispersion of FRGO promoted the formation of conductive networks throughout the WPU matrix. To enhance the electrical conductivity and EMI SE of the composites, silver nanoparticles (Ag NPs) were deposited on the FRGO surface. Subsequently, WPU composites were synthesized through the solution-mixing method. Their microstructure, morphology, electrical conductivity and EMI SE were investigated, and the results showed that, among Ag@FRGO/WPU composite with different values for the weight ratio of Ag NPs to FRGO, the Ag@FRGO/WPU composite with 5 wt% FRGO and weight ratio of Ag to NPs of 10:1 (10Ag@FRGO/WPU) exhibited the highest electrical conductivity (25.52 S/cm) and an EMI SE of 35 dB in the frequency range of 8.2–12.4 GHz (X-band).

Study of Modified LDHs as UV Protecting Materials for Polypropylene (PP)

Sajid Naseem[1,2, a)], Sunil P. Lonkar[1,3] , Andreas Leuteritz[1]

[1]*Polymer Processing Department, Leibniz Institute for Polymer Research Dresden, 01069, Germany*
[2]*Technical University Dresden, Germany*
[3]*Department of Chemical Engineering, The Petroleum Institute, Abu Dhabi, UAE*

[a)]Corresponding author: naseem@ipfdd.de

Abstract. The degradation and color loss of different polymer products which are being used in open environment is mainly due to Ultraviolet (UV) light. In this work the possibility of using different UV light absorbing Layered double hydroxides (LDHs) for the protection of polypropylene (PP) against UV light degradation is explored. First, possible transition metal containing LDH were synthesized and organomodification with ion exchange method were performed. Successfully, Zn-Ti LDH was synthesized and organically modified by Dodecylbenzene sodium sulfonate (SDBS) and Lauric acid (LA). Four combinations of Zn-Ti, Zn-Sn, Zn-Ti SDBS and Zn-Ti LA LDHs were used in PP composite preparation. Decomposition, UV absorption properties and photooxidation of all four composites and pure PP were determined. Most significant results were obtained by the addition of Zn-Ti-SDBS LDH in PP. Thermal stability of PP was increased by 70 °C by the addition of only 5 % Zn-Ti-SDBS LDH.

INTRODUCTION

Polymers degradation occurs due to exposure of materials to light, temperature and oxygen. UV light can produce reactive free radicals causing degradation and color loss of polymers and results in the decrease of mechanical strength [1-2]. The basic strategy to avoid degradation is to decrease the damage occurring due to radiation and suppress radical formation as shown in Fig. 1. There are different additives available in market to protect polymers against sunlight including inorganic and organic. UV absorbers mainly used nowadays are organic based materials such as cinnamates, triazines and benzophenones. Still, there are unsolved problems due to the use of organic UV-absorbers. They absorb the UV energy and transform it into heat and thus leading to the degradation of polymer matrix [2]. Sometimes the organic UV absorbers leach out from polymers and this phenomenon reduces their long term efficacy. Further, organic materials for light stabilization often show low temperature stability, thus limiting their use [2].

The inorganic fillers such as TiO_2 and ZnO are used mostly due their environmental friendly nature and thermal stability. As such particles are rather large, consequently the materials containing such particles are getting intransparent [3], and the particles have to be coated due to their ability to electrocatalytically degrade polymer structures [4-5]. Mixture of inorganic and organic UV absorber is advantageous over both organic and inorganic additives, the simplest examples are organic anions hosted by inorganic structures such as LDHs. LDHs have wide range of application and these materials are widely studied for different applications in last few years [6].

Polypropylene is most widely used polymer due to its several advantages. It is being used in many applications where it has to face the sunlight and it is susceptible for photodegradation [7, 11]. In the present study it is demonstrated that it is also possible to design the LDH structure as UV-scavenger and modify such structures in order to render them organophilic necessary for nanoscopic dispersion in polymer matrices.

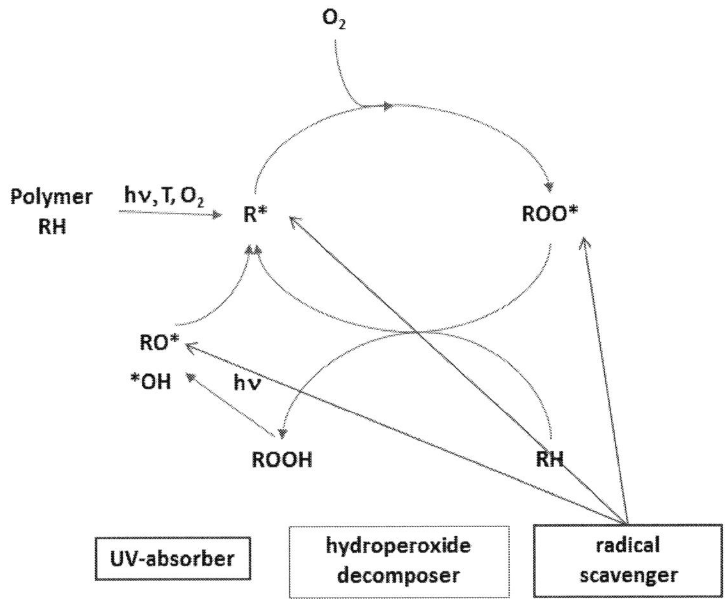

FIGURE 1. Polymer Protection Methods

EXPERIMENTAL AND CHRACTERIZATION DETAILS

All chemicals used for the synthesis and organic modification of different LDHs were purchased from Sigma-Aldrich. Zn-Ti, Zn-Sn LDHs were prepared by urea hydrolysis method [8]. A typical procedure is used in which 3:1 molar ratio of $Zn(NO_3)_2.6H_2O:TiCl_4$ is used. 0.22 ml of $TiCl_4$, 1.19 g of $Zn(NO_3)_2.6H_2O$ and 3 g of urea were dissolved in 100 ml of deionized water under vigorous stirring. The reactants were aged for 48 hour at 130 °C [8]. For Zn-Sn; 4:1 molar ratio of $Zn(NO_3)_2.6H_2O:SnCl_4$ was used. 0.04 M aqueous solution of $Zn(NO_3)_2.6H_2O$, 0.01 M $SnCl_4$ aqueous solution and 0.16 M urea solution mixed together and heated at 100 °C for 30 hour [8].

Zn-Ga, Zn-Cr and Cd-Cr LDHs were prepared by Co-precipitation method [9-11]. A typical synthesis procedure is as follows: solution of $Zn(NO_3)_2.6H_2O$ and $Ga(NO_3)_2.4H_2O$ in 2:1 molar ratio was added dropwise to 1 M Na_2CO_3 solution. The pH was maintained at 8 with the addition of NaOH solution. The precipatets were aged at 80 °C for 22 hours [9]. Solution of $Zn(NO_3)_2.6H_2O$ and $Cr(NO_3)_2.9H_2O$ in 2:1 molar ratio was mixed dropwise with NaOH and Na_2CO_3, pH was maintained at 9. Resulting preicipates were aged at 60 °C for 24 hours [10]. Solution of $Cd(NO_3)_2.4H_2O:Cr(NO_3)_2.9H_2O$ in 2:1 molar ratio was mixed dropwise with NaoH simultanesuly to maintain the pH at 8. The resulting mixture aged at 100 °C for 18 hours [11]. All these five type of LDHs were filtered, washed and dried in oven at 60 °C overnight.

Zn-Ti LDH was organically modified and organomodification was done with ion exchange method. A typical method was used in which 4 g of SDBS was dissolved in 100 ml of deionized water. 1 g of Zn-Ti LDH was mixed in SDBS solution and stirred for 24 hours at 80 °C under nitrogen environment. Modified Zn-Ti LDH were filtered, washed and dried in oven at 60 °C overnight [12].

X-ray diffractograms were recorded using a 2-circle diffractometer. UV-vis spectroscopy was carried out by Lambda 800 from Perkin Elmer.

RESULTS AND DISCUSSION

Several strategies are used to reduce the impact of UV-light irradiation on the degradation of polymeric materials based on organic UV absorbers, inorganic particles or both. LDH structures allow intercalating organic UV absorbers and thus enlarging the processing window for compounding. Generally, organomodification of LDH is necessary to achieve highly dispersed particles in the nanometer scale, thus not sacrificing the optical properties of the compound in visible light region. The possibility to design LDHs by changing the ratio of metal species as well as type of metal combinations for nearly the whole periodic table already had been discussed [13].

In order to be used as nano dispersible tunable intrinsically UV absorbing LDHs is investigated here in this study. Therefore five different types of LDHs structures were synthesized. Figure 2(a-d) describing results of UV-vis spectra and XRD spectra of different LDHs. In X-ray scattering, typically (001) reflections of LDH crystals are observed. From UV-vis graphs it is cleared that the different metal salts have different UV absorptions.

Figure 2(a) and Figure 2(b) representing the XRD spectra and UV-vis spectra of Cd-Cr and Zn-Cr LDHs respectively. It can be seen that there are three UV-bands of the resulting structures, one in the UV-range and two in the visible range. The WAXS curves indicate layered structures, but not necessarily an LDH. Figure 2(c) showing Zn-Ga LDH. It can be seen that there are no strong UV-band of the resulting structure, WAXS curve indicating three reflections, proving a layered structure. Figure 2(d) showing Zn-Sn LDH, It can be seen that there is strong UV-band in the UV rang of graph of the resulting structure, WAXS curve indicating three reflections, proving a layered structure. All attempts to organically modify these all four structures, indicated by a shift of the XRD reflection failed. Therefore no further investigation was done based on these systems.

Figure 2(e) showing Zn-Ti LDH. It can be seen that there is a strong UV-band in UV range of graph of the resulting structure, WAXS curve indicating three reflections, proving a layered structure of LDH. Organic modification of Zn-Ti LDH was successful with SDBS. Figure 3(a-b) showing the XRD spectra with comparison of SDBS modified and unmodified Zn-Ti LDH. It can be seen that there is a strong shift of WAXS curve to the left due to intercalation of SDBS in Zn-Ti LDH layers.

Typical reflections of crystals can be observed in Fig. 3(a) and in Fig. 3(b) at (003), (006) and (009). Peaks indicating the good crystallinity of LDHs. Reflection peak at (003) in Fig. 3(a) and Fig. 3(b) indicating the repeating unit of LDH and It is the sum of one metal hydroxide layers and spacing between two adjacent layers. In Fig. 3(b) the peaks shifted to the left which is indication of enhanced interlayer spacing and good intercalation of SDBS [14-15].

CONCLUSION

LDHs composed with Zn and Ti metal cations can be potential UV absorbing layered materials. Organic modification of Zn-Ti was successful and this can also leads to an easy processing of PP. The addition of these LDHs and organically modified LDHs in Polypropylene (PP) can reduce the degradation of Polypropylene (PP). The results appear to open a new field of research as UV absorbing LDHs for polymers, which may be used in different applications with combinations of polymers such as in solar cells, electrostatic devices, LED devices and other related application.

ACKNOWLEDGEMENT

The authors are thankful to the department of processing, Leibniz institute of polymer research Dresden, Germany for providing all the facilities to perform this research work.

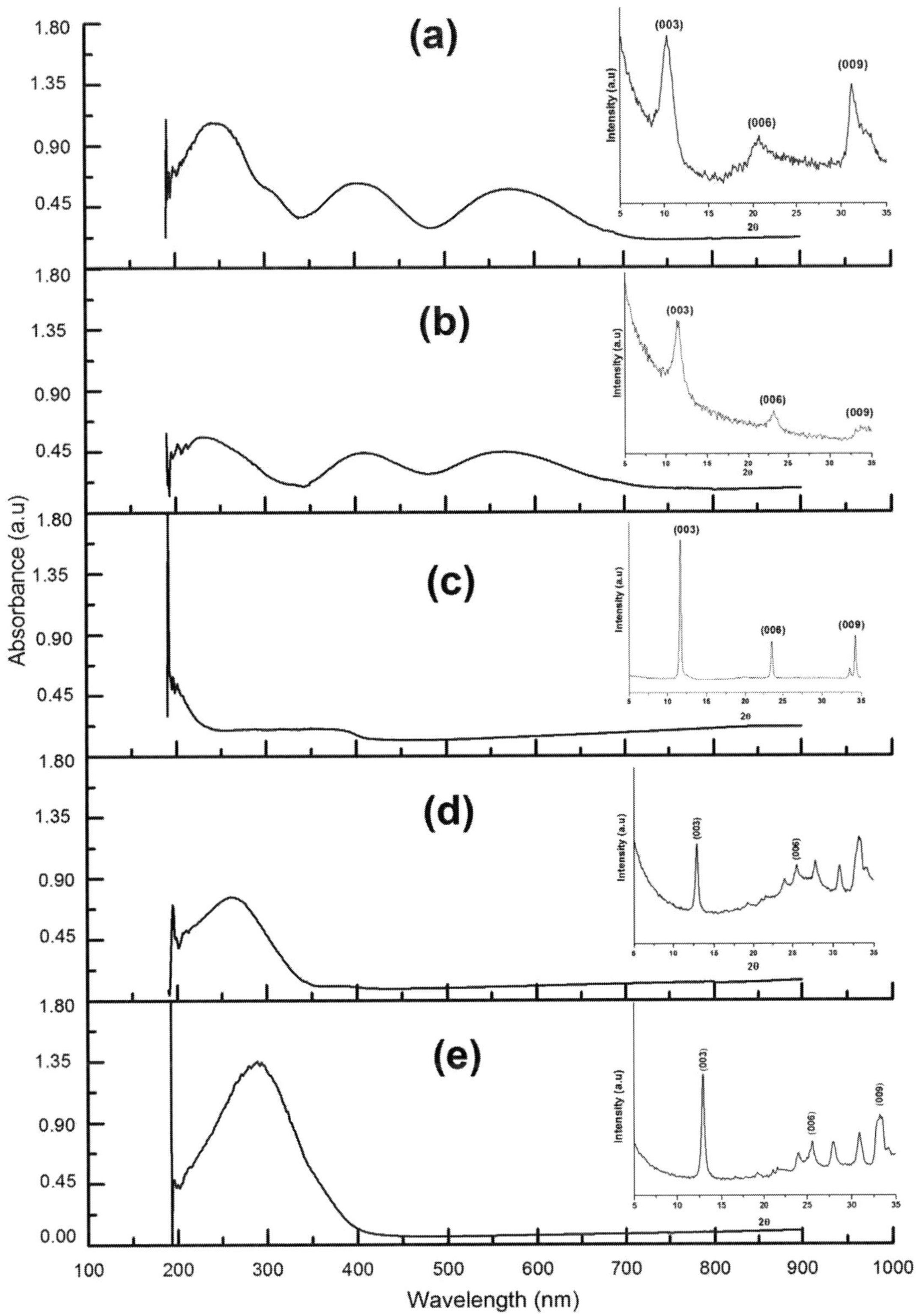

FIGURE 2. UV-Vis and XRD Spectra of (a) Cd-Cr LDH, (b)Zn-Cr LDH, (c) Zn-Ga LDH, (d) Zn-Sn LDH and (e) Zn-Ti LDH

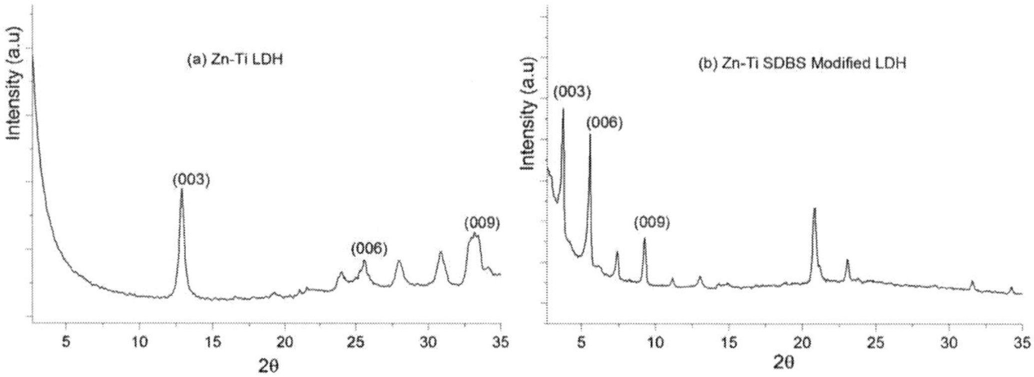

FIGURE. 3 XRD Spectra of (a) Zn-Ti LDH and (b) Zn-Ti SDBS modified LDH

REFERENCES

[1] G. Wang, S. Xu, C. Xia, D. Yan, Y. Lin and M. Wei, RSC Adv. **5**, 23708-23714 (2015).
[2] Y. Peng, W. Wang, J. Cao and Y. Huang, J. Appl. Polym. Sci. **134**, 44597 (2017).
[3] T.J. Turton and J.R. White, Polym. Degrad. Stab. **74**, 559-568 (2001).
[4] V. P. Gordienko and Yu. A. Dmitriev, Polym. Degrad. Stab. **53**, 79-87 (1996).
[5] P.O. Bussière, J. Peyroux, G. Chadeyron and S. Therias, Polym. Degrad. Stab. **98**, 2411-2418 (2013).
[6] D. Basu, A. Das, K.W. Stöckelhuber, U. Wagenknecht, G. Heinrich, Prog. Polym. Sci. **39**, 594 -626 (2014).
[7] Y. Gao, Y. Zhao, L. Qiu, Z. Guo, D. O'Hare and Q. Wang, Polym. Compos., DOI:10.1002/pc.23764 (2015)
[8] M. Shao, J. Han, M. Wei, D.G. Evans and X. Duan, Chem. Eng. J. **168**, 519-524 (2011).
[9] G.S. Thomas and P.V. Kamath, Solid State Sci. **8**, 1181-1186 (2006).
[10] C.G. Silva, Y. Bouizi, V. Fornés and H. García, J. Am. Chem. Soc. **131**, 13833-13839 (2009).
[11] Y. Guo, H. Zhang, L. Zhao, G.D. Li, J.S. Chen and L. Xu, J. Solid State Chem. **178**, 1830-1836 (2005).
[12] J. He, M. Wei, B. Li, Y. Kang, D. G. Evans and X. Duan, *Preparation of Layered Double Hydroxides*, Berlin, Heidelberg: Springer Berlin Heidelberg; 2006. p. 89-119.
[13] J.J. Bravo-Suárez, E.A. Páez-Mozo and S.T. Oyama, Quim. Nova **27**, 601-614 (2004).
[14] B. Kutlu, A. Leuteritz, L. Häußler, U. Oertel and G. Heinrich, Polym. Degrad. Stab. **102**, 9-14 (2014).
[15] B. Kutlu, P. Schröttner, A. Leuteritz, R. Boldt, E. Jacobs and G. Heinrich, Mater. Sci. Eng. C Mater. Biol. Appl. **41**, 8-16 (2014).

Layered Double Hydroxide Applications - Industry vs Research

Labuschagne Johan* [(1)]

[(1)] University of Pretoria - Gauteng - SouthAfrica

Research on Layered Double Hydroxides (LDH) polymer micro- and nano-omposites has been very active for close to 40 years now. Since the late 1970s, when the first research papers on LDH as heat stabiliser for first were published, the number of publications on LDH based polymer composites increased significantly. Many papers on the possible applications of LDH in polyethylene, polypropylene, ethylene vinyl acetate, ethylene vinyl alcohol, poly vinyl chloride, etc have been published. In these research papers based on LDH materials, functionalities such as thermal (heat) stability, acid scavengers, flame retardants, increase UV resistance, enhanced stiffness, increased impact strength, reduced heat distortion, increased cooling rates, reduced shrinkage, reduced warpage, reduced solvent uptake, increased barrier properties, increase bio-degradability, pigmentation, etc. Yet, only a hand full of companies currently sell any form of LDH (and mainly hydrotalcite) commercially. And the applications for these commercial LDH materials are only as PVC heat stabilisers and polyolefin acid scavengers. From this the question... Why such few commercial applications if research show so many possible applications. We think the answer is simple. Cost! And from here out interest in the commercially viable synthesis of LDH materials for polymer applications. The low cost, environmentally friendly synthesis techniques developed based on green chemistry based synthesis techniques are the focus of our research. We have already proved the comparable performance of our material to the typical materials synthesised in research papers. Some of these will be discussed in this presentation.

Polyamide GnP nanocomposites for permeation control

Leuteritz A. (1)*, Langner P. (1), Schult M. (1), Trampe M. (1), Boldt R. (1), Wagenknecht U. (1)

(1) Leibniz-Institut für Polymerforschung Dresden e.V. - Saxony - Germany

For an environmental sustainable future Hydrogen is an outstanding candidate as a fuel for aerospace application. It has the highest mass energy density compared to other fuels, is clean and eco-friendly. Further, hydrogen is the smallest stable molecule, which leads to rapid permeation through the walls of storage tanks. Therefore the main object of this work is the evaluation of different types of Graphene nano Platelets (GnP) as possible filler in different polyamides to limit permeation rate. The very high possible aspect ratio and the low weight make GnP to an excellent candidate in light weight construction. Based on rheology measurements the shear thinning effect shows that there is a good interaction between GnP and especially semi-aromatic polyamide, whereas aliphatic polyamide has less effect. Morphology development and permeation characterization depend on both processing conditions and filler – matrix combination was investigated, in order to enhance barrier effects achieved by melt compounded samples. This was shown in small scale (DSM X-plore, 15 ccm) and large scale Leistritz twin screw compounder combined with film extrusion.

Application of desert sands as reinforcing material for polymer concrete

Caba Stefan [1]*, Fiebig Christian [1], Koch Michael [1]

[1] Technische Universität Ilmenau - Thüringen - Germany

Low-cost houses are needed to face the demand for living space in the emerging markets in Africa and the Middle East. Desert sands cannot be used for common concrete. The lack of suitable sands and the high carbon footprint of cement have put the focus of research on new mixtures of polymer concrete. The application of desert sands for polymer concrete could help building up new cities creating local value added chains. Polymer concretes are used for special purpose e.g. sewage pipes for chemical plants or retention pots for chemicals. Benefits are the chemical resistance and the high strength compared to common concrete products. The composite material consists of 87 wt% quartz sands in predefined particle sizes mixed with a polyester resin. The resin is accelerated for cold hardening. Desert sands in a local production cannot be selected by particle size and particularly contain parts of salts and metals. These substances slow or even prevent the hardening reaction. The resin mixture of ground resin, hardener, inhibitor and accelerator must be adapted to the local sand composition. The content of metallic ions as well as the humidity of 12 sands were characterized. In rheometer experiments the slope of the viscosity during the hardening were examined. The resin recipe was alternated to find advantageous mixtures. A cobalt accelerator and a cobalt-amine accelerator were used. A multidimensional regressions analysis enabled to detect the substances slowing the reaction. The cobalt-amine hardener showed excellent reaction rates also at high metallic ion contents. Mechanical tests showed compression strengths equal to concrete used for houses. A system for building houses and a mobile production system were developed. Z-formed bricks can be produced locally and stacked by non-skilled workers creating local employment. A production mold with a breathing core was designed. Cycle times under 30 seconds providing enough bricks for 8 houses per day can be produced.

Investigation on the Thermoformability of Heat Conductive Plastics

K. Landsecker[a)] and C. Bonten

IKT, Institut für Kunststofftechnik, University of Stuttgart, Pfaffenwaldring 32, 70569 Stuttgart, Germany

[a)]Corresponding author: Kai.Landsecker@ikt.uni-stuttgart.de

Abstract. Thermoforming describes the production of three dimensional structures from flat sheet. In general, it is used for the production of simple shaped packaging or housing parts without any additional function. However, by the use of heat conductive plastic sheets, a heat-loss-function can be implemented to the thermoforming parts. Heat conductive plastics are polymer materials filled with thermally conductive fillers. The more filler is applied, the better the compound conducts heat. Though, with increasing filler content the stretching behavior of a sheet is changed too.

In this study, the influence of different filler geometries (plate-like and spherical fillers) and filler contents (0, 5, 10, 15, 25 vol.-%) on the thermoforming behavior of filled polystyrene material is determined. Therefore, the sheet's stretching behavior is analyzed by the IKT Thermoforming-material-characterization test. In addition, the influence of an altered stretching behavior on the thickness distribution of a thermoformed part is examined by thermoforming experiments.

INTRODUCTION

Thermoforming is commonly used for the production of areal parts with thin wall thicknesses [1]. Here the thermoforming process provides a cost-efficient solution for parts with high surface quality and low investment costs. The reason for the low investment costs can be found in the simple molding process. Thermoforming is normally performed with a single-sided mold, below melting temperature and at a low forming pressures of 0.1–14 bar [2]. However, the simple forming process leads to an increased dependency of a part's quality on the sheet's stretching behavior. Consequently, the introduction of new thermoforming material requires the characterization of the material's stretching behavior - especially if a material's properties are influenced by the stretching process. A new material for thermoforming processing is the group of heat conductive plastic materials.

STATE OF THE ART

Heat conductive plastics are polymer materials filled with heat conducting fillers [1,3]. The conductivity of these compounds depend on various factors. First, it is determined by the general composition of processed materials. The higher the heat conductivity of the single components, the higher becomes the heat conductivity of the whole compound. Furthermore, it is also dependent on the compound's inner structure. Non-spherical fillers lead to an anisotropic heat conductivity determined by the filler orientation. As a result, a change in filler orientation leads to a change in heat conductivity. [3,4]

It is known that material processing, for instance by injection molding [3] or extrusion [5], influences the orientation of non-spherical filler as well as the direction-dependent heat conductivity of a material. [6] and [7] show how non-spherical fillers are also orientated by the thermoforming process. As a result of an altered filler orientation, the through-plane heat-conductivity is reduced with increasing stretch ratio [6]. However, [6] also observes a strong influence of the filler type on the processing properties of a sheet. Still, the isolated role of filler geometry on the stretching behavior stays uncertain.

EXPERIMENTAL

In order to investigate the isolated influence of filler geometry, filler content and matrix material on the stretching behavior of sheets, sheets from the same basic materials were produced in a two-step manufacturing process. First of all, reference compounds were processed by a ZSK 25 twin-screw extruder from Coperion, Stuttgart, Germany. Two high-impact polystyrene materials, PS454N and PS486N from INEOS Styrolution, Frankfurt, Germany, were mixed with two graphite powders, GraphCOND® and GraphTHERM® from Georg H. Luh GmbH, Walluf, Germany. Altogether 18 different compounds were produced with different matrix/filler combinations and filler contents. Table 1 gives an overview of the materials produced as well as the nomenclature of the different compounds. While PS486N is a classic high-impact polystyrene material for thermoforming application with an MFI of 3,9 cm³/10 min (200 °C/5 kg), PS454N is a high-impact polystyrene with a reduced viscosity (MFI 14 cm³/10 min) for injection molding application. GraphCOND® is platelet graphite filler with D_{50} = 10–25 µm. GraphTHERM® is spherical graphite filler with a comparable fillersize of D_{50} = 21–25 µm. After the compounding process, the compounds were extruded to 0.8 mm thick sheet by a 30x25D single-screw extruder from Dr. COLLIN, Ebersberg.

TABLE 1. Overview of produced Compounds

Filler		Matrix	
Type	**Content**	**PS454N INEOS Styrolution**	**PS486N INEOS Styrolution**
-		PS454N	PS486N
GraphCOND®	5 vol.-%	PS454N5GC	PS486N5GC
	10 vol.-%	PS454N10GC	PS486N10GC
	15 vol.-%	PS454N15GC	PS486N15GC
	25 vol.-%	PS454N25GC	PS486N25GC
GraphTHERM®	5 vol.-%	PS454N5GT	PS486N5GT
	10 vol.-%	PS454N10GT	PS486N10GT
	15 vol.-%	PS454N15GT	PS486N15GT
	25 vol.-%	PS454N25GT	PS486N25GT

Material Characterization

The sheets were characterized according to their heat conductivity and stretching behavior. The thermal conductivity was analyzed by laser-flash-analysis (LFA) according to DIN EN ISO 22007-4 with a Netzsch, Selb, Germany, LFA 447 by in-plane as well as through-plane measurements. The density was measured according to DIN EN ISO 1183-1 (Verfahren A), while heat capacity was measured by DSC-analysis.

The stretching behavior of the sheets was characterized by the IKT TMC-test according to [8]. Figure 1 shows a scheme of the TMC-test setup. The specimen is placed on a piston that is coaxially arranged with a circular plug. Once the piston moves with a specified velocity, the specimen is deformed by a fixed plug and the resulting normal force is measured by a piezo-element. The force-deflection curves characterize the stretching behavior of a sheet. Since the whole apparatus is placed inside a heating chamber, the test can be performed at different temperatures. The sheets were characterized at a temperature of 130 °C using a hydraulic universal testing machine HUS 1 from MFL Systeme, Mannheim, Germany. The piston velocity was 200 mm/s and the test was carried out until the specimens failed by breakage.

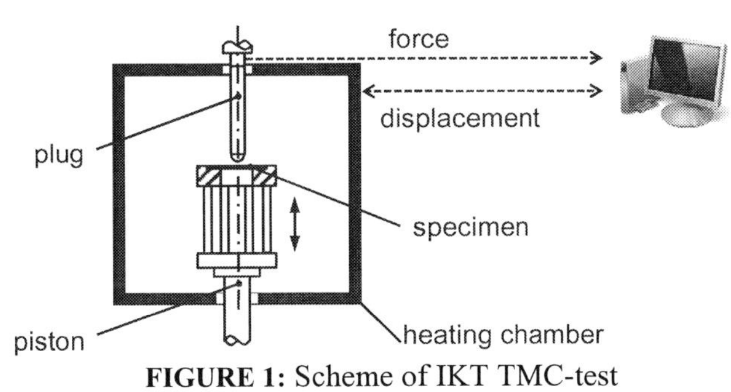

FIGURE 1: Scheme of IKT TMC-test

Thermoforming

In order to analyze the connection of TMC-measurements on the actual thermoforming behavior, thermoforming tests were performed using a laboratory thermoforming machine LDFG 23 b from ILLIG, Heilbronn, Germany. All sheets were thermoformed in a two-step forming process according to the same parameters shown in table 2. After the heating phase, the sheets were stretched by mechanical drawing with a positive mold and lastly, a vacuum was applied to give the part its final shape.

TABLE 2. Thermoforming parameters

Heater temperature:	inner loop: 450 °C
	outer loop: 500 °C
Heating time:	24 s
Mold temperature:	60 °C
Vacuum:	-900 mbar (rel.)
Time of vacuum:	10 s
Mold geometry:	box shaped positive mold

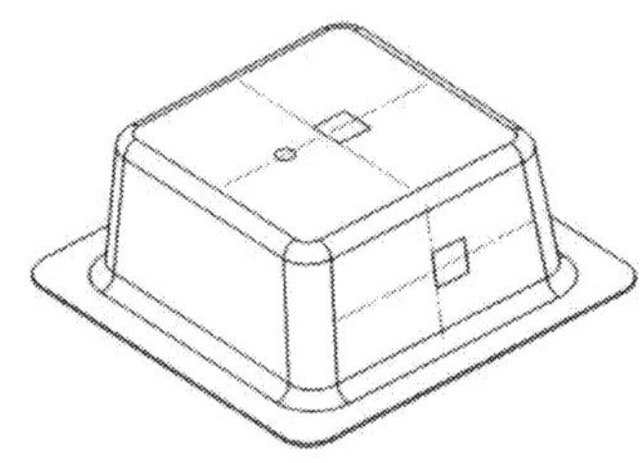

The thermoforming process was monitored by a 3D Digital-Image-Correlation system from Isi-Sys GmbH, Kassel, consisting of two Sony IMX174 cameras. The correlation was performed by Vic-3D 7 from Correlated Solutions, Irmo, USA. In addition, the thickness distribution of the thermoformed parts was measured with a hall-effect thickness gauge.

RESULTS AND DISCUSSION

The measurements of the sheets' heat-conductivity illustrate the increase in heat conductivity with increasing filler content. Figure 2 shows the heat conductivity of all materials in in-plane and through-plane measurements. Both matrix materials, PS454N and PS486N, show heat conductivities in a comparable range. The use of filler with a higher aspect ratio (GraphCOND®) leads to higher heat conductivity up to 3,2 W/mK in in-plane direction. Due to filler orientation the heat conductivity of the investigated sheets is higher in in-plane direction than in through-plane direction.

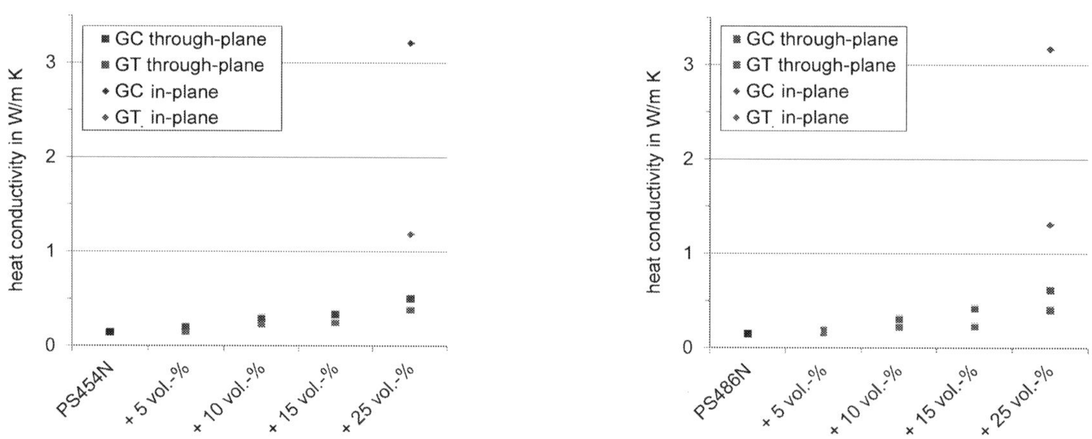

FIGURE 2: Heat conductivity of the PS454N sheets (left) and PS486N sheets (right)

The TMC-test describes the stretching behavior of a sheet under thermoforming conditions. Among other properties the force-deflection curves characterize the capability of a sheet to develop homogeneous sheet thicknesses in the thermoforming process. A monotonically growing force/deflection-curve characterizes a sheet with good forming properties. A distinct increase of force at high strain indicates strain-hardening behavior that facilitates a homogeneous sheet thickness distribution.

Figure 3 shows the force/deflection curves of the basic polystyrene materials and the heat conductive polystyrene sheets at a temperature of 130 °C and a piston velocity of 200 mm/s. The basic materials PS454N and PS486N show monotonically growing force/deflection curves. This shape describes materials with a good stretching behavior, leading to homogeneous sheet thickness distributions in thermoforming parts. With increasing filler content the force/deflection curves become more inhomogeneous, indicating a worse thermoforming behavior. At higher filler contents, a maximum is build up at a piston deflection of 20 mm and the sheets show necking behavior. With the increase in necking behavior, a decrease in thermoforming homogeneity is expected. Further, it can be observed that with increasing filler content the critical strain is reduced. The use of GraphCOND® has more influence on the force/deflection curves than the mixing with GraphTHERM®. The lower viscosity matrix PS454N is less influenced by the filler content or filler geometry than PS486N.

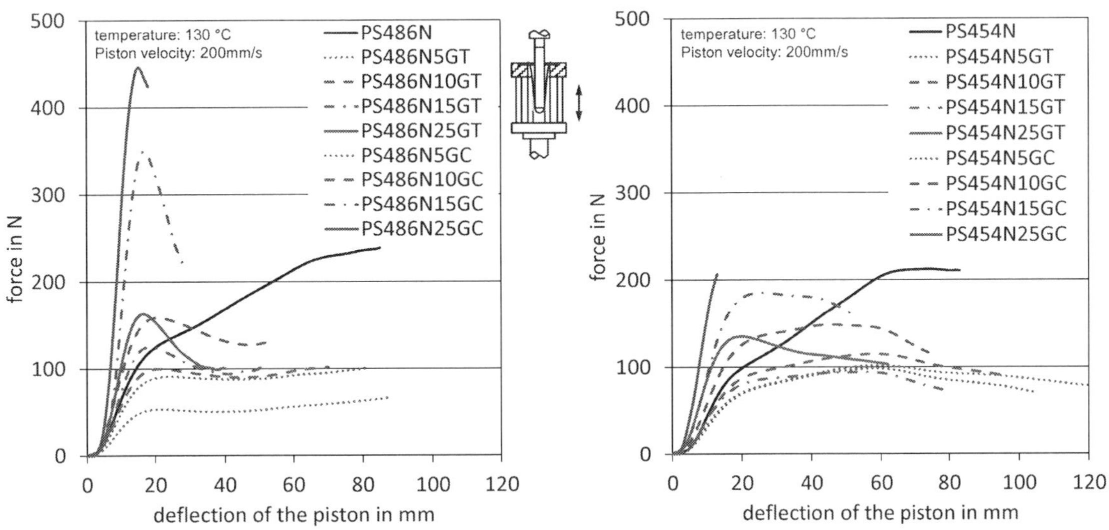

FIGURE 3: TMC measurements

The observed influence of different filler geometries and filler content as well as the matrix viscosity on the stretching behavior of sheets can be confirmed by thermoforming experiments. With an increase in filler content, the forming process becomes more inhomogeneous. In addition, with increasing filler content the sheets fail at lower strain. The GraphCOND® filler leads to more inhomogeneous stretching behavior than GraphTHERM®. Likewise in TMC-tests, it can be observed that the negative influence of increasing filler content is reduced by a lower viscosity matrix. Figure 4 shows the comparison of the main engineering strain of PS454N5GT and PS454N25GT. The Figure illustrates how increasing filler content leads to more inhomogeneous stretching.

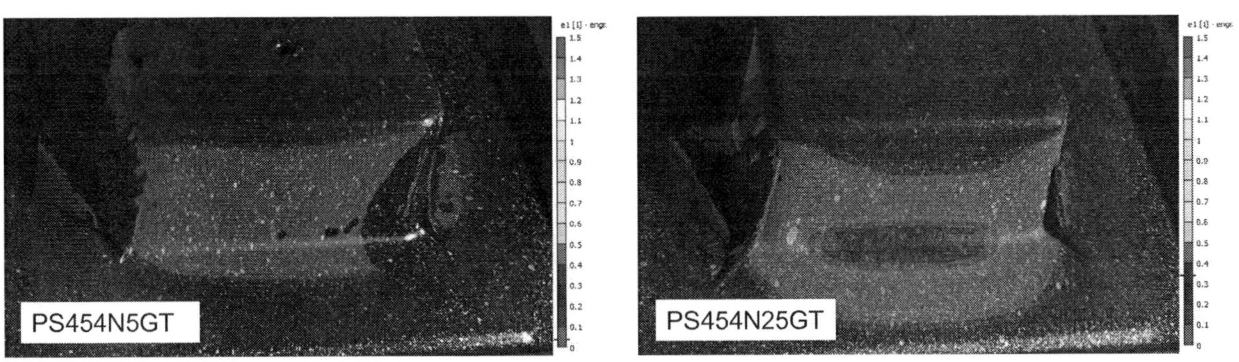

FIGURE 4: Thermoforming experiments analyzed by digital image correlation

Figure 5 shows the averaged thicknesses of the thermoformed parts. Sheets that failed in the forming process (due to breakage or incomplete forming) are excluded from the diagrams. It can be seen that with increasing filler content the sheet thickness distribution is more inhomogeneous. The spherical filler allow successful thermoforming even with higher filler content, while the low viscosity PS454N allows higher filler content than PS486N.

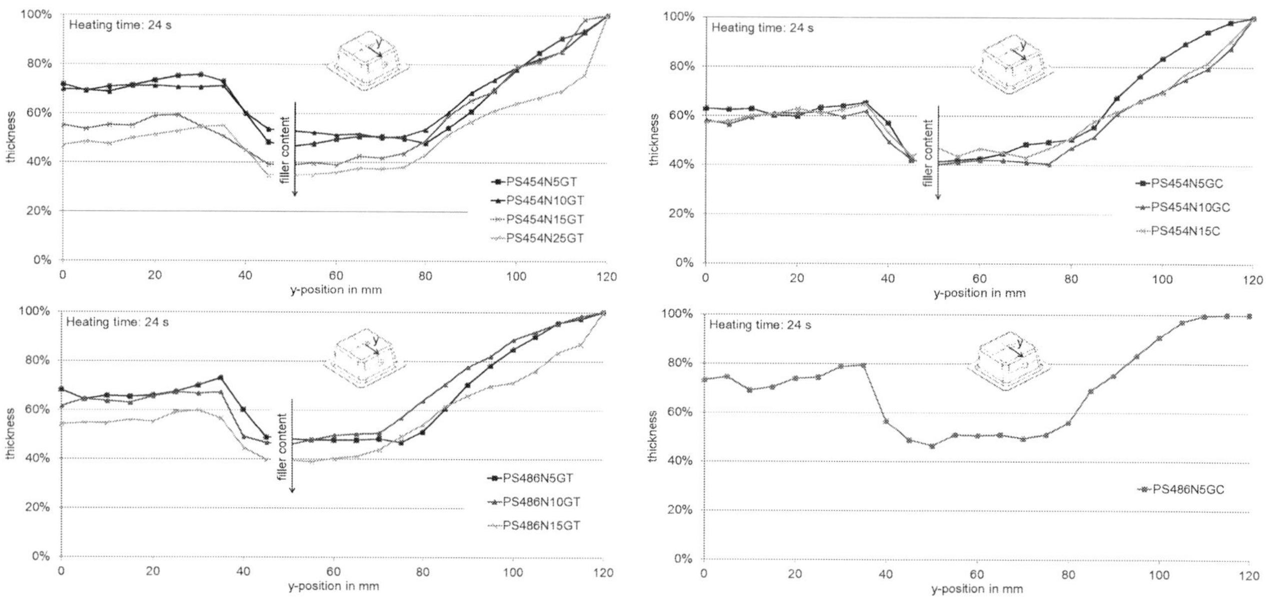

FIGURE 5: Thickness distribution of the thermoformed parts

OUTLINE

The TMC-test can be used for the evaluation of a sheet's necking and strain-hardening behavior. Since necking leads to a more inhomogeneous sheet thickness distribution in the thermoforming process, the TMC-test can be as well used for the prediction of a sheet's capability of forming a homogeneous thermoforming part. With increasing filler content, the stretching behavior as well as the resulting sheet thickness distribution become more inhomogeneous. Sheets with spherical fillers and a lower viscosity matrix show better thermoforming behavior than fillers with a higher aspect ratio in a higher viscosity matrix. However, this finding contradicts with a maximized thermal conductivity that is to be reached by filler with a high aspect ratio.

In future work the influence of different filler sizes should be investigated. The aim remains the selection of heat conductive materials with increased thermal conductivity and good thermoforming properties.

REFERENCES

1. C. Bonten, *Kunststofftechnik - Einführung und Grundlagen,* 2nd edition (Hanser, Munich, 2016).
2. L. Throne, *Technology of Thermoforming*. (Hanser, Munich, 1996).
3. C. Heinle, "Simulationsgestützte Entwicklung von Bauteilen aus wärmeleitenden Kunststoffen," Ph.D. thesis, Universität Erlangen-Nürnberg, Lehrstuhl für Kunststofftechnik, 2012.
4. S. Amesöder, "Wärmeleitende Kunststoffe für das Spritzgießen," Ph.D. thesis, Universität Erlangen-Nürnberg, Lehrstuhl für Kunststofftechnik, 2009.
5. O. Skrabala, "Wärmeleitfähige Kunststoffe: Verarbeitungsinduzierte Eigenschaftsbeeinflussung und deren numerische Vorhersage," Ph.D. thesis, Universität Stuttgart, Institut für Kunststofftechnik, 2016.
6. K. Landsecker and C. Bonten, "Untersuchung zur Thermoformbarkeit von wärmeleitfähigen Kunststoffen," in *25. Stuttgarter Kunststoff-Kolloquium 22.-23.03.2017* (Stuttgart, 2017).
7. C. H. Suh and J. L. White, "Particle and polymer chain orientation development in thermoformed talc-filled thermoplastics," in *Polymer Engineering and Science*, **36**(17), (SPE, 1996), pp. 2188-2197.
8. B. Hegemann, "*Deformationsverhalten von Kunststoffen beim Thermoformen*," Ph.D. thesis, Universität Stuttgart, Institut für Kunststoffkunde und Kunststoffprüfung, 2004.

The Effects of Surface Modification of MWCNTs with Hexanol on the Non-isothermal Crystallization Kinetics of Poly(butylene succinate)

Yarici Tugay [1], Kodal Mehmet [1], Ozkoc Guralp [1]*

[1] Kocaeli University - Kocaeli - Turkey

Research and development in biodegradable polymers has received increasing attention in recent years because of their wide range application in environmental friendly packaging and automotive applications. Among the biodegradable polymers, poly(butylene succinate) (PBS) is a promising polymer for various potential applications such as injection molded products and packaging films, sheets. However, insufficient gas barrier properties, higher liquid permeability and lower crystallization rate, are often not sufficient for various end-use applications. Carbon nanotubues (CNTs) are alternative candidates as nucleating agents for polymers. The aim of this study is to enhance the crystallization of PBS by using hexanol functionalized multi-walled carbon nanotube (Hf-MWCNT). A twin screw microcompounder was used to prepare the PBS/Hf-MWCNT nanocomposites. The modification process was characteriazed by FT-IR spectroscopy. The common kinetic models such as Avrami, Ozawa, Kissinger were used to evaluate non-isothermal crystallization behaviour of PBS and PBS/Hf-MWCNT nanocomposites by using differantial scanning calorimeter (DSC). Besides, a polarized optical microscope (POM) equipped with a hot-stage was used to examine the morphology during the crystal growth.

Keratin Fibres Recovered from Tannery Industry Wastes as Fire Retardant Agent on PLA composites

Sanchez-Olivares Guadalupe (1)*, Sanchez-Solis Antonio (2), Calderas Fausto (1), Alongi Jenny (3)

(1) CIATEC, A.C. - Guanajuato - Mexico, (2) Instituto de Investigaciones en Materiales, Universidad Nacional Autónoma de México - Ciudad de México - Mexico, (3) Dipartimento di Chimica, Università degli Studi di Milano - Piemonte - Italy

As it is well known, tannery leather process is a high pollutant emissor. During beamhouse stage, the wastes have a strong negative environmental impact because of the effluents and solids have a high content of organic matter. Hair skin derived from this stage, which is thrown in sanitary fills without any usual treatment. However, wasted hair is considered an important keratin fibre source with interesting applications in polymer materials. In this work keratin fibres (KFs) were recovered from tannery industry wastes. A specific treatment was applied to KFs in order to obtain polymer composites base on PLA, KFs and aluminium tri-hydroxide (ATH) by extrusion process. The effect of KF content and the joint action of KFs in combination with ATH as fire retardant for poly(lactic acid) -PLA- was studied by thermogravimetric analyses, UL94 classification, dynamical-mechanical, mechanical and rheological analysis. According to the collected results, the fire retardant potential of keratin fibres recovered from tannery industry wastes on PLA composites have been demonstrated. PLA composites were classified as V2 with only KFs (3, 5 and 10 phr). However, when KFs were added to 30 phr ATH, the composites reached V0 rating, the same classification obtained using 50 phr ATH. Therefore, by using low KF content (3 phr), it is possible to replace 20 phr ATH. In addition, KF-based composites exhibited higher mechanical properties than that containing only ATH. Flow behaviour under continuous simple and small amplitude oscillatory shear flows demonstrated that KFs decrease the viscosity of the PLA composites in a remarkable way improving the processability of PLA composites. Materials containing both KFs and ATH exhibited a higher elastic behaviour than those loaded with only KFs. The latter result represents an important advantage from an economical and environmental point of view in flame retarded materials.

Thermal Cycling effect on Mechanical properties of Multiwalled Carbon Nanotubes reinforced epoxy/carbon fiber composite laminates

***Chiang Chin-Lung** (1), **Chen Chin-Hsing** (2), **Chen Wei-Jen** (3), **Ma Chen-Chi M.** (4), **Shen Ming-Yuan** (3)**

(1) Hung Kuang University - Taichung - Taiwan, (2) Chinese Culture University - Taipei - Taiwan, (3) China University of Science and Technology - Hsinchu - Taiwan, (4) National Tsing-Hua University - Hsinchu - Taiwan

Carbon nanotubes (CNTs) possess special physical characteristics such as strength, stiffness, light weight, electrical conductivity, highly thermal conductivity and thermal stability, etc. Because of their high absolute strength and large surface area, multiwalled carbon nanotubes (MWCNTs) have high potential for improving the material properties of polymer-based composites. In this study, MWCNTs was used to reinforce epoxy/carbon fiber (CFRP) composite laminates to enhance their mechanical properties. The mechanical properties of MWCNTs / CFRP composite laminates including flexural and interlaminar shear strength (ILSS) under different thermal cycling conditions (i.e. 50, 100 and 200 times at -40 ~ 125 °C) were investigated, and the fracture surface of the specimen was also investigated using field emission scanning electron microscopy (FESEM) to determine the mechanisms of the MWCNTs in the CFRP composite laminates.

Preparation and properties of Antimicrobial PLA/PEG/Nano Ag nanocomposites

***Javadi Azizeh** (1)*, **Pazuki Mahmood** (1), **Khaki Amir** (1)*

(1) Amirkabir university of technology - Tehran - Iran

In this research, PLA/PEG blends containing 5, 10 and 15 w.% of PEG and also Nanocomposite of PLA/PEG (85/15 wt./wt.) with 0.05, 0.15 and .75 phr Nano Ag were prepared via melt mixing using an internal mixer. The mechanical and thermal properties of blends and nanocompositses, also their morphology were studied. DSC and DMTA results illustrated one Tg for PLA/PEG blends, which shows their miscibility in all studied compositions. Also the Tg and Tm of blends were decreased by increasing the PEG contents. The tensile and impact tests showed that the modulus and tensile strength of blends are decreased and their elongation at break and impact strength are increased by increasing the amount of PEG. In contrast, the addition of nano Ag to PLA/PEG 85/15 blend, led to increasing the tensile strength and modulus and decreasing the elongation at break and impact strength. Using EDAX, it was found that nano Ag particles are well dispersed in the nanocomposite samples. TGA showed that destruction temperature of composites are increased by increasing the amount of nano particles. Finally, the antimicrobial tests on nanocomposite sample containing 0.75 nano Ag, showed that it is sensitive to both gram-positive and gram- negative bacteria, but its antimicrobial property is more powerful in contact with gram- positive ones. Keywords: Poly(lactic acid), Poly (Ethylene glycol), Nano Ag, Antimicrobial, nanocomposites

Microencapsulation of Melamine Polyphosphate and Its Flame Retardance in Polyurethane Composites

Shen Ming-Yuan (1), Chen Wei-Jen (1), Kuan Chen-Feng (2), Kuan Hsu-Chiang (2), Yang Jia-Ming (3), Chiang Chin-lung (3)*

(1) China University of Science and Technology - Hsinchu County - Taiwan, (2) Far East University - Tainan - Taiwan, (3) Hungkuang University - Taichung - Taiwan

A new microencapsulated flame retardant containing melamine polyphosphate (MPP) and 4,4'-Oxydianiline-formaldehyde (OF) resin as core and shell material was synthesized by in situ polymerization technology. The structures and properties of microencapsulated melamine polyphosphate (OFMPP) were characterized by Fourier transform infrared spectroscopy, X-ray photoelectron spectroscopy, scanning electron microscopy, and water contact angle. Thermal behavior of OFMPP was systematically analyzed by thermogravimetric analysis. Flame retardant tests such as limiting oxygen index and UL-94 were employed to evaluate the effect of composition variation for the MPP and the OFMPP in polyurethane bridged polysilsesquioxane (Si-PU) composites. The results indicated that the microencapsulation of MPP with the OF resin led to an improvement of the hydrophobicity. Results also revealed that the flame retardancy of the Si-PU/OFMPP composite (LOI=32%) was better than that of the Si-PU/MPP composite (LOI=27%) at the same additive loading.

Friction and Wear Properties of Recycled Natural Fiber Reinforced Plant-Derived Polyamide 1010 Biomass Composites

Yosuke Nishitani[1,a)], Natsuki Sugawara[1], Kentaro Kawasaki[1], Shuto Oda[1], Jun Mukaida[1] and Tetsuto Kajiyama[2]

[1]*Department of Mechanical Engineering, Faculty of Engineering, Kogakuin University, 2665-1 Nakano, Hachioji, Tokyo 1920015, Japan*
[2]*Jonan Branch, Tokyo Metropolitan Industrial Technology Research Institute, 1-20-20 Minamikamata, Ota, Tokyo 1440035, Japan*

[a)]Corresponding author: at13152@ns.kogakuin.ac.jp

Abstract. To develop new engineering materials for sliding mechanical parts (referred to as tribomaterials in this paper) using only inedible plant-derived materials, this study aimed to investigate the frictional and wear properties of recycled natural fiber reinforced plant-derived polyamide 1010 (PA1010) biomass composites. Two types of recycled natural fibers: recycled jute fiber (R-JF) and recycled sisal fiber (R-SF) were used as the reinforcement fibers. PA1010 is made of sebacic acid and decamethylenediamine, which are obtained from plant-derived castor oil. In this study, R-JF and R-SF were surface-treated by two types of surface treatment: (a) alkali treatment by sodium hydroxide (NaOH) solution and surface treatment by ureidosilane coupling agent, and (b) alkali treatment by sodium chlorite ($NaClO_2$) solution and surface treatment by ureidosilane coupling agent before addition to PA1010. These fiber reinforced PA1010 biomass composites were then extruded using a twin extruder and injection-molded. Their frictional and wear properties were evaluated by ring-on-plate type sliding wear test. Their worn surface, the transfer film on the counterface, and wear debris were observed with a scanning electron microscope to understand the wear mode. It was found that the frictional coefficient and specific wear rate of the biomass composites improved when added with R-JF and R-SF whose surface have been treated by alkali treatment and silane coupling agent. In particular, the combination of $NaClO_2$ and ureido silane coupling agent was found to be the most effective surface treatment for improving the tribological properties of the biomass composites in constant load and velocity test. These results may be attributed to the change in the mode of wear mechanism according to the types of natural fiber used, surface treatment performed for realizing good dispersion of fibers in the composites, and good interaction adhesion between fiber and matrix polymer.

INTRODUCTION

Natural fiber-reinforced plant-derived polymer biomass composites are environment-friendly to a large extent and have unique performances. Many reports have been presented on the processing, characterization, rheological and mechanical properties of these biomass composites [1-3]. However, very few studies have been conducted on the use of biomass composites [4, 5] as tribomaterials, which are often applied for manufacturing mechanical sliding parts such as gears, bearings and seals. To develop polymeric tribomaterials, various properties need to be well balanced. In particular, not only sufficient mechanical properties and high moldability but also good tribological properties such as low frictional coefficient, low wear and high limiting pv (pressure x velocity) value are required. In our previous studies, we developed new polymeric tribomaterials based only on plant-derived materials [6-10]. In particular, we investigated the rheological, mechanical and tribological properties of natural fiber reinforced biopolymer composites such as hemp fiber (HF) reinforced plant-derived polyamide 1010 (PA1010) biomass composites. PA1010 is made of sebacic acid and decamethylenediamine, which are obtained from plant-derived castor oil [11]. As castor oil is not used for food, there is no competition with human food consumption. It was found that the mechanical and tribological properties of these biomass composites improve when they are filled with hemp fibers which have been surface-treated using silane coupling agent, and added with the plant-derived thermoplastic elastomer such as polyamide 11 elastomer. However, in order to further enhance the mechanical and tribological properties of these biomass composites, the effects of the type of natural fiber on the mechanical and tribological properties of these biomass composites were investigated in this study as not much research has been done in this area. In particular, recycled natural fibers such as recycled jute fiber (R-JF) and recycled sisal fibers are reused for the package bags of coffee beans, as reinforcement fibers have significant competitive advantages for these biomass composites because of low cost and environmental resistant performances. To develop such new tribomaterials using only inedible plant-derived engineering materials, this study aimed to investigate the frictional and wear properties of recycled natural fibers, namely recycled jute fiber (R-JF) and sisal fiber (R-SF), reinforced plant-derived polyamide 1010 (PA1010) biomass composites.

EXPERIMENTAL

The materials used in this study were recycled natural fiber (R-NF) reinforced plant-derived polyamide 1010 biomass composites (R-NF/PA1010). Plant-derived polyamide 1010 (PA1010, Vestamid Terra DS16, Daicel-Evonik Ltd., Japan) was used as the matrix polymer. Two types of recycled natural fibers: recycled jute fiber (R-JF, Takeishi Ltd., Japan) and recycled sisal fiber (R-SF, Takeishi Ltd., Japan) were used as the reinforcement fibers. These recycled natural fibers are reused to make the package bags of coffee beans and so on. R-JF and R-SF were precut into 5-mm-long pieces and surfaced-treated by two types of surface treatment: a) alkali treatment by sodium hydroxide (NaOH) solution and surface treatment by ureidosilane coupling agent (S3, 3-ureidopropyltrimethoxy silane, A-1160), and b) alkali treatment by sodium chlorite solution ($NaClO_2$) and surface treatment by ureidosilane. Two types of alkali treatment aqueous solutions such as sodium hydroxide solution (NaOH) and sodium chrolite solution ($NaClO_2$) were used as the mercerization agent. Alkali treatment by NaOH and by $NaClO_2$ was performed as follows: 5% alkali agent solution was placed in a stainless beaker. Chopped recycled natural fibers were then added to the beaker and stirred well. This was kept at room temperature for 4 h. Next, the fibers were washed thoroughly with water to remove the excess alkali agents sticking to the fibers. The alkali-treated fibers were dried in air for 24 h and in a vacuum oven at 80 °C for 5 h. The surface treatment of recycled natural fibers with the concentration of 1 wt.% ureidosilane coupling agent (S3) was carried out in 5 wt.% acetic acid aqueous solution in which the pH of the solution was adjusted to 3.5 and stirred continuously for 15 min. Then, the fibers were immersed in the solution for 60 min. After treatment, fibers were removed from the solution and dried in air for 24 h and in vacuum oven at 80 °C for 5 h. The volume fraction of fiber V_f was fixed at 20vol.% in this study. Various recycled natural fiber reinforced plant-derived PA1010 (R-NF/PA1010) biomass composites were extruded using a twin screw extruder and injection-mold. All the components were dried for 12 h at 80 °C in a vacuum oven beforehand until the moisture level was below 0.2%, then dry blended in a small bottle, and subsequently, the melt was mixed at 85 rpm and 220 °C on a twin screw extruder (TEX-30HSS, Japan Steel Works, Ltd., Japan). After mixing, the extruded strands of various R-NF/PA1010 biomass composites were cut into 5-mm-long pieces with a pelletizer and dried again at 24 h at 80 °C in the vacuum oven. Samples of various shapes were made by injection molding (NS20-A, Nissei Plastic Industrial, Japan) for various experiments. The molding conditions were as follows: cylinder temperatures of 220 °C, mold (cavity) temperature of 30 °C, and injection rate of 13 cm^3/s. To maintain the dry conditions of specimens for all measurements, they were stored in desiccators after molding in accordance with JIS K 6920-2 for at least 24 h at 23 °C.

The tribological properties were measured using a ring-on-plate type sliding wear tester (EFM-3-EN, Orientec, Co. Ltd., Japan). The tests were conducted using plate specimens (30 mm x 30 mm x 3 mm) with the surface (R_a=0.4 μm) finished by No. 240 and No. 800 polishing paper at room temperature under dry condition in accordance with JIS K 7218a (in which test is a type of full contact between the test specimen and metal counterface). A carbon steel (S45C) ring (ϕ20 mm x ϕ25.6 mm x 15 mm, apparent contact area A_a = 2 cm^2) whose surface (R_a=0.4 μm) was finished using the same polishing paper as that used for the metal counterpart. In this study, two types of tribological tests were carried out: one is the sliding wear test under constant normal load and constant sliding velocity, and the other is the limiting pv (pressure x velocity) value test based on the step load method. First, sliding wear test was conducted at the constant normal load P of 50 N (apparent contact pressure p_a =0.25 MPa), constant sliding velocity v of 0.2 m/s, and at the sliding distance L of 600 m. Tribological properties were evaluated by frictional coefficient μ and specific wear rate V_s. Frictional coefficient was calculated from the average values of frictional coefficient at the sliding distance between 400 m to 600m, which is a stable (steady state) distance range. The specific wear rate was calculated from the weight of sample (plate) before and after the test. The sliding surfaces of the test specimen, wear debris and counterface before/ after observation by a SEM with platinum (Pt) sputter coating. Secondly, the limiting *pv* (pressure x velocity) value was calculated using the step load method under a constant velocity v of 0.3 m/s at the initial normal load P_0 of 50 N (p_a =0.25 MPa) and step load P of 25 N every 3 min until the test piece fractured or melted. The step load just before the test piece fractured or melted was defined as the limiting load P_{lim}. This limiting load was divided with the apparent contact area A_a =2 cm^2, and the value obtained was taken to be the apparent limiting contact pressure p. The limiting *pv* value was calculated by multiplying this limiting contact pressure and the test velocity v.

RESUTS AND DISCUSSION

The tribological properties tested using a ring-on-plate type sliding wear tester under constant normal load, constant sliding velocity, and under dry conditions are discussed. Fig. 1 shows the relationship between specific wear rate (V_s) and frictional coefficient (μ) of various recycled natural fiber reinforced plant-derived polyamide 1010 biomass composites (R-NF/PA1010) investigated by constant normal load and sliding velocity test. The μ and V_s of neat PA1010 improved when added with recycled natural fibers (R-NF) such as recycled jute fiber (R-JF) and recycled sisal fiber (R-SF) which had been surface-treated using alkali treatment and ureidosilane coupling agent. However, the improvement effects of the type of R-NF used and surface treatment performed on the tribological properties differ with each property. The frictional coefficient of various R-NF/PA1010 biomass composites was found to decrease in the following order: PA1010 > R-JF ($NaClO_2$) > R-JF (NaOH) = R-SF (NaOH) > R-SF ($NaClO_2$). In short, the improvement effect of type of R-NF on the frictional coefficient of these biomass composites is stronger than that of type of surface treatment. In particular, R-SF improves the frictional properties of these biomass composites more than R-JF. This may be attributed to the difference in the self-lubricant properties of various natural fibers and the difference in the fiber dispersion in these composites according to the type of fibers. On the other hand, the specific wear rate of various R-NF/PA1010 biomass composites showed a different tendency from that of μ, decreasing as follows: PA1010> R-SF (NaOH) > R-JF (NaOH) > R-JF ($NaClO_2$) > R-SF ($NaClO_2$). The improvement effects of the type of surface treatment on the specific wear rate of these biomass composites are stronger than that of type of surface treatment. In particular, alkali treatment by $NaClO_2$ improves the wear properties of these biomass composites more than treatment by NaOH. These results may be attributed to the improvement in the interfacial adhesion between R-NF and matrix polymer. The interfacial interaction between R-NF and polymer matrix is thought to vary according to the type of alkali used for surface treatment. Therefore, the combination of R-SF and the surface treatment by $NaClO_2$ and ureidosilane has the best improvement effect on the tribological properties of R-NF/PA1010 biomass composites under constant normal load and constant normal velocity. These results may be attributed to the change in the mode of wear mechanism according to the type of R-NF and surface treatment performed for fiber dispersion in the composites and the good interaction adhesion between fiber and matrix polymer.

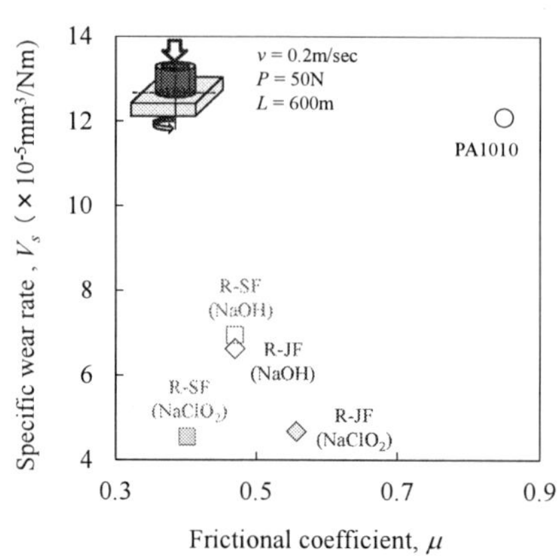

FIGURE 1. Relationship between specific wear rate and frictional coefficient of various R-NF/PA1010 biomass composites.

Thus, it is necessary to observe the wear debris, worn surface of materials and transfer films on the metallic counterface for understanding the mechanisms of tribological behavior, because these factors are strongly influenced by the characteristics of polymer composites [12-14]. Fig. 2 shows the SEM photographs of the wear debris of various R-NF/PA1010 biomass composites. The shape and size of wear debris vary with the type of alkali treatment but not with the type of R-NF. The shape and size of wear debris of R-SF (NaOH) (Fig. 2(a)) and R-JF (NaOH) (Fig. 2(c)) are long thin filamentary and big thick flaky particles. To the contrary, those of R-SF ($NaClO_2$) (Fig. 2(b)) and R-JF ($NaClO_2$) (Fig. 2(d)) are thin flakey ones. The tendencies mentioned above are thought to affect the tribological properties of these R-NF/PA1010 biomass composites.

Next, the step load test results of the limiting *pv* value of various R-NF/PA1010 biomass composites, which were more severe than the results of constant load and velocity test are discussed. Fig. 3 shows the limiting *pv* values of these R-NF/PA1010 biomass composites. The limiting *pv* values improved when added with R-NF whose surface has been treated using alkali treatment and ureidosilane coupling agent. In particular, in both R-JF and R-SF reinforced PA1010 biomass composites, the limiting *pv* values of alkali treatment by NaOH improved more than those of $NaClO_2$. These results suggest that the load carrying ability of R-NF/PA1010 biomass composites increases when added with R-NF which has been surface-treated by NaOH for improving the limiting *pv* values of various R-NF/PA1010 biomass composites.

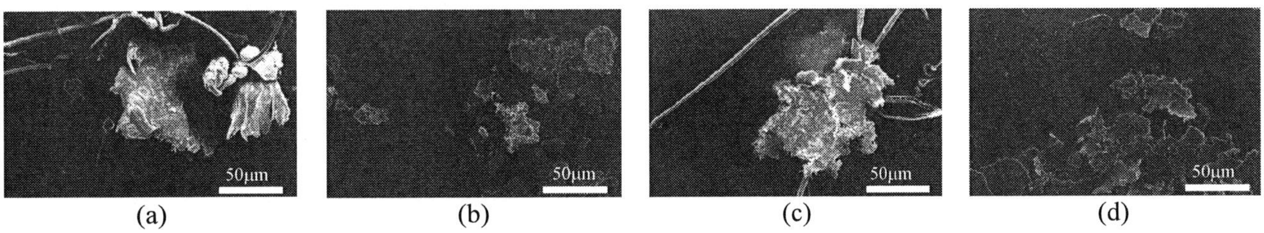

FIGURE 2. SEM photographs of wear debris of various R-NF/PA1010 biomass composites after sliding wear test: (a) R-SF (NaOH), (b) R-SF ($NaClO_2$), (c) R-JF (NaOH) and (d) R-JF ($NaClO_2$).

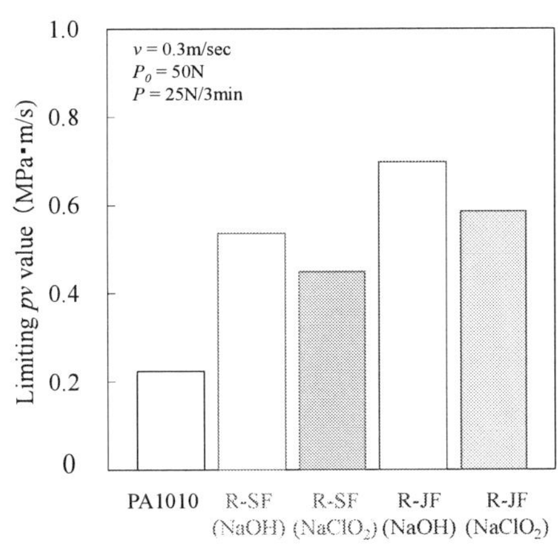

FIGURE 3. Limiting *pv* value of various R-NF/PA1010 biomass composites.

CONCLUSION

To develop new tribomaterials using only inedible plant-derived engineering materials, the aim of this study was to investigate the frictional and wear properties of recycled natural fibers reinforced plant-derived polyamide 1010 (PA1010) biomass composites. Two types of recycled natural fibers: recycled jute fiber (R-JF) and recycled sisal fiber (R-SF) were used as the reinforcement fibers. It was found that the tribological properties of plant-derived PA1010 improved when added with recycled natural fibers such as R-JF and R-SF which had been surface-treated by alkali treatment and ureidosilane coupling agent. However, the influence of the type of recycled natural fiber used and type of surface treatment performed on the tribological properties of recycled natural fiber reinforced plant-derived PA1010 biomass composites differs by property. In particular, the combination of R-SF and the surface treatment by $NaClO_2$ and ureidosilane improved tribological properties the most, such as frictional coefficient and specific wear rate. On the other hand, the limiting *pv* value improved when PA1010 was added with recycled natural fibers which had been surface-treated by alkali treatment and ureidosilane coupling agent. In particular, for both R-JF and R-SF reinforced PA1010 biomass composites, the limiting *pv* value of alkali treatment by NaOH improved more than that of $NaClO_2$.

ACKNOWLEDGMENTS

This work was supported by JSPS KAKENHI Grant Number 16K06750. We would like to thank the Functional Microstructured Surfaces Research Center (FMS, MEXT, Japan) of Kogakuin University and the Project Research of the Research Institute for Science and Technology of Kogakuin University for funding this study.

REFERENCES

1. O. Faruk, A. K. Bledzki, H.-P. Fink and M. Sain, Prog. Polym. Sci., **37**, 1552–1596 (2012).
2. L. Yu., K. Dean and L. Li, Prog. Polym. Sci., **31**, 576–602 (2006).
3. Y. Xie, C. A. S. Hill, Z. Xiao, H. Militz and C. Mai, Composites: Part A, **41**, 806–819 (2010).
4. N. Chand and M. Fashim, *Tribology of natural fiber polymer composites* (Woodhead Publishing Ltd., Cambridge, 2008), pp. 84–196.
5. U. Nirmal, J. Hashim, M. M. H. M. Ahmad, Tribol. Int., **83**, 77-104 (2015).
6. Y. Nishitani, M. Hasumi and T. Kitano, AIP Conference Proceedings, **1664**, 060007 (2015).
7. J. Mukaida, Y. Nishitani, and T. Kitano, AIP Conference Proceedings, **1664**, 060008 (2015).
8. Y. Nishitani, J. Mukaida, T. Kajiyama, Y. Yamanka, T. Yamanaka and T. Kitano, AIP Conference Proceedings, **1713**, 120007 (2016).
9. Y. Nishitani, J. Mukaida, T. Kajiyama, Y. Yamanka, T. Yamanaka and T. Kitano, AIP Conference Proceedings, **1779**, 060004 (2016).
10. Y. Nishitani, T. Yamanaka, T. Kajiyama and T. Kitano, "Thermal Properties of Hemp Fiber Reinforced Plants-Derived Polyamide Biomass Composites and Their Dynamic Viscoelastic Properties in Molten State" in *Viscoelastic and Viscoplastic Materials* edited by M. F. El-Amin (Intech, Rijeka, 2016), pp. 53–79.
11. T. Shimizu, JETI, **59**, 71–83 (2011) (in Japanese).
12. S. Bahadur, Wear, **245**, 92–99 (2000).
13. S. Yu, Z. Yu and Y.-W. Mai, Tribol. Int., **40**, 855–862 (2007).
14. Y. Nishitani, Y. Yamada, C. Ishii, I. Sekiguchi and T. Kitano, Polym. Eng. Sci., **50**, 100–112 (2010).

Modelling the Effects and Interactions of a Layered Double Hydroxide in a Flexible PVC Formulation

Fechter Reinhard (1)*, Sandrock Carl (1), Labuschagne Johan (1)

(1) University of Pretoria - --- - SouthAfrica

Developments by Labuschagne et al. (2006) in the manufacture of Layered Double Hydroxides (LDHs) have led to a patent describing a new effluent free synthesis method. To make the LDHs produced using the new synthesis method commercially viable requires that they have product applications. One such application is as a thermal stabiliser and fire retardant for poly(vinyl chloride) (PVC). The problem is that a PVC compound is a complex system that includes multiple additives that tend to interact strongly with each other. The aim of the investigation was to quantify and analyse the effect of an LDH on the important material properties of a flexible PVC compound. The effect of a Hydrotalcite-like LDH on the thermal stability, fire retardancy and basic mechanical properties was modelled using 2nd degree Scheffe polynomials. The empirical models for each response variable were determined using statistical experimental design. The particular models for each response variable, which are selected for predictive ability using k-fold cross validation, were interpreted using statistical analysis of the model terms. The analysis clearly shows and quantifies the significant effect that the LDH has on increasing the long term dynamic thermal stability and decreasing the smoke production of the PVC compound. The positive effects that the LDH has are offset by the negative effect it has on the elongation at break of the PVC compound. The analysis also reveals a number of interactions between the LDH and other additives. For instance, a strong synergistic relationship between the LDH and the plasticiser in terms of thermal stability and smoke reduction is evident, which to the knowledge of the authors has not been reported in the literature. There is also, somewhat surprisingly, evidence of antagonism between the LDH and the primary thermal stabiliser with regard to the thermal stability of the PVC compound.

Preparation and characterization of electrically conductive Thermoplastic polyurethane/Graphene-Graphene oxide nanocomposite porous films

Razeghi Mohammadali [1], Pircheraghi Gholamreza [1]*

[1] Sharif University of Technology - Tehran - Iran

Electrically conductive porous nanocomposite films have increasingly attracted interest because of the need for lightweight conductive materials in low filler content for potential application in EMI shielding. Graphene is an interesting candidate to prepare the conductive polymeric nanocomposites because of its excellent electrical conductivity, but obtaining a good dispersion of nonfunctional 2D nanomaterials inside the polar polymeric matrices is still a critical challenge. Instead, Graphene oxide nanoplatelets could almost properly be dispersed in polar polymeric matrices such as thermoplastic polyurethane (TPU) due to its functional surface groups. This study demonstrates a facile approach to synthesis graphene and graphene oxide nanoplatelets (GNP) simultaneously by electrochemical method and fabrication of TPU/GNP nanocomposite with non-solvent induced phase separation. Different methods were used to characterize the microstructure and properties of samples including optical microscopy, rheological measurements, SEM, DSC and electrical conductivity. SEM images confirmed the porous microstructure of nanocomposite films. While, rheological and DSC measurements indicate that incorporation of GNPs changes the microstructure of segmented polyurethane by interaction between hard segments and nanofillers in the solution state and final sample, respectively. Finally, although graphene oxide is electrically insulated, it improves the dispersion of conductive graphene inside the TPU matrix, observed in SEM & optical microscopy images. This state of dispersion and formed 3D network of graphene/graphene oxide lead to 5 order of magnitude increase in conductivity of TPU porous nanocomposite films observed with only 3wt% of nanofiller.

Universal and anisotropic simulation platform for the study of electrical properties of conductive polymer composites

Guanda Yang[1], Fritjof Nilsson[2], Dirk W. Schubert[1]

1 Institute of Polymer Materials, Friedrich-Alexander-University Erlangen-Nuremberg, Martensstr. 7, 91058, Germany

2 KTH Royal Institute of Technology, School of Chemical Science and Engineering, Fibre and Polymer Technology, SE- 10044 Stockholm, Sweden

Corresponding author: guanda.yang@fau.de

Abstract. In order to carry out a study on the synergistic relationship of variables which could affect the electrical properties of CPC, a universal and anisotropic platform of simulation, containing three simulation modules are explored. The simulation modules are: finite element modelling (FEM), percolation threshold modeling (PTM) and electrical networks modelling (ENM).

1. INTRODUCTION

Due to easy fabrication, tunable properties and a broad range of applications, the research area of conductive polymer composites (CPCs), consisting of insulating polymers and conductive fillers (carbon black (CB), carbon fibers (CFs), graphene, etc.) has increasingly gained more attention in the last decades.[1] CPCs with good flexibility exhibit a variety of sensing responses when subjected to external strain. The sensing mechanism is mainly based on the destruction and reconstruction of conductive networks. Therefore a variation of conductivity can be linked to the amount of exposure by external stimuli.

Recently, CPCs with good flexibility have been considered as satisfactory candidates to develop flexible strain sensors. Utilizing the advantages of the polymer matrices, CPCs can show excellent stretching ability and quick response to external stress.[2-4] Elastomers, such as vinylidene fluoride-hexafluoropropylene copolymer[5], polydimethylsiloxane (PDMS) [6], silicone rubber (SR) [7-9] or thermoplastic polyurethane (TPU) [10-12] are typically used as polymer matrices. However, the application of CPCs based strain sensors still face severe challenges, e.g., unstable signal output and insufficient reversibility. In particular, due to different demands for the strain sensing behavior in different applications, CPCs with tunable strain response patterns are urgently needed.

2. BACKGROUND

Besides the experimental investigation, simulations can provide an effective way of understanding the dynamic mechanism. A surplus of theoretical studies carried out by simulations have already been conducted in order to investigate the relationship between the geometry of fillers and electrical properties.[14-16] However, most of these studies assumed an idealized isotropic orientation of the fillers, represented by using just one parameter for the spatial orientation. If the geometrical factors are considered, as for example the aspect ratios of the fillers and the filler inclination, the influence on the electrical properties can be studied thoroughly.

Besides the experimental investigation, simulation can provide an effective way of understanding the dynamic mechanism. Many excellent theoretical studies via simulations have already been conducted[1-3] in order to investigate the relationship between the geometry of fillers and electrical properties. However, most of these studies assumed an idealized isotropic orientation of the fillers, and focused on only one variable of geometrical distribution. Considering the geometrical factors, i.e. fillers' aspect ratios and filler inclination might influence the electrical properties. Therefore, a systematic theoretical study on the influence of the geometrical distributions and filler dimensionality, especially for the 3-dimensional fillers, would still be scientifically necessary.

3. MODEL

For the simulation and modeling, the influence of different parameters on the electrical conductivity, particularly on the electrical percolation filler fraction was discussed in relation to experimental data. In order to carry out a study on the synergistic relationship of these variables, a universal and anisotropic platform of simulation, containing three simulation modules had been explored.[24] The simulation modules contain finite element modelling (FEM), percolation threshold modeling (PTM) and electrical networks modelling (ENM). After constantly updating, the simulation platform performs the following features:

- Anisotropy (orientation)
- High degree of flexibility
- High degree of automation
- High efficiency: Under the same magnitude of mesh and computing environment, our platform needs only 1/3 the time of other similar simulation program.
- High accuracy: All parameters and results can be accurate up to 12 decimal places.
- Very low error rate (less than 0.01%)

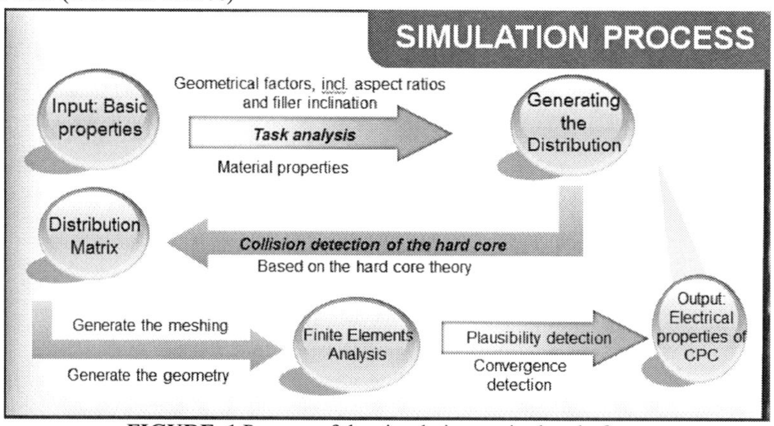

FIGURE. 1 Process of the simulation set in the platform

The mesh strategy is the key of simulation. For simulation with different settings and conditions, it is necessary to bring different mesh strategy when facing diverse models. In the platform of simulating CPCs, the mesh strategy is always optimized and updated.

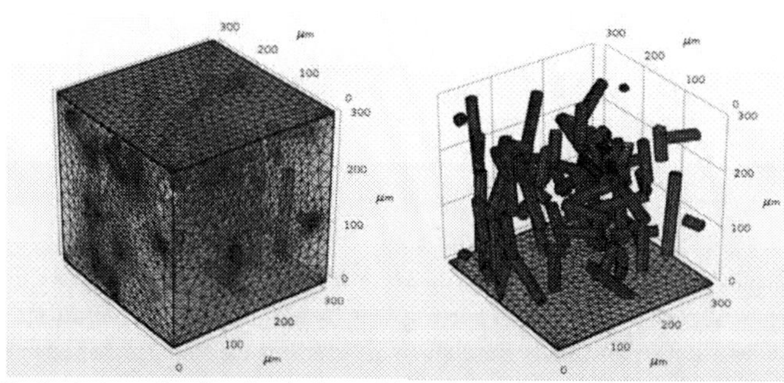

FIGURE. 2 (a) Mesh strategy for the simulation with 1~150 fillers, with the top and bot-voltage blocks a coefficient of geometry scale 0.3. The coefficient of the space and the fillers is 1.0. The mesh operation over the geometry is set to be free tetrahedral. (b) Mesh of the conductive filler, with the top voltage block and the space set to be hidden.

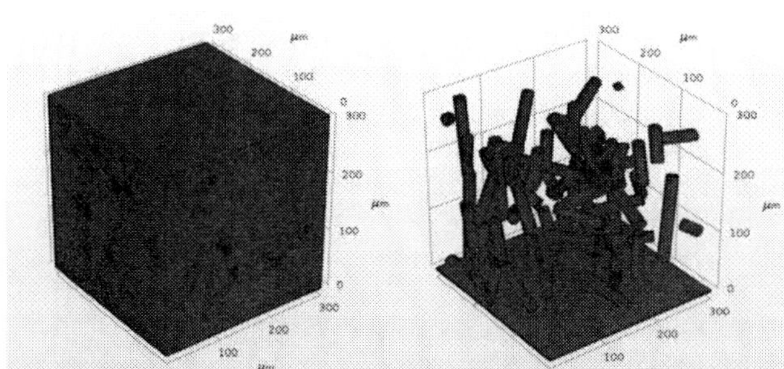

FIGURE. 3 (a) Mesh strategy for the simulation with >150 fillers, with the top and bot-voltage blocks a coefficient of geometry scale 2.0. The coefficient of the space and the fillers is 10.0. The mesh operation over the geometry is set to be free tetrahedral. (b) Mesh of the conductive filler, with the top voltage block and the space set to be hidden.

4. EXPERIMENTAL

The ordinate axis shows the number of elements, and the abscissa shows the ratio of current mesh quality and the smallest mesh quality. For the 3D model, it could also be approximated as the reciprocal of the ratio of the minimum element volume to the current element volume. This distribution is a practical tool for judging the quality of the mesh and the authenticity of the simulation results. A curve that satisfies the normal distribution represents a suitable meshing.

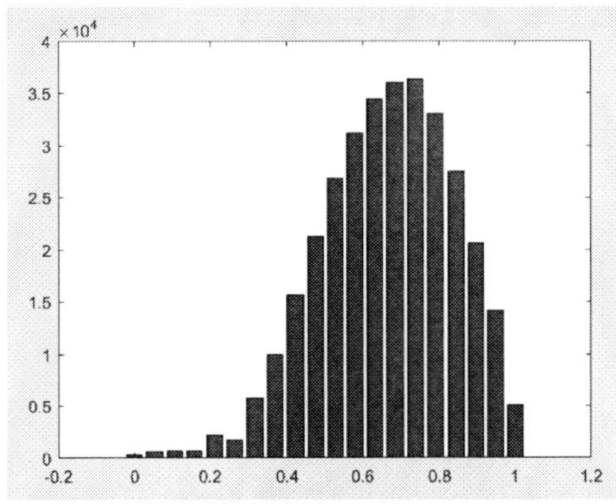

FIGURE. 4 Mesh quality control

After constantly updating and improving in the past years, the simulation platform performs high degree of flexibility, high degree of automation, high efficiency and high accuracy. Within the platform, all the factors could be controlled as either constant or variable, and a deeper understanding of the relationship between different factors is comprehended. An experimental consideration with multiple levels of variables is carried out, and the synergistic relationship of the variables could be systematic studied. The trend of change in each variable under different environmental settings is clearly displayed and systematically studied by the simulation platform.

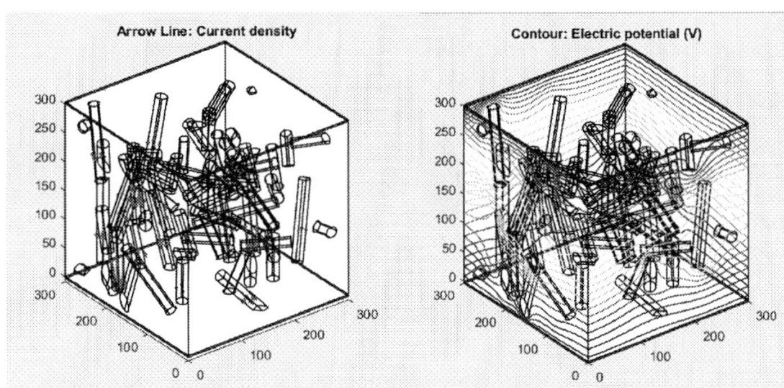

FIGURE. 5 Output of the current density and the electric potential

The arrow line of current density shows automatically the path built by the conductive fillers, the denser the red arrows, the greater is the current flow, also indicates the less resistance of the path. This output is significant at showing the electrical flow within the matrix according to the distribution of the fillers. After collecting more than 1000 groups of data and calculating the median value out of it, a McLachlan Fit has been done according to the function:

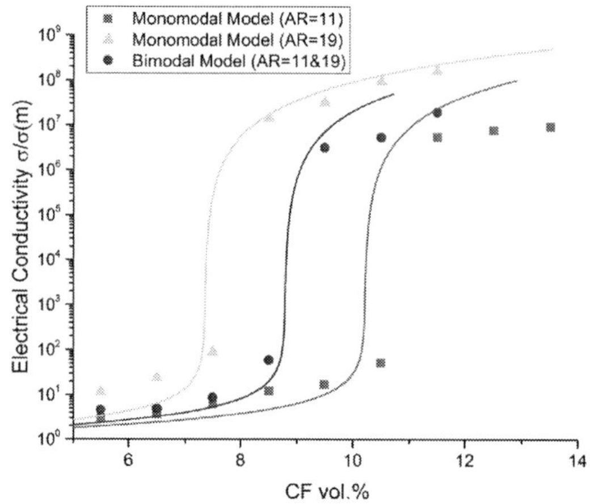

FIGURE. 6 Logarithm of conductivity vs CF volume fraction for the composites from different aspect ratios, as well as the bimodal model. The filler fraction of the both kinds of CFs in the bimodal model is set to be 50%. With the change in composition, the conductivity of the CPCs as well as the percolation threshold will also show a corresponding change.

The figure.5 shows the McLachlan Fit of the data, and the mono modal with aspect ratio 19 has a lowest percolation threshold of 7.6%; the group with AR=11 has the highest percolation threshold of 10.2%. With a bimodal model of a certain ratio between the both kinds of fillers (1:1), the percolation threshold could be adjusted to an intermediate value 8.4%. With more data from the simulation platform, a preparation method with a fixed percolation threshold will be determined.

CONCLUSIONS

A significant research aiming at the difficulties of simulating CPCs is carried out as the first phase of the simulation. The platform will be systematically updated for a better performance and efficiency, especially the percolation threshold method (PTM) module and the "high conductive layer" feature. Through the whole working phase, all the factors with high research value will be controlled as either constant or variable, and a deeper understanding of the relationship between different factors will be comprehended. An experimental consideration with multiple levels of variables will be carried out, and the synergistic relationship will be systematically studied. Combined with experimental data, the trend of change in each variable under different environmental settings will be clearly displayed by the simulation platform.

ACKNOWLEDGEMENTS:

The author would like to acknowledge financial support of the Europa Akademie e.V. and Cetrico Industrial Consultant GmbH.
Special thanks for the theoretical support and guidance on the McLachlan Fit from M. Sc. Muchao Qu.
Dr. Monika Voigt is thanked for the instructive ideas and kind modifications. (LSP, University of Erlangen-Nuremberg)

REFERENCES:

1. Modelling the relative permittivity of anisotropic insulating composites, F.Nilsson. Author links open the author workspace.U.W.Gedde. Author links open the author workspace.M.S.Hedenqvist, Composites Science and Technology, 2011
2. Simulating the effective electric conductivity of polymer composites with high aspect ratio fillers, Fritjof Nilsson, Johannes Krückel, Dirk W. Schubert, Fei Chen, Mikael Unge, Ulf W. Gedde, Mikael S. Hedenqvist, Composites Science and Technology, 2016
3. I. Balberg, J. App. Phys. 2012,112, 066104.
4. L. Berhan and A. M. Sastr,. Phys. Rev. E 2007,75,041120.
5. J. Krückel, Z. Starý, C. Triebel, D. W. Schubert and H. Münstedt, Polymer, 2012, 53(2), 395-402.
6. J. Krückel, Z. Starý and D. W. Schubert, Polymer, 2013, 54(3), 1106-1113.
7. J. Krückel and D. W. Schubert, Eur. Polym. J., 2014, 53 (1), 50-57.
8. X. Liu, J. Krückel, G. Zheng and D. W. Schubert, ACS Appl. Mater. Interfaces, 2013, 5 (18), 8857-8860.
9. S. D. Bondt et al. Journal of materials science 1992; 27: 1983-1988.

10. Chippendale,et al. The 8th International Conference on Computation in
11. Electromagnetics., 2011.
12. H. Deng, L. Lin, M. Ji, S. Zhang, M. Yang and Q. Fu, Prog. Polym. Sci., 2014, 39(4), 627-655.
13. C. S. Boland, U. Khan, G. Ryan, S. Barwich, R. Charifou, A. Harvey, C. Backes, Z. Li, M. S. Ferreira, M. E. Möbius, R. J. Young and J. N. Coleman, Science, 2016, 354(6317), 1257-1260.
14. H. Liu, W. Huang, X. Yang, K. Dai, G. Zheng, C. Liu, C. Shen, X. Yan, J. Guo and Z. Guo. J. Mater. Chem. C, 2016, 4, 4459-4469.
15. T. Sekitani, Y. Noguchi, K. Hata, T. Fukushima, T. Aida and T. Someya, Science, 2008, 321,1468-1472.
16. M. Kujawski, J. D. Pearse and E. Smela, Carbon, 2010, 48, 2409-2417.
17. L. Chen, G. Chen and L. Lu, Adv. Funct. Mater., 2007, 17, 898-904.
18. Z. Dang, M. Jiang, D. Xie, S. Yao, L. Zhang and J. Bai, J. Appl. Phys., 2008,104, 024114/1-6.
19. L. Bokobza, Silicon, 2009, 1,141-145.
20. H. Liu, J. Gao, W. Huang, K. Dai, G. Zheng, C. Liu, C. Shen, X. Yan, J. Guo and Z. Guo, Nanoscale, 2016, 8, 12977-12989.
21. H. Liu, Y. Li, K. Dai, G. Zheng, C. Liu, C. Shen, X. Yan, J. Guo, and Z. Guo, J. Mater. Chem. C, 2016, 4, 157-166.
22. Q. Fan, Z. Qin, S. Gao, Y. Wu, J. Pionteck, E. Mäder and M. Zhu, Carbon, 2012, 50(11), 4085-4092.
23. S. B. Kharchenko, J. F. Douglas, J. Obrzut, E. A. Grulke and K. B. Migler, Nat. Mater., 2004, 3(8), 564-568.
24. Balberg and N. Binenbaum, Phys. Rev. A, 1987, 35(12), 5174.
25. Balberg, J. App. Phys. 2012,112, 066104.
26. L. Berhan and A. M. Sastr,. Phys. Rev. E 2007,75,041120.

Electrical conductivities of melt spun PMMA/aligned carbon fibers/carbon black fiber composites

Qu Muchao [1]*, Nilsson Fritjof [2], Schubert Dirk W. [1]

[1] Friedrich-Alexander-Universität Erlangen-Nürnberg - Bavaria - Germany, [2] KTH Royal Institute of Technology - -- - Sweden

In this study, percolation thresholds of ternary composites of PMMA/CF/CB as well as PMMA/CF and PMMA/CB composites were investigated [1]. A conductivity contour plot for PMMA/CF/CB composites is presented for the first time. Moreover, a novel equation for predicting the electrical conductivity of composites containing two kinds of conductive fillers was proposed, showing results close to corresponding experimental data. Our work comprise also systematic measurements of tensile strength for the PMMA/CB/CF composites. A novel and significantly improved effective-media equations for conductive three-component composites is given.

Effect of ammonium and aminosilane montmorillonites organo-clays on the curing kinetics of unsaturated polyester (UP) resin nanocomposites

vargas maria angeles [(1)]*

[(1)] tecnologico de mexico - edo. mexico - Mexico

The curing kinetics of UP nanocomposites prepared by incorporating different amounts of twokinds of organo-montmorillonite (organo-MMT): trimethyloctadecacylammonium chloride (TMOA) andaminopropyl-triethoxysilane (APTES) were studied by non-isothermal differential scanning calorimetry(DSC) experiments. small angle X-ray scattering (SAXS) was used for measuring the d-spacings in themodified organo-clays, and no intercalation UP into these clays was observed for the nanocomposites.HRTEM images showed dispersed and agglomerated platelets in UP/APTES 2 and 10 wt%. DSC analysisshowed two peaks in UP resins and UP/organo-MMT, and a decrease in the exothermal peaks tempera-ture (Tp1and Tp2) for nanocomposites with the heating rate as compared with those of neat UP system;thus, the higher the heating rate, the higher the curing reaction rate. This effect was more clearly onthe UP/MMT-APTES nanocomposites. The effective activation energies (Ea) were determined with themode-free isoconversional Starik's method. Sesták–Berggren model was chosen to simulate the reactionrate with a good match achieved. Thermal gravimetric analysis showed that the cured UP/APTES at highconcentration were slightly more stable than UP and TMOA.

The effect of Nanosilica on Rheology, Crystallinity and Mechanical properties of High Density Poly Ethylene (HDPE), Ethylene,1- Octene Copolymer (EOC) and their blends

Javadi Azizeh (1), Karimi Samira (1), Nazockdast Hossein (1)*

(1) AmirKabir University of Technology - Tehran - Iran

In the of present work, 2, 5 and 7 wt.% of the Nanosilica were added to High Density Poly Ethylene (HDPE) and Ethylene,1- Octene Copolymer (EOC) respectively. Nanocomposites of HDPE/EOC blends with different compositions, containing 5 wt.% of Nanosilica were prepared too. All nanocomposites were prepared via melt mixing using an internal mixer at 180 and 60 rpm. The rheology, crystallinity and mechanical properties of all samples were studied. It was found that the addition of 5 wt.% of Nanosilica, to HDPE, leads to formation a network in it, which was confirmed by a viscosity upturn and a plateau in storage modulus against frequency diagrams. In other hand, even by addition 7 wt. % of nanoparticles to EOC matrix, there was no evidence of a network formation. DSC results showed that the addition of Nanosilica does not effect on crystallization temperature in both HDPE and EOC matrices, which is due to low surface area of this nano particle. In other hand, the amount of crystallinity is much lower in HDPE nanocomposite in comparison with neat polyethylene, which is due to the formation of network as confirmed by rheology too. The rheological data on HDPE/EOC blend nanocomposites, also DSC results of these samples showed that nearly all nanoparticles are placed in EOC phase. Tensile tests confirmed that the tensile strength and modulus of HDPE and EOC are increased by increasing the amount of nano particles and the elongation at break is decreased. It was also found that the blends of HDPE/EOC have lower mechanical properties in comparison with neat PE, but the addition of nanoparticles improve the strength of these blends. Keawords: HDPE, EOC, Nanosilica, Nanocomposite, rheology, crystallinity

Preparation and characterization of polyimide reinforced carbon fiber composites by vacuum assisted resin infusion

Jang Si-hoon (1), *Sur Young-sek* (1), *Choi Jun* (2), *Park No-hyung* (2)*

(1) Korea Institute of Industrial Technology - Ansan, Kyunggi-do - Korea, (2) Korea Institute of Industrial Technology - Ansan, Kyunggi-do - Korea

Carbon fiber (CF) was widely used for advanced composite fields such as transportation due to their various advantages of high strength and stiffness, etc. Also, polyimide (PI) was high performance polymer having dimensional stability, heat resistance and mechanical properties etc. In this study, CF non-woven was prepared by wet-laid process. And the composite materials based on CF non-woven with multilayer structure were prepared by vacuum assisted resin infusion (VaRI) molding technique with the use of PI to improve its mechanical and thermal properties. Two different molecular weights of PI have been used in the PI-CF composites. The thermal stabilities and mechanical properties of PI reinforced CF composites were investigated using several methods. Field emission scanning electron microscopy (FE-SEM) and fiber volume fraction of composites were adopted to characterize the morphology and structure properties. The obtained results showed that mechanical properties of composites using CF non-woven with multilayer structure were improved better strength than that of single or few layer composites. The morphology of PI resin in CF non-woven matrix indicated good degree of dispersion.

EFFECT OF MODIFIED MONTMORILLONITE IN THE FLEXURAL AND TENSILE PROPERTIES OF POLYLACTIC ACID NANOCOMPOSITES

***Perez-Fonseca Aida Alejandra* (1)*, *Martin del Campo Alan Salvador* (1), *Robledo-Ortiz Jorge Ramón* (1), *López-Naranjo Edgar José* (1), *Arellano-Martínez Martin Rigoberto* (1), *Jasso-Gastinel Carlos Federico* (1)**

(1) Universidad de Guadalajara - Jalisco - Mexico

Polylactid acid (PLA) is at the present time the most used biopolymer due to its high mechanical strength and easy processability (Arrieta et al., 2014). However, PLA shows disadvantages such as brittleness, sensitivity to high temperature and humidity, low impact strength and high cost. One alternative to modify PLA properties is by the addition of low amounts of nanoclays. Although thermal properties of PLA based nanocomposites have been studied extensively but there is still a lack of information about their mechanical properties. In this work, the effect of a modified nanomontmorillonite (NMN) added to PLA over its flexural and tensile properties was studied. The nanocomposites were prepared with PLA 3251D (Nature Works) and 1.34 NMN modified with methyl dihydroxyethyl hydrogenated tallow ammonium (25-30 wt. %) (Sigma Aldrich). The nanocomposites were obtained with NMN contents of 1, 3 and 5 % wt in a twin-screw extruder followed by an injection molding process. The nanocomposites were characterized by tensile and flexural tests as well as dynamic mechanical thermal analysis (DMTA) and X-ray diffraction (XRD) analysis. The results obtained showed increases in tensile and flexural modulus with NMN content, up to 26% in the case of flexural modulus and 19% for tensile modulus. The flexural strength results were increased independently of the NMN content (24%) indicating a good interaction between PLA and NMN. Also the tensile strength was increased by 10% with NMN addition. XRD tests indicated that the nanoclay was exfoliated with the extrusion process. DMTA results showed that storage modulus of nanocomposites increased with nanoclay addition; however the thermal stability was not modified displaying a significant drop in storage modulus around 60 °C (Tg of PLA) for all materials. Figure 1. Flexural and tensile properties of PLA nanocomposites References 1. M.P, Arrieta, F. Parres, J.López, A. Jiménez, Journal of Analytical and Applied Pyrolysis 101 (2013) 150–155.

On the effect of using a hybrid of nanoparticles containing clay and CaCO3 on the viscoelastic response of PP/EVA blends

Pourabdollah Khadar Mohammad Mahyar (1), Seyfi Javad (1), Khonakdar Hossein Ali (2)*

(1) Islamic Azad University of Shahrood - Semnan - Iran, (2) Leibniz-Institut für Polymerforschung Dresden - Saxony - Germany

On the effect of using a hybrid of nanoparticles containing clay and CaCO3 on the viscoelastic response of PP/EVA blends Mohammad Mahyar Pourabdollah Khadar1, Javad Seyfi1, Hossein Ali Khonakdar2* 1 Department of Chemical Engineering, Shahrood Branch, Islamic Azad University, Shahrood, Iran 2 Leibniz-Institut für Polymerforschung Dresden e.V., Hohe Strasse 6, D-01069 Dresden, Germany Abstract In this study, polymer blends based on polypropylene (PP) and ethylene-vinyl acetate (EVA) were employed to fabricate nanocomposites containing calcium carbonate (CaCO3) and clay nanoparticles via the melt-mixing method. The main aim is to study the effect of nanoparticles' hybrid on the rheological properties of PP/EVA blends. Both static and dynamic conditions were utilized for rheological investigation of the prepared samples. Results showed that nanoparticles had different effects on the viscoelastic response of the polymer blend. Those samples containing a fixed amount of clay (5 wt.%) and various amounts of CaCO3 showed a stronger elastic response. The most pronounced enhancement was observed when the content of clay increased from 1 to 5 wt.% upon which the elastic modulus increased from 70 to 750 Pa. The results showed that upon increment of clay content from 1 to 5 wt.%, the dynamic viscosity was also increased from 3000 to 17600 Pa.s, which could hinder the flow-ability of the polymer blend. However, a shear-thinning behavior was detected indicating the sudden drop of viscosity upon increasing the shear rate. In order to support the rheological data, scanning electron microscopy and hardness tests were also performed. Keywords: Viscoelastic response; nanocomposite; polymer blend; clay; calcium carbonate

Effect of ammonium and aminosilane montmorillonites organo-clays

vargas maria angeles (1)*

(1) tecnologico de estudios superiores - edo. mexico - Mexico

The curing kinetics of UP nanocomposites prepared by incorporating different amounts of two kinds of organo-montmorillonite (organo-MMT): trimethyloctadecacylammonium chloride (TMOA)and aminopropyl-triethoxysilane (APTES) were studied by non-isothermal differential scanning calorimetry DSC) experiments. small angle X-ray scattering (SAXS) was used for measuring the d-spacings in the modified organo-clays, and no intercalation UP into these clays was observed for the nanocomposites. HRTEM images showed dispersed and agglomerated platelets in UP/APTES 2 and 10 wt%. DSC Analysis showed two peaks in UP resins and UP/organo-MMT, and a decrease in the exothermal peaks temperature (Tp1 and Tp2) for nanocomposites with the heating rate as compared with those of neat UP system; thus, the higher the heating rate, the higher the curing reaction rate. This effect was more clearly on the UP/MMT-APTES nanocomposites. The effective activation energies (Ea) were determined with the mode-free isoconversional Starik's method. Sesták–Berggren model was chosen to simulate the reaction rate with a good match achieved. Thermal gravimetric analysis showed that the cured UP/APTES at high concentration were slightly more stable than UP and TMOA.

Mechanochemical Synthesis of Layered Double Hydroxides

Bester, K.M.[1,a)], Focke, W.W.[1,b)] and Labuschagné, F.J.W.J[1,c)]

[1] *Institute of Applied Materials, Department of Chemical Engineering, University of Pretoria, South Africa.*

[a)] Corresponding author: u12009271@tuks.co.za
[b)] walter.focke@up.ac.za
[c)] johan.labuschagne@up.ac.za

Abstract. Hydrotalcite-like layered double hydroxide was successfully synthesised from metal oxides and hydroxides using a horizontal bead mill and was identified as a mixture of hydrotalcite and meixnerite. The final thermally aged material that was milled for three hours has a particle size distribution in the range 0.5 μm to 12 μm. XRD and FTIR both suggest that the synthesised material is phase-pure LDH but the amorphous halo present in the XRD results suggests that an amorphous compound is present as a contaminant.

INTRODUCTION

Layered double hydroxides (LDH) are a class of anionic clays with many uses and applications, they are made up of a series of layered mixed metal hydroxides and an interlayer of anions. LDH materials find use in industry as flame retardants, catalyst packing, reaction catalysts, medicine delivery systems, adsorbers of halogen free radicals, waste water purifiers and many other useful products. [2, 1021-1022]

LDHs are specifically useful as flame retardants, smoke suppresants, reinforcing fillers, and acid scavengers in polymer compounds. The LDHs are non-toxic and heavy metal free, green alternatives to many additives currently used in the polymer industry. The problem with LDH synthesis is that all of the possible synthesis techniques either involve large amounts of waste water and metal salt effluent or require large amounts of energy to occur when large scale industrial production is concerned [1]. To determine if the mechanochemical synthesis is a viable synthesis technique, the existing techniques for LDH synthesis need to be understood.

Co-precipitation Method

The first technique discussed is the co-precipitation method, where aqueous solutions of the metal salts used to synthesise the desired LDH are added together in specific ratios at high stirring speeds, constant temperatures and high pH values using an inorganic base such as NaOH. The precipitate must then be hydrothermally aged to increase the crystallinity of the mixture using the process of Ostwald ripening. Advantages of this technique include the production of small particles before aging with high crystallinity after aging. The disadvantage of this technique is the large volume production of salt containing effluent and the size trade-off experienced during aging. [2, 1027–1028]

Urea Method

The urea method is very similar to the co-precipitation method, with the only difference being the use of urea as the base instead of an inorganic chemical base such as NaOH. This reaction takes place more slowly and as such it can be controlled more accurately to produce more closely distributed particle sizes, and does not require as severe washing due to the lack of the base metal and salt contaminants. [2, 1028-1029]

Dissolution Precipitation Method

The next method discussed is the dissolution precipitation method - or two powder synthesis method - in which the metal oxides or hydroxides are used as the reagents for the LDH synthesis. Labuschagné, Giesekke, and Van Schalkwyk [3] used this method to develop a one-step synthesis process with zero effluent for the synthesis of hydrotalcite. The method suggests a temperature within the range from 140 °C to 200 °C is held for between 4 and 6 hours at a pressure between 10 bar to 14 bar in order to produce hydrotalcite. This is the first method with no effluent, but the reaction conditions suggest that the process would be energy intensive in order to keep the reaction vessel at that high temperature for a long period of time.

Mechanochemistry

There have been some promising results in literature showing that LDH materials can be created using mechanochemistry. Mechanochemistry is the use of mechanical energy to affect a chemical reaction [4] and includes using a multitude of mechanical processes such as crushing, grinding, thermal shock and many other machining techniques to affect polymorphic transformations, solid surface property activation, crystal structure alteration, and solid state reactions on varying types of materials from metals to pharmaceuticals [5]. For the purpose of this investigation, only the use of grinding or milling and thermal treatment are considered.

A patent assigned to Süd-Chemie [6] describes a method for the synthesis and manufacture of hydrotalcite whereby one or both of the starting compounds are not used in the form of a solution. The mixture is subjected to intensive wet grinding, and further processing can lead to either a crystalline phase or an amorphous phase of a high quality hydrotalcite. The patent suggests starting compounds including unreactive MgO, $Al(OH)_3$ and CO_2 either in the form of carbonated water, gas or dry ice. It was found that where the hydrotalcite would be incorporated into a polymer, it is best if milling occurs until such a time that the material is milled down to a D_{90} cut size of 3 μm. [6]

Isupov, Chupakhina, and Mitrofanova [7] used a high-energy planetary-type activator to perform mechanochemical synthesis of magnesium-aluminium double hydroxides starting with a mixture of magnesium hydroxide and aluminium salts [7], where the interlayer spacing and species were determined by the anion in the aluminium salt. Khusnutdinov and Isupov [8] used a planetary mill to create a phase-pure magnesium-aluminium LDH from magnesium hydroxide, aluminium hydroxide and sodium bicarbonate. The interlayer of the LDH consisted of both carbonate and bicarbonate ions. The mill type allowed for 40 g samples to be synthesised at one time, and the milling was carried out for approximately 15 minutes before the samples were washed with water.

Ferencz *et al.* [9] used a two step mechanochemical process with a minute and controlled amount of water to produce almost phase pure Ca-Al LDH. The mill used for the synthesis was a mixer mill using 20 mm diameter stainless steel grinding balls. As this synthesis is water limited, the same results are not expected for the current synthesis investigation but are used as reference.

EXPERIMENTAL

All materials used for synthesis were industrial grade, so as to ensure that any successes achieved would not only be limited to the laboratory environment. $Mg(OH)_2$ of grade FR-20-200 was sourced from Dead-Sea Periclase, $Al(OH)_3$ of grade RPF-14 was supplied by Chemical Initiatives. Pure hydrotalcite, PURAL MG 63 HT, sourced from Sasol was used as a control sample for test work comparison. All the above materials are of high quality with mostly pure specifications. The $Mg(OH)_2$ was coated with 2 % by mass stearic acid, which was removed from before use in the experiment by calcining at a relatively low temperature of 800 °C for 30 minutes to allow the organic materials to burn off. The calcine step converted the $Mg(OH)_2$ into a mixture of coating-free $Mg(OH)_2$ and light-burned MgO [14].

A Netzsch LME1 horizontal bead mill using 2 mm yttrium stabilised zirconia beads was used for the milling of all materials. The mill uses a Watson-Marlow peristaltic pump to pass slurry through the bead mill.

RESULTS AND DISCUSSION

When the calcined $Mg(OH)_2$ (brucite) and $Al(OH)_3$ (gibbsite) materials were combined and milled for 3 hours, the particle size distributions measured at each time interval were generated as shown in Figure 1. Milling of the LDH precursor led to a size distribution over time curve that reached a minimum after 60 minutes and a new maximum after 3 hours of milling. While there is a contamination contribution to the size growth visible from the grey colouration of the white starting materials, it is hypothesised that this size growth is more prominently due to the formation of the LDH crystallites of hydrotalcite and meixnerite. The milled material was hydrothermally aged for 48 hours at 80 °C while CO_2 was bubbled through the mixture. The particle size distribution of this aged product is also shown in Figure 1.

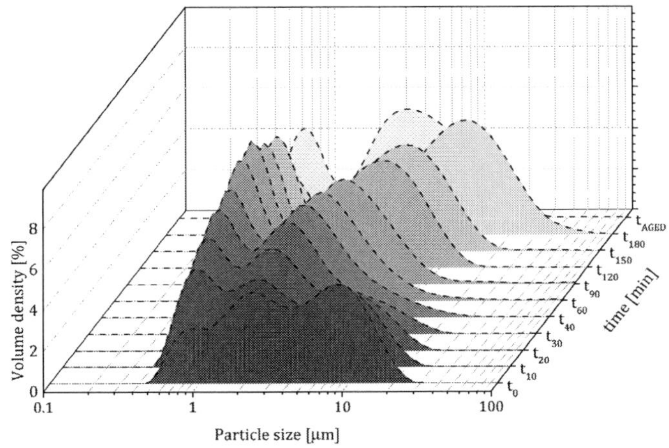

FIGURE 1. Particle size distribution cut sizes of the Mg-Al synthesis precursor after milling over time

From the FT-IR spectra shown in Figure 2, it is clear that a change in chemical species did occur over time. The peak at 3700 cm^{-1} associated with the OH stretch in brucite [10] disappeared within 30 minutes of milling. The peaks within the OH stretch region 3400 cm^{-1} to 3700 cm^{-1} that are associated with the gibbsite lose definition after approximately 60 minutes of milling and form one broad peak in the same range. The peaks within the OH bending region of 1000 cm^{-1} disappear after the same amount of time. In addition to the disappearance of all these peaks, a new peak in the range 1300 cm^{-1} to 1400 cm^{-1} formed as early as 30 minutes and grew in intensity over the duration of the milling and also during the hydrothermal aging step. It is unclear whether this peak is caused by a vibration in the hydroxyl group or the carbonate functional group; both groups have peaks in this region but spectra recorded by Lopez *et al.* [11] , Coates [12] and Palmer, Nguyen, and Frost [10] all seem to suggest that this is the peak of the intercalated carbonate present in hydrotalcite.

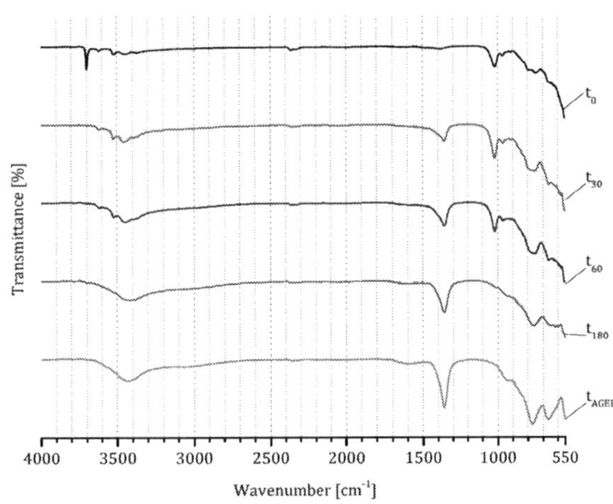

FIGURE 2. IR spectra for the milled Mg-Al precursor over time

All of the above changes in spectra suggest that an LDH was indeed formed, and comparison to the control sample of LDH shown in Figure 3 shows very little difference in FT-IR peak location. As the control LDH was a synthetic hydrotalcite it is believed that the synthesised product is also hydrotalcite.

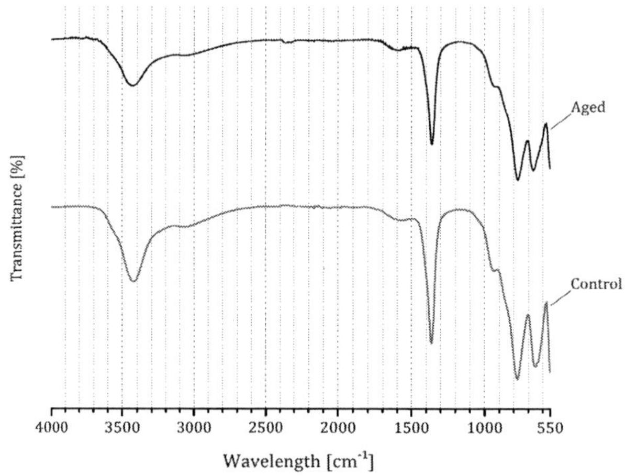

FIGURE 3. IR spectra comparison of the synthesised LDH (aged) and an industrially supplied LDH (control)

The disappearance of the brucite hydroxide could indicate the partial dissolution of an Mg^{2+} ion to substitute in water molecules and Al^{3+} ions. The disappearance of the gibbsite hydroxide peaks is a clear indication of the amorphatisation and insertion of the aluminium ions into the brucite crystal lattice, and the overall interaction of new double hydroxide complex could explain why the hydroxide peaks become less defined into the final broad hydroxide envelope. The presence of the intercalated carbonate or hydroxide peak also suggests that some type of layered crystal lattice has formed.

The milled samples were analysed with XRD scanning for crystal structure, crystallinity and LDH presence. The results are shown in Figure 4. It can be seen that an LDH crystal is present before milling even began, indicated by the [003] and [006] basal peaks at 13.5° and 27°, but that milling increased the intensity of the LDH peak, which can be associated with an increase in crystallinity and in amount. Another clear indication of the formation of LDH is the presence of the [110] and [112/3] peaks visible in the range of 70° to 80° [13].

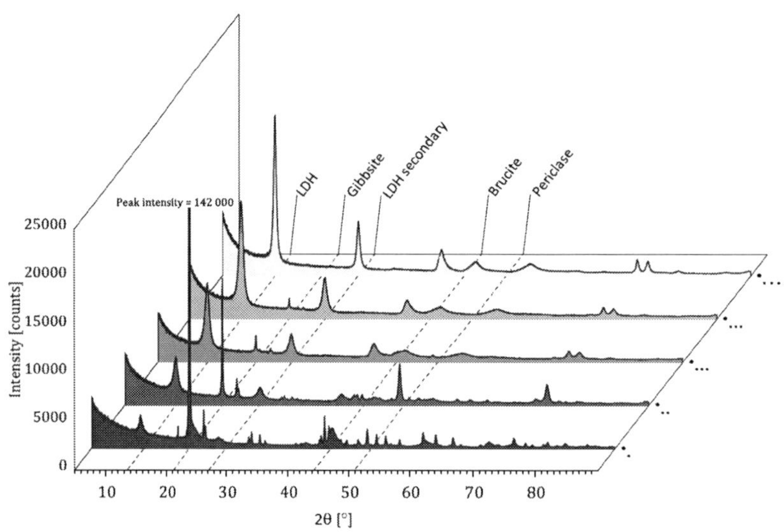

FIGURE 4. XRD patterns of the milled synthesis precursor material

The synthesis does appear to be successful, but further analysis of the XRD patterns show that after three hours of milling, there are still brucite and gibbsite crystals present in the mixture. After hydrothermal aging, no gibbsite was detected in the scans but it is hypothesised that the added heat and mixing effects allowed the aluminium to amorphatise completely, making it invisible to the XRD. This is supported by the increase in the broad deviation from the baseline of the XRD pattern (amorphous halo), associated with amorphous material present in the mixture.

CONCLUSIONS AND RECOMMENDATIONS

Based on the results obtained from the milling of magnesium hydroxide and aluminium hydroxide, mechanochemical synthesis of a magnesium aluminium layered double hydroxide was successful, hydrotalcite and meixnerite were both formed in the synthesis. Analysis of the synthesised layered double hydroxide shows that the conversion of the reaction was high but not complete and shows the presence of an amorphous material as well as contamination from the horizontal bead mill itself. The final particle size distribution of the synthesised material after thermal aging is in the range 0.5 μm to 12 μm. This size distribution is not suitable for shear-intolerant polymer matrix incorporation but after 60 minutes of milling the particle size distribution is in the range 0.5 μm to 7 μm which is more acceptable for polymer matrix incorporation by heuristic standards of polymer processing.

It is recommended that Mg-Al synthesis experiments be carried out for less than three hours to determine if a smaller particle size distribution and a more LDH-like morphology can be achieved after aging. This is because of the favoured particle size achieved at 60 minutes of milling, however it is not known whether the synthesis is complete after 60 minutes of milling and so further analysis on material at this point in milling is necessary. The total LDH yield should be investigated with Rietveld refinement analysis after autoclaving the synthesised material to crystallise all materials. The optimisation of the Mg-Al mechanochemical synthesis should be investigated to determine if a phase-pure hydrotalcite can be synthesised at a lower cost than the current synthesis methods.

REFERENCES

[1] H. P. Venter, "Optimisation of the synthesis of Mg-Al-CO_3 LDH and the partial substition of Mg/Ca-based LDHs," (2013).

[2] C. Forano, T. Hibino, F. Leroux, and C. Taviot-Guého, in *Handbook of Layered Materials*, Development in Clay Science, Vol. Vol. 1, edited by F. Bergaya, B. K. G. Theng, and G. Lagaly (Elsevier Ltd, 2006).

[3] F. J. W. J. Labuschagné, E. W. Giesekke, and J. D. Van Schalkwyk, Production of hydrotalcite, International patent 2006/123284 A2 (2006).

[4] P. G. McCormick and F. H. Froes, journal of Materials **50**, 61–65 (1998).

[5] I. J. Lin and S. Nadiv, Materials Science and Engineering **39**, 193–209 (1979).

[6] M. Eisgruber, J. Ladebeck, J. Koy, H. Schiessling, W. Buckl, and H. Ebert, Method for producing hydrotalcites, U.S. Patent No. 7,211,235 B2 (2009).

[7] V. P. Isupov, L. E. Chupakhina, and R. P. Mitrofanova, Journal of Material Synthesis and Processing **8**, 251–253 (2000).

[8] V. P. Khusnutdinov and V. P. Isupov, Inorganic Materials **44**, 263–267 (2008).

[9] Z. Ferencz, Á. Kukovecz, Z. Kónya, P. Sipos, and I. Pálinkó, Applied Clay Science **112-113**, 94–99 (2015).

[10] S. J. Palmer, T. Nguyen, and R. L. Frost, Coordination Chemistry reviews **253**, 250–267 (2009).

[11] T. Lopez, P. Bosch, M. Asomoza, R. Gomez, and E. Ramos, Materials Letters **31**, 311–316 (1997).

[12] J. Coates, Encyclopedia of Analytical Chemistry 10815–10837 (2000).

[13] F. Cavani, F. Trifiró, and F. Vaccari, Catalysis Today **11**, 173–301 (1991).

[14] W. R. Eubank and W. Roderick, Journal of the American Ceramic Society **34**, 225–229 (1951).

Temperature Effects on the Dissolution-Precipitation Synthesis of Hydrocalumite

B.R. Gevers[1,a)] and F.J.W.J. Labuschagné[1,b)]

[1] *University of Pretoria, Lynnwood Rd, Hatfield, Pretoria, 0002*

[a)]Corresponding author: u13040261@tuks.co.za
[b)]johan.labuschagne@up.ac.za

Abstract. The influence of temperature on the balance of formation between katoite and hydrocalumite (HC) was investigated. The conversion to HC could be increased by providing enough thermal energy to the reaction. Favourable reaction conditions for HC formation were reached at 60 °C. The rate of reaction to HC and katoite could be increased at 80 °C, although the formation of katoite seemingly remained low in comparison to HC formation. At 90 °C an increase in katoite and decrease in HC formation was observed indicating the decomposition point of HC.

INTRODUCTION

Layered double hydroxides (LDHs) are substances with physical and chemical properties similar to clay minerals [1, 1021]. Their general chemical formula is given by

$$[M_{1-x}^{II}M_x^{III}(OH)_2][X_{x/q}^{q-} \cdot nH_2O] \quad (1)$$

$[M_{1-x}^{II}\ M_x^{III}\ (OH)_2]$ describes the composition of the LDH layers consisting of trivalent (M^{III}) and divalent (M^{II}) metal cations, while $[X_{x/q}^{q-} \cdot nH_2O]$ represents the composition of the anionic interlayer [1, 1021].

HC is a hydroxide, carbonate or chloride intercalated Ca-Al LDH. For example, the carbonate intercalated form can be synthesised using dissolution-precipitation of CaO and $Al(OH)_3$ in H_2O with addition of a carbonate source. [2] While investigating the formation of katoite – a HC precursor – with cheaper calcium sources ($Ca(OH)_2$) readily available in South Africa, Wiid [2] observed the formation of small amounts of a HC phase as by-product of the required katoite phase without introduction of a carbonate source and in an inert environment. It is expected that this phase is OH^- intercalated or that CO_2 contamination might have led to the formation of HC as previously observed by Sánchez-Cantú *et al.* [3]. Synthesis of this phase in a one-step fashion from cheaper reactants and without a carbonate source would have large economic benefits and will, therefore, be investigated in this paper.

When forming HC, several different synthesis routes – which include the sol-gel method, urea hydrolysis, co-precipitation, hydrothermal treatment or dissolution-precipitation synthesis [1, 1027-1038] – can be used. The following is known about the formation of HC via a dissolution-precipitation mechanism:

- Wiid [2] investigated the formation of katoite from $Ca(OH)_2$, $Al(OH)_3$ and H_2O in inert conditions and found traces of HC when reacting the mixture below 90 °C using the dissolution-precipitation mechanism. Katoite was found at 120 °C.
- Sánchez-Cantú *et al.* [3] attempted to form a HC-like phase by reacting $Ca(OH)_2$ and AlOOH in water at a Ca:Al molar ratio of [2:1] and following a dissolution-precipitation mechanism; reaction occurred in an inert environment and at 80 °C. HC was formed.
- A HC-like phase has been synthesised by Buttler, Glasser, and Taylor [4] making use of supersaturated solution synthesis – a form of dissolution-precipitation synthesis – in an inert environment.

Formation of HC through dissolution-precipitation requires the desired metal hydroxides (an aluminium and calcium source) to be reacted in water [2] and subjected to conditions in which a dissolution of their ion constituents is favoured. Dissolution of the metal hydroxides is necessarily a function of the agitation speed due to a maximisation of fluid movement around the solid particles. Precipitation will occur if one or more of the products are insoluble [5, 120-124]. Furthermore, temperature and pH (which is a function of the amount of dissolved metal hydroxides in solution) are believed to play the largest roles during katoite and HC formation as they influence the dissolution behaviour of the metal compounds [2]. The HC phase is stable below temperatures of 85 °C and to break down into katoite and calcite at higher temperatures [6]. LDH decomposition could be ascribed to a decreasing pH with an increase in temperature. The optimum reaction conditions for katoite formation were identified to be 90 °C at a reaction time of 2 h [2].

Literature on HC synthesis from $Ca(OH)_2$, $Al(OH)_3$ and H_2O is very limited and no information is available regarding which, and the way in which, reaction parameters influence the reactions in the system. This is especially true with respect to the balance between katoite and HC formation, the extent of reaction, and the product crystallinity. This paper serves to discuss the influence of temperature on the balance between katoite and HC formation.

EXPERIMENTAL

Materials

Experiments were conducted using $Ca(OH)_2$ and $Al(OH)_3$ from ACE Chemicals (Gold Line) in chemically pure quality. Distilled and dissolved gas free water was used for all reactions.

Method

Experiments were performed in a bench top reactor set up using hotplate stirrers and glass beaker reaction vessels. An inert N_2 environment was maintained in all experiments. 100 ml distilled water was preheated to 80 °C while stirring slowly, allowing dissolved gases to vaporise, and afterwards brought to the desired reaction temperature. 20 g of dry reactant powders in stoichiometric ratio were weighed off according to Equation 2

$$4Ca(OH)_2 + 2Al(OH)_3 + cH_2O \longrightarrow Ca_4Al_2(OH)_{12}A_{2/n} \cdot xH_2O + (c - x)H_2O \tag{2}$$

and incorporated into the preheated water while stirring with a magnetic stirrer. The mixture was kept in suspension by magnetic stirring and reacted for 2 h at the desired temperature. The temperature span was chosen to commence at 30 °C – such that a reaction profile can be obtained – and end at a value at which HC should have exceeded its stability limit (85 °C) as previously found by Wiid [2]. The samples were filtered dry using vacuum filtration and analysed as a wet paste within 15 min.

Characterisation Techniques

X-ray diffraction analysis (XRD) and Fourier transform infrared analysis (FTIR) were used to identify and confirm phases present in the respective samples. XRD measurements were performed on a Panalytical XPert PRO X-ray diffractometer in $\theta - \theta$ configuration, using Fe filtered Co-Kα radiation (1.789 Å), an XCelerator detector and variable divergence- and fixed receiving slits. The data was collected in the angular range of $5° \leq 2\theta \leq 90°$ with a step size and time of 0.008 ° 2θ and 13 s respectively. Phases were identified using XPert Highscore plus software. Molar fractions of phases present were determined using Rietveld refinement. ATR-FTIR spectra were obtained using a Perkin Elmer 100 Spectrophotometer. Samples were pressed in place with a force arm. Spectra were obtained in the range of 550 - 4000 cm^{-1} each with 32 scans at a resolution of 2 cm^{-1}. Reactant standard FTIR scans were obtained and overlaid with the experimental data as explained by Galván-Ruiz *et al.* [7] in order to identify absorption peaks.

RESULTS AND DISCUSSION

Five reaction temperatures were tested to find a relationship between temperature and the favoured phase formation, an indication of a reaction profile and to test the influence of temperature on the crystallinity of the product. The conversion of reactants to HC and katoite could be shown through both XRD and FTIR analysis. Figure 1 shows a temperature reaction profile for the reaction obtained from XRD results of each sample.

Three interesting observations could be made from this plot which will be discussed below:

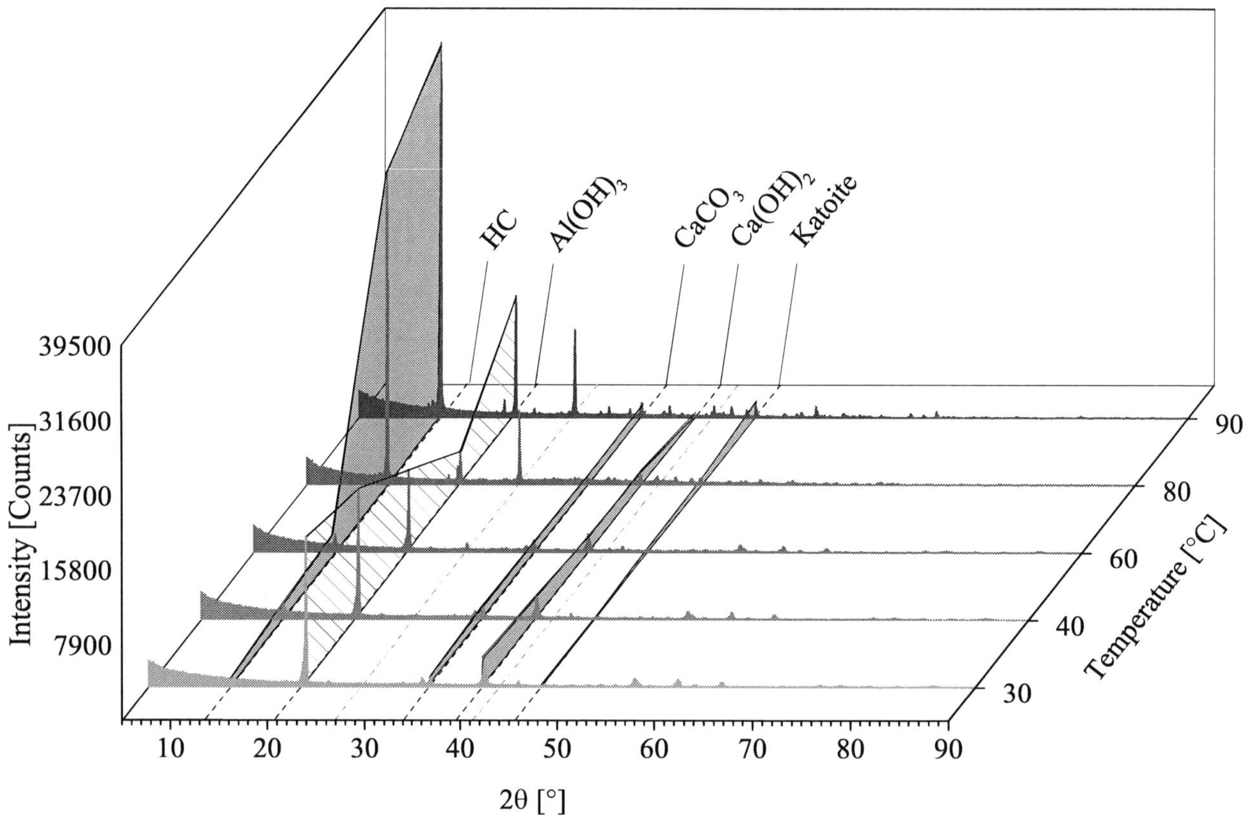

FIGURE 1. Plot of XRD data for the comparison of katoite and HC formation as well as reactant depletion with temperatures ranging from 30 °C to 90 °C. An increase in colour intensity indicates an increase in reaction temperature. Black dotted lines represent the primary peaks of HC, katoite, calcium carbonate and calcium hydroxide. Light grey dotted line represent the secondary and tertiary peaks of HC.

- The relationship between HC and katoite formation and depletion of the reactants with an increase in temperature.
- A sharp increasse in the intensity of the aluminium hydroxide peak at 90 °C after following the reactant depletion trend up to 80 °C.
- A drastic increase in HC peak intensity with temperature.

Balance Between Katoite and HC Formation

From Figure 1, it is clearly visible that enough thermal energy must be provided in order to start the reaction to form katoite and HC. With an increase in temperature an increase in HC and katoite and a decrease in $Ca(OH)_2$ could be observed. It is clear, that favourable reaction conditions for HC formation are reached at 60 °C.

Changes in extent of reaction with temperature could be fully ascribed to the energy required to start the reaction – enough thermal energy must be supplied to favour dissolution of the reactants to form katoite and HC. If not enough energy is supplied, very little conversion of reactants to products will be observed. No increase in katoite formation could be observed until the temperature increased to 90 °C; at this point the rate of HC formation seemingly slowed.

The balance of katoite versus HC was seemingly only influenced when favourable reaction conditions to form katoite exist; at the point where HC decomposition is thought to commence and katoite formation increases. Observations made by Wiid [2] – a small portion of HC forming at low temperatures – could be validated. It is expected that only katoite exists at 120 °C [2]. An onset of reduction in HC formation was already visible at 90 °C from the obtained results, seemingly confirming previous work in this field.

Aluminium Hydroxide Peak Intensity

It is assumed that the backward reaction from HC to katoite starts influencing the yield of the Ca-Al LDH at temperatures greater than 80 °C. Three explanations for the increase in aluminium hydroxide could be proposed:

1. The solubility of $Al(OH)_3$ decreases with an increase in temperature and more will therefore remain in the sample at 90 °C compared to 80 °C.
2. With an increase in temperature more energy is available for reaction and it might be possible that the excess in energy shifts the reaction equilibrium from LDH formation to LDH decomposition into katoite and $Ca(OH)_2$ and finally to a decomposition of katoite into gibbsite or a hydrated alumina phase.
3. The increase in peak intensity is observed through an increase in crystallinity of the remaining phase while the amount of the phase decreased.

Explanation 1 could be confirmed through auxiliary tests. Explanation 2 was also found to be plausible due to the rate of formation of HC seemingly slowing from 80 °C to 90 °C and an increase in katoite peak intensity could be observed. Explanation 3 could also be plausible, seeing that the solubility of aluminium hydroxide decreases and aluminium in solution could re-precipitate in a ore crystalline form.

To test the validity of these explanations, FTIR analysis was performed (Figure 2). Figure 2 shows the reactant

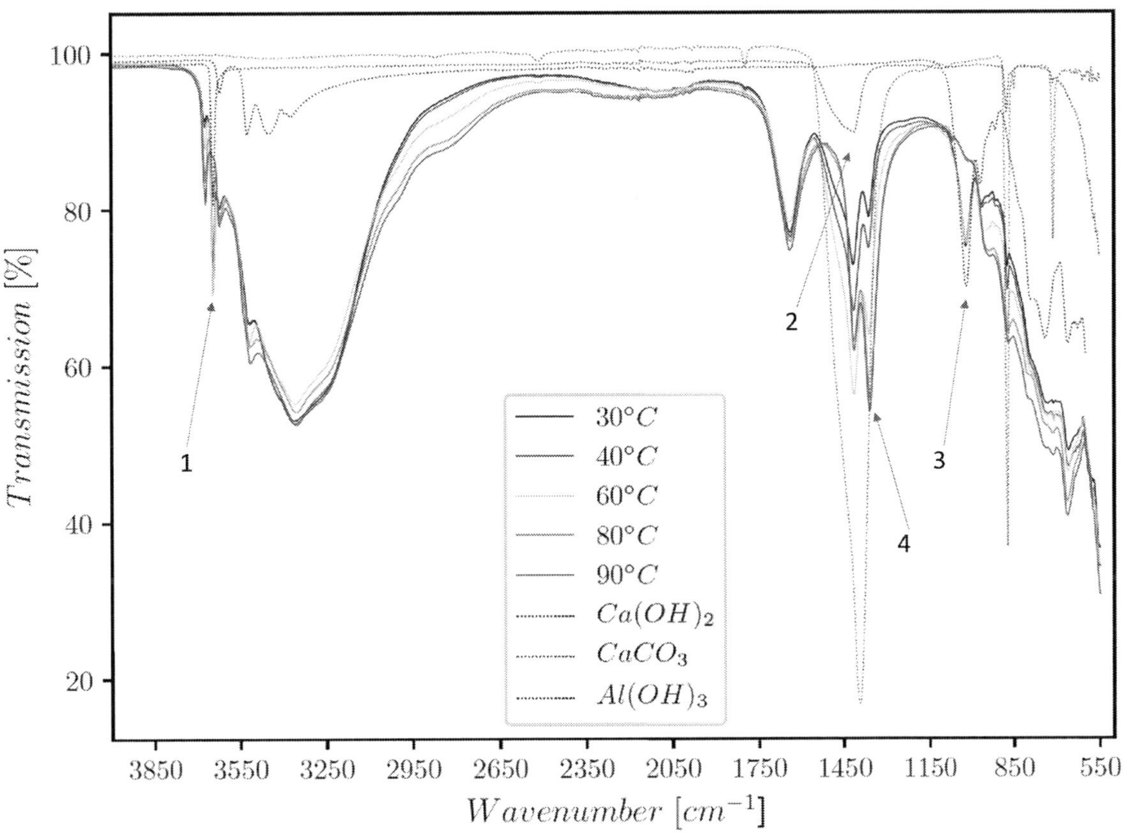

FIGURE 2. Comparative plot of FTIR scans for the temperature test series. Point 1, 2: $Ca(OH)_2$ depletion, Point 3: $Al(OH)_3$ depletion, Point 4: HC/ katoite phase.

depletion as already seen in Figure 1 (Point 1-3). However, here, the sharp increase in $Al(OH)_3$ is not visible. Although

FTIR is not a quantitative analysis technique, a difference was expected in the case of a large amount of $Al(OH)_3$ being present in the 90 °C sample. Instead, the plots for 80 °C and 90 °C lie almost on top of each other at Point 3. A large difference is however visible at this point between the lower temperature runs and the just-mentioned ones. This could lie in favour with the explanation of the HC phase decomposing and $Ca(OH)_2$ and $Al(OH)_3$ being used to form katoite. Decomposition of HC to katoite at this temperature confirms previous observations by Wiid [2]. Further clarification of the exact sample composition could clear this up (e.g. Rietveld refinement). However, techniques such as Rietveld refinement only give reliable information if the peak patterns fit the identified structure exactly and no unidentified peaks remain. With new or undefined structures this is a very difficult task to achieve.

Phase Crystallinity

It remains unclear at this point in time, whether the observed increase in peak intensity results from an increase in the amount of the phase or because of an increase in crystallinity. Other large changes in peak intensity can be observed in the HC peak at 80 °C. This increase in peak intensity can be ascribed to the increase in amount and crystallinity of the phase itself and possibly also to the analysis technique used. It is expected that platelet alignment can influence the intensity of an XRD pattern due to constructive or destructive interference. The higher the constructive interference, the higher the peak count observed in a pattern for the respective crystal structure [8]. This is also observed especially in platelet structures such as clays due to preferred orientation [8]. However, although this might have influenced the measured peak intensities, a relative increase with respect to the other phases is definitely observable, leading to the conclusion that this should only alert to the fact that the samples obtained at high temperatures do not necessarily consist of mainly HC.

CONCLUSIONS AND RECOMMENDATIONS

It was observed that HC formation increased with an increase in temperature. The data obtained also showed that the decomposition of the LDH phase possibly starts occurring between 80 °C and 90 °C. It was reasoned, that this shift in reaction pathway might occur due to a change in favoured katoite decomposition mechanism, due to more energy being available. Further proof needs to be found for this. It is expected that katoite remains as the stable phase at temperatures exceeding the tested temperature span.

Although no quantification of results can be given at this point in time, the correlations between the temperature and an increase in HC formation are quite strong and not unexpected. It would be interesting to further investigate the difference in yield of HC against katoite between 80 °C and 90 °C through quantitative means.

REFERENCES

[1] C. Forano, T. Hibino, F. Leroux, and C. Taviot-Guého, Developments in Clay Science **1**, 1021–1095 (2006).

[2] A. Wiid, *Exploratory Study of Single and Multi-step Dissolution-Precipitation Syntheses of Carbonate-containing Hydrocalumite*, Master's thesis, University of Pretoria (2015).

[3] M. Sánchez-Cantú, S. Camargo-Martínez, L. M. Pérez-Díaz, M. E. Hernández-Torres, E. Rubio-Rosas, and J. S. Valente, Applied Clay Science **114**, 509–516 (2015).

[4] F. G. Buttler, D. Glasser, and H. F. W. Taylor, Journal of the American Ceramic Society **38**, 4–9 (1958).

[5] J. C. Kotz, P. M. Treichel, and J. R. Townsend, *Chemistry & Chemical Reactivity*, 8th ed. (Brooks/Cole, United Kingdom, 2014).

[6] T. Matschei, B. Lothenbach, and F. P. Glasser, Cement and Concrete Research **37**, 1379–1410 (2007).

[7] M. Galván-Ruiz, J. Hernández, L. Baños, J. Noriega-Montes, and M. E. Rodríguez-García, Journal of Materials in Civil Engineering **21**, 694–698 (2009).

[8] D. Data and O. The, Introduction to XRay Powder Diffraction **2**, 1–12 (2009).

Effect of phosphorus-containing modified magnesium hydroxide on the mechanical properties and flammability of PLA/MH composites

Pengcheng Zhao[1, 2] Zhiqi Liu[1, 3], Michael Gehde[2], Ines Kuehnert[1] and Andreas Leuteritz[1, a)]

[1]*Polymer Processing Department, Leibniz Institut fuer Polymerforschung Dresden e. V., Dresden, Germany*
[2]*Faculty of Mechanical Engineering, Technical University Chemnitz, Germany*
[3]*Qinghai Institute of Salt Lakes, Chinese Academy of Sciences*

[a)] Corresponding author: *Andreas Leuteritz, leuteritz@ipfdd.de*

Abstract. In this study, a novel modifier (PPD-PTES) for Magnesium Hydroxide (MH) was successfully synthesized from (3-Aminopropyl) triethoxysilane and phenylphosphonic dichloride aiming to improve not only the compatibility of MH with polymer but also its efficiency. Fourier transform infrared spectroscopy (FTIR), ^{1}H, ^{31}P-nuclear magnetic resonance (NMR) were carried out to characterize its structure. The flammability, thermal degradation and mechanical properties of the PLA/MH composites were investigated by limiting oxygen index (LOI), microscale combustion calorimetry (MCC), thermogravimetric analysis (TGA) and tensile test. The results showed that the introduction of MH can significantly reduce the peak of heat release rate (PHRR) and total heat release (THR). Additionally, the LOI of modified MH reinforced PLA has also increased compared to the unmodified MH reinforced PLA. Further, results of tensile test proofed that the organic modification could improve the compatibility of MH with the polymer, but only slightly decrease brittleness.

INTRODUCTION

Poly (lactic acid) (PLA) is a biodegradable polymer derived from renewable resources. Nowadays it has become a promising material to replace some conventional petrochemical polymers because of not only its high strength and stiffness but also environmental concern. Currently, although PLA are mostly used in biomedical application and packaging industries, its other application are increasing because of its excellent properties.[1] Unfortunately, like other plastics its inflammability restricts the application of PLA in some important fields such as electrical industry, automotive parts and household and so on.[2,3]

Magnesium Hydroxide (MH) is a widely used, commercialized product which can also be used as flame retardant.[4] In the halogen free flame retardant systems of polymers, MH have been widely used due to its advantages such as low cost, environment friendliness and release of water in fire. However, the low efficiency affords high loading to achieve desired flame retardancy, and its poor compatibility with polymer is a big challenge, when used in brittle matrix such as PLA. Therefore, a surface modification that can enhance the interfacial interaction between MH and polymer is necessary.

In this work, a novel phosphorus containing modifier for MH was synthesized to realize the surface modification of MH. The structure was characterized by Fourier transform infrared spectroscopy FTIR, ^{1}H and ^{31}P nuclear magnetic resonance spectroscopy. The flammability was investigated using microscale combustion calorimetry

(MCC) and limiting oxygen index (LOI). Further, the mechanical properties and thermal decomposition behavior of composites were also studied.

Experimental

Material: Magnesium Hydroxide (Magnifin H5) was purchased from MAGNIFIN Magnesiaprodukte GmbH & Co KG; 3-Aminopropyltriethoxysilane (APTES) was purchased from TCI Deutschland GmbH; Phenylphosphonic dichloride (PPDCl, 90%) was purchased from Sigma-Aldrich Corporation and Poly (lactic acid) (PLA, 6202D) was purchased from NatureWorks LLC.

Synthesis and characterization of the modifier: The modifier of MH was synthesized from PPDCL and APTES. APTES was dissolved with dichloromethane in a three-necked flask which was putted in ice bath. Subsequently, an appropriate amount of PPDCL dissolved in dichloromethane was added dropwise. The temperature was kept at 0-5 °C for 2h and continued to react at room temperature for 5h. The mixture was filtered and washed with water. Then the raw product was separated from the liquid by reduced pressure distillation. The desired structure was proofed by ^{1}H NMR and FTIR.

Preparation of PLA and PLA/MH composites: The MH powder was mixed with 3 wt.-% of modifier using high speed mixer to realize surface modification. All the materials were dried under vacuum at 80 °C overnight. Then a twin-screw brabender mixer was used to prepare PLA/MH composites. Parameter settings follow: screw speed, 60 rpm; temperature 180 °C; duration 10 min. Then the resulting composites were cutted into pellet and molded at 190 °C by injection molding.
Characterization: The oxygen index meter (FTT) was used to measure the LOI value, microscale combustion calorimetry (Stanton Redcroft) was used to study flame retardancy of the composites, thermogravimetric analysis with a heating rate of 10 K/min on a Q5000 (TA Instruments) was carried out to study thermal properties and tensile test (Zwick) was used to study mechanical properties.

Results and discussion

Flammability: In this section, the effect of MH and m-MH on flammability was studied by limiting oxygen index (LOI) and microscale combustion calorimetry (MCC). Table 1 listed the results of MCC and LOI tests for pure PLA and PLA/MH systems, peak of heat release rate (PHRR), heat release capacity (HRC), total heat release (THR), temperature at the peak of heat release rate (T_{max}) and the LOI value. The results showed that the THR of samples decreased with increasing loading of MH. It decreased from 17.6 kJ/g of pure PLA to 11.0 kJ/g of PLA30MH. Based on MCC results, the effect of THR reduction is only slightly larger, than the reduction of polymeric phase in the composite (62 % vs. 70 %). In contrast, LOI measurement results only in very little effect at low loading, but reasonable effect with 30 % MH. That showed the potential of MH as flame retardant. A comparison of mMH composites and MH composites results in similar values of LOI and a further reduction in THR values. With only 3 wt% of surface modification, the THR value decreased further to 9.7 kJ/g. However, the PHRR of PLA with modified MH was higher than these with unmodified MH. Reason for that could be, that the low molecular weight modifier has participated in the combustion and contributed to heat release, subsequently the silane structure on the modifier has crosslinked other and built a protection layer, which can reduce the heat release and decomposition of the polymer. The LOI results showed that, at lower loading the composites had lowered LOI value. It might be due to the prevention of dripping caused by the filler MH which reduces the heat withdrawal by molten polymer, and due to the flammability of the modifier, it is explainable that the LOI value of PLA/mMH were higher than PLA/MH.

Table 1: MCC and LOI Results of PLA, PLA/MH and PLA/mMH Composites

Samples	PHRR[W/g]	HRC sum [J/g*K]	THR [kJ/g]	T_{max} [°C]	LOI [%]
Pure PLA	323	456	17.6	369	21.2
PLA10MH	359	360	14.3	270	20.2
PLA10mMH	455	531	14.3	275	20.7
PLA20MH	291	448	12.3	255	20.6
PLA20mMH	303	416	12.1	264	21.5
PLA30MH	155	309	11.0	251	24.3
PLA30mMH	230	347	9.7	256	23.5

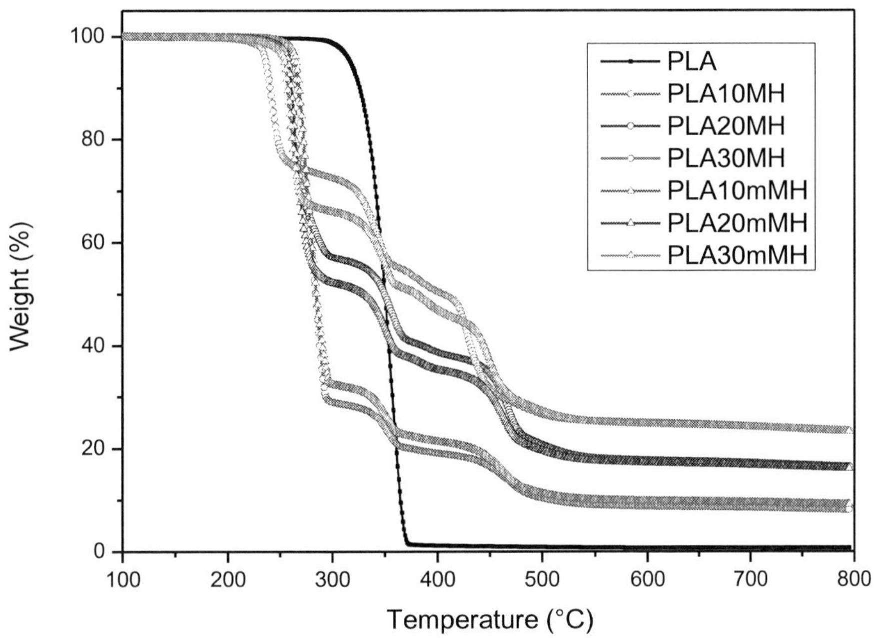

Figure 1: TGA curves of PLA/MH and PLA/mMH

Thermal decomposition behaviors: In this section, the effect of MH on the thermal decomposition of PLA was studied by TGA. Figure 1 shows the TGA curves of PLA/MH and PLA/mMH composites. It is shown that, the introduction of MH accelerated the decomposition of PLA. With increasing loading of MH the composites decomposed earlier. Reason for that could be, that under heating MH may release water which would lead to both initiation and acceleration of decomposition of PLA. As shown in the Figure 1, the starting temperature of decomposition of PLA10MH, PLA20MH and PLA30MH are 230 °C, 220 °C and 199 °C, and these of PLA10mMH, PLA20mMH and PLA30mMH are 234 °C, 222°C and 215 °C respectively. That indicates the organic surface modification of MH has formed a protection layer which can delay the decomposition of MH and subsequently that of PLA by creation of a crosslinked structure. The TGA results in that respect show similar trend as the fire test.

Both TGA and flammability test show limited advantage of using MH. Further, the organic modification is leading to a shifted decomposition at higher temperatures, but do not compensate the earlier degradation of the PLA matrix.

Tensile properties: Table 2 shows the tensile strength, strength and at break of pure PLA and MH based composites. It is observed that the tensile strength had an increasing trend as the concentration of MH increased. However, the inorganic filler had strongly impact to the toughness of PLA due to its poor compatibility with polymer. After the organic surface modification, MH should have a better dispersion in polymer as the interaction between matrix and filler is increased. Hence, the elongation at break of PLA with modified MH was slightly improved due to organic modification of mMH.

Table 2: Tensile test results of PLA, PLA/MH and PLA/mMH composites

Samples	E_t [MPa]	σ [MPa]	ε [%]
Pure PLA	3695±191	46.56±0.56	10.45±0.5
PLA10MH	3814±304	57.78±0.59	1.84±0.2
PLA10mMH	3535±224	52.48±0.5	1.72±0.36
PLA20MH	5121±525	49.23±18.23	1.15±0.45
PLA20mMH	4653±166	51.60±1.77	1.33±0.01
PLA30MH	5640±492	19.49±3.4	0.33±0.05
PLA30mMH	5584±285	49.98±2.59	1.08±0.08

Conclusion

In this work, the impact of Magnesium hydroxide to the properties of PLA was studied. At the loading of 30 wt % MH can increased the LOI value of PLA from 21.2 to 24.3. At the meantime, pHRR and THR value has significantly decreased from 323 W/g and 17.6 kJ/g to 155 W/g and 11.0 kJ/g. The surface modification seems to have only slight contribution on the flame retardancy of the PLA/MH composites. However, according to the results of tensile test, the presence of MH has strongly aggravated the mechanical properties of the PLA. It can be presumed that MH accelerated the decomposition of PLA at high temperature, which can again be slowed down by the surface modification with the FR active component. This contradictory effect limits the use of classical FR MH and mMH, showing that effective FR for bio-based PLA is still needed. At the meantime, it seems that modifying MH with organic component to enhance its interaction with polymer is a promising method, which could be used in other polymer system in order to improve both efficiency and compatibility with matrix of MH.

Acknowledgment

We want to thank to the China Scholarship Council (CSC) for the entire support of the project.

References

1. M. Jamshidian, E. A. Tehrany, M. Imran, M. Jacquot and S. Desobry, Comprehensive Reviews in Food Science and Food Safety, 5:552-571 (2010)
2. Y. Song, D. Wang, X. Wang, L. Lin and Y. Wang, Polym. Adv. Technol., 22, 2295 −2301, (2011)
3. X. Zhao, F. R. Guerrero, J. Llorca, and D. Wang, T. Wang, ACS Sustainable Chem. Eng., 4, 202 −209, (2016)
4. S. Fang, Y. Hu, L. Song, J. Zhan and Q. He, Journal of Materials Science, 43:1057-1062 (2008).

Assessment of poly(lactic acid) crystallization behavior in the presence of a plasticizer and nanoparticles

Monshizadeh Mehrnoush (1), *Khonakdar Hossein Ali* (2)*, *Seyfi Javad* (3)

(1) Central Tehran Branch, Islamic Azad University - Tehran - Iran, (2) Leibniz-Institut für Polymerforschung Dresden - Saxony - Germany, (3) Islamic Azad University of Shahrood - Semnan - Iran

Assessment of poly(lactic acid) crystallization behavior in the presence of a plasticizer and nanoparticles Mehrnoush Monshizadeh1, Hossein Ali Khonakdar2*, Javad Seyfi3 1 Chemical Engineering Department, Central Tehran Branch, Islamic Azad University, Tehran, Iran 2 Leibniz-Institut für Polymerforschung Dresden e.V., Hohe Strasse 6, D-01069 Dresden, Germany 3 Department of Chemical Engineering, Shahrood Branch, Islamic Azad University, Shahrood, Iran Abstract Poly(lactic acid) (PLA) is a biobased and compastable thermoplastic polyester that has been under investigation tremendously, and the main aim of those researches has been the introduction of a commodity material replacing the oil-based plastics [1]. The major obstacle on the way of extending the applications of PLA is the control of its crystallinity which is crucial on its degradation rate, thermal resistance, mechanical and also barrier properties [2]. In this work, nano-scale sized layered double hydroxide (LDH) was utilized in order to generally boost the properties of PLA as well as improve its crystallization extent and rate since they could act as nucleating agents. However, the growth rate was attempted to increase by incorporation of a proper plasticizer such as triethyl citrate (TEC). In this way, the combined use of LDH and TEC could synergistically increase the crystallization of PLA causing the application window of PLA to be further extended. DSC non-isothermal experiments were conducted and the above claim has been confirmed. Moreover, the POM results are also investigated in this research. Keywords: Poly(lactic acid); Layered double hydroxide; Crystallization References: [1] R. Auras, B. Harte, S. Selke, Macromolecular Bioscience, 4 (2004) 835–864. [2] S. Saeidlou, M.A. Huneault, H. Li, C.B. Park, Progress in Polymer Science, 37 (2012) 1657–1677.

PROPERTIES OF POLY (HYDROXYBUTYRATE-CO-VALERATE) COMPOSITES REINFORCED WITH ALFA FIBER AND ITS CELLULOSE

BOUKERROU Amar (1)*

(1) University of Bejaia, Algeria - Bejaia - Algeria

Currently the most viable way toward ecofriendly composites, with a less expensive price, is the use of natural fibers as reinforcement. This situation makes necessary the use of low cost fillers as a way of reducing the cost of the end product. The partial or total substitution of PHBV by lignocellulosic fiber could be of interest. PHBV can be reinforced with natural fiber in order to enhance its mechanical properties and to reduce the cost of final product. Although, there are a few hydroxyl and carboxylic groups on ends of PHBV molecular chain some degree of possible dipolar interactions among moderately polar (C=O) and polar groups (-OH and -COOH) of PHBV and hydroxyl group (-OH) of cellulosic fiber can provide an opportunity for an interfacial interaction between the fiber and the matrix. In this study, alfa fiber and its cellulose obtained after extraction were used as reinforcement for the development of totally biocomposite materials. This work characterization of tensile properties and a comparison between alfa fiber and its cellulose effect on poly (hydroxybutyrate-co-valerate) (PHBV) composites. Owing to higher production cost of PHBV as compared to traditional polymers, we investigated the possibility to replace a quantity of PHBV by lignocellulosic filler without damaging the properties of the polymers. The results show a narrow increasing trend in the tensile strength; this could be attributed to the compatibility between the filler and PHBV. [2] The tensile modulus of PHBV based composite was increased compared to neat PHBV. The diffraction profile of PHBV matrix is very similar of that presented for composites. Both PHBV, PHBV/alfa fiber and PHBV/cellulose shows reflections at the same values as for the neat biopolymer, indicating that addition of reinforcement does not alter the unit cell after incorporation of alfa fiber and its cellulose. Keywords: Cellulose,biopolymer, biocomposites, thermal properties.

USE OF NATURAL POLYMERS EXTRACTED FROM AGROINDUSTRIAL RESIDUES OF Agave tequilana Weber var. azul IN THE SYNTHESIS OF SILVER NANOPARTICLES AND THEIR ANTIBACTERIAL ACTIVITY

López-Naranjo Edgar José (1), Pérez-Fonseca Aida Alejandra (1), Robledo-Ortiz Jorge Ramón (1), Hernández-Rosales Irma Paz (2), Bueno-Durán Adela Yolanda (2), González-Ortiz Luis Javier (1), Manzano-Ramírez Alejandro (3)*

(1) Universidad de Guadalajara - Jalisco - Mexico, (2) Universidad Autónoma de Nayarit - Nayarit - Mexico, (3) CINVESTAV Unidad Querétaro - Querétaro - Mexico

Agave tequilana Weber var. azul (AT) fructans are fructose-based polymers that constitute the main raw material for the production of tequila. Chemically, they are β-(2,1) and/or β-(2,6)-linked fructose polymers anchored to a glucose residue found in the vacuoles of many plants including the Agavaceae family [1, 2]. In conventional tequila production, only AT "heads" are used, whereas the leaves are considered useless; regardless their high total reducing sugars (RS) content (up to 16.1%) [3]. Green synthesis of silver nanoparticles (AgNPs) encompasses the use of plants extracts containing natural reducing compounds (e.g. RS) for an eco-friendly and economically sustainable production of AgNPs [4]. In the present work, we demonstrate the use of an extract obtained from a major Mexican agroindustrial residue (i.e. AT leaves) for the synthesis of AgNPs. Additionally, the antimicrobial activity of the green-synthesized AgNPs was evaluated against human pathogenic bacteria. Results indicated that AT-extract could be used as a reducing agent to obtain AgNPs as those shown on Fig 1a. The as-synthesized AgNPs significantly inhibited bacterial growth against Staphylococcus aureus, Pseudomonas aeruginosa and Escherichia coli as exemplified by the inhibition zones presented on Fig 1b. 1. Arrizon, J., et al., Comparison of the water -soluble carbohydrate composition and fructan structures of Agave tequilana plants of different ages. Food Chem, 2010. 122: p. 123-130. 2. Narváez-Flores, M., et al., Ultrasound assisted extraction modeling from agave (Agave tequilana Weber var. Azul) at different temperatures and ultrasound powers. Food Bioprod Process, 2015. 96: p. 232-239. 3. Iñiguez-Covarrubias, G., et al., Utilization of by-products from the tequilana industry. Part 2: potential value of Agave tequilana Weber azul leaves. Biores Tech, 2001. 77: p. 101-108. 4. Meshram, S.M., et al., Green synthesis of silver nanoparticles using white sugar. IET Nanobiotechnol, 2012: p. 1-5. Fig 1. (a) SEM micrograph of green synthesized AgNPs Agave tequilana Weber cultivar azul extract, (b) bacterial growth inhibition zones displayed by AgNPs against Escherichia coli

Trends and perspectives of bead foam processing

Altstädt Volker (1)*, Neumeyer Thomas (1), Schreier Peter (1), Fafara Michael (1), Gensel Julia (1), Standau Tobias (2), Hädelt Bianca (2), Zhu Yejun (1)

(1) Neue Materialien Bayreuth GmbH - Bayreuth - Germany, (2) Department of Polymer Engineering, University of Bayreuth - Bayreuth - Germany

Expanded polystyrene (EPS) and expanded polypropylene (EPP) are known for tremendous potential for light weight design. New demands were raised up by the industry during the last few years, such as process combination, functionalization, and new material development [1]. First, the combination of automated tape laying technology and in-situ steam chest molding will boost the development of sandwich structures with organic face sheet and a bead foam core. Automatized unidirectional tape laying provides high flexibility for face sheet design, including geometry and fiber orientation. After in-situ foaming with a bead foam as a core material, sandwich structures with tailored organic face sheets and bead foam core can be achieved. With optimized face sheet design, high mechanical properties can be reached. Second, we present a novel approach for polymer bead foams with improved flame retardancy properties, by surface pre-modification of single beads and subsequent steam-chest molding. Since a good fusion of beads is necessary in order to achieve favourable mechanical properties, the effect of incorporated functional components on both the welding strength and the flame retardant properties was investigated. Third, in order to provide higher service temperature, novel core materials made from expanded polybutylene terephthalate beads (E-PBT) will be presented [3]. Chemical modification by chain extender overcomes the process difficulties for E-PBT bead foam production. At 85 °C, E-PBT foam provides 30 % better compression properties at 50% compression compared with EPP with density of 155 g/l. Furthermore, water absorption test and chemical resistance test will reveal the potential for practical applications.
Reference [1] Winterling, H., Sonntag, N., Kunststoffe 101, 10 (2011):18-21. [2] Altstädt, V., Mühlbacher, M. und Beck, J., Lightweight Design 7,4 (2014): 56-62 [3] Köppl, T., Raps, D., and Altstädt, V., Journal of Cellular Plastics 50, 5 (2014): 475-487.

Modelling the Stretching of Polypropylene Films by a Nonlinear Spring Dashpot Model

Reinders Frauke (1)*, **_Schöppner Volker_** (1)

(1) Paderborn University Kunststofftechnik - North Rhine Westphalia - Germany

The stretching of semi crystalline materials such as isotactic polypropylene leads to an increase of the material characteristics (tensile strength, E-Modulus etc.). The trend towards using wider and faster film production lines causes the need to develop a better understanding of the complex correlation between molecular structure of the raw material, the processing conditions, the morphology of the cast film and the final film properties. In this investigation a viscoelastic model for the simulation of the stretching process is selected. In this study the definition of the suitable material parameters is supported by further characterization of the material. In modern biaxially oriented Polypropylene (BOPP) lines film production speeds of up to 500 m/min can be achieved. The lines have dimensions of up to 30 m (width) times 100 m (length). This means that taking samples out of a running line is difficult and morphological changes cannot be controlled online. The model, which is presented in this paper shows a good compromise between physical understanding and correlation of simulation and experimental data. The model uses simple spring dashpot connections with a physical understanding of the isotactic polypropylene material. With a better understanding of the physical background of the morphological changes and a suitable model of the stretching process it will be possible to improve the product characteristics of BOPP material. Keywords: Films, BOPP, Stretching, Spring Dashpot Model.

Development Of Polyolefin Bicomponent Filaments As Controlled Release Devices Of Volatile Repellents For Use In Malaria Vector Control

Mthokozisi Sibanda[1,a)], Andreas Leuteritz[2,b)], Harald Brünig[2,c)] and Walter Focke[1,d)]

[1]*Institute of Applied Materials, Department of Chemical Engineering, University of Pretoria, Private Bag x20, Pretoria, 0028*

[2]*Leibniz Institute for Polymer Research Dresden (IPF), Hohe Straße 6, 01069 Dresden*

[a)]Corresponding author: u29709874@tuks.co.za

[b)]andreas.leuteritz@ipfdd.de

[c)]bruenig@ipfdd.de

[d)]walter.focke@up.ac.za

Abstract. The work presented here describes the development of controlled release and affordable mosquito repellent bicomponent polymer filaments. These filaments may be used in the production of alternative malaria vector control interventions aimed at reducing outdoor malaria transmission in resource-limited communities. The approach was to develop a bicomponent polymer filament with a core containing the volatile repellent active and an outer membrane layer that reduces the rate at which a volatile active is released to the atmosphere. The bicomponent filament was produced by a simple melt extrusion and drawing process. The core polymer used was poly(ethylene-co-vinyl acetate) (EVA) and the sheath polymer was high density polyethylene (HDPE). 40 wt.% N,N-Diethyl-*m*-toluamide (DEET) was incorporated into the EVA via a simple absorption process at a temperature of approximately 81 °C. Scanning electron microscopy (SEM) studies indicated that a bicomponent core-sheath structure was successfully formed by the melt extrusion and drawing process. Raman studies showed that a concentration gradient developed across the bicomponent filament after the manufacturing process. The Raman results suggest that the release of the repellent from the bicomponent filament is a diffusion controlled process, a characteristic of controlled release. With these findings, the study should proceed to the upscale of repellent impregnated bicomponent yarn production and fabric knitting. Such a fabric can be further studied for effectivenes in reducing outdoor malaria transmission in laboratory and field trials.

Key words: bicomponent, controlled release, malaria, mosquitoes, repellent

INTRODUCTION

Malaria is a debilitating parasitic disease that is transmitted by mosquitoes. Malaria mostly affects sub-Saharan Africa. Of the 394200 deaths recorded in Africa in 2015, 292000 were in children under the age of 5 [1]. Currently, there are no adequate interventions that can be used to mitigate outdoor mosquito bites. Further successes in malaria reductions in Africa will rely on the development of alternative vector control [2]. It is clear that targeting outdoor biting mosquitoes will have a significant impact in reducing malaria transmission. Topical repellents may be used to prevent or at least reduce mosquito bites in typical outdoor settings. N,N-Diethyl-*m*-toluamide (DEET) is the most widely used topical repellent active [3]. The main problem with topical repellents is that they have a residual efficacy of a few hours, which is very short [4]. To maintain effective residual repellent efficacy, these topical repellents must re-applied repeatedly. This makes them expensive for use in poor communities where malaria is most prevalent. The three most dominant malaria vectors in Africa i.e. *An. arabiensis, An. Gambiae s.s.* and *An. Funestus* have been shown to have a strong preference of feeding close to the ground level particularly targeting lower limbs, ankles and feet. This is regardless of whether they are indoors or outdoors [5]. Furthermore, it was shown that if the subjects were lying down they were randomly bitten on lower edges of parts of the body in contact with the ground. Another important detail is that if the lower limbs were protected i.e. mid calf and below, these vector mosquitoes did not attempt to fly higher in search of alternative biting sites but instead preferred to seek another victim. These findings present a simplified problem of reducing malaria transmission by protecting selected body sites. In this study, we report on the development of controlled release and affordable mosquito repelling bicomponent polymer filaments. These filaments can be knitted or woven to produce textiles used in the production of affordable personal protection clothing items such as insect repellent socks or ankle covers. The approach was to develop a bicomponent polymer filament with a core polymer impregnated with a volatile repellent active and a sheath layer that reduces the rate at which the volatile active is released to the atmosphere (figure 1). This proposed technology may offer significantly cheaper and more effective repellent fabrics than currently available repellent fabric technologies.

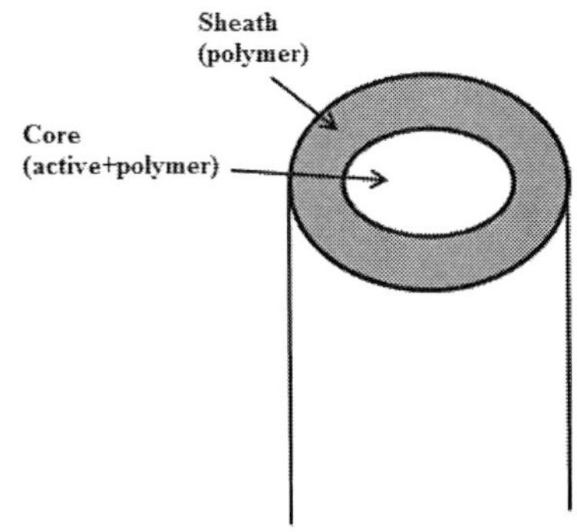

FIGURE 1. Schematic of envisaged bicomponent filament

EXPERIMENTAL

High density polyethylene (HDPE) (Borealis VL9500) with a melt flow index (MFI) of 34 at 190 °C/2.16 kg was used as the sheath polymer. Poly(ethylene-co-vinyl acetate) (EVA) (19 – 21%) (Evatane 2020) with an MFI of 21 at 190 °C/2.16 kg and a VA content of 19-21% was used as a core polymer. DEET with 97% purity, a density of 0.998 g/cm^3 and a normal boiling point is 288 °C was obtained from Sigma Aldrich. Bicomponent filament spinning trials were done using an in-house developed extrusion and melt drawing equipment. The equipment consisted of two heated cylinders with a plunger, one extruding the core and the other extruding the sheath. Scanning electron microscopy (SEM) images of produced filaments were recorded using a Zeiss Ultra Plus field emission spectroscope. The samples were embedded in an epoxy resin and frozen to -175 °C using nitrogen and cut with a diamond knife. The samples were then coated with gold before viewing. The concentration of DEET as a function of depth was studied using a RAMAN Imaging System WITec alpha300R. The spectra were taken using a 785 nm laser with an intensity of 75 mW. The objective used was 20x (Zeiss). The integration time was 0.5 s and accumulation was 200 acc. Total DEET content in the filaments was characterised using thermogravimetric analysis (TGA). The sample was placed in a Perkin Elmer TGA 4000 and the temperature was ramped up at a rate of 10 °C/min. The purge gas was nitrogen flowing at 50 mL/min. The mass loss was tracked as a function of temperature. The mechanical testing of the spun filaments were conducted using a Zwick/Roel Z05 universal test machine.

RESULTS

Extrusion and filament spinning

DEET was first incorporated into EVA pellets via absorption at 81 °C over a 24 hour period. The bicomponent melt spinning was a fairly simple process. The final melt temperature used was 160 °C and the maximum filament draw down speed was 800 m/min. The core-sheath melt throughput was fixed at 0.5 g/min respectively.

Thermogravimetric analysis

Figure 2 shows TGA traces of the produced bicomponent filament. The first mass loss event is attributed to DEET evaporation from the filament. The results indicate that a maximum of *ca.* 20 wt.% (based on total filament weight) was impregnated into the bicomponent filament.

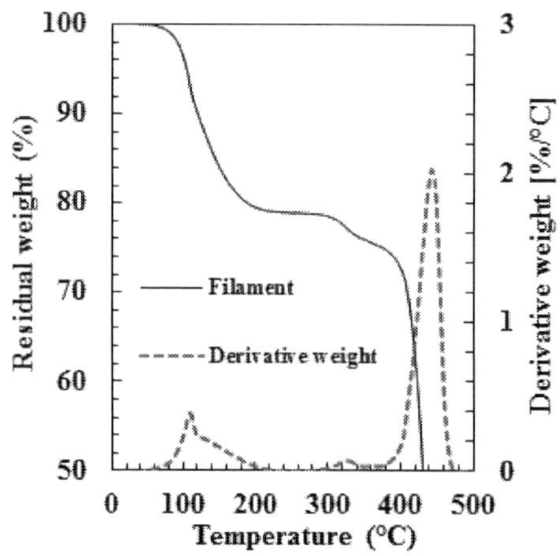

FIGURE 2. TGA traces of the lower and higher DEET content bicomponent filaments

Scanning Electron Microscopy

SEM cross-sectional micrographs of neat and DEET impregnated bicomponent filaments are shown in figure 3. All the filaments were produced at an equal core-sheath throughput of 0.5 g/min at 160 °C and a filament winding speed of 50 m/min. The micrographs show a perfectly formed core-sheath structure.

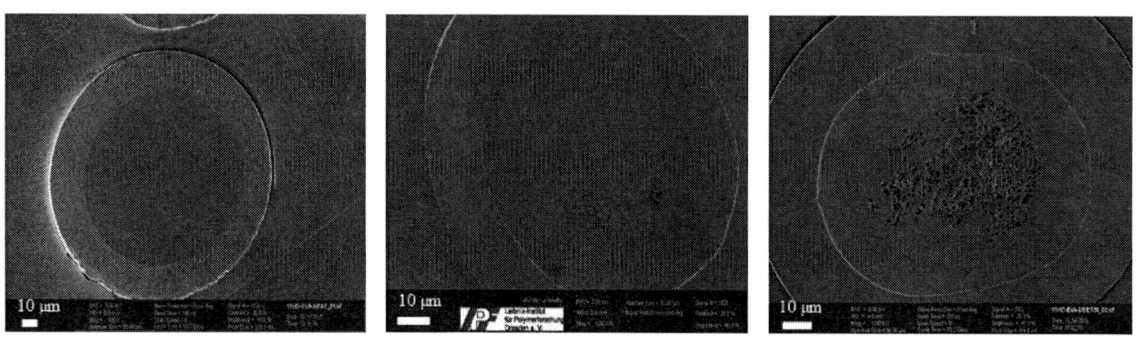

(a) (b) (c)

FIGURE 3. SEM micrographs of HDPE/EVA bicomponent filaments. (A) neat filament, (B) filament impregnated with *ca.* 10 wt.% DEET and (C) filament impregnated with *ca.* 20 wt.% DEET

DISCUSSION

The process of bicomponent filament spinning involves the drawing and cooling of melt. This aligns the polymer chains in the axial direction increasing the tenacity of the filament. At the same time, both polymer phases crystallize diminishing the amount of amorphous matrix available for dissolving the repellent active. The solubility of the active in the crystallized polymers drastically reduces creating a state of supersaturation. This causes blooming of the active relatively quickly in an effort to return to equilibrium [6]. DEET is a spatial repellent. It has to be available as a vapour in the immediate atmosphere surrounding the filament in sufficient quantities to effectively repel mosquitoes. The repellent in the vapour phase must constantly be replenished by the liquid repellent that has diffused to the surface of the filament. This implies unsteady diffusion of the liquid repellent from the core of the filament (EVA) to the surface via the HDPE sheath. If this was the case we would see a concentration gradient built up across the filament cross-section up to the centre point [7]. To test whether a concentration gradient had developed, we carried out a concentration profile study with increasing filament depth. In this study a Raman laser beam was focused within the filament at different depths to determine the respective DEET concentrations. Figure 4 shows Raman spectrograms at different depths for the filament containing *ca.* 20 wt.% DEET. The absorption at *ca.* 1000 cm^{-1} is attributed to DEET.

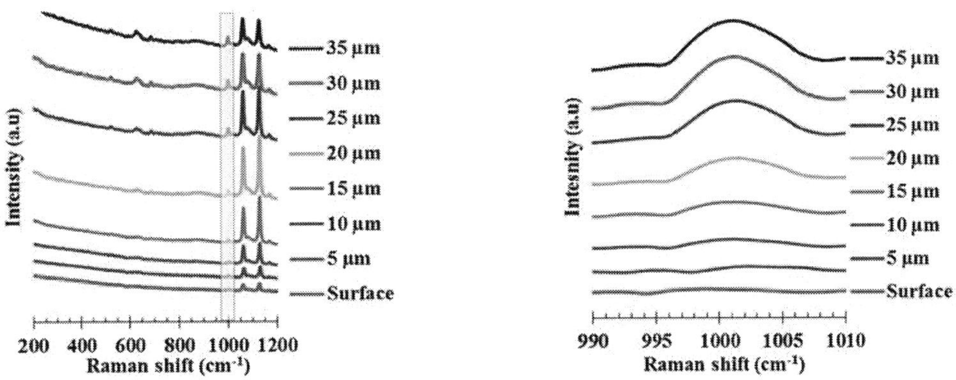

(a) (b)

FIGURE 4. Raman spectrograms of bicomponent filament containing 20 wt.% DEET based on total mass of fibe. (A) shows the spectra generated for the bicomponent filament. (B) shows zoomed in spectrograms attributed to DEET at *ca.* 1000 cm^{-1}

This absorption band gradually increases in size and in area as the filament is penetrated by the laser from the surface to the centre. This observation clearly shows a build up of a concentration gradient indicating the controlled release of DEET from the bicomponent filament. The development of the bicomponent filament has been done using HDPE and EVA. These polymers are relatively cheap and this will make the final product affordable. DEET was used as the volatile active. The reason we used DEET is because it is the most widely used and most effective repellent on the market [4]. In future studies, we should consider incorporating alternatives to DEET into the bicomponent filaments. IR3535 and picaridin are effective synthetic alternatives. P-menthane-3,8-diol may be considered as a potential natural repellent alternative.

CONCLUSIONS AND RECOMMENDATIONS

Increasing the residual effectiveness of repellents can make a significant impact in reducing outdoor malaria transmission. This work has proven the concept of trapping a liquid repellent and slowly releasing it into the environment using bicomponent filaments in laboratory scale studies. These filaments can be manufactured cost effectively by a simple melt extrusion and drawing process. This technology may offer a longer lasting residual repellent effectiveness compared to microencapsulation whilst offering a cost advantage. A wide range of repellent actives should also be considered for impregnation into the bicomponent filaments as alternatives to DEET. Further work needs to be carried out in upscaling the production of bicomponent yarn impregnated with repellent. This yarn can be knitted into a fabric that can be subjected to laboratory and field repellence testing. These studies will shed light on the effectiveness of this technology in reducing malaria transmission.

ACKNOWLEDGEMENTS

We would like to thank Mathias Häschel and Norbert Smolka for technical assistance during the bicomponent filament production. Funding from the Institute of Applied Materials, National Research Foundation (NRF) and the Institute for Polymer Research (IPF) is also acknowledged.

REFERENCES

1. WHO, (World Health Organisation, Geneva, 2016).
2. R. W. Snow and K. Marsh, The Lancet **376** (9735), 137-139 (2010).
3. M. Debboun, S. Frances and D. Strickman, in *Principles, methods, and uses* (Taylor & Francis group, Boca Raton, 2007), pp. 465.
4. E. Lupi, C. Hatz and P. Schlagenhauf, Travel Medicine and Infectious Disease **11** (6), 374-411 (2013).
5. L. Braack, R. Hunt, L. L. Koekemoer, A. Gericke, G. Munhenga, A. D. Haddow, P. Becker, M. Okia, I. Kimera and M. Coetzee, Parasites and Vectors **8** (1) (2015).
6. W. W. Focke and W. Van Pareen, Journal of Polymer Engineering **31** (6-7), 521-529 (2011).
7. W. Beek, K. Mutzall and J. van Heuven, *Transport Phenomena.* (Wiley & Sons, Chichester, 1999).

Performance of jute/PLA all-biodegredable overmolded eco-composites

alchekh wis abdulmounem (1)*, Ozkoc Guralp (2)

(1) kocaeli university - kocaeli - Turkey, (2) Department of Chemical Engineering Kocaeli University 41380 Kocaeli Turkey - kocaeli - Turkey

Development of ecofriendly biocomposites to replace non-biodegradable synthetic fiber composites is the main objective of this study. the plain woven jute fabric (WJF) reinforced poly(L-lactic acid) (PLA) composites were prepared by the hot press molding method. The mechanical and thermal properties of the composites were investigated .The influence of woven structure and direction on the mechanical properties i.e. DMA and flexural properties was investigated. The results show that the properties of the composites are better than those of PLA. When the content of the fiber is 30%, the composites can get the best mechanical properties. The dynamic mechanical analysis results show that the storage moduli of the PLA/jute composites increase with respect to the PLA. In the present study jute fibers were surface treated using two methods: i) alkali (NaOH) and ii) alkali followed by silane (NaOH+Silane) separately. Effects of surface treatments on jute fibers surface were characterized using fourier transform infrared spectroscopy (FT-IR) ,thermo gravimetric analysis (TGA) and (DSC) . The results from this investigation indicate that a 30 min dipping of the fibers in 0.5% alkali solution followed by 24h 1.5% Silane treatment leads to an increase in the tensile strength of up to 23%. The increase appears to be due to fiber separation and removal of non-cellulosic materials, which, in turn, resulted in an increased crystallinity. Blended of thermoplastic polyurethane (TPU) and (PLA) polymers at different ratios were made via twin screw extrusion and injection molding techniques. Then the blend injects to PLA/jute composite. (FTIR) confirmed the existence of the two components in the blends; differential scanning calorimetry (DSC) and dynamic mechanical analysis (DMA) confirmed the immiscibility between the TPU and PLA.

In-Situ Observation of Multiple-Necking Behavior in Continuous Cold Drawing of PET Filaments in Ethanol

Dongwoo Go[1], Wataru Takarada[1], Arun Aneja[2], and Takeshi Kikutani[1, a)]

[1]*Department of Materials Science and Engineering, School of Materials and Chemical Technology, Tokyo Institute of Technology, 2-12-1 S8-32, O-okayama, Meguro, Tokyo, Japan*
[2]*Department of Engineering, East Carolina University, Greenville, NC 27858, USA*

[a)]Corresponding author: kikutani.t.aa@m.titech.ac.jp

Abstract. In our earlier research, the batch-type drawing of undrawn poly(ethylene terephthalate) (PET) filament was carried out in ethanol at room temperature. The formation of multiple necking, reduction of drawing stress, infusion of ethanol into filament along with dye, and the occurrence of solvent-induced crystallization were observed. To explore the possibility of applying this unique behavior to accomplish a process of industrial significance, continuous drawing of PET fibers in ethanol was performed. Low oriented amorphous PET filaments were prepared. Continuous drawing of the filaments was carried out in ethanol using a set with rollers of different rotation speeds exposed to ethanol flowing between two glass blades (channel) between the two rollers. Through continuous of monitoring of necking behavior the formation of multiple necks was observed during the drawing process. The number of necks was found to increase with the increase of the rotation speed of feed roller. The location of initiation and propagation of neck shifted up stream with increased draw ratio and decrease of the rotation speed of feed roller. Through continuous monitoring of the drawing tension, it was found that the periodic occurrence of the formation of a group of multiple necks was accompanied by oscillation of drawing tension a source of instability of thread line with reduced yarn quality. The installation of a drawing pin was discovered to be useful for the stabilization of the drawing process.

INTRODUCE

Poly(ethylene terephthalate) (PET) is one of the most widely used polymer material for the synthetic fibers. Our earlier research focused on studies for drawing low oriented amorphous PET fibers in room temperature using organic solvent. It is well known that multiple-necking behavior occurs when tensile drawing is applied to PET fibers in an organic solvent [1]. In addition, infusion of an organic solvent was demonstrated that occurs when a fiber is drawn in an organic solvent [2]. Crystallization of molecules by organic solvents infused into fibers has also been reported [3]. In our previous studies [1-3], we have performed batch-type drawing using ethanol as an organic solvent. It has been confirmed that batch-type drawing using an ethanol, multiple-necking phenomenon occurred, yield stress and drawing stress were decreased, and the natural draw ratio was increased in drawing process. It was also confirmed that the solvent induced crystallization of the molecules occurred due to the influence of the ethanol infusion. It should be noted that neither infusion of ethanol into PET fiber nor solvent induced crystallization occurs if PET fiber is simply immersed in ethanol.

Various applications of infusion phenomenon occurring in the organic solvent drawing process are expected. Typical applications include low temperature dyeing of fiber and transport of material to yield desired functional properties. Infusion of both solvent and dye, and solvent induced crystallization occur in a single step drawing process. This is a cold drawing process, and the crystallization occurs at room temperature, with no additional annealing required. To explore the possibility of applying this unique behavior to industrial process, continuous drawing of PET fibers in ethanol was performed in this current research and drawing behavior was investigated in detail.

EXPERIMENTAL

Materials

Low oriented amorphous poly(ethylene terephthalate) (PET) filament with a diameter of 105 μm was used in this study. The PET filament was prepared by extruding of PET polymer at 300 °C from a spinneret with a single circular hole of 1 mm diameter. The filament was quenched with cross flow air at room temperature (25 °C). The throughput rate was maintained at 5 g/min, and filament was taken-up on a bobbin using a winder at a velocity of 400 m/min. Ethanol was used as the organic solvent in the solvent drawing process.

Filament Continuous Drawing

Continuous drawing of the PET filament was carried out in ethanol using a set of rollers operating at different rotation speeds. The circumference of rollers was 317 mm. Ethanol flowing in glass channel of the length of 300 mm was placed between the two roller. The schematic illustration of continuous drawing with ethanol channel is shown in Fig. 1. Drawing of as-spun filaments was carried out at various feed roll rotation speeds of 1 RPM to 10 RPM and draw ratios (DR) of DR 2x to DR 5x.

In-situ Observation of Continuous Drawing Process

Observation of multiple-necking behavior of PET filament in the drawing line was carried out using a high-speed camera (Phantom V9.0, Vision Research Inc., USA) during continuous solvent drawing process. Tension of PET filament in the draw line was also measured using a tension meter (TTM-201, Toray Engineering Co., Ltd., Japan) simultaneously.

RESULTS

Number of Necks

Photographs of PET fibers captured from the continuous drawing line in ethanol of various feed roller rotation speeds (from 1 RPM to 10 RPM) and draw ratio (DR) of 2x are shown in Fig. 2a. It can be seen that the number of necks per unit length increases as the rotation speed of the feed roller increases. It is speculated that the increase rate of the fiber velocity was larger than that of the propagation speed of the individual necks, thereby increasing the number of necks. Fig. 2b shows a photograph taken at fixed location in ethanol channel according to the distance from the entrance of ethanol bath. All photos are same drawing process of DR 2x and feed roll rotation speed of 1 RPM. As the filament moves toward the downstream side, the drawn part increases, the undrawn part decreases. And some undrawn part disappears with the pair of necks at each end of the undrawn part are merged and vanished. The number of necks decrease as the filament moves towards downstream side.

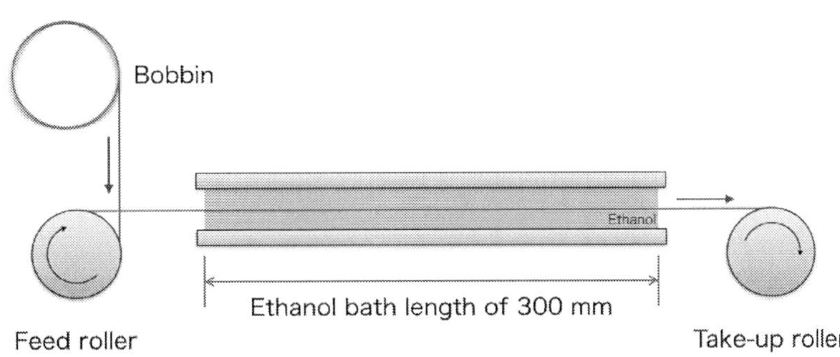

FIGURE 1. Schematic of continuous cold drawing system in ethanol channel.

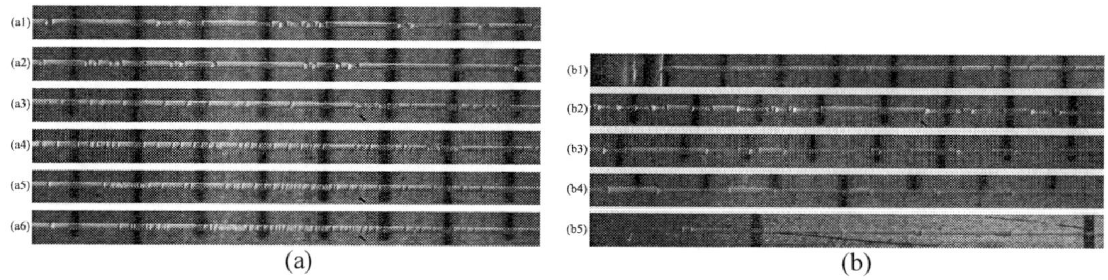

FIGURE 2. (a) is the photos of PET fiber under the drawing process in ethanol at a position near the entrance of ethanol channel. DR 2x, and feed roll rotation speeds were (a1) 1 RPM, (a2) 2 RPM, (a3) 4 RPM, (a4) 6 RPM, (a5) 8 RPM, and (a6) 10 RPM. (b) is the photos of PET fiber under drawing process in ethanol. Distance from the entrance of ethanol channel were (b1) 0 mm, (b2) 33 mm, (b3) 66 mm, (b4) 100 mm, and (b5) 200 mm. Feed roll rotation speed was 1 RPM and draw ratio was 2x.

Progression Rate of Elongation

The velocity of filament was measured for each drawing condition and position in ethanol channel. The drawing condition was varied from 1 RPM to 10 RPM at the feed roll rotation speed and for a draw ratio from DR 2x to DR 4x at each drawing speed. The distance from the entrance of the ethanol channel of 0 mm, 33 mm, 66 mm, 100 mm, and 200 mm, respectively, were measured with a high-speed camera. Fig. 3 shows the results of velocity measurements of the feed roll rotation speed of 1 RPM and 10 RPM, and the draw ratios of 2x, 3x and 4x. The moving velocity of the fiber at the entrance of ethanol bath is equal to the speed of the feed roll. The fiber velocity increases upon and elongation, and finally wound at the speed of the take-up roller. The filament velocity in the ethanol channel showed higher acceleration at slower drawing speed and higher draw ratio. The moving velocity of filament increases at an early stream of the ethanol bath as the acceleration is higher. As the drawing speed was faster and as the draw ratio was lower, the acceleration also lower. That is, as the moving velocity acceleration was higher, the drawing progression rate was large at the upstream side. In addition, when the DR 5x at the feed roll rotation speed of 1 RPM, the neck propagating position of the fiber was moved to the upstream side extremely due to the slow drawing speed and the high draw ratio, and finally it was observed to be escaped out of the ethanol channel.

Tension Measurement

In order to evaluate the stability of the continuous drawing process, the tension of draw line was measured during the drawing process under each drawing condition. Fig. 4a shows the result of measuring the tension on drawing line of DR 2x at each drawing speed, and Fig. 4b shows the result of measuring the tension of each draw ratio at the feed roll rotation speed of 1 RPM. As a result of measuring the tension on the draw line while drawing, it confirmed that the draw tension applied to the drawing line increased as the drawing speed increased. It was also confirmed that the draw tension increased as the draw ratio increased. The tension showed a specific periodic fluctuation depending on drawing condition.

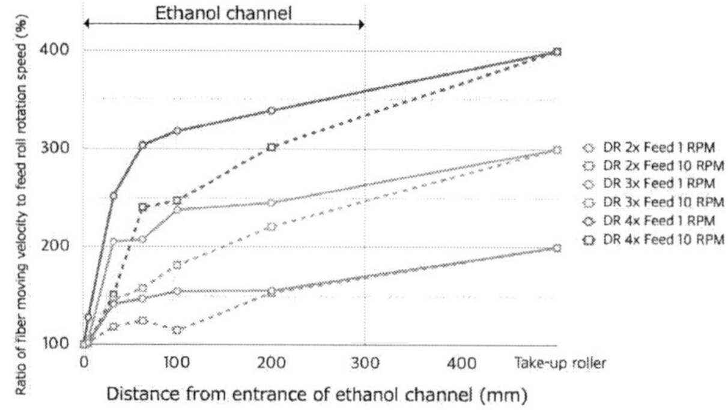

FIGURE 3. Ratio of fiber in ethanol channel velocity to feed roll rotation speed as a function of distance.

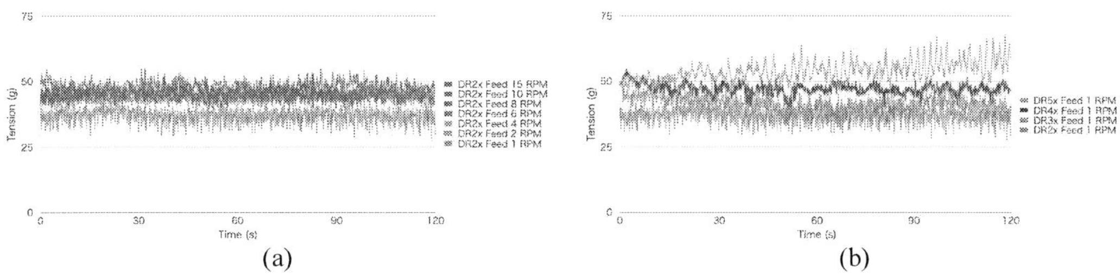

FIGURE 4. Draw line tension trace. (a) draw tension at DR 2x at each feed speed, and (b) feed roll rotation of 1 RPM at each draw ratio.

A snub pin was placed in the ethanol bath to stabilize the tension and induce initiation of neck. Schematic illustration of continuous drawing system equipped with pin in an ethanol bath is shown in Fig. 5. Since the tension after the pin installation appears different from the upstream versus the downstream around the pin, the tension was measured at both locations between the feed roller and the ethanol bath, and between the ethanol bath and the take-up roller. On comparing the tension at the upstream side and the downstream side, respectively, it was confirmed that the upstream side showed lower tension than the downstream side. In addition, the shown in Fig. 6, when the draw tension with the pin is compared with the draw tension without the pin, the draw tension of both sides around the pin is lower than the draw tension without the pin installation. The reason why the tension on the upstream side with the pin was the lowest is because the tension is induced on the downstream side due to the friction between the pin and the fiber. Nevertheless, it was speculated that the reason why the drawing tension on the downstream side with the pin is lower than that without the pin is that as the shown in Fig. 7, the number of necks is increased and the drawing stress is decreased. At various DR (Fig. 8). This should yield a higher quality yarn. The installation of the pin suppressed the fluctuation of draw tension significantly.

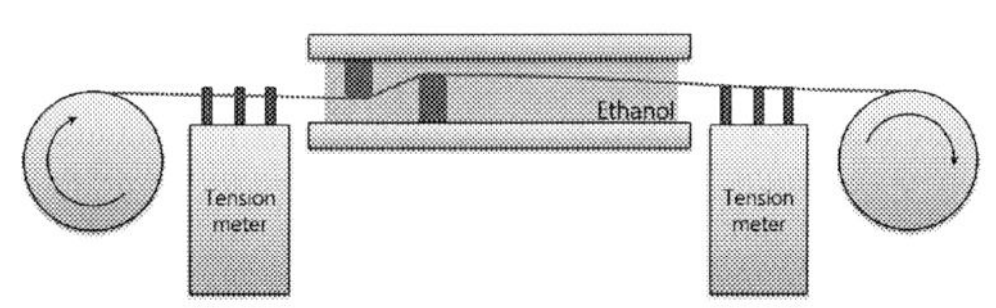

FIGURE 5. Schematic illustration of continuous drawing system equipped with pin in an ethanol bath. Tension meter was mounted on the upstream side and the downstream side, respectively.

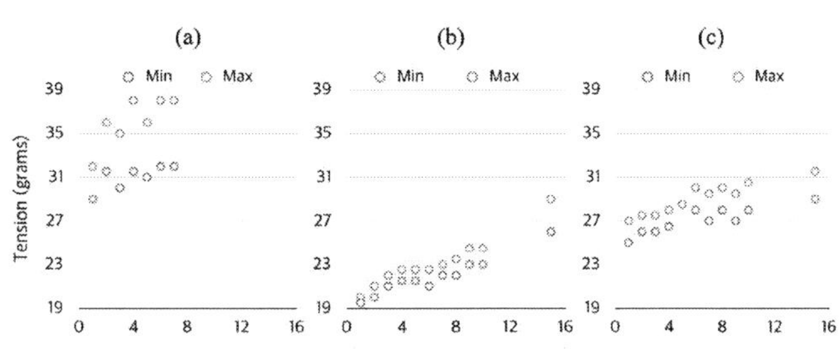

FIGURE 6. Impact of snub pin location on draw line tension. (a) no snub pin (b) tension upstream of pin (c) tension downstream of pin.

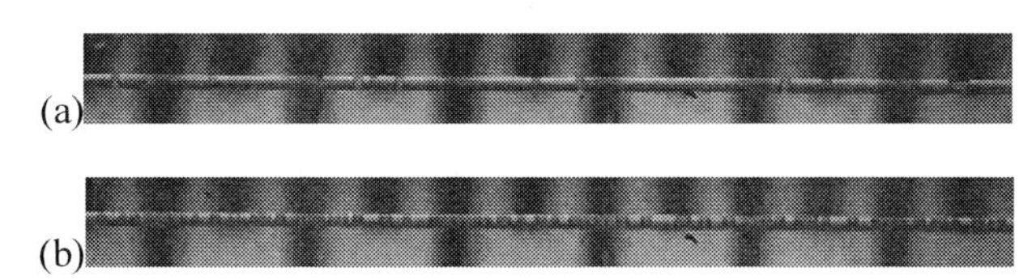

FIGURE 7. Photos for comparing the number of necks according to pin installation. (a) is when a pin is not installed and (b) is when a pin is installed

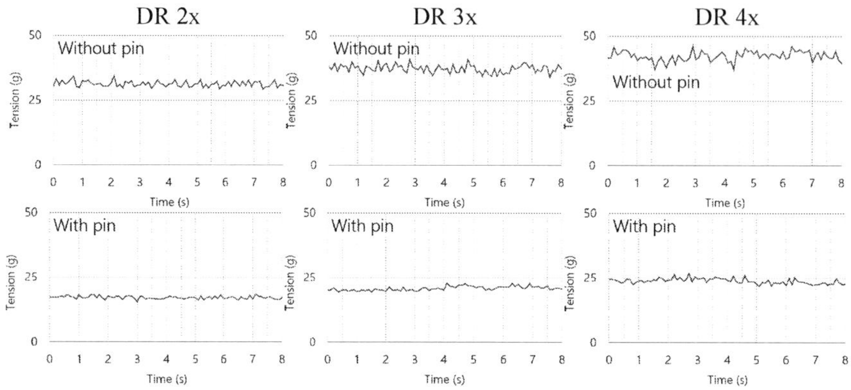

FIGURE 8. Result of draw tension on drawing line according to pin installation.

Uniformity of Drawn Fiber

In order to produce a commercial fiber product, the fiber must maintain uniform quality. In order to obtain drawn fiber a uniformity in the organic solvent continuous drawing process, it should be satisfy the following two conditions: (1)all necks should be propagated in ethanol channel, and (2)the filament should be fully drawn. With the draw ratio of 5x and the feed roll rotation speed of 10 RPM, the neck propagated in the ethanol channel and the drawing was completed. However, this condition is very unstable and difficult to maintain continuous process due to high frequency of breaks from broken due to the excessive tension applied on the drawing line. When drawing at feed roll rotation speed of 1 RPM, the draw tension on drawing line was stable, but draw with DR 5x at feed roll 1 RPM, position of the neck was moved out from the ethanol channel. Thus, the drawing is carried out at a feed roll rotational speed of 1 RPM and the draw ratio of at an interval of 0.1 from 4.5x to 5.0x to find a drawing condition in which the filament be able to fully drawn and also all of necks propagated in ethanol channel. As a result, when the draw ratio was less than 4.7x, the undrawn part remained and if that above 4.8x, the fiber was fully drawn. When the draw ratio was over 4.9x, we observed that the position of neck was moved out of the ethanol channel. At the draw ratio of 4.8x, it was possible for all necks to propagate in ethanol and the fiber was fully drawn. Therefore, it has been confirmed that the draw ratio of 4.8x was natural draw ratio in this experimental condition and it was possible to produce uniform drawn yarn consistently.

CONCLUSION

At the conclusion of this study, it was observed that the multiple neck occurred in the continuous drawing process. The influence of the drawing speed and the drawing ratio on the variation of the area of neck propagation was clarified. It is confirmed that the stability of the drawing process was improved by introducing a neck pin in the ethanol bath for the initiation of neck. Finally, the existence of a range of drawing conditions in which the drawing could be completed in the ethanol bath was confirmed.

REFERENCES

1. R. Khanum, D. Go, W. Takarada, A. Aneja and T. Kikutani, *SEN'I GAKKAISHI* **71**, No. 9, 273-283 (2015).
2. R. Khanum, D. Go, W. Takarada, A. Aneja and T. Kikutani, *Polymer* **70**, 19-29 (2015).
3. R. Khanum, W. Takarada, A. Aneja and T. Kikutani, *Polymer* **59**, 26-34 (2015).

Structural response of poly(3-hydroxybutyrate) fibers to heat and stress

Hufenus Rudolf [(1)*], Reifler Felix A. [(1)]

[(1)] Empa - St. Gallen - Switzerland

Poly(3-hydroxybutyrate) (P3HB) is a thermoplastic polyester produced by bacteria. Developing an upscalable melt-spinning process, we found that a controlled crystallization process is crucial for producing P3HB fibers: Primary crystallization has to be completed during the drawing step, while secondary crystallization has to be hindered to prevent brittleness and poor tensile properties. By modifying melt-spinning and drawing, we achieved fibers dominated by longitudinally oriented lamellae rather than spherulitic structures. To better understand the development of macromolecular orientation during fiber forming and annealing, we performed tensile stage and hot stage experiments in combination with wide-angle X-ray diffraction (WAXD). P3HB fibers were subjected to various tensions and temperatures in-situ. In the equatorial 2Theta scan of the WAXD patterns of these fibers, we observed a series of local maxima. In previous literature, the diffraction signals in this region are commonly described as one reflection, assigned to the so-called "beta-form" of P3HB. In contrast we postulate a highly oriented non-crystalline phase P(nc), which is kinetically trapped between the aligned lamellae of the crystalline alpha-phase. In our model, the oriented crystalline phase partially morphs into an oriented P(nc) under stress only. Cyclic change of tension reveals a high degree of reversibility for this transformation. Under heat only, the entrapped, oriented P(nc) gains mobility and thus irreversibly turns into an oriented crystalline phase. When combining stress and heat, the applied tension prevents such a transformation of the oriented P(nc), but it promotes an increase in orientation of the crystalline phase. On the fiber level, annealing under tension results in a higher modulus, while annealing without tension increases plasticity. This is in correspondence with the proposed structural model.

Effect of co-plasticizer type on properties of edible starch films

Ozkoc Semin Ozge (1)*, Sungur Seval (1), Kaval Melike (1), Ozkoc Guralp (1)

(1) Kocaeli University - Kocaeli - Turkey

An edible film is a biodegradable thin layer, made of bipolymers, coated on a food to set a barrier between food and the environment and can be consumed with food. To produce edible films, agricultural polymers are used and starch is one of them. Edible films obtained from native starch have some drawbacks (poor processing ability, high brittleness, etc.) in meeting the requirements of a packaging material. Therefore, thermoplastic starch films are obtained to meet most of the requirements. During this process, plasticizers are used to achieve ease of processing, increase polymer chain flexibility and molecular mobility and in some studies co-plasticizers were used to prevent retrogradation of thermoplastic starch. In this study, different esters such as tributhyl citrate (TBC), triisodecyl citrate (TIDC) and ethyl benzoate (EB) were used as co-plasticizers to obtain edible thermoplastic starch films. Mechanical properties, processing characteristics, clarity and hydrophylicity measurements were done to compare films formulated with different co-plasticizers. Compatibility between plasticizer and polymer has major significance in terms of effective plasticization. The chemical structure of the plasticizer, including chemical composition, molecular weight and functional groups, determines the degree of polymer plasticization. It is seen that the most compatible plasticizer was found to be TBC. The films produced using TBC elongates much higher than the others in tensile test, whereas their tensile strength and elastic modulus were lower. The mechanical tests showed that EB was incompatible with thermoplastic starch. The surface hydrophobicity of the thermoplastic starch films increased when co-plasticizers were used. The citrate esters having different molecular structures can be used as useful plasticizers to tune the properties of the thermoplastic starch edible films. Keywords: co-plasticizers, edible film, hydrophylicity, mechanical properties, thermoplastic starch

Biopolymer Foam Blends of Poly(lactic acid) and Novatein® Thermoplastic Protein

Walallavita Anuradha Sammani [(1)]*

(1) University of Waikato - Waikato - NewZealand

A batch processing method was used to fabricate foams comprising of a blend of poly(lactic acid) (PLA) and Novatein, a protein-based thermoplastic. Various compositions of Novatein/PLA were prepared with and without a compatibilizer, PLA grafted with itaconic anhydride (PLA-g-IA). These were foamed at varying temperatures above the material's Tg (80 – 160 °C). This batch foaming technique has successfully been implemented to foam both semi-crystalline and amorphous PLA using an environmentally benign blowing agent, subcritical CO2, however, batch foaming semi-crystalline Novatein is more difficult due to the constraining effects of the nature of the material. Pure Novatein cannot form a cellular structure at a foaming temperature of 80 °C, however in a blend with 50 wt. % of PLA, microcells formed with smaller cell sizes (3.36 µm) and higher cell density (8.44 x 1021 cells/cm3) compared to pure PLA and blends with higher amounts of PLA. The incorporation of 50 wt. % Novatein in PLA increased the crystallinity of the blend and the matrix stiffness of Novatein was high enough to restrain cell coalescence and cell collapse. Furthermore, the introduction of a crystalline phase from Novatein has created additional interfacial area for nucleation (interface between crystalline and amorphous phases). At a foaming temperature of 140 °C, NTP30-PLA70 showed a unique interconnected porous morphology with a pore size of 3.29 µm and cell density of 6.72 x1022 which can be attributed to the CO2-induced plasticization effect. Foaming of multi-phase systems offers a great opportunity to enhance the properties of porous materials, however, controlling the foaming process can be challenging. The outcome of this study can lead to a wider understanding about the foaming process and resulting morphologies of semi-crystalline and amorphous polymer blends.

Dynamics and Structure Development for Biaxial Stretching PA6 Films

Toshitaka Kanai [1,a)*] , Yoshimune Okuyama[2)], Masao Takashige[3)]

*[1] *KT Poymer, 5-7-14 Kuranamidai, Sodegaura, Chiba, 299-0245, Japan*
toshitaka.kanai@ktpolymer.com
[2]*Graduate School of Natural Science & Technology, Kanazawa University, 920-1192, Japan*
[3] *Idemitsu Unitech Co.,Ltd, Kamiizumi, Sodegaura, Chiba, 299-0193, Japan*

[a)]Corresponding author: toshitaka.kanai@ktpolymer.com

Abstract: The biaxial stretching film of polyamide 6 (PA6) is widely used for food and industrial products packaging, because it has good physical properties, such as high toughness and gas barrier properties. It is expected to improve film formability and performance for a larger market. So, in this research, both the stretching methods and PA6 blends were investigated in terms of dynamics and structure development for biaxial stretching films. There are three different stretching methods which can produce polyamide 6 films, namely the simultaneous biaxial stretching, the sequential biaxial stretching, and the double bubble tubular process. There are no reports which describe the difference among the three stretching methods and film properties, so it was not clarified why the films have big differences in their properties. For this reason, the higher order structure and stretching behaviours of the three stretching methods were studied. As a result, the sequential biaxial stretching shows strong stretched effect and molecular orientation in the transverse direction. In the simultaneous biaxial stretching, the molecule orientation was equal both in the machine direction and in the transverse direction of the stretched plane. The double bubble tubular process produced well balanced film and equal orientation in any directions, in terms of the phase difference in the plane of film. Blends of PA6 to meta-Xylene diamine (MXD6) improved the properties of easy straight line cut and high barrier, but did not have good stretchability. For this reason, they were also studied to develop the biaxial stretching film of polyamide which improved stretchability and physical properties. Furthermore, blend effects of MXD6 to PA6 and dynamics analyses were investigated.

1. INTRODUCTION

The biaxial stretching PA6 film is widely used for food and industrial products packaging, because it has good physical properties, such as high toughness, high tensile strength, hard wearing properties and gas barrier properties[1]. It is expected to improve film formability and performance for a larger market. So, in this research, both the stretching methods and PA6 blends were investigated in terms of dynamics and structure development for biaxial stretching films.

There are three different stretching methods which can produce polyamide 6 films, namely the simultaneous biaxial stretching[5], the sequential biaxial stretching[6], and the double bubble tubular process[1-4,8,9]. There are no reports which described the difference among the three stretching methods and film properties, so it was not clarified why the films have big differences in their properties. Also, as blends of PA6 to meta-Xylene diamine (MXD6) improved the properties of easy straight line cut and high barrier[10], they were studied to develop the biaxial stretching film of polyamide which has good stretchability and good physical properties. Furthermore, blend effects of MXD6 to PA6 and dynamics analyses were investigated. For these reasons, our study is to investigate the higher order structure and stretching behaviors of the three stretching methods and to develop the biaxial stretching film of polyamide which has good stretchability and good physical properties.

2. EXPERIMENT

2.1 Experimental Equipment

The film samples of PA6 were used to study the biaxial stretchablity and the super structure formation by using a biaxial stretcher. The biaxial stretcher SDR-527K, which was designed by our group and made by Etoh Corporation, was used[11].

The schematic view of this equipment is shown in Fig.1. Various data could be obtained at the same time, such as the stress strain curve (S-S), refractive indexes in three directions, light scattering data during the biaxial stretching and retardation distribution of the biaxial stretched film. The stretching conditions are shown in Table 1. The distribution of phase contrast at whole rotation angles for the biaxial stretched film was measured by RETS 100 Otsuka Electronics Co., Ltd.

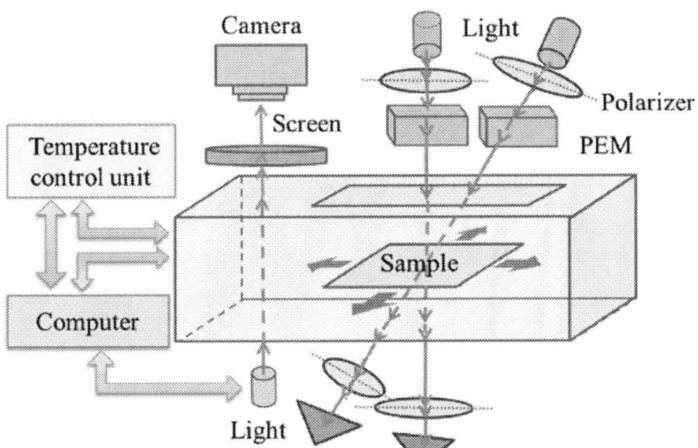

Fig.1 Schematic diagram of biaxial stretching machine

Table 1 Stretching conditions

Stretching mode	Sequential or Simultaneous biaxial stretching
Stretching speed	10mm/s
Stretching ratio	MD 3 times × TD 3 times (Standard condition)
Stretching temp.	110℃
Preheating time	2min

2.2 Materials

PA6 and MXD6 were used for our research. PA6 was Ube Nylon 1023FD with a mean molecular weight of 23,000 and the relative viscosity of η_r=3.5 in 98% sulfuric acid as a solvent. MXD6 was Mitsubishi Gas Chemical MX Nylon 6007 with a mean molecular weight of 25,000 and the relative viscosity of η_r= 2.7 in 96% sulfuric acid as a solvent. The chemical structure of both Nylon6 and MXD6 are shown below.

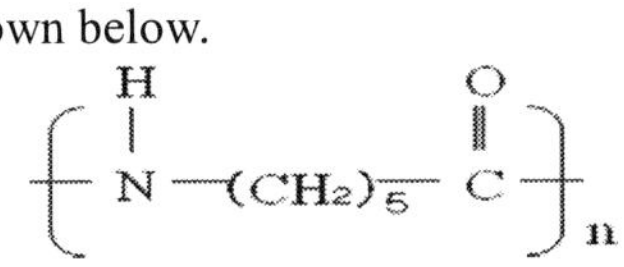

Polyamide 6 (PA6)

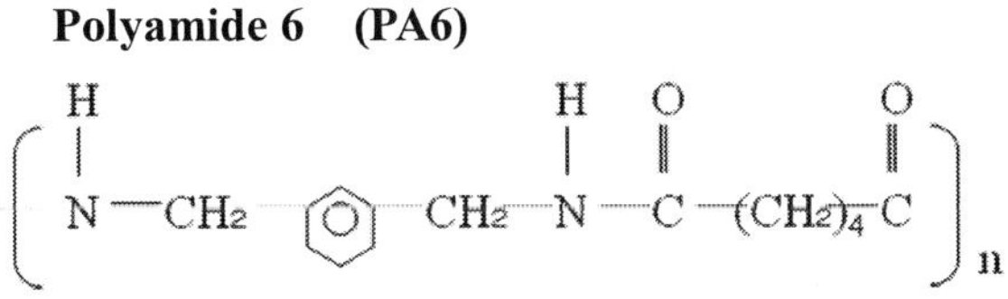

Poly (m-xylene adipamide 6) (MXD6)

2.3 Experimental Conditions

The 50μm thickness PA6 film samples were produced using an extruder with a die extruded at a resin temperature of 265℃ and chilled water bath of 18℃ to prevent crystallization. By using a 40 ϕ mm extruder (L/D=24) with a circular die of the diameter of 75mm and the lip clearance of 1mm and with a water-cooling ring having the diameter of 90mm, non-stretched film was produced at a resin temperature of 265℃ and a blow-up ratio of 1.2.

The film samples were stretched using the biaxial stretcher under the conditions of preheating time 2 minutes, standard stretching temperature 106°C, standard stretching speed 10mm/s and various stretching ratios. In-situ measurement of stress and strain curves and birefringence were obtained. The stretching ratio conditions of the sequential stretching were MD×TD 1.5×1, 2×1, 2.5×1, 3×1, 3×1.5, 3×2, 3×2.5, 3×3, respectively. The stretching ratio conditions of the simultaneous stretching were MD×TD 2×2, 2.5×2.5, 3×3, respectively.

The double bubble tubular film process was also produced in order to compare with tenter stretching films. Non-stretched films were produced under the same extrusion conditions as the above conditions. While passing through the part of apparatus for second blowing, which is composed of two pairs of pinch rolls and a heating furnace (a far infrared radiation heater is self-contained), this raw film was stretched simultaneously in the machine and transverse directions by using internal bubble air. The stretching ratio condition of the double bubble tubular stretching using a production line was MD×TD 3×3. Each sample was also measured by the polarizing microscope and three dimensional birefringences.

Furthermore, 80μm thickness films of the blends of MXD6 to PA were used as a base film for stretching. The stretching conditions were preheating time 2 minutes, stretching temperature 110°C, stretching speed 10mm/s and stretching ratio MD×TD 3×3 by the simultaneous biaxial stretching.

3. RESULTS AND DISCUSSION

3.1 PA stretching film

The photographic images through the polarizing plate for both the simultaneous biaxial stretching film and the sequential biaxial stretching film are shown in Fig.2. The photographic image through the polarizing plate of the simultaneous biaxial stretching film shows very uniform biaxial stretching, but the photographic image through the polarizing plate of the sequential biaxial stretching film shows ununiform stretching (Kink Stretching).

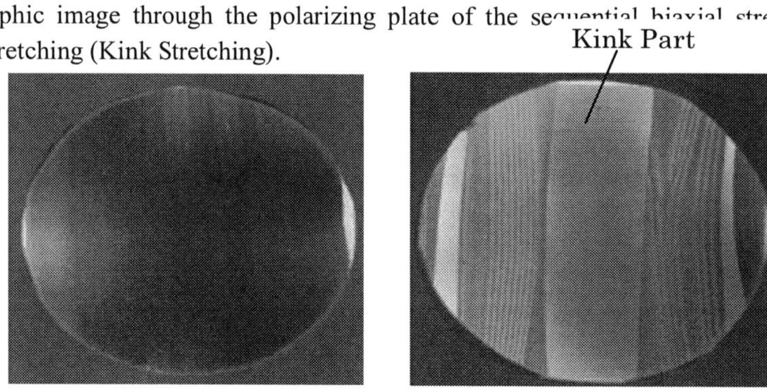

Simultaneous biaxial stretching　　　Sequential biaxial stretching

Fig.2 Pictures of PA6 films through polarizing plate

At first the sequential biaxial stretching results are shown. Fig.3 shows the stress and strain curve and retardation as a function of total strain ratio (surface ratio) under the simultaneous biaxial stretching MD×TD 3×3. The simultaneous biaxial stretching shows almost the same stress of X and Y and keeps very low retardation.

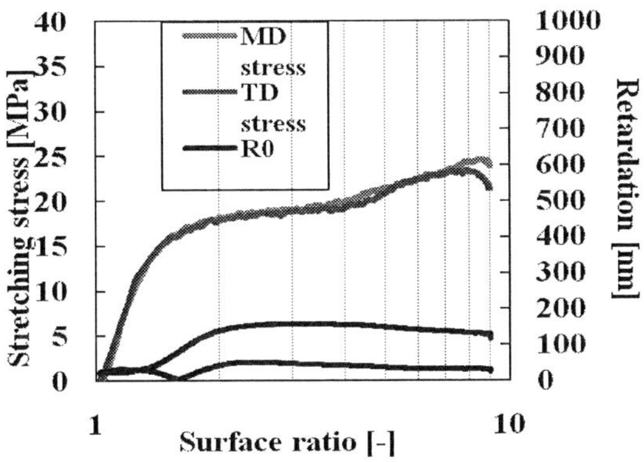

Fig.3 Stress-total stretching ratio curves and retardation-total stretching ratio curve of the simultaneous biaxial stretching

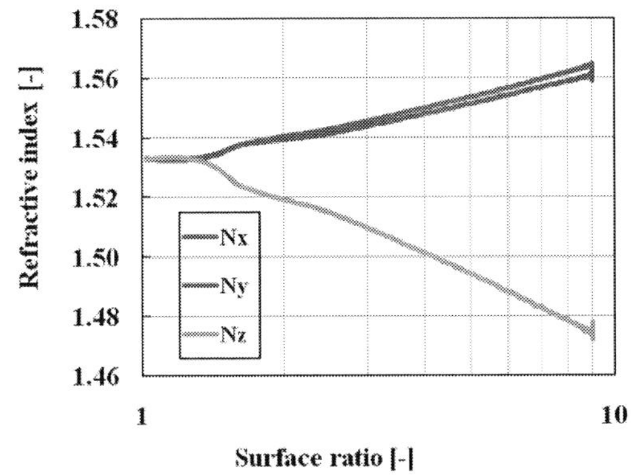

Fig.4 Refractive index and total stretching ratio curve of simultaneous biaxial stretching

Fig.4 shows three dimensional refractive indexes as a function of total strain ratio under the simultaneous biaxial stretching. The refractive indexes of Nx and Ny increase with increasing total strain ratio. It means that the simultaneous biaxial stretching increases both the MD and TD molecular orientations in the parallel to film plane and gives low retardation.

Next the sequential biaxial stretching results are shown. Fig.5 shows the stress–strain curve and retardation under the sequential biaxial stretching and stretching ratio MD×TD 3×3. Y stress (TD stretching stress) becomes larger than X stress (MD stretching stress). Furthermore, the retardation increases with increasing MD stretching ratio and then decreases with increasing TD stretching ratio.

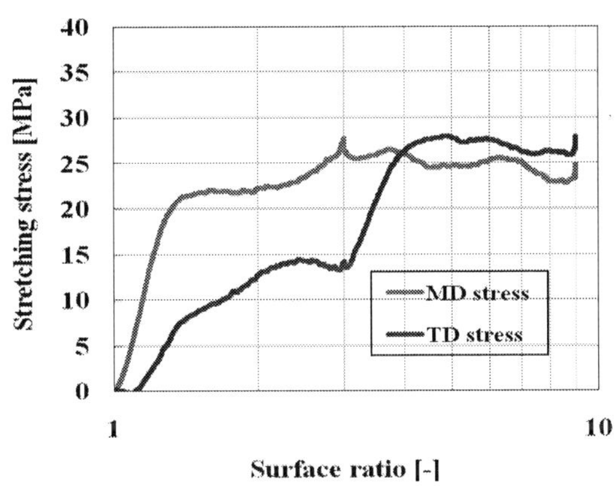

Fig.5 Stress-area stretching ratio curves of sequential biaxial stretching

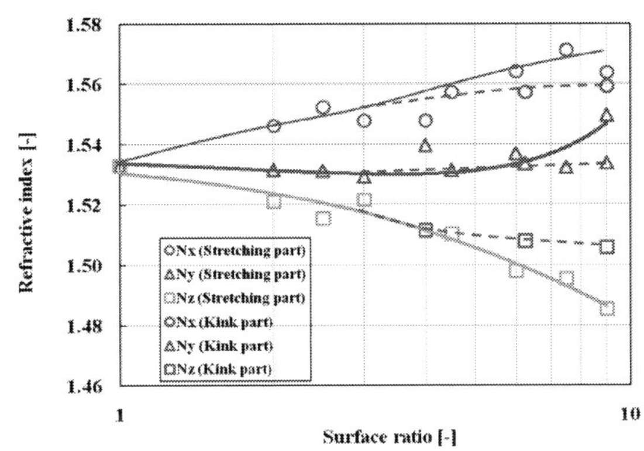

Fig.6 Refractive index and area stretching ratio curve of sequential biaxial stretching

Fig.6 shows three dimensional refractive indexes as a function of total strain ratio of the sequential biaxial stretching. MD stretching increases only Nx which means molecular orientation in MD. TD stretching increases Ny which means TD molecular orientation proceeds. In the unstretched zone, the refractive index during TD stretching does not change and keeps orientation constant.

From these results, the sequential biaxial stretching shortens the distance between the hydrogen bonds and makes the hydrogen bonds strong, and then increases TD stretching stress. Owing to this strong TD stretching stress, it is considered that it is difficult to stretch uniformly.

Fig.7 shows the distribution of the phase contrast at three different stretching methods. These results show that orientation balance at any angle is the best for the double bubble tubular film and the worst for the sequential biaxial stretched film.

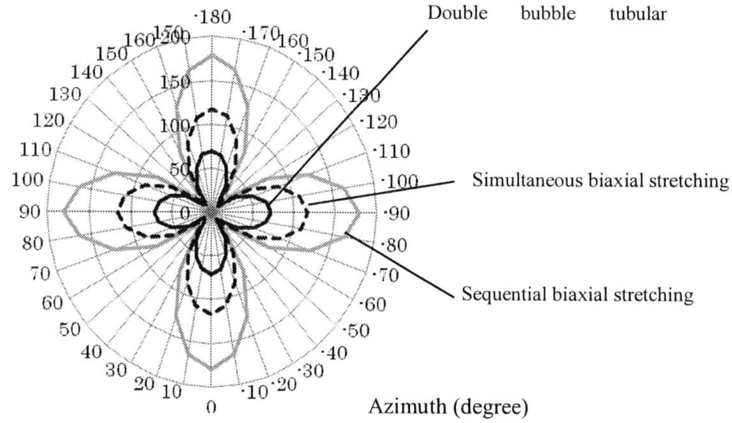

Fig.7 Phase contrast distribution of at whole angles produced by different stretching methods

3.2 PA/MXD6 blend stretching film

The blend of MXD6 to PA6 was studied to improve the gas barrier and easy tear properties. It was stretched and the stretching stress of the blend was evaluated, compared with PA 6. The results are shown in Fig.8.

Fig.8 shows blending of MXD6 to PA6 makes PA6 stretching stress decrease. This result means the hydrogen bond caused by PA6 was reduced by blending MXD6. Furthermore, the stress behavior is also changed after the stretching stress passes the yield stress. PA6 has larger stress build up than PA6/MXD6.

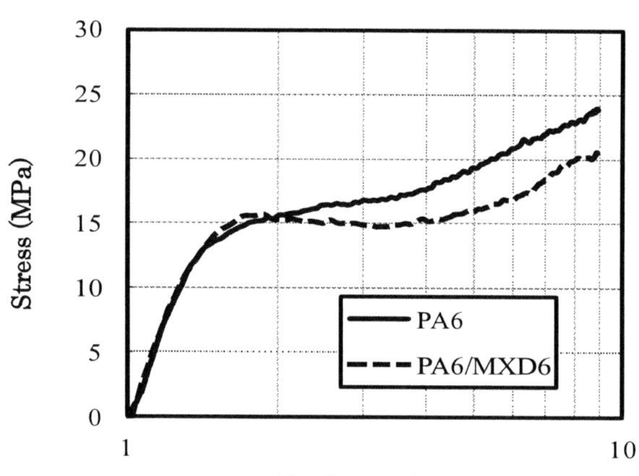

Fig.8 Stress-surface ratio curves of PA6, PA6/MXD6 blends at 110°C

3.2.1 Effect of Stretching Temperature

The stretching temperature was changed at 90°C, 100°C, 110°C, 120°C and 130°C. The maximum stretching stress as a function of stretching temperature was plotted in **Fig.9** in order to evaluate stretchability and stretching stress. The larger slope means the stronger stretching stress dependent on the temperature.

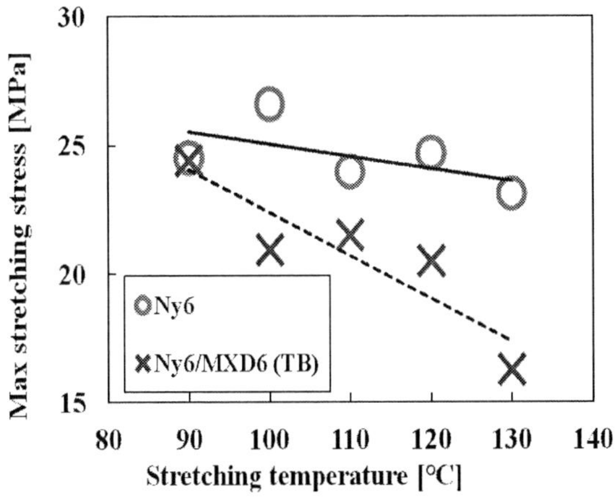

Fig.9 Relationships between maximum stress and stretching temperature

It was found that both PA6 and PA6/MXD6 blend decrease with increasing stretching temperature. PA6/MXD6 blend had much stronger temperature dependence than PA6. From this result, it was considered that the stretching temperature window of PA6/MXD6 was narrower than the one of PA6.

3.2.2 Effect of Stretching Ratio

Fig.10 shows the stretching stress and stretching ratio of PA6/MXD6 blends in order to evaluate the stretchability by changing stretching ratio of MD and TD. The degree of orientation in the plane of the film for PA6 and PA6/MXD6, which was calculated by three dimensional refractive indexes shown below, was plotted as a function of stretching ratio shown in Fig.11.

The degree of orientation in the plane of the film = [(Nx + Ny) / 2] - Nx

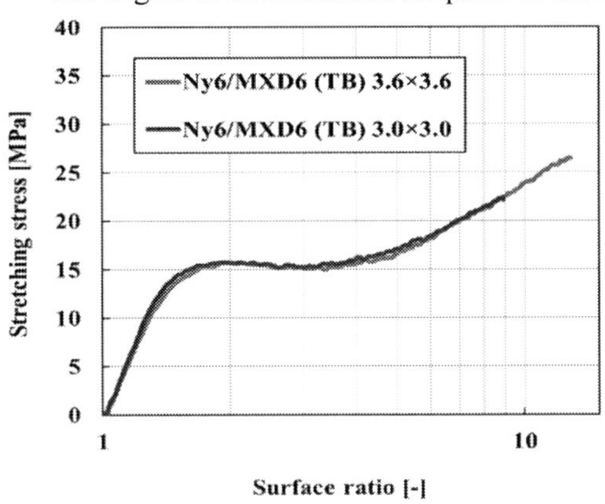

Fig.10 Stress-surface ratio curves of PA6/MXD6 blends at several different surface ratio

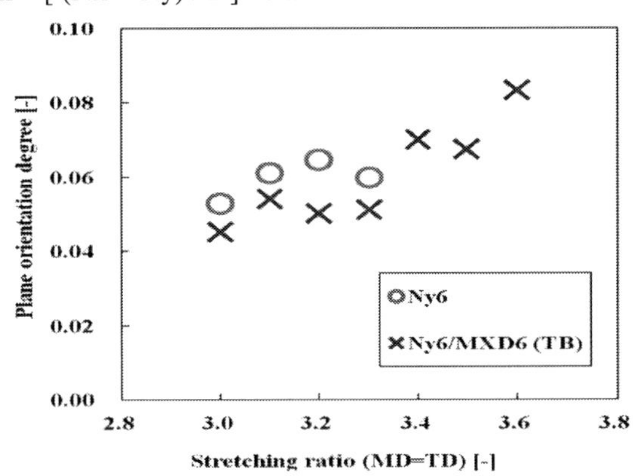

Fig.11 Degree of orientation-stretching ratio curves of PA6 and PA6/MXD6 blends at several different stretching ratios

Fig.10 shows the stretching stress increases with increasing the stretching ratio. Fig.11 shows the degree of orientation in the film plane of MD and TD for both samples increase with increasing the stretching ratio.

Table 2 shows the summary of the maximum stretching ratio, the maximum stretching stress and the degree of orientation. It shows the blending of MXD6 to PA6 reduces stretching stress, compared with PA6. So the blend can increase the stretching stress by increasing higher stretching ratio than PA6 without film break during the biaxial stretching.

Table 2 Maximum stretching ratio and maximum stress of PA6, PA6/MXD6 blends

	Maximum stretching ratio (-)	Maximum stress (MPa)	Degree of orientation (-)
PA6	3.3	29.0	0.060
PA6/MXD6	3.6	26.5	0.083

As the blending of MXD6 to PA6 prevents the hydrogen bond of PA6 from forming, the stretching stress is reduced. It was observed that the stretching stress between 25~30MPa causes film break during the biaxial stretching. PA6/MXD6 blend can be stretched at a higher stretching ratio without film break than PA6. From these results, PA6/MXD6 blend can produce the film having high orientation and high physical properties by increasing stretching ratio.

3.3 Film Uniformity

The retardation was measured and then the standard deviation of retardation was calculated to evaluate the film thickness uniformity. The result is shown in Table 3. It was found that the standard deviation of the retardation is closely related to the film thickness uniformity.

Table 3 shows PA6 has better uniformity than PA6/MXD6 at the stretching ratio 3 as PA6 has lower standard deviation than PA6/MXD6. This result comes from the stress build up ratio (maximum stretching stress / stretching stress at the yield point) after the yield stress passes. The higher stress build up of the stress and strain shown in Fig.12 improves thickness uniformity.

Table 3 Stretching ratio and standard deviation of retardation of PA6, PA6/MXD6 blends

	Stretching ratio (-)	Standard deviation of retardation (-)
PA6	3.0	7.6
PA6	3.3	6.1
PA6/MXD6	3.0	8.3
PA6/MXD6	3.6	5.8

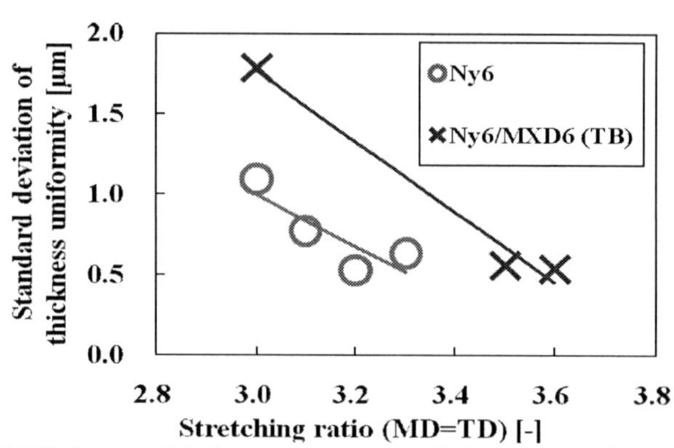

Fig.12 Relationship between standard deviation of thickness uniformity and surface ratio

In other words, as higher stress build up means the thicker film part is stretched much more than the thinner film part, PA6 which has higher stress build up and better thickness uniformity was produced, compared with PA6/MXD6.

Furthermore, the standard deviations of both samples decrease with increasing the stretching ratio and are almost the same at the maximum stretching ratio. This means that the film thickness uniformity of PA6/MXD6 is almost the same level as PA6, because as PA6/MXD6 has a lower stretching stress and can be set at a higher maximum stretching ratio.

4. CONCLUSION

1) Three different stretching methods, which can produce polyamide 6 films, namely the simultaneous biaxial stretching, the sequential biaxial stretching, and the double bubble tubular stretching, were investigated.

① Sequential Biaxial Stretching Process

Both the stretched part and the unstretched part were existed owing to the hydrogen bond during TD stretching. The stretched part shows high TD orientation. The unstretched part shows low TD orientation and high MD orientation, because of very low TD stretching effect. The ununiform part causes retardation distribution and the film balance to worsen

② Simultaneous biaxial stretching process

The polymer chain was oriented in MD or TD and almost the same orientation in MD and TD, so the orientation balance is good

③ Double bubble tubular process

Uniform stretching stress and orientation in every direction and good balanced film in every direction were obtained.

2) Low stretching temperature, slow stretching speed and high stretching ratio are important to obtain highly oriented film with high physical properties.

3) Blending of MXD6 to PA6 gives high gas barrier, easy tear properties and low stretching stress, but poor thickness uniformity. To increase stretching ratio increases stress build up and improves film thickness uniformity and produces high orientation film.

REFERENCES

1. Takashige, M., Kanai, T., Int. Polym. Process. (1990) 5, p. 287
2. Takashige, M., Kanai, T., Int. Polym. Process. (2005) 20, p. 100
3. Takashige, M., Kanai, T., Int. Polym. Process. (2006) 21, p. 86
4. Takashige, M., Kanai, T., J. Polym. Eng. (2008) 28, p. 179
5. Unitika, Japan Patent 44-2598
6. Toyobo, Japan Patent 48-80672
7. Takashige, M., Kanai, T., Yamada, T., Int. Polym. Process. (2003) 18, p. 368
8. Kang, H. J., White, J. L., Polym. Eng. Sci. (1990) 30, p. 1228
9. Kang, H. J., White, J. L., Cakmak, M., Int. Polym. Process. (1990) 1, p. 62
10. Takashige, M., Kanai, T., Yamada, T., Int. Polym. Process. (2004) 19, p. 147
11. Kanai, T., Film Processing Advances, Kanai, T.,Campbell,G.(Eds.) (2014) Hanser, Munich

The effect of Nano calcium carbonate on mechanical properties, crystallinity and Foam ability of PLA/Starch/ Epoxidized Soybean Oil (ESO) blends

Javadi Azizeh [(1)]*, *Vaziri Nasab Elham* [(1)], *Garmabi Hamid* [(1)]

[(1)] Amirkabir university of technology - Tehran - Iran

The blends of PLA/Starch containing Epoxidized Soybean Oil (ESO) as a compatibilizer, were made using an internal mixer, while the compositions of Starch and ESO were kept at 12.5 wt.%. Then 1, 4 and 7 wt.% of Nano Calcium Carbonate (NCC) were added to the mentioned blend respectively. The mechanical and thermal properties, as well as crystallinity, morphology and foamability of neat PLA, its blend with starch and nanocomposites were studied. It was found that ESO is a good compatibilizer for PLA/Starch blend and the elongation at break, also the impact strength of neat blend and nanocomposites are improved by using ESO. Also the results showed that using NCC leads to improvement in mechanical properties like tensile strength and modulus of the samples. In the other hand, NCC can play the role of nucleating agent and increase the crystallinity of PLA. Finally the neat blend and nanocomposites were foamed in an autoclave using high pressure Nitrogen gas. It was shown that the PLA/Starch/ESO/NCC foams have smaller cell size and higher cell density in comparison with neat blend. This is due to NCC, which acts as a nucleating agent for foaming process. So the heterogeneous nucleation via NCC, is happened and the homogeneous morphology and cell size are achieved. Also it was found that the cell density in nanocomposite samples is much smaller than neat blend. For example, in nanocomosite foam sample containing 4 wt.% NCC, the average cell size is decreased from 115 to 30 ìm.

Structural properties of high and ultrahigh strength polyacrylonitrile-based, ultrahigh modulus mesophase pitch-based and high ductility pitch-based carbon fibers studied by Raman spectroscopy, WAX and SAX techniques

Al Aiti Muhannad (1)*, Göbel Michael (1), Jehnichen Dieter (1), Fischer Dieter (1), Brünig Harald (1), Heinrich Gert (1)

(1) Leibniz Institute of Polymer Research Dresden - Sachsen - Germany

The structural properties of carbon fibers with different tensile modulus have been investigated. Carbon fibers used in this study are: 1- high ductility impregnated pitch based carbon fibers with the tensile modulus 110 and 155 GPa, 2- standard tensile modulus (high tenacity) polyacrylonitrile-based carbon fibers with the tensile modulus 240 GPa and 265 GPa, 3- intermediate tensile modulus (290 GPa), 4- high tensile modulus (390 GPa; 425GPa) and 5- ultrahigh tensile modulus (520 GPa; 630 GPa;720 GPa; 880 GPa; 923 GPa). Utilizing the Raman spectroscopy, the characteristic ID/IG ratio and the existence of the G'-band have been analyzed. Via the wide-angle x-Ray diffraction patterns, the crystal parameters, crystal thickness (Lc, La||, La), the crystal aspect ratio (La/Lc) and the apparent crystallinity, the graphitizing degree and the density have been characterized. The orientation factor of the crystals has been calculated for the different fibers from the azimuthal distributions of the respective (002) reflex. Conclusions regarding correlation of the Raman spectroscopy characteristics to the aspect ratio (La/Lc) and the orientation degree of the crystals have been drawn. Additionally, the nature of the pore structure of the carbon and graphite fibers has been studied utilizing the ESAX, SAX and MSAX techniques. The pore size and shape have been correlated to the Raman aspect ratio ID/IG.

Recycling of Composites – A New Approach Minimizes Downgrading

Jochen Wellekötter[1, a)] Stephan Baz[2], Johannes Schwingel[3], Götz Theo Gresser[2], Peter Middendorf[3] and Christian Bonten[1]

[1] IKT, Institut für Kunststofftechnik der Universität Stuttgart, Pfaffenwaldring 32, 70569 Stuttgart
Tel.: + 49 (0) 711 685 62864, Fax: + 49 (0) 711 685 62066.
[2] DITF, Deutsche Institute für Textil- und Faserforschung Denkendorf, Körschtalstraße 26, 73770 Denkendorf
Tel.: + 49 (0) 711 934 00, Fax: + 49 (0) 711 934 0297
[3] IFB, Institut für Flugzeugbau der Universität Stuttgart, Pfaffenwaldring 31, 70569 Stuttgart
Tel.: +49 (0) 711 685 62402, Fax: +49 (0) 711 685 62449
[a)]Corresponding author: jochen.wellekoetter@ikt.uni-stuttgart.de

Abstract. Fiber reinforced plastics (FRP) hold considerable potential for the construction and design of light-weight parts. FRP are rapidly gaining in importance in various industrial sectors. Although this group of materials combines high strength and stiffness while simultaneously being lightweight, major difficulties still arise when it comes to high volume production and the return of manufactured parts back into the cycle of materials at the end of their lifecycle. A new approach follows the idea of producing structural parts through the forming and back-injection molding of preforms with a large portion of recycled fibers and recycled matrix, without drastically compromising the mechanical properties. To achieve this, carbon fibers retrieved out of a recycling process are blended with recycled polyamide fibers and spun into a hybrid yarn. Subsequently these yarns are processed into preforms by means of Tailored Fiber Placement. This process allows for a precise fiber positioning in accordance with the load path. Additionally, back-injection molding allows functional integration into the parts, aiming for weight reduction and lowering installation effort. Recycled carbon fibers and recycled matrix granules are also utilized for the back-injection molding process using an Injection Moulding Compounder to investigate their influence on mechanical properties of the parts.

INTRODUCTION

When designing lightweight parts, fiber-reinforced-plastics (FRP) have proven to demonstrate outstanding potential due to their exceptional mechanical properties [1–4]. This is particularly important when trying to preserve our environment because of the possibility to reduce fuel consumption through light-weight construction in the transportation sector. The e-mobility sector is also in desperate need of light-weight solutions [5,6].

Although FRP seemingly display exclusively positive characteristics, their utilization still comes at a price. Especially issues like reutilization, reparation and recycling have not yet been thoroughly examined. This particularly affects FRP with thermoset matrices. [2,7]

The process of recycling FRP normally comes with the undesirable side-effect of "downgrading", meaning that the properties of the material gradually worsen with each recycling cycle [1,2,7], The reinforcing fibers cannot be reused in traditional fiber composite technologies, but are rather reutilized as additives in different plastic forming processes or in textile form with no orientation, e.g. as a non-woven. [7–9]

PRESENT RECYCLING CONCEPTS FOR FRP

Many studies (e.g. [7,10]) suggest that there will be a significant increase in demand for FRP in the near future. This is why it is of utmost importance to face the challenges of recycling of such materials early on. In general, FRP cannot simply be reused after their life cycle. The difficulty lies therein, that waste cannot easily be separated into „good“ and „bad“ parts, since structural damages like delaminations are not necessarily visible from the surface. Thermal recycling, i. e. combustion, and thereby recovery of stored energy is obviously possible, albeit extremely unsatisfactory from an ecological and economical standpoint [7,9]. The reuse of parts made of thermoplastic matrices or SMC as filling material in lesser strained plastic parts is well established, though.

Any other recycling process requires a high degree of purity of fibers and matrices. Currently, pyrolysis (combustion under exclusion of oxygen) is already utilized industrially to recover carbon fibers. In doing so, the matrix material is completely burned and energetically recovered. The carbon fibers stay mostly intact during the pyrolysis and keep their properties, although, due to being processed before, with shortened fiber-lengths. Possible alternative processing solutions are the solvolysis or the hydrolysis, among others [2,7,9].

Since the fibers are no longer of infinite length, but rather of varying shortness depending on the parts they were built into, most of the conventional textile processing methods cannot be applied anymore. Further processing is limited to being reused as an additive in an injection molding process or as fibers for non-woven fabrics [8].

A NEW APPROACH TO RECYCLING OF FRP

All of the recycling concepts presented here have in common that they result in a downgrading of the material, which effectively means that its properties after being processed are worse when compared to a respective new part. This is primarily caused by the significant reduction of fiber length during the process. Thomason [11] already proved in 2002, that different fiber lengths greatly impact the mechanical properties. Fibers of at least 10 mm can result in noticeable improvement of properties like stiffness, strength and impact resistance. (Figure 1). This demonstrates how necessary the preservation of fiber length during recycling is.

At the same time, due to the anisotropic properties of FRP, it is advantageous to optimize fiber orientation according to the load path of the part to further help in saving resources [1–4,12]. Especially the latter can prove to be difficult to implement when utilizing non-woven fabrics due to their isotropic structure.

Another possibility of improving resource efficiency presents itself when adding functional integration by adding force initiation elements and bars to the part. [1–4,12]

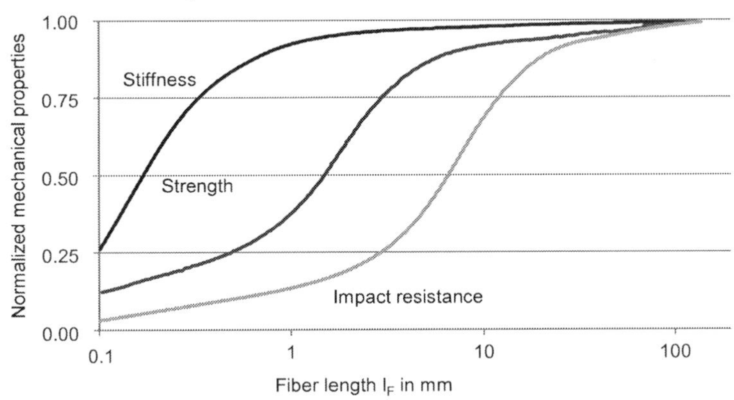

FIGURE 1. Standardized properties depending on different fiber-lengths of FRP [11]

Preservation of fiber length, optimization of fiber orientation and functional integration are being implemented in a new recycling concept by the Institut für Kunststofftechnik (IKT) of the University of Stuttgart, the Institut für Flugzeugbau (IFB) of the University of Stuttgart and the Deutsche Institute für Textil- und Faserforschung, Denkendorf, to thus receive drastically improved recycled parts. (Figure 2).

Recycled Staple Fiber Hybrid Yarn

After recycling the carbon fibers through the before described process, the pure carbon fiber flocks can be processed via textile processing. Compared to common processing of recycled carbon fibers to poor oriented semifinished products (e.g. non-woven), the shown concept utilizes the alignment of the fibers without radical fiber length reduction and transforms them into a hybrid yarn. For this purpose the carbon and matrix fibers are opened and blended by means of a textile opener from the Dilo Group, Eberbach, Germany. The used matrix fibers are spun out of recycled polyamide (rPA) and afterwards texturized and cut to staple fibers (Figure 3).

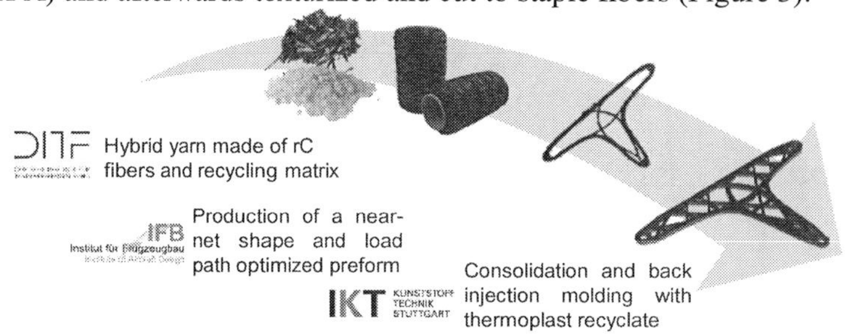

FIGURE 2. The Recycling Concept by IKT, IFB and DITF

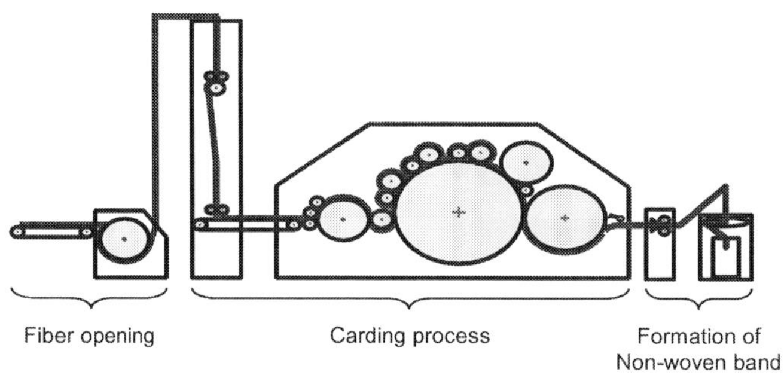

FIGURE 3. Carding Process

Central topic of the yarn production is the sliver formation by means of a card from the Maschinenfabrik Memmingen, Germany. The pre-opened and blended carbon and matrix fibers are fed into the card with a defined area weight by a feeding chute. The card performs the opening of the fiber flocks to single fibers, the fiber blend and the fiber orientation. At the delivery of the card the formed card web is merged into a sliver.

During the subsequent spinning process at a spiral covering spinning machine, the sliver is aligned in process direction and drawn to the final yarn count. After the draft, the thin sliver is entwined by an rPA filament in the spinning zone. The resulting staple fiber hybrid yarn out of rCF is shown in figure 4.

FIGURE 4. Recycling staple fiber hybrid yarn

Load Path Optimized Design Through Tailored Fiber Placement

Tailored fiber placement has proven to be an efficient method when producing hybrid preforms. It is beneficial that the fibers can be placed in any desired orientation since extremely small radii can be realized. Fiber placement strategies are optimized in order to realize parts that are completely load path compliant. This minimizes material waste. The explicit details of the optimization are not content of this paper, though. The produced samples utilized for research are manufactured on a 4-headed stitching machine by the Tajima Group, Japan.

Functional Integration and Consolidation Through Back-injection Molding

Technically, the produced hybrid preforms are processable into finished parts by a heated press. The preforms are put into a simple mold and consolidated under pressure and temperature. Injection molding offers high amounts of freedom of design. Hence, significantly more complex parts with functional integration can be obtained by means of injection molding instead of just consolidating the preforms. To achieve this, the hybrid preform is inserted into the injection molding tool of an injection molding machine Arburg Allrounder 520S 1600-400 by the Arburg GmbH & Co KG, Loßburg, Germany. In a back injection molding process, the added matrix consists of the same recycled thermoplastic material as the matrix fibers.

Recycling of Polyamide

To guarantee recycling material of consistent purity, raw polyamide (PA6: Ultramid B3S and PA66: Ultramid A3W by BASF SE, Ludwigshafen) is recycled. Firstly, tension rods of type 1A according to DIN EN ISO 527-2 are manufactured on the aforementioned injection molding machine. These are then shredded to ground up material in a cutting mill of type FZK 280 by the Fellner & Ziegler GmbH, Frankfurt am Main. This process is repeatedly conducted up to six times. The recycled material is then mixed with a percentage of raw material and processed into new granules in a twin-screw extruder of type ZSK 25 by the Coperion GmbH, Stuttgart. To increase the mechanical properties of the recycled material, carbon and glass fibers are compounded into the melt during the process.

RESULTS OF CURRENT RESEARCH

The hybrid yarns produced are made out of carbon fibers that are extracted from production waste by ELG, Coseley, England and rPA6-fibers (Ultramid B3S by BASF SE, recycled at the IKT). The amount of carbon fibers in the hybrid yarn amounts to 58 vol.-% while featuring a yarn fineness of 800 tex.

The average fiber length of the carbon fibers is about 22 mm. (The entire process results in a shortening of the fibers by approximately 50 %) Through the spinning process, all of the fibers are nearly parallel inside the core of the yarn (Figure 5). The tensile strength of 130 N of the yarn is sufficient for further processing in the TFP-process.

FIGURE 5. Recycling staple fiber hybrid yarn; detailed view

Characterization of processing and application properties

Alongside the textile and technical characterization of the yarn, its processing and application properties for the TFP are also investigated. Laminar embroidery designs are stitched and pressed into plates on a plate press by the Maschinenfabrik Lauffer GmbH & Co. KG, Horb am Neckar. Out of these plates, tension rods according to DIN EN ISO 527-5 in angles of 0°, 45° and 90° to the fiber orientation are milled. These tension rods exhibited an average fiber volume content of 47.7 % (relative standard deviation: 0.01 %). Table 1 shows the resulting mechanical properties. The obtained values are lower than the maximum possible values of over 200 GPa for FRP with carbon fibers but are still significantly above the usual values for long fiber reinforced thermoplastics as well as common aluminum alloys (approx. 75 GPa)[1].

Consequences of Recycling the Matrix

The recycling steps and their effect on material degradation of the matrix are mechanically and rheologically examined. Figures 6 and 7 show that the viscosity, the elastic modulus and the tensile strength remain almost constant even when going through several recycling iterations. These results can be confirmed with the compounds made out of recycled and raw material. Obviously, the addition of glass fibers greatly increases the mechanical properties. The mixing ratio of recycled and raw material as well as the amount of glass fibers are varied for further analysis. Figure 8 show that the mechanical properties for every analyzed material combination only marginally decrease depending on the ratio of fibers. Apart from the elasticity modulus and the strength, the impact strength and flexural modulus are also examined. Since, however, these show similar behavior they will not be further discussed in this paper.

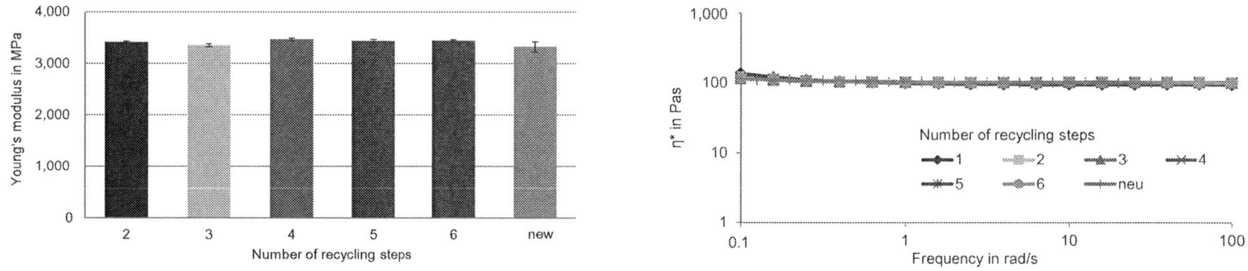

FIGURE 6. Dependency of the Young's Modulus (a) and the viscosity (b) on recycling steps

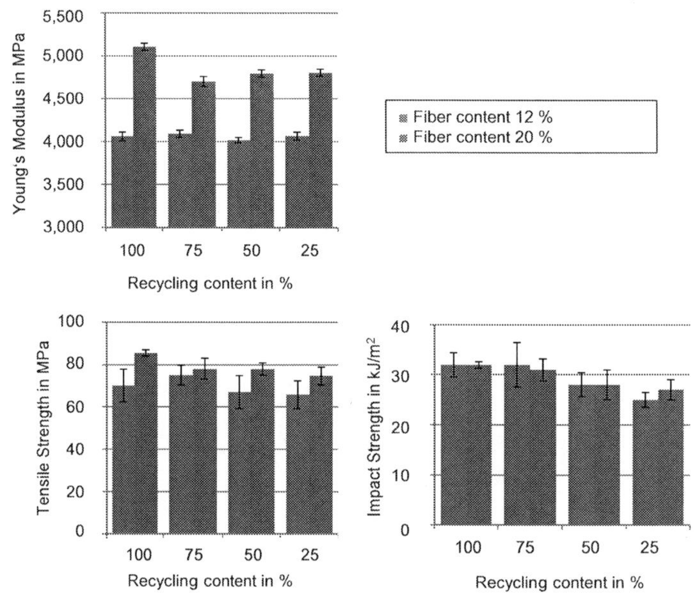

FIGURE 8. Young's Modulus, Tensile strength and Impact strength in dependency of recycling and fiber content

CONCLUSION AND OUTLOOK

Recycling will play a key role in the advance of FRP. A new and promising approach to recycling is the processing of recycled carbon fibers into a near-endless staple fiber hybrid yarn, the placement to structurally optimized and load path conform preforms by means of TFP and, finally, the back-injection molding in a subsequent consolidation step facilitates functional integration. First results indicate a promising concept in which the carbon fibers can be optimally oriented inside the parts while maintaining high fiber lengths. The matrix material can also be made out of recycled materials and its properties can be improved through the compounding of fibers. For this purpose a one-step processing chain is particularly suited due to its characteristic of carefully working the fibers into the material.

Until now, these promising results could not be tested under real-life conditions. Hence, currently a demonstrator to showcase the technology is being developed. This demonstrator will be examined in regards to its mechanical properties, especially compared to alternative recycling processes. A comparison to new parts made of non-recycled materials will also yield information about the performance of the materials. In order to save resources during manufacturing, the entire production process will be designed to run fully automated, including the production of hybrid yarn, TFP and the placement into the tool for the back-injection molding process. Furthermore, additional research for the production of hybrid yarn, the optimization of the fiber structure inside the parts and the aging and recycling behavior of the plastic matrix is still pending.

ACKNOWLEDGMENTS

This research and development project is funded by the Baden-Württemberg Stiftung. Special thanks also go to the IFB and the ITV for the successful collaboration and to the BASF SE for providing test materials.

REFERENCES

1. H. Schürmann, *Konstruieren mit Faser-Kunststoff-Verbunden*.2nd ed. (Springer, Berlin, Heidelberg, 2005).
2. G. W. Ehrenstein, *Faserverbund-Kunststoffe. Werkstoffe - Verarbeitung - Eigenschaften*, 2nd ed. (Hanser, München, 2006).
3. C. Bonten, *Kunststofftechnik. Einführung und Grundlagen*, 2nd ed. (Hanser, München, 2016).
4. C. Cherif, *Textile Werkstoffe für den Leichtbau. Techniken - Verfahren - Materialien - Eigenschaften*, 1st ed. (Springer, Berlin, Heidelberg, 2011).
5. M. Müller and M. Stegelmann, "Elektrisierender Leichtbau. Chancen und Herausforderungen im ressourceneffizienten Leichtbau für die Elektromobilität," Kunststoffe **16** (2), 22–28 (2016).
6. H. E. Friedrich, *Leichtbau in der Fahrzeugtechnik*, 1st ed. (Springer Vieweg, Wiesbaden, 2013).
7. S. Pimenta and S. T. Pinho, "Recycling carbon fibre reinforced polymers for structural applications: Technology review and market outlook," Waste Management (31), 378–392 (2010).
8. M. Hofmann and B. Gulich, "CFK-Recycling. Verarbeitung von rezyklierten Carbonfasern für die Herstellung von Verbundbauteilen," Lightweight Design **2013**, 20–23 (2013).
9. G. Gardiner, "Closing the CFRP lifecycle loop. Recycled carbon fiber update," Composites World, 28–33 (2014).
10. *Composites-Marktbericht 2015*.
11. J. L. Thomason, "The influence of fibre length and concentration on the properties of glass fibre reinforced polypropylene. 5. Injection moulded long and short fibre PP," Composites Part A: Applied Science and Manufacturing (33), 1641–1652 (2002).
12. M. Neitzel, P. Mitschang, and U. Breuer, *Handbuch Verbundwerkstoffe. Werkstoffe, Verarbeitung, Anwendung*, 2nd ed. (Hanser, München, 2014).

Radiation-based pre-foaming of expandable polystyrene beads (EPS)

Fafara Michael (1)*, Goerl Johannes (1), Schreier Peter (1), Neumeyer Dr.-Ing. Thomas (1), Lucht Miriam (2), Vetter Joerg (2), Altstaedt Prof. Dr.-Ing. Volker (1)

(1) Neue Materialien Bayreuth GmbH - Bavaria - Germany, (2) FOX Velution GmbH - Bavaria - Germany

For decades the particle foam EPS (expanded polystyrene) contributes to many scopes of application, e.g. packing for fragile goods and groceries, wall or perimeter thermal and acoustic insulation etc. However, the steam-based production processes along the EPS processing chain has been no subject of major changes in the last decades. Blowing agent loaded polystyrene microgranules are still exposed to steam for a default pressure and time. This results in an expansion of the microgranules and leads to foamed EPS beads [1, 2]. In this work, a novel radiation-based pre-foaming process is presented. Compared to the conventional process described above this technology reveals several advantages. Steam-based pre-foaming, according to the state of the art, requires an extensive peripheral equipment, which is cost-intensive in its purchase and operation because consisting of the supply of natural gas, steam generation, steam storage and supply lines. However the novel radiation-based process only requires a usage of dedicated power supply. The interaction of the radiation with the blowing agent loaded polystyrene microgranules is investigated and conclusions on the resulting effects of the beads regarding density and morphology are drawn. In detail it can be presented that during the novel process a smaller cell structure inside the beads can be achieved compared to state of the art beads exhibiting the same density (30 g/l). Furthermore, investigations on the energy consumptions indicate that the energy saving potential is up to 30 % compared to the state of the art steam based process. The weldability of the radiation based EPS has been verified, too. Thus mechanical and thermal trials have been conducted to compare both processes. First results revealed that due to the smaller cell morphology the compressive strength could be increased by approx. 22 % and also the thermal conductivity was reduced by approx. 5 %. Reference: [1] Cube, H.Pohl, K. Die Technologie des schäumbaren Polystyrols. Madison : Hüthig, 1965. [2] Schips, C. Verfahren zur Herstellung von expandierbaren thermoplastischen Partikeln durch Nachimprägnierung. 2011

Near-net-shape Rigid Foam Cores For CFRP Sandwich Composites Made From Polyethylene terephthalate Using Thermoforming

Jan Schütte[1,a)], Stefan Müller[2,b)] and Klaus Drechsler[3,c)]

[1]*BMW Group, Technical Planning, Knorrstr. 147, 80788 München, Germany*
[2]*Sahlberg GmbH, Friedrich-Schüle-Str. 20, 85622 Feldkirchen, Germany*
[3]*Institute for Carbon Composites, Faculty of Mechanical Engineering, Technische Universität München, Boltzmannstr. 15, 85748 Garching b. München, Germany*

[a)]Corresponding author: jan.schuette@bmw.de
[b)]stefan.mueller@sahlberg.de
[c)]drechsler@lcc.mw.tum.de

Abstract. In recent years, carbon fibre reinforced components have been introduced into the automotive industry in a mass production scale. A promising approach for reducing production steps while increasing the complexity of carbon fibre reinforced car body parts is manufacturing sandwich components using wet moulding. Equipping these wet moulding parts with high performance sandwich cores fabricated by milling, as implemented in the aerospace industry, does not comply with automotive requirements in terms of cost-efficient high-volume production. Recent investigations show that is possible to use near-net-shape rigid foam cores made from polyethylene terephthalate using thermoforming as an alternative core. Expanded polyethylene terephthalate is utilized for cladding or insulating in the marine industry or for wind energy applications, where foam parts are manufactured by simple bending after a heat treatment. In order to fabricate near-net-shape sandwich cores, an advanced thermoforming procedure for expanded polyethylene terephthalate needs to be developed. This contribution aims at describing the impact of the thermomechanical behaviour of the foam material and the resulting steps of the advanced procedure. Initial trials suggest that the occurrence of two major effects during thermoforming and subsequential wet moulding can result in faulty cores or sandwich parts: Firstly, the gas pressure inside the foam cells, which tend to collapse under forming but do not fracture and, secondly, the entropy elastic behaviour of polyethylene terephthalate above the glass transition temperature. In combination with external forces, both can result in delamination in the sandwich part. The development of an advanced thermoforming procedure aims at circumventing the negative impact of both effects. After the initial forming, smaller parts of the rigid foam core are reheated shortly above the melting point in order to locally destroy the foam cell structure. This results in rigid foam cores that are suitable for the wet moulding of sandwich parts without causing delamination or other failures.

INTRODUCTION

A new approach for using liquid composite moulding (LCM) methods in the automotive industry has been presented recently: Using the wet moulding process to fabricate carbon fibre reinforced plastic (CFRP) sandwich car body parts [1]. Manufacturing sandwich components using wet moulding combines the benefits of a high function integration into complex parts with the advantages of robust compression moulding methods. The wet moulding process for sandwich car body parts could be implemented in a fully automated production cell where textile handling, resin application, part transfer, and demoulding is carried out by robots [1].

Sandwich parts can be defined as a composite material consisting of high-tensile and rigid top layers, and a shear rigid core [2]. While the top layers can consist of various fibrous materials or even metallic panels, this investigation focuses on the core material. There are various kinds of core types that are typically used in CFRP components: foam cores, textile or wire cores, fold cores or honeycomb cores. They are made from metals, natural materials, and plastics, including polyethylene terephthalate (PET) [3, 4, 5, 6]. PET foam has reasonable mechanical properties in comparison to high performance foams [7], while being commercially available at a much lower price. Due to the complexity of the car body parts and the weight requirements, plastic foams are used for their geometrical adaptability and low density. Furthermore, foam cores offer a homogeneous supporting effect for the top layers, in contrast to other only partially supporting cores such as honeycomb or wire cores.

Since the machining of foam blocks into complex shapes is time and cost consuming, and therefore not applicable for mass production of car body parts, another technology needs to be implemented for fabricating foam cores of CFRP sandwich components. The thermoforming of PET foam blocks was introduced as a possible solution in a recent study [1]. Thermoforming of plastics is widely used for sheet or film materials [8], whereas thermoforming of foams commonly occurs during the manufacturing of all thermoplastic sandwich systems [9, 10, 11]. Shaping of PET foam via thermoforming has been described before [12]. However, that investigation only deals with a simple bending motion. The present contribution aims at presenting a process for fabricating PET foam cores via thermoforming that are suitable for the integration into complex CFRP sandwich parts.

MATERIAL PROPERTIES & BEHAVIOUR

PET is a semi-crystalline polymer with the ability to recover from non-elastic deformations. The recovery time decreases with rising temperature, possibly down to a few minutes at 40-50 °C above the glass transition temperature [13]. Furthermore, closed-cell PET foam shows the characteristic behaviour of elastic-plastic foams under a compressive load [14]. The behaviour of elastic-plastic foams under a compressive load is described by Ashby: The stress-strain-curve shows an initial linear-elastic phase, followed by a nearly constant stress plateau up to a strain of $\epsilon \approx 0.65$ [15]. For higher strain rates, the stress-strain-curve rises exponentially up to the densification area. In 2007, Mills [16] described the load on a closed-cell foam, when applying a compressive force, with the following formula:

$$\sigma_T = \sigma_0 + p_0 \frac{\epsilon}{1 - \epsilon - R},$$

where σ_T is the total stress, σ_0 is the portion of stress due to the polymeric material, p_0 is the initial pressure inside the foam cells and $\epsilon/(1 - \epsilon - R)$ is the gas volumetric strain. The formula is only valid on the assumption that the Poisson's ratio of the foam is $\nu = 0$. Here, R represents the ration between the foam density ($\tilde{\rho}$) and the density of the solid primary material (ρ_s)): $R = \tilde{\rho}/\rho_s$. Furthermore, Mills combined his formula with the ideal gas law, which results in the following equation that enables the calculation of the internal gas pressure inside the foam pressure p:

$$p = p_0 \left(1 + \frac{\epsilon}{1 - \epsilon - R}\right),$$

It is necessary to remark that this second formula is only valid on the assumption that the polymer portion is incompressible.

Furthermore, since closed-cell elastic-plastic foams, such as PET, fail by buckling of the cell edges [17] and remain sealed even under larger deformation, the gas temperature inside the cells should change with increasing compression. There are opposing statements on this topic: Gibson focusses on isothermal compression [17], others favour a model based on adiabatic compression [18]. Mills investigates the topic in detail and presents the following differentiation: A compressive load on a foam block with a cell size > 1.0 mm results in a temperature increase inside the cells up to 140 °C. For a cell size < 0.1 mm, the internal temperature increases only up to 10 °C, since the emerging heat can dissipate via the polymeric cell walls [16]. The latter applies to most commercially available foams, including the batch of PET foam blocks used for this investigation. Therefore, an isothermal compression process is assumed for the following discussions. The probed PET foam blocks (Airex AG, Sins, Switzerland) have a median density of 100 kg m^{-3}, a compressive strength of 1.5 N mm^{-2}, and a compressive modulus of 100 N mm^{-2}.

RESULTS & DISCUSSION

Thermoforming - Basic process

The basic thermoforming process, which is used for the production of PET foam prototype parts, is known for a long time and has been used in mass production of sheets and films frequently [8]. The main process steps are displayed in Figure 1: After heating a raw foam block to 220 °C, typically inside a convection oven, the block is transferred to the mould, which is closed immediately. This constitutes the forming process step. After a subsequent cooling period of

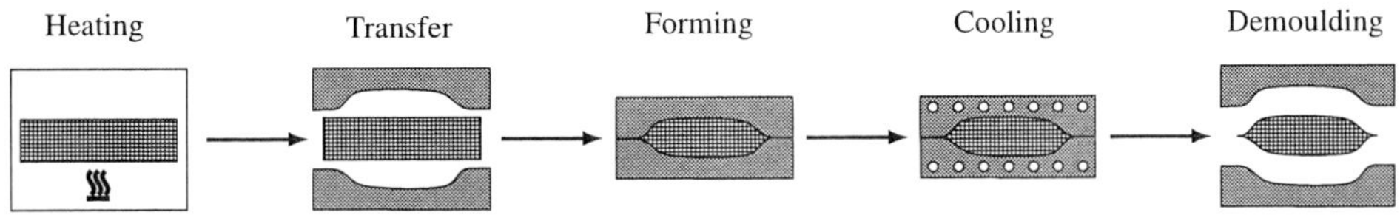

FIGURE 1. Process flow chart: basic thermoforming process for PET foam prototype parts.

180 s at 90 °C tool temperature, the shaped foam part is demoulded, finalising the thermoforming process. Afterwards, the shaped foam is used as core inside a CFRP sandwich component, which is fabricated via wet moulding in a second tool at 120–140 °C.

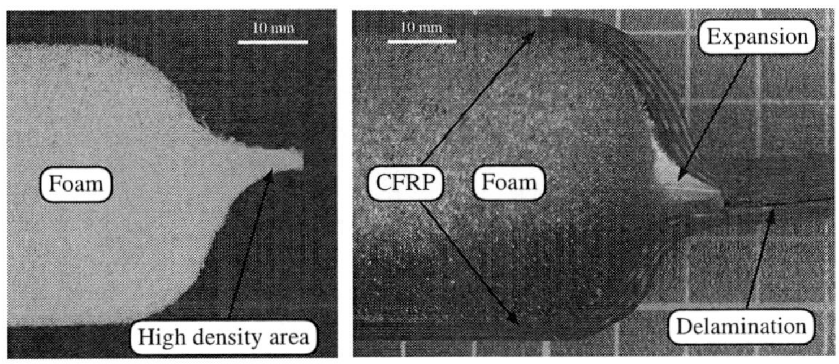

FIGURE 2. Section cut of a finished foam prototype part (Left) and section cut of a failed CFRP prototype part (Right).

Figure 2 shows the finished foam prototype part on the left. At the tip of the foam, a permanent high density area of compressed cells is formed. This high density area expands during the subsequent manufacture of a CFRP prototype part, clearly visible on the right side of the picture. Furthermore, the carbon fibre layers are separated from each other when the part is demoulded. This delamination could possibly be attributed to two main causes: Firstly, the gas pressure inside the condensed foam cells, and secondly, the entropy elastic behaviour of PET above its glass transition temperature. The temperature of the foam core rises during the manufacture due to the heated tool wall and the exothermal reaction of the resin [1], and thereby exceeding the glass transition temperature of the amorphous (67 °C) and the semi-crystalline phase (81 °C) of PET.

In order to determine the impact of the individual causes, the strain recovery of PET foam blocks is examined. Foam samples with an initial thickness of 40 mm are heated up to 220 °C and subsequently compressed, emulating the basic thermoforming process. The samples are divided into three groups and compressed to 32.5 mm (V1-3), 22.5 mm (V4-6), and 12.5 mm (V7-9), respectively. After a cooling down period, all samples are measured in thickness, and then reheated again in a convection oven at 150 °C for 15 min while free expansion being allowed. After a second complete cooling down period, the thickness is measured again to determine the recovery. The results displayed in Table 1 show that when reheated, thermoformed PET parts tend to recover disproportionally with increasing compression. The more the foam is initially condensed, the larger is the relative recovery (up to 24%).

TABLE 1. Strain recovery of PET foam samples with an initial thickness of 40 mm, after thermoforming at 220 °C and a cooling down period, and free expansion during reheating at 150 °C, ([14])

Thickness (40 mm ≙ 100%)	**V1-3** [%]	**V4-6** [%]	**V7-9** [%]
After thermoforming at 220 °C	83.33 ± 0.59	60.75 ± 0.54	34.58 ± 0.12
After reheating at 150 °C	88.58 ± 0.24	71.67 ± 2.36	58.75 ± 1.77
Recovery	≈ 5.3	≈ 10.9	≈ 24.2

The contribution of internal gas pressure and polymer behaviour needs to be quantified. Hence, 40 mm PET foam samples are compressed again, this time varying the foam temperature. Figure 3 shows typical stress-strain-curves for compression tests at 23 °C (Top) and 220 °C (Bottom). In both cases, total stresses are measured (black

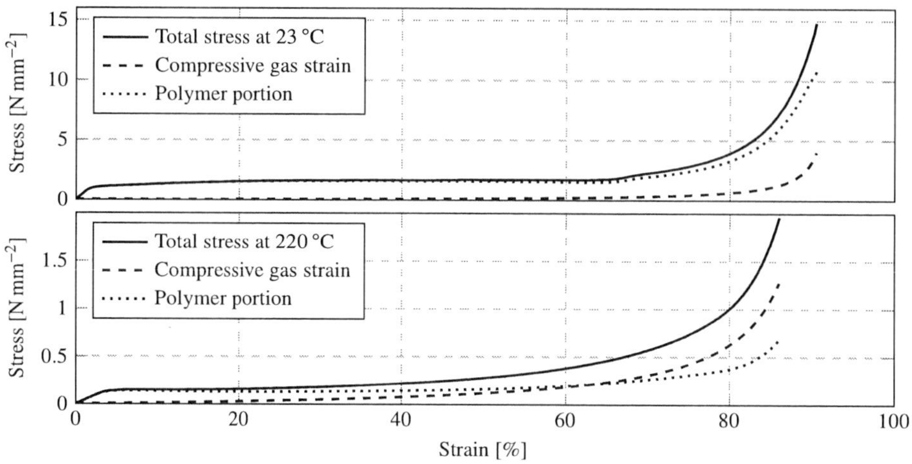

FIGURE 3. Stress strain curve for PET foam blocks at 23 °C (Top) and 220 °C (Bottom), ([14]).

line), then Mills' 2nd formula is used to calculate the contribution of the gas inside the foam cells (dashed line), which is subsequently subtracted from the total stress resulting in the contribution of the polymer (dotted line). At 23 °C the total stress is dominated by the polymer portion. Unsurprisingly, the total stress is significantly lower at 220 °C. However, the share of internal gas pressure increases remarkably. Since the foam cells remain undamaged with increasing compression, the disproportional recovery of the high density area (Figure 2) can be attributed to the rising share of the internal gas pressure at higher temperatures. Admittedly, the model seems to be inconclusive for very high compression rates. Gas has to leak from the cells eventually, when compressing the foam much as possible, i.e. at the tip of the prototype core. In these areas, a non-quantifiable combination of internal gas pressure and material behaviour is responsible for the expansion during reheating.

Thermoforming - Advanced process

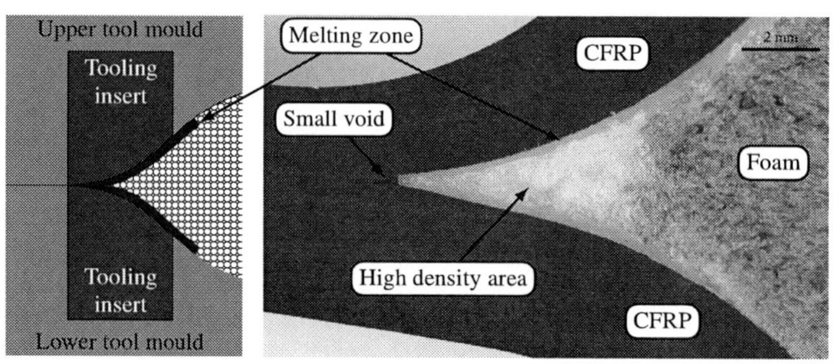

FIGURE 4. Advanced thermoforming with local heating (Left), CFRP sandwich part with a PET foam core manufactured via advanced thermoforming (Right).

A possible solution for the expansion failure of the CFRP sandwich part is the local destruction of the foam cell structure in high density areas in order to deplete any kind of internal tension. Therefore, the production method of the foam core needs to be adapted, i.e. the development of an advanced thermoforming process is necessary. The advanced method is almost identical to the basic one - with one major deviation: After the thermoforming mould is completely closed, the PET foam is locally reheated above its melting point (255 °C) by heatable tooling inserts inside the mould (Figure 4, left). After melting the foam locally at its tips, the outer face of the core re-solidifies during the subsequent cooling period. Again, the shaped foam fabricated with the advanced process is used as a core inside a CFRP sandwich part manufactured by wet moulding. After the fabrication of the sandwich part, the section cut on the right side of Figure 4 shows that solely a little void occurs at the tip of core, in contrast to the delamination in the previous tests (Figure 2). These voids typically do not exceed 1.0 mm in size. Furthermore, the difference between the foam, the high density area, and the melted and re-solidified outer face of the core is clearly distinguishable.

CONCLUSION

Thermoforming is a viable process for the fabrication of shaped PET foam cores. Using a thermoforming basic process, foam expansion and delamination of the textile layers occur in high density areas when fabricating a CFRP sandwich part including the basic core. Further examinations of the foam behaviour under compression load reveal that up to 60–70 % compression, the disproportional recovery of the foam can be attributed to the rising share of the gas pressure inside the foam cells. In high density areas, a non-quantifiable combination of internal gas pressure and material behaviour is responsible for the expansion during reheating. Using PET foam cores derived from the advanced thermoforming process – with local melting and re-solidifying of high density areas – results in CFRP sandwich parts with very small voids, but no delamination.

REFERENCES

[1] J. Schütte, J. Hoffmann, E. Ladstätter and K. Drechsler, "Manufacturing CFRP sandwich parts using wet moulding," in *Proceedings of the 17th European Conference on Composite Materials ECCM 17* (München, 2016).

[2] M. Flemming, *Faserverbundbauweisen: Halbzeuge und Bauweisen*, 1st ed. (Springer, Berlin, Heidelberg, New York, 1996).

[3] R. Juntikka and S. Hallström, Cellular polymers **23**, 263–297 (2004).

[4] M. A. Roth, "Strukturelles Nähen: ein Verfahren zur Armierung von Krafteinleitungen für Sandwich-Strukturen aus Faser-Kunststoff-Verbund," Ph.D. thesis, Technische Universität Kaiserslautern 2005.

[5] Y. Klett and K. Drechsler, "Design of multifunctional folded core structures for aerospace sandwich applications," in *1st CEAS European Air and Space Conference* (Berlin, 2007).

[6] D. Lutz, "Anwendungen und Innovationen bei Kernwerkstoffen für den Leicht- und Sandwichbau," in *Internationale AVK-Tagung* (Stuttgart, 2011).

[7] G. Kopp, "Auslegung und Dimensonierung von grossflächigen polyurethanbasierten Sandwichbauteilen unter Berücksichtigung von konzeptionellen und fertigungstechnischen Einflüssen," Ph.D. thesis, Universität Stuttgart 2015.

[8] C. Hopmann and W. Michaeli, *Einführung in die Kunststoffverarbeitung*, 7th ed. (Carl Hanser Verlag, München, 2015).

[9] U. Breuer, M. Ostgathe, and M. Neitzel, Polymer Composites **19**, 275–279 (1998).

[10] O. Rozant, P.-E. Bourban, and J.-A. E. Månson, Composites Part A: Applied Science and Manufacturing **32**, 1593–1601 (2001).

[11] J. Grünewald, P. Parlevliet, and V. Altstädt, Journal of Thermoplastic Composite Materials **30**, 437–464 (2016).

[12] L. Sorrentino, M. Aurilia, and T. Iannace, "Effects of the thermoforming on the cell morphology of a thermoplastic core," in *17th International Conference of Composite Materials* (Edinburgh, 2009).

[13] A. Pegoretti, A. Guardini, C. Migliaresi, and T. Ricco, Polymer **41**, 1857–1864 (2000).

[14] S. Müller, "Prozessentwicklung: Thermoformen von Hartschäumen," Master thesis, Technische Universität München 2016.

[15] M. F. Ashby, "Cellular solids - scaling of properties," in *Cellular Ceramics*, edited by M. Scheffler and P. Colombo (Wiley-VCH, Weinheim and Chichester, 2005), pp. 1–17.

[16] N. J. Mills, *Polymer foams handbook: Engineering and biomechanics applications and design guide*, 1st ed. (Oxford, 2007).

[17] L. J. Gibson and M. F. Ashby, *Cellular solids: Structure and properties*, 2nd ed. (Cambridge University Press, Cambridge, 1999).

[18] J. L. Throne and R. C. Progelhof, Journal of Cellular Plastics **20**, 437–442 (1984).

Open-celled foams based on diblock copolymers: Influence of melt elongational properties on foam morphology

Schulze Maria (1), Handge Ulrich A. (1)*, Abetz Volker (1)

(1) Helmholtz-Zentrum Geesthacht - 21502 Geesthacht - Germany

The preparation of polymer foams is strongly influenced by the rheological properties of the polymer matrix. Strain-hardening polymers are usually used for preparation of closed-cell foams. On the contrary, strain-softening in melt elongation favours the rupture of cell walls and consequently promotes the preparation of open-celled foams. In this work, we focus on diblock copolymers in the microphase-separated state. Our recent works have shown that diblock copolymers are associated with strain-softening in melt elongation and thus are ideally suited for fabrication of open-celled foams [1,2]. In this study, we elucidate the potential of a polystyrene-block-poly(4-vinylpyridine) and a polystyrene-block-poly(4-methylstyrene) diblock copolymer for preparing open-celled foams which were synthesized using the technique of sequential anionic polymerization. The properties of these foams are compared with foams of polystyrene and poly(4-methylstyrene). The technique of batch foaming using water and carbon dioxide as blowing agents is applied for foam preparation. In shear, the steady-state viscosity of the diblock copolymers of this study decreases with shear rate (structure-viscous behaviour). Our elongational experiments show that the diblock copolymers are associated with the phenomenon of strain-softening. The transient extensional viscosity in simple elongation agrees with the threefold of the shear viscosity. Our batch foaming experiments reveal that open-celled polymer foams using water and carbon dioxide as blowing agents can be prepared at appropriate processing conditions. On the contrary, the use of a strain-hardening polystyrene leads to a closed-celled foam. [1] M. Schulze, U.A. Handge, S. Rangou, J. Lillepärg, V. Abetz, Polymer 70, 88-99 (2015). [2] M. Schulze, U.A. Handge, V. Abetz, Polymer 108, 400-412 (2017).

Extrusion parameters optimisation for foaming of hydroxypropyl methylcellulose

Karlsson Kristina (1)*, Kádár Roland (1), Stading Mats (2), Rigdahl Mikael (1)

(1) Chalmers University of Technology - Gothenburg - Sweden, (2) Chalmers University of Technology and SP Food and Bioscience - Gothenburg - Sweden

Kristina Karlsson1, Roland Kádár1, Mats Stading1,2 and Mikael Rigdahl1 1Department of Materials and Manufacturing Technology, Chalmers University of Technology, SE-412 96, Gothenburg, Sweden 2SP Food and Bioscience, Box 5401, SE-402 29 Gothenburg, Sweden Foamed polymeric materials have many advantages such as being lightweight, strong with regard to its low weigh, versatile and relatively inexpensive. The majority of the foams are however produced from fossil-based resources. The two major drawbacks of such foams are their non-renewable nature and the difficult recycling process. Research on processing parameters of renewable candidates therefore comes of importance. Cellulose derivatives can be a promising material for foaming applications. They are in a sense lightweight, abundant, renewable and susceptible for modifications in order to be tailored for certain applications. Here, the processing parameters for laboratory-scale extrusion foaming of hydroxypropyl methylcellulose (HPMC) was investigated. Added water was here used as the blowing agent. Previous experiments were used to select suitable grades of HPMC for foaming, in terms of molecular weight, methoxy content and hydroxypropyl content. Initially, a capillary viscometer served the purpose of a batch extruder to study the selected aqueous HPMC-compounds. The foaming ability at different processing conditions was evaluated by this technique and assisted in the establishment of a general processing window in terms of flow rate, temperature and moisture content. The established processing window later served as a basis for the continuous extrusion foaming. The foaming ability was evaluated at different extrusion conditions and the produced HPMC-foams were characterized with regard to their apparent density and by scanning electron/optical microscopy. Key words: cellulose derivatives, extrusion, processing window

Continuous foam extrusion of high impact polystyrene (HIPS): Effects of processing parameters and blowing agent type and content

Emre Demirtaş[a,b], Hakan Özkan[b], Mohammadreza Nofar[a,*]

a. Department of Metallurgical and Materials Engineering, Chemical and Metallurgical Engineering Faculty, Istanbul Technical University, Maslak, 34469 Istanbul, Turkey
b. ARCELİK A.S. Central R&D Department. Materials Technologies, Tuzla 34950 Istanbul, Turkey

Abstract. This paper presents the optimization of a continuous foam extrusion process to manufacture high impact polystyrene (HIPS) foams through a twin-screw extruder. The effects of chemical blowing agent type and content as well as; process parameters: e.g., die temperature profile and screw speed (RPM) were investigated. After determination of the optimum die temperature profile, effect of chemical blowing agent type and content (i.e., 1, 3, 5 wt. %) and screw RPM (i.e., 100, 150, 200, and 250) were investigated. Results show that; the best void fraction (around 22%), with the highest cell density (around 6 x 10^5 cells/cm^3) obtained at die temperature profile of 150-145°C. It was also observed that 100 RPM, revealed the highest void fraction (around 32%) with cell density of around 3 x 10^5 cells/cm^3. The increase of chemical blowing agent, increased the void fraction from around 10% to 31% where the cell densities were also increased, respectively, from around 8 x 10^4 cells/cm^3 to 2 x 10^7 cells/cm^3.

Keywords. HIPS, Foaming, Chemical Blowing Agent, Twin Screw Extruder, RPM

INTRODUCTION

Polystyrene (PS) is an amorphous thermoplastic with a glass transition temperature around 100 °C and is widely being used in commodity applications due to its reasonable physical and mechanical features. [1]. Extruded PS foams are also rigid polymeric materials that could be produced in various densities according to its applications [2]. Park *et al.* [3-4] for the first time introduced the microcellular foaming of HIPS through extrusion foaming using CO_2 as physical blowing agent. In order to achieve desired foam expansion, the processing parameters are the most critical factors such as system pressure, die pressure, pressure-drop rate, die temperature, and blowing agent content. Furthermore, foaming strategies need to be identified and then applied in foaming process [5-6]. The effect of the injected gas content on cell density was also studied by Park and Suh [7]. According to these studies, when the injected gas content increased, the cell density also increased. When the injected gas amount was 10 wt%, the extruded HIPS foam structure was microcellular and this is the maximum soluble amount of CO_2. Up to this amount, all the injected gas will dissolve in the polymer. This dissolved gas amount also affects the cell density of the extruded HIPS foam. Extruded microcellular PS foams also shows the similar results [8]. The nucleation starts at the die section because of the thermodynamic instability and the pressure drop rate and to control of the die temperature is necessary. When the die temperature is too high, the gas loss via the hot skin layer of the extrudate is promoted due to the increased diffusivity of the blowing agent [9]. At higher die temperatures, the gas tends to escape through the polymer skin layer rather than to expand the foam; thus, it is very difficult to produce low density foams. Cell coalescence is also accelerated very seriously throughout the cell structure and the volume expansion decreases due to the enhance gas loss by coalesced cells at higher die temperatures [10]. However, if the die temperature is too low, the polymer melt becomes too stiff to be expanded by gas [9]. Moreover, increasing screw speed caused slow increment of the pressure in the die. According to Lee *et al.* [11] studies, with same die geometry, increasing die pressure increases the pressure drop and better void fraction and cell density are obtained with higher pressure drop rate. Nam *et al.* [12] also shows where void fraction increased slowly as they increased the screw speed in extrusion foaming of PP.

EXPERIMENTAL

1: Materials

High Impact Polystyrene (HIPS) was supplied from Versalis SPA. This HIPS shows the environmental stress cracking resistant (ESCR) behavior. The MFI for this material is 3.5 g/10min; the recommended melt temperature is 200 °C; and the specific gravity is 1.04 (g/cm³). The CO_2 based endothermic chemical blowing agents (CBA) used in these experiments were Hydrocerol produced by Clariant Inc. CBA 1 and 2 are different grades of Hydrocerol about decomposition temperatures. Both CBAs were used at the contents of 1, 3 and 5 wt%. Figure 1 and 2 show the DSC and TGA curves of these chemical blowing agents, respectively.

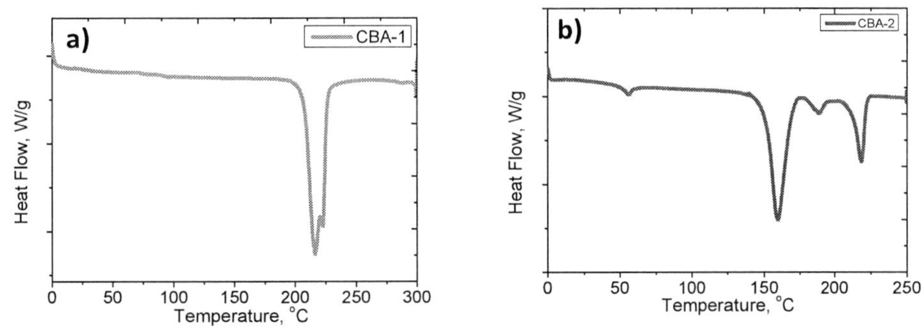

FIGURE 1. DSC Heating thermograms of the a) CBA-1 and b) CBA-2.

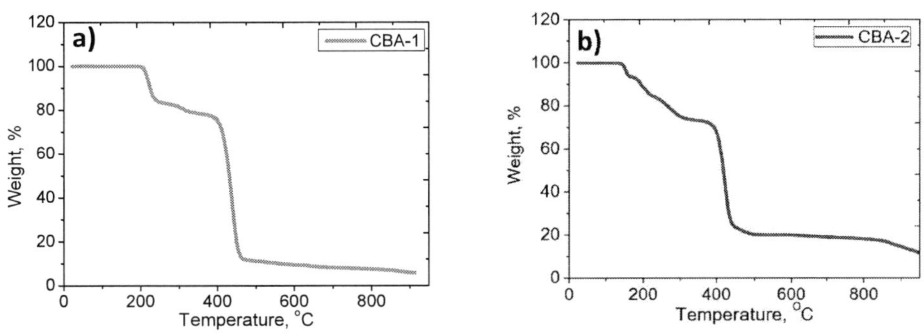

FIGURE 2. TGA curves of the a) CBA-1 and b) CBA-2.

2: Foaming Procedure and Characterization

Foaming was conducted in a twin screw extruder having 26 L/D ratio at a processing temperature range of 200-230° C. The filamentary die having 2 mm diameter was used in order to produce foamed filaments. The five different temperature of die was set as 130 to 180°C. After determination of the most proper die temperature for CBA – 1 and CBA – 2 used samples, the content of CBA (1,3, and 5 %wt.) was investigated. After the CBA content investigation, four different screw RPM (100, 150, 200 and 250) was set and the effect of RPM was observed on the foamed HIPS samples with different chemical blowing agents.

The foams densities were measured according to Archimedes principle using Mettler A1201 density measurement device based on ASTM D792-00. Void fraction values were then calculated using equation 1. Cell density and cell distributions of samples were observed using ZEISS scanning electron microscopy (SEM). Samples were monitored in the magnifications of 200X The cell density of samples were measured using equation 2. In these equations, n is the number of cells and A is the area of image. ρ_{foam} and $\rho_{polymer}$ is the density of the foam and unfoamed polymer.

$$\text{Void fraction} = 100\,.\,(1 - \frac{\rho_{foam}}{\rho_{polymer}}) \quad (1)$$

$$\text{Cell Density} = \left(\frac{n}{A}\right)^{3/2} . \frac{\rho_{polymer}}{\rho_{foam}} \quad (2)$$

RESULTS AND DISCUSSIONS

1: Effect of Die Temperature

Figure 3 shows the effect of die temperature on void fraction and cell density values of the foamed samples. It is clearly shown that, increasing the die temperature from 130 to 180, increased the void fraction and cell density for CBA – 2 used samples. On the other hand, the best void fraction (around 22%), with the highest cell density (around 6 x 10^5 cells/cm^3) obtained at die temperature profile of 150°for CBA – 1 used samples. Figure 4 also shows the SEM images of how various die temperatures influenced the cell morphology for foamed HIPS with different chemical blowing agents.

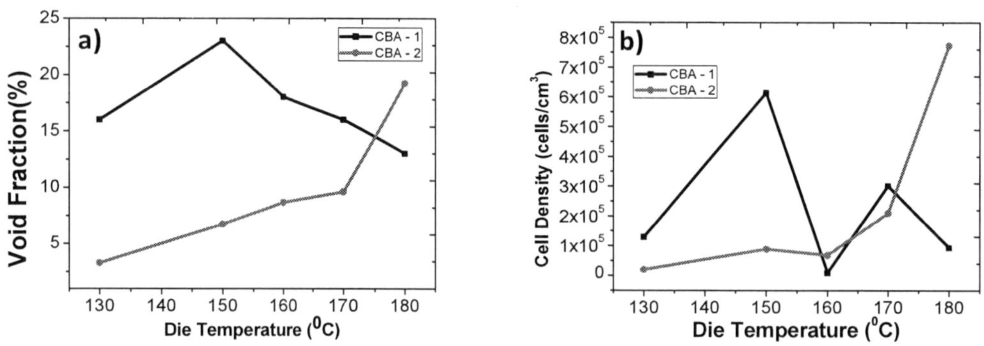

FIGURE 3. Effect of die temperature on (a) void fraction and (b) cell density

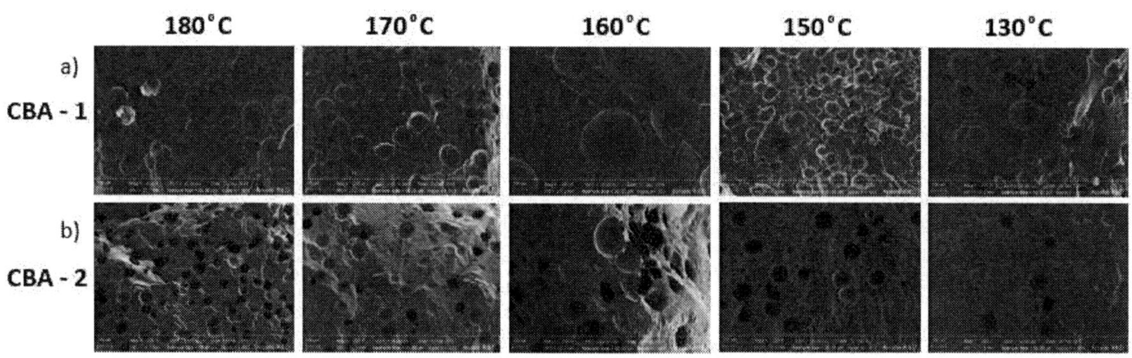

FIGURE 4. SEM images of the HIPS foams at different die temperatures for a) CBA-1 b) CBA-2 used samples.

2: Effect of CBA Content

Figure 5 shows the effect of chamical blowing agent content on void fraction and cell density values of the foamed samples. It is indicated that, increasing the void fraction from 1 to 5 wt %, increased the void fraction and cell density for both CBA – 1 and CBA – 2 used samples. The best void fraction (around 32%), with the highest cell density (around 2 x 10^7 cells/cm^3) obtained for 5 % wt. CBA – 1 used samples. On the other hand, the best void fraction (around 21%), with the highest cell density (around 8 x 10^5 cells/cm^3) obtained for 5 % wt. CBA – 2 used samples. It should be noted that, the cell density increased with increasing the chemical blowing agent content from 1 to 5%wt due to the increased the internal system pressure. Figure 6 also shows the SEM images of how increasing chemical blowing agent content influenced the cell morphology for foamed HIPS with different chemical blowing agents.

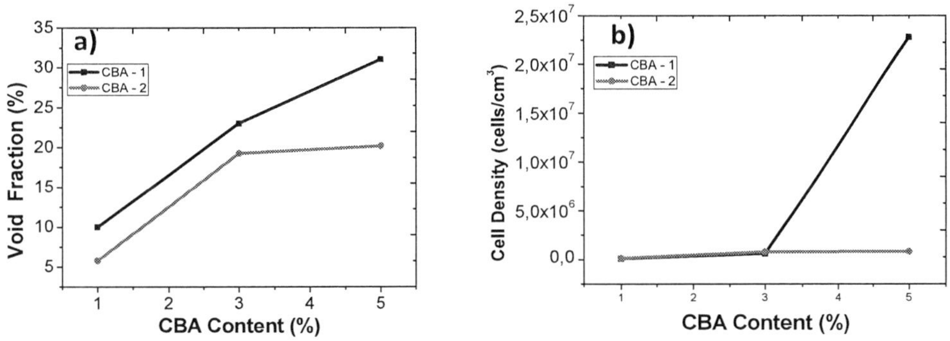

FIGURE 5. Effect of CBA content on (a) void fraction and (b) cell density

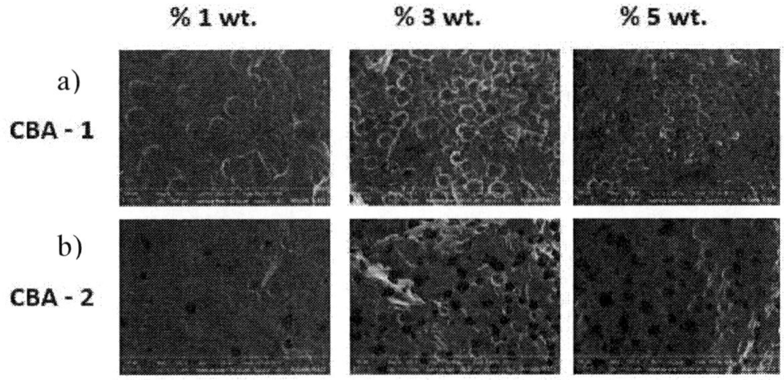

FIGURE 6. SEM images of the HIPS foams with various chemical blowing agent contents for a) CBA-1 b) CBA-2 used samples.

3: Effect of RPM

Figure 7 shows the effect of screw RPM on void fraction and cell density values of the foamed samples. It is seen that, the best void fraction (around 30 %) were obtained at the lowest screw RPM (100) for both CBA – 1 and CBA – 2 used samples. On the other hand, the highest cell density (around 2 x 10^7 cells/cm^3) was obtained at 150 RPM for CBA – 1, and 100 RPM for CBA – 2 used (around 10^6 cells/cm^3) samples. It should be noted that, the cell density and void fraction increased at the lower RPM values due to the better mixing and higher residual time. Figure 8 also shows the SEM images of how various die temperatures influenced the cell morphology for foamed HIPS with different chemical blowing agents.

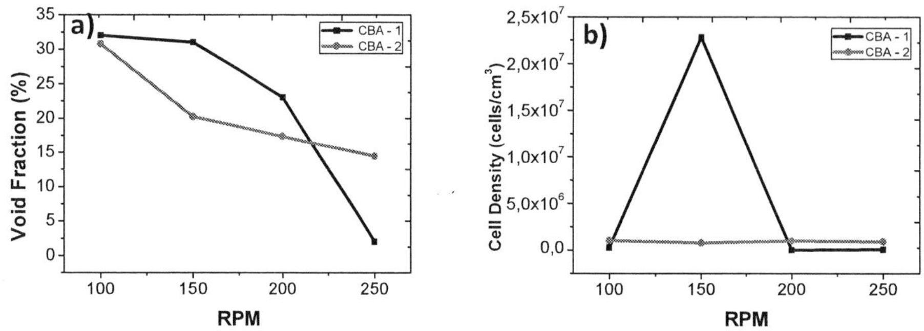

FIGURE 7. Effect of RPM on (a) void fraction and (b) cell density

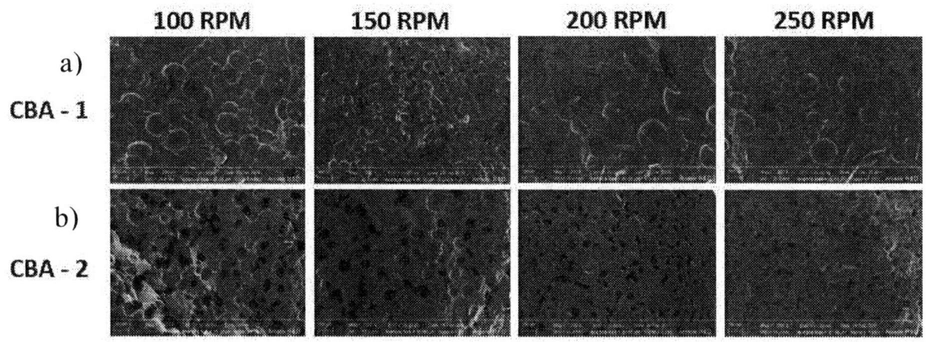

FIGURE 8. SEM images of the HIPS foams at different screw RPMs for a) CBA-1 b) CBA-2 used samples.

CONCLUSION

In this study, the extrusion foaming of HIPS through a twin-screw extruder was studied and the effects of process parameters: e.g., die temperature profile and screw speed (RPM); and chemical blowing agent type and content were investigated. According to results; the best void fraction (around 22%), with the highest cell density (around 6 x 10^5 cells/cm^3) obtained at die temperature profile of 150 °C; and 100 RPM, revealed the highest void fraction (around 32%) with cell density of around 3 x 10^5 cells/cm^3 for CBA -1 used samples. On the other hand, the best void fraction (around 20%), with the highest cell density (around 8 x 10^5 cells/cm^3) obtained at die temperature profile of 180 °C; and 100 RPM, revealed the highest void fraction (around 32%) with cell density of around 10^6 cells/cm^3 for CBA -2 used samples. The increase of chemical blowing agent from 1 to 5 % wt., increased the void fraction and the cell densities for both CBA -1 and CBA -2 used samples.

REFERENCES

1. S.T. Lee, C.B. Park, and N.S. Ramesh, *Polymeric Foams*, Taylor & Francis (2007)
2. K. W. Suh, Polystyrene and Structural Foam, in Handbook of Polymeric Foams and Foam Technology edited by D. Klempner and K. C. Frisch, Chapter 8, pp 151- 186, Hanser Publishers, (1991)
3. Park, C, B, Baldwin, D.F. and Suh, N.P., "Effect of the Pressure Drop Rate on Cell Nucleation in Continuous Processing of Microcellular Polymers", Polymer Engineering and Science, Vol. 35, No. 5, pp. 432-440, 1995.
4. Park, C, B, and Suh, N.P., "Filamentary Extrusion of Microcellular Polymers Using a Rapid Decompressive Element", Polymer Engineering and Science, Vol. 36, No. 1, pp. 34-48, 1996.
5. S.T. Lee, C.B. Park, Foam Extrusion Principles and Practise, chapter 14, pp. 465- 472, (2014)
6. M. Nofar, C.B. Park, Poly(lactic acid) foaming, Progress in Polymer Science, Vol. 39, No. 10, pp. 1721-1741, 2014

7. C. B. Park, PhD Thesis, Massachusetts Institute of Technology, Cambridge, MA, 1993.
8. M. Shimbo, K. Nishida, S. Nishikawa, T. Sueda and M. Eriguti, In: *Porous, Cellular and Microcellular Materials*, V. Kumar, ed., ASME, New York, 93, 1998.
9. Naguib, H.E., Park, C.B. and Reichelt, N., "Fundamental Foaming Mechanisms GoverningVolume Expansion of Extruded PP Foams," Journal of Applied Polymer Science, Vol. 91,No. 4, pp. 2661- 2668 (2004)
10. Lee J. W.S., Wang K., Park C.B., "Challenge to Manufacture of Low-Density Microcellular Polycarbonate Foams Using CO_2," Ind. Eng. Chem. Res. (2004)
11. J. Lee and C. Park, "Use of Nitrogen as a Blowing Agent for the Production of Fine-Celled High-Density HDPE Foams," *Macromolecular Materials & Engineering,* vol. 45, no. 23, pp. 1233-1244, 2006.
12. G. J. Nam, J. H. Yoo and J. W. Lee, "Effect of long-chain branches of polypropylene on rheological properties and foam-extrusion performances," *Journal of Applied Polymer Science,* vol. 96, no. 5, pp. 1793-1800, 2005.

The Evaluation of Biaxial Stretchability of Polypropylene Films Using a Newly Developed Test Machine

Kentaro EGOSHI [1,3,a)], Toshitaka KANAI [2,b)], Kazuhiro TAMURA [3,c)]

[1]*Eto Co., Ltd : 3403-1 Shake, Ebina city, Kanagawa, 243-0424 Japan*
[2]*KT Polymer : 2411-3 Kuranamidai, Sodegaura, Chiba, 299-0243, Japan*
[3]*Kanazawa University: Kakuma-machi, Kanazawa, Ishikawa, 920-1192 Japan*

[a)]k.egoshi.@eto.co.jp
[b)]toshitaka.kanai@ktplymer.com
[c)]tamura@se.kanazawa-u.ac.jp

Abstract. The evaluation method for a biaxially oriented film was developed using in-situ measurement during the stretching process. It could obtain basic data such as stress-strain curve, birefringence, light scattering, three-dimensional refractive indexes and birefringence distribution as functions of time, temperature and percentage in elongation, the information of superstructure could be obtained by measuring the birefringence and light scattering during the biaxial stretching process with a small piece of polymer sample. From the experimental results, stretchability, thickness uniformity and spherulite size could be evaluated simultaneously. Furthermore, as the thickness uniformity is closely related to the standard deviation of birefringence distribution, the thickness uniformity can be also evaluated by the retardation distribution

Keywords: biaxial stretching; in-situ measuring; birefringence distribution; light scattering; three-dimensional refractive indexes.

INTRODUCTION

The biaxial stretching films are widely used for both industrial products and our daily necessaries. Recently, the films with thin and uniform thickness in the production of stretching films are required from the reduction of raw materials and high quality films. In the production lines of the stretching films, one needs to understand not only the stretchable resin property, but also the optimum stretching conditions for the film resin. Besides, the evaluation of stretchability of the films becomes important in the development of the stretching film. The stress-strain curve and crystallization of crystalline polymer are keypoints for evaluation of stretchability during the biaxial stretching. The stress-strain behavior of the biaxial stretched film has been examined[1]. The relationship between the stretching force and stretching ratio of various samples was studied[2]. However, the relationship of both the stresses and retardation behavior as a function of the stretching ratio during the stretching process and the deformation of spherulite and stretching ratio have not been shown. The crystallization behavior and the higher order structure formation of crystalline polymer have been measured by the light scattering, but the deformation of spherulite by the stresses during the stretching process has not yet been done.

Moreover, the simultaneous measurement of stresses, birefringence and three dimensional molecular orientations are also essential for evaluation of stretchability. However, the simultaneous measurement data for the stresses, birefringence, spherulite and three dimensional molecular orientations during the biaxial stretching process have been not shown.

On the other hand, the in-situ measurement of the stress and retardation during the uniaxial stretching by using Photo Elastic Modulator(PEM) was previously proposed[3]. This method is superior in terms of high sensitivity and short measurement time. With the advantages of PEM, the stress and retardation in simultaneous measurement during the uniaxial stretching process can be measured even if the stretching speed is fast[1].

Above all, there are no reports of the simultaneous measurement of stress, birefringence, three dimensional refractive indexes and light scattering during the biaxial stretching process. In this paper, we describe on the results of crystalline polypropylene sample during the sequential biaxial stretching process obtained by a newly developed test machine.

SYSTEM CONSTRUCTION

FIGURE 1 illustrates the schematic diagram for the biaxial stretching test machine. The apparatus was constructed to carry out the simultaneous and the in-situ measurement of stress, birefringence, three dimensional refractive indexes and light scattering during the biaxial stretching process. The test machine was equipped with stretching unit and XY mapping controlled and driven by a computer, two load cells for the measurement of stresses of the biaxial stretching film, double birefringence measurement system using Photo-Elastic Modulator (PEM) for the measurement of retardations: the vertical and inclined incidence angle of laser beams respectively, and three dimensional molecular orientations of the biaxial stretching film, and the light scattering system for monitoring the spherulite conformation and spherulite size. The light source axis on the double birefringence and light scattering measurement system was adjusted to overlap in the center of the biaxial stretching film on the stretching unit.

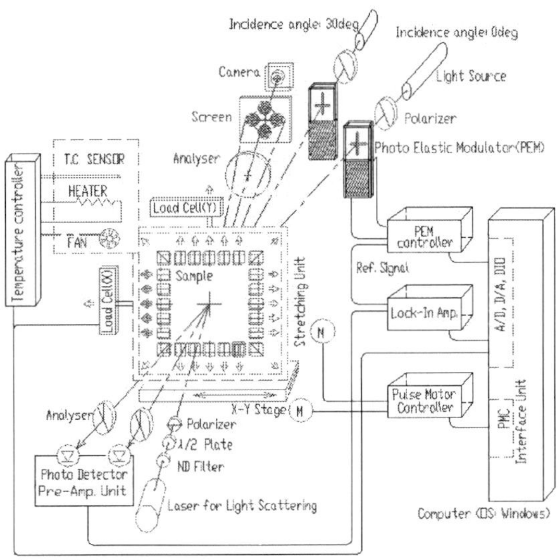

FIGURE 1 A diagram of a newly developed test machine

SAMPLE

Polypropylene is one of crystalline polymers which have high performance such as high formability, heat-resistance, high transparency and cost performance, so polypropylene film is widely used for food packaging and industrial films. Polypropylene is also the crystalline polymer which has highly ordered arrangement of molecules structure.

The resin of iPP sample, having MFR=3g/10min, Mw=3.6×105g/mol, Mw/Mn=5.0 and Tm=160℃, was produced by Prime Polymer(Japan). The non-stretching sheet with 500 micrometers as a film thickness was cut to size at 85 square millimeter of piece in use. The stretching conditions were fixed with the chamber temperature at 162℃, stretching speed at 25 mm/sec and preheating time at 120 sec before the biaxial stretching.

EXPERIMENTAL RESULTS AND DISCUSSION

Measurement of stresses and retardations

The stresses and retardations were simultaneously measured during the sequential biaxial stretching with stretching ratio at MD5×TD6.7 times as shown in FIGURE 2(a). As the results, the MD stress(blue line) and TD stress(pink line) are highly increasing in the beginning of MD stretching, and then they changed the increasing behavior where the MD yield point appears due to the neck-like deformation occurs during the MD stretching process. In the beginning of TD stretching, TD stress is dramatically increasing due to the neck-like occurs during the TD stretching process. Finally, MD and TD stresses are increasing again for the even-stretching in the last of TD stretching. The retardations of vertical incidence and inclined incidence angle of laser beams show the increasing in MD stretching and the decreasing in TD stretching process due to the sequential biaxial stretching. The variation of retardation in TD stretching is greater than the variation of retardation in MD stretching process at the middle stage of TD stretching. It means that the polypropylene has a high molecular orientation at the end of the sequential biaxial stretching.

Refractive indexes behavior during sequential biaxial stretching process

The refractive indexes which are information of molecular orientation were calculated from the results of retardations shown in FIGURE 2(b). Here, n_x and n_y are refractive indexes of MD and TD stretching direction, and n_z is refractive index of film thickness direction. The relationship between refractive indexes and stretching ratio shown: In the MD stretching, n_x is increasing due to the increasing of MD stress, and n_z is decreasing due to the decreasing molecular orientation in the film thickness direction. Next, in the TD stretching, n_y is increasing due to the increasing of MD stress, and n_z is still continuously decreasing due to the decreasing of film thickness during TD stretching process.

From the results, it found that the degree of n_y became higher than n_x where the variation of retardation in TD stretching is larger than in MD stretching. As well as from the increasing behavior in difference value of n_x and n_y, it was found that the stretched film is in a plane orientation according to the stresses and the sequential biaxial stretching ratio.

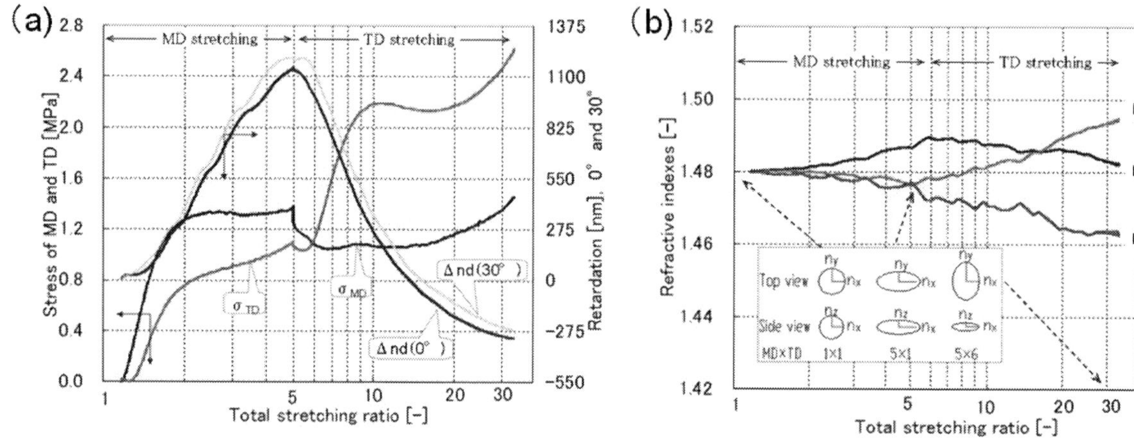

FIGURE 2

(a) Simultaneous measurements of stresses and retardations during the sequential biaxial stretching process

(b) Three dimensional refractive indexes behavior during the sequential biaxial stretching process

Spherulite deformation of stretched film at each sequential biaxial stretching ratio

The light scattering image of the stretched film was observed at each sequential biaxial stretching ratio shown in FIGURE 3(a). As the results of light scattering images as four-leaf clover pattern, it was found that the spherulite was deformed by the increasing of the sequential biaxial stretching ratio. And it was also obtained that the spherulite was dramatically deformed in the beginning of MD stretching region, and then it was starting to break up at MD2×TD1 stretched ratio where the MD yield point appears. Lastly, it was already disappeared at MD5×TD3 stretched ratio where the TD neck-like deformation finished before the TD stress increases again in the last of TD stretching.

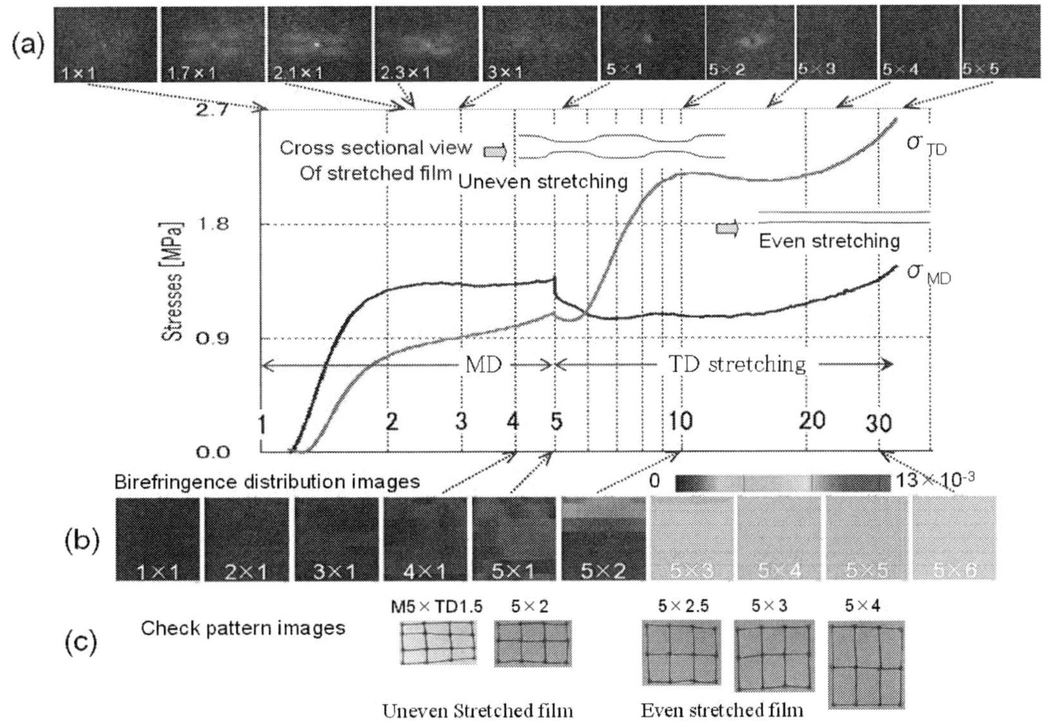

FIGURE 3

(a) Light scattering images at several stretching ratio of sequential biaxial stretching process

(b) Birefringence distribution images at several stretching ratio of sequential biaxial stretching process

(c) Stretched film pattern images at several TD stretching ratio

Birefringence distribution of stretched film at each sequential biaxial stretching ratio

The birefringence distributions at the center area of the stretched film were measured by XY mapping controlled system, and the standard deviation (SD) of birefringence distribution was calculated as shown in FIGURE 3(b).

As the results, it was found that the birefringence distribution images showed unevenly distributed results due to the neck-like until the spherulite had already disappeared at the stretched ratio MD5×TD3, and they showed good distribution from the stretching ratio where the MD and TD stresses were increasing behavior again in the last stage of TD stretching.

Neck-like deformation during sequential biaxial stretching process

Furthermore, the 5 mm^2 check pattern was drawn visually to check the stretched film pattern at several stretching ratios as shown in FIGURE 3(c). From the results of check pattern images, the uneven stretching occurred in the beginning of TD stretching (5×1.5 times) and then finally the stretching pattern became evenly after the TD yield point appeared at the stretching ratio where MD and TD stresses increased again in the last of TD stretching. It means that the stretched film had the neck-like stretching during the sequential biaxial stretching before the even stretching in the last stage of TD stretching.

Relationship between SD of birefringence distribution and thickness uniformity

Furthermore, in order to investigate the thickness variation at each sequential stretching ratio, the thickness average at 150 mm^2 of the stretched films was measured by dial gauge (Mitsutoyo Teclock PG-01J), and then the thickness uniformity could be calculated from equation (1) as follows.

$$\text{Thickness uniformity}[\%] = (\text{maximum thickness} - \text{minimum thickness}) / (2\times\text{average thickness})\times 100 \qquad (1)$$

The results of thickness uniformity and standard deviation of birefringence distribution of the stretched film at several TD stretching ratios are shown in TABLE 1 and FIGURE 4. As the results, it was found that the standard deviation of birefringence distribution and the thickness uniformity are improved with the increasing of sequential biaxial stretching ratio. It means that they are closely correlated in the TD stretching process.

TABLE 1 Results of standard deviation of birefringence distribution and thickness uniformity at several TD stretching ratio

Sequential stretching ratio	5×2	5×2.5	5×3	5×4	5×5	5×6
SD of birefringence distribution(×10^{-3})	0.544	0.426	0.195	0.042	0.021	0.018
Thickness uniformity [%]	60.7	48.1	29.7	18.4	12.7	11.9

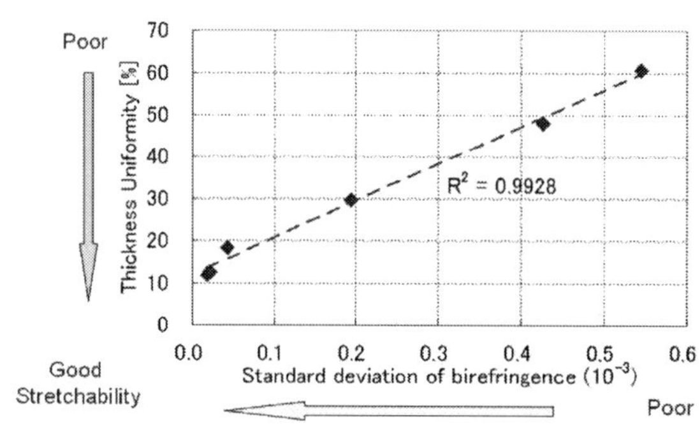

FIGURE 4 Relationship between standard deviation of birefringence distribution and thickness uniformity

CONCLUSIONS

We can summarize the stretchability of crystalline polypropylene film during the sequential stretching process obtained by a newly developed test machine and make conclusions as follows:

1. A newly developed test machine can simultaneously measure the data for evaluation of stretchability.
2. The even stretching was begun at MD5×TD3 where the spherulite was broken up and the neck-like stretching finished, so the thickness uniformity and the SD of birefringence became good.
3. In order to obtain a good stretchability of the stretched film, the proper stretching ratio should be above MD5×TD5.
4. As the thickness uniformity is closely related to the SD of birefringence distribution, the thickness uniformity can be easily evaluated by the retardation distribution.

REFERENCES

1. Kanai T, Campbell G A, *Film processing advances*, Hanser Publishers: Munich, 2014.
2. Tamura S, Kuramoto I, Kanai T. Polymer Engineering & Science, Vol. 52, Issue6, 1383-1393 (2012)
3. Egoshi K, Mochida Y. Proc. SPIE 2873, International Symposium on Polarization Analysis and Application to Device Technology, 1996, 58, doi:10.1117/12.246180

Preparation Of Intelligence Barrier Films Based On Low Density Poly Ethylene And Evaluation Of Its Properties

Bahari-Aban Babak [1], Mehranpour Milad [1]*, Oromiehie Abdulrasoul [2], Atabaki-Pasdar Pouria [1]

[1] Islamic Azad University Science & Research branch - Tehran - Iran, [2] Islamic Azad University South Tehran branch - Tehran - Iran

In this article, nanocomposite films based on low density polyethylene (LDPE), organoclay (OC) and ethylene vinyl alcohol (EVOH) copolymer were prepared and characterized through using scanning electron microscopy(SEM), transmission electron microscopy (TEM), X-ray diffraction (XRD), differential scanning calorimetry (DSC), oxygen permeation measurements, and tensile tests. The oxygen barrier properties of the LDPE/EVOH/OC film were better than those of the LDPE/OC film. Furthermore, from the TEM micrographs, the O2 permeability through the films was evaluated, which illustrated that adding both OC and LDPE to EVOH leads to a remarkable increase in the barrier properties of EVOH films. Also, a new colorimetric indicator sensor based on methyl red was sensitive to total volatile basic nitrogen (TVBN) released from red meat during storage which has been fabricated for real-time monitoring of meat quality. The methyl red as a freshness sensor worked based on pH increase as the basic spoilage volatile amines produced gradually in the package headspace, and subsequently, the colour of the sensor will change for spoilage indication, which is easily visible to the naked eye. The results reveal that the sticker sensor could be used to determine the degree of red meat cut freshness, as the relationship between the colour change of methyl red as a sensor response and the meat freshness follows a similar trend. Results have indicated that sensor response correlated well with microbial load of red meat, thus enabling the sensor for real-time monitoring of red meat spoils.

Crystals in-situ induced by supercritical carbon dioxide as bubble nucleation agents on foam structure controlling of poly(L-lactic acid)

Liao Xia [(1)]*, Li Junsong [(1)], Li Shaojie [(1)], Li Guangxian [(1)]

[(1)] Sichuan University - Sichuan Province - China

Semi-crystalline polymer like poly(L-lactic acid) (PLLA) is generally regarded as a multi-phase system where crystalline domains disperse in the amorphous matrix. The crystallization rate of PLLA is very slow under atmosphere. However, the crystallization process can be accelerated under compressed CO2, because the dissolved CO2 induces the swelling of polymers and decreases the energy barrier for the molecular retraction and fold. Furthermore, the intense interaction between the electron-donating carbonyl oxygen of PLLA and the electron-accepting carbon of CO2 greatly promotes the CO2 dissolution and then reinforces the plasticization effect. In the present work, CO2-induced spherulites were successfully utilized to regulate the morphology and structure of poly(L-lactide) (PLLA) foam with cells size ranging from nanoscale to microscale. The unique transitions from nanocells to microcells through need-like cells and bimodal cells in the spherulites formed under CO2 were investigated in detail. According to the characteristics of the DSC curves and the cell morphology, the saturation temperature was divided in three regions: Regions I (60-80 oC), II (90-110 oC) and III (115-120 oC). Results showed that cells could nucleate and growth in between the interlamellar amorphous layers in the spherulites formed under CO2. It was concluded that the crystallization morphology of PLLA treated under CO2 played a significant role in the cell nucleation and growth.

Assessment of Particles Interface Welding Quality in Microporous Structures Fabricated by Sintering Process

Salari Meysam [(1)], Pircheraghi Gholamreza [(1)*]

[(1)] Sharif University of Technology - Tehran - Iran

Sintering is one of the most practical methods for fabricating the porous polymeric microstructures through controlling the interfacial welding and mutual chain diffusion of polymeric particles. Porosity of sintered samples can be estimated using an offered simplified equation, considering sintering neck radius. Particles welding taking place through global or segmental relaxation and diffusion of polymeric chains at particles interface, exhibits a severe sensitivity to processing temperature, especially on the borderline between solid and melt states. Easier relaxation of chains was attributed to destruction of some crystalline regions and more freely motion of chains due to a transition to amorphous state as temperature increases. In this work, in order to simulate and control the microstructural porosity, the coalescence rate of two separated particles has been studied, using hot stage microscopy (HSM). Consequently, the necessary time for particles welding and coalescence with special neck radius to achieve determined final porosity, is estimated. The microporous structure has been confirmed by means of SEM images; and finally based on DSC and shear punch mechanical test results, interfacial microstructural changes and welding quality of particles in sintered porous samples have been discussed.

SHOCK DAMPING PROPERTIES OF OUT-OF-PLANE REINFORCED COMPOSITES WITH FUNTIONALLY GRADED SYNTACTIC FOAMS

Anwer Ahmed A. W. [(1)]***, Naguib Hani*** [(1)]*

[(1)] University of Toronto - Ontario - Canada

In addition to withstanding structural stresses during an impact event, there is an increasing demand to reduce force transferred to the receiver/structure during an impact event. Reduced load transfer is of special importance in personal protection, automobiles, and in aerospace industries, where high load transfer could cause personal and structural damage. Although most material structures cans absorb load without failing, these structures are not capable of reducing force transferred to adjacent structures and bodies. The high stiffness of most impact resistant material structures results in an isotropic high rate of shock transfer, and the coupled effect of low damping properties of these material structures, results in a high load transfer. In this study, Polyurethane resin was used as a matrix material and Ultrahigh molecular weight polyethylene braids were used for out-of-plane reinforcements of structures. Effects of fibrous reinforcement and functionally graded polyurethanes on transferred loads will be investigated with a customized ASTM D5420 drop weight set up. Correlation between processing parameters, material gradients, and orientations to the transferred loads will also be presented.

Controlled foam processing and cell structure development of gamma irradiated polypropylene/elastomer blends

GHOSH ANUP KUMAR [(1)*], Dutta Anindya [(2)], Singh Srishti [(3)]

[(1)] Head, Centre for polymer sciece & engineering, Indian Institute of Technology - NEW DELHI - India, [(2)] Research Scholar - Indian Institute of Technology Delhi, New Delhi - India, [(3)] Indian Institute of Technology - New Delhi - India

Currently a high demand in automotive industries motivated to make a comparative study between the gamma irradiated polypropylene/elastomer blends along with the effect of compositional variation and processing parameter dependence on foamability of polypropylene. The effects of gamma irradiation of blends directed towards the crosslinking of elastomer (Vamac Ultra IP) approach using gamma irradiation, prior to the preparation of blends with polypropylene as matrix on final foam morphology of different blends has been described. The introduction of crosslinked elastomer into the polypropylene matrix improved the foamability of polypropylene. Foaming was done in a high pressure autoclave at three different temperatures using supercritical CO_2 as a blowing agent. The x-ray diffraction, tensile, surface morphological and density properties of blends and their foams were studied by means of WAXD, universal testing machine, SEM (both secondary electron and back scattered electron detectors) and water displacement method respectively. The density of foamed blends decreased with increasing elastomer content, especially with foams achieved at higher foaming temperature. With the increase in elastomer content, domain size of crosslinked elastomer increased in blends and thus led to the bigger cells during foaming process, in spite of having its high gas entrapment capacity for its interconnected chains. Cell size of foams increased with increasing foaming temperature. Foams obtained at 145 °C showed good balance of microcellular structure along with higher cell density. Specific tensile properties of foams showed a promising improvement and can be correlated with the foam morphology.

Development of a Bead Foam based on the Engineering Polymer Polybutylene Terephthalate

Tobias Standau[1,a)], Bianca Hädelt[1)], Michael Fafara[2)], Volker Altstädt[1,2b)]

[1]*Lehrstuhl für Polymere Werkstoffe, University of Bayreuth, Universitätsstraße 30, 95447 Bayreuth, Germany*
[2]*Neue Materialien Bayreuth GmbH, Gottlieb-Keim-Straße 60, 95448 Bayreuth*

[a)]Corresponding author: tobias.standau@uni-bayreuth.de
[b)]altstaedt@uni-bayreuth.de

Abstract. In this work bead foams from a thermoplastic engineering polymer, exhibiting a higher service temperature than the established bead foam polymers polystyrene and polypropylene, were developed. For the first time bead foams from polybutylene terephthalate (PBT) with a density below 160 kg/m^3 were produced and welded. In particular, the influence of a multifunctional epoxy chain extender on the rheological behavior, the foam morphology as well as the weldability was investigated. Due to the reaction of the chain extender during the processing the viscosity could be increased and strain hardening was observed, leading to improved foaming behavior. An optimum chain extender concentration regarding density reduction and cell morphology was determined as 1 wt.-%.

INTRODUCTION

The most common bead foam polymers are large production-volume polymers like polystyrene, polypropylene, polyethylene and combinations thereof. However, all of these materials lack of thermal resistance. For instance, the heat deflection temperature of EPS (expandable polystyrene) and EPP (expanded polypropylene) is approximately 80 °C and 110 °C, respectively, and this drawback limits the application of these materials. Especially for under-the-hood applications bead foams that can withstand a permanent exposure to heat are demanded. The interest in engineering polymers such as polybutylene terephthalate (PBT) is growing since they could enable further applications for bead foams. In the earlier work of Köppl et al. [1] the continuous production of expanded polybutylene terephthalate (E-PBT) bead foams by foam extrusion coupled with underwater granulation (UWG) was reported. The work pointed out the main issues of PBT foaming, which are (i) the poor rheological properties of this polyester, namely a low melt strength that limits the foam expansion and (ii) the narrow processing window for foaming. Another drawback was (iii) that even though the production of E-PBT bead foams was possible, the foam beads failed to fuse under the conditions used in the steam chest molding (SCM) process.

Jeong et al. remarked on the strong impact of the rheological and crystallization behavior on the foaming behavior and the resulting morphology in the extrusion process [2]. A certain viscosity was needed, which was achieved by chemical modification with triglycidylisocyanurate. This modification led to an increase in the molecular weight and chain branching, resulting in higher viscosity [3]. To the best of our knowledge only a limited number of publications about the foaming PBT have been published and the reported densities are rather intermediate.

Certain ways to chemically modify polyesters have been described, including PET [4, 5] and PLA [6–8]. These works show an effective increase in molecular weight due to reaction of chemical modifiers with the polymer in the melt resulting in an increased melt strength, which is advantageous for foaming.

Our current study investigates the addition of a chemical modifier - namely a multifunctional epoxy chain extender – to enhance the rheological properties during the foam extrusion and thereby allowing increased density reduction and improvement of cell morphology. Furthermore, improved fusion technology and process control allows applying the necessary high steam pressures to weld the beads in a stable steam chest molding process.

EXPERIMENTAL

Chemical Modification during Continuous Bead Foam Production

Bead foams can be produced continuously by means of foam extrusion combined with underwater granulation. In this study PBT POCAN B1300 from Lanxess (Cologne / Germany) was used. The foam extrusion was carried out using a Dr. Collin tandem line (Ebersberg / Germany), which is a combination of (i) a twin screw extruder (L / D = 42, D = 25 mm), in which the blowing agent CO_2 (2 wt.-%) is added and (ii) a single screw extruder (L / D = 30, D = 45 mm), where the gas-loaded melt is cooled down successively. At the end of the single screw extruder an underwater-pelletizer UWG LPU (Gala / Eagle Rock, VA / USA) is attached, cutting the foamed strand with a rotating knife resulting in spherically shaped bead foams.

To establish certain rheological behavior a commercial multifunctional epoxy chain extender (CE) was added as dry blend in the foam extrusion process in various concentrations. TABLE 1 lists the PBT / chain extender combinations used during the trials.

TABLE 1. Overview of the foamed samples and the used CE concentration

notation	CE concentration (wt.-%)
PBT (neat)	0
PBT 0.5CE	0.5
PBT 1CE	1
PBT 2CE	2
PBT 4CE	4

Rheological Characterization

Bead foams were cryo-grinded (Retsch GmbH / Haan / Germany) and melt pressed (P/O/Weber / Remshalden / Germany) for rheological investigations. Plate-plate measurements were carried out on a RDA III (Rheometric Scientific / Piscataway, NJ / USA) to determine the shear properties.

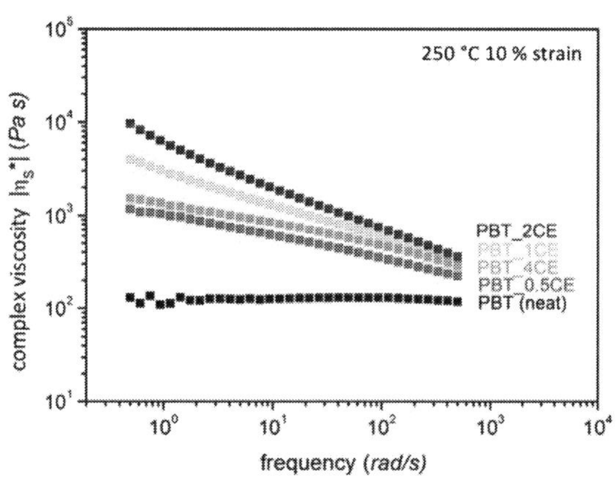

FIGURE 1. Frequency sweeps (at 250 °C / 10 % strain) of PBT modified with different concentrations of CE

As can be seen from the viscosity curves in Fig. 1 the incorporation of the CE leads to (i) an increase of the viscosity and (ii) a different shape of the curves. It is evident, that the addition of chain extender leads to an increased viscosity compared to the neat material, which shows a constant shear viscosity of 100 Pas over the measured frequencies, resulting from the chain extender reacting in the extrusion process with the polymer leading to an increased molecular mass. Up to 2 wt.-% chain extender, the viscosity is increasing with the chain extender concentration. However, PBT modified with 4 wt.-% CE shows a lower viscosity than PBT with 0.5 wt.-% CE. Thus it is presumed, that a too high amount of CE leads to higher shear during the processing resulting in chain breakage and consequently in a lower viscosity.

Furthermore, for the modified samples it can be derived from the increase of the viscosities at lower frequencies that the chain topology is not linear as it is for the neat PBT, which shows a Newtonean plateau. This is due to the multifunctionality of the chain extender that promotes branching and / or crosslinking.

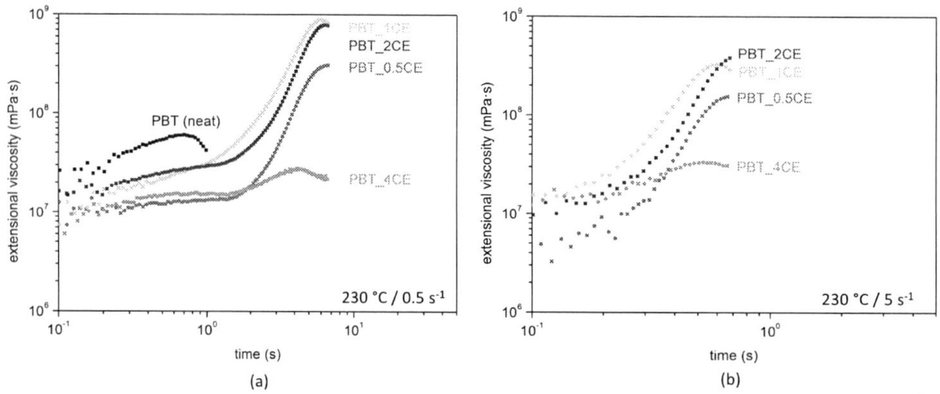

FIGURE 2. Extensional rheology measurements (at 230°C) for PBT with different strain rates of (a) 0.5 s^{-1} and (b) 5 s^{-1}

Additionally, the elongational viscosity was determined with an Anton Paar MCR702 (Graz / Austria) equipped with universal extension tool (UXF) at 230 °C and varying strain rates of 0.5 s^{-1} and 5 s^{-1}, as higher strain rates can be expected during foaming [9]. Rupture of the neat PBT sample occurred at the lower strain rate of 0.5 s^{-1} and could not be measured at higher strain rates.

From Fig. 2 it can be seen that the curves of the neat PBT modified with CE show a steep increase in the elongational viscosity over time at both strain rates, which is an evidence for strain hardening. Analogous to the shear measurements PBT modified with 4 wt.-% CE shows behavior divergent to the other modified materials. Here, the measurements were difficult and the obtained data revealed less pronounced strain hardening, which confirms the assumption from the shear experiments that a too high CE concentration leads to breakage of the chains by the high shear resulting in unintentional melt properties.

Steam-chest molding process was performed on a Teubert TVZ162/100PP (Blumberg / Germany) equipped with a specialized steam generator from HaeCo II from UniBell (Hwaseong/ South Korea) capable of providing unconventionally high pressures of up to 25 bar. The custom-made aluminum mold with dimensions of 20 cm x 20 cm x 6 cm was manually filled with 3.2 litres of foamed beads. A two-step steaming process was performed applying pressures between 7 to 15 bars for 30 to 60 s. All modified PBT bead foams could be fused using this process.

Characterization of the Foamed and Welded Part

In Fig. 3 SEM images of the modified and fused bead foams are shown. Depending on the CE concentration, beads have (i) different cell morphology and (ii) different welding behavior. The cell morphology is more homogeneous for the beads modified with 1 or 2 wt.% CE. Neat PBT (not shown here) and PBT with 4 wt.-% CE show very coarse morphologies. Also, PBT with 0.5 wt.-% CE actually shows large internal cavities in the bead, which get compressed during the steam chest molding and are not visible in the SEM images. Also, the welding quality is much better for PBT modified with 1 and 2 wt.-% CE, as it can be concluded from the absence of higher content of voids between the beads and a better surface appearance (not shown here).

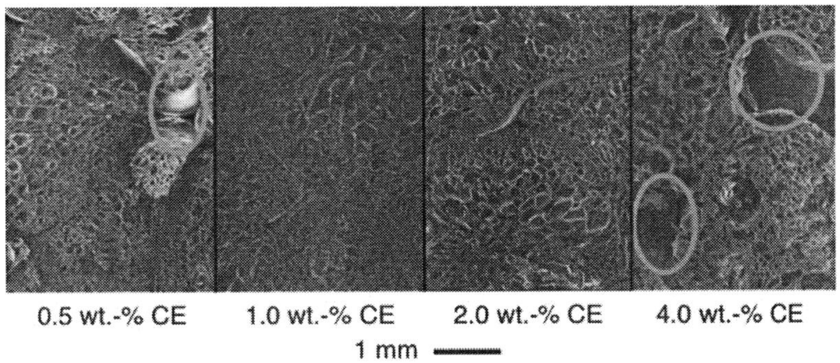

FIGURE 3. SEM images of the morphology of the modified PBT bead foams after welding

Figure 4 shows the densities of the bead foams and the welded parts. It is clearly visible that for the 1 and 2 wt.-% of CE the density is the lowest for the beads and the welded part. Interestingly, the welded parts show a further, slight decrease in density compared to the beads, resulting in the lowest achieved density of 155 kg/m3. It is worth noting that the PBT modified with 0.5 wt.-% and 4.0 wt.-%, respectively show increased densities when welded, what is a result of the poorer rheological properties during foaming and heating and compaction during the steam chest molding process leading to the melting and collapse of the foam structure.

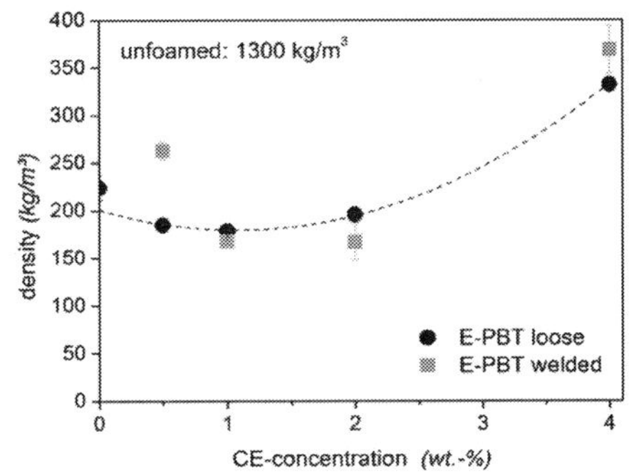

FIGURE 4. Overview of the densities measured for the loose beads (black dots) and the welded part (orange squares)

DMA analysis was performed on a Gabo Eplexor 500N (Ahlden / Germany) in compression mode on a sample of the welded part (diameter 25 mm, height 20 mm) with a load of 150 N, a frequency of 1 Hz and heating from 25 °C to 220°C at a heating rate of 1 K/min. Representative curves of PBT foamed beads (155 kg/m^3) and EPP (220 kg/m^3) are shown in Fig. 5. Even though both materials have different densities the shape of both curves is similar. In contrast to EPP the complex E modulus only starts to significantly decrease above 200 °C.

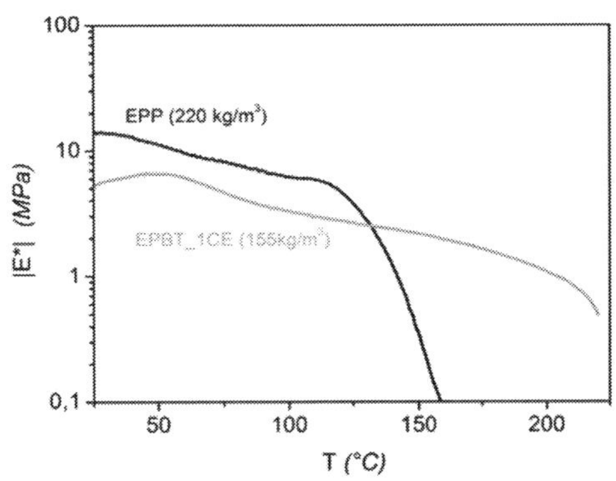

FIGURE 5. DMA curves of E-PBT (1 wt.-% chain extender) and EPP (reference)

CONCLUSION

It was shown that by reacting with a multifunctional epoxy chain extender incorporated during the foam extrusion process it was possible to achieve polybutylene terephthalate bead foams with low-medium densities that can be fused in a steam chest molding process with high steam pressures. The reaction of the polymer with the chain extender leads to changes in the chain topology resulting in a viscosity increase and strain hardening behavior, lower densities and more homogeneous morphologies. A chain extender concentration of 1 wt.-% was identified as optimum in terms of rheological behavior, density reduction and foam morphology. The parts made from PBT bead foams modified with 1 wt.% chain extender had densities down to 155 kg/m^3 with fine and homogeneously distributed cells and less voids.

AKNOWLEDGEMENT

This work was public funded by the German Research Foundation (DFG: AL474 / 28-1). The authors would like to thank all technicians that were involved creating the results. Special thanks to Jacqueline Uhm (University of Bayreuth) for proof reading the manuscript.

REFERENCES

1. Köppl T, Raps D, Altstädt V (2014) E-PBT—Bead foaming of poly(butylene terephthalate) by underwater pelletizing. J Cell Plast 50:475–487. doi: 10.1177/0021955X14528524
2. Jeong B (2006) Extrusion Foaming Behavior of PBT Resins. J Cell Plast 42:165–176. doi: 10.1177/0021955X06060952
3. Jeong BJ, Xanthos M (2007) Reactive Modification of Pbt With Applications in Low Density Extrusion Foaming. Polym Eng Sci 47:244–253. doi: 10.1002/pen.20699
4. Awaja F, Daver F, Kosior E (2004) Recycled poly(ethylene terephthalate) chain extension by a reactive extrusion process. Polym Eng Sci 44:1579–1587. doi: 10.1002/pen.20155
5. Awaja F, Daver F, Kosior E, Cser F (2004) The effect of chain extension on the thermal behaviour and crystallinity of reactive extruded recycled pet. J Therm Anal Calorim 78:865–884. doi: 10.1007/s10973-004-0454-z
6. Di Y, Iannace S, Di Maio E, Nicolais L (2005) Reactively modified poly (lactic acid): Properties and foam processing. Macromol Mater Eng 290:1083–1090. doi: 10.1002/mame.200500115
7. Corre Y-M, Maazouz A, Duchet J, Reignier J (2011) Batch foaming of chain extended PLA with supercritical CO2: Influence of the rheological properties and the process parameters on the cellular structure. J Supercrit Fluids 58:177–188. doi: 10.1016/j.supflu.2011.03.006
8. Al-Itry R, Lamnawar K, Maazouz A (2012) Improvement of thermal stability, rheological and mechanical properties of PLA, PBAT and their blends by reactive extrusion with functionalized epoxy. Polym Degrad Stab 97:1898–1914. doi: 10.1016/j.polymdegradstab.2012.06.028
9. Spitael P, Macosko CW (2004) Strain hardening in polypropylenes and its role in extrusion foaming. Polym Eng Sci 44:2090–2100. doi: 10.1002/pen.20214

Crack Arrest In Finger Jointed Thermoplastic Polyethersulfone Film Interleaved Carbon Fibre Reinforced Composites

Anthony David Benbow (1)*, Bacarreza Nogales Omar R (1), Shaffer Milo Sebastian Peter (1), Bismarck Alexander (1), Robinson Paul (1), Pimenta Soraia (1)

(1) Imperial College London - London - UnitedKingdom

Interleaved composites have shown promise in altering the failure mode(s) of composites and offer a facile approach to modify traditional carbon fibre reinforced composites (CFRCs). The addition of thermoplastics has been shown to enable easy repair [1], as well as controlled alteration of the stiffness upon heating [2]. The addition of a tough thermoplastic interleave in a CFRC is expected to improve shear properties in tension in matrix dominated failures/configurations [3]. Pre-cut unidirectional carbon fibre prepreg (M21/194/34%/T800S) composites were tested in tension with a 20 mm overlapped finger joint architectures, following previous methodology [4]. In between the overlapping finger jointed region the effect of introducing polyethersulfone (PES) interleave film is investigated. The PES interleave spans 10 mm, centrally, within this overlapped region. All of the carbon fibres in the finger joint arrangement were cut, such that no fibre(s) span the full length of any sample. The properties arising from these configurations are therefore strongly dependent on the interply overlap regions. Samples with the addition of a thick PES interleave arrested the initial crack which formed at the pre-cut site. The strain to-failure of the thick PES interleaved samples was over 3.2%, an increase of 85% compared to the baseline samples, and catastrophic failure was delayed in the majority of instances. The specific use of this embodiment could be to toughen regions in composites which require the overlap of prepregs and with further developments the approach could be exploited to yield a high performance yet (pseudo)ductile response under tension. [1] P. Robinson, 17th ECCM, Munich 2016, ISBN 978-3-00-053387-7 [2] H. A. Maples, Compos. Sci. Technol., 128 (2016), DOI:10.1016/j.compscitech.2016.03.028 [3] G. Grail, Compos. Sci. Technol., 106 (2015), DOI:10.1016/j.compscitech.2014.11.006 [4] G. Czél, Compos. Sci. Technol., 106 (2015), DOI:10.1016/j.compscitech.2014.10.022

Effect of Extrusion Parameters and Nanofillers on Mechanical Properties of PPSU Tufting Yarns

Cormac McGarrigle [1], Marcin Wegrzyn[1], Alistair McIlhagger[1], Eileen Harkin-Jones[1] and Edward Archer[1]

[1] Engineering Research Institute, School of Engineering, Ulster University Jordanstown, Northern Ireland, BT37 OQB

[a)] *Corresponding author: mcgarrigle-c2@email.ulster.ac.uk*

Abstract. Yarn diameters ranging from 160µm to 720µm of unfilled PPSU and PPSU nanocomposites with 1 wt.% of either carbon nanotubes (CNT) or carbon nanofibers (CNF) were prepared using a twin screw compounder. The tensile properties of these yarns generally improve with the addition of CNT or CNF at higher values of screw speed-to-haul off ratio. This effect is correlated with the yarn draw down ratio and attributed to nanofiller orientation and crystallinity induced in thermoplastic matrix. The fibres have as much as a 23% increase in Ultimate Tensile Strength (UTS) for the same parameter set when loaded with CNT, while the greatest increase observed for larger particle size fillers CNF is just 3.77%. Depending on filler and processing parameters set, yarns varied in UTS from 24.84MPa to 206.23MPa for unfilled PPSU and PPSU-CNT, respectively.

INTRODUCTION

Carbon fibre (CF) reinforced polymer composites offer significant improvement of material strength-to-weight ratio, when compared with materials traditionally used in aerospace and automotive industries.[1] However, delamination of composite structures is a considerable drawback that can be potentially solved by a through-plane stitching of CF fabrics before a resin impregnation step. Such a simple method of improving resistance to delamination requires identification of a suitable tufting yarn material and optimal yarn characteristics. Tufting arrests the displacement caused by crack opening displacement in mode I loading and crack sliding in mode II. The tufting thread functions as a pin, retaining the adjacent laminate regions following the transit of a delamination crack front through said region.[2] Preliminary testing has shown that polyphenylsulfone (PPSU) is a potential candidate for such a tufting yarn due to its excellent mechanical and physical properties which can be further modified by the control of processing conditions and the addition of nanofillers. In particular CNTs have shown great success with increasing the mechanical properties of polymers even at extremely low percentage volumes.[3]

EXPERIMENTAL

Materials

R-5500-NT grade Polyphenylsulfone (PPSU) from Solvay Advanced Polymers was used for all extrusions.

PR-24 carbon nanofiber (CNF) supplied by Pyrograf as heat treated materials with increased inherent conductivity had average fiber diameter 100nm and length ranging between 50-200um (aspect ratio 500-2000). The CNFs were homogeneously mixed in a charge bag with milled PPSU in order to give 1.0wt.% volume of CNF to PPSU.

The NC7000 multi-walled carbon nanotubes (CNT) supplied by Nancy as industrial grade (95% pure) had average fiber diameter 9.5nm and average length 1.5um (aspect ratio c.a. 160). The Multi-walled carbon nanotubes (MWCNTs) were homogeneously mixed in a charge bag with milled PPSU in order to give 1.0wt.% volume of MWCNTs to PPSU.

Polymer Preparation

PPSU pellets of ≈3mm were placed into a vessel with liquid nitrogen for 5 minutes in order to make them more brittle before being fed into a Fritsch Pulverisette 14 rotor mill, rotating at 15,000rpm.

Prior to the mixing with nanomaterials/extrusions the milled PPSU was dried in an over at 150°C for a period of 2 hours to expel any moisture within the material.

Extrusion

A Thermos Scientific HAAKE Rheomex PTW 16 OS twin-screw extruder combined with a Thermoscientific Type L-002-0061 pelletizer, modified to be used asa haul off unit, were used for all of the PPSU extrusions. The modified pelletizer was used as a continuous rate haul-off system. A ChargePoint PharmaSafe Powder Transfer Contanement Valve system was used for the extrusion of the PPSU nanocomposite yarns. The desired concentration of nanomaterials and milled PPSU were mixed in a Charge Bag. This Charge Bag is subsequently attached to the Pharma Safe Charge Point butterfly valve and both then are attached to the extruder feeding mechanism.

The extrusion barrels temperature profile began at 330°, in the zone closest to the hopper, and gradually heating up to 390°C before gradually cooling to 350°C at the die.

Optical Microscopy

A SZ-PT Olympus optical microscope was used to examine the shape and surface topography of the particles before and after milling. Image-J software was then used to take average sizing measurements from each pellet/particle

Scanning Electron Microscopy

In this study a JEOL JSM-IT100 InTouchScope SEM™ was used for the SEM imaging. The fibre samples were cut into sections of 2cm and these placed into an Emitech K500x sputter coater and the vacuum was set at 8*10-2 millibars of pressure for 2 minutes and 30 seconds at 3 milliamps (mA). The fibres were then turned over and the process repeated. The coated samples were mounted on stubs and placed into the SEM. 10kV at a magnification of *45 was employed giving a clear image of the surface topography of the fibres while still allowing enough of a field of view to perform diameter analysis on the fibres. The measure function within the SEM was exercised 5 times on each fibre sample, measuring the diameter at 500μm intervals.

Tensile Testing

An Instron 3344 tensile testing machine, equipped with a 500N load cell and Series 2710-200 Screw Action Grips, was employed throughout the fibre tensile testing. ASTM D3039/D3039M-15 "Standard Test Method for Tensile Properties of Polymer Matrix Composite Materials" and ASTM D3217-07 "Breaking Tenacity of Manufactured Textile Fibres in Loop or Knot Configurations" were followed for the tensile testing. Paper tabs were bonded to either ends of the polymer yarns, with SP106 resin, to avoid the grips causing damaged to the fibres. The gauge length between the grips in this case is set to 1 inch (25.4mm). The samples were tested at a constant speed of 20mm/min until failure.

RESULTS AND DISCUSSION

PPSU pellets and milled PPSU can be seen in Figure 1 below:

(a) (b)

FIGURE 1: (a) PPSU Pellet (b) Milled PPSU

The average diameter was reduced by 90.36% (3.47mm) from 3.84mm to 0.37mm. This should aid heat transfer ensuring a homogenous melt within the extrusion process, assisting the production of a more stable extruded yarn.[4] It will also aid the mixing of PPSU and nanomaterials.

This increase in extrusion stability is evident from the minuet error bars in Figure 2 where yarn diameters at the varying processing conditions are shown. The minute error The first number, followed by an "E" denotes the screw speed, in rpm, and the second, which is followed by a "P" represents the haul-off speed in rpm.

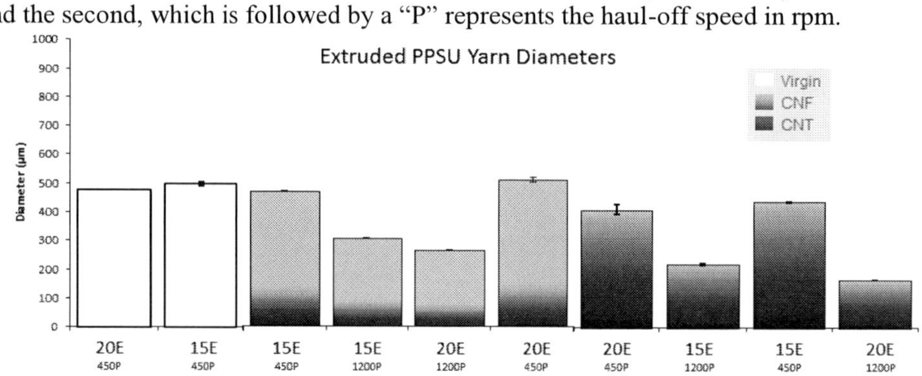

FIGURE 2: Extruded PPSU Yarn Diameters

Previous experimental work by Domenech et al .has shown structuring of extruded nanocomposites can be optimized by working under higher specific mechanical energy (SME) until a critical limit. In this case processing at a higher SME was limited by the maximum torque the machine was capable of achieving.[5]

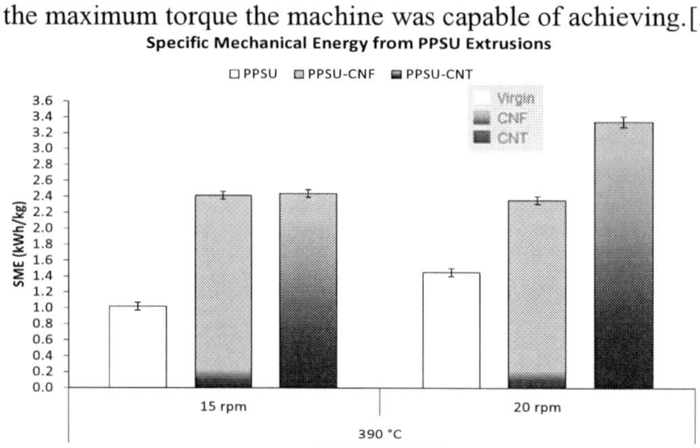

FIGURE 3: Specific Mechanical Energy from PPSU Extrusions

Figure 4 displays the ultimate tensile strength (UTS) for the PPSU and PPSU nanocomposites. It can be seen that little or no advantage in straight UTS was obtained from the edition of CNFs, the higher screw speed of 20rpm actually caused an 10.7% decrease in UTS. At the lower screw speed the CNF reinforced samples' UTS increased by an insignificant 3.77%. However, once carbon nanotubes (CNTs) are introduced as much as a 23.31% increase was observed at the same screw speed compared with the lower speeds.

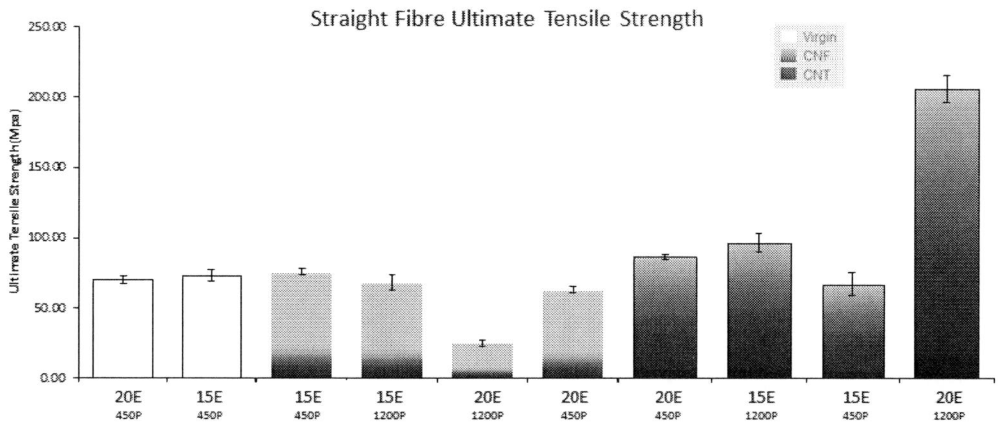

FIGURE 4: Straight Fibre Ultimate Tensile Strength

The UTS of each of the yarns calculated from the looped tensile testing are show in figure 5 below. The smooth surface topography allows the yarns to be pulled apart, when looped, without any risk of initial cutting or initial damage.

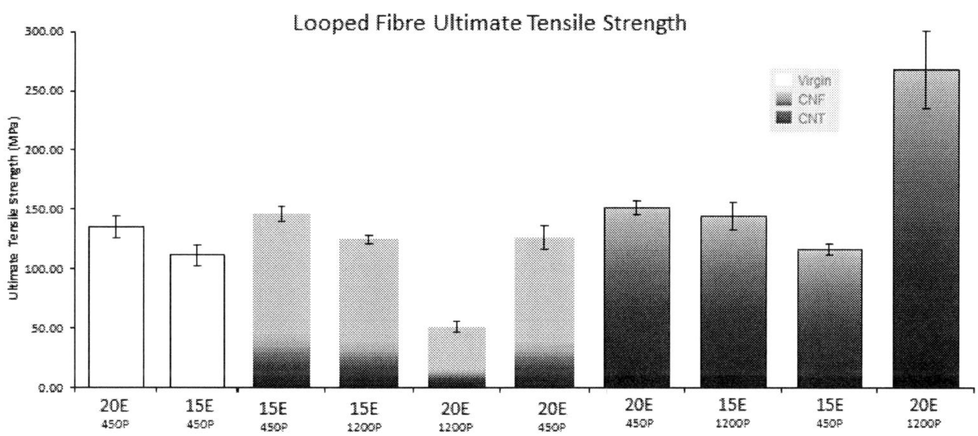

FIGURE 5: Looped Fibre Ultimate Tensile Strength

It is evident from both Figures 4 and 5 that a higher SME aids the increase of UTS of CNT reinforced PPSU, whereas lower SME increases the UTS of CNF reinforced PPSU.

In both the straight fibre and looped fibre tensile tests one particular parameter set exceeded the UTS substantially: screw speed of 20rpm and a pelletizer speed of 1200rpm. Compared to the other CNT reinforced extrusion at 20rpm the increase in haul-off resulted in a 138.5% increase in straight tensile UTS. This could be partially attributed to both the increased alignment of the nanotubes, as a result of the increased haul-off speed and as a result of the increase in SME.[5]

CONCLUSION

CNF reinforcement did not provide any significant improvements in mechanical properties and in certain situations actually resulted in a reduction in mechanical properties. CNTs did provide an increase in UTS under all tensile testing configurations. In both the straight fibre and looped fibre tensile tests one particular extrusion parameter set for the CNT reinforced PPSU exceeded the UTS substantially: screw speed of 20rpm and a pelletizer speed of 1200rpm. This could be partially attributed to both the increased alignment of the nanotubes, as a result of the increased haul-off speed and as a result of the increase in SME.

To date, some of the virgin PPSU yarns from this testing have been successfully tufted with little issue. Diameter consistency, where the diameter of the yarn varied across its length posed some minor slippages or yarn breakages. Testing with pellets and milled PPSU revealed that it was possible to get a more consistent diameter with the lower sized milled PPSU than the PPSU pellets. This could be as a result of both the removal of the air pockets within the pellets and by the smaller particles having a better heat transfer within the extruder, giving a more homogenous melt.

ACKNOWLEDGMENTS

This work was carried out under EPSRC grant number EP/L02697X/1 and was part funded by Department for Employment and Learning (Northern Ireland Executive).

REFERENCES

[1] Aktas A, Potluri P, Porat I, Group TC. Multi-needle stitched composites for improved damage tolerance. 17th Int Commitee Compos. 2009;

[2] Hartley JW, Kratz J, Ward C, Partridge IK. Effect of tufting density and loop length on the crushing behaviour of tufted sandwich specimens. Compos Part B Eng [Internet]. 2016;(December). Available from: http://linkinghub.elsevier.com/retrieve/pii/S1359836816331432

[3] Gojny FH, Wichmann MHG, Fiedler B, Schulte K. Influence of different carbon nanotubes on the mechanical properties of epoxy matrix composites - A comparative study. Compos Sci Technol. 2005;65(15–16 SPEC. ISS.):2300–13.

[4] Domenech T, Peuvrel-Disdier E, Vergnes B. The importance of specific mechanical energy during twin screw extrusion of organoclay based polypropylene nanocomposites. In: Composites Science and Technology [Internet]. Second Edi. Hanser; 2013. p. 7–14. Available from: http://www.hanserpublications.com/SampleChapters/9781569904596_9781569904596_Extrusion of Polymers 2E_Chung.pdf

[5] Domenech T, Peuvrel-Disdier E, Vergnes B. The importance of specific mechanical energy during twin screw extrusion of organoclay based polypropylene nanocomposites. Compos Sci Technol. 2013;75:7–14.

Effect of trisamide based nucleating agents on the morphology and mechanical properties of isotactic polypropylene foams

Demir Merve (1)*, Mörl Michaela (1), Schmidt Hans-Werner (2), Altstädt Volker (1)

(1) Polymer Engineering - Bavaria - Germany, (2) Macromolecular Chemie I - Bavaria - Germany

Having outstanding functional properties make isotactic polypropylene (i-PP) foams quite attractive to be applied in broad fields in polymer industry (e.g., acoustic and thermal insulators, automotive components) [1]. Nevertheless, obtaining foams with a fine cell structure is restricted due to the weak melt strength and melt elasticity of i-PP [2]. To address this, new generation benzene-trisamide based nucleating agents (NAs), which can be dissolved completely in i-PP melt at the operated processing temperature, have been used. The dissolved NAs have a tendency to crystallize upon cooling in very finely dispersed, one-dimensional, highly ordered nano-fibrillar objects which are formed by intermolecular hydrogen bonds between the stacks composed of trisamide based molecules. Self-assembly of these nanofibrils into three-dimensional objects, which act as nucleating sites for the foam cells, takes place [3]. In this study it was aimed to investigate the influence of 1,3,5-benzenetrisamide NAs with different molecular structures and concentrations on the nucleation behavior and foamability of extruded i-PP foams. Several testing methods were employed to characterize the morphological and mechanical properties of the foams. In the presence of NAs, i-PP exhibited excellent foaming performance, yielding smaller mean cell sizes and narrower cell size distributions than those of neat iPP. Depending on the molecular structure and the concentration of the NAs, significant enhancement in the compression modulus of the foams in comparison with a reference foam including talc was achieved. References: [1] R. Liao, W. Yu, and C. Zhou, Polymer, vol. 51, pp. 6334–6345, 2010. [2] K. Wang, F. Wu, W. Zhai, and W. Zheng, J Appl Polym Sci, vol. 129, pp. 2253–226, 2013. [3] N. Mohmeyer, B. Müller, N. Behrendt, J. Hillenbrand, M. Klaiber, X. Zhang, P. Smith, V. Altstädt, G. M. Sessler, and H.-W. Schmidt, Polymer, vol. 45, no. 19, pp. 6655–6663, Sep. 2004.

Foam injection molding of magneto sensitive polymer composites

Volpe Valentina [1)], Marco D Auria [2)], Luigi Sorrentino [2)], Davino Daniele [3)], Roberto Pantani [1), a]

[1]*DIIn - University of Salerno - Salerno - Italy.*
[2]*Institute of Polymers Composites and Biomaterials IPCB - CNR - Napoli - Italy.*
[3]*Department of Engineering University of Sannio - Benevento - Italy*

[a)]Corresponding author: rpantani@unisa.it

Abstract. The purpose of this work is obtaining magneto sensitive foamed parts made by iron microparticles distributed in a flexible polymer matrix. The material adopted is a compound of a thermoplastic elastomer with 20% by weight of iron microparticles. In particular, a masterbatch containing 50% by weight of iron microparticles was obtained by melt compounding in a twin screw extruder and was then diluted with the pure polymer directly in the hopper of the injection molding machine. A conventional injection molding machine was adapted to carry out the microcellular injection molding process. The cylinder of the injection molding machine was modified to host a system for controlling the quantity of gas injected during the batching step. The screw was modified introducing a mixing section in order to improve the dissolution of the gas inside the melt. An aluminum mold was designed and built in order to host an electro-magnet able to apply a magnetic field (up to 0.2 Tesla) during the injection and foaming steps. The samples obtained by foam injection molding achieved a density reduction up to 40% when compared with the corresponding non-foamed samples. In the case of samples obtained in the presence of magnetic field, the iron particles aligned in the direction of the field and organize into fibrillar structures of variable length. The mechanical tests have shown that the samples in which the iron particles are aligned in the direction of the magnetic field have a higher modulus compared to samples in which the particles are randomly dispersed. Furthermore, it was verified that the magnetic field aligned particles induces an anisotropic structural reinforcement. To the best of our knowledge, this work is the first attempt to obtain an alignment of filler particles by a magnetic field applied during foam injection molding

INTRODUCTION

Magneto-rheological (MR) materials represent a specific class of smart materials which can respond to an external magnetic field (MF) exhibiting a fast and reversible change of their viscoelastic properties. MR solids (such as elastomers, gels and foams) can be easily shaped and, unlike MR fluids, retain their shape after the molding process. In contrast to MR fluids, particles in magneto-sensitive (MS) solids cannot freely diffuse within the polymeric network and thus their average position is set during the molding and consolidating process. Therefore, depending on whether or not the particles are aligned by an applied MF during the consolidation of the polymer, the resulting spatial distribution of the particles can be either isotropic or anisotropic, respectively [1]. Several MS solids based on polymers and aligned magnetic particles have been proposed [2], but the high viscosity of the polymer precursors during the consolidation process implied the need for very high MFs (up to 1 T) to optimally distribute the particles and to exploit their magneto- elastic response, thus reducing their potential application fields.
The idea exploited in this work is to use foam injection molding to obtain lightweight magneto-sensitive solids. Among polymer processing technologies, injection molding for thermoplastic foams allows large scale productions of internally porous 3D articles. The polymer viscosity is kept low by the presence of the gas, and thus in principle the magnetic particles can be aligned by the application of a low intensity MF

.

MATERIALS

The magnetoparticles adopted in this work are Carbonyl Iron Powder produced by BASF with the trade name of CIP SQ, having a spherical shape, with an average diameter of about 5micron.

The polymeric material was a polyolefin elastomer, Engage 8402 supplied by DuPont Dow Elastomers (Midland, Michigan, USA).

ESPERIMENTAL

The custom electromagnet (Figure 1a), attached to the stationary platen of the injection molding machine (Figure 1b), consists of a laminated core with standard M235-35A ISO sheets and two coils of copper wires (diameter 1.4 mm, tolerance 2 mm) with 1200 windings each connected to a power supply. The electromagnet allowed to obtain a magnetic field of about 0.2T in the middle of the air gap.

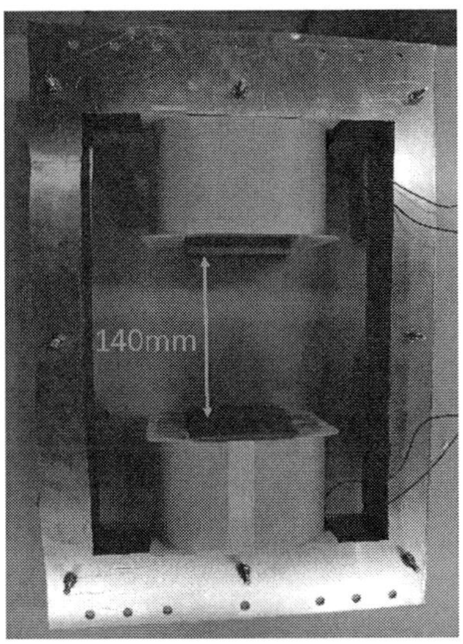

a

b

FIGURE 1. a) Custom Electromagnet; b) aluminum mold hosting the electromagnet

The cavity with a dumbbell geometry (Figure 2a) was obtained on a circular insert mounted on the moving half of the mold. The rotation of the circular insert allows to orient the dumbbell sample at different angles with respect to the magnetic field lines (Figure 2b). In this work, the dumbbell samples were molded at 0 degree, 45 degree and 90 degree with respect to the magnetic field orientation.

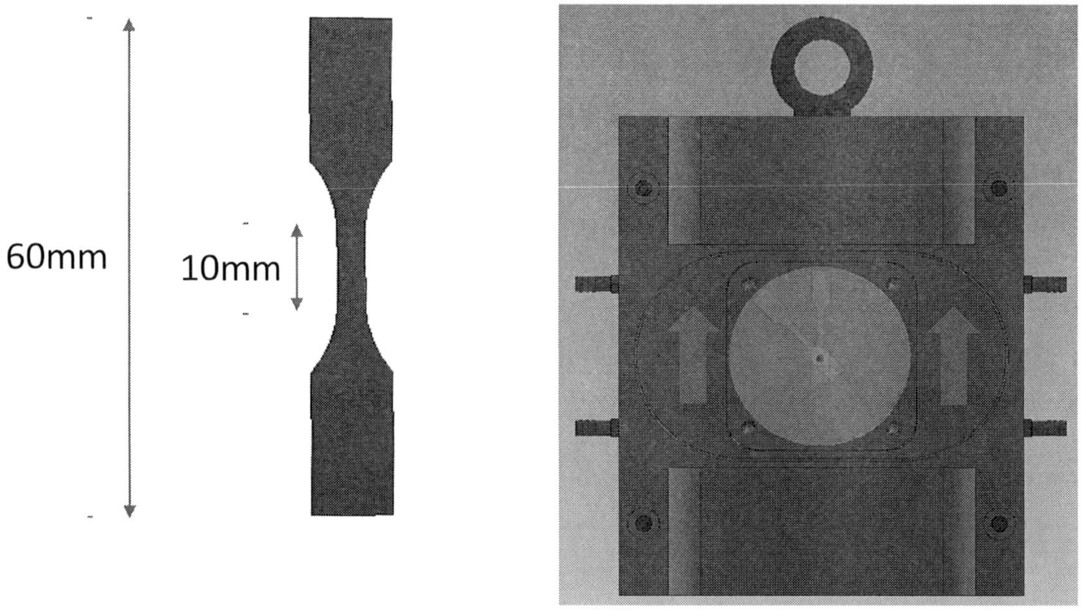

a b

FIGURE 2. a) Geometry of the molded parts; b) placement of the dumbbell geometry inside the moving half of the mold; the arrows indicate the direction of the magnetic fiels

Foam injection molding tests were performed on neat elastomers and their reinforced foams (containing 20% by weight of iron particles, corresponding to about 3% by weight) by solubilizing nitrogen, which was injected at a pressure of 150 bar into the barrel of the injection molding machine (a 70 ton CANBIMAT 65/185, from Negri-Bossi SpA, Italy) [3, 4]. The injection system measures the mass of gas conveyed to the barrel by monitoring the values of pressure and volume before and after each gas injection.

The samples have been tested in traction mode with a Lloyd LRXplus 5kN machine.

RESULTS AND DISCUSSION

The density reduction (R) with respect to the unfoamed part was calculated according to equation (1), where ρ_0 is the density of the unfoamed polymer (namely the same material, molded in the same conditions, but without nitrogen) and ρ_f is the density of the foamed part.

$$R = \frac{\rho_0 - \rho_f}{\rho_0} \qquad 1$$

The results of the density reduction are reported in figure 3. It can be noticed that a considerable reduction of the density was obtained for all the samples (more than 30% in all the conditions). A slight increase in density reduction is found for the samples containing microparticles. This is probably due to the nucleating effect of the microparticles, which enhances the foaming with respect to the pure polymer. The sample orientation with respect to the magnetic field did not affect the density reduction.

The results of the mechanical testing are reported in figure 4. On comparing the measurements carried out on the pure and the filled samples without magnetic field, it is clear that the filler itself increases the elastic modulus of the material

.

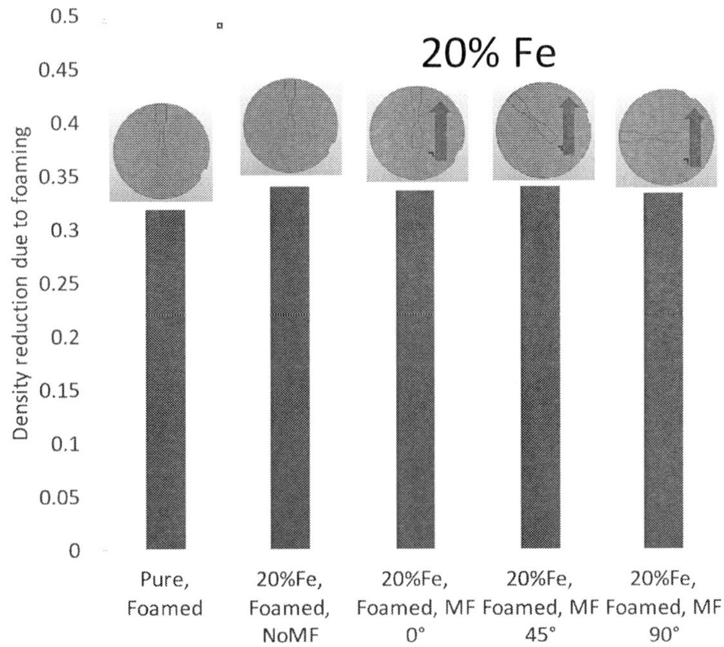

FIGURE 3. Density reduction of foamed samples with respect to a compact sample of the same material molded in the same conditions. The arrow indicates the direction of the magnetic field during molding. If the arrow is not present, the sample was molded without magnetic field.

The results of the mechanical testing are reported in figure 4. On comparing the measurements carried out on the pure and the filled samples without magnetic field, it is clear that the filler itself increases the elastic modulus of the material.

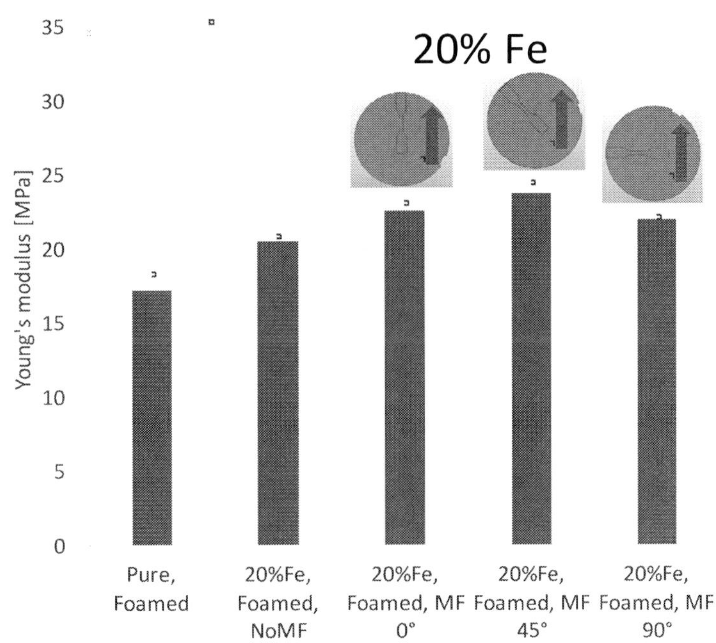

FIGURE 4. Young's modulus of the samples. The arrow indicates the direction of the magnetic field during molding. If the arrow is not present, the sample was molded without magnetic field

The samples molded under the magnetic field present a higher modulus with respect to those molded without the magnetic field. The orientation of the polymer flow with respect to the magnetic field is within the experimental uncertainty, even if it seems that the samples molded with the magnetic field orthogonal with respect to the injected flow direction present a lower modulus with respect to the other two configurations. Indeed, the samples with aligned particles shows a further 20% Young's modulus increment with respect to the filled sample with no MF applied. Among aligned samples, 0 and 45° perform better than 90° and this is in agreement with the results from [5] for open cell foams based on polyurethane matrix.

Further preliminary tests over the aligned samples (Figure 5) showed that they own an "active" character [6] and, if employed as cantilever beams, they visibly deform under constant magnetics fields.

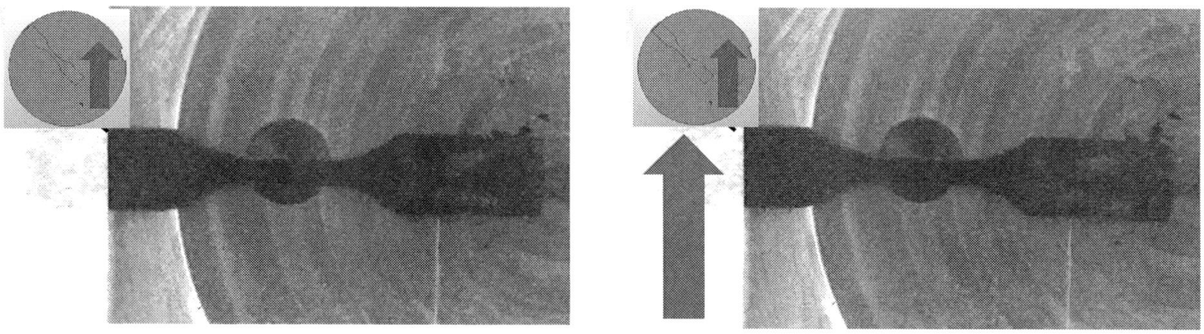

a b

FIGURE 4. The dumbbell sample placed into the electromagnet airgap, after production. The up left insets show the conditions of production while the red arrow show the magnetic field direction. a) The sample without magnetic field; b) The sample bent by the 0.2T magnetic field.

CONCLUSIONS

Composite lightweight materials based on a thermoplastic polymeric matrix with embedded magnetic micro-particles were produced by injection molding.

A magnetic field was applied to the mold during the polymer injection and sample cooling. This approach induced the alignment of magnetic particles along the MF lines and formed reinforcing chain-like structures. A special circular insert in the injection molding machine allowed the orientation of dog-bone samples at different angles with respect to the magnetic field lines. The mechanical characterization on reinforced samples showed that the use of MF to induce an anisotropic particle distribution in a thermoplastic polymer was an effective approach to increase the Young's modulus of injeciton molded foams, and that the best performing samples were produced at 45°.

Further preliminary tests over the aligned samples showed that they own an "active" character and, if employed as cantilever beams, they visibly deform under constant magnetics fields

ACKNOWLEDGMENTS

The research activity was supported by the Italian Ministry of Education and Research through the EPOCAM project (prin2012jwpmn9).

REFERENCES

1. D'Auria M, Davino D, Pantani R and Sorrentino L, Smart Mater. Struct. **25** 055014 (13pp) (2016).
2. Chertovich A V, Stepanov G V, Kramarenko E Y and Khokhlov A, Macromol. Mater. Eng. **295** 336–41 (2010).
3. R Pantani, V Volpe, G Titomanlio, Journal of Materials Processing Technology **214** (12), 3098-3107 (2014)
4. Volpe, V., Pantani, R., Journal of Applied Polymer Science **132** (48), 42612 (2015).
5. Sorrentino L et al, Journal of Applied Polymer Science, **119**, 1239–1247 (2011).
6. Davino D et al, IEEE Trans on Magnetics **48** (11), 3043-3046 (2012).

New fibers from PCM using the conventional melt spinning process

Nguyen Hoai An Tran[a,*], Martin Kirsten[a], and Chokri Cherif[a]

[a]*Technische Universität Dresden (TU Dresden), Institut für Textilmaschinen und Textile Hochleistungswerkstofftechnik (ITM), 01062 Dresden, Germany*

[*]Corresponding author: nguyen_hoai_an.tran@tu-dresden.de

Abstract. *Due to the increasing demand for technical textiles, particularly regarding functional textiles for clothing and home textiles, further developments have mainly focused on the investigation and fabrication of functional chemical textile fibers. The Institute for Textile Machinery and Textile High-Performance Materials (ITM) is currently developing a new type of functional fiber made from Phase Change Materials (PCM) using the simple and conventional melt spinning method on a standard industrial-scale melt spinning machine. PCM had originally been considered one of the most effective thermal energy storage systems for buildings. However, we recently have successfully incorporated innovative PCM pellets in the form of bi-component core and sheath multifilament yarns with a PCM content of 32 volume percent; they offer high latent heat up to 22.2 J/g, which is almost three times as high as that of the state-of-the-art fibers.*

Keywords: Phase Change Material, Bi-component Fibers, Melt Spinning

INTRODUCTION

The integration of Phase Change Materials (PCM) into textiles has attracted considerable attention recently, not only in industry [1, 2], but also in academic research [3-5]. There are various methods available for incorporating PCM into textiles such as coating, lamination, finishing, injection molding, foam techniques, and fiber technologies (dry, wet, and melt spinning) [5-7].

Among these methods, melt spinning is considered an economical and environmentally friendly process because of its high productivity and the absence of additional toxic substances. However, only few works have successfully incorporated PCM into textiles using the conventional melt spinning process due to high processing temperatures. Recently, Iqbal and Sun have developed multifilament polypropylene yarn containing PCM by mixing PCM pellets with Polypropylene (PP) [7]. The latent heat of this yarn was 8.08 J/g. Unfortunately, the PCM in these blend fibers can be easily removed by washing, wiping, or abrasion.

Within this research project, to avoid these disadvantages, we used bi-component fibers with core-sheath structures for the melt spinning process, in which PCM as core component is enclosed within the fibers by the sheath PP component. Furthermore, a high latent heat capacity of 22.2 J/g was obtained, which is almost three times as high as to that of state-of-the-art fibers [7].

EXPERIMENTAL

Phase Change Material as a commercial product was kindly supplied by our research partner, S-Textil Tech, Hof, Germany. Polypropylene (PP) Moplen HP561R, was kindly provided by LyondellBasell, Netherlands. Before melt spinning, PCM was dried at a temperature of 30 °C for 8 h.

The thermal behavior of PCM and PP was determined using Difference Scanning Calorimetry (DSC) and Thermogravimetric Analysis (TGA). All DSC and TGA analyses were carried out in a nitrogen atmosphere using a DSC Q2000 calorimeter (TA Instruments, Delaware, USA) and a TA Instruments Modulated TGA Q500 Thermal Analyser, respectively.

The melt spinning was performed on an industrial-scale bi-component melt spinning machine used for the fabrication of synthetic fibers at the Institute of Textile Machinery and High Performance Material Technology (ITM) with a 60-holes spinneret. The throughput is 60 g/min (at a ratio of core and sheath 1:1) with a take-up velocity of 1000 m/min. The sheath component is PP and the core component is a mixture of PCM and PP. The PCM content of the core component was varied with various volume percentage fractions of 22, 36, 50 and 64 % (PCM22, PCM36, PCM50, and PCM64).

The cross-section of fibers was investigated using a light microscope Carl Zeiss Axio Imager M1m with a digital colour camera AxioCam MRc5, Germany. The samples were observed at 1000 x magnification. The fiber surfaces was analyzed using a Scanning Electron Microscopy (SEM) (DSM 982 Gemini Carl Zeiss AG, Jena, Germany). All samples were observed at 1000 x magnification. The sample discs were prepared by sputtering a thin layer of 3 nm platinum.

The textile-physical properties of melt spun PCM-fibers were investigated using a Zwick/Roell Z 0.5 Universal Tester and controlled with TestXpert software (Zwick/Roell, Ulm, Germany) according to DIN EN ISO 2062. The fineness of the fibers (T_t) was measured by weighing 100 meters of fibers.

RESULTS AND DISCUSSION

Thermal behavior of PCM

Figure 1 shows a TGA curve and the thermal degradation of PCM. It reveals that PCM slowly started to lose its mass at around 150 °C, The loss in mass became substantial at 200 °C. The decomposition temperature of PCM is 234.5 °C, where PCM exhibits 5 % mass loss. At just above 250 °C, PCM can lose up to 10 mass %.

Figure 2 shows the isothermal thermogravimetric curves of PCM at different temperatures of 200, 210, 220, and 230 °C. The thermogravimetric (TG) curves in this Figure 2 were obtained by TGA measurement starting at 30 °C with a heating rate of 100 K/min up to the desired temperatures of 200, 210, 220, and 230 °C. The samples were then maintained at isothermal temperatures for 50 minutes. Figure leads to the conclusion that that PCM loses 5 % of its mass (decomposition point) at around 6.8 to 11.5 minutes depending on the isothermal temperature.

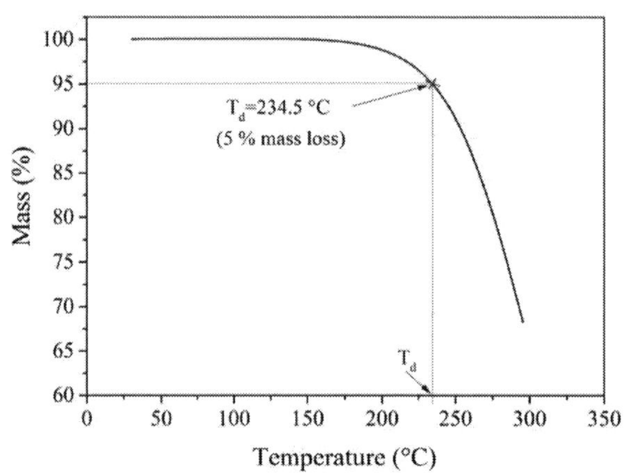

FIGURE 1. Mass loss vs. temperature in the TGA measurement.

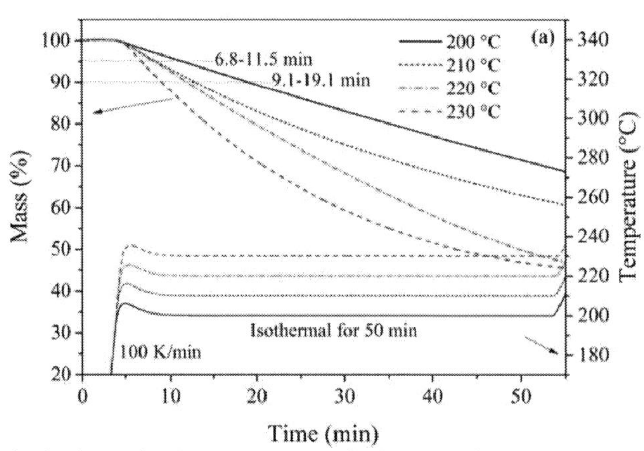

FIGURE 2. Mass loss vs. time in the isothermal TG measurement of PCM at various temperatures of 200, 210, 220, and 230 °C

Morphology and textile-physical properties of PCM bico fibers

Figure 3 presents a microscopic image of cross-section (left) and an SEM image of the surface morphologies of PCM64 bicomponent fibers (PCM64 bico fibers). It is seen that the PCM/PP blend is safely enclosed within the core component by the PP sheath component (left), which exhibits a smooth and featureless appearance (right). The filament has a diameter of about 35 μm. The fineness of the multifilament yarn (60 filaments) measured by weighing is T_t=62.9±0.2 tex, which is slightly more than the calculated fineness of 60 tex.

FIGURE 3. A microscopic image of cross-sections (left, scale bar:20 μm) and an SEM image of surfaces (right, scale bar:100 μm) of melt spun PCM64 bico fibers

Figure 4 illustrates the tensile strength and elongation properties of PCM bico fibers for different proportions of PCM in PCM bico fibers. With the increase in PCM content of the core component from 22 (PCM22) to 64 vol. % (PCM64), the tenacity σR(cN/tex) (the maximum specific stress) decreases from 18.2 to 6.8 cN/tex suggesting that the use of PCM led to a significant reduction in tensile strength of PCM bico fibers. Unlike the tenacity, the elongation of fibers showed a slight, linear increase with the increase in PCM content from 11 to 25 vol. %. However, with a high amount of PCM (PCM64 with 64 vol. % PCM in the core component), the elongation of fibers rapidly dropped from 306 (PCM50) to about 260 %, which is less than that of PCM11 bico fibers (Figure 10b). It could be due to PP amount in PCM64 bico fibers dropped to a lower limit value, which is equal to 36 vol. % PP in the core component. It was also found that, when the PCM content is higher than 64 vol. % (or the PP content is below 36 vol. %), the melt spinning process is no longer reproducible nor stable. Recently, Iqbal and Sun [15] have showed that PCM can be integrated into fibers with up to 8 % to develop multifilament yarn containing PCM. They utilized the melt spinning process of PCM and PP blends with their newly constructed spinneret. These results described above allow for conclusion that the incorporation of PCM into fibers leads to a reduction in tensile strength, but it results in an increase in elongation, if the PCM amount does not exceed certain limits.

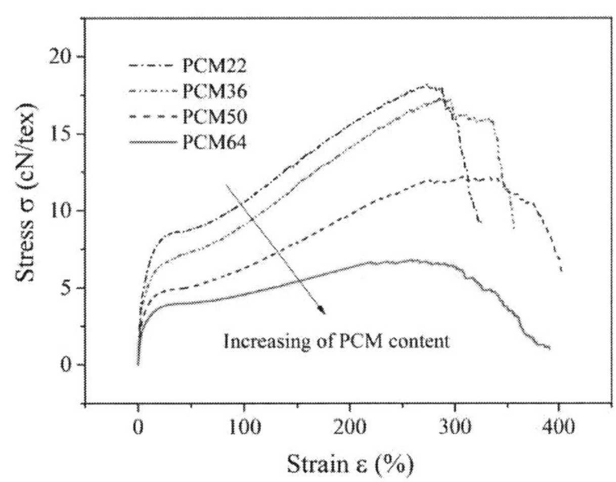

FIGURE 4. Stress-strain curves of PCM bico fibers with various PCM contents

Latent heat of PCM and PCM bico fibers

Figure 5 shows the DSC curves of PCM pellets and PCM64 bico fibers obtained using DSC measurement at a heating rate of 10 K/min from 0 °C to 220 °C. Table 1 summarizes the experimental and theoretical heat enthalpy (melting enthalpy ΔH_m and the cold crystallization ΔHc) of melt spun PCM bico fibers with 64 vol. % of PCM in the core component (32 vol.% in bi-component fibers). It is seen that the calculated and the measured latent heat of PCM64 bico fibres are different. The loss rates for melting and cooling enthalpies are 18.7 and 16.4 %, respectively. These differences in heat enthalpy are due to the thermal degradation of PCM, which occurs during the melt spinning process under the high process temperature. The melting enthalpy of melt spun PCM64 bico fibers is approximately 22.2 J/g, which is almost equal three times to that of state-of-the-art fibers [7].

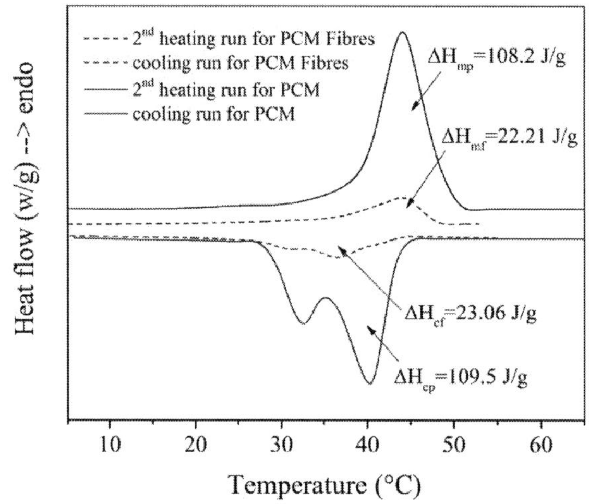

FIGURE 5. DSC curves of PCM pellets and PCM64 bico fibers

TABLE 1. Measured and calculated heat enthalpy of PCM bico fibers

Heat enthalpy	Experimental	Theoretical		Loss rate (%)
Melting enthalpy, ΔH_m (J/g)	22.2	27.3	(with 32 vol. % of PCM in	18.7 %
Cold crystallization enthalpy, ΔHc (J/g)	23.1	27.6	bi-component fibers)	16.4 %

ACKNOWLEDGMENTS

The authors thank the Deutsche Forschungsgemeinschaft (DFG) and the Dresden University of Technology (TU Dresden), and the Excellence Initiative "Support the Best", for funding of the bi-component melt spinning plant, and thus enabling this research project. We would also like to thank Mrs. Martina Dziewiencki, Mrs. Tina Anders, Mr. Michael Posselt (ITM, TU Dresden) for their technical support.

REFERENCES

1. M. H. Hartmann, J. B. Worley, M. North, *Cellulosic fibers having enhanced reversible thermal properties and methods of forming thereof.* 7790283, U.S. (2010)
2. B. Hagström, *Temperature regulating textile fibers containing large amounts of phase change material,* in '49th Dornbirn Man-made Fibers Congress/Chemical Fibers International'. Dornbirn, Austria, **60**, 221 (2010).
3. H. Yoo, J. Lim, E. Kim, *Effects of the number and position of phase-change material-treated fabrics on the thermo-regulating properties of phase-change material garments.* Textile Research Journal, **83**, 671-682 (2013).
4. G. Bartkowiak, A. Dabrowska, A. Marszalek, *Analysis of thermoregulation properties of PCM garments on the basis of ergonomic tests.* Textile Research Journal, **83**, 148-159 (2012).
5. W. L. Cheng, N. Liu, W. F. Wu, *Studies on thermal properties and thermal control effectiveness of a new shape-stabilized phase change material with high thermal conductivity,* Applied Thermal Engineering, **36**, 345-352 (2012).

6. N. Sarier, E. Onder, *Thermal characteristics of polyurethane foams incorporated with phase change materials*, Thermochimica Acta, **454**, 90–98 (2007).
7. K. Iqbal, D. Sun, *Development of thermal stable multifilament yarn containing micro-encapsulated phase change materials*, Fibers and Polymers, **16**, 1156–1162 (2015).

Experimental Study of Fiber Length Reduction of Highly Filled Long Glass Fiber-Reinforced Polypropylene in a Simple Shear Flow

Goris Sebastian [(1)]*, Simon Sara [(1)], Bechara Abrahan [(1)], Brands Dave [(2)], Yanev Angel [(2)], Osswald Tim [(1)]

[(1)] University of Wisconsin - Madison - Wisconsin - USA, [(2)] SABIC, Global Technology Automotive - Limburg - Netherlands

Fiber attrition during processing of long glass fiber-reinforced polypropylene at industry relevant fiber concentrations was studied experimentally to obtain a fundamental understanding of the physics in fiber breakage. The developed experimental setup of a Couette rheometer allows to study fiber breakage under highly controlled conditions in a simple shear flow. In combination with a novel fiber length measurement technique, the outcome of this work provides repeatable, accurate and robust measurement data quantifying the fiber length reduction at various process conditions. The experimental study aims to establish a theoretical link between processing conditions and fiber breakage of long glass fiber-reinforced thermoplastics. The impact of fiber concentration, initial fiber length, residence time, melt temperature and processing speed was quantified and studied. Even for the gentlest processing conditions, the results indicate that the residual fiber length after processing is less than 50% of the initial fiber length. For the most severe processing conditions, the results suggest a reduction to less than 10% of the initial fiber length, which highlights the challenges that fiber breakage poses for processing of long glass fiber-reinforced thermoplastics. The design of experiments aimed to isolate the effect of process variables on fiber breakage and their relative significance was determined. Lastly, the obtained data were used to test the Phelps-Tucker fiber breakage model, which assumes a buckling type failure of the fibers during processing. The continuum model was applied for all Couette rheometer experiments and the three empirical model parameters were fitted to the measurements. While it was possible to match simulation and measurements for all runs by adjusting the model parameter, the model parameters failed to show a material dependency of the parameters and it was not possible to compute them a priori.

Study of infrared sheet heating for thermoforming fibre-reinforced thermoplastic composite parts

Längauer Manuel [(1)]*, Zitzenbacher Gernot [(1)], Kneidinger Christian [(1)]

[(1)] University of Applied Sciences Upper Austria - Upper Austria - Austria

Lightweight structures made from fiber-reinforced thermoplastic semifinished parts gain more and more importance in the automotive and the aviation industry due to their excellent mechanical properties, their ability for functional integration, their low specific mass und their recyclability. Prior to forming, the sheets must be heated to soften the polymeric matrix material. This is mostly achieved via infrared radiation, convection or contact heating. Infrared heating is by far the most prominent method. A pilot experimental heating station was constructed to investigate this topic. It is equipped with ceramic heating elements for one-sided heating and it is designed to examine the behavior of heating methods and their influencing factors. Those include the temperature of the ceramic heaters, the distance from the heaters to the semifinished part and the number of heaters and their power. A simple clamping frame is used to hold the parts in place while four thermocouples on the top and the bottom of the sheets are included to monitor the temperature development. The experiments are conducted using twill-woven glass fibre reinforced polyamide 6.6 and polypropylene matrix systems of 0.5, 1 and 2 mm thickness. The measurements deliver sufficient data to enable the modeling of the heating behavior, especially the heat conduction through the thickness of the sheets. It can be shown that the heating behavior of thermoplastic composites is highly dependent on the used material and the experimental setup. In general, the surface that faced the heater exhibited a larger temperature distribution than the opposite side, which is an indicator for a high lateral thermal conductivity. The temperature difference between the heated and the opposite side had the tendency to be higher for PA 6.6 matrix systems in comparison with PP of equal thickness.

Silver Modified Polyarylonitrile-Activated Carbon Composite Fibers via Electrospinning Process

Thinnakorn Magdee[1, a)], Darunee Aussawasathien[2, b)], Sa-Nguansak Sriphalang[1, c)] and Kittipong Hrimchum[2, d)]

[1]*Department of Chemistry, Ubon Ratchathani Rajabhat University, 2 Ratchathani Road, Mueang District, Ubon Ratchathani, 34000, Thailand*
[2]*Plastics Technology Lab, Polymer Research Unit, National Metal and Materials Technology Center, 114 Thailand Science Park, Phahonyothin Road, Khlong Nueng, Khlong Luang, Pathum Thani 12120, Thailand*

[b)]Corresponding author: daruneea@mtec.or.th
[a)]nui24993@gmail.com
[c)]sangaunsak@yahoo.com
[d)]kittipoh@mtec.or.th

Abstract. This work involved the preparation of composite fibers from a mixture solution of 8% (w/v) of polyarylonitrile (PAN) containing silver (Ag) particles via a redox reaction and activated carbon (AC) using dimethyl formamide (DMF) as a solvent. The AC content was varied at concentrations of 0, 1, 2, 3, 4, and 5% (w/v). The electrospun composite fibers (ECFs) were prepared by using the electrospinning technique. The surface morphology and content of Ag particles of ECFs were determined by using field emission scanning electron microscopy (FE-SEM) and X-ray fluorescent spectroscopy (XRF) methods, respectively. Moreover, the viscosity and conductivity of spinning solutions were investigated as well. The viscosity of spinning solution increased from 45 to 54 Pa. s with the addition of AC particles from 0 to 4 % w/v and then suddenly dropped to 36 Pa. s at 5 % w/v of AC. The conductivity of spinning solutions decreased from 540 to 139 μS cm^{-1} when the amount of AC increased from 0 to 5% (w/v). ECFs had rough and porous surfaces with fiber diameters in the range of 200-250 nm. The size of fibers decreased with increasing of bead fibers as the dosage of AC increased.

INTRODUCTION

Electrospinning is a well-known technique to produce continuous fibers from various polymers with sub-micron to nanometer sizes [1, 2]. Various factors can affect the production of electrospun fibers during the electrospinning process: (a) the solution properties, such as viscosity, elasticity, conductivity, and surface tension; (b) the process parameters, such as the gap distance between the needle tip and the collector, applied voltage, and hydrostatic pressure in the solution container; and (c) the environmental effects, such as temperature, humidity, and air flow. A higher solution viscosity results in a larger fiber diameter [1, 3, 4]. Fong and Reneker [5] recognized that higher polymer concentration resulted in fewer beads. The shape of the beads changed from spherical to spindle like, when the polymer concentration varied from low to high levels. The charge ions in the polymer solution greatly influence fiber formation. As the charges carried by the jet increase, higher elongation forces are imposed on the jet under the electrical field, resulting in smaller bead and thinner fiber diameters [7].

Recently, Ag-containing electrospun nanofibers have been of great interest as a novel form of anti-microbial material in various applications including wound dressings [8], tissue scaffolds [9], chemical and biological protective materials [10], and medical devices and biotextiles [11, 12] due to the combination of the high specific surface area and fineness of electrospun nanofibers with the biocidal activity of Ag nanoparticles. AC is a highly porous material and extensively used to remove a variety of pollutants such as dyes, heavy metals, pesticides, and gases owing to its large specific surface area and high adsorption capacity [13-15]. It is useful in many industrial sectors such as food, pharmaceutical, chemical, oil, mining and in the treatment of both waste and drinking water [16]. However, ECFs containing both Ag and AC particles prepared via electrospinning process have not been reported.

In this study, polyacrylonitrile (PAN) was chosen as the polymer matrix since it is inexpensive and feasible to make fibers via the electrospinning technique. The Ag modified PAN particles were prepared through the in-situ reduction of silver nitrate using hydrazinium hydroxide as a reducing agent. The Ag-PAN-AC spinning solution containing different AC concentrations were used to prepare composite fibers by electrospinning process. The effect of AC contents on the viscosity and conductivity of spinning solutions and the fiber morphology were investigated.

EXPERIMENTAL

Materials

Polyacrylonitrile (PAN, average molecular weight of 150,000) supplied by Sigma-Aldrich, USA, was used as a base material. AC (CGC-11A 200C) was purchased from Gigantic Carbon Co., Ltd. with surface area and average particle size of 1000-1100 $m^2\ g^{-1}$ (N_2 BET method) and 74 μm, respectively. N,N-dimethylformamide (DMF) purchased from Labscan was utilized as a solvent for the preparation of Ag-PAN composites and spinning solutions. Silver nitrate ($AgNO_3$), hydrazinium hydroxide (N_2H_5OH), and acetone obtained from Analar Normapur, Carlo Erba, and Labscan were used for synthesis of Ag nanoparticles.

Preparation of Ag-PAN-AC composite fibers

1.0 g of PAN and 0.64 g of $AgNO_3$ were admixed in 50 g of DMF and then stirred for 1 h at 60°C until yellow clear solution was obtained. 50.0 g of DMF containing 1.132 g of N_2H_5OH was slowly added to the $AgNO_3$/ PAN solution under magnetic stirring at 10°C in ice bath. After that, 500 ml of acetone was added to the solution followed by filtering so that the free Ag nanoparticles were separated rapidly from the PAN-protected Ag nanoparticles (Ag-PAN nanoparticles) [57]. The latter part was dried in the vacuum oven overnight at 60°C. The composition of Ag-PAN-AC spinning solutions was presented in Table 1. The mixtures were stirred at 60°C for 24 h prior to electrospinning. The dark-brown solution was held in a needle with the gap distance from the needle tip (diameter ≅ 0.1 mm) to the rolling collector covered with aluminium foil (speed ≅ 220 rounds min^{-1} (rpm)) at about 30 cm under a voltage of 27 kV. The flow rate of spinning solution was 0.05 ml min^{-1}.

Table 1. The composition of Ag-PAN-AC spinning solutions.

Sample	Ag-PAN nanoparticles (% w/v)	AC particles (% w/v)
Ag-PAN	8.0	0.0
Ag-PAN-AC-1	8.0	1.0
Ag-PAN-AC-2	8.0	2.0
Ag-PAN-AC-3	8.0	3.0
Ag-PAN-AC-4	8.0	4.0
Ag-PAN-AC-5	8.0	5.0

Characterization

The conductivity of spinning solutions was measured at 25°C before the electrospinning process using a waterproof multiparameter (pH/ conductivity/ TDS/ salinity/ temperature) tester (PCSTestr 35, Okaton Instruments, USA). The probe was immersed in 50 ml of spinning solution. The average conductivity of each sample from three measurements was recorded. The viscosity of spinning solutions was determined at 25°C by a viscometer (RT49427, Brookfield AMETEK, USA). 20 ml of spinning solution was filled in the cylinder and spun with the speed of 0-200 rpm. The average viscosity of each sample from three measurements was recorded. The surface morphologies of Ag-PAN-AC composite fibers was characterized using a Schottky field-emission scanning electron microscope (SEM, SU5000,

Coax Group Corp. Ltd., Thailand). Prior to characterization, the sample on Al foil was attached onto an SEM stub using conducting tape and then sputter coated with a thin layer of platinum. The amount of Ag in Ag-PAN-AC composite fibers was determined using an X-ray fluorescence spectrometer (XRF PW2404, Philips, Thailand). All XRF analysis was performed using an X-ray beam collimator, an accelerating voltage of 30-60 kV, and an electron beam current of 66-125 mA.

RESULTS AND DISCUSSION

Spinning solution properties and Ag content

FIGURE 1 (a) presents the viscosity of spinning solutions versus the AC content. The viscosity of spinning solutions tended to increase approximately from 45 to 54 Pa. s with the addition of AC particles from 0 to 4 % w/v and then suddenly dropped to 36 Pa. s at 5 % w/v of AC. The viscosity of spinning solution increased when the AC particles was added in the solution due to the dispersion of AC particles in the polymer matrix. However, poor distribution of AC particles in polymer solution occurred at high AC concentration owing to the agglomeration of AC particles, causing the reduction of spinning solution viscosity. Moreover, the agglomerate structure may penetrate between the polymer chains, facilitating the polymer flow due to the reduction of chain entanglements.

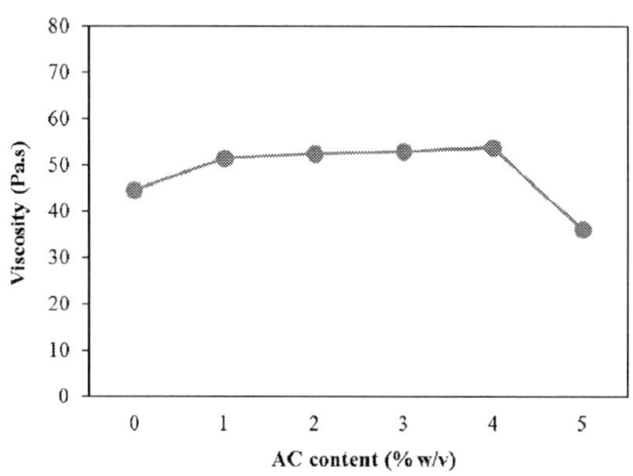

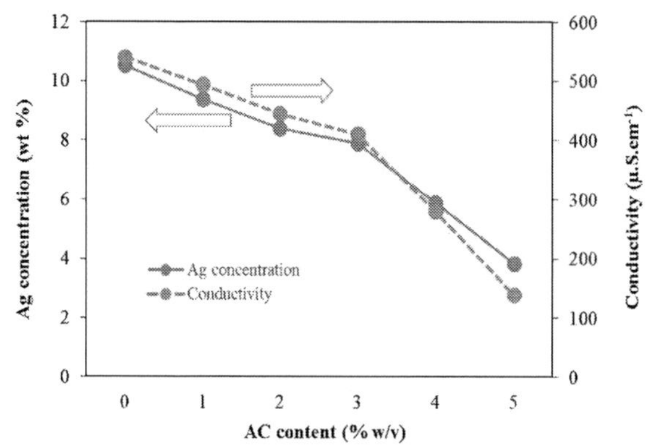

FIGURE 1. (a) Viscosities of spinning solutions and (b) conductivities of spinning solutions and Ag concentrations of Ag-PAN-AC composite fibers at different AC contents.

The Ag concentration obtained from XRF analysis decreased from 11 to 4 % as the loading of AC increased from 0 to 5 % w/v as shown in FIGURE 1 (b), since the proportion of Ag-PAN to AC decreased with the increase of AC content. The conductivity of spinning solution decreased from 540 to 139 $\mu.s.cm^{-1}$ when the AC concentration increased from 0 to 5 % w/v due to the reduction of Ag-PAN to AC ratio, corresponding to the Ag concentration results. Furthermore, AC has lower electrical conductivity than Ag in nature, therefore the conductivity of spinning solution decreased with the increment of AC loadings.

FIGURE 2 shows SEM micrographs of Ag-PAN and Ag-PAN-AC composite fibers with different AC dosages. The fiber diameter of ECFs was slightly influenced by the amount of AC. The composite fibers had rough and porous surfaces with some AC particles adhered on the fiber surfaces (see Figure 3). The average fiber diameter was approximately in the range of 200-250 nm. However, the bead fibers were obtained at high AC contents because of low viscosity of spinning solutions and AC accumulation, causing the difficulty in fiber stretching during the electrospinning process. This result corresponded to the previous work [1, 5]. In electrospinning, Solution concentration decides the limiting boundaries for the formation of electrospun fibers due to variations in the viscosity and surface tension [6]. For solution of low viscosities, surface tension is the dominant factor and just beads or beaded fibers are formed while above a critical viscosity or concentration, a continuous fibrous structure is obtained and its morphology is affected by the concentration of the solution. In addition, lower elongation forces were imposed on the jet under the electrical field at high AC dosage, e.g. 4 and 5% (w/v), since the polymer solution had low electrical conductivity, resulting in larger bead fibers. It has been found that with the increase of electrical conductivity of the solution, there is a significant decrease in the diameter of the electrospun nanofibers whereas with low conductivity of the solution, there results insufficient elongation of a jet by electrical force to produce uniform fiber, and beads may also be observed [7, 17].

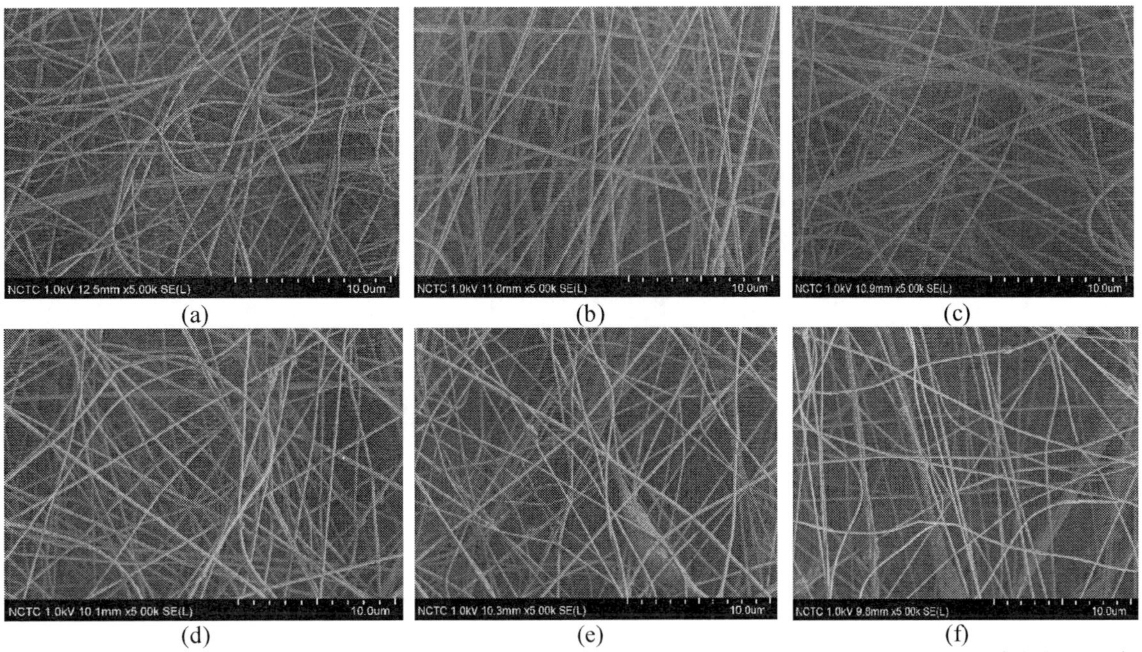

FIGURE 2. SEM micrographs of composite fibers with various AC loadings (% (w/v)): (a) 0, (b) 1, (c) 2, (d) 3, (e) 4, and (f) 5.

FIGURE 3. SEM micrographs of composite fibers at high magnifications: (a) no bead fibers and (b) with bead fibers.

CONCLUSIONS

In this work, ECFs were prepared from a mixture solution containing Ag-PAN and AC particles in DMF by using the electrospinning technique. The effects of AC contents on the viscosity and electrical conductivity of spinning solutions as well as fiber formation and morphology were investigated. The viscosity of spinning solution increased as the content of AC increased from 0-4% (w/v) and then suddenly decreased at 5% (w/v) of AC loading as a result of poor AC dispersion in the polymer matrix. Moreover, the AC agglomeration might occur and penetrate between polymer chains to reduce polymer chain entanglement and facilitate the polymer flow at high AC loadings. The conductivity of spinning solutions decreased when the concentration of AC increased from 0 to 5% (w/v) due to the reduction of Ag proportions, corresponding to XRF results. In addition, the size of fibers decreased with increasing of bead fibers as the dosage of AC increased. The obtained ECFs had a mean diameter of about 200-250 nm with rough and porous surfaces. The applications of as-prepared ECFs in methylene blue adsorption and antibacterial activity against Gram-positive bacteria (Staphylococcus aureus) and Gram-negative bacteria (Escherichia coli) will be further studied.

ACKNOWLEDGMENTS

The authors would like to express our appreciation to National Science and Technology Development Agency (NSTDA) for the senior project scholarship 2015 under Young Scientist and Technologist Program (YSTP).

REFERENCES

1. J. Doshi and D. H. Reneker, J. Electrostat. **35**, 151-160 (1995).
2. M. Bognitzki, W. Czado, T. Frese, A. Schaper, M. Hellwig, M. Steinhart, A. Greiner and H. J. Wendorff, Adv. Mater. **13**, 70-72 (2001).
3. H. Fong, I. Chun and D. H. Reneker, Polymer **40**, 4585-4592 (1999).
4. P. K. Baumgarten, J. Colloid Interface Sci. **36**, 71-79 (1971).
5. H. Fong and D. H. Reneker, J. Polym. Sci. B: Polym. Phys. **37**, 3488-3493 (1999).
6. J. M. Deitzel, J. Kleinmeyer, D. Harris, N. C. Beck Tan, Polymer **42**, 261-272 (2001)
7. X. Zong, K. Kim, D. Fang, S. Ran, B. S. Hsiao and B. Chu, Polymer **43**, 4403-4412 (2002).
8. P-O. Rujitanaroj, N. Pimpha and P. Supaphol, Polymer **49**, 4723-4732 (2008).
9. Z-C. Xing, W-P. Chae, J-Y. Baek, M-J. Choi, Y. Jung and I-K. Kang, Biomacromolecules **11**, 1248-1253 (2010).
10. K. H. Wu, P. Y. Yu, Y. J. Hsieh, C. C. Yang and G. P. Wang, Polym. Degrad. Stabil. **94**, 2170-2177 (2009).
11. F. Furno, K. S. Morley, B. Wong, B. L. Sharp, P. L. Arnold, S. M. Howdle and et al., J. Antimicrob. Chemother. **54**, 1019-1024 (2004).
12. G. Yuan and R. Cranston, Text. Res. J. **78**, 60-72 (2008).
13. C. K. Singh, J. N. Sahu, K. K. Mahalik, C. R. Mohanty, B. R. Mohan and B. C. Meikap, J. Hazard. Mater. **153**, 221-228 (2008).
14. J. Acharya, J. N. Sahu, B. K. Sahoo, C. R. Mohanty and B. C. Meikap, Chem. Eng. J. **150**, 25-39 (2009).
15. J. Acharya, J. N. Sahu, C. R. Mohanty and B. C. Meikap, Chem. Eng. J. **149**, 249-262 (2009).
16. R. H. Hesas, A. Arami-Niya, W. M. A. W. Daud and J. N. Sahu, Bioresources **8**, 2950-2966 (2013).
17. N. Bhardwaj, S. C. Kundu, Biotechnol. Adv. **28**, 325-347 (2010).

Synthesis of porous polymers by emulsion templated step-polymerization

BARBARA Imane [(1)]*, **Deleuze Hervé** [(1)]

[(1)] ISM University of Bordeaux - Bordeaux - France

PolyHIPEs are cellular materials obtained by polymerization within High Internal Phase Emulsions (HIPEs). The continuous phase of the emulsion is constituted of monomers and the volume of the dispersed phase (water) is superior than 74 % in respect to the total volume of the emulsion. The polymerization occurs in the continuous phase to create the polymer matrix and the dispersed phase induces the porosity. The polymers obtained are macroporous with pore sizes greater than 10 μm. Due to their low density and their high interface, polyHIPEs are used in several applications, such as drug encapsulation, supports for catalysts, and supports for the immobilization of bio-molecules, etc. PolyHIPEs are generally obtained by free-radical polymerization; that restricts the choice of monomers mainly to styrene and its derivatives and to acrylates. The majority of high performance polymers are not obtained by free-radical polymerization. Therefore, it is of great interest to extend the variety of monomers used to synthetize polyHIPEs using polycondensation.The objectives of this work are to synthetize and characterize polyHIPEs obtained by step growth polymerization: polycondensation or polyaddition in concentrated emulsion. Achieving a step polymerization without destabilizing the HIPE is a great challenge. Step polymerization requires extreme reaction conditions such as high temperatures and long reaction durations. Additionally, upon polycondensation a by-product is produced that has to be removed from the reaction media to achieve good yields. In order to successfully perform a polycondensation/polyaddition within the emulsion, we first focused on the synthesis of porous polyurethanes. Monomers were selected according to their Hydrophilic Lypophilic Balance (HLB) and synthesized. A stable emulsion was formulated and subsequently the polymerization was initiated by adding a catalyst. Macroporous polyurethanes were obtained and the homogeneity of the morphology was optimized.

Strain-induced Crystallization of PLA and PP in Extrusion

Tabatabaei Alireza* [1], *Park Chul B.* [1]

[1] University of Toronto - Ontario - Canada

Crystallization has been reported as one of the important phenomena when foaming of semi-crystalline thermoplastics. In this study, the effect of processing conditions on crystallization kinetics of polylactic acid (PLA) and polypropylene (PP) was investigated in a tandem extruder by means of direct visualization. It was observed that crystallites formed for both PLA and PP by decreasing the cooling extruder barrel temperature. Using the concept of the axiomatic design approach, the effect of cooling and flow-induced crystallization was decoupled during extrusion. It was achieved by processing the melt at a low flow rate (identical cooling) followed by a sudden increase in the flow rate (higher strain rates). A significant crystallization enhancement was visualized and quantified when the molten polymer experienced higher strain rates while maintaining the identical cooling history. In addition, the extrusion foaming behavior of PLA and PP was investigated in terms of verifying and elaborating the role of induced crystals on the cellular structure. A correlation was made to better understand the effect of crystals on foaming by comparing the images in-situ taken from the crystallites to the ex-situ characterization of the foam samples.

Fine-tuning the properties of Expandable Polystyrene (EPS)

Rath Dieter (1)*

(1) Leistritz Extrusionstechnik GmbH - Nuremberg - Germany

If expandable Polystyrene (EPS) was a human being, it would have reached its retirement age these days. Invented and introduced in the early 1950s, it has served for more than 65 years as an efficient and versatile material for thermal insulation and packaging. But even after such a long adolescence there is still room for further improvement. There are two basically different ways to produce EPS beads. The traditional route is the autoclave process. In this case the blowing agent, usually pentane, is added in the final phase of the polymerization. The product is obtained in form of white pentane-loaded unexpanded spheres of approx. 1 mm diameter. The alternative is to use an extrusion process starting from general purpose polystyrene (GPPS). In the second half of the extrusion process the products are thoroughly homogenized and cooled. It is important to ensure that there is a single-phase mixture at the outlet of the extruder. Further cooling may be carried out in a subsequent melt cooler, which is a specially designed static mixer. Finally the EPS-beads are obtained by underwater pelletizing. The pelletizer is designed to work with pressurized water in order to prevent the beads from premature expansion at this stage. Properties and performance of extruded EPS are comparable to autoclave EPS in every respect. However, the extrusion process is in our view far more flexible and in the end more economical as it allows simultaneous incorporation of any additive and also the use of recycled PS. Leistritz has carried out studies in co-operation with Fraunhofer ICT. Results have shown that there is potential for higher cost efficiency in the extrusion process. EPS – 65 years and still going strong. The paper provides a view into the complex extrusion process of EPS manufacturing: From the complete production system consisting of both hardware, a high performance extruder with all its ancillaries to "software", i. e. state-of-the-art EPS recipes and process know-how.

Properties and Foamability of Styrene-Ethylene-Butylene-Styrene (SEBS)/ Polystyrene (PS) Blends

Banerjee Ritima [(1)]*, Sinha Ray Suprakas [(2)], Ghosh Anup K [(3)]

[(1)] Centre for Polymer Science and Engineering,Indian Institute of Technology Delhi - Delhi - India, [(2)] Council for Scientific and Industrial Research - Pretoria - SouthAfrica, [(3)] Centre for Polymer Science and Engineering, Indian Instituteof Technology Delhi - Delhi - India

Thermoplastic elastomers (TPEs) are highly coveted, because they possess the properties of thermoset rubbers but can be processed like thermoplastics. Foaming will enhance their sustainability by lowering material consumption. SEBS is a commercially important TPE. SEBS is seldom used as a neat polymer, because of inferior mechanical properties. One way of enhancing the mechanical properties of SEBS is by blending it with a homopolymer. In this work, the properties and foamability of SEBS/ PS blends have been studied. SEBS and its blends (SEBS-10PS, SEBS-30PS and SEBS-50PS) containing 10%, 30% and 50 wt% PS were compounded and moulded for further studies. SEM studies revealed that a portion of the added PS increased the size of the styrene domains of SEBS, while the remaining portion got phase separated as seen earlier [1,2]. In SEBS-30PS and SEBS-50PS, the phase separated PS was in the form of fibrils aligned in the flow direction, leading to a significant enhancement of tensile modulus. All blends showed a negative deviation of melt viscosity from the additive rule of mixtures, attributed to the facilitation of slippage of SEBS by the low viscosity PS. When foamed with carbon dioxide at a higher foaming temperature (100 °C), SEBS-30PS and SEBS-50PS produced stable foams of higher volume expansion ratio, attributed to higher values of complex viscosity and storage modulus. Foams of SEBS and SEBS-10PS shrank. At a much lower foaming temperature (35 °C), PS showed a nucleating effect. However, all foams shrank. A higher PS content led to a more prominent shrinkage, attributed to the selective foaming of the elastomeric phase, hindered by the stiff PS aggregates [3]. Thus, SEBS/PS blends of a higher PS content have superior mechanical properties, enhanced flow at higher shear rates and superior foamability at higher foaming temperatures, as compared to blends of lower PS content. References [1] Raha, S; Kao, N; Bhattacharya, SN. Polymer Engineering & Science, 52(2012); [2] Sharudin, R; Ohshima M. Journal of Applied Polymer Science, 128(2013); 2245 [3] Banerjee, R; Ray, SS; Ghosh AK. Journal of Cellular Plastics, doi: 10.1177/0021955X16652108

Cyclic testing of polyurethane foam of plastic jacket pipes for district heating

***Leuteritz Andreas* [1]*, *Below Heiko* [2]**

[1] Leibniz-Institut für Polymerforschung Dresden e.V. - Saxony - Germany, [2] IMA Materialforschung und Anwendungstechnik GmbH - Saxony - Germany

Combined power and heat generation is state of the art to enlarge the energy usage of fuels above the level of producing electricity only. Therefore, in any case district heating is a simple and effective energy saving tool and for distribution plastic jacket pipes are used. For even larger reduction of pollution and carbon dioxide emission integrating solar power stations into district heating systems is one option. Plastic jacket pipes for district heating consist of a steel medium pipe with rigid polyurethane foam as insulation material and polyethylene casing. These pipes have to withstand stresses caused by the thermal elongation during heating and cooling cycles. Such stresses have to be transferred to the surrounding, which is in the most cases soil as compacted sand bed. Due to the specifics of solar radiation a fast change in high and low power output can happen e.g. during cloud movement. This is directly transferred into increasing number of load cycles causing cyclic stresses to specifically the rigid polyurethane foam. There is little known about the behavior of such rigid polyurethane foams under cyclic stresses over time. First attempts to characterize this kind of foams by using both static and dynamic testing combined with thermal and climate ageing had been performed. The results of such tests will be discussed and the testing scenario for the respective plastic jacked pipes presented.

Precursor fibers from Lignin-Polyacrylonitrile blends regenerated from chloride based ionic liquid solutions

Al Aiti Muhannad (1)*, Wulff Lucas (1), Brünig Harald (1), Wagenknecht Udo (1), Heinrich Gert (1)

(1) Leibniz Institute of Polymer Research Dresden - Sachsen - Germany

Precursor fibers derived from pure BioChoice® softwood and Organosolv Alcell® hardwood Lignin / textile graded Homo- and Co-Polyacrylonitrile (PAN) blends were spun utilizing chloride based ionic liquids (1-butyl-3-methylimidazolium chloride (>99%)) as solvent by dry jet wet spinning. Spinning dopes with different polyacrylonitrile concentrations and Lignin concentrations have been developed at low temperature, 85 °C. Shear rheological measurements were performed with rotational rheometer under different conditions, including different processing temperatures in the range between 65 °C and 105 °C, and different concentrations of lignin and PAN. Utilizing the oscillation measurements, Carreau-Yasuda model has been used for describing the pseudo-plastic flow behavior of those blends. Zero shear viscosity, infinite viscosity and consistency and the flow behavior power number have been calculated for the spinning dopes. Higher lignin share shows characteristic frequency and temperature dependency of the complex viscosity. Increased Organosolv lignin content lowers the complex viscosity at lower frequencies. At critical frequencies, the complex viscosity has been found to be equivalent for all lignin shares. At higher frequencies and higher lignin shares the complex viscosity shows higher values. The plasticizing effect caused by lignin suggests the in-situ fractionation of lignin by the ionic liquids at the blending stage. The activation energy of the dopes has been calculated using the Arrhenius plots. Time sweep measurements showed that the dopes were found in rheological terms to be thermally stable at 85 °C and up to 104 seconds. After 10^4 the stogare modulus increases which indicate the crosslinking of the structure. Comparative TGA-FTIR analysis of the pure polymer components and the blends after the regeneration from the ionic liquids showed a significant increase of the carbon yield at 800 °C of around 10%. Different spinneret diameters with constant characteristic L/D ratio of 2 have been used for the developing of the fibers. It has been found that the higher the lignin share is, the lower are the mechanical properties of the fibers.

Novel flame-retardant polymer bead foams

Gensel Julia [(1)]*, Klug Lisa [(1)], Niemeyer Gerd [(2)], Neumeyer Thomas [(1)], Pawelski Christin [(3)], Schreier Peter [(1)], Teubert Wolfgang [(4)], Wienhold Uwe [(1)], Wörthwein Hans [(5)], Altstädt Volker [(1)]

[(1)] Neue Materialien Bayreuth GmbH - Bayreuth - Germany, [(2)] Polysil Coatings GmbH - Wolfsburg - Germany, [(3)] University of Bayreuth, Polymer Engineering - Bayreuth - Germany, [(4)] Teubert Maschinenbau GmbH - Blumberg - Germany, [(5)] Wörthwein-EPP technologies - Cleebronn - Germany

One of the key challenges in the field of polymer engineering is the development of lightweight materials with improved flame retardant properties. Expanded polypropylene (EPP) is a low density material (typically 20-200 kg/m3) with excellent thermal insulating properties, very good energy absorption properties and complex 3D geometries. Therefore, this class of materials is used in many industrial applications, e.g. automotive, packaging, furniture etc. However, in order to increase the application scope of EPP, added functionalities are in a great demand. Since the modification of the polymer foam matrix is a challenge regarding the bulk properties and the foamability of the polymer composites, there is a need for universal functionalization methods. Flame retardant coatings on foamed beads provide a versatile and elegant route towards achieving this aim: the composition of the coatings can be adjusted on demand, which is more flexible than polymer matrix modification, while the bulk properties of the bead are not influenced. Motivated by these benefits, we present a novel approach for the engineering of multifunctional EPP foams. Applying a tailored surface pre-modification and subsequent steam-chest moulding, functional components were included within bead interfaces. Depending on the coating, it was possible to produce EPP foams with improved flammability (smoke and heat release) and self-extinguishing behavior. Further, since a good fusion of beads is necessary in order to achieve favorable mechanical properties, we discuss the effect of incorporated functional components on the bead fusion in the resulting foams. It was found that both, the nature and the size of additives strongly affect the welding quality due to different interdiffusion stages. These findings provide a deeper knowledge about the influence of heterogeneous interbead layer on EPP properties and allow further development of novel materials with fine-tuned functional properties.

Preparation Of Intelligence Antibacterial Films Based On Low Density Poly Ethylene And Evaluation Of Its Properties

Atabaki-Pasdar Pouria [(1)], Mehranpour Milad [(1)*], Oromiehie Abdulrasoul [(2)], Bahari-Aban Babak [(1)]

[(1)] Islamic Azad University Science & Research branch - Tehran - Iran, [(2)] Islamic Azad University South Tehran branch - Tehran - Iran

In this article, antibacterial film based on low density polyethylene (LDPE), potassium sorbate (KS) and ethylene vinyl alcohol (EVOH) copolymer were prepared and characterized using scanning electron microscopy(SEM), transmission electron microscopy (TEM), X-ray diffraction (XRD), differential scanning calorimetry (DSC), and tensile tests. The oxygen barrier properties of the LDPE/EVOH/KS film were significantly better than those of the LDPE/KS film. In addition to barrier properties, the LDPE/EVOH/KS film also had better elastic modulus than their counterparts without EVOH. The antimicrobial film containing potassium sorbate decreased the growth rate and maximum growth of yeast, and extended the lag period before mold growth was apparent. Therefore, it can prevent or reduce the rate of microbial spoilage in low viscosity liquids and on the contacted surface area of solid food products and as the result may prolong the shelf life of food products when it is used as a packaging material. Also, a novel sticker sensor has been fabricated based on methyl red, and tests have been conducted to detect the freshness of meat by its color change in real time (when the red methyl red paper changed to yellow). These are prepared by entrapping within a pH sensitive dye that responds, through visible color changes to the spoilage volatile compounds that contribute to a quantity known as total volatile basic nitrogen (TVBN). Finally, the methyl red was successfully used as a sticker sensor for the real-time monitoring of meat freshness in ambient and chiller conditions.

Influence of Raw Material Properties on Swelling and Spinning Stability of UltraHigh Molecular Weight PolyethyleneMonofilaments

***Yang Suhan* (1), *Yan Yurong* (1)*, *Jiang Deyuan* (1), *Liu Xin* (1), *Hufenus Rudolf* (2)**

(1) South China University of Technology - Guangdong - China, (2) Empa, Swiss Federal Laboratories for Material Science and Technology - St. Gallen - Switzerland

Ultra high molecular weight polyethylene (UHMWPE) monofilaments areconsidered as futurematerial for high performance lightweightcomposites. This prospect is mainly based on their high specific tensile strength, whichis the resultof an optimized gel-spinning process and amaximizeddraw ratio.Key steps in gel extrusion and spinning fibers from UHMWPE arehomogeneousswelling and gelation,which aredetermined by the raw material properties. In this study, three UHMWPE powderswith different average molecular weights and particle morphology were studied regarding their influence on swelling and spinnability of gel-spun UHMWPE monofilaments.The monofilaments were prepared by gel-spinning on a twin screw extruder using paraffin as solvent, followed by solvent extraction in xylene, and ultra-high drawing on heated godets.The testing methods applied were thermal analysis, SEM, XRD analysis, and measurement of bulk density and swelling behavior, The results show that the optimalswellingtemperaturedepends onmelt temperature of the raw material and powder particlesize, the swelling speed on melt temperature and particle porosity, and the swelling ratio on average molecular weight, particle size and distribution, andcrystallinity of the UHMWPE powder. Swelling temperature, speed and ratio on their part influence the shish-kebab formation during spinning,as well as theultra-high draw properties of gel-spun UHMWPE monofilaments.

Thermoelectric properties of PEDOT:PSS modified with vitamin C

Gnanaseelan Minoj (1)*, samanta Sumanta (1), Pötschke Petra (1), Pionteck Jürgen (1)

(1) Leibniz-Institut für Polymerforschung Dresden e.V. - Saxony - Germany

Thermoelectric (TE) materials convert waste heat into electricity; hence, this is foreseen as a possible source of green energy. The transformation of energy by TE generators has several advantages, namely they operate without moving parts, do not create noise, are less bulky, environment friendly, and may be coupled with other modes of energy generation. Compared to traditional TE materials like bismuth telluride or other metal oxides, polymeric materials are much easier to be fabricated into devices. Thus, there is a growing interest in such materials. PEDOT:PSS (1:2.5 by wt.) is one of the commercially available conducting polymers. However, it has relatively low electrical conductivity (σ) and Seebeck coefficient (S) and, hence, low power factor (S2σ). Vitamin C (L-ascorbic acid) is a natural weak acid which is found in citrus fruits, vegetables, etc. Previously, the use of mineral acids like sulfuric acid, hydrochloric acid etc., and organic solvents like dimethyl sulfoxide, ethylene glycol etc. for improving TE properties has been investigated, however, these compounds are not environment friendly. In view of environmental safety, in this study, the effect of vitamin C concentration (0-10 wt%) on electrical conductivity and Seebeck coefficient of PEDOT:PSS is investigated. It is found that vitamin C increased the electrical conductivity without negatively affecting the Seebeck coefficient. Hence, it could improve the power factor (~0.2 μW/mK2) of the conducting polymer, eventually increasing the efficiency of TE generators.

Novel method to measure surface crystallinity of fibers and nonwovens

Lanyi Franz J. [(1)]*, Wenzke Nicolai [(1)], Kaschta Joachim [(1)], Schubert Dirk W. [(1)]

[(1)] Friedrich-Alexander-University Erlangen-Nürnberg - Bavaria - Germany

The production of fibers from a polymer melt is an important process in many industrial sectors where fibers and nonwovens are utilized as hygiene and medical products, as geotextiles or as filter materials. The properties of such highly oriented products, like nonwovens or films, strongly depend on their crystallinity. As fibers are exposed to high cooling rates and draw ratios during their production, crystallinity normally is not a bulk property but a function changing with the diameter of the fiber or film and therefore bulk properties often are not informative for end consumer characteristics like a soft feel or the migration of additives to the surface. Currently the crystallinity of polymers is determined via differential scaling calorimetry (DSC) or via X-ray diffraction (XRD). Even though DSC is an easy and fast way to determine crystallinity, it suffers several disadvantages. In DSC studies sample preparation of fiber material is difficult, results are not calibration free and heat is introduced in the measuring process which might change instable crystal structures in the polymer. XRD is able to overcome these problems but is time intensive and complicated in terms of sample preparation and evaluation of the measurements. Furthermore, it is not possible to determine surface crystallinity with both methods. This study shows a novel approach to utilize Fourier transform infrared spectroscopy in attenuated total reflection mode (FTIR-ATR) as a highly sensitive, non-destructive and fast way to measure surface crystallinity of fibers in a penetration depth of 0,5 – 2,5 µm. Samples without crystallinity gradient produced with different cooling rates from 1 – 3500 K/min were measured via XRD and FTIR-ATR to calibrate the method. Subsequently fibers were produced at different processing conditions on a pilot spunbond line to verify the method and show its applicability to measure surface crystallinity of fibers and nonwovens.

A comprehensive study on the properties of polypropylene fibers and nonwovens in dependence of processing and storing conditions

***Lanyi Franz J.* [(1)*], *Wenzke Nicolai* [(1)], *Schubert Dirk W.* [(1)]**

[(1)] Friedrich-Alexander-University Erlangen-Nürnberg - Bavaria - Germany

We encounter products made from polypropylene fibers and nonwovens in a number of different areas of our daily life, like in hygiene products or filter systems. During their production, parameters like the haul-off speed of the fiber, the processing temperature or the residence time have an impact on the physical properties like the (surface) crystallinity or the molecular weight and the mechanical properties like the Young's modulus or the stress at break. The complexity of industrial facilities for the production of nonwovens does not allow to change parameters quickly and independently. Therefore, it is hardly possible to reveal the influence of individual processing parameters and systematically build knowledge about these. A pilot spunbond line developed at the institute of polymer materials allows to separate single process steps in order to show the influence of different spinning parameters on the properties of the single fibers and nonwovens. Even though studies exist which look into heat treatments of fibers, degradation of polypropylene in dependence of residence time and temperature or mechanical properties as a function of processing conditions, no studies exist which cover the influence of industrially relevant storing (time and temperature) and processing conditions (haul-off speeds, calendering settings and process temperatures) of fibers and nonwovens in their entirety. This study gives a comprehensive overview over the influence of mentioned spinning parameters on mechanical properties, degradation behavior and surface crystallinity of the products.

Development of a model for the mechanical behaviour of polypropylene nonwovens for different test directions and strain rates

Leucker Karsten (1)*, Schubert Dirk W. (1)

(1) Friedrich-Alexander-University Erlangen-Nürnberg - Bayern - Germany

During their usage nonwovens have to undergo a variety of load conditions with different directions and strain rates. Although the mechanical properties of nonwovens have been researched extensively there is still no model equation describing the tensile behaviour of nonwovens thoroughly. One reason for this is the anisotropy of the fibre orientation and thus of the mechanical properties of the nonwoven along with the local variation of base weight, which are both induced during the production process and cause variations in the mechanical performance [1]. While there are a few studies either dealing with the dependency of the tensile behaviour on the testing speed or with its direction dependency, the combination of both has not been investigated and modelled yet. Thus the aim of this work is to develop a model for the mechanical behaviour of polypropylene nonwovens for different test directions and strain rates. The mechanical properties of thermally point-bonded nonwovens, which are determined by the properties of the bonding points and of the unbonded areas, have been studied on two different polypropylene spunbond nonwovens with base weights of 15 and 25 gram per square meter. To work out a detailed model for the mechanical behaviour of the nonwovens the orientation distribution function of the filaments in both materials was determined. By performing tensile tests with three different speeds and in seven directions from machine direction to cross direction, a profound understanding of the tensile behaviour could be achieved. It appeared, that due to the strong anisotropy of the material the test direction has a much higher influence on the mechanical strength than the strain rate. Furthermore the results were used to develop a model for the mechanical behaviour of the nonwovens, which makes it possible to predict the stress-strain curves of nonwovens for any testing direction as well as for different strain rates and base weights. References: [1] Albrecht, W.; Fuchs, H.; Kittelmann, W. Nonwoven Fabrics: Raw Materials, Manufacture, Applications, Characteristics, Testing Processes (2006)

The SPLINE polymer extruder concept - A novel grooved barrel single screw extruder

***Miethlinger Jürgen* [(1)*], *Roland Wolfgang* [(1)]**

[(1)] Johannes Kepler University Linz, Institute of Polymer Extrusion and Compounding - Upper Austria - Austria

Grooved barrel single screw extruders (SSE's) are the first choice for high molecular weight polyolefin applications such as the extrusion of blown film, pipe and blow molded article. Nowadays grooved barrel SSE's are also used for a lot of special polymers. The grooves are classically designed with a tapered profile with a length of three to five times the screw diameter. The HELIBAR® system developed by Grünschloß has extended grooves with a more gentle tapered profile, which are also present in the plastication zone and enables reduced pressure, frictional forces and torque respectively high output at lower melt temperatures, good process stability, reduced wear and reduced energy consumption. The present work is related to a new grooved barrel SSE-design with extended grooves in the feeding and plastication zone without any tapered profile. That is without any compression and with extended diameter after the grooves in order to reduce pressure drop for metering and mixing zone. This new grooved barrel SSE-design is named SPLINE extruder emblematic following the designation of splined coupling in mechanical engineering. Results Due to the splines the mass throughput fraction of the grooves as well as the total mass throughput can be increased. The overall pressure profile could be decreased due to the larger volume at the end of the splines which is a result of this construction causing also more melt pumping capacity for metering zone and less pressure drop for mixing elements. Furthermore, the axial pressure profile is more balanced and levelled. High mass throughputs can be achieved at low melt temperatures, preferably with barrier screw concepts, as a consequence of dispersed melting effects causing specific cooling of the melt channel in barrier screws by means of individual solid particles. In summary the SPLINE polymer extruder concept offers significant benefits not only compared to smoothed barrel SSE's, but also towards classical grooved SSE's.

Mechanical characterization of iPP injection molded samples on multiple length scale

S. Liparoti[a*], *A. Sorrentino*[b], *V. Speranza*[a], *G. Titomanlio*[a]

[a] *Department of Industrial Engineering, University of Salerno, via G. Paolo II° 132, 84084 Fisciano (SA) Italy*
[b] *Institute for Polymers, Composites and Biomaterials (IPCB-CNR), P.le E. Fermi 1, 80055 Portici (NA) Italy*

*sliparoti@unisa.it

Abstract

The mechanical characterization at different length scales, from macro to nanometer scale, represents a very intriguing possibility to analyze the effect of developed morphologies on the properties of a molded polymer material. In this work, isotactic polypropylene molded samples, obtained with different processing conditions, have been mechanically characterized on different lengths scales. Dynamic Mechanical Analysis (DMA), micro Indentation and HarmoniX Atomic Force Microscopy (AFM) tests have been carried out to characterize injection molding iPP samples from millimeter to nanometer scale. Optical microscopy and AFM analyses show a complex morphology composed ? by a thin layer of globular elements at the sample surface, spherulites in the sample core and fibrils in the intermediate zone. The well organized and oriented structures, i.e. the fibrils, are characterized by higher values of mechanical modulus, whereas the globules, less organized structures, show very small elastic modulus values, close to the characteristic values observed in quiescent amorphous phase. Intermediate values of elastic modulus are observed in the sample core, where well developed spherulitic structures are characterized by both higher crystallinity and lower level of orientation.

The elastic modulus obtained by HarmoniX AFM technique are in good agreement with the values obtained by Indentation tests in the same sample positions, confirming the complementarity of these techniques for a complete multi-scale sample characterization. The cross section modulus distributions nicely compare with DMA results.

Introduction

Within the last two decades, an increasing interest from scientific and technological point of view has been focused on materials having complex micro and nanostructured morphologies [1,2]. The growing application of these materials requires the development of accurate material properties databases and realistic structure-processing-properties relationships. To that purpose, precise multiscale characterization methods and validated material models are necessary [3–5].

Only limited approaches have been tried to couple characterization results across different length-scales, moreover, the relationship among different morphological structures, whose presence is determined by local thermomechanical history, and the local properties of a polymeric material have received only limited attention [6–8]. For instance, in conventional and micro injection molding, strong viscoelastic behaviors coupled with inhomogeneous thermo-mechanical histories produce very complex microstructural organizations, with micro layers of different thickness and morphology, each of them being expected to have different mechanical properties [9–11]. Dynamical Mechanical Analysis (DMA), which is the most used and accepted mechanical characterization technique, is not able to provide direct correlation among microstructural organization and mechanical properties. In contrast, indentation tests give information about local mechanical properties at micrometric scale [12–20]. Recently introduced, HarmoniX AFM mode [21] has allowed to obtain elastic modulus maps with nano-metric spatial resolution. This technique provides simultaneously and at the same position the sample surface topography (morphology) and the map of the mechanical properties allowing a direct correlation of the sample properties with its local morphology. Up to now, AFM HarmoniX has never been used to measure elastic modulus of polymeric injection molding samples which are known to have complex morphology, composed by regions with different crystallinity and molecular orientations [10,22].

In this paper injection molded samples, obtained under different processing conditions have been mechanically characterized at different length-scale. The bulk mechanical properties have been measured by DMA, whereas both indentation and HarmoniX AFM have been used to detect the distribution of mechanical properties along the sample thickness, at different spatial resolutions. The overall mechanical characterization of iPP samples obtained under different injection molding conditions is compared with the micrometric and sub-micrometric characterization of both mechanical behavior and morphological features obtained by HarmoniX AFM.

Experimental

Polypropylene (iPP, Montell T30G, Mw=376000, Mw/Mn=6.7, tacticity=87.6%)[23–26] injection molded samples were obtained with two different mold temperatures (namely, 25 and 42°C) and two different packing pressures (namely, 260 and 720bar).

Molded samples were produced by injection molding adopting a 70-ton Negri-Bossi reciprocating screw injection molding machine [9,27]. The gate thickness is of 1.2 mm and the rectangular cavity has a length L=110 mm, width W=12.7 mm, and thickness S=1.5 mm [28]. A volumetric flow rate of 2.8 cm^3s^{-1} and a melt injection temperature (T_i) of 220°C were adopted.

Slices (0.1 mm thickness) were cut from the central part of the molded slab by means of a Leica slit microtome, along flow direction (x) and parallel to the flow-thickness plane (x,z). These samples were firstly subjected to morphology analysis by means of a light microscope (LM) model Olympus BX41 equipped with a digital camera and then mechanically characterized. Bulk moduli of the injection molded samples were determined by means of a TA instrument (DMA 2980) in tensile mode. Indentation tests were performed with a Nano Test™ Platform (Micro Materials Ltd). A three-sided Berkovich pyramidal diamond tip with a radius of about 100 μm was adopted to indent the surface of the sample. Tests were carried out with an initial load of 0.02 mN, a load rate of 1 mN/s and a maximum load of 120 mN held for 60s. Distribution of hardness and elastic modulus were evaluated from each imprint, directly by Nano Test™ softaware, according to the Oliver and Pharr method [29–31]. Superficially etched samples [6,10,32], were analyzed by AFM. The analyses were performed by a NanoScope MultiMode V scanning probe microscope (Veeco, Santa Barbara, CA) equipped with HarmoniX tool. Tests were performed with HMX probe silicon cantilevers (Bruker, Billerica, MA) with nominal radii of c.a. 10 nm.

RESULTS

An example of the sample morphology found in this position is presented in Fig. 1. The overall morphology of the sample produced with 25°C mold temperature and 260 bar packing pressure is shown by the optical cross section micrograph and it is deeper explored by magnifications in selected points by AFM height maps.

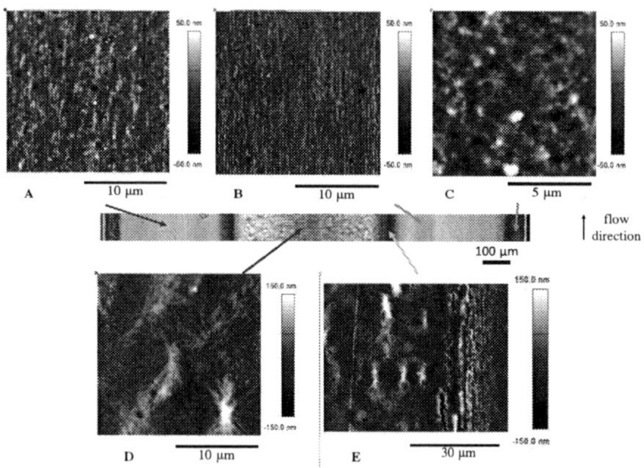

Fig. 1 Sample thickness morphology distribution detected by optical microscopy and AFM height maps. The positions where the analyses were performed are indicated by arrows pointing on the cross section optical micrograph. The sample was obtained with 25°C mold temperature and 260 bar packing pressure.

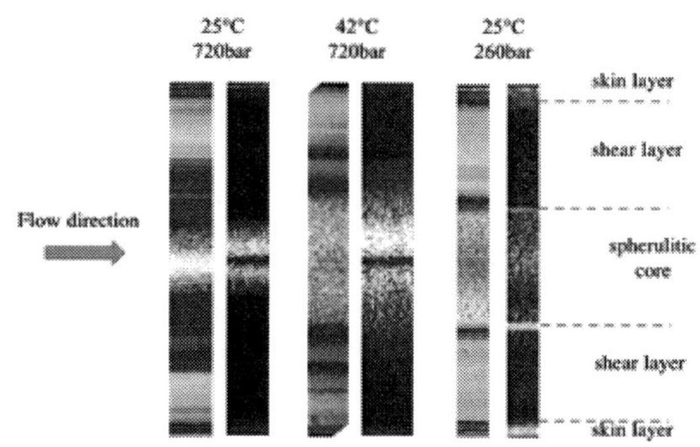

Fig. 2 Optical micrographs of the whole cross section (thickness-flow directions) of molded sample slices. The samples were obtained at two mold temperatures, 25°C and 40°C, and two holding pressures, 260 bar and 720 bar.

The sample morphology includes different areas: i.) a thin, oriented skin layer in the first 50 μm from the sample surface (Fig. 1C) composed by globular elements, ii.) highly oriented non-spherulitic zone (often referred as 'shear layer', Fig. 1A and B), that extends up to about 500 μm from the sample surface, composed by fibrillar structures, iii) a spherulitic core with small orientation (Fig. 1D). As evident from Fig. 1E, the morphology evolves gradually from fibrillar to spherulitic in the region named transitional [10]. In the literature, this type of structural organization is referred as skin-core morphology and is characteristic of injection molded samples [22,33–36]. The morphologies of the samples obtained selecting two packing pressures, 720 and 260 bar, and two mold temperatures, 25°C and 42°C, are compared in Fig. 2. Fig. 2 shows that the thicknesses of the different layers are strictly dependent on the injection molding conditions adopted. Indeed, the shear layer thickness, in which the frozen-in orientation is very high, is reduced by higher mold temperature and lower packing pressure [10,22,34].

HarmoniX AFM allows mapping the elastic modulus on the sample surface and, since it simultaneously detects morphology, it allows to identify a correlation between the morphology distribution along the sample thickness and the mechanical properties. The elastic modulus has been obtained from force-indentation curves by using Derjaguin-Muller-Toporov (DMT) model [31,37].

Fig. 3 shows the topography (height) and related maps of elastic modulus obtained by the HarmoniX AFM characterization of the samples produced with 260bar holding pressure and 25°C mold temperature. Below each topography (morphology) acquisition, the acquisition of the elastic modulus inside the same window and the elastic modulus profiles (overall elastic modulus scale of the sample taken along the white lines) are also reported.

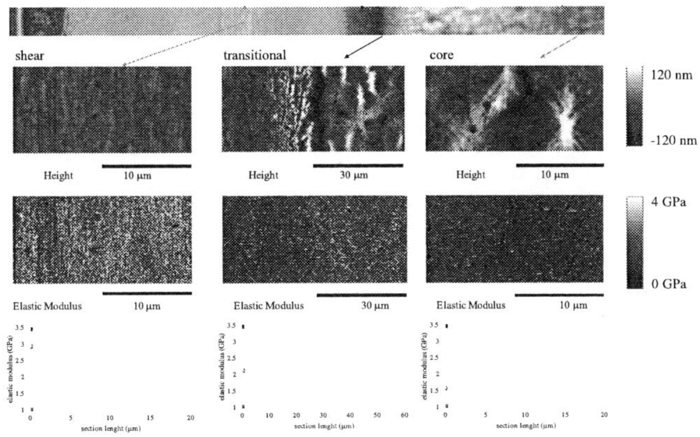

Fig. 3 Representative maps of Height (Morphology) acquisition on the top and, below each of them, corresponding elastic modulus obtained in the same window using HarmoniX on the sample molded with 260 bar holding pressure and 25°C mold temperature. The elastic modulus-profile (modulus vs. section length) represents an overall elastic modulus scale of the sample taken along the white lines shown in the image. Three representative positions along the sample thickness are shown: from left to right 300 μm (the shear) 500 μm (the transition) and 720 μm (the core) from the sample surface.

Inspection of Fig. 3 shows that, for the sample considered in the figure, the levels of elastic modulus are completely different among the three positions namely the shear, the transitional and the core layers (indicated in the figure). Indeed, the thin oriented fibrils present at 300 μm from the sample surface show higher moduli, whereas spherulitic core shows the lowest values of the modulus.

It would be interesting to compare the elastic modulus distributions obtained by AFM with the elastic modulus distributions obtained by elaborating the indentation data as described by Oliver and coworkers [38]. Thus, the moduli obtained by HarmoniX AFM analysis have been averaged on windows of 20x20 μm. These averaging windows are also comparable with the dimension of indentation imprints, which is about 60x60 μm.

Fig. 4 shows the effect of the processing conditions on the thickness distributions of the elastic modulus evaluated by both indentation and HarmoniX AFM analysis, the distributions of the number of fibrils per unit length (calculated from AFM height images) is also reported in the figure. The fibrillary density distribution on the cross section shows a good correlation with the modulus.

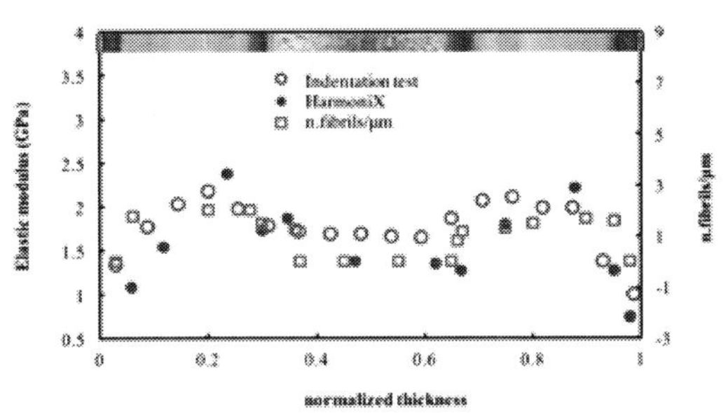

Fig. 4 a) Fibrillar density and elastic modulus distribution f evaluated by AFM. The elastic modulus values measured by indentation tests are also reported for comparison. All the plotted values are related to the sample produced with 25°C of mold temperature and 260bar of holding pressure.

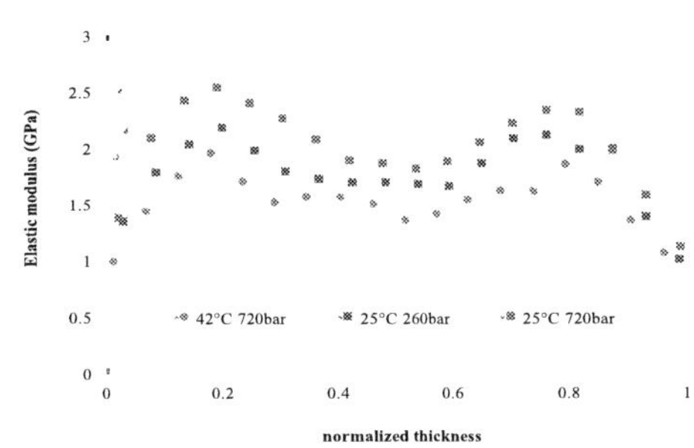

Fig. 5 Comparison of elastic modulus distributions obtained by indentation tests in samples produced with different holding pressures and mold temperatures.

In Fig. 4, the thickness distribution of elastic moduli taken on the cross section of the sample produced with 25°C of mold temperature and 260 bar of holding pressure nicely compares with the elastic modulus distribution obtained from indentation tests. This finding confirms the consistence of the results obtained at different scales. At the same time, these results demonstrate that HarmoniX AFM could be used as self-consistent analysis coupling morphology and modulus characterization in the same position.

In the area characterized by high fibrillar density, i.e. the shear layer, the elastic modulus values show a maximum. This maximum has to be generated by the cross section distributions of both cooling rate and molecular stretch. Lowest elastic modulus values are observed close to the sample surface, where fibrils are not present and globular elements are detected by AFM (Fig. 1C). The spherulitic core is characterized by constant values of elastic modulus, intermediate among the elastic moduli of shear layer and the elastic moduli close to the sample surface.

Fig. 5 shows the elastic modulus distributions obtained by Indentation tests on the other injection molded samples. The HarmoniX AFM analysis gives very similar results both for the modulus and for the hardness. In all the cases, the higher modulus values were detected in the fibrillary zone and the minimum values at the skin, consistently with the poor molecular organization observed in that area. In the central zone of the samples the molecules solidify in presence of a less intense flow field and lower cooling rate, therefore they present lower orientation level but the organization of the structure is more efficient than at the surface.

As for the effects of processing variables, the modulus increases by effect of an increase of packing pressure under constant surface temperature or by effect of a decrease of surface temperature under constant packing pressure; both effects are consistent with an increase of the molecular stretch determined by increase of packing pressure and decrease of surface temperature, respectively.

As further confirmation of the reliability of the previous experimental data, the bulk mechanical properties of the molded sample has been investigated by DMA.

Table 1 compares the bulk elastic moduli obtained by HarmoniX AFM and indentation with the bulk elastic modulus obtained by DMA analysis. The elastic modulus distributions obtained by HarmoniX and indentation (shown in Fig. 4) have been integrated along the thickness in order to obtain the corresponding bulk values.

Table 1 Bulk moduli evaluated by indentation, HarmoniX AFM and DMA analysis for the samples obtained with different packing and mold temperature conditions.

	HarmoniX AFM	*indentation*	*DMA (0.1Hz) 0.1% strain*
25°C 260 bar	1.69 GPa	1.73 GPa	1.70 GPa
25°C 720 bar	2.10 GPa	2.00 GPa	2.10 GPa
42°C 720bar	1.26 GPa	1.40 GPa	1.40 GPa

The bulk elastic moduli obtained from indentation tests are in good agreement with the values obtained by DMA.

DMA testing data cannot be strictly compared with AFM data due to the different deformation geometries and testing frequencies. Nevertheless, they should reveal similarities [8], as also shown in Table 1. This finding is obviously related to the good agreement of elastic modulus distribution obtained using AFM and indentation analyses shown in Fig. 4.

In literature only few results are present on the comparison among DMA, indentation and HarmoniX AFM [39]. The present work confirms a good agreement among all the mentioned analyses, at least for iPP molded samples. This is an interesting result especially considering the main differences among the proposed techniques, i.e. spatial resolution, stress field, deformation scales, dynamic or continuous load regime.

CONCLUSION

In this work iPP molded samples obtained by selecting various conditions of packing and mold temperature have been mechanically characterized at different length-scale by AFM HarmoniX, indentation and DMA techniques. A good agreement among HamoniX AFM and the other more conventional techniques for the evaluation of mechanical properties has been found. HarmoniX AFM has been proposed for the first time to analyze mechanical properties distribution inside injection molding samples, correlating local mechanical properties to the different morphologies formed along the sample thickness, which have been analyzed at nanometric scale. The morphological analysis reveals, more clearly with respect to the previous literatures, that, when the surface temperature is low, globular elements appear at the sample skin, their presence has been attributed to the fact that the cooling rate is too high to allow the formation of fibrillar structures characteristic of the adjacent shear layer. The increase of mold surface temperature, inducing a decrease of cooling rate, allows a certain organization increase, and the globular structures are replaced by the fibrillar shear. The changes in morphology determine parallel changes in mechanical modulus distribution. However, at the sample surface, where the cooling rate is the highest, modulus remains low since the molecules do not have time for reach high organization level. Spaced out from the surface, the cooling rate decreases, the molecules can reach a good organization level and, due to the stress induced by the flow, form compact structures characterized by higher moduli in the shear layer.

Taking into account the effect of orientation levels on the elastic moduli it is possible to explain the effect of mold temperature and packing pressure. High packing pressure and low mold temperature generally induce high orientation and organization levels that lead to the increase of elastic moduli; vice versa, the increase of mold temperature, inducing the decrease of relaxation time, lead to the production of less oriented structures with low elastic moduli.

References

1. A.M. Díez-Pascual, M.A. Gómez-Fatou, F. Ania, and A. Flores, Progress in Materials Science **67**, 1 (2015).
2. A. del Campo and E. Arzt, Chemical Reviews **108**, 911 (2008).
3. L. Guadagno, P. Longo, M. Raimondo, C. Naddeo, A. Mariconda, A. Sorrentino, V. Vittoria, G. Iannuzzo, and S. Russo, Journal of Polymer Science Part B: Polymer Physics **48**, 2413 (2010).
4. I. Puente Orench, F. Ania, E. Baer, A. Hiltner, T. Bernal, and F.J. Baltá Calleja, Philosophical Magazine **84**, 1841 (2004).
5. A. Sorrentino, G. Gorrasi, M. Tortora, and V. Vittoria, in *Polymer Nanocomposites*, edited by Z.-Z.Y. Yiu-Wing Mai (Elsevier, Cambridge England, 2006), pp. 273–296.

6. S. Liparoti, A. Sorrentino, G. Guzman, M. Cakmak, and G. Titomanlio, RSC Advances **5**, 36434 (2015).
7. A. Sorrentino, R. Pantani, and G. Titomanlio, Journal of Polymer Science Part B: Polymer Physics **45**, 196 (2007).
8. H. Gojzewski, B. Imre, C. Check, R. Chartoff, and J. Vancso, Journal of Polymer Science Part B: Polymer Physics **54**, 2298 (2016).
9. R. Pantani, V. Speranza, A. Sorrentino, and G. Titomanlio, Macromolecular Symposia **185**, 293 (2002).
10. S. Liparoti, A. Sorrentino, and G. Titomanlio, AiChE Journal **62**, 2699 (2016).
11. F. De Santis and R. Pantani, Journal of Materials Processing Technology **237**, 1 (2016).
12. L. Shen, I.Y. Phang, T. Liu, and K. Zeng, Polymer **45**, 8221 (2004).
13. Liu, I.Y. Phang, L. Shen, S.Y. Chow, and W.-D. Zhang, Macromolecules **37**, 7214 (2004).
14. F. Chavarria and D.R. Paul, Polymer **47**, 7760 (2006).
15. H. Lu, G. Huang, B. Wang, A. Mamedov, and S. Gupta, Thin Solid Films **500**, 197 (2006).
16. S.A. Shahdad, J.F. McCabe, and R.W. Wassell, Journal of Materials Science: Materials in Medicine **18**, 1599 (2007).
17. T. Uchiyama, M. Suyama, M.M. Alam, T. Asano, S. Henning, A. Flores, F.J. Baltá Calleja, and M.F. Mina, Polymer **48**, 542 (2007).
18. E. Wornyo, K. Gall, F. Yang, and W. King, Polymer **48**, 3213 (2007).
19. F. Ania, G. Broza, M.F. Mina, K. Schulte, Z. Roslaniec, and F.J. Baltá-Calleja, Composite Interfaces **13**, 33 (2006).
20. F.J. Baltá-Calleja, M.E. Cagiao, R. Adhikari, and G.H. Michler, Polymer **45**, 247 (2004).
21. O. Sahin, S. Magonov, C. Su, C.F. Quate, and O. Solgaard, Nature Nanotechnology **2**, 507 (2007).
22. S. Liparoti, A. Sorrentino, and G. Titomanlio, RSC Advances **6**, 99274 (2016).
23. R. Pantani, I. Coccorullo, V. Volpe, and G. Titomanlio, Macromolecules **43**, 9030 (2010).
24. R. Pantani, V. Speranza, and G. Titomanlio, Journal of Rheology **59**, 377 (2015).
25. R. Pantani, V. Speranza, and G. Titomanlio, Macromolecular Theory and Simulations **23**, 300 (2014).
26. A. Sorrentino and R. Pantani, Polymer Bulletin **70**, 2005 (2013).
27. R. Pantani, V. Speranza, I. Coccorullo, and G. Titomanlio, in *Macromolecular Symposia* (2002), pp. 309–326.
28. S. Liparoti, V. Speranza, R. Pantani, and G. Titomanlio, in *Proceeding of PPS 2015 Europe-Africa Polymer Processing Society Regional Conference - Graz (Austria)* (2015).
29. W.C. Oliver and G.M. Pharr, Journal of Materials Research **19**, 3 (2004).
30. G.M. Pharr, W.C. Oliver, and F.R. Brotzen, Journal of Materials Research **7**, 613 (1992).
31. W.C. Oliver and G.M. Pharr, Journal of Materials Research **7**, 1564 (1992).
32. H.M. White and D.C. Bassett, Polymer **39**, 3211 (1998).
33. S. Liparoti, A. Sorrentino, G. Guzman, M. Cakmak, and G. Titomanlio, in *PROCEEDINGS OF PPS-31: The 31st International Conference of the Polymer Processing Society–AIP Conference Proceeding* (2016), p. 170001.
34. S. Liparoti, A. Sorrentino, and G. Titomanlio, in *Polymer Processing with Resulting Morphology and Properties: Feet in the Present and Eyes in the Future: Proceedings of the GT70 International Conference - Salerno (Italy) 2015. AIP Conference Proceedings* (2015), p. 20052.
35. S. Liparoti, T.M. Hunag, A. Sorrentino, G. Titomanlio, and M. Cakmak, in *30th International Conference of Polymer Processing Society, Cleveland, Ohio, USA - AIP Conference Proceeding* (2014), p. 110015.
36. R. Pantani, I. Coccorullo, V. Speranza, and G. Titomanlio, Progress in Polymer Science **30**, 1185 (2005).
37. B. V. Derjaguin, V.M. Muller, and Y.P. Toporov, Journal of Colloid And Interface Science **53**, 314 (1975).
38. E.G. Herbert, W.C. Oliver, A. Lumsdaine, and G.M. Pharr, Journal of Materials Research **24**, 626 (2009).
39. M.E. Dokukin and I. Sokolov, Langmuir **28**, 16060 (2012).

AUTOMATION OF AN ULTRASOUND MIXING HEAD FOR LOW PRESSURE RESIN TRANSFER MOLDING

Schaefer Maximilian Kunibert (1)*, Raffelt Wolfgang (2), Wittmann Simon (1), Ehrenreich Stefan (1), Zaremba Swen (1), Drechsler Klaus (1)

(1) Technical University of Munich, Chair of Carbon Composites - Bavaria - Germany, (2) Dekumed Kunststoff- und Maschinenvertrieb GmbH & Co. KG - Bavaria - Germany

The automation of high volume production in fibre-reinforced polymer composite (FRPC) manufacturing is a key issue to lower manufacturing costs. In terms of injection technologies, the solvent free processing of fast curing resin systems is a key to reduce both labor work and waste production. Two types of injection machines with corresponding strategies are pursued today. Low-pressure injection machines are based on static or dynamic mixing devices that shear the material by their elements. This principle leads to high operation costs by using solvents or cleaning agents to flush the mixing chamber. High-pressure injection machines induce turbulent streams to mix the components. Enabling an efficient ejection of the reactive resin system, the necessary high pressure leads to complex and cost intensive plants. In this paper, the acoustic cavitation effect is applied to enable a solvent free Low Pressure Resin Transfer Molding (LP - RTM) production based on a moving ultrasonic horn. During injection, the micro scale turbulent streams enable a uniform homogenization of a three component fast curing resin system including an internal release agent at low mixing pressure. The use of an internal release agent eliminates the need of an external, manually applied release agent. Further, the poor contact principle enables a self-cleaning of the mixing chamber after the injection. The investigation of this Ultrasound Mixing Head is done in three steps: First, the integration, guidance and control of the ultrasound device during a LP - RTM injection are examined. Second, the process and material characteristic of parts produced with the Ultrasound Mixing Head and a Static Mixing Head are compared. Third, results of demolding tests performed with specimens produced with both Mixing Heads at various process conditions are presented.

A study on cooling performance in injection molding. Heat conductive mold materials versus conformal cooling channels.

Berger Gerald Roman (1)*, Friesenbichler Walter (2), Zorn David (2), Bevc Franz (3)

(1) Montanuniversitaet Leoben, Dpt. Polymer Engineering & Science - Styria - Austria, (2) Montanuniversitaet Leoben, Institute of Injection Molding - Styria - Austria, (3) MAHLE Filtersysteme Austria GmbH - Carinthia - Austria

The aim of this study was to optimize the manufacturing process of an oil filter housing (made of PA 66 GF35) for automotive applications. Mass accumulations in the design of this component increased the cooling time in injection molding considerably. To improve the cooling, a copper-beryllium alloy mold insert with drilled cooling channels was installed.While copper-beryllium reduced the hot spot significantly – due to its thermal conductivity λ of 105 Wm-1K-1 - its wear resistance was too weak for the polymer used. Thus we searched for a solution where cooling performance is equal and wear resistance is superior. By using simulation, we analyzed two approaches to build up the insert: A) We filled up the core by copper and shielded it by a 3mm steel skin. B) We designed conformal cooling channels (Ø3.5 mm) in a Selective Laser Sinter (SLS)-able steel grade (λ = 23 Wm-1K-1). Modeling mold and process in simulation software Sigmasoft v. 5.1, we calculated first the heat transfer coefficients in the cooling channels using the CFD module. In the original design, the HTC was mostly far above 10000 Wm-2K-1, however, at bore crossings and in a dead end, the HTC fell locally to 700. Both the copper core insert and the conformal channel insert showed values > 12000. Second, we simulated injection molding cycles. Although in the insert socket region the cooling performance of copper core design and conformal channel design was significantly better than of the original design, the (cooling time determining) finger-like insert tips were less affected as neither copper nor conformal channel reach that region. Thus at the moment of ejection (after 75 s), where the recommended ejection temperature is <210°C, the relevant hot spot temperature at the part surface was 141°C, 213°C, 225°C for the original design, the copper core design, and the conformal channel design, respectively. In conclusion: If conformal cooling channels do not reach the region of interest, they are useless.

Integrative Simulation of Weld Line Strength in Unreinforced Amorphous Thermoplastics

Christian Hopmann[1], Jakob Onken[1, a)], Lukas Gröning[1]

[1] *Institute for Plastics Processing at RWTH Aachen University (IKV), Seffenter Weg 201, 52074 Aachen, Germany*

[a)]jakob.onken@ikv.rwth-aachen.de

Abstract. In the production of injection moulded parts, weld lines are often unavoidable. They form in the confluence area of two melt fronts caused either by several gates or by different melt front velocities due to varying wall thicknesses. In the confluence or weld line area, interfacial disturbances and molecular orientation often lead to a significant reduction of the material strength. At present, there are no methods available for the product development of plastic parts that offer a reliable prediction of the weld line strength. The resulting uncertainty of the mechanical properties of the part in the weld line area often leads to quality defects as well as premature and unexpected component failure and loss of function. Therefore, an integrative simulation method for the prediction of weld line strength in unreinforced amorphous thermoplastics has been developed. Using data from injection moulding simulations, the location and orientation of the weld line is determined and a reduction factor for the material strength of each finite element in the weld line area is calculated. The model for the calculation of the reduction factor is based on an interdiffusion model by *Kim and Suh* which has been augmented to describe the influence of several process parameters such as melt temperature and holding pressure. For the calibration and validation of method and model, PC, PMMA and SAN tensile bars with defined weld lines were produced with varying process parameters using a double-gated mould. The tensile strength was determined and compared to the simulation results. The simulated weld line strengths show good accordance with the experimental results for the investigated materials.

INTRODUCTION

One of the most common phenomena in injection moulded plastics parts is weld lines, which form in the confluence of two melt fronts behind flow obstacles, due to the usage of several gates or due to melt front velocity differences caused by the part geometry (1,2). Due to boundary surface disturbances that occur in the confluence, a significant reduction of mechanical properties is often linked to weld lines. The reduced strength of weld lines in amorphous thermoplastics can be traced to three different effects: Insufficient molecular interdiffusion in the weld line, high molecular orientations parallel to the weld line and the formation of a v-shaped notch on the surface of the part (3-5).

At present, there are no commercially available methods for the product development of plastic parts that offer a reliable mechanical design including the weld line strength. The resulting uncertainty about the actual mechanical properties of the intermediate layers of the weld lines often leads to quality defects as well as premature and unexpected component failure. Therefore, an integrative simulation method for the prediction of weld line strength in unreinforced polycarbonate (PC) depending on process parameters such as melt temperature and packing pressure has been developed at the Institute of Plastics Processing at RWTH Aachen University (IKV) (Figure 1) (6,7). To validate and calibrate this simulation routine, weld lines in polymethylmethacrylate (PMMA) and styrene-acrylonitrile (SAN) where exanimated.

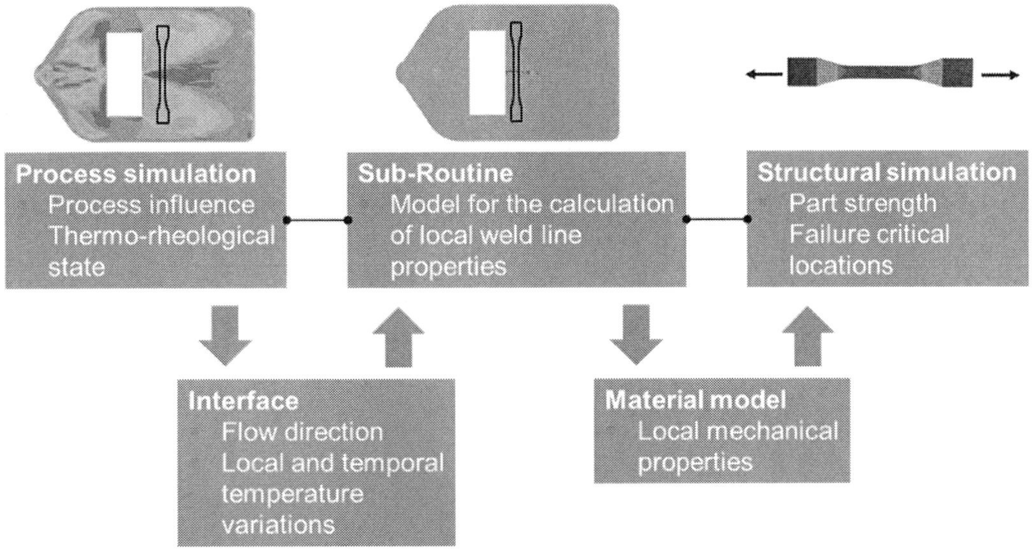

FIGURE 1. Schematic of the integrative simulation routine

SIMULATION

The integrative simulation chain uses process data created in an injection moulding simulation. Time dependent temperature and pressure from the process simulation is mapped on a FE-mesh. To calculate a reduction factor for the weld line strength based on that process data, a model based on the interdiffusion model by *Kim and Suh* is used (3,6,7). The reduction factor is defined as the breaking stress of the weld line divided by the breaking stress of reference samples without weld lines (Equation 1).

$$\text{Reduction Factor} = \frac{\text{Weld line strength}}{\text{Material strength}} \tag{1}$$

This model is implemented in a structural simulation subroutine that allows to predict the weld line strength. An example for the simulation routine and structural simulations results is shown in figure 2.

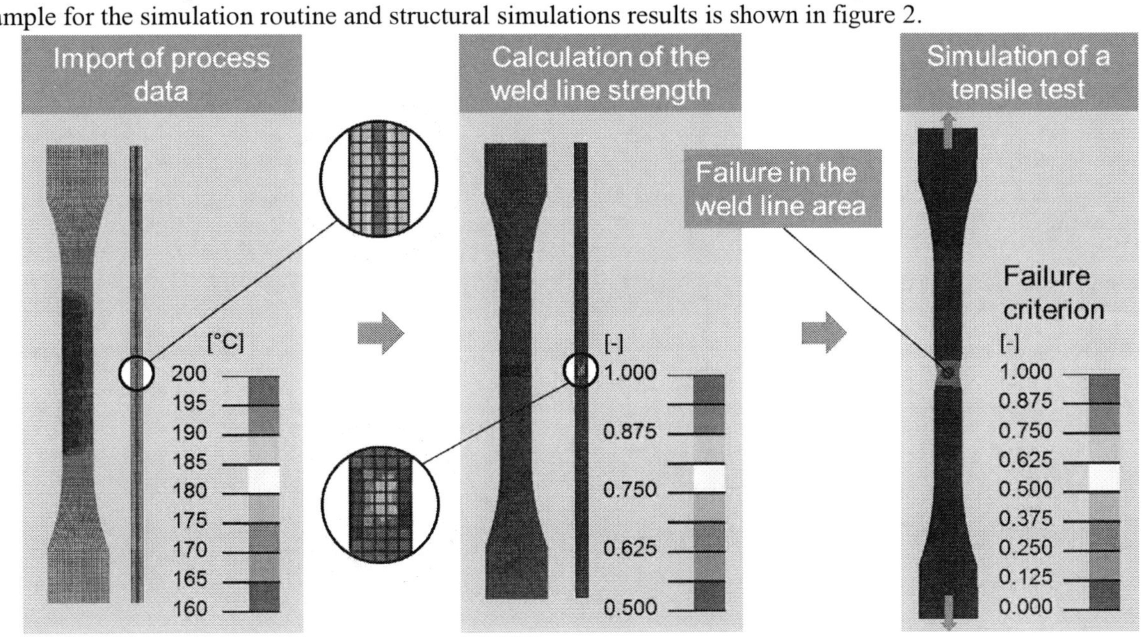

FIGURE 2. Integrative simulation routine for the prediction of weld line strength

EXPERIMENTAL

Methods and Materials

To examine the process parameter influence on the properties of weld lines, confocal microscopy and tensile testing was used. The materials investigated where type Luran 378P SAN by INEOS Styrolution Europe GmbH, Dormagen, Germany and type Plexiglas PMMA by Evonik Industries AG, Essen, Germany. Two different types of PMMA, type 6N and type 8N Plexiglas, were used to investigate the influence of different viscosities on the weld line strength in the same polymer.

As samples, type 1B tensile bars where manufactured using a KraussMaffei 160-1000CX injection moulding machine. To create stagnating weld lines in the tensile bars, a double gated mould was used. To produce reference samples without weld lines, a mould with a single bar was used. For the investigation of the process parameter influence, melt temperature and packing pressure where varied in three steps using a full factorial DOE. The following process parameters where used for the production of the samples (Table 1).

TABLE 1. Process parameters for the production of tensile bars with weld lines

Process Parameter	Factor level -1	Factor level 0	Factor level +1
Melt temperature [°C]	230	245	260
Packing pressure [bar]	500	800	1100
Injection volume flow [cm^3/s]		10	
Packing time [s]		10	
Mould temperature [°C]		80	

Confocal microscopy was conducted using a type VK-9700 laser confocal microscope by Keyence Deutschland GmbH, Montabaur, Germany. Tensile strength was tested using a type Z100 testing machine by Zwick GmbH & Co. KG, Ulm, Germany. Five samples were tested for each testing point. To measure the true strain und thus the true stress, the grey level offset of a speckle pattern on the tensile bars was recorded and analysed.

Results and Discussion

The confocal microscopy shows that the v-notches in the weld line have a maximum depth of 11 µm for both materials. The depth of these notches is bar far lower than the v-notch depths of 0.1-0.3 µm mentioned in the literature for polystyrene (PS) (4,8) but in a similar range as the notch depth of only up to 3.3 µm measured for polycarbonate (PC) previously at IKV (6).

The results of the tensile tests of PMMA Plexiglas 6N and 8N are shown in Figure 3. For Plexiglas 8N only the central point of the DOE is shown. The weld line strength is shown in Figure 3 as a reduction factor depending on melt temperature and packing pressure.

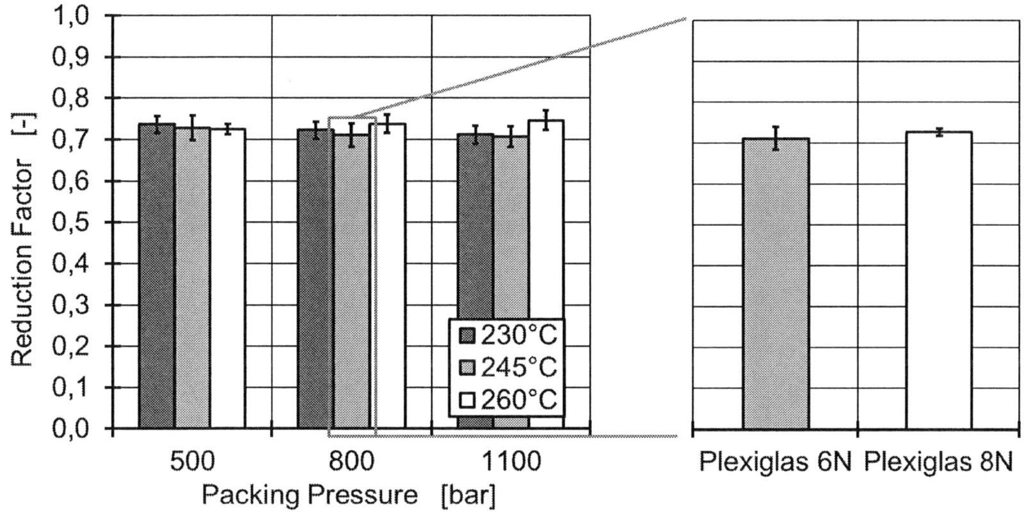

FIGURE 3. Tensile testing of samples with weld lines depending on melt temperature and packing pressure of PMMA Plexiglas 6N (left) und 8N (right)

The results show no significant influence of neither the process parameters nor the viscosity on weld line strength. This strongly disagrees with the correlations found in literature as several sources show a strong annealing of the weld line with rising melt temperatures (9-12). Previous investigations at IKV show an annealing of weld lines in PC with rising temperatures as well. Also, rising packing pressure has been observed to increase the weld line strength.

In Figure 4, the reduction factors in dependence of melt temperature and packing pressure for SAN are shown. Similar to the behaviour of PMMA, no influence of melt temperature and packing pressure on the weld line strength in PMMA can be observed.

An Explanation for this behaviour might be the thickness of the 1B tensile bars, which is 4 mm and therefore quite high compared to the thickness of most injection moulded parts.

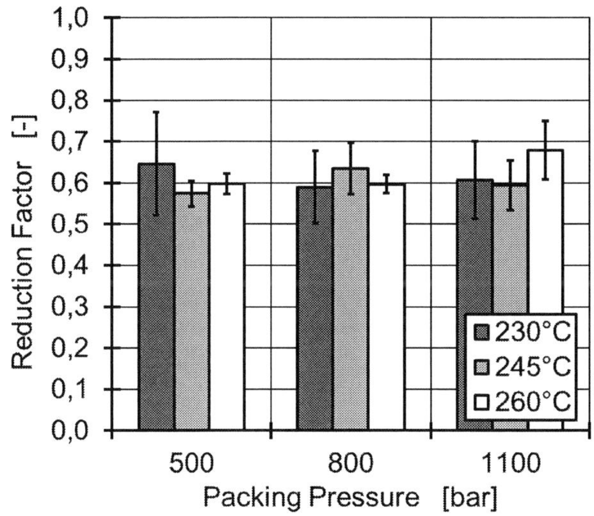

FIGURE 4. Tensile testing of samples with weld lines depending on melt temperature and packing pressure of SAN Luran 378P

CONCLUSION AND SUMMARY

The conducted investigations show no significant influence of melt temperature and packing pressure on weld line strength in PMMA and SAN. Therefore, the integrative simulation chain could not be calibrated and verified for these materials.

As the observed behaviour shows no accordance with the behaviour described in the literature, additional investigations have to be taken. To further investigate possible process parameter influences on weld lines in PMMA and SAN, different sample geometries and thicknesses should be tested.

ACKNOWLEDGMENTS

The research projects 18132 N and 19547 N of the Forschungsvereinigung Kunststoffverarbeitung were sponsored as part of the „Industrielle Gemeinschaftsforschung und –entwicklung (IGF)“ by the German Bundesministerium für Wirtschaft und Energie (BMWi) due to an enactment of the German Bundestag through the AiF. We would like to extend our thanks to all organizations mentioned.

REFERENCES

1. Nguyen-Chung T., Ph.D. thesis, Technische Universität Chemnitz, Chemnitz, 2001.
2. Seldén R., Effect of processing on weld line strength in five thermoplastics. Polym Eng Sci 1997; 37(1):205–18.
3. Kim S-G., Suh NP., Performance prediction of weldline structure in amorphous polymers. Polym. Eng. Sci. 1986; 26(17):1200–7.
4. Tomari K., Tonogai S., Harada T., Hamada H. et al., The V-notch at weld lines in polystyrene injection moldings. Polym. Eng. Sci. 1990; 30(15):931–6.
5. Fellahi S., Meddad A., Fisa B., Favis BD., Weldlines in injection-molded parts: A review. Adv. Polym. Technol. 1995; 14(3):169–95.
6. Onken J., Hopmann C., Prediction of weld line strength in injection-moulded parts made of unreinforced amorphous thermoplastics. International Polymer Science & Technology 2016; 43(11):T/1-T/8.
7. Hopmann C., Schöngart M., Onken J., Chen J., *Proceedings of the Polymer Processing Society Conference*, Graz, Österreich; 2015.
8. Nguyen-Chung T., Mennig G., Boyanova M., Fakirov S., Calleja F.J.B., Effect of an obstacle during processing on the weld line of injection-molded glassy polystyrene: Microhardness study. J. Appl. Polym. Sci. 2004; 92(5):3362–7.
9. Debondue E., Fournier J-E., Lacrampe M-F., Krawczak P., Weld-line sensitivity of injected amorphous polymers. J. Appl. Polym. Sci. 2004; 93(2):644–50.
10. Kovács JG., Sikló B., Experimental validation of simulated weld line formation in injection moulded parts. Polymer Testing 2010; 29(7):910–4.
11. Criens R.M., Moslé H.G., The influence of knit-lines on the tensile properties of injection molded parts. Polym Eng Sci 1983; 23(10):591–6.
12. Kühnert I., Ph.D. thesis, Technische Universität Chemnitz, Chemnitz, 2005.

SIMULATION OF THE MOLD STRUCTURE FOR REDUCING CAVITY FILLING DEVIATION IN MICRO INJECTION MOLDING

Lee Woo Il (1)*, Moon Sung Nam (2), Jung Jae Min (1)

(1) Seoul National University - Seoul - SouthKorea, (2) LG Chem - Seoul - SouthKorea

In micro injection molding process, it is important to make each cavity filling uniform. However, there are many factors that cause cavity filling deviation like mold temperature differential between each runner, diameter differential of gates or runners by mold manufacturing errors. In this study, we do numerical analysis for locating major factors cause cavity filling deviation and suggesting more robust design for micro injection molding. We used thermal characteristic values of EP-7000, one of polycarbonate series polymers used for injection molding. And the widely accepted Cross-Arrhenius model is adopted for the viscosity of polymer material. For characterizing specific heat (Cp) and thermal conductivity (k), we used piecewise-linear method. And volume of fluid method and piecewise-linear interface scheme was used for visualization of polymer flow. 6 runners were modeled and we simulated four cases (mold temperature, runner's diameter, gate's thickness and complex condition of mold temperature and runner's diameter). Numerical analysis has shown that case of mold temperature shows up to 28% difference in instant cavity filling rate and complex condition shows up to 33% difference. We suggest convergent runner and resin reservoir for reducing deviation. As a result, convergent runner design can reduce instant cavity filling rate deviation 20%, and resin reservoir design can reduce it 18%. Complex design of those two can reduce the deviation up to 22%.

Defects in Industrial Film Extrusion

Vlachopoulos John [1]*

[1] McMaster University - ON - Canada

Extrusion defects associated with the sharkskin/melt fracture occur at high wall shear rates and stresses. These have been extensively studied and discussed in numerous publications available in the open literature and there will not be part of the present presentation. However, there are other defects encountered in industrial extrusion operations (blown film, flat film), which have received very little attention. These include gels and flow streaks. They are due to a variety of causes which are not easily identifiable. Their common characteristic is that they involve some sort of degradation of the polymer melt. Frequently these defects occur in film extrusion lines which been running problem-free for a long time and may suddenly or gradually appear, especially when a resin with different processability (even minor) is fed to the extruder. While it is possible for gels to be present in the raw material, in this presentation we interested in gels formed in the extruder, due to excessive residence times, mainly because of flawed design of screw channels. They appear as black or brown specks. There is some evidence that the primary cause is the formation of secondary re-circulation flows known as Moffat eddies, in the fluid mechanics literature. Flow lines and streaks are due to flaws in the design of flow paths through adapters, screen changers, melt pipes and dies. Low wall shear rate regions tend to produce streaks of poor optical quality. In single layer extrusion they may be confused with sharkskin. In co-extrusion they might be confused with an interfacial instability known as zig-zag. Computer simulation of flow is used for pinpointing the regions where they originate.

New Approach for the Efficient Attainment of Flame Retardancy Using Multi Component Injection Molding

Dorothea Schneider[1,a),b)], Christof Hübner[2,c)] and Serge Bourbigot[3,d)]

[1]*A&E Produktionstechnik GmbH - Dresden - Germany*
[2]*Fraunhofer-Institut für Chemische Technologie ICT - Pfinztal - Germany*
[3]*École Nationale Supérieure de Chimie de Lille ENSCL - Villeneuve d'Ascq - France*

[a)]Corresponding author: d.schneider@a-e-produktionstechnik.de
[b)]URL: http://www.a-e-produktionstechnik.de
[c)]christof.huebner@ict.fraunhofer.de
[d)]serge.bourbigot@ensc-lille.fr

Abstract. The injection molding process yields many possibilities to create innovative plastic part solutions. Based on extensive research in distinct subject areas and the cross-disciplinary interconnection between different academic disciplines lately an interesting new approach for the efficient and effective attainment of flame retardancy could be identified: Multi component injection molding, where two or more plastic melts are injected next to or within each other during one cycle, can be used to produce innovative flame retardant plastic parts. These parts have an optimized skin/core structure with the core part consisting of the pristine polymer while the flame retardant polymer system is located in the skin. Such a structure proves to be ideal, because it allows for a faster protection dynamic.

This broadening of the field of application of standard injection molding machines is made possible by the use of intermediate plates with individually suitable modular technological inserts. The modular inserts required for this kind of application comprise hot runners including shut off and control functions to separate and join melt flows from different injection units into one cavity. These modules make for the combination of the sequential and simultaneous injection of melts.

In that account first trials which prove the concept to be viable have been conducted. Promising results will be shown as well as the principles of laboratory and industrial scale implementation of the underlying production process, while further research still is ongoing. In conclusion the paper will point out the advantages, possibilities and application potential given by this new approach. Cost savings will be given by the supersession of costly special machinery as well as achieved material savings and increased part quality and process stability.

INTRODUCTION

Small engineering companies as well as research institutions are facing innovation push and pull and its highs and lows all day every day since decades. Modern engineering always is a trade-off between efficiency and flexibility, since market requirements are becoming more and more volatile and still constraints are tight. Having the finger on the market steadily will consume a lot of valuable resources and simultaneously is a prerequisite for the generation of long term revenue. So market sensitivity should not come at any cost, just as well as efficiency should not cut short flexibility. Innovative engineering solutions have to strike the balance between sufficient market compliance and reasonable costs, and aim at so-called agility.

This paper will show in the first section the first insights from the development of a new approach for systematic innovation engineering to promote agility enhancement. The creation of such an approach is motivated by the necessary efforts to let market requirements be met by research and design processes. Further the paper will point out the particular market requirement for new flame retardancy solutions and the associated research challenges. In the third section an innovative solution to meet flame retardancy demands will be shown. Within the forth section the description of the experiments and results leading to this solution will be given before in the last section conclusions will be drawn as well as a short outlook given.

PLASTIC PART SOLUTIONS FOR AGILE INDUSTRY APPLICATION

An ongoing research project studies how it is possible to target industry relevant findings during research and creative processes proactively using given resources and core competencies to meet prioritized demands. The rise of modern methods and new standards allows for the use of the given data accumulated during engineering and research processes. The arising technological possibilities brought along with digital process support and data mining possibilities give way to match the wide range of innovation potential with actual necessities within the market. This allows for broadening the view and for moving away from the point of maximum efficiency and minimum requirements. This proactive approach may lead to a loss of efficiency and risen cost during the early phases of development. Meanwhile it will lift the developed solutions to the area of agile innovation proposition striking the balance between sufficient market compliance and reasonable efficiency.

As revenue equals application units multiplied by their value at the point where supply and demand meet, scaling the supply curve to fitting additional demands will nurture new fits and therefore let the revenue rise to the sum of those values. Scaling the innovation fits perfectly along the demand curve will even lead to revenue in the amount of the integral underneath the curve. To make this possible several methods of business informatics can be used to extract, collect and analyze data, emulating variability and evolutionary processes during product development and research processes [1, 2, 3].

The use of standard operations research methods, heuristics and optimization algorithms will then outline new fits of necessities stemming from market requirements with the given product scope. Of course as market requirements are volatile the scale of supply efforts leads to the necessity of modifications within the means of production. Modification expenses in plant machinery will be significantly higher if extensive changes are necessary within the core parts of the machinery itself. Interchangeable components yield less expensive flexibility potential and added inserts will do even more so. This is why in a lot of specialized application cases A&E technological inserts are used to equip standard machinery, transforming it into the perfect machine for the specialty method needed. One typical example for an added on specialty method which is often applied to standard machinery is sandwich injection molding while one way of the market changes to present themselves stand by as the altered requirements for flame retardancy.

FLAME RETARDANCY INNOVATION AS AN INDUSTRIAL REQUIREMENT

There is a steadily growing amount of electronics housed in plastics while also in sectors like transportation, building and furniture polymer materials can only be used when fulfilling fire retardant standards. The market volume of flame retardant chemicals is supposed to rise from 1.7 million tons in 2007 and 2.5 million tons in 2017 to over 3.7 million tons in 2022 [4].

Additionally new European regulations are progressively restricting the use of halogenated fire retardants, thus having an impact on the consumption of flame retardant additives due to obligations for manufacturers concerning health and environmental compliance. These facts have lead towards a pressing demand for a reduction of the fire hazard posed by highly combustible polymer materials. Therefore intensive efforts are being made to develop new safe and economical fire protection approaches [5, 6, 7].

The common approaches can be categorized into three types: The first type is the mechanical incorporation of flame retardant additives into the bulk polymeric matrix. The second possibility is to bind flame retardant chemicals to the matrix as integral part of the polymer chain forming functional groups. The third type is stated by surface modification. It stands to reason that this is a very convenient, economical and efficient way to protect substrates against fire, because of the 'paramount role played by the polymer surface during combustion' [5]. Placing the flame retardant chemicals at the surface proves to be ideal, because it allows for a faster protection dynamic. The heat penetration and diffusion to the gas phase occur at the polymer surface, so that the chemical and physical characteristics of the surface affect the ignition process predominantly.

During the recently conducted research project *PHOENIX* aiming at the development of compounds with new nanoparticle flame retardant additives one aspect of consideration was how to save valuable nanoparticle specimen material. Of course it was a reasonable attempt to use the flame retardant nanomaterial compound within the surface layer only, leaving the inner bulk material plain. This was realized by the formation of a layer structure during the injection molding process as described in the following section. Not only did it turn out to be advantageous because of the reduced amount of necessary flame retardant material but also resulted in a highly efficient way to protect the substrate which constitutes an industrially implementable highly productive one shot solution.

EFFICIENT REALIZATION USING MULTI COMPONENT INJECTION MOLDING

The variable configuration of standard injection molding machines using intermediate plates with individually suitable modular technological inserts was rendering possible the trials during the research project and at the same time gives way for the industrial use of the new approach. Sandwich injection molding is an injection molding process for the production of parts with a skin and core structure. The melt for the skin is injected first. Whereas the core material is injected subsequently, pushing forward the skin component. The skin material completely covers the core material in the final part if the determining injection parameters like material ratio, injection speeds, injection temperatures or mold temperature are properly set. Using a sandwich valve of two axially positioned pistons, which must be manufactured with very small tolerances, and ensuring the clean separation of the two different melt streams is essential [8].

Test series were conducted to minimize wear and optimize the process. The specimen production was realized using a 100 t clamping force two component injection molding machine which was equipped with a 2C intermediate platen as an added technological insert (working principle shown in figure 1).

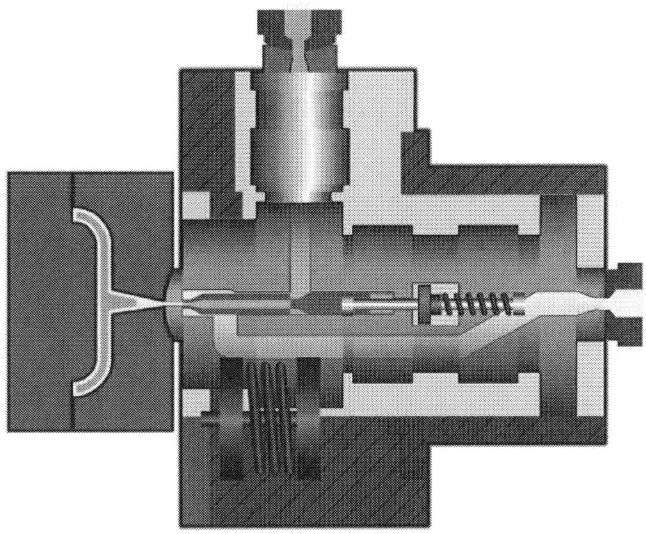

FIGURE 1. Scheme of the technological insert, an intermediate platen, used for the experiments

NEW APPROACH FOR ATTAINMENT OF FLAME RETARDANCY

Pursuing the intend to find effective flame retardant compounds trials were conducted concerning flammability parameters. During these trials very interesting findings were obtained: Flame retardant skin layers protected the specimen from burning nearly as well as the flame retardant protected the specimen if added throughout the complete plastic part. The core component was formed of pure Polypropylene (PP) *Moplen HP500N* while for the skin component various compounds were tested. Trials with a skin layer of a flame retardant composite with 25 % *Exolit 766*, as well as of a flame retardant PP copolymer with 25 % *Budit* (*Dinaplen PPC 1525*) in matrix of PP *Moplen HP500N* were conducted with various skin layer thicknesses in the range of 0.4 mm . . . 0.6 mm while the core layer thickness ranged from 1.8 mm . . . 2.0 mm. Additionally specimen of High Impact Polystyrene (HIPS) were prepared with skin layer thicknesses ranging from 0.5 mm . . . 0.8 mm containing an intumescent additive and nanoparticle flame retardant additives. The injection parameters and the mold temperature were varied in order to achieve thin but uniform skin layers. The most important process parameters besides the melt temperatures of the skin and the core component are the shot volumes of the two injection aggregates, the respective injection speeds, the injection time as well as the delay between the first and the second shot.

The samples were evaluated by mass loss calorimetry (MLC) at *ENSCL* using a Mass Loss Cone of 35 kW and 35 mm. The formulations are listed in table 1 and the results are presented in tables 2 and 3. The PP trials showed a reduction by 65 % of the peak heat release rate (pHRR) value of the sandwich samples compared to the pristine PP samples. The HIPS trials showed a reduction by 44 % resp. 54 % of the peak heat release rate value of the sandwich samples with a skin thickness of 0.5 mm resp. 0.8 mm compared to the pristine HIPS samples. The specimen with a flame retardant surface layer showed an equally satisfying performance as plates made from pure PP/*Exolit* composite confirming their applicability for flame retardancy requirements. The concept of multilayered plastic parts prepared by sandwich injection molding in order to form a fire retardant protective skin layer is therefore validated and should be considered for further applications.

TABLE 1. Formulations of sandwich injection molded samples

Specimen	Material abbreviation	Description	Result
Pristine	Ref PP	Neat PP plates made from *Moplen HP 500 N*	pHRR reference without flame retardant (FR)
Pristine with halogen free FR	3,3,1-2-1	Plates made of *Moplen HP 500 N* + 15 % Mg1	The incorporation of 15 % Mg1 in PP only permits a slight reduction of pHRR (24 %).
Sandwich with FR in the skin	4,5-4J	Sandwich plates made of *Moplen HP 500 N* as core and *PP copolymer Dinaplen PPC 1S25* with 25 % *Budit* as skin	The sandwich sample (4.5-4J) exhibits a 65 % HRR reduction and an intumescent layer is formed at the surface upon testing.
Pristine with halogen based FR	PPC Dinaplen	Plates made from PP copolymer *Dinaplen PPC 1S25* with 25 % *Budit*	pHRR is dramatically reduced (by 92 %) for the non-coextruded sample (PPC Dinaplen) thanks to the development of an intumescent coating

TABLE 2. MLC data of FR PP-based sandwich injection molded samples

Indicator	Unit	Ref PP	3,3,1-2-1	4,5-4HJ	PPC Dinaplen
t_{ign}	s	68	57	32	53
t_{fout}	s	704	877	1333	1642
$t_{flaming}$	s	636	820	1301	1589
$pHRR$	kW m^{-2}	388	296	136	31
t_{pHRR}	s	157	195	307	61
THR	MJ m^{-2}	95	90	79	10

CONCLUSIONS

During an interdisciplinary research project a new approach for the efficient and effective attainment of flame retardancy could be identified. The growing request for flame retardancy solutions which are safe and complying with health as well as environmental standards should be met by a proactive industry supply increase. A still progressing research attempt focuses on the targeted innovation development for such urging demand scenarios within research projects and small engineering companies including the economical potentials to be gained as a consequence. By the special multi component injection molding method of sandwich injection molding a skin and core layered structure can be created. Furthermore this processing method can be added to standard machinery by variable configuration using intermediate plates with individually suitable modular technological inserts giving way for the flexible industrial use of the new approach.

The results show a significant decrease of the amount of flame retardant chemicals necessary to obtain satisfying flame retardant behavior by pursuing the approach of a skin and core layered structure of the specimen. The PP trials showed a reduction by 65 % in the peak heat release rate (pHRR) value of the sandwich samples compared to the

TABLE 3. MLC data of FR HIPS-based co-extruded samples

Indicator	Unit	Ref HIPS	0.5 mm	0.8 mm
t_{ign}	s	115	88	135
t_{fout}	s	860	767	797
$t_{flaming}$	s	745	679	662
$pHRR$	kW m^{-2}	350	197	159
t_{pHRR}	s	212	241	294
THR	MJ m^{-2}	71	50	49

pristine PP samples. The HIPS trials showed a reduction by 44 % resp. 54 % of the peak heat release rate value in the sandwich samples with a skin thickness of 0.5 mm resp. 0.8 mm compared to the pristine HIPS samples.

Necessitating the sandwich injection molding process it is possible to obtain an industrial scale implementation of flame retardant surface modification at reasonable cost. Economical advantages will be given not only by the achieved material savings but also by the supersession of costly special machinery as well as an increased part quality and process stability. Following the possibility to produce the layered structure within one single production step with no additional coating machinery and handling needed, the repeatability within a stable production process increases part quality and safety while it decreases the risk of delamination as the skin layer forms an integral joint of the material within the part. In conclusion the presented solution comprises an affordable and satisfying response to the current flame retardancy challenges for standard injection molded parts.

Nevertheless further reaching trials should be conducted to find the best possible formulations and material combinations to optimize flame retardancy performance as well as relevant part characteristics like for example mechanical properties. Likewise the conceptualization of systematic research should be addressed in growing intensity, giving more weight to the anticipatory point of view targeting demand relevant and foresighted application development.

ACKNOWLEDGEMENTS

The research leading to these results has received funding from the *European Unions Seventh Framework Programme* for research, technological development and demonstration under grant agreement No. 310187 under project acronym *PHOENIX*. The authors would also like to thank the *European Social Fund* in Germany for financial support.

REFERENCES

[1] R. Braun and W. Esswein, WIT Transactions on Engineering Sciences 450–458 (2014).

[2] J. Agüero, M. Rebollo, C. Carrascosa, and V. Julián, Trends in Practical Applications of Agents and Multiagent Systems 9–17 (2010).

[3] N. Criado, S. Heras, E. Argente, and V. Julián, "Normative argumentation," in *Trends in Practical Applications of Agents and Multiagent Systems: 8th International Conference on Practical Applications of Agents and Multiagent Systems*, edited by Y. Demazeau, F. Dignum, J. M. Corchado, J. Bajo, R. Corchuelo, E. Corchado, F. Fernández-Riverola, V. J. Julián, P. Pawlewski, and A. Campbell (Springer Berlin Heidelberg, Berlin, Heidelberg, 2010), pp. 29–36.

[4] Clariant Produkte (Deutschland) GmbH, "The flame retardants market: https://www.flameretardants-online.com/flame-retardants/market," .

[5] G. Malucelli, F. Carosio, J. Alongi, A. Fina, A. Frache, and G. Camino, Materials Science and Engineering: R: Reports **84**, 1 – 20 (2014).

[6] S. Liang, N. M. Neisius, and S. Gaan, Progress in Organic Coatings **76**, 1642 – 1665 (2013).

[7] J. Alongi, Z. Han, and S. Bourbigot, Progress in Polymer Science **51**, 28 – 73 (2015).

[8] H. K. Volker Reichert, Dorothea Schneider, "Hotrunner intermediate plates for multicomponent injection moulding and hot runner defection plates for use in standard injection moulding processes," in *11th International Conference Advances In Plastics Technology APT'15 11-15 October 2015, Sosnowiec, Poland* (2015).

Multi-scale investigation and fundamental studies on interface/interphase in multilayered polymer systems towards better controlling coextrusion process

Lamnawar Khalid (1)*, Maazouz Abderrahim (1), Lu Bo (1), Bondon Arnaud (1)

(1) Université de Lyon, CNRS, UMR 5223, Ingénierie des Matériaux Polymères, INSA Lyon - Lyon - France

Multilayer coextrusion is a versatile technology to fabricate architectural multilayer polymers structures combining the properties of different materials in a single process. Understanding interfacial properties between neighboring layers is the key for suppressing the interfacial defects. Herein, we present some fundamental studies on interface/interphase in multilayered polymer systems for the better control of processing and final properties. Firstly, interfacial dynamics and rheology related to the diffuse interphase triggered from interdiffusion were studied with a model compatible polymeric system. The interdiffusion kinetics and the development of diffuse interphase were investigated by both rheological modeling and experimentally. Role of diffuse interphase at both nano- and macroscale in linear and nonlinear viscoelastic properties were elaborated. Secondly, the role of diffuse interphase in controlling the interfacial flow instability was studied in the model compatible multilayer system. Furthermore, the development of a reactive interphase with in situ formed copolymer and its effects on interfacial instabilities and the defects of coextruded films were further probed. The effect of the process parameters and the structure of the polymers on the generation of new defects called "grainy" were assessed. The macromolecular architecture of the in situ generated copolymer significantly impacted the interfacial roughness and the rheology of the multilayer stuctures. Relations between the relaxation process, elongation properties, the interfacial morphology and the copolymer architecture were established in correlation with the generation of the macroscopic grainy defect in coextrusion. Overall, our studies are aimed at uncovering the physicochemical properties of diffuse and reactive interphase to guide the better controlling of multilayer coextrussion. Lamnawar et al. Soft Matter 2016, J.Rheol. 2016, Macromolecules 2013, Encycl. of Polymer Science &Tech.2013

A Novel Method for the Prediction of Adhesive Strength for Two-Component Injection Molding of Thermoplastics with Thermoset Rubbers

W. Six[1, a)], G.J. Bex[2, b)], J. De Keyzer[3], F. Desplentere[1], A. Van Bael[2]

[1] *Materials Technology TC, KU Leuven Campus Bruges, Spoorwegstraat 12, 8200 Bruges, Belgium*
[2] *Materials Technology TC, KU Leuven Technology Campus Diepenbeek, Wetenschapspark 27, 3590 Diepenbeek, Belgium*
[3] *Sustainable Chemical Process Technology TC, KU Leuven Technology Campus Diepenbeek, Wetenschapspark 27, 3590 Diepenbeek, Belgium*

[a)]Corresponding author: wim.six@kuleuven.be
[b)]gertjan.bex@kuleuven.be

Abstract. Two component injection molding is a widespread technique to produce polymer products that consist of two materials. This technique is commonly used to combine various material properties, or functionalities in one product. The 'hard-soft' combination where a stiff material is over-molded by a soft layer, is one example, often seen in valve like products where the soft part is used as a seal. Thermoplastic elastomers (TPE's) are commonly used is this case. However, TPE's have only limited properties for chemical and temperature resistance. For such applications it would be beneficial to use a thermoset rubber as EPDM, NBR, NR, etc. Although injection molding of thermoset rubbers is not that uncommon, using it in an over molding process is rather rare and not much is known about the final properties of the over molded product. One of the most important parameters is the adhesion between the two materials. Recommendations for good adhesion can be found in literature for specific material combinations, but data on predicting the strength of adhesion in thermoset rubber over molding in function of material and process settings does not exist. This paper presents a novel, empirical method to predict the strength of adhesion in function of material and process settings.

INTRODUCTION

Numerical simulations for injection molding are commonly used for predicting process specific parameters and properties of the final molded product. This is done for single as well as for multicomponent injection molding. They predict flow patterns, clamping force, pressure during processing, fiber orientation, warp shapes, etc. Most often, numerical simulations are performed for the injection molding of thermoplastic materials. Using numerical simulations for two component injection molding of thermoplastic materials and thermoset rubbers is, as far as we know, uncommon. Moreover, commercial simulation software packages as Autodesk Moldflow have no straight-forward way of simulating this process [7]. Next to this, predicting the strength of adhesion between over molded materials is rather new. Nowadays, research is conducted in determining the strength of adhesion of overmolded organo sheets [1] and the healing capabilities of materials in weldlines [2]. Predicting strength of adhesion between a thermoset rubber and a thermoplastic material is new. This paper shows a possible workflow to determine the strength of adhesion between a thermoplastic material and an over-molded thermoset rubber via simulations. A combination of existing simulation software as Autodesk Moldflow, measurements and scripting was used to create novel results.

WORKFLOW

Previous to this research, measurements have been conducted to determine the strength of adhesion on tensile strips produced, produced in a 2K plate mold. This has been done for the HDPE M80064S and EPDM 005K material combination which is described by G-J Bex et al. [8]. The measurements describe the strength of adhesion for these tensile strips, as function of interface temperature between the rubber and thermoplastic material (1), curing of the thermoset material at the interface (2) and packing pressure during injection molding of the thermoset material (3) [3,4,8]. In the presented script, all the measured adhesion strengths are scaled from 0 to 1 (normalize data) and implemented as function of its specified parameters. To predict the adhesive strength for new designed over-molded products, numerical simulations are needed. In this case Autodesk Moldflow 2017 was used for the injection molding simulations. The first step is to implement the product and mold geometry in the simulation software. Next there is also a need for the material characteristics and the process settings. The test product for which the strength of adhesion will be predicted is shown in Figure 1. The product is a tube-like geometry over molded with a rubber ring for sealing purposes. The figure also shows the cooling set-up, consisting here of three cooling channels and a central bubbler.

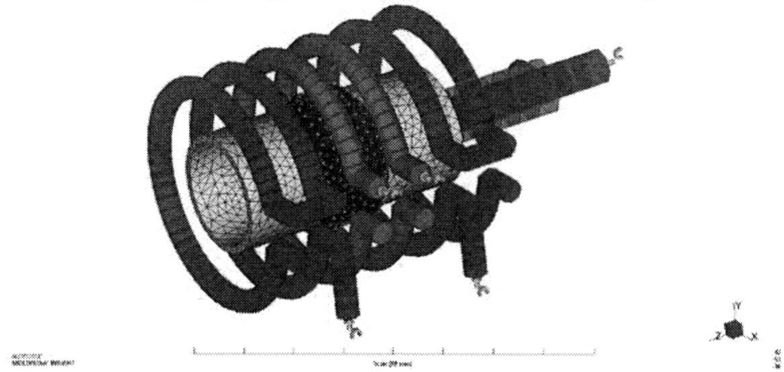

FIGURE 1. simulation case: tube with rubber ring

The product is manufactured using HDPE M80064S and EPDM 005K, of which the rheological and curing properties were measured [5]. To predict the final strength of adhesion, three important simulation results are needed as input for the presented script: the temperature of the thermoplastic part insert during rubber injection (i), the degree of cure as function of time (ii) and the internal pressure during the injection and packing phase (iii). Those three results are exported from Autodesk Moldflow as an XML format. As this file contains all nodal result data, which are used to produce new plots with combined results. Also the UDM-mesh file which contains all node locations is needed to reproduce the product for further calculations. Using this data, the developed script automatically finds the surface between the first component and the over molded rubber part. At this step, the script creates a new 2D mesh to plot and calculate the new results. For all nodes on the new created mesh, the right temperature, the cure and pressure data is found within the results data present in the XML files. Combining this data with the normalized input data, the script can produce novel plots which show the influence of a specific parameter on the strength of adhesion. The complete workflow is shown in Figure 2.

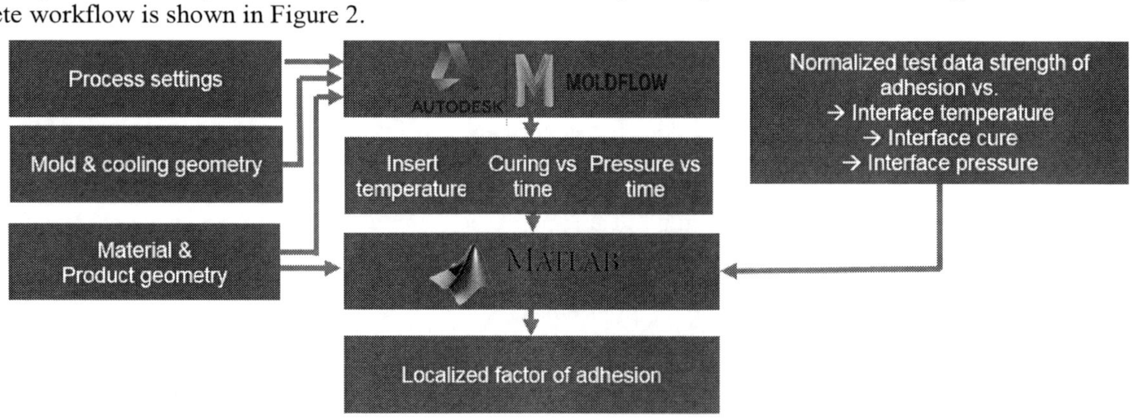

FIGURE 2. workflow to determine the localized strength of adhesion

NOVEL SIMULATION RESULTS

The presented script produces results which show the influence of individual parameters on the strength of adhesion, assuming the two other parameters have an optimal value at each point of the interface. It shows normalized values, between 0 (no adhesion) and 1 (optimal adhesion) for three individual parameters. The factor '$f_{temperature}$' shows the hypothetical influence of temperature on the strength of adhesion, 'f_{cure}' the influence of the degree of cure and '$f_{pressure}$' the influence of the local interface pressure. Although not shown in this paper, the script provides plots in function of time, so the development of adhesive strength can be monitored during the injection phase, packing phase and curing phase. These visualized time-steps are the same steps as available in the Autodesk Moldflow simulation result window.

Figure 3 on the left, shows the temperature at the surface of the interface. According to the simulations, the regions closest to the outer edges of the rubber ring, are the warmest. The surface temperatures lay between 128°C and 133°C. With these simulated surface temperatures, the script finds the corresponding adhesive strength, measured on the tensile strips. This measured tensile strength is normalized towards a value between 0 and 1 and plotted on the created mesh between the thermoplastic material and the thermoset rubber. Fig. 3 on the right, shows the $f_{temperature}$. As the interface temperatures are very near to the optimum, the simulated result shows an expected normalized strength between 0.8 and 0.99.

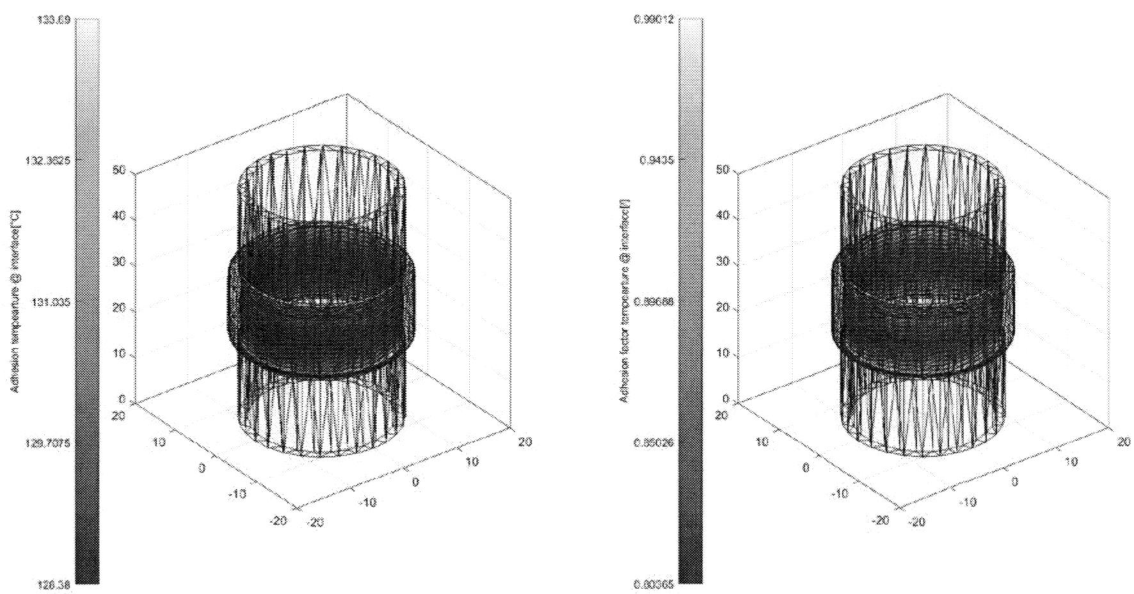

FIGURE 3. (left) Temperature at insert surface, (right) Simulated factor of adhesion vs. temperature: $f_{temperature}$

The second result is the factor of adhesion in function of the curing at the interface. The cure results in Figure 4 on the left, show the curing degrees at the end of the curing phase (1100s) at the surface of the rubber part, ranging from 0 to 0.86, where 1 is complete curing. Curing of the rubber ring is at its highest (0.7-0.86) at the middle section of the surface between the two product parts, whereas the edges show a lower degree of cure (less than 0.25) and even points where the rubbers is not cured at all. This has an effect on the adhesion as shown in Fig. 4 on the right. The factor of adhesion (between 0 and 1) varies between 0.11 and 0.96 (f_{cure}). At the edges of the product, a lower factor of adhesion is noticed compared to the calculated factor in the center, this due to a lower degree of cure at the edges in the rubber ring.

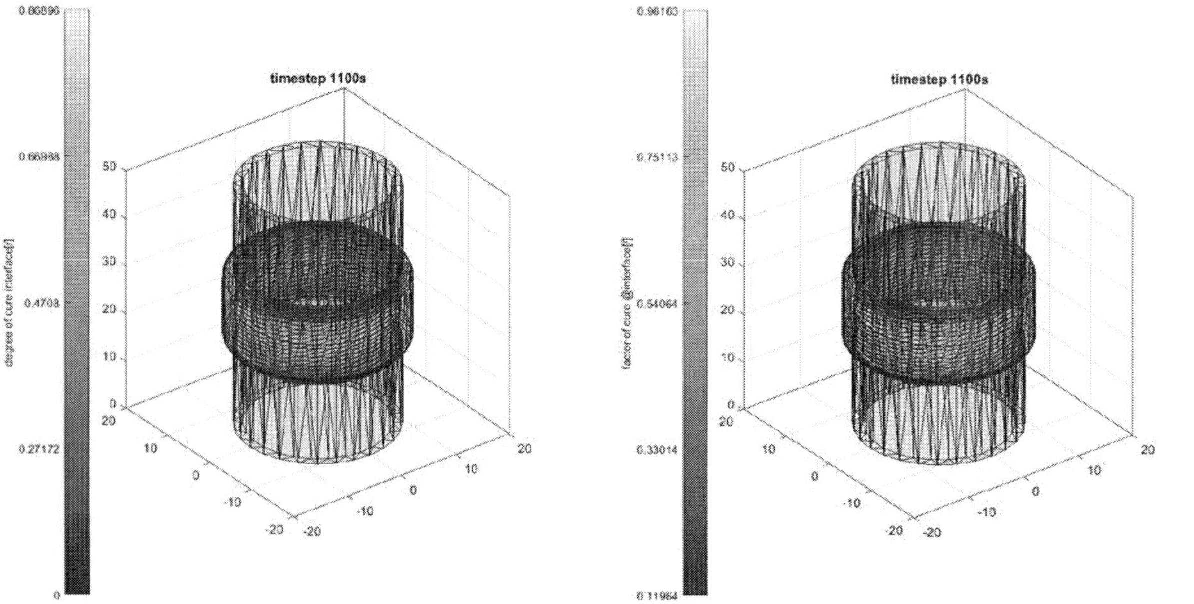

FIGURE 4. (left) Degree of cure at product surface, (right) Simulated factor of adhesion vs. degree of cure: f_{cure}

As last influencing parameter, pressure is displayed. Figure 5 on the left shows the pressure at the interface at the end of the packing phase (200.5s). Although pressure is equal on all locations during the packing phase, this does not mean that it results in an equal adhesive strength over the complete interface. Due to temperature differences, pressure has a localized effect. Fig. 5 on the right shows the $f_{pressure}$. This plot is very similar to the $f_{temperature}$ plot as the effect of pressure is temperature dependent. The $f_{pressure}$ shows an influence in a relative strength of adhesion between 0.69 and 0.98, where 1 is the maximum value.

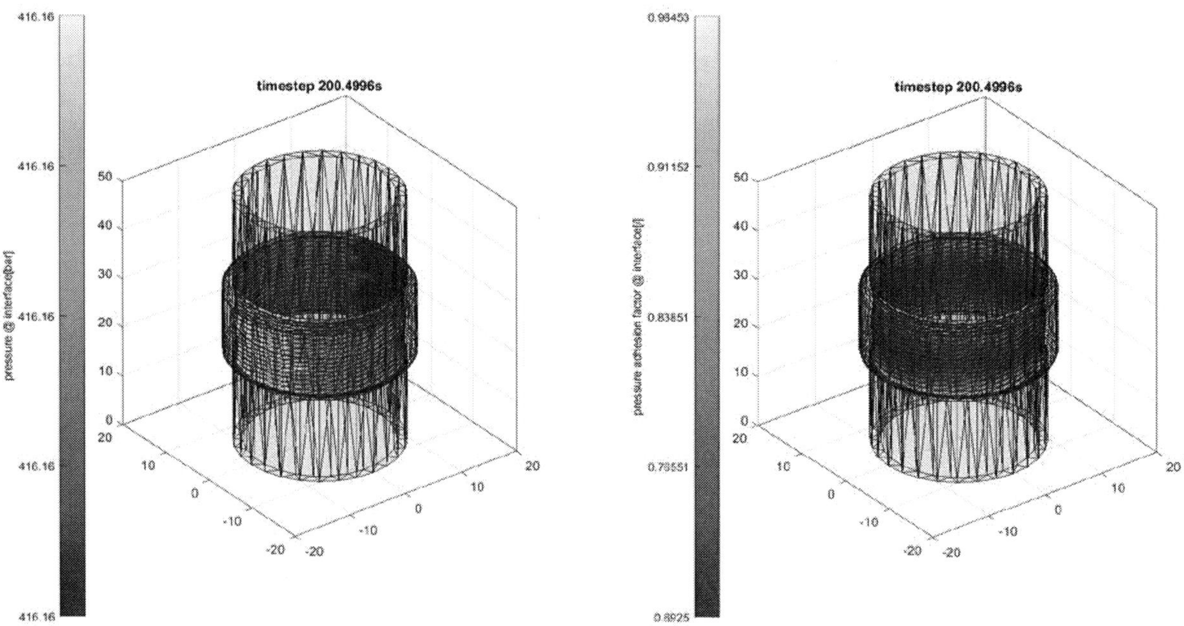

FIGURE 5. (left) Pressure at product surface, (right) Simulated factor of adhesion vs. pressure: $f_{pressure}$

CONCLUSION

The developed script uses measured adhesion strength data obtained by tensile testing of 2K molded test-strips obtained with well-defined processing parameters. With this data, the script is able to predict the individual influence of local values of temperature, cure and interface pressure on the expected local adhesion strength (assuming optimal values for the two other parameters). This approach can be used as initial assessment of the influence of injection molding process parameters, such as mold temperatures, injection speed, injection temperature and even mold design as they all affect the curing degree, insert temperature and packing pressure. In that way, it also can be used to optimize the product and mold geometry and even more important, to refine process settings. A more in depth design of experiments will be carried out to obtain a more realistic model that describes the adhesion strength as combined effect of local temperature, curing degree and packing pressure, in order to obtain more insight in the cross effects between these three parameters. A disadvantage of the proposed approach is that the adhesive behavior must be characterized in advance for each new material combination.

ACKNOWLEDGEMENT

The authors acknowledge the KU Leuven Impulsfonds for the financial support and the Company Hercorub NV for providing the uncured rubber material. G.J. Bex acknowledges the Flanders Research Foundation (FWO) for his PhD grant strategic basic research.

REFERENCES

[1] M.M. Bouwman T.G Donderwinkel, S. Wijskamp (2016) Overmoulding – An integrated Design Approach for Dimensional Accuracy and Strength of structural Parts. Proc. of the 3rd International conference & exhibition on thermoplastic composites, Bremen, Germany, 11-12 October 2016 (pp 37-40)

[2] J. Onken and C. Hopmann (2016) Prediction of weld line strength injection-moulded part made of unreinforced amorphous thermoplastics. Gummi Fasern Kunststoffe, 69, No. 9, 2016, pp.574-580

[3] Bex, G., De Keyzer, J., Desplentere, F., Van Bael, A. (2016). Two-component injection moulding of thermoplastics with thermoset rubbers: Process development. Proc. of the 32nd International Polymer Processing Society Conference. International Conference of the Polymer Processing Society. Lyon, France, 25-29 July 2016 (art.nr. S11-75)

[4] Bex, G., Six W., De Keyzer, J., Desplentere, F., Van Bael, A., Two-Component Injection Moulding of Thermoplastics with Thermoset Rubbers: the Effect of the Mould Temperature Distribution, presented at this conference.

[5] Six, W., Desplentere, F., Bex, G., De Keyzer, J., Van Bael, A. (2016). Simulation and optimization of vulcanization in the adhesion area in a 2K Thermoset-Thermoplastic product. Proceedings of the 7th biannuals internatinal PMI conference: polymer&mould innovations. PMI. Ghent, 21-23 sept 2016 (pp. 195-199).

[6] Sabic (2017,06 14) www.sabic-europe.com. cosulted on https://www.sabic.com/en/products/polymer/poly ethylene-pe/sabic-hdpe

[7] Moldflow (2017,07 08) https://knowledge.autodesk.com consulted on https://knowledge.autodesk.com/ support/moldflow-insight/learn-explore/caas/CloudHelp/cloudhelp/2018/ENU /MoldflowInsight-WhatsNew/files/ GUID-352A6BCD-9F36-42B2-9896-46F632116F9C-htm.html

[8] Bex, G., De Keyzer, J., Desplentere, F., Van Bael, A., Two-component injection moulding of thermoset rubber in combination with thermoplastics by thermally separated mould cavities and rapid heat cycling. International Journal of Advanced Manufacturing Technology

A new approach to model thermal expansion of semi crystalline polymers.

Ramakers-van Dorp Esther [1]*, Haenel Thomas [1], Reith Dirk [2], Hausnerová Berenika [3], Möginger Bernhard [1]

[1] Bonn-Rhein-Sieg University of Applied Sciences, Department of Natural Sciences, Rheinbach - NRW - Germany, [2] Bonn-Rhein-Sieg University of Applied Sciences, Department of EMT, Sankt Augustin - NRW - Germany, [3] Tomas Bata University, Faculty of Technology, Department of Production Engineering - Zlín - Czech

Extrusion blow molding is a successful processing method to produce hollow plastic parts for the packaging and automotive industry for more than 30 years. Unfortunately, it exist no adequate simulation software for the prediction of final part properties in blow molding in contrast to injection molding. The final properties of blow molded parts depend on the processing parameter such as melt temperature, extrusion speed and wall thickness. Especially shrinkage and warpage are important phenomena which affect the geometry and as a result the performance of the final part. Therefore, these phenomena have to be considered in part design and process development. The amount of shrinkage and warpage depend on the coefficient of thermal expansion (CTE). The classical method is a linear description of the CTE which is only valid in a certain temperature range and only for homogeneous materials. In reality, CTE depends strongly on local crystallinity and molecule orientation. Currently there is no non-linear description of CTE based on the anisotropic morphology of semi crystalline polymers. This study will propose a new non-linear CTE model for semi crystalline polymers based on crystallinity measurements and theoretical values of the direction depended young's modulus and CTE from literature for high density polyethylene HDPE. HDPE specimens were made with different mold temperatures and draw ratios to determine the influence on the anisotropic morphology of the parts. The thermal properties of HDPE were measured by differential scanning calorimetry (DSC) and thermomechanical analysis (TMA) in respect to the extrusion direction. These results were used to correlate and validate the new non-linear CTE model based on the measured crystallinity with the results of the thermal expansion. The new non-linear CTE model shows a good correlation with the results of the thermal expansion and gives the option to improve existing simulation software for blow molding in further studi

Design and Extrusion of High-Performance Polyetherimide Pipes with Variable Wall Thickness for Aircrafts

Michael Stegelmann[a)] and Peter Lucas and Niels Modler

Institute of Lightweight Engineering and Polymer Technology (ILK), TU Dresden
Holbeinstr. 3
01307 Dresden, Germany

[a)]Corresponding author: michael.stegelmann@tu-dresden.de

Abstract. In the aviation industry high-performance thermoplastic materials have started to replace metals in many fields. In this paper a novel pipe-extrusion process for high-temperature polyetherimide (PEI) is presented with special emphasis on the calibration and cooling of the pipes. Both the weight advantage over stainless steel pipes and the price advantage over titanium confirm a high potential of thermoplastic pipe systems in the aviation sector. The pipe systems installed in aircrafts are characterised by many variations in terms of geometries (diameter, wall thickness, etc.). The design diversity of thermoplastics can be used to exploit additional potentials in terms of mass savings, production efficiency and cost reduction. In order to address the forecasts of growing aviation transport capacity and aircraft production volume especially the production efficiency will be stronger focused in long term. This paper describes an approach to extrude PEI pipes with variable wall thickness. Thus, studies were conducted to analyse the effects of parameter changes on the pipe geometry and surface quality. In addition to that special emphasis was laid on the temperature profiles of the pipes inside the calibration system. An approach is presented to locally adjust the wall thickness of the pipes in longitudinal direction for defined geometry changes. The results of the simulation and the measurements led to an in-depth understanding of the process parameters needed to extrude PEI pipes with variable wall thickness.

INTRODUCTION

Freshwater, air-conditioning and waste disposal systems in commercial aircraft comprise several hundred meters of installed stainless steel and titanium pipes. An Airbus A320, for example, contains more than 100 running meters of freshwater piping. For the 150 to 200 aircraft delivered annually, this corresponds to 15,000 to 20,000 running meters just for the freshwater system of this one type of aircraft alone. The complex requirements that the aviation sector imposes on these pipe systems can also be met without major difficulty by using modern high-performance polymers, such as PEI.

A process for extruding variable-wall-thickness pipes made of PEI for freshwater and air-conditioning systems in aircraft has been designed, built and tested. The pipes are manufactured on a specially designed extrusion line and meet the complex and demanding requirements of the aviation sector. Thermoplastic piping systems offer significant advantages in terms of costs and weight. For example, PEI pipes just 0.6 mm thick and with an outer diameter of 25.4 mm (1") satisfy the pressure requirements imposed on freshwater systems as per ATA38 regarding the certification of water/waste systems for passenger aircraft. Comparing the mass of PEI pipes (0.6 mm thickness) with conventional aircraft pipes made of steel (0.3 mm thickness) the mass reduction is around 70 % or 45 % compared to titanium pipes (0.3 mm thickness) per meter, respectively.

With view on exploiting the advantages of thermoplastic pipe systems in the future, the presented extrusion process was specifically designed, built and tested for ruggedly extruding thin-walled aircraft-grade thermoplastic pipes with the option to vary the wall thickness online. This capability can be used to locally reinforce critical areas of the pipe-system in aircrafts. The outstanding issues of producing those variable high-performance polymer pipes are: the identification of stable process windows as well as the reproducibility of parameter variations to locally adjust the wall thickness. The studies that were conducted to answer these questions are presented in this paper.

MODULAR PIPE DESIGN SYSTEM

High-performance thermoplastics like PEI account for one percent of the plastics production. These polymers offer in comparison to standard and technical thermoplastics high mechanical properties, exceptional high temperature stability, flame retardancy, low smoke release rates and extraordinarily good media resistances [1-5]. The high market prices and the necessary special manufacturing equipment are the reason that these thermoplastics are mainly used in the aviation or other high-tech industries. PEI is typically used in aircraft cabins e.g. as tables, seat covers or cooking equipment. Polyetheretherketone (PEEK), having similar mechanical properties, was recently used to replace conventional steel or titanium drainage pipes in the new Airbus A350 XWB [6], leading to weight savings of up to 60 %. In comparison to the semi-crystalline PEEK, the amorphous PEI convinces with an attractive price (~10 €/kg compared to approximately 100 €/kg) but has a lower chemical resistance in general.

Based on these high performance polymers a modular pipe design-system has been developed (figure 1). The simplified system consists of straight, widened and bent pipes. Due to the sealing of the system-elements all pipe-endings need a groove that is suitable for common aircraft fittings. If the pipes are thermoformed using conventional straight pipes the wall thickness can lead to very thin areas that in worst case do not fulfil the mechanical requirements. If a pipe is widened from 1" to 1½" for example the wall thickness switches from 1,1 mm to 0,7 mm. One solution is to consequently use straight pipes with higher wall thickness that are dimensioned on the lowest wall thickness that might occur in the whole system. In that case the lightweight potential cannot be exploited. The solution presented in this paper is the use of pipes with locally adapted wall thickness. Therefore an extrusion process-concept with systematic parameter variations was developed.

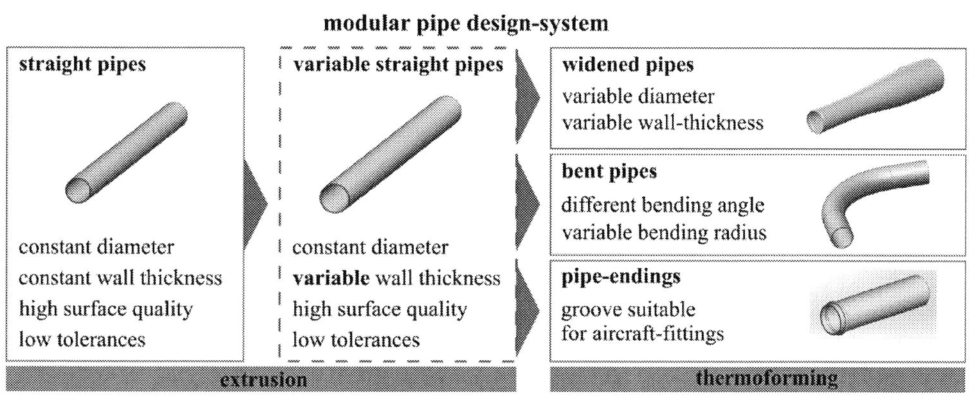

FIGURE 1. concept of a modular pipe design-system for aircrafts

VARIABLE PIPE-EXTRUSION OF HIGH PERFORMANCE THERMOPLASTICS

To extrude high performance thermoplastics a special manufacturing process was developed and optimised. The result is an installation for extruding high-quality, thin-walled PEI pipes. It contains a special high-temperature extruder, a nozzle and a high-temperature dry-calibration system that is specifically adjusted for the extrusion of PEI. Targeted parameter studies were carried out to identify suitable process windows and to draw conclusions from the extrusion parameters about the geometry-effects.

The extrusion line contains of a mechanical drive to generate the screw rotation, a special high temperature extruder, a spiral mandrel distributor, a dry calibration system, an ultrasonic measurement system, a cooling system, a haul-off unit and a cutting unit. All parts of the extrusion line are specially adjusted to extrude pipes fulfilling the high aeroplane standards using high temperature thermoplastics. In order to achieve the requirements for high surface qualities a dry calibration system with highly polished inner surfaces is used. Depending on the thermoplastic materials the calibration consists of one or two units that can be set on two different temperatures and two vacuum levels. Figure 2 shows the most important parameters of the extrusion process. From the extruder up to the calibration, 15 different temperature zones can be adjusted.

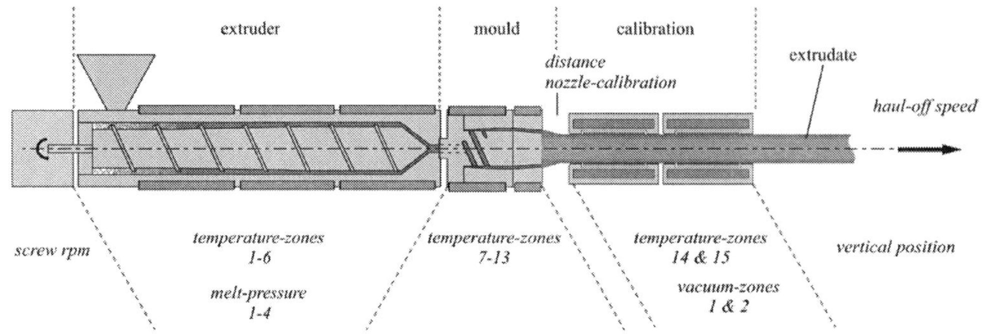

FIGURE 2. scheme of the parameters of the extrusion process

The melt pressure is utilized as an instrument to control the stability of the process. In connection with the screw rotation and the haul-off speed the stability of the process is primarily adjusted. While having a constant melt flow that is controlled by the screw rotation, the wall thickness of the pipe can be adjusted by varying the haul-off speed. One of the key parameters of the process stability is the distance between the nozzle and the calibration, ref. to figure 2. As shown in figure 3 (left) the deviation of the wall thickness is reduced when the distance is increased. In addition to that the limit of the maximum distance is related to the wall thickness. For instance, the maximum workable distance for a 1,1 mm pipe is at around 45 mm in comparison to 35 mm for a pipe with 0,8 mm wall thickness. Regarding the requirement of high surface qualities the screw speed has a significant influence. Figure 3 (right) shows the throatiness and fissures on the outer surface of the pipes when processing with a screw speed of 14 rpm. A higher screw speed of 35 rpm leads to a significant increase in the surface quality.

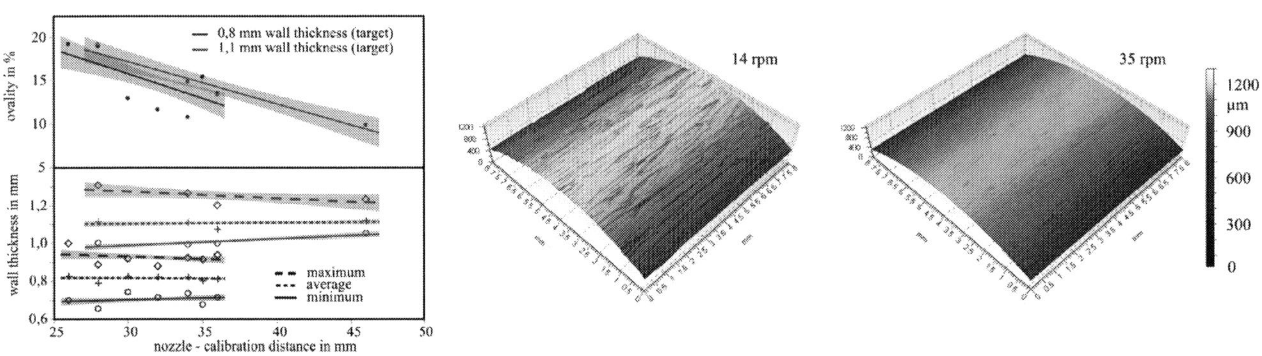

FIGURE 3. left: effect of the nozzle-calibration distance on wall thickness deviation measured online with an ultra-sonic system; right: surface roughness at different screw speeds measured with optical profilometry

To ensure the desired temperature conditioning of the extruded pipes a simulation model was developed. With the use of thermography before and after the calibration input parameters were generated and the numeric model was iteratively optimised as described in [7]. Figure 4 shows the calculated temperature profile for different positions inside the pipe-wall and on different processing speeds considering the haul-off speed. The drawn profiles illustrate that the cooling down on the outside of the pipe is quick, due to the direct contact to the wall. In contrast to that the inside of the pipe cools down slower due to a low heat conduction coefficient of the used polymers. After the calibration-unit the outside of the pipe gets heated up by the inside layers again, as can be seen in figure 4. When processing with slow speeds or thin pipes the polymer is cooled down rapidly in all layers. In that case a smaller calibration is enough for these processing conditions. As a result of these analyses a model was developed to forecast the cooling down behaviour of different pipe geometries and in the end to setup the right process parameters.

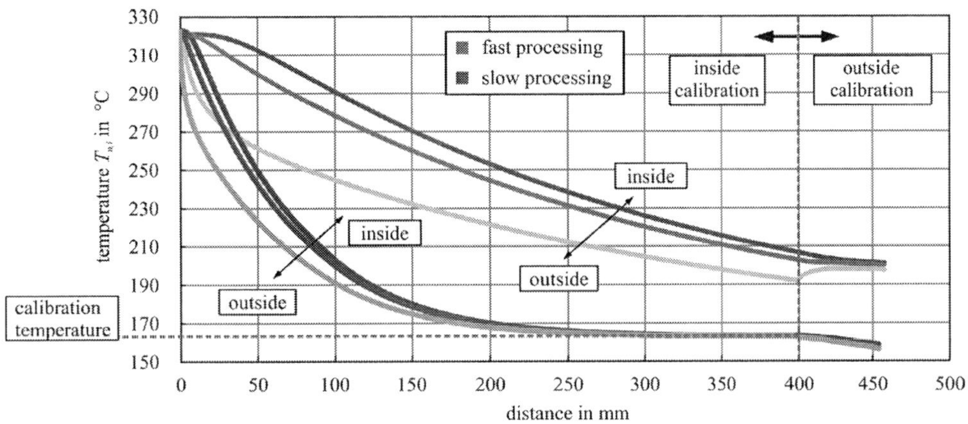

FIGURE 4. Calculated temperature profile during the pipe extrusion of PEI Ultem™ comparing slow and fast processing

In many process studies process parameter windows were identified for a reproducible production of PEI pipes with constant wall thicknesses reaching from 0.5 mm up to 1.4 mm and high quality surfaces. With these identified process windows a study was conducted to systematically vary the wall thickness in longitudinal pipe direction. Based on the geometry restrictions of the modular pipe design system, target pipe geometries were defined, including the change-length and the wall-thickness change. The defined pipe geometries were translated into process parameters. While a variation of the screw speed has a gentle influence on the process it was kept constant for the studies. The dimensional change was easily realised by simultaneously varying the haul-off speed and the distance between nozzle and calibration. In addition to that the direction of the dimensional change was analysed as well. Figure 5 shows representative measurements for two parameter changes. While changing the dimension of the pipe, the wall thickness was measured on varying positions using an ultra-sonic measurement system. It can be concluded that the variation from 0,8 to 1,1 mm and vice versa for different change-lengths is possible and the tolerance recommendations can be fulfilled.

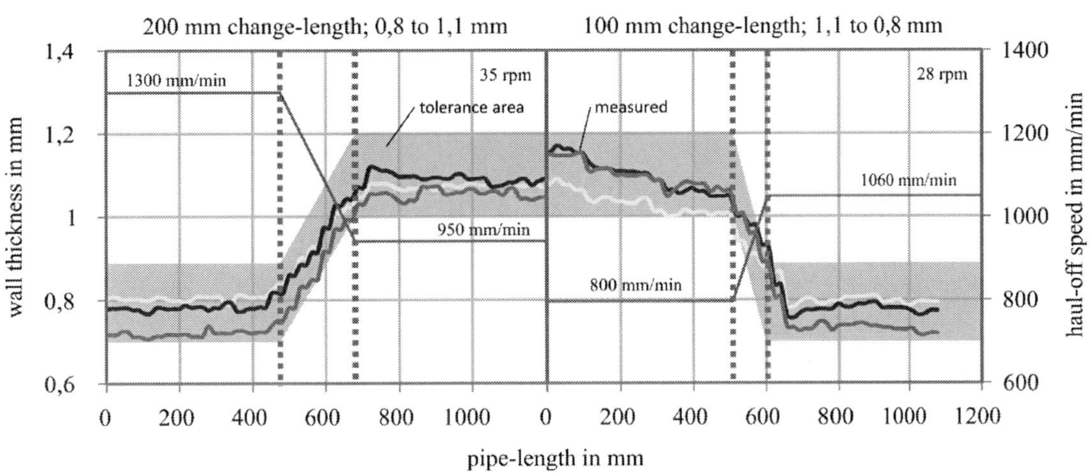

FIGURE 5. Measured wall thickness distribution during parameter variation for two setups

CONCLUSIONS AND OUTLOOK

It has been demonstrated that PEI can be extruded to high-quality pipes for use in commercial aircraft. Numerous parameter studies identified stable process windows for the extrusion of pipes with different wall thicknesses. The effects of the distance between the components nozzle and calibration on the pipe geometry were analysed and the influence of the screw speed on the pipe-surface quality was examined. A numerical model of the cooling down behaviour was developed to successfully calculate the temperature of the pipes inside the calibration for different processing conditions. Based on these analyses and the identified process windows a process setup for online varying the dimension of the extruded pipes was successfully tested and analysed.

The research results can now serve as a basis for further development for modular pipe systems in commercial aircrafts. The weight advantage over traditional stainless steel pipes, along with the price advantage over titanium additionally confirm the huge potential of thermoplastic pipe systems in aviation.

ACKNOWLEDGEMENT

Some of the research presented here was performed as part of the project "The technologies for an energetically self-sufficient, intelligent cabin (DIANA)", under the leadership of Diehl Aircabin GmbH, and was funded under funding code 20K1101K by the Aviation Research Program IV of the Federal Ministry for Economic Affairs and Energy, following a decision of the German Bundestag.

REFERENCES

1. Kaiser, W.: *Kunststoffchemie für Ingenieure: von der Synthese bis zur Anwendung*. Munich: Carl Hanser-Verlag, 2015.
2. Saudi Basic Industries Corporation. 09 Dec. 2015 *<https://www.sabic-ip.com/gep/Plastics/en/ProductsAndServices/ProductLine/ultem.html#>*.
3. N.N.: *ASTM E906-14*: Standard Test Method for Heat and Visible Smoke Release Rates for Materials and Products Using a Thermopile Method. ASTM International, West Conshohocken, PA, 2014, DOI: 10.1520/E0906-14, www.astm.org.
4. Stahl, P.O., Muurinen, P.: *Polyetherimid*. Kunststoffe. 85. (1995): 1611–1612.
5. Schobesberger, M.: *High-Tech für High-Performance*. Kunststoffe. 84. (1994): 759-761.
6. N.N.: *Rohre aus Victrex PEEK sparen Gewicht in der Luftfahrt*. 10 May 2012 K-Zeitung-online. 15 Feb. 2016 <http://www.k-zeitung.de/rohre-aus-victrex-peek-sparen-gewicht-in-der-luftfahrt/150/1195/38235/>.
7. Stegelmann, M.; Lucas, P.; Modler, N.; et.al.: *Extrusion of high-performance polyetherimide pipe systems for aircrafts*. Sampe Conference, Long Beach, CA, 2016

Bubble-free Polymer Devolatilization in a Simplified Extruder Model

Stefan Hirschfeld[1, a)] and Olaf Wünsch[1, b)]

[1]Institute of Mechanics, Chair of Fluid Mechanics, University of Kassel, Mönchebergstr. 7, 34125 Kassel, Germany
[a)]Corresponding author: stefan.hirschfeld@uni-kassel.de
[b)]wuensch@uni-kassel.de

Abstract. This paper presents an experimental study of bubble-free polymer devolatilization in a partially filled simplified extruder model. Systematic degassing experiments were performed with a model substance system at ambient temperature under variation of rotational speed and filling level. A stripping flow of nitrogen provides the necessary driving force for the mass transfer by reducing the partial pressure and prevents foam degassing. Furthermore, video analyses of the free surfaces and a theoretical reflection of the diffusion coefficient deliver more understanding of the process. Mass transfer is evaluated and assessed with models based on the penetration and surface renewal theory. A comparison with literature concerning experimental und numerical studies is performed. An explanation for known deviations between theory and results is given. The main outcome of this work is that intense mixing is crucial for a highly efficient degassing device.

INTRODUCTION

Polymer devolatilization is a widely used unit operation during polymer manufacturing and polymer processing. This separation process of volatile organic compounds (VOCs) from molten polymers is essential in order to obtain high quality polymer products and fulfill strict regulation limits for health and environment. There are two relevant mass transfer mechanisms: Diffusion-controlled film degassing and bubble-controlled foam degassing [1]. In rotating devolatilizers like screw extruders very low residual concentrations can be obtained and high-viscous polymers can be treated. For this case, film degassing is dominant. Only few published literature on the topic is available, which often focuses on a theoretical treatment without experimental validation. The expensive design of efficient degassing devices is still based more on empirical knowledge rather than a predictive theory [2].

THEORY

During polymer devolatilization mass transfer is usually rate controlled by the liquid-side mass transfer resistance. Assuming mass transport models based on the penetration [3] and surface renewal theory [4] leads to

$$kA = \frac{\dot{n}_i}{c_i - c_i^e(p_i)} = 2\sqrt{\frac{D}{\pi}}\underbrace{\left(\frac{A_{film}}{\sqrt{t_{film}}} + \frac{A_{pool}}{\sqrt{t_{pool}}}\right)}_{:=SRP}. \qquad (1)$$

The derivation of this equation is explained in more detail in [5]. kA is the total mass transfer coefficient k multiplied by the free surface A. The molar flux $\dot{n}_i$ of volatiles (i) thus is proportional to the molar concentration difference Δc_i, which is the driving force. Here c_i is the bulk concentration and c_i^e is the free surface concentration, which is coupled with the partial pressure p_i due to gas-liquid equilibrium (Flory-Huggins equation). In this paper, all the free surfaces A (*film*, *pool*) where degassing in rotating devices occur and their corresponding renewal times t are merged into the surface renewal parameter *SRP*. According to the model, a degassed fluid element is assumed to be completely mixed into the liquid bulk phase after exposure on the free surface. D is the diffusion coefficient.

EXPERIMENTAL

All degassing experiments were performed with the setup described in [5]. The model substance system consists of high-viscous polydimethylsiloxane (PDMS) (η_0 = $100\ Pas$) as polymer and 1,1,2-trichloro-1,2,2-trifluoroethane (CFC-113) as volatile. As mentioned in [5] the gas-side mass transfer resistance must be considered for a low mass flow of stripping nitrogen. Further experiments have revealed that it is negligible for $\dot{V}_{N2} > 15\ L_N/min$. Furthermore, a constant pressure level inside the devolatilizer is preferable. This was realized by using cold traps filled with Raschig rings as packing material to inhibit local plugging. Therefore, all of the following experiments were executed with $\dot{V}_{N2} = 20\ L_N/min$ and p = 1200 mbar.

To calculate kA by eq. (1) it is necessary to know $\dot{n}_i$, c_i and c_i^e. For that reason, respectively 10 samples were taken from liquid phase to measure a mean concentration c_i by thermogravimetric analysis for several points of time. A mean molar flux $\dot{n}_i$ is calculated directly from this data for each time interval. The free surface concentration c_i^e is determined with the partial pressure p_i calculated by Dalton's law, ideal gas law, measured gas phase data and the calculated value for $\dot{n}_i$. In this paper, all results are presented by mass fractions w_i, but the calculations are performed with molar concentrations c_i. Formulas for conversion of concentration units can be found in [5].

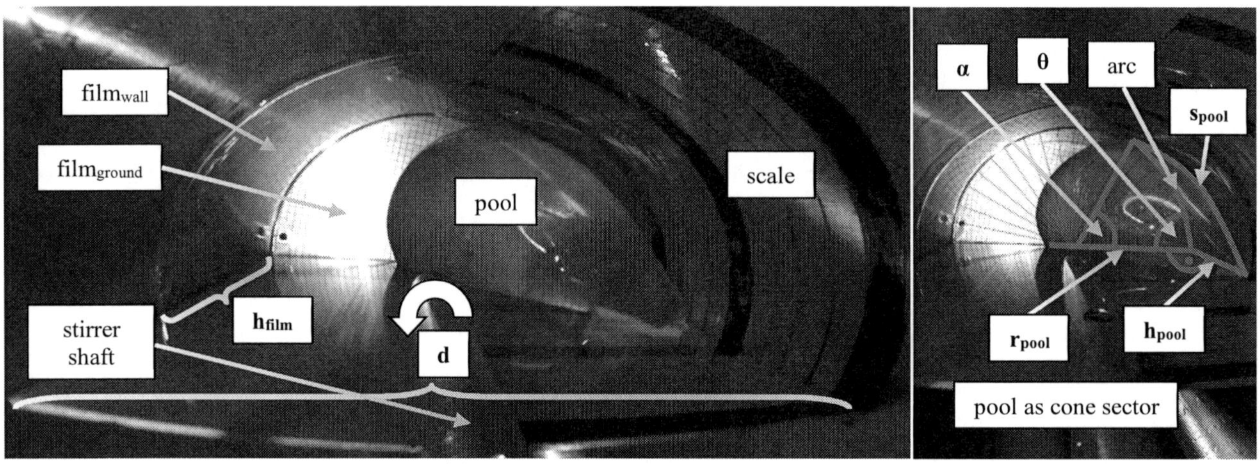

(a) (b)

FIGURE 1. (a) Simplified extruder model, d = 236 mm, V ≈ 2.2 L, n = 15 min^{-1}, (b) Parameters to calculate pool surface, V ≈ 1.1 L, n = 5 min^{-1}

To estimate free surfaces A and exposures times t in eq. (1) videos inside the simplified extruder model were recorded during degassing experiments with known fluid volume V. Fig. 1 (a) shows a photo of these recordings with the relevant degassing regions. A scale on stirrer and barrel wall helps to measure the relevant geometric parameters.

The film divides into a wiped melt film on the wall and the ground of the vessel. The surface of the wall film is calculated from the product of the film height h_{film} and the barrel circumference minus the covered pool area. The surface of the ground is equal to the base area minus the covered pool area. The exposure times t_{film} are directly measurable from the videos.

The rotating pool at the blade stirrer's flank is estimated using a simplification as a surface shell of a cone sector (cf. fig. 1 (b)) with the height h_{pool} $(= h_{film})$ and the radius r_{pool}. The surface line s_{pool} of the cone can be calculated with the pool contact angle α to obtain the circular arc on the base and h_{pool} by Pythagorean Theorem. With this information the free surface of the pool can be calculated as

$$A_{pool} = \pi\, r_{pool}\, s_{pool} \frac{\theta}{2\pi}, \qquad \frac{\theta}{2\pi} = \frac{V_{pool}}{\frac{1}{3}\pi\, r_{pool}^2\, h_{pool}}. \tag{2}$$

The missing central angle θ of the cone base sector can be determined by the known pool volume V_{pool} from mass balance of the sample weight. The exposure time t_{pool} is calculated with the surface line s_{pool} divided by the circumferential speed. The device consists of two chambers each with slightly asymmetric pool sizes due to fabrication tolerances. Consequently, all parameters were determined separately for both chambers resulting in six independent degassing regions each characterized with a surface value A and an exposure time t.

RESULTS AND DISCUSSION

Degassing results

Fig. 2 shows the raw data of the degassing experiments carried out for three different rotational speeds and three different filling masses of the device in dependence of the time. As expected, a higher rotational speed and lower filling level leads to faster degassing. Furthermore, it is obvious that a higher mass flux of volatiles leads to higher free surface concentrations, because the stripping flow of nitrogen is constant over all test runs. This is especially the case for $n = 15$ / min.

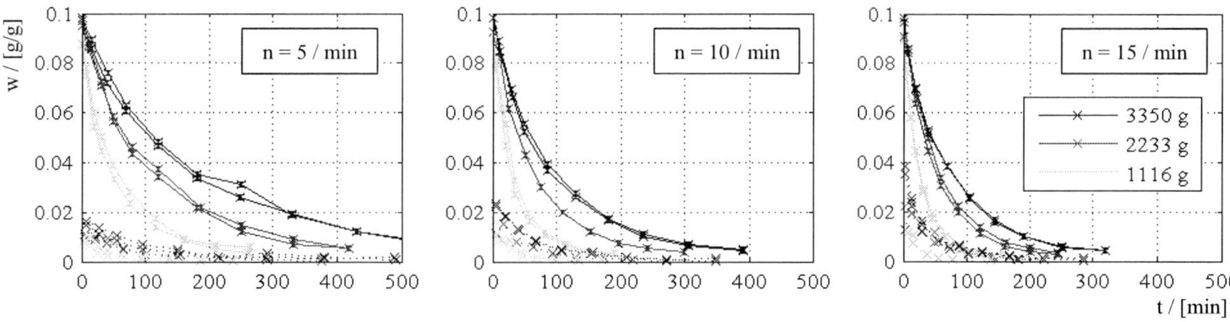

FIGURE 2. Raw data of degassing experiments: measured mass fraction w from samples (solid line), calculated mass fraction w_e for free surface from gas phase data (dotted line), two repeated measurements each

The aim of the investigations is to assess the limitations and restrictions of the mass transfer model. Therefore, the mass transfer kA is calculated by eq. (1) with data shown in fig. 2 for each time interval. The results are plotted in fig. 3 over mean concentration to ensure a uniform and comparable presentation. This is an inverse representation in comparison to fig. 2, since devolatilization starts on the right hand side and ends on the left hand side. The values in the gray area scatter significantly more due to measurement accuracy in combination with the low concentration differences towards the end of the degassing. As a result, this area is not taken into account for further evaluation.

Fig. 3 shows that kA decreases significantly over time. This is in contrast to the modeling. According to the model in eq. (1), kA should be constant during degassing unless either the surface renewal parameter SRP or the diffusion coefficient D or the molar flux $\dot{n}_i$ decreases due to reduced driving force Δc_i. All of these three possible reasons are analyzed in more detail in the following sections.

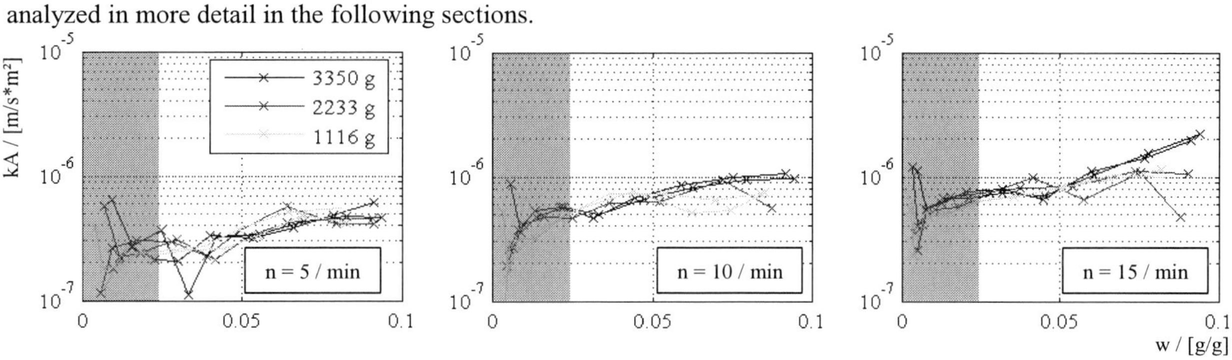

FIGURE 3. Results of degassing experiments: calculated mass transfer coefficient kA from data in fig. 2 with eq. (1)

Video results

Fig. 4 shows the surface renewal parameter SRP (cf. eq. (1)) determined by video analysis from the six mentioned degassing regions with the data of videos at the start and the end of devolatilization. The filling level and the rotational speed influence SRP. An increasing rotational speed leads to increasing free surfaces, because the height of the film and pool ($h_{film} = h_{pool}$) increase and the exposure times decrease. Both effects increase SRP. In contrast, an increasing filling level on the one hand as well leads to increasing free surfaces, but on the other hand also the exposure time for the pool (t_{pool}) increases. The second effect counteracts to some extent an increased SRP. Therefore, the influence of the rotational speed n is more crucial than the filling level. A second outcome is that *SRP* is nearly constant over each single degassing experiment (cf. the data "Start" with "End" in fig. 4). As a result, the first hypothesis, a decrease of *kA* (cf. fig. 3) due to *SRP,* is refuted by video analysis results.

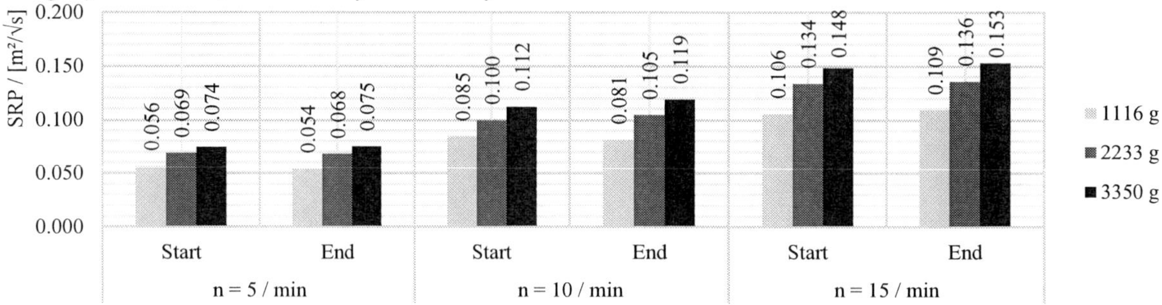

FIGURE 4. Surface renewal parameter *SRP* at the start and the end of devolatilization

Diffusion Coefficient

The diffusion coefficient D can be estimated on the one hand with a pure theoretical approach called the free volume theory [5] and on the other hand calculated from experimental results by eq. (1).

The free volume theory is an established approach to describe diffusion coefficients in polymer solutions. Several researchers developed this model (Cohen, Turnbull, Fujita, Vrentas, Duda) [6] assuming that the free volume between the molecules is the rate-limiting parameter. According to [6] all model parameters can be obtained from pure substance data and group contribution methods without any diffusion data. Diffusion occurs if a solvent molecule or a part of the polymer chain "jumps" into a "hole" of the free volume. Here ξ ratio of molar volumes of solvent to polymer "jumping units" is the relevant and most controversial discussed parameter [6]. The size of the polymer "jumping unit" is often about 1.5 times that of a monomer unit. Moreover, this value is larger for stiffer molecules than for more flexible ones. For PDMS 1.0 monomer "jumping unit" is assumed, because the glass transition temperature $T_g = 150$ K is very low for PDMS and thus the polymer chain is more flexible. In [7] this assumption leads to a good agreement with experimental data for PDMS. The data for CFC-113 were estimated as far as possible using pure substance data and group contribution models (Sugden in [6]). Some unknown parameters were estimated with the data for chemically similar solvents. Tab. 1 shows the results of this estimation for the relevant concentration range at $T = 298$ K. D_{min} and D_{max} are estimations with other chosen values for the unknown parameters, which lead to extreme low or high values to get an idea of how sensible those values act on the diffusion coefficient.

TABLE 1. Estimated D from free volume theory

w / [g/g]	D / [m²/s]	D_{min} / [m²/s]	D_{max} / [m²/s]
0.001	1.40E-10	9.25E-11	2.34E-10
0.01	1.44E-10	9.09E-11	2.56E-10
0.1	1.93E-10	7.52E-11	5.46E-10

TABLE 2. Calculated D from experiments

w / [g/g]	D_{min} / [m²/s]	D_{max} / [m²/s]
0.02	1.0E-11	2.0E-11
0.05	1.5E-11	6.0E-11
0.09	3.0E-11	1.6E-10

In eq. (1) the unknown diffusion coefficient D can also be calculated with the results from degassing experiments (*kA* in fig. 3) and video analysis (*SRP* in fig. 4) shown in tab. 2. D_{min} and D_{max} are the minimum and maximum values received for the corresponding concentrations. Regarding the change of these values during devolatilization, a decrease by a factor of about 10 is observed. In contrast, the values from the free volume theory decrease by a factor less than 2. The comparison of both approaches offers that the results in tab. 2 are not plausible. Consequently, the second hypothesis, a decrease of D could explain the results in fig. 3, is also refuted.

Driving Force and Mixing

As mentioned above, the molar concentration difference Δc_i is the driving force of the mass transfer. Eq. (1) takes the assumption of complete remixing of degassed fluid elements into the bulk phase after exposure at the free surface. Numerical simulations [8] revealed that this assumption is not valid. Because of laminar flow, outer parts close to the free surface of the rotating pool are not homogeneously mixed with central parts. Here the concentration is significantly higher. Consequently, the effective driving force *EDF* should be lower and therefore a reduced molar flux and kA-value are measured. EDF is calculated with $D = 2\text{E-}10\ \text{m}^2/\text{s}$ and normalized ($NEDF$) with the measured driving force:

$$EDF = c_{i,calc} - c_{i,calc}^{e}(p_i) = \frac{\dot{n}_i}{2SRP\sqrt{\frac{D}{\pi}}}, \qquad NEDF = \frac{EDF}{\underbrace{c_{i,\exp} - c_{i,\exp}^{e}(p_i)}_{\text{measurements presented in fig. 2}}}. \tag{3}$$

The results of eq. (3) are plotted in fig. 5. Especially the experiments with $n = 15$ / min reach nearly 100% normalized effective driving force $NEDF$ at the start of degassing. In the course of the experiment, $NEDF$ continuously reduces down to a factor of 3. This behavior fits to numerical simulations in [8]. A further comparison to experimental studies from [9] and [7] also confirms the thesis of incomplete mixing during devolatilization. Collins et al. [9] measured $ka = kA/V$ 3 times lower than theory and Gestring [7] predicts Sherwood numbers Sh 3 times higher than experimentally measured. In contrast to [9] and [7], this paper gives a further explanation for the measured behavior.

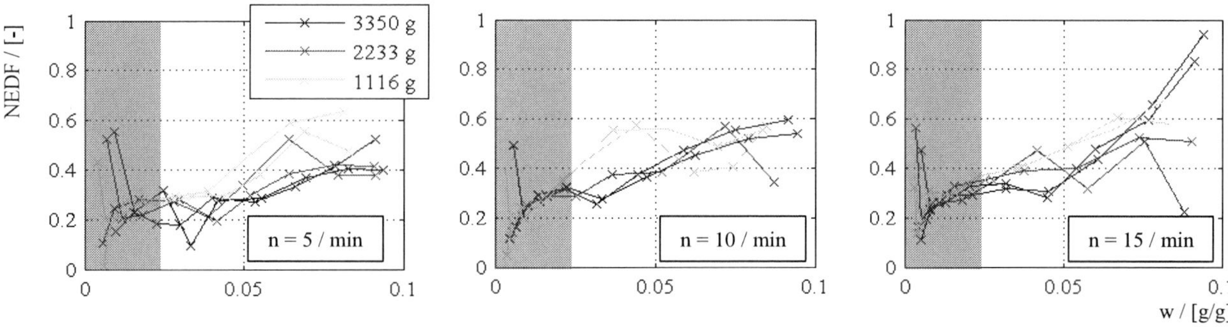

FIGURE 5. Results of degassing experiments: calculated normalized effective driving force $NEDF$ by eq. (3)

CONCLUSION AND OUTLOOK

This paper contributes a systematic experimental investigation of bubble-free film degassing with different methods and approaches. Limitations and problems of the penetration and surface renewal model to describe the mass transfer are considered and discussed. Degassing experiments revealed a decreasing mass transfer coefficient over time, which stands in contrast to theory and explains the described deviations between the theory and experiments in literature. Three possible hypothesis to explain this behavior are investigated. Video analysis was applied to determine a surface renewal parameter. This parameter is constant over each single experiment. Diffusion coefficients are estimated theoretically with the free-volume theory. In comparison with calculated values from experiments, these values are much higher than expected. It could be verified that incomplete mixing inside the liquid is the only plausible explanation for all mentioned deviations from theory. Numerical simulations confirm this statement.

The deviation of mass transfer measured in experiments and predicted from theory is not constant but increases during the degassing. Therefore, for technical applications it is recommended to use multiple shorter degassing zones coupled with mixing zones rather than only one long degassing zone. In future, more experiments on a twin-screw extruder are planned to prove findings from the simplified extruder model and transfer it to a technical device.

REFERENCES

1. R. J. Albalak, *Polymer Devolatilization* (Marcel Dekker, New York, 1996).
2. VDI-Gesellschaft Kunststofftechnik, *Aufbereitungstechnik 2006* (VDI-Verlag, Düsseldorf, 2006).
3. R. Higbie, AIChE J. **31**, 365-389 (1935).
4. P. V. Danckwerts, Ind. Eng. Chem. **43**, 1460-1467 (1951).
5. S. Hirschfeld, O. Wünsch, Proceedings of the 32nd International Conference of the Polymer Processing Society (PPS-32, Lyon, 2016), accepted conference paper for publishing.
6. J. M. Zielinski, J. L. Duda, AIChE J. **38**, No. 3, 405-415 (1992).
7. I. Gestring, "Entgasen von Polymeren", Ph.D. thesis, Universität Hannover, 2002.
8. O. Wünsch, M. Lübke, S. Hirschfeld, „Filmentgasung hochviskoser Polymere", Schlussbericht, IGF-Vorhaben 17298N der GVT, 2015.
9. G. P. Collins, C. D. Denson, G. Astarita, AIChE J. **31**, 1288-1296 (1985).

Simulation of the effects of using conformal cooling channels in SLM produced plastic extrusion calibrators.

K. Soete[1], F. Desplentere[1]

[1] *Materials Technology TC, KU Leuven Campus Bruges, Spoorwegstraat 12 8200 Bruges, Belgium*

[a)]Corresponding author: karen.soete@kuleuven.be

Abstract. In profile extrusion, one of the most crucial steps in obtaining a high quality, correctly sized profile is the calibration and accompanying cooling phase. In order to eliminate residual stresses and to induce adequate sizing of the profile, fast and uniform cooling of the extrudate is needed. By placing cooling channels as close as possible to the extrudates surface inside the calibrator, which is possible by using SLM(selective laser melting) produced calibrators, the temperature gradient along the profiles surface and its thickness, can be minimized resulting in better and faster cooling of the extrudate. The potential and advantages of using SLM produced calibrators with conformal cooling channels is even more apparent for profiles with small details which are currently unreachable using conventional mechanical processing techniques. In this paper the effect of using conformal cooling channels in extrusion calibrators is studied using Siemens NX simulation software. Temperatures of both cooling water and of the polymer extrudate are studied. These simulations are used to optimize calibrator design and reduce cooling time. Based on these simulation, an estimation is giving of the advantages and potential of using additive manufactured calibrators in the polymer extrusion process.

INTRODUCTION

Nowadays conformal cooling is a technique which is more commonly used for injection moulding. But from the industry, interest is growing in using these techniques in extrusion calibrators. The idea for conformal cooling is that the cooling channels are placed as close as possible to the product wall, this by using metal selective laser melted (SLM) 3D printed moulds. Due to this technique it is possible to follow the contour of the product even where this is not possible with conventional techniques. Because of the ability to design the cooling channels in this way, the temperature control of the product can be optimized. Using conformal cooling channels in mould design has several advantages. For example the design freedom which is possible because of SLM 3D printing techniques. Because of the better and more even cooling of the mould, lower internal stresses, less warpage and less sink marks in the polymer product can be achieved [1]. For injection moulding, reduced start-up times and shorter cycle times are two of the advantages as well, but for extrusion this can lead to an increased extrusion speed of the polymer profile. Some of the disadvantages of conformal cooling are the complex design processes that need to be taken into account. Also the production of the mould trough SLM implement a higher cost compared to the conventional production techniques used nowadays for mould making and is a much slower production process. This is mainly due to the high surface roughness of the printed mould which implements that several post-treatments are required like annealing and hardening and milling, eroding and polishing to meet the needed surface quality of the mould [2]. Also the limited number of producers of SLM printed moulds and the high costs of producing these moulds is a downside of this technique.

When looking at the possibilities for SLM in extrusion, this technique can be used for making both moulds and calibrators. In extrusion mould design, studies are done on how to optimize topology of the die for the polymer flow within extrusion [3]. Using this the SLM technique, additional cooling or heating channels can be implemented in the extrusion die as well as a combination can be made for adding additives needed for the extrusion process like waxes and lubricants. But in this study the focus will be on using the SLM technique to optimize the cooling in extrusion calibrators using conformal cooling channels. Again also in the calibrator the SLM gives the opportunity to implement also vacuum and if needed adding additives to the extrusion process, but this is out of the scope of this study.

METHOD

To be able to study the effect of using conformal cooling channels in SLM produced plastic extrusion calibrators a reference calibrator for an existing profile has been chosen to start from. Based on the external geometry of the profile a new design using conformal cooling channels has been designed.

Current situation

For the current situation of the calibrator, the cooling channels show to have different shapes, showing more or less the shape of the details in the calibrator. For example in figure 1 the shape of the cooling channel in the detail has an irregular shape to be able to cool the detail in the profile.

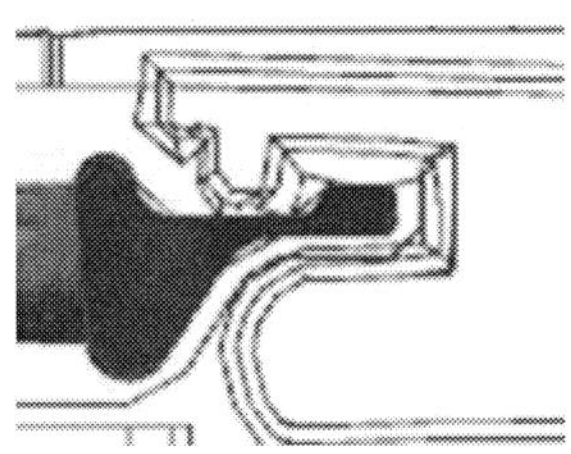

FIGURE 1. Detail of cooling channel inside detail of the profile in the calibrator

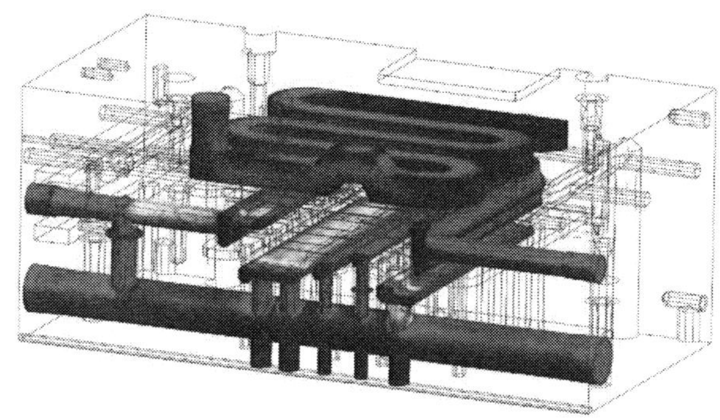

FIGURE 2. Original cooling channel design in the calibrator

For the cooling of the top of the profile, the calibrator is equipped with a zigzag cooling channel (see figure 1) . The rest of the cooling channels are straight cooling channels.

New calibrator design

For the new calibrator design, the technique of conformal cooling is implemented (see figure3). By placing a square shaped cooling channel with the cooling channel centerlines put as much as possible equidistant to the profile to be extruded, the intention is to create a more uniform cooling of the complete profile. The dimensions of the square shaped channel are 17 mm by 0.8 mm. Cooling channels are separated in two separate cooling circuits. One cooling the top an right side of the product, and one cooling parallel the left side and the bottom side of the profile. All cooling channels follow a zig zag pattern.

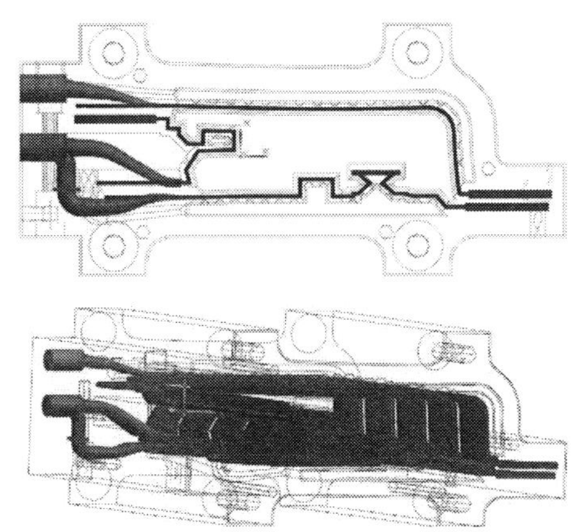

FIGURE 3. Design of new cooling channel geometry in calibrator

Simulation set-up

Simulations on both the existing case as for the new calibrator design are performed in siemens NX. For extrusion a coupled thermal/flow simulation combined with a solid motion effects simulation is needed. The disadvantage of combining the thermal flow simulation with the solid motion effects is that this implements a complex numerical model which requires high computational efforts. To reduces the calculation time, a simulation with solid motion effects is performed on a representative extruded profile with a height of 2mm and a width of 10 mm. The polymer temperature was set on 190°C while the calibrator was cooled on a temperature of 20°C. The extrusion speed was set on 0.05m/s (3m/min) and a heat transfer coefficient between the polymer and the calibrator of 500 W/m²C is used. Based on this simulation the heat flux from the polymer with a thickness of 2mm applied on the calibrator was found. By using exponential curve fitting on the results from the simulation the equation is found for the heat flux given by following formula:

$$Heat\ Flux\left[\frac{W}{mm^2}\right] = \frac{-0{,}5824\, e^{-60{,}35\, x} - 1{,}594\, e^{-8{,}337\, x}}{10} \qquad (1)$$

This formula can now be used in the simulations to apply the heat flux of the polymer melt to the calibrator and thus simplifying the complexity of the simulation model and reducing calculation time. For both simulation models (original and new) a water volume flow rate of 2 l/min at 20°C is used for the cooling channels.

RESULTS AND DISCUSSION

Flow profile in cooling channels

The flow profile inside the cooling channels for the original profile, for example in the detail show in figure 2, shows large deviations in flow velocity between the smaller (225 mm/s) and the larger (776 mm/s) cross section of the cooling channel geometry resulting in an uneven flow velocity profile. This will have a significant influence on the cooling potential of the calibrator for that section. Because of the rectangular shape in the new calibrator design, the flow velocity profile is uniform (± 1100 mm/s).

Pressure loss in cooling channels

An important disadvantage of the conformal cooling channels, because of their small dimensions, is the large pressure drop along the cooling channels. While for the original geometry the maximum pressure drop is 70 Pa, the pressure drop for the new cooling channel design is 60 000 Pa. Both for a flow rate of 2 l/min. This will be one of the major challenges need to be taken into account in conformal cooling channel design in extrusion calibrators.

Cooling water temperature

In the original design, there are seven outlets for the cooling water. In the new design only three. To be able to compare water temperatures, the cooling cannels for the original design are grouped into the three zones corresponding to the outlet points for the new geometry. When comparing all temperatures, it can be seen that for the original calibrator, there is a large variation in fluid temperature. This will have an influence on the temperature of calibrator surface and thus on the profile. For the new design, the cooling water temperature shows smaller differences for the

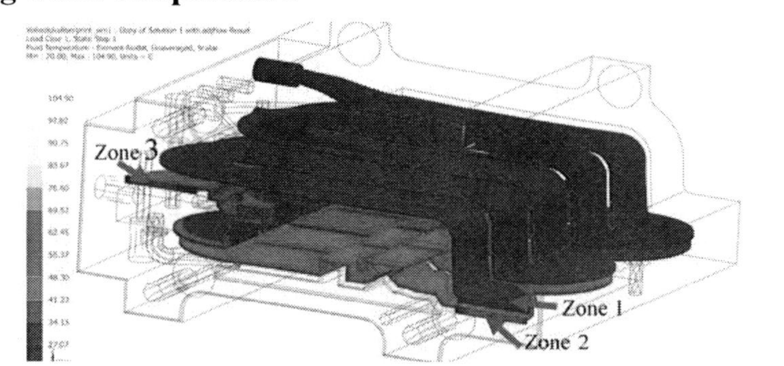

FIGURE 4 three outlet points for cooling channels in the calibrator design

three different zones which show that a more even cooling is achieved with the conformal cooling channels. This effect will be translated on to the surface of the calibrator and thus on the profile.

TABLE 1. Cooling water temperature for the different cooling channels in the calibrator for both original and new situation

Cooling zone	Temperature cooling water [°C]	
	Original Calibrator	New Calibrator
Zone 1: Top of profile + right side	45.1	30.5
	35.6	
Zone 2: Bottom of profile	75.8	42.3
	29.0	
	46.7	
	30.8	
Zone 3: Left side (detail) of profile	42.1	32

Temperature of calibrator surface

When looking at the temperature of the calibrator surface, there are two areas of interest (see figure 5). These areas are located at the two crucial details that need to be cooled in the profile. Because of the design, a lot of heat is applied on the surface of these bulges which needs to be disposed via the cooling water. Because these areas are hard to reach with conventional mould making techniques, because of the near entrance to the detail, the surface temperature of these two details is higher for the original design than for the new design. This is shown in table 2. Especially for surface 2 the temperature at the inlet is lower for the new calibrator design. But even more important is that the outlet temperatures for both location is significant lower for the new calibrator design showing that a more uniform and better cooling is achieved using the conformal cooling technique in extrusion calibrator. And this will have its effect on the temperature of the profile.

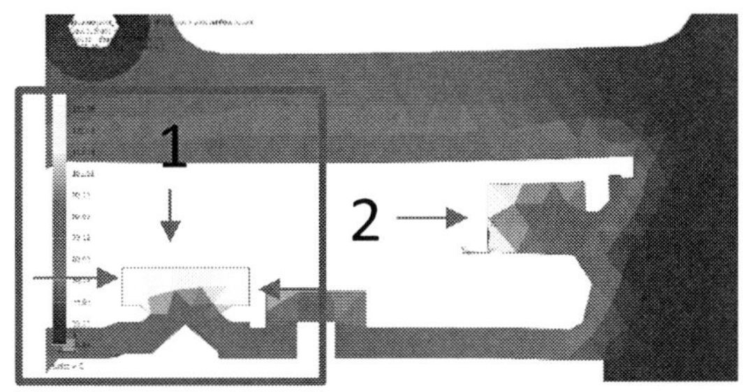

FIGURE 5. Location of the surfaces where temperatures are recorded for both the original and new calibrator design

SUMMARY AND CONCLUSION

Overall, it can be concluded that according to the first simulation, 3D printed calibrators with conformal cooling channels has several advantages compared to the conventional produced calibrator. This research shows that the contour of the surface to be cooled can be followed. The cooling channel centerlines are put as much as possible equidistant to the profile to be extruded. Because of the uniform cross section of the cooling profile, a more uniform flow profile can be achieved, which will have a large influence on the cooling potential of the cooling channel. These aspects result in a more uniform surface temperature of the calibrator which was shown as a result from the simulations. Besides the advantage, some remarks can be made for this technique. For example because of the limitation in flow channel shape and dimension, for example when small geometries are needed to be able to reach certain zones, large pressure drops need to be taken into account. Also the influence of wear, both inside the cooling channels as on the calibrator surface which comes into contact with the profile, need to be studied further.

Based on the simulations, first results seem promising, so in a next step the actual 3D printed calibrator will be tested in the production line and the results will be compared with the simulations.

ACKNOWLEDGMENTS

The author would like to thanks the Flemish innovation network for the support of the TETRA-project AM4XT innovative tooling concept for polymer extrusion using additive manufacturing. We would also like to thank Vives for the collaboration in this project.

REFERENCES

1. W. Six, R. Appermont, F. Desplentere, J. De Keyzer and A. Van Bael, "Design and simulation of conformal cooling channels In SLM produced injection molds," proc. International Conference of the Polymer Processing Society, pp. S13-250, 2016.
2. C. Hopmann, N. Yesildag, S. Bremen, K. Wissenbach and S. Merkt, "Surface quality of profile extrusion dies manufactured by Selective Laser Melting," RTejournal ¬ Forum für Rapid Technologie, 2016.
3. C. Hopmann, N. Yesildag, C. Windeck, S. Bremen, K. Wissenbach and S. Merkt, "opportunities and challenges of profile extrusion dies produced by additive manufacturing processes," in PPS32- 32nd World Congress of the Polymer Processing Society, Lyon, 2016.

Manufacturing and characterisation of lidocaine films prepared by hot melt extrusion.

Nair Karthik [(1)]*, Coates Phil [(1)], Whiteside Ben [(1)], Kelly Adrian [(1)], Korde Sachin [(1)], Wellings Simon [(1)], Paradkar Anant [(1)]

[(1)] University of Bradford - Bradford - UnitedKingdom

The aim of this study was to evaluate the influence of the hydrophilic carriers such as, PEO (polyethylene oxide), HPC (hydroxypropyl cellulose), and PVA (polyvinyl alcohol) on lidocaine mucoadhesive films manufactured via hot-melt extrusion. Polymer films loaded with 5% lidocaine processed by Xplore® (conical twin-screw extruder) were investigated for solid state characterisation, surface roughness, drug-polymer miscibility and molecular interactions, bioadhesion study, mechanical properties, drug release characteristics, and stability studies. The Hansen solubility parameters predicted a high miscibility between lidocaine and all three polymers. DSC and PXRD confirmed the conversion of the drug to the amorphous state. Lidocaine was found to increase the surface roughness of all the films except for PVA. Tensile testing showed the addition of lidocaine led to an increase in percentage strain at break, and a decrease in tensile strength, reflecting higher flexibility and ductility. FT-IR analysis of the extruded films showed no shifts in lidocaine bands in case of PEO whereas a total absence of lidocaine peaks was found in HPC and PVA due to the low incorporated amount of the drug (5%). All three extruded films demonstrated good adhesion ability, with PEO having the highest value. They also achieved immediate drug-release and were stable at 40°C/ 75% RH for one month with no detected recrystallisation of lidocaine. Overall, this study signified the suitability of a new Xplore® (pharma grade twin screw) extruder in producing thin and stable lidocaine mucoadhesive films.
Keywords: hot-melt extrusion, lidocaine, mucoadhesive films and Xplore®

Single Screw Extrusion with Low Pressurized Blowing Agents

M. Langlotz[1, a)] , M. Düngen[1] and M. Koch(†)[1]

[1]Department of Plastics Technologies, Ilmenau

[a)]Corresponding author: martin.langlotz@tu-ilmenau.de

Abstract. For the continuous production of foam products, the extrusion process is commonly used. Typical extrusion systems include an extruder for melting and mixing the thermoplastic material(s) as well as an extruder for cooling the melt. Both tasks can also be performed in one extruder. Between these extruders or extruder zones the blowing agent (e.g. CO2, N2) is injected into the melt with relatively high pressure. This study presents a foam extrusion process utilizing a two-stage-screw in which the blowing agent is injected at a low-pressure range and extruded as a strand. The study uses PP as well as nitrogen and carbon dioxide as blowing agents which are widely used for foaming. The screw concept makes it possible to produce a blowing agent loaded polymer melt with the present pressure of a standard gas bottle (or other gas storage devices) thus eliminating the need for expensive gas cooling, dosing and pumping technology. The influence of melt temperature and gas injection pressure on foam quality (expansion ratio, cell structure) is investigated. The produced foam has a regular cell structure.

INTRODUCTION

Foam extrusion is considered to be the method for generating continuous foam products as it is possible to generate remarkable specific material characteristics. Yet, costs for dosing devices are quite high, as it is necessary to inject gas at a very high pressure. This high pressure is not particularly necessary for generating the foam itself but needed to dissolve gas in the polymer melt [1]. In this work focus lies therefore on a validated concept that makes it possible to solve the gas and produce foam just by using the bottle pressure provided by the gas container.

LOW PRESSURE CONCEPT

Foam is produced by injecting a blowing agent into a polymer melt. When gas is injected, a two-phase-system is present. The gas then dissolves in the melt under pressure which results in a one-phase-system. When the pressure drops below a certain solubility pressure, nucleation takes place and bubbles start to form and expand. This pressure drop for initiating bubble expansion occurs in the extrusion die. [1]

It was already shown in earlier works by YUSA [2] that a high injection pressure of the blowing agent isn't needed for foaming. In that paper a screw design with a low pressure zone after the metering zone was implemented into an injection molding process and a gaseous blowing agent was added just by using the bottle pressure in an unpressurised process zone.

There is a basis for simulating and designing of devolatilization single screws by POTENTE [3]. These screws consist of six basic zones: feed, compression, metering, decompression/degassing, compression, metering. The low pressure concept of this study utilizes this setup to inject the gaseous blowing agent at relatively low pressures into the decompression zone of such a screw. In the first three zones plastic is melted and homogenized. After the decompression zone the pressure of the melt in the screw channel increases again.

The blowing agent is added into the melt in the decompression zone. Due to the low pressure in this zone the blowing agent continues to be in a gaseous phase as part of a two-phase-system until the pressure required for fully dissolving the gas in the melt is reached. This pressure is gained by another compression in the second zone in order to obtain the one-phase-system of blowing agent and polymer melt. A rapid pressure drop in the die then initializes nucleation and cell expansion. By utilizing the presented calculations provided by POTENTE [3] it is possible to design the screw for various operating points. These operating points preferably include low pressures for the gas injection zone. **FIGURE 1** shows the calculated pressure profile along the screw. In the operating points, rotational speed and counter-pressure were varied. It is shown, that for all speed and counter-pressure ranges there is always an unpressurised zone. The calculated screw was manufactured and contains the following process zones:

- 3-zone screw for plasticizing the polymer
- Shear element for ensuring a fully molted material and sealing the first stage from blowing agent
- Mixing element for improved mixing of polymer and additives
- Second 3-zone screw for decompression, recompression and pumping the melt to the die

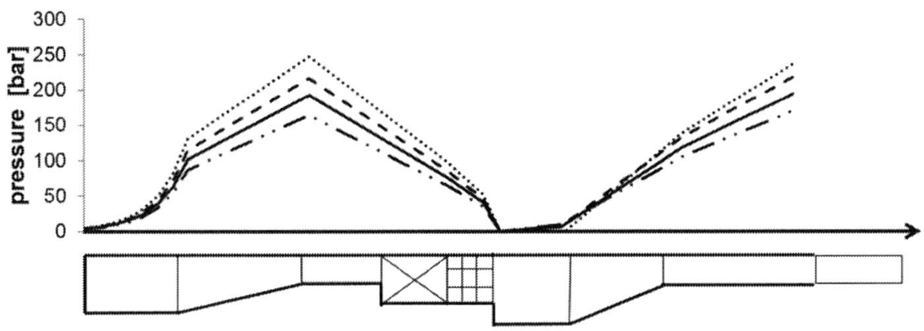

FIGURE 1. Pressure profile simulation along the screw for different rotational speed and counter pressures; conveying direction from left to right; screw design indicates the screw channel heights

EXPERIMENTATION

Experimental Setup

For the experiments at the University of Technology Ilmenau a customized laboratory extruder is used. The system is shown in **FIGURE 2 (a)**. The original extruder is suitable for a 45 mm screw and has a total length of 23D. By extending the device by another 26D, it is possible to implement a total length of 49D. At 25D the blowing agent is added to the melt. For this purpose, the gas cylinder was connected to the barrel via an adapter. In order to track and collect data from the process parameters, the barrel extension is equipped with several pressure sensors. This is shown in **FIGURE 2 (b)**. A sensor protruding into the melt for determining the melt temperature is located between screw and die. A strand die with diameters of 2 mm and 3 mm is used in order to vary the pressure level at the screw tip.

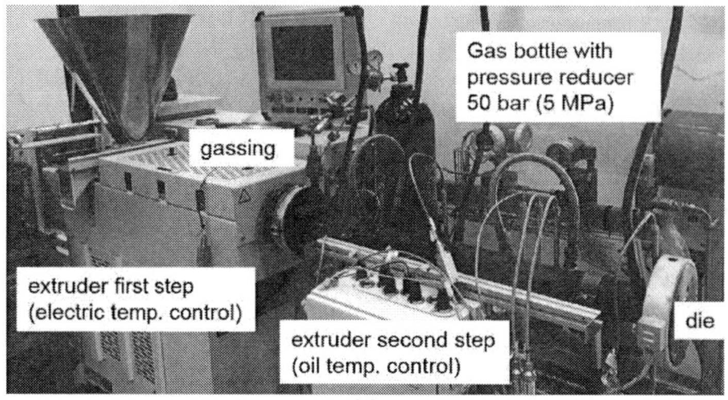

(a)

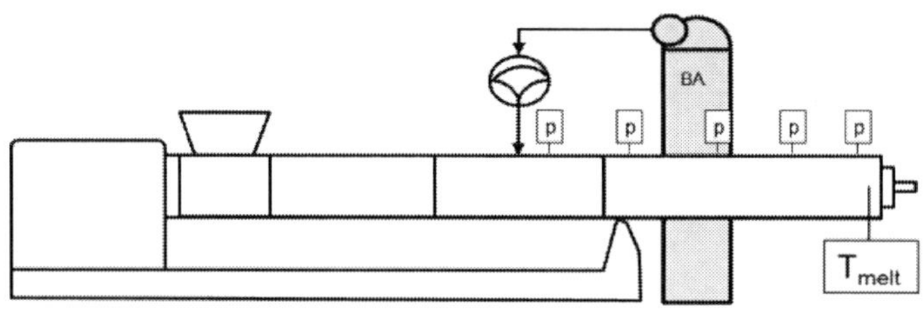

(b)

FIGURE 2. Experimental setup (a) 49 L/D laboratory extruder, (b) schematic experimental setup

Validation of Low Pressure Blowing Agent Injection Process

First experiments were carried out to test the calculated and manufactured screw. A total number of two experiments were conducted, one experiment with and one without blowing agent. The results are shown in **FIGURE 3**.

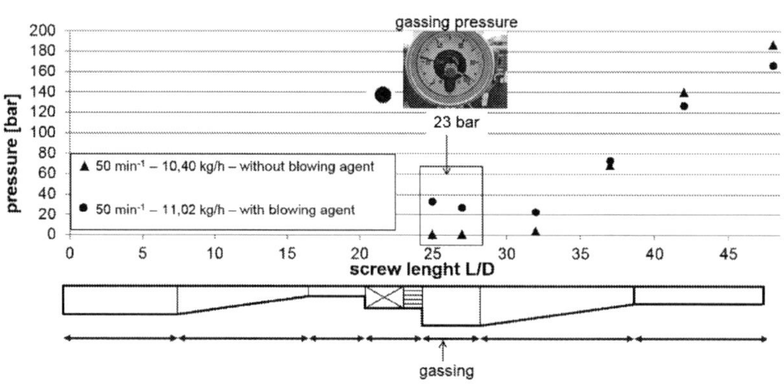

FIGURE 3. Validation of the low pressure gassing concept, measured pressure along the screw for tests with blowing agent and without blowing agent

As can be seen in the diagram for the experiment without blowing agent, no pressure was detected in the gas injection zone. Only afterwards an increase in pressure occurs. In the experiments with blowing agent, an increased pressure was measured in the gas injection zone, which equals the applied bottle pressure. The following pressure increase enables the gas to dissolve in the polymer melt. This is indicated by a decreased pressure profile compared to the first experiment, as an increased concentration of blowing in the melt leads to reduction of viscosity [4].

Experimental Design

The experiments are carried out with polypropylene with an MVR of 1,8 g/10min (2,16kg/230°C) and 4% nucleating agent in order to prove, that it is possible to dissolve gas into the melt without the need of expensive dosing systems. NAGUIB identifies the pressure before entering the die, gas content and melt temperature as significant factors [5]. Because gas injection is done entirely by the cylinder pressure, measuring the gas content is not possible with the given setup. Therefore, only the type of the blowing gas is varied. Carbon dioxide (CO_2) and Nitrogen (N_2) are tested as blowing agents with a bottle pressure of 45 bar. Counter pressure is varied by switching between the two die plates and the temperature is altered by cooling the barrel. Expansion rate is selected as response factor, which represents the ratio of density of the polymer melt to the density of the melt with blowing agent (equation 1).

$$\exp ansion \quad ratio \quad [\%] = \left(\frac{desity \quad without \quad blowing \quad agent}{density \quad with \quad blowing \quad agent} - 1 \right) * 100 \qquad (1)$$

A test series is characterized by the die and the blowing agent used in the experiments. Starting at a high melt temperature, blowing agent is added to the process and cooling of the extruder is initiated. This continuously reduces the melt temperature. For several temperatures a sample is collected. The density of the samples is measured and cross section cuts are microscopically examined.

RESULTS AND DISCUSSION

Experiments using CO_2 and the 3 mm die could not be conducted due to too little counter-pressure that prevented dissolution of the gas in the melt. Results of the remaining three test series are shown in **FIGURE 4**. As expected, the pressure rises with decreasing melt temperature (**FIGURE 4 (a)**). Higher melt temperatures than shown are also not possible with the factor combination of N_2 and the 3 mm die, as counter pressure is not sufficient to generate a 1-phase-system. **FIGURE 4 (b)** shows the development of the expansion rate at decreasing melt temperature. As can clearly be seen, CO_2 yields lower expansion rates than N_2 while decreasing significantly at falling melt temperatures.

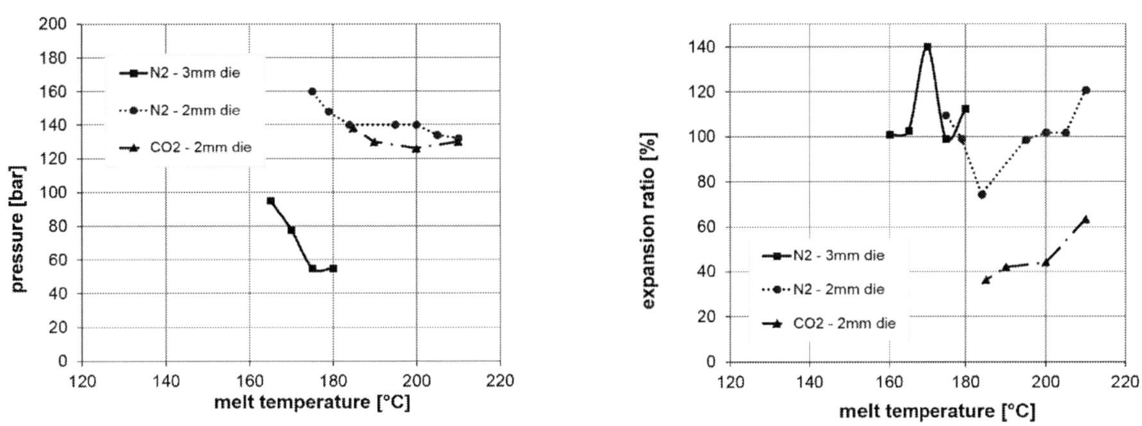

(a) (b)

FIGURE 4. (a) pressure in front of the die for different melt temperatures, (b) expansion ratio for different melt temperatures

After a certain temperature is reached, the gas dissolution is improved and a higher foaming ratio is achieved. Unfortunately, this increase does not occur with CO_2 as die pressure was too high and further cooling of the melt was not possible. After the melt temperature falls below a certain value and the melts resistance against the expanding gas increases, only little more cell expansion can take place which leads to the production of foams with higher densities.

FIGURE 5 shows cross sections for a varied die pressure at similar melt temperatures and the same blowing agent while the pictures of **FIGURE 6** show the cell morphology resulting from a constant die pressure and melt temperature but different blowing agents. It can be seen, that low counter-pressure leads to bigger cells (**FIGURE 5 (a)**) while the use of CO2 results in fine cells (**FIGURE 6 (a)**) that barely expand.

(a) (b)

FIGURE 5. cross-section with the same blowing agent N_2 and different counter pressure
(a) 3mm die, T_{melt} = 170°C, ρ = 0,385 g/cm³ / (b) 2mm die, T_{melt} = 172°C , ρ = 0,434 g/cm³

(a) (b)

FIGURE 6. cross-section with different blowing agent (BA) and the same die
(a) BA = CO_2, 2mm die, T_{melt} = 185°C, ρ = 0,674 g/cm³ / (b) BA = N_2, 2mm die, T_{melt} = 184°C , ρ = 0,510 g/cm³

CONCLUSION

A low pressure blowing agent injection process which enables the injection of blowing agents into a polymer melt and foaming it without further dosage devices was successfully implemented and tested. Additional experiments require the knowledge of the exact amount of gas injected. Furthermore, the use of dies, that allow an improved control of cell expansion, is recommended.

ACKNOWLEDGMENTS

The authors would like to acknowledge the members of the department plastics technologies of University of Technology Ilmenau and the recently deceased professor Michael Koch who was a major contributor to this technology.

REFERENCES

1. V. Altstädt and A. Mantey, *Thermoplast-Schaumspritzgießen* (Hanser, München, 2011).
2. A. Yusa, S. Yamamoto and H. Goto, Polymer Engineering and science. Letters **57**, 105-113 (2017).
3. H. Potente, *Kunststofftechnisches Seminar – Rechnergestützte Extruderauslegung* (University GH Paderborn, Paderborn, 1989).
4. L. Gerhardt, J Polymer Sci B Polymer Phys. Letters **35**, 524-534 (1997)
5. H. E. Naguib, C. B. Park and N. Reichelt, Journal of Applied Polymer Science. Letters **91**, 2661-2668 (2004).

Online analysis of melt viscosity during injection moulding with a hot runner rheometer

Prof. Dr.-Ing. Ch. Hopmann[1], Dr.-Ing. M. Theunissen[1] and M.Sc. J. Heinisch[1, a)]

[1]*Institute of Plastics Processing (IKV) at RWTH Aachen University, Seffenter Weg 201, D-52074 Aachen, Germany*

[a)]Corresponding author: Julian.Heinisch@ikv.rwth-aachen.de

Abstract. Injection moulding is one of the most productive methods in plastics processing allowing the production of parts with complex geometries in a single step. To stay competitive, part quality and its reproducibility remain a major issue for injection moulding manufacturers. Since modern machine technology allows sufficient reproducibility of machine variables like axis movements and pressures within the plasticising unit, the compensation of inevitable disturbances gains importance. Differences in the residual moisture of the plastics materials, the use of reground material, batch variations or interruptions in the production process may cause fluctuations in melt viscosity. These fluctuations affect the pressure distribution within the cavity resulting in warpage or even short shots or flash formation.
In this paper, a mould using pressure sensors within the hot runner manifold is used to analyse melt viscosity fluctuations during the injection moulding process. Similar to the principle of a high pressure capillary rheometer (HPCR), the viscosity can be analysed based on the pressure loss between two sensors. In contrast to steady rheometers, the transience of the injection moulding process has to be considered. Therefore, different indicators are analysed regarding their ability to distinguish polypropylene (PP) grades with different flowability. Finally, an indicator for the current melt viscosity is selected which enables an inline control of melt viscosity during injection moulding.

INTRODUCTION

Process control in injection moulding aims for a high and reproducible part quality. Measurement variables in production processes can be classified in machine variables, process variables and quality variables [1]. Today's injection moulding machines are usually controlled based on machine variables, such as injection speed or hydraulic pressure. Process variables correlate much closer to the desired part quality [2-4]. Inevitable disturbing effects, such as changing material properties, different environmental boundary conditions or a machine behaviour which is changing over time, can cause issues for the part quality with a process control based on machine variables [3, 5]. Due to the need for empirical quality models transferring the process variables into quality variables and the need for the ability to control process variables, a direct control of quality variables has not proven feasible for industrial injection moulding processes so far [6, 7]. Quality models based on the material behaviour could still provide a direct control of quality variables. For example, the material specific relationship between pressure p, specific volume v and temperature T has been successfully used as a quality model and an online control method for cavity pressure enabling a more stable process to thermal disturbances [3, 8]. Other material models, such as the rheological behaviour, are not considered for process control so far. However, especially changing melt viscosities are known to cause quality issues. Differences in melt viscosity may be caused by differences in the residual moisture content, batch variations, the use of regrind or regranulated material or interruptions of the production [9]. An increasing viscosity may cause underfilled cavities, whereas lower viscosities may cause overfilled cavities and consequently damage the parting plane of the mould. In any case, the pressure distribution will change with a different material viscosity causing a variation in the distribution of the shrinkage potential and part warpage [10]. Consequently, a process control considering the effects of different viscosities and realising a constant pressure distribution with the cavity would be highly desirable. Therefore, an online measurement of the current viscosity as a process variable is necessary. In this paper, a mould using cavity pressure sensors to measure a change in viscosity is analysed and a suitable indicator to distinguish materials with different viscosities based on the sensor information is derived.

RELATED WORK

Due to the important role of rheological properties, several measurement methods for melt viscosity have been developed. Commonly used measurement devices in a laboratory environment, such as high pressure capillary rheometers or melt index testers, are not suitable for online measurement during injection moulding processes [10]. For control purposes a measurement or at least an evaluation of viscosity changes has to be facilitated at the machine during the ongoing production. On the one hand, material viscosity can be evaluated based on the analysis of machine parameters. Therefore, indicators with a relationship to the current viscosity have to be identified. For example, indicators based on the pressure integral in the plasticising unit during the injection process or the plasticising torque during the plasticising phase can be used for that purpose [11]. For reason of machine dependency and better accuracy, viscosity evaluation based on process variables is more desirable. For this, sensors within the cavity of the mould are used to evaluate viscosity. A single cavity pressure sensor can be used to identify changing material viscosity based on indicators like the maximum value or the integral over time [12]. However, indicators based on one sensor only provide a stable evaluation basis if no other disturbances, such as thermal disturbances, affect the process. Therefore, concepts using two sensors have been developed. For example, a system uses a cavity pressure sensor near the gate and a temperature sensor at the end of the flow path to estimate the current melt viscosity based on the overall pressure value and the flow time between the two sensors [13]. Another possibility is to use the pressure difference between two cavity sensors along the flow path [14]. However, since the second sensor is reached at the end of the filling process, these concepts only provide the possibility for an inline viscosity measurement at the end of each cycle. Rheometer nozzles enable an online measurement of the current melt viscosity during the injection process. However, they are mainly applied for the generation of material data without the use of a mould [15].

VISCOSITY MEASRUREMENT IN A HOT RUNNER

For hot runner moulds, the placement of pressure sensors into the hot runner channel is a logical step since the hot runner is closer to the cavity and therefore the viscosity in the hot runner manifold correlates closer to the actual viscosity in the cavity. In [16] a mould using piezoresistive pressure sensors in the hot runner manifold to measure viscosity changes has been conceived and designed. The hot runner and the produced parts are shown in Fig. 1. In this paper, the mould is analysed regarding a suitable indicator for the viscosity evaluation. The advantage of the mould concept is that it allows an inline and an online measurement without overlaying cooling effects. The geometrical boundary conditions and the transience of the injection process may not enable an accurate overall measurand for the viscosity, but for control purposes it is enough to identify viscosity changes. Thus, the effects of changing material viscosities can be implemented into a control concept using a material model as quality model and a constant pressure distribution within the cavity can be realised by keeping the viscosity indicator on a reference level.

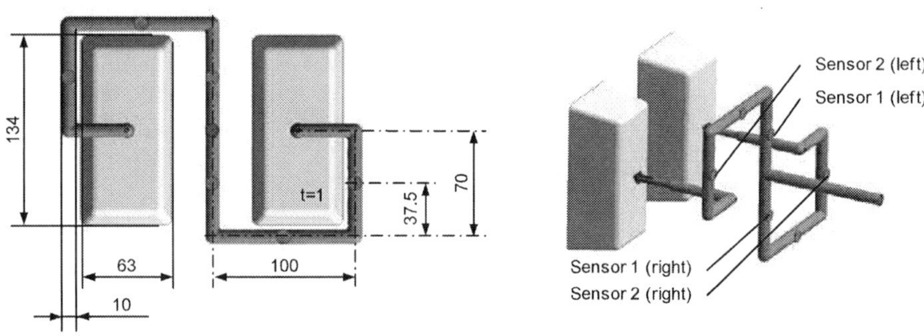

FIGURE 1. Dimensions of moulded parts and hot runner (left) plus sensor positions with notation (right) [17]

The mould shown in Fig. 1 was used for experiments on an injection moulding machine of type Allrounder 370 A by the manufacturer Arburg GmbH & Co. KG, Loßburg, Germany. The machine is equipped with a plasticising unit with a screw diameter of 30 mm. To measure the pressures in the hot runner at the depicted positions (Fig. 1) four piezoresistive pressure sensors of type 4021 B from the manufacturer Kistler AG, Winterthur, Swiss, with a diameter of 10 mm were used. Two polypropylene (PP) grades with different melt viscosities of the manufacturer Sabic Deutschland GmbH, Duesseldorf, Germany, were used. Grade PP 505P has a melt flow rate (MFR) of 2.0 g/10min, whereas grade PP 579S has an MFR of 47.0 g/10min (both materials at 230 °C and 2.16 kg; ISO 1133). The injection moulding parameters were selected according to the recommendations of the material supplier.

For comparative viscosity measurements, a HPCR of type HKR 2002 of the manufacturer Göttfert Werkstoff-Prüfmaschinen GmbH, Buchen, Germany, was used. For the HPCR measurements, the entrance pressure drop was considered by the *Bagley* correction [10]. For the calculated viscosities in the hot runner manifold no correction needs to be applied, since both measurement points have a sufficient long distance to changes in the cross section.

RESULTS AND DISCUSSION

In Fig. 2 a characteristic pressure course of a sensor pair in the hot runner manifold is shown. After the injection start, the pressures start to rise and after approximately 0.2 s a significant pressure difference between two sensors along the flow path can be measured. Due to the compressibility of the melt and the inertia of the screw and the melt, the melt flow establishes with a delay and the resulting pressure difference between the two sensors is delayed accordingly. At the start of the holding pressure, the pressures at both sensors drop to a constant pressure level under the set holding pressure of 300 bar. In the holding pressure phase, a reduced pressure difference occurs between the two sensors due to the negligible volume flow into the cavity. After the pressure holding phase, the hot runner pins close and the back pressure in the plasticising phase can be tracked.

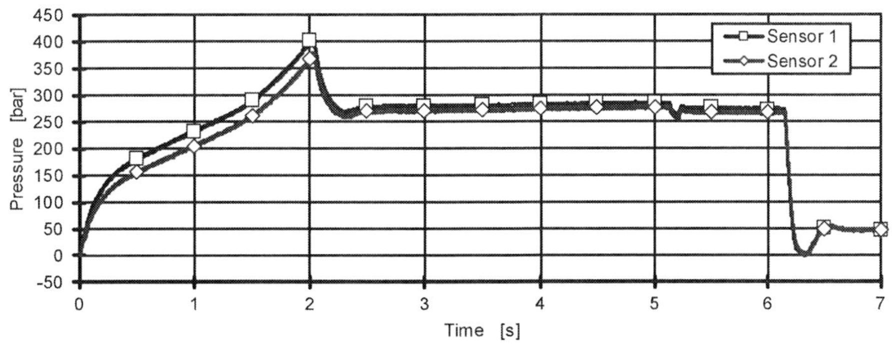

FIGURE 2. Exemplary course of pressure of sensor 1 and sensor 2 in the hot runner manifold (injection volume flow: 25 cm³/s)

For a comparison with HPCR measurements, the occurring viscosity in the hot runner was calculated based on a simple pressure flow in a pipe. Additional losses due to the redirections of the flow channel were neglected. According to *Schümmer,* the representative shear viscosity η_{rep} can be calculated based on the pressure loss Δp between two sensors which are positioned in the hot runner channel with the radius R in a distance l from each other [17]:

$$\eta_{rep} = \frac{\Delta p \cdot \pi \cdot R^4}{8 \cdot \dot{V} \cdot l} \quad (1)$$

Since the injection process is not stationary, a mean value of the pressure difference in the time from 0.2 s after the start to the end of the injection was used as Δp. As volume flow $\dot{V}$ half of the volume flow set in the injection moulding machine was used due to the two cavities. The representative shear rate in $\dot{\gamma}_{rep}$ the hot runner channel was calculated according to [18]:

$$\dot{\gamma}_{rep} = 0.815 \cdot \frac{4 \cdot \dot{V}}{\pi \cdot R^3} \quad (2)$$

In Fig. 3 the calculated viscosities in the hot runner and in the HPCR are shown. The overall level of the viscosities for both measurements is similar. The viscosity in the hot runner is a bit lower. This can be explained by the instationary injection process, due to which the melt flow is not fully developed during the measurement process. The mean standard deviation is lower than 2 % and cannot be depicted in the diagram. The viscosity degrades in both measurements with higher temperatures and higher shear rates. The temperature effect is lower in the hot runner since only the hot runner temperature and not in the plasticizing unit was changed to the specified level. Therefore, the overall melt temperatures presumably do not reach the setting point of the temperature in the hot runner manifold. Due to the hot runner channel diameter of 10 mm higher shear rates than 1,000 s^{-1} could not be achieved. The shear thinning effect occurs in both rheometers. It seems that due to the inertia of the system, the assumption that the volume flow in the hot runner can be calculated based on the injection speed, is not correct. However, this is not a problem for control purposes since not the set melt flow but the actual melt flow is decisive.

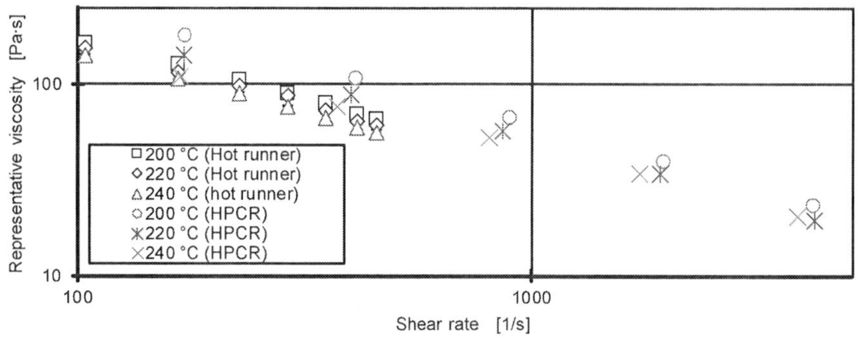

FIGURE 3. Comparison of viscosities measured in a hot runner and in high pressure capillary rheometer (HPCR)

Different possible indicators (Fig. 4) were analysed by a change of material grades with different viscosities regarding their suitability to inline identify different material viscosities. Since single point measurement tend to be affected by measurement noise, mainly integral indicators were chosen. Therefore, ten parts were produced at a stationary process with each a PP 579S and a PP 505P. The injection time during all experiments was 0.7 s. The relative change of different indicators and their coefficient of variation during the grade change are shown in Fig. 4. To evaluate the suitability of the indicators, the indicator quality as quotient of the relative change and the variation is additionally shown. It can be seen, that a change in viscosity can be detected reliably by the maximum value of a single sensor (indicator 1). This value however is highly dependent on the switch over point and is not useful in combination with a cavity pressure control. For a deeper analysis of viscosity changes, indicators based on data of at least two sensors are more reasonable. Indicators based on a single measuring point in time such as the difference between the maximal values (indicator 3 and 4) offer comparatively poor performance due to their high variation. Consequently, indicators based on integral values over time seem promising. Thereby, the integration boundaries have a significant effect on the indicator quality due to the instationary flow (see indicators 8-11). A variation of the integration boundaries has shown, that a delay time of about 0.2 s and an additional integration time of approximately 0.1 s offered good indicator performance (indicator 11).

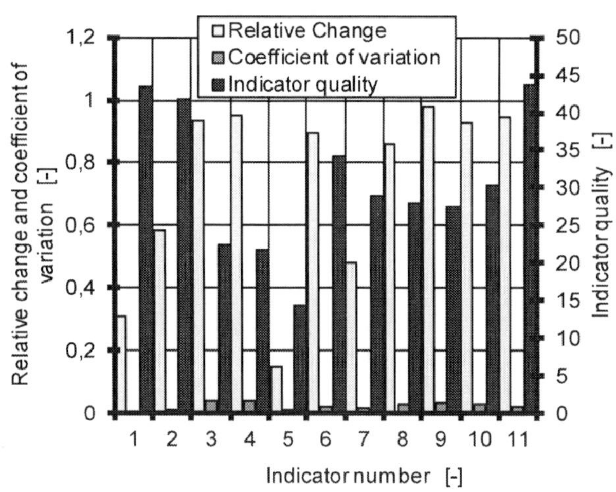

Indicator Number	Indicator name
1	Maximum of sensor 1
2	Maximum of sensor 1 (0 - 0.7 s)
3	Difference between maximum of sensor 1 and sensor 2
4	Difference between maximum of sensor 1 and sensor 2 (0 - 0.7 s)
5	Integral of sensor 1
6	Difference between integral of sensor 1 and sensor 2 (whole cycle)
7	Integral of sensor 1 (0 - 0.7 s)
8	Difference between integral of sensor 1 and sensor 2 (0 - 0.7 s)
9	Difference between integral of sensor 1 and sensor 2 (0.5 - 1 s)
10	Difference between integral of sensor 1 and sensor 2 (0 - 1 s)
11	Difference between integral of sensor 1 and sensor 2 (0.2 - 0.8 s)

FIGURE 4. Comparison of different indicators to detect a change in material viscosity

Therefore, in the future indicators based on the integrated pressure loss over time will be used to set up an inline control method for the compensation of changing rheological material properties. Additionally, an online control based on the current pressure loss during the injection process can be implemented.

CONCLUSION AND OUTLOOK

A mould with pressure sensors in the hot runner manifold has been used for an online measurement of material viscosity during an injection moulding process. Based on the pressure loss between the two sensors, the shear viscosity is calculated and compared to a laboratory measurement with a high pressure capillary rheometer. The measurement values show a similar course while the measurement in the hot runner shows slightly lower values due to the instationary measurement process. Furthermore, suitable indicators for a detection of changing material viscosities have been analysed with two polypropylene grades with different viscosities. Indicators based on the integrated pressure loss between the sensors along the flow path are well suitable. The inertia of the system should be considered with a delay in the integration.

In the future, the mould and the derived indicators can be used to build a control concept compensating the effects of different material viscosities to part quality. Therefore, inline and online control loops can be used to set up a control with hot runner temperatures and injection speed as manipulated values.

ACKNOWLEDGMENTS

The depicted research has been funded by the Deutsche Forschungsgemeinschaft DFG as part of the Cluster of Excellence 'Integrative Production Technology for High-Wage Countries'. We would like to extend our thanks to the DFG. We give thanks to Arburg GmbH & Co. KG (Loßburg, Germany) and Sabic Deutschland GmbH (Duesseldorf, Germany) for supplying the injection moulding machine and plastics material used in the experiments.

REFERENCES

1. U. Thombanse, et al., *The Road to Self-optimising Production Technologies*, in *Integrative Production Technology for High-Wage Countries*, C. Brecher, Editor. 2012, Berlin, Heidelberg: Springer. p. 793-848.
2. K. Schiffers, S. Kruppa, and S. Moser, *The Right Changeover Point for Each Shot.* Kunststoffe International, 2014. **104**(11): p. 26-30.
3. A. Schreiber, *Regelung des Spritzgießprozesses auf Basis von Prozessgrößen und im Werkzeug ermittelter Materialdaten.* 2011: RWTH Aachen.
4. C. Hopmann, et al., *Self-optimizing injection molding based on iterative learning cavity pressure control.* Production Engineering Research and Development, 2017. **11**(2): p. 97-106.
5. D. V. Rosato, D.V. Rosato, and M.G. Rosato, *Injection molding handbook.* 2012, New York: Springer
6. S.L.B. Woll and D.J. Cooper, *Pattern-Based Closed-loop Quality Control for the Injection Molding Process.* Polymer Engineering and Science, 1997. **37**(5): p. 801.
7. Y. Yang et al., *Injection Molding Process Control, Monitoring, and Optimization.* 2016, Munich,Vienna: Carl Hanser Verlag.
8. J. Wang and Q. Mao, *A Novel Process Control Methodology Based on the PVT Behavior of Polymer for Injection Molding.* Advances in Polymer Technology, 2013. **32**(1): p. 474-485.
9. F.A. Heinzler, C. Mielicki, and J. Wortberg, *A viscosity model for technical polymers describing chemical and physical aging.* Polymer Engineering & Science, 2015. **55**(7): p. 1628-1633.
10. T.A. Osswald and N. Rudolph, *Polymer Rheology.* 2015, Munich, Vienna: Carl Hanser Verlag.
11. R. Schiffers, *Verbesserung der Prozessfähigkeit beim Spritzgießen durch Nutzung von Prozessdaten und eine neuartige Schneckenhubführung.* 2009: Universität Duisburg-Essen.
12. J. Wortberg, R. Schiffers, and C. Ferriere. *Identification of Raw Material Condition by Process Monitoring and Derived Control Strategies in Injection Moulding.* in *PPS-24 Conference Proceedings.* 2008. Salerno.
13. C. Bader and E. König, *Quality control: taking control.* Kunststoffe International, 2012. **102**(6): p. 39-43.
14. C. Hopmann, M. Theunissen, and J. Heinisch. *Compensating Viscosity Fluctuations in Injection Moulding.* in *PPS-32 Conference Proceedings.* 2016. Lyon, France.
15. A. Fernandez, M. Muniesa, and C. Javierre, *In-line rheological testing of thermoplastics and a monitored device for an injection moulding machine: Application to raw and recycled polypropylene.* Polymer Testing, 2014. **33**: p. 107-115.
16. C. Hopmann, T. Fischer, and J. Heinisch. *Online melt viscosity measurement during injection molding for new control strategies.* in *Antec 2017 Conference Proceedings.* 2017. Anaheim, USA.
17. P. Schümmer, *Zur Darstellung der Durchflußcharakteristik viskoelastischer Flüssigkeiten in Rohrleitungen.* Chemie Ingenieur Technik, 1969. **41**(18): p. 1020-1022.

Influences of Processing Parameters on the Melt Temperature Homogeneity in Injection Moulding

Straka Klaus (1)*, Praher Bernhard (1), Steinbichler Georg (1)

(1) Johannes Kepler University Linz - Upper Austria - Austria

In polymer injection molding the melt homogeneity in terms of thermal, optical and mechanical quality is of great importance to produce high quality parts. While the influence of optical (e.g. color mixing) and of mechanical quality (e.g. fiber or filler distribution) is evident, the role of the thermal quality is often less obvious. Nonetheless, keeping in mind that axial temperature gradients in the screw chamber lead to similar gradients in the cavity, resulting in residual stresses in the molded part. Moreover, the melt temperature has significant influence on important molding process parameters as injection pressure and cooling time. The aim of the presented work is to show the influence of process parameter variations on the temperature of the melt cushion in the screw chamber of a reciprocating screw plasticizing unit. The temperature is measured with an ultrasonic measurement system directly in the screw chamber, as well as with a thermographic imaging system when the melt is leaving the nozzle during an air shot. It will be shown that although the maximum melt temperature measure is in good accordance for both type of measurement systems, the temperature drop in the screw chamber towards the screw tip is underestimated by the thermal imaging system. Results are given for two different polymeric materials.

Effects of different mold surface coatings on the ejection force in micro injection molding

Sorgato Marco **(1)***, ***Masato Davide*** **(1)**, ***Lucchetta Giovanni*** **(1)**

(1) University of Padova - Padova - Italy

The effect of different mold coatings on the ejection force in micro injection molding was experimentally investigated for a selection of thermoplastic polymers, at different molding conditions. Diamond like carbon (DLC), chromium nitride (CrN) and chromium titanium niobium nitride (CrTiNbN) coatings were selected in accordance with their potential affinity with the considered injected polymers. A specific mold cavity was designed, manufactured by micro milling and characterized (both topography and geometry) in order to isolate the effect of polymer adhesion on the ejection force. The obtained results were then correlated with the adhesion work for all the combinations of injected polymers and surface coatings. The wettability analysis was performed by measuring the interface contact angle at the melt temperature. The results indicated how an optimum mold coating, which minimizes the ejection force for a specific polymer, should be selected considering its wettability properties.

Enhancing the Weld Line Strength of Injection Molded Components

A. Geyer[a)] and C. Bonten

IKT, Institut für Kunststofftechnik, University of Stuttgart, Pfaffenwaldring 32, 70569 Stuttgart, Germany

[a)]Corresponding author: alexander.geyer@ikt.uni-stuttgart.de

Abstract. Weld lines often generate mechanically weak areas, because of the unfavorable perpendicular orientation of the macromolecules, fibers and/or other fillers to the direction of the main flow. In order to reduce these weak points, a special ram technique for injection molding, which can be integrated into the mold, was investigated. The results from the injection molding simulation of fiber-filled plastics prove that this fiber reorientation is possible. Tests with a prototype mold led to higher weld line strength and showed a correlation between these results of the process parameters.

INTRODUCTION

Through the use of injection molding, three-dimensional parts can be manufactured with little to no finishing work, thus exceeding the design possibilities of metal processing. In addition, the ongoing development of thermoplastics has created new fields of application so that metallic components can be increasingly replaced by plastics. Due to high strength requirements, the plastics are often reinforced with fibers in order to achieve higher specific strengths. [1, p. 311] However, during the molding process certain phenomena can occur that reduce the potential of plastics, such as orientation-dependent shrinkage, anisotropic strength properties and weld lines. Moreover, the influence of fibers on the weld line strength is a major challenge. For a long time, theory and practice have been looking for improvements. If these problems can be mastered, the range of applications of plastics will be expanded and the basis for a further, continuous growth of the plastic sector will be laid.

Weld Lines

Weld lines occur after the frontal fusion of two melt fronts in the mold and the subsequent solidification of the plastic in this state. The emergence of more than one melt flow may have various reasons. Besides obstacles, several sprues or gates, also different wall thicknesses can also lead to a split-up of the melt. [2]

As soon as the plastic melt comes into contact with the cold wall, it solidifies immediately. A thin edge layer is created without uniform orientations. Due to the high velocity gradient, the underlying region experiences a strong shear. In this zone, the strongest deformations occur and thus the highest orientations of molecules or fibers also arise in this area. As a result, fibers are aligned in the direction of flow. In the middle of the plastic melt, shear is low. At the melt front, the melt does not have any direct resistance. Therefore, the fibers and molecules are stretched flow by the fountain flow and are pushed towards the edge regions (Fig. 1, left). [3]

When two melt fronts meet, the orientations are frozen and the fibers and molecules remain aligned perpendicular to the direction of flow (Figure 1, right). In this region, the fiber orientation is parallel to the boundary surface over the entire structural thickness, which results in a force that is mostly transmitted by the intermolecular forces of the matrix material. This leads to a weakening of the part in this region, because of the anisotropy of fiber-filled plastics.

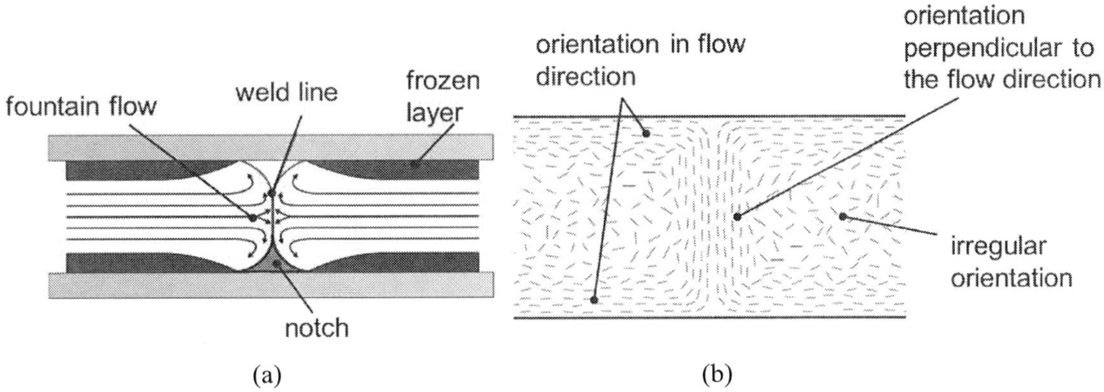

FIGURE 1. (a) Behavior of the melt in the mold. (b) Schematic fiber distribution in the weld line.

Materials with fibers that are oriented parallel to the load direction have a much higher stiffness and strength than perpendicular ones, because external stresses are transmitted mainly by the matrix material. The fibers, which are supposed to improve the mechanical properties, further increase the negative effects, since they impede the interdiffusion of the macromolecules, leading to an additional decrease in the weld line strength. In addition, air traps often occur, because the air cannot escape between the melt streams. An additional crack initiation can occur in the weld line area due to a notch which can be formed if the two melt streams freeze too fast at the mold wall.

In practice, there are hardly any differences in the tensile stress between unfilled plastics with and without a weld line. With fibers, however, the tensile stress and thus, also the weld line factor, decrease significantly, for example, for a PBT with 50 wt.-% glass fiber content only 30 % of the initial tensile strength is present in a tensile test. [4] This confirms the negative relationship between weld lines and the strength of the fiber filled plastics.

Possibilities for Improving the Weld Line Strength

The first attempt in designing new plastic parts should always involve preventing or relocating weld lines into areas with lower mechanical requirements. This is often possible with the use of a smart sprue design and can be predicted by using injection molding process simulation tools.

In order to achieve a sufficient force transmission over the boundary surface, it is important to integrate covalent bonds. This is only possible if molecules or fibers "migrate" across the weld line. This interdiffusion and "sewing" over the weld line is done by means of the Brownian molecular motion caused by thermal energy. The macromolecules can interfere with each other and form a mechanical bond over the boundary surface [5]. Thus, negative mechanical as well as optical properties in weld lines of unfilled plastics can be reduced or avoided by increasing the temperature [6]. However, the situation is different in the case of filled plastics, particulary for fiber-filled. Fibers cannot be displaced by a temperature variation alone. [7] The alteration of other process parameters also leads to minor improvements. Therefore, other means are required to significantly increase the strength. The aim here is to achieve a fiber reorientation by means of mechanical force, which is mostly realized with a flow through the weld line. Many processes have already been developed in the past, but they often involve a lot of effort. One example is the so-called push-pull-injection molding, which adjusts the through-flow of the weld line with the aid of two injection units. By oscillating injecting and pressing, but with the opposing stroke, a "reciprocating motion" of the still liquid melt is created in the cavity and thus, an orientation of the molecules and fillers into flow direction is achieved. However, this method entails many limitations such as the plurality of injection-molding units, a complex control system, special part design requirements and often a larger sprue, therefore requiring longer cooling times as well. [8] An easier way to improve the weld line strength is using a so-called overflow, where the cold weld line would be transformed into a hot weld line, because the melt can flow on. Depending on the position of the overflow, a slight improvement in the fiber orientation can be achieved.

The aim of this study is to implement an effective technology in order to increase the weld line strength, but with much less effort and a broader application range than for example the push-pull, cascade or SCORIM injection molding. A process that can be integrated directly into the injection mold will be analyzed in order to locally reduce the weakening of the weld line.

For this purpose, a ram is integrated in the area of the weld line. During the filling process, the overflow created by the ram in the retracted position is filled. At the end of the injection process, the ram pushes back the inflated melt into the main cavity which is shown schematically in Fig. 2. Due to this movement, an additional reorientation of the fibers should be occurring. Furthermore, due to the position of the ram, an additional ventilation of the weld line is possible and air traps can be removed.

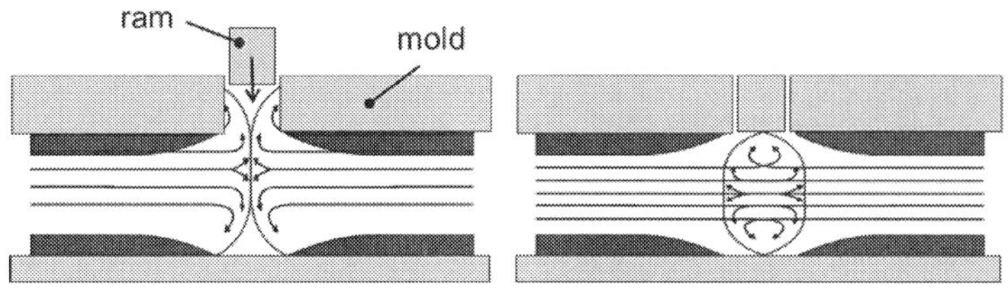

(a) (b)

FIGURE 2. Principle: (a) Ram in the retracted state. (b) Melt was pushed back into main cavity by ram movement.

EXPERIMENTAL

Based on simulation results, which showed the feasibility of the technology [4], two prototype molds were designed: a two-cavity mold with two 80 mm x 80 mm plates each with an obstacle (Fig. 3a) and a tensile bar design with two gates. A 10 mm x 10 mm ram is positioned in the weld line area in both cases, which can be controlled with a hydraulic cylinder MBZ160 from AHP Merkle GmbH, Gottenheim, Germany. The cylinder is controlled directly by the injection molding machine of the type Allrounder 570S from the company Arburg, Loßburg, Germany. After injection molding, 80 mm x 10 mm rods are milled from the plates for further tests. During the injection molding, the process parameters, timing of the ram movement as well as the temperature are varied.

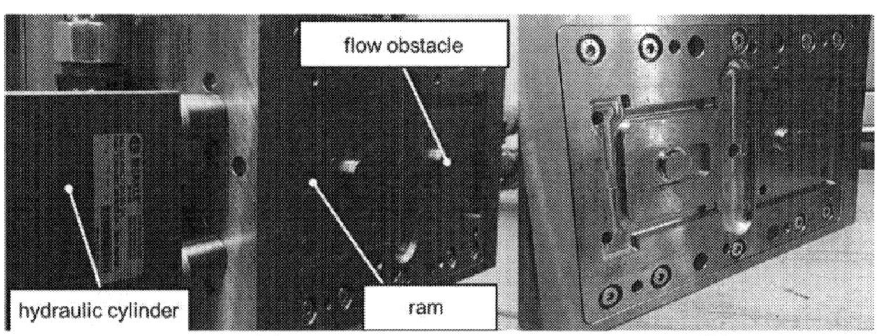

(a) (b)

FIGURE 3. Mold designs with plates (a) and tensile specimen (b)

Two tests are performed in order to analyze the results: a qualitative analysis by means of a macroscope type M420 from WILD, Heerbrugg, Germany, observing the fiber orientation and a quantitative assessment of the mechanical properties through tensile tests.

Two 30 wt.-% glass fiber reinforced polyamides are used in this study, an Ultramid B3WG6, BASF, Ludwigshafen, Germany (PA_1) and an Ultramid B27 E, BASF, compounded with 30 wt.-% glass fibers.

RESULTS AND DISCUSSION

By carrying out tensile tests with specimens, the influence of the process properties can be observed. The first process parameter to be analyzed is the timing of the ram movement. In Fig. 4a the tensile strength for a common weld line without modification is shown on the left. By pushing back the ram during the filling time and the switch-over, the highest tensile strengths can be reached. If the melt is pushed back too late, a drop in the strength occurs due to the freezing effect, which can damage the structure.

In the next step, the influence of the injection speed on the tensile strength was determined. Using a low injection volume rate of 50 cm^3/s (Fig. 4a), little improvement of the weld line strength can be reached. By doubling the injection speed, significantly higher tensile strengths can be achieved (Fig. 5b).

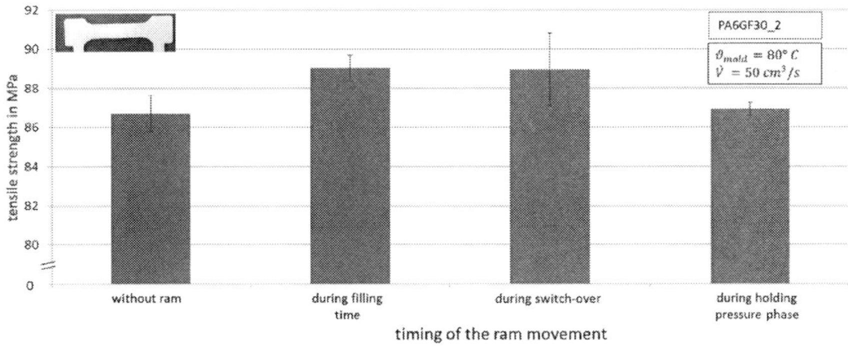

(a)

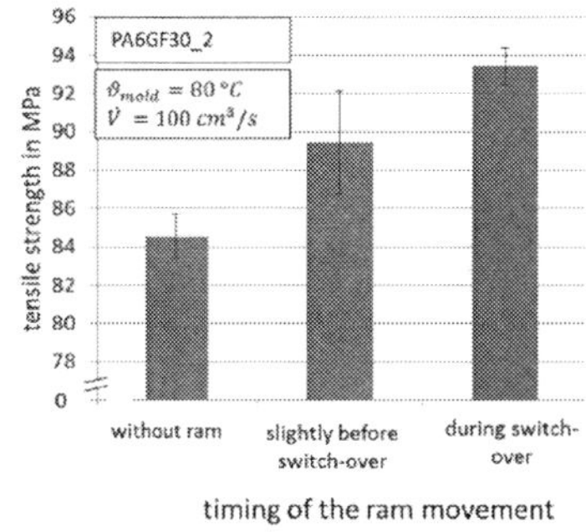

(b)

FIGURE 4. (a) Weld line strength depending on the timing of the ram movement for low injection speed. (b) Weld line strength for higher injection speeds

However, the variance of the results is high (Fig. 5a). In order to minimize this variance a special holding pressure profile was used (Fig. 5b). When the ram is moving, the holding pressure is kept low at 25 bars. After the melt is pushed back into the main cavity, the normal holding pressure is used. By means of this profile, slightly higher tensile strengths could be achieved and the variance of the results is minimized.

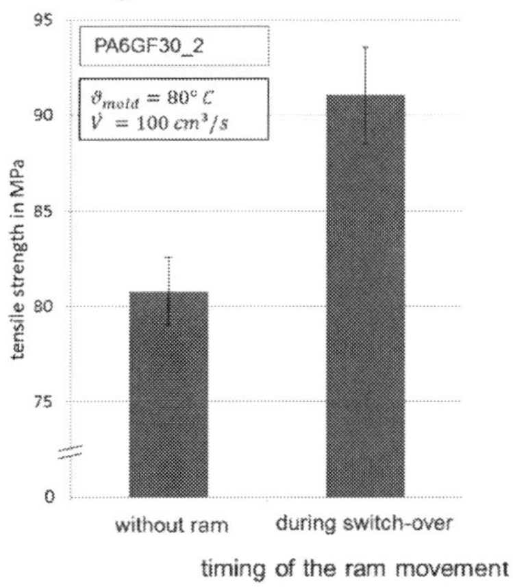

(a)

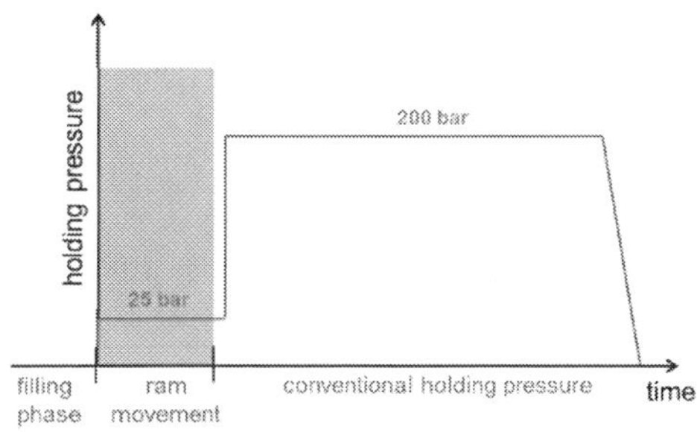

(b)

FIGURE 5. Weld line strength (a) with constant holding pressure (blue) and with special holding pressure profile (grey, b)

In addition to the tensile tests and to further understand why an increase of the strength could be achieved by means of the ram technology, macroscopic images are used for the qualitative evaluation of the fiber orientations. Figure 6 shows the images of the tensile specimen. It can be seen that the expected fiber orientation is formed at the left and right edge of the image, caused by the fountain flow. A large part of the fibers lie in the cross-sectional plane and are oriented in the flow direction. In the middle, which corresponds to the weld line area, the fibers are oriented perpendicular to the cutting plane. This deviation of the fiber orientation leads, as described, to a premature failure of the component. Figure 7, on the other hand, shows a recording of the weld line area after the application of the ram technique. All other machine parameters were kept constant. The ram is positioned on the top of the image. It can be observed that an area has been formed in which the majority of the fibers lie in the cross-sectional plane. This reorientation is clearly due to the ram movement and leads to the higher tensile strengths.

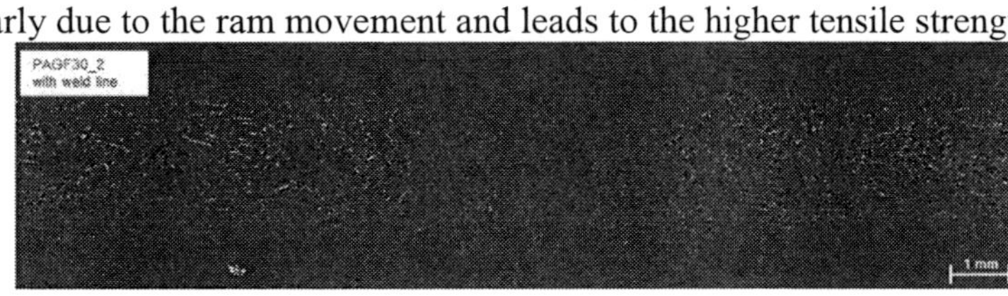

FIGURE 6. Macroscopic image of a common weld line.

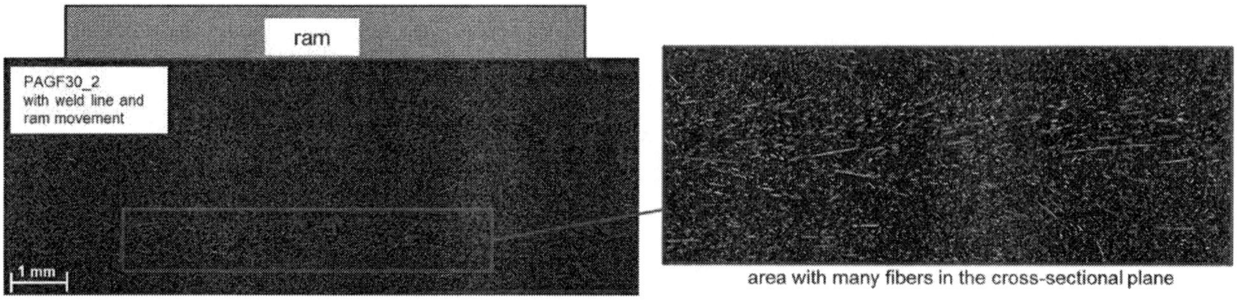

FIGURE 7. Macroscopic image of a modified weld line

CONCLUSION

The results presented in this study show a fundamental feasibility of the technology which uses a ram in order to improve the weld line strength. On the basis of various process simulations, two prototype molds were designed. Tensile tests with the injection molded specimens reveal that the process parameters have a significant influence on the technology and thus on the resulting weld line strengths, which showed a significant improvement in the mechanical properties. In terms of process parameters, the timing of the ram movement is an especially significant factor in the process. The best results could be achieved using the push back during the switch-over. In addition, it can be shown that higher injection speeds lead to higher tensile strengths and that the right holding pressure, as well as a special holding pressure profile, minimize the fluctuation in the results.

Moreover, the macroscopic examinations proved the predicted simulation results, since an area occurs where fibers are reoriented into flow direction and thus increase the weld line strength.

For an optimal use of the technology, further investigations have to be carried out, particular the behavior of the melt in the overflow, including the freezing effects, have to be analyzed and could potentially be improved by using higher temperatures in this area. Also, sensors should be integrated into the mold in order to monitor the melt movement and to determine the optimum time of the ram movement. In addition to that, further materials and/or fiber contents should be observed, as well as further ram geometries.

REFERENCES

1. C. Bonten, *Kunststofftechnik: Einführung und Grundlagen*, 2nd ed. (Hanser, Munich, 2016).
2. I. Kühnert, "Grenzflächen beim Mehrkunststoffspritzgießen", Ph.D. thesis, University of Chemnitz, 2005
3. D. J. Coyle, J. W. Blake and C. W. Macosko, *The kinematics of fountain flow in mold-filling,* in *AIChE journal*, 1987, Nr. 7, p. 1168-1177
4. A. Geyer and C. Bonten, *Werkstoffcharakterisierung im Spritzgießprozess*, in 25. Stuttgarter Kunststoffkolloquium, Stuttgart, 2017
5. C. Jaroschek, "Spritzgießen von Formteilen aus mehreren Komponenten", Ph.D. thesis, University of Aachen, 1993
6. L.-S. Turng and H. Kharbas, *Effect of process conditions on the weld-line strength and microstructure of microcellular injection molded parts*, in Polymer Engineering & Science, 2003, Nr. 1, p. 157-168.
7. R. Selden, *Effect of processing on weld line strength in five thermoplastics*, in Polymer Engineering & Science, 1997, Nr. 1, p. 205-218
8. H.-C. Ludwig, "Gegen-Takt-Spritzgießen - ein neues Verarbeitungsverfahren zur Optimierung der Werkstoffeigenschaften von verstärkten und unverstärkten Thermoplasten", Ph.D. thesis, University of Stuttgart, 1998

The Use of Micromechanical Models to Predict Fiber Reinforced Plastics

F. Willems[a] and C. Bonten

IKT, Institut für Kunststofftechnik der Universität Stuttgart, Pfaffenwaldring 32, 70569 Stuttgart
[a]Corresponding author: Fabian.Willems@ikt.uni-stuttgart.de

Abstract. The reliable prediction of lightweight structures of fiber reinforced plastics based on coupled process- and structural simulations are gaining importance in the computer aided design process. A high accuracy of prediction with acceptable complexity should be achieved. Usually, a large database and several iterations are required to calibrate these models. In this study, a reverse engineering method is used to calibrate the micromechanical model for the simulation. Based on coupled simulations, the influence of the input parameter of a micromechanical model is shown.

FIBER REINFORCED PLASTICS

Fiber reinforced composites are gaining importance due to an increasing influence of socio-political factors. In comparison to metals, the processing of plastics is far less energy consuming and allows for a higher design freedom [1]. Major challenges arise when substituting metallic structural parts because of high mechanical requirements and also varying ambient conditions. If a certain material type proved to meet the requirements, it has to be ensured that these material properties can also be achieved by the respective part. For fiber reinforced plastics, to reach a high accuracy within the design process, the process induced anisotropy caused by local fiber properties like the fiber orientation, the fiber length and the fiber concentration has be considered.

In general, this is possible by means of coupled simulations, which means transferring calculated fiber properties to the local elements of the structural simulation as material properties. This requires that the respective models are implemented into the solver as well as that the calibration parameters can sufficiently represent the processing. Furthermore, to successfully utilize micromechanical models (calculation of compound properties) the mechanical properties of each individual constituent have to be known. Also, an extensive database of experimental properties of the compound has to be available to reliably calibrate the model. Unfortunately, hardly any information about the microstructure exists because the determination of fiber orientation and length in general requires sophisticated analysis. Therefore the calibration of the fiber orientation model within the injection molding simulation and the required experimental data to each individual material or the material composites poses a significant challenge.

STATE OF THE ART

The resulting anisotropic mechanical properties inside injection molded, fiber reinforced plastic parts can vary heavily according to the local fiber properties. For example, for tensile bars which have been cut out of an injection molded plate, a reduction of stiffness and strength up to 30–50% has to be expected perpendicularly to the flow direction [2, 3] (figure 1a). This effect can also be transcribed to the influence of the fiber concentration and length. Usually, in industrial utilization, the properties of the fiber microstructure have to be assumed since this data is in general not available in material databases. That includes assuming the local varying fiber concentration to be a constant value according to the specifications of the manufacturer. Further, for short glass fiber reinforced plastics an often occurring fiber orientation within the tensile bar ($a_{11} = 0{,}8$ $a_{22} = 0{,}15$ $a_{33} = 0{,}05$) [4] is assumed as well as an aspect ratio of the fibers in between the following range: $a = 10\text{–}30$ [5].

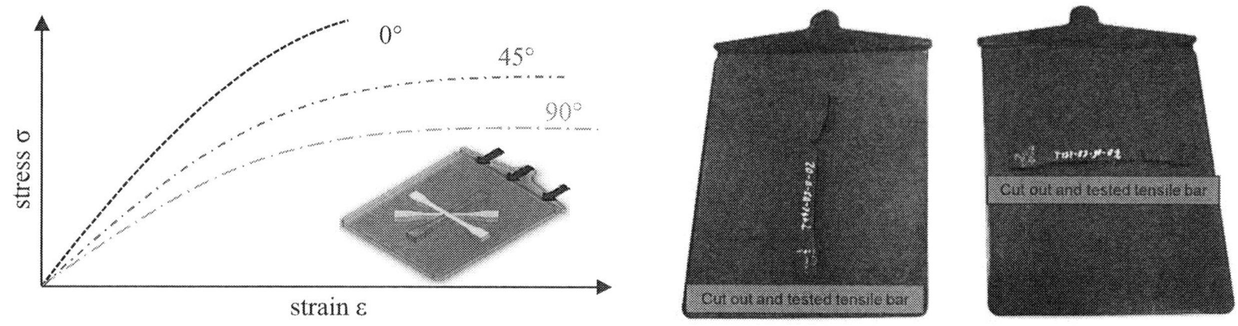

(a) (b)

FIGURE 1. (a) Stress-strain curve as a function of fiber orientation (b) Specimen parallel and perpendicular to the flow direction

The consideration of the anisotropy of fiber reinforced plastics in general follows the subsequent fundamental principle: First, the fiber properties are determined by process simulation. Then, based on micromechanical models, the mechanical properties are calculated and homogenized. Afterwards the distribution of fiber orientation based on the results of the simulation is taken into account. Lastly, everything is transferred to the structural mesh of the part (mapping). Numerous alternative individual methods already exist. However, they are mostly based on a combination of the already mentioned principle. Aside from the analytical methods there are also increasingly more numerical homogenizations being used due to the ever increasing computing power of modern computers.

Nowadays, to reliably predict the influences of processing, individual simulation methods have been established for each processing technology. To simulate the injection molding process, the three-dimensional base equations regarding continuity, impulse and energy conservation are solved. To calculate the fiber orientation almost exclusively the second order orientation tensors according to Advani and Tucker [6] are used. The related calculation models are mostly based on an extension of the Jeffery model [7] by Folgar and Tucker. To reduce computational effort, the models to predict the mechanical properties of inhomogeneous materials are based on a homogenization of the microstructure to the effective properties on the macroscopic level [8]. Often, only some of the numerous existing models like Halpin-Tsai [9] or Tandon-Weng [10] are used. These calculations result in the engineering constants of a volume element with a unidirectional orientation of fibers inside the matrix [11]. The input values for the calculation are the aspect ratio and the volume fraction of the fibers, and also the respective stiffness, shear moduli and Poisson's ratios of the fiber and the matrix. The orientation averaging [6] thereby calculates the stiffness matrix while taking the existing or predicted fiber orientation tensors into account. This results into an anisotropic stiffness matrix of the material compound.

The mesh topology for process and structure simulations often greatly varies because of different requirements for the solver. Due to this, it is in general necessary to translate the calculated results from the process simulation onto the structural mesh. This translation can be made by comparing both of the computational meshes by means of the Delauney-triangulation as described in [12]. The information about local fiber orientations only rarely exists since their determination is only possible through sufficient methods like a CT-analysis, ultrasound or the grinding pattern analysis [13, 14].

EXPERIMENTAL SET-UP

For the experiments, a polyamide 6.6 type Ultramid 13WG6 HRX with a glass fiber content of 30 wt.-% by BASF SE, Ludwigshafen, Germany, was used. The test specimens are of type BA in accordance to DIN 527-2 and were cut out of injection molded plates parallel and perpendicularly to the flow direction according to DIN 294-5. The chosen plate and the cut out test specimen are depicted in figure 1b. For the mechanical characterization, tensile tests according to DIN 527-1 with a universal tensile testing machine with five repetitions on conditioned specimens were carried out at standard atmosphere (25 °C, 50 %). The fiber lengths and concentrations within the specimens were determined through DIN EN ISO-3451 (image 2a). The remaining single fibers were examined on their number and weight averaged fiber length in accordance to the method described in [15]. The local fiber orientation was determined in the middle area of the tensile bars through grinding pattern analysis by applying the weighting function according to [16]. The evaluation of the ellipses was done with Matlab® by MathWorks® Natick, USA, using fiber orientation tensors. These were clustered over the entire cross-section. To ensure sufficient precision, a minimum of ten fibers per cluster was met.

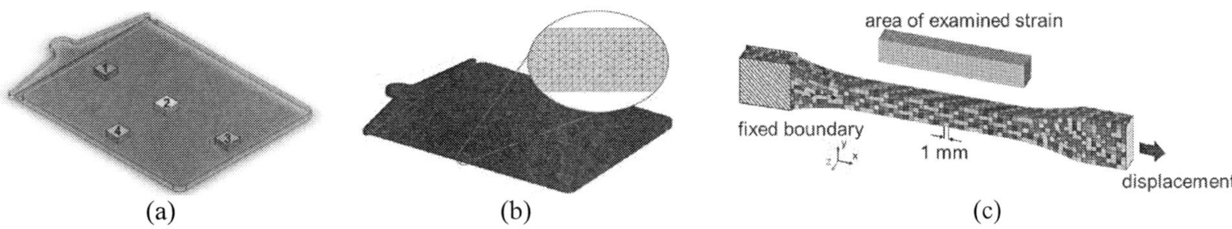

FIGURE 2. (a) Measure points to determine fiber lengths (b) Process simulation mesh with more than 15 elements along the thickness (c) Structural mesh and boundary conditions

The simulation set-up includes a coupled process and structural analysis. The meshes of the plate with more than 15 elements along the thickness as well as the tensile bar with six elements are depicted in figure 2b and 2c. For the process simulation the software Moldex3D® by CoreTech System Co., Ltd., Chupei City, Taiwan was used. The necessary parameters for the iARD-RPR fiber orientation model were determined through comparing the predicted fiber orientation tensors with their respective experimentally determined tensors. After successfully calibrating the process simulation, the micromechanical model is calibrated according to the developed approach shown in figure 3. The novel approach was developed in Matlab®. Here, the experimental stiffness (in parallel and perpendicular) were reproduced for the composite by reverse engineering procedure to determine the initial stiffness of the composite constituents for the applied micromechanical model. In the presented approach, the determined fiber orientations on the cross-sectional area of the specimen were averaged over the entire area. Since in this case every input value except for the unreinforced matrix stiffness was either determined or taken out of data sheets, merely the matrix stiffness remains to calibrate the model. Finally, the results of the calibrated process model as well as the calibrated micromechanical model are used to calculate the mechanical properties of the part (here, the plate).

The coupling is done by using a mapping routine [11] to transfer results onto the tensile bar or rather the corresponding mesh for the structural simulation in the form of an APDL Code by the solver ANSYS MAPDL by Ansys Inc., Cannonsburg, Pennsylvania. The structural simulation is set-up with a fixed boundary and a displacement. The reactionary forces and averaged elongation within the linear measurement range of the specimens is evaluated in accordance to the experimental tensile tests (figure 2c).

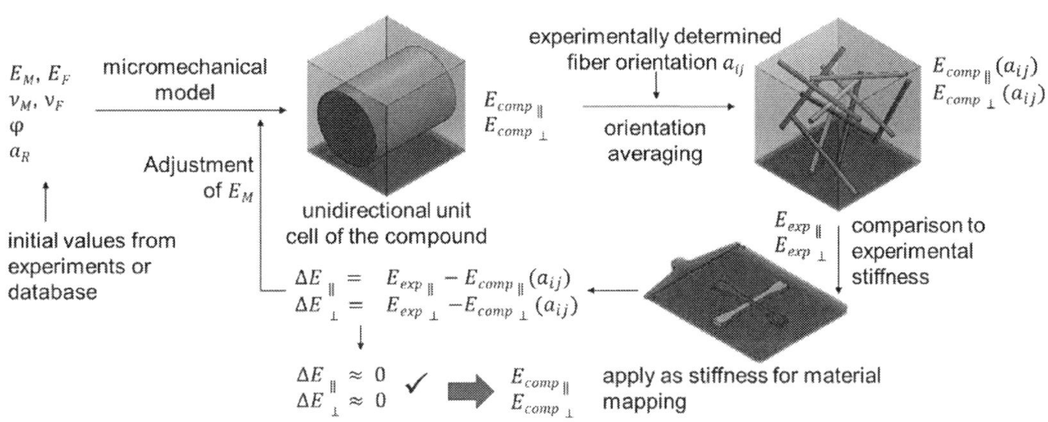

FIGURE 3. Iterative approach to calibrate a micromechanical model based on experimental values

Results and discussion

In table 1 the experimental results of the tensile tests for stiffness and strength as a function of the testing direction (parallel and perpendicular to flow direction) were summarized. A reduction of properties by nearly 25 % perpendicular to flow direction is observable. The fiber length distribution and also the fiber concentration were determined at four positions (figure 2a). The weight averaged fiber lengths and fractions at the measure points are shown in table 1. The number averaged fiber length of all measure points equals $\overline{L}_N = 343\ \mu m$ with an averaged fiber mass fraction of $\psi = 29{,}7\ \%$. The averaged diameter of the fibers was determined to 9.5 µm with a standard deviation of 1.5 µm. While taking into account the averaged fiber length, the resulting aspect ratio of the fibers equals to 36.

TABLE 1. Mechanical properties of tensile tests and averaged fiber length and fiber mass fraction of the plate

Mech. properties	Values in MPa	Fiber properties at measure point	1	2	3	4
$E_{\parallel} / E_{\perp}$	7610 / 5878	Number averaged fiber length L_N in µm	344	347	342	339
$\sigma_{m\parallel} / \sigma_{m\perp}$	152 / 112	Weight averaged fiber length L_W in µm	397	404	394	390
		Fiber mass fraction ψ in %	30,4	29,4	29,4	29,6

The determined fiber orientation tensor is shown by the component a_{11} in figure 4a and 4b with false-colors (blue = 0, red = 1). Here, the 1-direction is oriented perpendicularly to the cross-section surface so that the results parallel and perpendicular to the flow direction solely interchange by the orientation of the first and second component. This can also be shown by the diagonally averaged values (a_{11}, a_{22}, a_{33}) parallel (0.70, 0.27, 0.03) and perpendicular (0.43, 0.55, 0.02) to flow direction. It becomes evident that the determined fiber orientations of both specimens vary from each other. Especially the orientation in the core layer differs. An explanation for this is that even ideally perpendicularly oriented fibers obtain an angle of inclination through the method of fitting ellipses despite utilizing the weighing functions. The real fiber orientation within the specimens with its respective orientation will therefore be a value between those two results. That is why in the following an averaged orientation (0.63, 0.35, 0.02) will be assumed. Subsequently, with following inputs $E_M = 1.1\ GPa$, $E_F = 73\ GPa$, $\nu_M = 0.35$, $\nu_F = 0.22$, $\rho_M = 1.13\ g/cm^3$, $\rho_F = 2.54\ g/cm^3$, the micromechanical model was used to determine the unreinforced matrix stiffness according to the novel approach in figure 3.

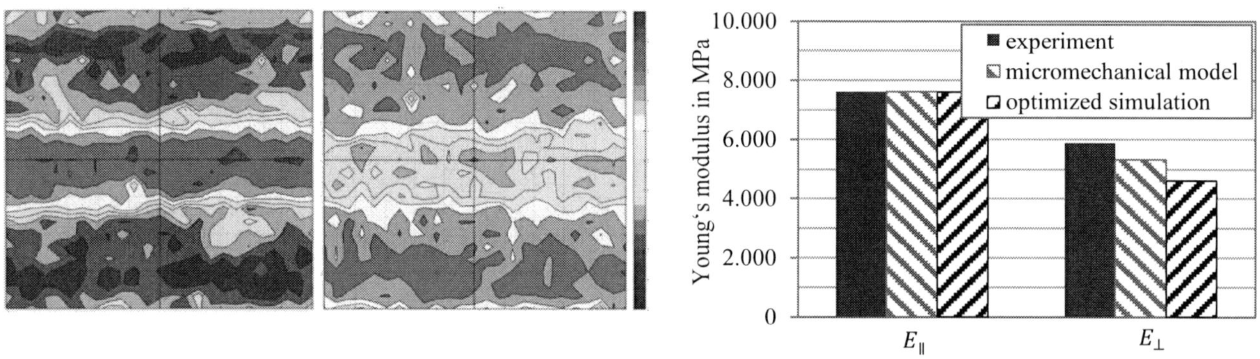

(a) (b) (c)

FIGURE 4. (a) Fiber orientation a_{11} parallel and (b) perpendicular to flow. (c) Young's modulus of experimental tensile test, fitted micromechanical model and simulation parallel and perpendicular to flow direction

The Young's-modulus of the unreinforced matrix was thus determined to be 2,610 MPa through the iterative calibration loop. As a result, further values like $E_{\parallel} = 7{,}610\ MPa$ and $E_{\perp} = 5{,}318\ MPa$ can be determined for the fiber reinforced material. Hence, the identified matrix stiffness for unreinforced PA66 is in between the range, tested in conditioned (1,100 MPa) and dry (3,000 MPa) state. Any further improvement of the perpendicular stiffness through adjustment of the Poisson's ratio of the matrix and/or the fibers is only limitedly reasonable and the achievable improvement remains minimal. On the basis of the used Poisson's ratio a deviation of 9.5 % from the experimentally determined values could be evaluated for the micromechanical model. That is why in principle, the here presented method of calibrating the micromechanical model is generally applicable, while a comparison with coupled simulations can provide additional information.

Based on the calibrated micromechanical model the simulation shows the influence of the predicted fiber orientation by process simulation on the resulting stiffness (figure 4c). The Young's modulus parallel to flow direction can also be predicted accurately by the whole simulation procedure, whereas the stiffness perpendicular to flow direction is showing a deviation of 21.5 %. This underestimation is caused by the predicted fiber orientation of the process simulation. Experimentally, the fiber orientation was determined to (0.63, 0.35, 0.02) by diagonally averaged values. Within the process simulation the fiber orientation in the second direction a_{22} (perpendicularly to flow) was highly underestimated (0.64, 0.23, 0.13). It can expected, that the remaining difference between the values in parallel and perpendicular direction can only be reduced through further optimizing the predicted fiber orientation by process simulation closer to the experimental values.

CONCLUSION

It could be proved that through the developed method (namely a reverse engineering approach) the used micromechanical model can be calibrated on the basis of experimental data (fiber orientation and compound stiffness). Thereby, the obtained deviation of the predicted compound stiffness perpendicular to flow direction is less than 10 %. The application of a micromechanical model through coupled simulation could also be illustrated. The achieved deviation of the prediction quality in parallel direction is less than 1 %, whereas the underestimated fiber orientation of the process simulation is leading to a deviation of nearly 22 % in the direction perpendicular to flow. The results show that the calibrations of the individual models have a strong influence on the resulting outcome. This means that especially the fiber orientation and also the aspect ratio of the fibers, which are rarely available as measured values, are a decisive parameter for the resulting accuracy of prediction.

Coupled simulations based on a correctly predicted fiber orientation may be carried out to further validate the developed calibration method. Likewise, through the consideration of all experimental input values for the calibration of the models, the remaining model induced deviation is also determinable. Another further improvement of the method can be achieved by considering the fiber orientation distribution instead of an averaged value. If these improvements enable a reliable calibration of the micromechanical model without requiring many simulation loops the entire process can be significantly accelerated. Lastly, when considering the increasing applications for long fiber usage, the fiber length distribution may also be taken into account in future calculations.

REFERENCES

1. C. Bonten, *Kunststofftechnik. Einführung und Grundlagen,* 2nd ed., (Hanser, München, 2016).
2. M. Stommel, W. Korte and M. Stojek, *FEM zur Berechnung von Kunststoff- und Elastomerbauteilen*, (Hanser, München, 2011).
3. S. Günzel, „Analyse der Schädigungsprozesse in einem kurzglasfaserverstärkten Polyamid unter mechanischer Belastung mittels Röntgenrefraktometrie,“ Ph.D. thesis, Technical University Berlin, 2012.
4. J.F. O'Gara, P.H. Foss and J.P. Harris, "Reinforced Plastic Design: Microstructure and Stiffness of Tensile Bars," in *SPE/ANTEC 2001 Proceedings*, (CRC Press, Dallas, Texas, USA, 2001), pp. 2803–2807.
5. G. Menges, E. Haberstroh, W. Michaeli and E. Schmachtenberg, *Menges Werkstoffkunde Kunststoffe*, (Hanser, München, 2011).
6. S.G. Advani and C.L. Tucker, "The Use of Tensors to Describe and Predict Fiber Orientation in Short Fiber Composites," in *Journal of Rheology*, (1987), pp. 751–784.
7. G.B. Jeffery, "The motion of ellipsoidal particles immersed in a viscous fluid," (Proc. Royal Soc., London A102, 1922), pp. 162–166.
8. D. Gross and T. Seelig, *Fracture mechanics. With an introduction to micromechanics.*, 2nd ed., (Springer, Berlin, 2011).
9. J.C. Halpin and J.L. Kardos, "The Halpin-Tsai equations: A review," Bd. 5. in *Polymer Engineering and Science*, (1976), pp. 344–352.
10 G.P. Tandon and G.J. Weng, "The effect of aspect ratio of inclusions on the elastic properties of unidirectionally aligned composites," in *Polymer Composites*, pp. 327–333.
11. F. Wolf, „Modellierung der anisotropen Eigenschaften faserverstärkter Kunststoffe für die Strukturberechnung auf Grundlage der Prozesssimulation,“ (2016).
12. E. Laurien and H. Oertel JR., *Numerische Strömungsmechanik. Grundgleichungen und Modelle - Lösungsmethoden - Qualität und Genauigkeit*, (Springer Fachmedien, Wiesbaden, 2013).
13. R.S. BAY and C.L. Tucker III., „Fiber orientation in simple injection moldings. Part II: Experimental results," in *Polymere Composites 13*, (1992), pp. 332–342.
14. S.G. Advani and C.L. Tucker III., "The use of tensors to describe and predict fiber orientations in short fiber composites," in *Journal of Rheology 31*, (1987), pp. 751–784.
15. F. Willems and C. Bonten, "Influence of processing on the fiber length degradation in fiber reinforced plastic parts," in *Proceedings of the Regional Conference Graz 2015 – Polymer Processing Society PPS*, (2016).
16. G. Fischer and P. Eyerer, "Fiber orientation and thermal expansion of short fiber reinforced thermoplastics," in *SPE Tech. Pap* 31, (1986), pp. 532–538.

AUTHOR INDEX

AUTHOR INDEX

AUTHOR INDEX

AUTHOR INDEX

AUTHOR INDEX

AUTHOR INDEX

AUTHOR INDEX

AUTHOR INDEX

AUTHOR INDEX

AUTHOR INDEX

Polymer Processing Society
810 E. 10th St.
Lawrence, KS 66044
USA

ISBN 978-1-5108-5487-1

CURRAN ASSOCIATES INC.
proceedings.com

Europe and Africa Regional Meeting of Polymer Processing Society (PPS2017)

Dresden, Germany
26 - 29 June 2017

Volume 2 of 2

Europe and Africa Regional Meeting of Polymer Processing Society (PPS2017)

Dresden, Germany
26 - 29 June 2017

Volume 2 of 2

ISBN: 978-1-5108-5487-1

Printed from e-media with permission by:

Curran Associates, Inc.
57 Morehouse Lane
Red Hook, NY 12571

Some format issues inherent in the e-media version may also appear in this print version.

Printed by Curran Associates, Inc. (2018)

For permission requests, please contact Polymer Processing Society
at the address below.

Polymer Processing Society
810 E. 10th St.
Lawrence, KS 66044
USA

Phone: +1-800-627-0326
+1-785-865-9403
Fax: +1-785-843-6153

pps@allenpress.com

Additional copies of this publication are available from:

Curran Associates, Inc.
57 Morehouse Lane
Red Hook, NY 12571 USA
Phone: 845-758-0400
Fax: 845-758-2633
Email: curran@proceedings.com
Web: www.proceedings.com

TABLE OF CONTENTS

VOLUME 1

S01 – MIXING AND COMPOUNDING

S02 – POLYMER BLENDS AND ALLOYS

S03 – RHEOLOGY AND PROCESS SIMULATION

S04 – FUNCTIONAL, NANO AND BIO COMPOSITES

S05 – FIBERS, FILMS AND FOAMS

S06 – INJECTION MOLDING AND EXTRUSION

VOLUME 2

S07 – ELASTOMER MATERIALS AND PROCESSING

S08 – NANOCARBON BASED COMPOSITES, WITH CNPCOMP2017

S09 – POLYMER MODIFICATION WITH IONIZING RADIATION

S10 – NANO- AND MICROSTRUCTURED SURFACES AND FILMS

S11 – IN-LINE ANALYTICS AND PROCESS MONITORING

S12 – WELDING AND JOINING TECHNOLOGY

S13 – POLYMER MATERIALS FOR MEDICAL APPLICATIONS

S14 – ADDITIVE MANUFACTURING

Experimental Investigation and Improved Modeling of the Melting Process in Single-Screw Extruders with a Grooved Plasticizing Barrel

J. Kettemann[a)], J.A. Avila-Alfaro and C. Bonten

IKT, Institut für Kunststofftechnik, University of Stuttgart, Pfaffenwaldring 32, 70569 Stuttgart, Germany

[a)]Corresponding author: Jochen.Kettemann@ikt.uni-stuttgart.de

Abstract. In the field of extrusion, nowadays, the main goals are reduction of the manufacturing costs as well as improvements in output and product quality. A very promising high performance extruder fulfills those requirements and is already used in the European industry. This new type of single-screw extruder was introduced to the European plastics industry in 1999 and consists of a combination of a grooved plasticizing barrel with a barrier screw. Although the exceptional performance of this extruder system has already been investigated comprehensively, the melting mechanisms are still not fully understood. Therefore, in the past both an analytical Newtonian model and a numerical non-Newtonian model using the finite difference method (FDM) had been developed at the IKT to predict the melting rate along the screw. In the present work, an enhanced analytical model is presented and compared to experimental investigations. The experimental study includes a qualitative method using optical analysis of the cross sections along the barrel. Hence, the melting process can be further characterized and assumptions of the previous analytical model are validated and partially extended. The enhanced mathematical model is based on the theory of melting in conventional single-screw extruders, the previous model and the experimental results. The model iteratively calculates the melt film thickness δ_0 while considering the melt to be a shear thinning fluid with temperature-dependent viscosity. Finally, the mathematical model is compared with the experimental results, the previous analytical model and a numerical simulation using the finite difference method (FDM). It could be shown that the simulations are in good agreement with the experimental results.

INTRODUCTION

Single-screw extrusion is still one of the most widely used processing technologies in the plastics industry [1]. Among high performance extruders, the single-screw extruder with a grooved plasticizing barrel is a very promising system with exceptional performance. It was developed at the Institut für Kunststofftechnik (IKT) at the University of Stuttgart and patented in 1999 [2]. This new type of high performance extruder consists of a barrier screw and a grooved barrel not only in the feeding zone but also in the melting zone. Therefore, in comparison with a grooved bush extruder the single-screw extruder with a grooved plasticizing barrel is characterized by efficient melting, increased specific throughput, consistent pressure build-up and lower processing costs. Although the benefits of this system have already been proved experimentally, the melting mechanisms are still not fully understood. Furthermore, todays design is still based on experiences and costly try-and-error experiments. Consequently, the IKT has investigated experimentally the melting zone and further developed mathematical models to predict the melting rate within a single-screw extruder with a grooved plasticizing barrel.

PRIOR INVESTIGATIONS OF THE MELTING ZONE

B.H. Maddock [3] was the first who experimentally investigated the melting zone in single-screw extruders. Based on Maddock's numerous cooling experiments in 1959, he published a qualitative melting mechanism. Although Maddock's mechanism was confirmed in many consecutive investigations, in the literature are also two different mechanisms presented by Klenk [4] and Dekker [5]. When extruding PVC, Klenk observed that the melting pool occurs at the opposite, passive flight of the screw. Decker, on the other hand, extruded Polypropylene (PP) in a 90 mm extruder and no melt pool could be seen. Rather, the solid bed was surrounded by melt. However, since the two deviating models represent exceptions, the mechanism according to Maddock has established in the literature.

Based on Maddock's qualitative melting mechanism, Tadmor [6] was the first who developed a mathematical model for an extruder using a conventional extruder screw. In his classical model, Tadmor considered a Newtonian fluid and a constant melt film thickness as shown in fig. 2 (a). Since then, several other models and improvements have been developed, which will not be presented in detail here. However, Rauwendaal [7] presented an advanced mathematical model considering a non-Newtonian fluid with temperature-dependent viscosity and a varying melt film thickness (fig. 2 (b)). Thus, he takes up Tadmor's models [6, 8] and the extensions made by Vermeulen et al. [9] to analytically calculate the melting rate.

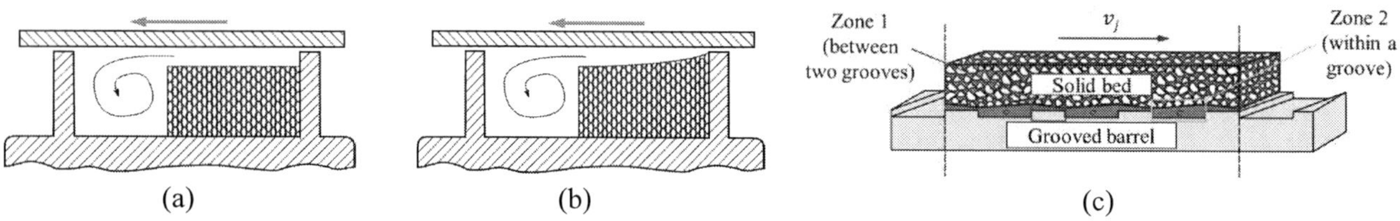

Figure 2. (a) Tadmor's classical model, (b) Rauwendaal's advanced mathematical model and (c) theoretical melting model for a single-screw extruder with a grooved plasticizing barrel (inspired by [10, pp.57]).

In a single-screw extruder with a grooved plasticizing barrel, to calculate the total melting rate two zones are distinguished from one another and the melting rate is calculated independently for each zone according to the classical melting theory. Subsequently, the two individual melting rates are combined by means of a geometric factor [11]. As can be seen in fig. 2 (c), the two zones to be distinguished are the geometrical area between two grooves (zone 1) and the area within a groove (zone 2). Based on both a theoretical model representation developed at the IKT and the drag induced melting mechanisms available in the literature, an analytical calculation using a Newtonian fluid and a numerical non-Newtonian simulation had already been developed [10, 11]. In the following, an enhanced mathematical model is presented, based on the theory of melting in conventional single-screw extruders, the previous model and the experimental results.

ENHANCED ANALYTICAL MODEL

The enhanced analytical model is based on the existing model at the IKT [11] as well as Tadmor's [6] classical melting model and is inspired by Rauwendaal's [7] mathematical descriptions. It considers a shear thinning fluid with temperature-dependent viscosity using a power law model and an iteratively calculated melt film thickness at the beginning of each section. The model enhancements and main assumptions are described in more detail below.

Assumptions

1. Screw channel and barrel are considered unwound and the curvatures of the screw, the grooves and the screw flights are neglected.
2. Melting only takes place at the interface between the barrel and the solid bed.
3. The solid channel is fully filled with solid material, up to the channel end and the solid bed is considered incompressible and homogeneous. The grooves are fully filled with molten material.
4. In the mathematical derivation, convection in the melt film is initially neglected. However, in zone 2 an approximated convection term according to Tadmor and Klein [12] will be added subsequently.
5. Heat transport occurs only one-dimensional in the normal direction to the interface.
6. The melt transport is laminar and takes place by means of a drag-induced melt removal with no wall slip.
7. There is no pressure gradient in the melt film and an incompressible shear thinning fluid with temperature-dependent viscosity is considered.
8. The melting mechanism is considered stationary and gravity forces as well as inertia forces are neglected.

9. For decoupling the energy equation and momentum equation, initially a linear temperature profile in the melt film is assumed.

Modeling

The mathematical description of zone 1 and 2 is based on the energy equation at the interface, developed by Tadmor [8]. To solve the energy equation, the temperature profile in the melt film as well as the temperature profile in the solid bed is needed. Therefore, the general conservation equations (energy equation, momentum equation and conservation of mass) are simplified and finally solved. The shear thinning behavior of the melt is described using the extended power law model by Ostwald and de Weale. Furthermore, equation (1) illustrates the assumed initial temperature profile with the melting temperature (T_m), the barrel temperature (T_b) and the melt film thickness (δ).

$$T(y) = T_b - (T_b - T_m)\left(\frac{y}{\delta}\right) \tag{1}$$

By solving the conservation equations, the melting rate in zone 1 (Ω_1') is given by equation (2) with the following variables: Melting density (ρ_m), relative velocity (v_j), width of zone 1 in the direction of v_j (W_1'), thermal conductivity of the melt (k_m), heat of fusion (λ), specific heat capacity of the solid material (c_s), initial temperature of the plastic material (T_f), melt film thickness in zone 1 (δ_1), consistency index (m_0), power law index (n) and the temperature sensitivity factor (α). However, for an exact derivation of the equations, reference is made to [10].

$$\Omega_1' = \rho_m v_j \left(\sqrt{\frac{\Phi_1}{v_j \rho_m} W_1' + \left(\frac{\delta_0}{2}\right)^2} - \frac{\delta_0}{2} \right) \tag{2}$$

with

$$\Phi_1 = \frac{k_m(T_b - T_m) + B_{31}}{\lambda + c_s(T_m - T_f)}, \quad B_{31} = \frac{k_m\, B_{21}\, \delta_1^2}{K^2}\left(K - e^K + 1\right), \quad B_{21} = -\frac{m_0}{k_m}\left[\frac{v_j\, K}{\delta_1\,(e^K - 1)}\right]^{n+1} \quad \text{and} \quad K = \frac{1}{n}\alpha(T_b - T_m)$$

The procedure to calculate the melting rate in zone 2 (Ω_2') is almost identical to that in zone 1. Only the assumptions of a constant melt film thickness in zone 2 (δ_2) and the consideration of an approximated convection term according to Tadmor and Klein [11] differentiate both zones, using the specific heat capacity of the melt (c_m) and the mean temperature in the melt (T_{AV}). However, the factors B_{31}, B_{32} and K are calculated identically.

$$\Omega_2' = \frac{\Phi_2\, W_2'}{\delta_2} \tag{3}$$

with

$$\Phi_2 = \frac{k_m(T_b - T_m) + B_{32}}{\lambda + c_s(T_m - T_f) + c_m(T_{AV} - T_m)}$$

The initial melt film thickness for each segment (δ_0) is dependent on many influencing factors and cannot be determined exactly mathematically. Nevertheless, according to the experimental observation that the solid channel is filled with solid material up to the end, an iterative method for calculating δ_0 has been developed. To do so, δ_0 is varied until either less than 5 % solid fraction is present at the end of the solid channel or the minimum ($\delta_0 = 0$) or maximum (max. distance between screw and barrel) value for δ_0 has been reached.

EXPERIMENTAL INVESTIGATION

Material

For the experimental investigations, a Polypropylene-Copolymer (PP) provided by Basell (RB-307-M0) was used. This material was chosen because of its industrial importance and its semi-crystalline structure. In order to be able to distinguish between the solid phase and the melt phase during the evaluation, a dry blend made of 97 % of an uncolored basic component and 3 % of a black masterbatch of the same material was extruded. For the subsequent comparison of the experimental data with the simulation, the material was completely characterized and measured in the analytical laboratories at the IKT.

Procedure

To visualize the melting mechanism, cooling experiments similar to Maddock [3] were performed at the IKT on a 50x34D axially grooved single-screw extruder. During extrusion of the dry blend, the machine was stopped abruptly and immediately afterwards the air-cooling system switched on. After the screw has completely cooled down to room temperature, the axially grooved barrel was pulled off as shown in fig. 3 (a). Subsequently, the cooled plastic was removed from the screw then sliced into small sections, polished and finally scanned. Fig. 3 (b) on the left side shows an example of a sliced sample and on the right side a scanned image. Within the framework of a fractional factorial design of experiments, the parameters rotational speed and die pressure were examined.

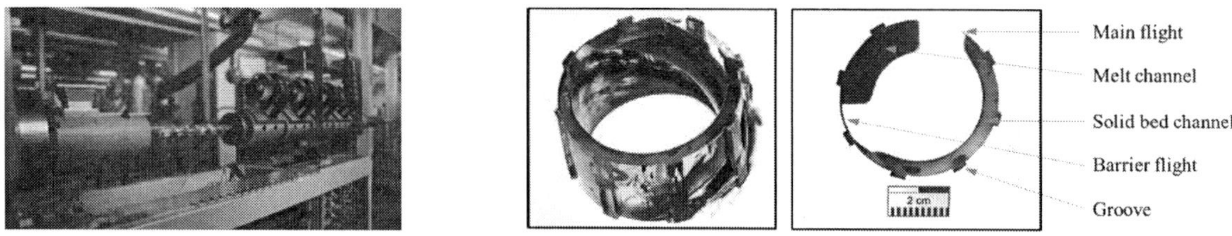

(a) (b)

Figure 3. (a) Pull out procedure at the IKT and (b) sample preparation of the cross sections.

RESULTS AND DISCUSSION

Figure 4 exemplary represents nine scanned cross sections along the barrier screw in the melting zone of the axially grooved barrel. The behavior observed in the shown test samples applies to all conducted experiments. The white color in the solid channel (fig. 4) proofs that it is fully filled with solid material up to the channel end. Thus, this observation confirms assumption 2 and forms the requirement for the iterative calculation method of δ_0.

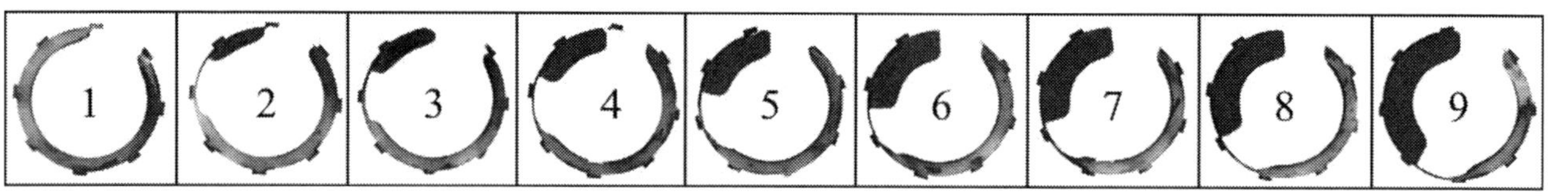

Figure 4. Cross sections along the barrier screw ($n = 250$ rpm, $p_W = 300$ bar and $\vartheta_b = 190$ °C)

Figure 5 (a) illustrates the calculated melting rates using different simulation approaches. It can be seen that the curve progressions of the different simulations are very similar. Especially at lower rotational speed, the curves are almost congruent. Nevertheless, with increasing rotational speed the difference increases slightly. As expected the Newtonian case has the highest melting rate and the analytically non-Newtonian case lies between the Newtonian case and the numerically non-Newtonian model. In this comparison, δ_0 was set to a constant value.

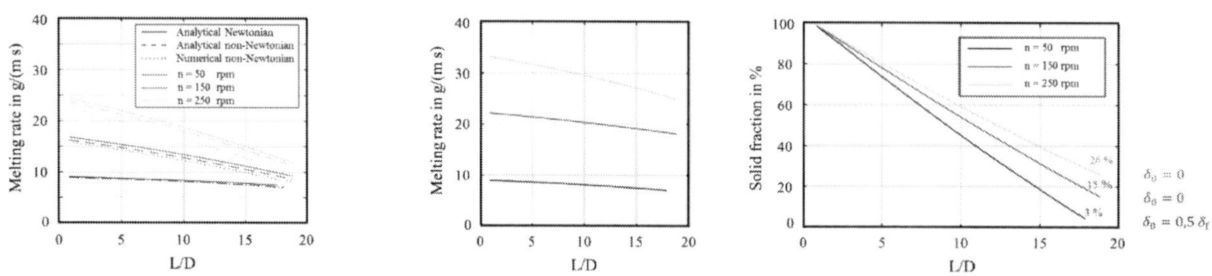

(a) (b)

Figure 5. (a) Comparison of the melting rate for three different simulation methods using a constant value for δ_0 and (b) melting rate and solid fraction calculated with the presented model and iteratively calculated δ_0.

However, to compare the experimental data with the presented model, δ_0 was calculated iteratively. The predicted melting rate as well as solid fraction is shown in fig. 5 (b) for three different rotational speeds. The calculated solid fractions at the end of the melting zone (fig. 5 (b) right) correspond with the experimental observations in fig. 6. A qualitatively comparison between simulation data and experimental results shows the same tendencies for all three rotational speeds. At a low rotational speed (50 rpm) a solid fraction of 3 % is calculated at the end of the melting zone, which is confirmed by the black melt in fig. 6 (a). At a higher rotational speed (150 rpm and 250 rpm), however, a higher solid fraction is calculated. This is also qualitatively confirmed in fig. 6 (b) and (c) by the significantly higher proportion of white material at the end of the melting zone.

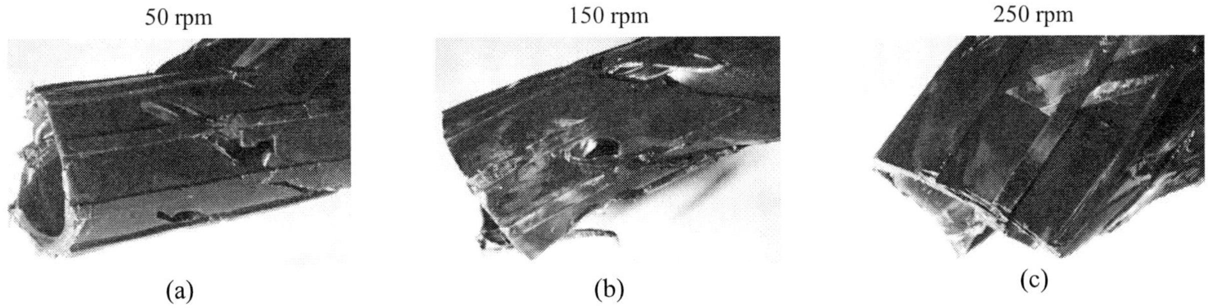

Figure 6. End of the melting zone at different rotational speeds with a die pressure of 300 bar and a barrel temperature of 190 °C.

CONCLUSION

Modeling of the melting zone allows increase in productivity in single-screw extrusion with a grooved plasticizing barrel. In this work an enhanced analytical model was presented to simulate the melting zone of single-screw extruders including a non-Newtonian fluid with temperature-dependent viscosity and an iterative calculation of the melt film thickness at the beginning of each segment (δ_0). Furthermore, cooling experiments according to Maddock were performed to better understand the melting mechanisms and validate the simulation results. Both, the experimental and simulative investigations improve the process knowledge of the melting mechanisms and the simulation is in good agreement with the experimental results.

In further steps, the model has to be validated with numerous experiments and different materials. Moreover, the experimental investigations will be extended to analyze an axially dividable helical grooved barrel. Furthermore, to compare the experimental results with the simulation data in more detail, the experimental solid fraction at the end of the melting zone should be determined quantitatively. In addition, the model is to be further improved in order to take into account the transport processes from the solids channel into the melt channel as well as the implementation of a pressure flow.

REFERENCES

1. C. Bonten, *Kunststofftechnik – Einführung und Grundlagen*, 2nd edition (Hanser, Munich, 2016).
2. E. Grünschloss, German Patent No.19928870-C2 (24.06.1999).
3. B.H. Maddock, SPE Journal **5**, 383-389 (1959).
4. P. Klenk, Rheologica Acta **7**(1)**,** 75-78 (1968).
5. J. Decker, Kunststoffe **66,** 130-135 (1976).
6. Z. Tadmor, Polymer Engineering and Science **6**, 185-190 (1966).
7. C. Rauwendaal, *Polymer Extrusion*, 5th edition (Hanser, Munich, 2014).
8. Z. Tadmor, I. Duvdewani, I. Klein, Polymer Engineering and Science **7**(3), 198-217 (1967).
9. J.R. Vermeulen, P.M. Gerson and W.J. Beek, Chemical Engineering Science **26**, 1445-1455 (1971).
10. J.A. Avila-Alfaro, Der Aufschmelzvorgang in Einschneckenextrudern mit genutetem Plastifizierzylinder, Ph.D. thesis, University of Stuttgart, Institut für Kunststofftechnik 2016.
11. J.A. Avila-Alfaro, E. Grünschloss, S. Epple and C. Bonten, International Polymer Processing **30**(2), 284-296 (2015).
12. Z. Tadmor and I. Klein, *Engineering Principles of Plasticating Screw Extrusion* (Van Nostrand Reinhold, New York, 1970).

Direct Visualization of Cavity Filling and Demolding Phenomena through Textured Glass Block

Wang Chenyang (1)*, **Owada Shigeru** (1), **Yokoi Hidetoshi** (1)

(1) The University of Tokyo - Tokyo - Japan

The surfaces of plastic products are transcribed with various textured patterns. During the transcription process, defects such as inhomogeneous transcription often occur, damaging the product exterior. In this research, several textured patterns were transcribed on the surface of glass by corrective reproduction. From the filling process, pressure holding and cooling process to demolding process, visualization of contact conditions between resin and textured surfaces was realized using a high speed video camera, glass-inserted mold, and new illuminator inside the mold. In the experiments, two types of textured surfaces were transcribed. During the filling process using matt finish textured patterns, the luminance of black resin contacting the textured patterns increased temporarily after the flow front passed dents on the surface due to the degassing of trapped air inside the dents, which caused the resin to float. Observation of the product surface using laser microscope indicated that this degassing phenomenon causes streaks on the product surface. In the inclined demolding process, demolding delay was observed. This demolding delay on the confronting side of the ejector pin was found to cause whitening defect on the surface of molded products. During the filling process using leather textured patterns, after the flow front of black resin passed the dents of the textured surface, the resin was found to continue filling into the dents asymmetrically due to the counter force of trapped and compressed air. In the inclined demolding process, indistinct demolding border contour profile was observed and confirmed to cause whitening and blushing defect. The demolding delay of cooled resin along the side of demolding direction is observed under a long distance microscope. The cooled resin deformed and was stretched conspicuously during demolding from the dents of concave-convex textured surface. The results of the above visualization experiments demonstrated the validity of this new visualization analysis method.

Mechanical Property Model for Fiber Filled Thermoplastics Incorporating Surface Tension

Matthias Bruchmüller[1, a)], Julius Geis[1, b)] and Michael Koch[1]

[1]*Technische Universität Ilmenau Fachgebiet Kunststofftechnik, Gustav-Kirchhoff-Str. 5, 98693 Ilmenau, Germany*

[a)] Corresponding author: matthias.bruchmueller@tu-ilmenau.de
[b)] julius.geis@tu-ilmenau.de

Abstract. This study presents an improved VOIGT law based model which incorporates the surface tension between filler and matrix material. VOIGT law based models require perfect adhesion between fibers and plastic material, which not always occurs. This study deals with the adhesive bonding conditions between matrix and fibers. The harmonic arithmetic mean equation of surface tension by WU is incorporated into a VOIGT law based model. For the experimental validation of this model different fiber reinforced compounds and test specimens are produced. Young's modulus, tensile strength as well as fiber length, orientations and surface tensions are measured.

INTRODUCTION

Short fibers and filaments as fillers are used to influence mechanical properties of thermoplastic materials. The most commonly utilized fillers are fibers made out of glass or carbon. Various authors investigate fibers as reinforcing element[2-11]. LIU revealed that basalt fibers (BF) could be a valid alternative to glass fibers (GF). BF are produced in a similar manufacturing process, but with less energy consumption[11]. The inorganic fiber originates from a volcanic mineral, thus it is one of the most common materials[11]. In addition, it is incombustible. Furthermore, it is free of health hazards and ecologically harmless[13]. BF have a vast potential, especially in combination with thermoplastics as composite materials.

PREVIOUS RESEARCH ON MODELING MECHANICAL PROPERTIES

Based on VOIGT, CHAMIS developed a mathematical model to describe compound young's modulus of endless reinforced fibers and matrix materials[1,3]. Different modifications are set to improve the prediction accuracy and consider short fiber reinforced plastics. COX included a reduction ratio η_1 to regard the fiber geometry and especially the length of the fiber[3]. KRENCHEL found out, that the fiber orientation also effects young's moduli. KRENCHEL created η_0 to consider orientation angles of fibers according to the tension direction[10]. Deviations in orientation lead to decreased young's modulus and a low compound stiffness. Considering these effects on the compound stiffness, specimens are produced to prove the accuracy of this model. The model and its modifications require perfect adhesion between both materials.

$$E_V = E_m \cdot (1 - \phi_f) + E_f \cdot \phi_f \cdot \eta_1 \cdot \eta_0 \tag{1}$$

NEW APPROACH CONSIDERING ADHESION

Regarding a compound in a tensile test, the stress need to be absorbed by the matrix material. Stresses can be transmitted from the matrix into the beginning of a fiber, through the fiber and back to the matrix again at the end of the fiber. The load transmission between fibers and matrix materials depend on their adhesion. Thus, adhesion between both materials need to be researched.

Three types of adhesion are known, diffusion, mechanical adhesion and surface tension. All three types can affect the adhesion between fibers and matrix materials.

Diffusion is known as interconnection of molecules of both phases. Regarding PP and BF or GF, both fiber types melt at lager temperatures than polypropylene or even polyamide can resist[14,15]. Thus, no diffusion effects are expected.

Mechanical adhesion require rough surfaces of e.g. a rigid part so that the melt can link to the fiber like a zip. In FIGURE 1 pulled-out basalt fibers with plain surfaces are shown, thus no melt can link to the fiber by mechanical adhesion.

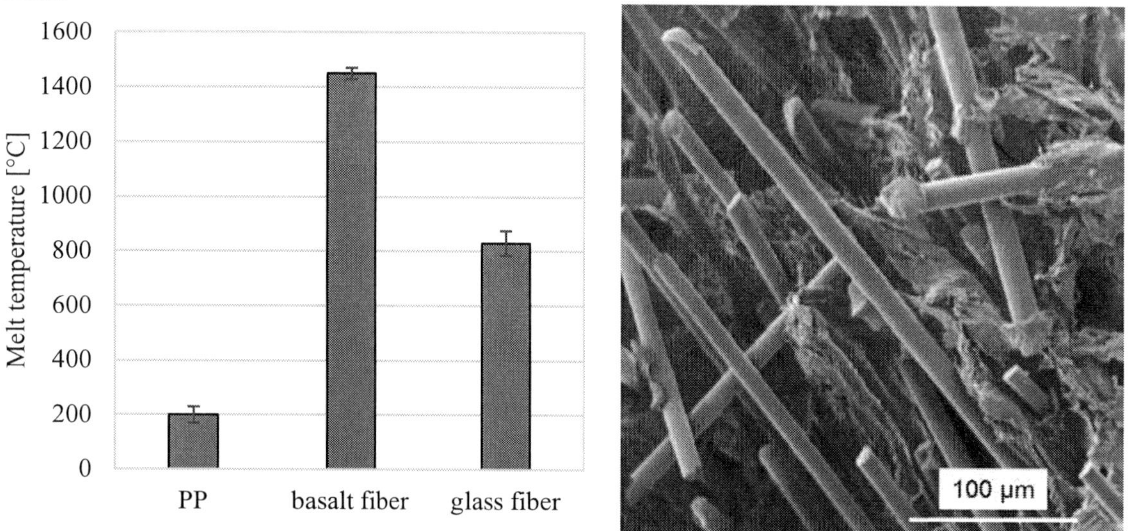

FIGURE 1. Comparison of melt behavior of compounded components[14,15] (r), pulled-out basalt fibers after a tensile test (SEM)

The continued research deal with adhesion between fibers and matrices considering surface tension. An existing mathematical model is used to combine the presented model with an extension regarding surface tension. WU created a formula to describe dependencies between material specific surface tensions and adhesion strengths.

$$\gamma_{12} = \left(\sqrt{\gamma_1^d} - \sqrt{\gamma_2^d}\right)^2 + \left(\sqrt{\gamma_1^p} - \sqrt{\gamma_2^p}\right)^2 \quad (2)$$

In combination with the maximum strength a correlation factor for surface tensions is formulated.

$$\boldsymbol{\eta_{adh}} = \eta_{st} = exp\left(-\frac{\alpha}{\gamma_2}\gamma_{12}\right) \quad (3)$$

If the disperse and polar components of the combined materials match, no attenuation between both occur and η_{adh} is 100 %. The value alpha is set by WU and need to be determined by measurements. The resulting formula includes the attenuation factor η_{adh} by the performance ratio of the fibers.

$$E_V = E_m \cdot (1 - \phi_f) + E_f \cdot \phi_f \cdot \eta_1 \cdot \eta_0 \cdot \boldsymbol{\eta_{adh}} \quad (4)$$

Both models, the former and the new one, are investigated and compared by experiments.

EXPERIMENTS

To investigate the models, specimens are produced using two types of matrix material and two types of reinforcing fibers. Mixtures are made and processed into test specimens. The specimens are tested in different measurement systems according to standards of the German Institute for Standardization.

Materials

A Polypropylene (PP) is used, which is suitable for injection molding. The density of the PP Borealis BC612WG is 0.90 g/cm³. Its nature mechanical properties are 25 MPa of tensile strength and 1.10 GPa of young's modulus. Additionally, a Polyamide 6.6, Durethan A30S is used, with a young's modulus of 3.8 GPa.

Endless basalt fiber rovings with 1,200 and 2,400 tex are used for the compounding process. All filaments of the rovings have diameters of 13 µm and a density of 2.68 g/cm³ by a tensile modulus of circa 90 GPa. Further, glass fibers are compounded into PP matrix to increase the varieties. The used glass fiber rovings have 2400 tex, diameters of 13 μm and a density of 2.60 g/cm³.

Compounds and Mixing Processes

BF and PP are compounded in a co-rotating twin-screw extruder with an L/D rate of 38 by a screw diameter of 40 mm. The temperatures from hopper to die are actively varied at nine different zones to adjust the melt temperature at the die. Compounds with PP are processed by 210 °C and with PA 6.6 are processed by 300 °C melt temperature at the die. Regarding the pre-tests, screw speeds set low between 50 and 80 rpm. The stages of mixing and forming are sketched in FIGURE 2.

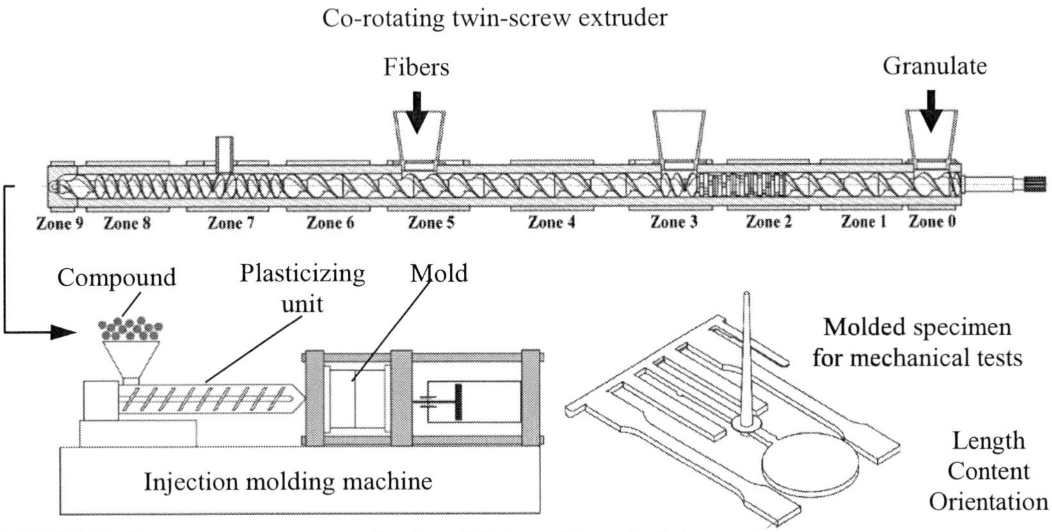

FIGURE 2. Steps to create specimen for the validation of the calculation models

Mechanical properties test methods

After Compounding, the fiber reinforced granulate is dried, prepared for fiber length measurements and additionally molded into specimens. The specimens are tested for their young's modulus.

To calculate the young's modulus, specific standards are used to measure fiber contents and length on the produced specimens as results of processing conditions. First, specimens are weighted and applied to calcination processes according to DIN 53568. Thus, the specimens are freed from plastic matrices and resultant fiber weights could be measured to calculate the fiber content. The lengths of the loose fibers are measured using a microscope camera.

The new attenuation factor need the polar and disperse surface tensions of the compound components. Therefore, two tensiometers with the following measurement methods are used, Wilhelmy-Method and the sessile drop method. Both are able to measure the contact angle between known liquids and unknown solid materials. Using the Method by Owens, Wendt, Rabel & Kaelble (OWRK-Method) the material specific surface tension can be measured and separated into disperse and polar components.

RESULTS

The surface tension results show significant differences in polar and disperse contents, shown in FIGURE 3.

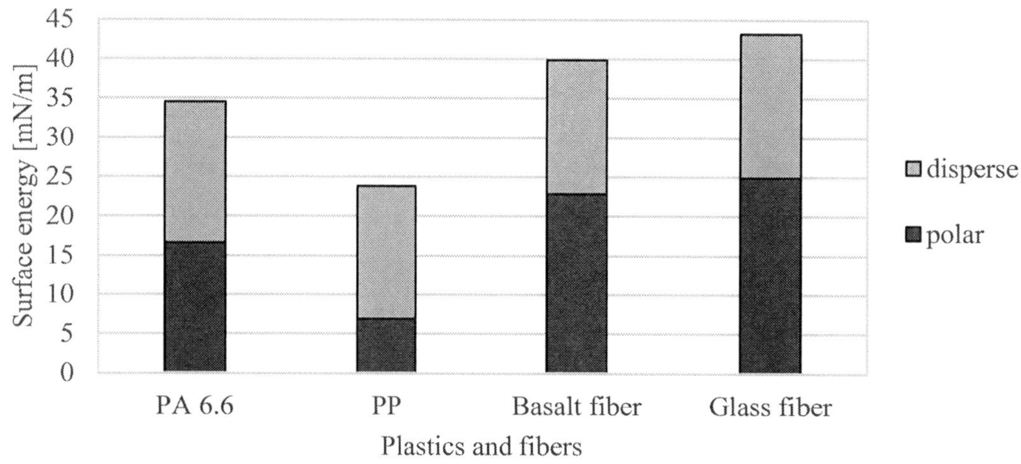

FIGURE 3. Comparison of the surface energies of the used plastics and fibers

Especially the difference between polyamide 6.6 and polypropylene is significant. In comparison to the used fibers a combination with polyamide 6.6 should cause an increase of adhesion and measureable Young's moduli.
The experiments to find alpha are done, using the results of basalt and glass fiber reinforced specimen. To find an accurate alpha, it was varied and correlated to the average young's moduli of both compounds.

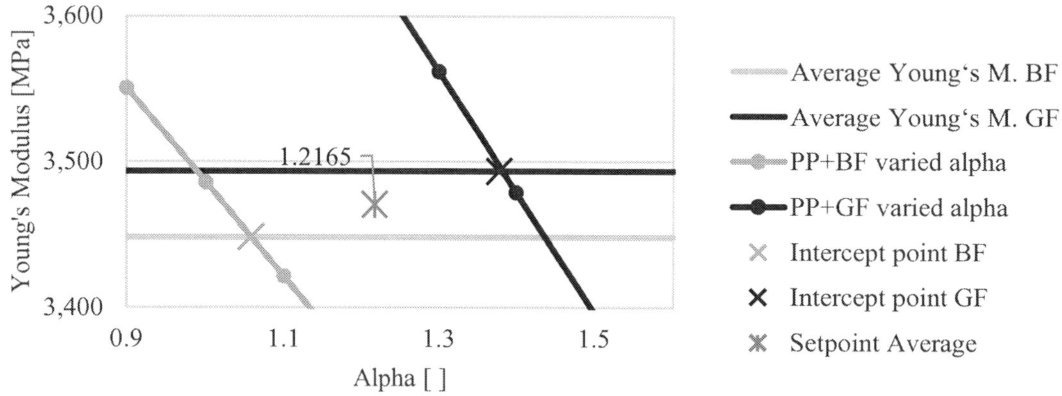

FIGURE 4. Method to define one alpha value for all fiber types

It is possible to compare measured and calculated results directly, by using the results of the experiments. Additionally the former model and new approach is compared.

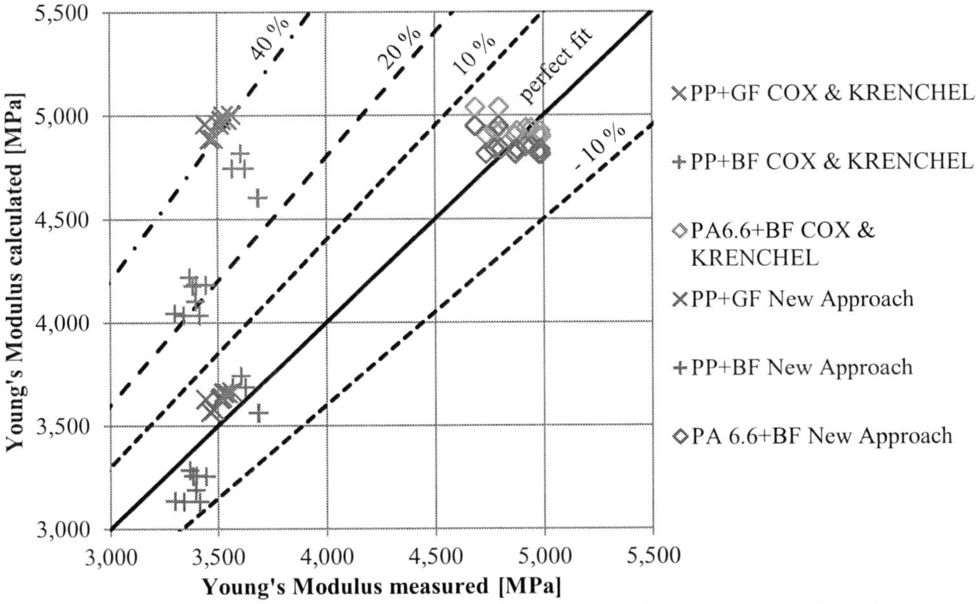

FIGURE 5. Comparison of the former model (red and orange) and the new approach (green)

In FIGURE 5, the results of the measurements and calculation are plotted. Regarding the orange results (PA 6.6 + BF), the former model calculate the young's moduli accurate. Using PP as matrix component, the results deviate up to 42 %. These results are caused by disregarding the influence of surface tension, which differs significantly, shown in FIGURE 3.

The new approach consider the surface tension and is able to minimize the deviance to 8 %. Even the accuracy of the results of PA 6.6 + BF are improved from 8 % to 6 % deviance.

CONCLUSION

The new approach allows a prediction of compound stiffness by a significant increased accuracy and is able to show the reduced stiffness by an unbalanced fiber and matrix material combination. The new approach is able to increase the accuracy of simulation tools and enables to find economic and technical compounds for the specific product requirements.

Further experiments should prove the possibility to predict adhesion agents and contents to improve stiffness of compounds.

ACKNOWLEDGMENTS

The results presented are obtained in the context of the research project KF2184742EB4 sponsored by the Bundesministerium für Wirtschaft und Energie (BMWi). The autors thank the BMWi for their financial support for the research, authorship, and publication of this article.

REFERENCES

1. Voigt, W.: Über die Beziehung zwischen den beiden Elastizitätskonstanten isotroper Körper, Wiedmanns Annalen der Physik und Chemie 38, pp.573-587, Verlag von Johann Ambrosius Barth, Leipzig 1889
2. Brast, K.: Verarbeitung von Langfaserverstärkten Thermoplasten im direkten Plastifizier-/Pressverfahren. Aachen: RWTH Aachen, 2001
3. Cox, H.L. The elasticity and strength of paper and other fibrous materials. Teddington, Middlesex: The National Physical Laboratory, 1951

4. Jianguo Zhang: The reinforcing effect of basalt fiber and polyamide-6 on the mechanical properties of polytetrafluoroethylene composite. Journal of Thermoplastic Composite Materials 2015
5. Hassan,* R. Yahya, A. H. Yahaya, A. R. M. Tahir: Tensile, Impact and Fiber Length Properties of Injection-molded Short and Long Glass Fiber-reinforced Polyamide 6,6 Composites Journal of reinforced plastics and composites, Vol. 23, No. 9/2004
6. Fu, S.-Y.; Lauke B.; Mai, Y.-W.: Science and engineering of short fibre reinforced polymer composites Woodhead Publishing 6th July 2009
7. Schmid, B.: Spritzgießen von langfaserverstärktem Polyamid 6.6 unter besonderer Berücksichtigung der Faserlängenreduktion während des Formgebungsprozesses Düsseldorf: VDI-Verlag, 1997
8. Stelzer, G.: Zum Faser- und Eigenschaftsabbau bei Verarbeitung und Recycling diskontinuierlich faserverstärkter Kunststoffe. Kaiserslautern: Universität Kaiserslautern, 2002
9. Yihe Zhang, Chunxiao YU, Paul K Chu, Fengzhu Lv, Changan Zhang, Junhui Ji, Rui Zhang, Heli Wang: Mechanical and thermal properties of basalt fiber reinforced poly(butylene succinate) composites. Materials Chemistry and Physics 133 (2012)
10. Krenchel , H.: Fibre reinforcement ; Akademisk Forlag, Copenhagen 1964
11. Liu, J.: Untersuchung von Verbundwerkstoffen mit Basalt- und PBO-Faser-Verstärkung. Dresden: Hanser, 2008
12. Tamás Deák, Tibor Czigány: Chemical Composition and Mechanical Properties of Basalt and Glass Fibers: A Comparison, Textile Research Journal Vol 79(7): 645–651
13. Thieltges, H.P.: Faserschädigung beim Spritzgießen verstärkter Kunststoffe. Aachen: RWTH Aachen, 1991
14. Durapact, Technisches Merkblatt: Alkaliresistente Glasfaser, Haan, 2013
15. Incotelogy, Basalt Rovings: Technisches Datenblatt, Pulheim 2015

Tailored microstructures and properties of microinjection moulded isotactic polypropylene/poly (ethylene terephthalate) blends

Zhao Zhongguo (1)***, Sun Hongwen*** (1)***, Wu Pingping*** (1)***, Yang Qi*** (1)*

(1) Sichuan university - Sichuan - China

In this study, the morphology evolution, crystallization behaviour, oriented analysis and mechanical performance of micropart and macropart were investigated in detail. Compared with conventional injection moulding, during the filling of melt into the mould cavity, MIM own the higher shear rate and quickly cooling rate. There are the significant differences in the morphology distribution between micropart and macropart due to the significant sample dimension discrepancy. For macropart, the dispersion of PET phase and the interfacial combination are poor and no orientation occurs. For micropart, however, the dispersed phase merges to form continuous and well-defined PET microfibrils along the flow direction in both skin layer and core layer (especially in PPET10 micropart). There are also the significant differences in the crystallization behavior between micropart and macropart. It is observed for the first time that the combined effects of a strong shear flow field and PET microfibrils facilitate the formation of bundle-like β-crystal epiphytic on the surface of PET microfibrils under microinjection moulding. When Tmould is 80 oC, the degree of orientation of the micropart increases and the crystalline structures are highly compact. However, the DSC and WAXD results also show that evaluating the PET content can increase the content of β-crystal and the orientation degree in micropart. In addition, it is also found that relative to macropart, micropart has the increase the formation of β-crystals and lamellar branching capability. The remarkably enhanced shear force field generated under microinjection

Investigation of the Melt Quality of a Double Wave Screw in a Grooved Plasticizing Zone

P. Thieleke[a)], T. Räpple and C. Bonten

IKT, Institut für Kunststofftechnik, University of Stuttgart, Pfaffenwaldring 32, 70569 Stuttgart, Germany

[a)]Corresponding author: philipp.thieleke@ikt.uni-stuttgart.de

Abstract. Double wave screws promise a good homogeneity at low melt temperatures. A patent of the IKT combines the double wave screw with a grooved plasticizing zone to enhance the single-screw extruder performance. In this work, trials with a double wave screw in a barrel with a grooved plasticizing zone, known under the trade name Helibar®, as well as a barrel with a grooved bushing are conducted and evaluated. In addition, these results are compared to former ones obtained by using a barrier screw in a grooved plasticizing zone. The evaluated parameters are melt temperature, pressure build-up, energy consumption and melt homogeneity. Microtome sections of the plastic strands of the trials are used in order to classify the melt homogeneity. The wave screw in combination with the grooved plasticizing zone shows enhancements regarding the pressure build-up along the screw and the homogeneity. With respect to the low melt temperatures, the double wave screw achieved good homogeneity results.

INTRODUCTION

Generally, an extruder must fulfil certain requirements simultaneously regarding mass throughput, homogeneity, pressure build-up and melt temperature. As not every parameter can be improved simultaneously, screw and barrel need to be designed for an optimum compromise. Wave screws are applicable for high performance extruder systems. They were developed from single wave screws to superior double wave screws [1]. However, they have always been used in combination with a smooth barrel in the plasticizing zone. Based on a patent [2] the performance of a double wave screw in combination with a grooved plasticizing zone will be investigated in this work. The patent's hypothesis is that the increased material transfer through the grooves leads to an improved melt homogeneity [2]. A grooved plasticizing zone shows promising enhancements regarding pressure build-up, melt capacity and the energy transfer into the solid material [3]. The combination of a double wave screw and a barrel with a grooved plasticizing zone enables the molten material to flow over an undercut barrier flight between different wave channels as well as through the grooves in the barrel, as shown in Fig. 1. Due to the permanent reordering of the melt, a double wave screw increases the heat transfer between molten and unmolten material and therefore promises a good homogeneity and keeping a low melt temperature at the same time [4].

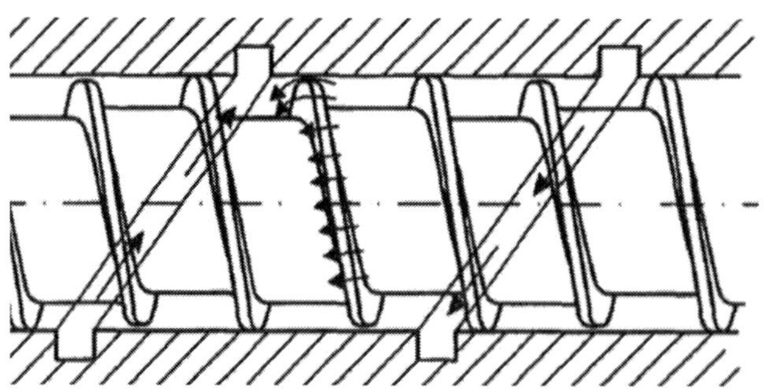

FIGURE 1. Double wave screw in a barrel with a grooved plasticizing zone [2]

EXPERIMENTAL

A double wave screw (35x34D) is used in combination with two different barrels. One barrel comprises a grooved plasticizing zone where the grooves run from the feed zone till almost the end of the barrel. The other barrel is a conventional grooved feed bushing. Both barrels have an identical groove geometry in the feed zone. While the grooved bushing needed water cooling, a passive cooling was sufficient for the barrel with grooved plasticizing zone. As both barrels generate force feeding the specific throughput will be the same. The trials are conducted at screw speeds from 100–300 rpm and die pressures from 100–300 bar. Through electric heater bands the barrel and die are set to a temperature of 220 °C. Figure 2 shows the experimental setup for both barrels. The following parameters are recorded and evaluated: pressure along the barrel, mass temperature, energy consumption and homogeneity.

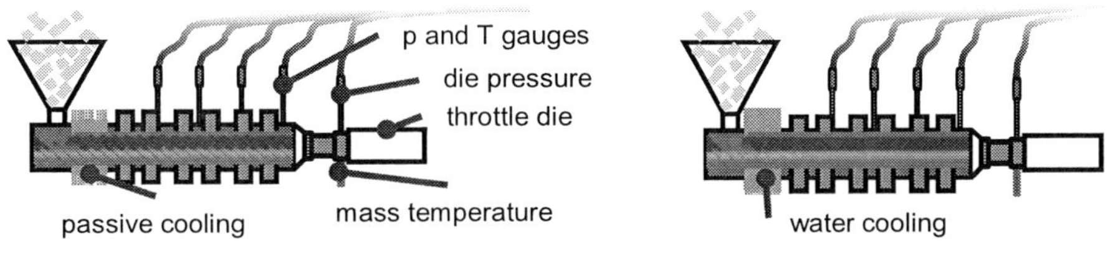

(a) (b)

FIGURE 2. Experimental setup of the barrel with grooved plasticizing zone (a) and grooved bushing (b)

Due to its industrial significance and good processing characteristics a high-density polyethylene (Hostalen ACP 5331 A, LyondellBasell, Netherlands) is chosen for the experiments. In order to classify the melt homogeneity 1 wt.-% of black masterbatch (Heucobatch-Schwarz 990 052, Heubach GmbH, Germany) is added to the PE-HD.

RESULTS

All the results obtained for the double wave screw (WS) in combination with a grooved bushing (GB) and a grooved plasticizing zone (GPZ) are compared to the results recorded by using a barrier screw (BS, 35x34D) in a grooved plasticizing zone at IKT [5]. The influence of the screw speed on the melt temperature is shown in Fig. 3. The melt temperature of the double wave screw decreases at 200 rpm and rises again at 300 rpm. The decrease is caused by the shortened residence time of the polyethylene in the barrel. Due to the rising effect of dissipation, the temperature increases for higher screw speeds. The level of the melt temperature for the grooved plasticizing zone is above the level for the grooved bushing at identical machine parameters. This is caused by the grooves in the plasticizing zone. The grooves lead to a thinner melt film since the film is produced newly every time the screw flight passes the grooves. In turn, a thinner melt film reduces the isolating effect between heated barrel wall and solid material. Therefore, the energy input is much more effective. With regards to the different set points of the heater bands for the double wave screw (220 °C) and the barrier screw (200 °C), the results clearly show that the wave screw lead to a lower melt temperature as predicted by literature. Neither for a grooved bushing nor for the grooved plasticizing zone the melt temperature exceeds the set point using the double wave screw. Unlike the double wave screw, the barrier screw enables a high melt temperature that slightly exceeds the set point.

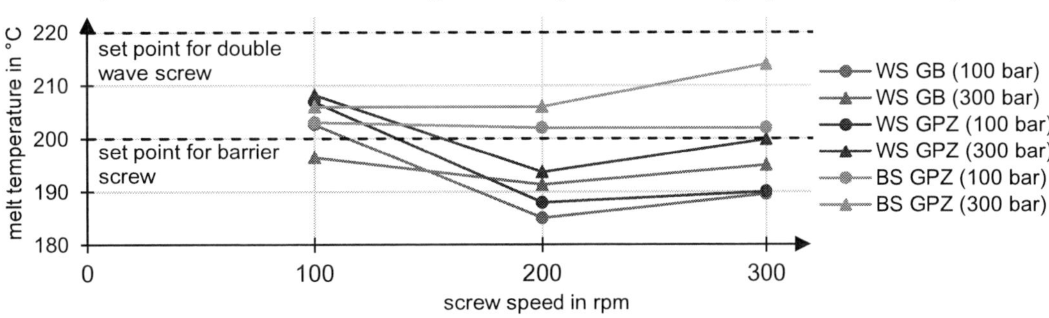

FIGURE 3. Melt temperature

Figure 4 shows the melt pressure along the barrel for die pressures of 100 and 300 bar at screw speeds of 100 and 300 rpm. The use of a grooved bushing leads to a higher pressure level compared to a grooved plasticizing zone. The highest pressure in a grooved bushing is reached in the feed zone. From then on the wave screw only consumes pressure till the end of the barrel. In contrast, the same wave screw in the grooved plasticizing zone is able to build-up pressure over a much longer distance. This reduces the absolute pressure level in the barrel significantly. Thus, the pressure in the feed zone is reduced. That means that the required screw torque will be decreased too. Moreover, wear of screw and barrel is reduced by a lower pressure in the feed zone. Comparing wave and barrier screw, the diagrams clearly show that the pressure level of the double wave screw is always higher than the level of the barrier screw. This can be explained by the dispersive mixing functionality of the wave screw. The early break-up of the solid bed and the squeezing of the melt over the wave crest splits up agglomerates on the one hand but leads to a high pressure level on the other hand.

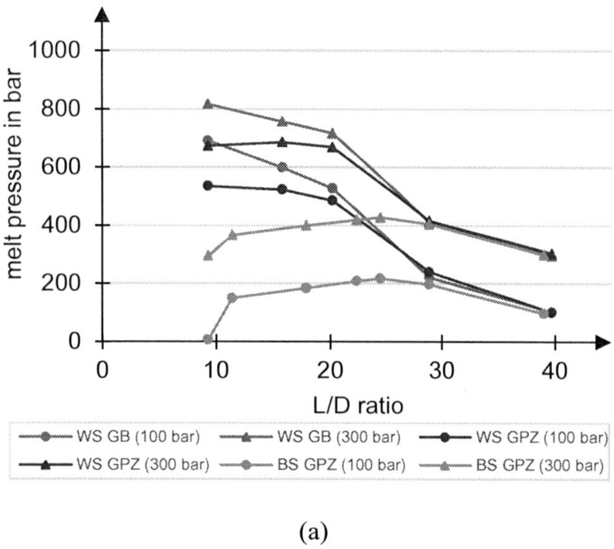

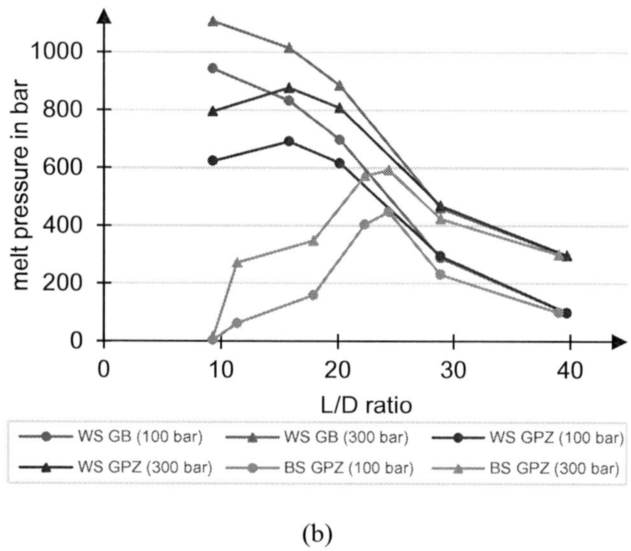

FIGURE 4. Pressure build-up along the barrel for 100 rpm (a) and 300 rpm (b)

The drive energy consumption is determined by the screw speed and the torque of the motor. Figure 5 compares the drive energy consumption for the barrel and screw combinations. The energy consumption rises as well if the screw speed is increased. The influence of the die pressure is not that clear. For the combination of double wave screw and grooved bushing as well as for the barrier screw with a grooved plasticizing zone more power is needed for rising die pressures. In contrast, the energy consumption decreases by changing the die pressure from 100 bar to 300 bar for the combination of double wave screw and grooved plasticizing zone. This can be obtained for all screw speeds. The lowest energy consumption can be obtained by the barrier screw. However, another wave screw design leads to a consistent energy saving of 3–9 % by using the barrel with the grooved plasticizing zone instead of the grooved bushing.

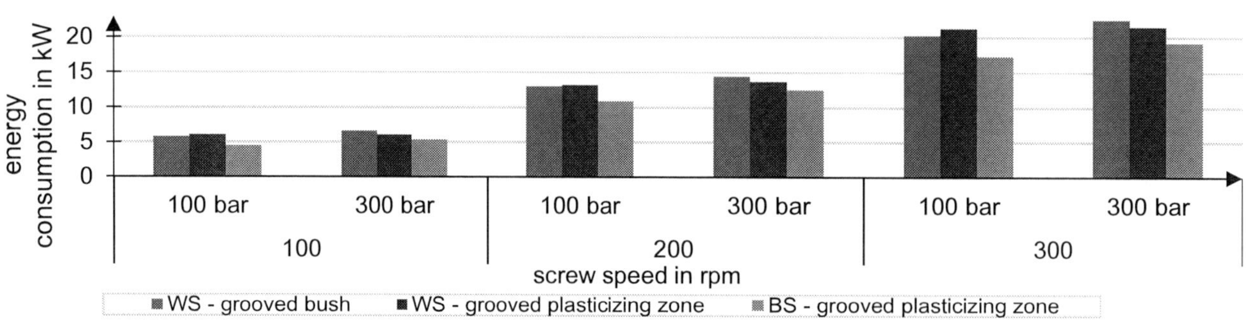

FIGURE 5. Energy consumption

The most promising impact of double wave screws is the enhancement of the homogeneity. The IKT developed an own approach in order to classify the homogeneity extrudates. A microtome cuts thin polymer films (40 µm) out of plastic strands. These polymer films are digitalized through a transmitted light scanner. Afterwards the homogeneity of the film is classified by the MATLAB® Image-Processing Tool-Box. There are ten classes. Zero represents the optimum and nine the worst homogeneity. To ensure the received class for an extrudate three polymer films are prepared and evaluated for each operation point. The results are presented in Table 2.

The best results are obtained for the barrier screw because the melt capacity of a barrier screw is obviously higher compared to a wave screw. Additionally, the barrier screw is used in combination with a standard mixing element. The mixing element comprises a Maddock shearing section and a Saxton distributive mixing section. It has to be emphasized that the wave screw is used without mixing element.

Table 2 also shows that the homogeneity of the grooved plasticizing zone is always better or equal compared to the grooved bushing. This enhancement can be explained by the higher melt efficiency through the grooves. This means that the solid material melts earlier and the time remaining for the mixing of the melt is increased. Obviously, there is also an improved material transfer through the grooves which helps to gain a better homogeneity. Although the homogeneity achieved by the double wave screw seems to be rather average, the low melt temperature must be considered, too. Better results regarding homogeneity by using the wave screw could be received with an additional mixing element.

TABLE 2. Classification of homogeneity

	grooved bushing		grooved plasticizing zone			
	double wave screw				barrier screw	
	microtome section	class.	microtome section	class.	microtome section	class.
100 rpm / 100 bar		**5**		**3**		**1**
100 rpm / 300 bar	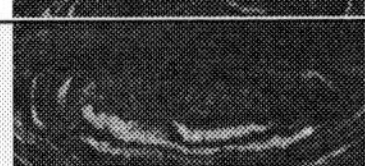	**5**		**4**		**1**
300 rpm / 100 bar		**6**		**6**		**3**
300 rpm / 300 bar		**6**		**5**		**1**

best ↑ 0 1 2 3 4 5 6 7 8 9 ↓ worst

CONCLUSION

The combination of a double wave screw in combination with a grooved plasticizing zone enhances the extruder performance. The grooves in the plasticizing zone lead to a pressure build-up over a longer distance which reduces the pressure in the feed zone. That means that the required screw torque will be decreased. Additionally, a lower pressure in the feed zone results in less wear of screw and barrel. As the double wave screw leads to low melt temperatures the screw is applicable to polymer processing for temperature-sensitive plastics. The hypothesis of the IKT patent is valid. A higher material transfer through the grooves enhances the homogeneity. It is also shown that the energy transfer from the heater bands into the solid material is improved through the grooved plasticizing zone due to a thinner melt film. In future, trials at high speed should be investigated.

REFERENCES

1. G. A. Kruder, U.S. Patent No. 4,173,417 (6 November 1979).
2. Epple, S., T. Erb and S. Geier, DE 10 2012 010 854 A1 (31 May 2012).
3. Bonten, C., *Kunststofftechnik*, 2nd edition (Hanser, Munich, 2016). ISBN 978-3-446-44674-8
4. G. A. Kruder and W. N. Calland, "Fundamentals of Wave Screw Extrusion," in SPE ANTEC Tech. Papers **36** (1990), pp. 74–85.
5. M. Makovac, "Study of Plastification Capability and Mixing in a High-performance Single Screw Extruder with a Grooved Barrel," Bachelor Thesis, University of Stuttgart, 2014.

Investigation of Residence Time Distribution of Different Screw-Barrel Concepts via Ultrasound in Extrusion

P. Thieleke[a)] and C. Bonten

IKT, Institut für Kunststofftechnik, University of Stuttgart, Pfaffenwaldring 32, 70569 Stuttgart, Germany

[a)]Corresponding author: philipp.thieleke@ikt.uni-stuttgart.de

Abstract. In today's extrusion plants, flexibility in production combined with a high productivity at the same time is an increasingly important requirement. Moreover, following strategies of minimized stock keeping and just-in-production, plastics converters must be able to produce small batch sizes economically. Due to the resulting high number of consecutive material and color changes on a production line, residence time in the extruder plays a decisive role. In this work, the residence time distribution is determined for three different extruder concepts using ultrasonic technology. A smooth bore, a grooved feed bush and a barrel with a grooved plasticizing zone, known under the trade name Helibar®, are investigated. A 35 mm barrier screw without mixing elements is used in each case. Calcium carbonate is applied as indicator material. A measuring flange between the extruder and the throttle die enables the recording of the amplitude in the inline measuring setup at different screw speeds and die back pressures. The results show that greater residence time distributions are obtained with the smooth bore extruder than with the two force feeding extruders which was to be expected. The results of the grooved feed bush extruder and grooved plasticizing zone extruder show similar process behavior.

INTRODUCTION

In the plastics industry, single-screw extruders are the core of an extrusion line. They generally have to meet various target requirements which depend on the respective application. One of the functions of a single-screw extruder is to provide the subsequent extrusion die with a polymer melt comprising defined characteristics. At the same time, the resistance of the extrusion die must be overcome [1]. The polymer melt of a single-screw extruder is characterized by the four target parameters melt homogeneity, mass temperature, mass pressure and throughput [2]. The four parameters are in a conflict of objectives since not every parameter can be improved without worsening another one.

From an economic point of view, further requirements of today's extruders are a low residence time distribution as they play an important role regarding costs at color and production changes. A narrow residence time spectrum increases the effective production time and decreases simultaneously scrap material. However, it implicates a lower melt homogeneity which is why the design of an extruder generally is a compromise of all requirements [3].

As there is still a gap between simulations of different extruder concepts and real test results by now, further experimental investigations are necessary to determine and explain procedures like residence time distribution.

EXPERIMENTAL

The test setup consists of a 35x34D extruder and a throttle die that can adjust the die pressure continuously. A measuring flange is implemented between extruder and throttle die in order to record the residence time distribution. Ultrasound is used as measuring method since calcium carbonate can be used as tracer material in the polymer melt with very low mass percentages. The test setup allows inline measurement during the extrusion process as illustrated in Fig 1(a). While the extruder runs, ultrasonic signals are transmitted from one ultrasonic probe. The ultrasonic signal runs through the polymer melt to the opposite ultrasonic probe which is able to detect the signal. If the material is altered where the ultrasonic signal runs through, a difference in amplitude or runtime can be recorded.

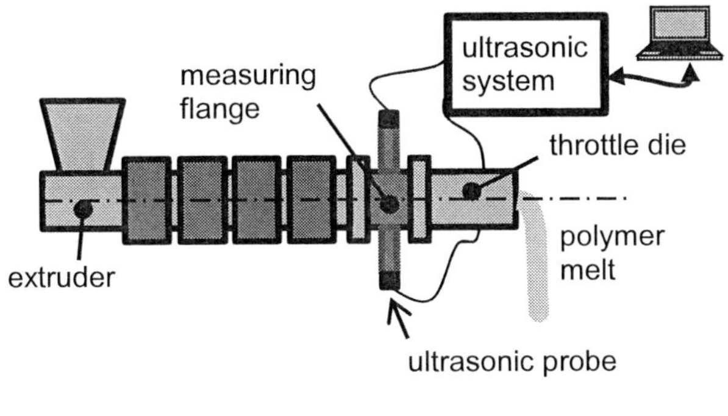

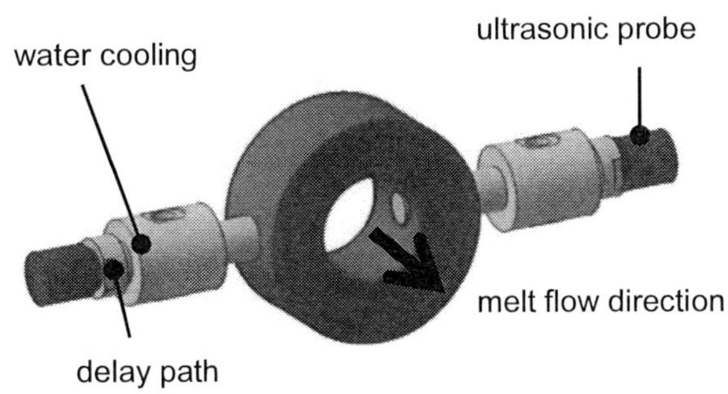

FIGURE 1. Test setup (a) and measuring flange (b)

In this work, different barrel concepts were investigated in combination with a barrier screw without mixing elements in each case. The inner diameter of the barrel is 35 mm with a process length of 34 L/D for every concept. The barrel concepts include a smooth bore barrel, a grooved bush barrel and a barrel with grooved plasticizing zone known under the trade name Helibar®. The grooved feed bush barrel and the barrel with a grooved plasticizing zone have the same feed zone geometry so that the specific throughput is the same for both barrel concepts.

Two different materials were processed for each barrel concept. A material overview with characteristic properties is shown in table 1. The PE-LD was chosen as a wall adhesive material. The PE-HD is a wall slipping material at a critical shear rate. The tests were conducted at rotational speeds from 50 to 300 rpm at die pressures of 150 and 300 bar.

TABLE 1. Material overview

Short Description	Trade Name	Process Temperature in °C	Viscosity in Pas	Bulk Density in g/cm³
PE-HD	Basell Hostalen ACP 5331A	200	83,000	0.52
PE-LD	Basell Lupolen 3020 H	200	12,828	0.52

The measurements of the residence time distribution were carried out by the impulse method. For each measurement, the extruder was set into steady state with a constant rotational speed and a constant die pressure. The feed hopper for the granules was run empty until the screw was only just filled with granules. At that moment, an impulse of 500 g including a small percentage of calcium carbonate was added while recording the ultrasonic signal. As soon as the impulse material was processed, virgin material was added to the feed hopper again. The result of the recorded ultrasonic measurement is shown in Fig. 2.

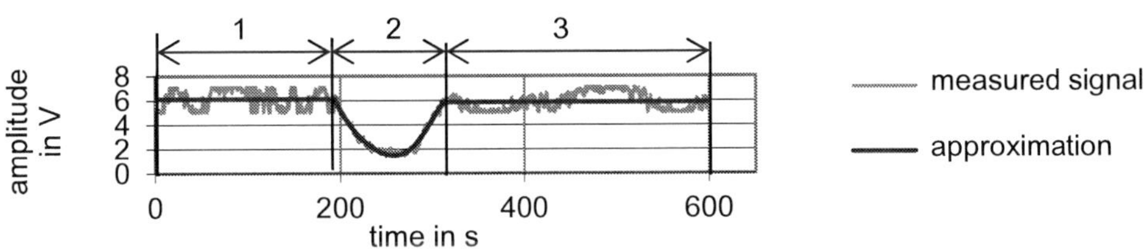

FIGURE 2. Amplitude measurement at an impulsive change in material by using a dry-blend

All trials were conducted in the same manner. The amplitude was measured over time. For a consistent evaluation, the measurement curves were separated into three sections. Average values were calculated for the section one and three because it is assumed that the amplitude is constant in these sections. A Gaussian function is used as approximation for section two. It is defined by formula (1):

$$f(t) = a_1 * e^{-\left(\frac{t-b_1}{c_1}\right)^2} + a_2 * e^{-\left(\frac{t-b_2}{c_2}\right)^2} \quad (1)$$

RESULTS AND DISCUSSION

Hereinafter, the results of the experiments conducted are presented. There are two options for the evaluation of the ultrasonic signal. One option is to determine of the runtime between the first waves that are sent out and the first rise of amplitude of the receiving signal by the opposite ultrasonic probe. In this case, the gradient of the first rise which stands for the starting point of the detection, can be adjusted. The ultrasonic measurement system records a frequency bandwidth, as the result strongly depends on disturbances and noise. Therefore, a wide range is recommended for a very precise measurement [4]. The more precise it is measured, the more time is required.

The transmission as well as the receiving signal of different materials are displayed in Fig. 3. For the shown diagram, the same material is continuously processed at steady state and the measurement signal is recorded. A closer look at the red framed period of time in Fig. 3(a) shows that the results are contradictory for a runtime recording. It is expected that the runtime of 10 wt.-% of compound and dry-blend has the lowest runtime as the Young's modulus is increased through a higher calcium carbonate content. A higher Youngs's modulus raises the velocity of the ultrasonic waves. However, this cannot be identified in general and the measurement of the runtime is too slow in order to record the impulse of the calcium carbonate tracer material.

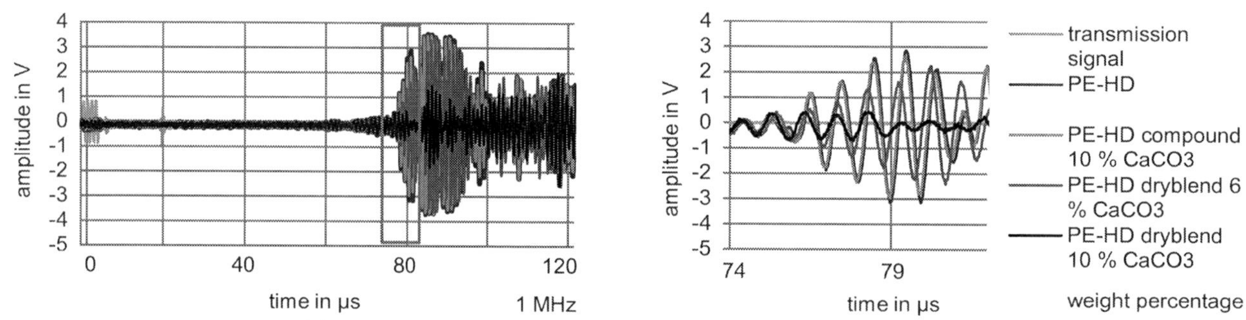

(a) (b) detailed view

FIGURE 3. Received ultrasonic signal for compounds and dry-blends

Alternatively, the amplitude of the receiving signal can be evaluated. With the software of the ultrasonic measurement system the amplitude of the receiving signal in a selectable period of time can be identified. By means of a Fast Fourier Transformation (FFT), the signal is converted from time to frequency domain. The amplitudes are plotted over frequency. If the frequency of the receiving signal equals the frequency of the transmission signal, the corresponding amplitude will be the proper output signal for the evaluation [5]. In addition, the result of the amplitude acquisition with the ultrasonic measurement system is very precise as the signal noise is already eliminated internally, so that the actual amplitude is recorded [4]. Figure 3 shows clearly that only dry-blends cause a reduction of amplitude of the receiving ultrasonic signal. Compounds cannot be used with the installed ultrasonic frequency of 1–2 MHz.

Figure 3 demonstrates the reason why compounds are not suitable for causing an amplitude attenuation. By processing the compound, the $CaCO_3$ particles are finely dispersed in the polymer melt whereas agglomerations of $CaCO_3$ particles occur with dry-blends. The agglomerations of the dry-blends are big enough to be encountered by the ultrasonic waves. This causes a dispersion and attenuation of the ultrasonic signal which reduces the amplitude of the receiving signal.

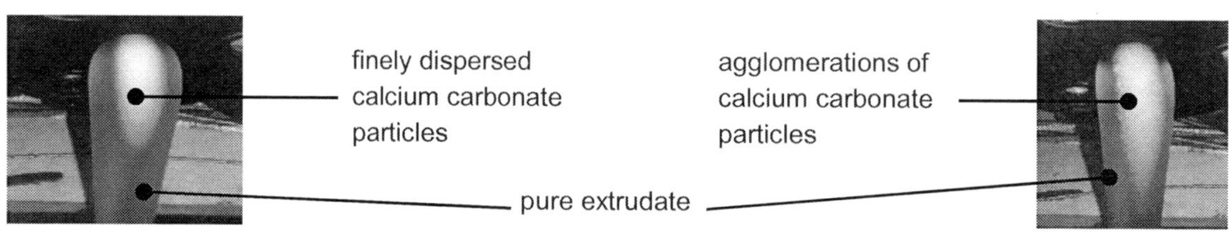

(a) Compound (b) Dry-blend

FIGURE 4. Extrudate of compound (a) and dry-blend (b) at die outlet

Referring to the different screw-barrel concepts, the results in Fig. 5 show that the smooth bore barrel in combination with the barrier screw reaches the lowest specific throughput. Additionally, the specific throughput varies dependent on the die pressure which was to be expected. In order to conduct color changes effectively, a high rotational speed should be chosen.

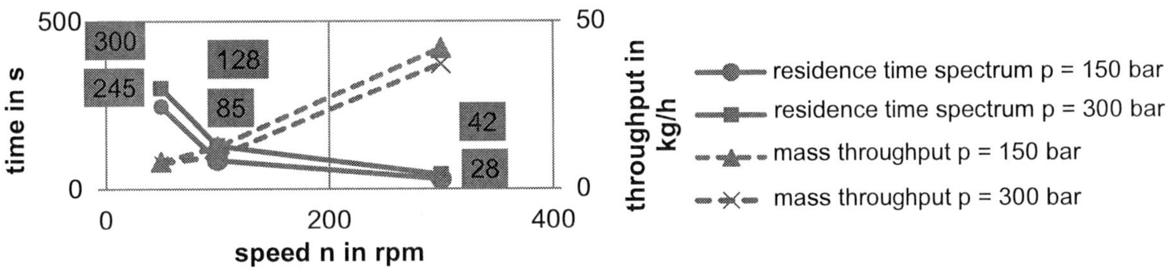

FIGURE 5. Residence time spectrum of the smooth bore barrel with PE-HD

For the grooved feed barrel, it is observed that the use of PE-HD and PE-LD result in a very similar specific throughput. For PE-HD, a higher counter pressure causes a narrower residence time spectrum as it is illustrated in Fig 6. Obviously, the viscosity increases with the counter pressure, which in turn leads to the effect that the velocity profile is more block-shaped. Generally, it is shown that a higher counter pressure at low speed has a positive effect on the residence time spectrum of a grooved feed extruder. At screw speeds over 100 rpm, the counter pressure has no significant influence on the residence time spectrum.

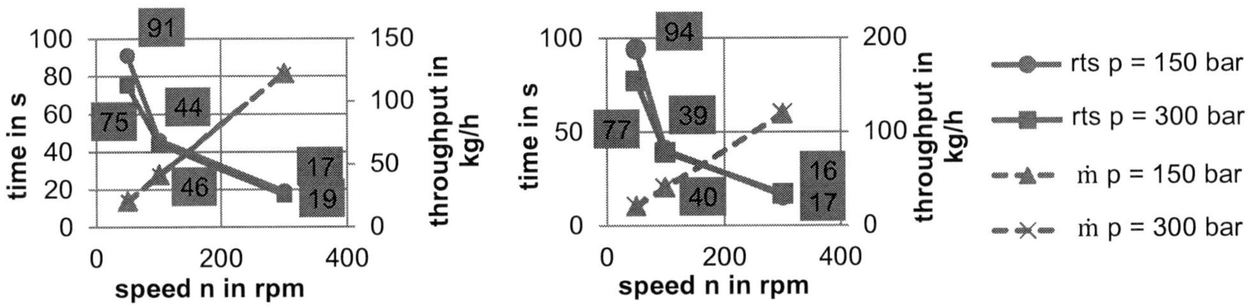

(a) PE-HD (b) PE-LD

FIGURE 6. Residence time spectrums (rts) of the grooved bush barrel with PE-HD (a) and PE-LD (b)

The specific throughput for the barrel with grooved plasticizing zone is approximately the same as for the grooved feed barrel. FIGURE 7 shows that at 100 rpm, a higher die pressure leads to a narrower residence time spectrum. Only at a rotational speed of 300 rpm the residence time spectrum becomes independent from the die pressure.

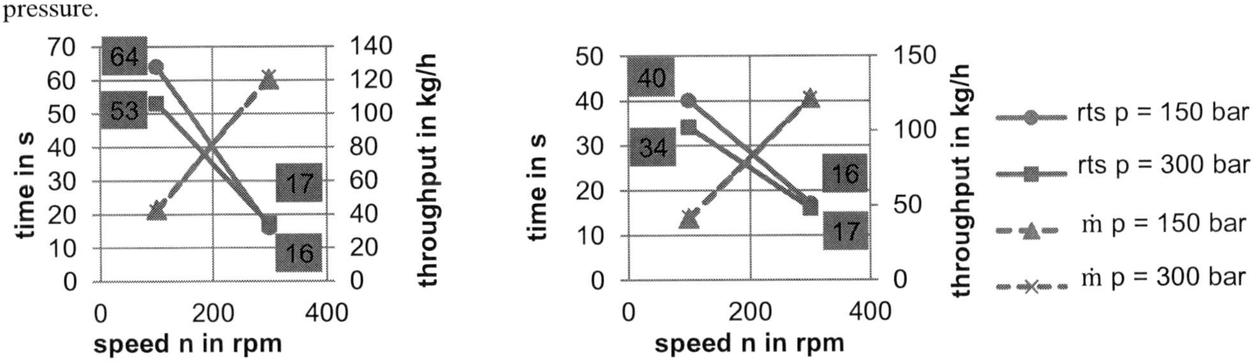

(a) PE-HD (b) PE-LD

FIGURE 7. Residence time spectrums (rts) of grooved plasticizing zone with PE-HD (a) and PE-LD (b)

When the different screw and barrel combinations are compared with each other, it turns out that the specific throughput is the dominating influencing factor on the residence time spectrum. For a smooth bore extruder, a considerably higher residence time is necessary for material or color changes. However, for the smooth bore extruder it is recommended that high screw speeds are used for color changes as materials can be saved.

The comparative studies of grooved bush barrel and barrel with grooved plasticizing zone show that they have almost similar residence time spectrums at equal process conditions. It is noticeable that at a screw speed of 100 rpm and the processing of PE-HD, the grooved feed extruder results in a narrower residence time spectrum. At higher speeds, the differences are negligible.

The difference in residence time distribution between PE-HD and PE-LD decreases with higher screw speed. Unfortunately, the test results hardly show definite trends. These inconsistencies complicate the evaluation of the test results. Hence, it is not clear in what extent wall slipping or different viscosities influence the residence time distribution.

CONCLUSION

The investigations show that the ultrasonic measurement system is successfully implemented into the extrusion process in order to record residence time distributions inline. The test results revealed that an evaluation of the amplitude is expedient. The investigations prove that for a given ultrasonic frequency, the agglomeration of calcium carbonate particles has a decisive influence on the ultrasonic signal. It should be possible to use compounds as tracer material as well if a higher ultrasound frequency is applied. The results also indicate that the receiving ultrasonic signal varies significantly at increasing screw speeds. The effect of the screw speed on the variation of the ultrasonic signal should be investigated in future work.

As both materials almost have the same bulk density, it is possible to evaluate the results without the influence of the specific throughput. However, no distinct influence of the wall slipping effect or differences in viscosity on the residence time distribution could be identified. Regarding the force feed extruders such as the grooved bush barrel and the barrel with grooved plasticizing zone, a higher counter pressure at low screw speed allows faster color changes. A further finding is that the residence time distribution becomes independent from counter pressure at high screw speed.

REFERENCES

1. Bonten, C., *Kunststofftechnik*, 2nd edition (Hanser, Munich, 2016). ISBN 978-3-446-44674-8
2. Michaeli, W.: *Einführung in die Kunststoffverarbeitung*, 6th edition (Hanser, Munich, 2010). ISBN 978-3-446-42488-3
3. Kast, O. and C. Bonten, „*Das Verweilzeitproblem : Mischqualität und Produktwechseleffizienz im Konflikt bei der Einschneckenextrusion*", in: K-Zeitung, 2014. **45**(23), pp. 13
4. Ritec Advanced Measurement System RAM-5000, *User Manual,* Version 1.0.86
5. Mühl, T., *Einführung in die elektrische Messtechnik*, 3rd edition (Vieweg+Teubner, Wiesbaden, 2008). ISBN 978-3-8351-0189-0

Effects of Pellet Characteristics and Feed Zone Design on the Output of Grooved-feed Extruders at High Screw Speeds

O. Kast[a)], M. Koch und C. Bonten

[1]*Institut für Kunststofftechnik, University of Stuttgart, Pfaffenwaldring 32, 70569 Stuttgart.*

[a)]corresponding author: oliver.kast@ikt.uni-stuttgart.de

Abstract. Rising requirements regarding throughput rates of modern extrusion processes have led to a high demand of high-speed single-screw extruders. However, whereas the throughput rate is directly correlated to screw speed for conventional extruders, this correlation decreases at higher screw speeds. As established design principles rely on this correlation, the calculation of throughput becomes imprecise for high-speed extrusion. There are approaches addressing this problem by assuming a partial filling of the screw, but these need experimental calibration for certain model parameters. Current research at the Institut für Kunststofftechnik at the University of Stuttgart focuses on the influence of pellet characteristics as well as the screw and groove design in the feeding section. The objective is to build up a more detailed knowledge on the single effects and implement this into throughput calculation models. For this, regression analysis is used to identify the different effects in regard to their size and relevance. Regression models for the total as well as the specific throughput are tested and their agreement with experimental data is evaluated. An approach with two separated screw speed ranges shows the most promising results. The application for different pellet shapes, e. g. recycled material, is not yet possible and has to be further investigated. In future works a fully integrated model should be developed, which implements the identified effects via their influence on bulk density and combines this with established models.

CHALLENGES TO MODERN EXTRUSION PROCESSES

High-speed extrusion with circumferential screw speed of above 1 m/s for increased production rates has become more and more important within the last two decades [1, 2, 3]. The most important advantages are high throughputs on small systems as well as a reduction of required motor power [4]. High-speed extruders have become widely available on the market since the mid-2000s [5]. However, current systems are limited to screw sizes of more than 60 mm [5, 6, 7]. Moreover, the high circumferential speeds affect the efficiency of the feed zone, which results in a reduction of the specific throughput. In previous works at the Institut für Kunststofftechnik (IKT), a high-speed extruder based on the HELIBAR® system was tested, for its advantages in regard to high throughput stability and a low melt temperature [8, 9]. It was shown that even with a 35 mm screw, throughputs were stable and as high as 300 to 400 kg/h [10, 11]. However, with increasing screw speed the specific throughput decreased, resulting in a non-linear correlation between screw speed and throughput (see Fig. 1 (a)). This phenomenon is well known [12, 13] and has been addressed by counter-measures like an increased diameter of the barrel bore [14] or a larger feed opening [13]. While these solutions have proven to be successful for conventional extruders, the problem of a reduced specific throughput remains for high-speed systems.

CLASSIC THROUGHPUT CALCULATION AND ADDITIONAL EFFECTS AT HIGH SCREW SPEEDS

A problem of the non-linearity between throughput and higher screw speeds is an appropriate consideration of this effect for the throughput calculation within screw design processes. Usually, the following approach is used:

$$\dot{m} = v_{ax} \cdot A_f \cdot \rho_{bulk} \ , \tag{1}$$

where $\dot{m}$ is the throughput, A_f the free cross-section of the screw channel and v_{ax} the axial speed component of conveying. For the calculation of v_{ax} a variety of methods exist, which all proclaim a linear correlation to screw speed. This results in an increasing off-set between calculated throughputs and those obtained in real processes.

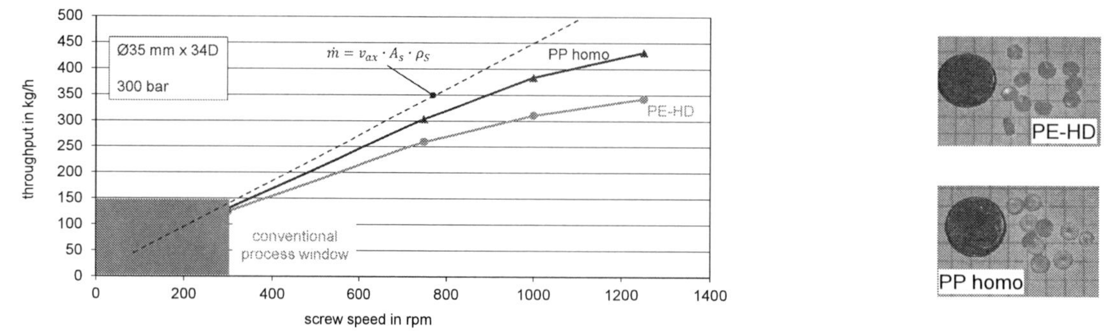

(a) (b)

FIGURE 1. Non-linearity of throughput with increasing screw speed

There are approaches to include this effect in the calculation. Schöppner states that the pellet flow from the hopper into the feed zone is physically limited. Thus, a partial filling of the screw channel occurs, when the conveying capacity of the screw exceeds the maximum pellet flow from the hopper [15]. He suggests calculating throughput curves with a horizontal asymptote at high screw speeds, which is given by this maximum pellet flow. This approach, however, requires an experimental calibration of the pellet flow, as it varies with polymer type and pellet shape. A current calibration of this model is given by Lessmann [16]. Another approach has been presented by Potente et al. [17] and uses a corrective factor <1 representing the filling degree of the screw. This also requires an experimental determination of functions for the filling degree.

Current works at the IKT aim to base corrective functions or factors for eq. (1) on actual, measurable pellet characteristics, such as bulk density, bulk flowability, pellet size and also pellet shape. Moreover, former findings from the investigation of a high-speed HELIBAR® extruder indicated an influence of the combination of different screws and barrels [19]. Therefore, in this study the influence of a variety of factors is investigated by means of statistical analysis.

REGRESSION ANALYSIS OF INFLUENTAL FACTORS ON THROUGHPUT

We use regression analysis to identify statistically significant influences of different pellet characteristics as well as geometrical factors of the screw and the barrel on the throughput. Other than a curve fit, which is used to replace single data points with a mathematical function, a regression is designed to find cause-and-effect correlations. For a given data set consisting of a dependent variable y and independent variables x_i, the regression coefficients β_i are determined via Ordinary Least Square (OLS) method. To test the regression's quality, statistical significance is evaluated with the so called t-test, which provides information about the probability of making a mistake, when a found correlation between y and x_i is regarded as true. This probability defines the so called significance level. In this work, a significance level of 5 % is used.

The data used for the regressions presented in this paper were taken from experimental tests published before [10, 11, 18]. A 35 mm single screw extruder consisting of a barrier screw and a continuously grooved barrel was used. Mass throughput was measured for different screw speeds and die pressures. The screw design and the design of the grooves in the barrel were varied in regard to channel height h_S, number of grooves N_G and also groove pitch S_G. A high density polyethylene (PE-HD, Hostalen ACP 5331A) and a polypropylene homopolymer (PP homo, Hostalen PP H2150) were used, both provided by Basell Polyolefine GmbH, Frankfurt (Fig. 1 (b)). The following pellet characteristics were measured: Average pellet weight g_P, average pellet diameter d_P, pellet aspect ratio a, bulk density ρ_{bulk} in dependence of channel height h and finally pellet flow time T from a hopper with a defined opening (as a measure for the flowability).

Different regression models were investigated, of which two are presented here. The first model describes the total mass throughput as a function of the presented pellet characteristics, the screw and the groove design, the screw speed n and also the square thereof (eq. (3)).

$$\dot{m} = \beta_0 + \beta_1 \cdot g_P + \beta_2 \cdot d_P + \beta_3 \cdot a + \beta_4 \cdot \rho_{bulk}(h) + \beta_5 \cdot T + \beta_6 \cdot h + \beta_7 \cdot N_g + \beta_8 \cdot S_g + \beta_9 \cdot n + \beta_{10} \cdot n^2 + \varepsilon \qquad (2)$$

Including n^2 inflicts the problem of multicollinearity; however, it also allows including a measure for the centrifugal force acting in the feed zone. For this regression, the whole data panel was used.

A second model describes the specific throughput, which is the quotient of total throughput and screw speed, as a function of the independent variables (eq. (4)).

$$\dot{m}_{spec} = \beta_0 + \beta_1 \cdot g_P + \beta_2 \cdot d_P + \beta_3 \cdot a + \beta_4 \cdot \rho_{bulk}(h) + \beta_5 \cdot T + \beta_6 \cdot h + \beta_7 \cdot N_g + \beta_8 \cdot S_g + \beta_9 \cdot n + \varepsilon \qquad (3)$$

In this second model there is no risk of multicollinearity as only n remains as an explaining variable, but not its square. For the regression, the data was categorized into two sub-panels for n < 500 min^{-1} and n > 500 min^{-1}.

RESULTS AND DISCUSSION

All variables not significant on the 5 % level were eliminated. The remaining variables were n, n^2, ρ_{bulk}, T and h. The regression coefficients were standardized for better interpretation and are shown in Fig. 2 (a). All results are plausible. The screw speed shows the biggest effect on the throughput. Also, the square of the screw speed turns out to be the second most influential variable and has a negative effect. This supports the assumption of a hindered pellet flow into the feed zone due to the centrifugal force at high speeds. T also has a negative effect, which shows that a poor flowability of the pellets also contributes to the under-proportional throughput. At the same time, channel height and higher bulk density show positive effects in the same range, which means the influence of T can be counter-acted. While the results do not hold any surprises, they show how the classic throughput calculations comes short when only taking bulk density into account but no other pellet characteristics and process influences.

When using the determined coefficients for the calculation of throughput, the curves depicted in Fig. 2 (b) are obtained. In comparison to the data points from the experiments, a good agreement is only given for PP homo and a flat screw channel as well as for PE-HD and a deep screw channel. The other two curves show big deviations at higher screw speeds

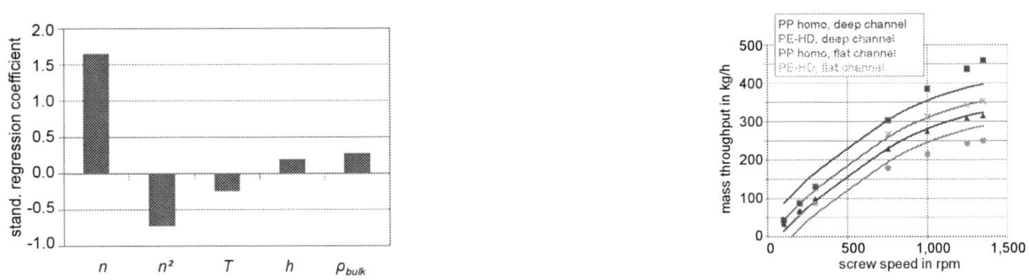

(a) (b)

FIGURE 2. (a) Standardized regression coefficients for the significant variables influencing throughput and (b) regression curves and data points for Model 1

For the second model, one more variable turns out to be significant on the 5 % level: Next to the variables from Model 1, the average pellet weight g_P is also included. This, again, is in line with the considerations about the centrifugal force, as the pellet mass contributes to it (see. eq. (5)). Apart from that, significant variables are the same as in the first model. Fig. 3 (a) shows the standardized regression coefficients for screw speeds below and above 500 min^{-1}. The different values clearly indicate a change of dominant factors for the different speed ranges. This is most pronounced for n and ρ_{bulk}. For low screw speeds, n does hardly influence the specific throughput, showing that the effect of the centrifugal force is low in this case. The bulk density is the most important determinant factor at low screw speeds. For higher screw speeds, however, the positive effect of the bulk density decreases and the negative effect of n increases. This is further supported by the fact that the influence of poor pellet flowability is also more pronounced at higher screw speeds, whereas the positive effects of pellet weight and channel height remain almost unchanged over screw speed.

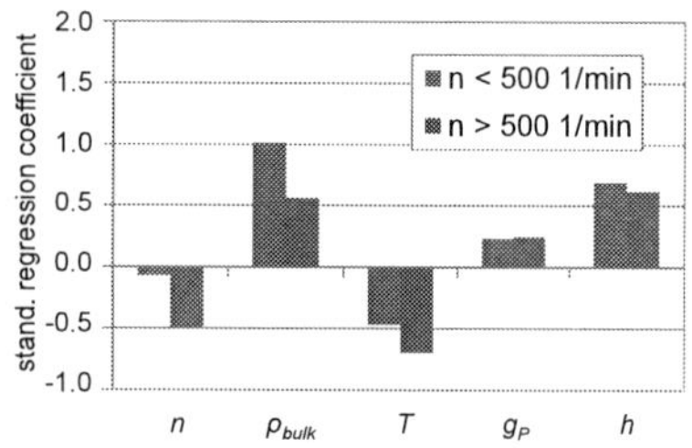

(a)

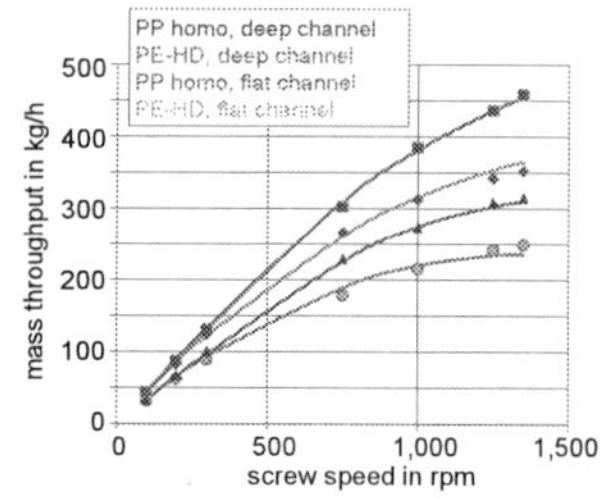

(b)

FIGURE 4. (a) Standardized regression coefficients for the significant variables influencing specific throughput and (b) regression curves and data points for Model 2

When using these results for calculating throughput curves as shown in Fig. 3 (b), the function has to be defined in sections and the intersection point, at which both sets of regression coefficients lead to the same specific throughput, has to be determined. The specific throughput is then multiplied with the screw speed to obtain total throughput. The results show a very good agreement with all data points, which shows that the change of dominant factors from low to high screw speed regime is important and has to be considered for throughput calculations.

Implications

The results show that including additional influences is important for a reliable throughput calculation and thus a successful design of extrusion screws and barrels. The classic theory has to be adapted to take this into account. However, the regression coefficients found in this work cannot simply be used for other materials and other screw and barrel combinations. Instead, the aim of this investigation was to identify which parameters have to be considered in the future and thus lay ground for a modification of eq. (1), which relies on physical mechanisms and measurable pellet characteristics, instead of using heuristic or empiric approaches. One of the main findings is that the results indicate the role of the centrifugal force F at high screw speeds.

$$F = m \cdot \frac{v_c^2}{r} = m \cdot 2 \cdot D \cdot (\pi \cdot n)^2 , \tag{4}$$

In eq. (5) m is the mass of the pellets (thus corresponding to g_P), v_c is the circumferential speed and r and D are the radius and the diameter of the screw, respectively. A possible modification for eq. (1) might then be to model the bulk density as a function of h, F and T, which would allow to include all factors found significant in this work into the classic throughput calculation, thus allowing for it to be applied even at high screw speeds:

$$\dot{m} = v_{ax} \cdot A_f \cdot \rho_{bulk} (h, F, T, ...) , \tag{5}$$

For the formulation of such a function for ρ_{bulk} will require further tests and investigations and broader data base.

CONCLUSION

The findings presents in this paper show that modern high-speed extrusion processes require consideration of more influential factors on throughput in order to reliably design screws and barrels for these applications. Next to bulk density, the flowability of the pellets as well as their average weight were found to have a significant influence on specific throughput. With increasing screw speed there is a change of regiment which parameters have the most important effects. The results strongly indicate that the centrifugal force becomes relevant at higher screw speeds and therefore should be considered for throughput calculations.

Other than expected, the channel height of the screw was the only design parameter found to be significant. However, previous experimental results imply an influence of the groove design, which has yet to be validated. That the groove parameters included in this work could not be show to be significant does not exclude an influence. A broader data base will be needed for further investigation. Hence, more factors should be regarded in the future, just as the pellet size distribution, which has been neglected by now. Fig. 4 shows an example of this for the two materials used in this study, showing clear differences between the two pellet types, which can be expected to further influence bulk density and throughput.

The findings presented here should be regarded as base point for a physically plausible modification of classic throughput calculation. The aim is to build a model taking into account all relevant parameters and allow a reliable calculation of throughput even at high screw speeds.

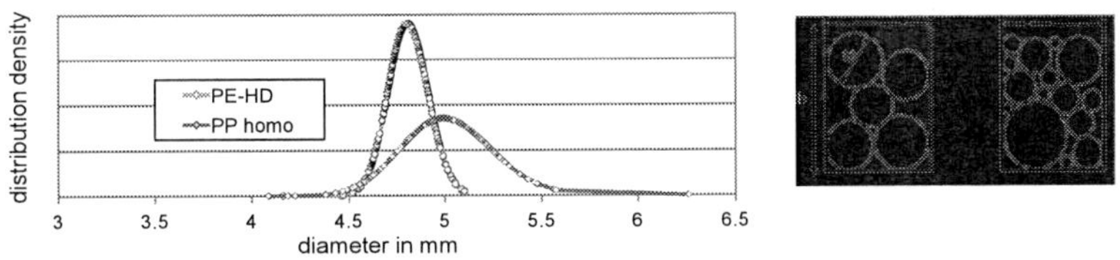

FIGURE 5. Pellet size distribution of PE-HD and PP homo and possible influence on bulk density

REFERENCES

1. T. Pohl, *Entwicklung schnelldrehender Einschneckensysteme für die Kunststoffverarbeitung auf Basis theoretischer Grundlagenuntersuchungen* (Shaker, Aachen, 2004).
2. J. Wortberg. and R. Michels, „Innovative Entwicklungen in der Einschneckenextrusion – Möglichkeiten und Grenzen in der Hochleistungsextrusion", in: *Der Einschneckenextruder : Grundlagen und Systemoptimierung*, (VDI, Düsseldorf, 1997).
3. W. Predöhl, *Technologie extrudierter Kunststofffolien* (VDI, Düsseldorf, 1979).
4. J. Wortberg and G. Karrenberg, „Neuentwicklungen in der Extrudertechnologie", in: *Der Einschneckenextruder,* Conference Proceedings (VDI, Düsseldorf, 2012).
5. M. Roth, „Schnelllaufende Einschneckenextruder : Entwicklungsstand", in: *VDI-Jahrestagung Extrusionstechnik,* Conference Proceedings (VDI, Bonn, 2010).
6. H. Hörmann, „Theoretische und experimentelle Betrachtung schnelllaufender Einschneckenextruder". Ph.D. thesis, University of Paderborn, 2014.
7. P. Gorczyca, „Analyse und Optimierung von Einschneckenextrudern mit schnelldrehenden Schnecken". Ph.D. thesis, University of Duisburg-Essen, 2011.
8. E. Grünschloß, „Helibar : Ein universelles Plastifiziersystem für Einschnecken-Extruder und Spritzgießmaschinen", in: *18. Stuttgarter Kunststoffkolloquium,* Conference Proceedings (Stuttgart, 2003).
9. C. Bonten, *Kunststofftechnik : Einführung und Grundlagen.* 2nd ed., (Hanser, München, 2016).
10. O. Kast and C. Bonten, Zeitschrift Kunststofftechnik/Journal of Plastics Technology, **11**(3), 180–203 (2015)
11. O. Kast and C. Bonten, "Influence of a Grooved Melting Zone on High-speed Single Screw Extrusion", in: *PPS Europe/Africa Regional Conference*, AIP Conference Proceedings (Graz, 2015).
12. H. Rahal, „Alternative Methoden zur Feststoffförderung und Plastifizierung in der Extrusionstechnik". Ph.D. thesis, University of Duisburg-Essen, 2008.
13. R. Michels, „Verbesserung der Verarbeitungsbandbereite und der Leistungsfähigkeit von Einschneckenextrudern". Ph.D. thesis, RWTH University, Aachen, 2005.
14. E. Grünschloß, Kunststoffe **83**(7), 309–311 (1993)
15. V. Schöppner, *Verfahrenstechnische Auslegung von Extrusionsanlagen* (VDI, Düsseldorf, 2001).
16. J. H. Lessmann, „Berechnung und Simulation von Feststoffförderprozessen in Einschneckenextrudern bis in den Hochgeschwindigkeitsbereich". Ph.D., University of Paderborn, 2016.
17. H. Potente et al., Kunststoffe **9**, 62–66 (2001).
18. O. Kast and C. Bonten, „Hochleistungs-Einschneckenextruder mit wendelgenuteter Plastifizierzone", in: *VDI-Jahrestagung Extrusionstechnik*, Conference Proceedings (Köln, 2016).
19. E. Grünschloss, „Feststoff-Förderung im Einschneckenextruder", in: *Der Einschneckenextruder – Grundlagen und Systemoptimierung* (VDI, Düsseldorf, 1997).

Integration of Electrochromic Devices in Plastic Parts by Injection Moulding

Sauer Viola [1], Schnieders Joachim [2], Heim Hans-Peter [1]*

[1] Universität Kassel - Hessen - Germany, [2] 3 Pi Consulting & Management GmbH - North Rhine-Westphalia - Germany

Aim of the BMBF funded research project "ProTeck" was the manufacture of electrochromic (EC) panes for automotive applications by back-injection moulding multilayer EC systems. A plastic substrate (consisting of polycarbonate, PC) is used as the basis of these systems, ITO as the conducting layer, TiO2 as the counter electrode, PEDOT:PSS as the electrochromic layer and a PU electrolyte to join both halves. To integrate the pre-manufactured electrochromic systems in three-dimensional plastic-based parts, e.g. for automotive applications, they are back-injection moulded with amorphous plastics like PC or PMMA. Eventhough the EC pane is cooled by the cold mould wall on the one side, on the other side the hot melt (T>200°C) inserts a big amount of thermal energy into the pane. Additionally the melt induces mechanical forces to the device, particularly shearing. Both influences, together with special properties of the EC devices and their single layers, can lead to different, partly interacting effects and the destruction of the device. Effects can be e.g. worse thermal conductivity due to the metal layers, vaporising of volatile components of the electrolyte and delamination of the device. For the experiments a specialised mould was designed, to achieve a homogenous temperature and pressure profile and a reduced shear loading of the EC system. The mould allowed the production of an application-oriented demonstrator, the size of a small side window, without over-moulding the contacts. The mentioned problems could largely be solved by adjustments of the mould, the process (e.g. mould and melt temperatures, injection pressure) or the EC system (e.g. electrolyte formulation for less humidity and gas content, better adhesion). The presentation will give summary of the experiments, mentioned phenomena and appropriate measures.

Smart Data Analysis for Optimized Manufacturing of Powder Coatings on Co-rotating Twin Screw Extruders

Sophie Pachner[1, a)] and Jürgen Miethlinger[1, b)]

[1]*Johannes Kepler University Linz, Institute of Polymer Extrusion and Compounding, Altenberger Straße 69, 4040 Linz, Austria*

[a)]*Corresponding author: sophie.pachner@jku.at*
[b)]*juergen.miethlinger@jku.at*

Abstract. This paper addresses smart data mining techniques for analyzing the processing of selected powder coatings on co-rotating twin screw extruders. The influences of different machinery and process parameters on optical quality criteria such as degree of gloss and color are analyzed. Experiments were performed on laboratory and production scale co-rotating twin screw extruders using different screw configurations. Decision trees are developed in order to analyze the process and to determine the essential influencing factors on the formation of the degree of gloss. For further analysis of the process and to enable optimal up-scale from laboratory to production scale analytic equations are evolved by using symbolic regression based on heuristic optimization algorithms.

INTRODUCTION

Powder coatings, in general consisting of binders (resins, hardeners and accelerators), pigments, fillers and additives are melted or chemically cross-linked by heat to form a close, well adherent coating. Figure 1 shows the main processing steps, this paper describing the extrusion step for homogenization of the single components in more detail. [1]

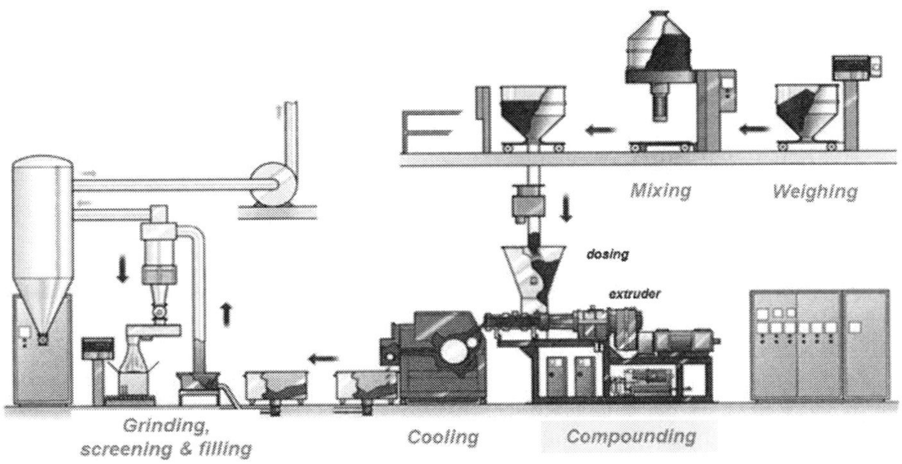

FIGURE 1: Industrial production of powder coatings, basic processing steps; [1]

In the production of powder coatings, a variety of quality criteria play an important role. In addition to mechanical properties such as abrasion, scratch and chemical resistance optical properties such as degree of gloss, color and surface roughness are among the most important quality criteria. [1] All of these quality characteristics are affected by countless influencing variables such as type of compounder, screw configuration, reactivity of the resin as well as process parameters. Processing is sensitive due to the high reactivity of the selected polymer system and therefore optimal process design as well as up-scale becomes more difficult, [2].

EXPERIMENTAL

The influence of different machinery and process parameters on the degree of gloss has been investigated. For this purpose, tests were carried out on co-rotating twin screw extruders of different types and sizes. Experiments were performed on laboratory scale as well as on production scale compounders, see Table 1.

TABLE 1: Characteristic dimensions of used co-rotating twin screw extruders:

Scale			Compounder 1 Laboratory	Compounder 2 Laboratory	Compounder 3 Production
Outer diameter	OD	mm	18	20	34
Ratio	OD/ID	-	1,55	1,6	1,8
Center distance	a	mm	15	20	26,2
Axial length	L	D	31,9	24	28

In addition also the influence of the screw geometry has been investigated by varying type, position and amount of kneading and mixing elements. Figure 2 shows the three tested screw geometries called 'soft', 'medium' and 'strong'.

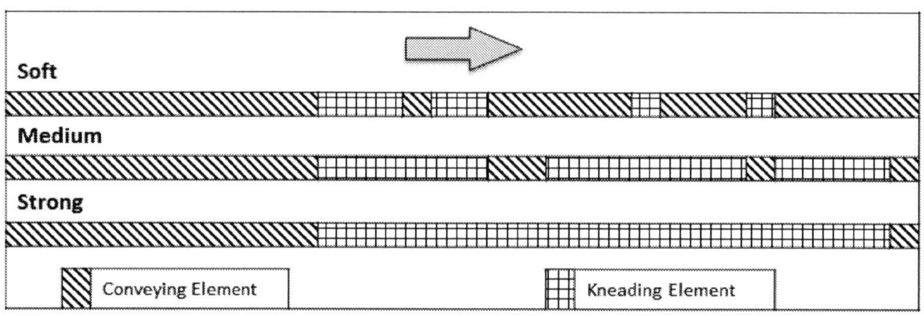

FIGURE 2: tested screw configuration; soft, medium and strong

Next to the determination of relevant process data during compounding such as melt temperature, torque and average residence time also optical properties of the coated sheets were determined. According to Howell [2] the degree of gloss was measured using a glossmeter, final color was determined using a spectrophotometer and the surface roughness was determined by laser confocal microscopy.

MODELING

Due to the high reactivity of the selected polymer system (polyester resin containing carboxyl groups) modeling of the process is very complex and therefore optimal process design as well as up-scale becomes more difficult. Figure 3 shows the essential steps of modeling starting from the parametric study up to the implementation of the developed models, [3, 4].

In addition to common analysis of the process we performed numerical calculations in order to describe the residence time characteristics and to calculate the degree of fill. For describing these complex procedures during processing we used smart data analysis and heuristic modeling whereby specific up-scale from laboratory to production scale becomes possible.

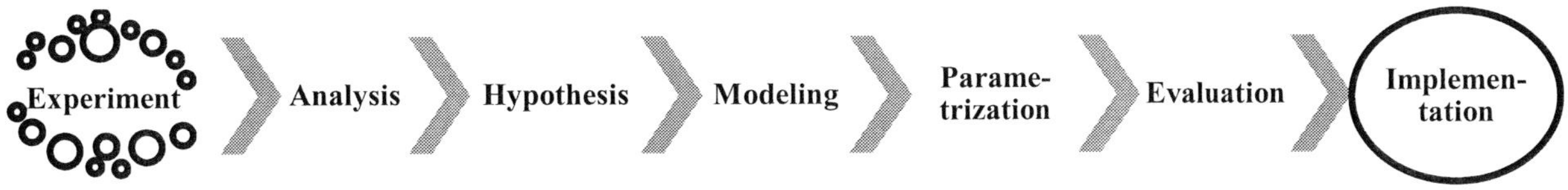

FIGURE 3: Smart data analysis for advanced production, [3, 5]

Symbolic Regression

Symbolic regression is the induction of mathematical expressions on data. Among modern data-based modeling techniques, symbolic regression [6] distinguishes itself by its ability to identify nonlinear white-box models without the need to specify the structure of the models This makes a big difference to other methods of regressions where a specific model is assumed and only the complexity of this model can be varied. So the main goal of this method is to determine the relationship of a target variable to a set of specified independent input variables [7] in terms of a mathematical expression.

Decision Tree

Classification is the assignment of data objects with unknown class membership to a best suitable class. Criteria for classification are e.g. the minimization of the classification error or the maximization of the degree of affiliation, [8]. Decision tree learners are commonly used in many practical applications of machine learning because they are fast and produce intelligible output that is quite accurate, [9]. The classifier is represented as a decision tree consisting of a set of nodes and edges where the model is constructed using training data with a known class membership, [8].

RESULTS AND DISCUSSION

In a first step, the influence of the different screw geometries on the degree of gloss was developed. Figure 4 shows the degree of gloss as a function of screw speed, throughput and the screw configuration for compounder 3 – production scale - and Compounder 1 – laboratory scale.

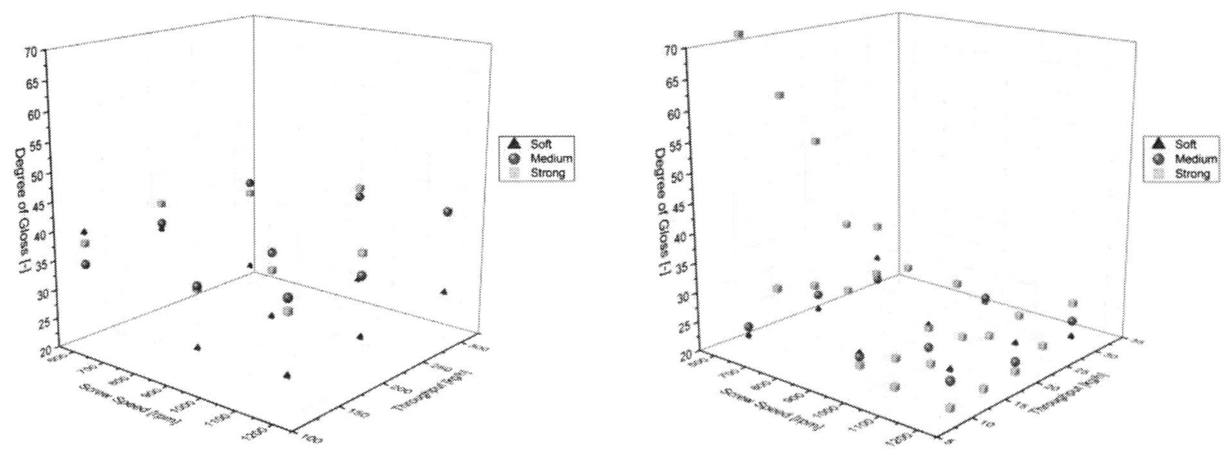

(a) (b)

FIGURE 4: Degree of gloss as a function of screw speed and throughput for different screw configurations soft, medium and strong; (a) Compounder 3 – production scale; (b) Compounder 1 – laboratory scale

These figures show that there are complex relationships between optical quality criterion degree of gloss, operation conditions and extruder set up which makes modeling of the operating behavior and also the up-scale hardly possible.

A decision tree was created in order to analyze the process and to determine the essential influencing factors on the development of the degree of gloss. Figure 5 shows the created decision tree for the lab scale compounder 1 operating with screw configuration strong. This tree shows the influence of melt temperature T_{Melt}, average residence time t_{res}, and dimensionless throughput π_V, calculated according to equation (2), on the surface quality criterion gloss.

$$\pi_V = \frac{\dot{V}}{N \cdot OD^3} \tag{1}$$

The dimensionless throughput π_V is defined in above mentioned equation (1) where $\dot{V}$ is the volume flow rate, N the screw speed and OD the outer diameter of the screw.

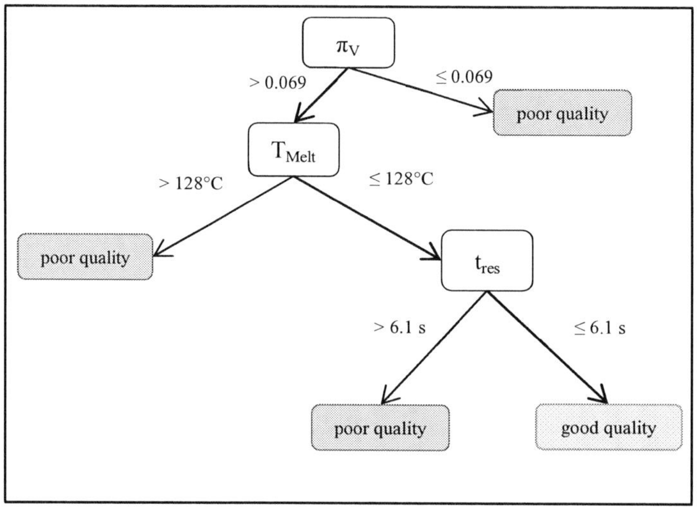

FIGURE 5: Decision tree for compounder 1 laboratory scale operating with screw configuration 'strong'; degree of gloss as function of dimensionless volume throughput π_V, melt temperature T_{Melt} and average residence time t_{res}

The evolved decision tree, illustrated in Figure 5, shows that the dimensionless throughput π_V is the most important influence factor for the quality criterion degree of gloss followed by the melt temperature T_{Melt} and the average residence time t_{res}. The developed model was validated by applying cross validation with 10 folds and additional test data from production where an overall classification performance of 88% could be determined.

Next to the development of a decision tree we developed analytic equations through the use of symbolic regression for the calculation of the degree of gloss as a function of processing conditions and geometry parameters which allow an optimal process design. The developed models thus also enable specific up-scale from laboratory to production scale. Equation (2) represents the scale-up model for the degree of gloss G:

$$G = \frac{k_0 \cdot OD \cdot \frac{OD}{ID} \cdot \left(k_1 \cdot \frac{OD}{ID} + k_2 \cdot N + a \cdot (k_3 \cdot \dot{m} + k_4 \cdot a)\right)}{k_5 \cdot a + k_6 \cdot N + k_7 \cdot \dot{m} + k_8 \cdot OD} + \cdots + k_{28} \tag{2}$$

where G is the degree of gloss, OD the outer diameter of the screw, OD/ID the ratio of outer and inner diameter, a the center distance, N the screw speed, $\dot{m}$ the throughput and k_0 to k_{28} are constants.

Figure 6 (a) shows the correlation of the experimentally determined values and the gloss values according to equation (2) for training and test data. Training data were used for learning and developing the model and test data were additional data from production just used for validation. The results of the calculated data according to the heuristic model are in good agreement with the experimentally determined training and test data. Coefficients of determination R^2 higher than 0,96 and mean relative errors MRE of less than 4,5% can be achieved, see Figure 6 (b).

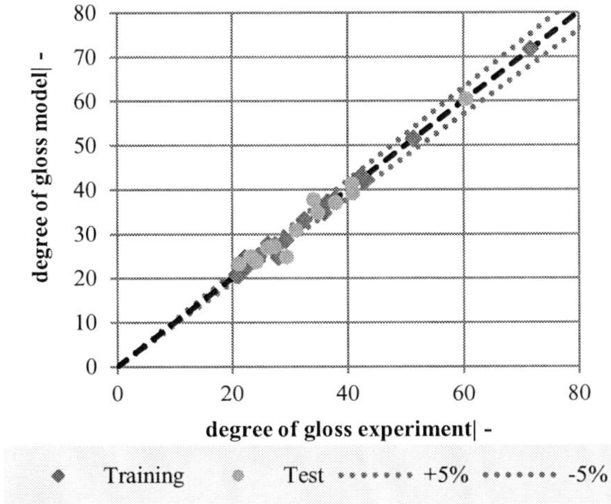

(a)

	Training	Test
R^2	0.991	0.967
MRE	3.0%	4.3%

R^2... Coefficient of determination
MRE... mean relative error

(b)

FIGURE 6: (a) Degree of gloss experimentally determined vs. degree of gloss determined according to the developed model; (b) results of error analysis with coefficient of determination and mean relative error

CONCLOUSION

We determined the influence of various processing parameters of co-rotating twin screw extruders on the quality parameter degree of gloss for the selected powder coating material. Experiments were performed using several types of co-rotating twin screw extruders as well as different screw geometries. Process data are analyzed by classification of the data due to the creation of decision trees. These decision trees show the importance of different influencing factors on optical quality parameters. Further analytic equations are determined for estimating the degree of gloss in dependency of the most important processing parameters by using heuristic optimization algorithms. These equations enable the optimal design of the process as well as specific up-scale from laboratory to production scale and can be used for close loop control of the process.

ACKNOWLEDGMENTS

Financial support by the Linz Center of Mechatronics GmbH (LCM) is gratefully acknowledged. LCM is funded by the Austrian Government and the Provincial Government of Upper Austria within the COMET program.

REFERENCES

1. J. Pietschmann, *Industrielle Pulverbeschichtung: Grundlagen, Verfahren, Praxiseinsatz,* 4th ed. Wiesbaden: Springer Vieweg, 2013.
2. D. M. Howell, *Powder coatings*. Chichester, New York: Wiley, op. 2000.
3. J. Miethlinger, "Zukunftspotentiale für Extrusion und Compounding Smarte Produktion, Selbstoptimierung, Multi-Technologie-Material-Mix," 2. Linzer Polymer Extrusion und Compounding Tagung, Dec. 2016.
4. S. Pachner and A. Lepschi, "Smart Data Mining und Prozessanalytik," 2. Linzer Polymer Extrusion und Compounding Tagung, Dec. 2016.
5. J. Miethlinger, W. Roland, S. Pachner, C. Marschik, and H.-J. Luger, "Investigations with new specific grooved barrel design, starved feeding and heuristic modeling for high capable single screw extruders," PPS Europe Africa Conference 2017, Dresden, Jun. 2017.
6. J. R. Koza, *Genetic programming,* 3rd ed. Cambridge, Mass. [u.a.]: MIT Press [u.a.], 1993.
7. M. Affenzeller, S. Wagner, S. Winkler, and A. Beham, *Genetic Algorithms and Genetic Programming: Modern Concepts and Practical Applications*: CRC Press, 2009.
8. H.-J. Lenz and R. M. Müller, *Business-Intelligence,* 1st ed. Berlin: Springer Berlin, 2013.
9. I. H. Witten, E. Frank, M. A. Hall, and C. J. Pal, *Data mining: Practical machine learning tools and techniques*. Amsterdam: Elsevier, op. 2017.

Investigations of starved-fed grooved barrel single screw extruders in industry-standard output ranges

MIETHLINGER JUERGEN (1)*, Marschik Christian (2)

(1) JKU - Johannes Kepler University - Linz - Austria, (2) Johannes Kepler University - Linz - Austria - Upper Austria - Austria

Motivation In industry, single screw extruders are frequently operated via flood feeding, that is, the material flows down from full material hopper by gravity providing a continuous material column over the screw channel and the extruder feeds in by gravitation as much as possible. Therefore, the output of the extruder is determined by the extruder screw speed. In case of flood feeding, the degree of filling is usually one, except for special cases of scrap or high-speed extrusion. Whereas co-rotating twin screw extruders are operated via starve feeding, where the material is metered into the extruder by a feeding device and the extruder screw is not fully filled with material. The output of the extruder is independent of the screw speed and determined by the feeder. Starve feeding for single screw extruders has not really been industrialized, although it has been successfully shown that it provides a wider processing window like improved mixing, reduced motor load or minimized wall-thickness variation and the investment for one starve feeding device is quite low. However, most of the investigations were carried out at a low output range. In addition, most of the studies was performed only for smoothed bore single-screw extruders. Therefore, the present work is focusing on high-output investigations of grooved barrel single screw extruders. Results The experimental studies confirms good extrusions stability also at industry-standard output ranges for grooved barrel single screw extruders. A reduction of axial pressure profile and motor load as known from experiments with smoothed bore single screw extruders could be also obtained for grooved barrel single screw extruders. Furthermore processing parameters like melt temperature, radial melt temperature profile, heat flow in the barrel or specific drive power consumption was determined as a function of the degree of filling.

Pelletize condition dependence of injection molded article of recycle plastic.

Tominaga Aya (1), Takenaka Nozomi (1), Yamasaki Natsumi (1), Nakano Ryoko (1), Sekiguchi Hiroshi (1), Michiue Tetsunari (2), Sugao Shunsuke (3), Yao Shigeru (1)*

(1) Fukuoka University - Fukuoka - Japan, (2) Eco-feel co., ltd - Hiroshima - Japan, (3) Eco-feel co.,ltd. - Hiroshima - Japan

Recently, we had cleared that the main cause of the worse physical properties of recycled resin is physical degradation, not chemical degradation. Moreover, we had found that by choosing a suitable compression condition we can improve the physical properties almost to that of virgin resin. However, in order to further increase the consumption of recycled resin, it is necessary to extend applicable molding methods. Among them, injection molding is particularly important. In this study, we pelletized the waste container and packaging plastics at various condition. And we measured mechanical properties of injection molded test piece by using those pellet. By this investigation, we found that the mechanical properties much depend on the pelletized condition and we can improve them by choosing the pelletized condition. This results suggest that if we had made a suitable pellet by choosing the pelletizer condition, we can apply various molding method to recycled plastics.

Mechanical Characterization of Frontal and Flowing Weld Lines in Injection-molded Short Fiber-reinforced Thermoplastics

M. B. Baradi[1, 2, a)], C. Cruz[2, b)], G. Régnier[1, c)]

[1]*PIMM, ENSAM, CNRS, CNAM, 151 bd de l'Hôpital, 75013 Paris, France.*
[2]*Robert Bosch GmbH, Corporate Sector Research and Advance Engineering, Plastics Engineering, 71272 Renningen, Germany.*

[a)]Corresponding author: mohamedbesher.baradi@de.bosch.com
[b)]camilo.cruz@de.bosch.com
[c)]gilles.regnier@ensam.eu

Abstract. Occurrence of weld lines (WLs) in injection-molded thermoplastic parts is often unavoidable and induces a significant reduction in the mechanical properties. In this work we present a mechanical characterization of the WLs induced in injection-molded PBT (Polybutylene terephthalate) wt. 30 % glass fiber-reinforced which constitutes the experimental reference for validating the respective prediction models. We considered frontal and flowing WLs by using different insert shapes, two flow types (parallel and divergent) and two thicknesses. We found that the influence of the process parameters and the insert shape (in the case of flowing WLs) on the quasi-static tensile properties of the WL is relatively low. Also we found that frontal WLs induce a more pronounced reduction of the tensile strength in comparison with the flowing WLs. Based on rheological tests performed on the unreinforced PBT matrix, we determined that the healing ability of the matrix at the WL is high. On the other hand, we observed that the unfavorable induced fiber orientation (FO) along the flowing WL does not vanish with the distance from the insert. The last results suggest that the FO is the dominant factor of the WL strength reduction, while matrix inter-diffusion does not play a significant role for this particular material system.

INTRODUCTION

Nowadays, one of the most commonly used technologies in the production of complex plastic parts in high production rates is injection molding. Warpage, voids or weld lines are typical defects of the manufactured articles. This work focuses on the influence of weld lines (WLs) in short fiber-reinforced thermoplastics (SFRT). WLs form during the filling phase at the zone where two melt fronts meet. Multiple injection gates, inserts inside the mold cavity or pins will lead to separation of the melt and WL formation. They are one of the most influential defects in injection molding due to their unavoidability and highly negative effects, which induce a diminution of strength that reaches up to 50%. We distinguish between two kinds of WLs; frontal WLs that form when two melt fronts meet in opposite directions and solidify immediately afterwards, and flowing WLs as the two melt fronts meet and flow through the mold until solidification. WL weakness is attributed to four main factors [1-5]:

1- ***Incomplete molecular diffusion (insufficient matrix healing):*** After intimate contact and wetting of the melt fronts, a diffusion process starts where the molecules reptate through the interface eliminating it. Though, due to the rapid cooling rates in injection molding the molecules might have less time than required to completely heal the WL.

2- ***Fiber orientation (FO):*** Fibers align parallel to the WL surface eliminating their role in strengthening the part in case of loading perpendicularly to the WL surface.

3- ***Molecular orientation:*** Molecules also orient parallel to the WL surface weakening the part. Nevertheless, Kim et al. [3] pointed out that molecular orientation plays a negligible role on strength reduction in SFRT.

4- ***V-notches:*** They act as stress amplifiers and form due to air entrapment or difference in shrinkage caused by the different molecular orientation. In this work we limited the influence of V-Notches by using ventilation channels and by applying high packing pressure.

Malguarnera et al. [6] researched the effect of different process parameters on the strength of frontal WLs using PP. A positive influence of increasing the mold and melt temperatures was evident. Frontal WLs of glass fiber-reinforced ABS had no influence on the tensile modulus but the strength reduction due to the WL reached about 50%. Moreover, increasing fiber content significantly amplified the strength reduction ratio [7]. The FO at a flowing WL was investigated using X-ray micro-tomography. Results showed that orientation behavior differed depending on the distance from the insert where a noticeable amount of fibers tended to orient in thickness direction just after the insert [8]. Kim and Suh [5] modeled the influence of poor healing and molecular orientation of a frontal WL with an amorphous polymer and the model was able to predict the strength of the WLs. A simpler model for reinforced materials gave the strength of the WL as the strength of the matrix reduced by the area taken by the fibers. It neglects the interfacial strength between fibers and matrix and assumes that the fibers are transversally oriented to the WL [9]. Based on the reptation theory of De Gennes, the healing quality of WLs was modelled by relating the polymer diffusion with the numerically computed thermomechanical history at the interface. The model showed good correlation with mechanical tests for unreinforced ABS. However, for reinforced PBT prediction was inaccurate, which was attributed to the inaccurate FO prediction [10].

Current state of the art indicates that both matrix healing and FO play a role on the quality of WLs in SFRT. However, a quantification of the role played by each mentioned weakness factor is still not clear especially in semi-crystalline materials. In this work, we present an experimental characterization, to be used as a reference for validating a new approach to quantify the contribution of the main weakness factors of WLs.

EXPERIMENTAL

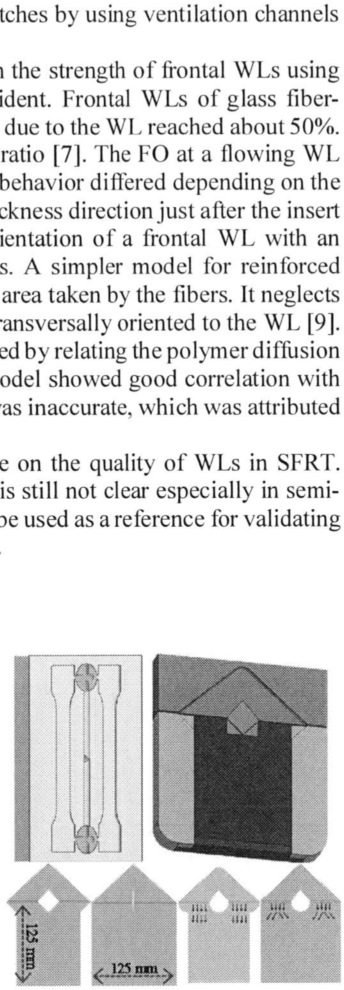

Figure 1. Frontal and flowing WL molds and plates

We investigated two materials: PBT reinforced with 30% wt. glass fiber (PBT-GF) and unreinforced PBT (PBT-Neat) of the same semi-crystalline matrix. The injection molding machine was a DK® CODIM 175/410. Ventilation channels were introduced to the frontal WL mold at the WL zone (10 mm width and 0.1 mm depth) to eliminate V-notches and voids. The flowing WL mold was designed with the assistance of rheological simulations using Moldflow® 2016. Results showed that the bigger the insert the bigger influence on the FO at the WL. We varied the thickness (1.5-3 mm), the insert shapes (prism, slit and cylinder) and the plate dimensions (125 x 125 and 125 x 95 mm) as observed in Fig. 1. The flowing WL tensile test specimens were machined by water cut at several distances from the insert tip (3, 14, 25, 47 and 69 mm). Quasi-static tensile testes were done using an Instron 5966 machine with an optical extensometer length of 10 mm and clamp speed of 10 mm/min. Rheological measurements were made using an Anton Paar 502 rheometer and an ARES-rheometer (TA-Instruments) in plate-plate configuration. SEM images to study the FO were made using a HITACHI 4800 II (15 KV voltage, 10 µA intensity and back scattered electron rate of 100%). Samples were cut with a diamond disk, then polished and coated with gold. Images were taken with magnification of x150 and stitched together using the open source software ImageJ. We studied four process parameters: packing pressure (10-60 MPa), injection time (0.8-3 s), mold temperature (PBT-GF: 60-100°C PBT-Neat: 40-80°C) and melt temperature (250-275°C).

RESULTS AND DISCUSSION

Mechanical Testing

Frontal WLs with PBT-GF showed a reduction of ~55% on strength and 30-40% on failure strain. Flowing WLs exhibited lower reduction in the strength (10-20%) and in failure strain (30-40%) Fig. 2. Contrarily, frontal and flowing WLs did not affect the nominal strength of PBT Fig. 2. In fact, failure did not take place at the WL region unlike in PBT-GF. Similar findings for PBT were reported by [3] evidencing the high healing ability of the PBT matrix and in turn supporting the assumption that the FO plays the dominant role in strength reduction with PBT-GF. Figure 3 presents the variations of frontal WLs strength with the packing pressure for PBT-GF. A slight increase in WL strength was evident by increasing packing pressure and injection speed for the 3 mm case. The variations of melt and mold temperatures did not affect the WL strength. The possible variations are in fact masked by the experimental noise. Similar behavior was followed by the flowing WLs. Average of WL strength over all positions after the insert with temperature variations are presented in Fig. 3. The packing pressure, injection time and melt/mold temperatures have negligible influence on flowing WL strength.

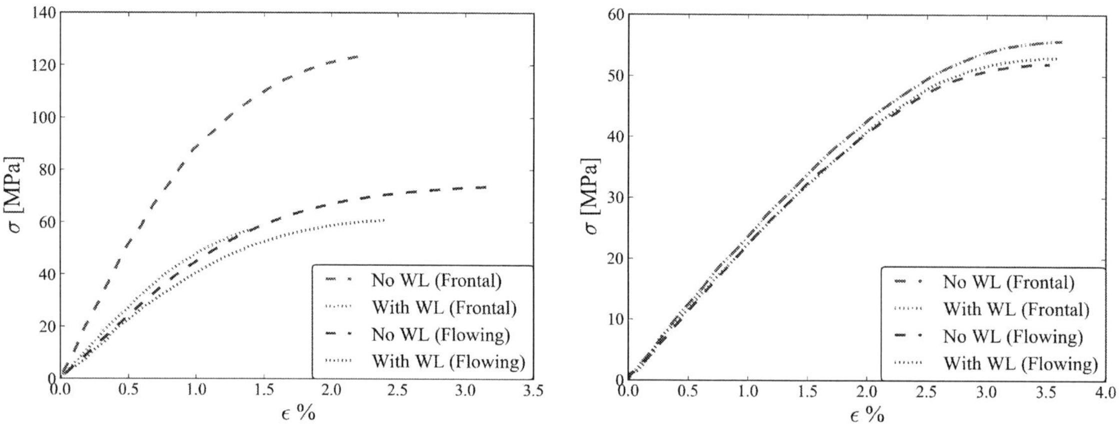

(a) (b)

FIGURE 2. Mechanical behavior of WLs (thickness of 3 mm) with PBT-GF (a) and Neat-PBT (b).

In fact, the biggest impact on strength comes from the section thickness, where the thinner frontal WL is the strongest and, on the contrary, the thickest flowing WL exhibit the highest strength, which points to an influence of FO. Another sign of FO role comes from the changes in mechanical response along the flowing WL, as shown in Fig. 4. Direct after prism and cylindrical inserts, the strength first decreases and then increases, which can be attributed to changes of the FO. After the slit insert the strength increases steadily with flowing distance. In all insert cases, the strain at failure recovers continually with the flowing distance. This probably indicates an interaction between the FO and matrix healing at the flowing WL and is subject of future quantification. Changing the insert shape had an influence on the mechanical properties immediately after the insert tip, but slight difference far from the insert tip.

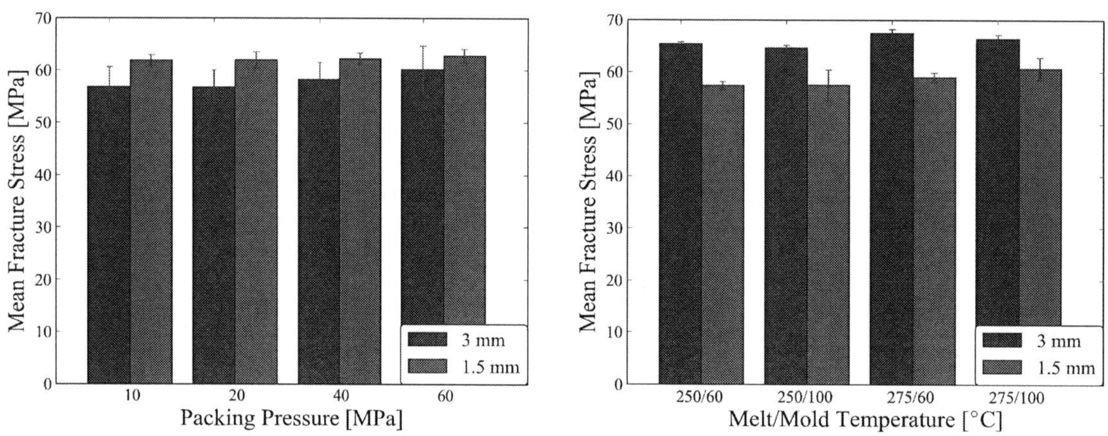

(a) (b)

FIGURE 3. Strength of frontal WLs (PBT-GF) with different packing pressure (a) and strength of frontal WLs (PBT-GF) with different mold/melt temperatures (b).

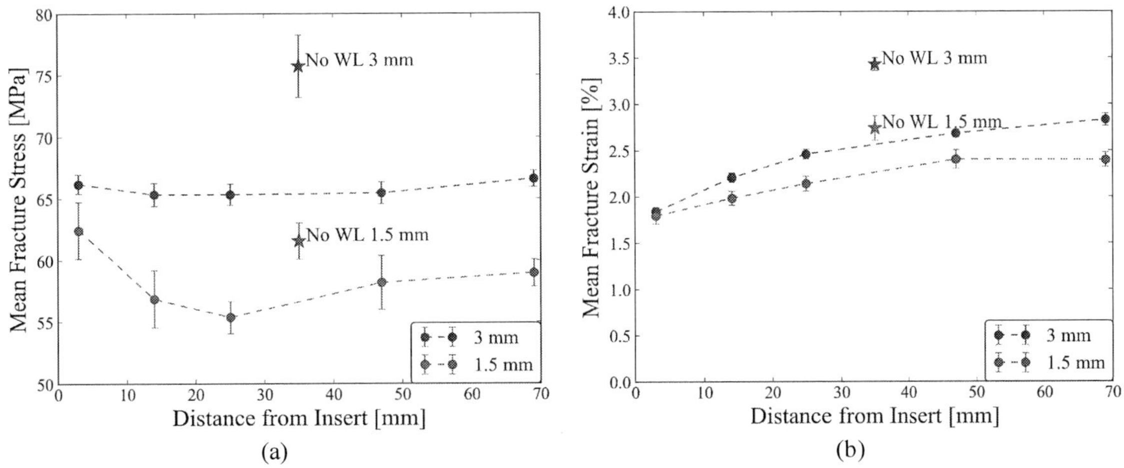

FIGURE 4. Changes in strength (a) and failure strain (b) along flowing WLs (PBT-GF, cylindrical insert).

Rheological Measurements

In order to verify the healing ability of the PBT matrix, we estimated the relaxation time t_{relax} of the molecular chains. At the WL interface, the relaxation time corresponds to the time required for a chain to reptate across the interface. As expected, t_{relax} decreases with increasing temperature. At the used injection temperatures, t_{relax} ranged between 0.001 s and 0.003 s, which is 1000 time faster than the injection time (Fig. 5). This allows to understand that the molecules have enough time to reptate across the WL interface explaining the absence of failure at the hypothetical position of the WL and proving the high healing ability for PBT used in this study.

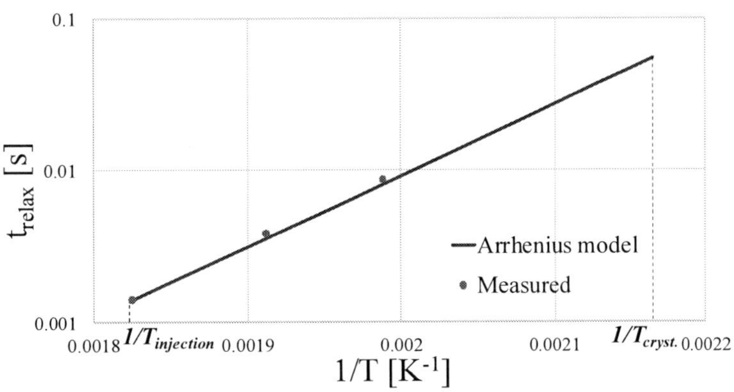

Figure 5. Rheological measurement and estimation of t_{relax} for Neat-PBT where t_{relax} is nearly equal to 0.001 s at injection temperatures

Analysis of Fiber Orientation (SEM)

In order to better understand the role of FO on both kinds of WLs a qualitative analysis was made using SEM. The SEM images were taken in both parallel and transversal directions to the WL. Results for frontal WLs show clearly the effect of fountain flow on FO. Two regions can mainly be defined, a region exactly at the WL with fibers orientated in thickness direction and a neighboring region where fibers orient transversally to the flow (Fig. 6). Flowing WLs were studied at the same positions of the mechanical tests. Results showed that at the WL region the core layer, which is usually oriented transversally to the flow, disappears along the WL and the fibers orient in the flow direction over the thickness. The distortion region in terms of FO at the WL extends up to about 2-4 mm from the WL plane depending on the thickness of the part. The characteristic FO at the WL is conserved along the whole

WL independently the distance from the insert (Fig 7). The FO exactly after the cylindrical insert FO is comparable to the one seen in frontal WLs.

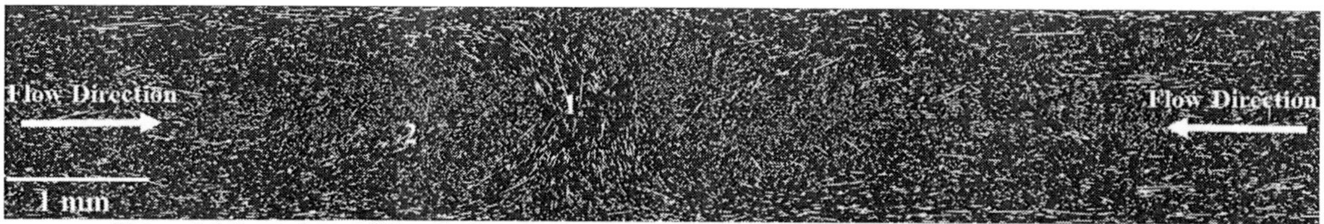

Figure 6. 1.5 mm frontal WL with fibers oriented in thickness direction (region 1) and transversal to flow (2)

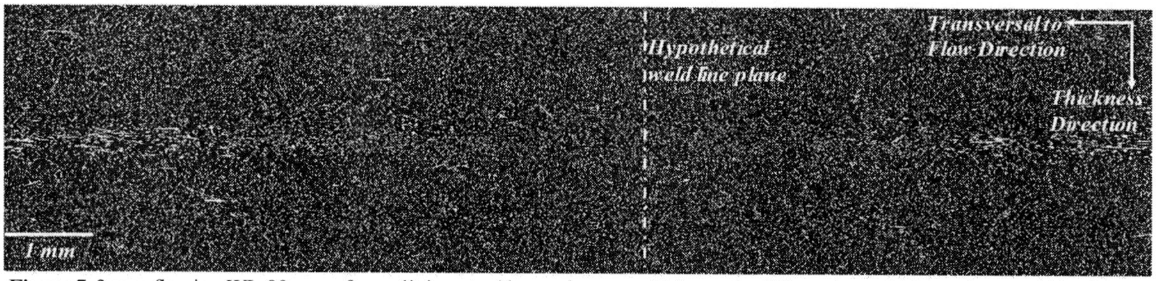

Figure 7. 3 mm flowing WL 90 mm after a slit insert with core layer vanishing at the WL region, which reaches a width of 6 mm

CONCLUSIONS

In this paper we have presented a characterization of frontal and flowing WLs using PBT-GF and PBT-Neat. Based on the relaxation time measured from rheological tests, we found that the PBT matrix shows high healing ability and this explains the absence of strength reduction in the case of WLs made of Neat-PBT. With PBT-GF, WLs induced a clear reduction of strength with frontal weld lines being more critical. The WL strength did not vary with the injection process parameters. In fact, the main factor affecting the WL strength is the part thickness. SEM analysis showed that the induced FO along the flowing weld line does not vanish and the influence zone reaches up to 8 mm width. The last results leads to assume that is FO the dominant factor of WL strength reduction. Future work will include a quantitative analysis of the FO to support a mean-field homogenization structural simulation at the WL in order to define the exact contribution of WL weakness factors.

ACKNOWLEDGMENTS

This work was financed by Robert Bosch GmbH. The experimental tests were done at the PIMM laboratory at ENSAM in Paris, France.

REFERENCES

1. G. Régnier and S. Le Corre, "Modeling of Thermoplastic Welding", in *Heat Transfer in Polymer Composite Materials,* Edited by N. Boyard (John Wiley & Sons, Inc., Hoboken, NJ, USA., 2016), pp. 235 – 268.
2. A. Vaxman, M. Narkis, A. Siegmann, and S. Kenig, Polym. Composites ***12***, 161-168 (1991).
3. J. K. Kim, J. H. Song, S. T. Chung and T. H. Kwon, Polym. Eng. and Sci. ***37***, 228-241 (1997).
4. J. Onken and C. Hopmann., Gummi, Fasern, Kunststoffe ***69,*** 574-580 (2016).
5. S. Kim and N. P. Suh., Polym. Eng. & Sci. ***26***, 1200-1207 (1986).
6. S. C. Malguarnera, A. I. Manisali, and D. C. Riggs., Polym. Eng. & Sci. ***21***, 1149-1155 (1981).
7. Hashemi S., Express Polymer Letters ***1***, pp. 688-697 (2007).
8. A. Ayadi, H. Nouri, S. Guessasma and F. Roger, 20th International Conference on Composite Materials (2015).
9. B. Fisa and M. Rahmani, Polym. Eng. & Sci. ***31***, 1330-1336 (1991).
10. Cruz, C. A., "Integrative simulation for assessing the mechanical performance of a weld line on injection moulded thermoplastic parts" in *WCCM* 2014, *ECCM* 2014 and, *ECFD* 2014, pp. 2334-2345

Visualization Analysis on Reciprocating Plastication Process of Glass Fiber Reinforced Resin by Glass-inserted Heating Cylinder

MA Sai [(1)]*, YOKOI Hidetoshi [(1)]

[(1)] The University of Tokyo - TOKYO - Japan

In the injection molding of long fiber reinforced resins, the greatest challenge to realize high strength molded products is minimizing the breakage of fibers during the plastication process inside the heating cylinder and during flow inside the mold. In the previous report 1), a melting model of 50wt% long glass fiber reinforced PP (L-GFPP) was proposed based on the visualization analysis of the continuous plastication process. Injection molding reciprocating plasticization is an unsteady process in which the relative positions of the screw and cylinder intermittently and periodically fluctuate corresponding to the molding cycle. In this study, a full flight screw and 50wt% L-GFPP were used to investigate the correlation between the reciprocating plasticizing process of L-GFPP and fiber length distribution during this process. In the plasticizing process, the formation of temporary unfilled area and the rotational behavior pellets moving from the orthogonal orientation to the parallel direction relative to the flight were observed. The results also showed that solid-bed fragments tend to generate easily in the compression zone. In addition, it was confirmed that the solid-bed ratio decreases temporarily in the middle of the plasticizing process and this is influenced by screw revolution and charge stroke. The weight average fiber length in the injected resin showed a temporary decreasing tendency in the middle stage of the injection process. Reference (1) S.Ma, H.Yokoi; Seikei-Kakou Symposia'13, 73(2013)

Improvement of the Reproduction Accuracy of a Defined Pole Length of Injection Molded Encoder Wheels

Katharina H. Kurth[1, a)] and Dietmar Drummer[1, b)]

[1]Institute of Polymer Technology, Friedrich-Alexander-Universität Erlangen-Nürnberg,
Am Weichselgarten 9, 91058 Erlangen, Germany

[a)] Corresponding author: kurth@lkt.uni-erlangen.de
[b)] drummer@lkt.uni-erlangen.de

Abstract. Magnetically sensitive sensors are increasingly used in different applications for the measurement of physical values, such as angle, position or speed. Depending on the application certain requirements for a high measurement accuracy have to be achieved, such as a high pole length accuracy and repeatability for the signal transmitter. These signal transmitter can be produced economically in the injection molding process using compounds consisting of hard magnetic particles embedded in a polymer matrix, so-called polymer bonded magnets. By integrating a magnetic field inside the mold an orientation of the particles in the polymer melt as well as a partial magnetization can be conducted, such that the second magnetization step is not required. This paper deals with the influence of the cooling conditions on the orientation of the particles as well as the final magnetic properties, such as the resulting pole length accuracy. The cooling conditions are varied by using different materials for the cavity wall. It can be shown that the cooling of the surface layer and resulting particle orientation has a major influence on the quality criteria of injection molded signal transmitter for sensor applications.

INTRODUCTION

Polymer bonded magnets are composite materials consisting of hard magnetic filler particles embedded in a polymer matrix [1]. Filler particles are primarily oxide ceramics, such as strontium ferrite or rare earth materials, such as neodymium iron boron [2]. The resulting compounds can be further shaped in conventional polymer processes, such as injection molding [3, 4]. Due to the injection molding process, the production of polymer bonded magnets in high quantities with narrow tolerances, as well as including integrated functional parts during the process is possible [2, 4, 5]. Compared to sintered magnets, a high freedom in design and magnetizing structure is feasible [2]. In order to utilize the complete potential of anisotropic hard magnetic particles, the particles have to be aligned during the process by an integrated magnetic field, which can be implemented by coils or permanent magnets [3, 6].

Polymer bonded magnets are mainly used in motor or sensor applications, as for example as multipolar encoder [4, 7]. For sensor applications, a sensor (e. g. hall sensor) measures the changing magnetic field of in order to evaluate speed, angle or position of machine elements [1, 8–10]. As multipolar encoder usually rings or strips consisting of a thermoplastic [2, 11] or elastomeric material [12] are used. Often, these encoder are magnetized after the processing [2]. However, the injection molding process enables the production of multipolar bonded magnets with anisotropic filler orientation and direct magnetization.

The properties of polymer bonded magnets are mainly influenced by the type, amount, distribution and orientation of the filler particles in the polymer matrix [13]. However, the particle orientation is affected in complex interactions of the flux density inside the cavity and the viscosity of the melt [1, 14, 15]. Further, the processing parameters, such as melt and mold temperature, holding pressure and melt flow behavior, e. g. influenced by the gating system influence the final magnetic properties [1]. Unfavorable melt flow can lead to different melt flow velocity, open jet formation or development of weld lines and lead to magnetic inhomogeneities [1]. The pole length accuracy of multipolar bonded magnets is mainly influenced of the gating system and, thus, flow behavior. Homogenous melt flow within one pole leads to good reproduction of the pole length without mayor influence of the processing parameters [16]. Melt flow along several poles lead to magnetic inhomogeneities and thus, increased pole length deviations, in particular close to the injection point and developing weld line [17–19]. Further, the pole length accuracy is highly influenced by the processing parameters and can be improved using higher melt and mold temperature [19].

These investigations deal with the influence of the frozen surface layer on the magnetic characteristics of multipolar rings. In order to vary the filler orientation in the surface layer inserts of different materials are used. The multipolar rings are then characterized regarding their magnetic properties, their particle orientation at different positions of the ring, as well as magneto-optically in order to identify magnetic inhomogeneities.

EXPERIMENTAL PROCEDURE

Processing

For these investigations a compound consisting of a polyamide 12 (Vestamid BS1636, Evonik Industries AG) and a strontium ferrite (SrFeO) powder (OP-71, Dowa Holdings Co., Ltd., provided by HAWA Magnetische und leitfähige Werkstoffe) is used. The materials are mixed with a filler volume content of 55 % using a twin screw extruder with a nozzle temperature of 220 °C.

As specimen multipolar rings (mold size: Ø 30.6 mm, height: 5 mm) with a pole length of 4 mm at the outer ring diameter, which corresponds to 24 poles, are used (compare to Fig. 2 a). The gating system is a pin-point gate system with two injection points at the side surface of the ring. The orientation of the filler particles, as well as the multipolar magnetization of the rings is realized by an integrated magnetic field in the cavity caused by 24 permanent magnets arranged radially around the cavity in the mold, such that the filler particles are oriented and partially magnetized during the injection molding process. No additional magnetization is conducted. In order to vary the cooling behavior of the rings' surface, the thermal conductivity of the mold insert between the permanent magnets and the cavity with a thickness of 0.4 mm is varied. The used materials are a steel with low permeability, as well as a glass fiber reinforced PEEK (PEEK-GF). The thermal conductivity, analyzed by using a Hot Disk device, amounts 13.9 W/mK for the used steel and 0.4 W/mK for the PEEK-GF material.

For injection molding the specimen, an injection molding machine Demag 25/280-80 of the company Sumitomo (SHI) Demag Plastics Machinery GmbH is used. All rings are produced using equal processing parameters, which include a melt temperature of 280°C, a mold temperature of 80°C, an injection speed of 40 mm/s and a holding pressure of 500 bar.

Characterization

The magnetic characteristics of the multipolar rings are evaluated, in particular regarding their maximum peak flux density as well as their pole angle. Therefore, the radial flux density B_R is measured in a distance of 0.5 mm from the outer ring surface using a hall sensor (KSY-44, Siemens AG) and is further correlated to the angle measured using a high-accuracy angular position encoder (RON 786, Dr. Johannes Heidenhain GmbH). The hall sensor is positioned flexibly with sliding contact on the outer ring surface, in order to adjust to any concentricity deviations of the ring. The resulting radial flux density curve is evaluated regarding the pole angle s, as well as the peak flux density $|B_{R,max}|$. The pole angle is defined as the angle between two adjacent zero crossings of the flux density, which is interpolated linearly. Figure 1 shows the evaluation of the flux density and pole angle in the mold, which is used as reference for the presented investigations. For the evaluation, the pole angle of characteristic poles, such as the poles with centrally located weld line (pole 6 and 18) or injection point (pole 12 and 24), are summarized. For the analysis of the process influences, the pole angle is analyzed regarding its reproduction accuracy in regard to the mold. Thus, the measurement device is adapted, such that the magnetic flux inside the cavity can be measured and evaluated accordingly.

Additionally, the rings are analyzed using imaging analysis methods. First, the characterization of the filler particle orientation is conducted using a scanning electron microscope (SEM) (Ultra Plus, Carl Zeiss AG) on polished surfaces at the ring's half height. The particle orientation is analyzed close to the surface of the ring, as well as at a distance of 800 µm to the ring surface at the pole center as well as the pole pitch. Second, the rings are analyzed magneto-optically (minimo, sensor type B, Matesy GmbH) on grinded surfaces at the ring's half height. The magneto-optic method can be used in order to analyze magnetic structures by using the faraday effect and the resulting rotation of linearly polarized light when interacting with a magnetic field. However, due to the non-calibrated device used, the shown grey values can be only evaluated qualitatively. The magnetization is analyzed at multiple locations, such as the pole with injection point and weld line, as well as two poles in between these two characteristic locations.

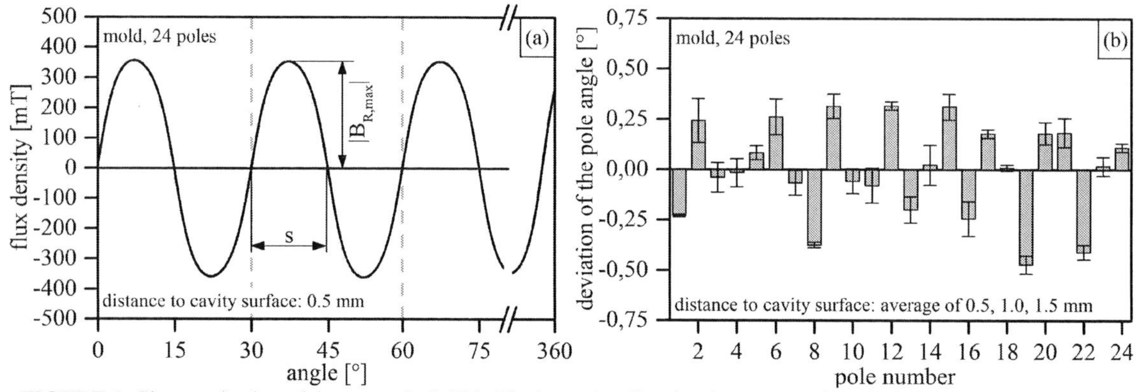

FIGURE 1. Characterization of the magnetic field inside the cavity: flux density curve and marked characteristic values (a), pole angle for each single pole (b)

RESULTS

Magnetic Characteristics

In the following, the influence of the intermediate layer as cavity surface on the magnetic characteristics of the injection molded multipolar rings is shown. Figure 2 shows the pole angle referred to the pole angle in the mold, as well as the average peak flux density of the poles. The pole angle of equal poles, such as poles with centrally located injection point, are summarized. For the multipolar rings with steel layer, the pole angle at the poles with centrally located injection points is smaller, the pole angle of the pole with centrally located weld line wider, compared to the pole angle of the mold, as it is already evaluated in previous research [16–18]. The poles in between these two characteristic locations are rather broader. However, a distinctive distribution cannot be identified. In contrast to this, the pole angle of the multipolar rings produced with the PEEK-GF intermediate layer show only small deviations compared to the mold. Furthermore, the peak flux density averaged for all poles of the multipolar rings produced with PEEK-GF intermediate layer is higher. This can be explained by the slower cooling of the surface layer due to the intermediate layer material and, thus, a higher degree of orientation of the filler particles (Fig. 3).

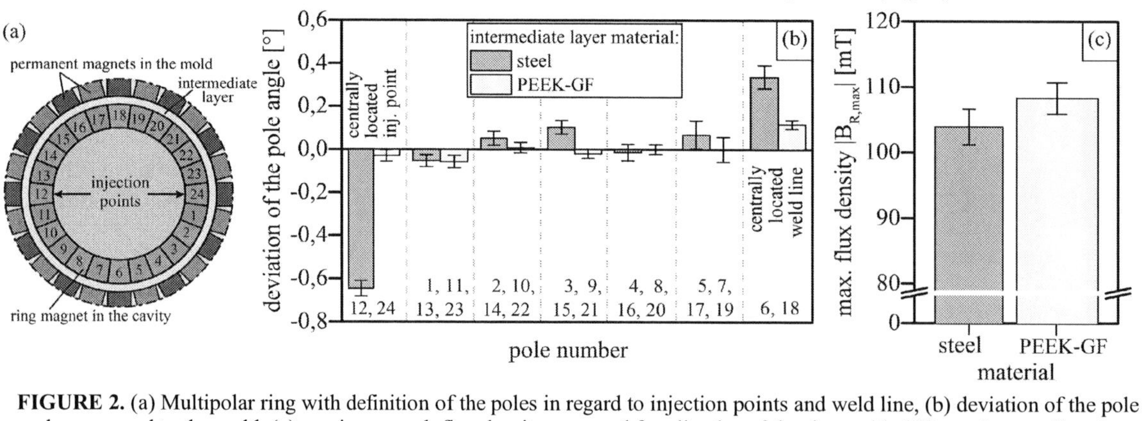

FIGURE 2. (a) Multipolar ring with definition of the poles in regard to injection points and weld line, (b) deviation of the pole angle compared to the mold, (c) maximum peak flux density averaged for all poles of the rings with different intermediate layer material

Imaging Analysis Methods

Imaging analysis methods are used in order to evaluate the filler particle orientation, as well as the resulting magnetization (Fig. 3, Fig. 4). The filler particle orientation is analyzed for the different surface materials at several locations of the ring (Fig. 3). The magnetic preferred direction of the anisotropic particles is perpendicular oriented to the large surface of the mainly platelet-shaped particles. For the parts produced with a steel intermediate layer no particle orientation can be identified at the rings' surface regardless the position relating to the magnetic field. However, at a distance of 800 µm from the surface a defined orientation is visible. In the pole center, the particles are aligned horizontally which corresponds to a vertical magnetic field. In the pole pitch this is reversed with a vertical particle orientation and corresponding horizontal magnetic field. However, the particle orientation for the rings produced with an intermediate layer consisting of PEEK-GF, the particle orientation at the surface and 800 µm from the surface is identical. Thus, the lower heat conductivity at the cavity surface leads to a slower cooling of the material at the surface and thus, a good particle orientation corresponding to the magnetic field. The higher degree of orientation additionally causes a higher flux density at the rings' surface, as can be seen in the magnetic characterization (compare to Fig. 2 c).

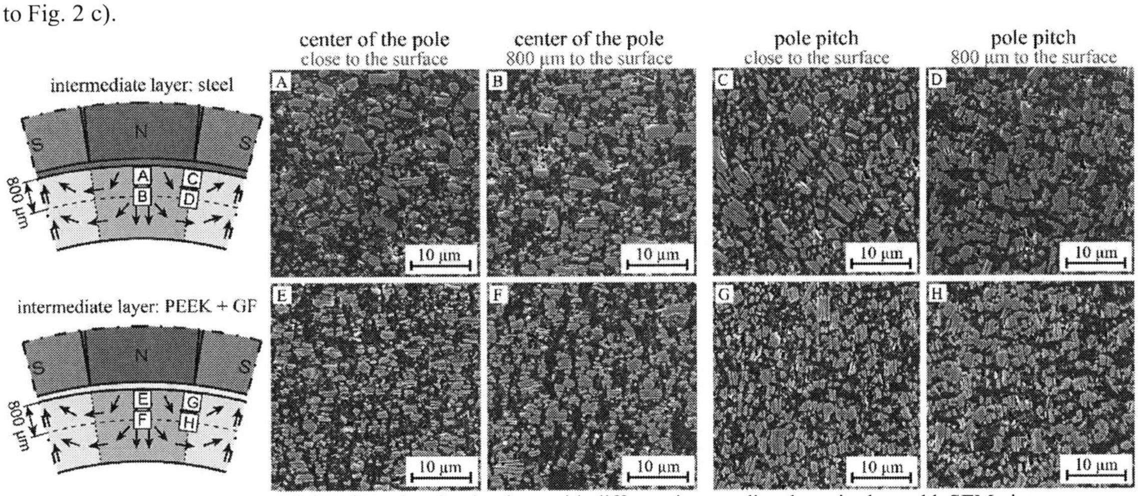

FIGURE 3. Filler particle orientation for the rings with different intermediate layer in the mold; SEM picture

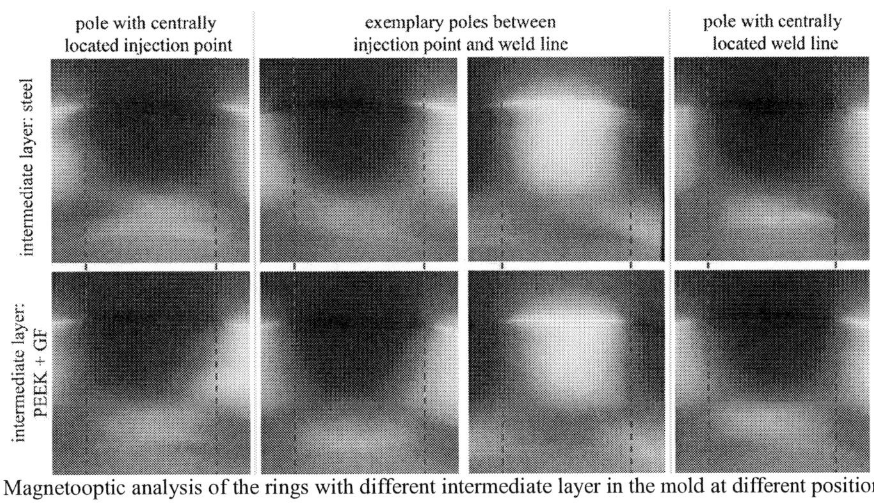

FIGURE 4. Magnetooptic analysis of the rings with different intermediate layer in the mold at different positions of the ring

Figure 4 shows the magneto-optic analysis of the multipolar rings with different intermediate layer in the mold for different positions at the ring. The alternating pole structure can be clearly seen for all poles. For the rings produced with steel intermediate layer, a surface layer with shifted magnetization opposed to the flow direction can be identified.

For the pole with centrally located injection point, the surface layer seems to be narrower compared to the inner part. In contrast to this, the pole with centrally located weld line appears broader. The poles between injection point and weld line show a clearly shifted magnetization at the surface layer. These effects, in particular the shifted magnetized surface layer, cannot be seen for the samples produced with PEEK-GF intermediate layer in the mold. Thus, the frozen surface layer with only partially oriented filler particles presumably leads to magnetic inhomogeneities, which influence the magnetic characteristics of the multipolar rings.

CONCLUSION

The investigations show that the pole length accuracy is highly influenced by the filler orientation in the surface layer. Using typical mold steel leads to a frozen surface layer and thus, partially non-oriented filler particles close to the surface. By varying the cooling rate of the rings' surface layer by varying the heat conductivity of the outer cavity surface, the filler particle orientation and the magnetic characteristics can be significantly influenced. Using the steel material leads to a high cooling rate in the surface layer with no filler particle orientation in regard to the magnetic field and resulting magnetic inhomogeneities and deviations in the pole length accuracy. A slower cooling rate, implemented by using a PEEK-GF material at the outer ring surface leads to considerable improvement of the filler orientation in the surface layer and, thus, a good representation accuracy of the pole length in the mold. Furthermore, the higher degree of particle orientation leads to higher peak flux densities.

ACKNOWLEDGMENTS

The authors thank the company Evonik Industries AG for providing the matrix material.

REFERENCES

1. D. Drummer, „Verarbeitung und Eigenschaften Kunststoffgebundener Dauermagnete," Ph.D. thesis, Universität Erlangen-Nürnberg, 2004.
2. M. Grönefeld, "Review on Bonded Magnets" in *Bonded Magnets*, edited by G.C. Hadjipanayis (NATO Science Series, 2004).
3. T. Sakai, K. Nakamura and A. Morii, International Polymer Processing **6** (1991).
4. T. Schliesch, Sonderdruck Industrieanzeiger, 35, 21-22 (2006).
5. A. Gardocki and D. Drummer, "Improvement of the Filler Orientability during Injection Molding of Multi-Polar SmCo-Magnets by Premagnetization" in *Conference Proceedings EDPC2012.*
6. S. Eimeke, A. Gardocki, S. Amesöder and E. Schmachtenberg, Kunststoffe international, 5 (2007).
7. D. Drummer, „Magnetizable Polymers," in *Proceedings 4th International Congress MID* 2000.
8. J. R. Brauer, *Magnetic actuators and sensors* (John Wiley & Sons, Inc., Hoboken, USA, 2014).
9. S. Gauthier, Motion System Design **54**, 10 (2012).
10. K. Reif, „Sensormessprinzipien," in *Sensoren im Kraftfahrzeug*, edited by K. Reif (Springer Vieweg, Wiesbaden, 2012).
11. A. Schmidbauer, in *Kunststoffgebundene Dauermagnete*, edited by G.W. Ehrenstein and D. Drummer (Springer VDI-Verlag, Düsseldorf, 2004).
12. T. Miyagawa, T. Takehara, T. Murakami, and T. Takajo, „A New High-Performance Plastic Magnetic Encoder," SAE Technical Paper 2009-01-1293, 2009, doi:10.4271/2009-01-1293.
13. E. Baur, S. Brinkmann, T. Osswald, N. Rudolph, and E. Schmachtenberg, *Saechtling Kunststoff Taschenbuch: 31. Ausgabe* (Carl Hanser Verlag, München, 2013).
14. S. Eimeke, *„Spritzgießen multipolarer, kunststoffgebundener Dauermagnete,"* Ph.D. thesis, Universität Erlangen-Nürnberg, 2008.
15. S. Eimeke, A. Gardocki, Z. Brocka, and E. Schmachtenberg, „Injection Molding of Polymer-Bonded Magnets with Multipolar Structure," in *Conference Proceedings SPE ANTEC 2008*, pp. 1508–1512.
16. K. H. Kurth and D. Drummer, International Polymer Processing **31**, 3 (2016).
17. K. H. Kurth and D. Drummer, Journal of Polymers **2017**, 23 (2017).
18. K. H. Kurth and D. Drummer, Journal of Polymer Engineering, doi: 10.1515/polyeng-2016-0306 (2016).
19. K. H. Kurth, D. Drummer and E. Grünewald, "Influences of the variation of process parameters on the pole length of multipolar bonded magnets", in *Conference Proceedings SPE ANTEC 2016.*

Capability database of injection moulding process-requirments for wider suitability and higher acceptance

Boorla Srinivasa Murthy [1]*, Eifler Tobias [2], Jepsen Jens Dines Obel [3], Howard Thomas J [1]

[1] Technical University of Denmark - Kgs.Lyngby, Copenhagen - Denmark, [2] Technical University fo Denmark - Kgs. Lyngby, Copenhagen - Denmark, [3] Technicla University of Denmark - kgs.Lyngby, Copenhagen - Denmark

Generally, there is little disagreement that an early consideration of dimensional accuracies achieved in production is conducive to the success of development of injection molding products. While different Process Capability databases (PCDBs) provide guidance for a meaningful estimation of the expected part variation, the adoption of corresponding guidelines and (proprietary) software tools seems to be, however, limited in industrial practice so far. This research paper addresses the gap between the available PCDBs and the requirements of designers in practice and investigates the key drivers for an improved applicability of corresponding database solutions in an industrial context. A survey of database users at all phases of product value chain in the plastic industry revealed that 59% of the participating companies use their own, internally created, databases, although reported to be not fully adequate in most cases. Essential influences are the suitability of the provided data, defined by the content such as material, tolerance types, etc. covered, as well as its accuracy, largely influenced by the updating frequency. Forming a consortium with stakeholders, linking database update to technology changes and connecting dimensioning standards to database offerings are proposed solutions.

Knowledge-Based Material Selection For Injection-Molded, Thermoplastic Parts

Alexander Porsch[1, a)], René Andrae[2, b)], Johannes Wortberg[1, c)], and Peter Köhler[2, d)]

[1]*University of Duisburg-Essen, Institute of Product Engineering, Department of Engineering Design and Plastics Machinery, Lotharstraße 1, 47057 Duisburg, Germany*
[2]*University of Duisburg-Essen, Institute of Product Engineering, Department of Computer Aided Engineering, Lotharstraße 1, 47057 Duisburg, Germany*

[a)]Corresponding author: alexander.porsch@uni-due.de
[b)]rene.andrae@uni-due.de
[c)]johannes.wortberg@uni-due.de
[d)]peter.koehler@uni-due.de

Abstract. Material selection is an important but often an arbitrary process at the beginning of the development of injection-molded parts. Due to the absence of a sufficient documentation, it cannot be retraced and optimized by other engineers. The use of expert systems, which apply a comprehensive knowledge base, is a possibility to reduce mistakes and make decisions more transparent. In this paper, a knowledge-based method for material selection is presented. Basis of the method is a knowledge base containing information about the properties of thermoplastics. This includes for example single- and multi-point-data, or facts like chemical resistances. Moreover, additional information concerning the modification of the properties, correlations, or the exact definitions are stored in the knowledge base. It also contains knowledge about production processes like standard injection molding, MuCell, or injection-compression molding restricting the number of suitable materials. Additional knowledge about the influence of part design for material selection completes the knowledge base. Considering the vast amount of available thermoplastics, the presented knowledge base includes selected examples. Based on this knowledge, a multistage method is developed using questionnaires to refine requirements and to exchange information between engineer and expert system. During the selection process, additional information, as it has been mentioned before, is given. At the end of the process, the user gets a ranking of possible materials. In addition, more detailed knowledge about their properties, advantages, and disadvantages are provided. For the development and the visualization of the method, System Modeling Language diagrams are used. These allow different views to the material selection method, the stages, and links between knowledge base and selection process. Finally, the results are discussed, and an outlook for development potential is given.

INTRODUCTION

Plastic parts are omnipresent in our everyday life. The range of products stretches from plastic bags or a PC mouse case to optical lenses and microinjection-molded medicine products. Moreover, the demand for plastic products increases steadily, which is documented by the continuous growing of the worldwide plastic materials production. Plastic products benefit from positive material properties (high quotient stiffness/density, chemical resistances, low electrical conductivity, etc.) and are able to replace metal products. Furthermore, new, innovative products in the areas of mobility, medicine, and sports are only enabled by the usage of customized polymer materials and corresponding manufacturing technologies. Present challenges are the increase of product requirements (smaller lot sizes and higher product diversity) and the simultaneous decrease of product life cycles (1.).

Additional, in comparison to metals, polymers possess a rather complex material behavior regarding the manufacturing process and the resulting part properties. For example, a high flow velocity may lead to increased mechanical properties in flow direction. Though, this influences also the shrinkage during the cooling of the hot material, so that geometrical tolerances may be not satisfied. The viscoelastic and highly temperature-depending material properties also affect the functional characteristics of plastic parts. All of these polymer specific properties are to be considered for the part design.

Of course, many literature resources regarding material science, process technology, and part design are available. Besides textbooks, journal and conference papers, especially, plastic materials producers, e.g., Quadrant Engineering Plastic Products (Reading, Pennsylvania, USA), provide their customers with detailed information in terms of brochures. Yet, the unexperienced designer has to gain and process the information by himself. Other tools supporting the designer are the commercial software CES Selector (Granta Design Ltd., Cambridge, UK), which is based on the research work of ASHBY (2.), and the web services Material Data Center (MDC, M-Base Engineering + Software, Aachen, Germany) and CAMPUS (Chemie Wirtschaftsförderungs-GmbH, Frankfurt, Germany), for example. The latter is freely accessible and in fact, only a database used in the CES Selector and the MDC. The MDC contains additionally polymer specific literature, a part database (pictures, used material and process, requirements, etc.), a register of trade names, and designing tools, e.g., for the estimation of the necessary wall thickness. Again, the user has to gain the information by himself. Also, the entered data or requirements are not checked for plausibility, and a ranking of the suitable materials is not created. Thereby, the designer gets no information how well the requirements are fulfilled by each material. In comparison, the CES Selector provides the user with a list of ranked materials. The review of the entered data is not conducted, and the supporting information mentioned before is also not provided to the user.

Because of the occurring challenges, methods for the knowledge-based manufacturing process selection and knowledge-based material selection were developed and implemented in a Knowledge-Based Engineering (KBE)-system. The presented results have been achieved in the framework of a research project.

In this paper, the focus is on the method for the material selection. As shown in (3.), it is performed in an early stage of the development process. The correction of adverse decisions becomes gradually pricier as the development process progresses (4.). Therefore, the usage of expert knowledge during the development process supports the designer. This results in a reduction of the required amount of iterations and therefore, of the development time, which leads to a decrease of the development costs.

STRUCTURE OF THE METHOD

Following ASHBY, the presented method contains four main steps, which are the specification of requirements, screening, ranking, and supporting information (2.). In comparison, the range of materials is limited to unfilled, thermoplastic polymers because of the scope of the research project. Other materials as metals, ceramics, or composites are excluded.

The first step of the method is the determination of the necessary requirements. Therefore, the KBE-system communicates with the designer using questionnaires. During the screening, the second main step, all materials are excluded, which do not fulfill the entered requirements. This selection bases on constraints as minimum and maximum values of density, for example, or on attributes as chemical resistances. Thereafter, a ranking of the remaining materials is created. Thereto, the user adds a weighting to each requirement, and based on a benefit analysis, the ranking is calculated. The result of the ranking is a list of suitable material types as Polyethylene, Polypropylene, or Polyamide. Because of the vast number of available polymers, which often have similar properties, - CAMPUS lists over 5300 Polymers- specific materials of material producers are not stored in the database. Nevertheless, over 250 different material types are available (5.). For the final material selection, the user gets additional supporting information, which is the fourth step of the method. Before each step is discussed in more detail, the developed KBE-system is presented.

KBE includes the systematic support of design engineers by providing methods and tools. Thereby, the methodological and IT-based design support is subsumed under the term of KBE-systems. In contrast to conventional programming, the rules and constraints are entered in a knowledge base in specified, mostly easily comprehensible syntax. This enables a straightforward maintenance and expansion of the knowledge base (6.).

In general, a KBE-system is closely linked to a CAD-system. In this project, it is implemented in Siemens PLM NX 10 (NX10, Siemens Industry Software GmbH, Cologne, Germany). By using a neutral knowledge representation of the relevant data and a coupled approach for the integration, the transferability to other CAD-systems or the usage as stand-alone program is guaranteed. The coupled approach for the integration is realized by the use of a web service and a database, as shown in Fig. 1, and the communication between each part of the KBE-system is implemented using the Hypertext Transfer Protocol (http).

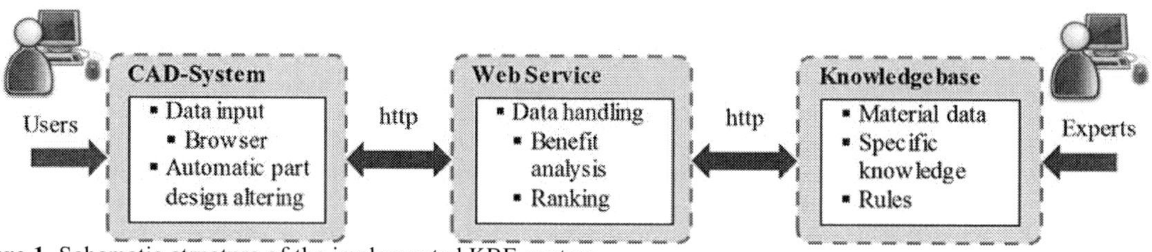

Figure 1. Schematic structure of the implemented KBE-system

The material selection is performed in the internal web browser of NX10. Queries are sent to the web service, which processes these and in turn, issues a MySQL request to the database. If a material type is finally selected, this may result in an automatic part design altering. However, this is not further discussed in this paper.

DEFINITION OF REQUIREMENTS

The first step of the material selection is the definition and input of requirements. Hereto, the KBE-system interviews the designer using three different forms. These are structured according to the top-down-methodology (7.). In the first form, a questionnaire, the user selects predefined groups of general requirements. For example, these are:

- Part geometry (maximum outer dimensions, part volume, minimum wall thickness, …),
- General requirements (tolerances, quantity, operational life, …),
- Thermal loads and thermodynamic properties (density, service temperature, application as heat insulator, …),
- Optical properties,
- Further processing (welding and bonding),
- Or economics (manufacturing costs and part costs).

Following the top-down-methodology, in the next questionnaire, the user is provided with a list containing detailed requirements. Naturally, these refer only to the selected groups of general requirements, which are selected in the first form. Now, additional information regarding the available properties are given. Besides the names, the standard, the unit, a description and the symbol are listed in the second questionnaire. The description is taken from the corresponding standard. Again, the user selects the necessary requirements.

Besides the predefined requirements, the user may create own ones by the combination of quantitative properties. Common used combinations as the specific stiffness (2.) (Young's modulus/density) are stored in the knowledge base. Based on all selected predefined and user defined requirements, the third and last form is generated.

In addition, there are also completed questionnaires implemented for selected applications. In these forms, only essential requirements are preselected, and the user needs to add further ones, which are necessary for the current application. Otherwise, preselected requirements can also be deselected. In the same manner, previous material selections can be loaded, and reviewed or adjusted.

Finally, values are entered for each requirement in the third and last form. These are minimum and maximum values for quantitative properties as e.g., for heat conductivity. Equally, it is also possible to enter only a minimum or a maximum value. Then, the opposite one is added automatically based on the corresponding extremum, which is stored in the database. Drop down menus enable the specification of qualitative requirements. Examples are chemical resistances or optical properties.

During the data input, all entered values are compared with the ones of the knowledge base. If a value exceeds the polymer characteristic boundaries for this particular property, the KBE-system will warn the user, and the extremum values are displayed. Moreover, the combination of requirements are verified.

The third form enables an automatic generation of a functional specification document. With this information and the results of the ranking, the user, respectively the purchasing department, may contact material producers.

For the ranking, the user adds a weighting to each requirement. Primarily, it is differentiated between must-have (weighting 5) and minimum requirements. According to (8.), the latter are divided into five groups, from unimportant to very important. Hereto, for the classification of the minimum requirements, an absolute scale - zero to four - is used. The KBE-system provides no assistance regarding the objectification of the weighting. Albeit, the influence of each requirement on the ranking is evaluated by an automatic variation of each weighting. In this way, requirements influencing strongly the ranking are detected. Now, the designer may recheck his decisions regarding these requirements.

Therewith, the first main step of the material selection method is completed. In the following, the entered requirements are aligned with the material data sets stored in the knowledge base.

SCREENING AND RANKING

The alignment is rule-based in form of IF-THEN rules. Firstly, the must-have and secondly, the minimum requirements are aligned with the knowledge base. Only, the material types fulfilling all requirements pass the screening and are transferred to the ranking, the next main step of the method. At the end of the ranking, an overview is generated regarding which requirement causes the exclusion of which materials. In the following, this list is called exclusion list and provides the user with additional information. The further usage of the exclusion list is described in a following section.

In the conclusive ranking, a benefit analysis is used ordering the remaining candidates. ASHBY uses so called material indices to quantify how well a material type can do the job. These are parameter combinations as the specific stiffness describing the main objective of the part. The better the material indices the higher is the rank (2.). In the developed method, the benefit analysis involves the user specified weighting for each requirement and the ability of the material types to fulfill this requirement in comparison to each other. This way, a score is calculated for each material type.

Supporting Information

As mentioned in a previous section, one group of general requirements is Economics. Because the material costs depend highly on the costumer (location, purchase quantity, discounts, etc.), the part costs are just a supporting information and do not influence the ranking. In this case, the part costs refer only to the material costs, and manufacturing costs are neglected. The part costs are calculated based on the part volume (provided by the CAD-system), the density (provided by the database), and a material price, that is available from different freely accessible websites as (9.). Nevertheless, these part costs are only a rough estimation of the real material costs because the part volume or more precisely the wall thickness is highly dependent on the used material. Therefore, based on given simplified, regular geometries, the influence of the material on the wall thickness and as a result on the part costs can be estimated. Exemplarily, some loading conditions, types of restraints, and regular geometries are implemented in the KBE-system. Further information about the equations and calculations are given in (3.).

Provided with the ranking and the supporting information, the user gets a substantive summary of material types fulfilling all requirements. In addition, it is specified how well a material type fulfills the requirements, and an estimation of the influence of the material on the part design and part costs is given.

If no material type fulfills the must-have and minimum requirements, the user has two possibilities. As mentioned before, it is saved which requirements lead to the exclusion of which material. Therefore, if an adjustment of the requirements is possible, the user may influence the resulting ranking. In the exclusion list, each entered value can be modified after the ranking, which is updated automatically. The second possibility is to modify the material properties. In general, most purchasable polymers contain already additives as pigments or heat stabilizers. The range of additives and their functionality spreads from glass fibers (improving mechanical properties) over UV-filters to antioxidants. Again, in the exclusion list, information about additives can be displayed. In comparison to the modification of the requirements, the updated ranking displays only the qualitative suitability of the polymers because the quantitative impact on each requirement is not precisely predictable.

At this point, the knowledge-based material selection method is completed. The user selects one or more material types preliminary based on all presented information. Subsequently, Finite Element Analyses or injection molding process simulations are to be performed to optimize the part design (3.). As mentioned before, the selected material influences the wall thickness, the cycle time of the manufacturing process, and in general, the resulting part costs. Moreover, unresolved, economical questions as the availability or exact material costs are to be clarified.

CONCLUSION

The presented method is intended for design engineers being unexperienced in the field of polymers. Due to the comparably complex material behavior, polymer specific knowledge is essential for the successful development of injection-molded parts. The KBE-system guides the user along the selection process. Moreover, the top-down-methodology helps specifying the necessary requirements. Because of the vast amount of available polymers, the selection method results in a ranking including material types. In addition, for the final material selection, supporting information regarding the part and processing costs is estimated and provided to the user. The presented KBE-system is based on a coupled approach for the integration into the CAD-system, and therefore, the transferability to other CAD-systems is ensured. Within the scope of the research project, only, the material parameters of selected material types are stored in the KBE-system. Likewise, the questionnaires do not include every upcoming requirement. For this reason, the KBE-system includes an interface for experts to add new material data and requirements.

In the current state of the method, the material data sets of the implemented material types are complete. The next step in the consistent further development is consideration of missing material parameters because each selected requirement influences the screening and the ranking significantly. One possible solution to this problem is the automatic generation of a functional specification document. Based on this document, the material producer can be asked for the missing information.

The benefit of the presented method is difficult to estimate. It depends on factors as the complexity of the project, the experience of the accountable design engineer, or the type of design (designing a new product or variants). Yet, two advantages are to be pointed out. The weighting of particular requirements enables the connection of several material indices possessing different importance to the designer. Additionally, the requirements of implemented projects can be saved and documented in a structured manner. Therefore, a company is able to store their specific knowledge.

ACKNOWLEDGMENTS

The German National Science Foundation (Deutsche Forschungsgemeinschaft, DFG), project no. WO 302/56-1 and KO 1620/15-1 supports this project, and we would like to thank the SIGMA Engineering GmbH for the support.

REFERENCES

1. N. N., Informationen aus dem kunststoffland NRW, Ausgabe 1/2016, 14-16 (2016).
2. M. F. Ashby, *Materials selection in mechanical design* (Butterworth-Heinemann, Burlington, 2011).
3. R. A. Malloy, *Plastic Part Design for Injection Molding, An Introduction* (Carl Hanser Verlag, Munich, 2011).
4. G. Reinhart, U. Lindemann and J. Heinzl, *Qualitätsmangement, Ein Kurs für Studium und Praxis* (Springer-Verlag, Berlin Heidelberg, 1996).
5. E. Baur, S. Brinkmann, T. A. Osswald, N. Rudolph and E. Schmachtenberg, *Saechtling Kunststoff Taschenbuch* (Carl Hanser Verlag, Munich, 2013).
6. N. N., "Knowledge management for engineering, Knowledge-based engineering (KBE)," in *VDI 2225 Blatt 3* (Beuth Verlag, Berlin, 2017).
7. J. Feldhusen and G. Lashin "Top-Down-Konstruktion auf Basis von 3D-CAD – Ansätze und Voraussetzungen für den Einsatz im Maschinenbau." in *Prozessketten für die virtuelle Produktentwicklung in virtueller Umgebung* (VDI-Verlag, Düsseldorf, 1998), pp. 277-301.
8. N. N., "Konstruktionsmethodik, Technisch-wirtschaftliches Konstruieren, Technisch-wirtschaftliche Bewertung," in *VDI 2225 Blatt 3* (Beuth Verlag, Berlin, 1998).
9. N. N., "KI Kunststoff Information Polymerpreise," in *https://www.kiweb.de/default.aspx?pageid=200* (Kunststoff Information Verlagsgesellschaft, Bad Homburg, 07-21-2017).

The Behavior of Bulk Solids in the Solids Conveying Zone of Smooth Barrel Single Screw Extruders: Friction, Bulk Density and Pressure Anisotropy

Kneidinger Christian [(1)]*, Zitzenbacher Gernot [(1)], Längauer Manuel [(1)], Schuschnigg Stefan [(2)], Miethlinger Jürgen [(3)]

[(1)] University of Applied Sciences Upper Austria - Upper Austria - Austria, [(2)] Department of Polymer Engineering and Science, University of Leoben - Styria - Austria, [(3)] Institute of Polymer Extrusion and Compounding, Johannes Kepler University - Upper Austria - Austria

This work deals with the material parameters of bulk solids in the solid conveying zone. New measurements show that the coefficient of friction of polypropylene-Homopolymer (PP-H) bulk material is clearly dependent on the frictional distance. Furthermore, a new mathematical model for the bulk density regarding the pressure and the temperature is presented. This model considers the melting point and the pvT-data and needs only three parameters to approximate the bulk density depending on pressure and temperature. The model fits experimental data from room temperature up to the melting point. There is a smooth crossover to the pvT-data at the melting point. These are clear advantages compared to existing models. The same approach approximates the pressure anisotropy sufficiently as well. Experimental results of the measurements of the external coefficient of friction, the bulk density and the pressure anisotropy of PP-Homopolymer (PP-H) utilizing model experiments are presented. The bulk density and the pressure anisotropy coefficient were measured in dependence of pressure and temperature using a special measuring device. The external coefficient of friction was measured dependent on frictional distance, pressure, temperature and the sliding velocity using a unique tribometer. Furthermore the influence of these dependencies is shown by applying these models to a segmented calculation of the solids conveying zone, regarding different models like that of Schöppner, Schneider, Hyun, Campbell as well as the classical one of Darnel and Mol. The results are compared to extrusion experiments.

Numerical analysis of the fountain flow instability

Smit Tim (1)*, Hulsen M. A. (2), Anderson P. D. (2)

(1) Dutch Polymer Institute, Technische Universiteit Eindhoven - Eindhoven - Netherlands, (2) Faculty of Mechanical Engineering, Eindhoven University of Technology - Eindhoven - Netherlands

Instabilities in polymer processing limit production rates and may influence the aesthetic or me- chanical properties of the final product. One example is the fountain flow instability, which takes place during the mold filling stage of an injection molding process. It has been shown experimen- tally that these instabilities manifest themselves in the form of a periodic oscillation of the flow front stagnation point between the two walls of the mold channels. The onset is determined by a critical Weissenberg number, the ratio of material response time scales to flow time scales. Different values have been predicted in previous work[1][2]. We use the finite element method and a number of stabilization techniques for flows of convective and viscoelastic nature (DEVSS-G/SUPG[3], LCR[4]) to do a full non-linear, transient simulation of a branched polymer melt propagating through a mold channel. We investigate the onset and behavior of the fountain flow instability with special focus on describing the contact point where the polymer is deposited on the mold wall. High stress and velocity gradients require special local treatment of the sharp interface method used to describe the free surface deformation. We observe the onset and behaviour of the flow during an instability. For low Weissenberg num- bers, we observe that the perturbation vanishes, indicating a stable system. For high Weissenberg numbers we observe oscillating behavior as the perturbation grows, as observed in experiments [6]. We investigate how varying material properties influences this critical Weissenberg number and the behaviour during unstable flow. This research forms part of the research programme of the Dutch Polymer Institute, project 788

In-situ X-ray analysis during Flash DSC experiments: study of isothermal crystallization and phase transformation of Polyamide 12

Paolucci Fabio (1)*, Govaert Leon E. (1), Peters Gerrit W. M. (1)

(1) Department of Mechanical Engineering, Eindhoven University of Technology - Eindhoven - Netherlands

Semi-crystalline polymers typically consist of crystalline and amorphous domain with dimension in the nanometer range. The final properties of such material depend on the volume fraction of the crystalline domains as well as on their size and structure. Polyamide 12, as well as many others polymer, shows a bimodal dependence of the crystallization rate as function of crystallization temperature. Depending on the processing conditions PA12 shows a complex polymorphism: □', □, α, α'. The main crystal form obtained by slow cooling from the melt at room temperature is the □ form. The □'-form can be obtained by fast quenching from the melt and it transforms in to □-form upon annealing. The α-phase is obtained only under specific conditions such as high pressure crystallization, casting from solution and drawing close to the melting point. The α'-form has been observed for crystallization at high temperatures. In order to study the isothermal crystallization of semicrystalline polymer in the whole temperature range between the glass transition temperature and the melting temperature, high cooling and heating rates are required to be sure that the crystallization takes place isothermally. In this research Flash DSC experiments were combined with in-situ wide angle X-ray scattering at high time resolution. We found that high super-cooling from the melt leads to only one strong reflection characteristic of the hexagonal structure of the □'-form and □-form. At lower supercooling, PA12 crystallizes into α'-form showing the two reflection peaks characteristic of a monoclinic structure. Most of these transformations are reversible during the heating and cooling, and thus we are able to investigate the structural changes of PA12 with in-situ X-ray experiment during fast heating/cooling. These results will provide insight in the complex phase behaviour of PA12 and the obtained data is used to develop a full crystallization model, based on well established knowledge present in our group, that will permit to predict the final structure of the material depending on the processing conditions.

Flow visualization of thin-wall injection molding and effect of mold coatings

***Masato Davide** (1)*, **Babenko Maksims** (2), **Lucchetta Giovanni** (1), **Whiteside Ben** (2)*

(1) Department of Industrial Engineering, University of Padova - Padova - Italy, (2) University of Bradford - West Yorkshire - UnitedKingdom

Injection molding of thin-wall parts is a challenge due to the raise of cavity pressure during the filling phase that can prevent the complete replication of mold geometry. The melt flow in thin-wall injection molding is characterized by high shear stress that can modify the no-slip boundary condition at the polymer-mold interface. Mold coatings can be used to reduce cavity pressure by exploiting a combination of thermal isolation of the flow and induction of the wall-slip phenomenon. In this work, the effects of different mold coatings (viz. diamond like carbon DLC, chromium nitride CrN, chromium titanium niobium nitride CrTiNbN) on the efficiency of the filling phase were investigated. The design of the mold allowed the high-speed visualization of the melt polymer flow through the 400 μm thickness of the cavity. The velocity of the polymer melt was evaluated by tracking the coordinates of some silica carbide particles dispersed in the polymer. The results indicated how mold coatings can be used to significantly reduce cavity pressure and thus improve the filling phase in thin-wall injection molding.

Thermoforming – Determining Temperature Profiles Across Sheet Thickness by Pyrometric Measurements

Benjamin Neubig[1, a)] and Christian Bonten

[1]*Institut für Kunststofftechnik, University of Stuttgart, Pfaffenwaldring 32, 70569 Stuttgart, Germany*

[a)]*Corresponding author: benjamin.neubig@ikt.uni-stuttgart.de*

Abstract. Thermoforming generally describes the shaping of a thermoplastic semifinished part above its softening temperature to a three dimensional part. The sub-step of heating the sheet can be seen as the very crucial process step. In most thermoforming machines, the sheet is heated by means of infrared radiation inside a radiation furnace. Depending on the thickness and the absorption characteristics of the sheet material as well as the working temperature of the radiators, different temperature profiles along the sheet thickness will occur. The knowledge of this temperature distribution is important to control the heating step since the sheet temperature directly influences process stability and part quality like sagging, burnings, orientations and wall thickness distribution of the part or accurate reproduction of surface details. In this paper a new approach by combining three different infrared sensors for determining the temperature distribution along the sheet thickness at any time of the heating phase will be presented. At first, the different amounts of radiation will be calculated in order to separate reflected and transmitted radiation from the actual measuring signals. At second the sheet will be divided into a certain number of layers and the amount of emitted radiation of each layer will be calculated and compared to measured signals in order to finally determine the real temperature distribution along sheet thickness. Calculated and measured results will be shown using polypropylene as sheet material.

INTRODUCTION

In thermoforming the temperature of the thermoplastic sheet can be seen as the most crucial process variable. The temperature directly influences the stretching behavior of the sheet and hence the wall thickness distribution of the final part as well as the accurate reproduction of the surface details [1]. In most thermoforming processes heating the sheet takes place inside a radiation furnace consisting of an upper and lower radiator field. In addition, the sheet temperature often is monitored simultaneously to heating inside the furnace. Therefore one or a small number of pyrometers are mounted inside mainly in the upper radiation field (see Fig. 1) to measure sheet temperature at specific points at its surface.

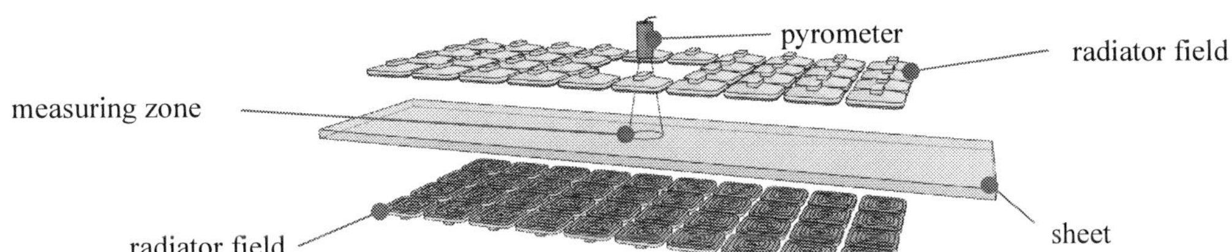

FIGURE 1. Measuring set-up in thermoforming machines

Pyrometers measure electromagnetic radiation and convert the measured signal into a corresponding temperature value. In the field of plastics processing this method of temperature measurement is established since decades [2]. However, applying this method to thermoforming processes with the set-up shown above some important aspects have to be taken into account in order to obtain reliable measurements: thermoplastics, especially unfilled materials, are partial transparent for electromagnetic radiation. Hence, with regard to Fig. 1 broad band pyrometers additionally measure contents of background radiation from the lower radiator field. These additional amounts in turn lead to displayed temperature values that normally are much higher than the sheet temperature really is.

The partial transparency also causes another effect. Most materials emit radiation as function of temperature only from a very thin layer underneath the surface. In contrast, thermoplastics emit radiation from a certain volume that means from deeper layers inside the material. Now, assuming not a homogeneous temperature but a temperature degree from sheet surface to sheet core for instance, then the displayed temperature values are merely mean values. Information about the real temperature degree does not exist.

Our investigations at IKT deal with both mentioned aspects: considering and particularly eliminating annoying background radiation from the actual measuring signal on the one hand as well as getting information about temperature degrees along sheet thickness on the other hand. Due to the shortness of this article, the following contents focus on the latter.

FUNDAMENTALS

As mentioned previously, pyrometers measure electromagnetic radiation, which each body with a temperature above absolute zero point (-273.15 °C = 0 Kelvin) emits. The emitted radiation, or more precisely the spectral radiant exitance, is proportional to the intrinsic temperature of each body [3]. Based on this correlation, pyrometers convert the measured radiation in a corresponding temperature value. For this purpose they consist of a collecting lens, a spectral band filter, a radiation detector and evaluation electronics [4], as illustrated in Fig. 2 (a). Different spectral band filters enable pyrometric measurements in different wavelength ranges and a classification in total radiation pyrometers, broadband radiation pyrometers and spectral pyrometers (Fig. 2 (b)).

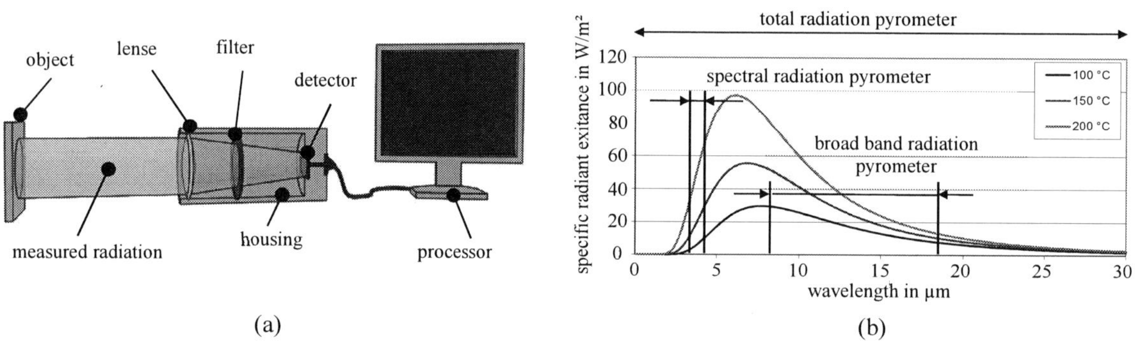

FIGURE 2. Basic components of a pyrometer (a) and measuring range of different pyrometer types (b)

Assuming the measuring set-up as shown in Fig. 1, partial transparency of the sheet material as well as using a broad band pyrometer the total measuring signal then consists of three components: the emitted radiation by the sheet $M_{sh,\lambda}$, the transmitted radiation through the sheet $M_{tr,\lambda}$ and the reflected radiation from the sheet's surface $M_{re,\lambda}$:

$$M_{to,\lambda_1-\lambda_2}(T^*) = \int_{\lambda_1}^{\lambda_2} M_{sh,\lambda}(T_{sh})\,d\lambda + \int_{\lambda_1}^{\lambda_2} M_{tr,\lambda}(T_{rf})d\lambda + \int_{\lambda_1}^{\lambda_2} M_{re,\lambda}(T_{rf})d\lambda. \qquad (1)$$

In equation 1 the T^* means the apparent temperature displayed by the pyrometer, index λ implicates the dependence on wavelength, T_{sh} is the sheet temperature and T_{rf} the temperature of the radiator field. With this radiation balance one is able to eliminate the annoying amounts of background radiation that distort the measured temperature. How we applied this equation for this purpose as well as we handle the calculation of transmitted and reflected radiation has been published for example in [5].

In the following we focus on how to calculate the emitted radiation from the sheet itself and on our approach how to get information about temperature degrees along the sheet thickness. For the first, we divide the sheet into a particular number of single layers according to [2] as illustrated in Fig. 3 (a). Then for each layer the emitted radiation can be calculated by:

$$M_{s_i,\lambda_1-\lambda_2}(T_{s_i}) = \int_{\lambda_1}^{\lambda_2} \varepsilon_{s_i,\lambda} \cdot M_{b,\lambda}(T_{s_i}) d\lambda. \tag{2}$$

In equation 2 ε_{si} is the emissivity for each layer thickness as function of wavelength. The determination of ε_{si} is also explained for example in [5]. $M_{b,\lambda}$ represents the radiation of a black body according to Planck's law

$$M_{b,\lambda}(T_{si}) = \frac{c_1}{\lambda^5 \cdot \left(\exp\left(\frac{c_2}{\lambda \cdot T_{si}} \right) - 1 \right)}, \tag{3}$$

wherein T_{si} represents each layer temperature according to a particular temperature profile across sheet thickness. Finally, the total amount of emitted radiation along the whole sheet thickness is obtained by accumulating the radiation from all layers of the number n:

$$M_{sh,\lambda_1-\lambda_2}(T_{s_i}) = \sum_{i=1}^{n} \int_{\lambda_1}^{\lambda_2} \varepsilon_{s_i,\lambda} \cdot M_{b,\lambda}(T_{s_i}) \cdot e^{(-k_\lambda \cdot (n-i) \cdot s_i)} d\lambda. \tag{4}$$

The last term in equation 4 corresponds to the Lambert-Beer law in order to consider only the transmitted portion of radiation of an inner layer, when covered by upper layers. The material property k_λ is the wavelength dependent absorption coefficient, which determines the decrease of radiative intensity inside the material (also see [5]). With the given equations above and the hints in [5] one is able to calculate the amount of radiation that only is emitted from the sheet itself when a certain temperature profile is assumed within the wavelength range of the pyrometer that is used. Now, due to the fact different temperature profiles obviously can lead to one and the same quantity of emitted radiation (e.g. profile 1 with higher inner and lower outer temperature can cause the same emitted radiation as profile 2 with lower inner, but higher outer temperatures), one central aspect of our approach is the usage of two broad band pyrometers, that measure radiation in different wavelength ranges (Fig. 3 (b)). In this approach we assume that the emission behavior of the sheet within the first wavelength range is different to the behavior in the second wavelength range. Then we postulate, if this requirement is fulfilled, that hypothetically there must be exactly one particular temperature profile so that equation 1 is fulfilled for the wavelength range of the first pyrometer as well as for the second pyrometer at the same time.

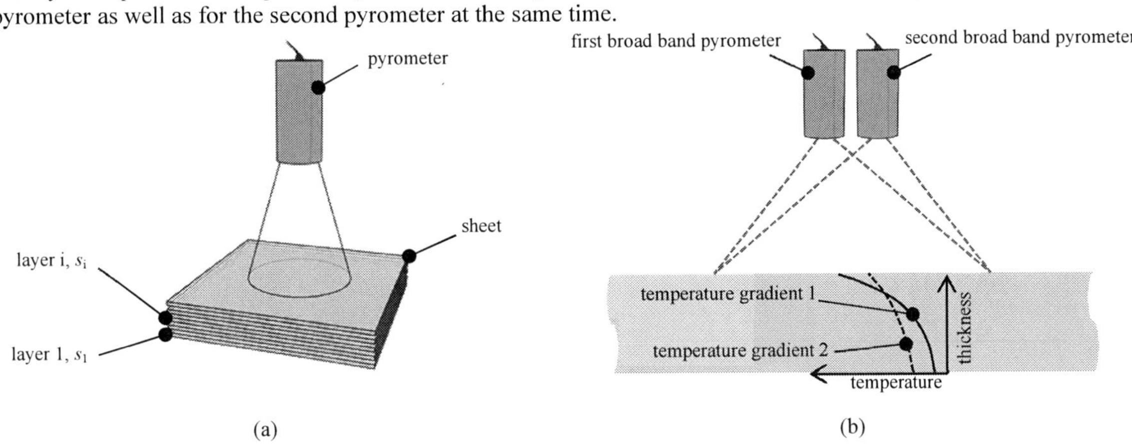

FIGURE 3. Illustration of layer model (a) and measuring set-up of two broad band pyrometers (b)

In order to conclude the temperature profile from the measurements of these two pyrometers at each point of heating we further assume an exponential degree of the temperature with the highest temperature being on the heated surface and the lowest temperature being either in the middle of the sheet or on the unheated surface in case of single sided heating. Mathematically this can be described as:

$$T_{gradient}(s_i) = a \cdot e^{b \cdot s_i} - c \cdot s_i. \tag{5}$$

Using equation 5 the shape of a temperature profile is determined by a constant value a and to variables b and c. The constant a represents the outermost temperature directly at the sheet surface. This temperature is measured by using a third (spectral) pyrometer that detects radiation exactly at a wavelength of 3.43 μm [6]. At this wavelength the most thermoplastics (thermoplastics with C-H-band) show a strong absorption band and this in turn means that emissivity is very high and no transmission of background radiation occurs. For the determination of the variables b and c we developed an algorithm that varies b and c with an optimization routine so that in a first step temperature profiles are obtained that fulfill equation 1 for the first broad band pyrometer with previously defined deviations. From these identified profiles in a second step the temperature profiles are selected that show the same preferably small deviation to the second broad band pyrometer, when equation 1 is applied again.

EXPERIMENTAL

The investigations presented in this article have been carried out using an unmodified Polypropylene (BASELL, Rotterdam/The Netherlands: Moplen HP400H). In a first step the material has been processed to sheets with a thickness of 1 mm using a laboratory flat film extrusion line (COLLIN, Ebersberg/Germany: ECS-30, screw diameter 30 mm, screw length 750 mm). In a second step these sheets have been compressed to sheets with a nominal thickness of 4 mm. Doing so thermocouples (type K) have been embedded between each interface in order to measure the temperature inside the compressed sheet at different points along the sheet thickness (see Fig. 4 (a)).

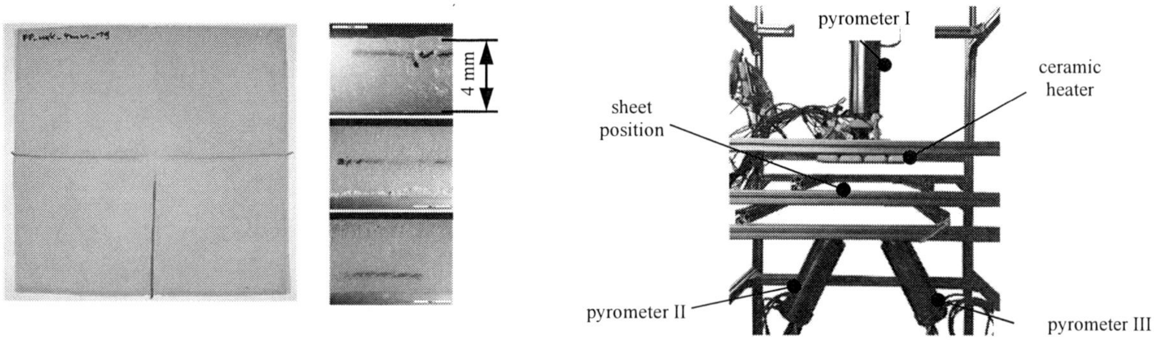

(a) (b)

FIGURE 4. Compressed sheets of PP (a) and experimental set-up (b)

The compressed sheets have been heated up using the experimental set-up shown in Fig. 4 (b). This set-up represents an infrared oven with an upper and a lower heating shield. Each shield consists of 16 single ceramic heater elements (ELSTEIN, Northeim/Germany: HTS/4). The operating temperature of the ceramic heater elements were set to 400 °C using a control system of type HASCO, Lüdenscheid/Germany: Z 1240/8/16/S. Furthermore, according to our approach three different pyrometers have been installed into the set-up: Pyrometer I (HEITRONICS, Wiesbaden/Germany: CT13.3i) is a spectral pyrometer, which detects radiation at 3.43 μm and hence measures the surface temperature of the sheet as described above. Pyrometer II (HEITRONICS, Wiesbaden/Germany: CT13.10) and Pyrometer III (HEITRONICS, Wiesbaden/Germany: KT15.81) are broadband radiation pyrometers and detect radiation within the range of 8 to 14 μm respectively 8 to 10 μm of wavelength.

The pyrometers can be arranged above or underneath the sheet so that the measurement is done either from the upper or the lower sheet surface. In this article only the case when pyrometer I is arranged above the sheet as shown in FIGURE 4 (b) is considered. The sheet has been heated single-sided from the upper sheet surface. For the measurements with pyrometer I the four center ceramic heater elements have been removed from the radiation shield so that the surface temperature of the sheet still can be measured.

RESULTS AND DISCUSSION

For determining the temperature degrees according to equation 5 we virtually divided the 4 mm sheet into 20 single layers. One requirement of our approach has been the different emission behavior of the sheet material within the measuring ranges of the broad band pyrometers. Therefore, Fig. 5 (a) shows the amount of calculated (equation 2 and 4) emitted radiation that comes from each layer to the sheet surface, when a uniform sheet temperature of 100 °C is assumed. The left side of Fig. 5 (a) corresponds to the upper sheet surface faced to pyrometer I, the right side to the lower sheet surface facing pyrometer II and III. From the diagram one can see that regarding to the sheet volume the most radiation comes from layer 11 to 20 that means from almost the half sheet thickness. Furthermore one can see the difference in the two pyrometer measuring ranges with lower emitted radiation in layer 19 and 20 and higher emitted radiation beneath layer 16 in the range of pyrometer II in comparison to the range of pyrometer III as well as the change in layer 17 and 18. Hence, the required condition above is fulfilled.

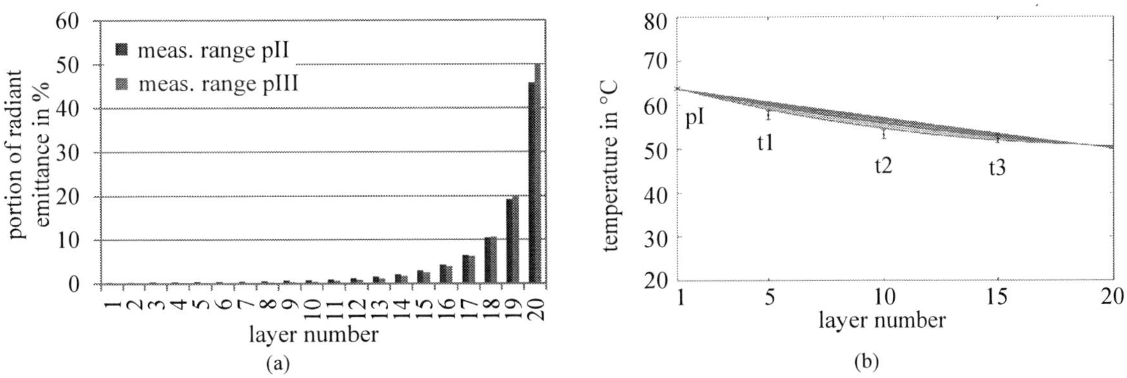

FIGURE 5. Calculated emitted radiation of each layer (a) and determined temperature degrees along sheet thickness (b)

Applying our algorithm and using equation 5 Fig. 5 (b) shows the determined temperature degrees along the sheet thickness at the particular time of 120 s heating. Again the left side corresponds to the upper and the right side to the lower sheet surface. Additionally Fig. 5 (b) contains the measured temperature values of the embedded thermocouples (t1 to t3) as well as the surface temperature measured by pyrometer I. All determined temperature degrees shown in Fig. 5 (b) result in a total deviation of only ±0.5 % to the calculated target value for pyrometer II and only -0.25 % to the target value for pyrometer III that were calculated by applying equation 1. A further differentiation of the presented temperature degrees has not been possible yet. However, the finding is a very tight band of temperature degrees that come very close to the reference temperature values (thermocouples) with the highest deviation of only 3 K.

CONCLUSIONS

The right determination of sheet temperature is still one of the key factors in thermoforming processes, if one aims to compare simulated with experimental results. This paper discusses the difficulties by measuring the sheet temperature inside a radiation furnace as well as an approach to determine the temperature distribution along the sheet thickness during the heating phase. On the example of single sided sheet heating it could be shown that with this approach a tight band of temperature gradients could be determined that are very close to the reference temperature values that have been measured by thermocouples embedded inside the sheet.

REFERENCES

1. J. L. Throne, Technology of Thermoforming, Hanser, New York, 1996.
2. W. Obendrauf, *doctoral thesis*, Leoben, 1996.
3. U. Grigull, *Wärmeübertragung durch Strahlung,* Teil 1, Springer, Berlin, 1988.
4. K. Stahl, G. Miosga, *Infrarottechnik*, Hüthig, Heidelberg, 1986.
5. B. Neubig, C. Bonten, Model-based Temperature Measurement for Thermoforming Applications, ANTEC, 2015.
6. K. D. Gruner, and H. Vogel, *Kunststoffe Synthetics*, Vol. 10, 26-30, 1995.

Influence of Hygro-Thermal Loads on the Durability of Thermoplastic-Polyurethane-Steel-Hybrids

Alejandro Puentes-Parodi[1,2], Leandro A. Santoro[1,3], Michael Gehde[2], Andreas Leuteritz[1], Ines Kuehnert[1,a)]

[1]*Leibniz-Institut fuer Polymerforschung Dresden e. V., Dresden, Germany*
[2]*Technische Universität Chemnitz, Chemnitz, Germany*
[3]*Technische Universität Dresden, Dresden, Germany*

[a)] Corresponding author: kuehnert@ipfdd.de

Abstract. The growing demand of polymer-metal hybrids as structural components in the automotive, aerospace, medical and military applications may require long-term stability under thermal stress. A full understanding of the long-term performance of bonded structures subjected to high temperature and humidity is required to use these products in engineering and industrial applications. In real applications, temperature will be higher at the proximity to the steel substrate, therefore, the boundary between steel and polymer suffers the highest hygrothermal stress and, therefore, a collapse of the structure starting at this boundary is most likely. In this work, kinetic parameters of hygrothermal decomposition of a thermoplastic polyurethane, TPU, were evaluated through the Ozawa-Flynn-Wall, Friedman and Chang methods after being subjected to atmospheres of 85 °C/40 % RH, and 40 °C/90 % RH for a period time of 1000 hours. Thermogravimetric tests were conducted at three different heating rates, 5, 10 and 15 K/min and different kinetic parameters were obtained, such as the activation energy, reaction order and the pre-exponential factor. These parameters were useful to predict the lifetime of the polymer that was in direct contact with the metal substrate, assuming a continuous operation work at certain temperatures.

INTRODUCTION

Investigation of durability of polymer-metal hybrids used in technological applications requires generally long-term experiments, especially when it is essential to evaluate thermal stresses and their effects on the lifecycle of the composite [1]. Natural aging tests are time-consuming [1–3], expensive and therefore not applicable for industrial development. On the other hand, thermogravimetric analysis (TGA) allows fast investigation of thermal decomposition with small probes in short time [6,7]. TGA can be helpful in predicting the lifetime of the hybrid part by evaluating the thermal decomposition phenomena at the polymer-metal joint. With regard to the long-term performance of thermoplastic-polyurethanes (TPU) jointed to metal substrates there is little information in the literature about its correlation with the thermal degradation process at the organic/inorganic interface [1,4,5]. In this work, small samples of TPU in the proximity of the steel substrate were used for investigation, as depicted in **Figure 1**, and kinetic parameters of solid decomposition were calculated to estimate the lifetime of the hybrid part.

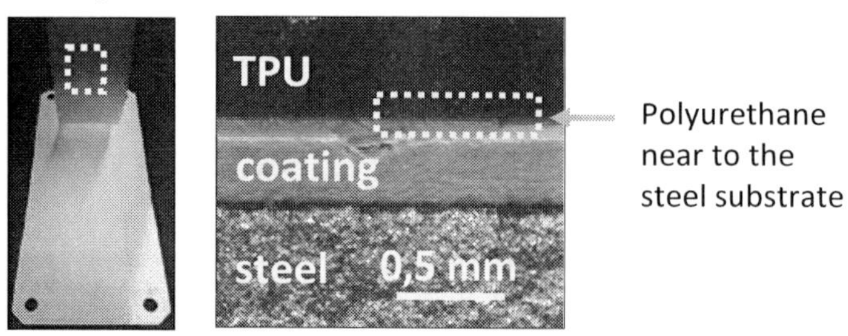

FIGURE 1. Illustration of TPU specimens in proximity to the steel substrate for TGA analysis.

MATERIALS

Phosphated Steel

A commercial low-carbon steel sheets, grade DC01, 2 mm thick, with a Zinc-phosphated superficial pretreatment were used as the material base for the hybrids. Steel sheets were obtained directly from an industrial partner.

Thermoplastic Elastomer Urethane

The commercial TPE is a polyester-based Thermoplastic Elastomer Urethane, TPU, ref. Elastollan C85A10 from BASF. It is characterized by its excellent tensile strength and high elongation at break, good damping characteristics, high resilience performance and excellent wear resistance [8]. The specific mechanical properties of TPUs are attributed to its multiphase chemistry with soft and hard segments [9]. Elastollan grade C85A10 TPU is formed basically from the combination of three components: the long chain polyester-based polyols, methylene diphenyl diisocyanates (MDI), and short diol chains. The linear polyurethane is formed by a polyaddition polymerization between the polyols and the short chain diols that react with the diisocyanates. The combination of a diisocyanate with the short chain diol produces the rigid component [10] as illustrated schematically in **Figure 2**.

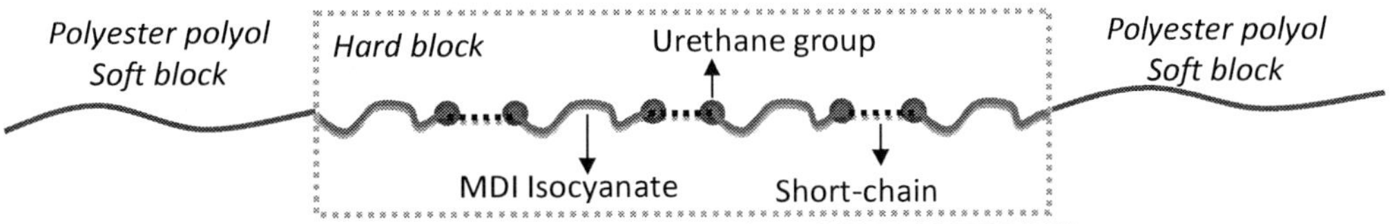

FIGURE 2. Schematic representation of C85A10 TPU [10].

Adhesive Promoter

An adhesive powder coating promoter, consisting of a commercially available low molecular weight polyester resin and an uretdione crosslinker, was used in this work. The powder coating was applied to the metal substrate by electrostatic charging gun, and it was baked in an oven at 150 °C for 15 min.

HYBRID PREPARATION

Manufacturing stages of polymer-metal hybrids consist of 4 steps. Plain steel sheets were cleaned and degreased with a solution of acetone, benzene and ethyl acetate to remove impurities. After cleaning, the adhesive powder-coating was applied with an electrostatic charging gun, and then the samples were placed in an oven at 150 °C for 15 min to crosslink the powder. Subsequently the TPU was overmolded on the pre-coated metal substrate in an Engel Victory 80 vertical injection unit. Finally, the samples were subjected to annealing in an air circulating oven at 100 °C for 20 hours. The fabrication process of the hybrid specimens is summarized in Figure 3.

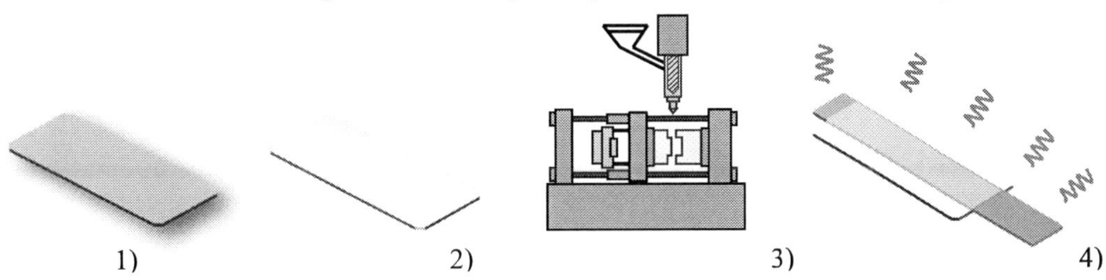

FIGURE 3. Manufacturing stages of polymer-metal hybrids. 1) Steel sheet preparation. 2) Powder-coating on substrate and crosslinking. 3) Overmolding of TPU. 4) Annealing.

EXPERIMENTAL

Artificially Aging

Artificially aging of hybrid parts was conducted at 85 °C/40 % RH and at 40 °C/90 % RH for 1000 hours in an environmental chamber (ESPEC-LHU from Japan). After aging samples were directly analyzed in the TGA device.

Thermogravimetric Analysis, TGA

Thermogravimetric analysis of TPU on direct contact with the steel substrate was performed by a TGA Q5000 of TA Instruments, from room temperature to 800 °C, under air at three different heating rates: 5, 10 and 15 K/min. The examination was conducted on fresh and hygrothermally aged polymers. From the TGA data, kinetic parameters of thermal degradation such as the activation energy (E_a), degradation order (n), and the pre-exponential factor (ln A) were calculated by the Ozawa-Flynn-Wall (OFW), Friedman and Chang methods [11–15].

Kinetic Parameters of Thermal Decomposition

Many experimental methods are used to determine the kinetic parameters and lifetime of the samples. Among the most common are the OFW, Friedman and Chang experimental method [11–15].
The OFW expressed in Equation (1), has been widely used to determine the E_A from thermogravimetric data. It is derived from the basic kinetic equation for heterogeneous chemical reactions and therefore has a extensive application in thermal decomposition investigations [11,7,17].

$$\frac{d\,log\,\beta}{d\frac{1}{T}} \cong \left(\frac{0{,}457}{R}\right) x E_A$$

1) Ozawa-Flynn-Wall method

Where β is the constant heating rate, T is the absolute temperature, R the gas constant and E_A the activation energy. The OFW method requires thermogravimetric data at different heating rates to calculate E_A in accurately way.
Then the Friedman and Chang methods, in Equation (2) and Equation (3) respectively, were used to compare the decomposition kinetics after the artificial aging of samples [12,18,19].

$$ln\left(\frac{d\alpha}{dt}\right) = lnA + nln(1-\alpha) - \frac{E}{RT}$$

2) Friedman method

$$\frac{ln\left(\frac{d\alpha}{dt}\right)}{(1-\alpha)^n} = lnA - \frac{E}{RT}$$

3) Chang method

Where $d\alpha/dt$ is the variation in the weight loss of the sample after a certain time, t, n is the decomposition reaction order A the pre-exponential factor, E_A activation energy, R the gas constant and T is the absolute temperature for a determined degradation system.

Table 1 exhibits the kinetic parameters for thermal decomposition of the TPU that was in direct contact with the steel sheet. These parameters were obtained from the methods referred above. E_A calculated by OFW and Chang methods were closer to the values found in the literature [2,11]; so lifetime estimations, displayed in Figure 4, were calculated from kinetic parameters obtained by the Chang experimental approach.
TGA data was extracted from specimens under air atmosphere in order to simulate normal service conditions.

TABLE 1. Different decomposition kinetics parameters calculated from OFW, Chang and Friedman for fresh and hygrothermally aged samples at 40 °C/90 %RH. Heating rate was at 10 K/min under air atmosphere.

Sample	OFW	Chang	Friedman
	E_A (kJ/mol)	E_A (kJ/mol)/n/ln A	E_A (kJ/mol)/n/ln A
Fresh	165	154/3/27	138/11,3/40
625h	162	151/3,2/27	104/8/22
1000h	78	147/3,1/26	148/12,3/28

Durability and Lifetime of TPU

Lifetime prediction of polymeric systems based on thermogravimetric data has been analyzed widely in the literature [11,15–17]. Lifetime can be predicted from the integration of the Arrhenius expression, Equation (4), as demonstrated elsewhere [17] and it is interpreted as follows:

$$t_f = \frac{1 - 0{,}95^{1-n}}{A(1-n)} exp\left(\frac{E_a}{RT}\right) \quad when\ n \neq 1$$ 4) Lifetime prediction

Decomposition kinetic parameters required for Equation (4) were determined experimentally following the methods described above. E_A calculated by OFW and Chang were closer to the values found in the literature for urethane-based polymers [2,11].

Table 2 exhibits the predicted lifetime of the fresh and pre-aged TPU near to the steel surface at certain continuous operation temperatures (COT). Estimations were obtained from equation 4), using kinetic parameters of solid decomposition from the Chang method, as indicated in Table 1. Specimens pre-aged for 1000 hours at the higher temperature (85°C/40% RH), showed shorter life spams when compared to the fresh and pre-aged samples at 40°C/90% RH. On the other hand, durability of fresh and pre-aged TPUs exposed to lower COTs, as in real engineering applications (T ≈ 50 °C), presented an hypothetical unlimited lifetime. The determined values of E_A show that durability of the TPU is definitely affected by temperature and humidity at the conditions described in this work. Lifecycle may be directly related to the kinetic parameters determined experimentally from TGA experiments, as detailed in table 1. E_A´s of pre-aged specimens, and especially for those subjected to the higher temperature conditions, were lower when compared to the fresh polymer. All three TGA methods evaluated showed the same tendency, meaning that lower E_A transfers into a shorter lifetime. The estimations for lifetime are not plausible as ageing for 1000 h at 90 °C/40 % RH leads to a predicted reduction of 100 years under similar thermal load. This demonstrates that further investigation for reliable prediction is still a requirement.

TABLE 2. Lifetime estimations (in years) of fresh and hygrothermally pre-aged TPU exposed to certain continuous operation temperatures.

	Lifetime Estimations, in years, of TPU exposed to Certain Continuous Operation Temperatures			
TPU Samples	**50 °C**	**100 °C**	**120 °C**	**150 °C**
Fresh		350 y	25 y	0.7 y
Pre-aged for 1000 h at 90°C/40%RH	100000+	250 y	18 y	0.4 y
Pre-aged for 1000 h at 85°C/40%RH		105 y	8 y	0.3 y

CONCLUSION

Specimens that were hygrothermally pre-aged for 1000 hours exhibited always a shorter lifecycle when they are subjected to different continuous operation temperatures. The lifetime of the TPU under thermal stresses may be correlated to the lower E_A determined by Chang and OFW methods. From the lifetime estimations, as presented in Table 2, there is evidence suggesting that durability of TPU is largely influenced by the combination of environments with high temperatures and certain degree of humidity. The combination of high humidity and low temperature (40 °C/90 % RH), which is normally found in tropical environments, resulted in less harmful effects in the durability of the polymer when compared to specimens subjected to a high temperature and low humidity (85 °C/40 % RH).

It is also worth to mention that accurate lifetime prediction for polymeric systems is, in most of the cases, a rather difficult task due to a number of variables involved in the service operational conditions, such as corrosion on the metal substrate, unexpected variations in humidity and temperature, quality of raw materials, among others. However, it was evidenced that TGA is a primary tool to predict lifetime, although results showed large deviations for some specific samples and further investigation is urgently needed for reliable values.

ACKNOWLEDGMENTS

The authors thank to the IPF team for all the support in the present work, especially to Martin Zimmerman, Andreas Scholze and Kristina Eichorn. The first author thanks to Colciencias in Colombia for the financial support to the research in Germany through the Scholarship "Beca Caldas 529".

REFERENCES

[1] Leuteritz A, Döring K-D, Lampke T, Kuehnert I. Accelerated ageing of plastic jacket pipes for district heating. Polym Test 2016;51:142–7.

[2] Ingenieurgesellschaft für Energie-technik und FernwärmeChemnitz mbH, IMA Materialforschung und Anwendungstechnik, Leibniz Institut für Polymerforschung. Zeitstandfestigkeit von Kunstoffmantelrohren. vol. 1. 2011.

[3] Boubakri A, Elleuch K, Guermazi N, Ayedi HF. Investigations on hygrothermal aging of thermoplastic polyurethane material. Mater Des 2009;30:3958–65.

[4] Hoikkanen M, Honkanen M, Frisk L, Vippola M, Lepistö T, Vuorinen J. Metal–thermoplastic urethane hybrids in environmental exposure. Int J Adhes Adhes 2012;35:21–6.

[5] Puentes-Parodi A, Gedan-Smolka M, Leuteritz A, Kuehnert I. Artificial Aging and Failure Modes of TPU-Metal Hybrid Composites. 32nd, Int. Conf. Polym. Process. Soc., Lyon, France: Polymer Processing Society; 2016, p. 113–4.

[6] Zong R, Hu Y, Wang S, Song L. Thermogravimetric evaluation of PC/ABS/montmorillonite nanocomposite. Polym Degrad Stab 2004;83:423–8.

[7] Gonçalves R, Rocco J, Iha K. Thermal Decomposition Kinetics of Aged Solid Propellant Based on Ammonium Perchlorate - AP/HTPB Binder. Appl. Calorim. a Wide Context - Differ. Scanning Calorimetry, Isothermal Titration Calorim. Microcalorim., INTECH; 2013, p. 325–42.

[8] Frick A, Rochman A. Characterization of TPU-elastomers by thermal analysis (DSC). Polym Test 2004;23:413–7.

[9] BASF. Thermoplastic Polyurethane Elastomers (TPU) Processing Recommendations 2011:28.

[10] BASF. Thermoplastic Polyurethane Elastomers (TPU). Elastollan-Material Properties 2011:1–44.

[11] Li X-B, Cao H-B, Zhang Y. Thermal degradation kinetics of rigid polyurethane foams blown with water. J Appl Polym Sci 2006;102:4149–56.

[12] Kalbende PP, Tarase M V, Zade AB. Preparation , Characterization and Thermal Degradation Studies of p- nitrophenol Based Copolymer. Hindawi Publ Corp 2013;2013:1–9.

[13] Flynn JH, Wall LA. A quick, direct method for the determination of activation energy from thermogavimetric data. Polym Lett 1966;4:323–8.

[14] Howard GT. Biodegradation of polyurethane: a review. Int Biodeterior Biodegradation 2002;49:245–52.

[15] Boubakri A, Haddar N, Elleuch K, Bienvenu Y. Impact of aging conditions on mechanical properties of thermoplastic polyurethane. Mater Des 2010;31:4194–201.

[16] Rychlý J, Lattuati-Derieux A, Lavédrine B, Matisová-Rychlá L, Malíková M, Csomorová K, et al. Assessing the progress of degradation in polyurethanes by chemiluminescence and thermal analysis. II. Flexible polyether- and polyester-type polyurethane foams. Polym Degrad Stab 2011;96:462–9.

[17] Roy PK. Thermal degradation studies of LDPE containing cobalt stearate as pro-oxidant. eXPRESS Polym Lett 2007;1:208–16.

[18] Brown ME. Kinetic Analysis of Isothermal Data. Introd. to Therm. Anal. Tech. Appl. 2nd ed., Springer; 2001.

[19] Yang X, Tallman D, Croll S, Bierwagen G. Morphological changes in polyurethane coatings on exposure to water. Polym Degrad Stab 2002;77:391–6.

Morphology-Property Behavior of Semi Crystalline Polymers in Injection Molded Parts

Y. Spoerer[1], C. Blanco[1,2], M. Zimmermann[1], M. Berger[1,3] and I. Kuehnert[1 a)]

[1]Department of Processing, Leibniz-Institut fuer Polymerforschung Dresden e.V., Dresden, Germany.
[2]Technical University Dresden, Dresden, Germany
[3]Hochschule fuer Technik und Wirtschaft, Dresden, Germany

[a)]Corresponding author: kuehnert@ipfdd.de

Abstract. During injection molding different types of interfaces (weld lines) are formed. The skin-core morphology in particular is affected by the cooling conditions (temperature difference between melt and mold) during the process. Inside the injection molded part the development of the structure depends on the melt flow conditions and the mold design, e.g. at the weld lines. Possible reasons for the development of weld lines are the division of the melt flow by flow obstacles or the use of multiple gates to create complex and/or large parts. This characteristic leads to optical irregularities and/ or mechanical weaknesses due to the structural orientation, which depends on thermodynamic and morphological behavior of the polymer chains. For a better understanding of the specific orientation of the molecular chains it is necessary to know how the structural development is affected by the processing conditions. Here an overview about the influence of the injection molding process on the crystalline morphology and the relation to mechanical properties for different semi-crystalline polymers is given. The crystalline morphology of the samples is visualized by using polarized light microscopy (POM) and the mechanical behavior is measured using a tensile test.

INTRODUCTION

Stringent requirements on the mechanical properties of injection molded parts make a better understanding of the fundamentals of the process – structure – property relationship necessary. Consequently several investigations on the thermal and crystallization behavior of semi-crystalline polymers [1-4] as well as on the processing behavior [5-9] have been conducted. The crystallization kinetics are well described in three phases: nucleation, crystalline growth and the optimization of crystalline structures [1]. Below the crystallization temperature the macromolecular chains will form a stable nucleus. During the crystalline growth the chains are folded into lamellae which eventually form spherulitic structures, where the formation of spherulites is slow in comparison to the nucleation and crystalline growth due to the inhibited mobility of the macromolecular chains [3]. The spherulites can be observed under polarized light due to their birefringent behavior. Parameters that influence the crystalline growth (in a process such as injection molding for example) are the chain length, temperature, pressure and nucleating effects [4]. The crystallization kinetics during the injection process were examined in further investigations [5, 6]. Fountain flow formation inside the cavity was described by Coyle [7], showing the formation of a v-profile at the flow front using tracers. Tadmor et. al [5] described the molecular orientation as a function of steady flow in which the orientation at the surface and the core of the sample is due to the elongation in the progressive front and the shearing behind the flow front of the macromolecules during injection, respectively. It is concluded that the macromolecular orientation of polymer materials during the injection molding process is complex due to the fact that it is a function of the temperature conditions and the rheological behavior of the material during the process. Especially high shear stress can occur in the formation of flow lines [8].

Due to the design of the mold used for the injection molding process the generation of interfaces e.g. weld lines is possible [10, 11]. Weld lines are formed if two melt streams come together because of the usage of flow obstacles inside the cavity or the use of multiple gates. They are evident as optical and/or mechanical blemishes. At the point where the melt streams come together they form a v-notch which is visible at the surface and mechanical weak because of the lower interdiffusion of the molecular chains in this area. Investigations on the tensile strength of amorphous, semi-crystalline and fiber-reinforced thermoplastics have shown that the strongest influence of the weld line compared to the reference without weld line occur in amorphous brittle and fiber-reinforced samples. These results are due to the less interdiffusion of molecular chains at the weld line and the orientation of fibers in the reinforced samples. The tensile behavior of the weld line sample compared to the reference without weld line can be described as weld line factor WF [12, 13]. Assuming that the weld line strength is lower than the strength of the reference, the WF will decrease as the difference between the tensile strength of the sample with weld line and the reference increase. The knowledge about the weld line factor is important for the mold design, e.g. in technical applications.

The study of the morphology-property behavior of semi-crystalline polymers in injection molded parts with and without weld lines is done using an optical investigation, the description of structural development after injection molding and an investigation into the effect on tensile behavior of the material.

EXPERIMENTS

Material and Processing

Two different types of semi-crystalline thermoplastics were investigated, namely, high-density polyethylene and polyamide 6.6 (Table 1). Tensile bars (Figure 1) with and without (reference) a weld line were injection molded from each material while the melt and mold temperature were varied according to the recommended processing temperatures.

TABLE 1. Materials and significant process parameters

Material	Trade Name	Melt temp.	Mold temp.
PE	HDPE 25055E	250 °C	80 °C
PA66	Ultramid A3K	300 °C	40°C/120°C

Optical Analysis and Testing

To analyze the morphology of the sample a small section was cut from the middle part of the sample (as shown in Figure 1). From this section thin cuts were prepared perpendicularly to the flow direction. Images of these thin sections were taken under polarized light using different levels of magnification.

A Zwick/ Roell universal tensile test bench was used to perform a standard tensile test according to DIN EN ISO 527-1 in a standard atmosphere. Testing speeds of 5 and 50 mm/min were used.

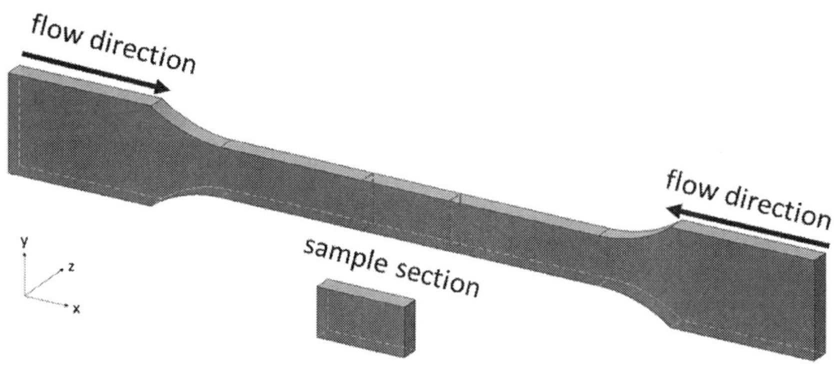

FIGURE 1. Tensile bar sample type 1A and middle section for the preparation of thin sections for polarized light microscopy.

RESULTS AND DISCUSSIONS

After the processing of PA6.6 [9] no significant change in the size of the v-notch was observed depending on the injection molding processing conditions. In Figure 2 a comparison of the v-notch influenced by the processing temperature is shown. The mold temperature for the sample in Figure 2a was 80 °C colder than for the sample shown in Figure 2b. With regard to the influence of processing temperature it was found that the skin area, without visible crystalline structures occurring under polarized light, becomes smaller with decreasing ΔT (where ΔT is defined as the difference between the melt and mold temperature i.e. $\Delta T = T_{melt} - T_{mold}$).

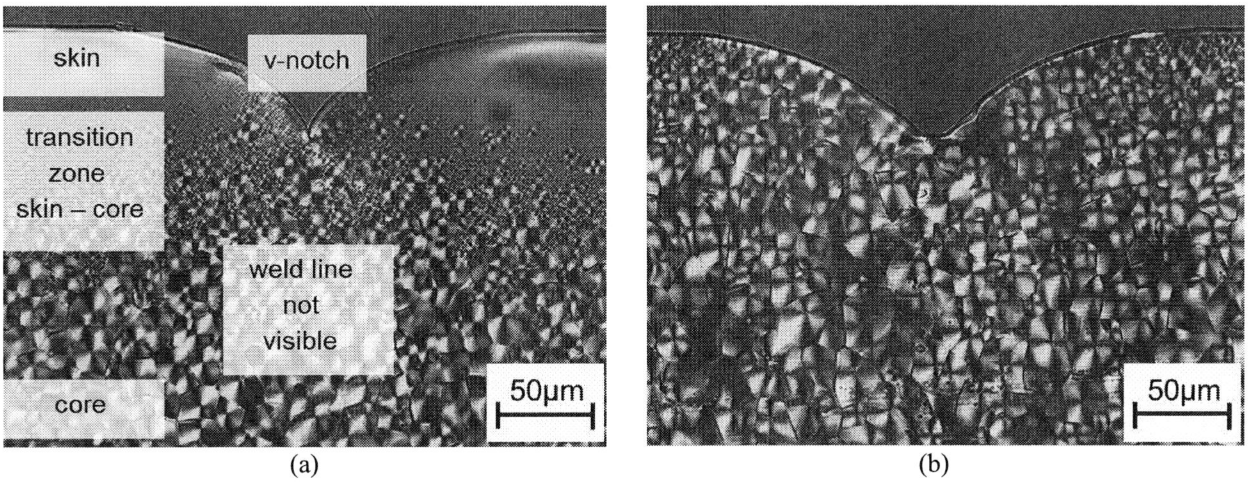

FIGURE 2. V-notch of a weld line sample under polarized light. (a) Material injected into a cold mold. (b) Material injected into a hot mold. [9]

Another effect on morphology behavior of the injection molded parts could be found in the HDPE samples (Figure 3). In the weld line samples the formation of stress induced crystalline flow lines was found due to the thermo-rheological history of the material during the filling phase. Visible is the symmetric formation of flow lines perpendicular to the flow direction. The weld line can be seen as mirror axis. Small spherulites are visible in the region of the flow lines while there are larger spherulites between the flow lines. These flow lines follow the formation as was found by Coyle [7]. With the use of the finite element method and tracers he described the kinematic inversion of the flow front. Considering the obtained morphology of HDPE it is accepted that the formation of flow lines is due to the inversion of the flow front during the filling process and the shearing between the forward moving and slowing down moving melt layers.

Figure 4 shows the tensile strength of PA6.6 and HDPE samples with weld line compared to the reference. The samples were prepared under optimized process conditions. For both materials the WF is approximately 0.9. No significant change in the tensile strength between the samples of each material was observed, like in previous investigations [13]. The negligible lower tensile strength of the weld line samples is maybe due to the different morphology at the v-notch and the crystalline structures depending on the formation of the weld line.

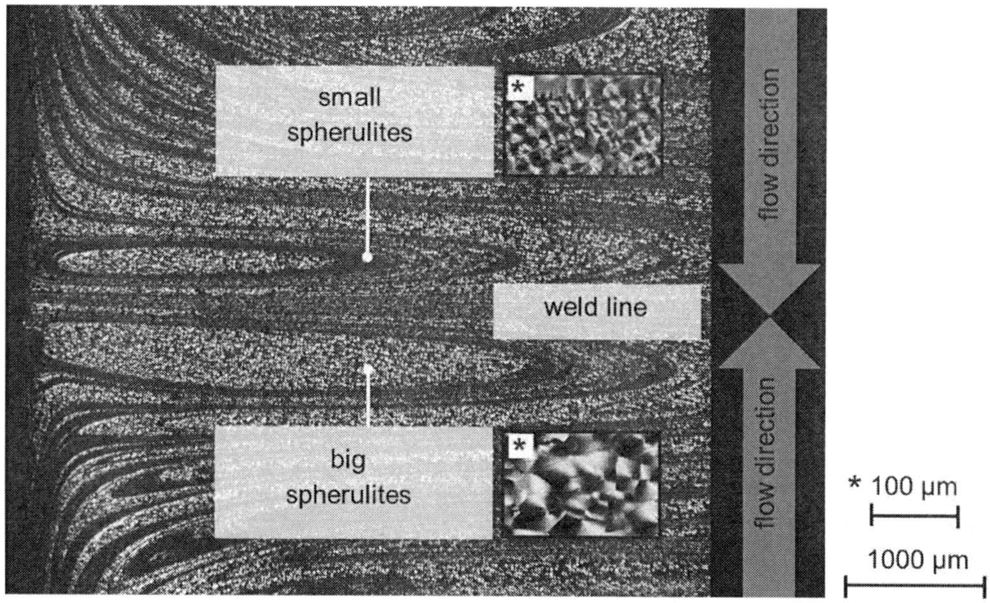

Figure 3. Morphology and flow line formation at a weld line of an HDPE sample under polarized light.

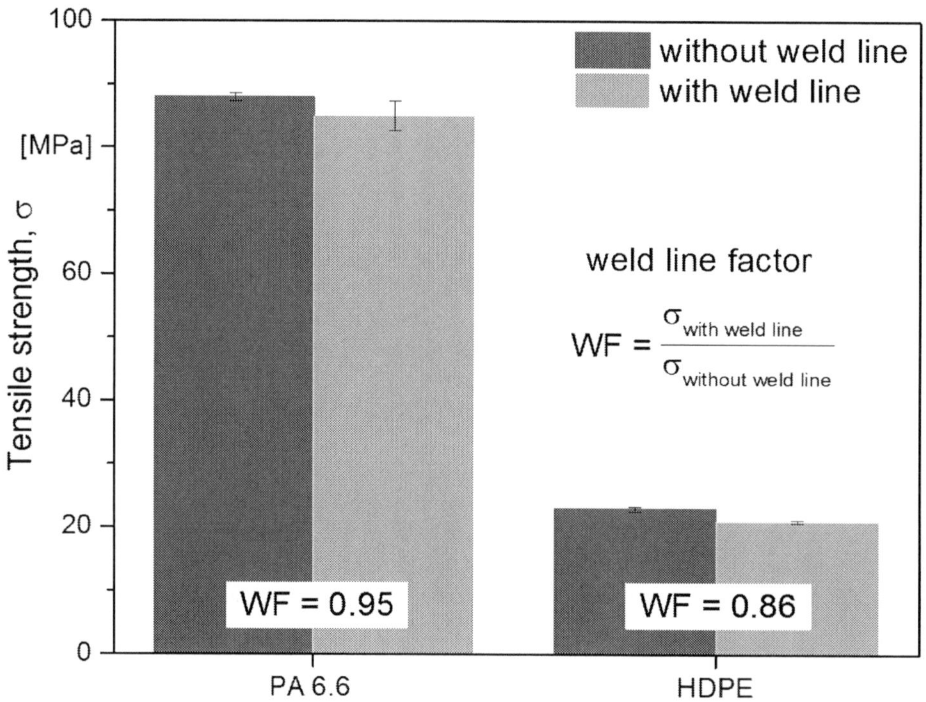

Figure 4. Maximum tensile strength of PA6.6 and HDPE for dog bone samples with and without weld line.

CONCLUSION

In the investigation on the morphology – property behavior of injection molded parts, the influence of the temperature conditions and the flow behavior during the filling phase on the structure development could be shown. Using polarized light it could be observed that the cooling conditions have a strong influence on the v-notch formation and the crystalline morphology at the surface of injection molded parts. The flow behavior during the injection process is also influenced by the processing temperature conditions. In the HDPE samples flow lines could be observed due to the inversion of the melt front during the filling process and the resultant shearing between the moving melt layers. The results of the tensile test have shown that in case of PA6.6 and HDPE after process optimization the effect of weld line on the tensile behavior is negligible. It is believed that the slight lower tensile strength of the weld line samples is due to the formation of a v-notch at the sample surface and the formation of crystalline structures along the weld line. Further investigations on the mechanical properties, e.g. on the elongation behavior, due to the influence of the crystalline formation during the process are part of future publications.

ACKNOWLEDGMENTS

The authors thank the colleagues at IPF for their support, in particular Mr. Scheibner for the mechanical testing, Mr. Scholze for the injection molding of the samples.

REFERENCES

1. Strobl, G.R., *The physics of polymers :concepts for understanding their structures and behavior*. 2. corr. ed., Berlin ; Heidelberg [u.a.]: Springer 1997.
2. Ehrenstein, G., G. Riedel, and P. Trawiel, *Thermal Analysis of Plastics - Theory and Practice*. Carl Hanser Munich: Carl Hanser 2004.
3. Sperling, L.H., *Introduction to physical polymer science*. 4. ed. ed., Wiley-Interscience Hoboken, NJ: Wiley-Interscience 2006.
4. Reiter, G., *Progress in understanding of polymer crystallization*. Berlin ; Heidelberg [u.a.]: Springer 2007.
5. Tadmor, Z., *MOLECULAR-ORIENTATION IN INJECTION MOLDING*. Journal of Applied Polymer Science, 1974. **18**(6): p. 1753-1772.
6. Kamal, M.R. and V. Tan, *ORIENTATION IN INJECTION MOLDED POLYSTYRENE*. Polymer Engineering and Science, 1979. **19**(8): p. 558-563.
7. Coyle, D.J., J.W. Blake, and C.W. Macosko, *THE KINEMATICS OF FOUNTAIN FLOW IN MOLD-FILLING*. Aiche Journal, 1987. **33**(7): p. 1168-1177.
8. Kamal, M.R., A.I. Isayev, S.-J. Liu, and J.L. White, *Injection Molding - Technology and Fundamentals*. Munich: Carl Hanser Verlag 2009.
9. Berger, M.,Untersuchungen zum thermomechanischen Verhalten und zur Kristallisation ausgewählter Thermoplasten für die Spritzgießverarbeitung. Bachelor Thesis, 2015, IPF-Dresden (Dr.-Ing. I. Kühnert) and HTW-Dresden (Prof. G. Lange).
10. Mennig, G., *Knit-line behaviour of polypropylene and polypropylene-blends* in *Polypropylene: structure, blends and composites*. Chapman & Hall: Chapman & Hall London. p. 205-226 1995.
11. Kuehnert, I. and I. Pompsch. *Morphology and strength of injection molded parts with interfaces*. Conference Proceedings in *ANTEC'2011*. 2011. Boston, USA.
12. Kuehnert, I., Y. Spoerer, and M. Zimmermann. *Weld Lines In Injection Molded Parts: Strength, Morphology and Improvement*. Conference Proceedings in *SPE ANTEC*. 2016. Indianapolis USA.
13. Kuehnert, I., M. Berger, Y. Spoerer, and M. Zimmermann. *Weld Line Behavior in Injection Molded Parts*. Conference Proceedings in *Polymer Processing Conference*. 2015. Korea.

Shrinkage, Warpage and Residual Stresses of Injection Molded Parts

Tristan Koslowski and Christian Bonten

Institute for Kunststofftechnik, University of Stuttgart
Pfaffenwaldring 32, 70569 Stuttgart.

tristan.koslowski@ikt.uni-stuttgart.de

Abstract. The shrinkage and warpage behavior of injection molded parts is very complex. By using a new measuring method it was shown for the first time that the warpage can be classified as warping and distortion. In general, warpage can only be reduced substantially by a sufficiently high volumetric compensation in the holding pressure phase.

INTRODUCTION

Injection molding allows to produce geometrically complex parts at low cost in a single process step. [1] During and after production, changes in the produced part´s shape can occur, caused by shrinkage, volume contraction and thus warpage. [2] In many publications shrinkage and volume contraction unfortunately are not distinguished. The term "volume contraction" describes the volume decrease while the term "shrinkage" describes the polymer chains´ reorientation and thus relaxation while the volume keeps constant. This paper uses "contraction" and "shrinkage" in the above mentioned manner.

The contraction and warpage behavior of the plastic parts has to be taken into account during the design and manufacturing phase of injection molds. Therefore, the mold shapes are generally made larger than the specified part dimensions. The prediction of contraction and warpage behavior is very complex. Nowadays contraction can already be predicted very well, but this is not true for warpage.

The influence factors on warpage are still far less understood. Thus, very cost-intensive mold changes are often necessary until the specified part shape is achieved.

STATE OF ART

The contraction and warpage of a plastic part are usually listed together in the literature. Contraction describes the dimensional deviation of a plastic part to the mold and is a normalized value, which can be determined for each plastic on standard specimen [3]. For warpage, however, the situation is different. In the German standard DIN 16742, warpage is defined as the sum of deviations in form, position and angle of the part due to a flexural warping, and torsional. An anisotropic contraction causes residual internal stresses in the part, which leads to warpage. [4] Due to the cooling of the part, a residual stress gradient as shown in Figure. 1 (a) occurs. After demolding, an anisotropic contraction over the wall thickness leads to a deformation of the part shape (Figure. 1, b). Thus a stress equilibrium is reached. [5, 6, 7]

If homogenous contraction is present in all spatial directions, it can be assumed that only a very small warpage and residual stress occurs. This makes clear that the effects on the contraction behavior have to be considered separately before investigating the warpage. In this work, the different influencing factors on the contraction are distinguished into material-dependent, process-dependent and shape-dependent contraction.

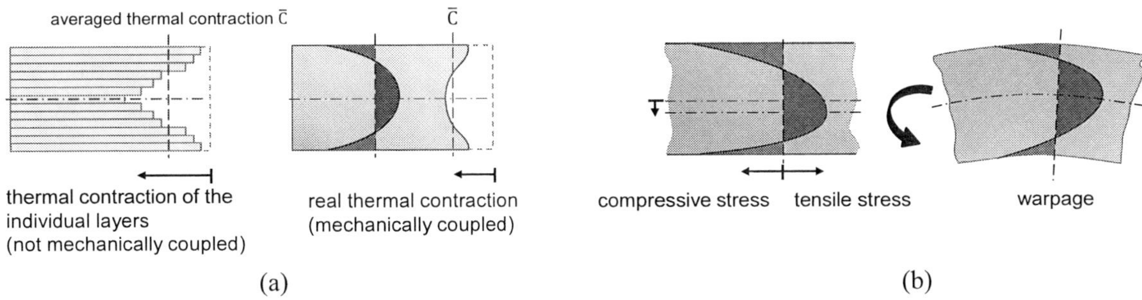

FIGURE 1. Residual stresses due to cooling effects. [5, 6]

The **material-dependent contraction** behavior can be shown by means of a pvT-graph. The contraction of semi crystalline thermoplastics is higher than that of amorphous plastics because of the crystalline phase. The reason is a lower free volume in the solid state, which is caused by the tight crystalline arrangement of the polymer chains [8].

The **process-dependent contraction** is influenced by a large number of different factors. In addition to the injection mold design and the positions and shape of the filling points and cooling channels, the process parameters have to be considered. Each of these factors influences the material characteristics during the injection molding process. [9, 10]

The shape of an injection-molded part has a decisive influence on the contraction behavior of the plastic. Depending on the part shape and mold concept, different dimensions and tolerances can be reached. In the withdrawn German standard DIN 16901 as well as in the current German standard DIN 16742, the **shape-dependent contraction** is divided into tool-specific dimensions and non-tool-specific dimensions [4]. The non-tool-specific dimensions must also be further differentiated into two types of contraction. Overall, there are three types of a shape-dependent contraction. The free contraction only occurs in the thickness direction. In this case the plastic can contract unhindered. Longitudinally and transversally to the flow direction of the plastic there is an internal contraction hindrance due to already cooled edge layers. The result is a hindered contraction. An additional external contraction hindrance is caused by a contraction of the plastic on mold cores or undercuts. The freezing material is fixed in the mold and cannot contract further. This results in so called prevented contraction. [2]

EXPERIMENTS AND MATERIALS

For the investigation of the contraction and warpage behavior, a semi-crystalline polypropylene of the type BJ368MO from Borealis AG, Vienna Austria, was used. A low viscosity and good nucleation behavior ensures a homogeneous contraction and a symmetrical stress profile caused by cooling effects. Two plates of 60 mm x 60 mm were used: In Figure. 2 the left plate used for determining the contraction according to DIN EN ISO 294-3 is shown. An identical plate with additional ribs at the edges, which cause a prevented contraction, is shown on the right. The experiment was carried out on an injection molding machine Arburg Allrounder of type 370S 700-100 / 70 from Arburg, Loßburg Germany, with a screw diameter Ø d = 25 mm. The used mold was designed as an injection compression mold. This allows a simple variation of the part thickness t. For each thickness the height position of the diaphragm runner was adjusted accordingly (0.75 x t). In addition, two pT-sensors (p1, p2) were integrated for process monitoring.

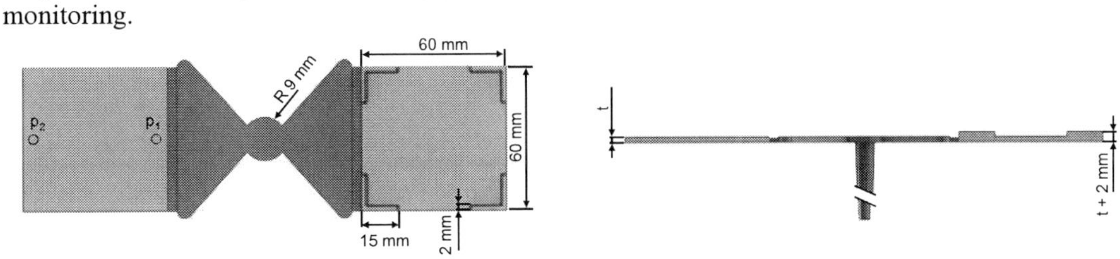

FIGURE 2. Used part shape.

The experiments include all relevant process parameters (Table 1). For this purpose a fractional factorial centrally composed design with 83 process variations was used. The injection volume flow of the machine (cm^3/s) was varied for each part thickness. This allows to realize the same melt front speed in the cavity for all part thicknesses.

TABLE 1. Used process conditions.

Parameter		Unit	-α	0	α
thickness	t	mm	1	2,5	5
melt temperature	ϑ_{me}	°C	210	235	260
mold temperature	ϑ_{mo}	°C	20	50	80
melt front speed	v_f	mm/s	150	500	850
holding pressure	p_h	bar	50	175	300
holding pressure time	t_h	s	1	4,5	8
ejection temperature	ϑ_E	°C	90	105	120

Measurement Methods

The contraction was determined according to the standard. A measurement of the warpage according to the standard is not possible yet for the lack of a standard. Till today, a standard measurement method for the warpage does not exist. Therefore, a new measurement method was developed with this work.

The height of the plate was determined tactilely at nine measuring points (see Fig. 3). Thus, a characterization of the warpage as the maximum difference in height is possible. In addition, the surface shape (eq. 1) or the surface profile in one direction (eq. 2) can be described with a polynomial function of second degree. As a reference for the tactilely determined warpage values, the warpage was additionally measured optically. For this purpose the plates were scanned with a 3D scanner ATOS Core 135 from GOM, Braunschweig Germany.

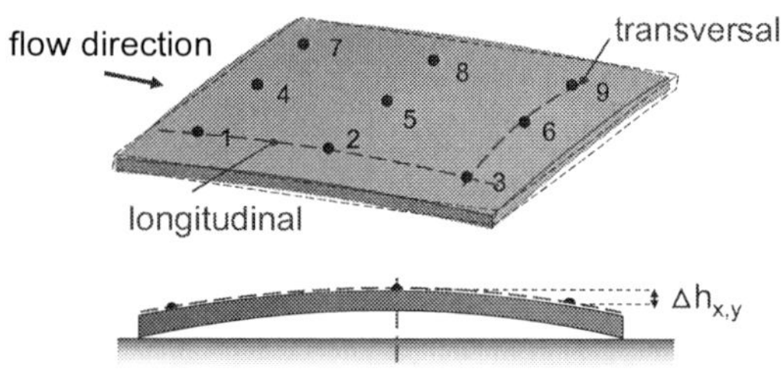

$$h_{(x,y)} = a_0 x^2 + a_1 xy + a_2 y^2 + a_3 x + a_4 y + a_5 \quad (1)$$

$$h_{(x)} = ax^2 \cdot bx + c \quad (2)$$

FIGURE 3. Warpage measurement method.

RESULTS AND DISCUSSION

Contraction

The contraction of injection molded plastic parts depends, as already mentioned, on a large number of influential factors. The effect analysis of the process parameters on the volumetric contraction of the two plates is shown in Fig. 4, a. The graph reads as follows: When changing the value of a parameter by 1, the dependent variable changes by the value of the respective effect coefficient. The total volumetric contraction of all process conditions consists of approx. 22–28 % longitudinal contraction, 18–24 % transversal contraction and 50–60 % thickness contraction. This means that the volumetric contraction is predominantly influenced by the thickness contraction. It is found that the parameters of the holding pressure phase – which affect the compensation of the volume contraction – have a significant influence on the volumetric contraction (Figure 4 a). The influence of the process parameters is nearly identical for both plate shapes. The hindered contraction is not fixated in the mold and therefore is influenced more strongly by the process parameters.

Warpage

The effect analysis of the maximum height deviation shows that it is influenced primarily by the parameters of the holding pressure phase, too (Figure 4 b). It can be assumed that the warpage mainly results from an insufficient volumetric compensation. The plastic melt freezes on the exterior walls immediately and forms a dimensionally stable frame. When the effect of holding pressure is insufficient (time and pressure), not enough plastic melt will flow into the mold cavity, resulting in the outer frame being deformed. Thus, increased warpage occurs. Unlike the findings on contraction presented above, it can be seen that even though volumetric contraction is similar, the effects on the height deviation of the two shapes differ clearly. The prevented contraction shows much higher process dependence than the hindered contraction. The plastic is additionally fixed longitudinally and transversally to the flow direction. This three-dimensional fixation of the plastic melt inside reacts more sensitively to the parameters of the holding pressure phase.

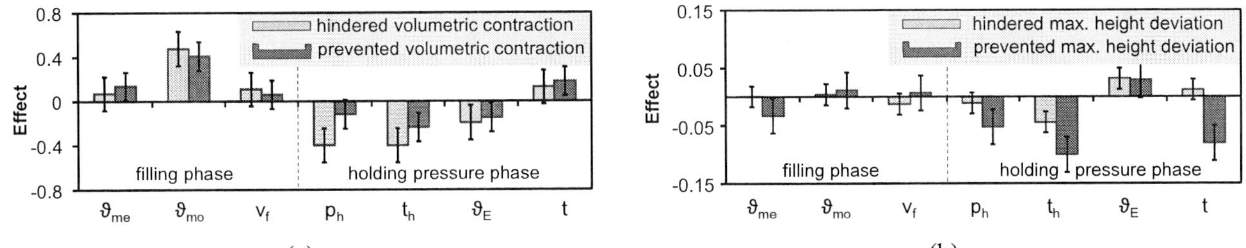

FIGURE 4. Process-dependent effect analyses to the volumetric contraction (a) and max. height deviation (b).

A more detailed analysis of the height warpage is possible with the new measurement method. The simplified function (eq. 2) was used to evaluate the warpage. This function comprises only half as many factors as the surface function. This new method allows a more precise classification of the warpage. a describes a flexural warping, which is arched and can be mapped with a square function with high accuracy. On the other hand, b describes a torsional warping which is expressed by a twisted surface pattern. The final parameter c is a thickness value and depends completely on the thickness of the part.

The measuring points of the central process settings and the respective calculated surface profile according to eq. 2 as well the surface scan of the optical measuring system are shown in Fig. 5,a. It can be seen that the simplified calculated surface profile describes the surface course very well. For the following investigations, an average of each of the three curves in longitudinal and transversal direction of the flow was used.

The new measuring method is explained in detail for the example of the holding pressure time at the central process settings of the experimental design. Figure 5 b, shows the maximum height deviation of the two plate shape at different holding pressure times. With an increasing holding pressure the height deviation decreases due to an improved volumetric compensation.

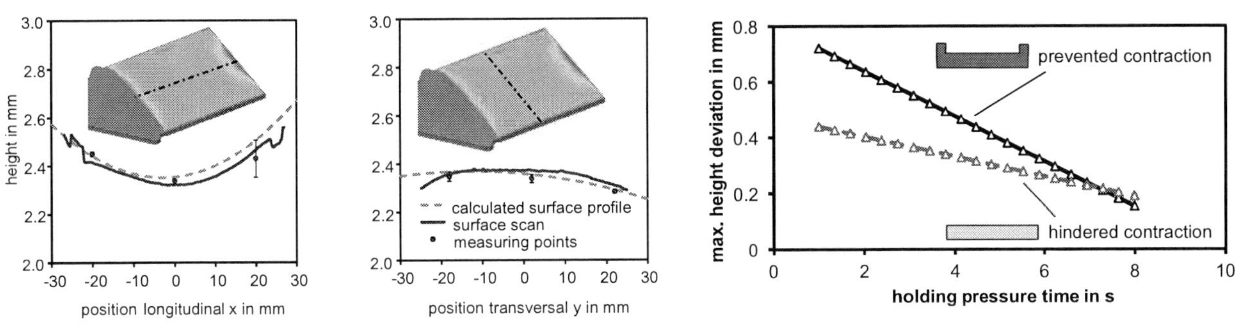

FIGURE 5. Comparison of the used measurement methods (a). Influence of the holding pressure time on the max. height deviation (b).

The parameters a and b (eq. 2) allow a more detailed view on the type of warpage (Fig. 6). In the case of hindered contraction the b-values longitudinal and transversal to the flow direction are almost zero (Fig. 6, b). Also the a-value transversal to the flow direction is very low. So the height deviation is caused by a parabolic warping (a-value) along the flow length. It is shown that an increasing holding pressure time reduces the value of a. Also the longitudinal contraction decreases due to the increased holding pressure time.

In the case of the prevented contraction Fig. 6, b shows very high b-values. Additional high a-values are shown in Fig. 6, a. This means that a flexural warping and additional torsional warping appearances. The influence of the holding pressure time can be clearly demonstrated. Furthermore it can be seen that a is negative. That means that the deflection of the plate is not concave, like before, it is convex. This shows that the warping of the hindered contraction occurs in opposite direction to the prevented contraction. This example shows the difficulty of the predictability of warpage. Despite a comparable contraction in both cases, the warpage behaves contrarily in its amount, shape and direction.

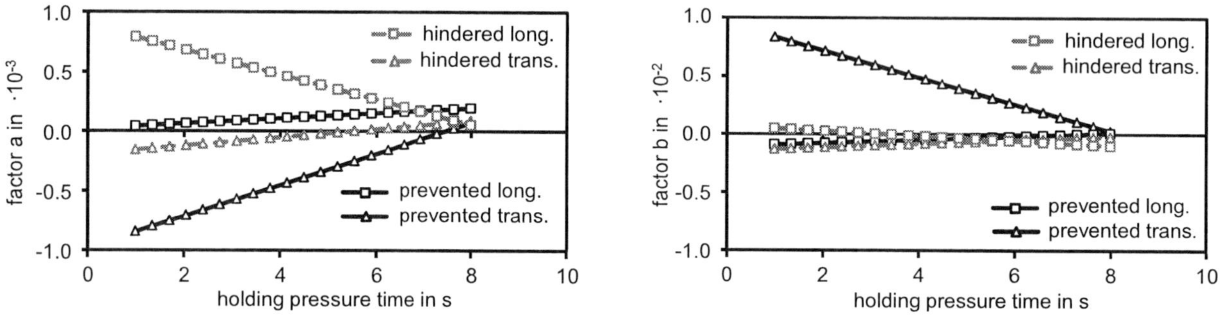

(a) (b)

FIGURE 6. Influence of the holding pressure time on the factor a (a) and b (b).

CONCLUSIONS

Within this work it was shown that the holding pressure phase has a significant influence on the contraction and warpage behavior, especially of unfilled semi-crystalline plastics. Generally, a slight warpage results from a slight contraction. However the type of injection mold has to be taken into account.

Depending on the mold concept the contraction can occur as a free contraction, hindered contraction or prevented contraction. In addition to the dimensional deviations, the shape and orientation of the molded part can be altered by the warpage. By using a new measuring method it was shown for the first time that the warpage can be classified as flexural warping and torsional warping. The prevented contraction, in particular, results in an increased warpage which depends on the process parameters markedly. In general, warpage can only be reduced substantially by a sufficiently high volumetric compensation in the holding pressure phase.

REFERENCES

1. C. Bonten Kunststofftechnik, „Einführung und Grundlagen“, Hanser (2016).
2. W.B. Hoven-Nievelstein, „Die Verarbeitungsschwindung thermoplastischer Formmassen“, Dissertation, RWTH-Aachen (1984).
3. ASTM D955-08, “Standard Test Method of Measuring Shrinkage from Mold Dimensions of Thermoplastics”, ASTM International, West Conshohocken PA, (2014)
4. Deutsches Institut für Normung e.V., „DIN 16742 Kunststoff-Formteile - Toleranzen und Abnahmebedingungen“, Beuth (2013).
5. S. Stitz, „Analyse der Formteilbildung beim Spritzgießen von Plastomeren als Grundlage für die Prozesssteuerung“, Dissertation, RWTH-Aachen, (1973).
6. P. Larpsuriyakul, H.-G. Fritz, *Journal of Polymer Engineering and Science*, **51**, 3 (2011).
7. T. A. Osswald, “International plastics handbook”, Hanser (2006).
8. G. Menges, P. Thienel, *Journal of Polymer Engineering and Science*, **17**, 10 (1977).
9. D. Kusić, T. Kek, J.M. Slabe; et. al., *Polymer Testing*, **32**, 3 (2013).
10. S. H. Tang, Y. J. Tan, S. M. Sapuan, et. al., *Journal of Materials Processing Technology*, **182**, 1-3 (2007)

A Concept of an Injection Compression Mould for Non-Invasive Ultrasound Tomographic Temperature Measurements

Jens Wipperfürth[1, a)] and Christian Hopmann[1]

[1] *Institute of Plastics Processing (IKV) at RWTH-Aachen*
Seffenter Weg 201, 52074 Aachen, Germany

[a)]Corresponding author: jens.wipperfuerth@ikv.rwth-aachen.de

Abstract. The temperature distribution is very important for the validation of models for the simulative part design due to the high influence on the moulded part, but is also very challenging to measure. Nowadays temperature measurements at the transition of the polymer melt to the mould or near the surface of the polymer melt are applicable and do not allow either a non-invasive analysis or a determination of the temperature field. In the approach of ultrasound tomography an ultrasound beam is emitted into the melt and the time-of-flight (TOF) is detected by a set of transducers, which are mounted around the melt. Subsequently the measurement is repeated from different directions. Using algebraic reconstruction techniques, a distribution of the ultrasound velocity can be calculated based on the TOF-dataset. With additional information about the polymer from pvT-behaviour, the distribution of the ultrasound velocity can be converted into a temperature field. To evaluate the applicability of ultrasound tomography an injection compression mould with a cylindrical-shaped cavity is currently designed, which fulfills the requirements for ultrasound tomography. Twenty ultrasound transducers are radially arranged around the cavity to reach a spatial resolution of 3.5 mm² per area element. The transducers are connected to buffer rods that provide a temperature shield for the ultrasound transducers and enable the possibility to scatter the ultrasound beam for fast signal detection. Filling simulations with Autodesk Moldflow 2015 were done to determine the temperature distribution in the mould at different time steps and are used to test the reconstruction algorithms. The results show a good agreement to the simulated temperatures. For accurate temperature measurements with ultrasound under real injection moulding conditions further research must be done, with respect to the influence of process parameters, e.g. cooling rate, and the behaviour of ultrasound in solidified polymer melts.

INTRODUCTION

The quality of injection moulded parts with respect to shrinkage and warpage is highly influenced by the injection moulding process [1]. In the last years, the increasing requirements for quality and accuracy of injection moulded parts lead to an increased usage of simulations to predict the part behaviour in an early phase of the part development. The simulations are based on physical models that need input parameters. Typically, these are either determined from laboratory experiments or simple assumptions.

In injection moulding, the temperature distribution in the melt during the process is an important input parameter, since it influences residual stresses and thus shrinkage and warpage. However, the temperature distribution is difficult to describe theoretically, due to high cooling rates, complex heat transfer mechanisms and the forming microstructure. Additionally, todays measurement techniques require either direct contact to the melt (e.g. thermocouple) or permit a non-invasive temperature measurement at or near the surface (e.g. pyrometry). These methods also deliver only punctual information and prevent a precise temperature measurement [2]. From this follows that new measurement techniques are required to resolve the temperature distribution precisely and improve the prediction of shrinkage and warpage via simulations.

In this context, recent approaches for the determination of the temperature distribution are based on ultrasound tomography [3-5]. This measurement technique allows a non-invasive and spatially resolved temperature measurement. Another advantage of this method is the stability against high cooling rates and the heat transfer mechanisms, since the acquired signal includes this information [4]. However, realising ultrasound tomography within the injection moulding process requires solving several constructional challenges. In this paper, we will describe a mould concept that allows performing ultrasound tomography during the process. The focus of the concept are highly space-resolved temperature measurements and fast acquisition times.

BRIEF REVIEW OF ULTRASOUND TOMOGRAPHY

The principle of ultrasound tomographic measurements for the injection moulding process is shown in Figure 1. An ultrasound transducer emits an ultrasound wave propagating through the melt (Fig. 1, (a)) along the direct path between emitter and receiver. By discretising the measurement region, the total arrival time can be determined by the sum of the arrival times in each areal element (Fig. 1, (b)). Depending on the temperature in each area element the arrival times differ since the ultrasound velocity changes. From this follows that dependent on the ultrasound velocity and the known length of the path in each element the total arrival time can be calculated (Fig. 1, (c)).

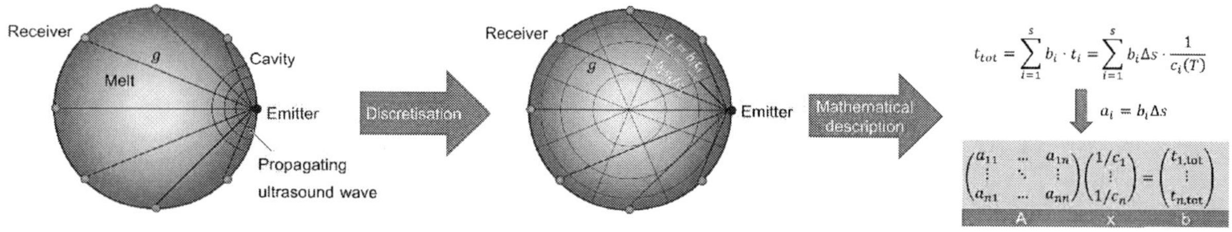

(a) (b) (c)

FIGURE 1. Ultrasound wave propagation on direct paths (a), discretisation of the measurement area (b) and the resulting linear set of equation that has to be solved for the ultrasound velocity (c).

Repeating the measurements from different direcetions, a set of linear equation can be obtained (Fig. 1, (c)) that has to be solved for the ultrasound velocity. Typically, this set of equation is overdetermined and cannot be solved directly. For this reason iterative reconstruction techniques must be used that are described in detail in the literature [6]. In a next step the temperature can be calculated from the reconstructed ultrasound velocities in each area element using equation 1.

$$c(x,y) = \left[\frac{1}{\rho(x,y)} \left(K + \frac{4G}{3} \right) \right]^{1/2} \quad (1)$$

In equation 1, ρ is the density, K is the adiabatic shear modulus and G the shear modulus, since the parameters are also temperature and pressure dependent. Therefore, these values have to be predetermined in laboratory experiments. The pressure in the measurement region is assumed to be constant [4]. Alternatively, the ultrasound velocity can be calibrated directly via run time measurements under defined temperature and pressure conditions, which requires the development of a suitable measurement device. A detailed description of the basics of ultrasound tomography can be found in [4].

MOULD DESIGN

To realise ultrasound tomography for the injection moulding process the following conditions must be considered:

1. Elimination of the pressure dependence
2. Temperature sensitive ultrasound transducer
3. Real behaviour of the propagating wave
4. Data acquisition time

A mould was designed to perform ultrasound tomography during the injection moulding process (Figure 2). The mould has a cylindrical shaped cavity with a radius of 15 mm and a height of 25 mm. The measurement region is in the middle of the axial height. There, 20 ultrasound transducers are radially arranged around the cavity. To ensure a homogenous pressure profile in the measurement region, the mould is designed as a compression mould.

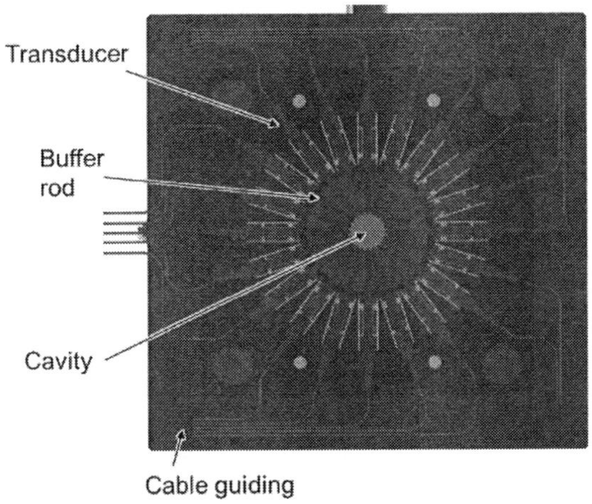

(a)

(b)

FIGURE 2. Layout of the buffer rods (a) and the ultrasound transducer (b).

The ultrasound transducers are not in direct contact to the melt since they are very sensitive towards high temperatures. Therefore, buffer rods of steel are used to shield the ultrasound transducer from the temperature (Figure 3). The buffer rods are also used to scatter the ultrasound wave in the melt, which is required since the emitted ultrasound wave is typically focused. This can be achieved by a semi-cylindrical cut at the end of the buffer rod that points towards the melt. At this curved interface the ultrasound behaves similar to light at a lens. Because of the different ultrasound velocities of the steel and the melt the ultrasound wave is scattered. To prevent circulation of ultrasound within the mould, the buffer rods are covered with PEEK.

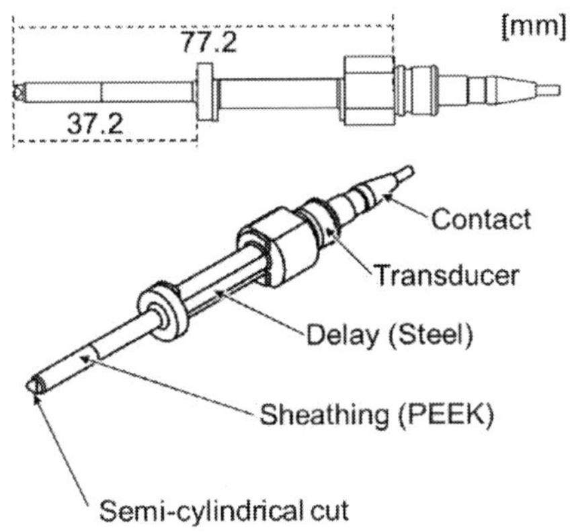

(a)

FIGURE 3. Layout of the buffer rods.

With the cavity radius used and the number of sensors mounted around the measurement region a maximal spatial resolution of 3.5 mm² per area element can be reached. However, using 20 ultrasound transducer requires 380 single measurements to acquire the necessary data for the reconstruction. Due to fast cooling of the melt, this data has to be acquired very fast. For this reason, a customised measurement device was developed (MACEAS GmbH, Bexbach, Germany) that is able to parallel measure and process the data. The measurement device has a time-resolution of 100 ns so that a complete dataset can be obtained within 80 ms.

EXPERIMENTAL

With the above described injection compression mould the complete dataset can be used to illustrate filling, post-pressure, cooling and ejection of the injection moulding process [4]. A typical signal measurement and the corresponding frequency spectrum is shown in Figure 4 for an opposing transducer-pair. The excitation frequency of the transducer is 2 MHz, which can be seen in the spectrum. Actually it is not fully understood, why 1 MHz and 0.5 MHz frequencies are measured, which are significantly larger than the excitation frequency. Since these frequencies can also be detected in offline measurements with steel or water, errors in the measurement device cannot be ruled out and possible process influences cannot be seen. However, this prevents a reconstruction of the temperature distribution, since the arrival time cannot be determined correctly by the internal signal processing. In a next step filter operations will be used that extract the correct arrival times from the measured data. Simultaneously the measurement device will be tested for its correct functionality.

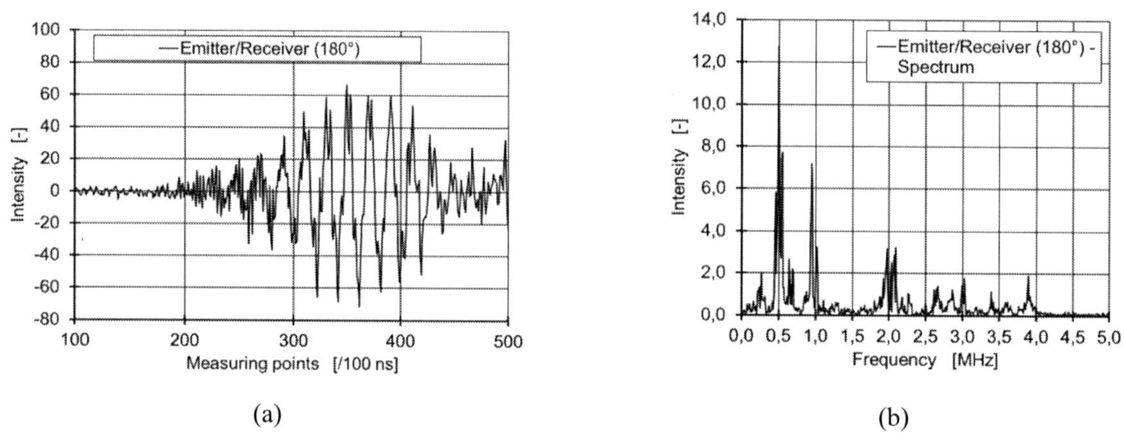

FIGURE 4. Measured arrival times (a) and corresponding frequency spectrum (b) of an opposing transducer-pair.

CONCLUSION

In this paper, a design for an injection compression mould is presented that allows measuring the temperature distribution using ultrasound tomography. The described design offers a spatial resolution of 3.5 mm² per area element and fast data acquisition times due to a customized measurement device. At the moment, unexplained effects lead to a difficult data processing. These effects are probably due to errors in the measurement device, but this has not yet been proven.

ACKNOWLEDGEMENTS

All presented investigations were conducted in the context of the Collaborative Research Centre SFB1120 "Precision Melt Engineering" at RWTH Aachen University and funded by the German Research Foundation (DFG). For the sponsorship and the support we wish to express our sincere gratitude.

REFERENCES

1. F. Ospald, Proceedings in Applied Mathematics and Mechanics **14**, 673-674 (2014).
2. K. Bendada, K. Cole, M. Lamontagne, Y. Simard, J. Opt. A Pure Apllied Opt. **5**, 464-470 (2003).
3. C. Hopmann, J. Wipperfürth, J. Appl. Mech. Eng. **6**, 1000264 (2017).
4. B. Praher, K. Straka, G. Steinbichler,. Meas Sci Technol. 24, 84004 (2013).
5. N. Halmen, C. Kugler, T. Hochrein, P. Heidemeyer, M. Bastian, J. Sensors Sens. Syst. **6**, 9-18 (2017)
6. A. C. Kak, M. Slaney, "Principles of Computerized Tomographic Imaging". Society for Industrial and Applied Mathematics (SIAM), Philadelphia, 2001.

Microinjection Molding of Polypropylene/Multi-walled Carbon Nanotubes Nanocomposites

Zhou Shengtai [(1)*], Hrymak Andrew [(1)], Kamal Musa [(2)]

[(1)] The University of Western Ontario - Ontario - Canada, [(2)] McGill University - Quebec - Canada

A series of polypropylene/multi-walled carbon nanotubes (PP/CNT) nanocomposite were prepared via melt dilution and then subjected to microinjection molding (μIM). A micro-part (μ-part) which has a three-step decrease in thickness along the flow direction was fabricated. The distribution of CNT within the μ-parts was evaluated by morphological observations and electrical resistivity measurements. To assess the influence of shear rates on electrical resistivity of subsequent moldings, compression molding was employed to prepare samples with the same filler concentrations as the injection molded μ-parts. Results indicated that the percolation threshold for the μ-parts shifted to higher filler concentrations when compared with the value obtained for their compression molded counterparts. This suggests that the microstructure of moldings could be significantly affected by the utilization of a different molding process. This is largely attributed to the variation of shearing conditions, which influence the orientation and quality of CNT dispersion.

Microinjection molding of carbon filled polypropylene nanocomposites: the effect of filler type on electrical and morphological properties of micro-moldings

Zhou Shengtai [(1)]*, ***Hrymak Andrew*** [(1)], ***Kamal Musa*** [(2)]

[(1)] The University of Western Ontario - Ontario - Canada, [(2)] McGill University - Quebec - Canada

Microinjection molding (μIM) is an important technology for fabricating micro-moldings (μ-moldings) used in various applications, such as microelectromechanical systems (MEMS) and other microsystems. Compared with conventional injection or compression molding processes, μIM involves higher thermal gradients, higher shear rates and shorter cycle times. The properties of μ-moldings are affected significantly by the molding conditions and materials employed. Presently, there is a trend towards manufacturing multifunctional components which could fulfill multiple-tasks simultaneously. Therefore, μIM of filler loaded polymer nanocomposites is receiving research attention. In this study, a series of polypropylene (PP) nanocomposites loaded with different carbon fillers, i.e. carbon nanotubes (CNT), carbon black (CB) and graphite nanoplatelets (GNP), were prepared by melt blending, followed by compression molding or μIM. Direct current electrical conductivity and melt rheology measurements were utilized to probe the percolated structure of the compression molded PP/carbon nanocomposites. For μIM, a mold insert which has a three-step decrease in thickness along the flow direction was adopted to study the effect of abrupt changes in mold geometry on the electrical and morphological properties of the μ-moldings. To facilitate characterization, the μ-moldings were mechanically sectioned into three different sections based on part thickness. Results indicated that the μ-moldings exhibited higher percolation thresholds when compared with their compression molded counterparts. This could be ascribed to the severe shearing conditions that are present in μIM. The morphology of the μ-moldings containing different carbon fillers was examined using scanning electron microscopy (SEM) for each cut section. The development of corresponding microstructure was found to be strongly dependent on the type of carbon filler used in μIM, which is critical for the enhancement of electrical conductivity of the μ-moldings.

Characterization of Bulk Materials for Single Screw Extrusion by Means of Fluidization Testing

M. Düngen[1, a)] , Z. Li[1] and † M. Koch[1]

[1]Department of Plastics Technologies, Ilmenau

[a)]Corresponding author: martin.langlotz@tu-ilmenau.de

Abstract. Conveying of bulk materials in the hopper and feed zone of single screw extruders can be challenging for materials which show a wide variation of particle shape and size like regrinded materials. Therefore this study presents an adapted fluidization testing method which is known for powders, e. g. SLS-materials. The test comprises a rotating barrel with transparent covers which is partially filled with bulk material. Due to the rotation the material moves in the barrel and forms characteristic surfaces which can be quantitatively described by the avalanche height and angle. The study deals with the design of such a device for typical particle sizes for single screw extruders. The impact of barrel geometry, barrel wall surface properties, rotation speed and bulk material properties are investigated in order to provide a test setup which yields reproducible and applicable results. The test setup is automated and features digital image processing in order to derive the avalanche height and angle for the characteristic sur-faces and movements.

1 INTRODUCTION

In single screw extrusion the thermoplastic bulk material usually flows freely from a hopper into the feeding zone of the screw and barrel. Especially for grinded materials like PET flakes originating from bottle recycling this process can show instabilities like clogging or a fluctuating throughput. The goal of this study is to provide a simple method to characterize the flow behavior of bulk materials. This flow behavior shall than be used as a material property which can be correlated to throughput in the hopper or feed zone of single screw extruders.

This study uses the principle of fluidization testing which is known for powder characterization [1-3]. The experimental setup for this testing method features a rotating drum which is partially filled with a bulk material. The flat top and bottom parts of the drum are made of a transparent material. A light source and a camera are used to take pictures of the moving bulk material. By image processing techniques the profile of the bulk material is extracted and the key figures avalanche angle and avalanche height can be calculated (see figure 1).

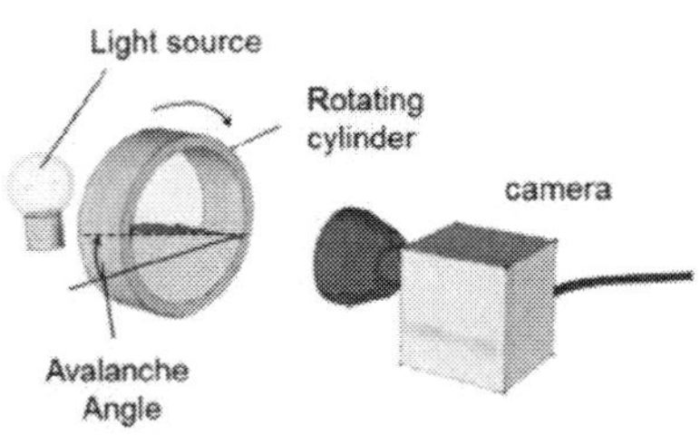

(a)

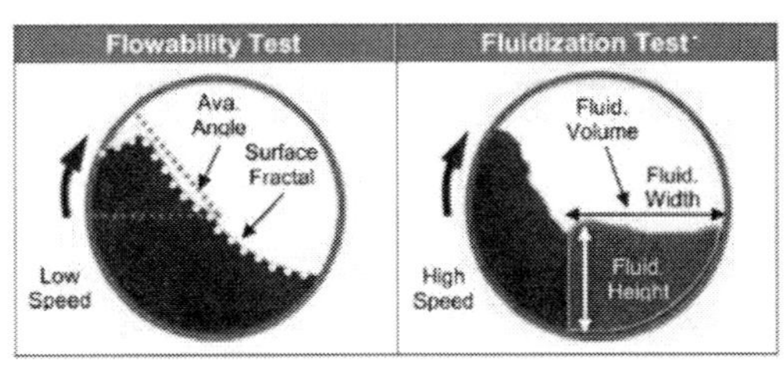

(b)

FIGURE 1. Principle setup of fluidization testing methods [1] and overview of resulting key figures from this method by evaluating the material profile in the drum [2]

2 EXPERIMENTAL

The experimental setup of this study is similar to fig 1. For this study the size of the drum is chosen larger since the particle size is significantly larger than in former studies (see table 1).

TABLE (1). Comparison of test setups for fluidization testing material material properties

	[1]	[2]	[3]	this study
particle diameter [µm]	0,5-150	22-115	4,4-12,7	>2000
drum diameter [mm]	50	35; 50	42; 50; 90	135; 195
drum rpm	0,3	0,6	40-200	0,4-6,2

The exact setup is given in figure 2. Two plates are included which are used to filter scattered light. The whole setup is boxed to exclude any other light sources. The cylindrical part of the drum has a sandpaper liner with a grit size of 240. The sandpaper prevents the bulk material from sliding along the cylinder wall. Thus the particle layer in contact with the cylinder wall follows the motion of the drum.

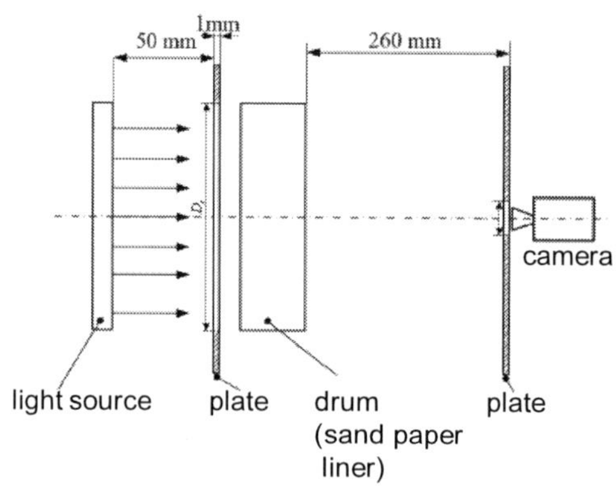

FIGURE 2. Experimental setup

The CCD camera produces simple black-white-pictures (figure 3) which are used to detect the characteristic profile line of the material. This line provides information about the width, height and angle of the material pile. The angle is only considered an avalanche angle, when it is the largest angle in a series of pictures (local maximum).

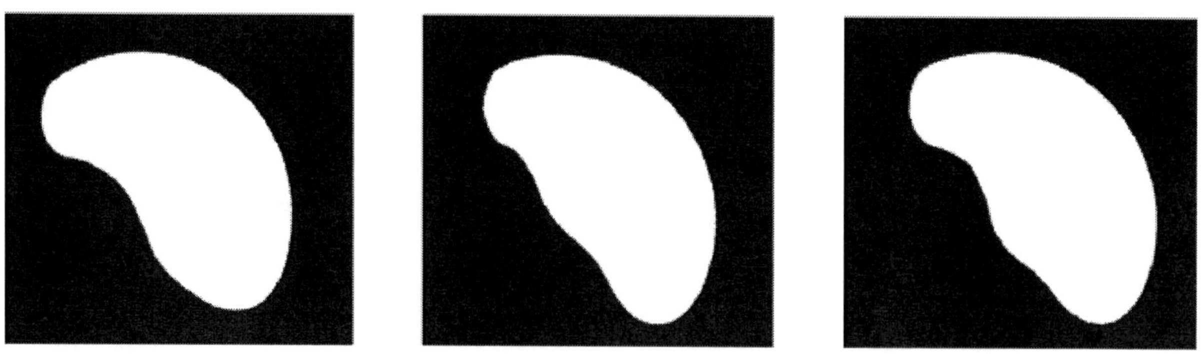

FIGURE 3. Images from the CCD camera showing the stages of material buildup (left), peek (center) and avalanche relief (right)

The useful measurement parameters are determined by preliminary tests using the virgin bulk material of LyondellBasell's PP Moplen RP 210 G. Figure 4a shows a decreasing avalanche degree for increasing drum diameters.

The avalanche degree is fluctuating for changing circumferential speeds. The standard deviation (SD) for this experiment given in figure 4b reveals a significant drop of the SD for increasing circumferential speeds. In order to ensure good reproducibility and small deviations the larger drum (195 mm diameter) and high circumferential speeds (i.e. 5 rpm) are chosen for further tests. The impact of the partial filling degree of the drum is tested for 30 %, 50 % and 70 % partial filling. For 50 % the dependency on rpm was the lowest yielding the best reproducibility. For 70 % partial filling the space for the bulk material to move is the smallest resulting in low avalanche heights, low avalanche degree and high standard deviations. Therefore 50 % partial filling is chosen for further experiments.

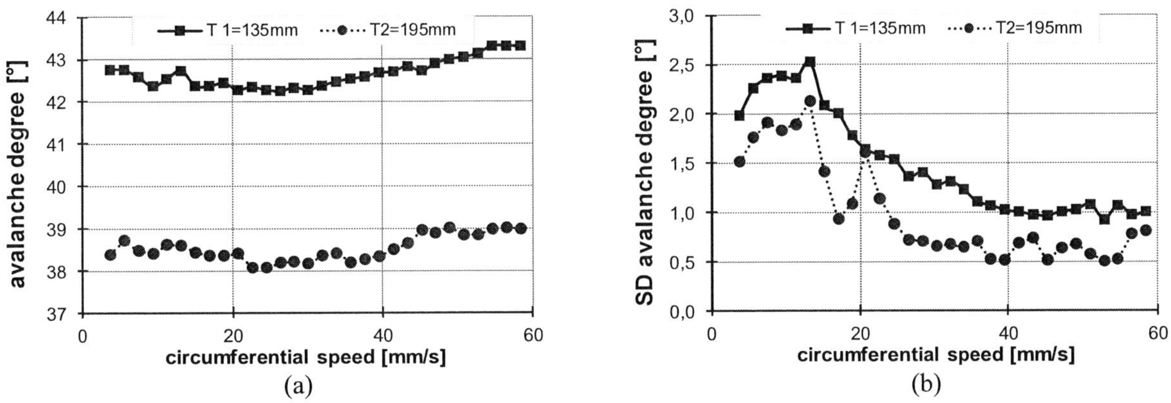

FIGURE 4. Parameter study of circumferential speeds and drum diameters (T1, T2); partial filling: 50%

All further avalanche degree measurements are based on at least 5000 pictures taken at a frame rate of 30 with 1.2 megapixels.

3 RESULTS

The adapted fluidization test method is used to evaluate the flow behavior of seven grinded PET materials and seven thermoplastic virgin materials. The results in figure 5 show several PET bottle flake materials. PET bottle flakes have the highest avalanche degree and standard deviation in this comparison. This was expected since the particles of these materials are irregular and have a broad size distribution [4]. PET Vitret V100 is a virgin PET type for bottle production and is included as a benchmark for the PET bottle flakes. The lower avalanche degree indicates a better flowability and the lower standard deviation indicates a more regular flow of the material.

PET flakes from the grinding of side cut material of a sheet extrusion line are also included in the test. In comparison to the bottle flakes the sheet flakes are flat and not crimped. The avalanche degree of the sheet flakes is lower than for PET flakes. For sheet flakes which had undergone a solid state crystallization process the avalanche degree is the lowest of all tested materials. This can indicate the importance of the particle shape and the coefficient of friction which is influenced by the degree of semi-cristallinity. The standard deviation is still higher than for virgin materials which can indicate the influence of the particle size distribution.

In general the results clearly distinguish the flow behavior of different thermoplastic bulk materials used in single screw extrusion.

The possible application of these results are shown by a simple experiment which features a standard hopper for injection molding machines with a diameter of 50 mm. By measuring the time it takes for 500 ml bulk material to flow out of the hopper a volume flow rate is determined. Figure shows a promising correlation of avalanche degree and volume flow rate.

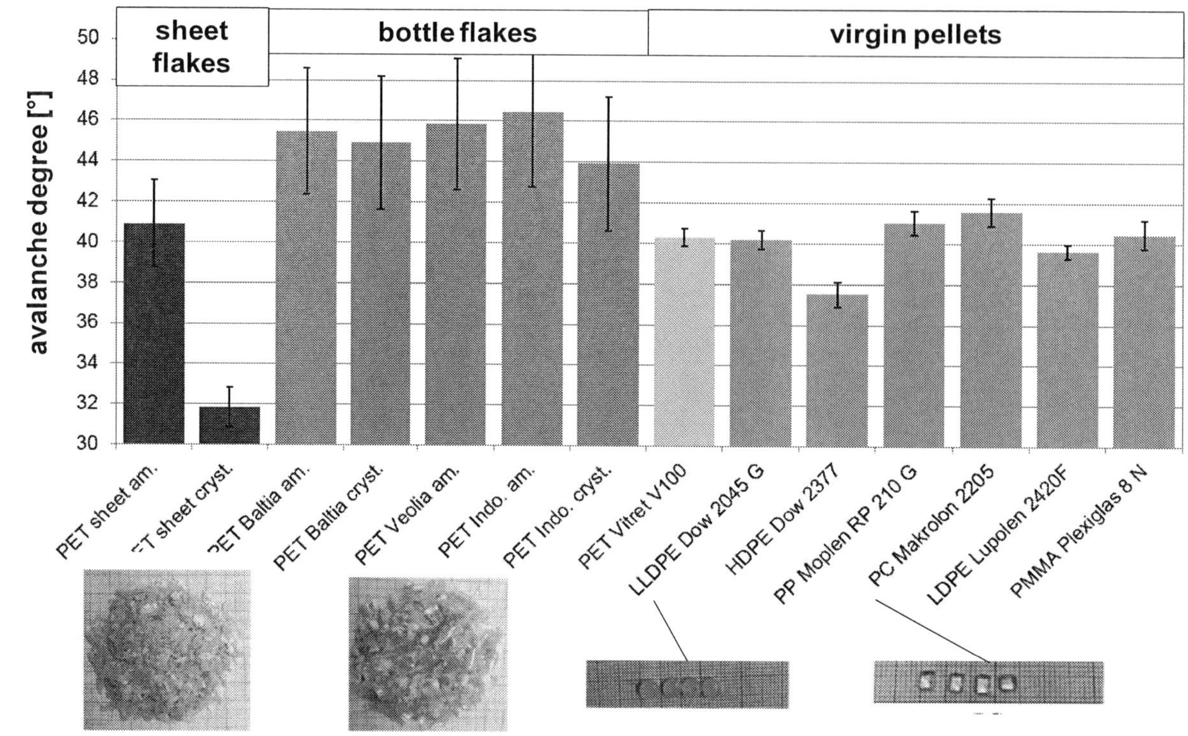

FIGURE 5. Avalanche degree and standard deviation for grinded and virgin thermoplastic materials

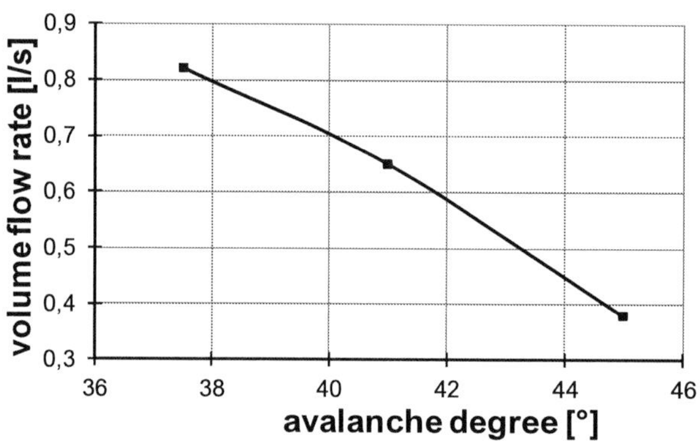

FIGURE 6. Correlation of volume flow rate through a hopper and the avalanche degree

4 CONCLUSION

The fluidization testing method can be adapted for thermoplastic bulk materials. The resulting key figures can be used to distinguish flow behaviors. First experiments also show promising applications of these results for forecasting hopper throughput. These results can also be useful to validate bulk material parameters in discrete element method simulation or to allow new modeling approaches for the solid conveying in single screw extruder feed zones.

5 ACKNOWLEDGMENTS

The authors would like to acknowledge the members of the department plastics technologies of the technical university Ilmenau and the recently deceased professor Michael Koch who was a major contributor to this topic.

REFERENCES

1. Amado, A, M. Schmid and K Wegener, "Advances in SLS powder characterization", Swiss Institute of Technology, Zürich (2011)
2. Schade, C. and R. Causton, "Development of a Maraging Steel Powder for additive manufacturing", Hoeganaes Corporation, Cinnaminson (2015)
3. A. Castellanos, J. M. Valverde and M. A. Quintanilla, Phys. Rev. E. **65**, id. 061301 (2002)
4. Friedländer, T., H. Z. Kuyumucu and L. Rolf, Aufbereitungstechnik **47**, 24-35 (2006)

Comparison of Different Ultrasonic Methods for Weld Line Characterization

*Fey Peter [1], Kreutzbruck Marc [1]**

[1] Universität Stuttgart - Baden-Württemberg - Germany

During injection moulding, thermoplastic materials are heated up to the molten state and then forced into a mould cavity at high pressures. For complex shaped components it is often necessary to use multiple sprues or to let the melt run around obstacles, which leads to the formation of weld lines. If two streams coming from opposite sides meet in the mould, the molecules of the two melt streams have different orientations and thus suppress effective crosslinking between the two melt fronts. Apart from being visible to the naked eye, such weld lines do reduce the strength of the part locally. This effect is even stronger when short fibre reinforced plastics are used. The fibres are aligned parallel to the weld line, while far from the weld line they are aligned parallel to the flow direction of the melt. This causes ad-ditional weakening of the part. Tensile testing specimens with and without weld line were pro-duced having glass fibre content ranging from 30 % to 50 % mass fraction. To characterize weld lines in short glass fibre re-inforced polymer, various ultrasonic methods were employed. Contact coupled ultrasound with longitudinal waves in through thickness direction was used to find the stiffness increase in-duced by the fibres being oriented parallel to the weld line. Addi-tionally, shear waves in through thickness direction were used to quantify the angular anisotropy of shear stiffness in the weld line as compared to regions further apart from it. The velocity of guided waves excited by air-coupled ultrasound transducers was measured perpendicular to the weld line and in pristine material. Air-coupled ultrasound in transmission visualized the weld lines in c-scan images. Results obtained from ultrasonic measurements were correlated to tensile modulus and strength measured destructively to understand, which ultrasonic method is able to substitute destructive testing.

Study of the factors affecting heating mechanism in ultrasound microinjection moulding

Babenko Maksims [(1)]*, Whiteside Ben [(1)], Gulcur Mert [(1)], Masato Davide [(2)]

[(1)] University of Bradford - West Yorkshire - UnitedKingdom, [(2)] University of Padova - Padova - Italy

Ultrasonic Injection moulding has emerged as an alternative production route for miniature and microscale polymeric components, where it offers some significant benefits over conventional microinjection processes, including reduced residence time at high temperature, lower required injection pressures, lower energy usage and easier cleaning. Heating of the polymer prior to injection is achieved entirely using applied ultrasonic energy using a 30 KHz source which is bought into contact with a predefined number of pellets in the sonication chamber. Heating of these pellets is generated from two distinct mechanisms, interfacial friction heating (caused by frictional heating at the contact points between the pellets) and viscoelastic heating (caused by heat generated by internal friction in the material, due to damping of the oscillations by the polymer itself). One challenge associated with these mechanisms is to ensure a repeatable moulding environment in the machine, because heating can be strongly influenced by the initial frictional heating, which itself is dependent on the size, shape and number of pellets and the resulting contact surface areas during the early phases of the heating process. The pellets are dropped into this chamber in an essentially random manner, so small changes in this initial configuration can affect heat generation and cause significant differences between consecutive moulding cycles. In this work we investigate the significance of these early events by controlling the initial contact areas using a range of defined particulate geometries including a stack of thin discs and uniform spheres. We quantify the uniformity of the temperature distribution for each shot and from shot to shot using a novel injection mould tool which allows direct viewing of the mould cavity using high speed conventional and infrared imaging. The uniformity of the resulting physical properties has been examined using small scale mechanical testing apparatus.

Evaluation of Physical Properties and Surface Microstructures Replication of Liquid Silicone Rubbers Obtained by Reactive Injection Molding

Ito Hiroshi [(1)]*, *Satoh Taisuke* [(1)], *Ishigami Akira* [(1)], *Nemoto Akihiko* [(1)]

[(1)] Yamagata University - Yamagata - Japan

In this study, the curing behavior, physical properties and surface microstructures replication of the crosslinked liquid silicone rubber (LSR) obtained by the reactive injection molding process were evaluated. Two types of LSR with different hardness were used in this study. The physical properties of the LSR were measured for optical transmittance and tensile test. In addition, the influence of the reaction process on the density of LSR and the molecular weight of a crosslinking points was evaluated. As a result of reaction injection molding experiment and physical property evaluation, the molecular weight between the crosslinking points inside the molded article decreased due to extension of reaction time and increase in mold temperature. As a result, it is considered that the elastic modulus in the tensile test increased. Furthermore, the replication of micro-lens allay shape was analyzed. As a result of evaluating the surface replication ratio, all samples showed a replication ratio higher than 90%. The effect of mold temperature and reaction time on replication ratio could not be confirmed. It was suggested that higher replication ratio can be obtained by raising holding pressure and injection speed during molding.

Development Of A New Measurement Concept For The Process-Oriented Characterization Of Rheological Material Properties

Michael Burgfeld[1, a)] and Johannes Wortberg[1,b)]

[1]*University of Duisburg-Essen, Institute of Product Engineering, Department of Engineering Design and Plastics Machinery, Lotharstraße 1, 47057 Duisburg, Germany*

[a)]michael.burgfeld@uni-due.de
[b)]johannes.wortberg@uni-due.de

Abstract. Today, plastics are used for varied applications and allow an economic production of complex geometries. Based on their thermal and mechanical properties more and more fields of applications accessible. This leads to an increasing demand of specific and function-oriented materials. For this purpose, different additives are used, which can lead to reactions between the metallic surface and plastic melt. The results are corrosive wear at the extruder components [1, 2] which then again result in a shorter lifetime of the machine and higher costs in case of failure. Furthermore, a change of material properties (for example the viscosity) during the processing time can lead to rejects of the product because of optical and mechanical weak points. In the following, a new measurement system will be presented, which allows testing new materials or standard materials regarding their processing stability and possible wear mechanisms. For this purpose, experiments with a standard LDPE will be shown.

INTRODUCTION

Packaging materials in the food technology include additives to absorb or desorb substances to increase the durability of the packed food. The used additives lead to chemical and physical reactions, which have an influence on the processing stability as well as a corrosive wear on the processing machine. According to this, the topics of process stability, process optimization, and reduction of rejects are in focus. Therefore, a cost-efficient and industrial-suited measurement system will be presented. This allows a process-oriented analysis of materials regarding the process stability. Besides thermal-rheological investigation, it is possible to detect an interaction between the metallic surface and the plastic melt. Such an interaction can lead to wear mechanisms or deposits on the metallic surfaces. Hereby, it is possible to detect wear or material instability before the material is used in the production.

In addition, the experiments should be conducted under processing conditions and variations of the residence time. Moreover, it is possible to change the shear rate in the range of $0\ s^{-1}$ to $150\ s^{-1}$ to test the material behavior on mechanical stresses. Furthermore, process parameters like processing temperature can be analyzed and optimized. Finally, it is possible to produce test films for further analyses.

The result is a cost-efficient testing method for new and existing materials, additives and coats against wear mechanisms. The test bench is intended to provide cost-effective potential for optimizing the process parameters during extrusion. In addition, the choice of materials as well as the selection of polymer blends and additives can be supported for error-free production of packaging materials.

LAYOUT OF THE TEST BENCH

The test bench based on three main components, a laboratory extruder for the plasticization of the material, a control volume to examine it, and a flat die to produce test films for further examinations. The complete measurement system as well as the heating zones for the control volume is schematically shown in the following figure

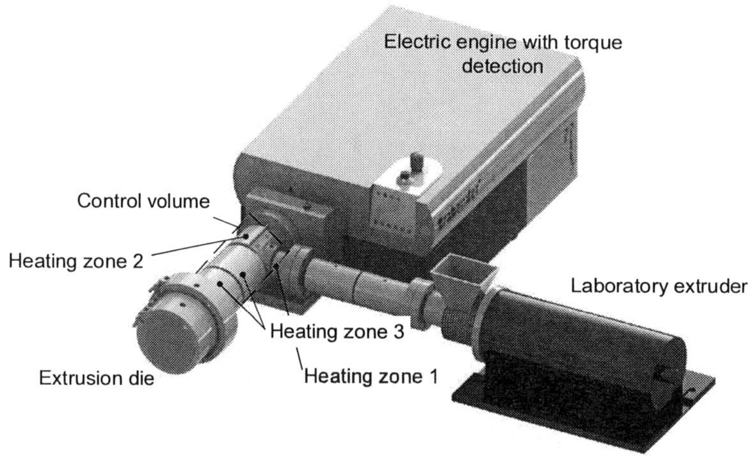

FIGURE 1. Measurement system

Through the use of a laboratory extruder, the test material is plasticized and extruded into a control volume, which is flanged orthogonally on the extruder. The short length of the extruder minimizes the residence time of the material and reduce the potentials of chemical changes before it reaches the control volume. To avoid reactions between plastic melt and metallic surface as well as wear of the barrel and the screw, a non-reactive and highly corrosive-resistant material (Inconel 625) is used.

Main component of the system is the control volume, based on a cylindrical test sample, which is surrounded by a heated barrel. In this case, the test sample corresponds to the wear part of the measurement system. To avoid chemical reactions, the barrel is manufactured of a non-reactive and corrosive-resistant material, too. Only for the test sample a standard material for extrusion dies (X36CrMo17 / 1.2316) is used. The result is that chemical reactions and wear only act on the test sample. Therefore, the geometry is kept simple to reduce the costs for the manufacturing. For the rotational movement, it is connected to an electric engine, which features a sensitive torque detection to measure a possible material change. With this, it is possible to make a statement about the material changing process. An increase/decrease of the viscosity during the test time, is consequently reflected by an increase/decrease in the measured torque. In addition to the torque detection, conclusions about reactions in the plastic melt can be drawn via additional pressure and temperature sensors at key positions in the system.

As already mentioned, a requirement of the measurement system is a reachable shear rate from 0 s^{-1} to 150 s^{-1} in the annular gap. To reach this requirement, the gap between test sample and barrel must be defined under consideration of the rotational speed. Referring to the two-plates-model of Newton the annular gap is designed for the calculation of the shear rate. For this the following equation is used.

$$\dot{\gamma} = \frac{dv}{dh} = \frac{\overbrace{2 \cdot \pi \cdot n \cdot R_i}^{circumferential\ speed}}{\underbrace{R_{barrel} - R_{test\ sample}}_{\Delta R}} \tag{1}$$

The internal diameter of the barrel (D_{barrel}) is set to 25 mm. Now, to reach the predefined shear rate in the gap, the shaft diameter of the test sample ($D_{test\ sample}$) is set to 22.6 mm. According to the dimensioning, the gap is 1.2 mm referred to the radius between test sample and inner surface of the barrel. Furthermore, to avoid a back flow of the melt to the electric engine, the test sample has grooves in the back area (Figure 1).

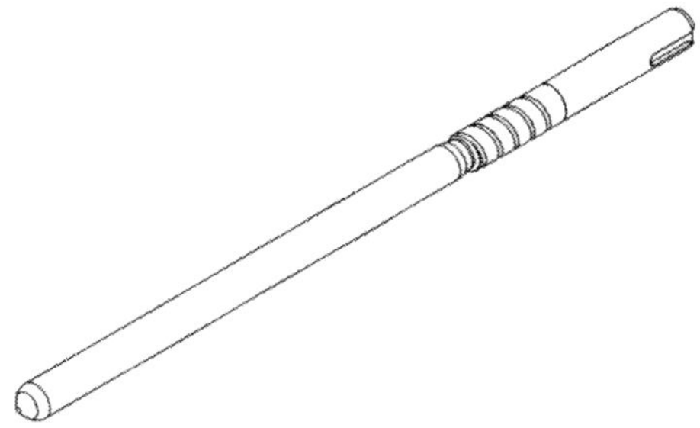

FIGURE 2. Schematic representation of the test sample

Due to the defined geometry and the limitation of the rotational speed of the drive, which has a range from 0 min^{-1} to 150 min^{-1}, a shear rate range in the annular gap between 0 s^{-1} and 147.92 s^{-1} has been achieved. A comparison between the reachable shear rate in the annular gap and associated rotational speed of the test sample show small deviations to each other. Through this, the desired shear rate can be specified directly via the rotational speed.

EXPERIMENTS

To test the function of the measurement system, the presented experiments are made with Sabic 2100N0, which is a standard LDPE. During the processing of this material a corrosive or abrasive wear is not excepted. In the following, the thermal sensitivity of the system and influence of the rotational speed such as processing temperature are investigated with a trial period of 3 hours. Furthermore, the control volume is closed with a blind flange instead of the extrusion die. For the recording of the torque and temperature, a sampling rate of 1 Hz is used. To reduce the variations of the measured torque values, the moving average is used. All following figures show the measure torque as smoothed.

At the beginning of every measurement, it is necessary to define the melt-respectively testing temperature. These temperatures correspond with the temperature of the barrel Afterwards the approach can be started. It must be stated that the process can be differentiated between a filled or not filled annular gap. If it is not filled, it is possible to start the trial directly when it is filled. In the other case, thermal and mechanical stressed material must be displaced with new material in the annular gap. In order to decrease the needed changeover time for the displacement process between the old and the new material, the test sample rotates at a maximum speed of 150 min^{-1} in synchrony with the extrusion process. This has the advantage that higher mass flow rates are possible because of a reduced viscosity in the annular gap. Subsequent for a time period of 5-10 minutes the extrusion process will be continued without a parallel rotating test sample. After that, the extrusion process is finished and the trial can be started. For this purpose, the test sample rotates with predefined rotational speed. From that point, the material is mechanically stressed over a defined period of time. Repeated measurements under the same test conditions exhibit satisfactory reproducibility. It can be seen in the following figure, that the curves of measured torque have the same characteristic curve shape.

Thermal Sensitivity

To test the thermal sensitivity of the measurement system, unoptimized parameters of the closed loop control of the band heaters are used. There are three in operation, which are relevant for the temperature control of the control volume. In the first trials, the temperature of the heating zones is set to 210 °C. Furthermore, the rotational speed of the test sample is 5 min^{-1} which correspond to a shear rate of 4.9 s^{-1} in the annular gap. During the test time, the measured temperature for heating zone 1 and 2 is constant at 210 °C. Only heating zone 3 (HZ3) shows a deviation to the set temperature. As shown in the following figure, the higher actual temperature of heating zone 3 leads to a change of the measured torque. This is due to the fact that a higher temperature leads to a decreasing viscosity.

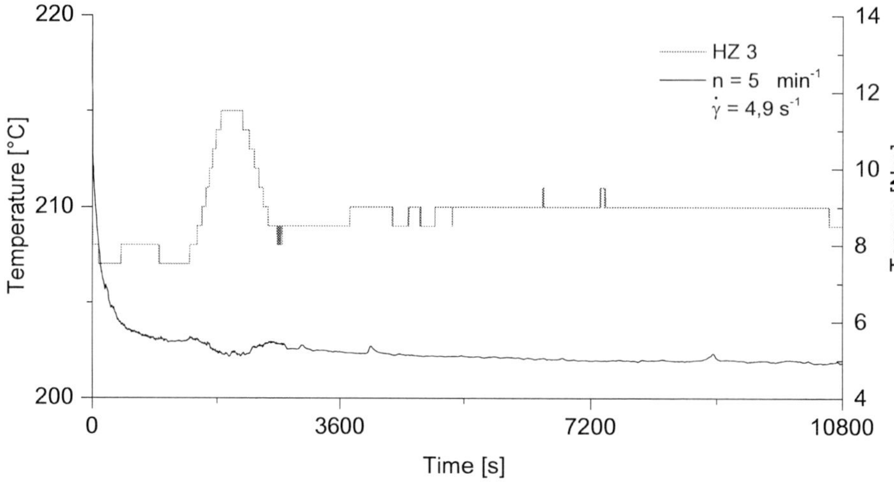

FIGURE 3. Measured torque and temperature over the test time

By means of this experiment it can be shown, that the measurement system has a sensitive torque detection, which shows a torque change by small deviations in temperature. It can be deduced that a material changing and the concomitant changes can be detected. Finally, used control parameters for the closed loop control were optimized. After this, the temperature profile of all heating zones is constant.

Influence of the Rotational Speed and Temperature on the Material Behavior

In the next trails, the influence of the rotational speed more specifically the mechanical stress on the test material is tested. Therefore, different experiments with rotational speeds of 5 min^{-1}, 25 min^{-1}, 50 min^{-1}, 100 min^{-1}, and 150 min^{-1} and a Temperature of 210 °C are conducted. As shown in the following figure 4, the material doesn´t change the viscosity over the test time because of a constant measured torque.

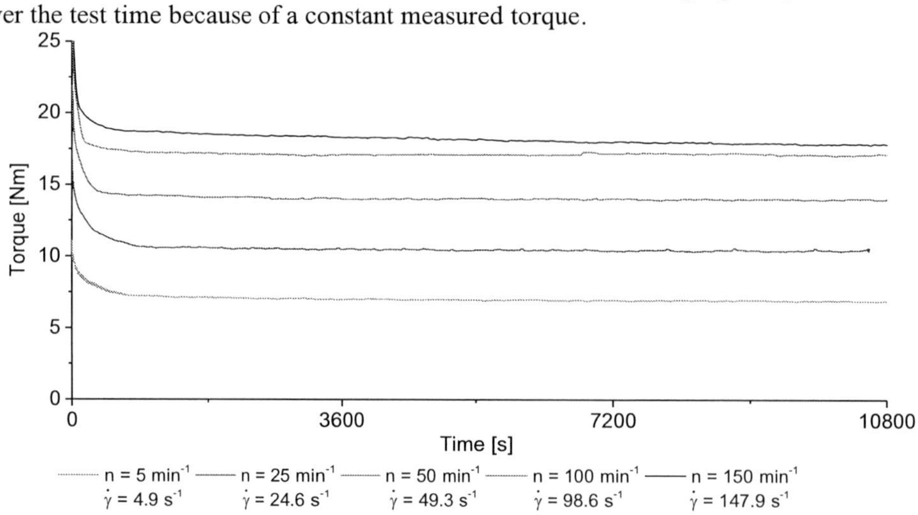

FIGURE 4. Influence of different rotational speeds on the material behavior of Sabic 2100N0 (210 °C)

It can be noted that an increasing rotational speed of the test sample leads to an increase of the measured torque. To explain this result, the following equation for a Searle-Rheometer [3] is consulted. This is due to the fact that the Searle-Rheometer and the presented measurement system have the same measurement principle.

$$T(n) = \eta(\dot{\gamma}) \cdot n \cdot \underbrace{F_{Geometry}}_{const.} \quad (2)$$

T: torque n: rotational speed $F_{Geometry}$: geometry factor

Out of this, it is obvious that there is a proportional correlation between the rotational speed and torque by a constant viscosity. This lead to a higher torque by increasing the rotational speed. Based on the shear thinning effect by plastics the torque has a degressive character. In the next step, the test temperature is increased up to 230 °C to test the influence of it on the material behavior. As before, the rotational speeds are 5 min^{-1}, 25 min^{-1}, 50 min^{-1}, 100 min^{-1}, and 150 min^{-1}. As anticipated, a higher temperature leads to a decrease of the measured torque. This is based on the known influence of the temperature on the viscosity. With regard to a possible material changing, it is shown that the temperature has no influence. Like the trials under 210 °C the detected torque is constant over the test time, which is shown in the following figure.

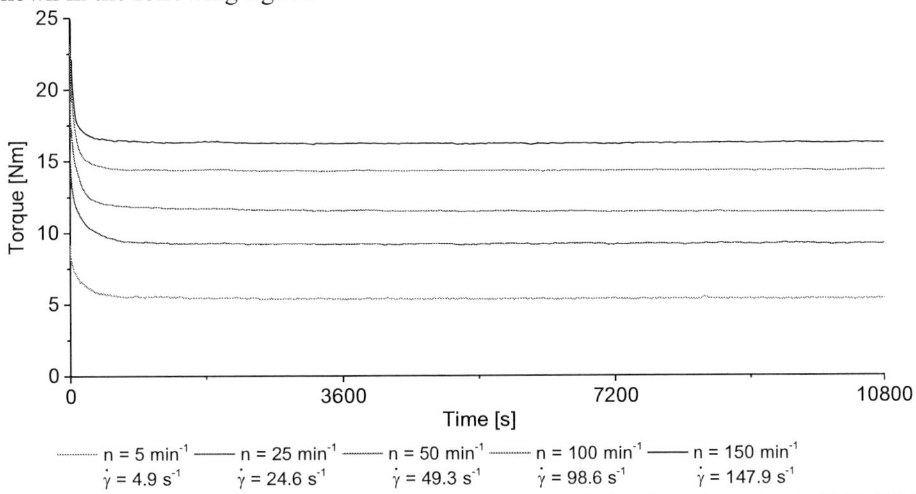

FIGURE 5. Influence of the Temperature on the material behavior of Sabic 2100N0 (230 °C)

CONCLUSION & OUTLOOK

By means of the presented test bench, versatile experiments with regard to the material behavior can be carried out under different test conditions. In addition to possible wear mechanisms, the material behavior can be investigated as a function of mechanical stress and residence time in the system. The presented trials show satisfactory results. The used torque detection is sensitive enough to measure minimal changes, which allows to make a statement about the material behavior. In the future, further materials should be tested and characterized by the measurement system. Possible Materials are HDPE, PP, PA, EVOH to analyze their material behavior. In the long term, a comparison between the measurements results of a rheometer and the presented system should be made.

ACKNOWLEDGMENTS

This study and the corresponding research project (ExtruStab, AiF-No. 18343 N) of the Industry Association for Food technology and Packaging (IVLV) was funded from the budget of the Federal Ministry for Economic Affairs and Energy (BMWi) through the Arbeitsgemeinschaft industrieller Forschungsvereinigungen "Otto von Guericke"e.V. (AiF) (German Federation of Industrial Research Associations). We cooperate in the project with the Fraunhofer IVV (Freising, Germany).

REFERENCES

1. De Kruijf, N. et al., *Active and intelligent packaging: applications and regulatory aspects*, Food Additives and Contaminants 19, 144-162, Taylor & Francis, 2002

2. Rooney, M.L: *Active Food Packaging*, 1st edn. Blackie Academic & Professional, Glasgow, UK, 1995

3. Grellmann, S., Seidler, S.: *Kunststoffprüfung*, 3rd edn. Carl Hanser Verlag München, 2015

Preparation and Microinjection Molding of Polyoxymethylene/Molybdenum Disulfide Nanocomposite

Chen Yinghong (1)*, Yang Sen (1), Yang Chaonan (1), Wang Qi (1)

(1) State Key Lab of Polymer Materials Engineering, Polymer Research Institute, Sichuan University - Sichuan - China

As an engineering plastic, polyoxymethylene (POM) possesses good comprehensive performance, such as good mechanical property, resistance to abrasion, fatigue resistance, resistance to chemicals and self-lubrication performance. One of the important applications of POM is in fabrication of the transmission parts, particularly the micro transmission parts e.g., microgears, etc. For some special fields such as high precision watch and aerospace industries, the micro transmission parts applied generally require the substantially improved self-lubrication performance and mechanical property. So, it is very necessary to conduct a modification on POM to enhance its properties. Compositing of molybdenum disulfide (MoS_2) with POM is a promising solution. However, the dispersion of MoS_2 fillers in polymer matrix could be a challenge. In this paper, a novel solid state shear milling technology was used to prepare the POM/MoS_2 nanocomposite and subsequently the microinjection molding investigation was also carried out. The results show that the nano-scale MoS_2 exfoliated platelets with perfect dispersion in POM matrix is successfully achieved. The obtained POM/MoS_2 nanocomposites exhibit the obviously enhanced tribological property and mechanical property. In addition, the microparts show four layers with different crystallization morphologies, where the transition layer mainly consists of shish-kebab structures. The mold cavity size and processing parameters (injection speed and mold temperature) were found to have dramatic influence on the orientation structure, morphology and mechanical performance. This work is supported by the Program for Youth Team of Sci & Tech Innovation Research in Sichuan Province and the Project of State Key Laboratory of Polymer Materials Engineering (Sichuan University) (sklpme2016-2-01).

Effect of screw design on fiber breakage and dispersion of FRTP in injection molding

SHIMOKUSUZONO So [(1)]*, INOUE Akira [(2)], TANAKA Tatsuya [(1)]

[(1)] DOSHISHA UNIVERSITY - Kyoto - Japan, [(2)] TOYO MACHINERY&METAL CO., LTD. - Hyogo - Japan

Reducing fiber breakage and improving fiber dispersion are the most important issue of FRTP in injection molding. The purpose of this study is to suggest an effective screw design for improving both residual fiber length and fiber dispersion. The factors that improve residual fiber length and fiber dispersion were investigated through the molded specimens by different design screws. We found that the screw design in the compression zone dominantly affected the residual fiber length, and that the choice of an appropriate design for the melting and mixing processes was the most important factor for improving both fiber dispersion and residual fiber length. Our experimental observations were compared with the results of flow analysis and consistency was observed. However, flow state of thermoplastic molten resin in heat barrel was not accessible to direct observation. So, we provided a device for visually confirming the flow state in barrel at a glance. As a result, we were able to grasp flow state of the resin in the barrel by using this visualization device and those made further optimization of screw design possible.

MEASUREMENT OF TWO-DIMENSIONAL TEMPERATURE DISTRIBUTIONS INSIDE FILLING MELTS ALONG CROSS-SECTION OF CAVITY USING INTEGRATED THERMOCOUPLE SENSOR

Yokoi Hidetoshi [(1)*], ***Ishida Shouma*** [(1)], ***Masuda Norimichi*** [(1)], ***Yamada Takehiro*** [(2)]

[(1)] The University of Tokyo - Tokyo - Japan, [(2)] Saitama Prefecture Industrial Technology Center - Saitama - Japan

Integrated thermocouple sensors are capable of high precision measurement of gapwise temperature distribution inside melts filling in the cavity, however they have such drawbacks as complicated measurement work, difficulty to change the sensor position, and poor sensor durability. In this study, the authors developed a unique measurement mold capable of sweeping the sensor to anywhere along the cavity width direction, with the sensor fixed at both ends of stationary and movable sides. Experiments were then conducted to clarify that the mold enables two-dimensional measurement of temperature distribution by sweeping a single sensor gradually from the center to the side of the cavity. In molding experiments using a rectangular caivty with a thickness of 3mm and PP, it was found that at low injection rate, the closer the sensor moves to the cavity side, the temperature drops more sharply due to cooling effect. On the other hand, at high injection rate, shear heating effect becomes more conspicuous as the sensor moves toward the side, resulting in a sudden increase in the temperature. Such dramatic temperature rise may cause the "ear-flow" phenomena along cavity sides. In addition, this mold was applied to measure gapwaise and two-dimensional melt temperature distributions during the cavity filling process of short-glass-fiber reinforced PP, and also for the first time to measue those during the super-critcal foam injection molding process of PP, which results in characteristic temperature change patterns for each moding process.

Investigation of the Relationship IR Heating and Injection Molding Parameters of Composite Laminates Which Shaped by Overmolding Technique

Gürcan Mert Gülsözlü[1,*], Yavuz Emre Yağcı[2], Burcu Girginer[2], M. Özgür Seydibeyoğlu[1]

[1] *Department of Materials Science and Engineering, Izmir Katip Celebi University*

[2] *Farplas Automotive Company*

ABSTRACT

Natural fiber-reinforced polymer (NFRP) composites pull attraction among lightweight structures in various studies both for academic and industrial research. The interest has risen due to its NVH (noise, vibration, harshness) and lightweight advantages in comparison to glass fiber and carbon alternatives. NFRP composites offer a surface structure on micro level that provides excellent bonding to polymer media without the need of sizing. Blends of carefully formulated kenaf, hemp, flax, jute, and other fibers in combination with appropriate thermoplastic matrix constitute composites laminates with performance characteristics such as formability, rigidity, and impact resistance.

This study comprised the investigation of IR pre-heating and injection molding process parameters of natural fiber-reinforced composites with innovative contribution of overmolding method. Fiber form technology enables to adapt injection overmolding method to combine the high processable thermoplastic melts with the thermoformable NFRP composite sheets for multi-functional hybrid part production. To understand the adhesion development through overmolding, a non-isothermal screening procedure for bonding molten polypropylene (PP) to an IR heated thermoformable NFRP sheets has been developed. The method has been implemented to optimize boundary process temperatures for good bond formation as a function of NFRP sheet composition while optimizing IR heating system and injection cycle. Specimens as output of optimized series of trials have been isolated from composites and tensile strength, impact resistance, and shear adhesion strength tests have been performed based on the ASTM D638, D7136, and D3163, respectively. This investigation allowed us to identify an ideal setup consisting of individual process and application temperatures for NFRP sheet and molten PP for ideally bonded composite-polymer hybrid part, results will inspire further studies on exploring boundaries of lightweight design.

Keywords: Injection Molding, Natural Fibers, Overmolding, Natural Fiber Reinforcement Polymers

*Corresponding Author: Izmir Katip Celebi University, Department of Materials Science and Eng., 35620, Cigli, Izmir, Turkey, T: +90 262 679 65 00, M: +90 539 829 44 88, E-mail: gurcanmertgulsozlu@gmail.com

1. Introduction

Automotive's demand for materials has evolved in the past decades. Environmental expectations have added new challenges next to well-known mechanical ones. An automotive part is expected to be multi-functional, cost effective, fastly produced, good in mechanical properties and lightweight. [5] Natural Fiber Reinforced Thermoplastic (NFRP) composites optimize the condition of low cost, light weight and functionality. NFRP Composites are made from cellulosic fibers such as jute, flax, kenaf, wood and different thermoplastic polymer matrices. The mechanical behavior of the composites depend directly on the matrix-fiber interface therefor production technique of a NFRP composite part is as important as the design and material selection. [1] To keep the mechanical properties of NFRP sheets and thermoplastic at desired points and achieve the expected hybrid bonding strength at junction point is vital. In our study we have used The Fiber Form. [9]

Technology developed by KraussMaffei in which one process gathers injection molding with the thermoforming of composite sheet i.e. sheet-like semi-finished materials with continuous or nonwoven fibers with various polymers.[6-9] In our case Isowood jut/kenaf-PP sheets were used in combination with thermoplastic PP in order to produce hybrid NFRP composites.

2. Materials and Processing

2.1. Materials

12 % talc filled polypropylene copolymer, with medium melt flow rate, high impact resistance and excellent UV resistance Hostacom CR 1171 G1A (Melt Flow Rate, 13 g/10 min. (230 °C/2.16 kg), ρ=0.99 g/cm^3 (23 °C)) was provided by Lyondell Basell Polymers. Natural fiber reinforcement polymer (NFRP) thermoplastic material which contains 50% bast fibers (flax, hemp, kenaf and mixtures), 50% polypropylene fibers in center layer with very high formability and very good crash-performance characteristics (Area weight, 1800 g/m^2, ρ=0.95 g/cm^3) was provided by ISOWOOD GmbH. Neat PP has been named C0, NFRP sheet C1.

2.2. Processing

Hybrid thermoplastic natural fiber composites were prepared with a Krauss Maffei Fiber Form line equipped with 650 Ton injection molding machine, infrared oven, a robot arm equipped with needles and grippers, drawer and centering station. An automation process was designed to prepare to desired hybrid thermoplastic parts.[4] Natural Fabric composites made of flux, hemp, kenaf and polypropylene were heated in integrated infrared oven, transferred to the injection mold when they were hot, thermoformed by closing the mold and overmolded with thermoplastic material. All fibers are wetted with polymer during the process. Injection molding parameters were adjusted in an appropriate way in order to molten PP to meet the NFRP mat in right temperature which is 220 °C. On the other hand NFRP mats were only heated up to 160-140/ 180-174/ 180-160 °C upper and bottom oven zone variations in order to keep natural fibers properties and wet PP polymer on surface. Necessary conditions were adapted as described for perfect polymer and NFRP bonding.[6] The thermoform and back injection process was applied to produce an automotive composite part.[9] The test specimens have been isolated from the part via a punch cutting tool(C2,C3 and C4 consecutively).

3. Characterizations

3.1. Mechanical Testing

The mechanical properties such as the bending strength, the value of elongation at break, and the value of Flexural Modulus of composite samples and neat PP, were characterized by 3 point bending testing. [2]The test samples were prepared according to the ASTM D790 standard. Then, the mechanical properties of samples were tested [7-8] by SHIMADZU AGS-X 3 point testing machine with a 5 kN load cell and testing speed of 1 mm/min.

3.2. Scanning Electron Microscopy

The morphologies of the NFRP (Natural Fiber Reinforcement Polymer) samples were studied by SEM using a Carl Zeiss Sigma 300 VP scanning electron microscope. SEM images were taken at an accelerating voltage of 15 kV at various magnifications. The samples were coated with gold by QUORUM Q150 coating device. [3]

4. Results and Discussions

The isolated test specimens mechanical performance actually mimic real life automotive scenarios just like crash resistance and form stability under load or thermal conditions.

Our chosen Isowood NFRP (Isowood Natural Fiber PP) meets the demands of automotive material standards for security parts in terms of Flexural Modulus and impact Resistance, on the other hand back injected plastic is also OEM approved thermoplastic with rigidity and impact resistant at the same time enabling trust worth performance on functional parts. [9] Our focus in depth was to investigate the bonding performance of the transition area where NFRP sheet and Thermoplastics meet. As bonded in temperatures close to their molten phases a successful bending performance has been measured supported by 3 point bending test and SEM pictures. Figure 1 shows the bending curve of NFRP composite C2 where IR heat conditions were (180°-174°)(Upper-Bottom Zone) with a bending strength is below the PP measured as 20.3065 N/mm^2 . Figure 2 shows bending curve of NFRP composite C3 with IR heat conditions 180°-160° which also has a lower bending strength measured 20.5945 N/mm2. Figure 3 and Table 1 for the comparison of bending result shows C4 hybrid composite sample with transition point in the middle achieving bending strength of neat PP (25 MPA).[Figure 3.] The infrared parameters show the milder temperature set is a preferred one, assuming higher setups would cause degradation or disorientation of fibers. [1] Figure 7.

and 8. SEM images for samples clearly demonsrate the the presence of fibers and polymer matrix(Figure 4-9) Figure 7-9 images show the surface and interface views of the perfect polymer and fiber interaction at the cross section showing the composite act as a homogeneous part where bonding occurs. Homogeneous bonding occurs was found while evaluating interphase of PP and Isowood bonding points. [3]

5. Conclusion

Our Study has shown fiber form technology and NFRP materials have huge potential for various automotive applications. [5-6] The design freedom and taking advantages of thermoforming and injection molding and not losing mechanical performance on joint sections expands the potential of applications. İnfrared technology is a fast and accurate way of preparing sheets to thermoform and composite process and variations of setup can influence cycle time and composite strength. Our future work will consist of using different combinations of sheets and thermoplastics and expand the test range towards finite elements analysis input to push the limits of material and technology for lighter and ecological automotive performance parts.

REFERENCES

[1] D.U. Shah, Developing plant fibre composites for structural applications by optimizing composite parameters: a critical review, J. Mater. Sci. 48 (18) (2013) 6083e6107.

[2] A. Jord_a-Vilaplana et. al. Development and characterization of a new natural fiber reinforced thermoplastic (NFRP) with Cortaderia selloana (Pampa grass) short fibers (2017)

[3] M.I. Misnon, M.M. Islam, J.A. Epaarachchi, K.-T. Lau, Potentiality of utilisingnatural textile materials for engineering composites applications, Mater. Des.59 (2014) 359e368.

[4] Mondo J., Wijskamp S. and Lenferink R., 2012, "Overview of Thermoplastic Composite ATL and AFP Technologies".

[5] August, Z., Ostrander, G., Michasiow, J. and Hauber, d., 2013, "Recent Developments in Automated Fiber Placement of Thermoplastic Composites".

[6] Henning, F.; Moeller, E.: Handbuch Leichtbau – Methoden, Werkstoffe, Fertigung (Engl. Lightweight construction handbook – methods, materials, engineering). Carl Hanser Verlag, 2011

[7] ASTM D638 Standard Test Method for Tensile Properties of Plastics

[8] ASTM D7136/D7136M – 12 Standard Test Method for Measuring the Damage Resistance of a Fiber-Reinforced Polymer Matrix Composite to a Drop-Weight Impact Event

[9] Greb, C-W.: Introduction in "Systematische Gestaltung von Fertigungsprozessen textiler Verstärkungs-strukturen für Hochleistungs-Faserverbundkunststoffe", edited by Greb, C-W, Shaker Verlag, Aachen, 2013

TABLE CAPTIONS

Table 1. The code of samples

Table 2. Mechanical properties of neat PP, ISOWOOD NF/PP and NFRP composites

Table 1. The code of samples

Samples	**Sample Codes**
PP	C0
ISOWOOD NF/PP	C1
NFRP Composite 180°-174°(Upper-Bottom Zone) IR Heating	C2
NFRP Composite 180°-160°(Upper-Bottom Zone) IR Heating	C3
NFRP Composite 160°-140°(Upper-Bottom Zone) IR Heating	C4

Table 2. Mechanical properties of neat PP, ISOWOOD NF/PP and NFRP composites

Sample	**Bending Strength (N/mm2)**	**Flexural's Modulus (GPa)**	**Elongation at break (%)**
C0	26	1.8	
C1	> 45	< 2	
C2	20.3065	0.0142	4.17386
C3	20.5945	0.0125	5.85185
C4	25.0475	0.0142	5.96815

FIGURE CAPTIONS

Figure 1. NFRP Composite specimen C2 (180°-174°) Bending Curve

Figure 2. NFRP Composite specimen C3 (180°-160°) Bending Curve

Figure 3. NFRP Composite specimen C4 (160°-140°) Bending Curve

Figure 4. NFRP Composite specimen C2 (180°-174°) Interphase SEM images

Figure 5. NFRP Composite specimen C2 (180°-174°) Surface SEM images

Figure 6. NFRP Composite specimen C3 (180°-160°) Interphase SEM images

Figure 7. NFRP Composite specimen C4 (180°-160°) Surface SEM images

Figure 8. NFRP Composite specimen C4 (160°-140°) Interphase SEM images

Figure 9. NFRP Composite specimen C4 (160°-140°) Surface SEM images

Figure 1. NFRP Composite specimen C2 (180°-174°) Bending Curve

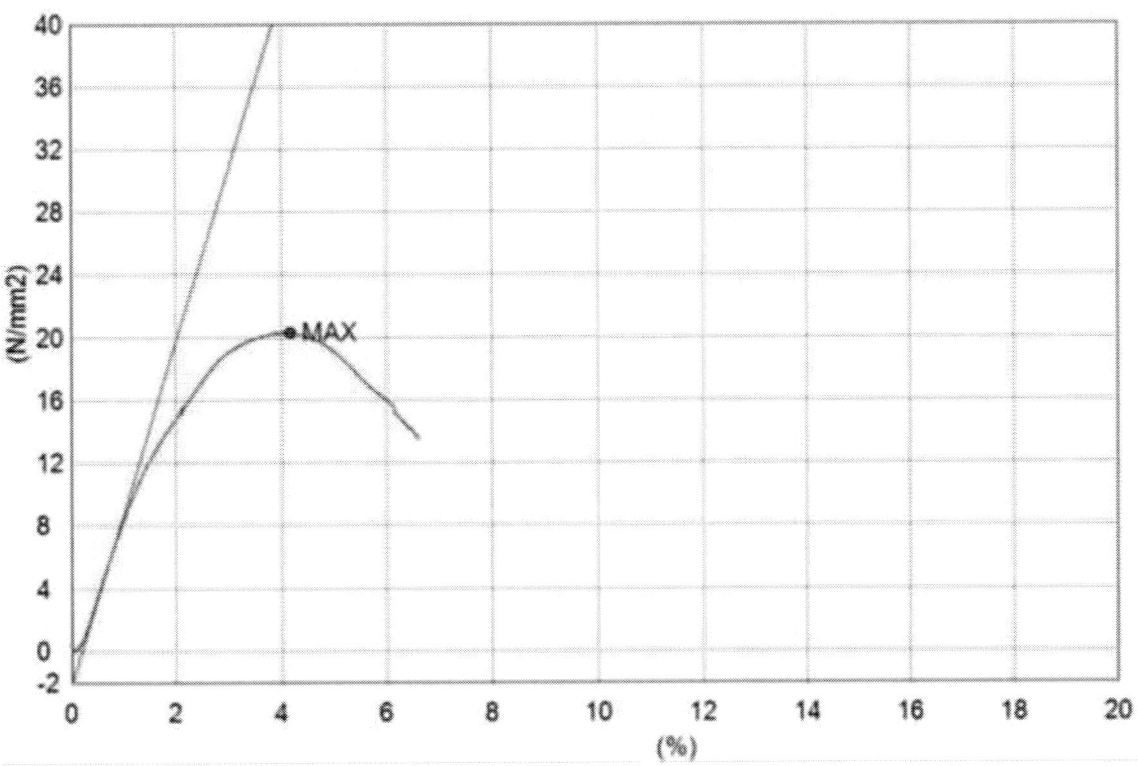

Figure 2. NFRP Composite specimen C3 (180°-160°) Bending Curve

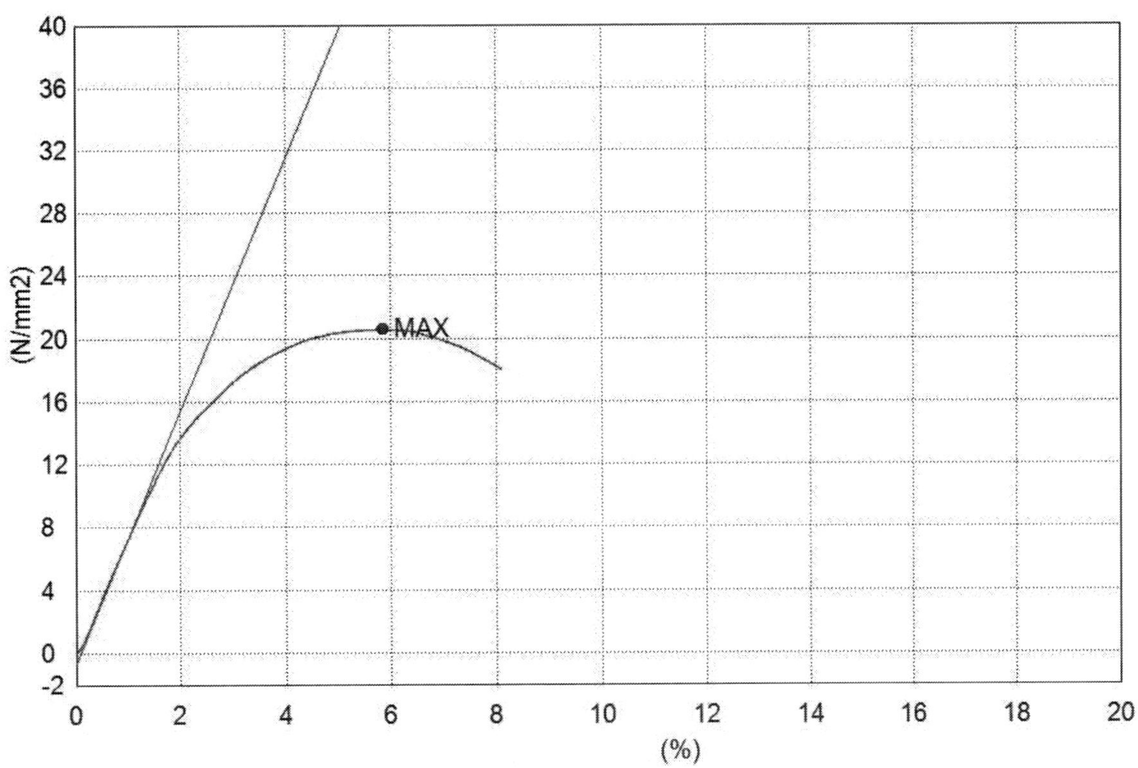

Figure 3. NFRP Composite specimen C4 (160°-140°) Bending Curve

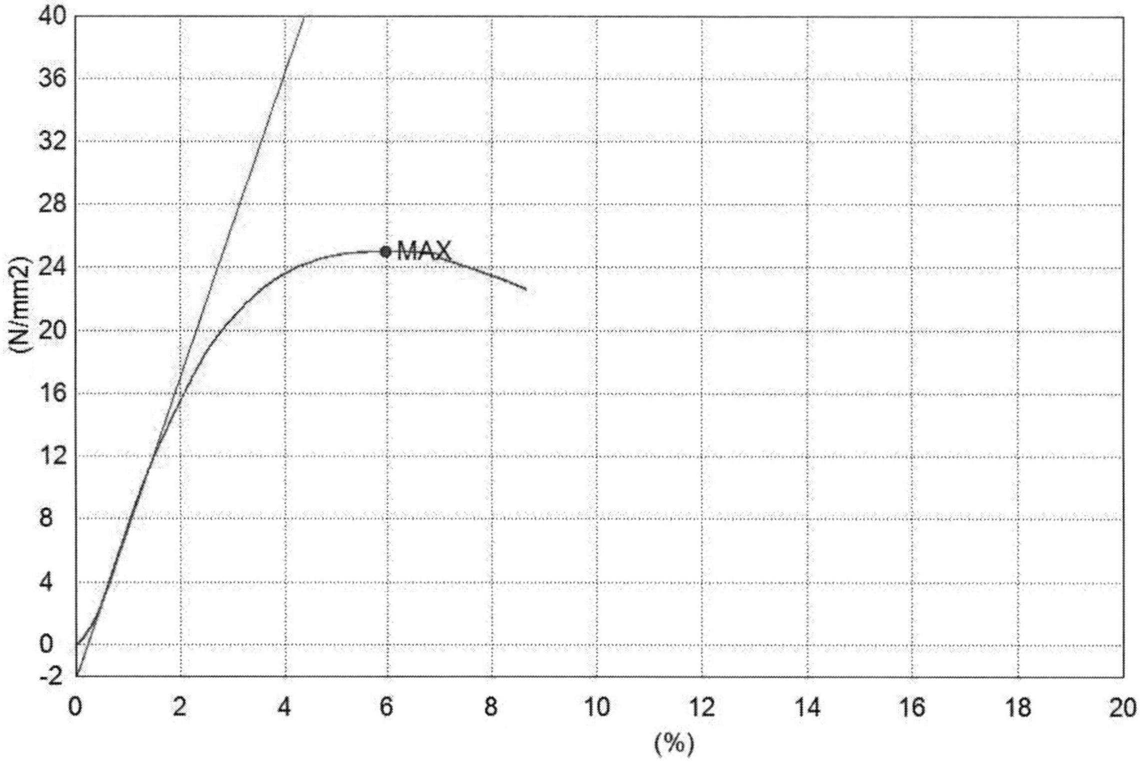

Figure 4. NFRP Composite specimen C2 (180°-174°) Interphase SEM images

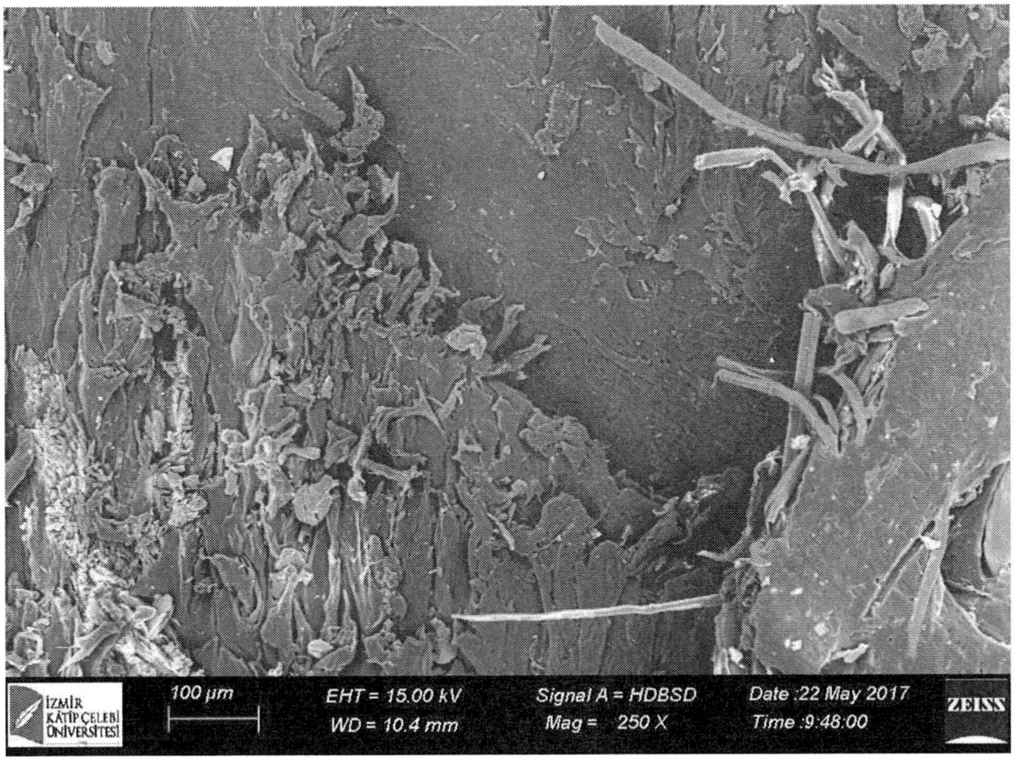

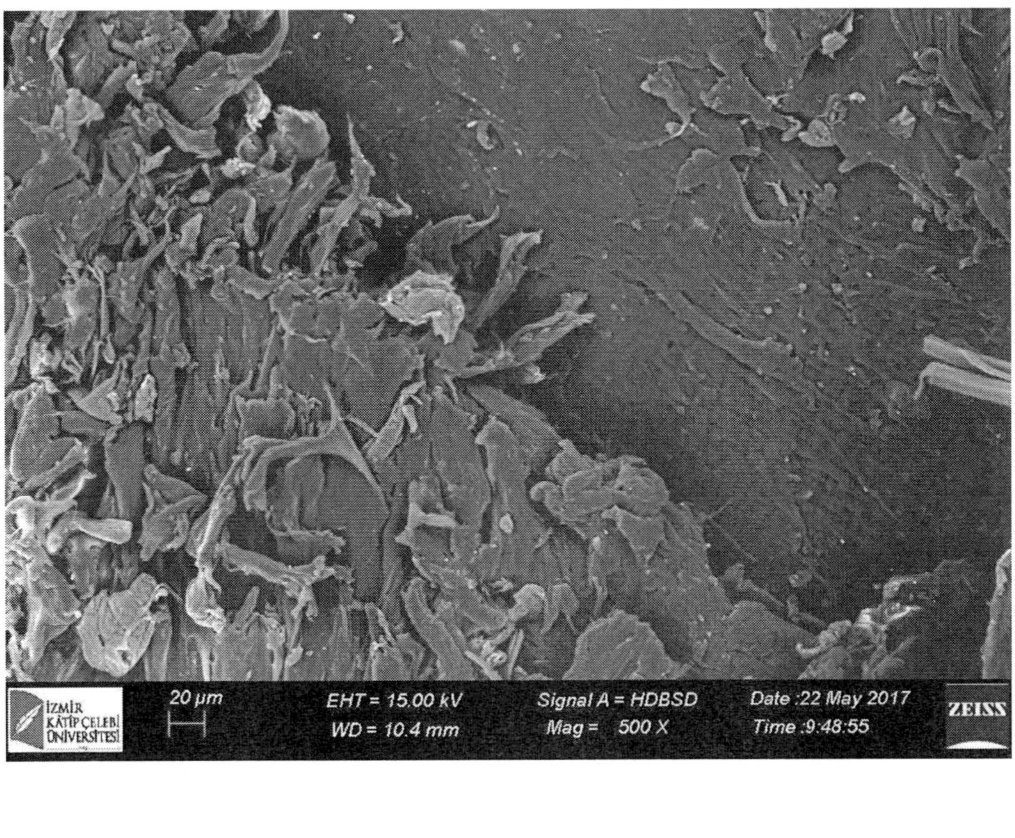

Figure 5. NFRP Composite specimen C2 (180°-174°) Surface SEM images

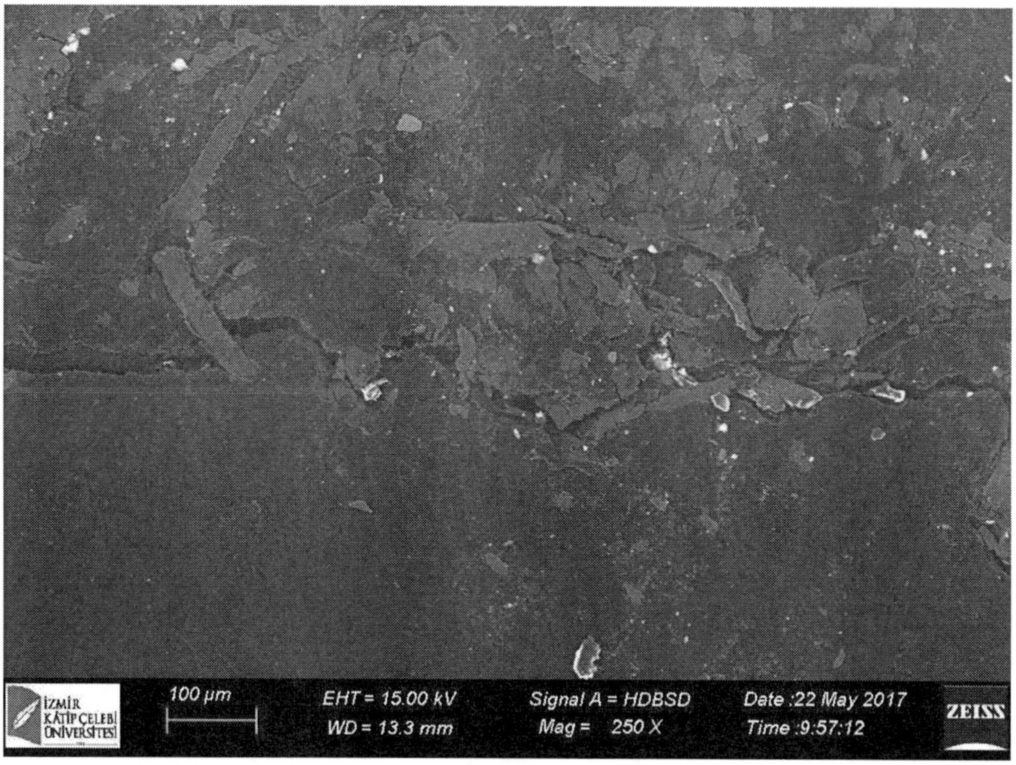

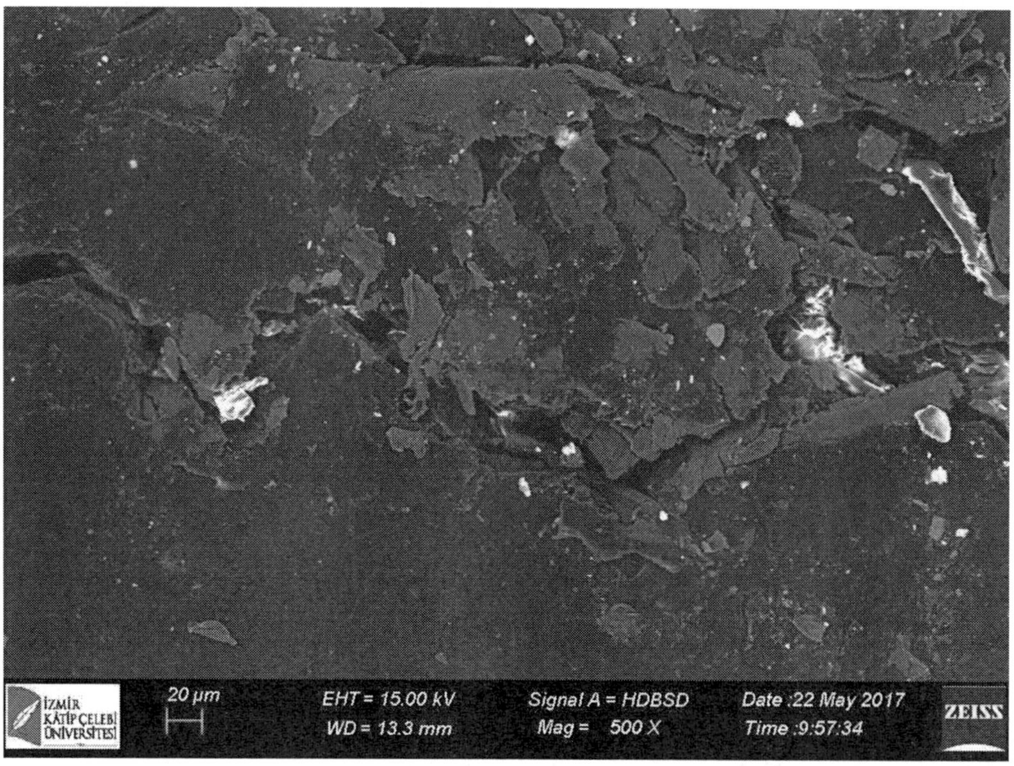

Figure 6. NFRP Composite specimen C3 (180°-160°) Interphase SEM images

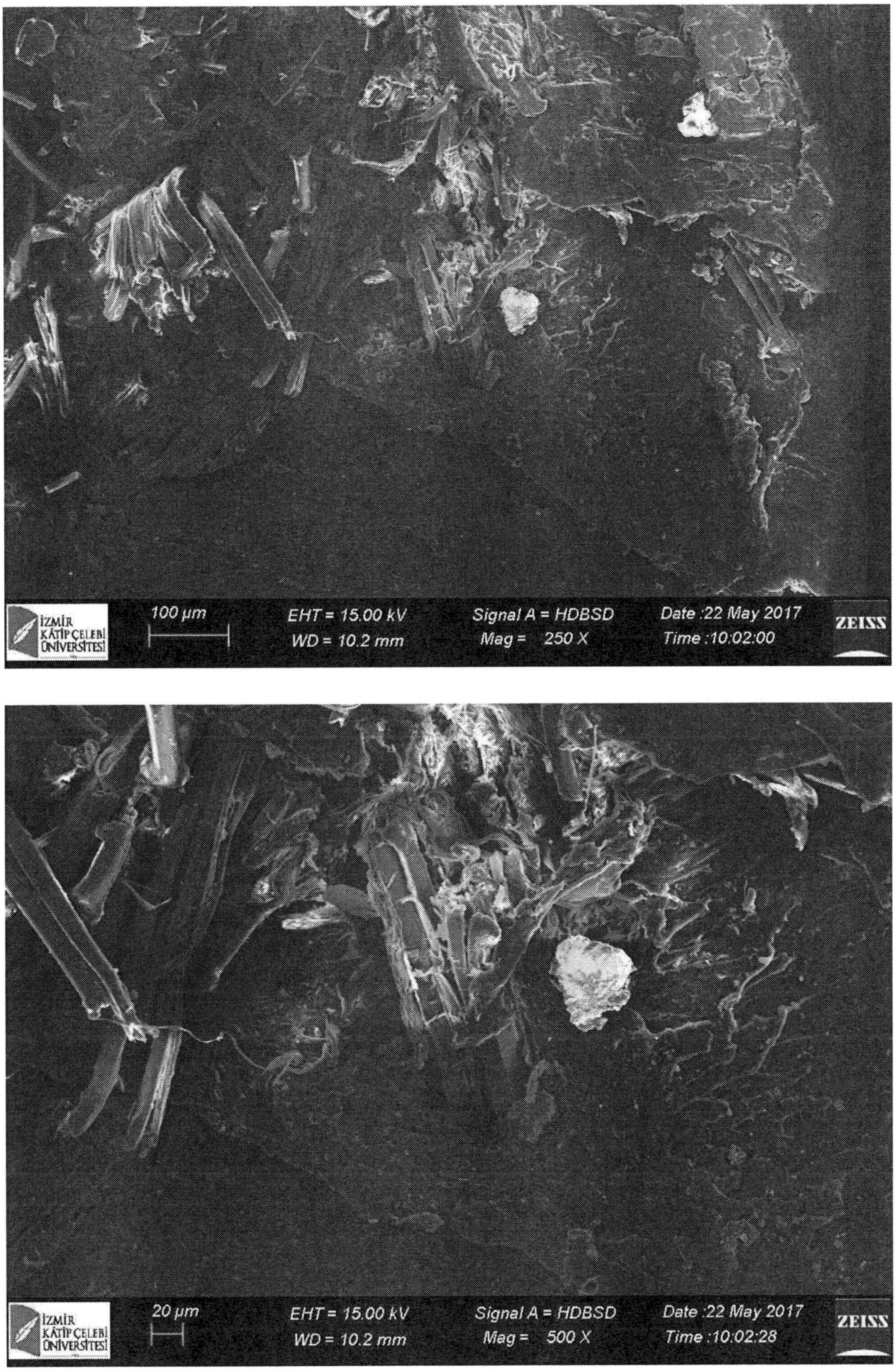

Figure 7. NFRP Composite specimen C4 (180°-160°) Surface SEM images

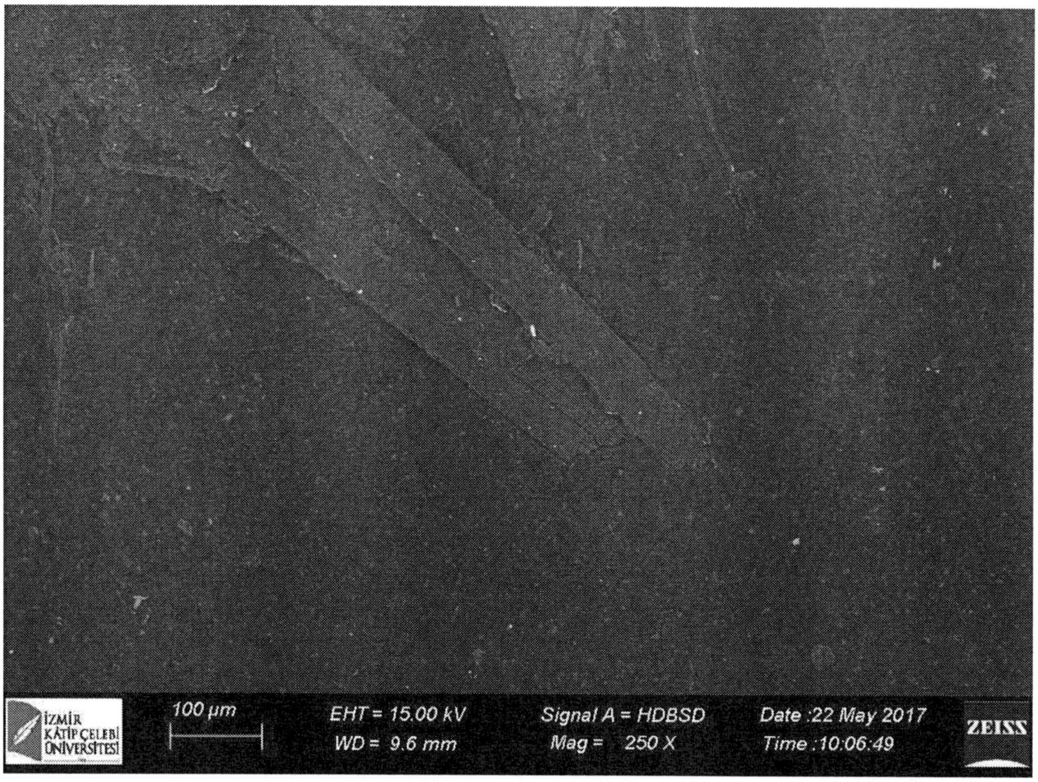

Figure 8. NFRP Composite specimen C4 (160°-140°) Interphase SEM images

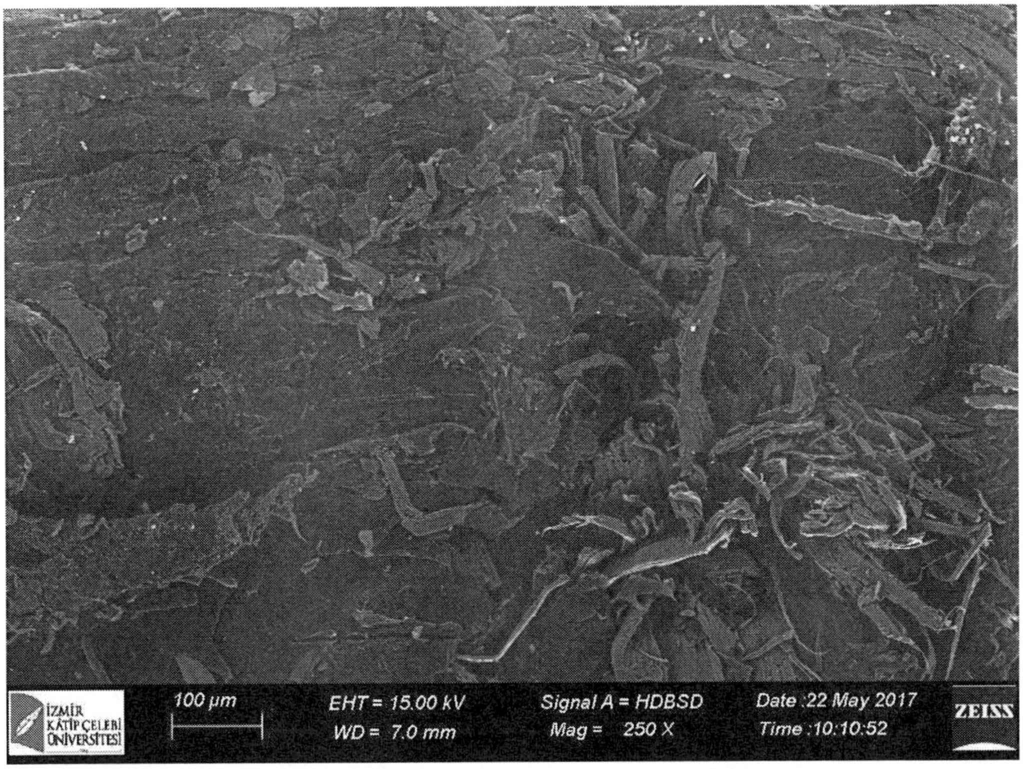

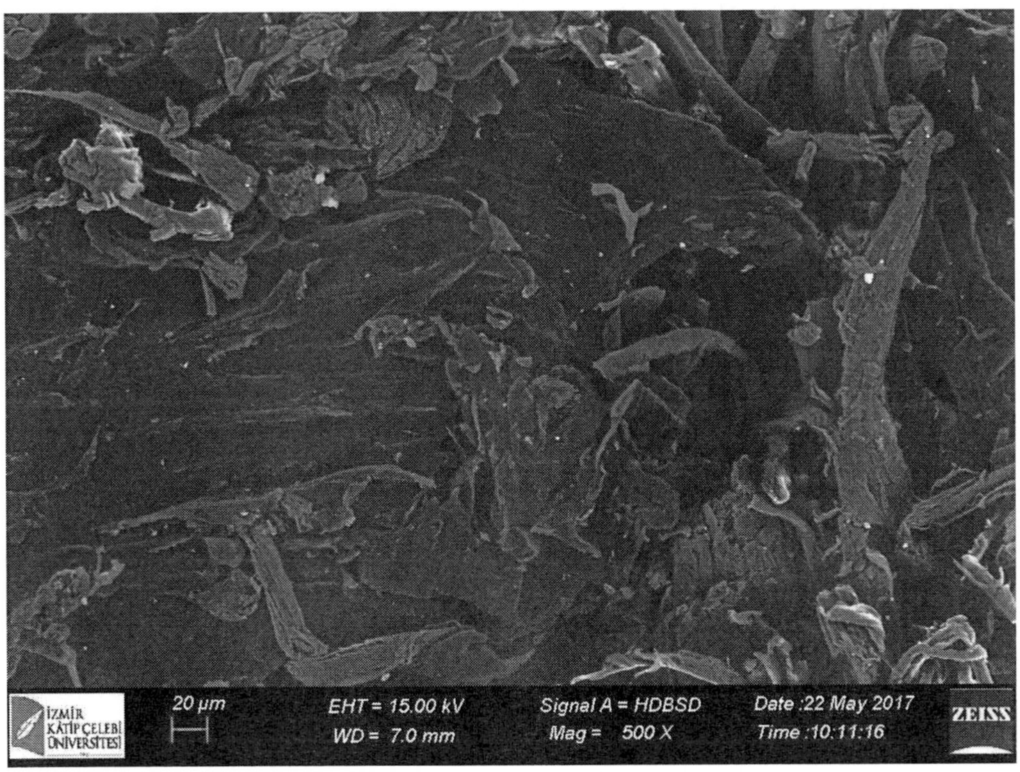

Figure 9. NFRP Composite specimen C4 (160°-140°) Surface SEM images

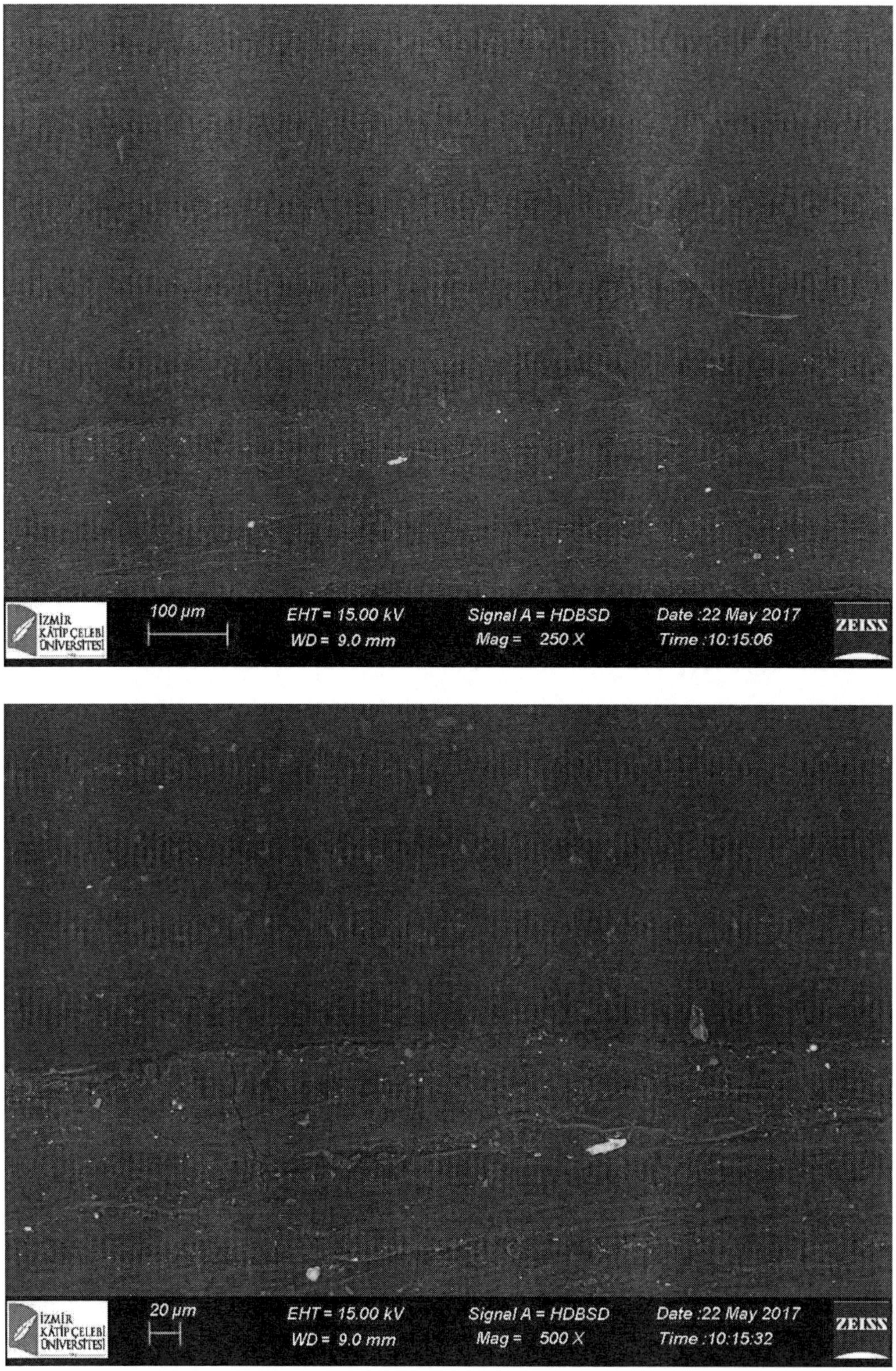

Construction of a pipe support with overmoulded metal insert

***Anger Karsten* [(1)*], *Lakemeyer Christoph* [(1)], *Lochbaum Ella* [(1)], *Haksteter Mark* [(1)]**

[(1)] Hadi-Plast GmbH & Co. KG - Hövelhof - Germany

Pipe supports contribute a big proportion to safe energy in modern buildings. Beside their function as the fastening of the pipeline, they have to insulate it from the environment. Conventional solutions consist of numerous components, of which every component fulfills one function. Because of this, conventional solutions are expensive and uncomfortable to install. The new developed pipe support provides a construction, that combines all functions in one component. The connection to the wall is given by an overmoulded metal insert. During the developing process, a with flame retardants filled polyethylene, that is foamed up with a chemical foaming agent, has been chosen as the material. It fulfills the mechanical requirements as well as the requirements to the insulation and fire prevention. Beside to the material, the focus of the developing process was on the construction of the pipe support. It was developed to the effect that the handling during the installation process is much easier and the pipe is well insulated. Moreover, the costs should decrease by being able to fasten different sizes of pipes with one pipe support. To reach this effect, the pipe support has a complex inner structure and pipe fastening, that promotes a good insulation by minimizing the contact surface between pipe and pipe support and using hollow chambers around the pipe. The pipeline is held by a minimal number of contact points, what leads to a contact surface, that is around 80% lower compared to conventional solutions. Different pipe sizes can be fastened with slats, that work as bending springs to fasten the pipeline. For installation or maintenance of the pipeline, the pipe support has snap-in hooks to open or close it as often as needed without additional tools. Finally, the injection molding tool and the process parameters are developed, whereby the challenge is to build a foam structure in some parts, and solid elements in other parts of the pipe support.

Fiber Length Reduction during Injection Molding

Elmar Moritzer[a], Gilmar Heiderich[b], Andre Hirsch[c]

[a]*Paderborn University, Kunststofftechnik Paderborn (KTP), Paderborn, Germany, elmar.moritzer@ktp.upb.de*
[b]*Paderborn University, Kunststofftechnik Paderborn (KTP), Paderborn, Germany, gilmar.heiderich@ktp.upb.de*
[c]*Paderborn University, Kunststofftechnik Paderborn (KTP), Paderborn, Germany, andre.hirsch@ktp.upb.de*

Abstract: When processing short fiber reinforced thermoplastics with screw machines (injection molding, extrusion), the fiber length is a resulting material property that is, to some extent, influenced by the process parameters. The extent of the influence and the effects of the process parameters have been investigated in many different studies, mainly by parameter variations. Nevertheless, the influence of the process parameters on the fiber breaking is described with regression models that are valid for the studied parameter set.

For a physical model of the fiber breakage during injection molding, simplified shear conditions are investigated. The pressures during plasticizing within the screw areas of injection molding machines are significantly lower compared to extrusion. For this reason the pressure flow within the screw channels can be neglected. The flow characteristic is basically influenced by the drag flow. To reproduce drag shear conditions, a test stand with a heated rotating piston and a stationary outer cylinder is used. In the gap between the piston and the outer cylinder the melted material faces the shear stress. The experimental results are used to find a correlation between the process parameters and the fiber breakage. This mathematical model is then transferred to the plasticizing process of injection molding. The validation of the model is done with a short fiber reinforced polypropylene.

Keywords: composite materials, fiber length reduction, injection molding
PACS: 81.05.Ni; 83.50.Uv

INTRODUCTION

The last few years have seen continuous growth in the sales of glass fiber-reinforced thermoplastics. The increasing use of these composites is due to their superior mechanical properties compared with unreinforced plastics. Their density is not much higher than that of unreinforced plastics so that if the mechanical data are related to the density, in other words if the weight-specific mechanical data are determined, the significant lightweight construction potential of fiber-reinforced plastics becomes evident.

Parts made of such composites must, however, not only have high lightweight construction potential, they must also be economical to produce. Fiber-reinforced thermoplastics have the greatest potential in this area because, compared with matrix polymers that have to be crosslinked, shorter cycle times can be achieved during the processing [1]. Fiber-reinforced thermoplastics are processed predominantly by injection molding with cycle times of less than one minute. Short and long fiber-reinforced granules are used as the base material, which have starting fiber lengths of 1–10 mm.

The strength and stiffness of the fiber-reinforced plastic are governed above all by the properties of the fiber. To obtain a benefit from the fibers, forces that act on the composite must be transferred to the fiber. The transfer of the forces takes place via the interface between fiber and matrix. As a function of fiber length, two different mechanisms occur when loading the composite to failure. If the fibers in the matrix material are of adequate length, failure of the fiber occurs. If the fiber length is inadequate, the fiber is pulled out of the matrix. Here, the strength of the fiber is not utilized, as a result of which the overall strength of the composite is reduced. To make full use of the strength properties of the fiber, a transfer of force must be ensured over the interface between the fiber and the matrix that leads to failure of the fiber. Assuming adequate adhesion of the composite, a minimum fiber length must be present, which is termed the critical fiber length l_c. The critical fiber length is estimated as follows with the aid of the fiber diameter d_f, interfacial shear strength τ_b and fiber breaking stress σ_{fB} [1]:

$$l_c = \frac{\sigma_{fB} * d_f}{\tau_b * 2} \qquad (1)$$

In order to shape the part in the injection molding process, the granules first have to be melted in the plasticizing unit of the injection molding machine. During the melting, the plastic or plastic melt is subjected to shear due to flow processes. This helps to achieve a disperse distribution of the fibers in the melt. At the same time, however, the shear also applies a mechanical stress to the fibers, causing fiber breakages [2].

During plasticization of both long fiber-reinforced and short fiber-reinforced thermoplastics, a shortening of the fibers takes place. Especially in the case of very short fibers, as are present in short fiber-reinforced thermoplastics, slightly increasing the fiber length is very effective and can lead to a significant increase in the mechanical data, as is shown by Fu and Lauke [3].

An optimization of the plasticizing process with regard to obtaining fibers that are as long as possible in the shot volume can be economically performed by simulations. For such simulations, modeling is needed. This is done by considering the physical boundary conditions. A few breakage models already exist, e.g. from Phelps [4], which describe the fiber breakage mathematically and physically in great detail. Through this detailed description, these models also need a variety of adjustment factors, resulting in a high degree of freedom. Furthermore, the adjustment factors must be determined separately by empirical means, necessitating a large amount of time-consuming experiments. The modeling approach described below is based on a simplified mathematical description of the physical conditions.

MODELING OF THE FIBER BREAKAGE

When plasticizing short fiber-reinforced thermoplastics in the injection molding process, a shortening of the fibers occurs from the moment of the first melt film formation. Prior to that, the fibers are in the solid plastics matrix, which protects them from damage. The flow processes that occur during the plasticizing process are predominantly dominated by the drag flow. In the melt film above the solid bed, in fact, only a drag flow is present. The shortening of the fibers is therefore examined as an example in preliminary tests within a test standard, in which a drag flow is generated through the rotation of a piston in a cylinder.

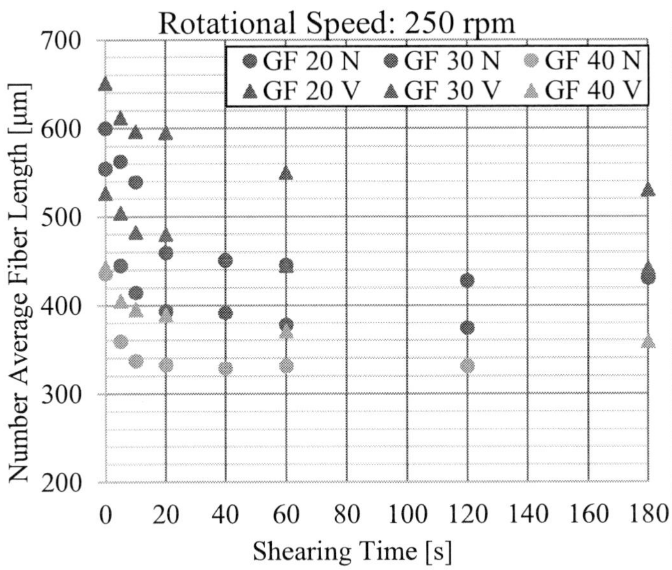

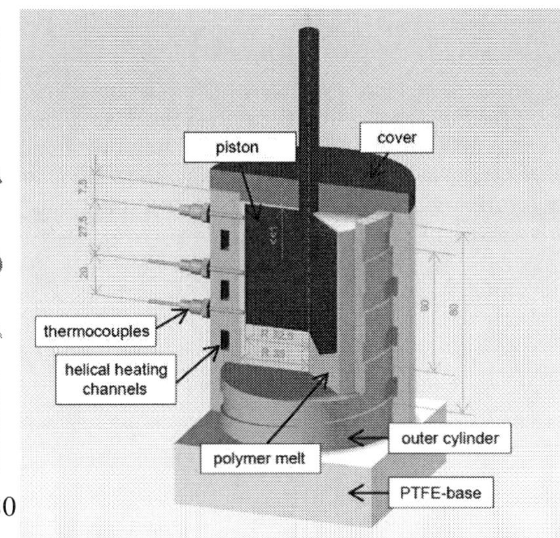

FIGURE 1. Fiber shortening vs. time when a drag flow is applied in a test stand

In this drag flow, different fiber-reinforced polypropylenes are subjected to shear. They differ in their matrix viscosity (high viscosity = N; low viscosity = V) and in their fiber contents (20% = GF20; 30% = GF30; 40% = GF40). It was found that a time-related regressive decrease in the weight-averaged fiber length always occurs, starting from the starting fiber length. In each case, a stationary end-value is reached, which is dependent on the shear stress $\eta * \dot{\gamma}$ occurring within the melt, the fiber weight content ψ and the stiffness of the fiber E. Through an examination by dimensional analysis, the following equation is established for the stationary fiber end-length L_∞ [5]:

$$L_\infty = ln\left(\frac{E^k}{(\eta * \dot{\gamma})^k * \psi^n}\right) * D \; where \; k = 1.43 \; and \; n = 13.71 \qquad (2)$$

Because similar flows occur during the plasticizing process as are generated in the test stand, a similar time-related decline in fiber length is to be expected. For this reason, the fiber length decrease shown in Fig. 1 for the melt area in the plasticizing process is assumed. Over the dwell time t_d, a regressive decrease in the fiber length L occurs. At time zero, the mean starting fiber length is L_0. This fiber length represents a boundary condition and corresponds to the mean fiber length in the granule. With advancing time, the fiber length approaches a stationary end-value, which can be determined by Equation 2.

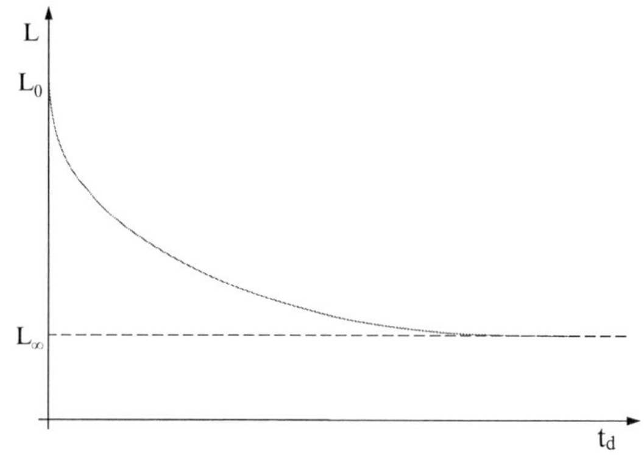

FIGURE 2. Mathematical description of the time-related shortening of the fiber

The time-related function curve between the two grid points L_0 and L_∞ is characterized by the flow conditions prevailing in the process and must thus be determined via the corresponding process values. The regressive time-related decrease from Figure 2 of the mean fiber length is described by the following equation, which Kloke [6] has also used to describe the decrease in fiber length in the twin-screw process:

$$\frac{dl}{dt_d} = -c \cdot l^2; \quad where\ l = \frac{L - L_\infty}{L_0 - L_\infty}\ for\ 0 < l \leq 1 \tag{3}$$

The solution of Equation 3 and the insertion of the boundary conditions produces the following descriptive equation

$$l(t_V) = \frac{1}{\frac{t_d}{t_0} + 1} \tag{4}$$

where t_0 is the time constant at which a halving of the fiber length occurs. The time constant is determined by an energy-related view of the fiber breakage, taking into account the flow processes during plasticization. Energy is used to break a fiber. This energy causes buckling in line with Euler's second buckling case. Through the buckling process, the fiber breaks in the middle, thus halving the fiber length. The energy needed for the buckling can only be introduced into the process via the screw motion in the form of dissipation. Generally speaking, the dissipation energy generated in the process can be described as follows:

$$E_{diss} = \eta \cdot \dot{\gamma}^2 \cdot V \cdot t_d \tag{5}$$

where V is the melt volume of the area being studied, η the viscosity of the plastic, $\dot{\gamma}$ the shear rate and t_d the dwell time.

The energy needed to break the fibers in the given melt volume can be expressed for Euler's second buckling case as follows, assuming a linear elastic behavior:

$$E_{break} = V \cdot \phi \cdot \frac{E \cdot \varepsilon_B}{2} \tag{6}$$

As described in [7], it is not only the viscous forces that are responsible for a breakage of the fibers, but also the forces that arise due to interactions of the fibers among each other. To take these into account, an interaction factor κ is introduced, which lowers the required breakage energy with increasing fiber content. In addition, this must be inversely proportional to the fiber volume content (Equation 7).

$$E_{break,int} = E_{break} * \kappa\ mit\ \kappa \sim \frac{1}{\phi} \tag{7}$$

The dissipation energy contributes not only to the fiber breakage but also brings about a temperature increase in the process. A dissipation coefficient ζ is thus introduced, which describes the proportion of the dissipation energy that contributes to the fiber breakage. This results in the following breakage condition:

$$\frac{E_{diss} * \zeta}{E_{break} * \kappa} = 1 \tag{8}$$

If this breakage condition from Equation 8 is met, breakage occurs in the middle of the fibers. Through this procedure, the time constant t_0 can thus be determined, because this describes exactly the time at which the length of the fiber is halved.

$$\frac{\zeta \cdot 2 \cdot \eta \cdot \dot{\gamma}^2}{\kappa \cdot \phi \cdot E \cdot \varepsilon_B} \cdot t_0 = 1 \tag{9}$$

Through reforming and insertion in Equation 4, we get the dimensionless fiber length l as a function of dwell time.

$$l(t_d) = \frac{\kappa \cdot \phi \cdot E \cdot \varepsilon_B}{t_d \cdot \zeta \cdot 2 \cdot \eta \cdot \dot{\gamma}^2 + \kappa \cdot \phi \cdot E \cdot \varepsilon_B} \tag{10}$$

Using the dimensioned factors L_0 and L_∞, the absolute mean fiber length in the process is determined (Equation 11).

$$L = l \cdot (L_0 - L_\infty) + L_\infty \tag{11}$$

The mean starting fiber length L_0 is determined experimentally by measuring the fiber length in the granule. The mean stationary fiber length L_∞ is the fiber length after which no further shortening of the fibers takes place.

MODEL IMPLEMENTATION

The application of the analytical fiber breakage model takes place within an existing 2D process simulation. This process simulation contains a melting simulation that makes it possible to determine the occurring melting rate and the solid bed geometry. Furthermore, the flows in the melt film and in the melt pool are calculated with the simplified basic equations of flow mechanics.

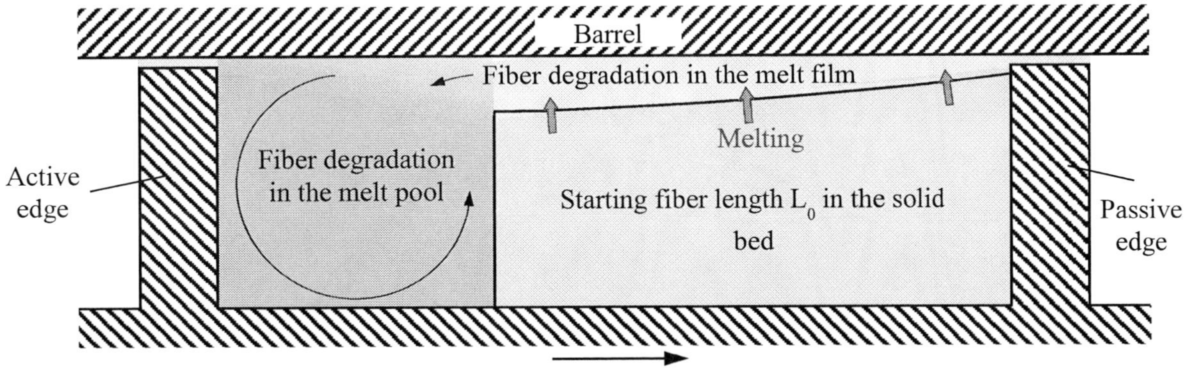

FIGURE 3. Areas of fiber damage during the plasticizing process

These values serve as input data for modeling the breakage. In the simulation, the screw is split up lengthwise into 100 segments for which the calculation of the fiber length is carried out individually. The fiber breakage in the melt film and melt pool are calculated separately and summarized via a weighting function to obtain a mean fiber length in the melt region of a channel segment. For the determination of the mean fiber length of an entire channel segment the mean fiber length in the melt region and the mean fiber length in the granule (as long as no complete melting has taken place) are entered into another weighting function.

EXPERIMENTAL VALIDATION

To validate the modeling, various process value variations were made on a Battenfeld injection molding machine (45 mm screw diameter, 3-zone screw).

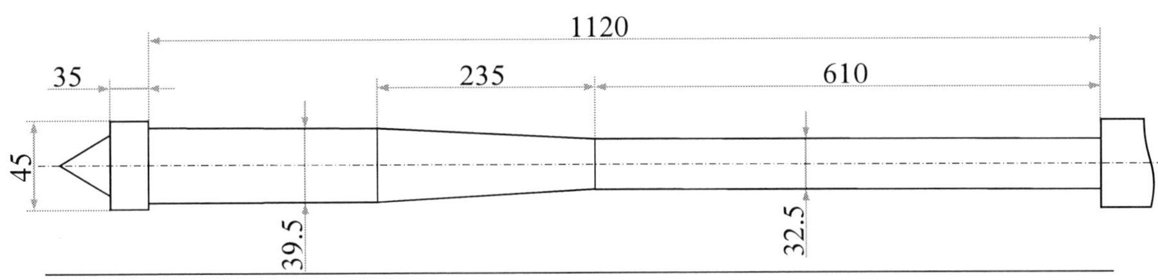

Injection molding parameter	Step 1	Central point	Step 2
Processing temperature [°C]	200	215	230
Metering velocity [mm/s]	100	250	400
Back pressure [bar]	20	60	80

FIGURE 4. Test setup and parameter settings

Use is made of cylindrical granules with three glass fiber contents, 20%, 30% and 40%, which are subsequently called GF20, GF30 and GF40. The matrix polymer of these materials is identical. The number average fiber length of GF20 is 623.82 µm, of GF30 is 525.54 µm, and of GF40 is 430.60 µm. The sampling is carried out within the metering volume. The samples are incinerated in a muffle furnace and the number average fiber length is determined with the help of an automated image analysis.

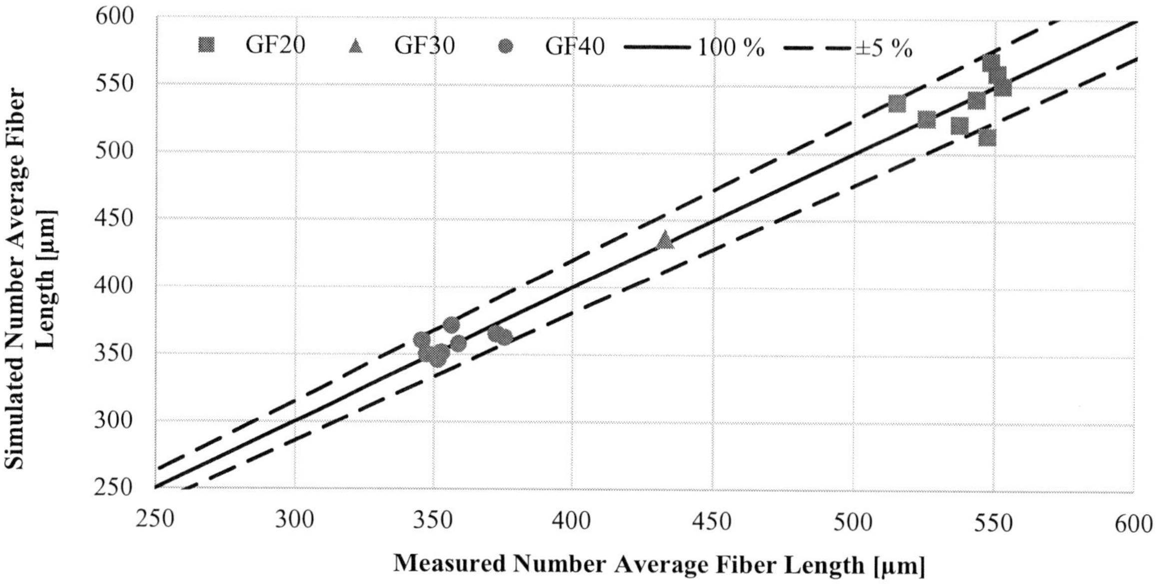

FIGURE 4. Comparison of measured and simulated fiber lengths after plasticizing (in the screw antechamber)

When comparing the experimentally measured number average fiber length with the modeled number average fiber length, only minor differences in the range of ±5 % were found. This means that the simulation quality of the model described in this paper as regards the shortening of the fiber during the plasticizing process is very good.

SUMMARY AND OUTLOOK

The modeling presented here is based on an energy-related view of the fiber breakage of short fiber-reinforced thermoplastics during the plasticizing process. To determine the starting values, use was made of a 2D process simulation. Through the melting simulation, both the shortening of the fiber in the melt film and in the melt pool can be determined. The simulated length reduction of the fiber describes the measured length reduction with a high degree of accuracy (maximum fluctuation 5%).

In future, the simulation will be validated for other matrix materials. Above all an examination will be made of whether the empirically determined factors are also valid for these materials. Should this not be the case, connections between the material properties and the shortening of the fiber will be examined by dimensional analysis and taken into account accordingly in the simulation.

REFERENCES

1. EHRENSTEIN, G. W.: *Faserverbundkunststoffe.* Werkstoffe - Verarbeitung - Eigenschaften. Carl Hanser Verlag, München Wien, 2006.
2. BAILEY, R.; KRAFT, H.: *A Study of Fibre Attrition in the Processing of Long Fibre Reinforced Thermoplastics. In:* International Polymer Processing, 1987, 2; S. 94–101.
3. FU, S.-Y.; LAUKE, B.: *Effects of fiber length and fiber orientation distributions on the tensile strength of short-fiber-reinforced polymers. In:* Composites Science and Technology, 1996, 56; S. 1179–1190.
4 PHELPS, J. H.: *Processing-microstructure models for short- and long-fiber thermoplastic composites,* 2009.
5. E. MORITZER; S. BROCKHAUS; G. HEIDERICH: *Fiber length degradation of Glass Fiber-Reinforced Polypropylene during Shearing.* 74th Annual Technical Conference of the Society of Plastics Engineers (ANTEC), Indianapolis (USA), 2016
6. KLOKE, P.: *Modellierung des Glasfaserlängenabbaus.* Dissertation, Paderborn, 2015.
7. THIELTGES, H.-P.: *Faserschädigung beim Spritzgießen verstärkter Kunststoffe.* Dissertation, Aachen, 1991.

Waste Heat Utilization for Granule Preheating in Extrusion Processes to Increase Energy Efficiency

Fabian Baumgarten[1, a)] and Johannes Wortberg[1, b)]

[1]*University of Duisburg-Essen, Institute of Product Engineering, Department of Engineering Design and Plastics Machinery, Lotharstraße 1, 47057 Duisburg, Germany*

[a)]fabian.baumgarten@uni-due.de
[b)]johannes.wortberg@uni-due.de

Abstract. Modern extrusion processes require 40% - 60% more energy than physically necessary. This excess of energy is mainly due to high heat radiation. One of the main sources of heat is dissipation.
In order to ensure stable thermal conditions, the barrel is tempered externally with heating and cooling elements and discontinuously operated radial fans. As a result, warm air is discharged unused to the environment.
In a new concept, the warm air is to be collected by a manifold and a central, continuously controlled, suction fan. This warm air is used for granule preheating. During the first development steps, the manifold, the barrel housing and the granulator funnel were already designed with the help of "Flow Simulation". Therefore, the flow simulation yields potentials of flow conditions caused by different flow geometries.
The next step is an adaption of the control technology for the control of the suction fan and the throttle valves in order to be able to control the temperature of each barrel zone.

OBJECTIVES OF THE PROJECT

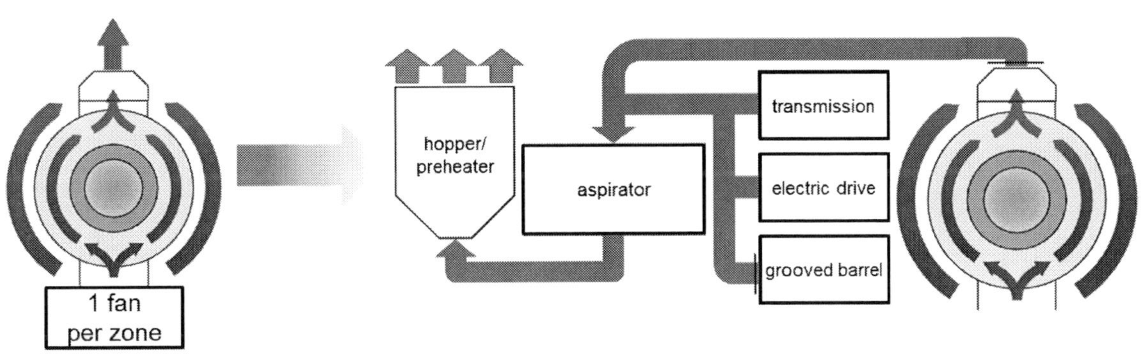

FIGURE 1. Schematic illustration of an extruder from the front. The left part of the illustration shows a conventional extruder and the right part of the illustration shows an extruder with additions for waste heat utilization.

Usually, in case of cooling the extruder, waste heat is lost to the environment by using cooling fans at each heating and cooling zone of the machine. Now, this waste heat shall be collected and used. Figure 1 shows a schematic illustration. All elementary components of the machine producing waste heat are shown here. Furthermore all components dissipating or using heat are shown either.

The primary objectives for the system to be developed are, tap accumulating waste heat, waste heat conduction with low energy losses, and efficient waste heat utilization for preheating materials. The secondary objectives are the tap accumulating waste heat of transmission, electric drive, and grooved barrel.

From these ambitions the new concept derives to realize continuous waste heat dissipation using a speed controlled aspirator and throttle valves.

THOUGHTS ON WASTE HEAT UTILIZATION

This project is about conducting the indicated heat flows for preheating to the granule. The added heat shall soften the granule to relieve the drive of the extruder. Following the torque shall be decreased and thus the demand of energy. However, these heat flows are tapable when the temperature of the melt is higher than the set point. The amount of energy of the waste heat flow being maximally available is depending on the volume flow of the aspirator and the temperature difference between the inflowing air and the barrel.

$$\dot{Q}_{\max} = \dot{V}_{\max_{air}} \rho_{air} c_{p_{air}} \Delta\vartheta_{\max_{air}} \tag{1}$$

ACCUMULATION OF WASTE HEAT

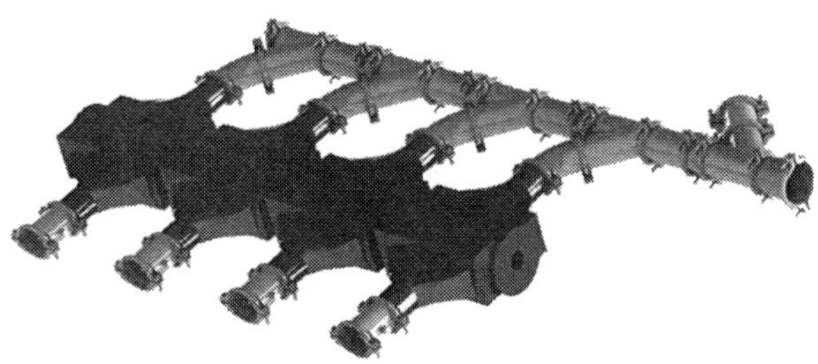

FIGURE 2. Throttle valves, barrel housing, and fan manifold for the accumulation of waste heat.

An elementary sketch of the alternate temperature control of the plasticizing unit, which has to be designed, is shown in Fig. 2. The central aspirator is connected to a hot air collector, the fan-type exhaust pipe and creates a vacuum in it. Branch pipes connect the hot air collector to every barrel zone, which has to be cooled, while every branch is spatially independent and possesses a hermetically sealed connection to the air outlet port of the heating/cooling combination unit. Due to the vacuum, the hot air is extracted in the respective barrel zone.

While operating in production, various zones demand different cooling capabilities; to reach the target temperature, an eduction of the heated air at every heating/cooling combination unit is needed. This is designed by the use of throttle valves, which control the flow rate of the cooling air. The opening up or closing down of the throttle valves is adjusted to the desired cooling power. Thus, also with respect to the frequency control of the aspirator, only the effectively required cooling power will be put into the system.

Barrel Housing

One part of the waste heat accumulation is the heat cooling combination consisting of a fan and heating and cooling tapes. Two different types of heating cooling tapes of the company WEMA are used. On one hand the Tape HK214 which allows heating and cooling of the barrel. It is made of a ceramic material and has a thermal conductivity of round about 16 to 28 watts per meter kelvin. On the other hand, the tape KE300 which allows cooling only. It is made of aluminum and has a much higher thermal conductivity with round about 200 watts per meter kelvin.

The most important requirement is the homogeneous temperature distribution in the barrel wall. Thereby it is distinguished if it is temperature distribution in peripheral or longitudinal direction. The distribution in peripheral direction has the higher priority. For homogenization of the temperature in peripheral direction an air baffle was installed which is narrowing the cross-section of the housing, in flow direction. For homogenization of the temperature in longitudinal direction, air baffles in the inlet areas have been installed for even distribution of the air in the housing. Further requirements are minimization of pressure losses and an effective dissipation of waste heat.

On the data basis which was prepared in the first working package of the project a volume flow of 100 cubic meters per hour and an exhaust air temperature of 100 degrees Celsius turned out to be a realistic scenario.

$$\frac{d_{outside}}{d_{inside}} = \frac{90\,mm}{45\,mm} = \frac{2}{1} \tag{2}$$

The proportion of cross-sections is necessary for the calculation of the heat flow density on the inner side of the barrel. From this information the heat flow of the exhaust air on the outer side of the barrel can be calculated.

$$\dot{Q} = \dot{V}\rho c_p \Delta\vartheta \tag{3}$$

The heat flow density derives from including the surface of the barrel.

$$\dot{q}_{outside} = \frac{\dot{Q}}{A} = \frac{\dot{Q}}{\pi d l} = \frac{\dot{V}\rho c_p \Delta\vartheta}{\pi d l} \tag{4}$$

Last the heat flow density on the inner surface can be calculated by using the cross-section proportion of the outer and the inner surface.

$$\dot{q}_{inside} = 2 \cdot \dot{q}_{outside} \tag{5}$$

All results are based on following simulation parameters: A volume flow of 100 m³/h, an exhaust air temperature of 100 °C, an ambient air temperature of 25 °C, and a heat flow density of 50000 W/m²

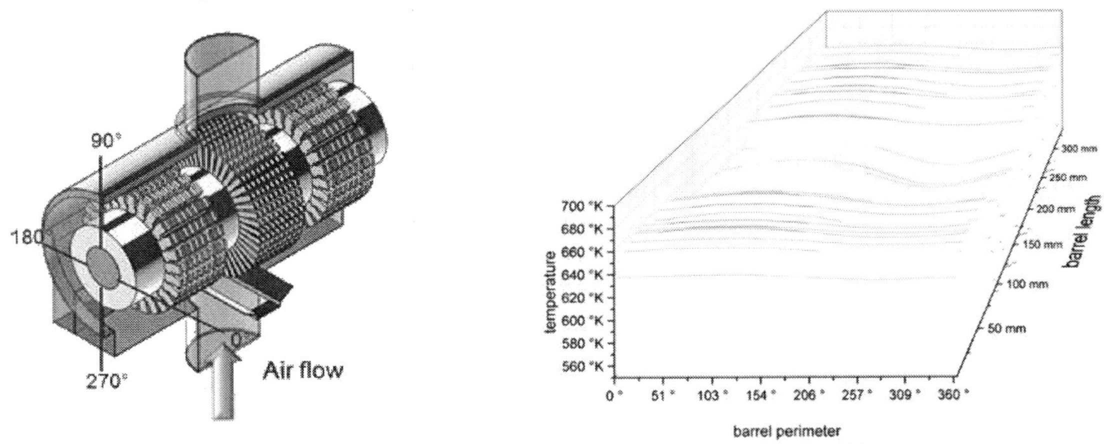

(a) (b)

FIGURE 3. Cutaway drawing of the first barrel housing concept (a) and 3D temperature diagram of the inner barrel surface (b).

The housing of the first concept has an air inlet on the 270° position and the outlet on the opposing side on the 90° position. Centrically a KE300 Tape and sidewise a HK214 Tape each is installed. In Fig. 3a you see the flow paths at a volume flow of round about 100 m³/h. The temperature values on the inner barrel wall have been calculated numerically in a simulation by using "SolidWorks". The temperature distribution on the inner wall is indicated in Fig. 3b. The temperature is laid on the z-axis, the barrel perimeter on the x-axis and the barrel length on the y-axis. Higher cooling in the area of the KE300 tape than of the HK214 Tapes can be seen. Furthermore, the homogeneity in peripheral direction in the center of the barrel is inferior to the margins. The heat is generally inferior compared to the other concepts.

The standard deviation is a measure for the homogeneity of the particular direction or rather the total homogeneity. The standard deviation in longitudinal direction is set between 41,93 and 48,53. The standard deviation in peripheral direction instead is in-between 1,45 and 10,05. The total standard deviation is 45,55.

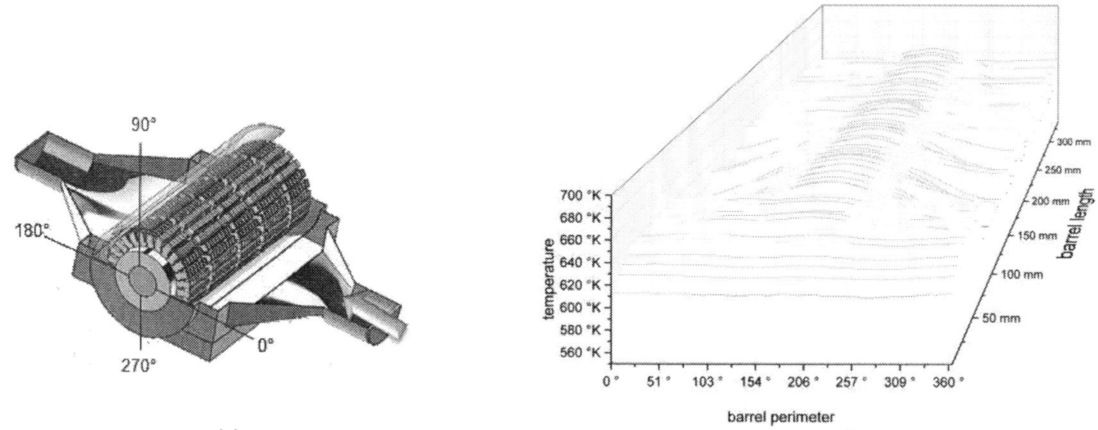

(a) (b)

FIGURE 4. Cutaway drawing of the second barrel housing concept (a) and 3D temperature diagram of the inner barrel surface (b).

Figure 4a is showing the housing of the second concept, which has the air inlet on the 0° position and the outlet on the opposing 180° position. Only heating cooling tapes of the type HK214 are attached.

The temperature distribution on the inside wall of the barrel can be seen in the 3D diagram (Fig. 4b). You can see in the central area – where there is no vortex formation and the streaming velocity has the highest values – the temperature distribution in peripheral direction is inferior to the margin areas. The temperature distribution in longitudinal direction instead is very acceptable. The heat dissipation is higher than in the first concept.

The standard deviation is a measure for the homogeneity of the particular direction or rather the total homogeneity. The standard deviation in longitudinal direction is in between 7,83 and 16,19. The standard deviation in peripheral direction instead is in between 0,79 and 20,86. The total standard deviation is 14,87.

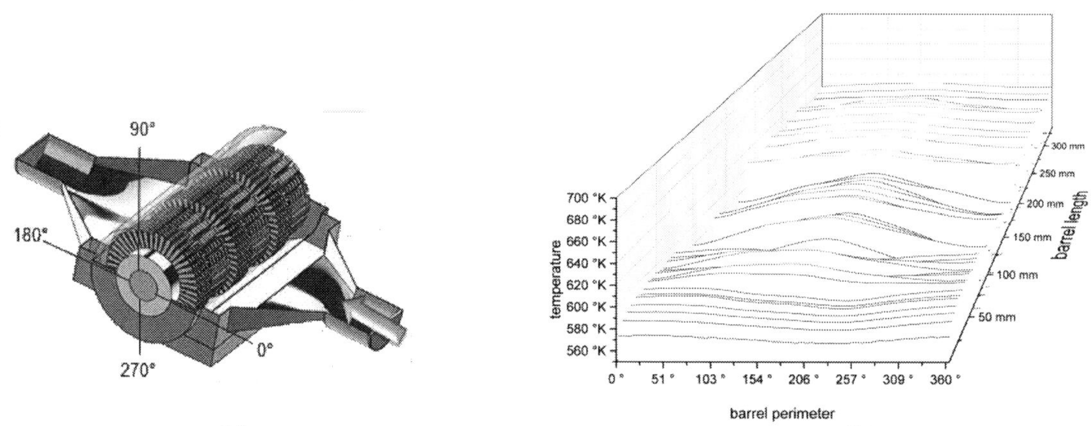

(a) (b)

FIGURE 5. Cutaway drawing of the third barrel housing concept (a) and 3D temperature diagram of the inner barrel surface (b).

The housing of the third concept has the air inlet on the 0° position and the outlet on the opposing 180° position. Three heating cooling tapes of the type HK214 and the cooling tapes of the Type KE300 are attached alternating to the barrel (Fig.5). The temperature distribution on the inside wall of the barrel can be seen in the 3D diagram. In the central area – where there is no vortex formation and the streaming velocity has the highest values – the temperature distribution in peripheral direction is inferior to the margin areas. In contrast to that, the temperature distribution in longitudinal direction is very acceptable. The heat dissipation in this concept is significantly the highest.

The standard deviation is a measure for the homogeneity of the particular direction or rather the total homogeneity. The standard deviation in longitudinal direction is in-between 11,21 and 22,36. The standard deviation in peripheral direction instead is in-between 1,40 and 17,71. The total standard deviation is 18,45.

WASTE HEAT UTILIZATION

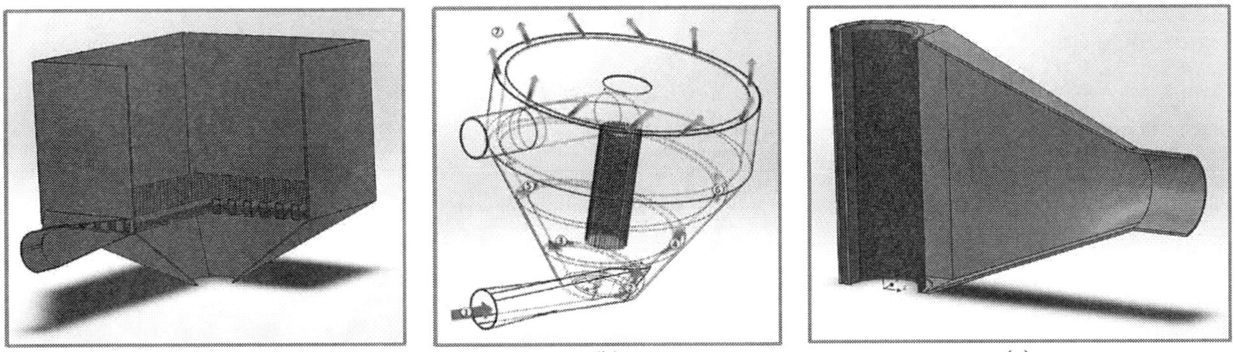

(a) (b) (c)

FIGURE 6. Drawings of the three concepts for waste heat utilization: mesh concept (a), slit pipe concept (b), ring concept (c).

Figure 6 shows the three concepts for the hopper with granule preheating which were elaborated during the project. The first two concepts are substituting the usual hopper, the third concept is an attachment.

The mesh concept (Fig. 6a) utilizes the warm exhaust air which is conducted through a diffusor in a mesh, brought in the hopper. The granule passes the mesh, the warm air rises because of the filling of the granule and preheats it therefore. The slit pipe concept (Fig. 6b) conducts the air through two openings in the hopper. The upper duct is being conducted through an arc pipe to the center of the hopper. The lower duct instead is being conducted cyclonically in a partition wall on the outer side of the hopper.

The ring concept (Fig. 6c) is an attachment which is being installed between the extruder itself and the hopper. The granule trickles through this attachment and is being heated by air, which flows in through a diffusor.

The mesh concept was chosen as the basis of the waste heat utilization.

Hopper – Mesh Concept

In the Mesh Concept the warm exhaust air is flowing in through a cylindrical pipe into a diffusor, at position number one. At position number two, this diffusor is increasing its cross-sectional area to decelerate the air and distribute it to all mesh struts. By perforations in these mesh struts, at position three, the warm air is being dissipated into the hopper. The warm air is rising up along the height of the hopper and heats the down falling granule. The height is a factor which yet has to be designed.

CONCLUSION AND OUTLOOK

Based on machine data simulation, parameters have been specified and an appropriate aspirator has been selected. Furthermore, barrel housings and fan manifolds have been designed. With the waste heat and CFD simulations, concept three was selected for the barrel housing, despite of the feeding zone. The arc pipe concept was selected for the fan manifold.

In addition to this, concepts for waste heat utilization have been elaborated and the mesh concept was selected as a basis for further investigation.

For the future, the integration of potential waste heat sources will be investigated, a prototype will be constructed and the simulation results will be verified by practical trials.

REFERENCES

1. Schneider, F.: *Energieverbrauch verringern.* In: Kunststoffe 10/2008, Carl Hanser Verlag, München, 2008
2. Körzel, K.W.; Michels, R.: *Neuentwicklung bei Zylindereinhausungen.* Beitrag zur 7. Fachtagung 'Neuigkeiten in der Extrusion', Süddeutsches Kunststoff-Zentrum, Würzburg, 05.-06. Juni 2002
3. Weiß, P.: *Energieeffizienz und Abwärmenutzung in der Extrusion.* Dissertation, Universität Duisburg-Essen, 2016

Investigation on the influences of injection moulding process parameters on the short-shot filling of polymeric micro features

Eladl Abdelkhalik Mohamed [(1)]*, ***Tosello Guido*** [(2)], ***Islam Aminul*** [(2)], ***Mostafa Rania*** [(3)], ***Soltan Hassan*** [(3)], ***Hansen Hans*** [(2)]

[(1)] DTU and Mansoura University - Faculty of Engineering - Denmark - Denmark, [(2)] DTU - Denmark - Denmark, [(3)] Mansoura University - Faculty of Engineering - Mansoura - Egypt

Micro injection molding is known to be one of the most important mass production technologies because of its high productivity with many applications in industries. Mould filling of micro-features is one of the most important process characteristics. This paper investigates the effects of four processing parameters on filling behavior presented by part mass and runner mass on micro finger test structure moulded with Acrylonitrile Butadiene Styrene (ABS) using DOE approach. In particular, the following factors are considered: injection speed, holding pressure, melt temperature and mould temperature. The study revealed that holding pressure and melt temperature are the most influential parameters increasing the magnitude of the part mass for all short shot stroke. Melt temperature is the most influential parameters decreasing magnitude of the runner mass for all short shot stroke and it the especially at 7 mm and 8 mm. From 9 mm to 13 mm short shot stroke holding pressure is the most influential parameters for increasing the runner mass.

Combining thermoset and rubber by injection molding - Characterization of factors determining compound properties

Ulrich Heyne[1, a)], Michael Gehde[1, b)]

[1]Technische Universität Chemnitz, Professur Kunststoffe, Reichenhainer Str. 70, D-09126 Chemnitz

[a)]Corresponding author: ulrich.heyne@mb.tu-chemnitz.de
[b)]michael.gehde@mb.tu-chemnitz.de

Abstract. Currently, there are no series applications of thermoset and rubber manufactured by multicomponent injection molding. For the reasons that the material behavior of thermoset and rubber is suitable for technical applications, mainly for high temperatures and chemical aggressive surrounding, as well as that their curing behavior is very similar in injection molding, the processing of multicomponent injection molding using reactive materials (thermoset and rubber) were analyzed. In present study, factors determining compound properties, the adhesive strength of the compounds, were investigated. The results show the practicality to create thermoset-rubber compounds with multicomponent injection molding technology and reveal important influencing variables for the compound properties and in the special case of the adhesive strength of the compounds.

INTRODUCTION

Nowadays there are high demands on the functionality and economy of products. One possibility meet these high requirements can be found in the multicomponent injection moulding technology. Main benefits of this technology is that it is possible to produce off tool parts as a way to reduce the production periods. Furthermore, it is possible to combine properties of different materials into one part, by that you can produce parts with high functionality. As a result, the number of parts needed can be reduced, which also reduces assembling costs and error sources [1, 2, 3].

The main material combinations in multicomponent injection molding technology are different kinds of thermoplastics, thermoplastics and rubber, or thermoplastic elastomers (TPE) and liquid silicone rubber (LSR) [4, 5, 6, 7, 8]. With these combinations it is possible to create parts with special profiles of properties such as hard/soft combinations. However, disadvantages of these combinations can be seen in the points that not all thermoplastics can be combined with rubber and in addition, that in general the thermal and chemical resistance of 'standard'-thermoplastics is low given the behavior of cured thermoset and rubber [9].

BACKGROUND

Particularly for parts in technical applications, stringent demands may be made on the heat distortion temperature, the inclination to creep, and the hardness and strength of the material, even at very high temperatures or chemical aggressive surroundings. With regard to the specified requirements, thermosetting plastics are generally better at fulfilling the specified properties than comparable thermoplastics such as PE, PP or PA.

Another interesting point when combining thermoset and rubber with multicomponent injection molding concerns the processing procedure. Both materials need heat for the reaction process and such the curing processes of each, and the reaction between both reactive material systems should take place simultaneously in one and the same heated tool by identical processing parameters by manufacturing. By that, the manufacturing can be very efficient in terms of energy, resources and economics. In addition, the processing behaviors of thermoset and rubber are very similar in injection molding. According to this the processing procedure could be simplified.

EXPERIMENT

As a first step of this study, a proof of concept was achieved. Subsequently, influencing variables for compound properties were investigated. Therefore, the study was divided into four main points of interest, shown in Table 1. Table 2 shows the selected test parameters for the different analyses. For the analysis of the influence of the process parameters the tested pairings of curing time and curing temperature chosen such that a complete chemical reaction of each component could be assured after manufacturing. The cross marks the tested pairings. For the tests of the influence of the material pairing, common, differently filled phenolic resins were selected. For the rubber component, materials for technical applications with a high resistance against technical fluids like gasoline or oil were selected. During the analysis of the process variants three possibilities were tested. Hereof, at first an uncured rubber was applied onto an uncured thermoset insert. Secondly, an uncured thermoset was applied onto a cured rubber and thirdly, very similar to the first, an uncured rubber was applied onto a cured thermoset. As a last point of the study, the influence of the surface roughness and the surface area of the inserts was investigated. Therefore, the surfaces of the inserts were modified by grinding with different sizes of abrasive grain before manufacturing.

Table 1. Main points of analysis

Test series	Background of analysis
1	Influence of processing parameters
2	Influence of material pairing
3	Influence of process variant
4	Influence of surface roughness and surface area

Table 2. Selected parameters and conditions for the different test series

Test series 1 - Influence of processing parameters - Selected process parameter pairings					
Curing time (min)	2	3	4	6	8
Curing temperature (°C)					
160					X
175			X	X	
190	X	X			

Test series 2 - Influence of material pairing - Selected material pairings			
Thermoset	PF1110 (45 GB, 35 GF -weight-%)	PF6510 (20 GB, 35 GF -weight-%)	X6952 (55 GF -weight-%)
Rubber	NBR (acrylonitrile butadiene rubber)	AEM (ethylene acrylate rubber)	ACM (acrylate rubber)

Test series 3 - Influence of process variant - Selected process variants		
Test runs	***Insert***	***Injected component***
1	(*) uncured thermoset	rubber
2	cured rubber	thermoset
3	cured thermoset	rubber

Test series 4 - Influence of surface roughness, roughness depending surface area - Selected sizes of grain for modification					
Test series	*1*	*2*	*3*	*4*	*5*
Paper grain size	P4000	P1200	P500	P320	P80
Averaged roughness [µm]	0,67	2,0	3,4	7,6	25,2

To manufacture the test specimens, the component injection molding was applied using inserts. To manufacture the test samples, the inserts have to be produced beforehand. Therefore, single tension bars were produced, selected and grinded on the surface. Subsequently, the insert had to be put into the tool. As a next step, the second component was injected. Was the curing time reached, the test sample could be demolded. Figure 1 shows the processing procedure of the test sample production. For each test series at least 5 test specimens were manufactured.

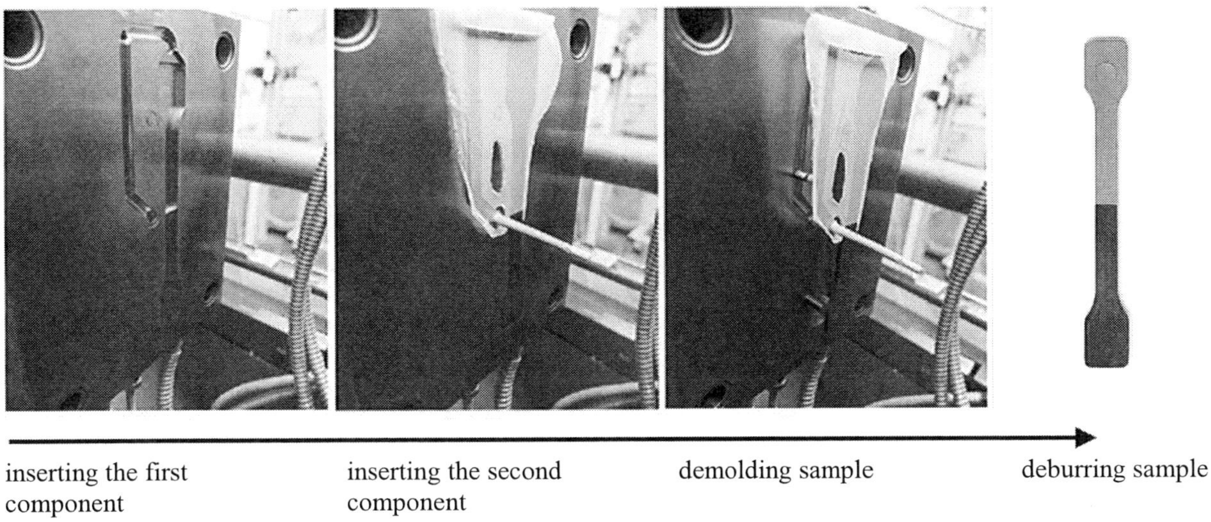

FIGURE 1. Manufacture of test samples using the produced inserts

RESULTS AND DISCUSSION

Influence of the processing parameters

Figure 2 shows influence of processing parameters curing time and curing temperature on the adhesive strength demonstrated on the combination of PF6510 with AEM or ACM. The figure shows that there is no significant influence of the tested parameter pairings on the adhesive strength across the whole tested area. The same behavior was found for all analyzed material pairings and process variants.

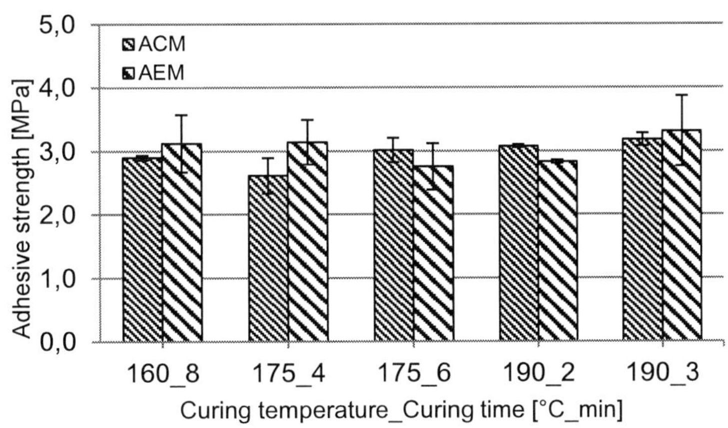

FIGURE 2. Influence of the processing parameters of the adhesive strength of the compounds

Influence of the material pairing

The left diagram in Figure 3 shows the tensile strength of the selected cured rubbers. The results in Figure 3 on the right side show the material pairings in dependence of adhesive strength. The figure shows that there is a significant influence of the material pairing on the adhesive strength. In general, the highest strength could be reached with a combination of AEM, and in special case with PF1110. In comparison, the figures show that over all the adhesive strength of the compounds reached about 40% of the tensile strength of the applied rubber component. This is indicated by the red dotted lines.

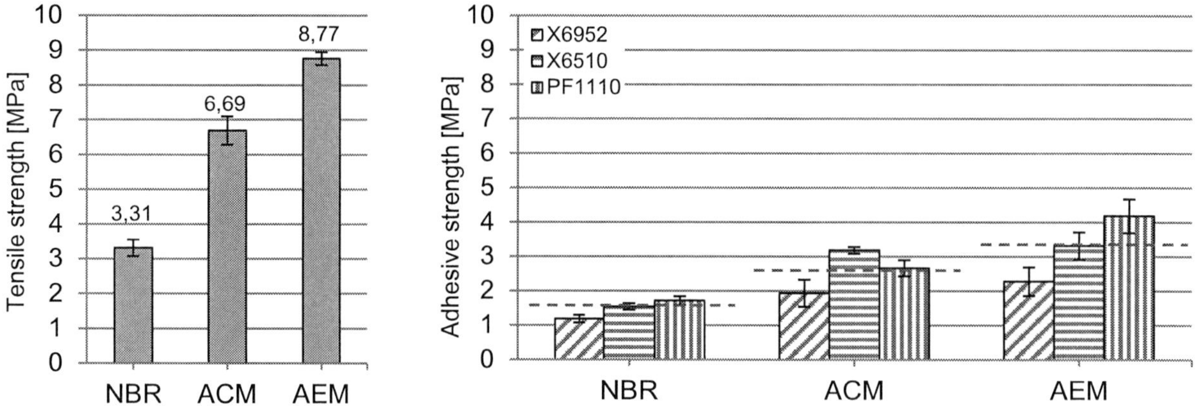

FIGURE 3. Influence of the material pairing on the adhesive strength of the compounds

Influence of the process variant

Figure 4 shows the influence of the process variant, demonstrated on selected thermoset-rubber pairings with PF1110 and X6952 as the thermoset (curing temperature 190°C; curing time 3 minutes). The figure shows that there is a significant influence of the processing variant depending on the material pairing. The figure with the pairing using PF1110 shows no differences between the different process variants. The results on the pairing with X6952 show notable differences between the different variants. In this context, the highest strength could be reached with uncured thermoset applied onto cured rubber. Similar results could be found for the pairing with PF6510. However, the influence of the process variant is considerably smaller than the influence of the material pairing.

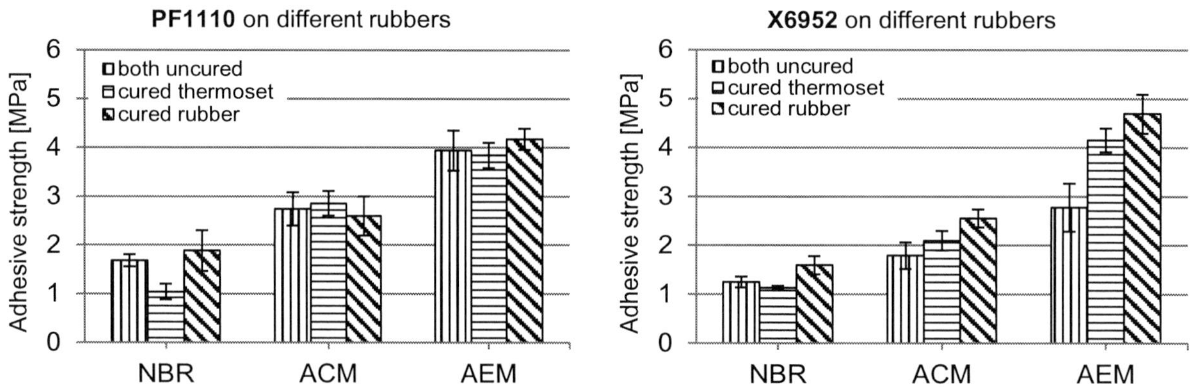

FIGURE 4. Influence of the process variant on the adhesive strength of the compounds

Influence of the surface roughness and surface area

To determine the roughness depending surface area, the inserts were analyzed with a Keyence digital Laser microscope VK9700 after modification. Table 6 shows the measured roughness depending surface areas. In Figure 5 (left side) the dependencies of adhesive strength and averaged roughness are shown, demonstrated on the combination of X6952 with AEM or NBR. For these and all other tested material pairings it was found that an increasing averaged roughness leads to an increase of the adhesive strength. Up to around 3 µm averaged roughness the increase grows strongly. Above about 3 µm averaged roughness the increase gets much weaker. Figure 5 (right side) shows the relationship between adhesive strength and the surface area. Although the surface area of the P80 grinded inserts is round 300% higher of the surface area of the P4000 grinded inserts, the adhesive strength of the test samples with the P80 grinded inserts increases only by round 100%. This behavior indicates that the coating and adhesion between the both components gets lower with increasing roughness.

Table 6. Roughness depending surface area (measuring surface: 351x351 μm)

Grain size	P4000	P1200	P500	P320	P80
Surface area [μm²]	169513	327569	468711	512750	654213
Surface area [%]	25,91 %	50,08 %	71,64 %	78,37 %	100 %

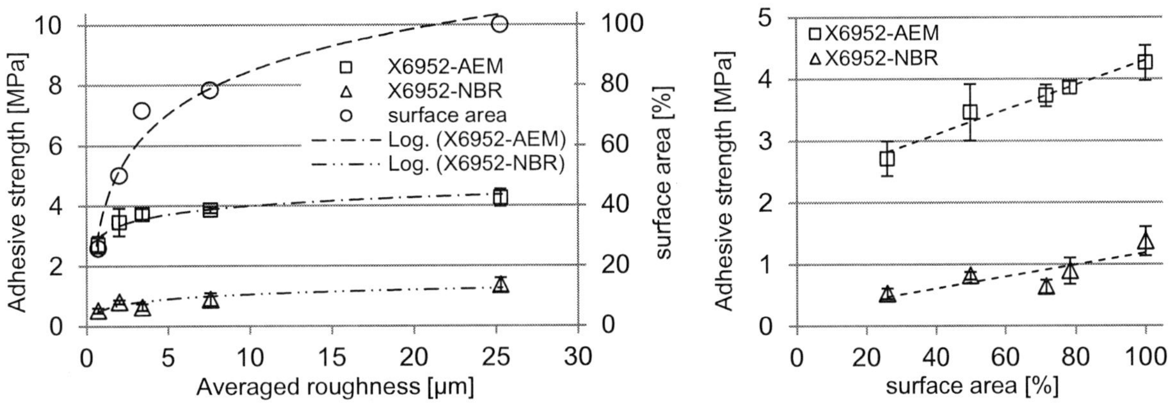

FIGURE 5. Influence of the averaged roughness and surface area on the adhesive strength of the compounds

CONCLUSIONS

The results of the study show that it is possible to produce thermoset-rubber compounds with high adhesive strength in one tool by using injection molding. Depending on the material pairing, strengths around 4,5 MPa can be reached, which complies to an elongation of the rubber component of around 300%.

While there was no notable influence of the processing parameters, a dependency was found for the different process variants, depending on material pairing, and the surface roughness. However, the biggest influence on the adhesive strength comes from the material pairing. This influence cannot be compensated by the other parameters analyzed here.

ACKNOWLEDGEMENTS

This project received funding from the research program "Zentrales Innovationsprogram Mittelstand". The authors thank the project management agency and the Willfried Mende GmbH for the support!

Supported by:

ZIM Zentrales Innovationsprogramm Mittelstand

Federal Ministry for Economic Affairs and Energy

on the basis of a decision by the German Bundestag

REFERENCES

1. R. Bauer, „Sequentieller Mehrkomponenten-Spritzguss," Kautschuk Gummi Kunststoffe, 06/2012
2. R. Bauer, „Der nächste Schritt in den Motorblock – 2K Pumpengehäuse aus Kunststoff," Kunststoffe, 05/2012
3. J. Milzer et al, „Eine einfache Alternative zu Gummi-Metall-Verbunden", Kunststoffe, 10/2007
4. A. Jaeger, „Kombinationen potenzieren das Leistungsvermögen", Kunststoffe, 10/2001
5. T. Stier, „Haftung im 2-K Spritzgussprozess – Verbesserte TPE-Haftung ohne Haftvermittler," Kautschuk Gummi Kunststoffe, 01/2015
6. M. Schuck, „Kompatibilitätsprinzipien beim Montagespritzgießen," Ph.D. thesis, Universität Erlangen-Nürnberg, 2009
7. T. Thust, „Rezeptur- und Prozesseinflüsse auf das Haftverhalten beim Mehrkomponentenspritzgießen von Thermoplast-Elastomer-Verbundbauteilen am Beispiel PA6.6-HNBR," Ph.D. thesis, Martin-Luther-Universität Halle-Wittenberg, 2014
8. I. Kühnert, „Grenzflächen beim Mehrkunststoffspritzgießen", Ph.D. thesis, Universität Chemnitz, 2005 A.
9. Oltmanns et al, „Dem Gummi auf der Spur", Kunststoffe, 03/2008

A new joint product made of PCM and plastic for application in water-powered heat storage systems manufactured by a coextrusion process

***Andretzky Michael* [(1)*], *Gardemann Ulrich* [(2)], *Wortberg Johannes* [(1)]**

[(1)] University Duisburg-Essen - NRW - Germany, [(2)] The fuel cell research center ZBT GmbH - NRW - Germany

In the research project "Development of a latent heat storage system with innovative PCM modules for the application in small and middle heat and power plants" an innovative heat storage system is to be developed. Modules with integrated phase change materials (PCM) shall be integrated into a heat storage system to make latent heat usable. Regarding this, the novel heat storage system presents an optimization towards conventional sensitive heat storage systems. One subsection of this research project is the development of an extrusion line for the inline inflation of extruded circular pipes with PCM. To design the extrusion line with regard to certain frame conditions the cooling behavior of the heat storage modules have to be known. Therefore, it is necessary to determine the exact product properties, which are depending on material properties and the final shape of the module. Taking the temperature-dependent specific volume of the phase change material into account the shape of module's shell is aimed to be the form of an elliptical prism. Considering the operating conditions of the heat and power plant and specific material properties a static mechanical analysis and an exact determination of material properties are made to design the final dimensions of the module's shell.

Influence of Processing Parameters on the Morphology of Co-Injected Foamed Parts

Wunderlich Katharina [(1)]*, Neumeyer Thomas [(1)], Altstaedt Volker [(2)]

(1) Neue Materialien Bayreuth GmbH - Bavaria - Germany, (2) Department of Polymer Engineering, University of Bayreuth - Bavaria - Germany

Soft-touch haptics currently gain importance in numerous applications, among them automotive interiors and various articles of daily use. A lot of parts which require soft touch haptics consist of two plastic components – a rigid structural component covered by a soft component providing the desired feeling. Besides common 2K injection molding, co-injection molding represents another approach to produce parts containing a hard and a soft component. In the process of co-injection molding (also called sandwich injection molding) two melts are injected into the mold sequentially: At first the skin material is injected and freezes at the cold mold wall partially. Secondly the core material is injected into the mold. The core material solidifies between the formed skin layers [1] [2]. To further reduce the material usage and thus the weight of such co-injected parts, foam injection molding can be employed [3]. In this work, co-injection molding with a foamed core is combined with a soft-touch skin layer based on TPE. Three grades of thermoplastic elastomers have been chosen. Different grades of polypropylene and polyamide are evaluated as structural component. The foaming of the core component is done using a MuCell® equipment and the physical foaming agent N2. The aim of this work is to understand the influence of the processing parameters on the morphology of the co-injected parts. A demonstrator part having a nearly circular cross-sectional area with a diameter of 40 mm is considered in this study. The processing parameters are systematically varied when producing demonstrator parts. The resulting morphology is investigated and characterized employing light microscopy. The skin to core-ratio as well as the foam morphology (cell size and cell size distribution) are evaluated. Crucial process parameters turned out to be injection speed, injection temperature as well as core content.

Hygrothermal aging and effect of corrosion in polymer-metal joints

Santoro Leandro Ariel* [(1)], *Puentes Parodi Jaime Alejandro* [(1)], *kuehnert Ines* [(1)]

[(1)] Leibniz Institut fuer Polymerforschung Dresden - Sachsen - Germany

In order to evaluate the effects of durability, in terms of corrosion, on a polymer-metal joint, hygrothermal aging at different conditions was conducted in a polymer-metal composite, which consists of a steel substrate, with and without corrosion protection; a polyester-based adhesive coating; and on top, an overmolded thermoplastic urethane. The hygrothermal aging test was carried out for 1000 hours at a relative high temperature and low humidity, 85 °C/40 %rH; at harsh conditions, 85 °C/85 %rH; and simulating a common tropical weather, 40 °C/90 %rH. After the artificial aging, the influence of temperature and humidity on corrosion and, subsequently, the failure modes of the composite, were investigated by digital microscopy, SEM/XDS, DSC/TGA and mechanical test.

Influence of reprocessing on the mechanical and thermal properties of polypropylene

Catelli Adriana Martinelli Catelli (1)*, Venâncio Amanda Marques (1)

(1) Centro Universitário FEI - São Paulo - Brazil

In this work, PP homopolymer was reprocessed through injection molding up to 6 processing cycles. The effects of reprocessing on the mechanical properties were analyzed by flexural, impact and fatigue tests. Significant differences in mechanical properties were assessed by analysis of variance (ANOVA). Thermal characterization was performed by differential scanning calorimetry tests (DSC). Melt flow index (MFI) tests was also performed. From the fourth processing cycle, MFI index increased with increasing processing cycles, indicating a possible reduction in PP molar mass. The percentage of crystallinity of PP increased and flexural strength and modulus decreased with increasing processing cycles. The impact strength value showed a decrease in the second processing cycle and did not showed significant differences with increasing of processing cycles.

Experimental analysis of conformal cooling in SLM produced injection moulds: Effects on process and product quality

T. Evens[a], W. Six[b], J. De Keyzer[c], F. Desplentere[b], A. Van Bael[a]

[a] KU Leuven Materials Technology TC, Technology Campus Diepenbeek, Wetenschapspark 27, 3590 Diepenbeek, Belgium
[b] KU Leuven Materials Technology TC, Technology Campus Ostend, Zeedijk 101, 8400 Oostende, Belgium
[c] KU Leuven Sustainable Chemical Process Technology TC, Technology Campus Diepenbeek, Wetenschapspark 27, 3590 Diepenbeek, Belgium

Corresponding author: tim.evens@kuleuven.be

Abstract. In an injection moulding process, the cooling step takes up to 50% or more of the total cycle time. Therefore, any reduction of the cooling time directly results into an increase of the production rate. Furthermore, products are often rejected because of defects such as warpage or sink marks. The cause of these defects is often an unevenly distributed cooling in the mould due to complex product or mould shapes. A solution for both problems is Conformal Cooling, a technique in which the cooling channels are designed to follow the contour of the mould cavity to obtain an optimal cooling. These often curved cooling channels are difficult or even impossible to produce with conventional techniques such as milling, drilling and EDM. The recent development of the Selective Laser Melting (SLM) 3D printing technology can be used to overcome these limitations. In a previous work [1] the cooling channels of an existing mould to produce sorting trays were assessed and redesigned to achieve a conformal cooled mould. A mould cavity and insert were provided with these cooling channels by means of the SLM technique. In this work the performance of the conformal cooled mould is compared with the original mould based on process characteristics and product measurements. In particular the start-up losses, mould temperature during the cycle, cycle time and product warpage are assessed.

Keywords: injection moulding, conformal cooling, SLM printing, process optimization.
PACS: S06-202

INTRODUCTION

The profit and competitiveness of an injection moulded product depends on the quality and the production rate. The cooling of a product has a great influence on both the quality and the production rate. Poor or unevenly distributed cooling directly results in higher cycle times and can even lead to defects in parts such as warpage or sink marks. Therefore Original equipment manufacturers (OEM's) are most interested in an optimal and evenly distributed mould cooling, known as Conformal Cooling. With conventional manufacturing techniques as drilling, milling and electrical discharge machining, it is difficult or even impossible to create an evenly distributed mould cooling which often follows the product shape. Selective Laser Melting (SLM) can be used as a solution for this problem. This SLM technique is based on a 3D printing technology in which the product is formed by selectively melting successive layers of metal powder by the interaction of a laser beam.

In a previous work [1] the cooling channels of an existing mould to produce sorting trays, were assessed using simulation software and redesigned according to design rules [2,3,4] to achieve a conformal cooled mould. Furthermore, simulations where conducted to predict the start-up time to reach a steady state situation, the cycle time based on the mould temperature and the difference in product dimensions. In this work, the product warpage, start-up losses and minimal cycle time without warpage are experimentally assessed using polypropylene and polyethylene material.

MATERIALS AND METHOD

Material

For this research two commodity thermoplastics were selected to analyse the effects of an SLM produced conformal cooled mould on process and product quality. The first material is a general purpose polypropylene (PP) copolymer (INEOS, grade 400-GA05) for injection moulding applications. The material has a shrinkage of ±1 %, a density of 946 kg/m³ (ISO 1183) and a melt mass-flow rate of 5 g/10 min (ISO 1133; 230 °C/2,16 kg). The second material is a high density polyethylene (HDPE) for injection moulding applications (Sabic, grade B6246LS). The material has a shrinkage of ±1,5 %, a density of 962 kg/m³ (ISO 1183) and a melt mass-flow rate of 5 g/10 min (ISO 1133; 190 °C/2,16 kg).

Method

Injection moulding is conducted on a Demag Ergotech IntElect 50/330-100 fully electric injection moulding machine. It has a maximum clamping force of 500 kN, a 22 mm screw with an L/D ratio of 20, a dry cycle (Euromap 6): 1,35 s – 231 mm and a plasticizing rate (PS) of 6,7 g/s.

The product (Figure 1) used in this research is a sorting tray with a wall thickness of 1.5mm. The inside of the tray is provided with four upstanding ribs. These ribs enhance the product stiffness and ensure that the trays can be stacked on top of each other. The mould is equipped with a cold runner, which creates a sprue at the bottom of the tray.

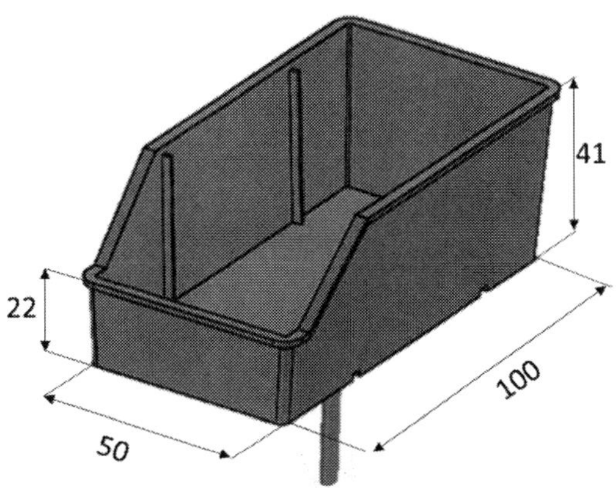

FIGURE 1. A representation of the sorting tray with a constant wall thickness of 1.5 mm. The outer dimensions are plotted.

The original mould (Figure 2) without conformal cooling has two conventional cooling channels with a diameter of ø8 mm and two bubblers in the core with diameter of ø16 mm. The mould is equipped with one cavity pressure sensor (Priamus 6008A) and three temperature sensors (Priamus 4008B).

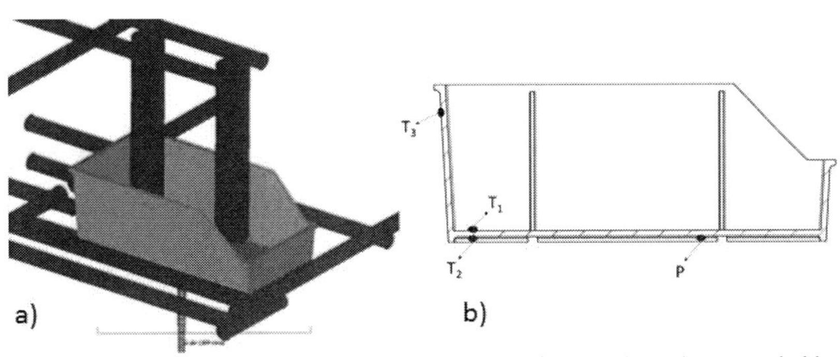

FIGURE 2. Representation of original sorting tray mould. (a) shows a sorting tray (green) surrounded by the conventional cooling circuit (blue), (b) shows the three fitted temperature sensors (T) and the cavity pressure sensor (P).

The redesigned sorting tray mould (Figure 3) with SLM printed conformal cooling channels with a diameter of ø4,5 mm, which follow the contour of the sorting tray closely. The mould is equipped with one cavity pressure sensor (Priamus 6008A) and two temperature sensors (Priamus 4008B). The mould is printed out of Maraging Steel (1.2709) metal powder, the recommended material for tooling applications [5].

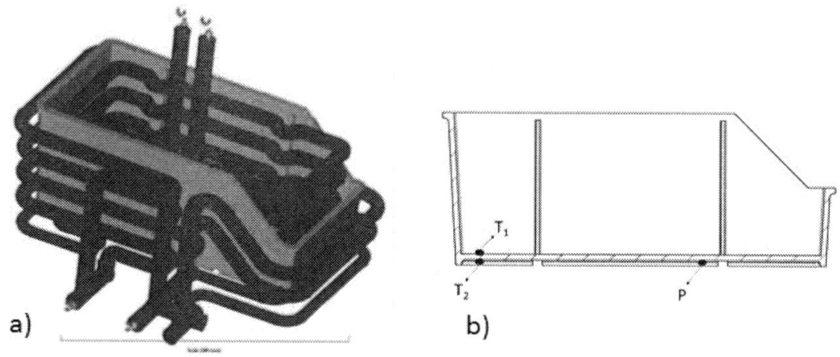

FIGURE 3. Representation of conformal cooled sorting tray mould. (a) shows a sorting tray (green) surrounded by the conformal cooling circuit (blue), (b) shows the two fitted temperature sensors (T) and the cavity pressure sensor (P).

The injection moulding parameters for the individual materials used in this research can be found in Table 1. The injection temperature and mould temperature were recommended by the material supplier. The injection speed was set as high as possible and was limited by the diesel effect. The holding pressure was set in a middle value, between an upper limit corresponding to flash and lower limit corresponding to sink marks.

TABLE 1. Process parameters for the individual materials

	HDPE	PP
Injection temperature	230 °C	230 °C
Mould temperature	30 °C	30 °C
Volumetric injection rate	98 cm^3/s	69 cm^3/s
Holding pressure	500 bar	300 bar

EXPERIMENTAL

Product warpage

The first criteria in which the conformal cooled mould was compared with the conventional cooled mould is the degree of product warpage. Therefore fifteen specimens where injection moulded with the conformal cooled mould and the conventional cooled mould. Both processes where set at the lowest possible cycle time of 12 s, limited by the injection moulding machine. The injection moulded specimens where measured with a flatbed scanner in combination with measuring software. The warpage is obtained by comparing the mould cavity dimensions with the inner width and inner length of the polymer product. The inner width and inner length of the specimens where measured at the centre of the product with a flatbed scanner, in combination with *SLTec Ltd Metric* measuring software. The resulting measurements are listed in in Table 2 for PE and Table 3 for PP.

TABLE 2. Product warpage with HDPE material

	Measured average width	Measured average length
Mould cavity dimensions	50 mm	101 mm
Product dimensions with Conventional cooling	46,62 ± 0,04 mm	99.96 ± 0,06 mm
Product dimensions with Conformal cooling	48,56 ± 0,02 mm	100,68 ± 0,05 mm
Product warpage with Conventional cooling	3,38 ± 0,04 mm	1,01 ± 0,06 mm
Product warpage with Conformal cooling	1,44 ± 0,02 mm	0,32 ± 0,05 mm

TABLE 3. Product warpage with PP material

	Measured average width	Measured average length
Mould cavity dimensions	50 mm	101 mm
Product dimensions with Conventional cooling	47,22 ± 0,03 mm	100,00 ± 0,06 mm
Product dimensions with Conformal cooling	49,12 ± 0,02 mm	100,89 ± 0,08 mm
Product warpage with Conventional cooling	2,78 ± 0,03 mm	1,00 ± 0,06 mm
Product warpage with Conformal cooling	0,88 ± 0,02 mm	0,11± 0,08 mm

From Table 2 and Table 3 it is clear that there is a large difference in product warpage between the sorting trays produced with the conventional cooled mould and the conformal cooled mould. This difference is due to a different cooling between the conventional and the conformal mould. Within the conventional cooled mould, there is a large difference of ±40°C between the mould core and the mould cavity (Figure 4a). This causes the conventional trays to warp in the inwards direction, making them impossible to stack (Figure 4b). On the other hand, the conformal cooled mould has an evenly distributed mould cooling, causing the products to have significantly less warpage.

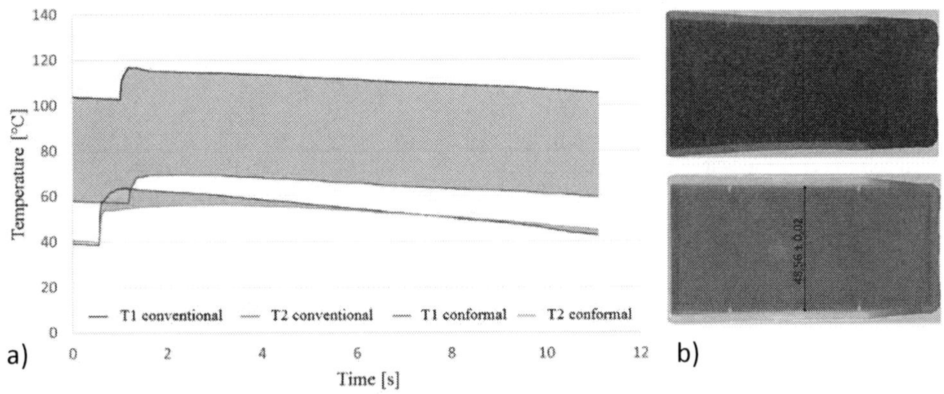

FIGURE 4. (a) Temperature variation within one injection moulding cycle. The conventional cooled mould (red) vs. the conformal cooled mould (blue). Temperature sensor T1 is placed in the mould core, Temperature sensor T1 is placed in the mould cavity (b) representation of a conventional cooled (blue) and conformal cooled (red) sorting tray.

Start-up losses

The second criteria in which the conformal cooled mould was compared with the conventional cooled mould are the start-up losses. When starting an injection moulding process, the product quality can only be guaranteed when the mould temperature before injection is stable. Minimizing the start-up time to reach this stable condition will result in a faster process start-up and a reduction of waste.

While the process is started-up, the temperature is measured within the conventional and conformal mould, as shown in Figure 5a for HDPE material. We can clearly see that the number of cycles to reach a stable mould temperature is ±20 cycles for the conventional mould and ±5 cycles for the conformal mould.

Minimal cycle time without warpage

The third and last criteria in which the conformal cooled mould was compared with the conventional cooled mould is the minimal cycle time to achieve a part without warpage in the product width. The injection moulding process is set to a very long cycle time by increasing the cooling time. Next, the cooling time is gradually reduced and the process is stabilized. At each of the different cooling times, the product width is measured as shown in Figure 5b for PE material. When using the conventional cooled mould, we can clearly see that the warpage decreases with an increasing cycle time and stabilises at a value of ±48,6 mm at a minimal cycle time of ±40 s. While using the conformal cooled mould, the product width ±48,6 mm is stable at all cycle times, even at the lowest possible cycle time of ±12 s. As a result the minimal cycle time to produce parts without warpage is reduced from ±40 s to ±12 s or the cycle time is reduced by 70%.

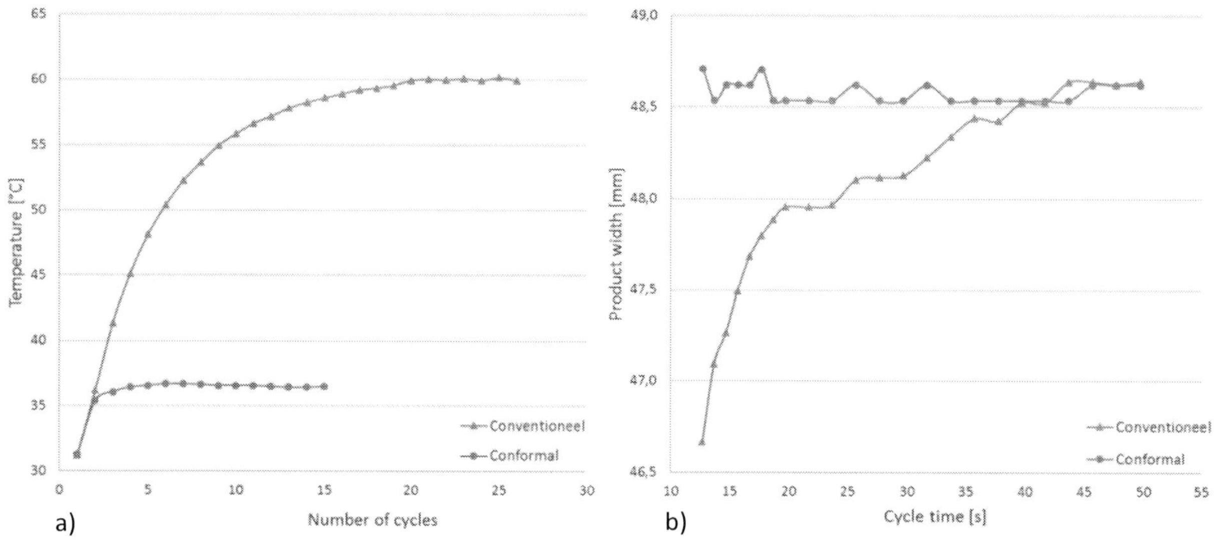

FIGURE 5. Start-up losses and the minimal cycle time for HDPE material. (a) Conventional cooled and conformal cooled mould temperature before injection, measured with sensor 2. (b) Product width using the conventional cooled and conformal cooled mould at different cycle times.

CONCLUSION

In this study the performance of a conformal cooled mould was compared with the original conventional cooled mould based on three criteria, product warpage, start-up losses and minimal cycle time without warpage. Because of an inefficient mould cooling, there was a difference in temperature between the mould core and mould cavity, therefore the trays warped in the inwards direction. With the conformal cooled mould the product warpage in width and length where reduced by ±2 mm and ±1 mm, making it possible to stack the sorting trays again. The start-up losses where decreased from ±20 cycles for the conventional mould to ±5 cycles with the conformal mould. This will result in a lower start-up time and a reduction of waste due to unpredicted quality during start-up. At last, the minimal cycle time to produce parts without warpage is reduced from ±40 s to ±12 s or the cycle time is reduced by 70%.

ACKNOWLEDGMENTS

The authors wish to thank the Flemish government for the funding in the IWT TETRA140320 project "*Conformal Cooling door middel van Selective Laser Melting*". The results discussed in the present paper where obtained within this project.

REFERENCES

1. W. Six, R. Appermont, F. Desplentere, J. De Keyzer, A. Van Bael (2016) Design and simulation of conformal cooling channels In SLM produced injection molds. *Proc. of the 32nd International Polymer Processing Society Conference S13-250.*
2. Y. Wang, K. Yu, C. Wang, Y. Zhang (2011) Automatic Design of Conformal Cooling Circuits for Rapid Tooling. *Computer-Aided Design 43:1001–1010.*
3. X. Xiaorong, S. Emanuel, A. Samuel (2001) The Design of Conformal Cooling Channels in Injection Molding Tooling. *Polymer Engineering & Science 41: 1265-1279.*
4. R. Mertens, S. Clijsters, K. Kempen and J.-P. Kruth (2014) Optimization of Scan Strategies in Selective Laser Melting of Aluminum Parts With Downfacing Areas, *Journal of Manufacturing Science and Engineering 136: 121-127.*
5. L. Thijs, J. Van Humbeeck, K. Kempen and J.-P. Kruth (2012) Investigation on the inclusions in maraging steel produced by selective laser melting, *Innovative Developments in Virtual and Physical Prototyping 5: 297-304.*

Highly filled suspensions with zirconia powders: Assessment and control of the surface chemistry

Auscher Marie-Camille (1)*, Cassagnau Philippe (1), Fulchiron René (1), Périé Thomas (2)

(1) IMP Lyon 1 - Villeurbanne - France, (2) Saint-Gobain CREE - Cavaillon - France

Processing of highly filled suspensions like ceramic in polymer matrix intended for injection molding or extrusion require specific properties, like a proper flow behavior and a good dispersion, that are often difficult to obtain in a first approach. The key is generally to control the particles-matrix' interface of the filled suspension, which is both a must and a challenge to reach the target properties. A simple method is here described to characterize and optimize the filler surface chemistry in the case of zirconia filled polymeric suspensions using stearic acid as a surfactant. It involves contact angle measurements by the sessile drop method between water and thermally treated zirconia powders mixed with stearic acid. It is shown that this is a powerful way to determine the optimum additive concentration. In addition, the effect of partially coated particles on the dynamic rheological behavior of the molten filled suspension is investigated. Besides, the kinetics of the hydrophobization of the powder surface is followed by contact angle measurements and can be a way to improve the powder preparation.

Extrusion-ultrasound on polyethylene terephthalate and polycarbonate composite from recycled materials

***Sanchez-Solis Antonio* [(1)]*, *Revilla Vianney* [(2)], *Corcuera Maria* [(2)], *Manero Octavio* [(2)]**

[(1)] UNAM - CDMX - Mexico, [(2)] IIM-UNAM - CDMX - Mexico

The concern about the environment and natural resources has been one of the most outstanding restlessness of contemporary worldwide society. For this reason, the polymer recycling materials emerges as an option that implies benefits and ecological improvement both in the reduction of waste and conservation of natural resources. The systems that have received most attention are those in which the constituents are reactive with each other, such as mixtures of polyethylene terephthalate (PET) and polycarbonate from bisphenol A (PC). In the present work, blends were developed from PET and PC from post-consumer residues formulated with the incorporation of a molecular chain extender additive, through the extrusion-ultrasound process, with the objective of revitalizing the recycled material through both polymers compatibilization. The study was divided into four stages: base formulations, percentage of suitable extender additive, adequate fill percentage, and single screw extrusion-sonication parameter process determinations. The structural molecular modifications were evaluated through tensile mechanical properties, differential scanning calorimetry, infrared spectroscopy, melt flow index and rheological measurements. It was found that the addition of a chain extender additive and extrusion–ultrasound process causes an increase in blend melt viscosity due at higher molecular weight produced. Also, it improved the compatibility between PET and PC, effect attributed to the copolymer generated by transesterification reaction. Additionally, this copolymer contributes to PET-PC miscibility observed by the glass transition temperature displacement. Blends with a filler material such as $CaCO_3$ leaded to highest values of Young's modulus because the excellent particles dispersion through entire polymer matrix by the effect of extrusion-ultrasound technique

Challenges in Rubber Processing: Learning from Game Theory and Complex Systems

Heinrich Gert (1)*, Stöckelhuber Klaus Werner (1), Wiessner Sven (1), Das Amit (1)

(1) Leibniz-Institut für Polymerforschung Dresden e.V. - Saxony - Germany

The performance of elastomeric materials, i.e. in car tires, is substantially determined by the used reinforcing filler system. Especially, the flocculation tendency of filler particles to form clusters and even network-like structures is significantly determining the mechanical properties of these elastomer materials and enhances especially their energy dissipation under periodic mechanical deformations. The industrial and even economic impact of the above discussed re-agglomeration of dispersed filler particles was even drastically grown when the tire industry has introduced the so-called "green tire technology" 20 years ago, i.e. the filler system in the tread compounds for high-performance passenger car tires was changed from carbon black to silica. Several technical challenges, e.g. the introduction of the tandem mixing approach, will be briefly mentioned only. Instead, we will demonstrate how several approaches and ideas from the science of complex systems help us to understand basic features of mixing highly filled rubbers. For example, it is well-known that successive iterations of the Baker transformation as a specific feature of complex systems is reflecting the filler fragmentation and, finally, the state of dispersion as a result of the mixing process. It is interesting to note that very recently Sumitomo Rubber Industries (SRI), Japan, demonstrated within their Advanced 4D Nano Design tool how the large-scale quantitative visualization of filler dispersion for further analysis and simulation of rubbers is realized successfully when using cutting-edge experimental facilities. In a simplified thermodynamic model, inspired by a segregation model from game theory, we describe fundamental mechanisms of filler structures formation in a polymer matrix as a result of the mixing process. We demonstrate how similar structures in society, nature or materials like rubbers emerge when supposing obvious cardinal mechanisms of structure formation in complex systems. The filler flocculation phenomenon itself is discussed with a simple kinetic model of autonomous (but forced by restrictions) saturation in an aggregated growth process which reflects the role of compl

NOVEL SILICONE THERMOPLASTIC ELASTOMERS WITH TAILORED PERMEATION PROPERTIES

Riess Gisbert [(1)]*, Berger Katrin [(1)], Andraschek Natascha [(1)]

[(1)] Department of Polymer Engineering and Science, Montanuniversität Leoben - Styria - Austria

Polysiloxanes such as polydimethylsiloxane (PDMS) combine several unique properties. They exhibit a low glass transition temperature, high thermal and oxidative stability, high UV-resistance, low surface energy and strong hydrophobicity as well as high permeability to many gases. These characteristics are independent over a wide temperature range. Our research deals with the synthesis and properties of thermoplastic silicone elastomers containing soft segments of dimethylsiloxane and hard urea segments. In order to determine structure-property relations, regarding to permeation properties, the chemical structure was varied. In this study thermoplastic silicone elastomers, based on polydimethylsiloxane oligomers with amino groups and different diisocyanates, were synthesized. The polymers were characterized by nuclear magnetic resonance (NMR) and infrared (FTIR) spectroscopy. The formation of hydrogen bonding of the hard segments was investigated by temperature dependent FTIR-spectroscopy, which demonstrated the thermal stability of the elastomer and defined the processing conditions of such polymers. Thin membranes of thermoplastic main-chain urea-siloxane copolymer were prepared and investigated with respect to the permeation properties of organic solvents. Vials were filled with a mixture of an aromatic (toluene) and an aliphatic hydrocarbon (iso-octane) and were sealed with the polymer films. The concentration of residual solvent mixture was measured by the relation of the peak area of the two solvents. This was investigated by the GC-MS-technique. A separation of the solvent mixture occurred. In conclusion, novel main-chain urea-siloxane thermoplastic elastomers were successfully synthesized with selective barrier properties. These polymers are highly permeable for organic solvents depending on the chemical structure. These siloxane polymers can be used as semipermeable membranes for separation of solvent mixtures in oil field applications.

Evaluation of the crosslinking density and the mechanical properties of silicone elastomers filled with bariumtitanates of different particle sizes as dielectric elastomers

Ziegmann Andreas [(1)]*, Schubert Dirk W [(1)]

[(1)] Institute of Polymer Materials, University Erlangen-Nürnberg - Bayern - Germany

In the present study, two commercially available silicone elastomers (high and room temperature vulcanizates) are filled with bariumtitanate with different particle sizes for the evaluation as a dielectric elastomer actuator. It could be shown that the particle size changes with the degree of filling. Therefore the resulting average molar mass between crosslinks and the mechanical properties are evaluated in dependence of the degree of filling and the particle size. Moreover it could be demonstrated that bariumtitanate acts as an active filler and three different bonding regmies are determined. Thus the stress and elongation at break could be increased up to more than 330 % and 160 % for the high temperature vulcanizate and up to more than 125 % and 145 % for the room temperature vulcanizate in comparison to the pure material, respectively. Furthermore a sweet spot of the mechanical properties is identified by analyzing the crosslinking density.

Analysis of the Extrusion Process of Silicone Rubber

***Verheyen Fabian* [(1)]*, *Heim Hans-Peter* [(1)]**

[(1)] University of Kassel - Hessen - Germany

High-consistency rubber (=HCR) is typically processed in the extrusion process into tubes or profiles. Because of the excellent properties in terms of flexibility down to low temperature, physiological harmlessness and the bride service temperature range from -80 to 250°C, there are a great number of applications for silicone rubber; especially in the field of automotive, food sector or medical engineering. During extrusion processing, there are some special phenomena, for example die-swelling. Die-swell means the swell of the extrudate after leaving the die. Due to the high requirements in terms of dimensional tolerance and the fact, that there is no calibration in the process line during extrusion of rubber, the manufacturer needs a lot of experience. With today's state of the art technology according to the used material, the process parameter are not changed to achieve the right dimensions, but the geometry of the die according to the used operating point will be customized. Those modifications are very time-consuming and because of that very expensive. The goal of this study is to identify the most influencing process parameters on the dimensional geometry of the extrudate by the help of design of experiment (=DoE) during the extrusion process of silicone rubber with a vertical extrusion line. It could be shown that the screw speed and as a consequence of that, the melt pressure and the vulcanisation temperature are the most influencing factors on the dimensional geometry of the extrudate. A profound knowledge of the whole process allows the manufacturer to change possible fluctuations in the dimension of the extrudate during processing by the change of chosen process parameters in an easy way. The running process does not have to stop or fundamental constructive changes at the die are not necessary.

Morphology-Property-Relationships of SEBS/PP Compounds for medical application

Christin Kiehle[1,a)], Stefan Roth[2,b)], Sven Wießner[3,c)]

[1] *Hospital Care, B. Braun Melsungen AG, Germany*
[2] *Hospital Care, B. Braun Melsungen AG, Germany*
[3] *Juniorprofessur für Elastomere Werkstoffe, Institut für Werkstoffwissenschaft, Technische Universität Dresden, Germany*

[a)] Corresponding author: christin.kiehle@bbraun.com
[b)] stefan.roth@bbraun.com
[c)] sven.wiessner@tu-dresden.de

Abstract: This work presents a comparative study on the morphology-property-relationships of SEBS/PP Compounds with varying PP content in modern medical engineering. A comparison of different characterization methods like thermal scanning stress relaxation (TSSR) and Hysteresis tests according to Eisele has been carried out. TSSR helps to describe an understanding about the phase morphology thereby. The works were completed by surveys of the morphology. Finally, the impact of modification of varying ratios of SEBS/PP and the influence of the PP morphology (random copolymer respectively homopolymer) on the material properties is presented.

INTRODUCTION

Since the discovery of the TPE in the 1960s, a continuous growth in applications has been taken place. Thus, TPE are very important materials in modern medical engineering. One typical application is medical tubing. Currently, this product is made of plasticized PVC. A growing demand for substitution of vinyl based medical tubings is however to be recognized. Thermoplastic styrenic elastomers (TPS) could be used as an alternative material but the material properties related to the intended application are not sufficiently known yet.

Thermoplastic styrenic elastomers for medical applications are primarily made of SEBS blended with PP and mineral oil. These types of TPE have been commercialized in the early 1990s. Its available in a hardness range of 30 Shore A - 53 Shore D [1]. Properties strongly depend on the combination of hard and soft phase. In the same manner, properties are dependent on the material morphology, which is evolving due to the blends composition and the process parameters in the compounding and molding process. Except for the works of Ohlsson et al. [2], Veenstra et al. [3] and Sengupta [4] comparative information on the SEBS–PP–oil systems are very limited up to now.

EXPERIMENTAL

Materials

In this study SEBS/PP Compounds with varying PP content are used. The samples where extruded and cut out in A5 test specimens according to ISO 527.

Method - Thermal scanning stress relaxation (TSSR)

Using the TSSR meter, both isothermal and anisothermic stress-relaxation measurements can be carried out. By means of isothermal stress relaxation (TSSR), mechanical properties such as the relaxation behavior of TPE and softer plastics can be characterized. During the test period, the specimen is subject to a relatively large constant elongation (predominantly $\geq$ 50 %) whereby the resulting force is recorded over time. Using these data, a stress-time curve is displayed at the end of the measurement.

In this work, a TSSR meter from Brabender Messtechnik GmbH was used. Tubing specimen were stretched by 50 % and heated at a constant elongation at 2 K/min until the resulting force has been settled to zero.

Figure 1 represents a typical result of a relaxation spectrum of an SEBS and SEBS/PP Compound. The glass transition point of polystyrene could be found as well as the melting area of polypropylene can be easily identified. The higher the PP content, the stiffer the material behaves and the higher the relaxation spectrum in the area of the glass transition point of PS is found.

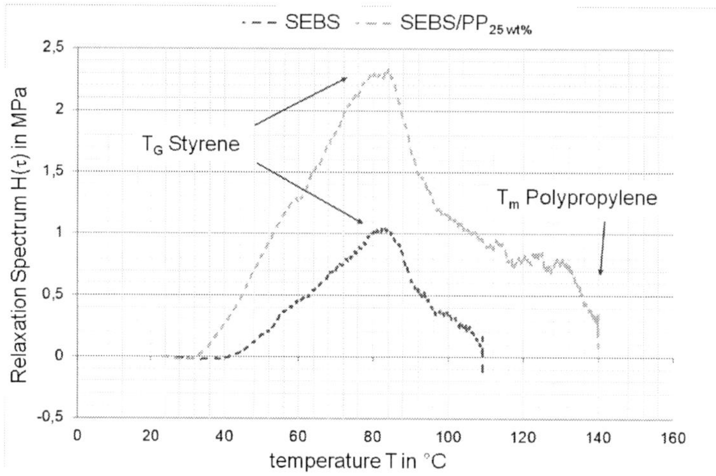

Figure 1: Example of anisothermal stress relaxation

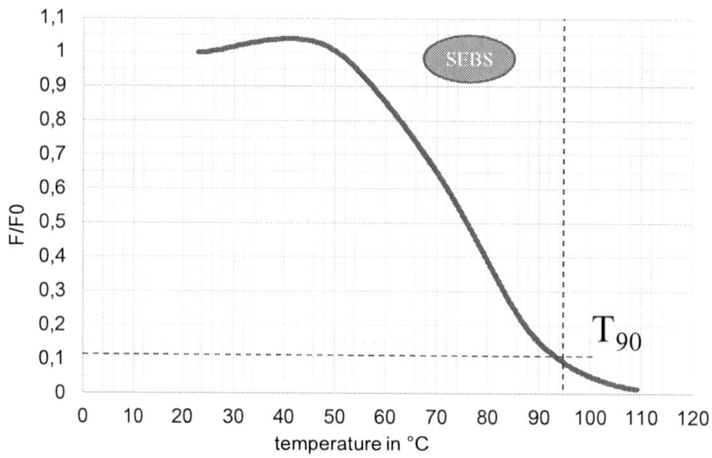

Figure 2: Example of nominal force in order to characterize the limited working temperature T_{90}

Figure 2 shows a typical result of the nominal force of an SEBS over temperature. As result the limited working temperature T_{90} can be determined. The limited working temperature T_{90} means that the material has lost 90% of the force in relation to the starting force.

Intermittent stress strain measurements to investigate the viscoelastic properties

In order to determine the deformation behavior, hysteresis tests were also carried out on a Zwick Z010 universal testing machine on the test specimen type A5. The remaining strain ε_{bl} was determined at room temperature and at a constant speed (50 mm/min) by deforming the specimen in a 20 % step up to a predetermined elongation ε_0 (400 %) and then the stress is removed from the sample at the same speed (see figure 3). As result the critical strain ε_{krit} can be found. The critical strain is defined as the point of intersection between the low strain region and the high strain region.

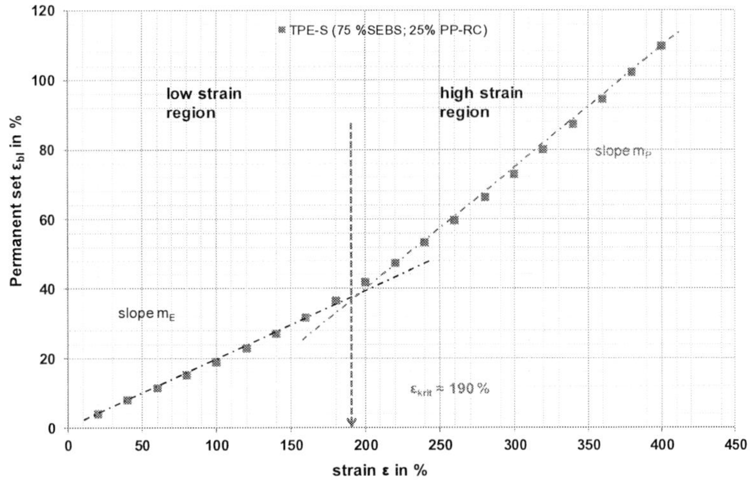

Figure 3: Hysteresis test for characterization the permanent strain $\varepsilon_{bl;}$, permanent set of an SEBS/PP-RC compound

Results and Discussion

Investigations have shown the influence of the morphology on the relaxation behavior and the elasticity of SEBS/PP compounds. The results of the relaxation spectrum showed the higher the PP content, the higher the stiffness of the SEBS/PP compounds as well as an increasing peak at approximately 150°C was found. Between 25 and 50 wt% PP the morphology of the compounds changes into a co-continuous phase morphology. The changing point were found within the increasing peak at the melting area of PP. In comparison of the different PP types the influence by the morphology are nearly the same. But the miscibility with the Homopolymer PP compounds are better as with the Random Copolymer PP (see figure 4(a)).

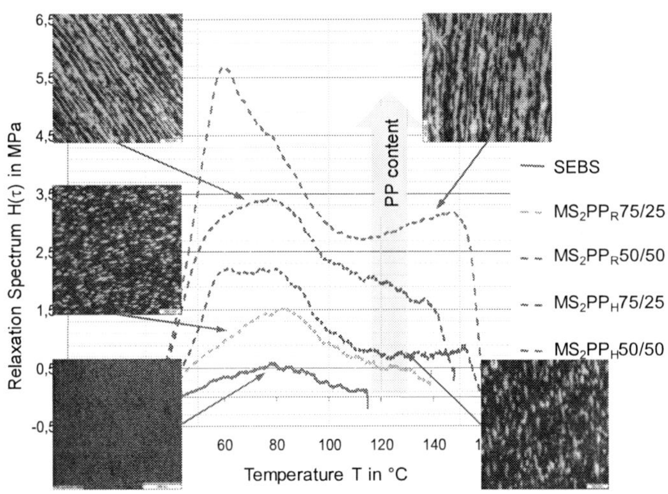

Figure 4(a)

The results of the hysteresis measurements according to Eisele shown also the influence of the PP content. The higher the PP content, the higher the remaining elongation after load subject and release (permanent set) and therefore the worse the elasticity of the compounds (see figure 4 (b)). The morphology is mainly dominated by the ratio of SEBS/PP and is not affected by the type of PP (PP-R or PP-H).

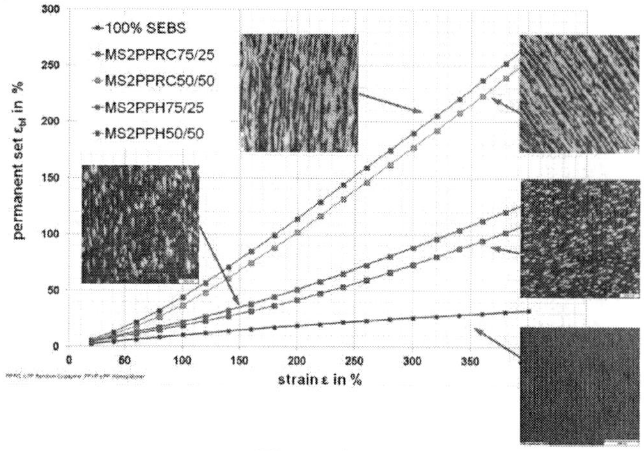

Figure 4(b)

Figure 4: (a) Relaxation Spectrum H(τ) with increasing PP content and comparison to morphology; (b) permanent set with increasing PP content and comparison to morphology

Figure 5 shows the comparison between the critical strain with increasing PP content. Both polypropylene types showed a change in the morphology at approximately 25 wt% PP. In this area, the morphology changes from dispersed into co-continuous. The critical strain of the SEBS/PP compounds with PP-H for example showed at 25 wt% PP a loose of elasticity. Because in this case the critical strain is approximately 50% less than the value of the SEBS/PP compound with 25 wt% PP-R.

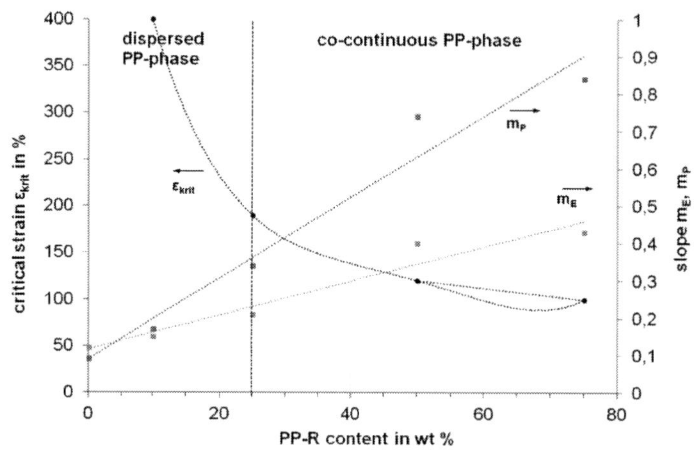

Figure 5(a)

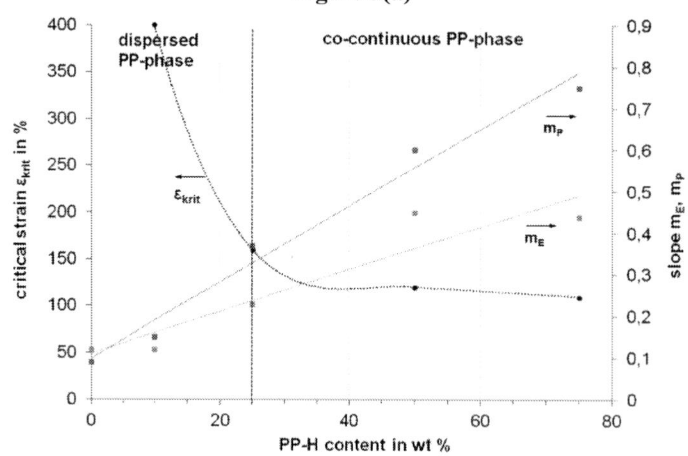

Figure 5(b)

Figure 5: (a), (b) change in morphology with increasing PP content

Figure 6 shows the resulting limited working temperature with increasing PP content. The results have shown the higher the PP content, the higher the limited working temperature. Also, here differences between the compounds with PP-H and PP-R are not really visible as long as the PP content is high enough. The limited working temperature T_{90} starts to change at approximately 18 wt% PP. At approximately 25 wt% PP the difference between PP-R and PP-H will be more distinct. Between 25 wt% and 50 wt% PP an increasing working temperature T_{90} is observed with increasing PP content. The compounds with PP-H shown also a better heat resistance in general.

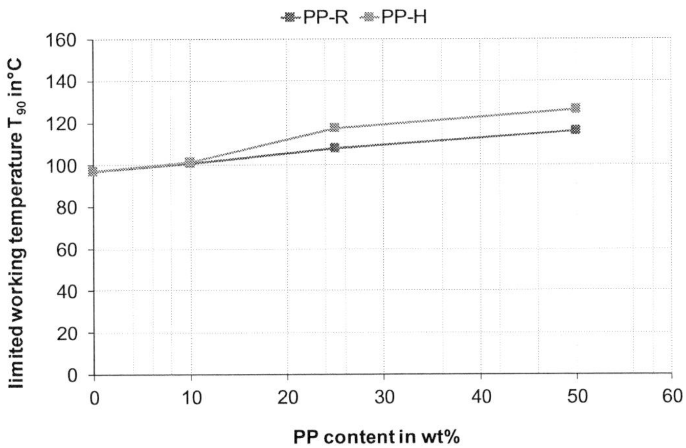

Figure 6: Comparison between critical strain and limited working temperature with increasing PP content

Conclusions

Investigations showed the influence of the morphology on the relaxation behavior and the elasticity of SEBS/PP compounds. Characterized were compounds with increasing PP content and different types of polypropylene. During this study two types of polypropylene were used, homopolymer and copolymer PP. The studies yield that TSSR measurement in the anisothermal test setup can be used in order to characterize the change of morphology from disperse into co-continuous state. Also, the glass transition point of polystyrene can be found by this method.

In addition, the intermittent stress strain measurement can be used in order to characterize the permanent elongation as a measure for the elasticity of SEBS/PP compounds. Differences between PP-H and PP-R in morphology have not be seen in this investigation. Both types of polypropylene showed in the compound with SEBS a spherical morphology at lower PP content. At 50 wt% PP the morphology changes into a lamellar structure. A higher working temperature in general was found for blends using PP-H compounds.

Further investigation of viscoelastic properties and relaxation behavior in comparison to morphology and its influence due to the manufacturing process and the possible applications in the medical field for TPS materials will subject of future works.

Acknowledgements

The authors gratefully acknowledge B. Braun Melsungen AG, in whose technical laboratory these investigations could be carried out. We also thank the manufacturers which provide us their materials.

References

1. W. Kaiser in *Kunststoffchemie für Ingenieure – Von der Synthese zur Anwendung*, Ed. 3; Carl Hanser Verlag, München, Wien, 2011; 444-445
2. B. Ohlsson et al in Polym. Eng. Sci., **36**, 1996; 501–510
3. H. Veenstra et al in Polymer **40**, 1999; 6661-6672
4. P. Sengupta, PhD Thesis, University of Twente, Enschede, The Netherlands 2004.

Simulative and Experimental Investigation of Rapid Heat Cycle Molding for Rubbers

T. Hutterer[1,a)] G.R. Berger[2)], C. Mager[1)], M.A. Fasching[3)], G.A. Pacher[1)] and W. Friesenbichler[2)]

[1]*Polymer Cometence Center Leoben GmbH, Roseggerstraße 12, 8700 Leoben, Austria*
[2]*Montanuniversitaet Leoben, Department of Polymer Engineering and Science, Injection Molding of Polymers, Otto Gloeckel-Strasse 2, 8700 Leoben, Austria*
[3]*SKF Sealing Solutions Austria GmbH, Gabelhofer Strasse 25, 8750 Judenburg*

[a)]Corresponding author: thomas.hutterer@pccl.at

Abstract. Injection molding of rubber compounds is a widely used method for manufacturing different rubber products, from consumer parts to seals for harsh environments. Yet its cycle time is tremendously higher than in injection molding of thermoplastics, as rubbers need several minutes of curing. Increasing the mold temperature would shorten the curing time, however, this leads to a premature start of the curing reaction which may cause severe quality and manufacturing problems. To avoid premature curing and to reduce filling problems, Rapid Heat Cycle Molding (RHCM) may be used to first, cool the mold in injection phase (until the cavity is filled) and second, to raise the mold temperature above what would be possible in isothermal conditions. In this study, this approach is evaluated. Employing design of experiments and injection molding simulation, the 3D transient temperature distribution in the mold is investigated. Injection molding experiments were conducted to determine the curing degree as a function of the temperature distribution at the cavity wall. Moreover, simulation accuracy was assessed. To summarize, RHCM can provide a way to achieve the desired degree of cure with higher temperatures than possible using standard temperature control.

INTRODUCTION

The timing of the start of the curing reaction is a critical factor in rubber injection molding and therefore has to be set precisely to ensure optimal processing and part quality. A premature start of the curing reaction can prohibit complete filling of the cavities due to a massively increased viscosity of the rubber, while delaying it after the cavities are already filled prolongs cycle times, thus reducing productivity. Rapid Heat Cycle Molding (RHCM), which is common in thermoplastic injection molding, is able to provide a solution for both problems [1]. The cavity temperature is set to be low during the injection phase, which suspends the reaction to start. Consecutively, curing mold temperatures can be set to higher set points when using RHCM causing the material to cure at rates which are often not possible under standard processing conditions. This can lead to reduced cycle times and avoid injection problems [2]. In this study, a special mold was developed and manufactured, which is ready for RHCM process control. The surface temperature distribution in the mold is simulated for various temperature control settings of a DOE and is compared to the conditions found in the actual mold. The spatial distribution of the degree of cure of parts manufactured with this mold is assessed both simulatively and with real experiments.

EXPERIMENTAL

Part and Mold Design

Figure 1a shows the geometry of parts produced using the RHCM mold. The design of the cavities enables manufacturing rubber parts which can be readily used for the following investigations:

- Compression set measurements [3] of different part areas as an industry-standard way to characterize the degree of cure
- Extraction of tensile test specimens [4] for measurements of mechanical properties (e.g. elongation at break) of rubber compounds
- Thin walled areas within the parts to show filling problems
- Thick walled sections, to characterize productivity improvements achieved by RHCM

In this work –which is a preliminary investigation to highlight the mold capabilities- only compression set measurements will be evaluated. Each of the cavities in Fig. 1 can be molded separately, to ensure better filling and reduce flashing. To enable high heating and cooling rates, the mold temperature control, depicted in Fig. 1b needs to be situated near the part surface. This was done by placing ceramic heating pads near to the cavity surface. Those are insulated against the outer mold to maximize the heat flow to the cavity. On the opposite side of the insulation, the cooling channels are situated.

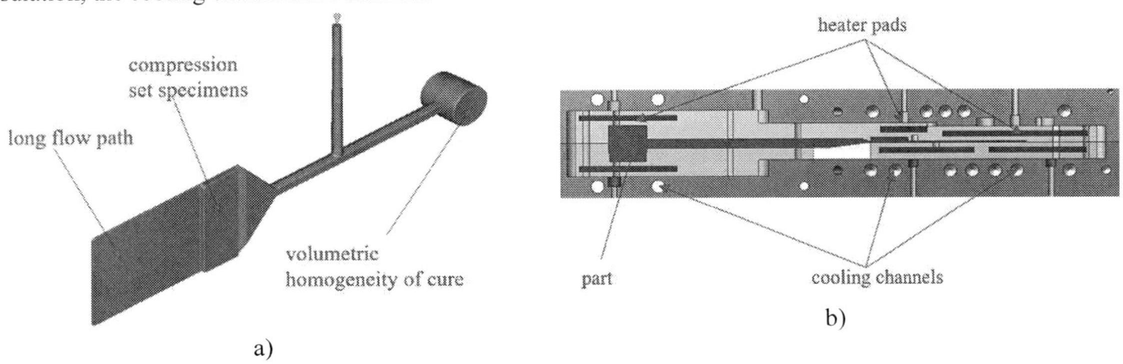

FIGURE 1. a) Part geometry of the RHCM mold. b) Layout of temperature control units in the RHCM mold

Design of Experiments

TABLE 1. Coded experimental settings of the 2^3 DOE

Experiment	T_m (°C)	T_c (°C)	v_i (cm³/s)
S1	110	160	5
S2	110	160	10
S3	110	200	5
S4	110	200	10
S5	150	160	5
S6	150	160	10
S7	150	200	5
S8	150	200	10
Center Point	135	180	7.5

To evaluate the accuracy of the simulated degree of cure and transient 3D temperature distribution in the RHCM mold, a 2^3 full factorial Design of Experiments (DOE) including a center point was carried out using an ENGEL injection molding machine (Engel Austria GmbH, Schwertberg, Austria) and simulated with Cadmould® 3D-F (Simcon kunststofftechnische Software GmbH, Wuerselen, Germany) respectively. Factors varied during the DOE were the mold temperature during the filling phase (T_m), the mold temperature during the curing phase (T_c) and the injection rate of the rubber compound (v_i). When filling is completed, the heat pads are heated to T_c until the end of curing, while the coolant temperature is set to T_m. As a material for performing the injection molding experiments, an industry standard NBR compound containing Sulphur based curatives is used.

RESULTS AND DISCUSSION

Homogeneity of Cure

Figure 2 a and b show the simulated degree of cure for experiments S1 and S8, which mark the extreme ends of each factor and are therefore representative of the maximum and minimum achievable degree of cure within the limits of the DOE. Fig. 2c and d show the calculated degree of cure based on compression set measurements as proposed by Friesenbichler et al. [5]. There is a strong correlation between the simulated and real crosslinking of the NBR samples, however, a strong spatial inhomogeneity across the investigated surface is also apparent. This inhomogeneous distribution of the local degree of cure across a rubber part is often caused by inhomogeneous tempering of the cavity wall. Thus, the simulated temperature field at the cavity wall is compared to observations made using a thermal imaging camera at the real mold. The temperature results of the simulation are shown in Fig. 3) and are very comparable to the measurements done on the actual mold. However, the real thermal conditions are slightly less evenly distributed (due to manufacturing tolerances or design restrictions) and the heat concentration at the edge between two wall thicknesses (Fig. 3, item 1) is more distinct than in the simulation. Measuring the actual surface temperature via thermal imaging is always prone to errors. To cope with reflections of the infrared radiation from other parts of the mold, the measurement area has to be shielded. Also, different emission coefficients of differently treated surfaces have to be taken into account. Hence, simulation provides a more straightforward way to estimate the amount of inhomogeneity in the cavity surface temperatures.

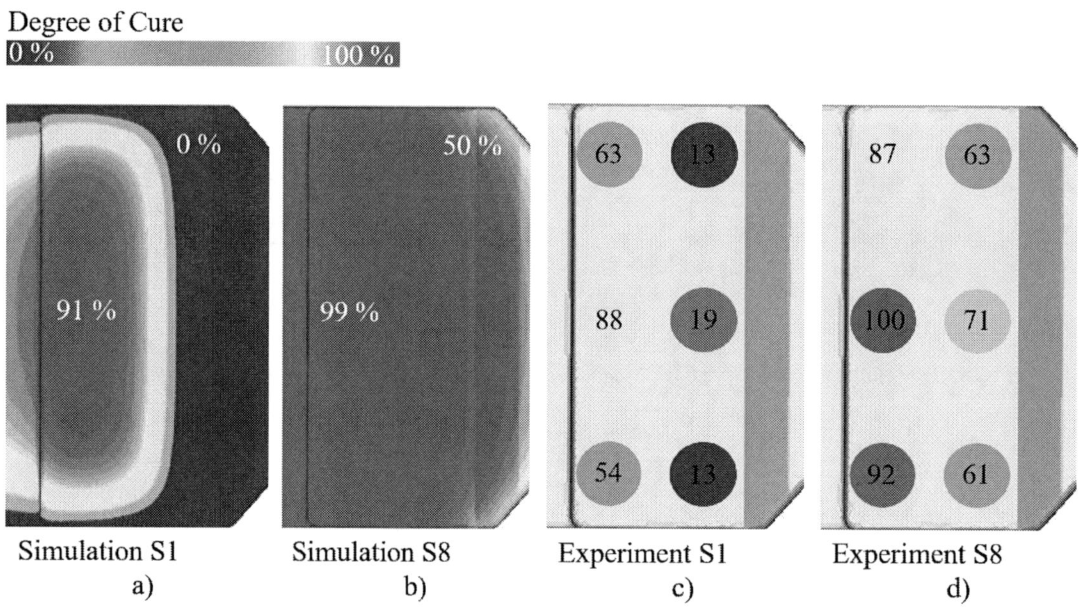

FIGURE 2. Spatial distributions of the degree of cure. a) simulation S1, b) simulation S8, c) experiment S1 d) experiment S8

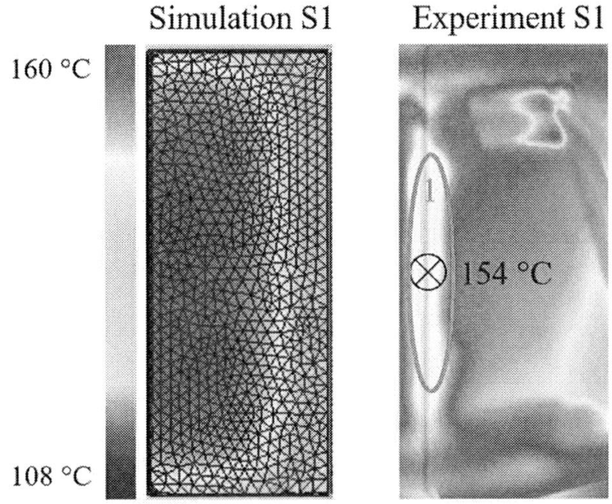

FIGURE 3. Temperature distribution of the cavity wall at the compression set section of the part

Regression Analysis

To analyze the curing behavior depending on the process settings, a linear regression model with an error level of 5 % was applied to the simulated degree of cure and compression set values obtained for NBR. Regression analysis was done using Minitab statics software (Minitab Inc, USA) Fig. 4) shows surface plots of the regression model for the simulated degree of cure and measured compression sets. The model is given in non-coded units for in Eq (1) for Fig. 4a) and in Eq (2) for Fig. 4b, With CD describing the degree of cure and CS describing compression set. Injection rate is not included in both models, as its effect was not statistically significant. Since the simulation does not take into account elongation flow right now, this is expected for the simulated degree of cure – elongation flow is the main mechanism leading to a temperature increase during injection [6, 7]. Injection rate not being significant when performing real experiments is noteworthy, however. This could be due to (1) the temperature increase in the rubber still being below the filling temperature or (2) the effects of T_m and T_c being much greater than any other influence, thus making v_i appear insignificant. The lower degree of determination of the model in fig. 4b is due to scattering of measured values, which is not present in simulations.

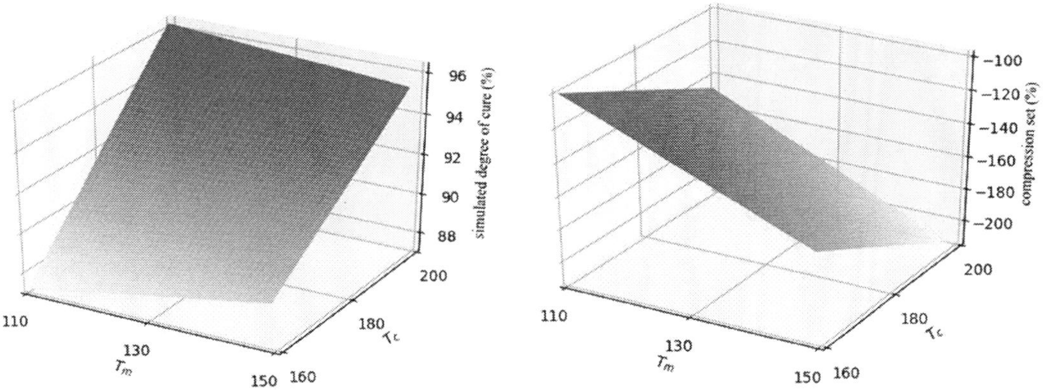

a) $R^2 = 95.48$ % b) $R^2 = 85.82$ %

FIGURE 4. Surface plots of the model obtained via regression. a) Simulated degree of cure. b) Compression set

$$CD = 9.000 + 0.374 * T_F + 0.452 * T_C - 0.002 * T_F * T_C \quad (1)$$

$$CS = 192.000 - 0.884 * T_F - 0.774 * T_C + 0.004 * T_F * T_C \quad (2)$$

CONCLUSION

3D transient injection molding simulation is able to adequately depict the real temperature distribution in a dynamically heated mold used for rubber injection molding. While infrared thermal images show a slightly less homogenous temperature distribution in the real cavity, real curing degree values calculated from compression set measurements and the simulated degree of cure correlate well. Regression analysis shows that for the RHCM mold, both T_m and T_c have to be taken into account when setting up the process. Due to the lower temperatures during filling (T_m), T_c can be set to higher values than normally possible when processing the investigated compounds. However, due to the nonlinear curing reaction kinetics, further investigations will include a design of experiments which is able to describe this behavior. Furthermore, it is important to improve the thermal homogeneity at the part surface and quantify the possible increase in productivity by rubber RHCM processing.

REFERENCES

1. M.H. Deckert, „Beitrag zur Entwicklung eines hochdynamischen variothermen Temperiersystems für Spritzgießwerkzeuge", Ph.D. thesis, TU Chemnitz, 2012
2. M. Fasching et al, Advanced mold concepts to improve part quality and productivity in rubber injection molding, Proceedings of the Regional Conference of the Polymer Processing Society, Graz 2015
3. Deutsches Institut für Normung (2010), Elastomere oder thermoplastische Elastomere – Bestimmung des Druckverformungsrestes (DIN ISO 815-1)
4. International Organization for Standardization (2012), Plastics- Determination of tensile properties – Part 2: Test condition for moulding and extrusion plastics (ISO 527-2:2012)
5. W. Friesenbichler et al., Methoden zu Bestimmung des Vernetzungsgrads an Kautschukformteilen, Proceedings of Rubber Meets Science Conference, Aachen 2015
6. D.M. Binding, J. Nonnewton Fluid Mechs. **27**, pp.173-189, 1988
7. D.M. Binding, J. Nonnewton Fluid Mechs. **35**, pp.27-42, 1991
8. J.S. Deng, A.I. Isayev, Rubber. Chem. Technol. **61**, pp. 340-361, 1988

Novel polyurethane elastomeric composites reinforced with Alumina, Aramid, and poly(p-phenylene-2,6-benzobisoxazole) short fibers, development and characterization of the thermal and dynamic mechanical properties

Pearson Adam [(1)], ***Naguib Hani*** [(1)]*

[(1)] University of Toronto - Ontario - Canada

Polyurethane is an extremely versatile material having seen use in a variety of applications. However, its relatively weak mechanical properties limit the scope of applicable functions. Recently the addition of various fillers to increase the functionality of polyurethane has been of great interest. This study outlines the development of novel polyurethane elastomers reinforced with Kevlar pulp, Alumina micro fibers, and poly(p-phenylene-2,6-benzobisoxazole) (PBO) fibers. One of the key goals in this study was to use an easily scalable elastomer fabrication process. To achieve this A high shear mixer was used for fiber dispersion followed be degassing and casting into a mould. The thermal and dynamic mechanical properties were characterized by thermogravimetric analysis (TGA), thermomechanical analysis (TMA), and dynamic mechanical analysis (DMA). Upon the addition of the reinforcing fibers an improvement in thermal stability as well as a significant reduction of 37% to the coefficient of linear thermal expansion was found. Additionally, an improvement of 97% to the storage modulus (G') was found with the inclusion of a small amount of PBO fibers, 1wt%, to the polyurethane matrix.

A New Class of Thermoplastic Elastomers via Atom Transfer Radical Polymerization and "Click" Chemistry

SINGHA NIKHIL KUMAR (1)*

(1) INDIAN INSTITUTE OF TECHNOLOGY KHARAGPUR, RUBBER TECHNOLOGY CENTRE - WEST BENGAL - India

Thermoplastic elastomers (TPE) are an important class of polymers which are processed like thermoplastics, but behave as elastomers. Apart from their interesting properties they have great advantage of recycling and reuse during processing. Conventional diene based TPEs (like SBS triblock copolymers) have poor resistance to weatherability, thermo-oxidative degradation and importantly poor resistance to oils and fuels. The inception of very versatile controlled radical polymerization (CRP) processes in late 1990s has led to a great opportunity to the development of new types of block copolymers which can have advantages over conventional TPEs. Acrylic block copolymers (BCP) have superior resistance to weatherability, oxidative/ozonolytic degradation and have much improved oil resistance properties. We have prepared different types of BCPs via transition metal catalyzed CRP, widely known as ATRP using poly(methyl methacrylate) (PMMA) or poly(furfuryl methacrylate) (PFMA) as hard segment and poly(hexyl acrylate) or poly(2-ethyl hexyl acrylate) as the soft segment. These BCPs were characterized by NMR, GPC, DSC and DMA analyses. Their morphology was analyzed by AFM, TEM and SAXS analyses. Interestingly, BCPs having furfuryl functionality were modified via Diels-Alder "click" reaction using bismaleimides. These modified BCPs showed self-healing characteristics. We have also modified SBS BCPs via thiol-ene reaction to improve their properties.

High Performance Thermoplastic Vulcanizates based on Carboxylated Acrylonitrile Butadiene Rubber and Polyamide 12

Chatterjee Tuhin (1)*, Basu Debdipta (2), Das Amit (3), Wiessner Sven (4), Heinrich Gert (5), Naskar Kinsuk (6)

(1) Senior Research Scholar at Rubber Technology Centre - West Bengal - India, (2) Scientist at IPF Dresden - Dresden - Germany, (3) Senior Scientist at IPF Dresden - Dresden - Germany, (4) Junior Professor at IPF Dresden - Dresden - Germany, (5) Professor @IPF Dresden - Dresden - Germany, (6) Associate Professor at Rubber Technology Centre - West Bengal - India

Thermoplastic vulcanizates (TPVs) are a special class of thermoplastic elastomers, in which dynamic vulcanization of the rubber phase takes place during melt mixing with a semicrystalline thermoplastic matrix phase at elevated temperature. High performance TPVs or so called "super TPVs" are a new generation of TPVs which exhibit high heat resistance as well as excellent oil resistance properties suitable for automotive under-the-hood applications. In the present work, a new Super TPV based on carboxylated acrylonitrile butadiene rubber (XNBR) and polyamide 12 (PA12) of different blend ratios has been prepared by using a fixed concentration of a novel cyclic monofunctional peroxide for initiating the dynamic vulcanization. The final morphology of TPVs varies from either a co-continuous to a dispersed one depending on the blend ratio. TPV with a wt% ratio of 50:50 XNBR-PA12 shows the highest mechanical properties as well as superior thermal stability among all other tested formulations. From differential scanning calorimetric (DSC) study, it can be clearly seen that the glass transition temperature (Tg) of XNBR has shifted to higher values in case of all TPVs as compared to that of the uncrosslinked blend system. Dynamic mechanical analysis (DMA) also demonstrates that tanδ values of all the TPVs are lower and the storage moduli are higher than the uncrosslinked blend system. The lowest tanδ peak of the TPV with a 50:50 wt% blend ratio of XNBR-PA12 indicates the highest degree of crosslinking and this is also supported by the swelling studies. Heat aging and oil resistance studies have also been carried out in details to characterize the behaviour of these high performance TPVs at service conditions. Key words: Thermoplastic vulcanizate, XNBR, PA12, Oil aging and heat aging.

Influence of ageing on the structure-property relationship of FKM, VMQ and EPDM rubber seals

Subramani Bhagavatheswaran Eshwaran [(1)*], Wiessner Sven [(1)], Jaunich Matthias [(2)], Wolff Dietmar [(2)]

[(1)] Technische Universität Dresden and Leibniz - Institut für Polymerforschung Dresden e.V. - Saxony - Germany, [(2)] Bundesanstalt für Materialforschung und -prüfung - Berlin - Germany

Elastomer seals are widely used as barrier seals in containers for low and intermediate level radioactive waste. In addition, they are also used in waste fuel storage and transportation as auxiliary seals. Depending on the area of use, the rubber materials have to demonstrate proper sealing performance with regard to mechanical, thermal and environmental conditions as well as irradiation during the entire operation period. A fundamental understanding of the structure property relationship can help to understand the performance of rubber seals in aged conditions. The present study focuses on the investigation of specific materials to study the structure-property and seal performance-correlation. Fluoro-elastomer (FKM), silicone rubber (VMQ) and ethylene propylene diene rubber (EPDM) were chosen as the prime candidates. Mechanical properties for these seals were achieved by careful control of base rubber formulation, i.e., by opting for the right filler and its concentration, amount of crosslinking chemicals, and with the assistance of process oils. Moreover, with the same rubber, similar mechanical properties were achieved by varying the filler concentration and crosslinking degree, which enables us to understand the structure property dependency of the seals in detail. The materials are investigated in the pristine and under aged conditions to evaluate the influence of ageing on e.g. the stress relaxation or recovery behavior. These values appear to be suitable parameters to judge the expected seal performance. Due to the electrically conducting nature of fillers used to reinforce rubbers (such as carbon blacks, carbon nanotubes (CNTs), Graphene, etc.) it is possible to obtain current information about mechanical and visco-elastic properties by monitoring the electrical conductivity or resistivity. A simple relation that bridges electrical resistance and the mechanical stress was derived and the stress values were theoretically predicted from the electrical resistance values, which showed good correlation with experimental results.

Hansen Solubility Parameters as a tool to evaluate the compatibility potential of Rubber Process oils (DAE, TDAE, MES, and NAP)

***Farshchi Negin* (1)*, *Abbasian Ali* (1)**

(1) Islamic Azad University Science and Research branch - Tehran - Iran

In order to improve the flexibility and processability of polymers, plasticizers are usually added to them. Highly aromatic oils [distillated Aromatic Extracts (DAEs)] are mostly used as process oils for rubbers in industry. As concerns for health and environment safety of DAEs due to their high content of polycyclic aromatic hydrocarbons (PAHs), rubber industry tends to consume process oils with lower content of PAHs such as, Treated distillate aromatic extract (TDAE), mildly extracted solvate (MES), hydro processed naphthenic oil (NAP). On the other hand, the more compatible the process oil and the rubber, the better the processability. "Solubility parameter" concept is a convenient way to determine the degree of solubility of materials in each other. In this study, inverse gas chromatography (IGC) method was used to calculate the solubility parameters and interaction parameters and depict solubility sphere for DAE, TDAE, MES, and NAP. The temperature dependence of these parameters was investigated as well. Results shows that in spite of the similarity in values of solubility parameter of DAE and TDAE at ambient temperature, increasing temperature leads to a significant difference in values. Despite other oils, TDAE shows a better compatibility with polar solvents. Also, interaction parameters show no specific dependence on the temperature for DAE, MES and NAP, except for polar solvents and TDAE. DAE has the highest compatibility with aromatic solvents . The compatibility trend with raising the temperature to values above 100◦C between oils and toluene is the same for all investigated except for TDAE which increase by increasing the temperature while others showed a reduction. The increase in alkyl chain is also effective in the way of increasing the compatibility of probes and their interaction parameters values as well. The solubility parameter δ2 of process oils determined to be 18.9, 18.9, 18.5, 19.0 (J/cm3)0.5 for DAE, TDAE, NAP, and MES at 25 °C.

Evaluation of the Influence of Crosslink Density and Penetrant Size on the Diffusion Properties of Silicone Oils into Silicone Elastomers

Jonas Daenicke[1, a)], Dirk W. Schubert[1, 4, b)], Mikael Hedenqvist[2)], Erik Linde[2)], Thomas Sigl[1)] and Raymund E. Horch[3)]

[1]*Institute of Polymer Materials, Friedrich-Alexander-University Erlangen-Nürnberg (FAU), Martensstraße 7, 91058 Erlangen, Germany*
[2]*Departement of Fibre and Polymer Technology, Royal Institute of Technology Stockholm (KTH), Teknikringen 56, 100 44 Stockholm, Sweden*
[3]*Plastisch- und Handchirurgische Klinik, Friedrich-Alexander-University Erlangen-Nürnberg (FAU), Krankenhausstraße 12, 91054 Erlangen, Germany*
[4]*Bavarian Polymer Institute, Dr.-Mack-Straße 77, 90762 Fürth, Germany*

[a)]Corresponding author: jonas.daenicke@fau.de
[b)]dirk.schubert@fau.de

Abstract. Driven by the continuing discussion on safety and quality of silicone breast implants, they have turned into focus of this study with respect to the diffusivity of low molar mass components from the silicone gel filling into the silicone breast implant shell. Therefore, the diffusivity of silicone oils into silicone elastomers were analysed by means of the crosslink density and the penetrant size. The study was focused on the diffusion of the cyclic siloxanes Octamethylcyclotetrasiloxane (D4), Decamethylcyclopentasiloxane (D5) and Dodecamethyl-cyclohexasiloxane (D6) due to their potential occurrence in silicone breast implants. The analysis of the diffusion behavior was carried out with silicone breast implant shells taken from explants and tailor-made silicone elastomer samples varying in crosslink density. Therefore, sorption experiments were performed. The subsequent evaluation of the sorption data yield to the corresponding diffusion properties. Based on the diffusion coefficient related to the crosslink density a model was developed to describe the material behavior.

INTRODUCTION

Quality and safety of silicone breast implants have been in focus of public media and research in the recent years due to the scandal of the French company Poly Implant Prothèse (PIP) [1]. Hence, this study deals with the sorption properties of cyclic siloxanes into the silicone breast implant shell and a commercial polydimethylsiloxane (PDMS) which is used as a reference material. The content of low molecular siloxanes in silicone breast implant gels is apparently small with about 1 – 2 % according to other studies [2, 3]. Nevertheless it is stated that PIP devices may contain higher amounts of certain cyclic siloxanes [2].

In the following a short summary is given on the current state of research on diffusion properties of low molar mass silicones (abbreviation: LM-silicones) through silicone breast implant shells into biological tissues. The diffusion of LM-silicones from silicone gel-filled breast implants in surrounding tissues and fluids is called gel-bleeding [4, 5]. Lykissa et al. have carried out studies on this topic in the late nineties [3]. The research was focused on diffusion of such polysiloxanes through implant shells and on the adsorption of polysiloxanes into biological imitating fluids. The diffusion of LM-silicones from intact silicone breast implants into several liquids was investigated. Moreover it is obvious that the cyclosiloxanes (D4 to D7) exhibit a constant release rate over time. The investigation of silicone gels indicates a content of about 1-2 % LM-silicones, containing cyclic and linear polysiloxanes varying from 3-20 siloxane units [3, 6, 7]. Considering the constant release rate and the fact that the examined silicone gels were taken from explants, the content of LM-silicones in new silicone breast implants might be higher than in the investigated ones [3].

Further studies were performed regarding the investigation on LM-siloxanes in biological tissues and fluids of women with silicone gel-filled breast implants. In those studies an increased content of cyclic and linear LM-silicones was detected in comparison to investigations on patients without silicone breast implants. This fact supports the hypothesis of a release of LM-silicones through the silicone breast implant shell [8, 9, 10].

In this report the diffusivity was related among other things to the penetrant size, herewith former studies of Gedde et al. in 1996 could be confirmed [11]. In regard to the relation between diffusivity and the crosslink density a similar behavior was obtained by Gent in 1982 [12]. In addition a model was introduced to explain the material behavior.

EXPERIMENTAL

Sample Preparation

Gel-filled silicone breast implants from several manufacturers were analyzed during the study. A total of nine explants of different shape, size, texture and implantation time were investigated. In addition reference silicone sheets with a deliberate adjusted degree of crosslinking were tailor made. Therefore Elastosil® LR 3003/30 A/B from Wacker Chemie AG, a two-component silicone, was used in different mixing ratios. The silicone elastomers were crosslinked in two-part molds for 16 hours at 120 °C. From both sample types square-shaped specimens were punched out with a side length of 10 mm. For the samples taken for the sorption tests an extraction in n-heptane was carried out. The experiments were performed in n-heptane of 99% purity obtained from ROTISOLV® of Carl Roth GmbH + Co. KG. The specimen were stored for 24 hours in n-heptane and afterwards dried under the hood. The same procedure of sample preparation was performed in n-hexane for the determination of the molar mass between crosslinks M_c. Therefore n-hexane of 95% purity was used which was again obtained from ROTISOLV® of Carl Roth GmbH + Co. KG.

Sorption Experiments

Sorption experiments were carried out to determine the diffusion properties of the cyclic siloxanes into the silicone samples. Therefore the prepared and extracted samples were stored in 3.5 g cyclic siloxane. The weight of the samples was measured for certain intervals in order to record an uptake diagram for the following diffusion coefficient calculations. After 48 hours the samples can be considered as saturated, that means no more mass-uptake of the swelling agent could be expected. All experiments were carried out in a ventilated oven at 37 °C. The used cyclic siloxanes Octamethylcyclotetrasiloxane (D4), Decamethylcyclopentasiloxane (D5), Dodecamethylcyclohexasiloxane (D6) were obtained from abcr GmbH.

Evaluation of the Molar Mass between Crosslinks

The average molar mass between crosslinks M_c can be determined using the data from the swelling experiments, which were carried out in n-hexane for 48 hours. This yields by calculation according to the Flory-Rehner theory to the desired parameter. M_c is defined by the following equation:

$$M_c = \frac{\rho V_m \left(v_B/2 - v_B^{1/2} \right)}{\ln(1 - v_B) + v_B + \chi v_B^2}, \quad (1)$$

where ρ is the density of the polymer, v_B describes the volume fraction of the polymer, χ is the interaction parameter and V_m represents the molar volume of the solvent. The density of the silicone elastomers was approximated by 1.09 g/cm³ and the interaction parameter χ between n-hexane and silicone rubber is 0.40 [13].

Determination of the Diffusion Coefficient

The diffusion coefficient of sorption was determined using a numerical calculation according to Fick's second law of diffusion leading to the concentration dependent diffusivity D(C). The derivation is adjusted for a plate geometry of the samples. The calculations were done by Matlab (Mathworks Inc., Natick, MA, USA) modeling, using the experimental uptake data of the sorption experiments.

The basis of the equation for the concentration dependent diffusion coefficient is defined as follows [14]:

$$\frac{\partial C}{\partial t} = \frac{\partial}{\partial x}\left(D(C)\frac{\partial C}{\partial x} \right) \qquad (2)$$

A more detailed derivation of the numerical calculation can be found in former papers of Gedde et al. and Hedenqvist et al. [4, 15]. The data evaluation results in the zero-concentration diffusivity D_{co}.

RESULTS AND DISCUSSION

Influence of Penetrant Size

The average value of the diffusivity for sorption experiments over all silicone breast implants is higher as the one of the reference material (PDMS). An overview of the values of the diffusion coefficient is listed in Tab. 1.

Table 1. Diffusion coefficient of the silicone breast implants in comparison to the reference material (PDMS)

Material	Diffusion coefficient for the different cyclic siloxanes in E-8 cm²/s		
	D4	D5	D6
Implants	42.2 ± 5.5	30.3 ± 6.0	20.8 ± 4.6
PDMS	18.7 ± 0.6	12.0 ± 0.4	7.8 ± 0.4

The difference is reasoned by the variation in the elastomeric network properties between the silicone breast implants and the reference material. Furthermore the silicone breast implant shells exhibit three distinguishing characteristics leading to a deviant diffusion behavior. At first a huge number of implants own a barrier layer, mainly consisting of phenylated silicone, preventing the diffusion process through the shell [5]. The barrier layer properties are neglected in this study due to the experimental setup. In addition only explants were used for the experiments that means that the implants already exhibit a prehistory which might handicap the material properties. One of the main factors is the texturized outer surface of the silicone breast implants which leads to an increase of the surface area and therefore enhances the diffusion process.

Figure 1 shows the diffusivity over the molar mass of the cyclic siloxanes. The plotted diffusivity values are synonymous with the zero-concentration diffusivity D_{co}. It is obvious that with an increase of molar mass of the diffusive agent the diffusion coefficient is decreasing. This fact is explained by the geometrical factor. The cyclic siloxanes own the same chemical backbone chain without any side groups affecting the sorption process. Only the space requirement in the cyclic arrangement of the penetrant is increased due to the number of siloxane units and has an evident influence on the diffusivity.

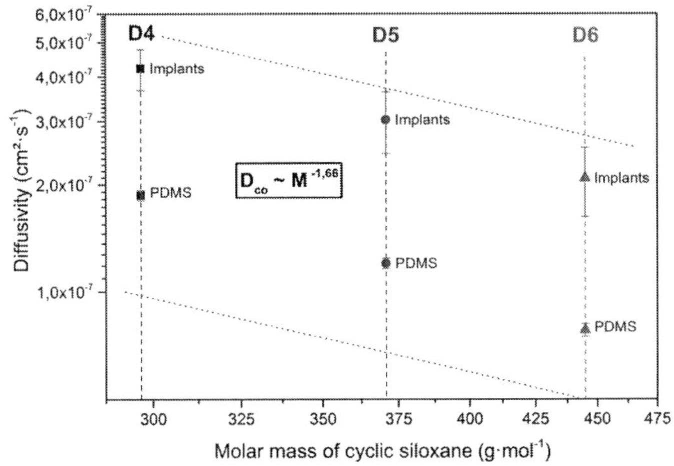

Figure 1. Diffusivity of the silicone breast implants (implants) and the reference material (PDMS) over the molar mass of the cyclic siloxanes – the doted lines indicate the empirical relationship of "$D_{co} \sim M^{-1.66}$"

An empirical relationship between the diffusion coefficient and the molar mass of "$D_{co} \sim M^{-1.66}$" can be defined according to the experimental data. An equal correlation was already determined by Gedde et al. for diffusive agents in the range of small molar masses in 1996 [11].

Influence of Crosslinking

This section considers the diffusivity related to the crosslinking properties of the elastomeric network. The average molar mass between crosslinks is used as a degree of crosslinking in this study. In Fig. 3 (a) it is shown that the diffusion coefficient of sorption is increased with increasing molar mass between crosslinks. The same observation was done by Gent and et al. in 1982 [12]. In this study the behavior is explained by a two phase model, considering two different states in the elastomeric structure (see Fig. 2 on the right), both assigned to own diffusion coefficients (D1 and D2). The difference of the two phases could be derived from the packing density of the polymer chains near or rather between the crosslinking points or it could be influenced by a filler degree, e.g. silicate, which may also differ in between the network. In Fig. 2 on the left the applied model is described including a parallel and a serial connection of the two phases in regard to the diffusion process [16]. Derived from these equations the used fitting equation is shown in the center of Fig. 2, including the parameters l and Φ. l specifies the application of parallel or serial connection and Φ is derived from the volume fraction, but contains the material parameter of the molar mass between crosslinks M_c, on whose dependency this relation is made, and a global fit parameter A. The subsequent data evaluation with the model is shown in Fig. 3 (a), where the lines indicate the fitting of the data. For the fitting step the global fit parameter A results in 7372 and the global fit parameter l results in 1, which is ambivalent to the parallel connection of the two phases. Figure 3 (b) illustrates the obtained diffusion coefficient for the two phases D1 and D2 over the molar mass of the diffusive agents. The D2 value is smaller than the D1 value, which means that in the area of the D1 value a lesser hindrance of the sorption process exists. Finally in Fig. 3 (b) the empirical relationship of "$D_{co} \sim M^{-1.66}$" is shown again for the modeled diffusivity data in relation to the molar mass of the diffusive agents.

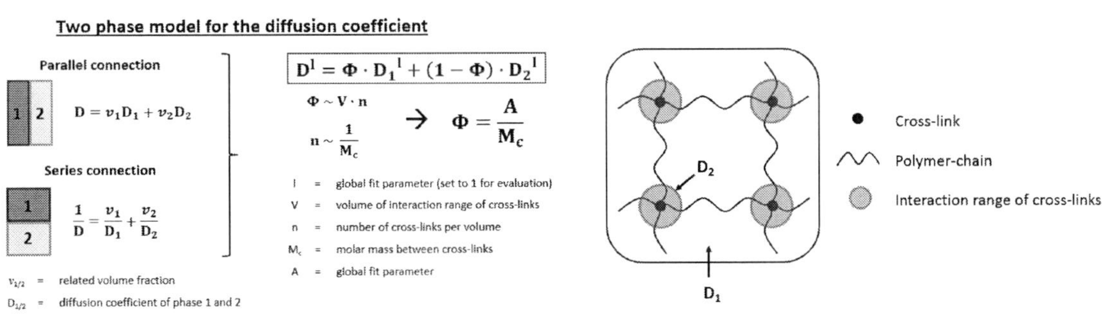

Figure 2: Description of the two phase model

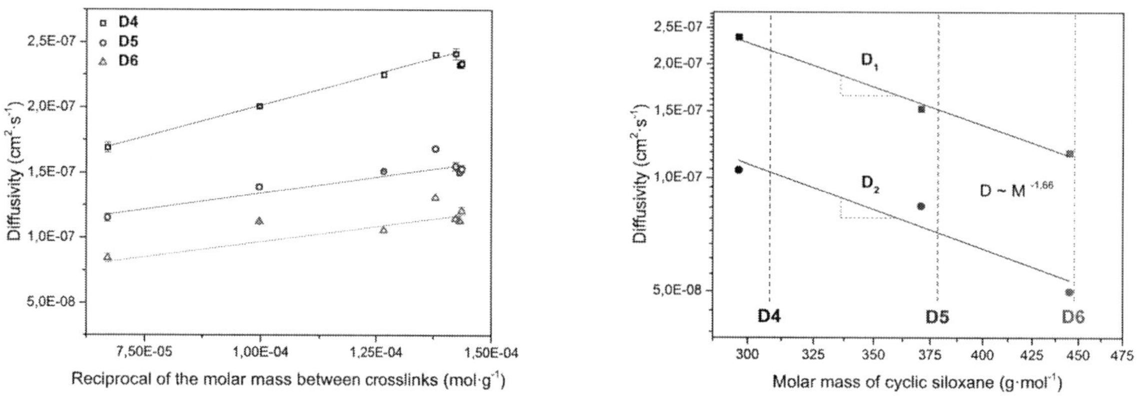

(a) (b)

Figure 3. (a) Illustration of the diffusivity over the reciprocal of the molar mass between crosslinks of the diffusive agents fitted with the two phase model and (b) illustration of the modeled diffusion coefficient over the molar masses of the cyclic siloxanes

CONCLUSION

An empirical relationship between the diffusivity and molar mass of the cyclic siloxanes was achieved, which shows a decrease of the diffusion coefficient with increasing molar mass of the diffusive agent. Furthermore the diffusion coefficient was related to the molar mass between crosslinks or rather to the crosslinking introducing a two phase model for the description. The applied model in this study has to be refined by using a larger number of different PDMS materials and by considering further material parameters, this refinement is planned for future studies.

ACKNOWLEDGMENT

The authors would like to thank Prof. Ulf W. Gedde especially for his help and advice regarding the experimental work in this study.

REFERENCES

1. D. W. Schubert *et al.*, Polym. Int. **63**, 172–178 (2014).
2. Scientific Committee on Emerging and Newly Identified Health Risks (SCENIHR), "Preliminary Opinion on the safety of Poly Implant Prothèse (PIP) Silicone Breast Implants," (European Commission, 2013 update).
3. E. D. Lykissa et al., Anal. Chem. **69**, 4912–4916 (1997).
4. S. Barr and A. Bayat, Aesthet. Surg. J. **31** (1), 56–67 (2011).
5. G.S. Brody, Silicone breast implant safety and efficacy, Medscape, 2017.
6. A. B. Birkefelda et al., Biomaterials **25**, 4405–4413 (2004).
7. G. Beretta, G. and M. Malacco, J. Pharm. Biomed. Anal. **78-79**, 75–82 (2013)
8. D. Flassbeck et al, Anal. Bioanal. Chem. **375**, 356–362 (2003).
9. S. V. Kala, et al., Anal. Chem. **69,** 1267–1272 (1997).
10. G. Beretta et al., J. Pharm. Biomed. Anal. **84**, 159– 167 (2013).
11. U. W. Gedde *et al.*, Polym. Eng. Sci. **36** (16), 2077–2082 (1996).
12. A. N. Gent and R. H. Tobias, J. Polym. Sci. **20**, 2317-2327 (1982).
13. L. R. G. Treloar, *The Physics of Rubber Elasticity* (Oxford University Press, USA, 1975).
14. J. Crank and G. S. Park, *Diffusion in Polymers* (Academic Press Inc., London and New York, 1968), pp. 1-37.
15. M. S. Hedenqvist and U. W. Gedde, Polymer **40** (9), 2381–239 (1999).
16. M. S. Hedenqvist and U. W. Gedde, Prog. Polym. Sci. **21**, 299-333 (1996).

Experimental Study on the Behavior Response for the CR / NR Hybrid Rubber Damper Subjected to a Cyclic Loading in Compression

Chiou Yung-Chuan (1)*

(1) National Chiayi University - Taiwan - Taiwan

The observation and analyses on the cyclic mechanic behavior responses of the CR / NR hybrid rubber damper subjected to a cyclic loading in compression have performed in this study. The waveform with a sine form has used in all performed cyclic tests in compression. Moreover, for the employed waveform, the average rate of cross head has controlled at the specified value of 30 mm/sec. Based on the observation on the experimental results, the facts have indicated that the response of the tested rubber damper subjected to a cyclic loading in compression is the closed form of a hysteresis loop; the measured maximum stress range decrease with decreasing the applied displacement range in compression; the stress range per cycle decreases with cycles. Moreover, the variation of absorbed strain energy per cycle, E_s , with cycles presents a gradually decrease and approach to a stable condition. The variation of strain energy per cycle with cycles depends on the magnitude of the applied displacement. A new cyclic performance indicator was proposed in order to estimate the correlation between the absorbed deformation energy and the transmitted force in per cycle by the tested rubber damper. Based on the calculated values, it was found that the magnitude of cyclic performance indicator decreases gradually to a stable value with cycles. The value of cyclic performance indicator increases with increasing the applied displacement range at the same number of cycle. In this paper, the Kelvin model was also employed to elucidate the variation of the visco-elastic properties of the tested rubber damper. According to the analyses on data, it was been found that the magnitude of cyclic equivalent damper coefficient increases to a stable value with cycles. On the contrary, the magnitude of cyclic equivalent stiffness decreases to a stable value with cycles. At the same number of applied loading cycle, both values of equivalent stiffness and damper coefficient increase with increasing the applied displacement range in compression.

Application of Functional Magnesium-based Nanomaterials in Room Temperature Vulcanized Silicon Rubber Composite

Liu Zhiqi (1)*, Li Lijuan (1), Chen Yunfa (2), Leuteritz Andreas (3), Wagenknecht Udo (3), Li Na (1), Ji Lianmin (1), Nie Feng (1)

(1) Qinghai Institute of Salt Lakes, Chinese Academy of Sciences - Qinghai - China, (2) Institute of Process Engineering, Chinese Academy of Sciences - Beijing - China, (3) Leibniz Institute of Polymer Research Dresden - Dresden - Germany

Abstract: The influence of the room temperature vulcanized silicon rubber(RTV) composites on mechanical properties, flame retardant and smoke suppression were studied by adding different proportion of superfine magnesium hydroxide(SMDH), nano magnesium hydroxide(NMDH), MgFe-LDH and MgFeCe-LDH into the RTV with dealcoholized one part room temperature vulcanized silicone rubber as the matrix. Tensile strength results indicate that the tensile strength of RTV/SMDH composite is the highest when the filler is 60 phr, the value is more than 3 MPa. Tensile results indicate that the elongation at break rate of RTV/MgFe-LDH composite is above 700%, which is two times than that of NMDH and the MgFe-LDH is suitable for preparation elastomer composites with RTV. Strength results indicate that strength of RTV composites grows gradually with the increasing of the fillers. Limited oxygen index(LOI) results indicate that the flame retardant property of the composites has improved a lot with the increasing of the fillers. Among these, the influence of SMDH on the composites is the most obvious, for example, the RTV/SMDH composite has a good flame retardant property with 60 phr of SMDH and the RTV/LDH has a slightly flame retardant property with the same filler of SMDH. Smoke density(SDR) results indicate that the SDR of composites are significantly lower with the same filler, which can greatly improve the security of RTV. Keywords: Magnesium hydroxide, Hydrotalcite-like compounds, Silicone rubber, Composites, Flame retardant, Mechanical property

Two-component injection moulding of thermoplastics with thermoset rubbers: The effect of the mould temperature distribution

G.J. Bex[a], W. Six[b], J. De Keyzer[c], F. Desplentere[b], A. Van Bael[a]

[a] *Materials Technology TC, KU Leuven Technology Campus Diepenbeek, Wetenschapspark 27, 3590 Diepenbeek, Belgium*
[b] *Materials Technology TC, KU Leuven Technology Campus Ostend, Zeedijk 101, 8400 Oostende, Belgium*
[c] *Sustainable Chemical Process Technology TC, KU Leuven Technology Campus Diepenbeek, Wetenschapspark 27, 3590 Diepenbeek, Belgium*

Corresponding author: gertjan.bex@kuleuven.be

Abstract. In a previous study [1] a novel two-component injection moulding process was proposed for combining thermoplastics with thermoset rubber. This process requires moulds in which the cavities for the thermoplastic and the rubber are thermally separated making it possible to use different mould temperatures for both materials. The process starts by first injecting the thermoplastic part. Afterwards the thermoset rubber is injected and vulcanised. The goal of two-component injection moulding is to combine materials within a single product without assembly. Therefore the adhesion strength between the two materials is an important property that determines the product quality for a two-component injection moulded product. A second important property for every injection moulding process is the cycle time. The cycle time for this specific process is mainly determined by the time required to sufficiently vulcanise the rubber. In this paper the effect of the mould temperature on both the adhesion strength and the vulcanisation time are investigated experimentally. Products are produced using different mould temperature distributions. Afterwards the adhesion strength is examined using tensile tests. The vulcanisation time is determined by using hardness measurements to verify the vulcanisation degree of the rubber. Results indicate that higher mould temperatures for the thermoplastic part lead to higher adhesion strength and lower vulcanisation time.

Keywords: Two-component injection moulding, thermoset rubbers, thermoplastics.
PACS: 81.20.Hy

INTRODUCTION

Two-component injection moulding is a technique used to combine polymers with different properties in a single injection moulding process. In previous research a novel two-component injection moulding process is proposed for combining thermoset rubber with thermoplastics [1, 2]. The main problem when combining these two materials is the difference in processing between thermoset rubber and thermoplastics. Thermoplastics are typically injected at a relatively high temperature (190 – 300 °C) in a cooled mould (20 – 100 °C) to solidify the product [5, 6]. Thermoset rubbers on the other hand are injected at a low temperature (80 – 100 °C) in a heated mould (160 - 180 °C) to enable the vulcanisation process [3, 4]. The proposed process uses moulds in which the mould cavities for the thermoplastic and the rubber part are thermally separated. In this way it is possible to use different temperatures for both cavities. Figure 1 shows the product used for this study. It consists out of two adjacent plates with a thickness of 2 mm and a length of 150 mm. Figure 2 shows the mould in which the cavities are thermally separated by air gaps and insertion of insulation material. The process starts by first injecting the thermoplastic material. Afterwards the thermoset rubber is injected and vulcanised. After the vulcanisation process it is important to cool the rubber part, otherwise the thermoplastic part is not solidified and will fail in the interface region upon ejection. Heating and cooling of the mould is done by hot water under pressure and cold water, similarly to rapid heat cycling in thermoplastic injection moulding [1, 2].The main goal of a two-component injection moulding process is to skip the assembly process. It is therefore important that the two parts, made out of different materials, are well connected to each other. Therefore the adhesion strength between the two materials is an important property that determines the product quality. In this work the effect of the mould temperature on the adhesion strength is discussed. The thermoset rubber needs to be vulcanised in order to gain its properties. This vulcanisation process takes a much longer time than needed for cooling the thermoplastic part, and therefore determines the total cycle time of the process. In this work the effect of the mould temperature distribution on the vulcanisation time is investigated for four material combinations.

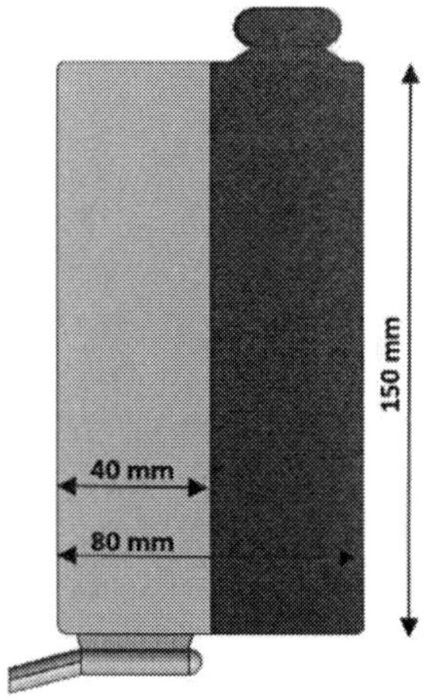

FIGURE 1. Geometry of the selected specimen with dimensions. The thickness of the specimen is 2 mm [2].

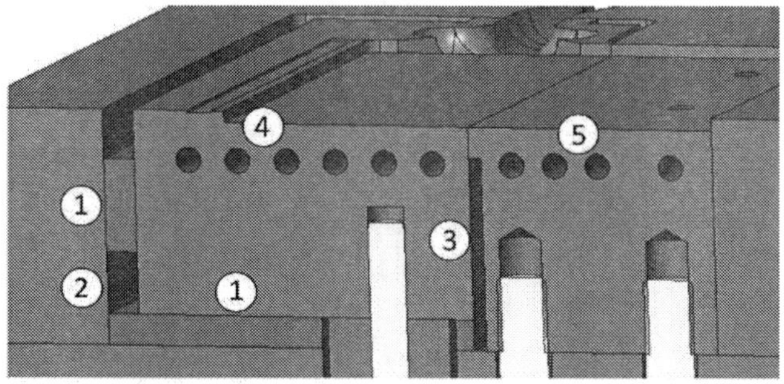

FIGURE 2. Design of the mould with thermally separated cavities [2].

MATERIALS AND METHOD

Material

Four different thermoplastic materials were selected for this research: two semi-crystalline apolar thermoplastics; i.e. a high density polyethylene (HDPE) (Sabic, grade M80064) with a melting point of 135 °C and a polypropylene (PP) (INEOS, grade 400-GA05) with a melting point of 164 °C; and two amorphous polar thermoplastics; acrylonitrile butadiene styrene (ABS) (INEOS, grade Novodur P2H-AT) with a glass transition temperature of 110 °C and polycarbonate (PC) (Covestro, grade Makrolon 2607) with a glass transition temperature of 143 °C. For these thermoplastics two compatible thermoset rubbers were selected: an apolar ethylene propylene diene monomer (EPDM) (Hercorub, grade 005K) compatible with HDPE and PP and a polar nitrile rubber (NBR) (Hercorub, grade 42G) compatible with ABS and PC.

Method

Injection moulding is conducted on an Engel ES330H/80V/80HL-F 2K injection moulding machine with a vertical rubber injection unit and a horizontal thermoplastic injection unit. The maximum clamping force of the machine is 1000 kN. The vertical injection unit is equipped with a screw with a diameter of 25 mm and an L/D ratio of 16. The screw for the horizontal unit has a diameter of 35 mm and an L/D ratio of 20. The mould and product used for this study have been discussed in the introduction.

The process parameters used for the injection moulding of the individual materials can be found in Table 1. For the thermoplastic part, these are determined for minimal shrinkage. The mould and injection temperature for the rubbers were recommended by the supplier. The injection speed was set as high as possible and limited by flash. The holding pressure was set to completely fill the product with minimal flash.

TABLE 1. Process parameters for the individual materials

	HDPE	PP	ABS	PC	EPDM	NBR
Injection temperature	230 °C	230 °C	220 °C	300 °C	80 °C	80 °C
Mould temperature	20 °C	20 °C	70 °C	80 °C	180 °C	180 °C
Volumetric injection rate	38 cm^3/s	38 cm^3/s	77 cm^3/s	87 cm^3/s	59 cm^3/s	59 cm^3/s
Holding pressure	790 bar	494 bar	346 bar	494 bar	198 bar	198 bar

The adhesion strength was evaluated by tensile tests on a Zwick Z050. Rectangular specimens with a width of 10 mm were punched out of the products. The interface is located in the middle of the sample The tests were performed at a speed of 200 mm/min. For each setting, five different samples were evaluated.

Since the cycle time depends on the necessary vulcanisation time it is necessary to evaluate the resulting vulcanisation degree. This is done by using Shore A hardness measurements according to ISO 7619 with a measuring time of 15 s.. The vulcanisation degree is evaluated at three different lines within the rubber as seen in Figure 3: one at the edge of the rubber part (c), one in the middle of the rubber part (b) and one at 1 mm away from the interface. Each hardness measurement is repeated on five different locations within the lines. The hardness measurements are then converted to vulcanisation degree using Equation 1, in which x is the hardness on a specific point, x_{min} is the hardness of unvulcanised rubber, and x_{max} is the hardness of 100% vulcanised rubber. In general, a vulcanisation degree of 90% is assumed to be a good value for ejecting a rubber product [2, 4].

$$vulcanisation\ degree = \frac{x - x_{min}}{x_{max} - x_{min}} \quad (1)$$

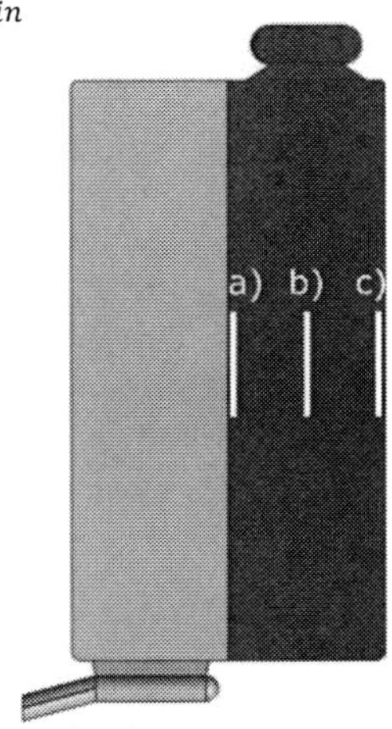

FIGURE 3. The lines are the different locations for the hardness measurements. (a) are the measurements at 1 mm away from the interface, (b) is for the measurements in the middle of the rubber part and (c) is for the measurements at 1 mm away from the edge [2].

EXPERIMENTAL

Effect of the mould temperature on the adhesion strength

In this section the effect of the mould temperature on the adhesion strength is investigated for multiple material combinations. Specimens are produced using different mould temperatures at the interface. Tensile tests were used to analyse the adhesion strength for each setting and thus the results are shown in the Figure 4. The vulcanisation times were chosen in order to ensure that the rubber is vulcanised for at least 90 % for each temperature setting. For the material combination HDPE-EPDM (Figure 4a) the vulcanisation time was 2600 s, for PP-EPDM (Figure 4b) a time of 800 s was used and for ABS-NBR (Figure 4c) the vulcanisation time was 1200 s. Figure 4 shows that the adhesion strength increases with increasing mould temperature at the interface. This effect is caused by an increased interdiffusion [2]. Figure 4a and Figure 4b show that there is a drop in adhesion strength when the temperature is too high for HDPE and PP. Both materials are semi-crystalline, and the drop in adhesion strength occurred when the mould temperature at the interface is higher than the melting point of the thermoplastic material. The amount of molten thermoplastic material at the interface is too high and therefore it can be pushed away, as a result sulphur gasses released during the vulcanisation process can penetrate the thermoplastic material at the interface. These gas bubbles lower the contact surface between both materials and therefore lower the adhesion strength [2]. For ABS, an amorphous material, shown in Figure 4c, no drop in adhesion strength was observed at the highest temperature. For this material the mould temperature was limited by the deformation of the thermoplastic part in bulk.

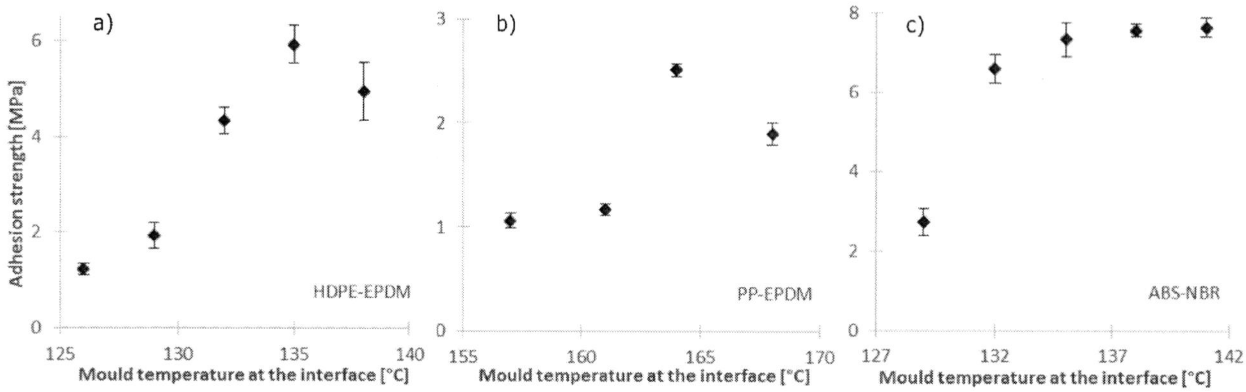

FIGURE 4. Adhesion strength in function of the mould temperature at the interface. In a) for the material combination HDPE and EPDM. In b) the results are shown for PP and EPDM and in c) ABS in combined with NBR. Error bars represent the 95% confidence interval.

Effect of the mould temperature on the cycle time

The vulcanisation reaction is critical in rubber processing because of its influence on the total cycle time. Figure 5 shows the vulcanisation reaction in the different areas of the rubber part for HDPE-EPDM using different interface temperatures. In Figure 5a a mould temperature at the interface of 126 °C was used while the mould interface temperature was 138 °C in Figure 5b. The mould temperature in the bulk of the rubber part is 174 °C. The temperature at the interface is lower due to the lower temperature of the thermoplastic cavity. It can be seen that the rubber in the middle and on the edge of the rubber parts are sufficiently vulcanised after 400 s. The rubber close to the interface however is only vulcanised after 1000 s when the temperature at the interface is 138 °C (Figure 5b). With an even lower interface temperature of 126 °C, the rubber is still not sufficiently vulcanised after 1400 s. The lowest possible cycle time is thus limited by the vulcanisation speed at the interface and thus by the mould temperature at the interface.

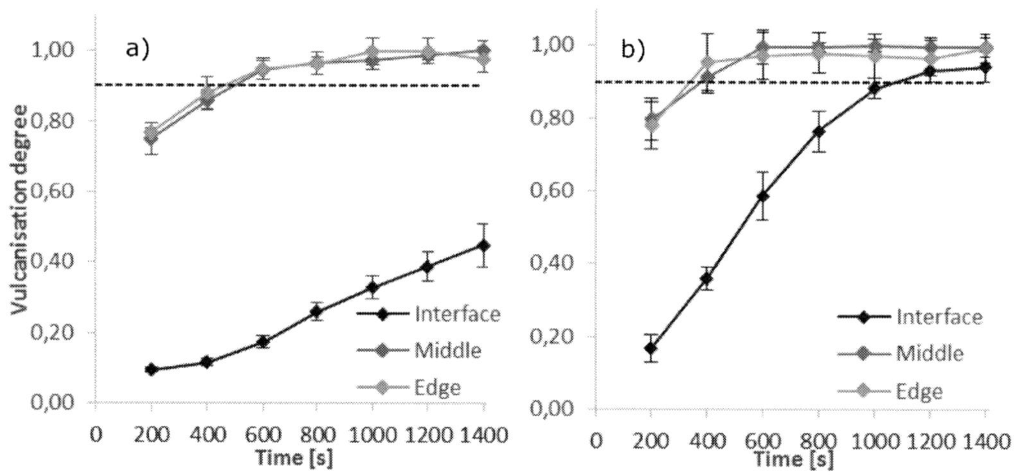

FIGURE 5. Vulcanisation process for specimens produced with a mould temperature at the interface of 126 °C in a) and 138 °C in b) for the material combination HDPE-EPDM. The error bars represent the 95 % confidence interval.

From Figure 5 it is clear that the vulcanisation time, and thus cycle time, depends on the mould temperature at the interface. As explained earlier the maximal mould temperature at the interface is limited by the amount of molten material at the interface for a semi-crystalline material and by the deformation of the thermoplastic part in bulk for an amorphous material. Both depend on the stability at high temperatures which can be estimated using the melting point for a thermoplastic material and the glass transition temperature for an amorphous material. In Figure 6a the vulcanisation reaction at the interface is shown for HPDE and PP using their maximum mould temperature at the interface. Because the melting temperature for the HDPE is lower (135 °C) than the melting temperature of the PP (164 °C), the maximum mould temperature is also lower. In this case the rubber is sufficiently vulcanised after 300 s while the rubber needs a vulcanisation time of 1000 s in combination with the HDPE. In Figure 6b the vulcanisation reaction is shown for PC with a glass transition temperature of 143 °C and ABS with a glass transition temperature of 110 °C, both in combination with NBR. The difference in glass transition temperature makes it possible to use a higher mould temperature for PC. In this case the rubber is sufficiently vulcanised after only 100 s while the rubber needs a vulcanisation time of 400 s when using ABS. In both cases the use of a more thermally stable thermoplastic material decreases the cycle time by ±70 %.

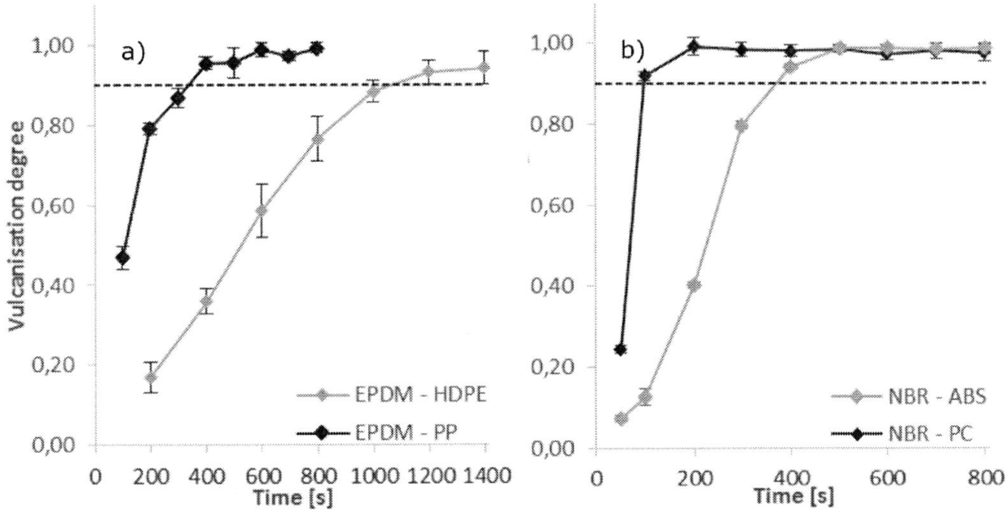

FIGURE 6. Vulcanisation process for specimens produced with HDPE and PP in combination with EPDM in a) and ABS and PC in combination with NBR in b). The error bars represent the 95 % confidence interval.

CONCLUSION

In this study the effect of the mould temperature on both the adhesion strength and the cycle time are investigated for two-component injection moulding processes of a thermoset rubber and thermoplastics. It was found that the adhesion strength increases with increasing mould temperature at the interface. However, the maximal possible temperature at this location is limited by the formation of gas bubbles in case of a semi-crystalline material and by deformation of the thermoplastic part in bulk in case of an amorphous material. Both phenomena depend on the stability of the thermoplastic material at high temperatures. This is related to the melting temperature for semi-crystalline materials and to the glass transition temperature for amorphous materials. The cycle time is limited by the vulcanisation speed at the interface, which also depends on the mould temperature at the interface. In this paper it was shown that it is possible to decrease the vulcanisation time by 70 % when using a more suitable thermoplastic material compared to a less temperature stable thermoplastic material. In further research the effect of other injection moulding process parameters, like the holding pressure and the injection speed, will be investigated.

ACKNOWLEDGMENTS

G.J. Bex acknowledges Research Foundation – Flanders (FWO) and the Impulsfonds project financed by the KU Leuven for funding his PhD grant Strategic Basic Research. The authors also acknowledge the company Hercorub NV for providing the rubber material.

REFERENCES

1. Bex G, De Keyzer J, Desplentere F, Van Bael A (2016). Two-component injection moulding of thermoplastics with thermoset rubbers: Process development. Proc. of the 32nd International Polymer Processing Society Conference, S11-75
2. Bex G, Desplentere F, De Keyzer J, Van Bael A (2017) Two-component injection moulding of thermoset rubber in combination with thermoplastics by thermally separated mould cavities and rapid heat cycling. International Journal of Advanced Manufacturing Technology
3. Ciesielski A (1999) An introduction to rubber technology. Rapra Technology Limited, Shawbury, pp 63–75, 140
4. Linday JA. (2012) Practical guide to rubber injection moulding. Smithers Rapra, London, p 14, 47–54, 69
5. Osswald TA, Turng LS, Gramann P (2008) Injection molding handbook. Hanser, Munich, p 65, 431–435
6. Kamal MR, Isayev A, Liu S-J (2009) Injection molding technology and fundamentals. Munich, pp 55–60

Nano silica from rice husk: How to solve an environmental problem through SBR

Raimonda Pablo [(1)]*, Mosca Daniel [(1)], Lesa Ana Ines [(1)], Mantero Carlos [(1)]

[(1)] Facultad de Ingeniería - Instituto de Ensayo de Materiales - Montevideo - Uruguay

It is well known that the mechanical properties of filled rubber materials are enhanced when compared with rubbers alone, as a result of the interactions between the filler and the rubber chains. The most common fillers are Carbon black and amorphous silica. Furthermore, rice is an important crop in Uruguay whose production rises to one million tons per year. For this reason, husk disposal is a serious problem. Regarding a solution to this environmental issue, the authors firstly evaluated the controlled burning of acid pretreated husk to yield reasonable results as SBR rubber reinforcement. In this paper, we demonstrate that we can obtain Nano silica in a very log energetic form, and it can be used as a reformate filler In SBR rubber with mechanical properties very similar to de commercial products. To obtain amorphous Nano silica with the smallest superficial area possible, we first boiled the rice husk during 3 hours and then burn it at controlled temperature (600ºC) to obtain a grayish white residue. This material (RH-A), is allowed to cool down. Afterwards we precipitated silica from caustic solutions of ash, with a mineral acid. The product was centrifuged and washed first with hydro clohidric acid and them with distillated water. The last stage of this process involves drying at 110° C for about 12 hours. The yield of silica in this reaction was 90%. The Nano silica (PNS) obtained was characterized by, XRD, SEM, TEM, EDS, FTIR, TGA, DSC and BET analysis. To produce the rubber compound we made a formulation based on SBR and other components. On this master without accelerant, we added the amount of Nano silica needed to obtain the ratio of 60 pHr. We did the same with a commercial product (CNS), to compare. Finally, we made comparative mechanical tests of all the products obtained by ASTM methods The Nano silica produced by this method and used as filler of SBR rubber showed similar mechanical properties compared to a commercial product.

The synthesis and characterization of silane coupling agents to increase silica filler content in styrene-butadiene rubber wet master batch for all-season automobile tires

Choi Jun [(1)]*, Choi Ji Hye [(1)], Lee Bo Ram [(1)], Kim Jung Soo [(1)], Jang Si-hoon [(1)], Park No-hyung [(1)], Kim Dong Hyun [(1)]

[(1)] Korea institute of industrial technology - Ansan - Korea

The synthetic rubber compound, which is widely used as the core material for automobile tire industry, is manufactured by mixing styrene-butadiene rubber (SBR) and organic / inorganic filler. It is known that the most important factor for the physical properties of rubber compound is the interaction between the filler and the rubber, which affects the rotational, braking and abrasion resistance. Silica filler has hydrophilic groups such as a silanol group on their surface which has a low affinity with hydrophobic rubbers. In order to solve this problem, researches on an efficient silane coupling agent (SCA) has been continuously carried out. In this study, methoxy type SCAs which are expected to show higher hydrolysis efficiency were synthesized. Their structures were analysed by gas chromatography-mass spectrometry (GC/MS) and nuclear magnetic resonance (1H-NMR) spectroscopy. In addition, their hydrolysis efficiency and condensation side reaction in SBR wet master batch were examined by fourier transform infrared spectroscopy (FT-IR) and gel permeation chromatography (GPC), respectively.

Mechanical properties of Cellulose Nanofibers- Graphene oxide Biocomposite

Mianehow Hanieh [(1)]*, **Berglund Lars** [(1)]

[(1)] Wallenberg Wood Science Center, KTH Royal Institute of Technology - Stockholm - Sweden

Graphene with its planar structure and exceptional mechanical and electrical properties has attracted a great deal of attention since 2004. Many researchers have tried to utilize this nanomaterial in different polymeric matrices to make high performance nanocomposites. However, the lack of hydrophilic functional groups on graphene surface and strong ð-ð interaction between graphene sheets restrict its dispersion in hydrophilic matrices, resulting in poor mechanical properties. The hydrophilic derivative of graphene, graphene oxide (GO), has high affinity to any hydrophilic matrices and improve mechanical properties of the resultant nanocomposite. On the other hand, cellulose nanofibers (CNFs) also have excellent properties such as high aspect ratio, high crystallinity, low density and high elastic modulus. Therefore, the objective of this work was to take advantage of superior properties of GO and CNF to make a new bio-based composite with exceptional mechanical properties. To study the mechanical properties of such composite, different samples containing 0.1 - 4 wt% of GO were prepared by vacuum filtration and characterized using Scanning electron microscopy (SEM), Atomic Force Microscopy (AFM), X-ray diffraction spectroscopy. Then, the mechanical properties of the resultant composites were investigated by tensile testing in 50% relative humidity. Results showed that by addition of only 0.1 wt% of GO to CNF, the modulus of the nanocomposite reaches to 16 GPa from 9,3 GPa for pure CNF and by adding more GO the modulus reaches a plateau and start to decrease. Comparison of microscopy and mechanical testing shows that the high modulus of the resultant composite is coming from the layered structure of the nanocomposite and the effective dispersion of GO that results in efficient stress transfer between GO and CNF.

Fatigue and Structural Health Monitoring of Delamination in Composite Materials

Kravchenko Oleksandr (1), Pedrazzoli Diego (1), Manas-Zloczower Ica (1)*

(1) Case Western Reserve University - Ohio - USA

Structural health monitoring of delamination in polymer based composite laminates is an important problem for ensuring safe operation of composite based structures. The use of carbon nanofillers dispersed in polymeric matrix has shown promise as a way of structural health monitoring by correlating changes in electrical resistance with the accumulated damage in the composite. In the present work we use a new approach employing carbon nanostructure (CNS) buckypapers (BP) to prepare glass fiber/epoxy composite laminates with enhanced resistance to delamination and damage monitoring capability. Selective reinforcement and structural health monitoring in composite laminates is achieved by introducing the CNS-BS at the interlaminar region using a method of frictional sliding. The changes in the compliance of short beam samples subjected to fatigue loading in the three point bending configuration due to delamination were correlated with the changes in electrical resistance. Delamination in the sample is manifested in increased electrical resistance caused by the breakage of the conductive interlaminar layer with carbon nanofillers. The electrical resistance results show correspondence with delamination crack nucleation, followed by stable and unstable crack growth. Crack propagation in the three point bending configuration shows a combination of opening and sliding modes of fracture. The sliding, (Mode II) resulted in partial closing of the delamination cracks during cycling. The mechanics of crack closing during fatigue and in the unloaded configuration was also studied using finite element analysis to help understand the experimentally observed partial recovery of the electrical conductivity. The proposed method of structural health monitoring offers a promising technique for detecting and examining delamination damage in composite laminates in fatigue life. Keywords: structural health monitoring, carbon nanostructures, delamination, electrical resistance, fatigue life

Orientation Development of Nanotubes During Processing of Polymer Nanocomposites

***Kenig Samuel* [(1)]*, *Tenne Reshef* [(2)], *Gomid Eyad* [(1)]**

[(1)] Shenkar College - Israel - Israel, [(2)] Weizmann Institue of Science - Israel - Israel

The orientation development of nanotubes (NTs) during flow of nanocomposite melts was studied in elongational as well as shear flows. Two types of NTs were included in the investigation: Multi-Wall Carbon Nanotubes (MWCNTs) and Tungsten di-Sulfide based Inorganic Nanotubes (WS2-INTs). INTs differ from CNTs with respect to its diameter (100 nm compared to 10nm) and length (1 micron compared to a few microns). Distinctively, pristine INTs surfaces are rich in sulfur and tungsten elements with a small fraction of oxygen. CNTs surfaces are composed of aromatic moieties that lead to agglomeration. The surface attributes of these two classes of nanotubes effect their rheological properties and dispersion in polymers. As the orientation of the NTs is the decisive factor determining the final nanocomposite properties (mechanical, thermal, electrical), it is of significant importance to study the mechanisms of orientation development of NTs containing polymer melts. The orientation state of the elongated NTs is governed by two principle flow fields that prevail in all processing methods namely: elongation and shear flows. Experimental results and derived modeling showed that the orientation development in elongational flow is enhanced in the case of INTs compared to CNTs as is the case in shear flow. A rigorous analysis of the time dependent orientation indicates that while the INTs demonstrate rigid like behavior the CNTs are affected by the relaxation of the host polymers. In addition, the formation of agglomerate structures in the case CNTs should be taken into account with respect to the orientation development. Based on the orientation state and resulting morphology a generic engineering design methodology and modeling could be formulated for the prediction of the final polymer nanocomposite properties.

Nanotubes in Immiscible Polymer Blends: Three Studies

Grady Brian (1)*

(1) University of Oklahoma - Oklahoma - USA

Our work in immiscible polymer blends has three parts. First, we have shown that it is possible to control which phase of an immiscible polymer blend a nanotube resides in by altering the surface chemistry of the nanotube. Second, we have shown that if a nanotube resides in the minor phase of an immiscible polymer blend, it is protected from breakage during mixing in a conical laboratory twin-screw extruder. In fact, even if the nanotube resides in the major phase, breakage is reduced. Third, we will present some preliminary results on the use of block nanotubes, that is a single nanotube having one block with one chemistry and another block with a second chemistry, on the modification of polymer blends.

Engineering Graphene with Polymers for the Preparation of Nanocomposites

Rubio Noelia [(1)*], ***Au Heather*** [(1)], ***Shaffer Milo S.P.*** [(1)]

[(1)] NanoHAC, Department of Chemistry, Imperial College London, London, SW7 2AZ - London - UnitedKingdom

Graphene has attracted increased attention in recent years due to its exceptional electronic, mechanical and thermal properties. The combination of graphene with existing polymer-based materials1 has afforded promising applications. Strong efforts have been made to produce single layer graphene with minimal framework damage. A useful approach would be to exfoliate and functionalise the graphene layers directly from graphite or few-layer graphene. Graphite intercalation compounds (GICs) are well-known precursors that afford isolated graphene layers avoiding framework damage2.In this work we have used this method in order to introduce charges into few-layer graphene starting material to create electrostatic repulsion. We have used different grafting techniques (grafting from and grafting to) that allow the introduction of well-defined polymers on the graphene layers. The results suggest that the grafting from approach results in an increase in the grafting ratio and solubility of the material in organic solvents. On the other hand, the grafting to approach showed an increase in the grafting ratio when using low molecular weight polymers; the solubility did not change significantly in this case. The conformation of the attached polymers was predicted using de Gennes' model3; the grafting from products would grow in a brush-like fashion according to this model, while the grafting to hybrids would follow a mushroom regime (where the polymer chain coils up on itself). Moreover, the functionalisation process is highly versatile and would allow the introduction of a wide variety of polymers and copolymers on graphene. The improved solubility of the material in acetone after functionalisation will allow better compatibility with an epoxy resin in order to prepare novel composites 1 Huang, X et al Chem. Soc. Rev.,2012,41,666. 2 Milner, EM, Skipper, NT, Howard, CA, Shaffer, MSP, J.Am.Chem.Soc., 2012,134,8302. 3 Degennes, PG Macromolecules,1980,13,1069.

Tuning the Phase Separated Morphology and Resulting Electrical Conductivity of Carbon Nanotube Filled Blends with a Random or Block Copolymer

Moldenaers Paula [1]*, Bharati Avanish [1], Cardinaels Ruth [2]

[1] KU Leuven - Leuven - Belgium, [2] TU Eindhoven - Eindhoven - Netherlands

The need for conducting polymeric materials in applications such as flexible electronics is on the rise. Immiscible blends containing conducting particles offer a generic route to tune the electrical functionality by means of tailoring their morphology combined with selective localization of the particles. The conductivity of biphasic blends consisting of PáMSAN and PMMA containing multiwall carbon nanotubes (MWNTs) was enhanced by tailoring the blend morphology with a PS-PMMA copolymer. The PáMSAN/PMMA blends undergo phase separation by spinodal decomposition during which the MWNTs preferentially localize in the PáMSAN phase. The effects of the architecture (block or random) and molecular weights of the copolymers on the morphology, linear viscoelastic moduli and the resulting blend conductivity were systematically investigated at various copolymer concentrations. Effective compatibilization is achieved with block copolymers independent of their molecular weight. Short random copolymers are not effective, whereas long random copolymers are as effective as the long block copolymers. An effective compatibilizer led to interfacial tension mediated suppression of coarsening of the PáMSAN domains, resulting in increased connectivity and refinement of the PáMSAN phase containing the MWNTs, as confirmed by optical microscopy and the linear viscoelastic response of the blends. A pronounced increase by 3 decades in magnitude of the low frequency electrical conductivity was achieved upon addition of only 0.25 wt% long random copolymer to blends with 0.5 wt% MWNTs. Furthermore, the copolymer concentration required to develop a percolating network of MWNT-MWNT and PáMSAN-MWNT are approximately the same, independent of the type and molecular weight of the copolymer, suggesting that the electrical and rheological percolations are comparable. In conclusion, the ability of the copolymer to refine and stabilize the PáMSAN domains governs the blends electrical conductivity.

Mechanical and electrical properties of carbon nanoparticle modified epoxy matrix with regard to the size effect

Leopold Christian [(1)], Augustin Till [(1)], Schwebler Thomas [(1)], Lehmann Jonas [(1)], Liebig Wilfried V. [(2)], Fiedler Bodo [(1)*]

[(1)] Hamburg University of Technology - TUHH - Hamburg - Germany, [(2)] Karlsruhe Institute of Technology - KIT - Karlsruhe - Germany

The influence of nanoparticle morphology and filler content on the mechanical and electrical properties of carbon nanoparticle modified epoxy matrix is investigated regarding the size effect. Three types of particles, representing spherical, tubular and layered morphologies are used. A clear size effect of increasing true failure strength with decreasing volume is found for neat and carbon black modified matrix. Carbon nanotube (CNT) modified matrix show high potential for stress increase, but dispersion and purity are critical. In few layer graphene modified matrix fibres, particles are larger than statistically distributed defects and initiate cracks, counteracting any size effect. Different toughness increasing mechanisms on the nano- and micro-scale depending on particle morphology are discussed based on scanning electron microscopy images. Electrical percolation thresholds in the small volume fibres are significantly higher compared to bulk volume, with CNT being found to be the most suitable morphology to form electrical conductive paths. Good correlation between electrical resistance change and stress strain behaviour under tensile loads is observed. The results show the possibility to detect internal damage in small volumes by measuring electrical resistance and therefore indicate to the high potential for the use of CNT modified polymers of fibre reinforced plastics as a multifunctional, self-monitoring material.

NEW INSIGHTS FROM THE OPTIMIZATION OF MELT MIXING PARAMETERS IN POLYPROPYLENE/ MULTI-WALLED CARBON NANOTUBE NANOCOMPOSITES

Lison Valérie (1)*, ***Hurtgen Marie*** (1), ***Kchit Nadir*** (1), ***Rul Alicia*** (1), ***Claes Michaël*** (1)

(1) Nanocyl - Sambreville - Belgium

In polymer nanocomposites containing carbon nanotubes (CNT), the critical challenge remains to manage the dispersion of CNT in most of thermoplastics resins, in terms of size and number of agglomerates. Indeed, it is quite obvious to achieve the percolation threshold above a critical filler volume fraction which coincides with the formation of a conduction network of filler particles in the continuous polymer phase. However, despite intensive studies on this field, the higher level of CNT's dispersion required by a growing number of applications in various industries has not yet been reached. It is typically the case, for example, in melt spinning process or automotive sector. In this work, the research will be focused on processing parameters during melt mixing in twin-screw extruder to obtain a disentanglement and uniform dispersion of CNT. It includes feed method of CNT (through a masterbatch or by direct melt mixing), feed position, screw profile, feed rate, screw speed, profile temperature, residence time. The relationship between these tailored parameters and the effective dispersion of CNT will be evaluated through an optical control system. This system detects on a processed film via a camera the number and the size of defaults by unit of surface. The strategy is to develop the expertise on a commodity thermoplastic with a high consumption and a large variety of applications. It's why the polypropylene was an ideal choice for this study. Its apolar behavior makes the objective even more likely complex. Different grades will be evaluated to identify the influence of resin's viscosity, correlated with melt mixing parameters. In this context, Nanocyl will be able to develop customized masterbatches dedicated to the final applications, with on the enhancement of CNT's dispersion.

Carbon nanotube fibres, fifteen years of progress with challenges and prospects for the future

Windle Alan [(1)]*

[(1)] Cambridge University - Cambs. - UnitedKingdom

There is no doubt about the strength of the carbon-carbon covalent bond, which coupled with the lightness of the atom, makes it a perfect subject for high strength materials. The initial strength hype over carbon nanotubes (CNT), now being repeated in some circles with graphene, arose from measurements of the strength of individual graphitic layers, and while the scientists were usually (but not always) professionally explicit as to what they had measured, the publicity machines assumed that these figures could also be readily realized in any material made from nanotubes. It is against a background of this overselling that we have watched the gradual improvement in strength of useful materials, especially yarn-like fibres made of carbon nanotubes. Fibres made by the direct spinning process have been used as the subject of this study. If CNT fibres are to displace carbon fibre on a world scale, it will be firstly on the basis of cost, as the direct spinning process for nanotube fibre is eminently scaleable. Secondly, it is the toughness of the fibre which is likely to be one of its major selling points, so while matching existing fibres in tensile properties, it promises to be much more resilient to damage and very much tougher in any case of non-axial loading. Strategies for further development are outlined, as will be the prospects for tuning properties for particular applications. Finally, there will be a personal view as to where it all might be leading.

Understanding the Enhancement of Young's modulus of Macroscopic Carbon Nanotube Fibers After Polymer Infiltration

Bartolomé Mas[1], Alfonso Monreal-Bernal[1], Hangbo Yue[1,2], Haiquin Zhang[3] and Juan J. Vilatela[1,a)]

[1]*IMDEA Materials Institute, Eric Kandel 2, Getafe, Madrid, Spain, 28906*
[2]*Departamento de Ciencia de Materiales, Technical University of Madrid, Madrid, Spain 28040*
[3]*Beihang University, 37 Xueyuan Rd, Haidian Qu, Beijing Shi, China, 100191*

[a)]Corresponding author: juanjose.vilatela@imdea.org

Abstract. Macroscopic fibres of carbon nanotubes (CNTF) are prone to polymer infiltration into their internal pores by capillary forces, often leading to an increase in fibre tensile properties. Here we study CNTFs infiltrated with different polymers *via* solution infiltration. The results show increases in specific modulus of up to 50 % after polymer infiltration. This increase occurs due to improved stress transfer between load-bearing elements in the fibres, and saturates for a fibre mass fraction of around 0.85. The modulus of the composites fibres can be described by a rule of mixtures considering an effective fibre modulus, higher than that of the dry fibre and dependant on the polymer matrix modulus. The dependence of this effective modulus on matrix properties can be estimated from *in situ* Raman spectroscopy measurements during axial tensile testing. A comparison of composite fibres with different polymer matrices shows that the rate of downshift of the 2D Raman band increases almost linearly with increasing matrix modulus. Their dependence provide a direct relation between Raman measurements and the modified rule of mixtures, and is thus close to enabling prediction of composite properties in arbitrary matrices.

INTRODUCTION

There is an ever increasing plethora of nanomaterials becoming available in sufficient quantities to integrate them into macroscopic structures, typically in combination with a matrix to form a nanocomposite. Examples of nanofillers include the widely-studied nanocarbons (CNTs, graphene) [1, 2, 3], metal nanowires [4], inorganic nanowires [5], organic whiskers [6], etc. Nanocomposites have been widely seen as a promising route to translate the exciting properties of nanomaterials into macroscopic materials with improved mechanical and electrical properties. In this quest, it has become apparent that dispersing nanoparticles as fillers becomes rapidly ineffective at volume fractions above 1 % due to aggregation and processing difficulties intrinsic to nanoparticle dispersion [1, 2, 7]. Instead, the controlled assembly of nanobuilding blocks into macroscopic structures with preferential alignment parallel to the nanocarbon axis of interest (nanotube/wire axis or in-plane for 2D material) has proven more effective. Continuous fibers and films made up of CNTs, graphene, and their combinations are now routinely produced by wet spinning [8, 9], drawing from arrays [10], and direct spinning from CVD [11, 12], some of them in semi-industrial quantities.

CNT fibers, under development for over 14 years, already rival or outperform reference high-performance materials in terms of: specific mechanical properties (similar to carbon fiber)[13], electrical conductivity (2.5 of copper per unit mass [14]), thermal conductivity (above copper [15]) and composite performance [16, 17, 18, 19, 20]. In addition, they have demonstrated outstanding behavior as actuators, sensors, electrodes, and in general in applications involving charge transfer/storage [21].

A unique aspect shared by these fibers and films is the combination of high-performance properties and an exceptionally high surface area and porosity. At around $250\,m^2\,g^{-1}$ [22], the specific surface of CNT fibers is 1000 greater than that of a traditional material such as a CF filament, with the implication that their interfaces with gases, liquids and solids is also exceptionally large and thus plays a critical part in the physical properties of the macroscopic ensemble. This porosity is reminiscent of the bottom-up assembly process and results in a hierarchical structure [23] made up individual nanobuilding blocks that associate in nanocrystals (e.g. bundles), which then form microfibrils and subsequently a filament. Such a hierarchical structure is perhaps similar to a natural fibre [24] or a cotton yarn but otherwise unprecedented as a synthetic material.

As a result of this yarn-like porous structure these materials have outstanding flexibility in bending [25] and can therefore be knitted and knotted, they can host multifunctional nanoparticles, and can readily be infiltrated with liquids [26], polymers [27] and polymer electrolytes [28]. The ingress of liquid and polymer molecules into the fibers has been shown to lead to sensing properties [29].

In the context of composite fabrication, capillary forces are strong enough to ensure infiltration of polymer molecules in solution or molten state into the fibres, as observed for example by inspection of FIB-milled composite cross-sections [18]. This relatively simple integration of CNT fibres with polymer matrices has resulted in composites with tensile strength and Young's modulus as high as $3.0\,GPa\,SG^{-1}$ and $234.4\,GPa\,SG^{-1}$ [17], respectively, without recourse to chemical treatments to improve fibre/matrix adhesion, although so far realized in very thin specimens with thickness in the micron range. In spite of these encouraging results, it is surprising how little is understood about the mechanical properties of these fibers when infiltrated with different polymers and liquids. Experiments show the presence of polymer inside fibre pores to produce increases in axial fibre properties compared to the dry (i.e. polymer-free) state, such as in Young's modulus, tensile strength and fracture energy [30, 31, 32, 33]. These effects highlight the fact that the properties of the fibres are strongly dependant on the matrix they are embedded in, which is unusual for high-performance fibres, whose tensile properties are well defined and identical with and without the presence of polymer at the filament level. Recent modelling work using coarse-grained molecular dynamics applied to 30nm-wide "fibres" successfully predicts large increases in tensile properties after infiltration. Yet, this effect is largely attributed to the polymer compressing the fibres and thus increasing inter-CNT stress transfer via a short separation between adjacent CNTs, although such radial fibre shrinkage is not observed in real CNT fibre composite systems. Overall, the strong effect of polymer infiltration contrasts with the lack of a framework to describe it, which is the more important considering its immediate implications for the use of CNT fibres in structural polymer composites.

In this work we first show the infiltration of polymers and liquids into CNT fibre arrays to be described by an empirial modified Lucas-Washburn expression, which confirms the filling of pores by a range of molecules. Then, we show that the introduction of polymer into the fibers enhances stress transfer between the load-bearing elements. By comparing the Raman spectra during *in-situ* tensile testing of specimens, we establish the role of the polymer as a stress transfer element. Results for CNTF-composites with different polymer matrices, hence having different bulk Young's modulus, demonstrates that stress transfer is not univocally determined by the fiber properties, but depends on the modulus of the embedding matrix. We then propose a simple model to determine fiber properties for arbitrary matrices based on a rule of mixtures and using an experimentally-determined correction factor for the infiltrated fibre based on *in-situ* Raman spectroscopy measurements during tensile testing. This would enable the subsequent use of standard composite theory to predict properties of large composite structures with multiple filaments, such as a non-woven unidirectional lamina, the basic element in laminate composites.

EXPERIMENTAL DETAILS

CNT fibers were produced by the direct spinning process, whereby a fibre is produced by drawing an aerogel directly from the gas-phase during growth by chemical vapour deposition [11]. Ferrocene, thiphene and butanol were used as catalyst, promoter and carbon source, using conditions leading to the formation of highly graphitic CNTs with around 3-5 layers, as described earlier [12].

Two types of samples were prepared: individual (around 10-micron diameter) CNT fiber filaments and films of mulitple filaments. In both cases samples were collected directly onto a plastic support frame, which enabled further densification with acetone without inducing undesired samples stretching. The dynamics of liquid spread into CNT fibre films were recorded with a high resolution digital camera placed just above the specimen. The advance of the liquid front could be monitored using standard image analysis software (ImageJ). The liquids used consisted of: benzyl alcohol (BA), N-methyl-2-pyrrolidine (NMP), Poly(ethylene glycol) (PEG), Epoxy(D.E.R.736) resin.

In order to produce low polymer content composite samples, films were infiltrated with polymer solutions by direct deposition of droplets with a high-precision micropipette. The content could be adjusted by varying the polymer concentration in the solution and with knowledge of injected volume. The thermoplastic matrices used were thermoplastic polyurethane (TPU), PEG and polystyrene (PS), which were dissolved in dichloromethane. The epoxy consisted of monocomponent aerospace resin RTM6, which was cured at room temperature after 24 hours. For infiltration experiments uncured epoxy was dissolved in acetone. For Raman measurements individual filaments were directly immersed in the polymer solutions. Electron microscopy was carried out with an FIB-FEGSEM Helios NanoLab 600i (FEI) at 5 kV. Tensile tests of individual filaments were performed with a Kammrath and Weiss microtensile machine and a MTS Nanobionix, with 1 and 0.5 N as maximum load, respectively. For film specimens, tensile tests were conducted using an Instron 5966 dual column tabletop universal testing system. Samples were tested directly on the PTFE frames with a gauge length of $35 \pm 1\,mm$. In all cases, the strain speed was $3\%\,min^{-1}$.

The Kammrath and Weiss microtensile machine was also used to apply a controlled elongation in a stepwise manner to the individual filaments of CNT fiber, in order to relate the 2D Raman shift to the applied strain. After each step Raman spectra were recorded with a Jasco NRS-5100 spectrometer, using a 20× objective lens and a 532 nm laser (2.33 eV) as excitation source (for 15 s and 3 accumulations). The laser power was kept below 0.6 mW to avoid increasing the temperature of samples and the possible damage of the polymer structure. The polarization of the laser was kept parallel to the fiber axis with the aim to maximize the signal. Three spectra were taken for each deformation step, and the spacing between them was 50 μm. A visual reference was used to locate the probing area after each deformation step.

RESULTS AND DISCUSSION

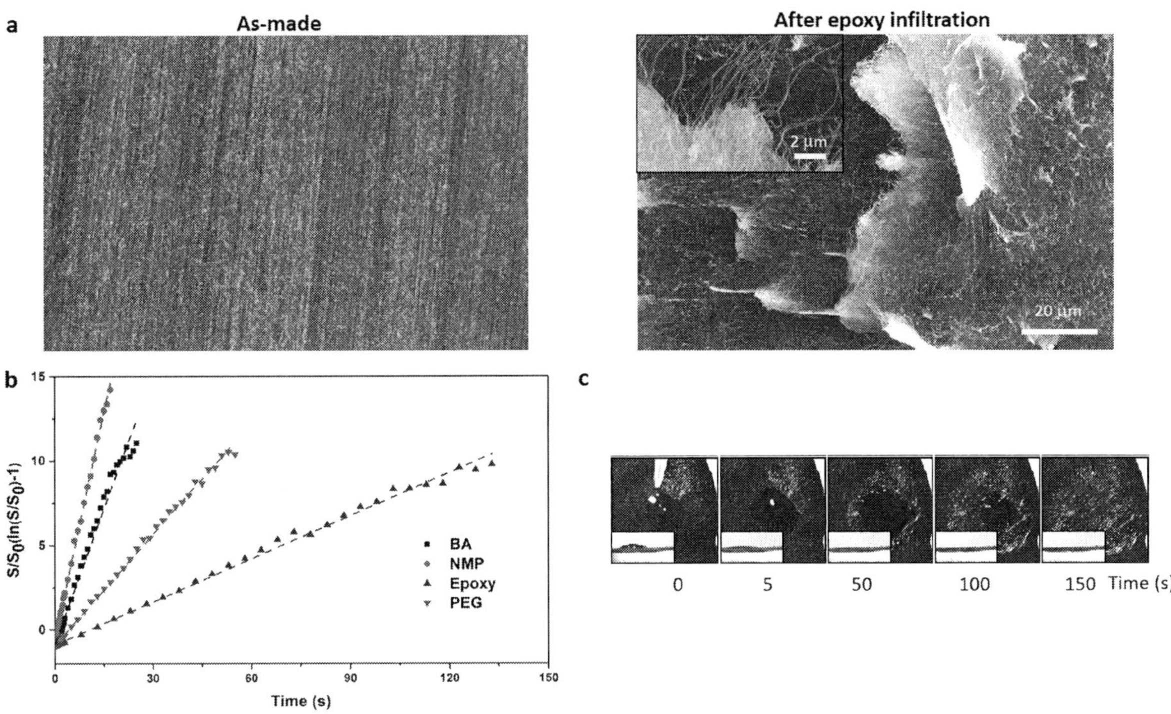

FIGURE 1. Wicking of liquids in CNT fibre films. a) Dynamics of liquid spreading obtained from image analysis using different liquids, with experimental data fitted to an empirical modified Lucas-Washburn equation. b) Examples of optical micrographs of epoxy spreading into CNT fibre films (inset shows side view). c) Electron micrographs of samples as made (left) and after epoxy infiltration (right).

The porous structure of CNT fibers and the consequent strong interaction with foreign molecules can be observed in Figure 1. SEM micrographs (Figure 1a-left) of a dry fibre (i.e. non-infiltrated) show the preferential alignment of CNTs parallel to each other and to the fiber axis which nevertheless are imperfectly packed and therefore give rise to a substantial porosity and surface area accessible to small molecules. This porosity provides strong capillary forces for the infiltration of liquids and polymers. Figure 1a-right shows extensive coverage of the internal structure of a

CNT fibre by epoxy, as also observed elsewhere [20]. To illustrate this effect, in Figure 1b we present data obtained from optical imaging of the spreading of a drop of epoxy resin and other liquids in a CNT fibre film. These liquids rapidly wick into the film and fill the internal pores of the fibers as a consequence of capillary forces arising from the fibre pore structure (Figure 1c). The wicking rate in this configuration can be accurately described by a modified Lucas-Washburn equation of the form [34]

$$\frac{S}{S_0} \cdot \left(\frac{\ln S}{S_0} - 1 \right) = -1 + \left(\frac{2\pi r \gamma \cos\theta}{3\eta S_0} \right) \cdot t \tag{1}$$

where S_0 is the initial wicking area covered by the droplet at time zero, S the wicking area at time t, r the effective capillary radius of the fiber, γ the liquid surface tension, θ the Young contact angle and η the liquid viscosity. Figure 1c presents a plot of $S/S_0 \cdot (lnS/S_0 - 1)$ against t for a range of liquids with different viscosities and contact angles. The experimental data can be accurately fit by equation 1, implying that the liquid flow can be treated as a conventional (meso)macroscopic flow. The value of r is determined from the slope of the linear fitting, which comes out about $15 \sim 40\,nm$ (Table 1), in agreement with measurements on indivdual CNT fibre filaments [35] and in the range observed in electron micrographs and obtained from gas adsoprtion measurements [23]. We note, however, that this calculated radius represents the *average* pore radius encountered by the liquid flow, but in reality the pores are highly irregular and reflect a wide distribution of interbundle separations. This implies that initial filling of macropores can proceed faster than the bulk wicking rate.

TABLE 1. Parameters of different liquids wicking CNT fibre films

Liquid	Contact angle (°)	Surface tension (mN/m)	Viscosity ($mPas$)	Capillary radius (nm)
Benzyl alcohol	64.4 [36]	36.8	5.47	43
N-methyl-2-pyrrolidine	10 [26]	44.6	1.67	14
Poly(ethylene glycol)	25 [37]	43.5	95.6	26
Epoxy(D.E.R.736)	42.5 [38]	50	51	16

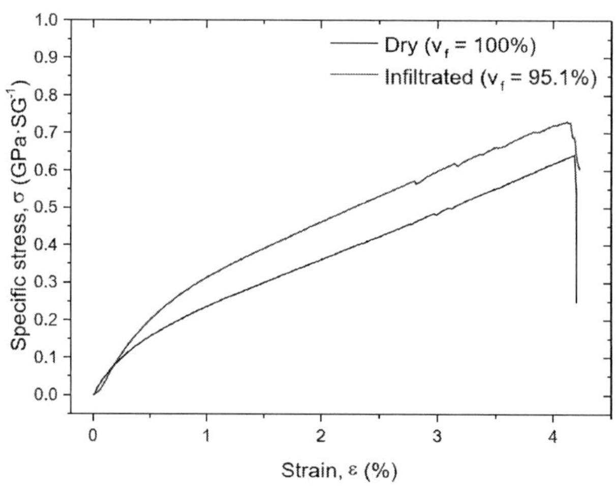

FIGURE 2. Stress-strain curves at 35 mm gauge length of dry CNT fibre film (i.e. non-infiltrated) (black) and epoxy-infiltrated CNT fibre film with fiber mass fraction (w_f) of 0.95 (blue).

The results above highlight the porous nature of CNT fibres as evidenced by their rapid uptake of foreign molecules. Furthermore, it thus seems inevitable that during composite fabrication polymer molecules penetrate into their structure. The first question then is how the polymer contributes to the mechanical properties of the infiltrated fiber. We study this by plotting specific stress-strain curves of CNTFs before and after infiltration and curing of an epoxy resin (Figure 2). The samples have been purposely infiltrated with dilute polymer solutions so as to produce composites with a low polymer volume fraction and thus fibre-dominated properties. The use of specific properties

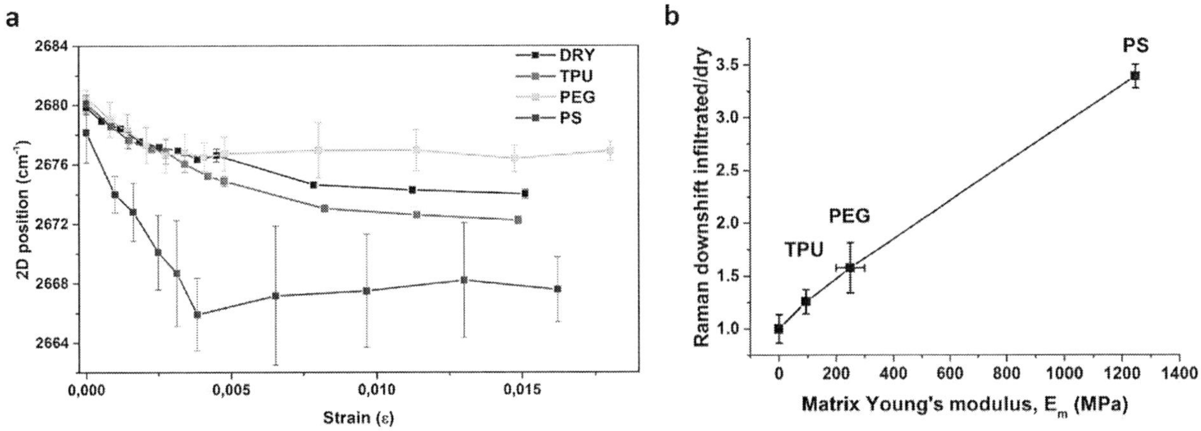

FIGURE 3. (a) 2D Raman peak position under tensile strain of dry individual filaments of CNT fibre (black), and embedded in TPU (red), PEG (green) and PS (blue). (b) Plot of magnitude of 2D Raman peak position downshift per unit tensile strain, versus Young's modulus of the infiltrated polymer.

requires measurements of sample linear density but avoids defining cross sections for the samples and thus improves accuracy. More importantly, the increase in specific modulus observed after infiltration provides a direct evidence of enhanced stress transfer, as opposed to an increment in the more common volume-normalised mechanical properties, which could simply reflect the filling up of voids inside the material. Most reports on improvements in tensile properties of CNT fibres upon polymer infiltration obviate this simple but important distinction.

Such enhancement could be directly measured by *in-situ* Raman spectroscopy measurements during tensile testing [39, 40, 41]. Under tensile (compressive) strain, Raman active modes are down (up)shifted from their original position as a consequence of changes in CNT lattice vibrational frequencies (i.e. phonon modes). The rate of Raman peak position shift with respect to bulk strain represents a measure of the effective modulus of the material.

Figure 3a presents a plot of the 2D Raman mode position against the fiber strain for both, the dry fiber and infiltrated with TPU and PS, as soft and hard matrices, respectively. All graphs preserve the elastoplastic behaviour observed in Figure 2 and confirm that the addition of polymer increases stress transfer to the CNTs, as indicated by the increase in the magnitude of the rate of Raman shift change. The calculated slopes magnitude in the initial part of the curve come out as 8.65 ± 1.17, 10.86 ± 1.25, 13.64 ± 3.2 and $29.35 \pm 3.35\, cm^{-1}\, \%^{-1}$ for the dry, TPU-infiltrated and PS-infiltrated, respectively. These values are in agreement with previous reports [30]. The Raman data show a clear correlation between the enhancement in fibre modulus and the modulus of the polymer matrix (Figure 3b). Further experiments are required to establish an analytical relation between the two parameters, but we note that the relation is nearly liner, in contrast with shear-lag models, where matrix stress transfer to the reinforcing fiber by shear scales with $\sqrt{E_m}$.

The higher values of fibre modulus (tensile and through Raman) observed after polymer infiltration indicate that the polymer can effectively transfer stress to regions of the fibre that otherwise shared little load in the dry state. Although Raman measurements correspond to surface bundles, this situation mimics what is found at the inner fiber portions, where adjacent bundles previously separated by pores are joined by polymer after infiltration. Overall, the infiltrated fibre can be seen as two interpenetrating phases. The enhancement of stress transfer extracted from Raman measurements would thus reflect the situation across the whole polymer/fibre interface, including its large component "inside" the fibre.

However, the absolute rates of Raman downshift must be taken with caution. In monolithic materials $\left(\frac{\Delta Pos(2D)}{\Delta\epsilon}\right)$ is linearly related with bulk modulus. For different grades of carbon fibre, for example, $E = \left(\frac{\Delta Pos(2D)}{\Delta\epsilon}\right) x \frac{1000TPa}{60cm-1percent}$ [42]. This expression would also apply to individual CNTs or graphene platelets, for which the local deformation can usually be determined. However, in macroscopic ensembles of CNTs (or graphene), the stress transfer throught the material need not be uniform, particulalry in the presence of polymer [39, 43]. Hence, local strains can differ from the measured bulk strain (fibre elongation). The implication is that the observed Raman shift rate would still provide a measure of stress transfer but would not necessarily reflect the bulk modulus of the material. This can be observed by plotting experimental values against fibre modulus obtained from tensile tests and comparing with CF values. These data are shown in Figure 4.

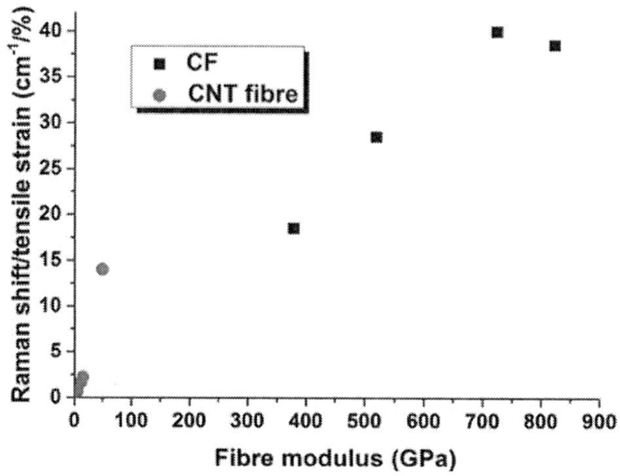

FIGURE 4. Magnitude of 2D Raman mode downshift with tensile strain, plotted agains fibre tensile modulus for CNTF and CF. Experimental values for CNTF deviate from the linear behaviour observed for CF because of non-uniform stress distribution in the material. Data for CF from [44]

An important implication of Figure 3 is that unlike traditional fibers (e.g. CF) which have defined properties that can be used to predict composite properties, CNT fibers have properties that depend on the choice of matrix and the extent to which it infiltrated the fiber porous structure. This situation is a general consequence of polymer infiltration into a porous structure which occurs in materials such as conventional Yarns [45] and macroscopic fabrics. We turn to the effect of the polymer mass fraction on fiber and composite properties to address this uncertainty and obtain a corrected fiber modulus that can be used to predict properties of systems of multiple fibers in a matrix such as a traditional unidirectional laminate. These experiments also help us determine the optimum fraction to enhance fiber properties.

Figure 5 presents a plot of the modulus of fiber film against fiber mass fraction. As shown also schematically, the system starts as a dry fiber ($w_f = 100\,\%$) but is gradually filled up with matrix as w_f decreases. The fiber/composite modulus increases as the polymer fills up the pores and binds bundles, with modulus peaking at a critical value around ($w_{fc} \approx 90\,\%$) and then decreasing as the matrix starts adding weight surrounding the fiber. The shape of E vs w_f in Figure 5,is also an indication of much the polymer inside the fiber is acting as an adhesive that requires only to bridge across bundles rather than filling up the entire pore to enhance stress transfer. This distinction is important because it provides a design parameter suggesting that the ideal structure corresponds to highly aligned bundles with a small separation between them sufficiently large only to host the adhesive polymer. There are reports on the improvement in fiber properties by adding polymer from solution at the point of densification of the array of nanotubes drawn from a forest [17, 46]. The observation in those references that the properties of the final composite fiber peak at intermediate v_f values is another example of the behavior observed in Figure 5.

Clearly, if the modulus E_{fc} at v_{fc} is known, then the material can be treated as a non-porous fiber and the properties of composite structures predicted using well established composite models. However, producing a similar plot for each fiber and matrix combination is not practical. Instead, as a first approximation we can use the Raman data to correct the dry fiber modulus and obtain an effective fibre modulus and its dependence on matrix properties.

The rule of mixtures can still be used, considering the effective fibre modulus after polymer infiltration (E_f^{eff})

$$E_C = E_f^{eff} v_f + E_m(1 - v_f) \tag{2}$$

With the E_f^{eff} given by the modulus of the dry fibre (i.e. polymer-free) and the additional contribution to the modulus arising from the polymer matrix transferring stress between elements in the fibre

$$E_f^{eff} = E_f^{dry} + aE_f^p \tag{3}$$

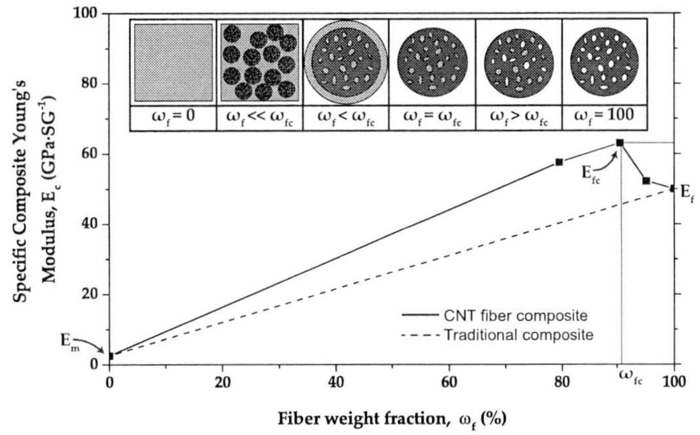

FIGURE 5. Plot of Young's modulus of an epoxy-CNT fibre film composite against polymer mass fraction (w_p). The dashed line shows the rule of mixtures that describes the behaviour of traditional composites. Continuous line shows the results obtained in this work. Where E_f represent de Young's modulus of non-infiltrated CNT fibre, E_m the neat polymer and E^* is the Young's Modulus corresponding at critical polymer mass fraction (ν_{pc}).

where E_f^p represents the modulus of the matrix-reinforced elements in the fibre. a is a constant that modulates the relative contribution of matrix reinforced elements relative to dry elements, and therefore related to fibre porosity and specific surface area. The stress transfer between two load-bearing elements connected by the polymer matrix is also governed by the matrix properties and thus the second term in equation 3 is related to the shear modulus of the matrix. This dependence is implicit in the rates of Raman downshift observed in Figure 3b. To a first approximation, we consider a linear dependence between $\frac{\Delta Pos(2D)}{\Delta\epsilon}$ and E_M, leading to

$$E_f^{eff} = E_f^{dry} + bE_M \tag{4}$$

where the constant b represents the increase in fibre modulus for a given matrix modulus, proportional to the enhancement determined from Raman.

$$b \propto \frac{\Delta \frac{\Delta Pos(2D)}{\Delta\epsilon}}{\Delta E_M} \approx 0.017 cm^{-1}/MPa \tag{5}$$

The rule of mixture for the composite modulus can thus be expressed in terms of (dry) fibre and matrix properties and the constant factor b.

$$E_C = E_f v_f + E_m(1 - v_f + b) \tag{6}$$

Equation 6 has a pleasant simplicity, and the merit of being suitable for the prediction of properties of larger composite architectures such as unidirectional laminae. Still, embedded in b is a complex dependence of the enhancement in fibre modulus on fibre structure, porosity and matrix shear modulus. A more detailed analysis of these factors is underway.

Conclusions

The results and discussion above rationalize the effects of polymer infiltration into porous CNT fibers and provide a first method to predict the mechanical properties of larger composite structures made up of them. But these results also reveal details about the deformation mechanism of the fiber, which are in principle applicable to other macroscopic assemblies of weakly interacting nanobuilding blocks imperfectly packed.

This simple model currently ignores fiber/matrix chemistry, degree of infiltration of large molecular weight polymers and requires to know the *in-situ* Raman data during tensile testing. Work is in progress to take these factors into account. In spite of these limitations, through the correction of the dry fiber modulus it is possible to obtain a first prediction of the properties of the fiber in a polymer matrix and thus engineer more complex structures with arrays of fibers and thus begin to exploit the properties of these nanostructured materials in structural applications.

ACKNOWLEDGMENTS

The authors are grateful to V. Reguero for fabrication of samples and acknowledge generous financial support from the European Union Seventh Framework Program under grant agreements 678565 (ERC-STEM), FP7-People-Marie Curie Action-CIG (2012-322129 MUFIN) and from MINECO (MT2012-37552-C03-02, MAT2015-62584-ERC, MAT2015-66888-C3-3-R, RyC-2014-15115).

REFERENCES

[1] J. J. Vilatela and D. Eder, ChemSusChem **5**, 456–78 (2012).
[2] J. N. Coleman, U. Khan, W. J. Blau, and Y. K. Gun'ko, Carbon **44**, 1624–1652 (2006).
[3] R. J. Young, I. A. Kinloch, L. Gong, and K. S. Novoselov, Composites Science and Technology **72**, 1459 – 1476 (2012).
[4] Y. Tang, S. Gong, Y. Chen, L. W. Yap, and W. Cheng, ACS nano **8**, 5707–5714 (2014).
[5] Y. Dou, K. Lin, and J. Chang, Nanoscale **3**, 1508–11 (2011).
[6] L. Tang and C. Weder, ACS applied materials & interfaces **2**, 1073–80 (2010).
[7] M. Shaffer and I. A. Kinloch, Composites Science and Technology **64**, 2281 – 2282 (2004).
[8] R. Cruz-Silva, A. Morelos-Gomez, H.-I. Kim, H.-K. Jang, F. Tristan, S. Vega-Diaz, L. P. Rajukumar, A. L. Elías, N. Perea-Lopez, J. Suhr, M. Endo, and M. Terrones, ACS nano **8**, 5959–67 (2014).
[9] Z. Xu and C. Gao, Nature communications **2**, p. 571 (2011).
[10] M. Zhang, K. R. Atkinson, and R. H. Baughman, Science **306**, 1358–1361 (2004).
[11] Y.-L. Li, I. a. Kinloch, and A. H. Windle, Science **304**, 276–278 (2004).
[12] V. Reguero, B. Aleman, B. Mas, and J. Vilatela, Chemistry of Materials **26**, 3550–3557 (2014).
[13] K. Koziol, J. Vilatela, A. Moisala, M. Motta, P. Cunniff, M. Sennett, and A. Windle, Science **318**, 1892–1895 (2007).
[14] K. L. Stano, K. Koziol, M. Pick, M. S. Motta, A. Moisala, J. J. Vilatela, S. Frasier, and A. H. Windle, International Journal of Material Forming **1**, 59–62 (2008).
[15] T. S. Gspann, S. M. Juckes, J. F. Niven, M. B. Johnson, J. A. Elliott, M. A. White, and A. H. Windle, Carbon **114**, 160 – 168 (2017).
[16] T.-W. Chou, L. Gao, E. T. Thostenson, Z. Zhang, and J.-H. Byun, Composites Science and Technology **70**, 1–19 (2010).
[17] X. Wang, Z. Z. Yong, Q. W. Li, P. D. Bradford, W. Liu, D. S. Tucker, W. Cai, F. G. Yuan, and Y. T. Zhu, Materials Research Letters **1**, 19–25 (2013).
[18] J. J. Vilatela, R. Khare, and A. H. Windle, Carbon **50**, 1227–1234 (2012).
[19] Y. Han, X. Zhang, X. Yu, J. Zhao, S. Li, F. Liu, P. Gao, Y. Zhang, T. Zhao, and Q. Li, Scientific Reports **5** (2015), 10.1038/srep11533, cited By 9.
[20] A. Mikhalchan, T. Gspann, and A. Windle, Journal of Materials Science **51**, 10005–10025Nov (2016).
[21] J. J. Vilatela and R. Marcilla, Chemistry of Materials **27**, 6901–6917 (2015).
[22] E. Senokos, V. Reguero, J. Palma, J. J. Vilatela, and R. Marcilla, Nanoscale **8**, 3620–3628 (2016).
[23] H. Yue, V. Reguero, E. Senokos, A. Monreal-Bernal, B. Mas, J. Fernández-Blzquez, R. Marcilla, and J. Vilatela, Carbon **122**, 47 – 53 (2017).
[24] T. Zimmermann, E. Pöhler, and T. Geiger, Advanced Engineering Materials **6**, 754–761 (2004).
[25] J. J. Vilatela and A. H. Windle, Advanced Materials **22**, 4959–4963 (2010).
[26] J. Qiu, J. Terrones, J. J. Vilatela, M. E. Vickers, J. a. Elliott, and A. H. Windle, ACS nano **7**, 8412–8422 (2013).
[27] J. Terrones, J. a. Elliott, J. J. Vilatela, and A. H. Windle, ACS nano **8**, 8497–8504 (2014).
[28] E. Senokos, V. Reguero, L. Cabana, J. Palma, R. Marcilla, and J. J. Vilatela, Advanced Materials Technologies **2**, 1600290–n/a (2017), 1600290.
[29] J. C. Fernández-Toribio, A. Íñiguez Rábago, J. Vilà, C. González, . Ridruejo, and J. J. Vilatela, Advanced Functional Materials **26**, 7139–7147 (2016).
[30] Carbon **88**, 60 – 69 (2015).

[31] E. Mder, J. Liu, J. Hiller, W. Lu, Q. Li, S. Zhandarov, and T.-W. Chou, Frontiers in Materials **2**, p. 53 (2015).
[32] J. Jia, J. Zhao, G. Xu, J. Di, Z. Yong, Y. Tao, C. Fang, Z. Zhang, X. Zhang, L. Zheng, and Q. Li, Carbon **49**, 1333 – 1339 (2011).
[33] C. Fang, J. Zhao, J. Jia, Z. Zhang, X. Zhang, and Q. Li, Applied Physics Letters **97**, p. 181906 (2010), http://dx.doi.org/10.1063/1.3511451 .
[34] A. Borhan and K. Rungta, Journal of Colloid and Interface Science **158**, 403 – 411 (1993).
[35] A. V. Neimark, S. Ruetsch, K. G. Kornev, P. I. Ravikovitch, P. Poulin, S. Badaire, and M. Maugey, Nano Letters **3**, 419–423 (2003), http://dx.doi.org/10.1021/nl034013x .
[36] X. Chen, M. Farber, Y. Gao, I. Kulaots, E. M. Suuberg, and R. H. Hurt, Carbon **41**, 1489 – 1500 (2003).
[37] S. Nuriel, L. Liu, A. Barber, and H. Wagner, Chemical Physics Letters **404**, 263 – 266 (2005).
[38] Z. Dai, B. Zhang, F. Shi, M. Li, Z. Zhang, and Y. Gu, Applied Surface Science **257**, 8457 – 8461 (2011).
[39] J. Vilatela, L. Deng, I. Kinloch, R. Young, and a.H. Windle, Carbon **49**, 4149–4158 (2011).
[40] R. J. Young and M. A. Montes-moran, Carbon **40**, 845–855 (2002).
[41] M. Montes-Morán and R. Young, Carbon **40**, 857–875 (2002).
[42] R. J. Young, I. a. Kinloch, L. Gong, and K. S. Novoselov, Composites Science and Technology **72**, 1459–1476 (2012).
[43] S. Li, X. Zhang, J. Zhao, F. Meng, G. Xu, Z. Yong, J. Jia, Z. Zhang, and Q. Li, Composites Science and Technology **72**, 1402–1407 (2012).
[44] C. Cooper, R. Young, and M. Halsall, Composites Part A: Applied Science and Manufacturing **32**, 401 – 411 (2001).
[45] O. a. Khondker, U. S. Ishiaku, A. Nakai, and H. Hamada, Journal of Polymers and the Environment **13**, 115–126 (2005).
[46] X. Wang, P. D. Bradford, W. Liu, H. Zhao, Y. Inoue, J.-P. Maria, Q. Li, F.-G. Yuan, and Y. Zhu, Composites Science and Technology **71**, 1677–1683 (2011).
[47] A. P. A. Raju, A. Lewis, B. Derby, R. J. Young, I. a. Kinloch, R. Zan, and K. S. Novoselov, Advanced Functional Materials **24**, 2865–2874 (2014).
[48] V. Pichot, M. Burghammer, S. Badaire, C. Zakri, C. Riekel, P. Poulin, and P. Launois, Europhysics Letters (EPL) **79**, p. 46002 (2007).
[49] H. Cox, British journal of applied physics **72** (1952).
[50] T. Filleter, R. Bernal, S. Li, and H. D. Espinosa, Advanced Materials **23**, 2855–60 (2011).
[51] J. Vilatela, J. Elliott, and A. Windle, ACS nano **5**, 1921–1927 (2011).
[52] R. J. Young, Journal of the Textile Institute **86**, 360–381 (1995).
[53] A. M. Beese, X. Wei, S. Sarkar, R. Ramachandramoorthy, M. R. Roenbeck, A. Moravsky, M. Ford, F. Yavari, D. T. Keane, R. O. Loutfy, S. T. Nguyen, and H. D. Espinosa, ACS Nano **8**, 11454–11466 (2014).
[54] S. Li, X. Zhang, J. Zhao, F. Meng, G. Xu, Z. Yong, J. Jia, Z. Zhang, and Q. Li, Composites Science and Technology **72**, 1402 – 1407 (2012).
[55] C. Fang, J. Zhao, J. Jia, Z. Zhang, X. Zhang, and Q. Li, Applied Physics Letters **97**, p. 181906 (2010).
[56] Y.-L. Li, I. A. Kinloch, and A. H. Windle, Science **304**, 276–278 (2004), http://science.sciencemag.org/content/304/5668/276.full.pdf .

Polyurethane Composite Sheets Incorporating BNNTs and CNTs at High Loading: Morphological, Mechanical and Electrical Characterization and Advantages of the Fabrication Method

***Martinez Rubi Yadienka* [(1)]*, *Ashrafi Behnam* [(2)], *Jakubinek Michael B.* [(1)], *Homenick Christa* [(1)], *Barnes Michael* [(1)], *Park Daesun* [(1)], *Kim Keun Su* [(1)], *Kingston Christopher* [(1)], *Plunkett Mark* [(1)], *Walker Steven* [(1)], *Simard Benoit* [(1)]**

[(1)] National Research Council Canada - Ontario - Canada, [(2)] National Research Council Canada - Ontario - Canada

Carbon nanotubes (CNTs) and boron nitride nanotubes (BNNTs) are promising candidates for the development of light-weight, nanoreinforced, multifunctional composites. Although BNNTs and CNTs share similar mechanical properties they can impart a different set of functionalities. While CNTs are electrically conductive BNNTs show transparency in the visible region, are electrically insulating, thermally more resistant than CNT and have the ability to shield neutron radiation. High nanotube (NT) contents are often necessary in order to achieve mechanical, thermal or electrical conductivity requirements for a particular application. However, it is still challenging to attain a simple and economical method that improves nanotubes dispersion and interfacial interaction with the matrix. We have developed a process for the fabrication of high-loading polyurethane (TPU)/NT nanocomposite sheets using a one-step solution method that is easy to scale-up. Through optimization of the processing conditions it was possible to control the nanotube dispersion and the morphology of the nanocomposite sheets. The fabrication of 30 x 30 cm2 NT/TPU composite sheets was demonstrated showing the scalability of the method. A comparative study of the influence of BNNTs and CNTs on the tensile properties of the developed NT/TPU sheets was undertaken. After incorporation of pristine nanotubes, superior tensile properties were obtained with CNTs (modulus, strength and failure strain of 1 GPa, 40 MPa and 100%, respectively) suggesting a stronger interaction of CNTs with the TPU than pristine BNNTs. After functionalization of the BNNTs a significant improvement was observed with 100% improvement in stiffness compared to CNTs. In the case of CNTs, electrical conductivities up to 7000 S/m have been obtained. The nanocomposite sheets are flexible, robust and can be integrated, similarly to other fiber fabrics, into conventional composite manufacturing processes to fabricate hierarchical structures.

Challenges to the industrial melt-processing of conductive plastics

Choong Gabriel Y.H.[1, 2 a)] De Focatiis Davide S.A.[1], Lidgett Mark J.[2], Thornton Matthew J.[2] and Clifford Mike J.[1]

[1]*Faculty of Engineering, University of Nottingham, Nottingham, NG7 2RD, UK*
[2]*Haydale Composites Solution, Loughborough, LE11 1QJ, UK*

[a)]Corresponding author: gabriel.choong@nottingham.ac.uk

Abstract. In this work, we investigate the relationship between the timescales available for polymer mobility during processing and post-processing and the electrical resistivity of melt-processed thermoplastics filled with carbon nanoparticles. Post-process annealing below the glass transition temperature was one avenue explored to uplift electrical conductivity. Detailed analysis of available literature on thermoplastics filled with either graphite nanoplatelets or carbon nanotubes, and of relevant processing data suggests that the required timescale for shaping process or post-processing to obtain conductive material needs to be sufficiently longer than that of the base polymer characteristic relaxation time τ_d. Four factors have been identified that promote the formation of a conductive filler network in thermoplastics: filler loading content, polymer molar mass, processing temperature and processing timescales.

INTRODUCTION

The preferred industrial processing route to incorporate graphite nanoplatelets (GNPs) or carbon nanotubes (CNTs) into thermoplastics is melt compounding. These compounds, made either by direct compounding or by masterbatch dilution, are subsequently formed into parts through further melt-processing. The bulk material properties, particularly the electrical properties, are strongly dependent on the nanocomposite morphology obtained in the melt and frozen-in by subsequent cooling to form a part. It is widely known that achieving satisfactory filler dispersion in melt-mixing is a major challenge. This is further complicated by the influence of filler geometry from various commercial carbon nanoparticle suppliers [1]. In addition, specific properties are associated with different filler network morphologies. For example, secondary agglomeration promotes electrical conductivity whereas good dispersion and distribution of the particles are required for mechanical property enhancement.

Despite these challenges, there is significant commercial interest in employing traditional melt processing equipment to produce GNP-filled thermoplastics with electrical conductivity uplifts due to both production cost and volume considerations. The two challenges faced in this work were: (1) rapid processing is unfavourable for electrical conductivity, and (2) certain parts may be unsuitable for long timescale melt processing due to factors such as intricate shapes and production cost.

Previous studies have focused on material-related (e.g. filler loading content and polymer type) and compounding equipment-related factors (e.g. extruder parameters) [2, 3]. Relaxation of the polymer is the dominant relaxation mechanism in nanocomposite rheology [4], and therefore here, we look at whether and how polymer mobility can be used to drive the re-organisation of the filler network as a post-processing procedure. Thermal annealing is one such method, and we explore this on injection moulded GNP filled amorphous thermoplastic as a possible route to enhance electrical conductivity. The focus of this work is to find a link between the timescales available for polymer mobility during processing, and the resulting electrical conductivity of processed parts. The aim is to identify a suitable route to develop carbon-filled conductive plastics by appropriate selection of both the materials and processes, with the goal of achieving savings in both cost and development time.

MATERIAL AND METHODS

All polymer materials used in this work are commercial grade plastics. Acrylonitrile-butadiene-styrene (ABS) and polypropylene (PP) were both filled with 10 wt% of unfunctionalised GNP. The materials were compounded using a twin-screw extruder and pelletised in a commercial environment.

Injection moulding (IM) was performed to produce plate specimens with dimensions of 100 mm × 80 mm × 3.2 mm at a temperature of 235 °C in a commercial setting. Compression moulding (CM) using flash moulds [5] was employed to produce disc specimens at a temperature of 200 °C in two sizes: Ø 25 mm × 0.5 mm and Ø 65 mm × 2 mm.

Annealing parameters were identified through thermal and rheological characterisation. Rheometry of unfilled ABS was performed with an Anton Paar MCR 302 equipped with a torsional shear geometry. Measurements were carried out at a frequency range 100 – 0.1 rad s^{-1} at 0.01% strain for a temperature T range of 100 – 135 °C. Isothermal frequency sweeps were shifted using an Arrhenius equation to determine the time-temperature dependence of unfilled ABS. The glass transition temperature T_g of ABS-GNP was determined as 138 °C using a Linkam DSC600 with a heating rate of 10 °C min^{-1}. Annealing of IM plaque was performed at $T = 130$ °C in a Memmert Universal oven; at this temperature, the relaxation time of unfilled ABS was $\tau_d = 9.7 \times 10^3$ s.

Volume resistivity ρ was measured using a Keithley 8009 resistivity test fixture with a Keithley 6517B electrometer as the voltage source. Specimens were electrified with 100 V for 60 s in accordance to ASTM D257-14 [6]. After set annealing times, plaques were removed from the oven and quenched in a water bath, followed by thorough drying with paper towels prior to each electrical measurement.

Time-temperature dependences of the unfilled ABS and PP in the melt state were determined using rheometry. Isothermal frequency sweeps were performed using a Ø 25mm parallel plate geometry with a 0.5 mm gap size in an air atmosphere. Unfilled materials were measured between 250 °C and 140 °C. The frequency range was 100 – 0.1 rad s^{-1} at 1% strain. Mastercurves were obtained by applying time-temperature superposition (considering horizontal shifting only) using a procedure detailed in [4]. The relaxation time τ_d was determined from the inverse of the cross-over frequency of each mastercurve shifted to the desired reference T.

RESULTS AND DISCUSSION

Ratio of processing timescale to τ_d

In order to assist on decision making for both matrix and processing timescales, we determined the ratio of the processing timescale t to τ_d of the unfilled polymer; this is subsequently referred to as the relaxation time fraction, t/τ_d, which is typically a number greater than unity in typical processes. The polymer melt undergoes a cooling profile during IM, and thus relaxes by a varying amount as τ_d also changes with temperature. The average mould temperature $\bar{T}$ is obtained from the Fourier series solution for 1-dimensional cooling of a plate, given by

$$\bar{T}(t) = T_e + (T_{\text{initial}} - T_e) \sum_{\text{odd } n} \frac{8}{n^2\pi^2} \exp\left(-\frac{n^2\pi^2}{4}\frac{\alpha t}{L^2}\right)$$

where T_e = mould T, T_{initial} = melt T, α = polymer thermal diffusivity, t = time, and $2L$ = plate thickness.

The total relaxation time fraction for a given cooling profile is determined by numerically integrating $\int dt/\tau_d$ over the cooling profile, and at each $\bar{T}$, τ_d was determined using Williams-Landel-Ferry (WLF) equation by shifting to the desired $\bar{T}$. The lower limit of $\bar{T}$ was set to the material's T_g. The number of terms of the Fourier series solution n was fixed at 49 to improve the description of the high temperature profile in the initial cooling stage.

In the case of CM, we calculated t/τ_d using only the isothermal holding period as this is dominant over the contributions of the heating and cooling periods. Fig. 1 shows a comparison of ρ as a function of the fractional relaxation time for the corresponding IM and CM melt-processes.

It is recognised that there are other factors that influence the electrical resistivity of an IM specimen, such as shear effects and filler orientation. However, it should be noted that the typical rapid processing timescales significantly limit the system's ability to rearrange the filler network to a state that favours electrical conductivity. Therefore, as a first approximation, the polymer's fractional relaxation time is used as an indicator of the feasibility of obtaining a conductive material.

Thermal annealing

Evolution of volume resistivity as a function of the fractional relaxation time of GNP-filled ABS (10 wt%) for different moulding processes and including post-processing annealing is presented in Fig. 1. Here the fractional relaxation taking place during annealing is added to that during processing with the same method. After 22 hours of annealing at 130 °C, ρ of IM ABS-GNP decreased by an order of magnitude relative to the same material without annealing. By changing the timescale of the shaping process from IM to CM, ρ of the identical ABS-GNP material decreased by 9 orders of magnitude. This suggests that the annealing procedure resulted in limited additional polymer mobility to modify the filler network to create more conductive pathways.

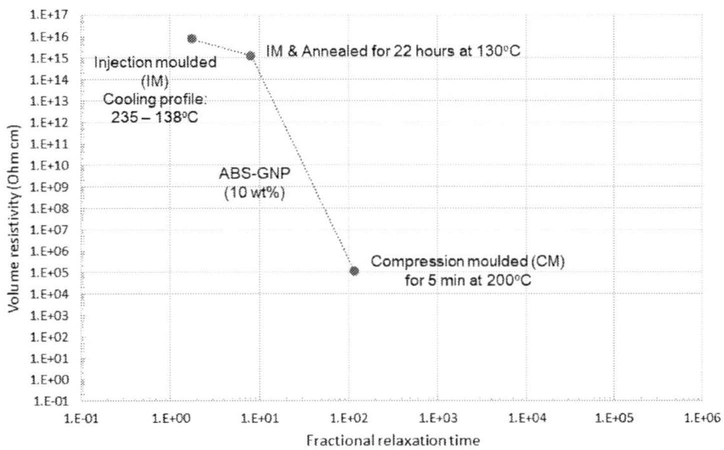

FIGURE 1. Electrical resistivity as a function of the fractional relaxation time of injection moulded, injection moulded and annealed, and compression moulded GNP-filled ABS (10 wt%). *Dotted lines are a guide to the eye*

Thermoplastics filled with carbon nanoparticles

The technique described in the previous section was employed to analyse literature results. Literature [7-15] was selected based on the availability of three components of required information: (1) the rheological response from which to identify the time-temperature dependence of the unfilled matrix and τ_d ; (2) sufficient specimen preparation parameters to obtain the temperature history (CM/IM); and (3) direct current volume resistivity measurements. In cases where the time-temperature dependence of the matrix was not directly available, it was obtained from published data on similar materials since this does not vary significantly between polymers of different grades [4].

Fig. 2 displays volume resistivity obtained from either GNP or CNT filled thermoplastics (semi-crystalline and amorphous) as a function of the fractional relaxation time on a wide range of literature studies. The plot suggests that it is easier to decrease ρ with CNTs at lower filler loading content; in addition, it can also be seen that, at least for the cases shown here, filled amorphous polymers also tend to achieve lower ρ when compared to filled crystalline materials.

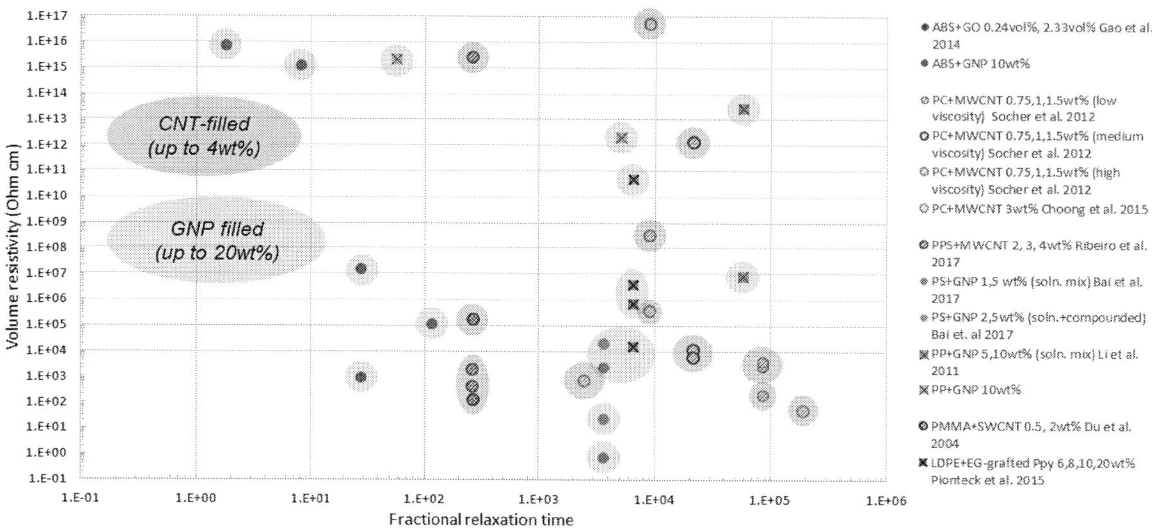

FIGURE 2. Volume resistivity as a function of the fractional relaxation time (multiple) for two different types of carbon nanoparticle fillers

From the literature, we identified four factors that contribute to the reduction of ρ, as displayed in Fig. 3: (1) increasing the filler loading content, (2) increasing the process timescales, (3) increasing the processing temperature and (4) decreasing the polymer molar mass. Factors (3) and (4) are supported by Zeiler et al.'s findings that the CNT network formation is enhanced with higher temperature and lower molar mass [15]. It is plausible to consider that the same factors apply to GNP-filled thermoplastics, although there is limited experimental evidence in published literature of this at present. In addition, factors (2) and (3) are both exposed to the risk of polymer degradation that can be detrimental to the nanocomposite bulk properties. Long processing timescales and temperature can also be constrained by external factors such as manufacturing costs, making such processes commercially unattractive.

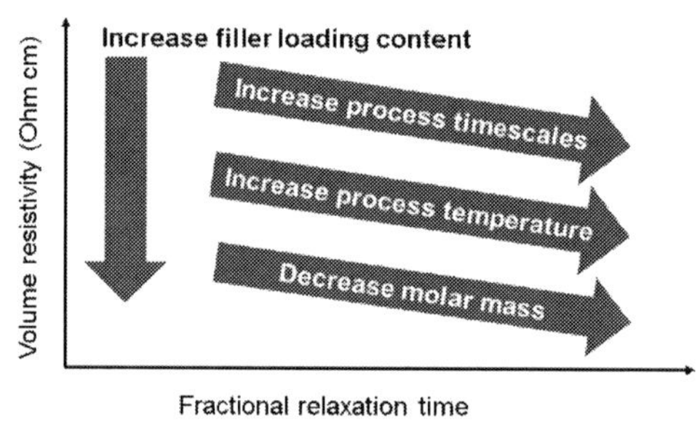

FIGURE 3. Material-related and process-related factors that promote electrically conductive filled thermoplastics

Fig. 4 illustrates the regions of interest for the development of conductive thermoplastics. There are 4 regions, labelled as: (1) Insulating, (2) Future target, (3) Likely viable commercially and (4) Not likely viable commercially. Region (1) is where the filled polymer systems have not yet reached the electrical percolation threshold. Region (2) includes conductive materials that can be processed rapidly – it is where one might aspire to be, but as yet has no reported results in the literature; Region (3) is where the bulk of the current conductive nanocomposite systems lie, depending on the conductivity, they are suitable for either electrostatic discharge or full electrical conductivity; Region (4) is similar to region (3) but timescales make these polymers less commercially viable due to long processing times and the risk of polymer degradation. Precise boundaries between the regions are somewhat arbitrary and should be tailored according to user requirements. Also, the diagram is not intended to identify materials that are commercially viable from the literature, but rather to aid the planning for the selection of both the material and the process that is most likely to be commercially relevant. It is worth mentioning that since Fig. 3 identifies the factors that influence ρ, it serves as an approximation tool and can provide a strategy to further decrease ρ when Fig. 4 is populated with the relevant data.

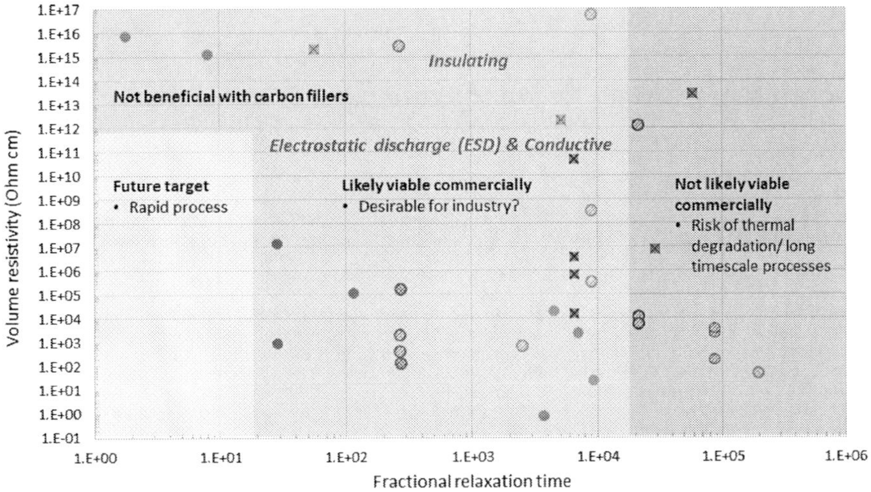

FIGURE 4. Volume resistivity of carbon nanoparticle-filled thermoplastics as a function of the fractional (multiple) relaxation time. The coloured segments indicate regions of interest to assist in both material and process selection when developing electrically conductive materials. *There is no intention to identify commercially viable materials from the literature*

CONCLUSIONS

This work has shown that uplifts in electrical conductivity of carbon nanoparticle based thermoplastic can be achieved if the timescales of the shaping or post-processing are sufficiently greater than the matrix polymer's relaxation time τ_d. In addition, four factors that promotes the formation of a conductive filler network in thermoplastics were identified from the literature: lower polymer molar mass, increased filler loading content, higher processing temperature, and longer process timescales.

ACKNOWLEDGMENTS

The authors would like to acknowledge Innovate UK for part-funding this work through the Knowledge Transfer Partnership scheme (ref. KTP 010109).

REFERENCES

1. Müller M.T., Hilarius K., Liebscher M., Lellinger D., Alig I., Pötschke P., *Materials*, **10**, 5, 545 (2017).
2. Khanam P.N., AlMaadeed M.A., Ouederni M., Harkin-Jones E., Mayoral B., Hamilton A., Sun D., *Vacuum*, **130**, 63–71 (2016).
3. Kalaitzidou K., Fukushima H., Askeland P., Drzal L.T., *J. Mater. Sci.*, **43**, 8, 2895 – 2907 (2008).
4. Choong G.Y.H., De Focatiis D.S.A., Hassell D.G., *Rheol. Acta*, **52**, 8-9, 801–814 (2013).
5. De Focatiis D.S.A., *Polym. Test.*, **31**, 4, 550–556 (2012).
6. ASTM, 'D257-14', Standard Test Methods for DC Resistance or Conductance of Insulating Materials (2014).
7. Gao C., Zhang S., Wang F., Wen B., Han C., Ding Y., Yang M., *ACS Appl. Mater. Interfaces*, **6**, 15, 12252–12260 (2014).
8. Socher R., Krause B., Muller M. T., Boldt R., Pötschke P., *Polym.*, **53**, 2, 495–504 (2012).
9. Choong G.Y.H., Lew C.Y., De Focatiis D.S.A., *J. Appl. Polym. Sci.*, **132**, 42277 (2015).
10. Ribeiro B., Pipes R.B., Costa M.L., Botelho E.C., *J. Comp. Mat.*, **51**, 2, 199–208 (2017).
11. Bai Q., Wei X., Yang J., Zhang N., Huang T., Wang Y., *Compos. Part A Appl. Sci. Manuf.*, **96**, 89–98 (2017).
12. Li Y., Zhu J., Wei S., Ryu J., Sun L., Guo Z., *Macromol. Chem. Phys.*, **212**, 18, 1951–1959 (2011).
13. Du F., Scogna R.C., Zhou W., Brand S., Fisher J.E., Wiley K.I., *Macromol.*, **37**, 24, 9048–9055 (2004).
14. Pionteck K., Valdez E.M.M, Piana F., Omastová M., Luyt A.S., Voit B., *J. Appl. Polym. Sci.*, **132**, 20 (2015).
15. Zeiler R., Handge U.A., Dijkstra D.J., Meyer H., Altstädt V., *Polym.*, **52**, 2, 430–442 (2010).

Optimal Performance of Nano-carbon Polymer Composite: Dimensionality Matters

Nadiv Roey [(1)]*, **Regev Oren** [(1)]

[(1)] Ben-Gurion University of the Negev - Israel - Israel

This study offers a comparative study on various nano-carbon reinforcements differing in their dimensionality (D): carbon nanotubes (CNTs) and graphene nanoribbons (1D), Graphene Nano platelets (GNPs) (2D) and Graphite (3D). A wide range of composite characteristics were investigated including mechanical, electrical and rheological properties. We found that the optimal nanomaterial concentration (ONC) yielding a maximal reinforcement is typically located in close proximity to the rheological percolation threshold, in which an abrupt increase in polymer matrix viscosity is found. Therefore, such increase may be utilized as an indicator for the ONC. The 1D filler, i.e., CNT, is more effective than fillers of higher dimensionality: it yields optimal reinforcement and percolates at lower filler loading. In terms of processability, the GNPs are advantageous in dispersion and demonstrate milder rheological effect. Moreover, the GNP offers valuable reinforcement in a wider range of filler loading. To provide a facile comparison, a figure of merit is introduced taking into account the essential aspects of composite performance including rheological effect, enhancement, and loading (including range). Based on the figure of merit values, we were able to manufacture a robust hybrid nanocomposite utilizing the virtues of both 1D and 2D nano-carbon fillers. References: 1. Roey Nadiv, Michael Shtein, Matat Buzaglo, Sivan Peretz-Damari, Anton Kovalchuk, Tuo Wang, James M. Tour and Oren Regev. Carbon, 2016/ 99: 444-50 2. Michael Shtein, Roey Nadiv, Noa Lachman, H. Daniel Wagner and Oren Regev. Composites Science and Technology, 2013. 87(0): 157-163. 3. Roey Nadiv, Michael Shtein, Alva Peled, Gleb Vasilyev, Eyal Zussman and Oren Regev. Composites Science and Technology, 2016, 133, 192-199.

Reactive Coagulation of Single-Walled Carbon Nanotubes for Tougher Composites – Solution Processing and Assembly

Adam J. Clancy[1, a)], David B. Anthony[1, b)], and Milo S. P. Shaffer[1, 2, c)]

[1] Department of Chemistry, Imperial College London, South Kensington Campus, London, SW7 2AZ, U.K
[2] Department of Materials, Imperial College London, South Kensington Campus, London, SW7 2AZ, U.K

[a)] Corresponding author: a.clancy11@imperial.ac.uk
[b)] d.anthony08@imperial.ac.uk
[c)] m.shaffer@imperial.ac.uk

Abstract. The injection of reduced single-walled carbon nanotubes into a coagulation bath of polyvinyl chloride (PVC) solution leads to the formation of nanocomposite fibres with polymer covalently bound to the nanotubes. The influence of PVC concentration and molecular weight, and the extrusion diameter on the nanocomposite fibre tensile properties and composition have been examined. The nanocomposite fibres produced have strengths as high as 480 MPa and modulus of 15 GPa, making them the strongest and stiffest PVC composites recorded to date.

BACKGROUND

Single-walled carbon nanotubes (SWCNTs) are ideal fillers for nanocomposites owing to their high intrinsic mechanical properties, low weight, and high aspect ratio.[1] However, SWCNTs dispersed at high weight loadings through shear based approaches tend to form bundled agglomerates in the final composites which act as defects in the resulting composite. Instead, reduction to 'nanotubide' anions[2] can be performed generating salts which are spontaneously soluble as individualised SWCNTs[3] at high concentrations. The approach relies on stabilization through Columbic repulsion, and avoids damage more typically induced during shear dispersion or acid oxidation. Nanotubides are highly reactive, opening the possibility of a family of functionalisation reactions, most commonly through addition of alkyl halides. The high concentration nanotubide solution can be used as a dope for coagulation spinning of pure SWCNT fibres,[4] though their mechanical properties are relatively poor. 'Reactive coagulation spinning', using a polymer coagulant that is reactive towards nanotubide, such as polyvinyl chloride (PVC), allows formation of stronger composites with matrix covalently grafted onto the filler. In this approach, the SWCNTs are wrapped in polymer, as evidenced by their lack of electrical conductivity and high composite dielectric constant, even at high loadings.[5] The composite mechanical properties can be modulated by SWCNT selection: longer SWCNTs give higher strain-to-failure and more crystalline nanotubes provide higher strength. Nevertheless, while the mechanical properties of these composites are higher than nanotubide-derived pure SWCNT fibres, their performance is limited by poor processing and has significant scope for improvement. Here, the effects of coagulant concentration, polymer molecular weight and extrusion cannula diameter are examined to improve the reactive coagulation fibre spinning of nanotubide.

EXPERIMENTAL

Materials

Sodium (99.99 % ingot), N,N-dimethylacetamide (DMAc, 99 %), naphthalene (99 %) and PVC were purchased from Sigma Aldrich Ltd. (U.K.). PVC was purchased at three different molecular weights: M_w 43 kDa (M_n 22 kDa),

M_w 80 kDa (M_n 47 kDa), M_w 233 kDa (M_n 99 kDa). Polymers are referred to by weighted average molecular weights throughout. Methyl isobutyl ketone (MiBK, 99.9 %), ethanol (96 %), and water (HPLC grade) were purchased from VWR (U.K.). SWCNTs (Elicarb, P929, Batch K108511/g) were provided by Thomas Swan (U.K.) and purified before use through a previously reported reductive route.[6]

Characterization

Thermogravimetric analysis (TGA) was performed on a Pyris 1 (Perkin Elmer, U.S.A.), under 60 sccm nitrogen flow on 0.5 - 1 mg samples. The samples were held at 100 °C for 90 min to dry the sample and degas the furnace, weights are normalized to the weight at, and are presented from, the end of this drying step. The samples were then heated at 10 °C min^{-1} to 700 °C. Single fibre tensile tests were adapted from BS ISO 11566 with fibres mounted to a card template using Araldite Rapid (Huntsman, U.S.A) for ease of handling, and tested at 15 mm gauge lengths at a cross head displacement of 15 µm s^{-1} on a TST350 Tensile testing stage (20 N load cell, Linkam U.K.). The fibre cross-sectional area was assumed to be circular at fracture site, with true stress provided accordingly, and standard error is provided from at least 10 samples tested for each sample set.

Processing

Nanotubide solutions were made using a previously reported technique;[7] in short, in a glovebox, sodium (2.3 mg, 0.1 mmol) and naphthalene (12.8 mg, 0.1 mmol) were dissolved in dry DMAc (3.4 mL) through stirring with a glass stirrer bar for 1 h to give a deep green solution, before dried SWCNT powder (12 mg, 1 mmol, 3.5 mg mL^{-1}) was added and stirred overnight to give a viscous black solution. Reactive coagulation spinning was performed according to previously reported procedure.[5] PVC was dissolved in boiling MiBK at a desired concentration and cooled to room temperature before use, poured into a crystallization dish on a rotating platform. The nanotubide dope was drawn into a syringe in the glovebox and capped before removal to limit exposure to non-ambient conditions before spinning. The cap was replaced by a z-shaped cannula allowing the dope to be injected parallel to the flow of the rotating bath and was injected via syringe pump at a gauge-dependent rate (Table 1) into a bath traveling at 20 cm s^{-1} at the point of injection. The extruded fibres were removed from the coagulation bath and dried under tension (0.49 mN) before soaking in fresh MiBK (5 min, 50 °C) to remove adsorbed PVC, and dried at room temperature.

TABLE 1. Injection rates for varying diameter cannulas, calculated to give flow velocity of 15 cm s^{-1}.

Cannula Gauge	Cannula Diameter (µm)	Cross-Sectional Area (m^2)	Injection Rate (mL h^{-1})
24	311	7.54×10^{-8}	41
27	210	3.46×10^{-8}	19
29	184	2.66×10^{-8}	14

RESULTS AND DISCUSSION

Extrusion Cannula Diameter

It is a common adage within the materials community that smaller is stronger: by decreasing the cross sectional area of a material, the probability of the material containing a weakening flaw is decreased. In coagulation spinning, decreasing the extrusion diameter will not only decrease the fibre diameter, but is also expected to increase the SWCNT alignment in the composite due to higher shear of the dope. Here, three cannulas (24 gauge, 27 gauge, 29 gauge) were used to a spin nanotubide dope into a 5 wt% $PVC_{(MiBK)}$ (43 kDa) coagulant. Final fibre diameter followed the same trend as the cannulae diameter (Table 2). As the fibre diameter reduced, strength (143 MPa to 234 MPa) and modulus (7.8 GPa to 17.3 GPa) increased substantially, whilst strain-to-failure (4.5% to 2.5%, Fig 1) decreased. This trend is consistent with increasing alignment of SWCNTs within the nanocomposite fibre. Increasing the initial alignment of SWCNTs increases their contribution to reinforcement along the fibre axis but reduces the capacity to realign under tension, subsequently reducing the strain-to-failure of the nanocomposite fibres. As the PVC content is similar between all samples (26.7 wt% to 32.9 wt%, Fig 1d) and independent of cannula diameter, the mechanical differences can be attributed to the fibre hierarchy. The consistency in composition also demonstrates that the fibre formation is not hampered by PVC diffusion which might have been expected to decrease PVC content for larger diameter fibres.

TABLE 2. Diameters and mechanical properties of SWCNT/PVC composite fibres coagulated using differing gauge dope injection cannulae. † Values in parentheses represent standard deviations. * SWCNT content calculated at 600 °C from TGA. ‡ Calculated average value from mean strength and mean diameter.

Cannula Gauge	Fibre Diameter (µm) †	SWCNT (wt%) *	Strength (MPa)	Young's Modulus (GPa)	Strain-to-failure (%)	Breaking Force (mN) ‡
24	34.0 (± 14.6)	32.9	143 (± 25)	7.8 (± 0.4)	4.5 (± 0.2)	519
27	31.2 (± 9.9)	26.7	163 (± 11)	10.2 (± 1.0)	2.8 (± 1.8)	498
29	24.7 (± 7.8)	26.8	234 (± 17)	17.3 (± 1.3)	2.5 (± 0.7)	448

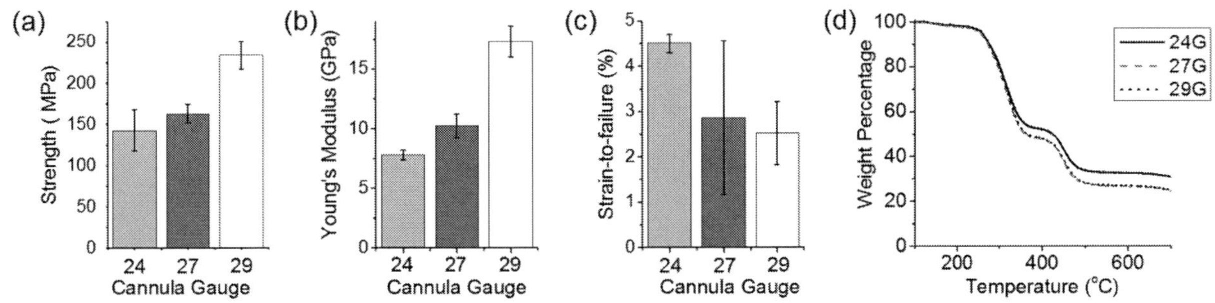

FIGURE 1. Left: Mechanical properties of SWCNT/PVC fibres created using Elicarb nanotubide dope and PVC solution (M_w = 43 kDa, 5 wt% in MiBK) as coagulant. Right: TGA of SWCNT/PVC composite fibres coagulated using differing gauge dope injection cannulae 5 wt%, 43 kDa PVC coagulant.

PVC Coagulant Control

To measure the influence of the PVC coagulant, three commercial PVC samples (M_w 43 kDa, 80 kDa, and 233 kDa) were dissolved in MiBK at two different concentrations: 1 wt% and 3.5 wt%. The 'high' concentration was decreased from the 5 wt% used for the previous experiments as 233 kDa PVC forms gels at ≥4 wt% at room temperature (Fig 2). Fibres were spun with 24 gauge cannula to increase handleability, as in spite of the increased mechanical performance from narrower cannulas/fibres, the absolute breaking force of the thicker fibres is still higher (Table 2).

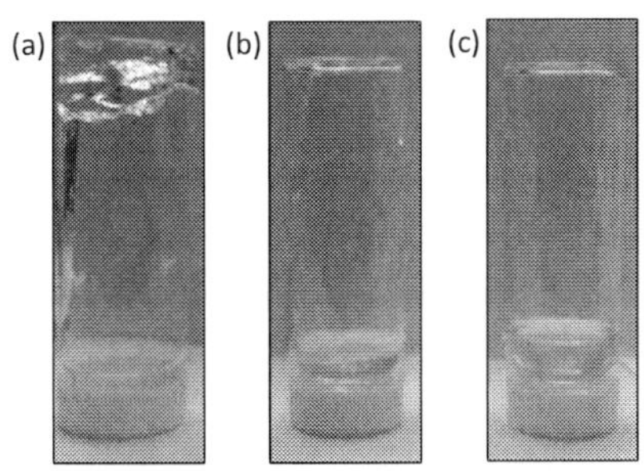

FIGURE 2. PVC (Mw 233 kDa) in MiBK at varying concentrations. 5 mL of solution at (a) 5.0 wt%, (b) 4.0 wt%, and (c) 3.5 wt% after inverting and leaving to stand for ~5 seconds.

The molecular weight of the PVC in the coagulation bath was inversely related to the percentage of polymer within the final composite fibre (Fig 3), *i.e.* the heaviest 233 kDa PVC leads to the lowest PVC content for both coagulant concentrations likely due to the increased diffusion times required for the polymer into the coagulating fibre, akin to the effects seen by Mercader *et al.*[8] for coagulation SWCNTs in polyvinyl alcohol solutions. Decreasing the concentration of the polymer decreased the loading of polymer in the final composite for all PVC molecular weights. While increasing the polymer concentration could be thought to increase the quantity of polymer which can diffuse into the coagulating fibre, as shown for varying gauge nanocomposite fibres, the composition is not diffusion controlled. Instead it is thought that the number of grafting sites for distinct PVC molecules along each SWCNT is increased with increasing polymer concentration. A single PVC molecule contains thousands of C-Cl bonds capable of reacting with a nanotubide, with each reaction depleting the finite number of reactive charges on the nanotubide.[9] If one PVC molecule grafts to a nanotubide, it will, given time, eventually wrap the SWCNT, covalently bonding at many sites along the surface, depleting charge and sterically occluding the grafting of additional PVC molecules. By increasing the polymer concentration, the number of distinct PVC molecules which can react to each nanotubide, before the surface is occluded/the charge depleted, is likely increased.

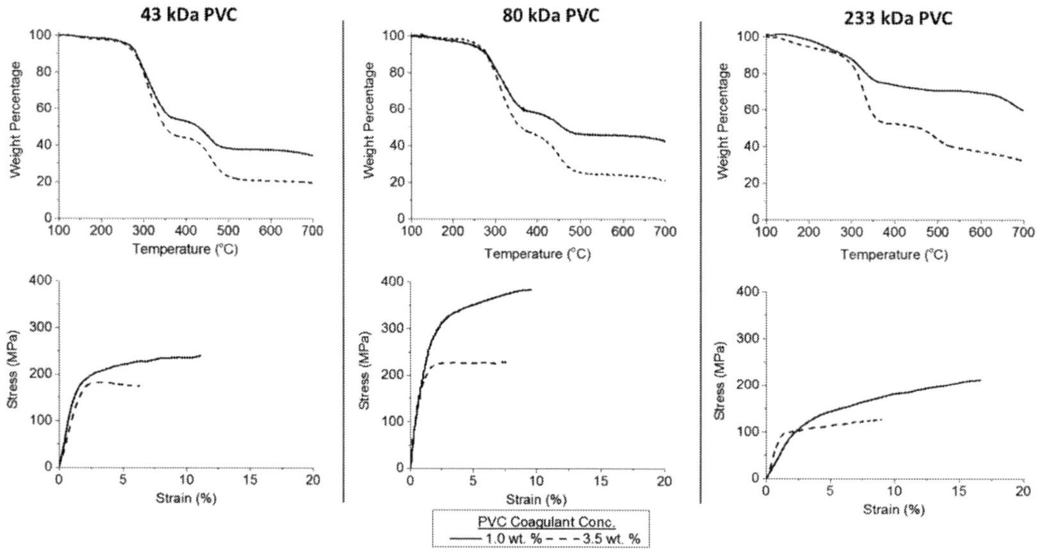

FIGURE 3. Thermograms (top) and representative stress-strain diagrams (bottom) for SWCNT/PVC nanocomposites fibres.

As previously reported,[5] SWCNT/PVC nanocomposite fibres behaved elastically to a yield point at 1% - 2% strain, however, after yield, the mechanical behavior was dependent on the SWCNT loading. Composites spun using 1 wt% PVC coagulant (*i.e.* with higher SWCNT loadings, Fig 3) displayed noticeable strain hardening. In the 3.5 wt% PVC samples, only the 233 kDa sample (which also indicated high CNT loading) showed a strain hardening response (Fig 3, right); the rate of strain hardening was noticeably higher for the 1 wt% PVC samples containing the higher SWCNT content (69.5 wt% vs 37.3 wt%). The dependence on SWCNT loading combined with the lack of strain hardening seen intrinsically for PVC leads to the conclusion that the strain hardening is a SWCNT effect, tentatively assigned to reorientation. It is worth noting that these trends are consistent with the 5 wt%, 24 gauge experiments (*vide supra*) which show lower strength, modulus, strain-to-failure (Fig 1) than either of the 1 wt% and 3.5 wt% PVC concentration coagulant based nanocomposite fibres.

TABLE 3. Properties of SWCNT/PVC composite fibres from varying coagulants * Calculated from TGA at 600 °C.

Mw (kDa)	PVC Coagulant (%)	SWCNT (wt%) *	Strength (MPa)	Young's Modulus (GPa)	Strain-to-failure (%)
43	1.0	37.7	217 (± 25)	11.0 (± 1.1)	8.0 (± 3.1)
43	3.5	20.7	183 (± 10)	10.0 (± 1.1)	4.6 (± 2.6)
80	1.0	45.7	295 (± 19)	15.0 (± 0.3)	8.9 (± 1.9)
80	3.5	23.8	253 (± 19)	15.0 (± 1.2)	7.7 (± 2.5)
233	1.0	69.5	211 (± 17)	5.98 (± 0.8)	12.3 (± 6.3)
233	3.5	37.3	122 (± 14)	8.90 (± 1.0)	7.5 (± 2.1)

The molecular weight of the coagulant polymer has a strong influence on the composite's mechanical properties (Fig 4). The 80 kDa PVC shows the best strengths (295 MPa for 1 wt% PVC, max 482 MPa measured) and moduli (15.0 GPa). The increase in strength versus the 43 kDa PVC composites can be either attributed to the higher SWCNT loading, or the higher inherent strength of longer polymers due to heightened intermolecular bonding. The 233 kDa PVC composites have the highest SWCNT loading but the lowest average strengths and moduli. Combined with their high strain-to-failures, it is thought that the SWCNTs are very poorly aligned leading to high reorientation under tension but poor SWCNT contribution to strength due to their off axis loading orientation.

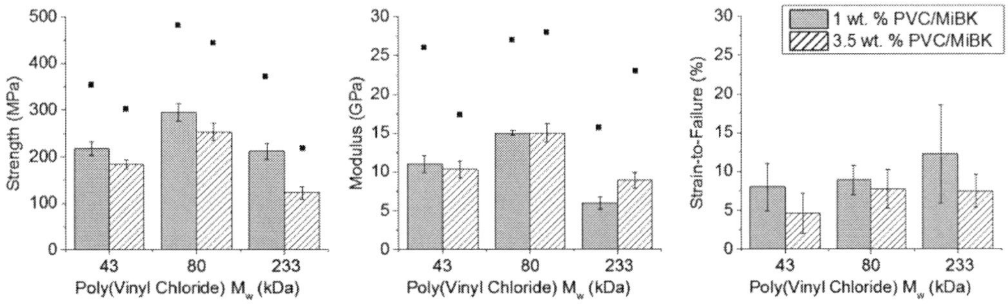

FIGURE 4. Tensile properties of Elicarb/PVC nanocomposite fibres with varied PVC coagulant concentrations and PVC molecular weights. Squares represent the highest measured values.

CONCLUSION

Reactive coagulation spinning of individualised, reduced SWCNT nanotubide ions into PVC solutions provides an effective new route to SWCNT composites. By examining the spinning parameters, the mechanical properties have been substantially improved, over the nanocomposite fibres reported initially. By decreasing the extrusion diameter, strength and modulus are increased, at the expense of strain-to-failure. By decreasing the concentration of the polymer in the coagulation bath, the loading of SWCNTs in the composite can be increased, subsequently increasing the strength and strain-to-failure of the nanocomposite fibres; at high SWCNT loadings, the toughness is further increased by introduction of a strain hardening mechanism. Finally, the matrix molecular weight has an important effect on the mechanical performance: an intermediate weight appears to maximise strength and stiffness by balancing the poor intrinsic properties of short polymers and the limited matrix loading obtained for larger polymers.

ACKNOWLEDGEMENTS

The authors would like to acknowledge the UK Engineering and Physical Sciences Research Council (EP/I02946X/1) for funding and Thomas Swan Ltd. for providing Elicarb SWCNTs. All underlying data to support the conclusions are provided within this paper. The authors declare no competing financial interest.

REFERENCES

1. J. N. Coleman, U. Khan, W. J. Blau and Y. K. Gun'ko, *Carbon*, 2006, **44**, 1624–1652.
2. A. Pénicaud, P. Poulin, A. Derré, E. Anglaret and P. Petit, *J. Am. Chem. Soc.*, 2005, **127**, 8-9.
3. S. Hodge, S. Fogden, C. Howard, N. Skipper and M. S. P. Shaffer, *ACS Nano*, 2013, **7**, 1769-1778.
4. C. Jiang, A. Saha, C. C. Young, D. P. Hashim, C. E. Ramirez, P. M. Ajayan, M. Pasquali and A. A. Martí, *ACS Nano*, 2014, **8**, 9107-9112.
5. A. J. Clancy, D. B. Anthony, S. J. Fisher, H. S. Leese, C. S. Roberts and M. S. P. S. Shaffer, *Nanoscale*, 2017, **9**, 8764-8773.
6. A. J. Clancy, E. R. White, H. H. Tay, H. C. Yau and M. S. P. Shaffer, *Carbon*, 2016, **108**, 423–432.
7. A. J. Clancy, J. Melbourne and M. S. P. Shaffer, *J. Mater. Chem. A*, 2015, **3**, 16708-16715.
8. C. Mercader, A. Lucas, A. Derré, C. Zakri, S. Moisan, M. Maugey and P. Poulin, *Proc. Natl. Acad. Sci. U. S. A.*, 2010, **107**, 18331-18335.
9. F. Hof, S. Bosch, S. Eigler, F. Hauke and A. Hirsch, *J. Am. Chem. Soc.*, 2013, **135**, 18385–18395

The Influence of Thermal Treatments on Dielectric Behaviors of Carbon Nanotubes-BaTiO3 Hybrids Reinforced Polyvinylidene Fluoride Composites

Fan Benhui (1), Bedoui Fahmi (2), He Delong (1), Bai Jinbo (1)*

(1) Ecole CentraleSupelec - Ile de France - France, (2) Université de Technologie de Compiègne - Compiègne - France

Incorporated carbon nanotube-BaTiO3 hybrids (H-CNT-BT) with core-shell structure, a good dispersion of CNTs can be achieved in a semi-crystalline polymeric matrix, polyvinylidene fluoride (PVDF). Carried by BT particles, CNTs are easy to mutually connect which helps to obtain an extremely low percolation threshold (fc). After thermal treatments, the dielectric constants (ε') of samples further increase which depends on the conditions of thermal treatments such as annealing temperatures, annealing durations and cooling ways. Thus, in order to study more comprehensively about the influence of thermal treatments on composite's dielectric behaviors, in situ synchrotron X-ray is used to detect re-crystalline behavior of PVDF. Results of wide-angle X-ray diffraction (WAXD) and small-angle X-ray scattering (SAXS) show that after the thermal treatment, the content of β polymorph (the polymorph with the highest ε' among all the polymorphs of PVDF's crystalline structure) has increased nearly double times at the interfacial region of CNT-PVDF, and the thickness of amorphous layers (La) in PVDF's long periods (Lp) has shrunk around 10 Å. The evolution of CNT's network possibly occurs in the procedure of La shrinkage, where the strong interfacial polarization may be aroused and increases ε' at low frequency. Moreover, an increase in the thickness of crystalline lamella may also arouse more orientational polarization and improve ε' at high frequency.

ULTRASONIC PROCESSING OF MULTIFUNCTIONAL EPOXY COMPOSITE REINFORCED BY LOW-COST MWNT NANOPAPER

Castro Jose M (1)*

(1) The Ohio State University - Columbus, Ohio - USA

This work presents a novel ultrasonic infiltration approach for epoxy and multiwall carbon nanotube (MWCNT) nanopaper prepreg. The prepared nanocomposites demonstrated promising performance in processing flexibility, abrasive resistance, scratch resistance and EMI shielding for potential applications in wind energy, electronics, and automotive industries.

Rheological and dynamic mechanical behavior of epoxy resins filled with multiwalled carbon nanotubes and graphene-based nanoparticles

Nobile Maria Rossella [(1)]*, Rosolia Salvatore [(1)], Raimondo Marialuigia [(1)], Naddeo Carlo [(1)], Guadagno Liberata [(1)]

[(1)] University of Salerno - Fisciano Salerno - Italy

Nanocomposites based on the tetraglycidyl methylene dianiline (TGMDA) epoxy resin using various kinds of nanofillers such as multiwalled carbon nanotubes (MWCNT) (1D filler) and graphene-based nanofillers (2D filler) are investigated in this paper. The 4,4-diaminodiphenyl sulfone (DDS) is used as curing agent. Despite very good mechanical performance of the TGMDA epoxy resin, one of the current limitations of this epoxy precursor is related to the high value of the viscosity. In order to optimize the dispersion of the nanofillers inside the epoxy-DDS mixture we have fluidified the initial epoxy precursor with an epoxy reactive monomer which has proven to be very effective to lower the viscosity of the nanoparticles/resin mixture and therefore in facilitating the nanofiller dispersion step. In this work the rheological properties of MWCNT/epoxy nanocomposites and of graphene/epoxy nanocomposites filled with two similar graphene-based nanofillers, differing essentially for the exfoliation degree and for the content of carboxylated groups, are investigated in the liquid state (with the inclusion of DDS) before curing. The rheological measurements were carried out on a Physica-MCR301 (Anton-Paar) rotational rheometer equipped with a parallel plate geometry. Small amplitude oscillatory shear measurements, within the linear viscoelasticity regime, on the liquid mixtures with different contents of MWCNT and graphene-based nanoparticles were performed at different temperatures. Dynamic mechanical properties of the cured samples were performed with a dynamic mechanical thermo-analyzer (Tritec2000 DMA-Triton Technology). Thermal analysis has been used to study the oxidative stability of the components and the formulated resins. The effect of CNTs and graphene-based nanoparticles on the thermogravimetric curves has also been analyzed in this paper.

Rapid quantitative mapping of multi-walled carbon nanotube concentration in nanocomposites

Fisher Sandra J. [(1)]*, ***Shaffer Milo S. P.*** [(1)]

[(1)] Imperial College London - United Kingdom - UnitedKingdom

Carbon nanocomposites have exciting potential but often display poor performance. The limitations are often attributed qualitatively to issues of distribution and dispersion within the matrix, factors that are sensitive to the processing technique and conditions employed. Unfortunately, there are relatively few methodologies available for the quantification of inhomogeneous carbon nanotube distributions. Understanding the relationships between microstructure and properties is crucial to developing the next generation of carbon nanocomposites; consequently, there is a growing need for new, quantitative approaches to characterise the distribution of fillers in nanocomposite materials. Here, Raman spectroscopy is used to quantify the concentration of multi-walled carbon nanotubes (MWCNT) in compression moulded epoxy composites across differing concentration regimes. While Raman is a powerful tool, scans of statistically relevant areas can be time consuming to obtain. Since MWCNT are strong optical absorbers, optical microscopy in transmission mode can also be used to quantify the density of MWCNT fillers and visualize the MWCNT concentration across the composite. Through correlating the intensity of the Raman G-band to the transmittance intensity, rapid and quantitative analysis of MWCNT distributions over large areas of the composite (>100 μm x 100 μm) can be achieved. This powerful tool enables quantitative mapping of the concentration of MWCNT, thus allowing quick verification of manufacturing processes with applications for fundamental development and quality assurance in a commercial context.

Electrically conducting CNT/polymer composites towards 3D-printed electronics

Andersson Emelie [(1)], Kiefer David [(1)], Lund Anja [(1)]*, Müller Christian [(1)]

[(1)] Chalmers University of Technology - Gothenburg - Sweden

3D-printing, in this case fused filament fabrication (FFF), shows potential as a versatile method to print simple electronic circuits attractive for the development of wearable devices, such as sensors and energy harvesting. We present the development of a nanocomposite - carbon nanotubes (CNT) in a polypropylene (PP) matrix - suitable for FFF. Both, short multi walled (MWNT), single walled (SWNT) and double walled (DWNT) nanotubes were evaluated. Firstly, masterbatches with up to 33 wt% CNT in the polymer matrix were produced via solution processing using a low Mw PP. The result was a porous structure of CNTs wrapped by the polymer, as observed by SEM. The highest resulting electrical conductivity c for our 33% composites was observed for the MWNT-masterbatch; cMWNT = 11.4 S/cm. This material was diluted by melt-mixing with an extrusion grade PP, with the aim to extrude the nanocomposite into a filament and finally print it using a commercial FFF printer.

Morphology and Mechanical Properties of Elemental sulfur Reinforced Nylon-12 / Multi-walled carbon nanotubes (MWCNTs) Composites

JENA KISHORE (1), ALHASSAN SAEED (1)*

(1) THE PETROLEUM INSTITUTE - ABUDHABI - UNITED ARAB EMIRATES

ABSTRACT: Multi-walled carbon nanotubes (MWNTs)/Nylon-12 nanocomposites with different MWNTs (1-5wt.%) loadings have been prepared by the simple melt-compounding approach at 200 °C for 20 mins with a rotation of 150 rpm for all the composites. A dark and homogeneous Nylon-12/MWNTs nanocomposite was collected from extruder after 20 mins melt-compounding. A fine and homogeneous dispersion of MWNTs throughout Nylon-12 matrix is observed by transmission electron microscopy (TEM). Scanning electron microscopy (SEM) observation on the fracture surfaces of the composites shows not only a uniform dispersion of MWNTs but also a strong interfacial adhesion with the matrix, as evidenced by the presence of many broken but strongly embedded carbon nanotubes (CNTs) in the matrix and by the absence of de-bonding of CNTs from the matrix. Mechanical testing by tensile tests as well as dynamic mechanical analysis shows that, compared with neat Nylon-12, the storage modulus and the tensile strength of the nanocomposite are greatly improved by about 113% and 134%, respectively, with incorporating only 3 wt % MWNTs. In addition to the above mechanical analysis, crystallization and melting behavior of Nylon-12/MWNTs nanocomposites are also observed and discussed by differential scanning calorimetry and X-ray diffraction. Keywords: CNT; Melt processing; Mechanical properties; TEM; SEM.

Mechanical Properties of Carbon Monoxide Reduced Graphene–Polyamide-6 Nanocomposites Prepared by Melt-mixing

Kok Peng Marcian Lee[1, a)], Michael Czajka[2], Robert Shanks[2], Fugen Daver[1, b)]

[1] School of Engineering, RMIT University, PO Box 71, Bundoora, Victoria 3083, Australia.
[2] School of Science, RMIT University, GPO Box 2476 Melbourne, Victoria 3001, Australia.

[a)] s3367109@student.rmit.edu.au
[b)] Corresponding author: fugen.daver@rmit.edu.au

Abstract. Carbon monoxide reduced graphene (CO-rG) was compounded with polyamide-6 (PA6) by melt-mixing to investigate the suitability of CO-rG as novel filler for enhanced mechanical properties in PA6-based composites. CO-rG was first exfoliated via probe ultrasonication in p-xylene prior to melt-mixing. Nanocomposites of differing compositions were then prepared and characterised for their elastic modulus (E), tensile yield strength (σ_{yield}), and strain at break (ε_b). Interactions between CO-rG and PA6 were characterised via Fourier-transform infrared (FTIR) spectroscopy and observed via scanning electron microscopy (SEM). Significantly enhanced mechanical properties were achieved even at low CO-rG content, with mechanical percolation observed at 1.0 %·w/w CO-rG. Beyond the mechanical percolation concentration, the addition of 6.0 %·w/w CO-rG resulted in overall enhancement in bulk composite properties with a measured E and σ_{yield} of 4107±225 MPa and 62±1 MPa, respectively. FTIR spectroscopy confirmed the existence of hydrogen bonding between CO-rG and PA6. SEM imaging indicated uniform dispersion of CO-rG in PA6.

Key words: graphene, polyamide 6, nanocomposite, mechanical properties, melt-mixing.

INTRODUCTION

Carbon monoxide reduced graphene (CO-rG) prepared by thermal reduction in a reductive CO atmosphere, has fewer defects than graphenes that have been reduced in inert atmospheres. CO-rG was successfully dispersed in poly(styrene-b-butadiene-b-styrene) (SBS) and poly(trimethylene terephthalate)-epoxy resin by utilising π-interactions between filler and matrix for enhanced dynamic mechanical properties (1, 2). While the preparation of CO-rG and its dispersion via π-interactions represent an efficient means to produce and apply high-quality graphene in aromatic polymer composite systems, its application as a filler in non-aromatic polymers, like polyamide-6 (PA6), via melt-mixing has not been investigated.

The aim is to demonstrate that CO-rG–PA6 composites can be effectively prepared via melt-mixing for enhanced mechanical properties. Melt-mixed CO-rG–PA6 composites were characterized by tensile mechanical testing, scanning electron microscopy (SEM), and Fourier transform infrared (FTIR) spectroscopy.

EXPERIMENTAL

Materials Processing and Composite Preparation

PA6, trade name Taulman 645, was supplied by Taulman3D (Missouri, USA) in the form of white pellets.

Graftech 220-50N (GT) expandable graphite was supplied by UCAR Carbon Company (Ohio, USA). GT in 1 g amount was expanded in a furnace for 30 s to create expanded GT. The furnace was preheated to 1000 °C and expansion was performed in air. Expanded GT was then further heated in a ceramic tube furnace at 1000 °C for 8 h in a 100 % carbon monoxide atmosphere to produce CO-rG (1, 2).

A Qsonica Q500 sonicator with 12.7 mm diameter probe was used to exfoliate CO-rG prior to melt-mixing. A Sonopuls RZ 2 rosette cell was used to encourage uniform ultrasonication of the dispersion. Amplitude of 100 % was applied for 10 min/g of CO-rG. All composites were ultrasonicated in 40 mL of p-xylene, with a maximum batch size of 0.5 g G for an energy density of 128 J/g of CO-rG. Ultrasonication was carried out by alternating between 1 min of sonication and 1 min of cooling in an ice bath.

CO-rG-PA6 composites, in approximately 50 g batches with varying CO-rG content were produced via melt-mixing with a Haake Rheomix 600 Batch Intensive Mixer with counter-rotating roller impellors. PA6 pellets were dried in a vacuum oven at 80 °C for 20 h prior to processing. Mixing was performed for a total of 15 min at 245 °C and at 100 rpm. PA6 was first added to the melt-mixer and allowed to disperse for 5 min. CO-rG suspended in p-xylene was then added to the melt-mixer in a steady stream and allowed to disperse for a further 10 min.

Compression moulding of all nanocomposites was performed with a heated press at 245 °C. The composite was first allowed to melt at contact pressure for 10 min. Pressure was subsequently ramped to 20 MPa over 1 min, and pressure was maintained for an additional 5 min. The mould was water-cooled at an average cooling rate of 35 °C/min to 60 °C before the composites was removed.

Characterisation

Uni-axial tensile testing was conducted with an Instron Universal Test Instrument, Model 4465, with 10 kN load cell according to ASTM D638-97, specimen type IV specifications. An Epsilon Model 3542 axial extensometer with a gauge length of 25 mm was used to measure strain. Elastic modulus (E), tensile yield strength (σ_{yield}), and elongation at break (ε_b) were evaluated at a cross-head speed of 5 mm/min at ambient temperature. Results presented are an average of at least three measurements.

FTIR spectroscopy was performed with a Perkin-Elmer Spectrum 2000 FTIR spectrometer and PIKE Technologies GladiATR. All spectra were scanned within the range of 400-4000 cm^{-1}, with a total of 24 scans and a resolution of 4 cm^{-1}. All spectra were normalised to the peaks corresponding to CH_2 rocking modes at ~730 cm^{-1}.

SEM images of the composites were taken using a FEI Quanta 200 environmental scanning electron microscope operating at 3 kV in high vacuum. Specimens were mounted to the specimen holder using conducting carbon-impregnated tape. Prior to imaging, a gold (Au) sputter coating was applied onto the specimen.

RESULTS AND DISCUSSION

The contribution of CO-rG to the E, σ_{yield}, and ε_b are presented in Fig. 1. As the ε_b of GN0.0 and GN0.1 had exceeded the measuring range of the axial extensometer, data presented for ε_b are relative values and should be interpreted qualitatively. Specimens are labeled from GN0.0 to GN6.0, with the numbers corresponding to the weight fraction of CO-rG in the composite.

The measured σ_{yield} was maximum at a critical composition of GN1.0 (66±3 MPa), subsequently decreasing as CO-rG content increased to 3.0 %·w/w (61±3 MPa). This critical composition is referred to as mechanical percolation threshold (3). Above the mechanical percolation composition, the reduced physical distances between individual graphene sheets can cause restacking due to attractive van der Waals forces. As a result, mechanical reinforcement will become limited by slippage occurring between the graphene sheets.

An increase in particle agglomeration with increasing CO-rG content beyond the mechanical percolation threshold is expected to be detrimental to the σ_{yield} and E of the composite. However, σ_{yield} and E generally remained unchanged between GN3.0 - 6.0, and between GN1.0 - 3.0 respectively. Furthermore, a significant increase in E was observed when CO-rG content was increased to 6.0 %·w/w. GN6.0 possessed significantly enhanced mechanical properties with measured E and σ_{yield} of 4107±225 MPa and 62±1 MPa, respectively. As such, the mechanical properties of CO-rG-PA6 composites were indicative of efficient CO-rG dispersion.

Empirical data were compared against Halpin-Tsai predictions to further evaluate the mechanical properties of the composite (4). Theoretical predictions underestimated the E of the composite when CO-rG content was low (Fig. 1b). The significance of these observations was further accentuated by the assumption of model parameters that can only be achieved in truly pristine monolayer graphene. The phenomenon had been attributed to the synergism of mechanical interlocking and hydrogen bonding interactions between the matrix and filler phases on mechanical reinforcement. The Halpin-Tsai model is a micromechanical model that was developed by fitting model parameters to empirical data. As such, its failure to take into account any synergistic interactions as a result of the length scale of the nanofillers may have contributed to its poor fit to the experimental data.

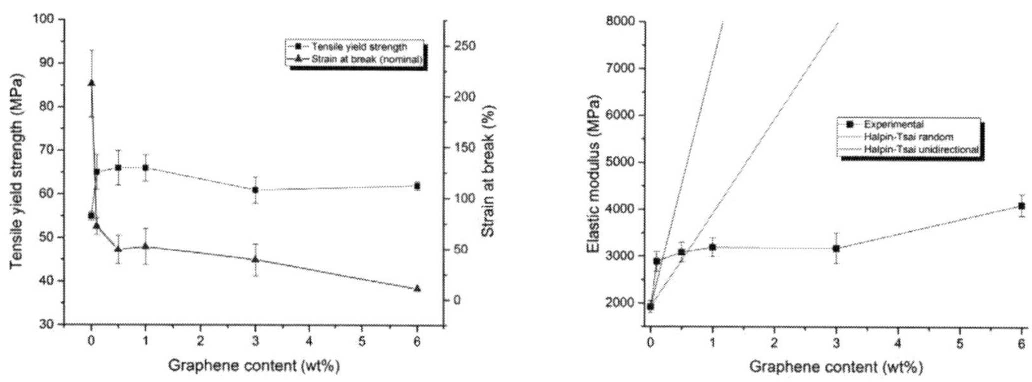

(a) (b)

FIGURE 1. a) Graphene modification of tensile yield strength and strain at break, and b) comparison of experimental elastic modulus with micromechanical models

SEM imaging was used to observe fracture surface morphology of G-PA6 nanocomposites. Neat PA6 displays relatively smooth fracture surface, with crack propagation in a generally linear direction (Fig. 2a). With the addition of CO-rG, a noticeable increase in surface roughness was observed (Fig. 2b). This is attributed to crack deflection due to the presence of CO-rG. Furthermore, it is observed that CO-rG was firmly embedded in the polymer matrix and still efficiently dispersed at 6.0 %·w/w CO-rG content (Fig. 2b).

An increase in surface roughness had been associated with enhancements to material toughness, and is indicative of good filler-matrix bonding (5).

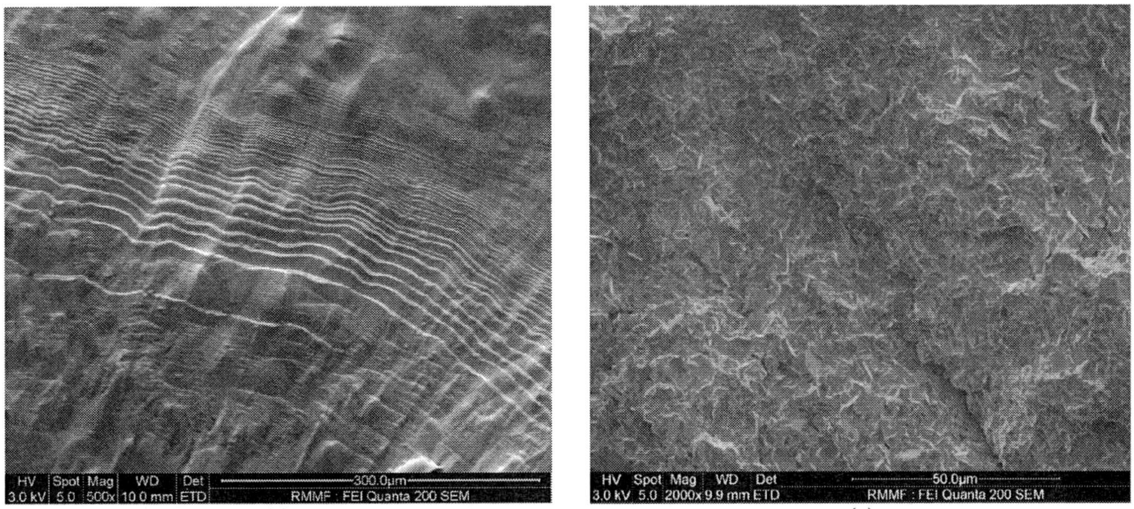

(a) (c)

FIGURE 2. SEM images of CO-rG-PA6 composite tensile fracture surfaces for (a) GN0.0, and b) GN6.0

FTIR spectroscopy was applied to identify interactions between CO-rG and PA6. The region of 1700-1500 cm^{-1} in the FTIR spectra is significant as the amide I and amide II bands have been associated with hydrogen bonding (6) (Fig 2b). A peak shift in the amide I band to a lower frequency is attributed to formation of ordered hydrogen-bonded C=O groups. The amide II band in neat PA6 splits into two components, with peaks at 1542 cm^{-1} and 1535 cm^{-1}. The intensity of the latter peak relative to the former increases with increasing CO-rG content,, culminating in a distinct peak at 1535 cm^{-1}. This can be attributed to an increasing component of ordered hydrogen-bonded N-H groups, resulting in a distinct peak at lower frequency. A general increase in amide I and amide II peak intensities as CO-rG content increases is indicative of the formation of hydrogen bonds between CO-rG and PA6.

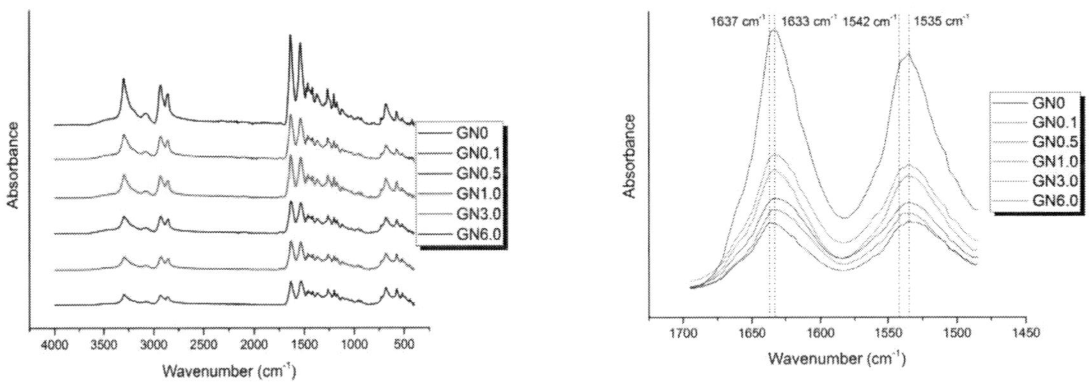

(a) (b)

FIGURE 3. a) FTIR spectra of CO-rG–PA6 composites and b) Close-up of FTIR spectra between 1700-1500 cm^{-1}

CONCLUSION

CO-rG can be effectively dispersed into PA6 by melt-mixing. At 0.1 %·w/w CO-rG content, the measured *E* of CO-rG-PA6 composites displayed synergistic extreme reinforcement by exceeding micromechanical predictions. This is a unique observation of such behavior in composites containing graphene with deliberately reduced functionalization. Uniform dispersion and matrix-filler interactions were observed on the tensile fracture surfaces of the composites. This is attributed to formation of hydrogen bonds between CO-rG and PA6, as indicated in the FTIR spectra. As a result, the detrimental effect of increasing graphene content beyond the mechanical percolation threshold, to bulk composite mechanical properties, was diminished. CO-rG was effectively dispersed in PA6 through melt-mixing for enhanced mechanical properties.

REFERENCES

1. Czajka M, Shanks R, Oldfield D, editors. Carbon monoxide reduced low-defect graphene nanocomposites with poly(styrene-bbutadieneb-styrene). Annual Technical Conference - ANTEC, Conference Proceedings; 2015.
2. Sarathchandran C, Czajka M, Chan CH, Shanks RA, Thomas S. Interfacial interactions of thermally reduced graphene in poly(trimethylene terephthalate)-epoxy resin based composites. Polymer (United Kingdom). 2016;106:140-51.
3. Zhao X, Zhang Q, Chen D, Lu P. Enhanced mechanical properties of graphene-based polyvinyl alcohol composites. Macromolecules. 2010;43(5):2357-63.
4. Halpin JC, Kardos JL. The Halpin-Tsai equations: A review. Polymer Engineering & Science. 1976;16(5):344-52.
5. Hou W, Tang B, Lu L, Sun J, Wang J, Qin C, et al. Preparation and physico-mechanical properties of amine-functionalized graphene/polyamide 6 nanocomposite fiber as a high performance material. RSC Adv. 2014;4(10):4848-55.
6. Navarro-Pardo F, Martínez-Barrera G, Martínez-Hernández AL, Castaño VM, Rivera-Armenta JL, Medellín-Rodríguez F, et al. Effects on the thermo-mechanical and crystallinity properties of nylon 6,6 electrospun fibres reinforced with one dimensional (1D) and two dimensional (2D) carbon. Mater. 2013;6(8):3494-513.

Effect of mixing sequence on the fibrillation of thermo tropic liquid crystalline polymer in the blend system in presence of graphene

ADHIKARY MUNESH CHANDRA (1)*, RATH SANJIT KUMAR (1), MH PRIYADARSHINI (1), DAS CHAPAL KUMAR (1)

(1) FAKIR MOHAN UNIVERSITY - ODISHA - India

The literature review revealed that the presence of nano fillers can alter the morphology of TLCP droplets by altering the flow properties of the blend system. However, the question remains unexplored, what is the effect of sheet like nano fillers on in-situ fibrillation behavior of TLCP in the polymer blend system. In this study, we have used the graphene as nano filler in the nylon-6 and TLCP blends. We have also studied the different sequences of mixing process so as to impart the maximum in-situ fibrillation of TLCP in order to achieve the better properties of the blends. Four types of blending techniques we have considered as follows. a. Entire graphene added to nylon-6 and then blended with TLCP b. Entire graphene is added to TLCP and then blended with nylon-6. c. 50% graphene is added to nylon-6 and 50% graphene is added to TLCP and then both master batches were blended together. d. Both virgin polymers were blended together first and then the entire graphene is added to the virgin polymer blend system. All the blends were then characterized rheologically, mechanically and spectroscopically to assess the extent of fibrillation behavior of graphene as well as its preferable location in the phases. The viscosities of TLCP containing blends are affected by two factors like interfacial adhesion and TLCP fiber formation and orientation. The orientation of TLCP fibers along the flow direction can enhance

Evaluation of the Fracture Resistance of Epoxy-Carbon Nano-Composites

***Anwer Muhammad A. S.* [(1)], *Naguib Hani* [(1)*]**

[(1)] University of Toronto - Ontario - Canada

Epoxy has long been recognized for its high strength, stiffness, and bond strength. It's high bond strength makes possible the transfer of high shear stresses. This makes it extremely useful for discontinuous and continuous fiber composites applications. However, epoxy is inherently brittle with low toughness. To enhance its fracture resistance, adding filler particles to cause extrinsic toughening is a common technique. Nanofillers like CNT and Graphene are popular candidates due to their extremely high intrinsic properties. However, they are expensive and hence pose difficulty for mass scale production. This work evaluates the effect of Carbon Nanofibers and Graphite nano-particles in the fracture toughness and tensile characteristics of epoxy. Both Carbon Nanofibers and Graphite nanoparticles are high performance fillers but at a fraction of the cost of other nanofillers. Detailed morphological characterization is also performed to elucidate the structure property relationships of the developed composites.

In-situ polymerization of poly (butylene terephthalate) in presence of graphene-related materials: effects of nanoparticles structure and defectiveness on crystallinity and thermal conductivity of the relevant nanocomposites

Colonna Samuele (1)*, Bernal Maria Del Mar (1), Monticelli Orietta (2), Gomez Julio (3), Novara Chiara (4), Saracco Guido (5), Müller Alejandro Jesus (6), Liu Guoming (7), Wang Dujin (7), Fina Alberto (1)

(1) Politecnico di Torino - Alessandria - Italy, (2) Università degli studi di Genova - Genova - Italy, (3) AVANZARE Innovacion Tecnologica S.L. - La Rioja - Spain, (4) Politecnico di Torino - Torino - Italy, (5) Istituto Italiano di Tecnologia, Centre for Sustainable Futures - Torino - Italy, (6) POLYMAT, University of the Basque country and Ikerbasque - Basque Country - Spain, (7) Beijing National Laboratory for Molecular Sciences, CAS Key Laboratory of Engineering Plastics - Bejing - China

Cyclic butylene terephthalate (CBT) were polymerized into poly (butylene terephthalate) (pCBT) in presence of different types of graphene-related materials, with variable size and structural defectiveness. Selected nanoflakes were annealed at 1700°C in vacuum, resulting in a reduced amount of defects, as consistently proven by several techniques [1]. Nanocomposites were prepared by pre-dispersing nanoflakes in CBT, followed by catalyzed ring-opening polymerization during extrusion. This technique allowed obtaining superior dispersion of nanoflakes compared to conventional melt processing methods, thanks to facile infiltration of CBT between the nanoflakes and further dispersion as a result of shear applied during mixing [2]. Thermal conductivity tests showed significant variability as a function of nanoparticle defectiveness, surface area and lateral dimensions. A dramatic increase in the thermal conductivity of the nanocomposite was observed in presence of annealed nanoflakes (up to 1.7 W/m·K with 10 wt.% loading), compared to the pristine nanoflakes, evidencing the importance of using low defective materials. The study of nanocomposite thermal conductivity before and after polymerization was also addressed, to gain insight in the modification of nanocomposites morphology upon polymerization [3]. Furthermore, the effect of different processing parameters (time, temperature, shear rate) when preparing nanocomposites was addressed [4]. The presence of nanoflakes was found to strongly affect crystallization of nanocomposites evidencing for nucleating effects and the formation of a new high-temperature melting phase depending on the nanoparticle grade. References M. Tortello, S. Colonna, et al. Carbon 109 (2016) 390-401 2 S. Colonna, et al., Polymer 102 (2016) 292-300 3 S. Colonna, et al, Sumbitted to European Polymer Journal 4 S. Colonna, et al, Materials & Design 119 (2017) 124-132

Effect of UV-enhanced Oxidation on Tensile Strength of Carbon Nanofiber Reinforced Polyamide 11

ENOMOTO Kazuki (1)*, INABA Takumi (2), TAKAHASHI Yukiya (2)

(1) Meijo University - Aichi - Japan, (2) Graduate School of Meijo University - Aichi - Japan

In our previous study (Ref.1), two-step oxidation by using mixed acid and hydrogen peroxide was performed on the CNFs, and tensile property of the composites reinforced with CNF was improved by two-step oxidation. This was because that the mild oxidation by using hydrogen peroxide in the second step was effective to introduce functional groups including oxygen from the results of XPS analysis. In this study, in the second step of the two-step oxidation, it was investigated whether the functional groups can be introduced more efficiently by irradiating with ultraviolet rays, moreover, optimization of UV-enhanced oxidation conditions was performed. From the results of the XPS analysis, by UV-enhanced oxidation, more functional groups including oxygen were introduced to the surface of CNFs compared to the conventional two-step oxidation. When the UV irradiation time was varied in the second step of the two-step oxidation, the oxygen content increased until 2 hours and then saturated. From the results of tensile tests, the tensile strength of the composited reinforced with CNF treated by UV-enhanced oxidation reached the maximum at 1 hour of UV irradiation and it decreased at the further treatment time. This is because that even in the mild oxidation using hydrogen peroxide, the structure of CNF was destroyed by UV irradiation and the strength of CNF itself was reduced. Ultimately, by applying UV-enhanced oxidation for 1 hour, the tensile strength was improved compared to the conventional oxidation without UV rays. References (1) K. Enomoto, Y. Takahashi: Oxidization of Carbon Nanofiber and Its Effect on Tensile Properties of PA11 Matrix Composite Reinforced with Carbon Nanofiber, Conference Paper of PPS-32(Lyon), (2016)

Stronger Cement

Regev Oren [1]*, Nadiv Roey [1]

[1] Ben-Gurion University of the Negev - Israel - Israel

The growing availability of nanotubes (NT) and the increased knowledge about their loading in polymers have prompted their incorporation in cementitious matrices. The main challenge towards achieving a significant enhancement in cement properties is an effective dispersion of the agglomerated NT. In my talk I will demonstrate how efficient NT (carbon or WS2) dispersion in cement results in substantial flexural and compressive strength enhancements at NT concentration below 0.15 wt%. The reinforcement by WS2 NTs remains significant after a variety of curing processes, suggesting a genuine nanoscale reinforcing effect. By employing a comprehensive fractography we show that the WS2 NTs inhibit crack propagation by bridging with a pullout failure mechanism.

Hybrid nanocomposite as strain sensor on the base of SBS/CNT/RGO

Haghshenas Samira [(1)], Mehranpour Milad [(1)*], Bazgir Saeed [(1)]

[(1)] Islamic azad university - Tehran - Iran

In the recent years conductive polymer composites (CPCs) as strain sensors has been widely investigated. A wide range of strain sensitivities and high repeatability are vital for different applications of strain sensors. Many of researches had done on the sensitivity of the strain and a bit researches were found on investigating the repeatability and stability with time of these sensors. In the present study hybrid conductive nanocomposites based on styrene-butadiene-styrene three-block copolymer (SBS), multi wall carbon nanotubes and reduced graphene oxide with different ratios of fillers was made by melt mixing method. Hybrid and non-hybrid nanocomposites were made by internal mixer and different sensitivity of strain sensor were obtained. Hybrid nanocomposites (SBS / MWCNT / RGO) with filler content of 4 wt% were selected as a high sensitivity sensor as well as good repeatability. Simultaneous the results showed that synergistic effect of two fillers, which causes a more stable conductive networks .To achieve such proper networks using individual filler in the polymer matrix was not possible.Morphology of the all samples were investigated by SEM.

Development of Polymer Composites Containing Nano-carbon Materials by High-pressure Fluid Mixing Method

Shin-ichi Kihara[a)], Yusuke Tsukuda, Masao Asada, and Shigeki Takishima

Department of Chemical Engineering, Graduate School of Engineering, Academy of Science and Technology, Applied Chemistry/Chemical Engineering Unit, Hiroshima University, 1-4-1 Kagamiyama, Higashi-hiroshima 739-8527 JAPAN

[a)] *Corresponding author: snkihara@hiroshima-u.ac.jp*

Abstract. It is well known that it is not easy to develop a polymer nanocomposite with nanocarbon materials, NCMs (carbon nanotube; CNT or graphite), which have several superior electrical, thermal and mechanical properties. The difficulty comes from the less-reactive high surface area of NCMs which is basically hydrophobic and interacts by the van der Waals (vdW) force makes strong bundled or clustered structure. To defibrate the bundle structures of NCMs in the hydrophobic polymeric matrix, we have developed a new melt mixing method under a high-pressure fluid situation; polymer melts mixing is carried out under unfilled state where the unfilled space is filled with supercritical fluids (SCFs). The SCF such as $SCCO_2$ makes polymers plasticized and enhances the mobility of polymers and also has a similar density of organic solvents with less interfacial tension, and can enhance polymer penetration into nano-spaces between bundles. In this study, we have performed the mixing under high pressure CO_2 and N_2 circumstances, we found that the mixing under $SCCO_2$ produced a well dispersed CNTs composites with less breaking CNTs around the shear stress of several kPa. To advance more finely dispersion of CNTs under high-pressure fluid mixing, we added some polar and non-polar solvents. However, it was not effective for dispersive mixing for high-polar solvent with high pressure of CO_2 or N_2.

1. Introduction

As one of nano-carbon materials (NCMs), CNTs are one dimensional, stable and strong materials and show several superior properties [1]. Researchers and engineers want to build in these properties in the conventional polymers. It is well known that polymers as standard soft materials have properties of lightweight, easy forming, insulation and flexible, but some of these polymeric properties show mutually exclusive characteristics to NCMs. However, by fusing these contradicting properties with a designated nano-hierarchical structure, it is expected that new functional soft materials will be produced. For sufficient performance of CNTs in the composites, it is required to spatially disperse CNTs in a sufficiently large aspect ratio without buckling and breaking. However, it is difficult to disperse CNTs in uniformly and finely because CNTs have a high cohesiveness energy comes from van der Waals (vdW) interaction of high specific surface area and form bundle structures naturally [2]. In this study, to develop a method to defibrate the bundles and disperse CNTs with less defects in polymer, we have carried out polymer melt mixing with high-pressure fluids that have affinity for polymers and have a property of high diffusivity for the nano-space.

2. Concept of high-pressure fluid mixing on dispersion of CNTs into polymer

The basic concept of High Pressure Fluid Mixing (HPFM) [3] for defibrating CNTs bundles and dispersing CNTs in polymer melts is as follows; This method is similar to melt blending (MB) but mixing under supercritical fluids (SCF). The disadvantages of dispersion process of MB [4] might be overcome by utilizing the characteristics of the supercritical carbon dioxide ($SCCO_2$) [5]. Firstly, swelling of polymer matrix enhancing the mobility of the polymer chain and lower the surface tension of the polymer by dissolving of $SCCO_2$. Secondary, enhancing penetrating of low molecular compounds into hydrophobic nano-space between CNTs bundles. Thirdly, the bundle structure stabilized by the vdW force will be separate by the shear stress on its ends like opening the tear cheese from the end. Shear stress above the amount of the vdW forces around the tip of the CNTs bundles makes the bundles opening and then polymer chains can penetrate the new gap by reptation motion while preventing reaggregation of CNTs. By repeating the defibrating of bundles from the tips, like the zipper effect, the HPFM process will be able to produce CNTs/polymer nanocomposites with less damage of CNTs.

3. Developed mixing apparatus and sample preparation conditions

Figure 1 shows our developed HPFM apparatus [3]. This is a batch type co-rotating twin screw mixer with neutral set of four two-tip kneading disks. The mixing elements are possible to be changed depending on the materials. This system consists of the pressurized gas supplying unit, the high-pressure cell unit and the gas trapped unit. The feature of the apparatus is that it is possible to mix polymer melts under unfilled state up to 16 MPa atmosphere. Sample preparing procedure is as follows: charging CNTs and polymer pellets into the cell and vacuuming the cell, heating up to the mixing temperature and left for 1 h, and then introducing gas and left for 1 h to set the constant pressure. After mixing under the constant pressure, depressurization the cell and collecting sample.

Samples used in the present study are followings; CNTs is kindlily provided from the Nippon ZEON Corp. The CNTs is composed of single wall produced by the super-growth chemical vapor deposition method [6] and the average diameter is 4 nm and length is about 400 μm. The initial average bundle diameter counted by 200 points of FE-SEM images (S-5200, Hitachi High-Technologies Corp.) is about 9 nm and the bundle is arranged at interval of about 50 nm like an afforested forest. Some CNTs are bended and crossing each other. Polymer used is the general polystyrene (M_w = 329 kg/mol, M_w/M_n =2.08, T_g = 100℃, PS Japan Co. Ltd). Sample preparation conditions in this study is shows in Table 1.

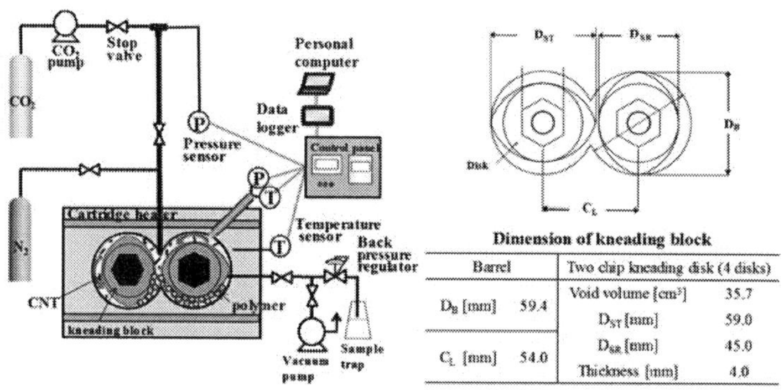

Figure 1. High-pressure fluid mixing (HPFM) apparatus.

Table 1 Experimental conditions.

Mixing method		High Pressure Fluid Mixing (HPFM)			Melt Blending (MB)		
Sample #		(a)	(b)	(c)	(d)	(e)	(f)
Organic solvent		-	toluene	NMP	-	toluene	NMP
Composition [wt%]	CNT	1.0	1.0	1.0	1.0	1.0	1.0
	PS	99.0	96.2	96.4	99.0	96.4	95.7
	solvent	-	2.8	2.6	-	2.6	3.3
Mixing temp. [℃]		120～240	180		150～240	180	
Mixing press. (gas) [MPa]		12～16 (CO_2) 12 (N_2)	12 (CO_2)		0.4 (N_2)		
Rotational speed [rpm]		60～180	60		60		
Mixing time [h]		2					

Evaluation of dispersibility of CNTs in the composite used in this study was the followings. The defibrating of bundles was estimated by measuring the SEM images of diameter randomly selected 200 points of CNTs bundles extracting from the composite. The extracting method was that dissolving the composites in a large amount of toluene over one day and filtrating the solution with PTFE membrane filter (pore size: 0.5 μm, Advantech), and then collecting CNTs on the membrane filter. The damage of CNTs was estimated by the intensity ratio G/D of the graphite band (G-band; 1590 cm^{-1}) and defect band (D-band; 1350 cm^{-1}) using Raman spectroscopy [7] of wave length 514 nm (HT64000, HORIBA Ltd.). The G/D ratio of the pristine CNTs is 4.52.

4. Result and discussion.

(a) Effect of mixing conditions: We compare the mixing ability on defibrating of CNTs bundles in polymer by the result of atmosphere gases (CO_2, and N_2), the temperature, and rotational speed of kneading. From Figure 2 (a), as increasing the temperature, the large bundles having larger than 1μm in diameter decreased independent of species of atmospheric gases. Above the temperature of 200℃, the HPFM and MB were almost the same performance in the point of the average diameter of bundled CNTs. The tendency of the mixing temperature on defibrating of CNTs bundles would reveal that the thermal fluctuation of CNTs strongly enhances the de-bundle process. However, $SCCO_2$ mixing can defibrate CNTs bundles at a lower temperature than MB. This result means that HPFM can prevent thermal decomposition of polymer in a long time of melt mixing. On the other hand, when CO_2 pressure increases up to 16MPa and temperature above 200℃, the integrated bundles of larger than 1 μm in diameter remained. The reason is that polymer in this mixing condition was excessively plasticized by CO_2

dissolution and the shear stress loaded in the mixing was smaller than the vdW forces between CNTs. To defibrate the bundles, it will be necessary to apply a shear stress above a certain critical value in the mixing. Figure 3 shows the effect of the shear stress on the defibrating of CNT bundles. The representative shear stress was estimated by the shear rate at the inter meshing zone of twin screw and the shear viscosity of PS of gas saturation at the shear rate. The upper value of the error bar in Fig. 3 (a) indicates the maximum bundle diameter and the lower value of error bar indicates the minimum bundle diameter observed. Regardless of the mixing under CO_2 and N_2 atmospheres, the shear stress ranges from 2 to 4 kPa was needed to defibrate the bundles, which was corresponding to the order of shear rate 10^6 s^{-1} at the jet milling in solution process (assuming 1 Pa·s). However, the higher the shear stress, the larger bundles remained but the smaller bundles were also produced. Therefore, by increasing the mobility of the polymer as much as possible as around the critical stress, the defibrating of bundles progresses uniformly. Fig. 3 (b) and (c) shows the index of the breaking of CNTs structure, G/D band ratio, against the shear stress and the average bundle diameter, respectively. From Fig. 3, mixing with $SCCO_2$ was superior to mixing under N_2 in spite of the average bundle diameter was almost same. In addition, the zero shear viscosity of the CNTs/PS nanocomposites at 180℃ that produced by HPFM at 200℃, 12MPa CO_2 , MB at 240℃, 0.4 MPa N_2 were 14.3 and 3.1 times larger than that of the pure polymer, respectively. Therefore, the HPFM with $SCCO_2$ will be able to disperse CNTs with larger *L*/*d* in polymer melts. Since maintaining the crystalline structure of CNT as possible as the pristine CNT is needed for advanced functional nanocomposites, the HPFM will be a superior method to prepare the CNTs/polymer nanocomposites.

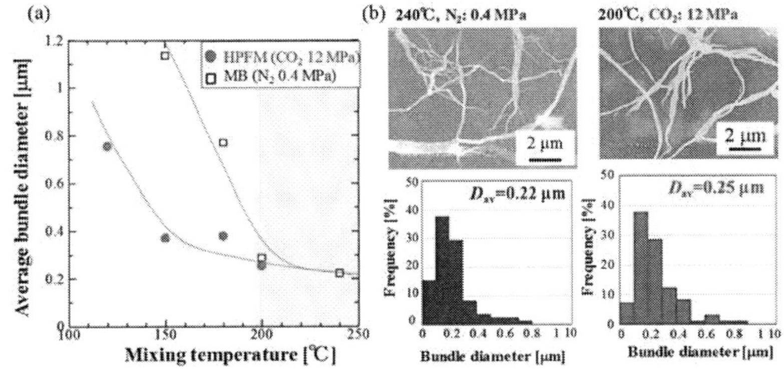

Figure 2. Effect of mixing performance on the mixing temperature. The sample preparation conditions were shown in Table 1 (a) and (d). (a) the average bundle diameter against the mixing temperature (lines are for the guide to tendency), (b) comparison of bundle diameter distributions between MB with 0.4 MPa N_2 at 240℃ and HPFM with 12 MPa CO_2 at 200℃.

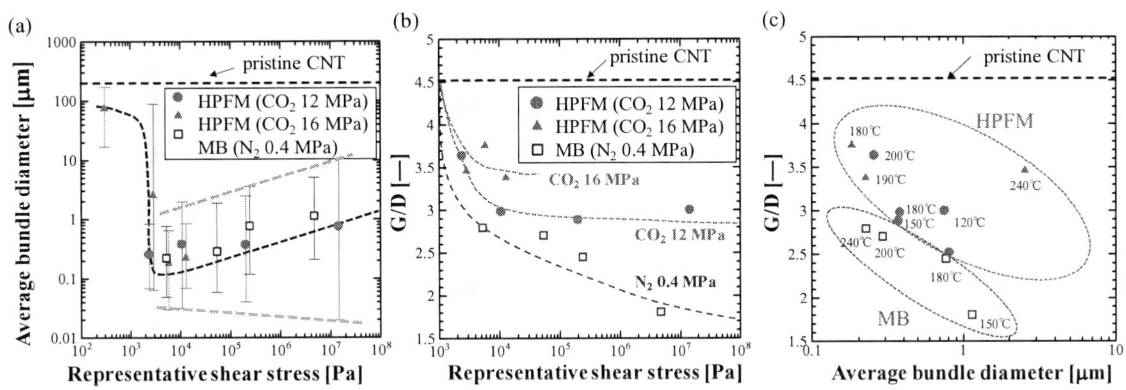

Figure 3. Comparison between HPFM with $SCCO_2$ and MB with N_2 on the CNTs dispersion. The sample preparation conditions were shown in Table 1 (a) and (d). (a) the average bundle diameter against representative shear stress, (b) the G/D against representative shear stress, and (c) the G/D against average bundle diameter. Lines were for the guide of tendency.

(b) Effect of organic solvent/SCF: In order to exam the effect of the organic solvent to enhance the dispersion of CNTs, N-methyl-2-pyrrolidone (NMP) which is compatible to NCMs and toluene which is not compatible to NCMs were added as shown in Table 1. These compatibilities were judged from the Hansen solubility parameter [8]. From the SEM images of figure 4, a slightly defibrating and de-bundle progress was observed for adding toluene, but not progress for case of adding NMP. In addition, toluene is soluble in $SCCO_2$ and it was removed only by de-pressuring process after HPFM with $SCCO_2$. In spite that NMP might be a good dispersant of NCMs and also dissolve PS, the bundle of CNTs was not defibrating because of partially solubility of NMP to SCO_2. Therefore, the wetting balance among CNTs and polymer/solvent under the constraint of the critical shear stress will be required to promote defibrating process with less damage of the smaller bundles that remaining in mixing condition with pure $SCCO_2$.

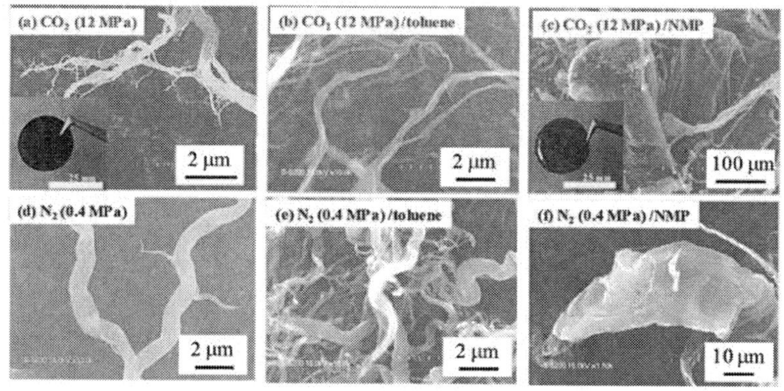

Figure 4. SEM images of the extracted CNTs. (a) and (d) were the sample prepared without organic solvent, (b) and (e) were of the sample prepared with toluene, (c) and (f) were of the sample prepared with NMP. The sample preparation conditions were shown in Table 1.

5. Conclusion

By comparing of defibration ability of CNTs bundles in PS between the high-pressure fluid mixing (HPFM) and melt blending (MB), we have studied the effect of $SCCO_2$ on the mixing. We found the followings; (1) the higher shear stress larger than the critical shear stress about 2kPa to 4 kPa was needed to defibrate the CNTs bundles. (2) HPFM with $SCCO_2$ around the critical stress was a superior mixing condition for dispersing CNTs with less damage in polymer melt. (3) some of CNTs formed strong bundles, and it was difficult to defibrate the CNT bundles less than 200 nm or less in diameter. (4) the effect of introducing an organic solvent which plasticizes polymers and is also compatible to NCMs more than pure $SCCO_2$ was not possible to be judged only by the wettability to CNTs.

In the future, we will further experimentally investigate the influence of organic solvent/SCFs and shear stress on dispersion and defibration of NCMs including semi-graphene, and will develop a model of the defibration processing for achieving finer NCMs/polymer composites by HPFM.

REFERENCES

1. (a) J. N. Coleman, U. Khan, W. J. Blau, Y. K. Gun'ko, "Small but strong: A review of the mechanical properties of carbon nanotube–polymer composites", Carbon, **44**, 1624-1652 (2006), (b) Z. Han, and A. Fina," Thermal conductivity of carbon nanotubes and their polymer nanocomposites: A review", Prog. Polym. Sci., **36**, 914-944 (2001)
2. R. S. Ruoff, J. Tersoff, D. C. Lorents, S. Subramoney and B. Chan, "Radial deformation of carbon nanotubes by van der Waals forces", Nature, **364**, 514–516 (1993).
3. S. Kihara, M. Asada, M. Haruki, and S. Takishima," Effect of the high pressure fluid mixing on preparation of polymer nanocomposites containing carbon nanotubes", 5402, 17th International Congress on Rheology 2016, 08 - 13 August 2016, Kyoto.
4. S. Ata, H. Yoon, C. Subramaniam, T. Mizuno, A. Nishizawa, and K. Hata, "Scalable, solvent-less de-bundling of single-wall carbon nanotube into elastomers for high conductive functionality", Polymer, **55**, 5276-5283 (2014).
5. S. G. Kazarian, "Polymer processing with supercritical fluids", Polym. Sci., ser. C, **42**, 78-101 (2000).
6. K. Hata, D. N. Futaba, K. Mizuno, T. Namai, M. Yumura, S. Iijima. "Water-assisted highly efficient synthesis of impurity-free single walled carbon nanotubes," Science, **306**, 1362–1364 (2004).
7. A. Jorio, M.A. Pimenta, A. G. S. Filho, R. Saito, G. Dresselhaus, and M. S. Dresselhaus, "Characterizing carbon nanotube samples with resonance Raman scattering", New J. Phys., **5**, 139.1-139. 17 (2003).
8. C. M. Hansen, "Hansen solubility parameters: a user's handbook", CRC Press. 2000.

Influence of carbon particle dimension on formation of the filler network in a polymer melt

Lellinger Dirk (1), Hilarius Konrad (2), Müller Michael Thomas (3), Pötschke Petra (3), Alig Ingo (2)*

(1) Faunhofer LBF - Darmstadt - Germany, (2) Fraunhofer LBF - Darmstadt - Germany, (3) Leibniz Institute of Polymer Research - Dresden - Germany

Electrical conductivity and rheological measurements were performed to study the influence of particle dimension on the formation of the filler network in a polymer melt. For the fillers carbon nanoparticles with one, two, three or fractal ($2 < d \leq 3$) dimensions were compared. For this purpose, multi-walled carbon nanotubes (CNT), graphite nanoplates (GNP) and carbon black (CB) were incorporated in a polycarbonate matrix. The insulator-conductor transition of the different composites was studied by dielectric spectroscopy. The electrical conductivity in the melt was measured in a rheo-electrical setup with the electrical field perpendicular to the shear plane and in the shear plane under steady shear conditions as well as in the quiescent melt. During a weak shear flow a breakdown of the electrical conductivity and steady state values of conductivity and viscosity were observed for all filler systems. This general feature can be explained by the interplay of destruction and built up of the filler network. After stopping the shear flow a reorganization of the filler network occurs, which leads to an increase in electrical conductivity. This increase is slower and less complete for GNP and GNP/CB as compared to CB and CNT. For fillers with platelet geometry (GNP and GNP/CB composites) the conductivity perpendicular to the shear plane was found to be higher than in shear direction. This anisotropic conductivity is preserved during recovery of the filler network in the quiescent melt, whereas the CB composites show an equalization of the conductivities of the two directions. This is explained by incomplete reorientation of GNP due to slow rotational diffusion or sterically hindered orientation and the faster translational diffusion for CB. In the case of CNT the CNT orientation is suppressed by agglomeration.

Strategies for Carbon Nanotubes Incorporation in Structural Composites

kchit Nadir [(1)]*

[(1)] Nanocyl - Namur - Belgium

Nanomodified structural composite is presented as the solution to overcome the fiber reinforced polymer FRP issues such as delamination crack propagation. In this contribution the nano-modification for FRP is implemented through nano-engineered thermoplastic polymer. Two approaches were proposed to introduce thermoplastic modified CNTs in multiscale structural composite. The first one is deposition of CNTs thermoplastic veils between the layers of a composite laminate and the second one is powder coating of conventional prepreg by CNTs thermoplastic powder. It is expected that thermoplastic modified CNTs will improve the resin toughness and reduce crack propagation increasing CAI strengths. Thanks to the combination of good electrical and mechanical characteristic of CNTs and toughness of thermoplastic it was shown that the damage tolerance as well as electrical conductivity of multifunctional carbon fiber reinforced polymer was improved compared to conventional composite.

Freestanding mesoporous three-dimensional graphene bulk prepared by chemical vapor deposition on sacrificial Cu nanopowder sinter templates

***Ma Yuxiao* [1]*, *Liu Jianhua* [1], *Yu Mei* [1], *Li Songmei* [1]**

[1] Beihang University - Beijing - China

With the development of energy storage devices, electrode with high specific surface area has attracted significant attention. Meanwhile, miniaturization of energy storage system requires not only high gravimetric but also high volumetric specific surface area. Three-dimensional graphene (3DG) and its application in electrodes has been widely researched. Fabricating reduced graphene oxide (rGO) sheets into a foam structure is a simple and low-cost method to prepare 3DG. However, the inter-sheet resistance between rGO sheets hinders electron transfer, and large amounts of defects in rGO reduces the electron conductivity. An effective way to avoid such drawbacks is to synthesize interconnected 3DG by chemical vapor deposition (CVD) of graphene on metal templates. Interconnected 3DG synthesized by CVD features high conductivity and stability. Traditionally, the metal templates used in CVD methods are metal foams with large pore size of hundreds of micrometers. Due to the hollow structure with large interior pores, it usually has quite low volumetric specific area and poor mechanical robustness. Herein we reports a methods of preparing freestanding mesoporous 3DG by using metal powder sinter as template. Copper powder with average particle size of ~50 nm is sintered into mesoporous copper bulk, meanwhile methane and hydrogen are introduced to carry out the CVD process. The Cu template is then etched by $Fe_2(SO_4)_3$ to obtain freestanding mesoporous 3DG bulk. By changing the parameters of the process, the morphology and structure of 3DG can be precisely controlled. The as-prepared 3DG bulk is with average pore size of ~7 nm, and the surface area of the sample is around 450 m2/g. Furthermore, it features high mechanical resiliency after pressed to high strain of 50%. As the pore size is minimized, the necessary thickness of graphene layer to achieve mechanical robustness can also be remarkably reduced, leading to high gravimetric and volumetric specific area. Such pore size distribution not only increases surface area and improves mechanical properties, but also provides unblocked diffusion path for electrolyte ions. Thus, it is favorable for energy storage devices.

Thermally Conductive Microcellular Polymer Nanocomposites of Graphene Nanoplatelets (GnP)

Hamidinejad S. Mahdi [(1)]*, Chu Raymond K.M. [(1)], Filleter Tobin [(2)], Park Chul B. [(1)]

[(1)] MPML, Department of Mechanical and Industrial Engineering, University of Toronto - ON - Canada, [(2)] Department of Mechanical and Industrial Engineering, University of Toronto - ON - Canada

A new class of thermally conductive microcellular polymer nanocomposites of graphene nanoplatelets (GnP) is reported. Foamed and solid high-density-polyethylene (HDPE)-GnP composites containing different GnP contents (0-16 vol.%) were injection-molded. Foamed composites were fabricated using dissolved pressurized nitrogen gas. The effects of foaming on the dispersion, exfoliation, orientation and inter-connectivity of platelets, and thermal percolation threshold were investigated. The introduction of microcellular foaming, significantly changed the orientation of platelets, enhanced their inter-connectivity via biaxial stretching of the matrix and further exfoliated GnPs in the polymer. Hence, foaming reduced the density of the injection-molded samples up to 30%, and enhanced the thermal conductivity up to two folds. The results revealed that lightweight, highly thermally conductive products with lower filler loading, can be fabricated using foam injection molding for various applications in miniaturized electronic devices, and electric motor systems.

Antistatic and Flame Retardant Rotomouldable Polymer Graphite Nanocomposites

Mhike Washington (1)*

(1) University of Pretoria - Gauteng - SouthAfrica

Two commercial expandable graphites were evaluated as precursors for the synthesis of graphite nanoplatelets. Microwave radiation treatment was shown to be more efficient in exfoliating expandable graphite than furnace heating. The expandable graphite with better exfoliating characteristics was selected. XRD results of this graphite showed that it was a high stage graphite intercalation compound. Graphite nanoplatelets with an average particle size of 13 μm and an estimated thickness of about 76 nm were prepared by microwave exfoliation and ultrasonication-assisted liquid phase exfoliation in isopropanol from the selected expandable graphite. Prior to the selection of isopropanol as the ultrasonication media, various exfoliation media that encompassed different solvents and water with various surfactants had been evaluated, on the basis of their acoustic cavitation characteristics. The graphite nanoplatelets were used as a functional additive to fabricate linear low density polyethylene (LLDPE) and poly(ethylene-co-vinyl acetate) (EVA) based nanocomposites using the rotational moulding (rotomoulding) process. The dry blending approach yielded surface resistivities within the static dissipation range (antistatic) at filler loadings as low as 0.25 wt.% (0.1 vol.%). However, even at this low graphite content, impact properties were significantly reduced compared to the neat polymers. Bilayer mouldings via the double dumping method proved to be a feasible approach to achieve both acceptable mechanical properties and antistatic properties. This was achieved by rotomoulding nanocomposites with a 1 mm outer layer containing the filler and a 2 mm inner layer of neat LLDPE. Excellent fire resistance, in terms of cone calorimeter testing, was achieved when the outer layer also contained 10 wt.% expandable graphite.

Functional Nanocomposites Based on Cellulose and Carbon Nanotubes

Hajian Alireza [(1)]*, Berglund Lars [(1)]

[(1)] KTH Royal Institute of Technology - Stockholm - Sweden

We hereby report a new method of making carbon nanotube (CNT) nanocomposites with one of the most abundant bio-based polymers (cellulose as a matrix). This is done by the assembly of CNTs in cellulose nanofibrils (CNFs) colloidal suspensions in water without any surface modification on CNTs or using surfactant solutions. This gives the capability to add the functionality of the CNTs to cellulose structures without alteration on the pristine CNT structure. Another important issue is the ability to use CNFs themselves as building blocks of various structures such as multifunctional aerogels and microfibers. Moreover, CNFs as well as CNF-CNT hybrids can be transferred into gels and arrested structures in relatively low concentrations. Then, this gel transitions can be done inside the microfluidic channels after hydrodynamic alignment of the nanoparticles. So the aligned particles are arrested in the gel form with high Herman orientation factor that results in high strength and conductivity of the microfibers in the fiber direction. Another important application of these gels is in the formation of aerogels with very low density, relatively high modulus, and conductivity. These aerogels can be formed via ice-templating the gels with a known concentration and density. This assembly opens a new route in CNT nanocomposites. Such conductive and strong composite structures can be used in various applications including sensors, electronic textiles, supercapacitors, etc.

Compatibilizing of PP/NBR Immiscible Blend Using Modified Nano Graphene Oxide

Khodabakhshi Hamid [1], Mehranpour Milad [1]*, Goharpey Fatemeh [2]

[1] Islamic azad university - Tehran - Iran, [2] Amirkabir university of technology - Tehran - Iran

Compatibilized polypropylene(PP)/acrylonitrile butadiene rubber(NBR) blend were prepared using a new method of compatibilizing based on modified graphene oxide nano sheets(MGO). in the beginning, graphene oxide nano sheets synthesised and modified on nucleophilic situation with methylenedianiline and then maleated pp was grafted on it. afterwards, NBR was blended with MGO by solution method as masterbatch. Finally, the compatibilizer masterbatch with PP and NBR were melt mixed with internal mixer. in each step accuracy of assertion were characterized using specific tests. For this purpose FTIR, EDXA, TGA and XRD tests were used in synthesis and modification steps of graphene oxide sheets. Additionally, the mechanical , rheological as well as morphological properties of compatiblized and reference blends were investigated by using DMTA, FE-SEM and RMS tests and the corresponding results were illustrated.

Polyethylene-graphite composites: Comparing cone calorimeter and micro-flammability test results

Focke Walter Wilhelm [(1)]*, **Mhike Washington** [(1)], **Mbonane Cebolenkosi** [(1)], **Mwakikunga Bonex Wakufwa** [(2)]

[(1)] University of Pretoria - Gauteng - SouthAfrica, [(2)] CSIR - Gauteng - SouthAfrica

Polyethylene (LLDPE) compounds containing 10 wt-% graphite fillers were rotationally molded into flat sheets. Flame retardancy was studied using cone calorimeter tests conducted at a radiative heat flux of 35 kW m-2. Only the expandable graphite, an established flame retardant for polyethylene, significantly reduced the peak heat release rate. Compared to the neat polyethylene, it was easier to ignite the LLDPE composites containing carbon black, expandable graphite and exfoliated graphite. However, rather unexpectedly, the inclusion of flake graphite increased the time to ignition by up to 80%. Simulations conducted with the ThermaKin numerical pyrolysis software suggest that increased reflectivity was mainly responsible for this effect. The behaviour of the compounds in micro-flammability tests did not reflect these results. The reasons for the discrepancy will b.e discussed

POLAR POLYMORPHS CHARACTERIZATION BY SYNCHROTRON X-RAY DIFFRACTION OF ANNEALED CNT FIBRE/PVDF COMPOSITES

Fernandez-Blazquez Juan P. ***(1)*, Monreal-Bernal Alfonso (1), Vilatela Juan J. (1)***

(1) IMDEA Materials Institute - Madrid - Spain

The use of polymers in microelectromechanical systems (MEMS) has raised exponentially, due to higher flexibility in comparision to standard inorganic materials and lower cost in terms of materials and processing. Polyvinylidene fluoride (PVDF) is the most commonly used polymer in MEMS. PVDF crystallizes into three different regular conformations. The regular packing of these conformations induces different crystal phases, named alpha, beta, gamma and delta, these last three are polar phases, so with potential application in MEMS. There are several methodologies to obtain these phases from the most stable one, being one of them the heterogeneous nucleation. Carbon nanotubes (CNTs) can act as nucleating agents. CNTs can accelerate the crystallization, and also provide the nucleation of different polymorphs. The challenge to disperse a high concentration of CNT in polymer matrix and to exploit the axial properties of CNTs can be faced by assembling them into a macroscopic fiber, with the tubes aligned parallel with the fiber axis.[1] The formation of polar phases in PVDF can be favored by nucleation on CNT. After isothermal crystallization at high temperatures, a quite different melting peak sequence is observed by DSC melting experiments. The monitoring of these melting processes by a sequence of X-Ray diffractograms every few degrees Celsisus may give a relevant information about the main polymorphs formed. This monitoring can be performed by X-Ray Synchrotron Radiation due to its high intensity,[2] being this study in CNT Fibre/PVDF composites the objective of this work. References. [1] V. Reguero, et al., Chem. Mater. 2014, 26, 3550. [2] J. P. Fernandez-Blazquez et al., Macromolecules 2007, 40, 1775.

Improving the Processability of Graphene Nanoplatelets in Polyamide 6 during Melt Compounding Extrusion

José Francisco Blanco-Villalba[1] Imanol Recio[2], Julio Gómez-Cordón[2], Christof Hübner[1, a)]

[1]*Fraunhofer Institute for Chemical Technology, Joseph-von-Fraunhofer-Straße 7, 76327 Pfinztal, Germany*
[2]*Avanzare Innovación Tecnológica S.L., Avda Lentiscares 4-6, 26370 Navarrete, Spain*

[a)]Corresponding author: christof.huebner@ict.fraunhofer.de

Abstract. In this paper, we consider the compaction of very low apparent density graphene nanoplatelets (GNP) to increase the maximal loading that can be melt compounded with polyamide 6 (PA6) using a twin-screw extruder. We use an effective medium model to describe the thermal properties of the composites as a function of filler dispersion, shape and orientation.

INTRODUCTION

Polymer nanocomposites are candidates to exploit the excellent electrical, thermal and mechanical properties of graphene and other related nanomaterials. Using thermoplastic polymers to this aim - e.g. by melt compound extrusion - is especially useful due to their versatility and reusability.

The low apparent density of some graphene nanoplatelet (GNP) powders (in the order of a few grams per liter) makes the melt compounding extrusion of GNP/polymer composites problematic, especially at high GNP concentrations. The high GNP/polymer volume ratio makes the premixture in the extruder hopper quite irregular and results in composites with badly distributed GNPs.

In this paper, we consider the compaction of very low apparent density GNP to increase the maximal loading that can be melt compounded with polyamide 6 (PA6) using a twin-screw extruder.

EXPERIMENTAL

Avanzare's GNP1 was used as filler for the composites. In Fig. 1, SEM and TEM images of GNP1 are shown. GNP1 are partially oxidized graphene nanoplatelets (4% oxygen by XPS) with a lateral size of 84 μm (D50) and an average thickness of 6 nm (approximately 18 layers). Polyamide 6 (Ultramid© B3K) was used as matrix.

A compaction process was performed to reduce the apparent density of the filler. The GPN1 powder was dispersed in water and mechanically stirred for 30 minutes at room temperature. The dispersion was vacuum filtered and the solid was dried at 80 °C for 24 hours.

A twin-screw extruder (Rondol 10 mm Microlab Extruder) was used for the melt compounding at 40 rpm and an average temperature of 240 °C. Disk-shaped samples (50 mm diameter, 2 mm thickness) were injection molded and their through-plane thermal conductivity was measured according to the ASTM E1530-06 standard.

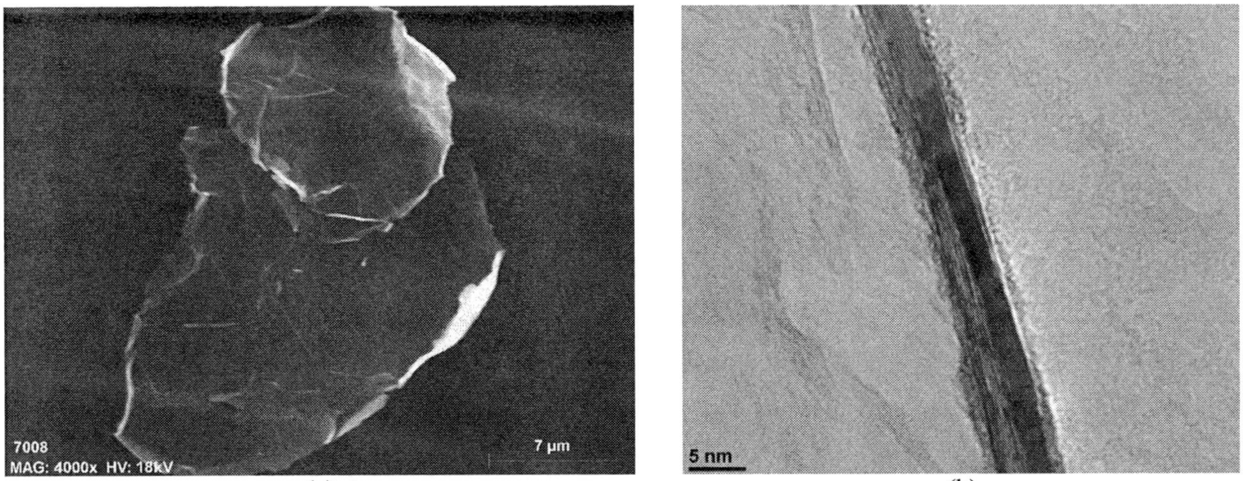

(a) (b)

FIGURE 1. SEM (a) and TEM (b) images of GNP1

The composites thus produced were compared with those produced with non-compacted GNP1 using the NanoDirekt process [1]. In this process, a GNP1 dispersion is introduced in an extruder. The dispersion media is then evaporated, as depicted in Fig. 2.

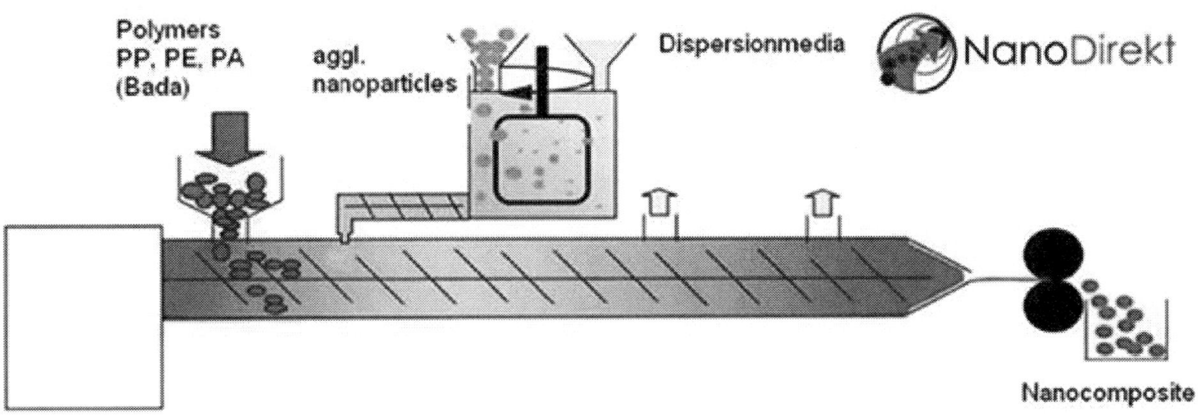

FIGURE 2. Description of the NanoDirekt process: nanoparticles are fed to the extruder in a dispersion

RESULTS

Before the compaction process, the relative density of GNP1 was 5 g/L. Therefore, the volume of GNP1 that was fed to the extruder was much bigger than the volume of polymer, especially for higher loadings, which in practice limited the maximal loading of GNP1 to 5 wt%. The relative density of the compacted GNP1 increased to 80 g/L, allowing loadings up to 20 wt%.

Figure 3, shows SEM images PA6/GNP1 composites produced in two different ways: using the NanoDirekt process with non-compacted GNP1; and by direct melt compounding of compacted GNP1. A good dispersion was observed for the composites produced with the NanoDirekt process, while the composites produced with compacted GNP1 consist of mainly GNP1 agglomerates.

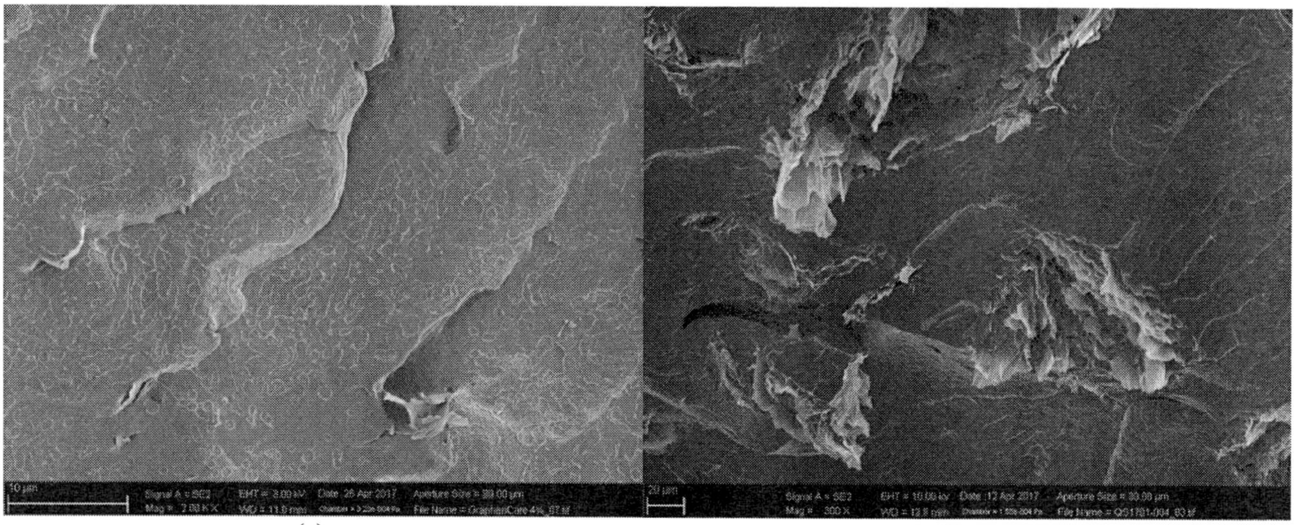

(a) (b)

FIGURE 3. SEM image of the cross section of a pellet from a PA6 + 2% GNP1 composite, produced: (a) Using the NanoDirekt process; (b) by direct compounding of compacted GNP1

The through-plane thermal conductivity of the composites compounded under identical conditions with compacted and non-compacted GNP1 was compared. No difference can be observed between both, suggesting that the compaction process did not hinder the thermal properties or dispersion of the GNP1 in the composite. A maximal enhancement of more than 900% was achieved at the highest GNP1 concentration.

An effective medium model, namely the Polder-van Santen-Böttcher model [2-4], was used to describe thermal conductivity of the composites (equation (1)). This model takes into account the concentration, shape and orientation of the fillers. The degree of dispersion was also considered, assuming that two kinds of fillers are present: perfectly dispersed GNPs, and agglomerates of GNP.

$$\begin{aligned}\lambda &= \lambda_{matrix} \\ &+ \left(\lambda_{GNP} - \lambda_{matrix}\right)\left[\frac{\lambda \cdot v_{GNP_x}}{\lambda + (\lambda_{GNP} - \lambda) A_{GNP_x}} + \frac{\lambda \cdot v_{GNP_y}}{\lambda + (\lambda_{GNP} - \lambda) A_{GNP_y}} + \frac{\lambda \cdot v_{GNP_z}}{\lambda + (\lambda_{GNP} - \lambda) A_{GNP_z}}\right] \\ &+ \left(\lambda_{Aggl} - \lambda_{matrix}\right)\left[\frac{\lambda \cdot v_{Aggl_x}}{\lambda + (\lambda_{Aggl} - \lambda) A_{Aggl_x}} + \frac{\lambda \cdot v_{Aggl_y}}{\lambda + (\lambda_{Aggl} - \lambda) A_{Aggl_y}} + \frac{\lambda \cdot v_{Aggl_z}}{\lambda + (\lambda_{Aggl} - \lambda) A_{Aggl_z}}\right]\end{aligned} \tag{1}$$

Where λ is thermal conductivity of the composite, λ_n is the thermal conductivity of the component n, while v_{n_i} and A_{n_i} are the volumetric fraction and the polarization factor (which is a function of the aspect ratio of the filler) of the component n in the i direction, respectively. The alignment of the n component in the x direction and the degree of dispersion of the composite can be quantified according to the model with following parameters:

$$\%Allignment_{n,x} = \frac{v_{n_x}}{v_{n_x} + v_{n_y} + v_{n_z}} \tag{2}$$

$$\%Dispersion = \frac{\sum_i v_{GNP_i}}{\sum_i v_{GNP_i} + \sum_i v_{Aggl_i}} \tag{3}$$

Where *GNP* and *Aggl* represent dispersed GNPs and agglomerates of GNPs, respectively. Thus, a composite with randomly oriented particles can be considered as the case in which the particle orientation is equally distributed in the three main directions.

The modeled was compared to experimental data, and an estimation of the GNP orientation and degree of dispersion was obtained from the best fitting. As a first approximation, the following assumptions were made: the GNP1 particles and agglomerates have an aspect ratio of 1:10000 and 1:20, respectively, and an isotropic thermal conductivity of 2000 W/mK.

In a more rigorous consideration, the anisotropic properties of GNP1 (3000 W/mK in-plane, 6 W/mK through-plane) and GNP1 agglomerates (2850 W/mK in-plane, 0.38 W/mK through-plane) were considered. For the GNP1, the thermal conductivity values of graphite were assumed. While for the agglomerates, the Polder-van Santen-Böttcher model was used, considering an agglomerate to be a composite of stacked GNPs with a 5% of air content.

The isotropic case provided a degree of dispersion of 1% and a 70% orientation of the GNP/agglomerates parallel to the surface of the sample (tangential to injection flow direction). For the anisotropic case, a degree of dispersion of 1% and an 83% orientation parallel to the surface of the sample was obtained. At this stage, these values are to be considered rough estimations, being their qualitative aspect more important than the quantitative one. Although the degree of dispersion was not quantified by other methods, the 1% value is compatible with the poor dispersion observed in Figure 3. The orientation was not measured in any way, but a certain degree of orientation tangential to the flow during the injection molding process is to be expected.

CONCLUSIONS

The compaction process of GNP1 allowed a higher loading in PA6, up to 20 wt%. Both SEM images and measurements of thermal conductivity suggest that the compaction process did not hinder the dispersion of the GNP1 or their thermal properties.

We propose an effective medium model to describe the thermal properties of the composite as a function of the filler dispersion, shape and orientation. This analysis can be used as a tool for a fast estimation of the quality of the dispersion and orientation of nanoparticles in a composite.

ACKNOWLEDGMENTS

This project (TheLink: http://thelink-project.eu) has received funding from the European Union's Horizon 2020 research and innovation programme under the Marie Skłodowska-Curie Grant Agreement no. 642890.

REFERENCES

1. I. Mikonsaari and A. Zanki, "NanoDirekt - Direktprozess zur Herstellung von Nanosuspensionen und Zudosierung in thermoplastische Matrices zur Produktion von Nanocomposites", in *Produktgestaltung in der Partikeltechnologie: 4. Symposium Partikeltechnologie-2008*, edited by U. Teipel (Fraunhofer-Institut für Chemische Technologie, Pfinztal, Germany, 2008).
2. D. Polder and J. H. van Santen, Physica **XII**, no 5 (1946).
3. W. Y. Hsu, T.D. Gierke and C. J. Molnar, Macromolecules **16**, 1947–1949 (1983).
4. G. Bánhegyi, Colloid & Polymer Sci **264**, 1030–1050 (1986).

Fabrication and properties of electrically conductive bicomponent fiber

Lellinger Dirk [(1)]*, Kunkel Daniel [(1)], Stock Sebastian [(1)], Weise Benjamin [(2)], Mroszczok Jens [(2)], Alig Ingo [(1)]

[(1)] Fraunhofer LBF - Darmstadt - Germany, [(2)] Institut für Textiltechnik of RWTH Aachen University - Aachen - Germany

There are several applications of tailored carbon fiber materials in the automotive, wind energy or aircraft industry. Electrically conductive fibers possess e.g. the potential for sensing or heating applications, e.g. by insertion into fabrics or lightweight constructions. An electrically conductive bicomponent fiber with core-sheath geometry has been developed. The core component consists of an electrically conductive carbon nanoparticle (GNP) based polymer composite covered by an insulating polymer sheath. Despite electrical insulation the sheathing with a polymer guarantees resistance against abrasion and the washing processes. A conductive polymer composite based on polycaprolactone (PCL) and containing a mixture of carbon black (CB) and multi-walled carbon nanotubes (CNT) was developed for the conductive core. The sheath consists of polyamide-6. Based on these materials a laboratory and pilot spinning process was developed for bicomponent melt spinning machines: After melting the two polymer components by extrusion, the core- sheath geometry was formed in the spinning nozzle. After crystallization the filaments are post-stretched and winded up. Due to the higher melting temperature of the sheath the conductivity of the core could be increased by thermal annealing. The conductivity increase of the CB/CNT filler network is due to the reformation of filler network after the process induced rupture. For electrical measurements several methods for contacting the conductive core were developed and tested. Electrical conductivity was measured for different CNP compositions and processing conditions. Furthermore, the electrical conductivity of the fibers was measured in a tensile test arrangement together with the strain-stress-curves. The possibility to scale-up the production of electrically conductive bi-component fibers has been shown. Future challenges are an improved stretchability of the fibers for subsequent weaving and an improvement of the spinning process.

Carbon Nanoparticle Surface Modification for Thermoplastic Polymers

Schumm Benjamin [(1)]*, Abendroth Thomas [(1)], Althues Holger [(1)]

[(1)] Fraunhofer IWS - Saxony - Germany

Polymers exhibiting an electrically conductive surface are crucial for being used in different products. The demands for conductivity depend on application: for antistatic properties a sheet resistance of 1e5 - 1e9 Ù, for electro-optical applications a sheet resistance 1e2 Ù is required. To achieve surfaces with electrical conductivity additives, such as surfactants and filler materials or coatings are applied. However, many restrictions need to be considered for choosing the best additive or coating for a certain polymer, as they directly influence the durability in arid conditions (surfactants), mechanical and optical properties (filler) and costs (coating). At Fraunhofer IWS a process was developed for refining polymeric surfaces in order to achieve electrical conductivity, which can be applied in conventional roll-to-roll (R2R) and injection molding processes. High aspect ratio nanomaterials forming a nanoscaled network were therefore deposited either on calender cylinder (R2R) or mold cavity (injection molding) and afterwards transferred and integrated into polymeric surface. In dependence of film thickness and nanomaterials being used different optical transparency and sheet resistances in the range of 1e2-1e6 Ù can be adjusted. Due to the nanomaterials only are integrated into polymeric surface, a high durability of electrical properties is provided and a very low material consumption is needed. The mechanical bulk properties of the refined element are not affected by the nanomaterials. By using this new technology, high transparent (>90 %) antistatic (~1e5 Ù) polymeric devices were produced by integrating a multi-walled carbon nanotube (MWCNT) network with an areal load of 25 mg/m2 either into PE and PET-foils or PC or PMMA elements. Easy implementation in already existing production processes and very inexpensive material costs (< 0,01 €) render this nanomaterial integration technology for manufacturing electrically conductive polymer foils and devices.

An Investigation on Cure Kinetics of Epoxy/MWCNTs Nanocomposites through Isothermal Calorimetric and Rheological Analyses

***Saeb Mohammad Reza* [(1)]*, *Rastin Hadi* [(1)]**

[(1)] Department of Resin and Additives, Institute for Color Science and Technology - Tehran - Iran

Cure kinetics of epoxy nanocomposites containing multi-walled carbon nanotubes (MWCNTs) was studied by differential scanning calorimetry and rheological analyses. A very low content of pristine and amine-modified MWCNTs were introduced into epoxy resin to study the agreement of calorimetric and rheological methods on cure kinetics assessment of epoxy/MWCNTs nanocomposites. In this way, Model-free and model-fitting methods were used to gives rise the more useful information about cure kinetics of studied samples. As expected, incorporation of unmodified MWCNTs into epoxy resin increase activation energy caused by physical hindrance, while for modified MWCNTs, it decline. This is mainly because of existing amine groups on the filler surface that accelerate the curing reaction. Isothermal rheokinetic studies reveal the crosslinking structure of epoxy resin in the presence of various amount of MWCNTs content. Accordingly, 0.3 wt.% MWCNTs impede the curing reaction as well as destruct the 3D crosslinking network.

Rheologically determined the role of nanoparticle geometry on the morphological evolution of ABA block copolymers

Hasanabadi Noushin [1], Lattuada Marco [2], Ranjbar Behnaz [1], Nazockdast Hossein [1]*

[1] Amirkabir University of Technology - Tehran - Iran, [2] Adolphe Merkle Institute, University of Fribourg - Fribourg - Switzerland

Block copolymer(BCP) containing nanoparticles has drawn much attention as novel advanced material in the recent years. However, some aspects of these nanostructured materials including the influence of nanoparticle geometry on the BCP structural development have not been fully understood. In this study, the role of nanoparticle geometry (Multi-Walled Carbon Nanotube/ graphene nanosheet) on the microphase separation of BCP based nanocomposite material composed of cylindrical forming polystyrene-b-poly(ethylene-butylene)-b-polystyrene (SEBS) has been investigated by means of rheological technique. The results of temperature sweep experiments indicated an enhancing effect of nanoparticles on increasing the BCP microphase separation temperature and accelerating its kinetic which could be explained in terms of confinement of BCP segments. The resulting confinement can lead to domination of segmental enthalpic energy over entropic one which will, in turn, force the microphase separation to occur at higher temperatures. This was evidenced by greater enhancing effect of graphene nanosheets with higher geometrical confinement on microphase separation compared to Multi-Walled Carbon Nanotube. Moreover, transient viscoelastic analysis implied the retarding effect of nanoparticles on the orientation and also restructuring of BCP domains as well as nanoparticles whose extent was dependent upon the concentration and also type of nanoparticles. Furthermore, DMTA results indicated that the addition of nanoparticles reduces the interphase area in the favor of increasing the extent of microstructural evolution. Keywords: Block copolymer, MWCNT, Graphene nanosheets, Rheology, Morphology development

Solid-state carbon-based textile supercapacitors for energy storage applications

Paleo Antonio J [(1)]*, Staiti P. [(2)], Ferreira F. N. [(1)], Rocha A. M. [(1)], Lufrano F. [(2)]

[(1)] University of Minho - Guimaraes - Portugal, [(2)] CNR-ITAE, Istituto di Tecnologie Avanzate per l´Energia "Nicola Giordano - S. Lucia - Italy

In this work, carbon-based conducting electrodes based on two different types of carbon nanofibers (CNF) have been produced by the dip and dry coating method onto cotton substrates. Furthermore, activated carbon (Norit A Supra Eur) and manganese oxide (MnO2) have been subsequenlty added to the CNF-based dip-coated cotton fabrics electrodes and asymmetric supercapacitors have been constructed and tested with the focus of obtaining devices with increased capacitive performance. In particular, the carbon-based active layer was prepared by spreading on the CNF-based electrodes a slurry containing the activated carbon (AC) material, graphite fibres, polyvinylidene difluoride (PVDF) as binder and N,N dimethylacetamide (DMA) solvent, whereas the MnO2 based active layer was prepared by spreading on the CNF-based textile electrodes a slurry formed by MnO2, carbon black, graphite fibers, PVDF and DMA. A solution of 1M Na2SO4 impregnated in porous paper separator (Nippon Kodoshi Corportion, Japan) was employed as neutral aqueous electrolyte. The supercapacitors were electrochemical investigated by cyclic voltammetry (CV), galvanostatic charge/discharge (GCD) and electrochemical impedance spectroscopy (EIS). The results indicated that with this particular combination of carbon and manganese oxide active layers on CNF-based cotton fabrics it was possible to obtain specific capacitance of 100 F/g and a high specific energy density of 10 Wh/kg. Acknowledgments This work was partly financed by FEDER funds through the Competitivity Factors Operational Programme - COMPETE and by national funds through FCT – Foundation for Science and Technology within the scope of the project POCI-01-0145-FEDER-007136. A. J. Paleo acknowledges the support of COST Action CA15107- Multi-Functional Nano-Carbon Composite Materials Network (MultiComp) by means of a short term scientific mission (STSM).

The effect of graphene flake diameter on the performance of anti-corrosion coatings

Istrate Oana M [1]*, Gibbon S. R. [2], Moghal J. [3], Budd P. M. [4], Kinloch I. A. [1]

[1] School of Materials, National Graphene Institute, University of Manchester - -- - UnitedKingdom, [2] AkzoNobel Research Development & Innovation - -- - UnitedKingdom, [3] CROWN Technology Wantage - - - UnitedKingdom, [4] School of Chemistry, University of Manchester - - - UnitedKingdom

The world of organic coatings has been quickly revolutionised by the use of graphene due to this material's unique properties (e.g. impermeability, high surface area, high aspect ratio and low cost) which make it an ideal candidate for anti-corrosive barrier coatings [1-6]. Polymer/graphene nanocomposite formulations have been prepared with commercially available graphene nanoparticles, i.e., XGnP® 5M, 15M and 25M (XG Sciences, Inc., Lansing, Michigan, USA) with flake diameters of 5 µm, 15 µm and 25 µm, and a thermosetting polymer matrix, i.e., BPA free coating, Gold Lacquer (AkzoNobel, Packaging Coatings Ltd., UK). The polymer/graphene formulations were coated onto electrolytic chromium coated steel (Crown Packaging UK Plc) using a spin coater. Electrochemical tests were performed in a 0.1 M NaCl solution, using a conventional three electrode system. Electrochemical impedance spectroscopy tests showed that the presence of graphene improved the corrosion resistance of the materials with flake diameter strongly affecting the performance of the coating.

Nanoprobe Investigations of Viscoelastic Behavior in Elastomeric Nanocomposites

Eaton Matthew David [(1)]*, Kolluru Pavan V [(1)], Collinson David William [(1)], Delgado David [(1)], Shull Kenneth R [(1)], Brinson L. Catherine [(1)]

[(1)] Northwestern Univeristy - Illinois - USA

The addition of nanomaterials into a polymer matrix at low concentrations (~0.2wt%) can lead to dramatic changes in the viscoelastic, mechanical, electrical, and optical properties of the material, enabling a wide variety of applications in structural, electronic, and biological implant materials and devices. Although the bulk properties of the composites can be easily measured through traditional means, much is still unknown on how the local polymer properties change approaching the interface of the nanoparticles, especially for polymers in the rubber regime. Previous work has used atomic force microscopy (AFM) to measure the local elastic modulus at this interphase, but time dependent viscoelastic properties of polymers present more challenges in capturing quantitative frequency/time resolved responses. A quantitative technique to measure viscoelastic properties at the nanoscale for these composites using a modified Quantitative Nanomechanical Mapping (QNM) AFM technique has been developed. An oscillating cantilever indents into an elastomer in a sinusoidal motion and the resulting force is recorded. From the phase lag between these two responses, the viscoelastic properties can be measured. The accuracy of this method has been compared to oscillatory data of the bulk material using dynamic mechanical analysis. The method can be applied to biphasal systems to measure how the viscoelastic properties change at the interphase. Model composite samples made from carbon-rubber "sandwich" structures have been investigated for quantitative interphase measurements. These structures minimize the geometrical complexity found in bulk nanocomposites, allowing the isolation of the interphase effect. Various geometries can then be explored to understand their influence on locally changing polymer properties. The goal is to apply these results to computational models and data mining tools to find guiding relationships in material physics and to enable targeted material design.

Advanced Electrically Conductive Adhesives for High Complexity PCB Assembly

P.E. Lopes[1,a)], D. Moura[1], D. Freitas[1], M.F.Proença[2], H. Figueiredo[3], R. Alves[3], M.C. Paiva[1, b)]

[1] IPC/i3N, Department of Polymer Engendering, University of Minho, Guimarães, Portugal; [2] Department of Chemistry, University of Minho, Braga, Portugal; [3] CM/MFT3 Bosch, Braga Portugal.

[a)]Corresponding author: pelopes@dep.uminho.pt
[b)] mcpaiva@dep.uminho.pt

Abstract. Electronic packaging, or assembly of packed electronic components on printed circuit boards, present challenges that require innovative solder pastes and electrically conductive adhesives to face the increasing complexity of PCB assembly, with denser board occupation and demanding thermal management during assembly. Our aim is to prepare carbon particle based conductive adhesives. The first step to achieve this goal was to prepare composites with epoxy resin and a variety of nano to micron scale carbon particle, produced by mixing on a three roll mill. The percolation threshold for each particle type was determined as well as the conductivity level reached after percolation.

INTRODUCTION

Electronic packaging has become one of the most important sectors in modern electronic industry. In fact, this interest rose from the miniaturization of electronic components, where the interconnection of chips on printed circuit boards (PCBs) is frequently made by soldering [1]. Traditionally, this process is achieved using lead-based or lead-free solders, which provides an electronic conductive pathway between PCB and components [2].

As an alternative to solder, electrically conductive adhesives (ECAs) have been produced, since they can be environmentally friendly, light weight and requiring less processing steps [3, 4]. There are a few commercial ECAs being used, most of them made of epoxy and high amounts of silver microflakes [5]. Despite all the progress with these ECAs, the complete replacement of the traditional solders has not been achieved yet because of unstable contact with non-metal finished components, silver migration, and poor electrical conductivity, which results from the small contact points formed in the percolated network [5]. To overcome this limitation, different strategies have been employed such as the incorporation of silver nanoparticles (NPs) or the usage of one dimensional materials (1D) like carbon nanotubes and metal nanowires, simultaneously with two dimensional (2D), graphene or graphite flakes, or three dimensional particles (3D) such as microparticles. NPs may present different behaviour, very much depending on their size. An example of that is the effect that NPs above 10 nm have on the electrical resistance above a certain NP content. The increase of NP contact points results in higher electrical resistance. In turn, for NPs smaller than 10 nm, a better contact between flakes is achieved and thus the number of contacts are reduced. This is possible through the sintering process that produces larger particles from its powders, based on atoms diffusion. In the particular case of metallic particles, the surface atoms are less bonded than the bulk atoms and thus may leave the surface when sufficient energy is given. Thus, in the case of the NPs, the large surface area per given volume provides a good source of atoms with weak bonds. Sintering of silver NPs has been reported for temperatures as low as 150 °C [4, 6].

Carbon materials such as carbon nanotubes (CNTs) and graphene are recognized their high electrical conductivity, nano size and high aspect ratio. These features allow them to easily establish an electrically conductive network for electron transport, achieving a percolation threshold at low nanoparticle contents. Nevertheless, some drawbacks still exist. For 1D materials such as CNTs the main problem is related with their dispersion in the polymeric matrix. After composite production, reagglomeration of CNTs is observed in low viscosity resins or melted polymers. Reagglomeration has been prevented through covalent or non-covalent surface functionalization [7]. The former leads to atomic rearrangement of the surface carbon structure, while the latter involves the non-covalent modification of the carbon nanoparticle surface with molecules such as pyrene derivatives, without disrupting the carbon structure of the nanoparticles and maintaining the CNT properties [8]. Another approach consists on the "decoration" of graphene or CNTs with silver nanoparticles [4, 9-11]. Tab. 1 presents a selection of results reported in the literature for carbon hybrid composites.

TABLE 1 Silver and carbon hybrid composites properties.

Silver flakes (wt %)	CNTs (wt %)	Graphene (wt %)	Bulk resistivity (Ω.cm)	Reference
66.5	0.27	-	10^{-3}	12
80	-	0.05	4.3×10^{-5}	13
79.5	1.5	-	4×10^{-6}	14
65	-	15	2.37×10^{-4}	15

In the present work it was intended to use mainly carbon particles to achieve a high electrical conductivity level and at the same time to improve the adhesive and mechanical properties of the materials using, to achieve that goal, the minimum amount of filler.

EXPERIMENTAL

Single wall carbon nanotubes (SWCNT) were Tuball from OCSiAl; multi wall carbon nanotubes (MWCNT) were NC 7000 from Nanocyl; Carbon nanofibers (CNF) were Pyrograf III from Applied Sciences Inc.; Chopped carbon fibers were KRECA Chop from Kureha; Graphite nanoplatelets were XGnP H5, and XGnP C from XGSciences and Micrograf from Nacional de Grafite. The epoxy was Biresin® CR83 and hardener Biresin® CR83-2, from Sika.

The dispersion of the carbon nanoparticles in the epoxy resin was carried out on a three roll mill equipment EXAKT 80E (EXAKT Advanced Technologies GmbH, Germany). The epoxy and nanoparticles were initially hand mixed and then fed to the rear rolls of the mill and collected from the front collection tray. The mixing process consists of passing the composite mixture through the mill several times, decreasing the gap between the rolls until the minimum gap, 5 µm, is reached, followed by several passages using force mode between the front rolls. The epoxy resin hardener was manually mixed after the dispersion process was carried out. The final mixture was degassed under vacuum and cured at 80 °C for 60 min.

Surface resistivity was measured on samples prepared by manually impregnating medium porosity filter paper (Prat Dumas 009210 Ø of 90 mm) with the composite, squeezing it with a glass rod onto filter paper over a glass base. "Dog bone" test samples were prepared with standard dimensions for tensile testing on the universal testing machine and were also used for volume electrical characterization. The samples were produced on a silicone mold.

Electrical surface resistivity of the epoxy composite-impregnated papers was obtained using a Keithley Picoammeter/Voltage Source Model 6487 with Model 8009 Resistivity Test Fixture. A Keithley System Source Meter (SMU) Model 2635B with Model 5809 Kelvin Clip Lead Set was used for volume resistivity measurements.
The measurements of surface resistivity were performed appling a potential of 10V to the sample, and measuring the current intensity. Electrical resistivity was calculated for the electrode geometry of the test fixture. I-V curves were obtained by sweeping the potential from -10 V to 10 V at 0.5 V increments. At each potential step, the corresponding current was measured. Volume resistivity was calculated from the slope of the I-V curve and geometry of the sample.

RESULTS AND DISCUSSION

The initial characterization of the electrical properties of the composites prepared was based on the surface resistivity of composite impregnated paper. The lowest surface resistivity measured for each filler combination used, and the filler content, are presented in Tab.2. The compositions tested included the electrical percolation, that is, a large change of the electrical properties within the range of filler contents used. The single filler composites with lower percolation threshold were those based on SWCNT, MWCNT and exfoliated graphite H5, and the corresponding percolation curves are presented in Fig, 1. Notice that in this set of composites those with SWCNT required ten times lower filler weight content to attain the same level of resistivity.

The hybrid composites (containing two types of fillers) also presented noticeable gain in electrical properties, in particular those with SWCNT, MWCNT and graphite XGnP H5.

TABLE 2 Fillers and amount ranges used, for preparation of epoxy composites, and lowest surface resistivity measured in the sample of paper impregnated with composite.

Filler	Filler Amount (wt %)	Filler	Filler Amount (wt %)	Lowest Surface Resistivity (kΩ.sqr)
No filler				$1.36x10^{11}$
SWCNTs	0.001 – 0.12			$7.05x10^{1}$
MWCNTs	– 1.9			$9.56x10^{1}$
Graphite H5	0.001 – 10			$3.07x10^{5}$
Exf. Graphite H5				$9.47x10^{1}$
Graphite C	0.02 – 1			$7.92x10^{8}$
Graphite NG	0.79 – 5.83			$4.66x10^{2}$
Carbon nanofiber (CNF)	0.35 – 13			$1.34x10^{6}$
Carbon fiber (CF)	1.16 – 34			$3.59x10^{3}$
SWCNTs	0.27	MWCNTs	0.2	$2.32x10^{1}$
SWCNTs	0.12	Graphite H5	0.8 – 3.2	$3.90x10^{1}$
SWCNTs	0.12	Graphite C	0.4 - 1	$1.18x10^{2}$
MWCNTs	0.1 – 1.7	Graphite H5	0.1 – 2.5	$1.13x10^{2}$
MWCNTs	0.1	Graphite C	0.5 – 3.5	$5.39x10^{5}$
MWCNTs	0.1 – 0.3	CNF	5	$4.44x10^{3}$
MWCNTs	0.1 – 0.3	CF	7 – 28	$2.53x10^{3}$
SWCNTs and MWCNTs	0.005 – 0.26 0.1 – 1	Graphite H5	1.4 – 3.1	$1.60x10^{1}$

All the ternary systems studied present electrical conductivity. The presence of graphite nanomaterials as filler up to 3.5 wt % in combination with carbon nanotubes did not change significantly the surface electrical properties, as shown in Fig. 1 b)**Error! Reference source not found.**. The presence of graphite, however, has the advantage of an apparent viscosity reduction of the epoxy composite.

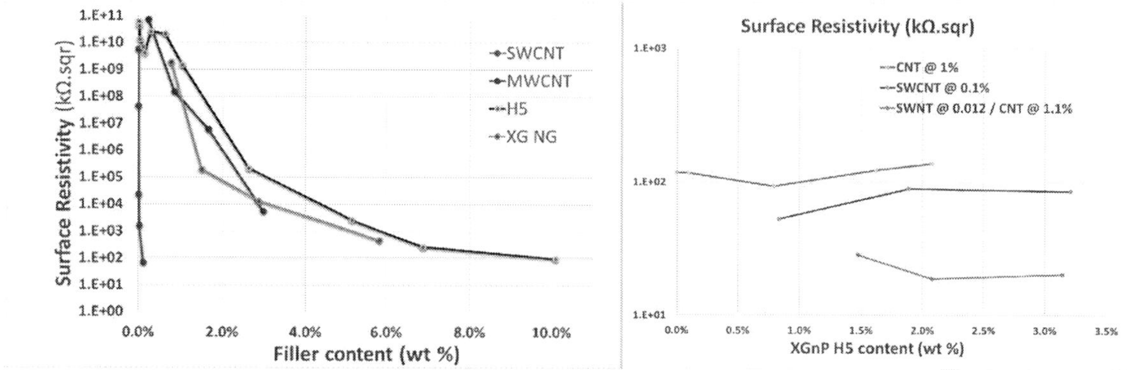

a) b)

FIGURE 1. Surface resistivity measured on the paper impregnated composite samples a) Percolation curves for selected fillers, and b) Influence of XGnP H5 content on the electrical characteristics of SWCNT and MWCNT composites

FIGURE 2. presents the I-V curves and percolation curve of SWCNT samples with filler contents of 0.05,. 0.1, 0.2 and 1 wt %, where the resistivity values obtained at the higher filler content are in the range of a few Ω.cm with percolation at 0.25 wt% of SWCNT.

Composites with 0.5 and 1 wt % of SWCNT and 5 wt % of graphite H5, were prepared and characterized. They presented a reduction of the electrical resistivity, from 51.5 and 8.45 Ω.cm without graphite, to 11.1 and 4,82 Ω.cm with 5 wt % of graphite.

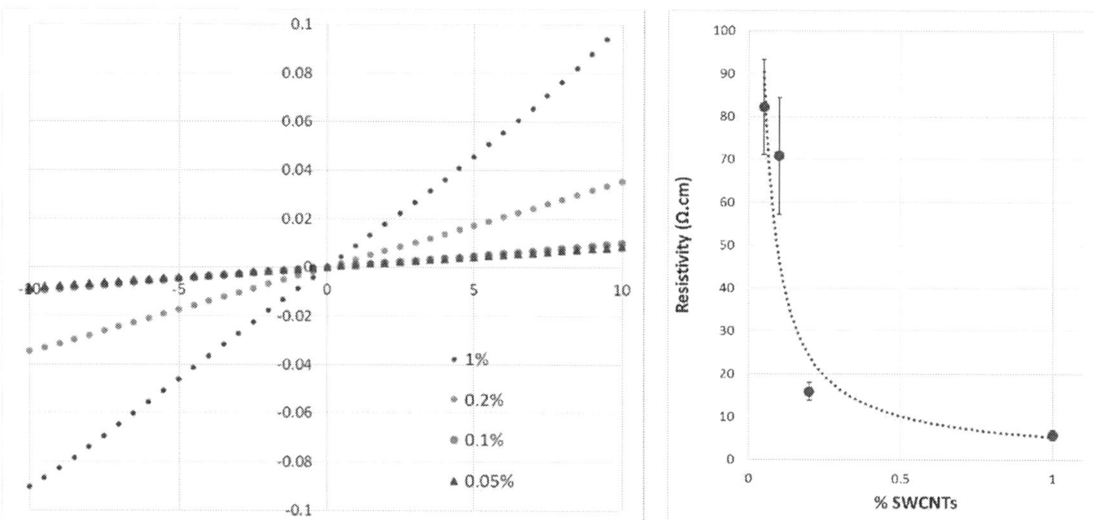

FIGURE 2. I-V curves (left) obtained with SWCNT composites using the dog bone samples and volume resistivity (right) determined from the slope of the I–V curves, accounting for the geometry of the samples.

CONCLUSIONS

Dispersion of the carbon nanoparticles in epoxy resin was achieved using a three roll mill, and percolation was observed in four composite materials with filler loads below 5 wt %. Also, hybrid composites with two nanomaterials with total loadings around 5 wt %, presented electrical conductivity.

Addition of graphite up to 3.5 wt % to SWCNT and/or MWCNT epoxy composites had no impact on the electrical properties, and at 5 wt % with SWCNT an increase of the bulk electrical conductivity was observed. Conductivity values achieved are still insufficient for the wide application on electronic soldering. Different materials combinations and carbon surface modifications will be tested in the future, to improve these properties.

ACKNOWLEDGMENTS

This research is sponsored by the Portugal Incentive System for Research and Technological Development. Project in co-promotion n° 002814/2015 (iFACTORY 2015-2018).

REFERENCES

1. H. S. Zhao, T. X. Liang and B. Liu, International Journal of Adhesion and Adhesives **27** (6), 429-433 (2007).
2. J. Y. Kim, S. J. Jeong, Y. J. Cho and K. S. Kim, Journal of Cleaner Production **67**, 249-257 (2014).
3. Y. Li, K. S. Moon and C. P. Wong, Science **308** (5727), 1419-1420 (2005).
4. B. M. Amoli, J. Trinidad, A. Hu, Y. N. Zhou and B. X. Zhao, Journal of Materials Science-Materials in Electronics **26** (1), 590-600 (2015).
5. B. M. Amoli, A. M. Hu, N. Y. Zhou and B. X. Zhao, Journal of Materials Science-Materials in Electronics **26** (7), 4730-4745 (2015).
6. H. J. Jiang, K. S. Moon, Y. Li and C. P. Wong, Chemistry of Materials **18** (13), 2969-2973 (2006).
7. L. Vaisman, H. D. Wagner and G. Marom, Advances in Colloid and Interface Science **128**, 37-46 (2006).
8. R. H. Baughman, A. A. Zakhidov and W. A. de Heer, Science **297** (5582), 787-792 (2002).
9. P. C. Ma, B. Z. Tang and J. K. Kim, Carbon **46** (11), 1497-1505 (2008).
10. P. C. Ma, N. A. Siddiqui, G. Marom and J. K. Kim, Composites Part a-Applied Science and Manufacturing **41** (10), 1345-1367 (2010).
11. H. P. Wu, X. J. Wu, M. Y. Ge, G. Q. Zhang, Y. W. Wang and J. Z. Jiang, Composites Science and Technology **67** (6), 1116-1120 (2007).
12. K. Y. Chun, Y. Oh, J. Rho, J. H. Ahn, Y. J. Kim, H. R. Choi and S. Baik, Nature Nanotechnology 5 (12), 853-857 (2010).
13. X. Peng, F. T. Tan, W. Wang, X. L. Qiu, F. Z. Sun, X. L. Qiao and J. G. Chen, Journal of Materials Science-Materials in Electronics 25 (3), 1149-1155 (2014)
14. K. H. Liu, L. Liu, Y. F. Luo and D. M. Jia, Journal of Materials Chemistry 22 (38), 20342-20352 (2012).
15. Y. Li, D. Lu, C. P. Wong. Electrical Conductive Adhesives with Nanotechnologies; Springer: New York, New York, USA, 2010; p 210

Continuous production of carbon nanotube-grafted quartz fibres: Effect of carbon nanotube length on fibre/matrix adhesion

De Luca Hugo (1)*, Anthony David B. (1), Greenhalgh Emile S. (1), Bismarck Alexander (1), Shaffer Milo S. P. (1)

(1) Imperial College London - London - UnitedKingdom

The mechanical properties of fibre-reinforced composites depend on the properties of the fibre matrix interface, where stress concentrations prevail. In situ growth of carbon nanotubes (CNTs) on the surface of silica fibres produces "hairy" (or "fuzzy") fibres, incorporating both micrometre and nanometre reinforcement length scales. Simulations of neighbouring CNT-grafted-fibres carrying CNTs with different lengths, perpendicular to the fibre surface, showed a reduction in radial and shear stresses at the fibre matrix interface. Stress concentrations are predicted to shift from the fibre matrix interface to the end of grafted CNT forest, effectively increasing the fibre diameter, resulting in higher stress concentrations in the resin rich region between the hierarchical fibres for longer CNTs [1]. While this mechanism has yet to be demonstrated experimentally, the introduction of grown CNTs has imbued thermal and electrical conductivity [2], and improved interfacial shear strength, albeit with reduced primary fibre tensile strength [3]. Here, the continuous production of CNT-grafted quartz fibres was performed in an open chemical vapour deposition (CVD) reactor with continuous in line catalyst deposition. Highly graphitic CNTs with controllable lengths ranging from 0.1 to 20 μm were grown on the quartz fibre surface by adjusting the reduction and growth times, with shorter fibres growing homogeneously and longer CNTs growing in a splayed "Mohawk" manner. The effect of CNTs length (and thus microstructure) upon the mechanical properties of CNT-grafted-quartz-fibre/epoxy composites was investigated through single fibre pull-out and tensile tests. This study not only allows direct comparison to the computational predictions on hairy fibres, but provides a route towards continuously grown CNT-grafted quartz fibres. [1]Romanov et al., Comp. Sci. Technol., 2015, 114, 79 [2]Qian et al., J. Mater. Chem., 2010, 20, 4751 [3]Qian et al., Comp. Sci. Technol., 2010, 70, 393

Breaking the nanofiller loading/dispersion dichotomy in polymer nanocomposites

Santagiuliana Giovanni (1), Picot Olivier T. (1), Rubinic Luca (2), Crespo Maria (1), Porwal Harshit (1), Li Yan (1), Zhang Han (1), Pugno Nicola M (3), Peijs Ton (1), Bilotti Emiliano (1)*

(1) Queen Mary University of London - -- - UnitedKingdom, (2) Department of Civil, Environmental and Mechanical Engineering, University of Trento - -- - Italy, (3) d Centre for Materials and Microsystems - -- - Italy

Graphene has shown to possess a host of extraordinary properties, such as high carrier mobility (15,000 cm2V−1s−1 [1]), high thermal conductivity (~5000 W/mK) [2], high mechanical properties (1 TPa Young's modulus and 130 GPa tensile strength [3]) and non-permeability to standard gases [4], which promises to revolutionise many fields from optoelectronics to energy storage and generation, from sensors to transport (polymer composites for aerospace). And yet, the intrinsic properties of nanoparticles like graphene, are difficult to exploit due to major challenges in achieving precision, control and material design strategies that work for nanomaterials. In fact, we face a typical dilemma: the smaller the material's dimensions, the more appealing its intrinsic physical properties, but also the more limited is our control over them. Herein we present a simple yet effective way, to produce nanocomposites with a well-defined micro and nanostructure based on a pastry making technique. The process, which we called "pressing and folding" (P&F), is an iterative top-down production technique whereby a filler and a binder are combined by successive pressing and folding steps. Not only is this process scalable, it also enables producing composites films with various compositions (up to 35 vol.%) and degree of organisation. In this work we study the processing of low density polyethylene, a widely used plastic, with graphene nano-platelets. We investigated the impact of the processing for various compositions on the final properties and established a theoretical model to predict the final properties based on the nano-filler dispersion level.

The Effect of Modified Graphene Variants on Physical, Mechanical and Thermal Properties of PP Composites

Moschopoulou Eleni (1)*, Gough Tim (1), Kelly Adrian L. (1)

(1) Polymer IRC, Faculty of Engineering and Informatics, University of Bradford - West Yorkshire - UnitedKingdom

The development of polymer/graphene-based nanocomposites has opened an interesting area in materials science in recent years and has already found use in industry. However, there are still challenges that must be addressed for these composites to reach their full potential in a wider range of markets. Similar to other nanofillers, dispersion of graphene nanoplatelets is one of the key issues which has a direct effect on the properties of the composites. A comparative study was carried out to investigate the effect of surface treatment/modification of several different commercial forms of graphene nanoplatelets (with or without surfactant, functionalized or not) on mechanical, thermal and morphological properties of polypropylene composites. Surface functionalization techniques were used to provide surface functionalities which enhanced polymer compatibility. Compounds with 1% wt. graphene loading were fabricated by melt compounding in a twin-screw extruder. The efficiency of the graphene addition in polypropylene matrix was verified through measurements of rheological, mechanical and physical properties (rotational Rheometry, DSC, TGA & DMA) as well by morphological and distribution studies (Raman spectroscopy, SEM & optical microscopy).

Multifunctional properties of carbon nanotube fibres

Fernández-Toribio Juan Carlos (1)*, Ridruejo Álvaro (2), Vilatela Juan José (1)

(1) IMDEA Materials Institute - Getafe Madrid - Spain, (2) Department of Materials Science, Polytechnic University of Madrid - Madrid - Spain

Carbon nanotube (CNT) yarns are formed by the macroscopic assembly of nanotubes well aligned along a fibre axis. Hence, due to this anisotropic assembly, CNT fibres exploit efficiently the excellent axial properties shown by nanotubes. However, fibres do not show by far the outstanding physical behaviour of individual nanotubes. This fact can be explained in terms of the complex hierarchical morphology of CNT yarns. Thus, CNTs are closely packed into bundles joined each other by Van der Waals forces; these bundles, in turn, are entangled in a high anisotropic network. Therefore, the fully transference of properties from single tubes to the macroscopic assembly is clearly diminished. In this work CNT fibres have been produced by the direct spinning method. This route consists on mechanically drawing CNT aerogel from the gaseous reaction zone of vertical CVD furnace, and then directly winding it onto a rotating rod. This is a really versatile method which enables to control the morphology of the samples by simply varying process parameters. In this regard, mechanical properties of CNT fibres obtained are in the range of high performance yarns. In fact, CNTs can be thought as rigid polymer chains. That is, both nanotubes and polymers are macromolecules consisting of long chains of repeating molecular units, and have large values of aspect ratio (length/diameter). This is why a classic analytical model used for polymer fibres has been successfully applied to this case. Through this model a close relationship between mechanical properties and morphology was successfully established. For this purpose, different morphology samples were submitted to synchrotron source X-Ray diffraction experiments in order to quantify the changes of morphology. Finally, it has been strongly reported the use of nanotubes as chemical sensors due to changes of electrical conductivity in the presence of certain chemicals. In this work it has been demonstrated the scalability of this mechanism up to fibre. In fact, a novel model based on electrochemical doping of CNT fibres has been developed. According to it, CNT yarns were successfully used as composite manufacturing s

Lightweight metal composites reinforced by surface-decorated nanocarbon

Li Qianqian [(1)], Nasiri Samaneh [(2)], Zaiser Michael [(2)]*

[(1)] Department of Aeronautics, Imperial College London - London - UnitedKingdom, [(2)] Institute of Material Simulation WW8, University of Erlangen-Nuremberg - Fuerth - Germany

The interest to use nanocarbon such as CNTs to reinforce light metals has increased in recent years. Our previous research showed that by adding 0.1wt% CNTs, the mechanical properties of the Mg composites have all been improved up to 40%. However, the results are still below the theoretically predicted value [1-2] due to the CNT agglomerates still existing on the fracture surface of the composites. In this paper we present for the first time using metal coated CNTs to produce such Mg AZ91 composites. In order to effectively disperse CNTs in the Mg melt, we have coated metals such as Pt on the surface of CNTs. TEM and EDX confirm that atomic Pt has been successfully deposited onto CNTs. Different from coating metal layers on CNTs in literature [3], surface decoration of CNT with metals as in this study may provide further benefits. Discrete metal nanoclusters on the CNT surface can help to prevent agglomeration by the simple means of acting as geometrical 'spacers'. This effect is well known in the context of graphene where decoration of exfoliated graphene sheets with Pt nanoclusters was shown to prevent face-to-face aggregation of the sheets [4]. The dispersion ability and stability of CNTs have been improved after the Pt coating. The mechanical properties are all improved compared to the pure AZ91 and to the uncoated CNTs/AZ91 composites. Individual pulled out CNTs can be observed on the fracture surface of the Pt-coated CNT/AZ91 composites under SEM. We attribute this to a better dispersion and a possible improved wettability of CNTs in Mg melt facilitated due to the Pt coating. Simulation of the interactions between different metals and CNTs has also been carried out and the results agree with the experimental data. The simulation results of different metal coatings are used to guide the experimental work. [1] Q. Li et al.Comp.Sci.Tech.69(2009). [2] Q. Li et al,Euromat 2013. [3] N. Du et al.J. Phys.Chem.C 113(2009). [4] N. Park et al App.Phys.Lett.94 (2009)

Alignment study of CNT veils and the influence on their composites.

Brandley Eileen [(1)]*, Greenhalgh Emile S. [(1)], Shaffer Milo S.P. [(1)], Li Qianqian [(1)]

[(1)] Imperial College London - London - UnitedKingdom

Carbon nanotubes veils are sheets of non-woven entangled carbon nanotubes (CNTs) which have exhibited substantial property improvements, such as Young's modulus and electrical conductivity, when uniaxial strains are applied to the veil causing the CNTs to align within the veil [1, 2]. Due to the importance that the alignment plays in the performance of the veils, it is critical to quantify the alignment of CNTs in veils. This work develops a method of applying a two-dimensional Fourier transform (2D-FFT) on scanning electron micrographs to calculate the orientation of CNTs in pre-formed arrays such as veils. From this 2D-FFT an orientation distribution function and the associated Herman parameter were calculated. The method was used to evaluate the orientation of veils that had been pre-strained to 5%, 10% and 16%. The algorithm developed showed the increased strain in the orientation distribution function and calculated an increasing Herman's parameter with increased pre-strain. To further validate the algorithm the measurement was compared to polarized Raman spectroscopy, a common method used to calculate orientation of CNTs [3, 4]. The orientation maps produced by both methods were in good agreement with both identifying areas of higher orientation in the same locations of the mapped area. Following the alignment of the CNT veil, the veil was impregnated with an epoxy composite to produce the CNT veil epoxy composite with improved mechanical properties from increased alignment of CNTs in the array. References: 1. A.E. Aliev et. al; Carbon, 2007, 45: 2880-2888 2. L. Zhang; G. Zhang; C. Liu; S. Fan; Nano Lett, 2012, 12(9): 4848-4852 3. W. Liu et al.; Carbon, 2011, 49(14): 4786–4791 4. X. Wang et. al.; Mater. Res. Letters, 2013, 1(1): 19–25

Functionalization of carbon nanotubes (CNTs) by using different oxidative agents and characterization with regard to functionalization efficiency

***Kröning Katharina** (1)*, **Fiedler Bodo** (1)*

(1) Hamburg University of Technology – TUHH - Hamburg - Germany

Carbon nanotubes (CNT) combine high electrical conductivity with good chemical and mechanical properties, making them interesting for many applications, such as reinforcement in polymer composites. In order to better adapt the tubes to specific applications, they can be modified in various ways. Our research focuses on the functionalization of single-walled and multi-walled carbon nanotubes with different oxidation agents. By adding carboxylic groups to the surface defects of CNT, the interactions between the tube surface and the epoxy resin can be improved. This results in a stronger CNT-matrix bonding and thus improves the mechanical properties of the composite material. The CNT are synthesized by chemical vapor deposition and afterwards functionalized in different steps. The first step is the purification of the tubes to remove catalytic particles and amorphous carbon. This is followed by oxidative functionalization by using different oxidative acids like nitric acid, sulfuric acid and mixtures of the acids to include carboxylic groups at the tube surface as the second step. The reactions will be carried out by reflux and sonication with varying the parameters like reaction time and temperature, in order to investigate the influence of the reaction parameters on the degree of functionalization and its effects on the tube surface structure. The influence of the functionalization method on the dispersive behavior of the functionalized CNT in epoxy is investigated as well as mechanical properties of the resulting nanocomposites by fracture toughness and tensile tests. The characterization of the acid treated tubes is carried out using different analysis methods. The effect of functionalization on the CNT-matrix bonding is investigated by fractography of the nanocomposite fracture surfaces in SEM. With this approach, the potential of CNT functionalization for improving physical properties of nanocomposites and for use in different applications is analyzed.

Universal control of pyro-resistive behaviour for flexible self-regulating heating devices

Liu Yi [(1)], *Porwal Harshit* [(1)]*, *Peijs Ton* [(1)], *Bilotti Emiliano* [(1)]

[(1)] Queen Mary University of London - -- - UnitedKingdom

Conductive polymer composites (CPCs) are considered intelligent materials to be utilized in smart devices and have attracted numerous interest from both academia and industries for decades as they combined the merits of polymer matrices and conductive filler [1]. Their capability of detecting and responding to external stimuli has offered a range of promising applications, in particular a practical choice for pyro-resistive applications such as heating devices [2]. The safety and efficiency requirement on smart heating devices has promoted the rapid growth for novel self-regulating heating devices. In order to design a good self-regulating heater, at least two equally important and fundamental aspects need to be considered: (i) good positive temperature coefficient (PTC) performance that self-regulates the heater; (ii) good Joule heating property to fulfill the heating purpose. Unfortunately, most of the existed self-regulating systems are experiencing a competitive effect between those two fundamental properties.[3] It is generally acknowledged that the low aspect ratio filler system led to a clear PTC effect, while efficient Joule heating behaviour requires large aspect filler system [4]. These two competing effects have limited novel design based on CPC hence smart design is worthy developing to achieve the wider range of applications for self-regulating heating devices. From application point of view, a more flexible heater is also desired since it could fulfill more versatile application. Here, an innovative and simple design has been developed for flexible heating devices with capability of having a universal control over the pyro-resistive behavior, which led to unlocking some of the current typical compromises in materials performance. For the first time, an heating device with large PTC intensity and good Joule heating performance, together with physical flexibility is shown, based on series connections of different conductive polymer composites.

Process development for industrialization of CNT Electrodes for Electro Active Polymers Actuator

KOLARIC Ivica [1], Glanz Carsten [1], Bodny Florian [1]*

[1] Fraunhofer-Institut für Produktionstechnik und Automatisierung IPA - -- - Germany

Electroactive polymer (EAP) actuators based on carbon nanotubes (CNTs) are in the field of programmable materials and have found great interest in research. Applications of the materials are seen in the field of soft robotics, biomedicine, bio-inspired implants. They provide many advantages such as a low driving voltage, a noiseless movement and are lightweight. However, up to now ionic CNT actuators are not ready to be integrated into products that are ready-to-market. This is due to their moderate performance and their low reproducibility. In this talk the focus is on the reproducibility and the manufacturing speed of screen printed ionic CNT EAP actuators. The manufacturing of solution processed actuators can be divided into two steps. Firstly, the dispersion for the separator and the electrodes has to be made. Secondly, these materials have to be processed with different coating technologies. Commonly the dispersion is made by the use of a magnetic stirrer which in general is a slow dispersing method. In contrast an ultrasonic horn is used to reduce the time that is required to make the dispersion for the electrodes. Casting into a mould is mostly used to make the separation layers and the electrodes. With this technique thick layers can be realized, again however, solidification using this method is time consuming due to the large amount of solvent which has to evaporate. To reduce the manufacturing time of actuators and to improve the reproducibility, screen printing is used to print the electrodes of ionic CNT EAP actuators. During the development all process steps are connected via manufacturing service bus software with industry 4.0 to reduce the cycle time for different batches and to ensure the scalability of all processes.

Development of a Colored GFRP with Antistatic Properties

Hauke Meeuw[1, a)], Markus Radek[1] and Bodo Fiedler[1]

[1]Institute of Polymer Composites, Hamburg University of Technology, Denickestraße 15, D-21073 Hamburg, Germany

[a)]*Corresponding author: hauke.meeuw@tuhh.de*

Abstract. This study shows that a modification with ultralow filler content of novel single wall carbon nanotubes (SWCNT) converts an intrinsic insulating GFRP into one with antistatic properties. These properties remain even by adding pigments for customizing without affecting the wanted bright coloring (e.g. signal color). We developed a bright colored and antistatic glass fiber reinforced polymer (GFRP) by addition of carbon nanoparticle and pigments. Novel, in industrial scale available SWCNT dispersed in polyester resin with low content of volatile organic compounds (VOC) show an ultra-low percolation threshold of 0.005wt.%. This ultralow filler content leads to the required conductivity as well as a given transparency of the nano composite. In a next step, we transferred these properties into a GFRP, manufactured by infusion process. The addition of pigments lead to the individual coloring of the GFRP. Both, the SWCNT modified and SWCNT colorized GFRP fulfilled the required electrical resistances for ESD protection.

INTRODUCTION

The development of novel multifunctional materials, in particular by modification of polymers with carbon-based nanoparticles, became one of the main research topics in the last decades. Especial interest lays on SWCNT due to their unique material properties. By adding this filler type, the intrinsic insulating polymer becomes electrical conductive. This modification provides the possibility for industrial applications requiring electrostatic discharge. In the 1880s, the topic of electrostatic discharge began to attract attention, when nearly 80 explosions occurred in German laundries due to electric discharge [1]. Recent statistics published in Europe and the USA show that, even today, electrical discharges are the cause for explosions in 35% of all cases [2]. In general, charge formation occurs due to body separation. There are three typical cases to distinguish [3]:

1. Two bodies in contact are separated by sliding each other (e.g. unwinding tape)
2. Repetitive separation (e.g. walking on a carpet)
3. Quasi-continuous contact and separation (e.g. flow of liquids in pipes)

The possibility of static electricity to be a source of ignition rises, if following conditions occur [4]:

- Charge generation is higher than dissipation rate
- Discharge occurs in ignitable atmosphere and exceeds ignition energy

Following equation calculates the discharge energy W in dependence of the generated potential U and the capacity C of the material:

$$W = \frac{1}{2} C_{HB} U^2 \tag{1}$$

A person walking across a carpet generates a potential of about 15kV. According to human body model, the capacity of the human body C_{HB} differs between 100pF to 300pF [5]. This leads to a generated energy dissipation of 10mJ up to 35mJ. Britton et al. classified the discharge energies according to the human reaction, whereby more than 15mJ leads to an unpleasant shock [4]. In comparison, the generated energy exceeds the lowest minimum ignition energy by a factor 100 for common vaporous solvents (e.g. 0.19 for acetone at a concentration of 4.97vol.% in air) [6]. The dissipation of generated charge avoids high charge formations. Antistatic materials provide an electrical conductivity that is sufficient to dissipate the charge in short time at accompanying hazard-free electrical current flow preventing a generation of a sparks. For handling electrostatic sensitive devices (ESD), three types of resistances influence the charge dissipation: The Surface resistance, measured between two electrodes, the volume resistance measured between top and bottom surface and the dissipation resistance, measured between top surface and ground. According to ASTM F 150-06 the static dissipative region lays between 1MΩ (safe dissipation) to 1GΩ (sufficient fast dissipation) [7]. Figure 1a sums up the classification of conductors and insulators in respect to their surface resistance. By incorporation of carbon nano fillers, e.g. SWCNT, into an intrinsic insulating polymer it becomes electrical conductive. We described in 2016 the possibility to adjust the electrical properties of carbon filled epoxy systems. Figure 1b shows the dependency of filler content on the electrical conductivity for different carbon fillers [8].

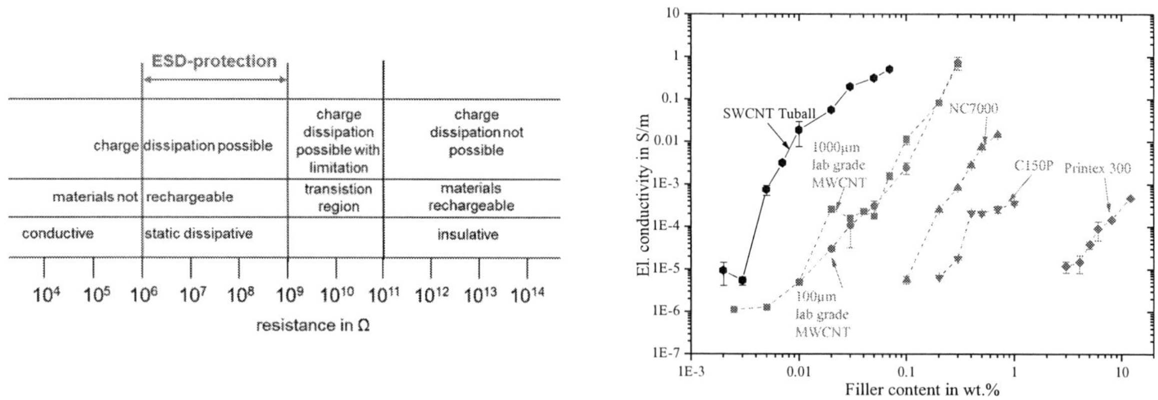

(a) (b)

FIGURE 1. a) Classification of materials in respect to their surface resistance in style of [9] b) Dependency of filler content on the electrical conductivity for different carbon fillers (SWCNT results were generated in this study) [8]

The fact of adjustable electrical properties combined with the outstanding mechanical properties normalized on fiber-reinforced carbon nano filler modified polymers' density leads to promising novel industrial applications and opens further markets. Furthermore, possibility of coloring fulfills the customers' demands for individualized products. In the past carbon black was mostly utilized to achieve conductivity in polymers. If color was a requirement, then the addition of carbon-based fillers cannot be used. With the development of next generation carbon nanotubes, especial SWCNT, ultra-low filler addition leads a sufficient conductivity of the polymer (fig. 1b). The aim of this work is the development of a signal colored and at the same time ESD protective GFRP flooring by utilization of ultra-low filler addition of SWCNT and color pigments.

MATERIALS AND METHODES

Unsaturated polyester resin AROPOL FS7993M-25, in combination with a cobalt accelerator and a peroxide catalyst ME50LY, Romar Voss, Netherlands, was used as polymer matrix. This resin provides a low volatile content. Dicyclopentadiene serves as replacement for styrene for environmental safety purpose, flame-retardation and improved fiber wetting [10]. A low viscosity of 100-150mPas makes it suitable for vacuum infusion processing [11]. Incorporation of SWCNT Tuball, OCSiAl, Russia, enables the required electrical conductivity for ESD protection of the material. They feature a SWCNT content of more than 75wt.%, nearly no amorphous carbon (less than 1wt.%), a G/D ratio over 70, diameter of 1.8nm, a length of 5µm and a market price at industrial scale 50 times lower compared to other SWCNT suppliers [12]. Addition of coloring paste PU5011 lemon yellow provided by Altropol, Germany, leads to signal color of the composite. It contains of organic pigments yellow PY74, a hydrocarbon based binder and a disperse agent based on alkylammonium salt and silicic acid to stabilize the dispersion. For manufacturing of GFRP by infusion processing unidirectional non-crimp glass fiber fabric (91%warp, 9%weft, 600g/m^2) in $[0]_6$ configuration was used. In a first step, the percolation behavior of the SWCNT/polyster resin composite (without coloring and glass fiber) was investigated. Incorporation of SWCNT into the polymer by three-roll milling process with parameters listed in **tab.1** leads to homogenous dispersion. After dispersion 1wt.% catalyst and 0.5wt.% accelerator were added, the dispersion stirred under vacuum and vacuum infused into a mold (170mm x 81mm x4mm).

TABLE 1. Process parameters for dispersion via three-roll mill

Steps	Gap sizes [µm]	Roll speed [rpm]
1	13/5	300/100/33
2	13/5	300/100/33
3-5	5/5	300/100/33

After curing for 25h at 30°C, rectangular specimens (80mm x 10mm x4mm) were cut and electrical contacts applied on the edges. The DC resistance was measured with source meter Keithley 2606 according to DIN EN ISO 3915 [13] (referring fig. 2a). Chosen source voltage was 10V. With the gained results the most adequate filler content out of the percolation curve was picked for vacuum assisted infusion process of the SWCNT modified GFRP plate aswell as the colored SWCNT modified GFRP plate. Six areas were cut out from the manufactured plates and the volume and surface resistance was determined according to ASTM D257-07 [14] with combination of Keithley resistivity test fixture 8009 and Keithley source meter 2606 (fig. 2b and c), respectively. Accompanying analysis of the dispersion quality via light microscopy as well as determining the influence of particle modification on glass transition temperature via differential scanning calorimetry were conducted.

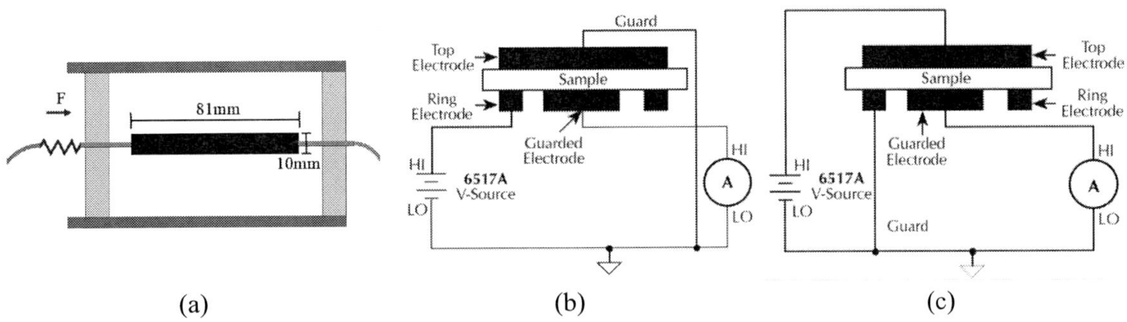

FIGURE 2. Measuring setup according to: a) volume resistance DIN EN ISO 3915; b) surface and c) volume resistance ASTM D257-07 [15]

RESULTS AND DISCUSSION

The SWCNT/resin composite shows an expected decrease of the resistance with increasing filler content (see fig 3a). Starting with a filler content of 0.007wt.% the material fulfills the requirements for ESD protection. For the further material development a filler content of 0.01wt.% was chosen, because an increase of resistance results from addition of coloring paste and color pigments. Figure 3b visualizes the influence of either 5wt%. coloring paste or 3wt.% color pigment addition on the electrical resistance.

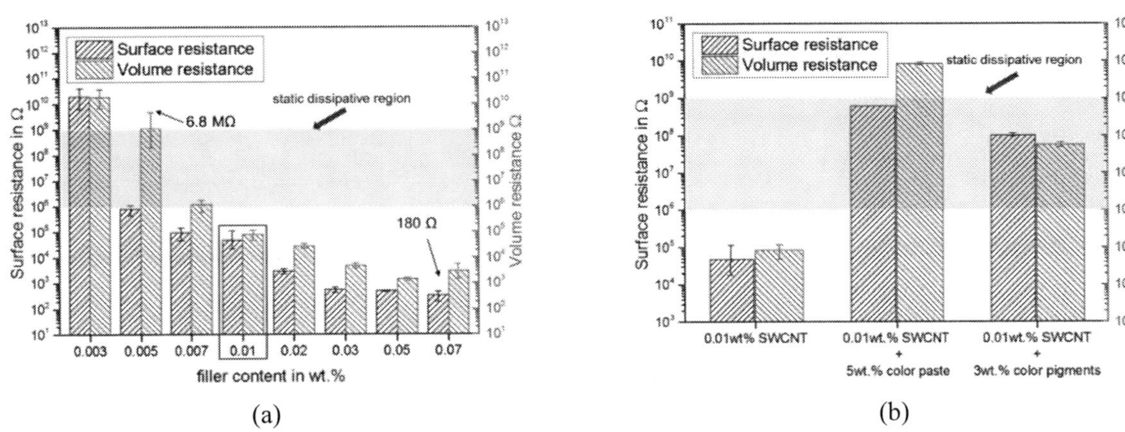

FIGURE 3. a) Dependency of the surface and volume resistance on filler content and b) influence of color additives on the resistances

As expected, the resistance increases with the addition of both materials. Pigment addition fulfils the required demands. For the coloring paste addition, the increase in resistance exceeds the for ESD application required range. Conducted DSC measurements show, that color paste lowers the glass transition temperature slightly. Sumfleth [16] as well as Ma et al. [17] investigated the influence of multi filler systems on the electrical resistance change of epoxy resin dispersions. Globular, high conductive carbon black shows a synergetic effect in combination with CNT regarding the electrical properties. Carbon black particles settle at the free ends of the CNT. There they function as bridges between the CNT and benefit the formation of the percolating electrical network by reducing tunneling distances. Transferred to this study, the globular color pigments settle as well at the free ends of the SWCNT. These intrinsic electrical isolative pigments disturb the tunnel mechanism and therefore the electrical resistance. Taken micrographs show a homogeneous distribution of particles and no percolating network (see fig. 4).

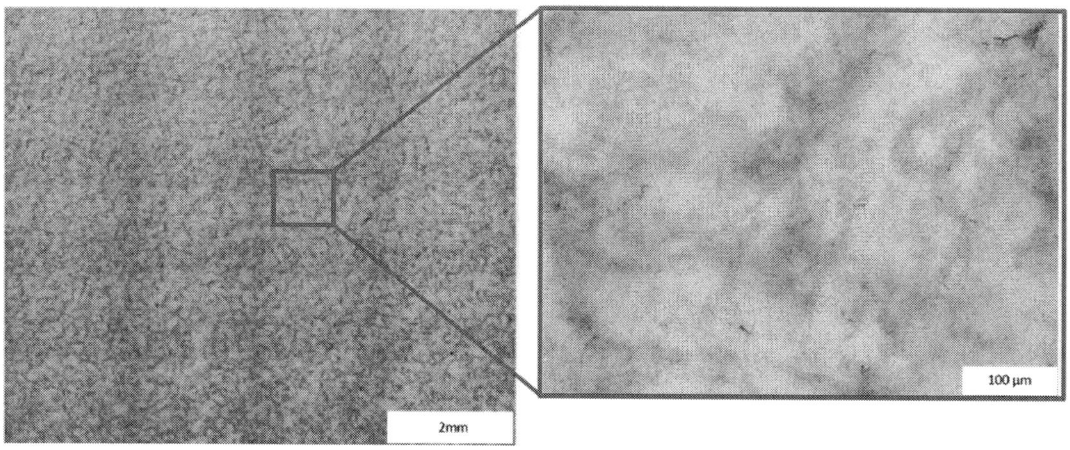

(a) (b)

FIGURE 4. Micrograph of polyester resin modified with 0.01wt.% SWCNT and 5wt.% PU5011 lemon yellow

Alkylammonium salt and silicic acid act as stabilizers and disturb the formation of the percolating network. Ma et al. [17] and Yue et al. [18] concluded that additional particle stabilize dispersions. Different particle geometries prevent re-agglomeration. The composition of the of the coloring paste results in significant higher resistances compared to the addition of pristine pigments. Thus, the SWCNT modified resin with the pristine pigment features ESD protection these parameters are transferred into the GFRP. Figure 5 shows the results for the only SWCNT modified (a) as reference and for the colored GFRP (b).

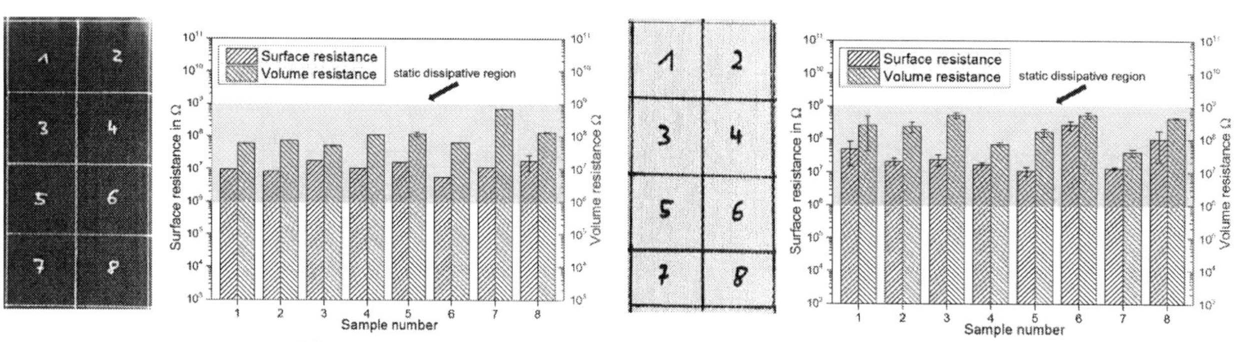

(a) (b)

FIGURE 5. Surface and volume resistances for a) 0.01wt.% SWCNT and b) 0.01wt.% SWCNT and 3wt.% color pigment yellow PY74

In comparison, there occurs a slight increase in volume resistance for the colored GFRP, but the developed GFRP fulfills the requirements for ESD protection.

CONCLUSION

In this study we demonstrated, that the new generation SWNT feature an electrical percolating network at ultra-low filler loadings (0.01wt.%). Within the development process, ESD protection in the nano composite was achieved. Transferring the used modification parameters into the fiber composite leads to the desired antistatic properties. Even with coloring, the GFRP does not lose the ESD protective properties. Ultra-low filler addition SWCNT allow even of signal coloring.

ACKNOWLEDGMENTS

The authors thank the Ministry of Science, Research and Equalities Hamburg (BWFG, project LFF-FV 05) for financial support of this project. Moreover, the authors are grateful to OCSiAl for providing the SWCNT and Altropol for the great support with the coloring materials.

REFERENCES

1. Pidoll U von. Helmut Krämer memorial lecture: Electrostatic assessment of products and processes – A view backwards and forwards. Journal of Electrostatics. 2013;71:586–90. doi:10.1016/j.elstat.2012.10.006.
2. Perrin L, Laurent A, Falk V, Dufaud O, Traoré M. Dust and electrostatic hazards, could we improve the current standards? Journal of Loss Prevention in the Process Industries. 2007;20:207–17. doi:10.1016/j.jlp.2007.03.006.
3. Drake N. Polymeric materials for electrostatic applications: A report from Rapra's Industry Analysis and Publishing Group. Shrewsbury, Shropshire, UK: Rapra Technology Ltd; 1996.
4. Britton LG. Avoiding static ignition hazards in chemical operations. New York, NY: Center for Chemical Process Safety of the American Institute of Chemical Engineers; 1999.
5. American National Standards Institute (ANSI), American National Standards Institute. ESDA/JEDEC Joint Standard for Electrostatic Discharge Sensitivity Testing - Human Body Model (HBM) - Component Level 2017.
6. Wypych G, Pionteck J, editors. Handbook of antistatics. Toronto: ChemTec Publishing; 2016.
7. F06 Committee. Test Method for Electrical Resistance of Conductive and Static Dissipative Resilient Flooring 2013. West Conshohocken, PA: ASTM International. doi:10.1520/F0150.
8. Meeuw H, Viets C, Liebig WV, Schulte K, Fiedler B. Morphological influence of carbon nanofillers on the piezoresistive response of carbon nanoparticle/epoxy composites under mechanical load. European Polymer Journal. 2016;85:198–210. doi:10.1016/j.eurpolymj.2016.10.027.
9. proHPL Fachgruppe Dekorative Schichtstoffplatten. Elektrostatische Ableitfähigkeit von Dekorativen Schichtstoffen (HPL): HPL nach EN 438. Frankfurt am Main; Juni 2008.
10. Thomas Allspach. DCPF-UP-Harze. Kunststoffe. 1999;89:117–8.
11. Romar Voss. Datasheet: AROPOL FS7993M-25; 2003.
12. TUBALL™ Single Wall Carbon Nanotubes. http://inanocomm.org/tuball/. Accessed 23 Jul 2017.
13. DIN. Messung des spezifischen elektrischen Widerstandes von leitfähigen Kunststoffen 1999. Berlin: Beuth Verlag GmbH.
14. D09 Committee. Test Methods for DC Resistance or Conductance of Insulating Materials 2007. West Conshohocken, PA: ASTM International. doi:10.1520/D0257-07.
15. Keithley Instrumentes. Volume and Surface Resistivity Measurements of Insulating Materials Using the Model 6517A Electrometer/High Resistance Meter; 2001.
16. Sumfleth J. Rheological and electrical characterisation of single- and multi-filler polymer nanocomposites. Hamburg: TuTech-Innovation GmbH; 2010.
17. Ma P-C, Liu M-Y, Zhang H, Wang S-Q, Wang R, Wang K, et al. Enhanced electrical conductivity of nanocomposites containing hybrid fillers of carbon nanotubes and carbon black. ACS Appl Mater Interfaces. 2009;1:1090–6. doi:10.1021/am9000503.
18. Yue L, Pircheraghi G, Monemian SA, Manas-Zloczower I. Epoxy composites with carbon nanotubes and graphene nanoplatelets – Dispersion and synergy effects. Carbon. 2014;78:268–78. doi:10.1016/j.carbon.2014.07.003.

Nanostructured Glass Fibre Sensor Yarns for Structural Health Monitoring

Müller Michael Thomas (1)*, Pötzsch Florian Hendrik (1), Wiegand Niclas (1), Gohs Uwe (1), Heinrich Gert (1)

(1) Leibniz Institute of Polymer Research Dresden - Saxonia - Germany

Interphases in fibre-reinforced composites are vital regions for the stress transfer between fibres and surrounding matrix. Adding an electrical percolated network into the interphases allows the detection of mechanical stress strain behavior by monitoring its electrical response. To create an electro-mechanical-response sensing ability of fibres by a continuous surface nanocoating, highly conductive nanomaterials are needed. Therefore, a comparison of the electro-mechanical-response behavior of different carbon modification, such as fibre like (carbon nanotubes), sheet like (graphene) and spherical (carbon black) modifications, were performed. The fillers where stabilized in aqueous polypropylene (PP) emulsion. Due to the additive excess of the initial PP emulsion, the respective carbon fillers can be easily stabilized by using an ultrasonic supported dispersion unit. Subsequently, the carbon filled PP emulsions were used for the in-line nanostructuring of the glass fibres during the melt drawing process. In order to achieve a homogeneous electrical conductive interphase, the average coating thickness must be above 450nm. Additionally, the temperature and the drawing speed of the coating process have a high impact on resulting electrical resistivity. For the determination of the strain-sensing behavior of the nanostructured glass fibres in PP composites, a mechanical 3-point bending test was chosen. In a static test assembly, the online resistance measurement shows a clear correlation between the measured electrical signal and the applied mechanical deformation.

Characterization of the nucleation efficiency of carbon nanotubes for polymer crystallization

Schawe Juergen E.K. (1)*

(1) Mettler Toledo GmbH - Switzerland - Switzerland

The nucleation efficiency of carbon nanotubes in a polypropylene (PP) matrix has been investigated by fast scanning calorimetry (FSC). The influence of the nanotubes on the crystallization behavior was studied in isothermal and non-isothermal crystallization experiments. For the latter case the critical cooling rate to ensure a fully amorphous structure was determined as a function of the nanotube content. From this function, the nucleation activity efficiency parameter is derived, which describes the efficiency of nucleation agents independent of temperature and filler content. Isothermal measurements show that the carbon nanotubes accelerate the á-phase crystallization significantly, but do not influence the mesophase formation. The acceleration influence of the nanotubes on the crystallization rate of the á-phase formation crystallization due to nanotubes was found to be related to the (i) number of additional heterogeneous nuclei per nanotube and (ii) the filler influence on the growth rate of the crystalline structure.

Morphology and Mechanical Properties of Elemental sulfur Reinforced Nylon-12 / Multi-walled carbon nanotubes (MWCNTs) Composites

JENA KISHORE KUMAR [(1)*], Alhassan Saeed M [(2)]

[(1)] THE PETROLEUM INSTITUTE - ABU DHABI - UAE, [(2)] THE PETROLEUM INSTITUTE - ABUDHABI - UAE

ABSTRACT: Multi-walled carbon nanotubes (MWNTs)/Nylon-12 nanocomposites with different MWNTs (1-5wt.%) loadings have been prepared by the simple melt-compounding approach at 200 °C for 20 mins with a rotation of 150 rpm for all the composites. A dark and homogeneous Nylon-12/MWNTs nanocomposite was collected from extruder after 20 mins melt-compounding. A fine and homogeneous dispersion of MWNTs throughout Nylon-12 matrix is observed by transmission electron microscopy (TEM). Scanning electron microscopy (SEM) observation on the fracture surfaces of the composites shows not only a uniform dispersion of MWNTs but also a strong interfacial adhesion with the matrix, as evidenced by the presence of many broken but strongly embedded carbon nanotubes (CNTs) in the matrix and by the absence of de-bonding of CNTs from the matrix. Mechanical testing by tensile tests as well as dynamic mechanical analysis shows that, compared with neat Nylon-12, the storage modulus and the tensile strength of the nanocomposite are greatly improved by about 113% and 134%, respectively, with incorporating only 3 wt % MWNTs. In addition to the above mechanical analysis, crystallization and melting behavior of Nylon-12/MWNTs nanocomposites are also observed and discussed by differential scanning calorimetry and X-ray diffraction. Keywords: CNT; Melt processing; Mechanical properties; TEM; SEM.

Nanoscale chemical imaging as an advanced tool for carbon nanotube fibre nanocomposite engineering

Mikhalchan Anastasiia (1)*, Banas Agnieszka (2), Banas Krzysztof (2), Borkowska Anna (3), Breese Mark B.H. (2), Kwiatek Wojciech M. (3), Tay Tong-Earn (1)

(1) Department of Mechanical Engineering, National University of Singapore - Singapore - Singapore, (2) Singapore Synchrotron Light Source - Singapore - Singapore, (3) Laboratory of Spectroscopic Imaging, Institute of Nuclear Physics, Polish Academy of Sciences - Krakow - Poland

With recent advances in the direct Floating Catalyst CVD synthesis and its on-going commercialisation [1,2], the significant quantities of 'kilometre'-long CNT fibres could reach the market in the nearest future. This expands their applicability in cost-effective light-weight engineering composites with enhanced toughness and damage tolerance [3,4]. During the composite processing, the internal (inter-bundle) nanoscale pores of a CNT fibre undergo infiltration by a monomer that could be subsequently cross-linked or polymerised. The thin polymer interphase emerging within this hierarchical structure is particularly important for better stress transfer between neighbouring CNT bundles, with significant impact on the overall mechanical performance of a composite. In this work, we present the first-of-this-kind analysis of the chemical structure of CNT fibre-reinforced nanocomposites by means of AFM-IR spectroscopy [5]. The morphology and polymer distribution within the infiltrated CNT fibre could be unravelled with sub-40 nm spatial resolution, including nanoscale chemical imaging of the polymer interphase formed along each single CNT bundle within the nanocomposite. In addition, the complementary mapping of local stiffness could be reconstructed across the area with CNT localization. The results and established methodology reveal new possibilities to analyse reinforcing potential of CNT fibres and intricate CNT-polymer nanocomposites. Acknowledgements NUS Strategic Funding (R-265-000-523-646), malopolska Regional Operational Programme (MRPO.05.01.00-12-013/15), and NUS Materials Science group for providing CNT fibres. References: [1] T. Gspann, F. Smail, A. Windle, Farad Discuss, 2014, 173:47-65. [2] C. Hoecker, F. Smail, M. Pick [et al.], Chem Eng J, 2017, 314:388-395. [3] J. Vilatela, R. Khare, A. Windle, Carbon, 2012, 50:1227-1234. [4] A. Mikhalchan, T. Gspann, A. Windle, J Mater Sci, 2016, 51(22):10005-10025. [5] A. Dazzi, C. Prater, Chem Rev, 2016.

Highly filled PP/graphite composites: Study of electrical and thermal conductivity as well as adhesive joining

Rzeczkowski Piotr [(1)]*, Lucia Martin [(1)], Knapp Andre [(1)], Müller Anett [(1)], Cohnen Andreas [(2)], Hopmann Christian [(2)], Hickmann Thorsten [(3)], Pötschke Petra [(1)], Krause Beate [(1)]

[(1)] Leibniz-Institut für Polymerforschung Dresden - Dresden - Germany, [(2)] Institute of Plastic Processing at RWTH Aachen University - Aachen - Germany, [(3)] Eisenhuth GmbH & Co. KG - Osterode am Harz - Germany

Highly filled polymer composites with carbon-based nanofillers (e.g. graphite, carbon black, carbon nanotubes) are considered as very suitable thermal and electrical conductive materials. Therefore they find applications in energy industry as bipolar plates in fuel cells or heat exchanger's housings. In both cases it is their task to carry away the heat and in case of fuel cells they need to conduct electrical current, providing efficient operation conditions in the cells. In this work, expanded graphite and polypropylene was melt mixed in a lab scale co-rotating twin-screw extruder and then compression molded to plates. The aim of the investigation was to study the influence of the filler content on thermal and electrical conductivity of the composites. For polypropylene (PP) based composites containing up to 86 wt.-% of different kinds of graphite values of thermal conductivity (laser flash analysis) up to 22.2±0.5 W/(m·K) and electrical volume resistivity as low as 0.17±0.02 Ohm·cm were obtained. Regarding the construction of a demonstrator from several parts, an assembly and sealing method of PP/graphite composite plates was investigated. Adhesive joints seem to be more efficient, easier and faster to apply than traditionally sealing used in those applications, e.g. pressed gaskets. An optimal adhesive system should exhibit good thermal stability (up to 80 °C), good water and chemical resistance, sufficient mechanical properties and good adhesion to PP/graphite. Epoxy, polyurethane and acrylate adhesives were put into consideration and examined to join such highly filled polymer materials. Keywords: Electrical conductivity, thermal conductivity, polymer composites, adhesive joint, fuel cells

POLYPROPYLENE NANO-GRAPHITE COMPOSITES OBTAINED BY AN ULTRASOUND TECHNIQUE AND MELT EXTRUSION

Manero Octavio [(1)]*, Sanchez Solis Antonio [(1)]

[(1)] National Autonomous University of Mexico - Distrito Federal - Mexico

In this work, the effect of inclusion of nanographite particles in a polypropylene matrix (PP) is studied. Nanographite particles were obtained through ultrasound exfoliation from graphite upon using a water-based HASE associative polymer as a surfactant. Nanographite was blended with polypropylene through two extrusion processes: twin-screw and single-screw, the latter includes the coupling to a static-mixer head, to generate extensional flows. Concurrently, ultrasonic waves are applied to the molten flow through ultrasonic transducers attached to the mixing head, which induces high particle dispersion and good particle distribution in the polymer matrix. It was found that at HASE concentration of 5% by weight and sonication time of 14 days (period of the exfoliation process) optimum tensile properties of the compound were achieved. Also, with respect to the PP matrix, the rate of thermal degradation decreased from 2.1 (PP) to 1.9 (% /°C), melt temperature ranged from 442 (PP) to 396°C, melt index decreased from 7.4 (PP) to 6.2 (g/10 min). Raman spectroscopy confirmed the exfoliation process, rendering sizes ranged from graphite particles of few graphene layers to micron- size particles. Rheological measurements of the compounds revealed that the extrusion-ultrasound process influence the viscosity, storage and loss moduli. The dispersion and distribution of nanoparticles improved the electromagnetic radiation shield (approximately 35%). The dielectric constant changed from 2.21 (pristine PP) to 9.02 for the compounds, which enables a good level of electrostatic charge dissipation.

Effect of functionalization on the thermal conductivity of multilayer graphene nanocomposites

Tsai Jia-Lin [(1)*], **_Tseng Po-Ying_** [(1)]

[(1)] National Chiao Tung University - Hsinchu - Taiwan

This research investigated the effect of functionalization on the thermal conductivity of multilayer graphene nanocomposites using molecular dynamics simulation. The functional groups grafted on the graphene surface are carboxyl and amine groups. The interfacial thermal conductance (ITC) between the graphene and the surrounding epoxy matrix was examined and the effects of the functional groups on ITC were explicated using vibrational density of state (VDOS). It was found that the VDOS mismatch between the surrounding epoxy and the outmost layer of graphene could be lessened by the functional groups. Nevertheless, the VDOS mismatch between the outmost layer and the inner layers within multilayer graphene was increased. Thus, the functional group grafted on the surface can effectively improve the ITC in the interface, but decrease the thermal conductivity of multilayer graphene. Moreover, the effect of layer number in the thermal conductivity of multilayer graphene was characterized through the Fourier's law. Results indicated when the layer number increased, the thermal conductivity of multilayer graphene increased accordingly.

Rate Effects in Thermogravimetric Analysis of Carbon Nanotubes

Grady Brian (1)*

(1) University of Oklahoma - Oklahoma - USA

Thermogravimetric analysis is often used as a means to measure properties of nanotubes, for example the amount of chemical modification of tubes. We will present results that show conclusively that, unless very slow heating rates are used (e.g. less than 1 K/min), the measured degradation curve depends on the density of the nanotubes. Rate effects in TGA are well-known, what was surprising to us was how slow the rate had to be in order to eliminate these effects.

Synergistic effects of graphite mixtures with different particle sizes and morphology on the electrical and thermal conductivity of highly filled polypropylene composites

Krause Beate [1]*, *Cohnen Andreas* [2], *Rzeczkowski Piotr* [1], *Facklam Martin* [2], *Hickmann Thorsten* [3], *Bangert M. D.* [2], *Theodor A.* [2], *Hopmann Christian* [2], *Pötschke Petra* [1]

[1] Leibniz-Institut für Polymerforschung Dresden - Dresden - Germany, [2] Institute of Plastic Processing at RWTH Aachen University - Aachen - Germany, [3] Eisenhuth GmbH & Co. KG - Osterode am Harz - Germany

Despite the established commercial polymer composites available for producing bipolar plates for fuel cell stacks with suitable conductive properties, there are still research activities necessary to develop bipolar plates with maximized electrical and thermal conductivity. The common pathway to develop such materials is to enhance the content of conductive fillers (e.g. graphite) in the composite. Other research activities focus on the use of expanded graphites that have high intrinsic electrical and thermal conductivity. The approach of using synergistic effects of bimodal filler systems based on two graphites of different types and particle-size distributions is not deeply explored in literature. There is a need to develop recipes with maximized conductive behavior by holding down the proportion of conductive filler to enhance the mechanical properties and plastics processing. In this study, composites based on polypropylene (PP) and different filler systems are melt-mixed using a small scale microcompounder as well as a lab scale co-rotating twin-screw extruder. The measurements of the electrical and thermal conductivity as well as mechanical properties of the composites are performed on pressed plates. The influence of synthetic/synthetic and synthetic/expanded graphite combinations and their different mass ratios in the composite on the conductive and mechanical behavior are investigated. The results show that a composite with 80 wt% of a bimodal filler system of two synthetic graphites (mass ratio: 4:1) and 30 wt% ethylene-propylene-diene-rubber-monomer in the PP component has a lower volume resistivity (0.17 Ω·cm) than the most electrical conductive composite with monomodal synthetic graphite (0.31 Ω·cm) at the same thermal conductivity of 15.1 W/(m·K). Additionally, this combination shows a lower volume resistivity and same thermal conductivity as compared to a composite containing 80 wt% of highly structured expanded graphite (0.19 Ω·cm, 15.5 W/(m·K)).

Low percolation thresholds of electrical conductivity and rheology in PC/SAN/MWCNT nanocomposites

Bizhani Hasti [(1)]*, ***Nazockdast Hossein*** [(1)], ***Jalali Arani Azam*** [(1)]

[(1)] Amirkabir University of Technology - Tehran - Iran

The role of multiwalled carbon nanotubes (MWCNT) on morphology and electrical conductivity of PC/SAN with the same blend ratio, but varying CNT content was studied by means of a combination of rheology and scanning electron microscopy (SEM) techniques. A pronounced low frequency solid-like behavior of molten nanocomposited blends suggested a fine state of dispersion and a strong 3D network formed in the samples with the concentration about 0.4 %wt. CNT localization was found to be mainly determined by the thermodynamic parameters relative to kinetic parameters. Thermodynamic calculations predicted a preferential localization of CNTs in PC phase and this study shows that order of feeding doesn't play a significant role in the nanotubes distribution in the blend constituent. An electrical percolation threshold about 0.2% wt. was found in the samples prepared by direct feeding, which is closed to the rheological percolation.

Influence of a Supplemental Filler in Twin-Screw Extruded PP/CNT Composites using Masterbatch Dilution

Michael Thomas Müller[1,b)], Beate Krause[1], Bernd Kretzschmar[1] and Petra Pötschke[1,a)]

[1] *Leibniz-Institut für Polymerforschung Dresden e.V., Hohe Straße 6, 01069 Dresden, Germany.*

[a)] Corresponding author: poe@ipfdd.de
[b)] müller-michael@ipfdd.de

Abstract. In this study commercially available multiwalled carbon nanotubes (2-8 wt.%) were incorporated in polypropylene (PP) by direct powder feeding or by a masterbatch dilution procedure using a twin-screw extruder. The influence of a supplemental, electrical non-conductive talc or electrically conductive carbon black (CB), filler on the resulting composite properties was investigated. In comparison to the direct carbon nanotube (CNT) incorporation the masterbatch dilution step resulted in improved CNT macro dispersion. The use of the supplemental fillers CB or talc does not show a significant influence on the CNT dispersion state. When compared to direct CNT incorporation, the second compounding process involved in masterbatch dilution leads to higher electrical resistivity of injection molded samples. On the other hand, the supplemental fillers talc or CB decreased the electrical resistivity values. With the addition of talc or CB an increase of the Young's modulus due to the reinforcing effect of the second filler was achieved. However, no synergistic effect between the used supplemental fillers and the CNT on the mechanical properties was obtained.

INTRODUCTION

The CNT dispersion state has a significant impact on quality and performance of the resulting nanocomposite. An optimized CNT dispersion process during melt extrusion results in better property profiles in the final application. However, the suitable distribution and dispersion of the CNT powders of as-produced agglomerated and entangled tubes during the melt processing are challenging. If the state of CNT dispersion is insufficient, the full potential of the CNT material, like electrical resistivity and mechanical properties, is not accessible. When adapting the masterbatch (MB) approach, the primary CNT agglomerates are subjected to shear stresses twice, which is expected to result in better dispersion as compared to direct CNT incorporation. In addition to the typical double extrusion process using the MB dilution, the use of maleic anhydride (MA)-grafted polyolefin as compatibilizer or ultrasound supported CNT-polymer solution mixing are described in the literature [1-4] to improve CNT dispersion. Furthermore, the addition of organic additives [5, 6] or secondary fillers like carbon black [7, 8] or talc [9, 10] are shown to be suitable to support the dispersion processes of rupture and erosion acting onto the primary CNT agglomerates towards agglomerate-free composites.

The aim of this research was the study of the MB dilution effect and the effect of feeding supplemental fillers in twin-screw extrusion on the CNT dispersion and the electrical and mechanical properties of the composites. Therefore, direct melt compounding and MB dilution with and without additional dispersion of supporting fillers, namely CB or talc, were examined. In order to reduce the CNT shortening, which was shown to occur always during the melt mixing processes [11], in some experiments a low viscosity PP was added replacing the higher viscosity PP matrix. The lower viscosity and lower shear stresses acting on CNT agglomerates are expected to result in lower nanotube shortening. Hence, the higher CNT aspect ratio can lead to a lower electrical percolation threshold. Furthermore, higher retained CNT lengths are supposed to influence the mechanical properties positively.

EXPERIMENTAL

Polypropylene HD120MO (Borealis, Vienna, Austria) with a melt flow index (MFI) of 8 g/10 min (ISO 1133) was used as the matrix polymer and additionally, as low viscosity PP Novolen 1100UC (BASF, Ludwigshafen, Germany) with a MFI of 101 g/10 min was applied. A commercially available nonfunctionalized multiwalled CNT

(MWCNT) material was employed namely Nanocyl™ NC7000 (Nanocyl S.A., Sambreville, Belgium). Nanocyl™ NC7000 nanotubes have a mean diameter of 10.0 nm, a mean length of 1341 nm [12], and a bulk density of 66 kg·m^{-3} [13]. A highly electrical conductive carbon black (CB), namely Printex XE2B (Orion Engineered Carbons) with a mean particle diameter of 30–35 nm and a bulk density of 100–400 kg·m^{-3} was used as the second conductive filler. Talc Luzenac A7 (Imerys Talc, Toulouse, France) with a mean particle diameter D50 of 2.1 µm and a tapped bulk density (ISO 787/11) of 0.26 g·cm^{-3} was used as a nonconductive filler.

For twin-screw extrusion, an intermeshing co-rotating twin screw extruder ZE 25 (Berstorff, Germany) having a distributive mixing screw configuration with L/D length of 48 (similar to screw SC-5 in [14]) was used. The polymer granules (PP high viscosity and low viscosity) were fed gravimetrically in the hopper, whereas the filler powders (2, 5, and 8 wt.%) were fed gravimetrically in the side feeder attached at 14 D. In the case of a supplemental filler additionally a second side feeder was attached at 26 D. For all mixtures, an increasing temperature program from 180°C to 200°C, a screw speed of 500 min^{-1}, and a throughput of 10 kg/h were used. For the MB production 10 wt.% MWCNT were fed gravimetrically in the side feeder at 14D. This MB was diluted with virgin PP by feeding both in hopper whereas any additional filler was fed at the 14D side feeder. In one series 20 wt.% of the added PP was replaced by the low viscosity PP. The talc loading was related to the whole composite [9].

The compounded pellets were compression molded (CM) at 200 °C, 50 kN, 2 min into circular plates (60 mm diameter, 0.3 mm thickness) using a hot press (Model PW 40 EH, Paul Otto Weber GmbH, Germany). Injection molded (IM) plates (80 mm x 80 mm x 2 mm) for electrical testing or dogbones of the type A1 (DIN EN ISO 527-2) for tensile tests were produced using an injection molding machine Ergotech 100/420-310 (Demag, Germany) with a two-cavity mold, fed by a flash gate, at a melt temperature of 240 °C and a mold temperature of 40 °C.

To determine the state of filler macro dispersion in the composites transmission light microscopy (LM) investigations using microscope BH2 combined with a camera DP71 (Olympus Deutschland GmbH, Germany) were performed according to the standard ISO-18553 on thin sections (5 µm thickness) which were cut perpendicular to the length direction of the extruded strands. The electrical volume resistivity of CM and IM plates was determined according to the standards ASTM-D257 and ASTM D 4496. At least three CM samples were measured to get the arithmetic mean value with the associated standard deviation of resistivity. The measurements were performed through the plates using a Keithley 8009 Resistivity Test Fixture (resistivity > 1E7 Ohm) or along strip cut from the plates using a 4-point test fixture with gold electrodes (resistivity < 1E7 Ohm), both combined with a Keithley electrometer 6517A (Keithley Instruments Inc., USA). The tensile properties of neat PP and filled PP composites were tested according to DIN EN ISO 527-2/1A/50 on IM dogbone tensile bars. Young's modulus (elastic modulus) E_t, maximum stress (tensile strength) σ_M and elongation at break ε_B were determined on 10 samples each. The unnotched Charpy impact strength at 23°C (acU+23) was measured according to ISO 179/ /1eU on 10 IM bars.

RESULTS AND DISCUSSION

The performance (e.g. electrical and mechanical) and the quality (e.g. surface roughness) of CNT composites are significantly depending on the state of filler dispersion. Therefore, the CNT macro dispersion was examined using LM on thin sections of the extruded strands (Figure 1). In all compositions, the MB dilution procedure leads to a decreased number and size of remaining primary CNT agglomerates. Better dispersion can be also observed in both cases, direct CNT incorporation and MB dilution, if the CNT content is lower (2 wt.% vs. 5 and 8 wt.%). An exception is the sample by direct CNT incorporation of 2 wt.% CNT and 30 wt.% talc, which shows largest CNT agglomerates. The supporting effect of the supplemental fillers CB and talc on the CNT dispersion during direct CNT incorporation is not so strongly pronounced for the extrusion process like it was described in the field of discontinuous melt compounding [7, 9]. Actually, in the case of the addition of 2 wt.% CNT and 30 wt.% talc in PP the CNT dispersion gets worse in comparison to the sample without talc. A slight dispersion improvement by the supplemental filler is only obtained during the MB dilution of 5 wt.% CNT using 30 wt.% talc or 5 wt.% CB.

The effects of the addition of the supplemental filler and the MB dilution process on the electrical volume resistivity are summarized in Figure 2a, b. In general, CM prepared samples show lower percolation (> 2 wt.%) and resistivity than IM samples (percolation between 2.0 and 5.0 wt.%). The measurements on CM plates show that the MB dilution process leads to lower resistivity values as compared to direct incorporation. On the other hand the values measured on IM plates show the opposite effect. The reduced values in CM samples assumed to have random CNT orientation follow the trend in nanotube dispersion. In IM samples, nanotube orientation along the flow direction may be assumed. In addition with stronger CNT shortening after twofold extrusion in MB dilution such more aligned and shorter nanotubes obviously percolate at higher loadings (starting 5 wt.%) and create networks

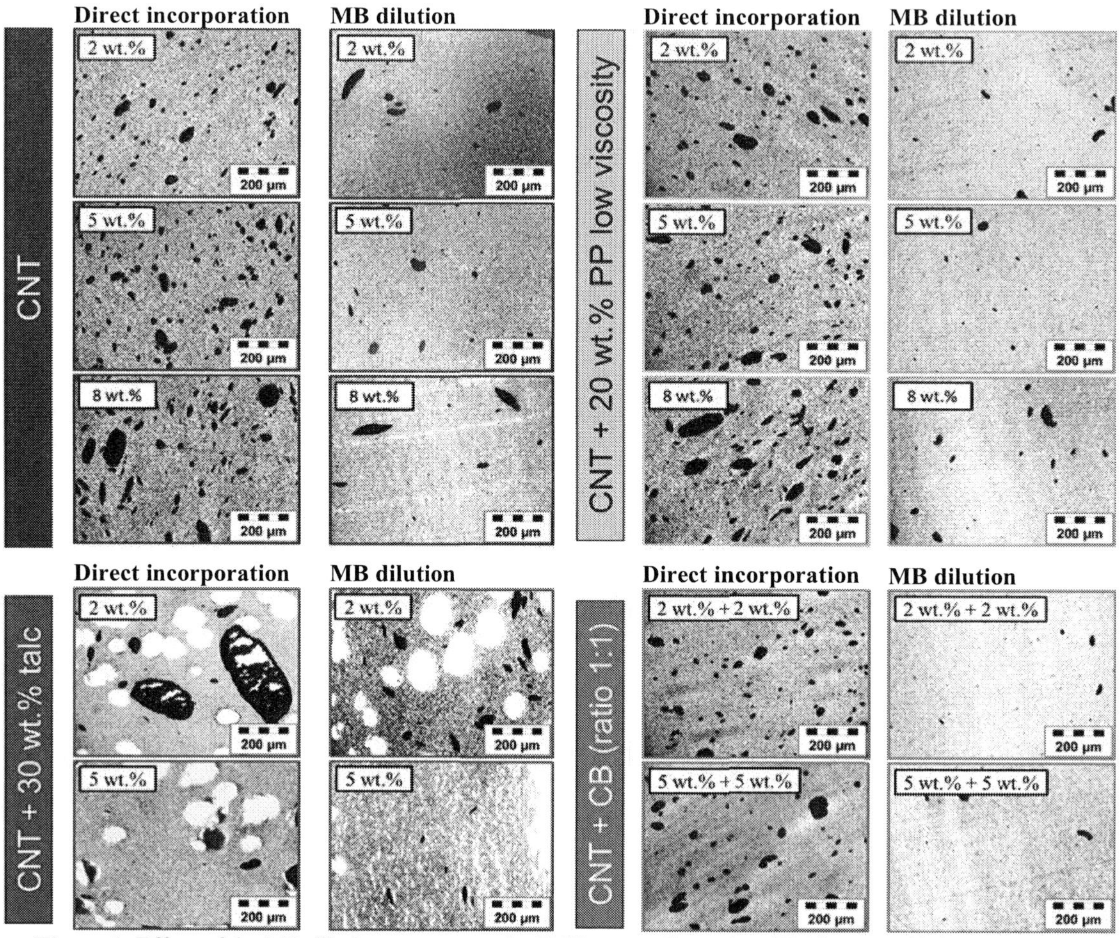

Figure 1 Effect of the MB dilution process on the CNT macro dispersion (LM images, thickness 5µm)

with lower conductivity. The replacement of PP by 20 wt.% low viscosity PP has only a slight impact on the volume resistivity measured on CM plates for both direct incorporation and MB dilution processes. This is in accordance with literature results which state that mainly the percolation threshold can be reduced due to the use of a lower viscosity matrix but values at higher loadings are comparable [11]. Looking at values of IM samples, the low viscosity PP reduced the volume resistivity only at 5 wt.% CNT using direct incorporation.

In comparison to the single filled composites, the hybrid filled composites with 30 wt.% of talc shows reduced electrical resistivity due to the volume exclusion effect [9]. Among the IM samples that with 2 wt.% CNT prepared by direct incorporation shows significantly reduced resistivity compared to the corresponding IM samples. The addition of the conductive supplemental filler CB shows only a slight resistivity reduction. However, there is no advantage against talc by comparing the electrical values, although the state of CNT dispersion is better (Figure 1).

The influences of different CNT incorporation procedures on the tensile behavior and the impact strength of IM samples are shown in Figure 2c-f. Young's modulus and tensile strength increase with CNT loading, whereas the elongation at break reduces. This major deterioration in strain at break, from round about 500% of virgin PP to values of 20-2% for filled PP, is probably related to the presence of the relatively large residual CNT agglomerates whose size and number of the agglomerates increased with CNT content (Figure 1). Large CNT agglomerates acts as stress concentrators and reduce the possibility of the typical PP matrix deformation.

To achieve higher Young's modulus and tensile strength the addition of 20 wt.% low viscosity PP was favorable only for direct incorporation of 5 wt.% CNT. For all other CNT loadings, the addition of low viscous PP is not

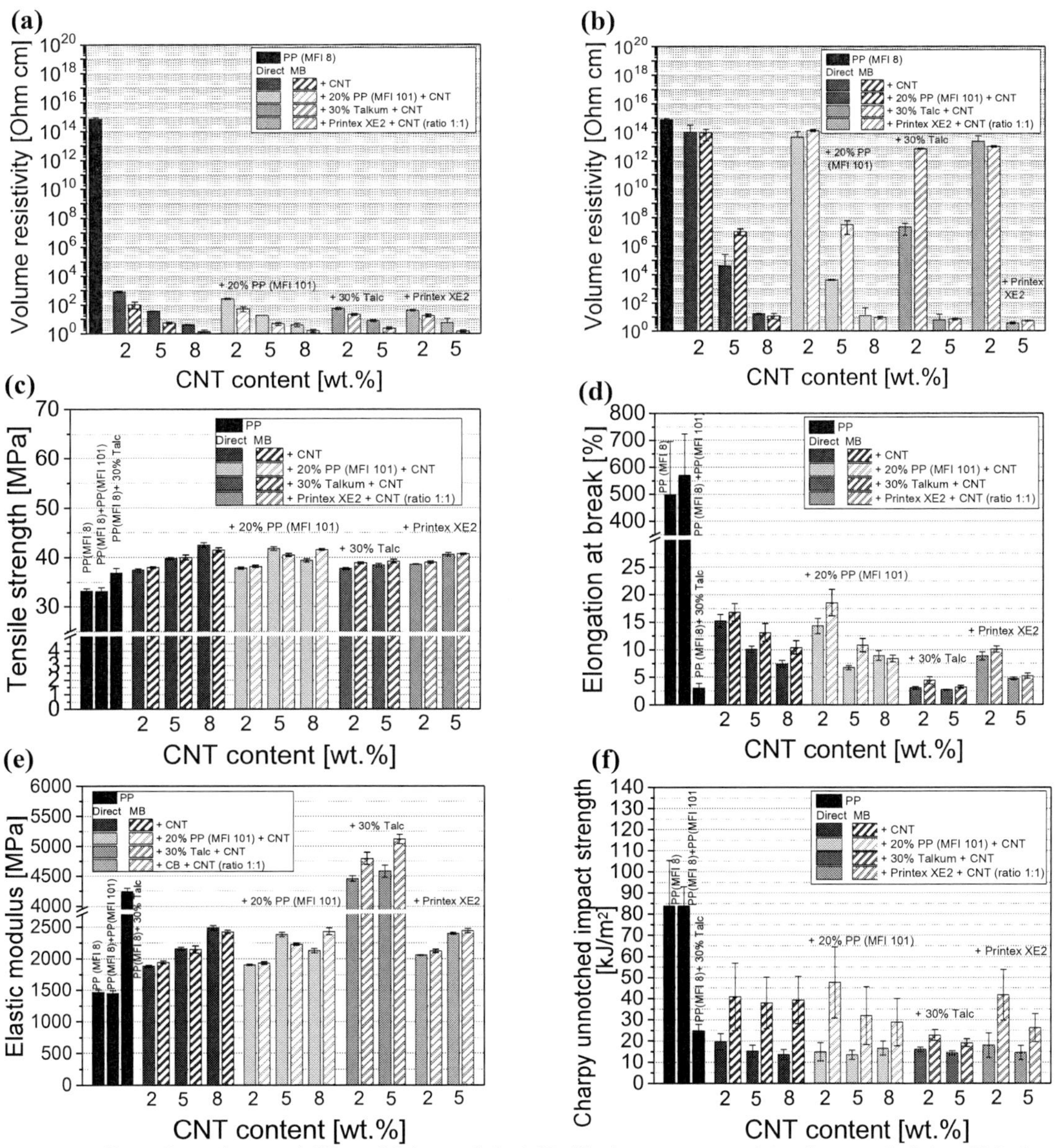

Figure 2 Effect of supplemental filler addition and the MB dilution process on (a) volume resistivity, CM plates, (b) volume resistivity, IM plates, (c) tensile strength, (d) elongation at break, (e) elastic modulus and (f) Charpy unnotched impact strength (values of (c)-(f) measured on IM parts)

suitable to improve these values. The use of supplemental fillers did not affect the tensile strength, however increased the Young's modulus, due to the added stiffness of the second filler. No synergistic effect between the used supplemental fillers and CNT on the mechanical properties was obtained.

When comparing MB dilution with direct incorporation, no clear trend in Young's modulus and tensile strength was observed whereas the elongation at break is always higher using the MB procedure. The situation is quite different when using hybrid fillers in MB dilution, where improvement in Young's modulus, tensile strength and strain at break was found. One reason is probably the improved CNT nano dispersion. The impact strength is significantly higher when using MB dilution compared to direct CNT incorporation. This can be attributed to the improved CNT dispersion after the MB dilution step (Figure 1). This improvement is lowest for the talc addition. All modifications do not improve impact strength.

SUMMARY

In this investigation electrically conductive PP based composites containing commercially available MWCNTs with electrical non-conductive talc or electrical conductive carbon black were prepared using a twin-screw extruder. The MWCNTs were incorporated in the PP matrix directly at the side feeder or via a MB dilution procedure at the hopper. The resulting electrical and mechanical properties were measured on IM parts. In general, the MB dilution step leads to an improved CNT dispersion meaning that number and size of remaining primary CNT agglomerates are decreased. The use of the supplemental filler CB and talc does not lead to significantly improved state of CNT macro dispersion. The supplemental fillers increase the Young's modulus due to the reinforcing effect of the second filler. Further, the electrical resistivity values of the CNT-composites can be reduced by adding these two kinds of fillers. However, no synergistic effect between the used supplemental fillers and CNT on the mechanical properties could be found. The additional compounding process involved in the MB dilution did not show a clear trend to improve Young's modulus and the tensile strength whereas the elongation at break and the impact strength are improved. In comparison to direct CNT incorporation, the achievable electrical resistivity, of IM samples, is higher. One reason is probably the CNT shorting during the double extrusion process, whereby the reduced nanotube length leads to different percolation and orientation behavior during the injection molding process. Additionally, the effect of 20 wt.% replacement of PP by a low viscosity PP was studied. For the CNT direct incorporation process, a slight impact on the volume resistivity and on the mechanical properties was observed only when the CNT content was in the percolation threshold range. However, when using the MB dilution process these improvements disappeared. Finally, it can be concluded, that an excellent state of carbon nanotubes dispersion is an important prerequisite to enhance the composite properties and hybrid filler systems can lead to further enhancements.

ACKNOWLEDGMENTS

The authors thank Mrs. Kristina Eichhorn and Mr. Holger Scheibner for mechanical testing and German Federal Ministry of Education and Research (BMBF) for financial support within project no. 03X0054B (CarboTube).

REFERENCES

1. M. Pöllänen, S. Pirinen, M. Suvanto and T. T. Pakkanen, *Composites Science and Technology* **71**, 1353-1360 (2011).
2. K. Stoeffler, Perrin-Sarazin, F., Ton-That, M.-T., Bélanger, C., Denault, J., *Journal of Nanostructured Polymers and Nanocomposite* **6**, 128-135 (2010).
3. B. X. Yang, K. P. Pramoda, G. Q. Xu and S. H. Goh, *Advanced Functional Materials* **17**, 2062-2069 (2007).
4. J. F. Vega, J. Martínez-Salazar, M. Trujillo, M. L. Arnal, A. J. Müller, S. Bredeau and P. Dubois, *Macromolecules* **42**, 4719-4727 (2009).
5. M. T. Müller, B. Krause and P. Pötschke, *Polymer* **53**, 3079-3083 (2012).
6. M. T. Müller, P. Pötschke and B. Voit, *Polymer* **66**, 210-221 (2015).
7. R. Socher, B. Krause, S. Hermasch, R. Wursche and P. Pötschke, *Composites Science and Technology* **71**, 1053-1059 (2011).
8. E. Bilotti, H. Zhang, H. Deng, R. Zhang, Q. Fu and T. Peijs, *Composites Science and Technology* **74**, 85-90 (2013).
9. M. T. Müller, J. Dreyße, L. Häußler, B. Krause and P. Pötschke, *Journal of Polymer Science Part B: Polymer Physics* **51**, 1680-1691 (2013).
10. H.-D. Bao, Z.-X. Guo and J. Yu, *Polymer* **49**, 3826-3831 (2008).
11. R. Socher, B. Krause, M. T. Müller, R. Boldt and P. Pötschke, *Polymer* **53**, 495-504 (2012).
12. B. Krause, R. Boldt and P. Pötschke, Carbon **49**, 1243 –1247(2011).
13. B. Krause, M. Mende, P. Pötschke and G. Petzold, *Carbon* **48**, 2746-2754 (2010).
14. T. Villmow, B. Kretzschmar and P. Pötschke, *Composites Science and Technology* **70**, 2045-2055 (2010).

The influence of graphene nanoplatelets on mechanical and shape memory properties of thermoplastic polyurethane/ acrylonitrile butadiene styrene nanocomposites

***Memarian Farzaneh* [(1)]*, *Fereidoon Abdolhossein* [(1)], *Khonakdar Hossein Ali* [(2)]**

[(1)] Faculty of Mechanical engineering, Semnan University, Semnan, Iran - Semnan - Iran, [(2)] Leibniz Institute of Polymer Research, D-01067 Dresden, Germany - Dresden - Germany

The influence of graphene nanoplatelets on mechanical and shape memory properties of thermoplastic polyurethane/ acrylonitrile butadiene styrene nanocomposites Farzaneh Memarian1*, Abdolhossein Fereidoon1, Hossein Ali Khonakdar2 1Faculty of Mechanical engineering, Semnan University, Semnan, Iran 2Leibniz Institute of Polymer Research, D-01067 Dresden, Germany Corresponding author: Farzaneh Memarian (Email: farzane.memarian@gmail.com) Abstract shape memory and mechanical properties of a new polymer blend nanocomposites based on crystalline thermoplastic polyurethane (TPU) and acrylonitrile butadiene styrene (ABS) reinforced with graphene nanoplatelets were investigated. The TPU/ABS polymer blend which was prepared by melt blending technique is a thermo-responsive material which benefits the glass transition of TPU as a driving force to be recovered. Two types of graphene nanoplatelets were used, single or few-layered graphene (SLGS) and multi-layered graphene sheet (MLGS). Loading only 0.5 wt% MLGS in to the blend improved tensile strength about 55%. Also, Young's modulus of the blend was increased 28% by loading 1.5 wt% MLGS. In comparison, TPU/ABS/SLGS nanocomposites showed greater Young's modulus but lower tensile strength. The presence of 0.5 wt% SLGS caused up to 42% and 66% increment in tensile strength and Young's modulus, respectively. Moreover, a significant increase of 180% in Young's modulus is observed, by adding 1.5 wt% SLGS. Shape recovery of nanocomposites reinforced with SLGS was higher than nanocomposites reinforced with MLGS. By the presence of only 1 wt% SLGS, shape recovery reaches to more than 99%. Furthermore, shape fixity of both types of nanocomposites was comparable and in the most cases it was larger than 97%, even after three cycles. Keywords: Thermoplastic polyurethane; Acrylonitrile butadiene styrene; shape memory property; mechanical property; Single layered graphene, Multi layered graphene

Monitoring polymer diffusivity in nanocomoposites by NMR

Nadiv Roey (1)*, Fernandes Ricardo M. F. (2), Dai Jing (2), Buzaglo Matat (1), Varenik Maxim (1), Furó István (2), Regev Oren (1)

(1) Ben-Gurion University of the Negev - Israel - Israel, (2) KTH Royal Institute of Technology - Sweden - Sweden

Nano fillers are known to dramatically affect the rheological properties of polymer systems. Understanding and manipulating the rheological behavior of carbon nanotubes (CNTs)-loaded polymer system is of high importance for achieving optimal performance. The remarkable effect of CNTs on the polymer system rheology is mainly associated with their high aspect ratio, enabling them to percolate and form a network structure. The NT network inhibits the polymer chain mobility already at very low filler concentration (~0.1 vol%). In this study, we explored the mobility (diffusivity) of epoxy chains when loaded by CNTs using NMR diffusion experiments. At the molecular level, we detect the self-diffusion coefficient of CNT-bound epoxy chains and correlated their concentration to bulk rheological properties (e.g., viscosity and storage modulus). The concentration of CNT-bound epoxy molecules scales with the rheological percolation, specifically, with the crossover volume fraction in which the polymer transforms for liquid into solid-like behavior (G'= G''). We therefore suggest that diffusion NMR can be employed as an indicative characterization tool for rheological and interface properties of nanocomposite systems. Reference: 1. Roey Nadiv, Michael Shtein, Alva Peled, Gleb Vasilyev, Eyal Zussman and Oren Regev. Composites Science and Technology, 2016, 133, 192-199.

The relationship between microstructure and the vapor sensing behaviors of multi-walled carbon nanotubes filled polycarbonate/polystyrene blends

Li Yilong [(1)], *Pionteck Jürgen* [(1)], *Pötschke Petra* [(1)*], *Voit Brigitte* [(1)]

[(1)] Leibniz-Institut für Polymerforschung Dresden - Sachsen - Germany

Vapor sensor is an important application for conductive polymer composites (CPCs) chemiresistors which is based on the typical percolation behaviors of CPCs. When towards organic vapor, the conductive networks in CPCs would be changed, which results in the electrical resistance changes. Based on this novel property, they can be used as e-nose, human health monitor, environment detector etc. In this study, we explore the microstructure of polycarbonate/polystyrene/multi-walled carbon nanotube (PC/PS/MWCNT) blend composites with different composition prepared by melt mixing. SEM exhibits a sea-island morphology for blend composites containing 0.75 wt.% MWCNT with a PC/PS weight composition of 70/30 (C70S30M0.75), while the weight composition of 50/50 at same MWCNT content (C50S50M0.75) forms a co-continuous structure. Based on surface tension parameters of components, it is assumed that MWCNT preferentially locate within the PC phase. This can also be proven by selective extraction experiment: a clear solution can be obtained using cyclohexane, which is used for PS extraction. Vapor sensing performance is related with the swellability of the PC phase with MWCNT network. Vapor of good solvents for PC (dichloromethane and acetone) show a poor reversibility for all CPCs due to the strong interaction between vapor and PC, but C70S30M0.75 exhibits the highest sensitivity among the three CPCs. Vapors with moderate PC solubility (ethyl acetate and toluene) display similar sensing response as good solvent vapors, however, C50S50M0.75 shows an excellent reversibility due to large interface which facilitates the evaporation of vapors. Finally, the PC/PS blend CPCs are also suitable for sensing poor solvent vapors for PC (e.g. cyclohexane) due to the forced MWCNT network changes by the swelling of PS phase, even if the MWCNT are located within the PC phase. Such novel vapor sensor provides a new possibility for sensing vapors with wide ranges of solubility parameters.

Comparative study of electrical percolation of single-, multiwalled and branched carbon nanotubes melt mixed in different thermoplastic matrices

Krause Beate [(1)]***, Barbier Carine*** [(2)]***, Kunz Karina*** [(1)]***, Pötschke Petra*** [(1)*]

[(1)] IPF Dresden - Saxony - Germany, [(2)] Université de Bretagne-Sud - Lorient - France

Electrical conductive thermoplastic composites filled with carbon nanotubes (CNTs) have been widely researched during the last decade. One of the aims of current efforts is to reach high electrical conductivity at low filler content. Next to the properties of the polymer matrix, mainly the CNT characteristics, like type of CNT, its length or active surface area and chemistry, determine the electrical conductivity of polymer/CNT composites produced by melt mixing. In this contribution, three different types of CNTs were incorporated in polypropylene (PP), polycarbonate (PC) and polyvinylidenefluoride (PVDF) using a conical twin screw microcompounder Xplore DSM15 and compression molded plates were prepared. The commercial CNTs were singlewalled CNT Tuball® from OCSiAl Ltd. (Russia), multiwalled CNT NC7000 from Nanocyl, S.A. (Belgium) as well as branched and entangled MWCNTs having surface modification of polyethylene glycol (CNS-PEG) from Applied NanoStructured Solutions (USA). The composites were characterized concerning their electrical and thermal properties and filler dispersion. The electrical percolation thresholds for composites of PP/Tuball, PP/CNS-PEG and PC/CNS-PEG were found already at 0.1 wt%. For PVDF/Tuball, PVDF/CNS-PEG and PC/Tuball percolation starts between 0.1 and 0.25 wt%. At 2 wt% filler all composites with CNS-PEG show lower volume resistivity values than the other fillers. Values of about 3 Ohm·cm were achieved in PVDF, 7 Ohm·cm in PP and 11 Ohm·cm in PC based composites. The highest percolation thresholds were always measured for composites with NC7000. The main reason for this may be the significantly lower aspect ratio of NC7000 compared to the other kinds of CNT. A homogeneous dispersion was observed especially for composites containing CNS-PEG. For all CNTs a nucleation effect was found in differential scanning calorimetry. At the same CNT content the highest values of crystallisation temperature were found for CNS-PEG and Tuball.

Melt mixed composites of polypropylene with singlewalled carbon nanotubes for thermoelectric applications: switching from p- to n-type behavior by additive addition

Petra Pötschke[1,a)], Beate Krause[1], Jinji Luo[1]

[1] *Leibniz-Institut für Polymerforschung Dresden e.V., Hohe Straße 6, 01069 Dresden, Germany.*

[a)] *Corresponding author: poe@ipfdd.de*

Abstract. *Composites were prepared with polypropylene (PP) as the matrix and singlewalled CNTs (SWCNTs) of the type TUBALL from OCSiAl Ltd. as the conducting component by melt processing in a small-scale twin-screw compounder. In order to switch the typical p-type behavior of such composites from positive Seebeck coefficients (S) into n-type behavior with negative Seebeck coefficients, a non-ionic surfactant polyoxyethylene 20 cetyl ether (Brij58) was used and compared with a PEG additive, which was shown previously to be able to induce such switching. For PP-2 wt% SWCNT composites Brij58 is shown to result in n-type composites. The negative S values (up to -48.2 μV/K) are not as high as in the case of previous results using PEG (-56.6 μV/K). However, due to the more pronounced effect of Brij58 on the electrical conductivity, the achieved power factors are higher and reach a maximum of 0.144 μW/(m·K²) compared to previous 0.078 μW/(m·K²) with PEG. Dispersion improvement depends on the type of SWCNTs obtained by using varied synthesis/treatment conditions. Solution prepared composites of PEG with SWCNTs also have negative S values, indicating the donation of electrons from PEG to the SWCNTs. However, such composites are brittle and not suitable as thermoelectric materials.*

INTRODUCTION

Electrically conductive polymer composites (CPC) based on insulating matrices and carbon nanotubes (CNTs) have been developed for diverse applications, such as for the dissipation of static electricity. Recently, such CPCs are also being studied as potential thermoelectric (TE) materials to recover waste heat into electric energy. The thermoelectric effect (also called Seebeck effect) is described as an electrical potential (voltage ΔU) induced by a temperature difference (ΔT) between the two sides of a material. High Seebeck coefficient (S) and high electrical conductivity (σ) are favorable for high power factor ($PF = \sigma \cdot S^2$), whereas a simultaneous low thermal conductivity (κ) ensures a high TE efficiency determined by ZT ($ZT = \sigma \cdot S^2 T/\kappa$) The advantages of polymer based thermoelectric materials over traditional materials, like bismuth telluride alloys, are not only their cost efficiency, but also the ease of processing, flexibility, low density, and intrinsic low thermal conductivity.

An efficient thermoelectric generator (TEG) requires both high performance p-type (positive S) and n-type (negative S) TE materials. To avoid problems caused by different thermal expansion coefficients or corrosion effects of two different materials, it is desired to combine similar p- and n-type materials for the device fabrication. An ideal scenario is to apply the same base material, which can be doped both into p-type and n-type.

Using the approach of melt mixed composites of insulating thermoplastic polymers with conductive fillers like carbon nanotubes (CNTs) normally p-type materials can be obtained. This was shown on the example of the polymer matrix polypropylene (PP) with industrially available singlewalled CNTs (SWCNTs) [1]. SWCNTs of the type TUBALL™ from OCSiAl were employed and melt mixed into the PP matrix using a small-scale compounder. In addition, copper oxide (CuO) representing a nontoxic compound with high Seebeck coefficient was applied together with the SWCNTs to enhance S and PF. The effect of SWCNT and CuO content on σ, S, and PF was studied [1]. The SWCNTs form an electrical conducting network already at very low amounts of ~ 0.1 wt% [2] and melt mixing conditions only play a marginal role on the composite properties [3]. When adding an ionic liquid (IL) to composites with 2 wt% SWCNT, the maximum S value reached was 63.8 μV/K, resulting in a power factor of 0.26 μW/(m·K²) [3]. In a recent work, we showed that such melt processed p-type PP/SWCNT/CuO composites could be converted into relative stable n-type by using a processing additive during the melt mixing. At the investigated SWCNT concentrations (0.8 wt% and 2 wt%) and a fixed CuO content of 5 wt%, the addition of polyethylene glycol (PEG, M_n 10,000 g/mol) during melt mixing converted the p-type composites into n-type.

Values of S up to -56 μV/K were achieved using 2 wt% SWCNT and 10 wt% PEG, resulting in a power factor of 0.078 μW/(m·K^2) [4].

This contribution investigates whether this switching by PEG also occurs without the CuO additive, for different SWCNT modifications and when replacing the so far used PEG by a non-ionic surfactant polyoxyethylene 20 cetyl ether (Brij 58). Brij58 contains a hydrophilic PEG part and a hydrophobic aliphatic hexadecyl segment connected by an ether group. On one hand side, it is assumed that the hydrophobic part of Brij58 leads to a better wettability of PP due to similar polarity of PP and hexadecane. On the other side, it is expected that the wettability of SWCNTs, depending on the SWCNT surface properties (e.g. functionalization and resulting polarity), by Brij58 having hydrophilic and hydrophobic parts is better compared to hydrophilic PEG. Despite their very low melt viscosities which could reduce the shear stresses acting on the primary CNT agglomerates during the nanotube dispersion, PEG as well as Brij 58 as additives were shown to improve dispersion and increase electrical conductivity. This was proven for composites based on linear low density polyethylene with multiwalled CNT (Nanocyl NC7000), whereby Brij58 was more effective than PEG [6]. In addition, to understand the effect of PEG, composites of PEG with SWCNTs were prepared by solution mixing and studied concerning their Seebeck coefficients.

EXPERIMENTAL

As polypropylene matrix, Moplen HP400R (LyondellBasell Industries) was applied. Three different types (named A, B and mA) of SWCNTs Tuball^{TM} provided by OCSiAl Ltd) (> 75% purity, diameter 1-2 nm and length > 5 μm [2]) were used. Thereby, SWCNTs type A and B represent as-synthesized SWCNTs prepared at different synthesis parameters. SWCNTs of type A were plasma oxidized to produce the mA type SWCNTs. Polyethylene glycol (PEG) flakes, provided by Sigma-Aldrich (M_n 10,000 g/mol) were used. In order to compare with previous investigations using PP-SWCNT composites with 5 wt% CuO [4], an amount of 10 wt% PEG was selected. In comparison, polyoxyethylene 20 cetyl ether (Brij58, Sigma–Aldrich) with the molecular formula $HO(CH_2CH_2O)_{20}C_{16}H_{33}$, a molecular weight M_n of ~1124 g/mol and a melting temperature of ~38°C was applied as switching additive. This Brij58 has a waxy solid structure and is commonly used in biochemical applications. Based on previous investigations using Brij58 for improving CNT dispersion [6], an amount of 4 wt% was selected.

For melt mixed composites, a conical twin-screw microcompounder (DSM Xplore 15) with a capacity of 15 ccm was used. Mixtures of grounded PP powders (diameter <1 mm) and 2 wt% SWCNT and 4 wt% Brij58 or 10 wt% PEG pre-shaken in glass vial were fed into the running microcompounder. The standard processing conditions were fixed to a melt temperature of 210°C, a rotation speed of 250 rpm and a mixing time of 5 min. For samples containing Brij58 also trials with the mixing temperature of 230°C were performed to see the temperature effect. Afterwards, the composites were pressed at 210°C under a pressure of 50 kN into circular plates with a diameter of 60 mm and a thickness of ca. 0.3 mm using a press PW 40 EH (Paul Otto Weber GmbH, Germany). Rectangular samples were manually cut from the plates and used for measuring the electrical conductivity and Seebeck coefficient.

For PEG/SWCNT composites, solution mixing was applied due to the extremely low viscosity of the PEG melt: 0.5 g PEG was dissolved in 6 mL water by magnetic stirring (600 rpm, 5 hours). Certain amounts of CNTs were added into the PEG-water solution and the resulting mixtures were sonicated with a probe sonicator (UP400S, Hielscher). Composite mixtures were then drop-casted on the top of a polyimide substrate. The brittle composite films were then carefully detached from the polyimide substrate and used for the Seebeck coefficient characterization.

For the electrical conductivity characterization, a 4-electrode test fixture combined with a Keithley electrometer E6517A or Keithley multimeter DMM2000 was used and eight values were averaged. The Seebeck coefficient at room temperature was determined using a measurement platform SRX [5] and the results are the average of at least 3 measurements for each sample.

The study of CNT macrodispersion was done on thin sections (5 μm) which were cut from the pressed plates at room temperature using a Leica microtome RM2265 and a diamond knife. The transmission light microscopy (TLM) investigations were performed using a microscope Olympus BX53M and a camera Olympus DP71. The quantification was done by calculating the agglomerate area ratio A_A (%) which represents the quotient between the area of agglomerates A and the area of the complete image A_0. Details are given in [7]. At least 15 cuts were used to determine a mean value and standard deviation.

RESULTS AND DISCUSSION

Obtained values of electrical conductivity σ, Seebeck coefficient S and the resulting power factor PF of PP based composites containing the 3 different types of SWCNTs without or with PEG or Brij58 are summarized in Table 1.

TABLE 1. Thermoelectric properties of PP based composites with 2 wt% SWCNT containing 10 wt% PEG or 4 wt% Brij58.

Material	Electrical conductivity [S/cm]	Seebeck coefficient [μV/K]	Power Factor PF [μW/(m·K²)]
PP-A-210°C	0.17	23.1	0.009
PP-A-Brij-230°C	0.78	-43.1	0.144
PP-mA-210°C	0.08	35.6	0.099
PP-mA-Brij-210°C	0.04	-48.1	0.008
PP-mA-Brij-230°C	0.50	-48.2	0.115
PP-mA-PEG-210°C	0.11	-45.7	0.023
PP-B-210°C	0.24	43.2	0.045
PP-B-Brij-210°C	0.59	-35.9	0.076
PP-B-Brij-230°C	0.78	-37.9	0.111
PP-B-PEG-210°C	0.63	-48.0	0.145

As it is clearly seen, not only PEG but also Brij58 switches the Seebeck coefficients of the composites into negative values. This occurs for all 3 SWCNT variations as well as for both melt mixing temperatures. Among the SWCNT variations, highest electrical conductivity and Seebeck coefficient resulting in highest PF are observed for type B (p-type material). In nearly all cases, the addition of Brij58 or PEG enhances electrical conductivity. The achieved negative S values range between -36 μV/K and -48 μV/K which is lower compared to our previous results (-56.6 μV/K [4]). However, due to a more pronounced effect of Brij58 and PEG on the electrical conductivity, the achieved power factors are higher and reach a maximum of 0.144 μW/(m·K^2) compared to 0.078 μW/(m·K^2) in previous work [4]. Increasing the mixing temperature from 210°C to 230° under otherwise identical conditions enhances the electrical conductivity and leads to slightly higher negative S values.

Morphological investigations (Figure 1) show inconsistent influence of PEG or Brij58 addition on the CNT macrodispersion. For PP-SWCNT-Brij58 composites, black SWCNT agglomerates are visible in a gray matrix concluding that PP and Brij58 are miscible in this scale. The Brij58 addition results in PP-A composites in a significant better CNT macrodispersion as indicated by lower numbers of agglomerates and agglomerate area ratio A_A. For PP-mA composites, the Brij58 addition leads only to a marginal decrease of A_A independent of the mixing temperature. In contrast to PP-A and PP-mA composites, PP-B-Brij58 composites show a worse CNT macrodispersion, especially at higher mixing temperature, though this composite without additives had the best dispersion and highest conductivity in comparison to the other composites without additives.

In the light microscopy images of composites containing PEG, a two-component system is visible, indicating two non-miscible materials (Figure 1, right). The PP matric appears gray and PEG is visible as white areas. The PEG addition leads to a significant decrease of A_A in the PP-mA composite whereas in the PP-B composites the CNT macrodispersion becomes worse. It can be concluded that the different surface properties of SWCNTs (mA, B) play an important role for the efficiency of PEG to improve the CNT macrodispersion. SWCNTs of type mA are expected to have more functional groups on their surface and thus higher surface polarity and to interact in a better way with these additives.

In addition, SWCNT filled composites with only PEG as the matrix were prepared using SWNT type mA by solution mixing and characterized to elucidate the electron donating characteristic of PEG. The negative Seebeck coefficients of all these PEG/SWCNT composites with different SWCNT concentrations (from 1 wt% to 6 wt% SWCNT) shown in Table 2 confirm that PEG donates electrons to the SWCNTs and generates n-type materials. Thereby the amount of SWCNTs does not influence the S value in a systematic way; starting at 2 wt% the values are comparable at around -40 μV/K. However, the PEG/SWCNT composites are very brittle (see Figure 2). Similar to n-type PEI/SWCNT [8], PEG/SWCNT composites melt at low temperature due to the low melting temperature of the PEG matrix and they can be easily dissolved/influenced by water. Thus, they are unsuitable as TE materials. In comparison, melt mixed n-type composites, with PP as matrix and PEG as additive, are highly flexible as shown in the photo in Figure 2 right. In addition, the weight loss for PP-SWCNT-PEG strips after 1 day immersion in water was only 2 wt%, which confirms that most of PEG is locked inside the PP matrix.

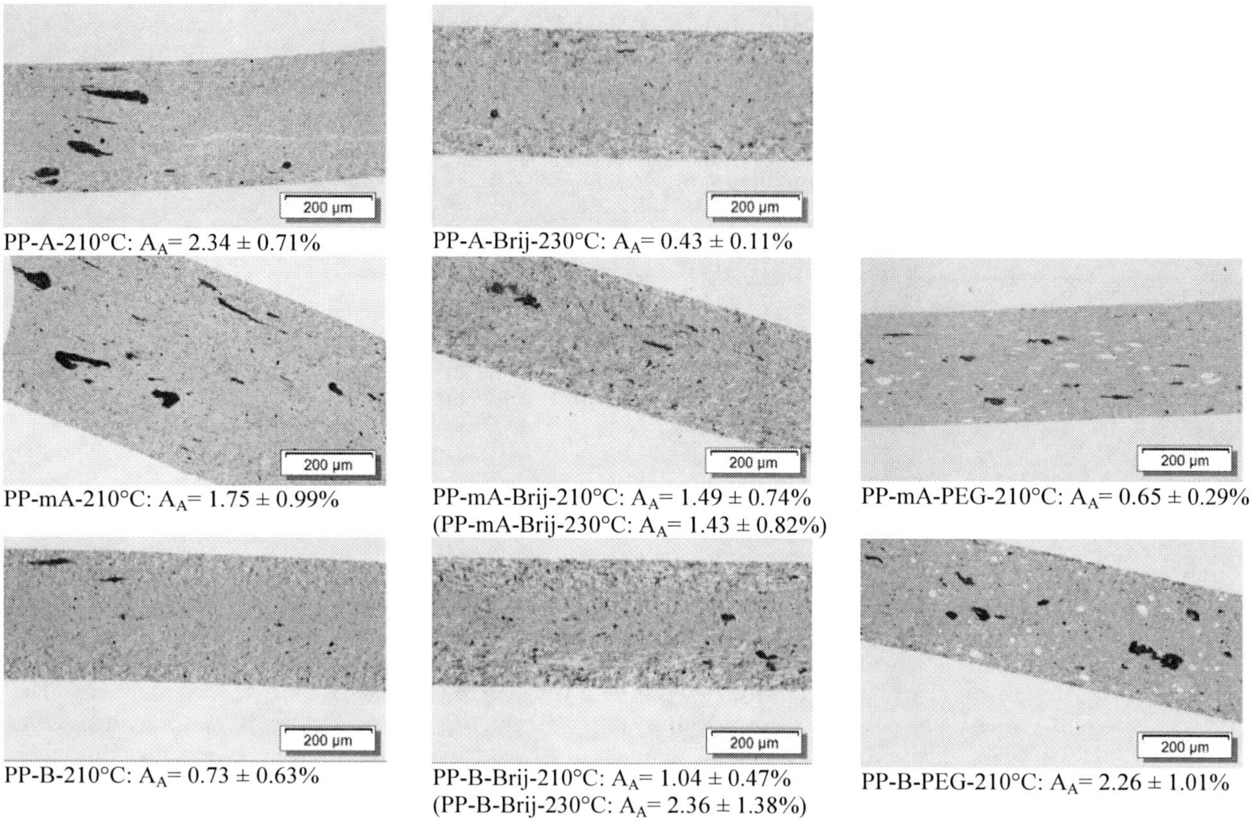

FIGURE 1. TLM images of PP based composites with 2 wt% SWCNT containing 10 wt% PEG or 4 wt% Brij58.

TABLE 2. Seebeck coefficient of solution prepared PEG/SWCNT (mA) composites.

SWCNT content mA [wt%]	Seebeck coefficient [μV/K]
1	-32.5
2	-40.7
4	-38.2
6	-38.9

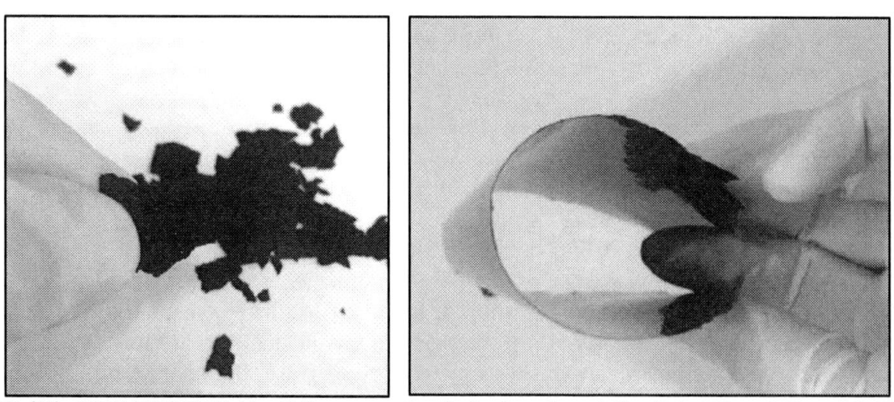

FIGURE 2. Brittle PEG/SWCNT (4wt%) composite (left photo) and flexible PP-mA(2 wt%)-PEG(10 wt%) composite, prepared at 210°C (right photo).

SUMMARY

In this investigation, electrically conductive PP based composites containing SWCNTs were studied concerning their thermoelectric (TE) behavior. The composites were produced by melt mixing, which represents an industrial viable method. In order to generate materials with p- and n-type character from the same polymer base, additives switching the typical p-type behavior into n-type were investigated. Next to PEG, a surfactant containing both a PEG part also a hydrophobic aliphatic hexadecyl segment connected by an ether group (Brij58) was investigated aiming for a higher power factor.

It was shown that also the additive Brij58 switches the material behavior to be n-type, whereby Seebeck coefficients between -35.9 and -48.2 μV/K were achieved when adding 4 wt% to composites with 2 wt% SWCNTs. Addition of Brij58 and PEG in most cases enhances the SWCNT macrodispersion, however, the improvement depends on the SWCNT variation. In composites with SWCNT type B having the best starting macrodispersion, additive addition worsens the macrodispersion. Increasing the mixing temperature from 210°C to 230° under otherwise identical conditions enhances electrical conductivity and leads to slightly higher negative S-values. Best values of Seebeck coefficient was -48.2 μV/K for the composite of PP with 2 wt% SWCNT type A and 4 wt% Brij58 prepared at 230°C. Highest power factor of 0.145 $\mu W/(m \cdot K^2)$ was obtained for PP with 2 wt% SWCNT type B and 10 wt% PEG prepared at 210°C.

Seebeck coefficient measurements on solution prepared composites of PEG with SWCNT (1-6 wt%) confirmed that PEG donates electrons to the SWCNTs and generates n-type materials. Nevertheless, these composites with only PEG and SWCNTs are too brittle to be used in thermoelectric generators, where flexible materials show advantages.

This investigation confirms the feasibility of using melt mixing to fabricate both p-type and n-type materials from the same base materials by the simple use of a switching additive, which donates electrons to the nanotubes and can at the same time enhance their macrodispersion. More investigations are planned to further enhance Seebeck coefficient, electrical conductivity and power factor of such melt mixed composites based on insulating matrices which are still much lower than for traditional TE materials or intrinsically conductive polymers, like PEDOT:PSS and their composites [9,10].

ACKNOWLEDGMENTS

The authors thank OCSiAl Ltd for providing the different kinds of SWCNTs Tuball™ and Dr. Michael Thomas Müller for helpful discussion.

REFERENCES

1. J. Luo, B. Krause, P. Pötschke, *AIMS Mater. Sci.* **3**, 1054-1062 (2016).
2. B. Krause, P. Pötschke, E. Ilin, M. Predtechenskiy, *Polymer* **98**, 45-50 (2016).
3. J. Luo, B. Krause, P. Pötschke, PPS32 Conference, 2016, Lyon, France, Proc. S02-34.
4. J. Luo, G. Cerretti, B. Krause, L. Zhang, T. Otto, W. Jenschke, M. Ullrich, W. Tremel, B. Voit, P. Pötschke, *Polymer* **108**, 513-520 (2017).
5. http://www.ipm.fraunhofer.de/en/services/contract-measurements/thermoelectric-measurements.html
6. M. T. Müller, P. Pötschke, B. Voit, *Polymer* **66,** 210-221 (2015).
7. R. Socher, B. Krause, M. T. Müller, R. Boldt and P. Pötschke, *Polymer* **53** (2), 495-504 (2012).
8. Y. Nonoguchi, K. Ohashi, R. Kanazawa, K. Ashiba, K. Hata, T. Nakagawa, C. Adachi, T. Tanase, T. Kawai, *Sci. Rep.* ***3*,** 3344 (2013).
9. O. Bubnova, Z.U. Khan, A. Malti, S. Braun, M. Fahlman, M. Berggren, X. Crispin, *Nat. Mater.* **10**, 429-433 (2011).
10. C. Cho, B. Stevens, J.H. Hsu, R. Bureau, D.A. Hagen, O. Regev, C. Yu, J.C. Grunlan, *Adv. Mater.* **27**, 2996-3001 (2015).

Melt compounding of Poly(Lactic Acid) with carbon-based fillers: effect on crystallization kinetics

De Santis Felice [(1)]*, D'Urso Luciana [(2)], Acocella Maria Rosaria [(2)], Pantani Roberto [(1)], Guerra Gaetano [(2)]

[(1)] University of Salerno, Department of Industrial Engineering - Fisciano - SA - Italy, [(2)] University of Salerno, Department of Chemistry and Biology - Fisciano - SA - Italy

Biodegradable polymers are raising an increasing interest as a promising alternative to petroleum-based materials since they can be produced from renewable resources at reasonable costs. Among all the possible biodegradable polymers, the most commonly studied is poly(lactic acid) (PLA). However, PLA has some processing limitations such as poor thermal and mechanical resistance and slow crystallization kinetics, if compared to other thermoplastic resins, which prevent its complete access to industrial sectors. Carbon materials have gained a new significant attention as a filler to enhance polymer properties: in this work graphite, carbon black, and an ad-hoc functionalized carbon black application is investigated. These carbon-based materials were melt compounded in a twin screw extruder with different mixing times and characterized by differential scanning calorimetry (DSC), rotational rheometry and GPC tests. Experimental results show that the compounds exhibit a lower crystallization time, if compared to the neat resin, and a good thermo-mechanical stability. In particular, the functionalized carbon black shows the fastest crystallization kinetics, probably due to a specific nucleation effect of the functional tails.

High Performance Electromagnetic Wave Absorber Derived from Multi Layered Assembly of PC/PVDF Blends Containing Different Functional Conducting and Magnetic nanomaterials

***Biswas Sourav* [1]*, *Bose suryasarathi* [2], *Panja Sujit Sankar* [1]**

[1] National Institute of Technology Durgapur. India - West Bengal - India, [2] Indian Institute of Science, Bangalore - Karnataka - India

A novel ultra-thin three-layered assembly of different nano-structured polymer blends is designed here for attenuating electromagnetic (EM) waves by 99.999%. All the key parameters1; intrinsic wave impedance matching, intense absorption of the incoming EM radiation and surface reflection of incident EM waves from the shield material were targeted to design the final assembly using inclusion of chemically modified nano materials in an immiscible bi-component blends of PC (poly carbonate) and PVDF (polyvinylidene fluoride). Acid functionalized MWCNTs contained blend was used as a front layer of the assembly which guided the maximum incident EM wave into the assembly due to the intrinsic impedance matching. In the middle layer α-MnO_2 doped MWCNTs and ferrite-doped cross-linked GO framework contained blend facilitated the enhanced attenuation of incident EM wave by absorption due to the high dielectric and magnetic loss parameters. Surface reflection phenomenon was utilized here at the final layer of the assembly, which directed the penetrated waves to the middle layer of the assembly for secondary interaction, when only MWCNTs were present in the blend. Surely, this unique study offers amendable replacement of the existing solution related to EM absorption.

Electrical conductivity of PVDF composites melt mixed with MWCNTs having different morphologies

Kunz Karina [(1)]*, Krause Beate [(1)], Kretzschmar Bernd [(1)], Pötschke Petra [(1)]

[(1)] Leibniz-Institut für Polymerforschungschung Dresden e.V. - Saxony - Germany

The study investigates the electrical conductivity of melt mixed PVDF/MWCNT composites. The experiments compare sheets produced by extrusion using a flat die with compression molded plates. Such samples differ significantly in nanotube orientation. Whereas compression molded plates are expected to be nonoriented, during the film extrusion strong flow-induced orientation of the MWCNTs occurs, which can cause differences in electrical volume conductivity depending on the measurement direction (in-plane x- and y-direction, through-plane). The purpose of the investigation is to show the relationship between the morphology of different MWCNTs and the electrical properties of plates and films of composites (0.1-7.5 wt. % loading). Among long and short MWCNTs two types namely branched and linear MWCNTs were chosen. First experiments for melt mixed composites prepared using a small scale conical twin- screw compounder (DSM Xplore) showed much lower electrical percolation threshold and much higher plateau conductivity for long branched MWCNTs (CNS flakes from ANS (USA), 0.1 wt.%, 1.9 S/cm @3wt.% respectively) compared to short linear MWCNTs (Nanocyl NC3150 (Belgium), 0.25 wt.%, 0.12 S/cm @3wt.% respectively), which correlates very well with the different aspect ratio of the CNTs. Furthermore, no differences in electrical volume conductivity could be observed between results obtained for compression molded plates of PVDF/CNS composites produced by laboratory twin-screw extruder (Berstorff ZE 25) and microcompounder. For PVDF/2 wt. % CNS composites only low differences in in-plane electrical volume conductivity were measured between extruded films (100 µm thickness: 0.2 and 0.7 S/cm (x and y-direction, respectively); 180 µm thickness: 0.2 and 0.5 S/cm) and pressed plates (240 µm thickness, 0.9 S/cm). It is assumed that the conducting network formed of long branched CNTs is less sensitive to flow-induced orientation during the film extrusion compared to nonbranched tubes.

Synergetic Effect of Carbon Nanotubes and Graphene Nanoplatelets on the Fatigue Crack Growth Rate of Epoxy Composites

Jen Yi-Ming [(1)]*, Zheng Kun-Yang [(1)]

[(1)] National Taiwan Ocean University - Taiwan - Republic of China

The fatigue crack growth rate of carbon nanotube (CNT)/graphene nanoplatelet (GNP)/epoxy composites were experimentally analyzed to study the synergetic effect of the two nano-fillers on the fatigue crack behavior of the studied composites. In the experimental program, The content of two fillers of all the nano-composite specimens was controlled to be 0.4 wt%, and five allocations between the two fillers (CNTs : GNPs =1:9, 3:7, 5:5, 7:3, and 9:1) were considered in the preparation of specimens. The compact-tension (CT) specimens were employed to investigate the fatigue crack growth rates of the composites with various filler allocations. Experimental results show that the specimens with the CNT/GNP allocation of 1:9 have the lowest fatigue crack propagation rate than those with different filler allocations, only one-tenths of that of neat epoxy. Examining the fracture surfaces of the specimens reveals that the composites with the CNT/GNP allocation of 1:9 have the roughest surfaces than those with the other filler allocations. The fillers with three-dimensional structures composed by integrating the two nano-fillers were found to deflect the direction of fatigue crack fronts effectively, reducing the crack propagation rate significantly.

Bipolymer-Carbon Nanotube Composites

Sójka Anna (1), Krause Beate (1), Pötschke Petra (1), Elias Anastasia (1)*

(1) Leibniz Institute of Polymer Research Dresden - Saxony - Germany

Polymers derived from biological sources are attractive materials from an environmental perspective. Like most polymers, their properties may be modified through the inclusion of micro- and nanoscaled fillers. Polyhydroxybutyrate (PHB) is a biopolymer that is both produced and enzymatically-degraded by bacteria. This material has excellent biocompatibility and stability in the body, and conductive PHB-based composites could be used in a variety biomedical applications, including tissue engineering, biosensors, and neural interfaces. In general, conductivity may be imparted to (insulating) polymers through the introduction of conducting fillers, such as carbon nanotubes (CNTs). While a few solution-processed PHB-CNT composites have been described in the literature, conductivity has not been demonstrated. In this contribution, PHB-CNT composites are formed by melt-mixing PHB with multiwalled CNTs. The morphology of the resulting material is explored by scanning electron microscopy (SEM), and well-distributed bundles of CNTs are observed within the polymer matrix. The CNT macrodispersion in the composites was characterized by transmission light microscopy showing some remaining CNT agglomerates. Despite this, a relatively low percolation threshold (~ 0.5 wt%) is achieved,with volume resistivity of 6 Ω·cm at 5 wt% loading. Data describing the thermal and mechanical properties of the composites as a function of CNT loading will also be presented.

Nanofilled epoxy adhesives: optimization of the rheological properties of the nanofilled formulations

Nobile Maria Rossella ***[1]*****, ***Rosolia Salvatore [1]****,* ***Vertuccio Luigi [1]****,* ***Guadagno Liberata [1]***

[1] University of Salerno - Fisciano Salerno - Italy

In this study nano-modified epoxy adhesives using carbon nanofillers have been investigated. The epoxy mixture used in this work is based on the tetraglycidylmethylenedianiline TGMDA and the epoxy monomer 1-4butanedioldiglycidyl ether (BDE), that acts as a reactive diluent. The 4,4-diaminodiphenyl sulfone (DDS) is used as curing agent. These components were supplied by Sigma Aldrich. Moreover, the liquid rubber carboxyl-terminated butadiene acrylonitrile copolymer (CTBN Hycar-Reactive Liquid Polymers, Mn = 3600 containing terminal carboxy groups) is also employed. The epoxy adhesive was prepared by adding 12.5 phr of CTBN and 10 phr of triphenylphosphine to the epoxy TGMDA in the liquid state. The epoxy mixture is then mixed with BDE monomer at a concentration TGMDA/BDE of 80%:20% wt/wt. This particular epoxy formulation has proven to be very effective for improving nanofiller dispersion due to a strong decrease in the viscosity. The MWCNTs (3100 Grade purchased from Nanocyl S.A.) were incorporated into the TGMDA-CTBN-BDE-DDS blend by using an ultrasonication Hielscher model UP200S-24 kHz. The rheological measurements in the liquid state, before curing, were carried out on a Physica-MCR301 (Anton-Paar) rotational rheometer equipped with a parallel plate geometry. Small amplitude oscillatory shear measurements, within the linear viscoelasticity regime, on the TGMDA-CTBN-BDE-DDS blend and on the mixtures with 0.1, 0.5 and 1 wt% of MWCNT content were performed at different temperatures. The inclusion of 0.5 wt% MWCNT increases the complex viscosity of the uncured dispersions of about 2 orders of magnitude at low frequency while the storage modulus G' at low frequency becomes nearly independent of frequency indicating the formation of a percolation network in the nanocomposite.

FUNCTIONALIZATION OF FLAX FIBRES BY RADIATION-GRAFTING

Sonnier Rodolphe [(1)], Otazaghine Belkacem [(1)], Rouif Sophie [(2)*]

[(1)] C2MA, Ecole des Mines d'Alès - -- - France, [(2)] IONISOS - Auvergne-Rhône-Alpes - France

The use of natural reinforcements in composite materials has considerably increased due to numerous advantages. However, some drawbacks limit their industrial applications such as their hydrophilic character, their low thermal stability and also their flammability. Among different strategies to functionalize natural fibers, the radiation grafting appears promising. In this work, it was used to improve the flame retardancy and the hydrophobicity of flax fabrics. As an example, the procedure to radiation-graft phosphorus compounds on flax fabrics is presented. The influence of various experimental conditions on the grafting efficiency was assessed.It is possible to control precisely the grafting efficiency and the location of phosphorus in order to achieve the desired level of flame retardancy. Self-extinguishing behavior of fabrics can be achieved [1]. Hydrophobic fibers were also prepared according to similar procedure. This work is partly included in the French project called BIONICOMP, headed by IONISOS, with a consortium gathering of academic and industrial partners, in order to improve the performances of composites reinforced by natural fibers fabrics.

Controlling of Radiation-induced Grafting at Nanoscale for the Preparation of Advanced Functional Membranes

Güven Olgun (1)*

(1) Hacettepe University - - - Turkey

Radiation-induced grafting has been developing in three main directions: i) polymeric membranes, ii) polymeric sorbents, and iii) graft polymers for medical applications. In conjunction with these directions membranes play key roles in environmental, industrial and health-care applications. Various types of membranes have been developed for use in reverse osmosis, nanofiltration, electrodialysis, solid polymer electrolytes, Li-ion battery and fuel cell applications, sensors and tissue engineering[1,2]. The base materials used in these applications are mostly fluorinated polymers to meet very stringent chemical, mechanical, thermal properties required for demanding applications. Grafting of monomers on fluorinated polymers which are known with their chemical inertness is possible by using ionizing radiation as the tool for generating radicals. Until quite recently however, radiation-induced grafting has been accomplished by conventional free radical polymerization method which lacks control over the molecular weight and dispersity of graft chains. Close control of graft chain lengths with very narrow molecular weight distributions opened up new opportunities for environmental, industrial and health-care applications[3]. In this context our recent research interest has been focused on controlling the molecular weight and distribution of radiation-induced grafted chains to obtain nanosize polymeric brushes via 'Reversible Addition-Fragmentation Chain Transfer' RAFT process. The application of this process for the preparation of polymer thin film electrodes to detect lead ions in aqueous systems at ppb levels[4], radiation-induced deposition of copper nanoparticles in the nanochannels of PET track-etched membranes[5], nanoscale grafting of poly(N-isopropyl acrylamide) on cell culture dishes for tissue engineering and preparation of highly conductive ETFE membranes for fuel cells[6] will be highlighted in this presentation.

The Influence of Gamma Irradiation on the Chemical Structure and Crystallinity of PHBV/PLA/Cloisite 30B Nanocomposites

Idris Zembouai[1,2)], Mustapha Kaci[1,a)], Stéphane Bruzaud[3]

[1] *Laboratoire des Matériaux Polymères Avancés (LMPA), Faculté de Technologie, Université de Bejaia 06000, Algeria*

[2] *Département de Génie des Procédés, Faculté de Technologie, Université Saad Dahlab- Blida 1, Blida 09000, Algeria*

[3] *Institut de Recherche Dupuy de Lôme (IRDL), FRE CNRS 3744, Université de Bretagne Sud, Rue de Saint Maudé, 56321 Lorient Cedex, France*

[a)] *Corresponding author: kacimu@yahoo.fr*

Abstract. The objective of the work was to investigate the effects of gamma irradiation at absorbed dose of 75 kGy on the chemical structure and crystallinity of PHBV/PLA blend (50/50 w/w) without and with Cloisite 30B (C30B) incorporated at 3 wt %. IRTF spectra showed that gamma irradiation induced changes in the chemical structure of PHBV/PLA blends due to oxidation reactions involving the transformation of ester groups to hydroperoxides. However, the addition of C30B in the polymer blend had no influence on the nature of the degradation process. GPC data showed that the value of the average number of chain scission (n_t) of irradiated PHBV/PLA/C30B sample is lower than that of the virgin blend highlighting the role of C30B as a retardant of the chain scission mechanism. WAXS analysis indicated an increase in the crystallinity of all irradiated samples at 75 kGy.

Keywords: PHBV, PLA, Cloisite 30B, Polymer blend, Nanocomposites, Gamma irradiation,

INTRODUCTION

The interest in polymer blends is constantly increasing because this is a good approach to adjust the functional properties of materials [1-4] to make them suitable for a variety of applications [5]. Indeed, blends of poly(3-hydroxybutyrate-*co*-3-hydroxyvalerate) (PHBV) and polylactide (PLA) were explored as an alternative way of achieving novel materials with desired properties [6]. Such blend materials are biodegradable and biocompatible and provide economical benefits with combined properties [7]. Further, the incorporation of organomodified montmorillonite in polymer materials often leads to improved properties. On the other hand, medical and pharmaceutical devices as well as food packaging require sterilization by ionizing radiation to guarantee microbiological safety [7,8]. Therefore, the paper aims to investigate the changes under gamma irradiation in the chemical structure and crystallinity of PHBV/PLA blend: 50/50 (w/w) with and without C30B (3 wt %) at absorbed dose of 75 kGy.

EXPERIMENTAL SECTION

Materials used

PLA was supplied in pellets form by Nature-Works under the trade name 7001D. The polymer is a semi crystalline one having the following main properties: density = 1.25 g/cm^3, MFI = 6 g/10 min (210°C, 2.16 kg), T_g = 60°C and T_m = 160°C. PHBV was manufactured by Tianan Biological Materials Co. Ltd. (China) and commercialized in pellets form under the trade name ENMAT Y1000P. According to the manufacturer, PHBV has the following properties: density = 1.25 g/cm^3, T_g = 8°C and T_m = 165°C. Cloisite 30B (C30B) is an organically modified montmorillonite, which is commercially available and supplied by Southern Clay Products (Texas). C30B is a montmorillonite modified with bis-(2-hydroxyethyl)methyl tallow alkyl ammonium cations.

Sample preparation

Prior to use, all materials were dried under vacuum at 60°C for 24 h. The mixing temperature, screw speed and residence time were set at 180°C, 50 rpm and 8 min, respectively. The extruded materials were ground into pellets. Prior to compression molding, the pellets were dried over 12h under vacuum at 60°C and compression molded in a Carver® hydraulic press at 180°C under a pressure of 40 bar for 3 min to obtain thin films of about 150μm of thickness.

Gamma irradiation Conditions

Gamma irradiation was carried out on films in the form of circular bands of almost 20 cm diameter, in ^{60}Co industrial equipment at the Nuclear Research Center of Algiers (Algeria). The samples were exposed to 75 kGy at a dose rate of 1.92 $Gy.h^{-1}$ in the presence of air at room temperature.

TECHNICAL CHARACTERIZATION

Fourier Transform Infrared Spectroscopy (FT-IR)

FT-IR spectra of various +ilm samples were recorded by an FT-IR spectrometer (Shimadzu 8400 M) using 4 cm^{-1} resolution and 40 scans. All spectra were recorded in absorbance mode in the 4000-400 cm^{-1} region.

Wide angle X-ray scattering (WAXS)

The WAXS measurements were performed at room temperature using advance diffractometer equipment (BRUKER AXS D8), wavelength $\lambda = 0.154$ nm for 40 kV and 40 mA. The 2θ scan range was used from 2 to 40° at a scan speed of 40 $steps^{-1}$.

Average number of chain scission (n_t) determined by SEC

SEC was used to determine the average number of chain scission (n_t) of PHBV/PLA and PHBV/PLA/C30B samples at 50 and 75 kGy through the measurements of the weight and number-averages molecular weight. The apparatus was equipped with a set of three columns: two ResiPore and one PL gelMixed C (Polymer Labs.). The detection system was composed by a refractometer and a UV detector. Chloroform was used as an eluant with a flow rate of 0.8 mL/min. The elution profiles were analyzed using the Empower SEC module software (Waters). Calculations were based on calibration curves obtained from polystyrene standards ranging from 200 up to 6×10^6g/mol.
The average number of random chain scissions per unit mass (n_t) was calculated according to the following Eq. **[9]**:

$$n_t = \frac{1}{Mn_t} - \frac{1}{Mn_0}$$

Where, Mn_t and Mn_0 are the number-average molecular weight at a given dose and non-irradiated samples, respectively.

RESULTS AND DISCUSSION

FT-IR analysis

FTIR analysis was used to show the changes in the chemical structure of the irradiated materials. In this respect, Fig.1(a) and (b) display the FT-IR spectra of PHBV/PLA/C30B nanocomposite before and after 75 kGy exposure recorded in the regions 3600-3200 cm^{-1} and 1900-1600 cm^{-1}, respectively. In Fig.1(a), the FTIR spectra of PHBV/PLA/C30B sample exhibits before gamma irradiation, three broad absorption bands at 3568, 3505 and 3438 cm^{-1}, which are attributed to OH groups present in alcohols, hydroperoxides and carboxylic acids **[6]**. At 75 kGy, all the absorption bands increase in intensity. In Fig.1(b), the FTIR spectra of PHBV/PLA/C30B reveal the presence of a large absorption band in the carbonyl region assigned to ester groups before exposure **[1]**. After gamma irradiation, this band becomes wider, while its intensity is lower compared to the control sample. This could be explained as a result of oxidation reactions involving the decomposition of the ester groups of PHBV/PLA blend leading to the formation of various oxidative species and hydroxyl groups contained in hydroperoxides. Similar results are observed in the case of unfilled PHBV/PLA blend (FTIR spectra not shown here) suggesting that the presence of C30B in the blend matrix has no effect on the nature of degradation process.

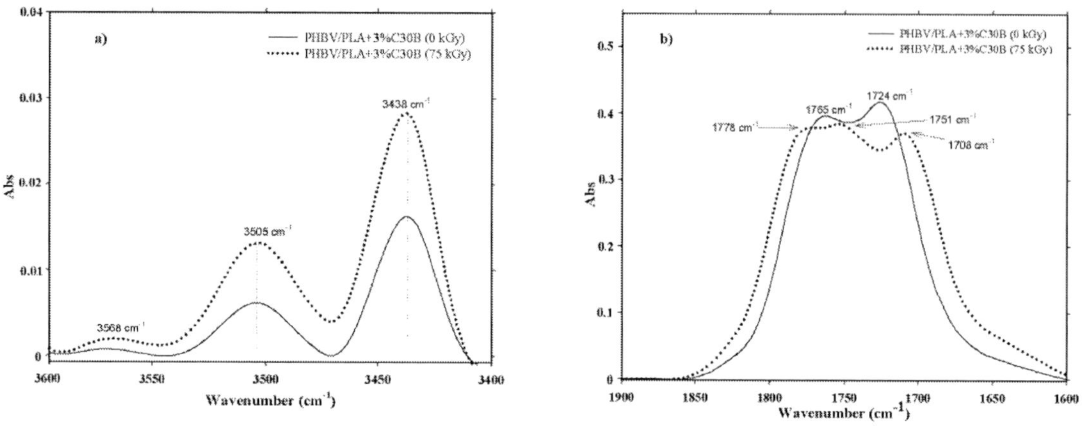

FIGURE 1. FT-IR spectra of PHBV/PLA/C30B nanocomposite before exposure and after 75 kGy. **(a)**: 3600-3400 cm^{-1} and **(b)**: 1900-1500 cm^{-1}.

WAXS Analysis

Fig.2(a) and (b) show WAXS patterns of PHB/PLA blend and PHBV/PLA+3%C30B nanocomposite, respectively before and after absorbed dose of 75 kGy. As presented in Fig.2(a), the WAXS spectrum of PHBV/PLA blend reveals before exposure, the presence of two strong diffraction peaks assigned to (020) and (110) planes of the orthorhombic unit cell, and small peaks assigned to the (021), (101), (111) and (121) planes **[1]**. At 75 kGy, no new diffraction peak is observed in the WAXS pattern, while the angular position of the peaks remains unchanged. However, Fig.2(a) shows clearly an increase in intensity for the main peaks. This results from an increase in the crystallinity of the irradiated material. Fig.2(b) shows the WAXS patterns of PHBV/PLA/C30B before and after 75 kGy. Similar trend is also observed characterized by an increase in the crystallinity of the nanocomposite sample after gamma irradiation. According to the literature **[7,8]**, the increase in crystallinity of the irradiated samples could be attributed to nucleation phenomenon, i.e., the great crystallites are randomly destroyed by gamma irradiation, followed by the appearance of small and innumerable crystallites **[10]**.

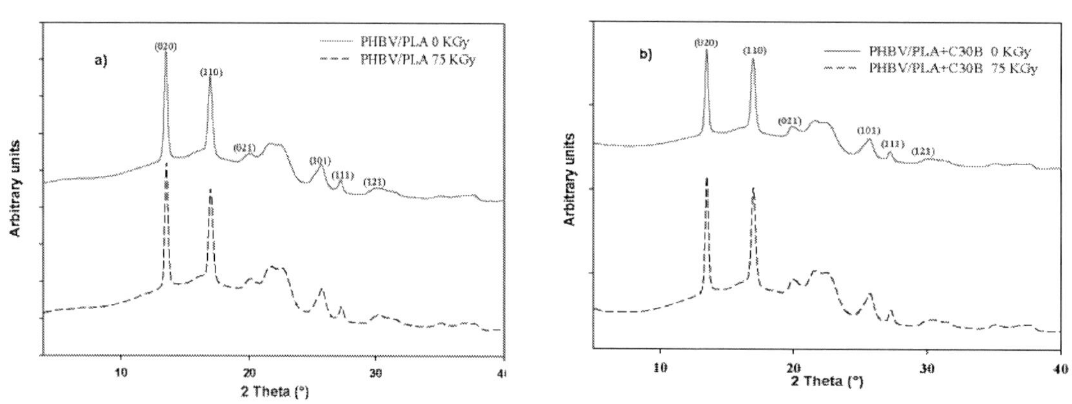

FIGURE2. WAXS patterns of **(a)**: PHBV/PLA blend and **(b)**: PHBV/PLA+3%C30B before and after 75 kGy.

Evolution of the average number of chain scission (n_t)

Fig.3 shows the changes in the average number of chain scission (n_t) per unit mass at 50 and 75 kGy for PHBV/PLA blend and PHBV/PLA/C30B nanocomposite samples. From Fig.3, it is observed the linearity of the plots during gamma irradiation for all the samples with an increase in the value of n_t at 50 kGy, however, more pronounced at 75 kGy. The linearity of the plots implies that random chain scission is occurring in the irradiated samples. Further, the solubility test indicates that the irradiated samples have no gel fraction at 75 kGy and they are totally soluble in chloroform. Another important feature is the evolution of n_t value which is lower for the nanocomposite sample at 75 kGy compared with that of the virgin blend indicating the role of C30B as a retardant of the chain scission mechanism of PHBV/PLA blend.

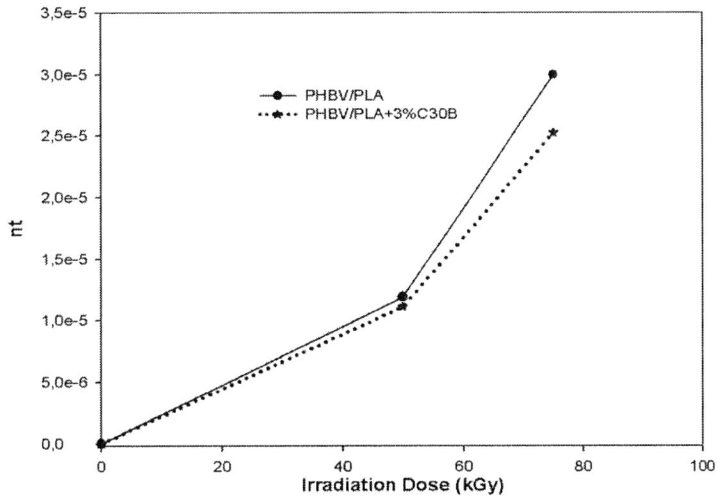

FIGURE3. Evolution of the average number of random chain scission (n_t) per unit mass for PHBV/PLA blend and PHBV/PLA+3%C30B nanocomposite at 50 and 75 kGy.

CONCLUSION

In this paper, the effects of gamma irradiation on the chemical structure, the average number of chain scission as well as crystallinity of both PHBV/PLA blend and PHBV/PLA/C30B nanocomposite was investigated at 75 kGy. It can be concluded that although the incorporation of C30B into PHBV/PLA blend has no influence on the nature of the degradation process, which is dominantly the chain scission mechanism. However, the presence of C30B in PHBV/PLA blend results in the reduction in the value of the average number of chain scission conferring better resistance for the nanocomposite material against gamma irradiation. This result is important regarding the potential applications of such materials in food packaging and biomedical. Moreover, the irradiated materials exhibited higher crystallinity at 75 kGy.

ACKNOWLEDGMENTS

The authors would like to thank the International Atomic Energy Agency through the Research Contract n°17692/R1 and they are also grateful to EGIDE through the TASSILI program 12 MDU 871 for its financial support in this collaborative project.The authors are pleased to express their grateful acknowledgements to the Nuclear Research Center of Algiers (CRNA).

REFERENCES

1. I. Zembouai, M. Kaci, S. Bruzaud, A. Benhamida, Y.M. Corre, Y. Grohens, Polym.Test**32**,842-851(2013).
2. I. Zembouai, S. Bruzaud, M. Kaci, A. Benhamida, Y.M. Corre, Y. Grohens, A. Taguet, J.-M. Lopez-Cuesta, J. Polym. Env**22**,449-459(2014).
3. I. Zembouai, S. Bruzaud, M. Kaci, A. Benhamida, Y.M. Corre, Y. Grohens, J.-M.Lopez-Cuesta, Polym. Eng. Sci**54**,2239-2251(2014).
4. I. Zembouai, S. Bruzaud, M. Kaci, A. Benhamida, Y.M. Corre, Y. Grohens, J. Polym.Env**22**,131-139(2014).
5. T. Gerard, T. Budtova, Eur. Polym. J**48**, 1110-1117 (2012).
6. I. Zembouai, M. Kaci, S. Bruzaud, I. Pillin , J-L. Audic , S. Shayanfar , S-D. Pillai Polym.Degrad. Stab **132**, 117-126(2016).
7. I. Zembouai, M. Kaci, S. Bruzaud, L. Dumazert, A. Bourmaud, M. Mahlous, J.-M. Lopez-Cuesta, Y. Grohens, Polym.Test**49**,29-37(2016).
8. L. Zaidi, S. Bruzaud, M. Kaci, A. Bourmaud, N. Gautier, Y. Grohens. Polym.Degrad. Stab**98** ,348-355(2013).

9. M. Deroiné, A. Le Duigou, Y-M. Corre, P-Y. Le Gac, P. Davies, G. César, S. Bruzaud. Polym. Degrad. Stab **108**, 319-329(2014).
10. L-M. Oliveira, E.S. Araujo, S-L. Guedes, Polym. Degrad. Stab **91**, 2157-2162(2006).

Influence of Gas Atmosphere on Electron-Induced Cyclization of Polyacrylnitrile Powder at Elevated Temperature

Müller Michael Thomas [1]*, Brünig Harald [1], Fischer Dieter [1], Gohs Uwe [1], Häussler Liane [1], Kirsten Martin [1], Malanin Mikhail [1]

[1] Leibniz-Institut für Polymerforschung Dresden e.V. - Saxony - Germany

The oxidation/thermal stabilization of polyacrylnitrile (PAN) precursor fibers is a highly exothermic and time-consuming process. In several reports, the reduction of thermal stabilization time was shown for an irradiation of PAN at room temperature in different atmospheres. Since electron induced chemical reactions strongly depends on the polymer chain segment mobility, the present investigation reports the electron treatment of PAN homopolymer and copolymer powder in air and nitrogen atmosphere at an average temperature of 130 °C for the first time. This irradiation temperature was precisely selected in order to study the electron induced changes at temperatures above the glass transition temperature, but below the initiation temperature of thermal induced cyclization. The average dose rate of all electron beam (EB) treatments amounted to 0.77 kGy/s in order to keep the temperature fluctuation within the range of ± 5 K. In addition, several analyses were done in order to characterize the type of radicals, the crosslinking behavior, the functional groups, the exothermic behavior as well as the thermal stability of virgin and EB modified powder samples. Based on the calculated values of extent and degree of cyclization some correlations between functionalization and crosslinking were drawn with respect to cyclization and maximum residual mass at 800 °C.

Melt spinning and characterization of matrix blend based on polypropylene and ethylene-octene copolymers modified by high energy electrons

Al Rahhal Samer [(1)]*, ***Bruenig Harald*** [(1)], ***Gohs Uwe*** [(1)], ***Heinrich Gert*** [(1)]

[(1)] Leibniz-Institut für Polymerforschung Dresden e.V. - Saxony - Germany

Polypropylen (PP) is known as a thermoplastic material with lot of advantages as low cost, low density, chemical resistance, well processability, etc. The simultaneously melt spinning of PP and glass filaments (Online Hybrid Fiber Spinning) opens the possibility to produce hybrid yarns for glass fiber reinforced composites using textile processes. However, the applications of PP in electrical and automotive industries as matrix material for endless fiber reinforced composites are limited because of its low stiffness, low tensile strength, and insufficient impact strength. Improving the poor mechanical properties of PP to get at least a low cost high performance material that benefits the requirements of modern reinforced light weight constructions seems to be a challenge of our time. In the presented work, PP and blends of PP and ethylene-octene copolymer (EOC) were prepared by electron induced reactive processing (EIReP) using various blend ratios at low electron doses (up to 18 kGy) and fixed electron energy of 1.5 MeV. Melt spinning trials of the EIReP-modified PP and PP/EOC blend materials were carried out by means of an industrial near extruder spinning equipment to investigate the spinnability and to get optimum spinning parameters as well as the effects of blend components and electron beam irradiation to the mechanical properties of as-spun fibers. It is well known, that blending PP with 2.5mass-% EOC increase the Charpy impact strength but has no or lowering effect to E-modulus. The treatment with electron beam allows further improvements. The presentation shows, that under well balanced conditions, significant improvement of adhesion and impact strength as well as modulus for the modified PP-matrix material can be reached.

INLINE EDGE LAYER MODIFICATION OF THREE DIMENSIONAL SHEET MOLDING COMPOUNDS

Zschech Carsten [(1)]*, Pech Mathias [(1)], Gedan-Smolka Michaela [(1)], Gohs Uwe [(1)], Heinrich Gert [(1)]

[(1)] Leibniz-Institut für Polymerforschung Dresden e.V. - Saxony - Germany

Incomplete curing of sheet molding compounds (SMC) as well as the post-curing and out gassing during the industrial coating process are the two main reasons for the coating problems of SMC Class-A automotive parts. Furthermore, their surface properties as contact angles and surface roughness strongly depend on processing parameters of thermal curing and coating as well as used cleaning and pickling methods. Consequently, cost and energy intensive additional process steps are used in the automotive industry in order to reduce or avoid above mentioned problems during the industrial coating process of SMC-Class-A automotive parts. A cost efficient robust procedure is required in order to overcome these disadvantages as well as to fulfil the requirements of industrial coating processes. Based on electron beam curing and surface functionalization by low energy electrons a novel pre-treatment procedure was developed and tested. For this purpose, two different compact electron emitters have been coupled to a robot in order to validate this novel inline pre-treatment procedure. In one additional step, post-curing and surface functionalization of the three dimensional SMC-parts is achieved. Presently, this novel pre-coating technology is tested for different SMC-parts.

Development and Investigation of High Performance Fire Retardant Polypropylene Composites

***Xiao Dan* (1)*, *Gohs Uwe* (1), *Wagenknecht Udo* (1), *Voit Brigitte* (1), *Wang De-Yi* (2)**

(1) Leibniz-Institut für Polymerforschung Dresden e.V. - Saxony - Germany, (2) IMDEA Materials Institute - Saxony - Spain

Polypropylene (PP) has excellent mechanical and chemical properties. Thus, it is used in a wide range of applications. However, high flammability of PP limits its application in remarkable flame-retardant standards fields [1]. Some of halogenated flame retardants (FRs) are restricted by the European Community directives ROHs, WEEE and REACH. Now metallic hydroxides are widely used in industry, but the high loading (~ 60 wt.%) seriously destroys the mechanical properties of flame retardant PP. Intumescent flame retardant (IFR) is established in polymeric materials since it is regarded as a safe and environmentally friendly flame retardant [2-3]. However, some components of traditional IFR are made from low molecular mass components which might undergo migration and thermal degradation during processing. In this work, a multifunctional flame retardant (MFR) was prepared based on ammonium polyphosphate (APP). In addition, we used electron beam (EB) technology for inducing of chemical couplings between MFR and PP. The results showed that MFR was excellent flame retardant crosslinker for PP. Further, MFR exhibited outstanding flame retardancy. In comparison to the traditional APP, the LOI value of PP/MFR composites (with 20~35 wt.% MFR) were more than 30 and effective melt drop resistance in UL-94 test; in cone calorimeter test, reduction of the peak heat release rate (~ 41 %), total heat release (~ 53 %) and smoke production (85 %). Finally, the thermal stability and tensile properties of these designed flame retarded PP composites were improved by EB treatment. Based on these results high performance fire retardant polymer nanocomposites might be developed for industrial applications such as insulated material of wire, cable, etc.

Polymer electrolyte membranes from pre-irradiation induced graft copolymerization on ETFE

Ke Xi (1), Li Xin (1), Drache Marco (1), Gohs Uwe (2), Beuermann Sabine (1)*

(1) TU Clausthal - - - Germany, (2) Leibniz Institut für Polymerforschung Dresden e.V. - - - Germany

Polymer electrolyte membranes (PEM) are a key component in many fuel cells and in redox flow batteries (RFB). So far in most applications Nafion membranes are utilized. In search of alternate membrane materials various approaches are pursued. One of them is the radiation-induced activation of polymer films and subsequent graft copolymerization [1]. In this contribution the preparation of PEMs in a three step process is reported: Firstly, commercially available fluoropolymer films, e.g., poly(ethylene-co-tetrafluoroethylene) (ETFE), are activated via electron beam treatment. Secondly, graft copolymerizations with functional methacrylates, e.g., such as hydroxyethyl methacrylate and glycidyl methacrylate, are performed. Then, the functional groups are sulfonated yielding the final PEM. Alternately, copolymerizations may be carried out with acrylic acid and hydroxyethyl acrylate, and subsequently the material is phosphonated to yield proton conductivity. The impact of the ETFE activation process, in particular the adsorbed dose, on the grafting process, and consequently the performance of the PEM is addressed. In order to reduce cross-over in vanadium redox flow batteries the polymerizations may be carried out in the presence of crosslinkers. The electrical performance of the resulting membranes is evaluated by electrochemical impedance spectroscopy as well as in fuel cell and in VRFB tests. In the latter up to 300 charge and discharge cycles were carried out, lasting for nearly a month. Stable electrochemical performance directly indicates membrane stability under typical VRFB conditions [2]. [1] L. Gubler, Adv. Energy Mater. 2014, 4, 1300827 [2] X. Li, A. R. dos Santos, M. Drache, X. Ke, U. Gohs, T. Turek, M. Becker, U. Kunz, S. Beuermann, J. Membr. Sci. 2017, 524, 419

DC High Power Electron ELV Accelerators for Polymer Treatment

Kuksanov Nikolay (1)*, Salimov Rustam (1), Fadeev Sergey (1), Nemytov Petr (1), Golubenko Yuri (1), Korchagin Alexey (1), Semenov Alexey (1), Lavruchin Alexander (1), Kogut Dmitryi (1), Domarov Evgeny (1), Vorobiev Denis (1), Cherepkov Victor (1)

(1) Budker Institute of Nuclear Physics - Novosibirsk - Russia

Budker Institute of Nuclear Physics simultaneously with investigation in high energy Physics develops and manufactures high-tech equipment for industrial and research applications. One kind of such equipment is electron accelerators. The main application of these machines is treatment of the polymers by E-beam. BINP is research institute and has limited power for manufacturing the industrial accelerators. Nevertheless beginning from 1973 we had manufactured over 150 accelerators. Additionally we have partners in South Korea and China. BINP also has collaboration with different research institutes for polymer like IPF. BINP has a special installation for the beam experiments for all potential users of accelerators in BINP. Different models of ELV accelerators are covering now the energy range from 0.2 to 2.5 MeV with maximum beam current 100 mA or beam power 100 kW. For environmental application there was developed and manufactured accelerator with maximum power 400 kW and beam power 0.5A. ELV accelerator can be equipped with a set of underbeam equipment that improves the quality of EB treatment and increases the application of EB. There are: - double window extraction device that allow to increase a linear beam current density; - 4-side irradiation system, which decreases the nonhomogeneity of azimuth dose distribution during cable irradiation; - ring irradiation system that allow irradiate during 1 pass the cables with external diameter up to 60 mm: - the system for transportation of wires, cables and pipes through irradiation area; - the system for grain irradiation; - increasing of homogeneity of absorbed dose on the width of irradiated films, band and tapes. Up to now there did not find application in industry the device with extraction into atmosphere the focused beam. 100 kW electron beam is extracted through the 1 mm holes in diaphragm. This device allows to operate with extra high beam power density.

INFLUENCE OF REACTIVE PROCESSING ON NANOMECHANICAL PROPERTIES OF POLYAMIDE 6 AND ITS BLENDS

Banerjee Shib Shankar [(1)]*, Janke Andreas [(1)], Gohs Uwe [(1)], Heinrich Gert [(1)]

[(1)] Leibniz-Institut für Polymerforschung Dresden e. V - saxony - Germany

The presentation reports about the influence of reactive processing on some nanomechanical properties of polyamide 6 (PA6) and its blends with fluoroelastomer (FKM) using PeakForce Quantitative Nanomechanical Mapping (PF-QNM) atomic force microscopy (AFM) technique. PA6 was modified by electron beam (EB) technology within a selected dose range (25-200 kGy) where branching was generated instead of crosslinking. DMT modulus (based on the Derjaguin-Muller-Toporov model) of PA6 was found to increase (~ 131 %), whereas, adhesion force (between tip and sample) and dissipation energy (energy loss) decreased (~ 40 % and ~ 66 %, respectively) with increasing dose (25 and 200 kGy). The changes of above properties with dose were correlated to electron-induced branching, restricted segmental mobility and significant reduction of water absorption. Finally, dynamic vulcanization was used to prepare PA6/FKM Thermoplastic Vulcanizates. Using PF-QNM AFM, DMT modulus and adhesion force were analyzed from the continuous matrix phase to the dispersed domain through the nanometric interface (4 - 5 nm). The effect of dynamic vulcanization on the nanomechanical properties of the interface and the dispersed rubber phase of this blend was also discussed. Systematic investigations are in progress to predict the effect of electron-induced reactive processing on nanomechanical properties of PA6/FKM blends.

Process-Induced Morphology and Mechanical Properties of HD-PE

Boldt Regine [(1)]*, Gohs Uwe [(1)], Wagenknecht Udo [(1)], Stamm Manfred [(1)]

[(1)] Leibniz-Institut für Polymerforschung Dresden e.V. - Saxony - Germany

Based on results of previous work high-density polyethylene has been modified with high energy electrons either before or after injection moulding in order to compare the influence of sequences of processing (irradiation – forming). Electron treatment of polyethylene was performed under inert atmosphere at different states of aggregation. The polymer was irradiated in the solid state as well as in the molten state under stationary and dynamic conditions. The process induced morphological and mechanical properties have been investigated using light microscopy, transmission electron microscopy as well as thermal analysis and tensile tests. Changes of molecular mass were measured with gel permeation chromatography. The results have shown that it has to distinguish between process-related properties affected by different thermal preconditions like cooling rate or influenced by electron treatment for instant absorbed dose from properties dependent on the process order.

RADIATION CROSS-LINKING TECHNOLOGY FOR EXTRUDED PRODUCTS

***Rouif Sophie* [(1)*], *Mitschak Henry* [(2)]**

[(1)] IONISOS - Auvergne-Rhône-Alpes - France, [(2)] IONISOS GmbH - Saxony - Germany

The IONISOS Group acquired a facility for electron beam cross-linking in Germany in November 2016. This location in Salzenforst near Bautzen and close to Dresden was built in 2009 by LEONI AG. Presently, two accelerator lines (Energy 1.5 – 3.0 MeV, output 220 kW) are used to treat reeled products such as cables, pipes and heat shrink tubing. In the process, the long polymer chains of the Polyolefins (mainly polyethylene PE) are cross-linked by introducing new carbon-carbon bonds into the molecular structure. Cross-linking leads to considerable improvement in the heat performance, mechanical and chemical properties of the materials, so melting, for example, is not possible anymore. Particularly, in case of polyethylene, radiation cross-linking (PEX C material) has more advantages and less constraints compared to the chemical cross-linking with peroxydes (PEX A) or vinylsilanes (PEX C). The actual processing of materials takes place in a shielded concrete vault with a speed of up to several 100 meters per minute. The goods are unreeled and exposed to the accelerated electrons according to clearly defined process parameters. The electrons penetrate the materials homogenously and lead to the desired crosslinking effect. The electron beam equipment includes hi-tech in-line monitoring of the cross-linking process thus ensuring the high quality requirements of our customers can be reproducibly achieved. The facility currently processed approximately 100 million linear meters of customer product each year and has capacity available to cater for market growth over the foreseeable years.

Superhydrophobic Polymer Coatings: Effect of Composition and Thermodynamics on the Surface Morphology

Mead Joey (1)*, Nahum Tehila (1), Dodiuk Hanna (2), Kenig Samuel (2), Panwar Artee (1), Barry Carol (1)

(1) University of Massachusetts Lowell - Massachusetts - USA, (2) Shenkar College of Engineering Design and Art - Ramat Gan - Israel

Durable superhydrophobic coatings have been prepared from a system of silica nanoparticles (NPs) to provide nanoscale roughness, fluorosilane to give hydrophobic chemistry, and three different polymer binders: urethane acrylate, ethyl 2-cyanoacrylate, and epoxy. The coatings using NPs in various concentrations with the three binders showed different superhydrophobic behavior when applied to both glass and polycarbonate (PC) substrates. Coating composition and substrate surface characteristics and wettability affected the superhydrophobic characteristics of the coatings. Interfacial tension and spreading coefficient parameters (thermodynamics) of the coating components were used to predict the localization of the NPs for different binders and concentrations. The thermodynamic analysis of the NPs localization was in good agreement with the experimental observations. Using thermodynamic analysis and the experimental data (scanning electron microscopy (SEM), X-ray photoelectron spectroscopy (XPS), profilometry, and atomic force microscopy (AFM)), it was determined that localization of the nanoparticles on the surface was critical to yield the needed roughness for superhydrophobicity. The durability was evaluated by tape testing for both glass and polycarbonate substrates. Several coating compositions retained their superhydrophobicity after the tape test. Of the three binders the epoxy was found to be the most durable. In summary, it was concluded that thermodynamic analysis is a powerful tool for predicting the roughness of the coating as a result of location of NPs on the surface and can be used in the design of superhydrophobic coatings.

Bio-inspired Textured Composite Surfaces with Abrasion Resistance and High Friction Properties

Anwer Ali O. (1), Bagheri Zahra S. (2), Ferni Geoff (2), Dutta Tilak (2), Naguib Hani (1)*

(1) University of Toronto - Ontario - Canada, (2) Toronto Rehabilitation Institute - Ontario - Canada

This study details the development of textured composite surfaces with evolution of the coefficient of friction on ice with varying fiber types and matrices. Different types of fibers, such as metallic and carbon based, were melt blended with Thermoplastic polyurethane (TPU) and Styrene-Butadiene Styrene (SBS) matrices in equal volume fractions. The texturing process involved a mechanical ablation technique to expose the fibrous phase from the surface of the composite using fiber de-bonding and pull-out. The surface properties of these textured composites were compared to their pure counterparts, and the effect of the coefficient of friction of these surfaces on ice was determined before and after simulated wear. Surface characterization techniques, including roughness testing and Scanning electron microscopy images, were conducted to determine the effect of abrasion on the frictional properties of the composite surfaces.

Physical coloring of multilayered elastomeric structures

Dagan Maayan [(1)], Naveh Naum [(1)]*

[(1)] Shenkar College of Engineering and Design - Ramat Gan - Israel

When two transparent polymers with different refractive indices are stacked alternatively in a multilayered structure, a preferred wavelength range is reflected, depending on the refractive index difference and layer thickness. Colors generated in such a way are referred to as "physical colors". Visualization of a color in the visible range requires a refractive index difference of about 0.1 and layer thicknesses in the order of 100 nm. Increasing the number of layers contributes to the level of saturation (intensity) of the color. Transparent multilayered films and sheets were studied. Two thermoplastic polyurethanes (TPUs) of refractive indices 1.50 and 1.60 were co-extruded; the extrusion temperatures of each grade were adjusted to equilibrate the viscosities and thus stabilize the layers. The extrusion system comprises two extruders and a series of multiplying units, each doubling the number of layers, such that 12 multiplying units render 212+1+1 (= 8193) layers. For a 0.8 mm-thick die, the resulting layers after die swell are ~ 120 nm thick. TPUs are highly elastic, and large stretching ratios allow for significant reduction in thickness. Physical coloring developed in a controlled, reversible manner. The visual color correlates with elongation, and can be offset by choice of initial thickness. Both reflected and transmitted wavelengths change throughout the whole visible range, the former decreases and the latter increases with stretch.

Determination of Influencing Factors on Adhesion and Surface Quality in the UV-based In-mold Coating Process of BMC

Larissa Kutscha, M.Sc. M.Sc.[1, a)], Dipl.-Ing. Stefan Apelt[1, b)], Carsten Koch, M.Sc.[2] and Prof. Dr.-Ing. Christian Hopmann[2]

[1]*Robert Bosch GmbH, Robert-Bosch-Campus 1, 71272 Renningen, Germany*
[2]*Institute of Plastics Processing at RWTH Aachen University, Seffenter Weg 201, 52074 Aachen, Germany*

[a)]Corresponding author: larissa.kutscha@de.bosch.com
[b)]stefan.apelt@de.bosch.com

Abstract. The in-mold coating (IMC) as a mass production process offers potential to generate low-cost, high precision surface qualities on polymeric materials for optical applications, such as mirrors and reflectors. The combination of photo-curable coating systems and the IMC process provides several advantages, especially short curing durations and reduced plant complexity. This work presents the generation of optical precision surface finishes on highly filled bulk molding compound (BMC) substrates by coating with UV-curable epoxy in a closed mold. The influence of process parameters as intensity, radiation dose and film thickness on surface finish and adhesion to the substrate is analyzed. Methods for adhesion improvement, such as surface pre-treatment, are taken into account as well. The curing behavior during the coating process is detected in-situ via dielectric analysis (DEA).

INTRODUCTION

In times of new applications regarding autonomous driving, appliance and process automation, sensor and optical detection systems and therefore optical components such as lenses and mirrors become increasingly important (1,2). Since such components are still cost drivers of optical systems due to the high requirements on the optical functionality, there is a high economical interest on cost and time-efficient mass production processes for polymer based optical components. In comparison to the intensive research on lens production during the last decades, there is still a low level of experience in manufacturing polymeric high-precision mirrors. Those mirror are still mainly produced out of thermoplastic materials, which have disadvantages regarding shrinkage and heat resistance. A new approach is the use of highly filled thermoset bulk molding compounds (BMC) as basic materials for mirror surfaces. BMC has nearly no shrinkage, very high mechanical stability and heat resistance and can be processed by injection molding or injection compression molding. However, the high filling grade results in low surface qualities, which need to be equalized by a separate coating step. In-mold coating (IMC) and therefore the exact replication of the mold insert surface quality enables the generation of nanoscale precision mirror surfaces after metallization (3). Due to its high curing speed, its processability at room temperature and the low plant complexity and energy consumption, the combination of the IMC process with UV-curable coatings is considered as most effective for the above-described application (4, 5). Additionally, the solvent-free, one-component, radical or cationic photo-curing resins offer advantages regarding environmental issues (6). Whereas such resins are already used as dental composites, in the graphic and building industries, and for temperature-sensitive electrical applications (6, 7, 8), there is currently no experience in the use of UV-curable resins as IMC coating materials and their adhesion on BMC substrates. Therefore, the aim of this research work is to determine process and material-based influencing factors on layer adhesion and surface quality of UV-cured in-mold-coated BMC surfaces fulfilling optical requirements.

MATERIALS

As substrate material for the following investigations a bulk molding compound (BMC) by Menzolit GmbH, Bruchsal, Germany, based on unsaturated polyester resin was used. According to its datasheet the doughy mass has a filler content of 70 - 80%, including 10 % textile glass fibers and a density of 2.02 g/cm³. It's very low molding shrinkage of 0.02% and high heat resistance of 200 °C are considered advantageous for optical applications in the automotive sector, such as headlights and reflectors, for which this material type is typically used. To determine the influence of a low-pressure plasma treatment on adhesion between coating and substrate, the described BMC is considered in a pure untreated (BMC23) and in a plasma-treated mode (BMC23 + Plasma) for these investigations.

As UV-curable coating material Catiowell 3057 by Wellmann Technologies GmbH, Friedelsheim, Germany was investigated. The resin is based on a cycloaliphatic epoxide (ECC), which is combined to an iodonium salt photo-initiator. The cationic polymerization reaction is inducible by UV and visible light between 350 and 400 nm wavelength and temperatures higher than 120 °C. Catiowell 3057 show nearly Newtonian flow behavior with a low viscosity of 345 mPas.

EXPERIMENTAL SET-UP AND ANALYTICAL METHODS

The UV-based in-mold coating process is performed in a closed mold as demonstrated in Fig. 1. The plate-shaped substrate specimens are placed into the bottom cavity and fixed by vacuum suction. The closed mold is evacuated to avoid blistering of the coating. The substrate is floated by photo-curable resin, injected through a film gate into the cavity. The coating is cured by UV irradiation, which transmits through an UV-transparent mold insert integrated in the top cavity. As UV source LED-plane radiator DeloLux 20 emitting irradiation of a wavelength of 400 nm is used, connected to the Delo-UniPro Light control unit for intensity and time setting (Delo Industrie Klebstoffe GmbH & Co. KGaA, Windach, Germany). The curing behavior is monitored in-situ by a dielectric analysis (DEA) sensor monotrode 4/3 RC (Netzsch Gerätebau GmbH, Selb, Germany), which is positioned in the film gate of the bottom mold. After cure, the specimen is ejected mechanically. Due to its simple structure, the set-up can be integrated into a tensile testing machine Zwick Z005 (Zwick Roell AG, Ulm Germany) for clamp force realization as well as for mold movement and demolding force measurement.

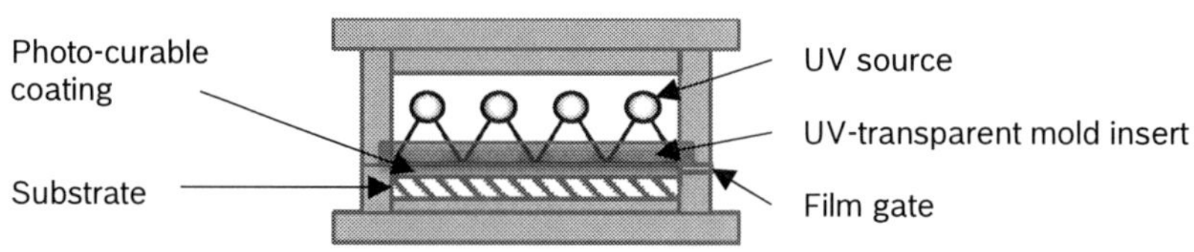

Fig. 1. Experimental set-up for the UV-based in-mold coating process.

To determine influencing factors as UV intensity, duration of irradiation, film thickness and substrate material, BMC23 and BMC23 + plasma were coated in the described UV-based in-mold coating process with varying intensities of 25 mW/cm², 137.5 mW/cm² and 250 mW/cm² and durations of 40 s, 80 s and 120 s. Layer thicknesses of 200 µm, 300 µm, 400 µm and 500 µm were considered. The curing behavior was measured in-situ via DEA and compared to conversion values from DSC measurements. The used low-pressure plasma treatment parameters were evaluated based on the surface energy calculated on contact angle measurement data according to the Owens-Wendt-Rabel-Kaelble method and XPS measurements of the element consumption on the BMC surface before and after plasma-treatment (9).

To analyze the quality of the generated coating surfaces a confocal laser scanning microscope Olympus Lext OLS 4100 (Olympus Europe SE & Co. KG, Hamburg, Germany) with a 50x objective resulting in a z-resolution of 8nm, was used. Roughness and waviness values were calculated out of 40 line-measurements in x- and y-direction at 5 positions on the surface. The adhesive strength between coating and substrate material was measured via pull-off test (ISO 4624:2016). All pull-off tests were processed by the adhesion tester PosiTest AT-A (DeFelsko Corporation, Ogdensburg, NY, USA), using test dollies of 20 mm diameter. The dollies were applied on the tested surface by a 2-component epoxide adhesive and pulled off in vertical to surface direction after 24 h of hardening time, measuring the force necessary.

RESULTS AND DISCUSSION

As described before, the specimens were cured at varying intensities and durations of irradiation. To analyze the influence of those parameters on the curing behavior, DEA measurements were processed and compared with the results of swelling and DSC measurements. Fig. 2 shows the ion viscosity measured by DEA at varying curing parameters. It is noticeable that the curing rate – represented by the curve gradient – increases with increasing intensity, whereas there is nearly no difference between the curing speed at 137.5 mW/cm² and 250 mW/cm² compared to 25 mW/cm². Considering the ion viscosity at 25 mW/cm² there is also an influence of the duration of irradiation on the curing behavior. If the irradiation stops after 40 s, the ion viscosity reaches a constant level earlier as for the irradiation of 120 s. This effect is not noticeable for the higher intensities. It could be concluded, that for lower intensities, there is still photoinitiator activatable after 40 s of irradiation which accelerates the cross linking reaction, whereas it is already consumed at higher intensities. This also leads to lower final ion viscosity values at low intensities indicating lower degrees of conversion and cross-linking as demonstrated in Fig. 3 and Fig. 4 (10).

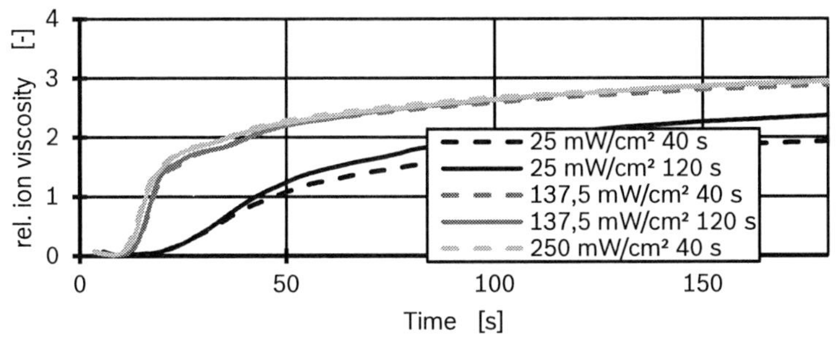

Fig. 2. Relative ion viscosity measured via DEA during cure at different intensities and durations of irradiation

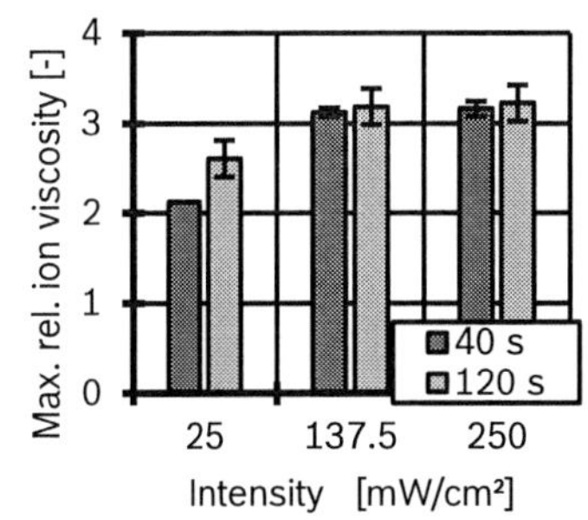

Fig. 3. Maximum value of the relative ion viscosity after 5 min after start of irradiation.

To investigate the influence of varying curing behavior of the coating on adhesion between coating and substrate material, the adhesive strength is measured by pull-off test after the in-mold coating step. Fig. 5 and Fig. 6 show the influence of intensity, duration of irradiation and layer thickness on adhesion. It could be considered that the curing parameters intensity and duration have no significant influence on the adhesive strength. In contrast, the adhesive strength is significantly lower at lower layer thicknesses of 200 and 300 µm than at higher layer thicknesses. This effect might be due to differences in shrinkage behavior influenced by thickness layer and degree of conversion and should be focused in further investigations.

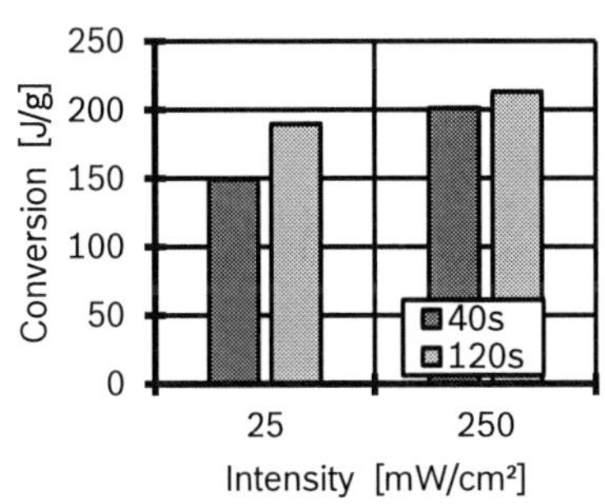

Fig. 4. Conversion measured via UV-DSC during cure at different intensities and durations of irradiation.

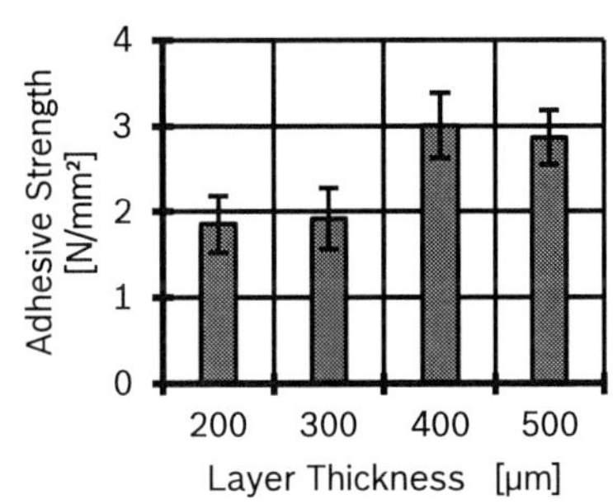

Fig. 5. Influence of different layer thicknesses on adhesive strength after cure at 137.5 mW/cm² and 120 s.

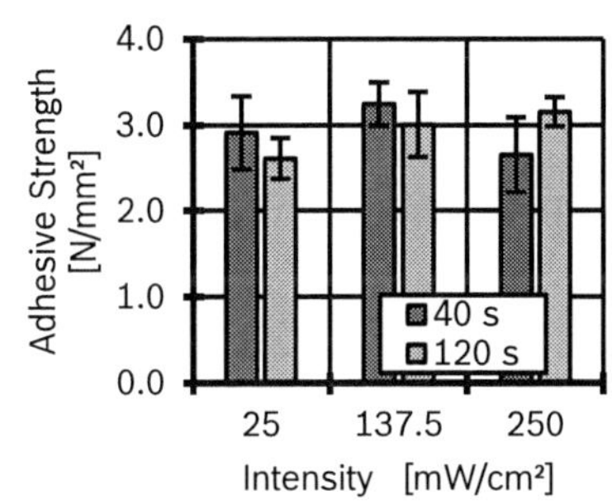

Fig. 6. Influence of different intensities and durations of irradiation on adhesive strength (layer thickness 400 µm).

To gain information about the influence of a low-pressure plasma treatment of BMC the surface energy in polar and disperse parts were calculated out of contact angle values considering different plasma parameters, as demonstrated in Fig. 7. It is noticeable that the surface energy increases due to the low pressure plasma treatment. In particular the polar part increases from 2.97 mN/m for an untreated BMC to 32.45 mN/m after a plasma treatment at 300 W for 1 min. Higher treatment durations or intensities do not result in further increase of the surface energy.

Additionally the results of XPS measurements of the pure and plasma treated surface demonstrated in Fig. 8 show that the carbon consumption decreases, whereas the amount of reactive oxygen groups at the surface increases during plasma treatment. For all further investigations the plasma parameters 300 W and 3 min were used.

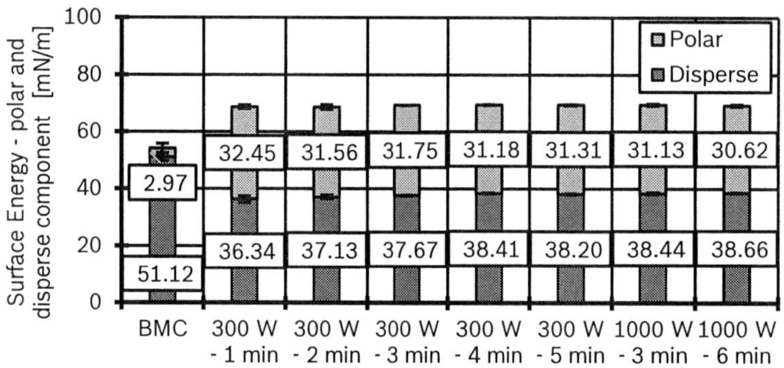

Fig. 7. Surface energy in polar and disperse components of the pure BMC surface and after low-pressure plasma treatment at different parameters.

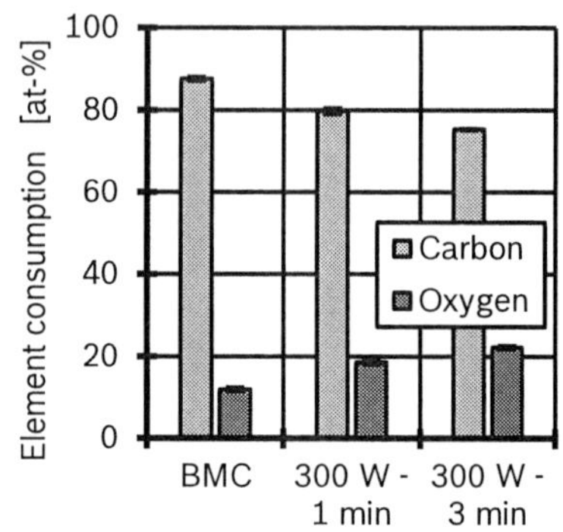

Fig. 8. Carbon and oxygen consumption on pure and plasma treated BMC at different parameters by XPS.

Fig. 9 and Fig. 10 demonstrate the results of adhesive strength measurements on pure and plasma-pretreated BMC for different curing parameters. It is noticeable that there is no significant improvement of adhesion between coating and substrate by the processed plasma treatment step. Considering the fact that surface polarity changes and reactive oxygen is bound to the BMC surface due to plasma treatment it could be concluded, that the polarity/wettability and reactivity of the surface doesn't affect adhesion dominantly. The value differences might be caused by overlaying effects between the described and other adhesion mechanism (e.g. mechanical anchoring). This should also be focused in further investigations.

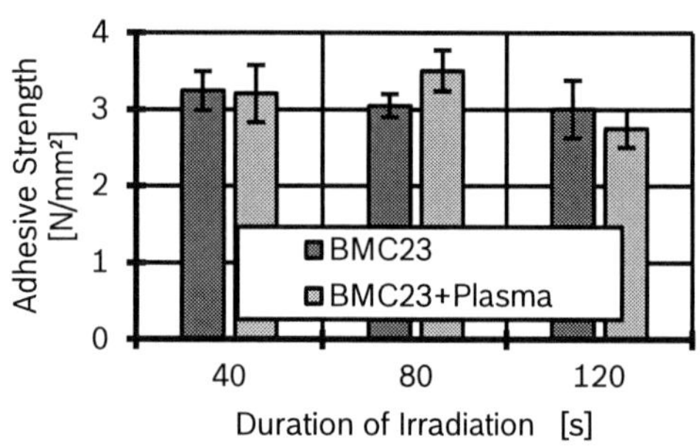

Fig. 9. Influence of plasma treatment of BMC on adhesive strength after cure with different durations of irradiation at 137,5 mW/cm and a layer thickness of 400 μm.

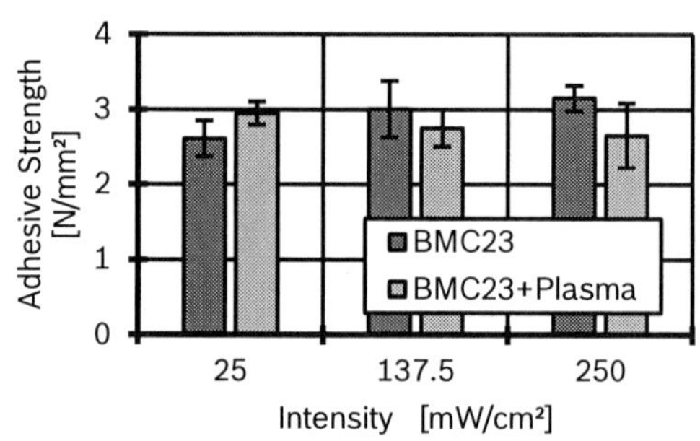

Fig. 10. Influence of plasma treatment of BMC on adhesive strength after cure at different intensities for 120s and a layer thickness of 400 μm.

Since the aim of the in-mold coating process is to generate high precision surface qualities, the roughness and waviness values of the coated specimens were analyzed. According to mirror manufacturers an averaged roughness value of less than 20 nm is targeted. Fig. 10 a) and b) demonstrate, that the UV-based in-mold-coating process is suitable to generate roughness values of less than 21 nm and waviness values of less than 27 nm, which corresponds to a surface quality improvement of more than 90% and nearly reaches the targeted value. Further roughness improvements might be achieved by higher mold insert surface qualities. The specific analysis on the influence of curing parameters on surface quality is not appropriate in this case, due to the z-resolution of 8 nm of the used confocal laser scanning microscope.

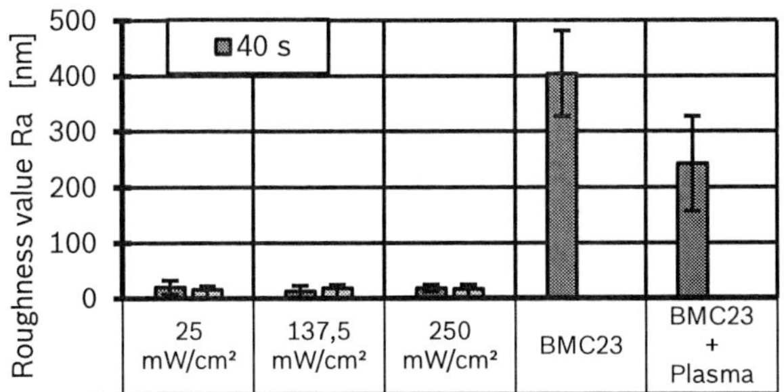

	25 mW/cm²	137,5 mW/cm²	250 mW/cm²	BMC23	BMC23 + Plasma
40 s	20.50	13.00	19.00	404.00	241.50
120 s	16.50	19.00	17.00		

Fig. 91a) Averaged roughness values Ra of the coated surface compared to the uncoated BMC surfaces.

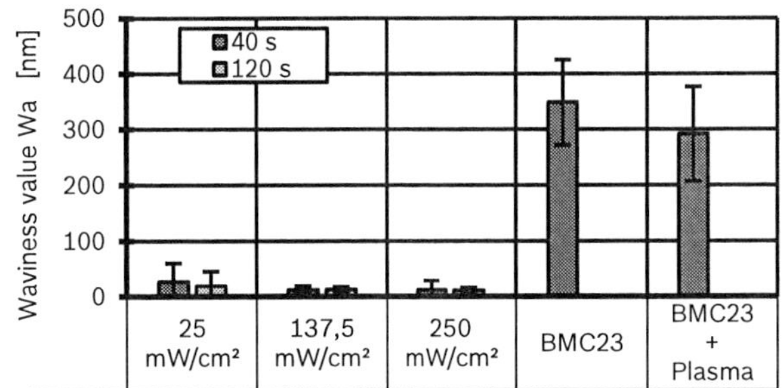

	25 mW/cm²	137,5 mW/cm²	250 mW/cm²	BMC23	BMC23 + Plasma
40 s	26.75	11.00	11.50	348.25	292.00
120 s	19.50	12.00	10.50		

Fig. 11b). Averaged waviness values Wa of the coated surface compared to the uncoated BMC surfaces.

CONCLUSION

The UV-based in-mold coating process of BMC was presented. It was shown, that this process is suitable to generate high precision surface qualities with roughness values less than 21 nm and waviness values less than 27 nm. The curing behavior was in-situ monitored by dielectric analysis, which clearly shows positive dependencies between intensity and duration of irradiation on curing speed and final conversion. This was verified in comparison to DSC values. Furthermore the adhesive strength between coating and BMC substrate was measured, showing no influence of intensity and irradiation duration. In contrast adhesion increases with increasing layer thickness. Aiming at adhesion improvement, the low-pressure plasma treatment of BMC was analyzed via contact measurement. Although polarity and oxygen consumption increased via plasma treatment, no significant influence on the adhesive strength was found.

ACKNOWLEDGMENTS

For supporting this research work with supply of materials and knowledge, we want to extend our thanks to Menzolit GmbH, Topas Advanced Polymers GmbH and Wellmann Technologies GmbH.

REFERENCES

1. Tsai, K.-M., Hsieh, C.-Y., Lo, W.-C., 2009. A study of the effects of process parameters for injection molding on surface quality of optical lenses. Journal of Materials Processing Technology 209, 3469–3477.
2. Michaeli, W., Hopmann, C., Heßner, S., Walach, P., 2011. Injection-compression moulding of aspherical plastic lenses. Journal of Plastics Technology 7, 1–22.
3. Apelt, S., Geise, S., Weiss, C., 2012. Method for manufacturing light reflecting device i.e. reflection mirror, for e.g. head-up display, involves applying coating material on surface of cavity of injection molding tool, and applying base body to coating material.
4. Golaz, B., Michaud, V., Leterrier, Y., Månson, J.-A.E., 2012. UV intensity, temperature and dark-curing effects in cationic photo-polymerization of a cycloaliphatic epoxy resin. Polymer 53, 2038–2048.
5. Jančovičová, V., Mikula, M., Havlínová, B., Jakubíková, Z., 2013. Influence of UV-curing conditions on polymerization kinetics and gloss of urethane acrylate coatings. Progress in Organic Coatings 76, 432–438.
6. Masson, F., Decker, C., Jaworek, T., Schwalm, R., 2000. UV-radiation curing of waterbased urethane–acrylate coatings. Progress in Organic Coatings 39, 115–126.
7. Takamizawa, T., Yamamoto, A., Inoue, N., Tsujimoto, A., Oto, T., Irokawa, A., Tsubota, K., Miyazaki, M., 2008. Influence of light intensity on contraction stress of flowable resins. Journal of Oral Science 50, 37–43
8. Behnsen, P., n.d. Einsatz der UV-Technologie zur Erzeugung von Papierbeschichtungen mit funktionalen Oberflächen (PTS-Forschungsbericht No. IK_VF110016).
9. Rabel, W., 1977. Flüssigkeitsgrenzflächen in Theorie und Anwendungstechnik. Physik Journal 33, 151–161.
10. Knappe, S., 2005. Cure Monitoring of Paints by Dielectrical Analysis - Even in-situ. www.pcimag.com, Paint & Coatings Industry 26–34.

Multifunctional and responsive surfaces using nanostrucrured polymer brushes

Uhlmann Petra (1)*, Münch Alexander (1), Adam Stefan (1), Rauch Sebastian (1)

(1) Leibniz-Institut für Polymerforschung Dresden e.V. - Saxony - Germany

Polymer brushes are used as a toolbox to build up coatings by using the buttom-up principle to create the desired surface function [1]. The used building blocks are (bio)functional linear block or comb polymers which ensure at one hand the chemical coupling to the surface and at the other hand the desired function. Such coatings can be composed by one or several different polymers to realize for instance switching (i.e. adaption to changing environmental conditions or active switching of surface properties)or to serve as optical sensor or bioactive substrate by incorporating nanoparticles or biomolecules into the layer [2,3]. Those hybrid layers are enormously be in principle applied to every substrate by adapting the grafting procedure to technological coatings processes [4]. Examples for those applications are anti-icing, anti-fouling, anti-soil or novel easy-to clean coatings [5].

Confinement Induced Ordering of Phase Segregated Domains in Polymer Blend Thin Film

Bhandaru Nandini (1), Mukherjee Rabibrata (1)*

(1) Indian Institute of Technology Kharagpur - west bengal - India

Meso scale structures obtained due to phase segregation of an immiscible polymer blends has attracted significant interest over the years as these surfaces can potentially be used in a wide variety of applications such as anti-reflection coatings in organic solar cells, as templates in tissue engineering, as templates for selective deposition, as multifunctional substrates for combinatorial studies and many more. Such polymer blends when spin coated on a flat substrate results into variety of random structures which largely depends on the composition of the two polymers, the film thickness as well as the nature of the substrate. With this work we investigate polystyrene (PS) / poly(methylmethaacrylate) (PMMA) blend thin films of various concentrations and compositions spin casted on flat as well as on different topographically patterned substrates. Myriad of nanostructures were obtained on varying parameters such as molecular weight of the polymers, the surface energy of flat and patterned substrates and nature of the topographic confinement. Three broad regimes of the phase segregated morphology could be identified based on the concentration (Cn); Regime 1: Cn <0.5%, at which the solution spin dewets, resulting in isolated polymer droplets aligned along substrate grooves, Regime 2: at an intermediate Cn where isolated threads are formed with aligned phase segregated domains along each substrate groove, and finally in Regime 3:above a critical concentration (Cn*) which depends strongly on ratio of the blend composition, the overall morphology of the film transforms forms discontinuous to continuous. The physical confinement imposed by the topographical patterns enabled ordering of these domains which are otherwise random and also significant miniaturization of the domain size was also achieved. While the extent of ordering of the domains gradually diminishes with increase in film thickness (vis-à-vis Cn), the size of the domains remains much smaller.

Morphological Stability of Nanoparticles Containing Polymer Blends and Bilayer Thin Films

Das Anuja (1)*, Mukherjee Rabibrata (1)

(1) Indian Institute of Technology Kharagpur - west bengal - India

Stability of nanoparticle filled multi-polymeric systems have attracted significant interest due to the fact that addition of such particles drastically changes the properties of the polymers and have huge potential in the field of optoelectronic devices, nanocoatings. With this work, we present here a fundamental study on morphological stability on addition of fullerene NP in poly(styrene) (PS)/ poly(methylmethaacrylate) (PMMA) blend thin films. PS and PMMA being immiscible polymer blends at room temperature laterally phase segregates leading to formation of meso scale surface structures. Atomic Force Microscope images of as coated PS-PMMA blend reveals that on increasing the concentration of NP, the lateral phase segregated domains reduces. Moreover, beyond a critical concentration, the phase segregation is completely arrested leading to a flat film. By ellipsometric studies, thickness of the polymer blend with nanoparticle before and after selective phase removal hinted vertical phase separation instead of a typical lateral phase separation. We can qualitatively infer that the nanoparticles while spin coating, migrates towards the polymer–polymer interfaces and impart stability to the interface, thereby arresting the lateral phase segregation process during the latter stages of spin coating. This interfacial stability led to a vertical phase segregation of the PS and PMMA forming a polymer–polymer bilayer. This technique can be used to coat ultra-thin multifunctional coatings without the need of additional coating processes. While the effect of addition of NP is well known for imparting stability to single layer of polymer thin film, for the case of polymer bilayer with NP, a reduction in dewetted length scale of top layer is observed between specific ranges of concentration and the dynamics of the instability becomes sluggish. Apart from stability, this provided an additional parameter to impart miniaturization in dewetted morphology.

Superhydrophobic and superoleophilic meshes coated by PDMS/ZnO nanocomposites for oil/water separation

Bolvardi Beleta (1)*

(1) IslamicAzadUniversity - Tehran - Iran

Separation of oil/water mixtures has been turned into a very interesting topic due to the environmental concerns from industrial oily wastewater and polluted oceanic water as well as oil leakage accidents [1]. One of the most efficient methods to separate oily materials from water is to induce superhydrophobic and superoleophilic properties to a certain substrate like sponges, meshes, fabrics, and etc. [2]. To achieve the superhydrophobicity (water contact angle > 150o), one should provide two conditions including proper level of roughness and low surface energy on the desired material. If the surface energy of the material is tuned in a way that is lower than that of water but higher than those of oil, the modified material repels water and absorbs oil. The main aim of this study is to impart superhydrophobicity and superoleophilicity onto the metal-based meshes with different pore sizes via coating the meshes by a polymer-based nanocomposite. To this end, polydimethylsiloxane (PDMS) is used as the matrix for the nanocomposite causing the surface energy of the meshes to be highly lowered. To increase the roughness, zinc oxide (ZnO) nanoparticles are utilized. Firstly, the wettability o the coated meshes is investigated by the water contact angle measurement. The surface morphology of the coated nanocomposite on the meshes' surface is characterized via scanning electron microscopy. References: [1] Feng L, Zhang Z, Mai Z, et al. (2004) Angew Chem Int Ed 43:2012-2014 [2] Piltan S, Seyfi J, Hejazi I, Khonakdar HA (2016) Cellulose 23:3913-3924

Polyethylene surface modification ability of side chain crystalline polymer and side chain crystalline block copolymer.

Miho Yuga [(1)]*******, ***Yao Shigeru*** [(1)], ***Nakano Ryoko*** [(1)], ***Sekiguchi Hiroshi*** [(2)]

[(1)] Fukuoka University - Fukuoka - Japan, [(2)] Fukuoka-University - Fukuoka - Japan

Side Chain Crystalline Block Copolymer (SCCBC) is composed with a side chain crystalline monomer unit and a functional monomer unit. It can be adsorbed to a polyethylene (PE) crystalline surface through crystalline supramolecular interaction. By using this interaction, we can modify the PE surface to various properties. The modification method is very simple, just dipping PE films into the SCCBC dilute solution or coating the dilute solution to the PE surface.@ In this time, we polymerized a new side chain crystalline polymer (SCCP). And, we compared the PE surface modification ability of this SCCP and SCCBC.

An Investigation of morphology and chemical composition of water-repellent films based on Poly(dimethylsiloxane)/titanium dioxide nanocomposites

Bolvardi Beleta (1)*

(1) IslamicAzadUniversity - Tehran - Iran

In this study, Poly(dimethylsiloxane) (PDMS)/titanium dioxide (TiO2) films were fabricated through simple solution casting technique via a modified phase separation process. The presented approach resulted in a remarkable reduction in the required amount of nanoparticles for achieving super hydrophobicity. Scanning electron microscopy (SEM) and 3D confocal microscopy were utilized to characterize surface morphology and topography of samples, respectively. An attempt was made to give an in-depth analysis on the surface rough structure using 3D roughness profiles. It was found that high inclusions of non-solvent and nano particles resulted in a stable self-cleaning behavior due to the strong presence of hydrophobic TiO2 nanoparticles on the surface. Quite unexpectedly, low inclusions of nanoparticles and non-solvent also resulted in super hydrophobic property mainly due to the proper level of induced surface roughness. XPS analysis was also utilized to determine the chemical composition of the films' surfaces. The results of falling drop experiments showed that the sample containing a higher level of nanoparticles had a much lower mechanical resistance against the induced harsh conditions.

Influence of processing method and parameters on the quality of composite films of montmorillonite clay (MMT) and poly(vinyl alcohol) (PVA)

***Liu Jingjing* [1], *Chavez Sonia* [1], *Shaw Montgomery* [1]*, *Sun Luyi* [1]**

[1] University of Connecticut - CT - USA

Composite films of montmorillonite clay (MMT) dispersed in a crosslinked poly(vinyl alcohol) matrix exhibit unusually favorable mechanical, flame-retarding and oxygen-barrier properties. For example, the oxygen permeability of a 22-micrometer PET film can be reduced by a factor of ~200 by adding a 0.6-micrometer coating. To optimize the processing and properties of these coatings we are examining factors that control the ordering of the clay nanoplates that are responsible for the favorable properties mentioned above. Among these factors are (1) the shear-stress magnitude and history during the coating process and (2) the interaction of the clay with the substrate, each other and the polymer "glue." Bench-scale coating processes considered in this study were dip, spin and doctor-blade coating. For ultra-low shear stress, Meyer bar coating was examined. Considering the clay platelets as isolated disks, these processes can give Péclet numbers from ~1 to several hundred. All but the spin coating can be used for continuous processing.

Investigating the effect of surface energy of chitosan/titanium dioxide nanocomposite films on their antibacterial behavior

Raisi Mahdieh (1), Khonakdar Hossein Ali (2), Seyfi Javad (3)*

(1) Islamic Azad University, Tehran Branch - Tehran - Iran, (2) Leibniz-Institut für Polymerforschung Dresden - Dresden - Germany, (3) Islamic Azad University of Shahrood - Semnan - Iran

Investigating the effect of surface energy of chitosan/titanium dioxide nanocomposite films on their antibacterial behavior Mahdieh Raeesi1, Hossein Ali Khonakdar2, Javad Seyfi3 1 Chemical Engineering Department, Central Tehran Branch, Islamic Azad University, Tehran, Iran 2 Leibniz-Institut für Polymerforschung Dresden e.V., Hohe Strasse 6, D-01069 Dresden, Germany 3 Department of Chemical Engineering, Shahrood Branch, Islamic Azad University, Shahrood, Iran Abstract Chitosan is the most important derivative of chitin, and is obtained by partial deacetylation of chitin in the solid state under alkaline conditions or by enzymatic hydrolysis. Since chitosan is biocompatible, biodegradable and non-toxic, it has found a vast range of applications within the biomedical applications such as tissue engineering and wound healing [1]. In order to improve some properties of chitosan like cell adhesion, different nano-scale inorganic materials have been incorporated into the chitosan matrix. It was shown that materials such as titania, zirconia and alumina could enhance the cellular behavior if they are used in nano-scale sizes [2]. One of the most important characteristics of chitosan is its antibacterial property which could be utilized especially in wound dressing applications. In this research, the thin films based on chitosan and titanium dioxide (TiO2) nanoparticles are first fabricated, and then, the effect of nano-TiO2 concentration on the bacterial adhesion is assessed. To this end, the surface energy is calculated by measuring the contact angles of different liquids, and then, an attempt is made to establish correlations between the surface energy of the films and the antibacterial property. Keywords: Chitosan; Antibacterial; Surface Energy References: [1] R. Jayakumar, R. Ramachandran, V.V. Divyarani, K.P. Chennazhi, H. Tamura, S.V. Nair, Int. J. Bio. Macromol. 48 (2011) 336. [2] P.A. Ramires, A. Romito, F. Cosentino, E. Milella, Biomaterials 22 (2001) 1467.

Assessment of the role of surface modification of nanosilica on the wetting behavior of thermoplastic polyurethane surfaces

Seyedani Niloofar (1), Otadi Maryam (1), Seyfi Javad (2), Khonakdar Hossein Ali (3)*

(1) Central Tehran Branch, Islamic Azad University - Tehran - Iran, (2) Islamic Azad University of Shahrood - Semnan - Iran, (3) Leibniz-Institut für Polymerforschung Dresden - Saxony - Germany

For over a decade, superhydrophobic surfaces have gained significant attention within the scientific community as well as different industries because of their unique water-repellent, self-cleaning, anti-fouling, anti-icing, anti-sticking, anti-contamination properties, and their promising potential for practical applications [1]. Among polymers, polyurethane (PU) is one of the most versatile materials which is widely used in applications ranging from foams and coatings to medical devices [2]. Surface modification of PU films and coatings could be very influential on the behavior of this material for biomedical applications especially where blood compatibilitiy is concerned. Therefore, PU/Silica thin films were prepared in this work, and the main aim is to study the role of surface modification of silica nanoparticles on the wetting behavior of PU surfaces. To this end, scanning electron microscopy is also used to observe the effect of different silica nanoparticles on the surface morphology of the films. Water contact angle measurement results showed that, depending on the used nanosilica type, a diverse range of wettabilities including superhydrophobic and superhydrophilic behaviors were attained. Keywords: Polyurethane; Nanosilica; Superhydrophobic References: [1] I. Hejazi, B. Hajalizadeh, J. Seyfi, G.M.M. Sadeghi, S.H. Jafari, H.A. Khonakdar, Applied Surface Science 293 (2014) 116. [2] S.A. Madbouly, J.U. Otaigbe, Prog. Polym. Sci. 34 (2009) 1283.

The polymerization of the side chain crystalline block copolymer and the study of super hydration

Fukaya Koki [(1)*], Sekiguchi Hiroshi [(1)], Nakano Ryoko [(1)], Yao Shigeru [(1)]

[(1)] Fukuoka Univercity - Fukuoka - Japan

Koki Fukaya, Hiroshi Sekiguchi, Ryoko Nakano, Shigeru Yao Dep. Chem. Eng. Fukuoka Univ. A block copolymer composed of a monomer having a long alkane side chain and another monomer having a functionality such as a hydrophobic property becomes a side chain crystalline block copolymer (SCCBC) which crystallizes at the side chain of the former. We had found the SCCBC has an adsorption function to polyethylene surface by crystalline supramolecular interaction strongly. By using the SCCBC, we can modify surface property of PE chemically and easily. Polytetrafluoroethylene (PTFE) is also known as a polymer material with super hydrophobicity and super-soil resistance, and used in many fields. In addition, the porous membrane is well known as a functional material having both super hydrophobicity and water vapor permeability. However, it is expensive and also a material that has problems in process ability and strength. In this time, we had polymerized a new chemical structural and triblock type SCCBC that is composed with a monomer having fluorocarbon side chain as a functional monomer by RAFT living radical polymerization method. We found that this Fluorinated SCCBC (F-SCCBC) has an ability that can modify PE surface to super hydrophobic. We also found that when applied the F-SCCBC to porous membranes, the porous membrane becomes to have both super hydrophobic and water vapor permeability like PTFE membrane.

Alignment of carbon nanoparticles in composites with electric fields

Morais Manuel (1)*

(1) Fraunhofer ICT - Pfinztal - Germany

Manipulation of carbon nanotubes and other carbon nanoparticles in liquid media through the application of alternate electric fields has been frequently addressed in literature. Even though various models have been proposed, the mechanisms guiding this phenomenon are up to the moment not completely understood, given the disparity of results found. For nanoparticles in liquid suspensions, Brownian motion and Van der Waals forces are usually regarded as the relevant driving forces in particle dynamics. With the application of an electric field, electrophoresis, dielectrophoresis and Coulombic interactions arise. There is, as reported in literature, a strong dependency on the electric field intensity and frequency. Experimental setups were developed for investigations of the effect of high frequency electric fields (up to 45 MHz) at electric field strengths up to 20 KV/m on different kinds of polymer/nanoparticle composite systems. Liquid systems, such as solvents, monomeric systems, and epoxy mixed with different types of nanoparticles are investigated. In addition, a setup coupled with a heating and cooling plate was developed for trials with electric field assisted annealing of molten polymer samples. In-situ current and voltage measurements are performed for studying the dynamic evolution of such electrical processes. A stereo-macroscope with transmission lighting was used for visual data collection.

Polyethylene plating by using the function of side chain crystalline block copolymer

Uchino Tomonori [(1)]*, Nakano Ryoko [(1)], Sekiguchi Hiroshi [(1)], Yao Shigeru [(1)]

[(1)] Fukuoka Univercity - Fukuoka - Japan

Tomonori Uchino Fukuoka Univercity Polyethylene (PE) is known to be a very convenient and high performance plastic because of an inexpensive cost and its hydrophobicity and chemical resistance property. Therefore, PE is used in many applications. On the other hand, PE is a very highly crystalline polymer and do not have any polarity on its chemical structure, it is known to be a very difficult material to modify the surface properties. Because of this reason, it has been considered that it is impossible to apply plating to the surface of PE. Recently, we have reported that the side chain crystalline polymers constructed with a monomer having a long alkane side chain shows a good adsorption property to PE due to the crystalline supramolecular interaction. Moreover, we have polymerized a Side-Chain Crystalline Block Co-Polymer (SCCBC) which is composed of the above side-chain crystalline monomer and a functional monomer (which can confer solvent-compatibility, polarity, etc.) and found that the SCCBC could modify the surface property of PE to variously. In this time, by using this function, we have tried electroless copper plating to the surface of PE.

Bio-inspired hierarchical nanocomposite foam for rapid oil/water separation

Cherukupally Pavani [(1)]*, ***Sun Wei*** [(1)], ***Wong Annabelle P.Y.*** [(1)], ***Ozin Geoffery A.*** [(1)], ***Bilton Amy M.*** [(1)], ***Park Chul B.*** [(1)]

[(1)] University of Toronto - Ontario - Canada

Note: We would like to deliver this work as both poster and oral presentations. If there is a poster competition please allow us to present a poster. Removal of micro-oil droplets from industrial effluents, such as oil spills, oil field wastewater, and petrochemical effluents, is a global environmental challenge. For example, Canada's oil sands industry has generated over 1 billion m3 of dissolved organic contaminated water resulting in severe damage to the environment, human health, and wildlife habitat. Unfortunately, the existing superwetting foams for oil/water mixture separation are ineffective for micro-oil droplet removal and are made from expensive, toxic, or non-scalable materials, such as metal oxides, carbon nanotubes, and graphene. One promising solution is to use polymeric foams for micro-oil and water separation due to their low cost, effectiveness, and multifunctional properties. In this work, a scalable, facile dip-coating method was employed to functionalize polyester polyurethane foam with decyl-capped nanocrystalline silicon (ncSi), resulting in complementary electronic and oleophilic properties. Within only 10 minutes, the novel hybrid foam is able to adsorb crude oil micro-droplets via electrostatic and van der Waals forces with over 97% efficacy. Moreover, due to its large free volume and surface area, an oil uptake capacity of 24 g/g was achieved in only 60 seconds. The nanocomposite foam also exhibited other favourable properties, such as its resistance towards acids and bases, low density and hydrophobicity. These preliminary studies indicate the low-cost ncSi coated polymeric foam has a great potential to remediate organic contamination in industrial effluents.

In-situ x-ray structure measurements of polymers during die-drawing

Coates Phil (1)*, Men Yongfeng (2), Caton-Rose Fin (1), Thompson Glen (1), Jiang Zhiyong (2), Lai Yuqing (2)

(1) University of Bradford - W Yorks - UnitedKingdom, (2) Changchun Institute of Applied Chemistry Chinese Academy of Science - Jilin - China

A novel small scale precision die drawing solid state orientation facility has been designed and constructed in the Polymer IRC Laboratories, and installed in the high resolution USAXS/SAXS/WAXD Xeuss 10m line in Changchun, with initial experiments on solid phase deformation of rectangular cross section polymer products. We combine microstructural analysis X-ray scattering skills with process engineering to provide new insights into process structuring mechanisms for property-enhanced (so added value) commercial PLAs and polyolefins aimed primarily at affordable healthcare technologies. We seek to elucidate factors affecting molecular and crystalline orientation during processing of these polymers, and to establish links between molecular parameters, processing conditions, hierarchical microstructures and the required properties in the final products, with an emphasis on enhanced materials for medical technologies. Initial results for polyolefins will be presented. In addition, an in-situ micro moulding facility is also being designed, to explore structure development in that process.

Inline melt homogeneity measurement in injection molding

Bernhard Praher[1, a)], Marvin Goldmann[1] and Georg Steinbichler[1]

[1]*Institute of Polymer Injection Moulding and Process Automation, Johannes Kepler University Linz, Altenbergerstrasse 69, 4040 Linz, Austria*

[a)]*Corresponding author:bernhard.praher@jku.at*

Abstract. Unwanted inclusions in the plasticized polymer melt (unmelted granules, wear particles, dirt, gas,...) can reduce the visual and structural properties of the resulting injection molded part. We developed a non-invasive ultrasound based measurement system working in the high-pressure and high-temperature environment of polymer processing for the detection and classification of such inclusions. The system is installed between the heating barrel and nozzle of the injection molding machine and it is based on ultrasound transmission measurements during the injection of the polymer melt into the mold. The novel system shows reliable capability for the classification of the part quality before it is demolded. Furthermore, we can detect changes in the polymer melt temperature and density due to problems and/or alterations in the plasticization process or polymer raw material. The developed low-cost inspection system can be used for process monitoring, prediction of the final part quality and optimization of the plasticization parameters in scientific and industrial applications.

INTRODUCTION

In an injection molding process typically solid polymer granules are plasticized in a heated barrel containing a rotating screw and stored in the screw ante-chamber in front of the screw tip. The polymer melt is injected under high pressure into the mold cavity using the axial moved screw. After the cooling and solidification period the mold opens to allow part removal.

The homogeneity of the plasticized polymer melt in injection molding is often inadequate due to mechanical/structural inclusions and/or temperature gradients in axial and radial direction. Typically mechanical inhomogeneities in the polymer melt are gas bubbles, unmelted granules, wear particle from the plasticizing unit, dust and filler material agglomerations. Temperature gradients of the polymer melt in the screw ante-chamber arise due to the complex thermal history of the polymer melt in the screw channels and especially in axial direction because of the reduction of the active screw length during the plasticizing process. Both types of inhomogeneities can reduce the mechanical and optical properties of the final part. Due to the measurement boundary conditions (high pressure, high temperature) the non-invasive detection of mechanical and temperature inhomogeneities during the process is a challenging task. Nevertheless, temperature gradients in the screw ante-chamber can be measured by ultrasound based measurements in radial direction [1] and axial direction [2] or alternatively using infrared sensors and thermocouple mesh grids [3]. Mechanical/structural inclusion measurements were performed for example using optical methods [4] (e.g. detetection of gas bubbles) and ultrasound based methods [5] (e.g. detection of unmelted granules).

We developed an ultrasound based system for the simultaneous in-situ measurement of mechanical inhomogeneities and temperature gradients. The system (consisting of measurement adapter, ultrasound probes, pulser/receiver unit and signal processing) is working within the high-temperature and high-pressure environment of polymer processing without expensive and sensitive laboratory equipment.

OVERVIEW MELT INHOMOGENEITIES

Figure 1 shows an overview and classification of different detectable (using the proposed ultrasound based measurement system) types of inhomogeneities in the polymer melt in the screw ante-chamber after the plasticizing process.

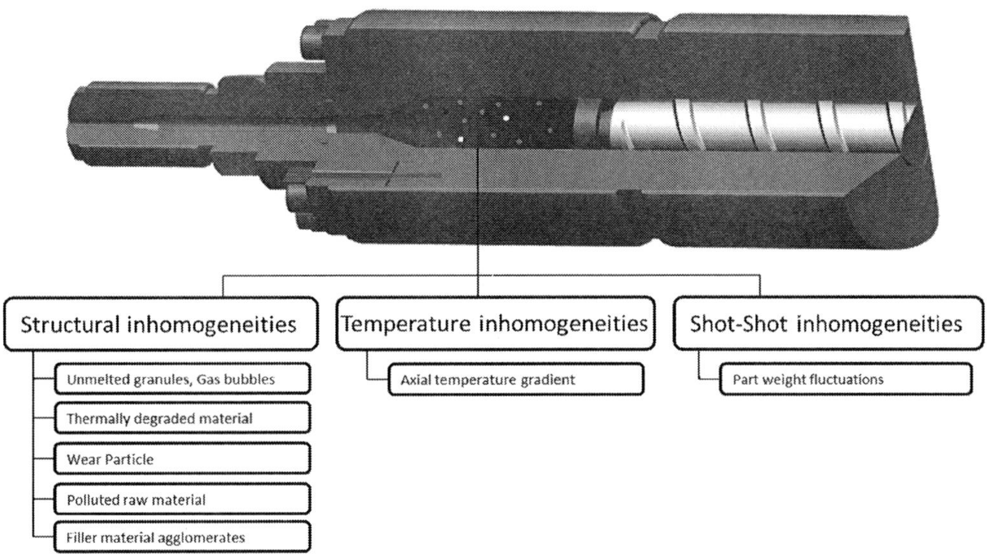

FIGURE 1. Overview and classification of different types of inhomogeneities in the polymer melt.

Strutural inhomogeneities

Mechanical/structural inclusions can reduce the optical and mechanical properties of the final part. Process or material depending inhomogeneities (unmelted granules, gas bubbles, thermally degraded material) can be avoided by suitable process parameter (especially barrel temperature, back pressure, dry raw material). For that reason a melt homogeneity sensor system can help to find the optimum process parameter for improving the final part quality and continuous monitor the injection molding process to find unavoidable inclusions like metallic wear particle from the plasticizing unit (non-return valve, screw flight and inner surface of the barrel) or filler material agglomerations.

Temperature inhomogeneities

Axial temperature gradients in the polymer melt are mainly due to the reduction of the active screw length during the plasticizing process. Typically, we can observe a decrease of polymer melt temperature from the nozzle to the tip of the screw. The resulting viscosity differences during the injection process can lead to flow irregularities in distribution systems and therefor to non-uniform filling of cavities. Furthermore, the temperature gradients can induce warpage effects at the final part due to different cooling conditions. The proposed ultrasound based system can evaluate the axial temperature gradients by the measurement of the temperature (and pressure) depending sound velocity in the polymer melt during the injection into the mold.

Shot-Shot inhomogeneities

Typically reasons for part weight fluctuations are shot-shot inhomogeneities due to a non-steady process (e.g. interruption of the process) or non-constant closing time of the non-return valve. By monitoring the sound velocity (which is closely related to the thermodynamic condition variables of the melt) the part weight uniformness can be predicted.

EXPERIMENTAL SETUP

Figure 2 (a) shows the schematic of the proposed ultrasound based measurement system. The ultrasound transmitter and receiver are mounted in a measurement adapter between the nozzle and barrel of the injection molding machine. The ultrasound transmitter requires a high-voltage (360 V) pulse generated by a pulser-receiver unit (Optel, OPBOX ver 2.0 - USB 2.0) to produce an ultrasound pulse with centre frequency 5 MHz. The electrical signal from the received ultrasonic pulse is processed by the analog-digital converter (100 MHz sampling rate, 10 bit) of the pulser-receiver unit. The maximum pulse repetition frequency rate is 1200 Hz. The control of the pulser/receiver unit and the signal processing of the received ultrasound signals were carried out in a single-board computer using Matlab.

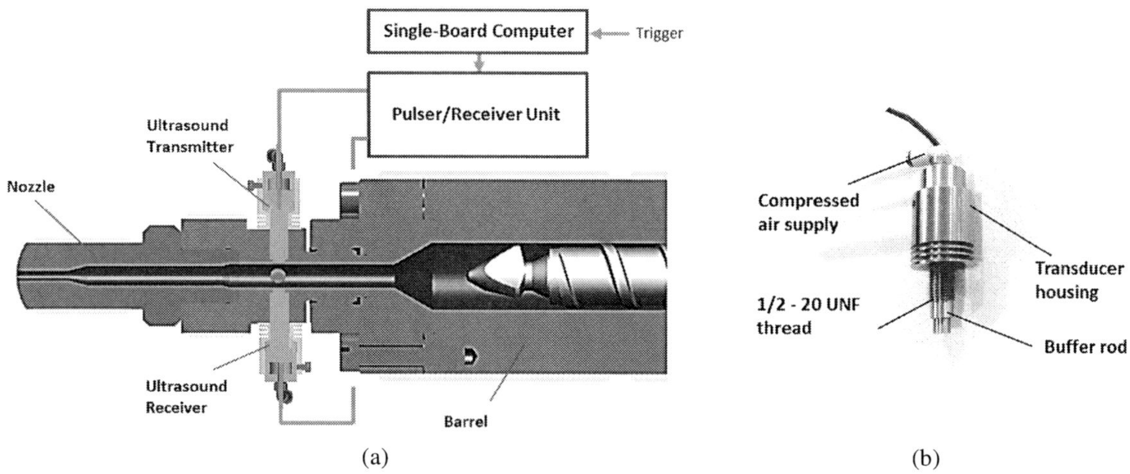

FIGURE 2. (a) Schematic of experimental setup and (b) developed high-temperature and high-pressure ultrasound probe.

Figure 2 (b) shows a photo of the (high-temperature and high-pressure) ultrasound probe we developed. A piezoelectric disc is placed behind a steel buffer rod and enclosed in an transducer housing. For high-temperature applications, the probe can be purged with cooled air via a flexible tube. The probe can be mounted by an ½ - 20 UNF thread in the measurement adapter.

All measurements were carried out on a 500 kN injection molding machine (ENGEL victory VC 200/50) at the University of Linz / Austria. A standard screw with a diameter of D = 30 mm and a length of 20D was used to process the polymer material (Polypropylene Borealis HD120MO).

DATA ANALYSIS

Sound energy is attenuated as it passes through the polymer melt because parts of it are absorbed, reflected, scattered, refracted or diffracted. Signal attenuation strongly depends on the type and diameter of inhomogeneity. For the detection of mechanical/structural inclusions we analyzed the intensity

$$I = \int |A(t_{inj})| dt \tag{1}$$

of the received pulse $A(t_{inj})$ during the injection process at the injection time t_{inj}. The melt temperature (gradients) can be determined by the simultaneous measurement of the longitudinal sound velocity $c_L(t_{inj})$ and the melt pressure $P(t_{inj})$ using a calibration function $T=f(c_L,P)$. Shot-Shot inhomogeneities can be detected by comparing the mean value $c_{L,mean}=mean(c_L(t_{inj}))$ between different injection mold cycles.

RESULTS AND DISCUSSION

Strutural inhomogeneities

For evaluation of the proposed system regarding the capability of the detection of structural/mechanical inhomogeneities in the polymer melt we contaminated the polymer granules in the feeder by different types of particles (dust, paper particle, PMMA granule, PA granule, ceramic powder, silicone particle). All specified inclusions were successfully detected. Figure 3 (a) shows the measured intensity $I(t_{inj})$ with 3 single paper particle (diameter > 0.5 mm) in the injected melt and the corresponding black spots in the molded test specimen (70 mm x 70 mm x 4 mm plate). The inclusions can be identified by the time resolved strong decrease of the intensity signal. In Figure 3 (b) the result of a Polypropylene melt contaminated by single PMMA granules is shown.

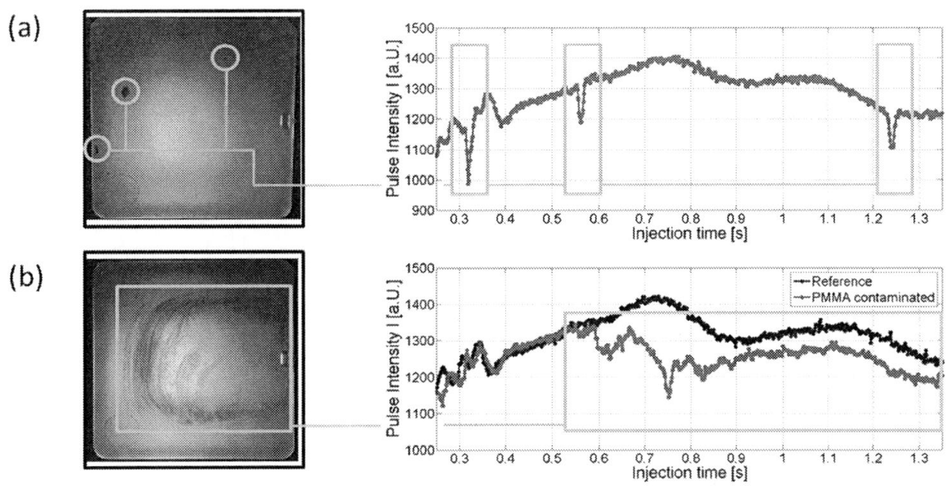

FIGURE 3. (a) Detected single paper particle in the polymer melt and resulting black spots in the molded test specimen. (b) Comparison of PMMA contaminated Polypropylene with reference measurement (pure Polypropylene) and resulting streaks in the molded test specimen.

In the molded test specimen streaks can be observed, the intensity signal (red line) is reduced in comparison to a reference measurement without inclusions (black line).

Temperature inhomogeneities

Figure 4 shows an example result of the axial melt temperature measurement during the injection molding process (using c_L and P).

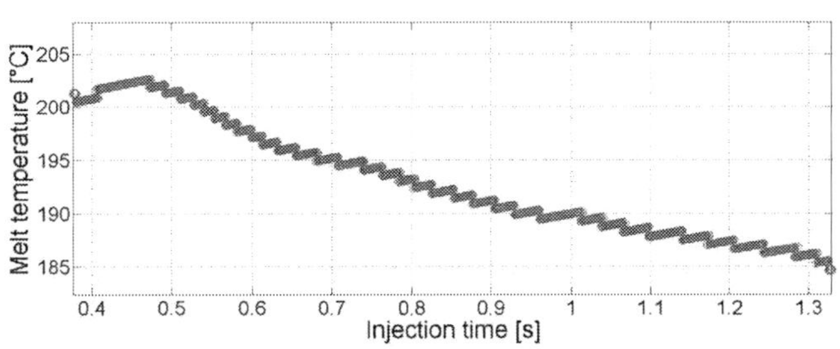

FIGURE 4. Example result of axial melt temperature measurement.

A temperature gradient $\Delta T > 15$ °C can be observed. The system can be used to optimize process parameter (for example back pressure) for reducing the temperature gradients in polymer melt.

Shot-Shot inhomogeneities

Figure 5 shows the possibility to detect part weight fluctuations by monitoring the mean sound velocity $c_{L,mean}$. In this experiment we changed the barrel temperature at cycle number 11 from $T_{barrel,max}$ = 180°C to $T_{barrel,max}$ = 200°C. At cycle number 19 the barrel reached the set-point temperature and at cycle number 25 we reached a steady state process.

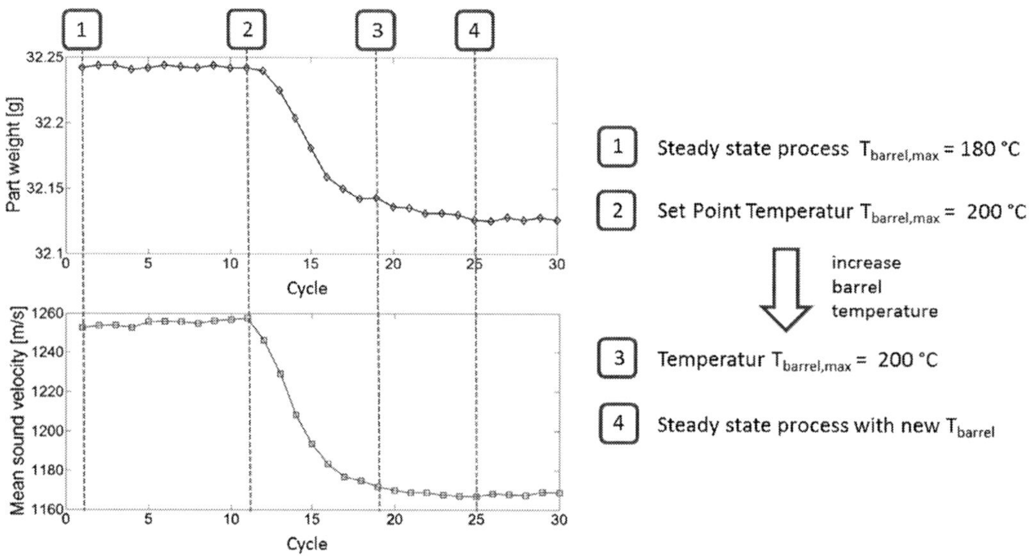

FIGURE 5. Monitoring of part weight fluctuations using the mean sound velocity $c_{L,mean}$.

Due to the close relationship of the sound velocity c_L with the density of the melt the proposed system can be used for the prediction of part weight fluctuations and detection of process parameter changes.

ACKNOWLEDGMENTS

We thank Michael Hettrich-Keller and Alfred Mayr for their support in constructing the measurement setup, and the tech2B GmbH for the Edison Silver Award 2017 for the project 'MoldSonics' presented to the Institute of Polymer Injection Moulding and Process Automation.

REFERENCES

1. B. Praher, K. Straka and G. Steinbichler, Meas. Sci. Technol. **24,** 084004 (2013)
2. K. Straka, B. Praher and G. Steinbichler, AIP Conf. Proc **1664**, 110010 (2015)
3. J. Vera-Sorroche, A.L. Kelly, E.C. Brown and P.D. Coates, Polym. Eng. Sci. **55**, 1059–1066 (2015)
4. M. Stephan, S. Große, M. Stintz and U. Blankschein, Plast. Rubber Compos. 35, 432 – 438 (2006)
5. B. Praher, K. Straka, J. Usanovic and G. Steinbichler, AIP Conf. Proc. **1593**, 159-162 (2014)

Nondestructive Testing of 3D-Hybrid Components Using Air-coupled Ultrasound

W. Essig [a] and M. Kreutzbruck

IKT, Institut für Kunststofftechnik, University of Stuttgart, Pfaffenwaldring 32, 70569 Stuttgart, Germany

[a]Corresponding author: wolfgang.essig@ikt.uni-stuttgart.de

Abstract. The increasing use of hybrid lightweight structures in the automotive industry results in new challenges in the area of multi-material interfaces. Insufficient bonding quality between two materials, especially in highly loaded structures, affects the component`s performance. Hence, it is very important to monitor the bonding quality during the multi-material component fabrication process obtaining a reliable save-life-design. Flat specimens, consisting of sheet steel and organic sheet layers were examined using air-coupled ultrasound transducers in reflection setup. By slanted incidence of the ultrasound on the profile`s surface, guided waves were excited in the specimen and delaminated areas could be visualized due to changed phase values in the received signal. Furthermore, a 3D-hybrid cap profile consisting of sheet steel, organic sheet layers and injection molded reinforcing ribs was examined using a special transducer setup. The received air-coupled guided wave signal reveals different signatures of bonded and delaminated organic sheets.

INTRODUCTION

At present, no quality assurance procedure has yet been implemented which may determine the material compounding and component quality of 3D-hybrid components. The non-destructive material and component testing (NDT) with air-coupled ultrasound is suitable for checking defects between joining partners [1]. The testing is carried out without direct contact, thus avoiding time consuming subsequent process steps such as coating, varnishing or gluing. In contrast to other methods of NDT, inline quality detection is also possible without extended safety precautions. However, the increasing complexity of the test objects will result in new challenges with respect to access, sufficient high signal-to-noise ratio and the involved defect recognition algorithm with adapted interpretation criteria. Consequently, the presented contactless ultrasound inspection approach has to pass through the standard validation procedure to prove its full capability and reliability within real 3D hybrid part testing scenarios.

HYBRID COMPONENTS

Hybrid components are compound systems that are functionally assembled from at least two different materials [2]. Hybrid components made of plastic and metal are primarily used in the automotive industry, communications technology and electrical engineering [3]. The metallic component in the interconnection system serves the strength and stiffness of the component, whereas the plastic component additionally stabilizes the component's cross-section and allows an extended function integration. This type of structure concept is partly used in serial vehicles to connect the two A-pillars. The form-locking connection of an aluminum support with a co-polyamide coating can support the dashboard, glove compartment and steering [4].

There are different hybrid types and structural concepts, which are predestined for particular structures with different requirements. Innovative hybrid structure concepts, in particular for flexurally stressed beams, are:

- Composite castings: hybrids made of sheet metal and metal castings
- Hybrids made of steel sheet and textile reinforced duroplastics
- Hybrids made of textile and long fiber reinforced thermoplastics
- 3D-hybrids

3D-hybrids are hybrid components made of metal sheet, organo-sheet and long-fiber reinforced thermoplastics (LFRP).

The metal sheet is planar reinforced with the organ sheet and stiffed by rips, made out of long fibre reinforced plastics. This results in a cross-sectional stabilization of the metal plate which is at risk of bending. Due to the load-adapted reinforcement, thinner metal plates can be used than with pure metal structures, which leads to a generally lighter component.

SPECIMENS

A 3D-hybrid component, which has been developed in the research project (SAB project) "3D hybrid technology", serves as a basis for the experiments. The component is a standardized bending beam of the construction space and mass specification P1. Figure 1 shows both sides of the component. The carrier mass corresponds to the normalized weight of 0.65 kg. The outer dimensions are also standardized with a length of 370 mm, a width of 85 mm and a height of 34 mm.

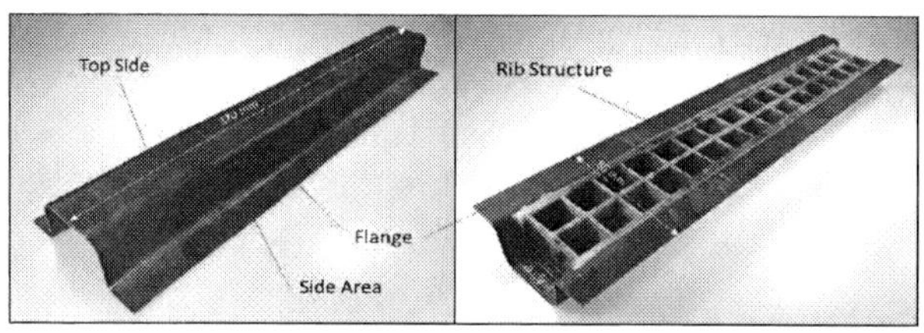

FIGURE 1. Standardized bending beam as a 3D-hybrid component

The steel plate has the designation "22MnB5" and is press-hardened. It is coated with an adhesive agent for the joining process. The adhesion agent is based on epoxy resin. The organo sheet has a polyamide matrix, the long-fiber reinforced thermoplastic (LFRP) is also based on polyamide. The ribs have an average distance of 20 mm along the component. The distance from the longitudinal axis is approximately 15 mm to 17 mm. The ribs have an average wall thickness of 3 mm.

In addition to the 3D-hybrid component, a preformed steel sheet was available for the examinations. This is also intended for a 3D-hybrid bending beam of the construction space and mass class P1. The semi-finished product is of the same material and has identical dimensions. The only difference is the lack of plastic. For a simulated detection of adherent plastic, a glass fiber reinforced polyamide tensile test rod with an epoxy resin adhesive was glued into the left side of the semi-finished steel product.

MEASUREMENT SETUP

Since the complex LFRP structure is only accessible on one side in in series production condition, the measurements are carried out in the re-emission setup. At first, the 3D-hybrid component is examined at the top. The measurement setup 1 is shown in Figure 2, left. The signal path runs in the x-direction. The flat flange of the component allows a further measurement setup. The measurement configuration 2 is shown in Figure 2,right. In this configuration, the Lamb waves are generated in the flange region and are detected on the upper side.

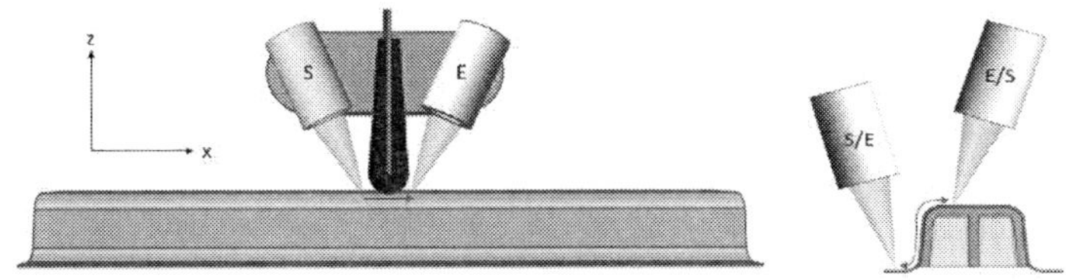

FIGURE 2. Measurement setup 1 measurment setup 2, right: measurement setup 3

The curves between the flange and the side wall have a radius of approx. 9 mm. The radius is about 7 mm between the side wall and the top of the component. Curved waveguides have different dispersion characteristics than plane waveguides. According to [10], curves can influence the wave propagation as soon as the ratio of the radius to plate thickness is less than 10:1. The ratios of the bending radius to the steel plate thickness (1 mm), respectively, to the composite thickness (3.5 mm to 5.5 mm) may therefore influence the wave propagation. Since the roundings are symmetrical and constant over the complete length of the component, their influence on the measuring results is neglected.

All shown measurements have been carried out with focussed transducers, produced by the ULTRAN Group ® and featuring a resonace frequency of 200 kHz.

RESULTS AND DISCUSSION

The C-scans generated are shown in Figure 3 and Figure 4 and show the tops of both test specimens. Figure 3 shows the 3D-hybrid component and Figure 4 illustrates the analogous C-scan of the semi-finished steel sheet, with the glass-fiber-reinforced polyamide being glued in the left half of the component. In both figures, a region of the middle rib features shows an increased amplitude along the x-direction (at $y \approx 50$ mm). A comparison with results from preliminary tests on steel plates reveals that this is a typical interference pattern due to edge reflections.

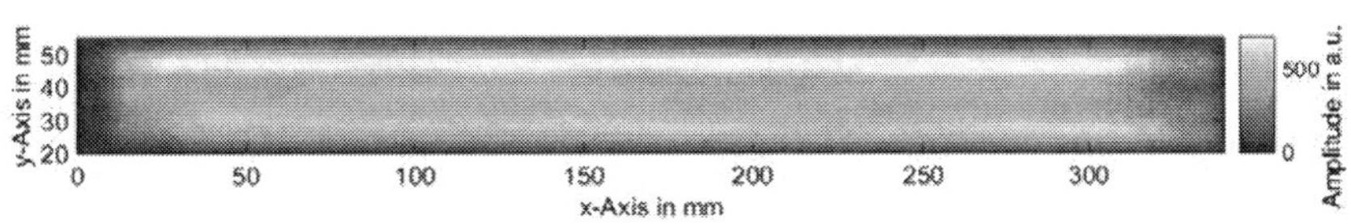

FIGURE 3. C-scan of the upper side of the 3D-hybrid component at measurement setup 1

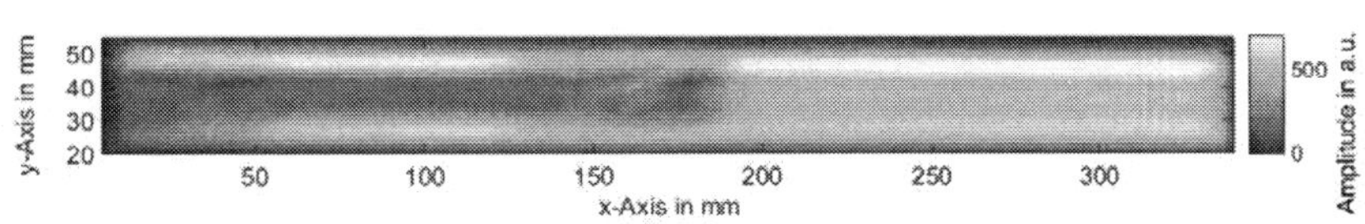

FIGURE 4. C-Scan of the upper side of the preformed sheet metal with the polyamide composite glued to the left side at measurement setup 1

The C-scan of the 3D-hybrid (Figure 3), especially the x-range between 100 mm and 150 mm, is similar to the C-scan of the pure semi-finished metal sheet. The similar gradients of the 3D-hybrid component and the semi-finished steel sheet suggest a primary influence of the steel component. The middle strip (around $y = 40$ mm) presumably does not show the middle rib of the component.

The sites with a low amplitude, for example at $x = 250$ mm and at $x = 50$ mm in Figure 3, could result from the damping of the adhering plastic. It is therefore assumed that large-area detachments could be found at the locations of increased amplitudes, such as in the areas where the interference strip is noticeable. This would also explain the significantly greater influence of the PA tensile sample on the amplitude profile.

Figure 5 shows A-scans of both test specimens taken with the measurement setup 2 (guided wave from the flange to the top of the profile). On the left, one can see the mode of the 3D-hybrid component and on the right side the mode of the semi-finished steel sheet. Hereby, the different amplitude heights and the reflected wave modes in the 3D-hybrid component are noticable.

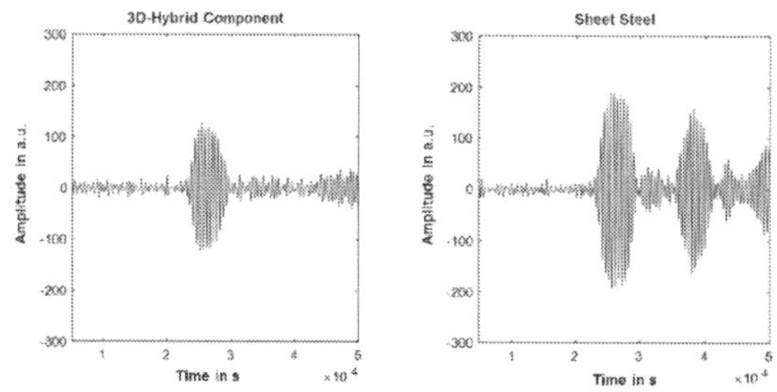

FIGURE 5. left: A-scan of the 3D-hybrid component right: A-scan of the semi-finished metal sheet bot at measurement setup 2

The corresponding C-scans are shown in Figure 6 and 7. The scan begins at x = 0 mm at the left end of the component. The transducer is moved to the vertical center of the specimen. The plastic portion of the 3D-hybrid has an influence on the amplitude profile (see alternating high and low amplitudes along the x-direction in Figure 6), when compared to an analogous C-scan of the pure steel sheet (Figure 7). The two C-scans were measured with identical parameters and the same angle of incidence immediately after each other.

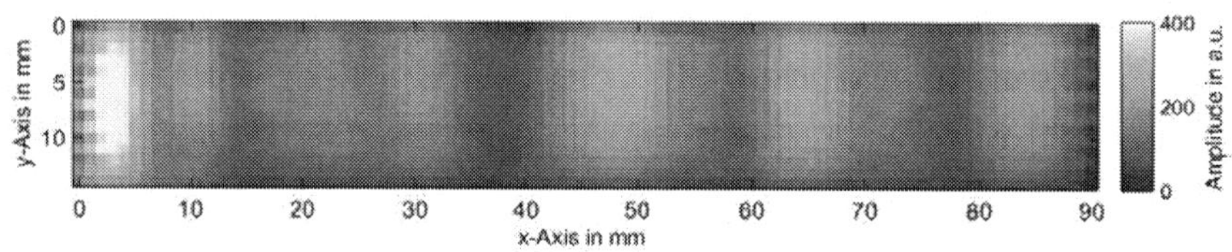

FIGURE 6. C-scan of the 3D-hybrid component at measurement setup 2

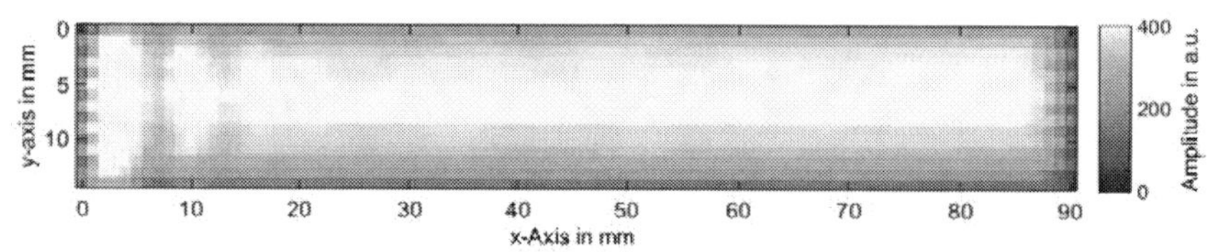

FIGURE 7. C-scan of the semi-finished sheet metal at measurement setup 2

Figure 8 shows the amplitudes of the semi-finished steel sheet and the 3D-hybrid components for y = 5 mm. The amplitude measured at the specimen with plastic content is weakened compared to the amplitude of the pure steel specimen. In both cases interference fringes appear due to edge effects. These can consequently also occur for strongly curved structures. In the case of the 3D-hybrid component, one can see a periodical pattern. The distance between the maxima suggests that these are the internal ribs.

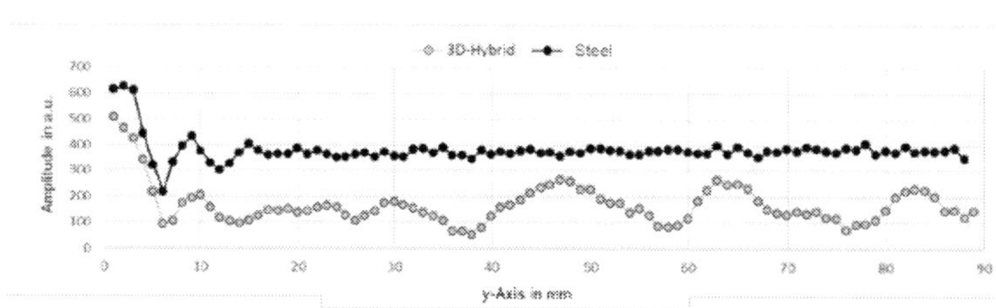

FIGURE 8. Amplitude comparison of both C-scans

CONCLUSION AND FUTURE WORKS

The experimental investigations conducted within this thesis propose a feasible approach to detect defects in metal-plastic compounds by means of air-coupled ultrasonic testing and guided waves.

On a 3D-hybrid component, in the form of a normalized beam carrier, several waves could be generated with the re-emission setup and different transducer configurations. A comparison with a steel half showed that the plastic components have an influence on the wave profile and the ribs are indicated in the C-scan.

At the same time, the investigations of the 3D-hybrid component revealed new challenges. The rib structure, for example, prevented double-sided accessibility for a two-sided examination. In addition, the highly curved roundings of the component are not suitable as coupling or radiation points. The less curved surfaces, which are suitable as potential points, are partly small-dimensioned. Focused transducers showed a larger potential than non-focused transducer and induced guided waves, which also can be detected in the strongly curved structures.

It has been shown that the dissipation of the guided waves in 3D hybrid structures is significantly lower than in steel. Nevertheless, the range could be suitable for a structure monitoring or an extensive point check.

It is critical to question whether the influence of the present phenomenon of amplitude damping at defective points is sufficient. In particular when defects are not present in large areas or are superimposed by other structural effects. The feasibility is currently investigated in other research activities. For this purpose, more suitable test objects with artificially introduced and defined damages were produced and tested.

ACKNOWLEDGMENTS

This research and development project is funded by the Federal Ministry of Education and Research (BMBF) under the concept "Innovations for tomorrow's production, services and work" (promotion code 02P14Z040 - 02P14Z049) and funds from the Energy and Climate Fund By the project sponsor Karlsruhe (PTKA).

REFERENCES

1. Kreutzbruck, M.: Herausforderungen an die zerstörungsfreie Prüfung im Trend moderner Leichtbaustrategien. In: 24. Stuttgarter Kunststoffkolloquium, Universität Stuttgart, Institut für Kunststofftechnik, 2015
2. Hybrid-Bauteile - Von allem das Beste. In: Kunststoff Magazin Online. URL: kunststoff-magazin.de/thermoplaste/hybrid-bauteile/von-allem-das-beste.htm Zul. Gepr.: 01.04.2012
3. Munde, A.: Hybridbauteile: Das Beste aus zwei Welten. URL: umformtechnik.net/Hybridbauteile-das-beste-aus-zwei-welten_21580_de/ Zul. Gepr.: 20.05.2016
4. Aluminium-Kunststoff-Hybrid geht in die Serienproduktion. In: Kunststoff Magazin Online. URL http://www.kunststoff-magazin.de/fvk-werkstoffe/aluminium-kunststoff-hybrid-geht-in-die-serienproduktion.htm Aktualisierungsdatum: 03.04.2013
5. Maaß, J.: 3D-Hybrid-Technologie - Höchstbelastete PKW-Karosseriestrukturen in Multi-Material-Design In: CARBON COMPOSITES MAGAZIN, 2nd edition, 2015
6. Ambrozinski, L., Piwakowski, B., Stepinski, T., Pieczonka, L., Uhl, T. Pitch-Catch Air-Coupled Ultrasonic Technique for Detection of Barely Visible Impact Damages in Composite Laminates. (2014)
7. Döring, D. Luftgekoppelter Ultraschall und geführte Wellen für die Anwendung in der Zerstörungsfreien Werkstoffprüfung [Dissertation]. Stuttgart, 2011: Universität Stuttgart, Institut für Kunststofftechnik
8. Gaal, M., Bartusch, J., Döring, J., Dohse, E., Lange, T., Hillger, W., Kreutzbruck, M. Air-Coupled Ferroelectret Ultrasonic Transducers Applied to Testing of Firber-Reinforced Polymers.
9. Gaal, M., Dohse, E., Bartusch, J., Köppe, E., Kreutzbruck, M., Hillger, W., Amos, J. Ultrasonic Testing of Adhesively Bonded Joints Using Air-Coupled Cellular Polypropylene Transducers.
10. Raddatz, F.; Flächige Analyse der Ausbreitung von Lambwellen an gekrümmten, anisotropen Strukturen, Seminar des Fachausschusses Ultraschallprüfung, Vortrag 6, 11. DGZfP Berlin, November 2013

Material Characterization within the Injection Molding Process

A. Geyer[a)] and C. Bonten

IKT, Institut für Kunststofftechnik, University of Stuttgart, Pfaffenwaldring 32, 70569 Stuttgart, Germany

[a)]Corresponding author: alexander.geyer@ikt.uni-stuttgart.de

Abstract. In injection molding, economic production can only be achieved with the highest possible process speeds, however the high quality of the final products has to be maintained. Therefore, not only the machines, but especially the materials, are continually further developed. These new or modified materials currently are mainly characterized in laboratory investigations. However, the transfer to the actual injection molding process is not well implemented. For this reason, the IKT researches new measuring methods, which allow the determination of various material properties directly in the injection molding process. This method would be a combined in-line viscosity measuring die and also an ejection force measuring mold in order to characterize the ejection behavior of various plastics and/or additives. In addition, the force on the mold core can be measured and these values allow us to draw conclusions about the shrinkage. The results provide means to further increase of the efficiency of the injection molding process.

INTRODUCTION

In plastic engineering, the injection molding process represents one of the most important processing techniques and allows manufacturing economically large amounts of parts with various intricate shapes and designs. However, to achieve a sufficient added value, the highest possible process speed must be guaranteed. At the same time, the quality of the parts must not be reduced. To reach a fair compromise between the two, not only the injection molding machines have to be improved, but also the materials. With the help of processing aids, or so-called additives, as well as material compounds, it is possible to either use new plastic materials in the injection molding process or to make existing materials even more effective. For this purpose, it is important to know how such new or modified materials behave during injection molding. This is usually estimated by various laboratory tests. However, laboratory tests often lack in practical relevance and do not predict exact information about the effects of additives on the materials. As a result, it is of great interest to be able to perform material characterizations directly within the injection molding machine. Thus application-related material data can be collected for comparison and the injection molding process can be further optimized so that it becomes even more economical.

STATE OF THE ART

This study focuses on the characterization of materials which were modified by means of internal and external lubricants as well as nucleating agents.

Processing Aids

In order to ensure high injection speeds without damaging the plastic, a low viscosity of the melts is required. This can be achieved by using so-called flow aids, also referred to as internal lubricants. These are low molecular weight compounds which have a similar polarity to the plastic itself, which makes them soluble in the melt. These compounds are interposed between the polymer chains and allow the molecular chains to slide off more easily, thus reducing shearing forces on the plastic melt as well as the friction heat. This leads to a better flow behavior and a better homogeneity of the melt, whereby shorter injection times can be realized. [1]

For a fast and clean ejection, it is important that the molded parts do not adhere to the cavity. In addition to correct injection, post pressures and draft angles, the lowest possible friction coefficients between the plastic and the mold steel are desired. This can be achieved by using external lubricants. These, in contrast to the internal lubricants, have higher molecular weights and an opposite polarity to the plastic, so that they are insoluble in the melt. During processing, they migrate to the hot interface between the plastic and the cavity, where they act as a release agent and thus optimize the ejection step. [2]

Furthermore it is important that the plastics have fallen below a certain temperature before ejection in order to reach a sufficient stiffness. Otherwise ejector marks or deformations are created which can cause component damage. The ejection temperature has a direct influence on the cooling time and is therefore a relevant factor in reducing the cycle time. Therefore, raising the temperature of the form stability would be of interest. In the case of semi-crystalline thermoplastics, increasing the crystallization temperature can be achieved by adding crystallization accelerators. With so-called nucleating agents, the degree of crystallization can be increased. This raises the crystallization temperature and shortens the cooling times. [3, S. 128]

Material Characterization

To be able to characterize the effect of additives on the plastics, various laboratory tests are performed. Rheometry is often used to characterize the flow properties. By means of rheometers, viscosity curves can be recorded and can show if internal lubricants do indeed lead to lower viscosities. The effects on the flow path length in the mold can be observed by using a so-called flow spiral. To determine the effects of external lubricants on the ejection step, friction values of the two friction partners (plastic/steel) are carried out at corresponding temperatures.

Crystallization effects are characterized with the aid of dynamic differential calorimetry (DSC). With a DSC analysis, it is possible determine melting- and glass-transition temperatures, degree of crystallization and crystallization temperatures. Thus, the effects of crystallization accelerators can be reproduced with this method. [4, S. 124]

All these methods require their own, partly costly and expensive laboratory equipment, samples and often special test specimens that require additional effort and time. At the same time, it is never possible to duplicate application-related conditions with laboratory tests.

Therefore, in this study a measurement methodology should be developed, which is integrated directly into the injection molding process. The aim is to optimize materials through additives in a way that the injection molding cycle is as short as possible while maintaining a high quality. In this paper, methods are presented which characterize the flow behavior of the plastic as well as the ejection behavior.

EXPERIMENTAL

Viscosity Measuring Die

With a so-called viscosity measuring die, the viscosity can be measured directly during injection. For the tests carried out, a die was developed and manufactured, which can be mounted on an Allrounder 520S 1600-400 injection molding machine from Arburg, Loßburg, Germany. The principle of this die is similar to that of a slit capillary rheometer. The constant volume flow of the melt is therefore achieved by the axial movement of the screw. The die has a rectangular duct with the dimensions 10 mm × 3 mm. Inside, three pressure sensors of the MDA series from Dynisco, Heilbronn, Germany, can be positioned with a measuring range of up to 1600 bars. The pressure sensors have a distance of 40 mm from each other. The structure of the nozzle is shown in Fig. 1 through a longitudinal cross section.

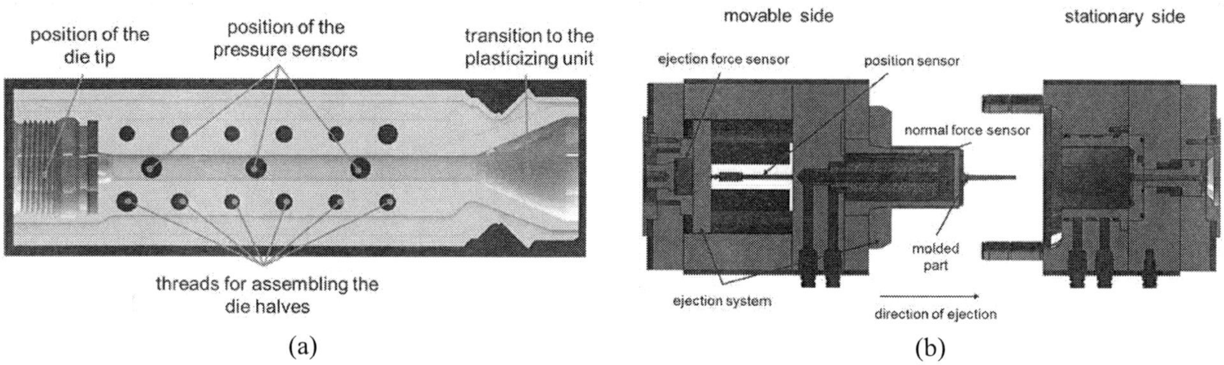

FIGURE 1. (a) Cross section of the viscosity measuring die. (b) Ejection force mold.

In order to characterize and predict the viscosity curves, the Carreau's approach [5] is used to describe unfilled thermoplastics.

$$\eta = \frac{a}{(1 + b \cdot \dot{\gamma})^c}. \qquad (1)$$

Here a represents the zero viscosity, parameter b is the inverse of the shear rate, and the dimensionless flow exponent c describes the gradient in the shear thinning region.

Ejection Force Measuring Mold

With the so-called ejection force mold developed by Koslowski [6] at IKT, the ejection force can be detected over the ejection path. It is a special cup mold without any draft angles and with an octahedral basic shape (Fig. 1b).

The ejection force can be determined by a DMS force sensor type 4577 A20 from Kistler Instruments, Winterthur, Switzerland. At the same time, the ejection path can be observed by a linear potentiometer of the LRW1 series from WayCon, Munich, which is mounted on the ejector unit. In addition to this, forces perpendicular to the mold core, which are the normal forces, are recorded by means of a miniature force sensor type 9211B from the company Kistler. The overall design of the mold is shown in Fig. 1b.

The characteristic parameters can be determined automatically during processing. The maximum ejection force should provide information about the stiction and adhesion. The gradient of the force over the path should give information about the kinetic friction. The normal force shows a renewed increase in the remaining cooling phase, which is due to shrinkage on the core and thus should provide information about the crystallization behavior.

Materials

In the present study, the used materials and additives can be seen in Table 1.

TABLE 1. Investigated materials

Name	Matrix	Additive
PP_A1	Polypropylene	0.5 wt.-% prototype copolymer wax 1
PP_A2	Polypropylene	0.5 wt.-% prototype copolymer wax 2
PP_A3	Polypropylene	0.5 wt.-% prototype copolymer wax 3
PP_A4	Polypropylene	0.5 wt.-% prototype polyester wax
PP_A5	Polypropylene	0.5 wt.-% prototype wax

RESULTS AND DISCUSSION

With validation trials the reliability of the measuring die could be shown, since it is possible to determine the viscosity curves for unfilled, as well as highly filled plastics. In addition, it is also possible to measure the viscosity for low shear rates when injecting without a mold and high shear rates when using a mold. [7]

In the following figures the results for some PP based modified plastics which were injected without a mold are shown. Figure 2a shows that the viscosity curves differ clearly from each other and thus the influence of the additives is measurable. For a better characterization, Carreau's law was attempted and analyzed. In Fig. 2b, the zero viscosity as well as the shear thinning, represented by the Carreau parameters *a* and *c*, are shown depending on the used additive.

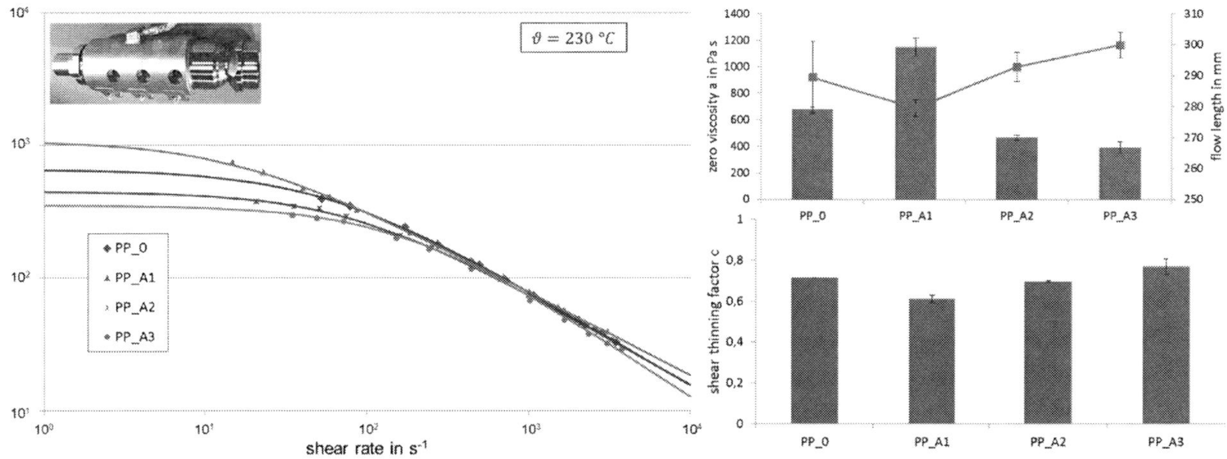

(a) (b)

FIGURE 2. (a) Viscosity curves for different modified PP. (b) Carreau parameter *a* and *c* in comparison to the flow path length.

In this case, additive A1 leads to the highest zero viscosity while having the lowest shear thinning. Additives A2 and A3, on the other hand, have lower zero viscosity values and a higher shear thinning factor. This result is reflected in the validation test using a flow spiral. The results show that additive A1 indeed has the lowest flow path length, while additive A3 has the highest one. This confirms that it is possible to characterize the effects of the additives on the flow behavior and it even allows us to predict the relative flow path length at the same processing conditions. An exact prediction of the flow path length is difficult, because additional factors, such as the temperature in the mold, have to be taken into account.

For characterizing the ejection behavior, the ejection force mold was used. The results of the maximum ejection force and the gradient of the force over the path can be seen in Fig 3. All the used additives lead to lower maximum ejection forces and, thus to lower stiction. Also, the negative gradient can be lowered by using those additives. This leads to the conclusion that all of the additives act as very good releasing agents.

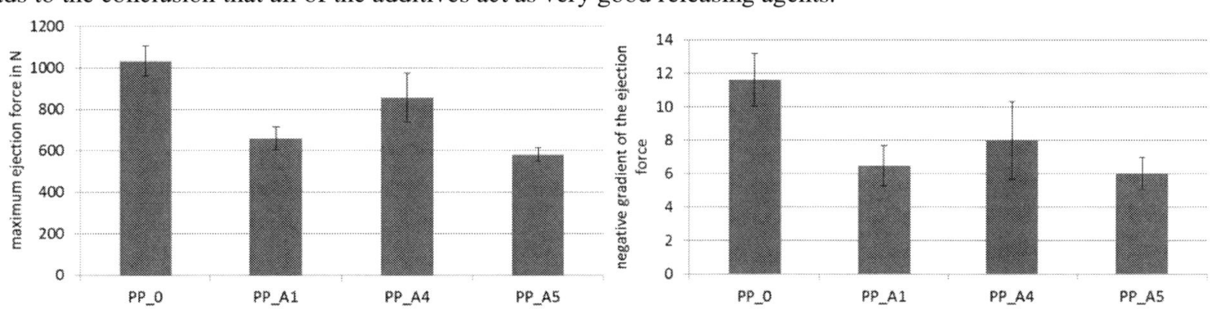

(a) (b)

FIGURE 3. (a) Maximum ejection foreces. (b) Negative gradient of the ejection force over ejection path.

As mentioned before, the pressure curve at the core shows a renewed increase in the remaining cooling time (Fig. 4a), which occurs due to shrinkage on the core. By looking at the maximum normal forces for the different compounds, there is hardly any difference between the natural PP and the modified types. Only additive A5 leads to a slightly higher maximum normal force. This could be explained by a higher degree of crystallization.

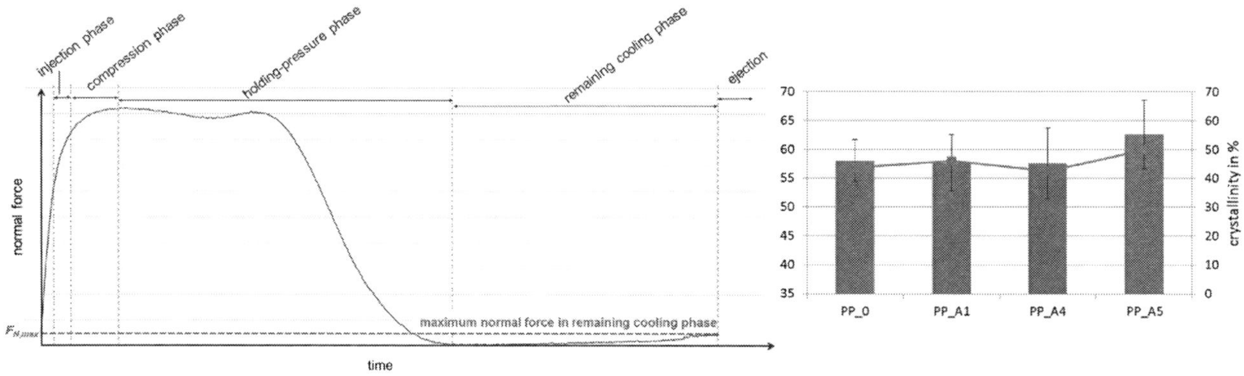

(a) (b)

FIGURE 4. (a) Normal force signal over one injection molding cycle. (b) Maximum normal force and crystallinity.

By means of differential scanning calometry (DSC), the crystallinity of the materials was calculated and once again, additive A5 presented slightly higher values. Thus, it seems reasonable that higher normal forces in the remaining cooling phase are a hint for higher degrees of crystallization of the material. Because the differences in these measurements were not significant, no definite statement can be made. In order to come to a definite conclusion, more materials have to be observed.

CONCLUSION

By means of the in-line measurement methods, various additives could be analyzed. On one hand, using an in-line viscosity measuring die could provide viscosity curves over a large range of shear rates. Furthermore, conclusions about the flow path length of the materials can be found.

The ejection behavior could be analyzed using a special cup mold. Determining the ejection force over the ejection path provides detailed information about the effect of processing aids on stiction and kinetic friction of the material. The results of a normal force sensor in the mold core provide information about the shrinkage and, as a result potentially even about the crystallization behavior. In order to validate this, more measurements are necessary with further materials. The main aim is to combine both measuring devices into a one-step additive characterization tool.

REFERENCES

1. J. Murphy, *Additives for plastics handbook* (Elsevier, Amsterdam, 2001).
2. L. Matthies and F. Preusser, *Mehr als nur ein Gleitmittel,* Kunststoffe 103 (2013) 11.
3. P. Eyerer; T. Hirth and P. Elsner, *Polymer Engineering: Technologien und Praxis* (Springer, Berlin, 2008).
4. C. Bonten, *Kunststofftechnik: Einführung und Grundlagen* 2nd ed. (Hanser, Munich, 2016).
5. P. J. Carreau, *Rheological equations from molecular network theories*, Transactions of the Society of Rheology, 1972, No. 1, p. 99–127.
6. T. Koslowski and C. Bonten, *Neue Inline-Messverfahren zur Bewertung von Verarbeitungshilfsstoffen im Spritzgießprozess*, in 24. Stuttgarter Kunststoffkolloquium (Stuttgart, 2015).
7. A. Geyer and C. Bonten, *Werkstoffcharakterisierung im Spritzgießprozess,* 25. Stuttgarter Kunststoffkolloquium (Stuttgart, 2017).

Non-contact Inline Monitoring of Thermoplastic CFRP Tape Quality Using Air-coupled Ultrasound

W.Essig [a)], P.Fey, S. Meiler and M. Kreutzbruck

IKT, Institut für Kunststofftechnik, University of Stuttgart, Pfaffenwaldring 32, 70569 Stuttgart, Germany

[a)]Corresponding author: wolfgang.essig@ikt.uni-stuttgart.de

Abstract. Thin (0.1 mm) thermoplastic continuous fiber CFRP tapes with a width of 100 mm were examined using air-coupled ultrasound. Transducers were arranged in reflection as well as transmission setup. By slanted incidence of the ultrasound on the tape surface, guided waves were excited in the material in fiber direction and perpendicular to the fiber direction. Artificial defects – fiber cuts, matrix cuts, circular holes, low velocity impacts from tool drop, and sharp bends – were produced. Experiments on a stationary tape showed good detectability of all artificial defects by guided waves. Also the effects of variation in material properties, fiber volume content and fiber matrix adhesion being the most relevant, on guided wave propagation were examined, to allow for quality assessment. Guided wave measurements were supported by destructive analysis. Also an apparatus containing one endless loop of CFRP tape was constructed and built to simulate inline testing of CFRP tapes, as it would be employed in a CFRP tape production environment or at a CFRP tape processing facility. The influences of tape conveying speed on detectability of artificial defects as well as material properties were elaborated and recommendations for implementation in production scale inline monitoring are given.

INTRODUCTION

Whilst most prepregs are manufactured from carbon-fiber and thermosetting matrix, latest developments tend to replace the thermosetting with a thermoplastic matrix. The most common method of production is the pultrusion process, impregnating the fibers with molten thermoplast pellets. After the pultrusion process the carbon-fiber reinforced prepreg with thermoplastic matrix (CFR-Tape) is wound on a spool and ready for further processing [1].

For the analysis of the tapes air-coupled ultrasound testing was chosen because it is a well-known technique, it does not need any radiation shielding or coupling agent and offers a range of possible experimental set-ups [2]. Besides that, non-destructive testing may be part of an inline-quality inspection, just as demanded within the upcoming concept of "industry 4.0" [3]. Recent research was performed on the electromechanical response of polypropylene ferroelectret transducers [4], air-coupled cellular polypropylene transducers as mentioned in [5], as well as the resonance-free measuring and excitation of ultrasound [6].

SIMULATION OF ULTRASOUND PROPAGATION

Consisting of two materials, carbon fiber and polyamide 6, with quite different elastic properties the tape shows anisotropic material behavior similar to other composite materials. For the simulation of ultrasound propagation, the computer program "Disperse" [7], developed at the Imperial College London was used. With the program only accepting a single set of homogenized material parameters, the elastic properties of both materials have to be combined, applying Puck's and Jones' laws [7].

TABLE 1: Calculated phase velocities for f = 500 kHz

	v_{a0} acc. to Puck in m/s	v_{a0} acc. to Jones in m/s
longitudinal	888.3	980.6

TAPE SPECIMENS

The tape with a thickness of 0.18 mm was manufactured from UD carbon fibers and the widespread polyamide 6 with an intended fiber volume ratio of 50 %. In order to achieve comprehensive results, five different specimens with specific properties were prepared. Besides the detection of defects, a second task was to find a correlation between ultrasound data and the actual fiber volume ratio of the tape. The purpose of the first specimen, named TP1, was to serve as a standardized sample with specific defects to test different experimental set-ups. Its schematic representation is given in Figure 1. The second specimen (TP2) featured an incorrect fiber volume ratio over entire length, the third specimen (TP3) showed a transition from correct to incorrect fiber volume ratio with the fourth specimen (TP4) being a flawless specimen. The fifth and last specimen (TP5) was manufactured as a flawless continuous tape with a length of 2 m.

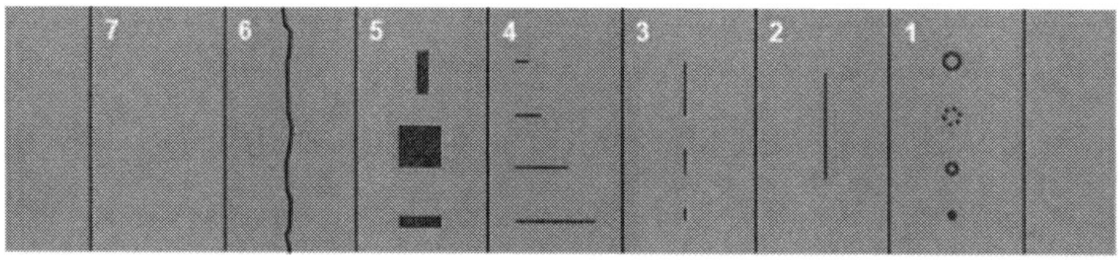

FIGURE 1: TP1, defect pattern

In the standardized specimen TP1, the first segment included holes with four different diameters, the second one featured one long cut perpendicular to the fiber orientation with a length of 40 mm. The third section displayed three cuts perpendicular to the fiber orientation with a length of 5 mm, 10 mm and 20 mm. The fourth segment showed four cuts parallel to the fiber orientation with a length of 5 mm, 10 mm, 20 mm and 30 mm, whereas the fifth segment featured three impact damages. The second to last segment included a crack perpendicular to the fiber orientation with the last section serving as a flawless reference zone.

ULTRASOUND TEST EQUIPMENT

For the experiments two ultrasound transducers manufactured by the Ultran Group were used. The first one, namely NCG500-S50, has an active area of 50 mm by 50 mm and served as the transmitter, whilst the other transducer, called NCG500-S25, with an active area of 25 mm by 25 mm was used as the receiver. Both transducers were driven by a Dasel Airscope TT, wired to a laptop which was used for the ultrasound user interface and for programming the control device of a scanning table. For testing the samples were mounted to an adjustable frame.

Measuring Methods

Two different alignment methods for the transducers have been tested within the analysis of TP1, revealing that single-sided ultrasound testing is unsuited for testing CFR-Tape. Consequently only two-sided alignment was applied to the other specimens with variation of transducer to surface angle from 27° up to 43° and variation of ultrasound coupling parallel or perpendicular to the fiber orientation.

Continuous Tape Test Rig

FIGURE 2: Experimental set-up for testing continuous tapes

A continuous tape test rig was developed in order to simulate the inline testing in the production process. The test rig was designed to accept specimens with a length of up to 1880 mm and to be run at speeds of up to 10 m min^{-1}. Aluminum profiles were used for the construction as well as acrylic glass elements. It features four cylinders directing the tape with one of them being driven by a stepper motor equipped with an encoder.

ULTRASOUND DATA EVALUATION

The data evaluation program based on Matlab was designed to function as a quality gate. After cutting off the interferences at the edges of the tape, the average value as well as the standard deviation for both the amplitude and the phase image is calculated. Moreover the relative standard deviation in the entire range of values is computed. In the program all pixels that are not in the range average value plus/minus standard deviation are marked as defect. After cleaning the image from very small defects using morphological processes, the ratio of area indicated per area examined is calculated. If the size of the indicated area exceeds a certain value, the tape also no longer matches the quality standards. Linking both methods, a good method for quality testing of CFR-Tapes was obtained. The whole process is shown in Figure 3.

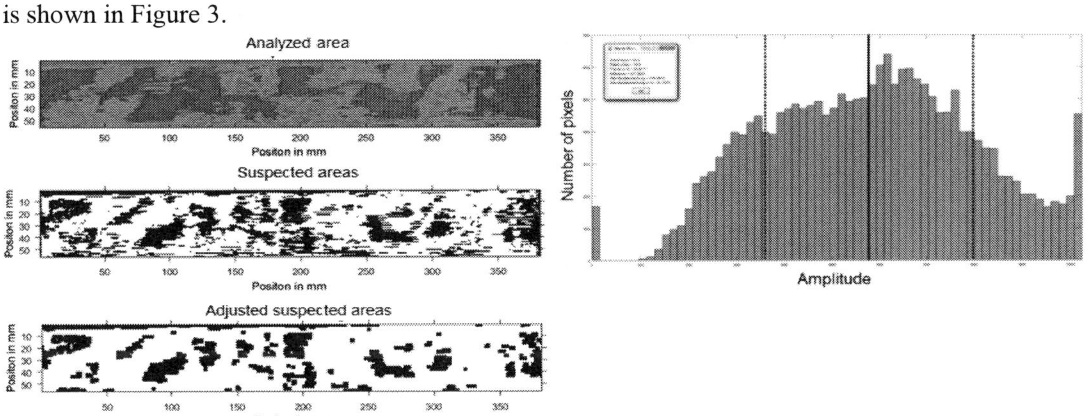

FIGURE 3: left: Amplitude image of TP2 with image processing shown to mark suspicious areas; right: Histogram of specimen TP2 with a separate window giving information about average value (solid line) and standard deviation (dashed lines) as well as suspected surface area.

Analysis of Fiber Volume Ratio

According to [8], it is possible to ascertain the fiber volume ratio by examining micrographs of the tape. After the microscopy, the micrographs are extracted from the surrounding microscope picture and the contrast of the image is increased. The fibers appear in white color whereas the matrix remains black. A program clears the image from smaller particles and calculates the total area of the cross-section, as well as the area of the fibers. With knowledge of both figures it is possible to give an estimate of the fiber volume ratio.

Maximum tape velocity

The maximum tape velocity was determined by observing the A-scan image of the specimen TP5 while increasing the tape velocity until gaps between two measuring points became visible on the user interface of the Airscope TT. Applying this method, the tape speed topped at 5.35 m min^{-1}. This limit is due to the used ultrasound system, it is not a physical limit.

RESULTS AND DISCUSSION

Artificial Defects and Decreasing Fiber Volume Ratio

Both, wave propagation parallel and perpendicular to the fiber orientation provided good data. As expected, defects with perpendicular orientation to the ultrasound wave propagation direction were detected more clearly than those oriented parallel to it. For the type of tape investigated the best results were gained with a transducer-to-surface angle of 27°, measured from the vertical direction. An example of an amplitude image of the specimen TP1 is given in Figure 4.

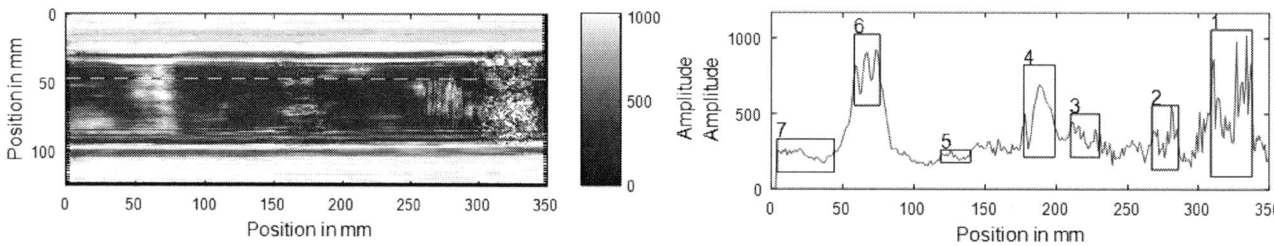

FIGURE 4: left: amplitude image of specimen TP1 measuring parallel to the fiber orientation; right: amplitude profile along line in c-scan

For the analysis of the tape TP3 with the transition from a good to a bad fiber volume ratio the tape was separated into eight sections and the amplitude values as well as the standard deviations were calculated for each sector. Within the flawless sections the amplitude values and the standard deviation remained steady. Regarding the sections displaying an incorrect fiber volume ratio, both the average values and the standard deviations varied as seen in Figure 5.

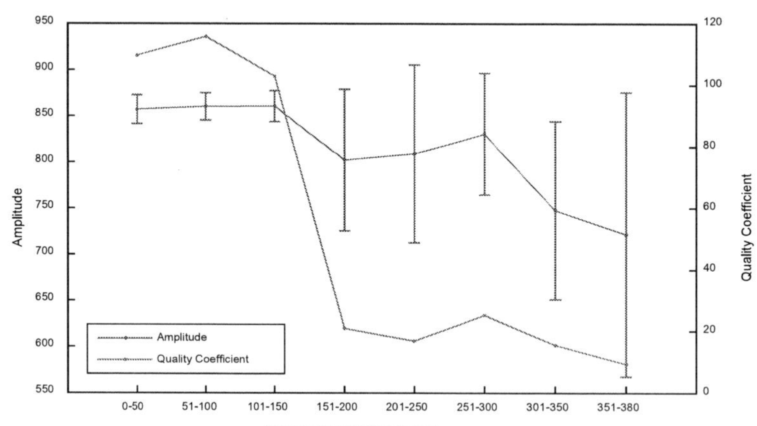

FIGURE 5: Specimen TP3; progress of the average amplitude value and standard deviation

With the amplitude values dropping and the standard deviation increasing for the sections with incorrect fiber volume ratio, the hypothesis about using the average values and the standard deviation as a quality gate was supported. This led to the definition of an empirical quality coefficient ***2mean/std***, which easily separates to flawless sections of the tape from those with incorrect fiber volume ratio.

Analysis of Fiber Volume Ratio and Simulated In-line Inspection

By overlaying a photograph of the second specimen with a transparent ultrasound image in an image processing software, it became apparent that the ultrasound image corresponds quite well with the visual pattern of incorrect fiber volume ratio.

Comparing the specimens with an incorrect fiber volume ratio to an optically flawless tape, the measurements showed exceptional variations. Consequently, no clear statement regarding a correlation can be made, although the calculated fiber volume ratio of the flawless specimen was measured in fact higher than the fiber volume ratio of the specimens with an obviously incorrect fiber volume ratio. According to [8], the method should be capable of exactly determining the fiber volume ratio, but further examination has to be done, in order to adjust the method to CFR-Tapes.

As the method of non-destructive testing with air coupled ultrasound should be evaluated as a quality assurance method, the process should be capable of adapting to different tape speeds. This fact was proven by running the tape at three different speeds. For measurements at 2.17 m min^{-1}, 2.97 m min^{-1} and 4.41 m min^{-1}, the values of the amplitude varied only by 0.4 % between measurements, whereas the values of the phase varied by 1.4 %. These results state clearly that the testing method results are independent of the tape production speed.

CONCLUSION AND FUTURE WORKS

Air coupled ultrasound is capable of being applied as a quality assurance element for thermoplastic CFR tapes. For future application it is essential to increase the spatial resolution by using focused or phased array transducers. Regarding the correlation between the fiber volume ratio and the measured ultrasound data, further examination needs to be performed.

REFERENCES

1. Lang, K. Zur Erweiterung des Spektrums von Faser-Kunststoff-Verbunden mit thermoplastischer Matrix. Einsatz industriell faserverstärkter Halbzeuge in hochbelasteten Strukturbauteilen des Automobil- und Flugzeugbaus. Ph.D. thesis. Darmstadt, 31. Jan. 2012.

2. Peters, J., V. Kommareddy, Z. Liu, D. F and D. Hsu. Non-Contact Inspection of Composites Using Air-Coupled Ultrasound. In: REVIEW OF PROGRESS IN QUANTITATIVE NONDESTRUCTIVE EVALUATION:Volume 22, S. pp. 973-980.

3. Hermann, M., T. Pentek und B. Otto. Design Principles for Industrie 4.0 Scenarios. A Literature Review. Dortmund.

4. Döring, J., V. Bovtun, M. Gaal, J. Bartusch, A. Erhard, M. Kreutzbruck and Y. Yakymenko. Piezoelectric and electrostrictive effects in ferroelectret ultrasonic transducers. In: Journal of Applied Physics, 2012, 112(8), S. 84505. Doi:10.1063/1.4759052

5. Mate Gaal, Elmar Dohse, Jürgen Bartusch, Enrico Köppe, Marc Kreutzbruck, Wolfgang Hillger and Jay Amos. Ultrasonic testing of adhesively bonded joints using air-coupled cellular polypropylene transducers.

6. Daschewski, M., M. Kreutzbruck, J. Prager, E. Dohse, M. Gaal and A. Harrer. Resonanzfreie Messung und Anregung von Ultraschall. In: tm - Technisches Messen, 2015, 82(3). Doi:10.1515/teme-2014-0020

7. Stellbrink, K.K.U. Micromechanics of composites. Composite properties of fiber and matrix constituents. Munich: Hanser, 1996. ISBN 1569902062.

8. George, A. Optimization of Resin Infusion Processing for Composite Materials. Simulation and Characterization Strategies. Dissertation. Stuttgart, 15. Mrz. 2011.

9. Blackstock, D.T. Fundamentals of physical acoustics. New York, NY: Wiley, 2000. A Wiley-Interscience publication. ISBN 978-0-471-31979-5.

Local Defect Resonance Excitation Thermography for Damage Detection in Plastic Composites

Markus Rahammer[1, a)] and Marc Kreutzbruck[1]

[1]*Institut für Kunststofftechnik, University of Stuttgart, Pfaffenwaldring 32, 70569 Stuttgart, Germany*
[a)]*Corresponding author: markus.rahammer@ikt.uni-stuttgart.de*

Abstract. Ultrasonic excitation thermography testing (UTT) is a well-established non-destructive testing (NDT) method that excels in its high defect selectivity in very short measurement times. The former is advantageous for providing clear and user-friendly results. Especially in the case of cracks and crack-like defects like delaminations, a defect signal can be expected after just a few seconds independently of the material tested. While common ultrasonic testing devices use high frequencies (MHz), UTT typically utilizes high power sonotrodes with kHz-frequencies adopted from the ultrasonic welding industry. Usually high electric and acoustic powers are necessary to provide sufficient energy to the defect. The principle of local defect resonance opens up new methods to significantly reduce the energy required and enables the employment of small piezo-based transducers. Defects are modelled as distinct volumes with a fixed mass and geometry that have their own resonance frequencies. Flooding the specimen with ultrasound of that frequency leads to a distinct excitation of the defect vibration while the residual specimen area is resting. Due to the small mass vibrating only a very small amount of excitation energy is needed. This energy-efficient excitation due to the resonant vibration can be used for UTT, because all common heating mechanisms (hysteresis, crack tip friction) respond. Experiments have shown that acoustic energies in the range of mW suffice, in order to produce temperature increases greater than 1 K. The largest hindrance so far has been the determination of LDR frequencies of unknown defects. However, the upper and lower limit of these frequencies depends on the material and the defect size and is known due to a large experimental base. Using a broadband sweep excitation in this frequency range this method can be applied even to unknown parts with delaminations of few mm size and bigger.

INTRODUCTION

Ultrasonic excitation thermography testing is well known for its defect selective, dark-field damage detection capability, yet it is currently only used with bulky high power equipment. In order to advance this method from its current lab status to a reliable and certified NDT-procedure a reduction in energy consumption combined with improved defect activation and signal processing is necessary. One way to achieve this goal is the combination of Ultrasonic excitation thermography testing with the concept of local defect resonance (LDR).

Local Defect Resonance

The concept of LDR has been first investigated and described in [1, 2]. Local defect resonances occur on the basis that the presence of a defect leads to a local decrease in stiffness for a certain mass of the material in this area. Using a frequency match of the driving ultrasonic source and the resonance frequency of the local defect area leads to a very efficient energy pumping into the defect area. This results in high amplifications of the defect vibration amplitude as compared to the residual sound specimen area. There is a distinct relation between the defect size and its resonance frequency. Eq. (1) was first derived in [2] for cylindrical defects and relates the defect geometry (residual thickness H, radius r) and the material parameters (Young's modulus E, density ρ and Poisson's number υ) to the LDR frequency f0.

$$f_0 \approx \frac{1.6H}{r^2}\sqrt{\frac{E}{12\rho(1-\upsilon^2)}} \qquad \text{i)}$$

Later, an equivalent solution (eq. (2)) for quadratic-shaped defects with side length s was found [3]:

$$f_0 \approx \frac{4\pi H}{3s^2}\sqrt{\frac{E}{6\rho(1-\upsilon^2)}} \qquad \text{ii)}$$

Recent contributions of other groups in the subject demonstrated the benefit of LDR for inspection of delaminations [4], impact damage [5] and kissing bonds [6].

Due to the high efficiency that LDR can provide, a low input power (< 1 W electrical power) is sufficient to activate the defects. This opens up applications with inexpensive excitation sources, e.g. simple piezo disc transducers, or very flexible solutions like vacuum attached piezo shakers.

While using laser scanning vibrometry is the most direct way for visualizing vibration patterns, other useful methods of visualizing local defect resonances exist. Interferometric techniques, for instance shearography and thermal techniques like Ultrasonic excitation thermography [7] can provide much faster and in particular more cost-efficient imaging approaches.

Ultrasonic excitation thermography using LDR

Ultrasonic excitation thermography, also known as acoustic thermography, vibrothermography, sonic IR or thermosonics, is a well-established NDT technique mainly used for finding cracks in various materials. Next to crack friction, when it comes to polymer-based materials the visco-elastic effect generates significant heat during vibrations. A sample application is shown in Fig. 1. The part tested is a CFRP rim with several small cracks and delamination due to destructive fatigue testing. The sound is introduced via a high power ultrasonic welding sonotrode seen in the center of the image. While the contact area has the highest temperature, several spots of heat due to crack friction are found as well.

FIGURE 1. Photograph (left) and UTT result (right) of a CFRP rim (image source: Greenteam Stuttgart)

Ultrasonic excitation thermography's main advantage is the fast and simple employment in conjunction with an unambiguous measurement result. Recent progress has shown that the use of high power ultrasonic equipment is not necessary, when the aforementioned principle of LDR is incorporated [8]. Low input powers of a few W electrical are often sufficient for activating the defect to vibration amplitudes high enough for production of measurable heat. Defect resonances are typically of a very high Q-factor. For this reason, even a small deviation from the resonance frequency will result in a significant drop of amplitude. Results from [7-8] illustrate this effect for different applications.

The principle of LDR in itself is a step forward towards improving the efficiency of ultrasonic thermography. Further research is done in advancing this method hardware-wise, for example by using guided waves to channel the ultrasonic energy more efficiently towards the defect area [9]. Defect resonance frequencies are usually unknown and require substantial time to determine via wideband excited vibrometry even with analytical estimates in place beforehand. Therefore, a method called Resonance Frequency Sweep Thermography (RFST) using broadband excitation signals was introduced in [10].

EXPERIMENTAL

All results shown in the following chapter use RFST as described in [10]. A broadband sound signal is introduced to the parts by a vacuum-attachable transducer. The signal is a sweep signal of low modulation frequency and is repeated once, in order to produce a periodic temperature rise. The thermal signal from the surface is recorded with an IRCAM Equus 327 pro thermal imaging camera (640 x 512 pixels, <20 mK NETD, 100 Hz framerate). The recorded signal is then evaluated with a Fourier transform at one over the sweep time. The testing parameters for all results were 50 V excitation amplitude and 5–50 kHz ultrasonic frequency.

RESULTS

In this chapter, two examples are given for the application of RFST to CFRP parts.

CFRP Part with Delamination due to Heat Damage

This quadratic (295 x 295 x 4.5 mm^3) CFRP plate of unknown layup includes nine delaminations of different sizes. They were introduced by local overheating. The first step was to investigate possible defect resonance frequencies. This was done with a vibrometry scan of the defect areas. The vibrometry result is shown in Fig. 2. The false color coding shows the vibration amplitude of the scanned areas from green (low amplitude) to red (high amplitude). The vacuum-attached transducer is also seen in the picture.

The corresponding amplitude spectrums for the frequency band of interest are shown in the left part of Fig. 2. They reveal that both the defects have different resonance frequencies, ca. 19 kHz for defect no. 2 and ca. 24.5 kHz for the 4th defect. This was to be expected, because the defect differ in size.

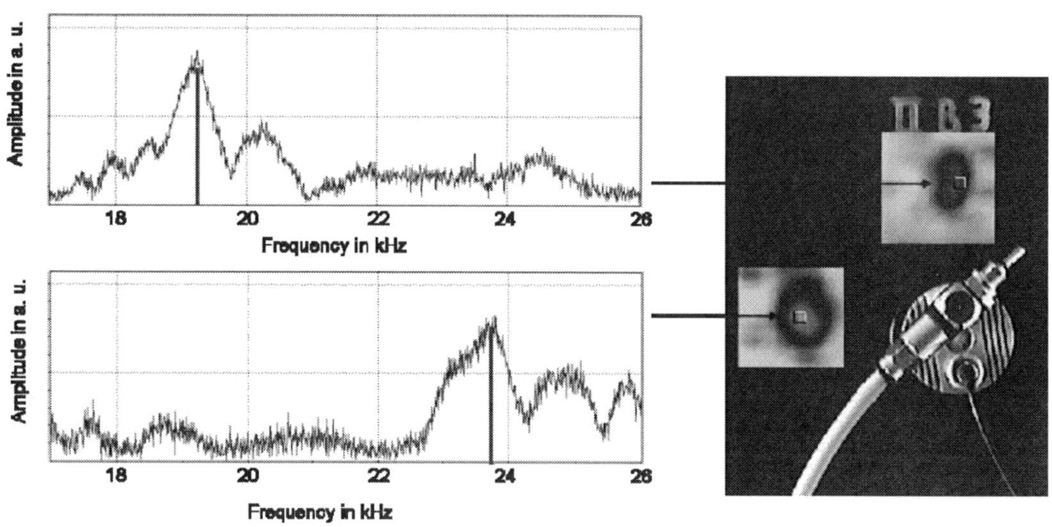

FIGURE 2. Laser Scanning Vibrometry result of two defects of the CFRP plate (right) and corresponding amplitude spectrums for the defects (left)

The amplitude and phase image results for an RFST measurement with the parameters described before are shown in Fig. 3. The total testing time was 30 s with three repetitions of a 10 s sweep with the frequency band shown in Fig. 3. The Fourier frequency was therefore 0.1 Hz. While the amplitude image reveals the defects clearly with very high signal-to-noise ratios (SNR), the phase image includes additional characterizing information. The different phase angels of the defects correspond to the resonance occurrence times with the spectrums. The 2nd defect (on the right in Fig. 4) has a smaller phase angle than the 4th, because its temperature raise happened earlier during the 10 s sweep excitation.

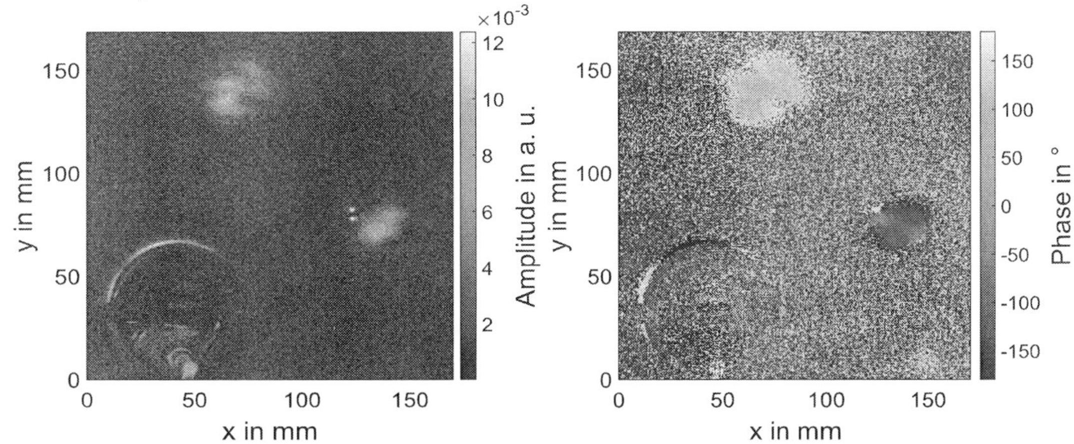

FIGURE 3. Amplitude (left) and phase (right) images of the RFST measurement of the CFRP plate

CFRP Maintenance Access Hatch

The CFRP maintenance access hatch from an aircraft is a real world sample application for damage detection by NDT. A photograph of the test setup with the transducer is shown in Fig. 4.

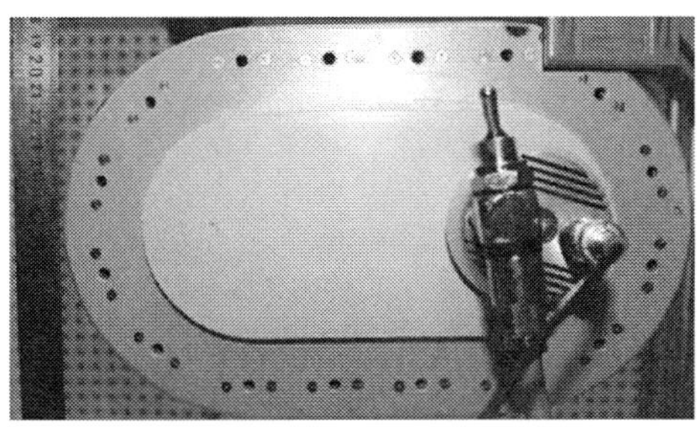

FIGURE 4. CFRP maintenance access hatch with vacuum transducer

This specimen has several large cracks and delamination. There extend cannot be determined from visual inspection. The results of the RFST testing are shown in Fig. 5.

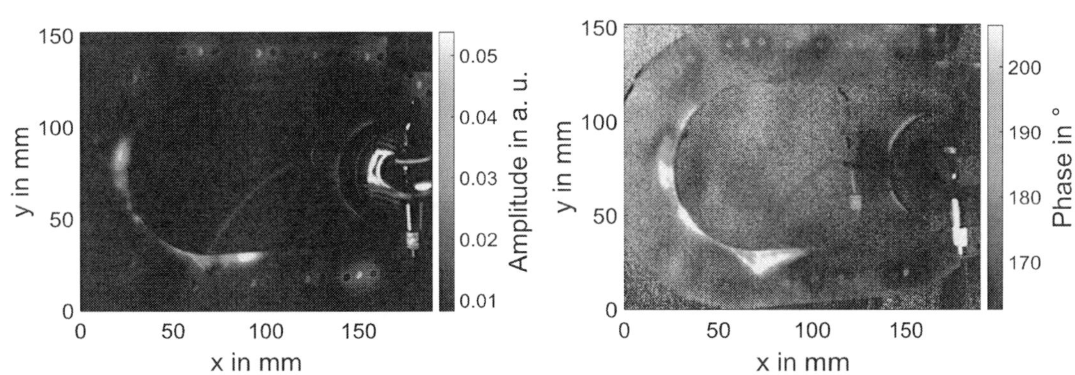

FIGURE 5. Amplitude (left) and phase (right) images of the RFST measurement of the CFRP hatch

As before, the amplitude image reveals the defects with very high defect selectivity and SNR. The phase image, while a little bit noisier, shows an even greater extent of the defect sizes.

CONCLUSIONS

In ultrasonic thermography heat is produced due to local energy dissipation. Crack friction and visco elastic dampening are the mechanism in plastic parts. The heat power is proportional to the excitation frequency and the square of the elongation, which is a function of the vibration amplitude. Local defect resonance strongly amplifies the vibration amplitudes and provides a means to reach measurable temperature signals for low input powers.

For unknown defects and defect resonance frequencies a broadband excitation is possible. A Fourier transform of the recorded temperature leads to amplitude and phase images with high SNR defect signals.

The method Resonance Frequency Sweep Thermography has been successfully applied to several academic and real parts and is viable for a fast and safe defect detection in fibre reinforced plastics.

REFERENCES

1. I. Solodov, J. Bai, S. Bekgulyan and G. Busse, "A local defect resonance to enhance acoustic wave-defect interaction in ultrasonic nondestructive evaluation," in *Applied Physics Letters,* **99**(21), (2011), 211911.
2. I. Solodov, J. Bai and G. Busse "Resonant ultrasound spectroscopy of defects. Case study of flat-bottomed holes," in *Journal of Applied Physics*, **113**(22), (2013), 223512.
3. I. Solodov, "Resonant acoustic nonlinearity for defect-selective imaging and NDT," in *AIP Conference Proceedings* 1685, (2015), 20003.
4. A. Klepka, L. Pieczonka, W. J. Staszewski and F. Aymerisch, "Impact damage detection in laminated composites by non-linear vibro-acoustic wave modulations. Damage Mechanics," in *Composites Part B: Engineering*, 65, (2014), pp. 99–108.
5. G. P. Malfense Fierro, D. Ginzburg, F. Ciampa and M. Meo, "Nonlinear thermosonics and laser vibrometry for barely visible impact damage of a composite stiffener panel," in *T. YU, A.L. GYEKENYESI, P.J. SHULL und H.F. WU, Hg. SPIE Smart Structures and Materials + Nondestructive Evaluation and Health Monitoring*, Las Vegas, Nevada, United States, (2016), 980419.
6. S. Delrue, M. Tabatabaeipour, J. Hettler and K. van den Abeele, "Non-Destructive Evaluation of Kissing Bonds using Local Defect Resonance (LDR) Spectroscopy: A Simulation Study" in *Physics Procedia*, **70**, (2015), pp. 648–651.
7. I. Solodov, M. Rahamer and N. Gulnizkij, "Highly-Sensitive and Frequency-Selective Imaging of Defects via Local Defect Resonance," in *Proceedings of European Conference on Non-destructive Testing (ECNDT 2014),* October 6–10. Prague, Czech Republic, (2014).
8. I. Solodov, M. Rahammer, D. Derusova and G. Busse, "Highly-efficient and noncontact vibro-thermography via local defect resonance," in *Quantitative InfraRed Thermography Journal*, **12**(1), (2015), pp. 98–111.
9. M. Rahammer, I. Solodov, W. Bisle, D. Scherling and M. Kreutzbruck, "Thermosonic Testing with Phase Matched Guided Wave Excitation," in *Journal of Nondestructive Evaluation*, **35**(3), (2016), pp. 1–7.
10. M. Rahammer and M. Kreutzbruck, "Fourier-transform vibrothermography with frequency sweep excitation utilizing local defect resonances," in *NDT & E International*, **86**, (2017), pp. 83–88.

IN-LINE ANALYTICS OF THE CONTENT, THE DISPERSION AND THE NANOSTRUCTURATION PROCESS OF POLYMER NANOCOMPOSITES DURING EXTRUSION

Fischer D. [(1)]*, ***Kretzschmar B.*** [(1)], ***Müller J.*** [(2)], ***Grosse S.*** [(2)], ***Kummer S.*** [(3)]

[(1)] Leibniz-Institut für Polymerforschung Dresden e.V. - Saxony - Germany, [(2)] TOPAS GmbH - Saxony - Germany, [(3)] Plastic Logic GmbH - Saxony - Germany

Extrusion is one of the most applied technologies for polymer processing. We use extrusion to create polymer nanocomposites for application in automotive, electrical and packaging industrial sectors. These nanostructured materials have advantages in comparison to traditional polymer materials. Mechanical properties, barrier and surface properties, electrical properties and flame retardancy will be improved. We show the determination of concentration, particle size, dispersion (degree of intercalation/exfoliation) and mechanical properties of polymer nanocomposites in the melt by Ultrasonic, Near Infrared spectroscopic and light extinction measurements during extrusion in real time. We correlated all in-line data with off-line rheological measurements, electron microscopy, Raman imaging and mechanical properties. The polymers used are polypropylene and polyamide 6 with different nanofillers (layered silicates) and nanoparticles. We studied the influence of different processing conditions on the morphology and on final product properties. So we could measure the nanostructure formation process in real time along the extruder on different measurement points by adaption of ultrasonic, near infrared, Raman and light extinction sensors at the end and at different positions along the extruder. It was possible to measure different dispersion states caused by different processing conditions. We could describe the nanostructuration process along the extruder for polymeric nanocomposites using the different in-line methods.

Structure evolution during film blowing: An experimental study using in-situ small angle X-ray scattering

Troisi Enrico [1]*, van Drongelen Martin [1], Caelers Harm J. M. [1], Peters Gerrit W. M. [1]

[1] Eindhoven University of Technology - Eindhoven - Netherlands

The blown film extrusion process is extensively used in the packaging industry to make large volumes of semi-crystalline polymer films. Due to the considerable amount of research dedicated to polymer processing in the last few decades (1), it has become well known that the conditions applied during the production process, together with the molecular features of the resin, determine in a large amount the final microstructure, mechanical and optical properties of the film. The final morphology of the blown film has been studied as function of processing conditions and material composition by means of numerous experimental methods (2-4). However, in these cases, measurements were carried out off-line on the final products and the optimal processing conditions could be determined only by trial and error, making the procedure time consuming and expensive. On-line measurements of the structure development by means of X-ray scattering techniques are the key requirement in order to fully understand the influence of processing conditions, such as blow-up and take-up ratio (BUR and TUR), and of molecular features such as long chain branching content (LCB) and molecular weight distribution (MWD). A blown film unit and extruder were installed on the DUBBLE beamline at the European Synchrotron Radiation Facility (ESRF-Grenoble, France). Since film-blowing is a stationary process, 2D small angle scattering patterns were recorded at different heights, ranging from the die exit to beyond the frost line, typically 15-20 cm above. Experiments were performed on 3 different materials (2 linear low density polyethylenes (LLDPE) with different amount of butyl branching and a blend of LLDPE with 10% of low density polyethylene) using 4 different sets of BUR and TUR. Using 1D autocorrelation function analysis of the SAXS patterns obtained we obtained the long period, linear crystallinity and lamellar thickness evolution along the bubble as a function of the processing variables. The orientation was also determined using the Herman orientation factor method. In this research, we used, for the first time, small-angle X-ray scattering (SAXS) to probe the evolution of lamellar thickness and molecular orientation, the main quantities governing mechanical properties, as function of both blow-up and take-up ratio for different material compositions.

Environmental Stress Cracking Of PE-HD Induced By Liquid Media – FNCT Testing

Schilling Markus (1)*, Böhning Martin (1), Oehler Harald (2), Alig Ingo (2), Niebergall Ute (1)

(1) Bundesanstalt für Materialforschung und -prüfung - BAM - Berlin - Germany, (2) Fraunhofer-Institut für Betriebsfestigkeit und Systemzuverlässigkeit LBF - Darmstadt - Germany

Due to the increasing demand for utilization and improvement of energy efficient materials, especially concerning requirements in lightweight construction and design, polyolefin materials are used extensively and in an increasingly broad range of applications. Although loaded with stresses under yield stress, preformed voids, inhomogeneities or notches can possibly lead to damage and unexpected failure induced by liquid media in polyolefin materials. For PE-HD, slow crack growth (SCG) as well as environmental stress cracking (ESC) are relevant damage mechanisms. SCG appears without influence of a surrounding medium or in "inert" media whereas ESC occurs in "active" media, which have decisive influence on damage mechanism and time to failure. To characterize the inherent resistance of the material against those damage mechanisms, the Full-Notch Creep Test (FNCT) is widely used. Therefore, the FNCT is of paramount importance particularly for the assessment of pipe and blow molding types of PE-HD – especially in the field of polymer development and improvement for high-performance applications. In this study, the FNCT – usually applied as a standardized testing method (ISO 16770) using a few universal model liquid media – is extended by investigation of selected relevant PE-HD materials with a variety of properties also in real media. Mainly topical fuels, customary in the market such as diesel and biodiesel are examined. Especially the influence of temperature, and the ESC behavior of PE-HD in media that are sorbed to a significant extent, are addressed. The investigations were performed using a novel FNCT-device with 12 individual sub-stations, each equipped with individual electronic stress and temperature control and continuous online monitoring of the specimen elongation. Additional to acquired results concerning time to failure and elongation behavior, imaging techniques, such as light microscopy (LM), laser scanning microscopy (LSM) and X-ray computed tomography (CT scan) were utilized to obtain data concerning crack propagation and media migration. Correlations of elongation behavior and data derived from imaging analysis were obtained.

Determination of the flow and curing behavior of highly filled phenolic injection molding compounds by means of spiral mold

Thomas Scheffler[1, a)], Michael Gehde[1, b)], Michael Späth[2, c)], Peter Karlinger[2, d)]

[1] Chemnitz University of Technology, Professorship of Plastics Engineering,
Reichenhainer Str. 70, GER - 09126 Chemnitz
[2]University of Applied Sciences, Fachhochschule Rosenheim, Hochschulstr. 1, D-83024 Rosenheim

[a)]Corresponding author: thomas.scheffler@mb.tu-chemnitz.de
[b)]michael.gehde@mb.tu-chemnitz.de, [c)]michael.spaeth@fh-rosenheim.de, [d)]peter.Karlinger@fh-rosenheim.de

Abstract. Due to the complex flow and curing behavior of thermosetting molding compounds, the quality of parts out of these materials depends to a high degree on the reactive and viscous characteristics during their processing. In the presented study a flow spiral was used to determine the viscosity directly during the injection molding process. In addition to the flow path, the flowability is determined by measuring the pressure loss over time at three different flow path lengths with two pressure sensors. To compare the measuring results of the sensors in the mold with the data obtained by the injection molding machine, the hydraulic pressure and the screw-driving torque were recorded. The results display a high accordance of the recorded machine parameters with the parameters measured in the mold. They will be discussed depending on the degree of filling, the content of water of the phenolic molding compound and the mold temperature.

INTRODUCTION

The reaction and the processing of thermosets depend highly on the material condition. The analysis of the effects of the condition changes (e.g. due to creeping reaction even despite relatively low temperatures or due to water absorption) require a continuous observation and adaptation of the process parameters during the processing of thermosetting molding compounds. Nevertheless, reject rates up to 30 % are possible. For example, volatile fuse and filling behaviors can be possible reasons for rejection. The water content has an impact on this behavior especially on the behavior of technical relevant phenol resins. These compounds have a strong tendency to absorb water which directly affects the viscosity of the material and constitutes a major element of uncertainty for the quality of the produced part [1, 2, 3,].

Among others it is described that the water content has no impact on the curing reaction and that a higher water absorption reduces the melt viscosity [4]. The degree of water absorption has several reasons and depends on the amount of filler, the kind of the filler and the degree of preconditioning of the resin. Additionally, the degree of preconditioning directly impacts the flowability of the polymer [5]. The described studies were conducted using the molding compound PF 31 which is filled with short-fibered organic wood flour. Compared to technically heavy-duty phenol resins with an anorganic filler content up to 80 wg.-% this molding compound displays completely different material properties. Thus, the insights of the mentioned studies cannot be applied to heavy-duty phenol resins sufficiently [6, 7]. Nevertheless, a few studies using highly filled technical thermosetting molding compounds to investigate the impact of water content on the processing and the resulting component quality also found that a higher water content increases the shrinkage and reduces the dimensional stability [8].

BACKGROUND

Thermosetting molding compounds are delivered with an extremely high tolerance range regarding the formulation of the molding compound. This may include differences in the filler content of different batches up to 15 % which influences distinctly the proportion of the fluid polymer. Additionally, the degree of preconditioning, this means the existing interconnectedness, the chain length in the initial situation and the amount of phenol nuclei, impacts the flowability of molding compounds. The degree of preconditioning can differ up to 20 % and depends directly on the respective process conditions [9]. The resulting process stability and the uniform production quality of thermosetting molding compounds impact massively the subsequent flowability of the material as an early curing can easily occur [10].

The aim of the production of molding compounds is a high pre-curing to enable a time and energy saving further processing until the complete curing of the material. The fact that the viscosity has to be still processable is a limiting factor for the prior interconnectedness of molding compounds used in injection molding.

Like previously mentioned, it is possible to deduce the viscosity from the injection pressure, however, there are clear dependencies regarding the moisture content, the temperature of the tool and the injection velocity. Further indirect methods to determine the viscosity constitute the measurement of the flow path length in the flow spiral with a predefined pressure or the determination of the viscosity with the measurement of the pressure difference. For thermosetting molding compounds the dependence of the area without wall contact on various processing parameters has been displayed in a flow spiral [11]. Only for the described molding compound PF31 data on viscosity within a flow spiral tool has been obtained and the dependence of the viscosity on tool temperature and injection speed has been pointed out [12].

EXPERIMENTAL

Three novolac based phenol formaldehyde resin molding compounds from Hexion GmbH were examined. The Bakelite® PF6510 (GB35 GF20), the Bakelite® PF6680 (GB35 GF30) and the Bakelite® PF1110 (GB35 GF45) are molding compounds highly filled with glass fibers (GF) and glass beads (GB). The results of the control at good issue in the continuous kneader (minimal torque (M_{min}), reaction time (T_R)) are displayed in Table 1. To investigate the influence of the water content, the materials were impinged with 98 % air moisture at the temperature of 30°C within a climate cabinet of Binder MKFT 115. The samples were removed after 1 h, 2 h, 4 h, 8 h, 24 h, 48 h, 96 h and 168 h. The times of removal result from independent pretests and are equivalent to Höer 2014 who ascertained a saturation of water absorption after 168 h under these storage conditions [9]. The water content was analyzed at 130°C using the Karl-Fischer-Titration.

TABLE 1. Initial values of the tested materials

Material	Number	minimal torque 140°C [Nm]	reaction time - DIN 53764 [s]
PF6510	FG5K510602	6.7	44
PF6680	FG5M510106	4.7	118
PF1110	FG5L571102	3.4	75

For the injection molding experiments a fully hydraulic injection molding machine with a horizontal two plate closing system (Type KM 160-750 CX, Control MC-6, Fa. Krauss Maffei) and a plasticizing cylinder for thermosets (cylinder diameter D= 45 mm, nozzle diameter 4 mm) was used. The plasticizing parameters were constant during the whole experiment with a back pressure of 25 bar, a screw speed of 30 RPM and a from the nozzle on declining cylinder temperature of 90°C – 75°C – 60°C. Different injection speeds (12 ccm/s, 24 ccm/s, 48 ccm/s) and the tool temperatures (155°C, 170°C, 185°C) were employed. The experiments were executed using a flow spiral tool with a cavity height of 4 mm and a cavity width of 10 mm. The maximum flow path length is 1385 mm. Six cavity pressure sensors were integrated over the flow path. Thereby three measurement lengths were created: measurement length one from 15 to 115 mm, measurement length two from 215 to 315 mm and measurement length three from 1170 to 1270 mm. The apparent viscosity was calculated according to the indicated formula using the pressure loss and the volume flow within the respective measuring length. Differing from the standard use of the flow spiral, the experiments were executed way controlled to detect backflow. The injection pressure limitation was set by 2300 bar.

$$\eta = \frac{\tau}{\dot{\gamma}} = \frac{\Delta P * K}{\dot{V} * K} \quad (1)$$

RESULTS AND DISCUSSION

Water Absorption

The initial water content of the molding compounds amounted from about 0.5 % (PF1110) up to about 0.9 % (PF6680). In general, thermosetting molding compounds are delivered with a water content up to 5 % [5]. These high water contents could not be detected in the inorganically filled molding compounds that were conditioned. The results of the material conditioning of the molding compound PF1110 and the PF6510 show a water absorption up to a water content of 2.4 % within the first 48 h. Moreover, no saturation could be detected after 168 h and the molding compound PF1110 featured a higher water absorption than PF6510. This leads to the conclusion that the water absorption does not only depend on the resin but also on the filler and its interface. A processing within the injection molding is only possible with a water content up to about 2.5 %, as a higher moisture content leads to significant fluctuations during the processing.

Influence of the Moisture Content

Figure 1 shows the influence of moisture on flow path length and injection pressure of the molding compounds PF1110 (a) and PF6510 (b).

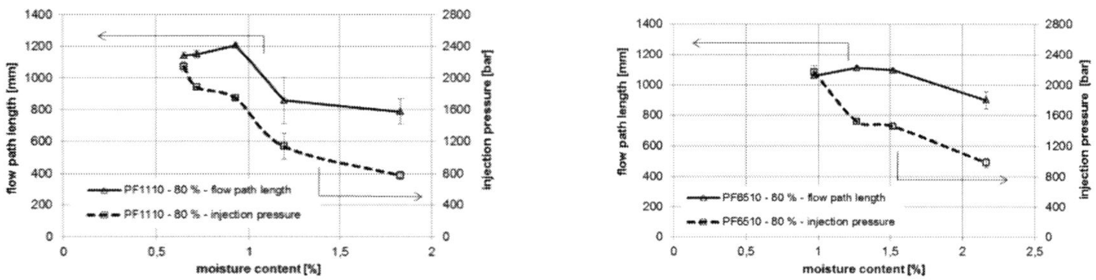

(a) (b)

FIGURE 1. Influence of moisture on flow path length and injection pressure of (a) PF1110 and (b) PF6510

Higher moisture contents lead to a decrease of the injection pressure to 63 % (PF1110) and to 54 % (PF6510). In addition, the flow path length increases slightly with the water content rising to 0.93 % (PF1110) and 1.27 % (PF6510) and decreases with a further increase of the moisture content by 35 % (PF1110) and 20 % (PF6510). The higher compressibility of the molding compound during the plasticizing leads to the fact that a greater amount of the molding compound can be injected with the same dispense volume explains the extension of the flow path. The following decrease of the flow path results from a higher backflow through the screw walks. As a result, it can be concluded that a light increase of the water content improves the flowability until it decreases significantly due to the excessive material moisture which results in a stronger process fluctuation and a higher backflow that diminishes the form filling.

To control the process further, the plasticizing torque was recorded. Figure 2 depicts the torques that have been measured and the screw path of the conditioned molding compound PF6510 in dependence on the water content during the plasticizing period.

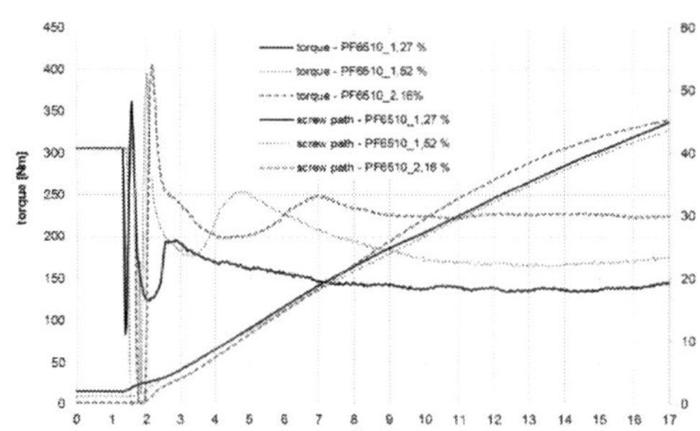

FIGURE 2. Influence of moisture on the plasticizing torque of the PF6510

It becomes obvious that the screw paths proceed identically. Regarding the plasticizing torque a clear increase can be detected with a rising water content. This can be explained by the stronger backflow and the resulting higher number of melted screw walks during the plasticizing. From this phenomenon it can be deduced that by recording the plasticizing torque the backflow can be detected and thus this parameter can be employed to control the processing of thermosetting molding compounds.

Another objective of this study was the determination of the apparent viscosity using the measurement of the pressure difference dependent in relation to the flow path length within the mold. Figure 3 shows the apparent viscosities of the molding compounds PF6510 dependent on the water content and the flow path length (measurement length one and two) during the injection period standardized on the first overflow of the second pressure sensor of the respective measurement length.

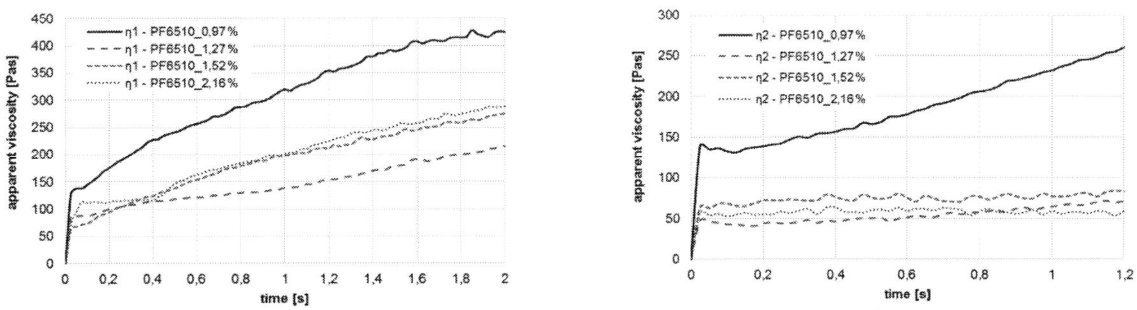

(a) (b)

FIGURE 3. Influence of moisture on flow path length and injection pressure of the (a) measurement length one and the (b) measurement length two

It illustrates that the unconditioned batch (PF6510_0.97 %) always displays the highest apparent viscosity. The conditioned batches feature a viscosity that is about 60 % lower over the flow path but with an similar course. Only the apparent viscosity of the molding compound that has been conditioned for four hours (PF6510_1.27 %) displays a different behavior which can be explained by the insufficient homogeneity of the water absorption of the molding compound due to the short conditioning duration. In addition, all viscosities of measurement length two are lower than at measurement length one which is due to a higher melt temperature caused by the heat transfer of the hot mold. As the shear rate is constant, the increase of the viscosity results from a rising pressure difference during the injection period and might be explained by a pressure dependence of the viscosity. This results from the constantly rising hydraulic injection pressure during the injection period.

Influence of the mold temperature

Figure 4 depicts the apparent viscosity of the molding compound PF6510 in dependence on the temperature of the mold and the length of the flow path (flow path length one and two) during the injection period standardized on the first overflow of the second pressure sensor of the respective measurement length.

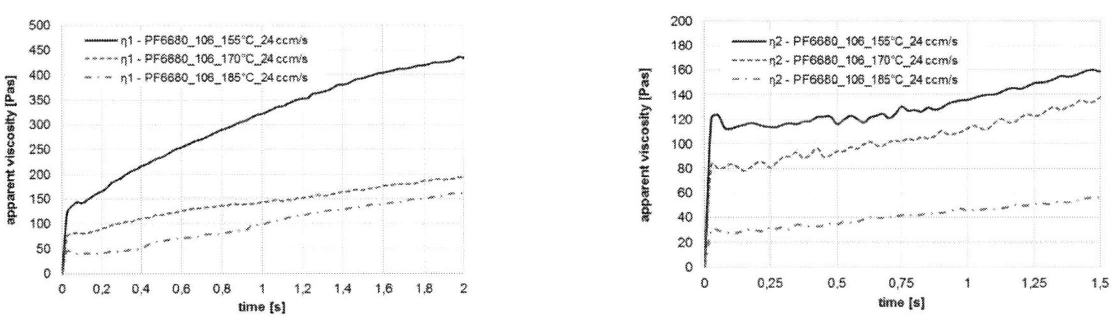

(a) (b)

FIGURE 4. Influence of the molding compound - apparent viscosity over the flow path (a) measurement length one and the (b) measurement length two

It becomes obvious that the apparent viscosity decreases by 60 % with the increasing temperature of the mold at the flow path length one and that this difference is almost constant over the whole injection phase. At the flow path length two, there is a lower apparent viscosity with increasing temperature of the mold and the difference decreases from 75 % to 60 %. Furthermore, a lower viscosity over the flow path and an increase in viscosity during the injection phase can be measured. The second phenomenon may be explained by the growing injection pressure during this phase.

CONCLUSION

The results show that the water content of the molding compound increases during the period of conditioning (30 °C, 98 % rel. humidity) with the highest increase occurring within the first 48 h. This demonstrates the necessity of a hermetic packaging of thermosetting molding compounds. The increasing water content leads to a sinking injection- and cavity-pressure. Furthermore, a higher backflow and a higher plasticizing torque during the whole plasticizing phase correlates with an increasing moisture content. Therefore, a process control of the plasticizing torque should be considered. All effects result from the lower viscosity stemming from a higher water content. A low pressure-dependent apparent viscosity over the flow path can be detected. At the second flow path length, the increased moisture content leads hardly to a decrease in the apparent viscosity. Thereby, it is likely that the minimum viscosities of the molding compound PF6510 lie within the range of 50 – 70 Pas at the mold temperature of 175°C. This minimum viscosity is also prevalent at the measurement length two for the molding compound (PF6680), see Figure 5 (b), at a tool temperature of 185°C with 30 Pas at the beginning of the injection period. In short, a clear reduction of the viscosity in dependence on the tool temperature can be found also over the flow path length. This contradicts a few theories which posit a higher hardening progress due to a higher tool temperature during the injection period. The investigation of this coherence is subject of further studies.

Further studies should investigate the hypothesis of the pressure-dependence of the viscosity taking the filler content into account. Additionally, these dependencies should be investigated with different shear rates, which can be set by using various injection speeds. Finally, the obtained knowledge should be compared to data that has been measured offline.

REFERENCES

1. W. Dahlhoff, “Beitrag zur Verarbeitung von duroplastischen Formmassen durch Spritzgießen,“ Ph.D. thesis, Universität RWTH Aachen,1969
2. M. Fischbach, “Prozessführung beim Spritzgießen Härtbarer Formmassen,“ Ph.D. thesis, Universität RWTH Aachen, 1982
3. M. Höer, G. Hülder, and M. Gehde, “Post-curing effects on thermo-mechanical properties of injection molded high performance phenolics,“ SPE-ANTEC Proceedings, Cincinnati, 2013
4. M. Rath, ”Online Ultraschalluntersuchungen an duroplastischen Formmassen zur Prozesskontrolle,” Ph.D. thesis, Universität Berlin, 2000
5. J. Weise, ”Technologische Untersuchungen zum Spritzgießen ausgewählter Duromerformmassen unter Verwendung von Kaltkanalwerkzeugen,” Ph.D. thesis, Universität Karl-Marx-Stadt, 1987
6. T. Barth, ”Das mechanische Werkstoffverhalten duroplastischer Formstoffe bei Feuchteeinwirkung, ” Fortschrittsbericht VDI-Zeitung Reihe 5 Nr. 69, Düsseldorf, 1983
7. S. Englich, T. Scheffler and M. Gehde, ”Specific mold filling characteristics of highly filled phenolic injection molding compounds, SPE-ANTEC Proceedings,” Cincinnati, 2013
8. T. Maenz, M. Höer, G. Hülder, C. Tüchert, and M. Gehde, “Influence of Material Batch and Moisture Content on the Processing Behavior and Dimensional Stability of Phenolic Molding Compounds,” SPE-Antec proceedings, Orlando, 2015
9. M. Höer, ”Einfluss der Material- und Verarbeitungseigenschaften von Phenolharzformmassen auf die Qualität spritzgegossener Bauteile,” Ph.D. thesis, TU Chemnitz, 2014
10. D. Kaylon and M. Hallouch, ”Compounding of Thermosets in Continuous Kneaders, ” Advances in Polymer Technology, Vol. 6, pp. 237 – 249, 1986
11. W. Michaeli, D. Hunold and T. Kloubert, “Formfüllung bei Spritzgießen von Duroplasten,” Plastverarbeiter 43. Jhrg, 1992, Nr.12
12. W. Michaeli and T. Gernot, “Rheologische Stoffdaten für Duroplaste,” Plastverarbeiter 45. Jhrg., 1994, Nr.3

CFRP Pipe Inspection by Means of Air-Coupled Ultrasound

Sebastian Joas[a)], Wolfgang Essig, Florian Fröhlich, Marc Kreutzbruck

IKT, Institut für Kunststofftechnik, University of Stuttgart, Pfaffenwaldring 32, 70569 Stuttgart, Germany

[a)]Corresponding author: sebastian.joas@ikt.uni-stuttgart.de

Abstract. Due to their high strength and stiffness at comparatively low weight, lightweight construction materials such as fiber reinforced plastics (FRP) are used in a variety of technical systems. The areas of application range from aerospace and renewable energies to the automotive sector, which is gaining increasing importance. The non-destructive testing (NDT) by means of air-coupled ultrasound is a decisive factor for the detection of production and operation induced defects. As a result, it is possible to examine components regarding possible defects without contact or coupling medium. By using carbon fiber reinforced plastic (CFRP) pipes with different geometries, it was possible to demonstrate the suitability of air-coupled ultrasound for defect detection. For this purpose, guided waves were contactlessly excited in the pipe, which circumferentially propagated along the components' cross-section. Based on the amplitude and phase characteristics of the waves, defects could be detected in different axial and circumferential positions within short inspection times of a few seconds. The developed testing method is automatable and hence offers a high potential for future quality control in serial production.

1. INTRODUCTION

Fiber reinforced plastics (FRP) are used in many different areas like automotive, aerospace, renewable energies, sports and medical engineering. Due to their high strength and stiffness at comparatively low weight, FRP are often the material of choice in lightweight construction. [1] An essential disadvantage of FRP is the relatively complex production process, which might lead to defects. For the detection of these production induced defects as well as for periodic component inspection, the non-destructive testing (NDT) by means of air-coupled ultrasound meets a decisive role. This contactless method enables the examination of objects without the need for an additional coupling medium. [2] One major challenge in air-coupled ultrasound testing is the high acoustical impedance difference in between air and the material of the ultrasound probe. For this reason, testing heads with a special adjusted membrane are utilized. Thereby, ultrasound probes based on cellular polypropylene have been successfully applied. [3, 4]

In the NDT using air-coupled ultrasound, there are different measurement arrangements. The so called slanted re-emission mode (SRM) is based on the excitation of Rayleigh- or lamb waves. These lamb waves are excited when the ultrasonic wave hits the components surface under a certain incidence angle, which causes a symmetrical or asymmetrical oscillation of the whole part. In dependence on the components thickness and the frequency, the oscillations may occur in different modes. [7, 8, 9] If the incidence angle becomes higher, the transverse wave is present as a surface or rather Rayleigh wave, which propagates along the components surface for a long time due to the weak coupling between specimen and air and gets damped or rather reflected by the surface roughness and imperfections. Guided waves propagate parallel to the components surface and can be detected at larger distances from the excitation point [3], which enables the air-coupled ultrasonic testing with unilateral accessibility. The measurement setup of the SRM is illustrated in Fig. 1. Changes in amplitude enable the detection of existing imperfections in the investigated specimen. In previous studies it was shown, that not only plates [10, 11] but also hollow cylinders, as illustrated in Fig. 1, can be examined. [5, 9]

However, the line scan and the following rotation of the pipe lead to a very time consuming test procedure of the entire surface area. The aim of the research project is to avoid this time consuming mechanical scanning process and enable not only the testing of straight pipes but also of more complex shaped structures using an industrial robot.

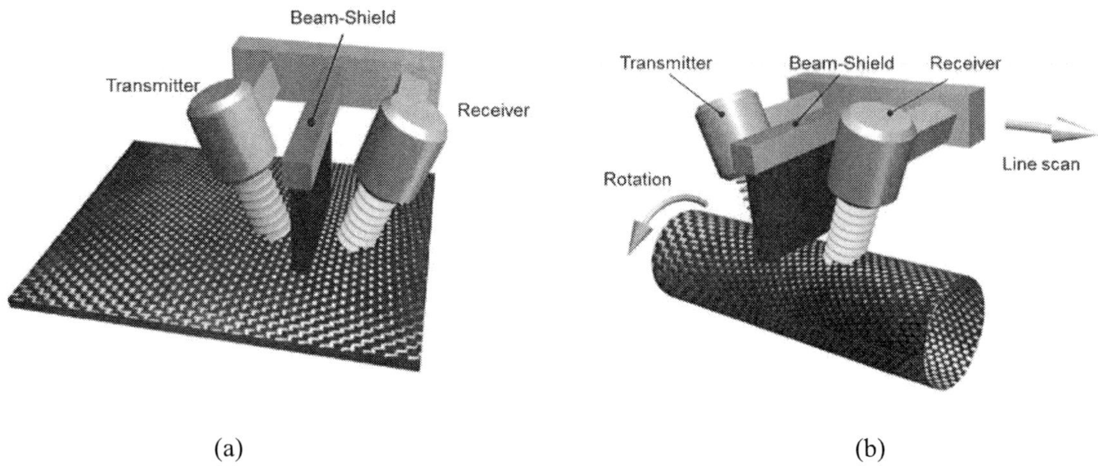

FIGURE 1. SRM for the investigation of a CFRP plate (a) and a CFRP pipe (b)

2. EXPERIMENTAL PROCEDURE

All measurements were performed using the non-focused ultrasonic probes AS200T (Air Star, Inc), which have a center frequency of 207 kHz. A straight CFRP pipe as well as a chainsaw handle bar is used as test object, as shown in Fig. 2 (a) and (b).

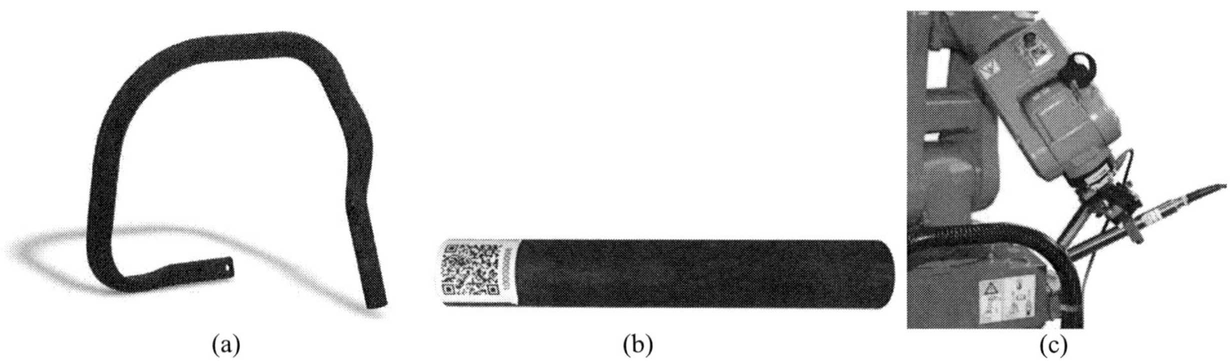

FIGURE 2. Straight CFRP pipe (a) and chainsaw handle bar (b) are tested using an X-Y scanning table as well as an industrial robot (c)

The straight, thin-walled CFRP pipe has a length of 200 mm and an outer diameter of 32.5 mm. The wall thickness is 1 mm and the resulting inner diameter 30.5 mm. The CFRP pipe is clamped on the front between two cones of plastic. The wall thickness of the chainsaw handle bar is not constant in the circumferential direction.

For all measurements, rectangular adhesive tapes were used in order to generate artificial defects on the outer surface of the pipe. These provide an attenuation of the ultrasonic waves, which changes the amplitude or phase course and thus, simulate real defect behavior. The dimensions of the artificial defects are varied both in the axial and in the circumferential direction. All artificial defects are always placed perpendicular to the measurement and circumferential direction. The pipes are tested in the axial direction (cf. measurement direction) by means of a line scan, as schematically shown in Fig. 3 using the straight CFRP pipe. An X-Y scanning table operated by a stepper motor is used to perform the linear scan of the CFRP hollow cylinder.

A guided wave, which is propagating in circumferential direction of the pipe, is used to draw conclusions on the presence of artificial defects and their exact dimension and location. The results are presented as an A-scan at each axial position of the pipe, whereby a B-scan can finally be generated (c.f. Fig. 3).

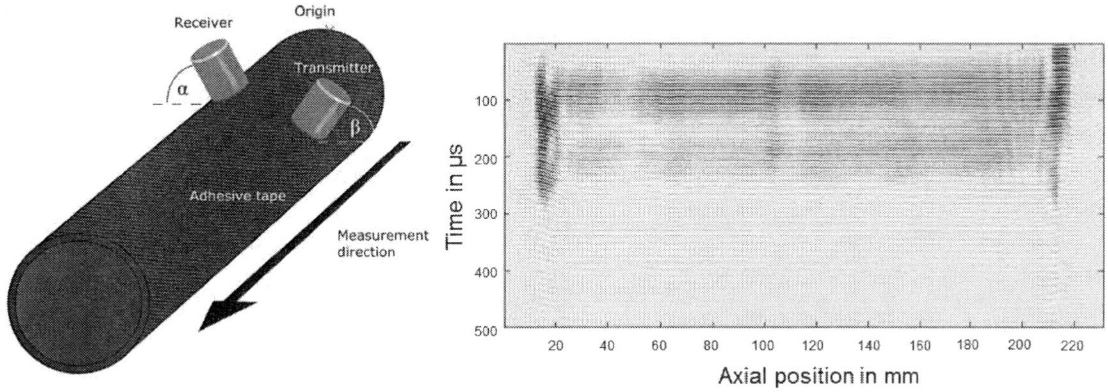

(a) (b)

FIGURE 3. Schematic representation of the sequence of a measurement (a) and results presentation as B-scan (b)

In the B-scan, the relevant time range needs to be identified and the difference between minima and maxima of the amplitude in the specific time range can finally be determined. Based on the resulting course, the aim is to examine if artificial defects in axial and circumferential positions can be detected and whether their exact size can be determined. In order to investigate the suitability of the test method for industrial practice, the chainsaw handle bar is investigated using the IRB 120 industrial robot (ABB Ltd.) as an example for complex pipe geometries.

For the automated measurements, a model of the chain saw handle bar is loaded into the virtual environment of RobotStudio. Afterwards, the program calculates twelve trajectory points, which enable an interpolation of the measuring distance with its total length of approximately 250 mm.

Transmitter and receiver are attached to an optical bench with a rotary table, which allows a precise adjustment of the angles, and are finally flanged to the robot arm. A photo of the measurement setup was previously shown in Fig. 2 (c). The MATLAB®-based UniversalUltraSchallSoftware (UUSS), developed at the IKT, is used for the evaluation of all measurements. In the conducted measurements, the guided wave replaces the need for a mechanical scan in circumferential direction and thus minimizes the required test time.

3. RESULTS AND DISCUSSION

3.1. Investigation of the Straight CFRP Pipe

An important aspect of the investigations by means of air-coupled ultrasound is the determination of the dimensions of the respective defect. While the exact defect size in circumferential direction cannot be determined, it is possible to draw very precise conclusions on the dimensions of the artificial defects in axial direction.

For instance, the amplitude information for an artificial defect having the dimensions 15 mm x 25 mm is shown in Fig. 4. It is apparent, that there is a local maximum at the first edge of the adhesive tape, which is followed by a local minimum at the rear edge of the adhesive tape. The gray area indicates the position of the defect, whereas the red lines show the beginning and the end of the pipe. This characteristic course can also be found in further experiments using artificial defects of different dimensions (10 mm x 15 mm, 15 mm × 15 mm and 25 mm × 15 mm) in the same circumferential position.

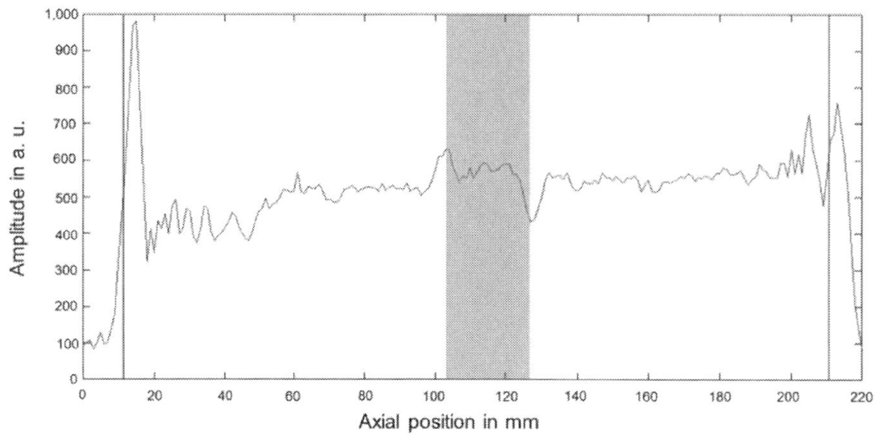

FIGURE 4. Amplitude profile for a defect (25 mm × 15 mm) in a circumferential position of 35 mm (time range 355-525 μs)

Even if the circumferential position of the defect is varied, the axial size of defects can be determined. In this case, three different adhesive tapes (15 mm x 15 mm, 20 mm x 15 mm and 20 mm x 15 mm) are applied as artificial defects. The resulting measurement signal enables the determination of the axial size of all adhesives, which result in 16 mm, 20 mm and 20.5 mm, respectively. The corresponding signal profile for the adhesive tape placed in a circumferential position of 52 mm with a size of 20 mm x 15 mm is shown in Fig. 5.

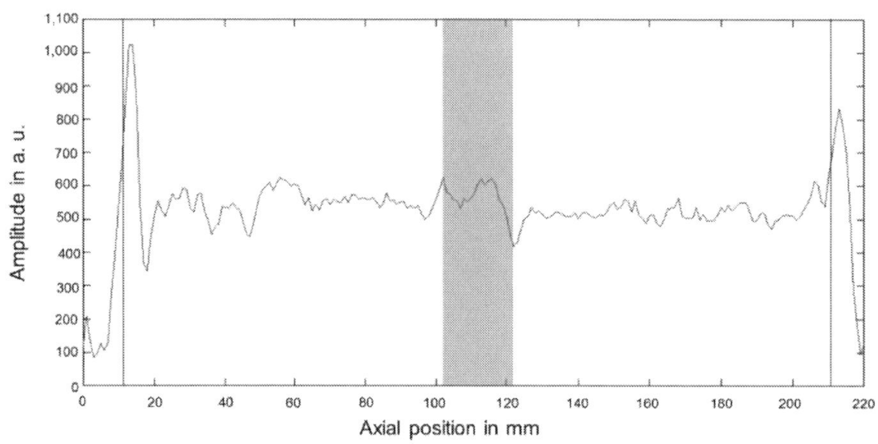

FIGURE 5. Amplitude profile for a defect (20 mm × 15 mm) in a circumferential position of 52 mm (time range 355-475 μs)

3.2. Automated Inspection of the Chainsaw Handle Bar

All previously performed measurements have been carried out for the simple case of the straight CFRP pipe, which has a constant diameter and a constant wall thickness. These simplifications are not valid for more complex components. Due to this, the following section deals with the testing of a chain saw handle bar as an example for a real application. At first, two measurements with the same test parameters are carried out with and without adhesive tape (15 mm x 15 mm) in order to perform an investigation regarding interferences and variation of the signal. It can be shown that both measuring signals, except for the area of the defect, are in good accordance and thus, it can be shown that the test parameters are also suitable for inspection. The adhesive tape can be detected by a very pronounced amplitude change, and its axial dimension can also be determined to be 14 mm, as shown in Fig. 6. In this case, the dimensions of the artificial defect can be determined by the difference of the axial values of the curves inflection points. This is in contrast to the procedure for the straight CFRP pipe, which could be due to the non-constant wall thickness of the pipe in circumferential direction.

Finally, three adhesive tapes with dimensions of 15 mm x 15 mm are positioned at the curved section of the chainsaw handle bar. The robot assisted measurement shows that all three artificial defects can be detected using the phase information by clearly visible phase steps. However, using the phase information, it is not possible to quantify the size of the defects in axial direction.

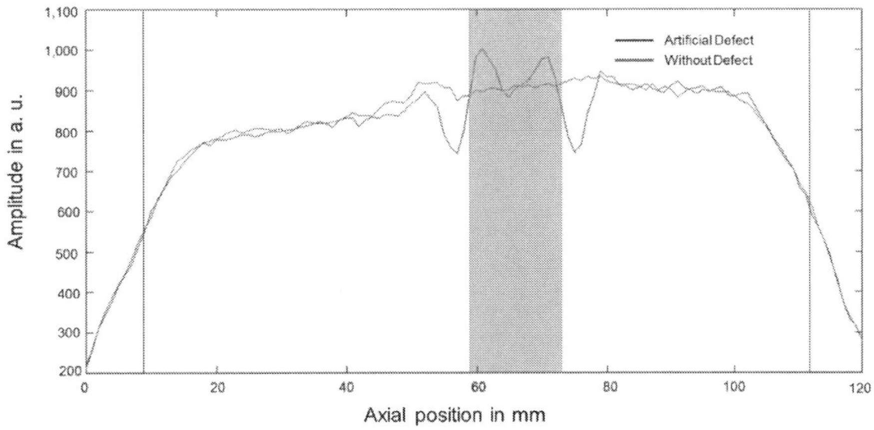

FIGURE 6. Amplitude profile for the test of the chainsaw handle bar with (red) and without (blue) artificial defect (time range 225-334 μs)

4. CONCLUSION

The presented results show that air-coupled ultrasound is a suitable test method for simple as well as complex CFRP pipes. A guided wave in circumferential direction combined with a movement in axial direction enables a detection of defects in different axial and circumferential positions by the analyzation of the phase and amplitude profiles. Furthermore, the detected flaws can also be quantified in axial direction for both of the investigated pipe geometries. The test of the chainsaw handle bar showed, that guided waves also propagate along non-circular outer contours.

The pipe inspection by means of air-coupled ultrasound thus offers a high potential for the inline-quality control. Hereby, it is possible to investigate simple as well as complex shaped pipes regarding defects in very short inspection times of a few seconds and with only unilateral accessibility. In addition, an automation of the measurement procedure is possible and enables the testing of curved pipes, which provides not only a reduced inspection time but also a higher probability of detection. Future research activities should focus on the validation of the measuring method on specimen with real defects such as delaminations, cracks and impact damages as well as on further automation.

REFERENCES

[1] C. Bonten, *Kunststofftechnik – Einführung und Grundlagen,* 2nd edition (Hanser, Munich, 2016).
[2] A. J. Rogocsky, Material Evaluation. **50**, 1491 (1991).
[3] M. Gaal, J. Bartusch, E. Dohse and M. Kreutzbruck, *AIP Conference Proceedings*, **33A**, 471 (2014).
[4] M. Gaal, J. Bartusch, J. Döhring, E. Dohse, T. Lange, W. Hillger and M. Kreutzbruck, *Proceedings of the 12th International Conference of the Slovenian Society for Non-Destructive Testing*, 43 (2013).
[5] D. Döring, *Luftgekoppelter Ultraschall und geführte Wellen für die Anwendung in der Zerstörungsfreien Werkstoffprüfung* (Ph.D. thesis, University of Stuttgart, Institut für Kunststofftechnik, 2011).
[6] J. Krautkrämer und H. Krautkrämer, *Werkstoffprüfung mit Ultraschall* (Berlin, Springer, 1986).
[7] K. Schiebold, *Zerstörungsfreie Werkstoffprüfung – Ultraschallprüfung* (Springer Vieweg, 2015).
[8] S. Steeb, *Zerstörungsfreie Werkstück- und Werkstoffprüfung: Die gebräuchlichsten Verfahren im Überblick* (Renningen: Expert-Verl., 2011).
[9] J. Prager, H. Gravenkamp, M. Rahman and E. Köppe, Technisches Messen. 251 (2012).
[10] M. Castaings and P. Cawley, J. Acoust. Soc. Am. **100**, 3070 (1996)
[11] D. N. Alleyne and P. Cawley, IEEE Transactions on Ultrasonics, Ferroelectrics, and Frequency Control. **39**, 381 (1992).

Degradation of Bimodal Distribution Polypropylene and Polyethylene during the Extrusion Process

Bayazian Hoda [1]*, Reinders Frauke [2], Goebel Christian [2], Schoeppner Volker [2]

[1] Paderborn university - North rhein westfalia - Germany, [2] Paderborn University Kunststofftechnik - North Rhine Westphalia - Germany - North rhein westfalia - Germany

The molecular weight distribution (MWD) of the polyolefins is an essential criterion to determine the properties of the polymer and its applications. Bimodal distribution is one of the most popular MWD in industry when melt strength besides convenient processing such as producing biaxial oriented film or pipes processing, is required. Bimodal grade is attributed to an increase in the number of entanglements and better network connectivity so that grad pretend polymer for sagging or fell down the melt strength. On the other hand during the processing of extrusion, chain scission is displayed, initially in this study polypropylene and polyethylene are used in multiple extrusion which it is causes the chain scission processes during thermo-mechanical degradation. Thus, HP525J as a bimodal polypropylene, and HyPE 35 P FA as a bimodal polyethylene grade are selected then these grades are degraded in high temperature profile and different screw speed in extruder. In the end for evaluating this phenomenon on these grades, high temperature gel permeation chromatography (HT-GPC) is employed. The MWD behavior has been measured after each turn in extruder and on the odd occasion intensity in the low molecular weight region shift to higher molecular weight region, and their MWD are compared with each other, also there are some left shifts to low molecular weight in MWD of polypropylene grade by it degradation.

DIELECTRIC ANALYSIS FOR ONLINE CURE MONITORING: CORRELATION WITH THERMAL AND RHEOLOGICAL PROPERTIES OF EPOXY RESINS

Bakis Gökhan (1), Demleitner Martin (1), Sanchez Aline (1), Pflock Tobias (2)*, Chaloupka Alexander (3), Altstädt Volker (1)

(1) Department of Polymer Engineering, University of Bayreuth - Bayreuth, Bayern - Germany, (2) NETZSCH-Gerätebau GmbH - Bayern - Germany, (3) NETZSCH-Gerätebau GmbH - Selb, Bayern - Germany

Fiber reinforced polymer composites have been increasingly used by various industries such as aerospace and automotive due to their high specific stiffness and strength, chemical resistance and thermo-mechanical properties. To achieve the best part quality and cost-efficient processing, online cure-monitoring is desirable. In the present work, the objective was to obtain a correlation between the results of the dielectric analysis (DEA) and several material properties such as degree of cure, glass transition temperature (Tg) and gel point of epoxy systems. Experiments were conducted on four different resin types, commonly used for industrial applications. Moreover, these findings were evaluated finally in the resin transfer molding (RTM) process. At first, modeling of the curing kinetics of the resins was done. The necessary parameters were studied via differential scanning calorimetry (DSC) to allow the prediction of the degree of cure and the development of the Tg during curing. A relation between Tg and conversion was established which was used for the construction of diffusion-dominated reaction kinetics in the later stage of curing enhancing the cure-prediction quality. Dielectric analysis was then conducted to study the curing of the epoxy resin systems. This technique allows in-mold monitoring of the crosslinking. The ion viscosity was used for correlation with the above mentioned material properties and good agreement could be found. To precisely determine the gel-point of epoxies, rheological experiments were conducted. The focus was the precise correlation of the gel point from rheological experiments with dielectric results. The results agreed well between within experimental error. Finally, the obtained results of two chosen systems were compared and verified during RTM processing. Successful flow front and cure monitoring were done with the dielectric analysis. Good agreement with theoretical cure prediction and rheological properties were observed.

Cavitation and Lamellae Fragmentation-Reconstruction of Isotactic Polypropylene during Uniaxial Drawing: the Influence of Annealing

***Chang Baobao* (1)*, *Schneider Konrad* (2), *Heinrich Gert* (2)**

(1) Leibniz Institute of Polymer Research Dresden - Dresden2014 - Germany, (2) Leibniz Institute of Polymer Research Dresden - Dresden - Germany

Plastic deformation of semi-crystalline polymer is a complicated process containing two processes: cavitation and lamellae fragmentation-reconstruction. The relationship between macroscopic yielding point and the two processes and the relationship between the two processes themselves have long been a hot topic for the past decades. In this study, the structural evolution of isotactic polypropylene (iPP) during uniaxial drawing is monitored by synchrotron small angle X-ray scattering (SAXS). The results suggest that before the yielding point, lamellae rotation could be observed because of the elastic deformation of the amorphous phase. On the contrary, lamellae tilting induced by crystallographic slip could be found only after the yielding point. And melting-recrystallization of the lamellae takes place afterward. Cavities can be triggered before or after the yielding point. The cavities are oriented mainly perpendicular to the loading direction. The development of nano-cavities into micro-cavities before the yielding point will lead to the fracture of the specimen. Otherwise, a stable neck could be formed if the nano-cracks coalesce into micro-cavities after the yielding point. The micro-cavities are oriented along loading direction. Generally, annealing shows no influence on the lamellae melting-recrystallization. The thickness of the newly formed lamellae is in linear proportion to the reciprocal of the drawing temperature, irrespective of the original lamellae thickness. However, annealing could greatly promote the formation of the cavities because of the enhanced connection between the amorphous phase and the lamellae.

Design of material adapted joints – Potentials and challenges in joining fibre reinforced thermoplastics to dissimiliar materials at the example of thermoclinching

Gude Maik [(1)]*

[(1)] Technische Universität Dresden, Institut für Leichtbau und Kunststofftechnik - Saxony - Germany

Design of material adapted joints – Potentials and challenges in joining fibre reinforced thermoplastics to dissimiliar materials at the example of thermoclinching

Molecular Structure of Reactive Polycarbonate – Amine Interfaces Characterized by IR-Spectroscopy and Differential Scanning Calorimetry

Cordelia Zimmerer[1, a)], Liane Häußler[2], Kerstin Arnhold[2], Lisa Ziegler[1], Gert Heinrich[1]

[1] *Leibniz Institute of Polymer Research Dresden e.V., Institute of Polymer Materials, Department Reactive Processing, Hohe Straße 6, 01069 Dresden, Germany*
[2] *Leibniz Institute of Polymer Research Dresden e.V., Institute of Macromolecular Chemistry, Department Analytics, Hohe Straße 6, 01069 Dresden, Germany*

[a)] Corresponding author: Zimmerer@ipfdd.de

Abstract. Solid-state reactions of Polycarbonate are quite complex and totally different to reactions in solution state. As various reaction pathways with different kinetics may occur, the in-depth characterization of molecular processes is not only of technological importance but also essential for understanding the reactive processing of Polycarbonate. Infrared (IR) spectroscopy is a molecular sensitive technique to characterize the formation of functional groups as well as their intermolecular and intramolecular interaction. The formation of low amounts of Bisphenol-A as a side product leads to the formation of mixed phases containing both, PC and Bisphenol-A. Bisphenol-A is soluble in Polycarbonate and serves as a plasticizer for the Polycarbonate phase. An intermolecular interaction between both compounds is observed. The maximum reduction of glass transition temperature of about 100 K is determined for Polycarbonate mixed with different amounts of Bisphenol-A.

INTRODUCTION

Bisphenol-A-based Polycarbonate (PC) is an often used engineering material with excellent mechanical and optical properties but with both, a limited hydrophilicity and biocompatibility. A functionalization or surface modification of PC offers an increasing field of applications.

Amine compounds are chemically well suited as modifiers because a covalent linkage is formed between carbonate groups of PC and amine. These covalent linkages are expected to be stable and permanent. Chemical reactions of PC with amines, e.g. Octadecylamine (ODA), result in block-co-polymer formation (c.f. Fig. 1) [1].

However, these chemical reactions are associated with an erosion process of PC reducing the mechanical stability of the material [2]. The understanding of interfacial reactions of PC with amino compounds offers the possibility to optimize reactive processing and to design interfaces with high adhesive strength [3].

For a long time it is known that the surface modification of PC substrates can be done by utilization of chemical reactions subsequently to polymer processing, e.g. in bath solutions [4]. However, subsequent surface modifications are time consuming, expensive, and solvents have to dispose. If keeping energy effective technologies in mind a process-integrated solvent free technology, for example injection molding with integrated surface modification, is worth striving for. To promote such technologies a fundamental understanding of interfacial chemical processes is crucial.

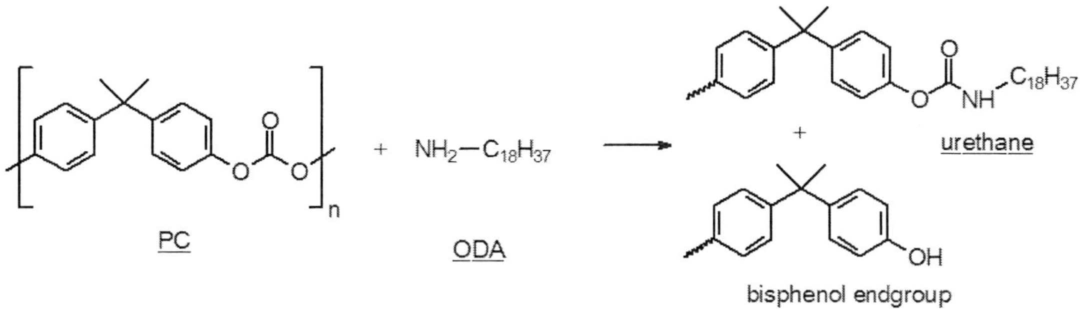

FIGURE 1. Reaction scheme of PC with ODA

For the development and optimization of process-integrated chemical reactions the different reactivities of compounds in solid-state, the reaction control under high temperatures and altered pressure conditions and, finally, a high-performance analysis of the resulting interfaces have carefully to taken into account. Hereby, the analysis will provide information about the molecular structure of the interface. Additionally, investigation of time-dependent alterations of the composition of the interphase is required to explain reaction mechanisms. IR spectroscopy is a non-destructive analytical method and useful to provide insights into the molecular structure of polymeric interfaces. It can be applied to evidence new formed covalent linkages as well as weak interactions, e.g. hydrogen bonding or other intermolecular interactions.

X-ray and thermal analysis methods must be applied to identify the formation of new products. The investigation of glass transition temperature measured by Differential Scanning Colorimetry (DSC) can be used to probe molar mass changes or phase miscibilities.

EXPERIMENATAL SECTION

Chemicals

PC (Makrolon 2207) was purchased from Bayer, Germany. Octadecylamine (ODA), grade 99% (Sigma-Aldrich, Germany) was recrystallized in chloroform to obtain a mp 50-52°C. Bisphenol-A, grade 97%, mp 158-159°C was supplied by Sigma-Aldrich, Germany and used as delivered.

Preparation of solid-state PC aminolysis

An equimolar ratio of PC and ODA was mixed in a glass tube. The mixture with a mass of approx. 1 g was homogenized and then transferred into a three-necked flask equipped with a stirrer, heated to 180 °C for 7 hours and purged with nitrogen, and under temperature control. Totally seven samples, each approx. 1 mg, were taken in an interval of 60 min from the PC-ODA mixture. Cooling time of approx. 5 min was applied. IR spectroscopic characterization was performed at room temperature.

IR spectroscopy

IR spectra from samples and from the reference substance were recorded in ATR mode using a FT-IR spectrometer Vertex 70 (Bruker Optik GmbH, Ettlingen, Germany) equipped with a horizontal ATR diamond crystal unit (Platinum ATR diamond unit, Bruker Optik GmbH, Ettlingen, Germany).

Spectra were analyzed using the Matlab package (Version 7.0 (Math Works Inc., MA, U.S.). Prior to data analysis, IR spectra were reduced to the region 950 cm^{-1} - 1800 cm^{-1}. A linear two-point baseline correction was performed by using the function msbackadj. Area normalization was carried out in order to scale all IR spectra to the same range. The algorithm for 2D correlation analysis was implemented in Matlab as described elsewhere [12].

DSC measurements

A DSC Q 2000 (TA Instruments, USA) was used for sample measurements. Under nitrogen, a temperature range of -80°C to 200°C was scanned with a rate of 2 K/min with temperature modulation of 0.31 K/40s for heating-cooling-heating cycles. Tg was calculated from the reversing heat flow signal.

CHARACTERIZATION OF AMINOLYSIS COMPOUNDS

IR spectroscopic characterization of reaction products of aminolysis of PC and ODA shows a complex mixture of different compounds. Evaluation of IR spectra reveals the formation of urethane groups but also of other compounds. The position of the strong stretching vibration band of carbonyl group indicates the structural units of different aminolysis compounds. Appropriated reference data were taken from literature [1,3] or from spectra of synthesized model substances. The IR spectra of the major reaction products are shown in Fig. 2. The spectra collection and implementation in an IR library represents an important tool for interfacial molecular characterization.

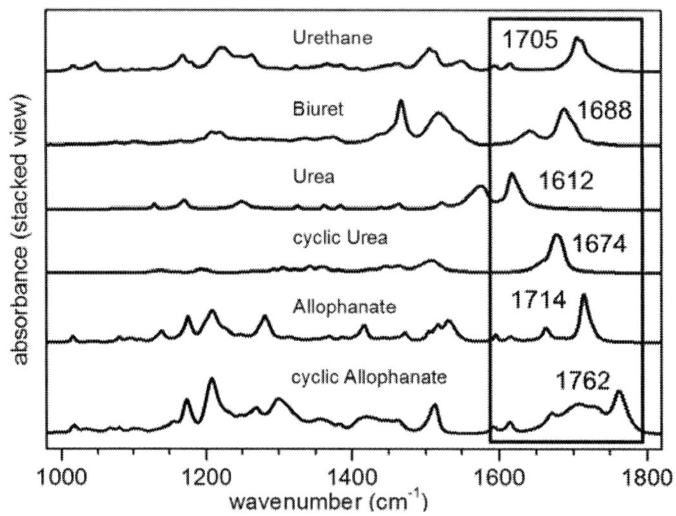

FIGURE 2. IR spectra of model aminolysis compounds. Only bands between 1000 cm^{-1} and 1800 cm^{-1}, relevant for the identification of reaction products formed through aminolysis of PC-ODA, are depicted. Especially, carbonyl stretching vibration range is highlighted and labeled.

INTERMOLECULAR INTERACTIONS BETWEEN COMPOUNDS FORMED THROUGH AMINOLYSIS AND POLYCARBONATE

The carbonyl absorption bands of PC - Bisphenol-A mixture (A) and PC-ODA mixture after thermal treatment (B) are shown in Fig. 3. While a pure PC exhibits only one band, located at 1774 cm^{-1}, the mixture of PC-Bisphenol-A shows a second band centered at 1750 cm^{-1}. The presence of this band indicates a molecular interaction of carbonyl groups of PC with hydroxyl groups of Bisphenol-A. A similar shift of the carbonyl band was reported for other carbonate - Bisphenol-A mixtures [5]. Therefore, the observed band shift of 24 cm^{-1} points to an interaction between C=O and O-H groups. The PC-ODA mixture after thermal treatment possess beside these two carbonyl stretching vibrations a band at 1712 cm^{-1} of a new formed urethane linkage and at 1683 cm^{-1} of another aminolysis compound.

INVESTIGATION OF REACTION MECHANISM

Processing conditions may vary, for instance in cycle times, pressure, or temperature profile. A time resolved characterization of the formation of reaction products points to changes of the molecular composition. A time resolved series of spectra was taken under identical conditions. 2D correlation analysis was applied to study the formation of new functional groups and intermolecular interactions (Fig. 4). The signs of the correlation coefficients define whether the signal is changed in the same direction (positive sign) or in opposite direction (negative sign).

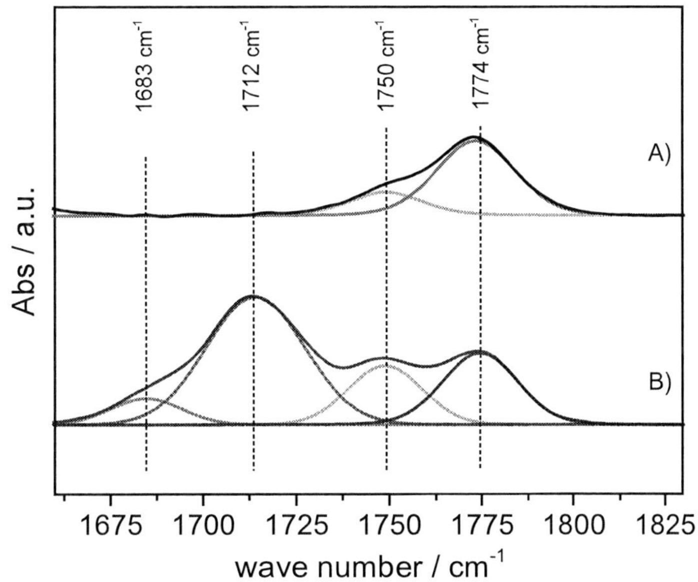

Figure 3. IR spectra in the range of carbonyl absorption bands of PC – Bisphenol-A mixture and curve fit analysis for two bands, depicted in gray tones (A) and of PC - ODA mixture (B) after heating to 180°C and cooling to room temperature. Curve fit analysis reveals 4 overlapped bands, depicted in gray tones. Interactions between C=O and O-H groups causes the shift of the carbonyl band from 1774 cm^{-1} to 1750 cm^{-1}.

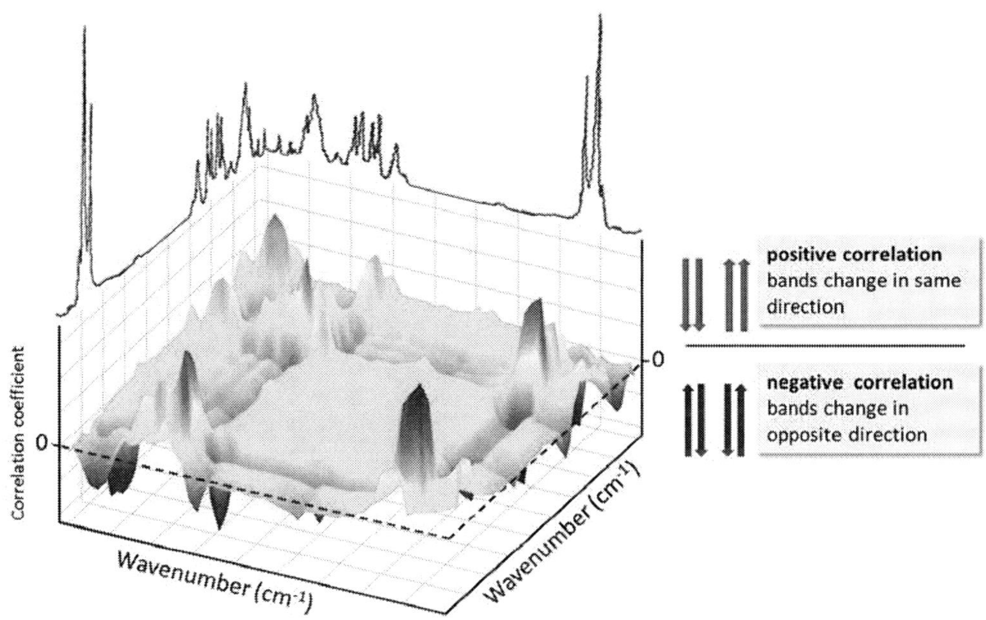

FIGURE 4. 2D map of the synchronous correlation coefficients

These signals indicate two parallel processes, namely the degradation of carbonyl groups and the formation of amide groups. In particular the temporal differences between degradation of carbonyl groups and the formation of amino and amide groups point to a multi-step formation and/or to different reaction products. It can be hypothesized that the solid state reaction route involves a number of steps including the formation of urethane as an intermediate product and different other reaction products, which were identified as Bisphenol-A, biuret and urea.

FORMATION OF INTERPHASES AND MIXED PHASES OF BISPHENOL-A AND POLYCARBONATE

The formation of new aminolysis products in the PC-ODA interface leads obviously to the formation of mixed phases with PC, amine, or with other reaction products as urethane. A new interphase occurs if no miscibility is obtained between the new formed aminolysis compound and the parent phases. In order to correlate the molecular structure of the PC-ODA interface with the adhesive strength and the final mechanical properties of the composite, this formed interphase has to be characterized with regard to its structure in dependence on time and on the corresponding processing parameters.The reversing heat flow - temperature diagram (see Fig. 5) shows a glass transition temperature of PC at 148°C. Bisphenol-A shows no signal before it melts at approx. 158°C. PC-Bisphenol-A mixtures show a drastic decrease of glass transition temperature to about 40-70°C depending on molecular ratio of PC to Bisphenol-A. This can be interpreted by the plasticizing impact of Bisphenol-A on PC. The exothermal peak in the curves of PC-BPA 50-50 and PC-BPA 20-80 are interpreted as a cold crystallization process of Bisphenol-A crystallites in the mixed PC-Bisphenol-A phase.

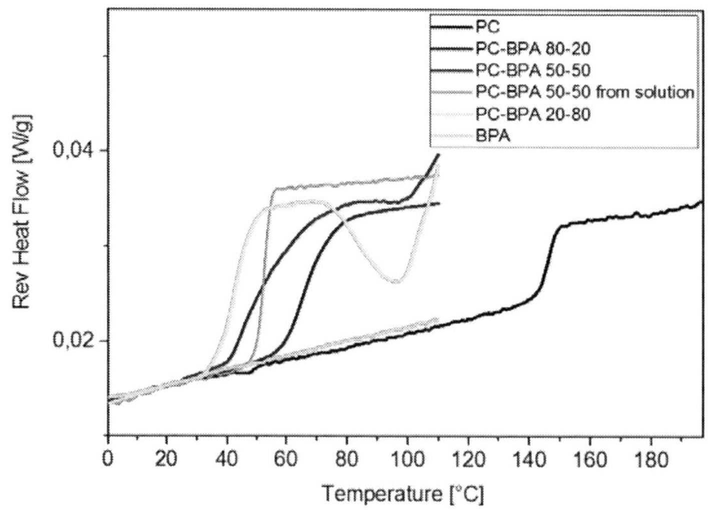

FIGURE 5. DSC curve of PC, Bisphenol-A and mixtures of PC-Bisphenol-A (PC-BPA) for different ratios

SUMMARY

The chemical reaction starting at a binary PC-amine interface comprises compositional changes between reactants and products and a possible phase separation of the formed interphase. The IR spectroscopic study of the solid state reaction reveals the formation of desired urethane groups as interfacial linker. The applied 2D correlation analysis demonstrates that urethane is only an intermediate product. Biuret and finally urea will be formed during the subsequent processing time. Moreover, Bisphenol-A as degradation product of polycarbonate is also formed. Consequently, a surface modification of PC with amine requires a defined reaction time in order to ensure an optimized interfacial link and to minimize the degradation of the materials. Nevertheless, the results of this study clearly demonstrate that surface modification of PC during injection molding can be considered as an effective advancement of this primary forming process technology.

REFERENCES

1. C. Zimmerer, J. Nagel, G. Steiner, G. Heinrich, Macromol. Mater. Eng. **301(6)**, 648-652 (2016).
2. P. Jankowski, D. Ogonczyk, A. Kosinski, W. Lisowski, P. Garstecki, Lab Chip **11**, 748-752 (2011).
3. C. Zimmerer, F. Wolff-Fabris, E.Koch, G.Steiner, G.Heinrich, Polymer **54(25)**, 6732-6738 (2013).
4. J. R. Caldwell, W. J. Jackson, Jr., J. POLYMER SCI.: PART C **24**, 15-23 (1968).
5. B. Fei, C. Chen, S. Peng, X. Zhao, X. Wang, L. Dong, Polym. Int. **53(12)**, 2092-2098 (2004).

Effects of Vibration welding process parameters on the weldability of the thermoplastics

DAS CHAPAL KUMAR [1]*, ADHIKARY MUNESH CHANDRA [1], FRIEDRICH S [2], GIHDE MICHEL [2]

[1] FAKIR MOHAN UNIVERSITY - ODISHA - India, [2] TU-Chemnitz - Chemnitz - Germany

Joining of Plastics is mainly done at the places where the product cannot be fabricated in a single shot manufacturing process. Plastic Welding Processes can be divided into 2 groups; one involves mechanical movement and the other requires external heating. In this study the vibration welding, which involves mechanical movement has been applied to join the two thermo plastics of different character. It is very easy to weld two crystalline plastics or two semi crystalline plastics but it is very difficult to join the amorphous thermoplastics either together or crystalline or semi crystalline plastics as another partner. In this case, either in-situ compatibilizer or externally applied compatibilizer is a must. We tried to join the PMMA-PBT system and PMMA- nylon 66 systems using various welding process parameters. Externally applied compatibilizers used are PMMA-PBT= 50-50%, 70-30%, 30-70%. Similar proportions were used for PMMA- Nylon-66 system. It has been observed that the viscosity play a vital role in vibration welding process which changes the process parameters in order to get the efficient welding. The various parameters like amplitude, welding pressure, Dwell Time and melt down depth (distance) were studied. It has also been found that in case of PMMA-PBT system the more welding pressure and higher amplitude is required. Melt down depth however is little bit lower may be because of higher viscosity of PBT. The weld line was characterized by FESEM and found to be folded down in the case of PMMA- Nylon-66 system. However for PBT system, the folding is in the same plane of weld line. Better weldability have been observed when the two types of externally applied compatibilizer are applied separately on to the two substrates and then vibration welded. Two types of solution compatibilizer in the ratio of 70-30% and 30-70% is the better option than the 50-50% variety. Keywords: Vibration welding, PMMA- PBT, PMMA-Nylon66, Welding Pressure, welding Amplitude.

Ultrasonic welding of polystyrene microinjected moulded components – physic-chemical and mechanical characterisation

Tuinea-Bobe Cristina Luminita [(1)]*, Smith Nigel [(1)], Whiteside Ben R [(1)], Coates Phil D [(1)]

[(1)] University of Bradford - West Yorkshire - UnitedKingdom

The plastics industry is a fast evolving sector, with customers requiring higher volumes of products, complex shapes, miniaturisation and added functionality, in an environment friendly manner at competitive cost. Ultrasonic welding is one of the processes used to assembly injected moulded components to fulfil the above requirements, in various industries: automotive, medical, electrical and electronic, communications, appliances, consumer products, toys, textile, packaging, etc. This economic process is fast, clean, efficient, repeatable, and can significantly increase production and lower assembly cost. Offering flexibility and versatility, high reliability, consistency, long life, and low-cost makes ultrasonic welding the preferred method of assembly. The ultrasonic created bonds are desired to be strong and integral, and retaining the properties of the pristine material. Microinjection moulding makes possible the production of fine details with high precision and accuracy, requirements extended to the assembly process. The bonding process needs to maintain the crystallinity, the optical and mechanical properties of the original component – attributes highly influenced by the heat generated from high-frequency mechanical motion. This study will show how the quality of welds varies due to the mechanical motion produced by a 20 kHz ultrasonic welding system, along with applied force that creates frictional heat at the plastic components joint area. The paper includes the manufacturing (microinjection moulding), the physico-chemical (DSC, confocal microscopy, SEM, microCT) and mechanical (tensile testing) characterisation of polystyrene components. Parameters such as: energy director shape, ultrasound amplitude and energy, weld time, hold time and pressure will be discussed in the assessment of welds resulting from this high-frequency mechanical motion process.

Preparation of Ultra-Thin Metallic Susceptor Structures for Electromagnetic Induction Heating of Reactive Polymer Interphases

L. Ziegler[1], E. Kampert[2], A. Janke[1], G. Heinrich[1], C. Zimmerer[1, a)]

[1]*Leibniz Institute of Polymer Research Dresden, Hohe Str. 6, 01069 Dresden, Germany*

[2]*Dresden High Magnetic Field Laboratory, Helmholtz-Zentrum Dresden-Rossendorf, 01314 Dresden, Germany*

[a)]Corresponding author: Zimmerer@ipfdd.de

Abstract. A single high magnetic field pulse of 60 T can be used to initiate chemical reactions between Bisphenol-A based Polycarbonate and Octadecylamine. Therefore, susceptor structures are needed for a localized heating of the interphase between the reactive polymer materials. As susceptors, stable gold structures were developed and produced by physical vapor deposition. The gold structures were characterized morphologically and electrically to optimize their heating efficiency.

INTRODUCTION

Induction heating is widely employed for the heating and bonding of metals. For non-conductive materials, susceptor structures are useful for this technique. In general, an alternating electromagnetic field induces an eddy current in a susceptor structure, often made from metals, which delivers energy in terms of thermal load to the surrounding medium [1].

Our aim is to apply induction heating to reactive interfacial bonding of polymeric materials. Therefore, we extent the above explained macroscopic principle of induction heating to nanoscopic dimensions by utilizing nanometer thick susceptor structures. Localized induction heating of reactive polymeric interfaces of Polycarbonate (PC) and Octadecylamine (ODA) leads to the formation of covalent bonds. Previous works showed that high magnetic field induction is a suited technique for the reactive bonding between these polymers using silver structures as susceptors [2]. In the present approach, silver was replaced by gold to avoid the energy loss caused by a spontaneous oxidation of silver. Requirements for localized interfacial induction heating are the application of ultra-thin metallic structures in order to reduce the heating of the bulk material and their careful placement in between both materials.

The new joining method is presented schematically in Fig. 1. Gold susceptor structures were deposited into the interface of both materials. The electromagnetic induction heating was achieved by applying a high magnetic field pulse of 60 T. Consequently, the susceptor structures heat the surrounding polymer materials. Afterwards new covalent bonds arise between the PC and ODA, which are detected and analyzed by Fourier Transform Infrared (FTIR) Spectroscopy.

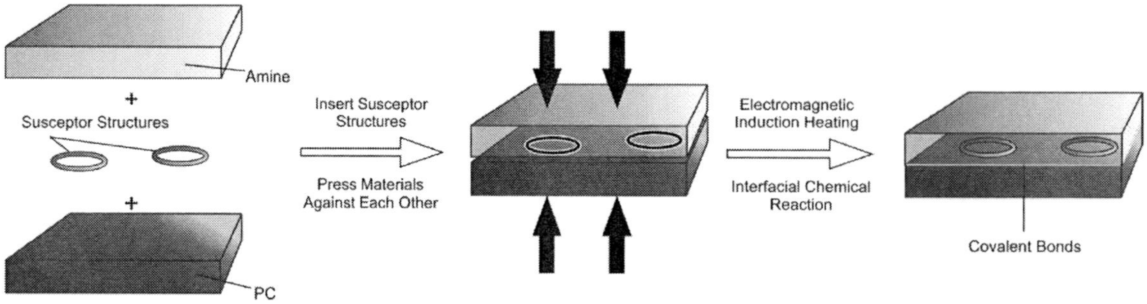

FIGURE 1. Preparation route for covalent bonding of PC and primary amine ODA by electromagnetic induction heating with susceptor structures.

PHYSICAL PRINCIPLES AND CALCULATION OF TEMPERATURE INCREASE

The simultaneous discharging of a number of 4 capacitor banks, with an electrical energy of 1455 kJ, to one copper coil creates an alternating high magnetic field pulse of about 70 T [3]. This alternating magnetic field of the copper coil is denoted as primary magnetic field. The polymer materials, including the susceptor structures, are placed to the core of the copper coil in the space of homogeneous field lines. The primary magnetic field induces a secondary magnetic field and, subsequently, an eddy current in the susceptor structures. Susceptor structures are heated where heat radiation causes a definite warm up of the surrounding polymer materials. As soon as the activating energy is exceeded through the increase of temperature in the polymer materials, the initiated chemical reaction results in a formation of new covalent bonds between both materials. The formation of covalent bonds is required to obtain high adhesion strength and permanent bonding between two polymer materials.

A high magnetic field pulse is able to transfer energy contactless to the susceptor structure and its application is independent from the polymer material or added organic or inorganic auxiliaries with polar molecular structure. To estimate the resulting temperature increase in the interphase of the materials, it is assumed that the electrical energy is completely transferred in form of thermal energy. After adjusting the equations from Ref. [2], the following equation for the temperature increase is obtained (1).

$$\Delta T = \frac{U^2 \cdot \Delta t}{R \cdot m \cdot c} \tag{1}$$

It demonstrates that the temperature increase ΔT (K) is restricted by the electrical resistance R (Ω), the specific heat capacity c ($J \cdot kg^{-1} \cdot K^{-1}$) and the mass m (kg). The greater these values, the lower the temperature increase. The electrical voltage U (V) and time t (s) also have an impact on the temperature but they are limited by the properties of the applied 70 T copper coil.

Figure 2 presents schematically the temporal path from a single high magnetic field pulse of about 60 T [3]. After 7 ms it has reached its maximum. After 19 ms the magnetic flux density has reached zero and the energy input into the material almost ends.

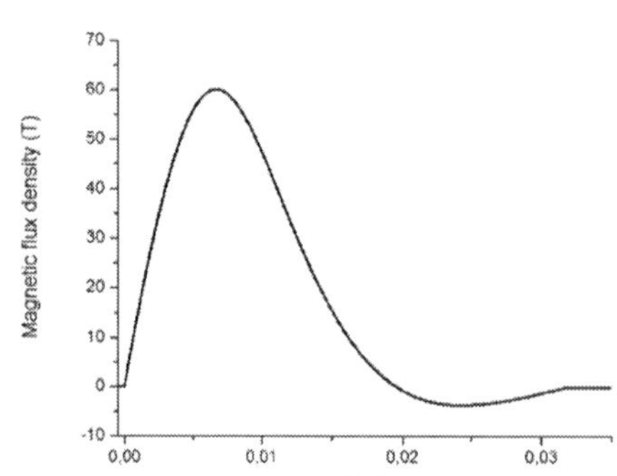

FIGURE 2. Temporal path of a high magnetic field pulse with a magnetic flux density of 60 T.

EXPERIMENTAL

Materials

PC Makrolon 2245 was received from Bayer, Germany. ODA (99 %) was obtained from Sigma Aldrich, Germany. Chloroform (99 +%) ($CHCl_3$) and Toluene were received from Acros Organics, Spain, and Ethanol absolute from VWR Chemicals, France. Gold (99.99 %) was obtained from SAXONIA Edelmetalle GmbH, Germany.

Sample Preparation

Silicon wafers with native SiO_2 (CZ prime, (100) orientation, 10 Ω·cm resistance, TU Dresden, Germany) were cleaned with chloroform and afterwards with ethanol each for 5 minutes in an ultrasonic bath at room temperature. PC solution (10 g/l in $CHCl_3$:Toluene, 1:1) was spin-coated on Si-wafers with a spin-coater Spin150i from APT GmbH, Germany, to yield PC films with thicknesses of about 50 nm. These samples were vaporized with gold (rate of 0.1 A/s) by Physical Vapor Deposition (PVD) with thermal evaporation (Creamet 300 V2 from Creavac GmbH, Germany). To achieve gold rings, the wafers were masked so that rings with an outer diameter of 6 mm and an internal diameter of 4 mm were produced.

Characterization

Resistance Measurements. The gold susceptor structures were cut once with a needle. On both sides of the cut two copper wires were fixed. With an ohmmeter (Model 350 from Lake Shore Cryotronics, USA) the resistance was measured of the susceptor structures.

Atomic Force Microscopy (AFM). AFM measurements of the gold surfaces were done in the peak force tapping mode by a Dimension ICON (Bruker-Nano, USA). We used silicon nitride sensors SCANASYST-AIR (Bruker, USA) with a nominal spring constant of 0.4 N/m, the tip radius is 2 nm.

RESULTS AND DISCUSSION

A chemical reaction between PC and ODA yields in a formation of urethane as linker compound. In the reaction with a single high magnetic field pulse, both substances (PC and ODA) are present as solid phases. Due to the low melting point of ODA, at 53°C, [4], the reaction is initiated on the interphase of the materials at temperatures in this range. The successful coupling of both materials will be analyzed by FTIR measurements.

From Equation 1 it can be concluded that the electrical resistance reciprocally affects the temperature increase of the susceptor structures. To reach an optimized heating effect of the susceptor structures, a low electrical resistance is highly desirable. Table 1 lists different elements which are used for electromagnetic induction heating and their specific electrical resistance. As it can be seen the electrical resistance from gold is quite high compared to silver and copper, but lower than nickel.

TABEL 1. Elements and their specific electrical resistances [4].

Material	Electrical resistance [Ω·m]
Ag	$1.467 \cdot 10^{-8}$
Cu	$1.678 \cdot 10^{-8}$
Au	$2.214 \cdot 10^{-8}$
Ni	$6.93 \cdot 10^{-8}$

Gold susceptor structures were prepared as rings with a thickness of 10-50 nm (Fig. 3a). Figure 3b presents the electrical resistance of the gold ring susceptor structures and the theoretically calculated values. The gold susceptor structures were also subjected to a thermal annealing (2 h at 200°C) to improve the electrical resistance.

For all gold ring structures an annealing results in a decrease of electrical resistance. For a 10 nm thick ring structure the electrical resistance is decreased to about 500 Ω but always 27 times higher than the calculated value of about 19 Ω. For the thicker gold ring susceptor structures this annealing effect could be confirmed, but is less prominent. The highest concordance between experimental and theoretical electrical resistivity was achieved for 50 nm thick gold rings.

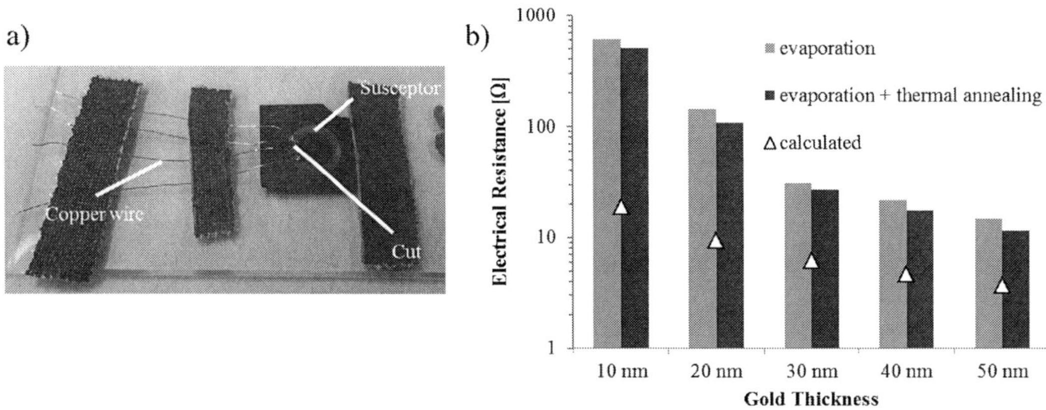

FIGURE 3. a) Gold susceptor structure on a PC coated silicon wafer contacted with copper wires for electrical resistance measurements, b) Results of the electrical resistance measurements of 10-50 nm thick gold structures and calculated data.

The surface measurements by AFM of the gold susceptor structures show that clusters are deposited. By that they do not present a homogeneous morphology in the thickness range of 10-50 nm. With increasing thickness of the gold layer, metal layer grows to a closed but not homogeneous layer, see Fig. 4a) - left. The clusters are not immobile but can move during a thermal treatment [5]. We suppose that the cluster like structure of Au layer causes the higher experimental resistance values.

Figure 4a) demonstrates the morphology of the gold susceptor structures with different gold thicknesses before and after the thermal treatment. From the AFM images the property of the surface coverage is calculated and related to ring thickness. The coverage at a thickness of 10 nm is 55 %, and 93 % were observed when the thickness was 50 nm (Fig. 4b). Through the thermal treatment larger regions of different height arise and the electrical resistance decreases. Figure 4b shows the theoretically achievable temperature increase, calculated from Equation 1, for the gold susceptor structures from measured electrical resistances applying a single magnetic high field pulse of definite flux density of maximal 60 T with the pulse shape from Fig. 2. The 10 nm gold susceptor structure behavior, even after thermal treatment, prohibit an induction heating application for PC-ODA system, because the temperature increase is too low. Since the melting point of PC or ODA is not reached with a temperature increase of 7 K. For 30 nm thick and annealed gold ring susceptor structure a temperature increase of 45 K is calculated. A temperature increase of 63 K with 50 nm thick gold ring structures is feasible for the application to bond both, PC and ODA.

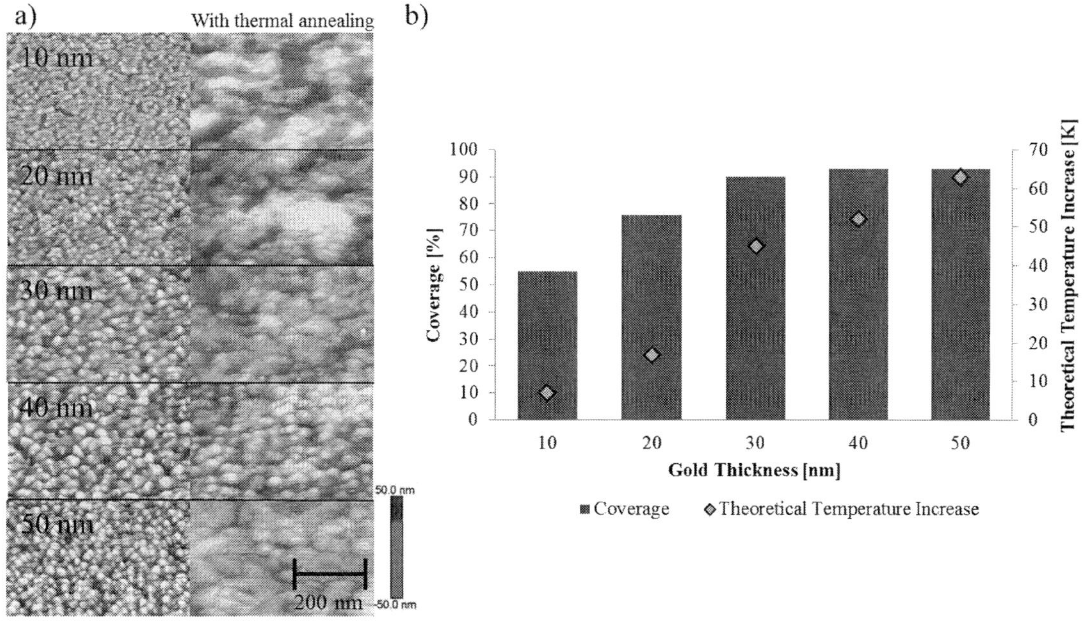

FIGURE 4. a) AFM images of gold ring structures before (left) and after (right) thermal treatment (200°C, 2 h), b) measured electrical resistance in relation to layer thickness and the theoretical temperature increase calculated from experimental values with a definite magnetic pulse path.

CONCLUSION

In this study a route is presented for the bonding of solid-state Polycarbonate with a solid-state primary amino group containing material by electromagnetic induction heating. The approach is focused on the preparation and improvement of a few nanometer thick gold susceptor structures. After thermal annealing, an appropriate electrical resistance of susceptor structures can be obtained for 50 nm or thicker ring structures. The chemical reaction of PC with amine can be initiated at a final temperature of 361 K, with 63 K temperature increase by induction heating of 50 nm thick gold ring structures and started at 298 K (room temperature).

ACKNOWLEDGMENTS

The authors would like to thank the support of the High Magnetic Field Laboratory, Helmholz Zentrum Dresden-Rossendorf.

REFERENCES

1. J. Chovan and M. Slodička, J. Comput. Appl. Math. **311**, 630 (2017).

2. C. Zimmerer, G. Heinrich, F. Wolff-Fabris, E. Koch and G. Steiner, Polymer **54**, 6732-6738 (2013).

3. E. Kampert (private communication).

4. D. R. Lide, *CRC Handbook of Chemistry and Physics*, 84th ed. (CRC Press, Boca Raton, FL, 2004) p 3-438 – 12-44.

5. J. Lee, P. Pandey, M. Sui, M.-Y. Li, Q. Zhang and S. Kunwar, Nanoscale Res. Lett. **10**, 494 (2015).

Joining Polycarbonate - Manufacturing and Evaluation of Transparent Joints Using an Innovative, Objective Test Method

Karoline Hofmann[1, a)] Michael Gehde[1)]

[1]*Chemnitz University of Technology, Institute of Materials Handling, Conveying and Plastics Engineering, Professorship of Plastics Engineering, Reichenhainerstr. 70, 09126 Chemnitz, Germany*

[a)] Corresponding author: karoline.hofmann@mb.tu-chemnitz.de

Abstract. Large-scale laminated safety glass is increasingly used in the public and private sector for example museums, jewelers or villas. Special safety is provided by these glasses in layered combination with polycarbonate sheets. Due to the low weight and its high tensile elongation comparing glass of the same thickness, highly transparent polycarbonate is suitable for safety-relevant components. Because of the dimensional limitation in the extrusion process, joining polycarbonate compounds are necessary for large-area joints of this laminated safety glass. According to the present state of the art, a transparent weld joint of polycarbonate is not possible. Within the scope of the project, polycarbonate joints were produced by infrared welding. The transparency of the joints was realized by varying the process parameters, for example the heating time. Additionally, the present study focusses on a new innovative light intensity measurement, because conventional optical test methods can only subjectively assess the transparency and quality of the part and the welding seam. The principle is based on the transmission of the welded and polished polycarbonate sample by means of a laser and the measurement of the current at the connected photoelectric cell, which is proportional to the intensity of the laser light. This enables a qualitative evaluation of the joining dimension and quality such as structural changes over the entire sample width. The results of this objective method correlate with the width of the stress distribution around the joining level, which are made visible with the photoelasticity using a polariscope and thus contribute to a better process understanding.

INTRODUCTION

Polycarbonate is used in a large variety of everyday applications where it has mostly been proven critical to the performance of the component. Such unique applications are found widely in the construction and building, the automotive and medical sector. For these industries, the special features of polycarbonate render it unique for its respective uses.

Especially in the construction sector, polycarbonate offers unique possibilities thanks to its properties. The combination of fire retardance, low weight, glass-like transparency and high impact resistance helps to reduce costs, develop new structures and improve energy efficiency [1]. In comparison with other amorphous thermoplastics, polycarbonate shows the highest elongation at break point, and compared to glass with the same thickness it saves up to 50 % of weight [2]. That makes it ideally suited for use in laminated safety glass panels, which are mainly used by private individuals or jewelry stores and museums. This laminated safety glass is made of polycarbonate sheets and glass in a layered combination and combines the positive properties from both materials. Therefore, polycarbonate safety glass can be UV- protective, lightweight, thin and soundproof, and it provides both active protection against thrown projectiles, smashing and gunshots, and passive safety thanks to its shatterproof properties [3].

ASSIGNMENT

Because of the increasing demand and growing design requirements regarding the shape and size of the laminated safety glass, it is necessary to develop strategies to meet these requirements. Due to the manufacturing process of the polycarbonate plates in the extrusion process, the possible dimensions of the plates are limited. Currently it is possible to produce plates dimensions from 2500 x 8000 mm^2 [3]. Larger solutions require a joining process of the polycarbonate sheets. This could be achieved with different welding processes or bonding.

The present study focuses on two main tasks. Firstly, finding a suitable method to produce welded joints with good mechanical and optical properties. That means that the initial transparency is to be restored. And secondly, the welded connections should be examined and evaluated regarding to their transparency. By using conventional optical measurements, only subjective gradations of transparency are possible. Objective statements about the degree of transparency to the base material cannot be made. Therefore, it is necessary to develop an innovative test method which would help to validate the welded seams objectively.

METHODOLOGY

The test material was a PC Lexan in different material thicknesses. Injection-molded plates with a thickness of 2 mm and 4 mm were used as well as extruded sheets with a thickness of 12 mm. The test material had been re-dried before joining, because polycarbonate is hygroscopic, i.e. it absorbs moisture from the atmosphere.

Regarding the high optical requirements, the infrared welding is a suitable method for joining polycarbonate. The welding tests were taken at a servo motoric horizontal heating element welding machine Type K2150 bielomatik with a modular infrared beam unit. The utilized emitters were short-wave quartz glass emitters with maximum power of 1250 W. Preliminary investigations were carried out to characterize the basic material. The test methods included the Karl- Fischer- Titration (KFT), the differential scanning calorimetry (DSC analysis) and mechanical tests such as bending and tensile tests to get reference values.

Additionally, heating tests were made to get a first impression about the melting behavior depending on the selected parameters.

Following this, the welding tests were made to determine the influence of the welding parameters to the optical properties (tab.1). Therefore, the radiator distance, type and power were kept constant over all welding tests. The heating time, was varied between 40 s and 60 s depending on the material thickness. A variation of the drying time was conducted to determine the influence of the material moisture to mechanical and optical properties. Another influencing parameter, to the transparency of the connection in the welding process, is the joining pressure which has been changed between 0.25 N/mm^2 and 2 N/mm^2.

TABLE 1. welding parameters for infrared welding of polycarbonate sheets

radiator distance [mm]	radiator type [W]/ power [%]	material thickness [mm]	heating time [s]	drying time [h]	joining pressure [N/mm^2]
25	1250/ 100	2,4,12	40,45,50,50,55,60	0,2,8,12,18,24	0.25,0.5,1.5,2

After cooling, the welded plates were sawn in test specimens for various mechanical and optical tests. The mechanical properties were determined by a standardized tensile and bending test at a universal testing machine type Z250 with a 50 kN loadcell (Zwick/Roell) and an INSPEKT table 10 with 10 kN loadcell (Hegewald & Peschke).

For determining the optical properties various laboratory devices were used, such as a polarization microscope type BX51 (Olympus), a polariscope (Zeiss), and an SLR camera (Sony). Furthermore, a new test method has been developed – the Light Intensity Measurement (LIM).

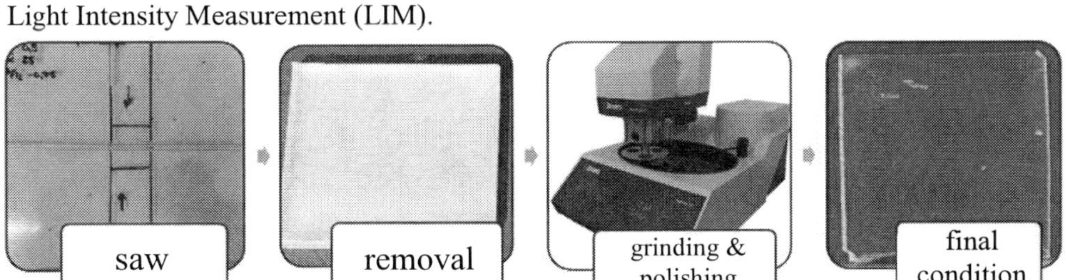

FIGURE 1. process of restoring the initial transparency

Before evaluating the optical properties of the joints, the specimens had to be prepared in a time- and material consuming process (fig.1). Firstly, smaller samples had to be sawn from the welded plates. Then the weld bead was to be removed on both sides using a belt sander to generate a flat surface. This was followed by a professional finishing of the samples' surface using an automatically grinding and polishing machine. This procedure needed several steps of grinding with different grain sizes and polishing procedures until it reached its initial transparency.

Thus, the test setup for the developed Light Intensity Measurement is shown in fig.2.

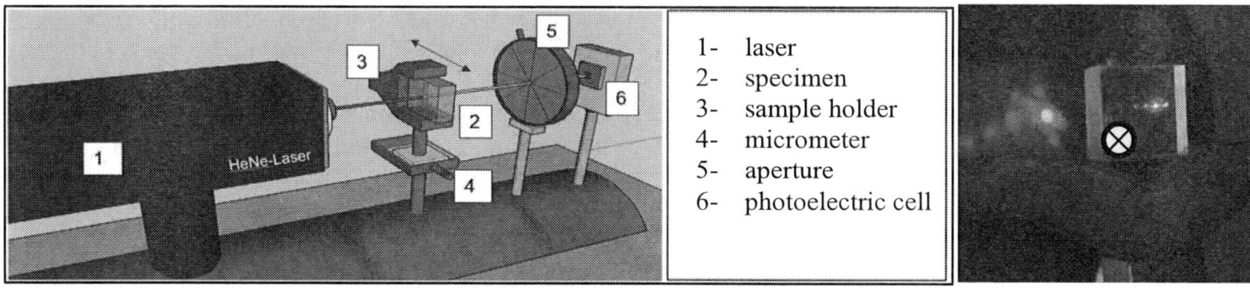

FIGURE 2. schematic representation of the measurement setup (left), 12 mm PC sample during the measurement

A new test method was found, using a Helium-Neon-Laser (632 nm) and a photoelectric cell. As illustrated in the picture, the polycarbonate sample is placed between a laser and a pinhole or aperture on a sample holder, which can be moved with the aid of a micrometer. Behind the pinhole is a photoelectric cell which uses the external photoelectric effect [4,5]. The laser light emits a constant laser beam with a defined light intensity to the sample. Depending on the surface and internal structural changes or defects in the sample, parts of the light are transmitted, reflected or absorbed [6]. The transmitted part of the light goes through the sample and hits the photoelectric cell. The remaining part of the light can be measured as current. The test begins at the edge of the sample and the values of the current are recorded every 100 µm over the entire width of the sample. If the measured data is at the same level, without any changes, the investigated specimen can reffered as transparent. But whenever the current alters, the sample isn´t completely transparent because structural changes at the surface or within the sample are affecting the laser beam. By this means the welded samples can be evaluated in an easy and objective way.

RESULTS AND DISCUSSION

Pre-tests

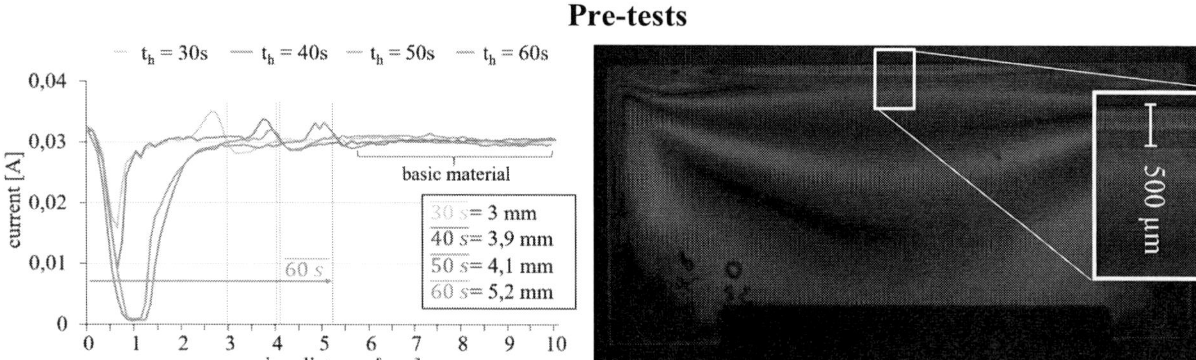

FIGURE 3. results of the heating tests using the LIM (left) and distance measurement using a polariscope at a PC sample with 60 s heating time (right)

The results from the heating tests are shown in fig. 3 for LIM (left) and for the photoelastic analysis from a PC sample with 60 s heating time (right). The plates were heated from 30 s to 60 s in 25 mm distance to the radiator with 1250 W at 100 % power. After the cooling process the heat affected zone, starting at the edge of the sample, was measured using the LIM. In fig.3 (left) are strongly declining values at the beginning of the course, followed by typical peaks in every heating time curve. The decreasing values can be explained by light scattering at the rounded edge of the samples after grinding. The peaks after these areas representing the influence of the heating time. The visualized results from the LIM show that the longer the heating time, the higher the melt layer thickness and the smaller the change in the current, compared to the basic material level. Furthermore, the comparison of the LIM and the photoelastic analysis show approximately the same distance of the heat-affected zone in the material.

Optical Results

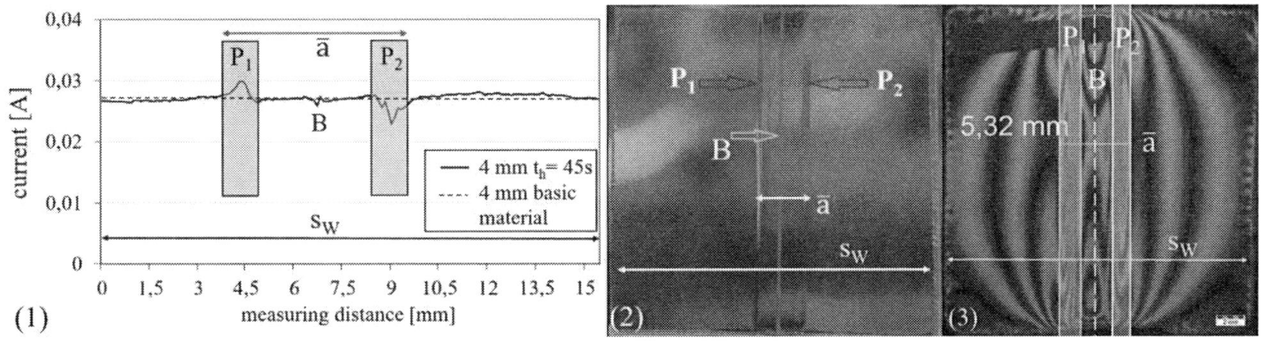

FIGURE 4. optical results of the different test methods - LIM (1), subjective evaluation by means of photography (2) and photoelasticity (monochromatic light) using a polariscope (3)

Figure 4 shows the correlation between the optical results using different test methods like LIM (1), the subjective evaluation using photography (2) and the stress distribution measurement using a polariscope (3). By investigating the sample macroscopically (2) there can be recognized up to 3 areas, some of them being more visible than others. These lines appear after the polishing process and affect the visual impression. The joining zone can be recognized in the middle as a sharp line B (2) additionally on both sides of the joining zone there are two blurry areas, which hadn´t been previously observed or investigated (fig.4 P_1, P_2). The section between P_1 and P_2 is shown as the distance $\bar{a}$ and s_W is equivalent to the entire measuring distance using the LIM.

Figure 4 (1) shows the typical progression of a curve for 4 mm PC with a heating time t_h = 45 s and a joining pressure of p = 0.5 N/mm^2. The three typical areas are also visible in this curve. Before P_1 and after P_2, a very flat course is detectable, which represents the basic material. Within the distance $\bar{a}$, the current changes so that a certain structural change takes place and the transparency of the sample differs, depending on the welding parameters. This distance $\bar{a}$ can be measured both with the LIM and with the evaluation of the polariscope photos, and the results are approximately equal.

Afterwards, it must be evaluated which welding parameters are responsible for the macroscopically visible areas. In addition to the humidity, the heating time and the joining pressure could have an influence. Figure 5 shows the influence of material moisture (left) and duration of heating time (right) to the optical properties.

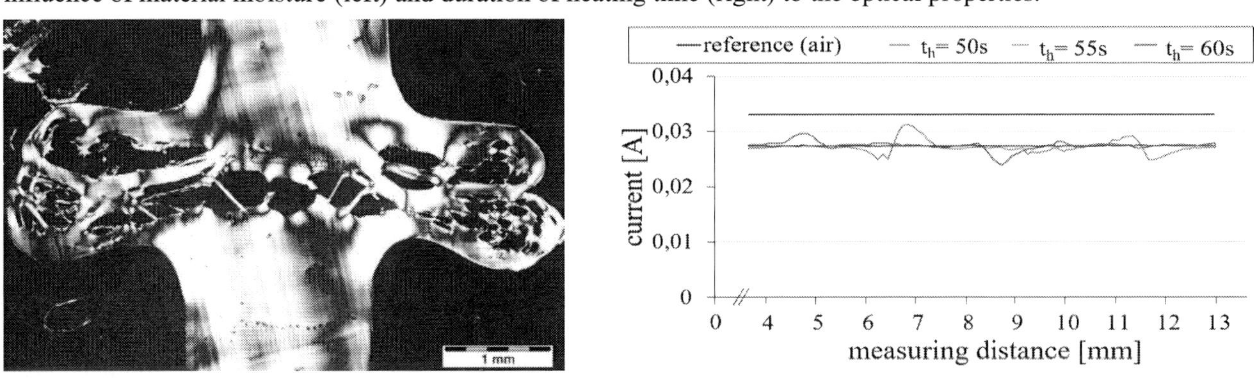

FIGURE 5. transmitted light microscopic image of moist PC d= 2mm, polarized, 25x (left) output signal using LIM shown for PC d= 2 mm with different heating times (right)

After a prolonged storage time after the extrusion process, the moisture content in the material increases. During the welding process the water will react with the molten polymer at the procession temperature. This leads to a formation of bubbles in the joining zone and both mechanical and optical weakening. Figure 5 (right) shows the results of different heating times for 2 mm PC using the LIM. As with the 4 mm and 12 mm thick material, the heating time has a decisive influence on the visual impression of the connection. The longer it is, the wider becomes distance $\bar{a}$

until it last in an even course of the curve as with the 60 s heating time and a transparency which correspond to the basic material.

Furthermore, the reference on air is the basis of the measurement and can be defined when nothing is placed between the photoelectric cell and the laser. The thicker the material, the lesser the measured light intensity (current) at the photoelectric cell, as the laser beam must pass through more material (see fig.6, left).

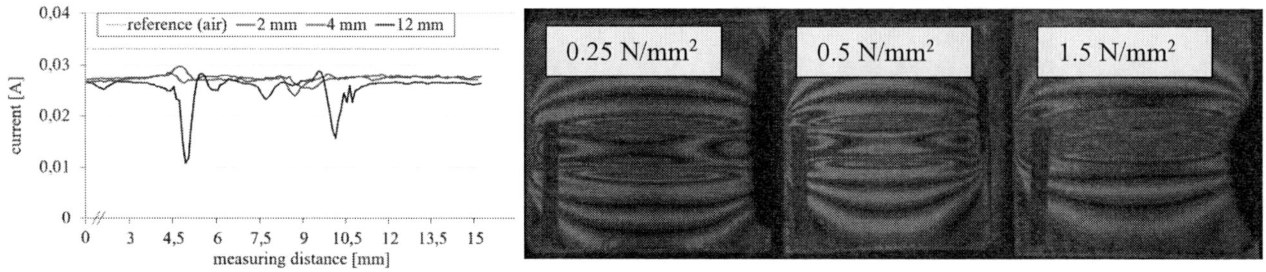

FIGURE 6. output signal using the LIM for different material thicknesses, t_h= 50 s and p= 0.5 N/mm^2 (left) and different joining pressures for PC d= 4 mm, t_h= 50 s, photoelastic analysis using monochromatic light (right)

Figure 6 (right) shows the influence of different joining pressures by means of the photoelasticity. An increase in the joining pressure causes a reduction in the distance $\bar{a}$, because the P_1 and P_2 areas get closer to each other. The decrease of this area $\bar{a}$ can also be seen macroscopically, which leads to a degradation of the subjective impression. Regarding the polariscope measurement, concentrated residual stresses can be seen around the joining zone, caused by the welding process. A tempering process was made to reduce the concentrated residual stresses in these areas. The thermal inducted residual stresses in the basic material of the sample were reduced, however, the P_1 and P_2 regions remained the same. With a rising joining pressure the concentration of the stresses in the P_1 and P_2 regions even increase. Thus, the joining pressure also has an important influence regarding to the transparency of the PC samples.

CONCLUSION

After the removing of the weld bead it is necessary to restore the initial transparency of the material, which was possible with the aid of several grinding- and polishing procedures. By means of the developed LIM, it is now possible to evaluate the transparency of the samples in an easy and effective way. Moreover, these can be compared with one another and the influence of the welding parameters is also predictable. The results of the LIM correlate with the results of the macroscopic subjective evaluation of the samples and with the results of the polariscope measurements.

PC needs to be dried before welding, because the moisture in the material has an influence on the optical properties.

The heating time has a decisive influence as well. The longer it is the better the results. Furthermore, the area $\bar{a}$ becomes wider with an increasing heating time, because a higher energy quantity is introduced into the material. Within the 2 mm material and a heating time from 60 s it was possible to produce a welded joint which corresponds to the basic material transparency.

Together with medium joining pressure (around 0.5 N/mm^2) it is possible to create an invisible connection.

Thus, the new test method must be validated and extended in further studies. Furthermore, the process of restoring the initial transparency is very time-consuming and must become more efficient. Only then large-area PC connections can be produced and further processed to laminated safety glass.

REFERENCES

[1] Lau, K.: Plasmagestützte Aufdampfprozesse für die Herstellung haftfester optischer Beschichtungen auf Bisphenol-A Polycarbonat, Martin-Luther-University Halle Wittenberg, Ph.D.Thesis. Halle 2006.
[2] N. N., Technische Kunststoffe im Überblick, ThyssenKrupp, 12/ 2009.
[3] Silatec Sicherheits- und Laminatglastechnik GmbH, Gelting, Germany, available at https://www.sicherheitsglas.de/fileadmin/user_upload/downloads/SILATEC_produktkatalog.pdf.
[4] Herrmann, K.H.: Der Photoeffekt- Grundlagen der Strahlungsmessung, Vieweg+Teubner Verlag, 1994.
[5] Naumann, H., Schröder, G., Löffler-Mang, M.: Handbuch Bauelemente der Optik, Carl Hanser Verlag, München Wien, 2014.
[6] Schröder, G., Treiber, H.: Technische Optik, Vogel Business Media GmbH & Co KG, Würzburg, 2014.

CLINCHING AND TORSIONAL ULTRASONIC WELDING - AN INNOVATIVE PROCESS COMBINATION FOR JOINING METAL-POLYMER HYBRID STRUCTURES

Eric Brueckner[1,a)], Wolf Georgi[2,b)], Michael Gehde[1,c)] and Peter Mayr[2,d)]

[1]Chemnitz University of Technology, Institute of Materials Handling, Conveying and Plastics Engineering, Professorship of Plastics Engineering, Reichenhainer Str. 70, 09126 Chemnitz, Germany
[2]Chemnitz University of Technology, Institute of Joining and Assembly Technology, Professorship of Welding Engineering, Reichenhainer Str. 70, 09126 Chemnitz, Germany
[a)]*Corresponding author:* eric.brueckner@mb.tu-chemnitz.de
[b)]wolf.georgi@mb.tu-chemnitz.de, [c)]michael.gehde@mb.tu-chemnitz.de, [d)]peter.mayr@mb.tu-chemnitz.de

Abstract. Due to the optimized local material adaption by using multi-material-structures, the final part properties can reach maximized performance and efficiency. This fact correlates with a rising demand for processes which join dissimilar materials (e.g. metal-polymer, thermoplastic-thermoset or incompatible thermoplastics) reliably. The present study deals with an innovative process combination of clinching and hot-melt adhesive bonding for joining fiber-reinforced polyamide (PA6-GF30) to steel. The energy input for the hot-melt adhesive bonding is performed by torsional ultrasonic welding. The study was aimed to analyze design variations of the ultrasonic weld seam to reach steady joint properties in spite of unsteady manufacturing conditions and to minimize interactions between clinching and ultrasonic welding. However, the hot-melt adhesive bonding process has always to keep the geometric preconditions for the following clinching process.

INTRODUCTION

The market offers various suitable techniques for joining metal-polymer hybrid structures, which can be classified in adhesive bonding, welding-based techniques, mechanical fastening as well as hybrid joining technologies like injection over molding or combinations of the individual techniques above [1,2]. One of the challenges is described by the reliable joining of thin-sheet metal-polymer hybrid structures, which means a thickness up to 3.0 mm. By studying the existing research with respect to welding-based techniques, the adhesion mechanism mainly based on mechanical interlocking. Consequently and exemplary for laser welding, specific surface modifications (e.g. laser micro-structuring) are often required to reach a sufficient joint strength [3,4]. Similarly for longitudinal ultrasonic welding, specific tie layers are needed to make joining even possible [5,6].

A recent and innovative procedure is the application of torsional ultrasonic welding for joining metal-polymer hybrid structures. In this case, the joint is characterized by a hot-melt adhesive zone between metal and polymer by using the polymer melt itself and no additional adhesive materials. Respectively for joining steel and short fiber-reinforced polyamide (PA6-GF30), Friedrich et al. [7] have recently presented a promising process combination of clinching and torsional ultrasonic welding. The study shows the applicability by increasing the joint strength in comparison to the usage of the single methods, especially regarding the combination of high properties for shear loading and T-peel loading. Furthermore, an improving of the failure behavior which means a temporally separated failure of the single joints (fail-safe-behavior) could be detected without the usage of any additional adhesive for torsional ultrasonic bonding. In fact, no times are needed for its handling and curing. [7]

Nevertheless, Friedrich et al. [7] have also disclosed several open questions. Regarding the clinching process, the long-term stability of the joint and the thermal durability (relaxation behavior) were still unknown. In addition, the forming behavior of the polymer below the melting temperature was not investigated extensively which is correlated with the avoidance of cracks during forming. Concerning the torsional ultrasonic welding process, the increasing of the process window and the avoidance of material overflow should be focused. Furthermore, the required ultrasonic energy and the joining time should decreased to optimize the economic feasibility and the material stress.

While Brueckner et al. [8] have carried out some innovative information about the relaxation behavior of metal-polymer hybrid structures which were mechanically joined by clinching, the present paper focusses on the optimization of the joint design for the torsional ultrasonic process. It was aimed to analyze the influence of different socket geometries regarding the required energy input, joining time, material overflow and joint strength. However, all optimizations are restricted by the demand to keep the geometric preconditions for the following clinching process which means the avoidance of any gaps between metal and polymer.

EXPERIMENTAL

The experimental based on materials which are primarily used in the automotive sector. A DC04 steel was selected as the metallic component. The test specimens with the dimension of 110 x 45 x 1 mm³ (length x width x thickness) were cut out of panels. These panels can usually be purchased on the market with a thickness between 0.25 mm and 3.0 mm in the uncoated condition. The plastic matching component was injection molded out of 30 % short glass fiber-reinforced polyamide (PA6-GF30) and had the dimension of 110 x 45 x 1 mm³ (length x width x thickness).

The torsional ultrasonic process was realized by using a TELSONIC Ultrasonic welding Press TSP 3000. The process was carried out path-controlled with a frequency of 20 000 Hz, a maximal amplitude of 80 µm, a joining force of 1100 N and a joining path of 0.4 mm. For joining metal to polymer, the overlapping area was 45 x 45 mm², fabricated as a tensile shear test specimen as shown in figure 1a. The overlapping area of the metal component was decreased and brushed. The joint was positioned central to the overlapping area.

The polymer specimen is also characterized by a socket geometry made as a circular ring which is placed in the center of the overlapping area. A different internal socket diameter of 20 mm, 25 mm and 30 mm leads to different joining areas. The present study based on joining areas of 648 mm², 471 mm² and 255 mm² with a constant socket height of 0.4 mm (see figure 1b). Additionally, the investigations were carried out with a completely flat polymer specimen (without socket geometry). The contact area of the horn was 708 mm² also made as a circular ring area.

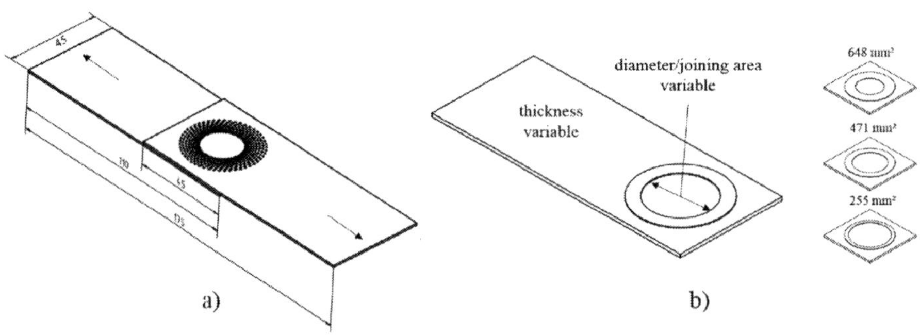

FIGURE 1. a) Experimental setup – Tensile shear test specimen, b) Design of the polymer specimen with a various joining area

The mechanical joint properties were analyzed by using a universal testing machine of Zwick/Roell with a test speed of 10 mm/min. The microscopic analysis was done by reflected light microscopy (device: polarizing microscope Olympus BX51). The specimens were polished to the vertically center of the joint (full section view).

RESULTS AND DISCUSSION

Joint strength and optical quality

The experimental was planned to have a joining path of 0.4 mm which correlates exactly with the height of the socket geometry. It was aimed to reduce the plate thickness as low as possible during joining. However, a sufficient melt generation is always presupposed to have an adequate adhesion. Figure 2 shows the reachable shear force at failure in dependence on the joining area. In contrast to flat polymer specimen (contact area of the horn: 708 mm²), the shear force at failure could be increased by using defined joining areas. In maximum, it is

5302 ± 99 N by using a joining area of 471 mm². That means an increasing of 18 % by having a clearly lower standard deviation compared to a joining area of 708 mm². Nevertheless, it can be also pointed out that a joining area of at least 471 mm² is required to have a sufficient adhesion between metal and polymer as a consequence of the melt generation. Whereas the fracture behavior was typically in the base material of PA6-GF30 at the changeover of metal to polymer, the fracture appearance for a joining area of 255 mm² is characterized by an adhesive failure as shown in figure 2 which indicates a too low melt volume and melt viscosity as well. Regarding the joint strength, the joining area of 255 mm² also shows a clearly minor shear force at failure. Another advantage is the optical quality of the joints which were created by using a socket geometry. In dependence on the joining area, the material overflow, which is characteristically for flat specimens, could be avoided. Figure 2 shows no flash material for joining areas of 471 mm² and 255 mm². Consequently, the joining area of 471 mm² seems to be most appropriate considering joint strength and optical quality. Furthermore, possible unsteady manufacturing conditions of the mating parts in the industrial environment could be compensated easily.

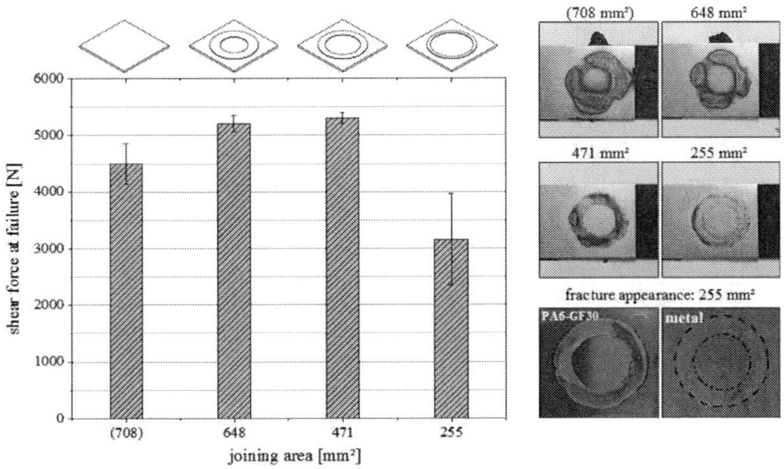

FIGURE 2. Shear force at failure as a function of the joining area, optical quality of the joint, joining path 0.4 mm, amplitude 80 µm, joining force 1100 N

Ultrasonic energy and joining time

Due to the annealing colors shown at the surface of the metal component in figure 2 and figure 3, the level of energy input could be estimated comparatively. With a decreasing joining area the energy input seems to decrease as well. Therefore, figure 3 also represents the required ultrasonic energy and joining time which was displayed by the torsional ultrasonic machine.

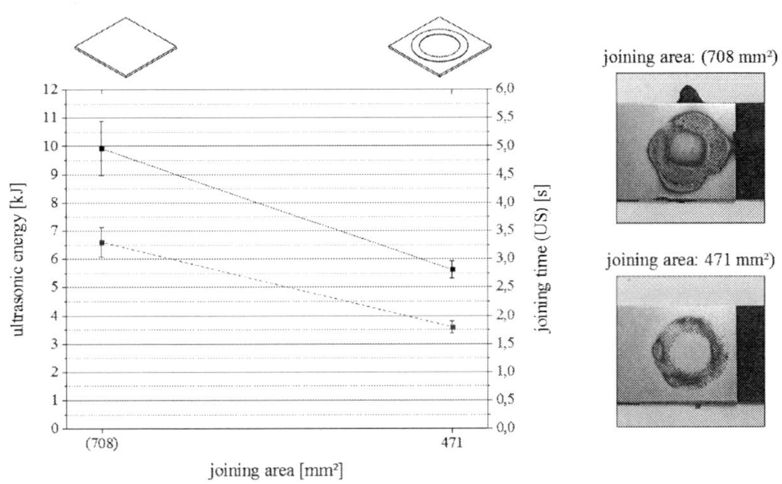

FIGURE 3. Ultrasonic energy and joining time as a function of the joining area, joining path 0.4 mm, amplitude 80 µm, joining force 1100 N

In fact, using a joining area of 708 mm² an ultrasonic energy of 9.9 ± 1.0 kJ was detected. However, using a joining area of 471 mm² the maximal energy decreased to 5.6 ± 0.3 kJ which means an energy saving of 43 %. Additionally, the maximal joining time of 3.3 s (joining area 708 mm²) decreased to 1.8 s and correlates with a time saving of 45 %. Due to this fact, advantages regarding the service life of the machine parts and the material´s stress during joining could be expected.

Geometric pre-conditions

The approach to join thin-sheet metal polymer hybrid structures by using the torsional ultrasonic welding process has always to be noticed in combination with the subsequent mechanical clinching process. Consequently, the ultrasonic joining process has to keep the geometric pre-conditions for this method. That means, a complete form closure without any gaps has to exist between metal and polymer. Using a socket geometry, the challenge is to reach a joining path which correlates preferably exactly with the socket height. In this case, no gaps occur and no thin-out of the polymer specimen arises. A thin-out during joining would mean a reduction of the cross section and consequently a decreasing of the strength. However, as shown in figure 4, a complete form closure between metal and polymer could be detected exemplarily for a joining area of 471 mm². In fact, independent of the socket design, the geometric pre-conditions could be fulfilled. Additionally, a joining path of 0.4 mm ideally preserve the initial thickness of the polymer plate of 1.0 mm (see figure 4).

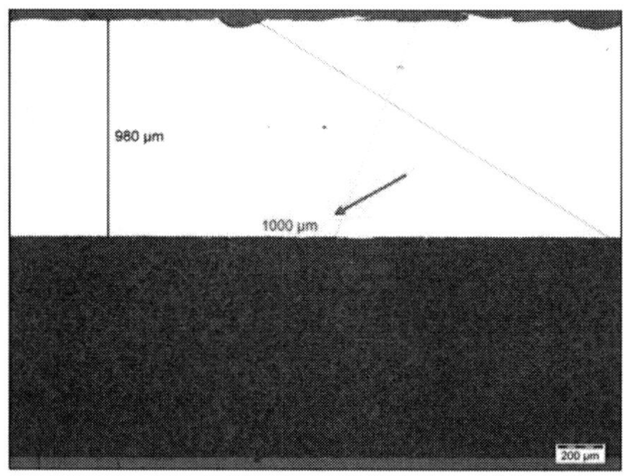

FIGURE 4. Microscopic analysis of the adhesive bonding between metal (DC04) and polymer (PA6-GF30) component, joining path 0.4 mm, amplitude 80 µm, joining force 1100 N

CONCLUSION

The present study focusses on joining of thin-sheet metal-polymer hybrid structures by using the torsional ultrasonic welding process. The ultrasonic joining is only one sequence of a process combination with the subsequent clinching process. Consequently, the geometric pre-conditions for the clinching process have always to be considered. The results clearly show that it is possible to join short glass fiber-reinforced polyamide (PA6-GF30) and steel reliably by using the torsional ultrasonic process. Since the adhesion mechanism seems to be mainly based on mechanical interlocking, high shear forces at failure of about 5300 N are reachable by using polymer specimens with a thickness of 1.0 mm. In contrast to further investigations, the present study focusses on polymer specimens which were modified with a special socket design. As shown, the socket geometry can ensure steady joint properties without any material overflow. Additionally, a decreasing required ultrasonic energy and joining time could be detected with a decreasing joining area. However, the usage of a socket geometry ensures an increased joint strength compared to flat polymer specimens as a consequence of the preserved cross section of the specimen. By having a socket, the process can carried out path-controlled. An aligned joining path avoids a thin-out of the polymer and keeps the geometric pre-condition for the following clinching process as well. The results above show a minimal joining area of at least 471 mm² to create a sufficient melt volume and to have an adequate adhesion between metal and polymer. Now it is possible to compensate unsteady manufacturing conditions of the mating parts in the industrial environment more easily.

ACKNOWLEDGMENTS

This study was promoted from budgetary fund of the German Research Foundation (DFG). We would like to record our appreciation of this support. Furthermore we thank the companies TOX® Pressotechnik GmbH & Co. KG and Telsonic AG for the technical and advisory support.

REFERENCES

1. S.M. Goushegir, *Friction Spot Joining of Metal-Composites Hybrid Structures*, Ph.D. Thesis, TU Hamburg-Harburg, 2015, ISSN: 2191-7833
2. S.T. Amancio-Filho, J.F. dos Santos, *Joining of Polymers and Polymer-Metal Hybrid Structures: Recent Developments and Trends*, Polymer Engineering and Science, p. 1461-1476, DOI: 10.1002/pen
3. A. Roesner, S. Scheik, A. Olowinsky, A. Gillner, U. Reisgen, M. Schleser, *Laser Assisted Joining of Plastic Metal Hybrids*, Physics Procedia, 12 (2011), p. 370-377
4. P. Amend, S. Pfindel, M. Schmidt, *Thermal joining of thermoplastic metal hybrids by means of mono- and polychromatic radiation*, Physics Procedia, 41 (2013), p. 98-105
5. D. Flock, E. Haberstroh, K. Bobzin, T. Schläfer, T. Warda, P. Kutschmann, *Ultrasonic welding of hybrid metal-plastic components with flame spraying of adhesion layer,* Journal of Plastics Technology, Carl-Hanser-Verlag, 7 (2011)
6. G. Ramarathnam, M. Libertucci, M.M. Sadowski, T.H. North, *Joining of Polymers to Metal*; Welding Research Supplement, December 1992, p. 483-490
7. S. Friedrich, W. Georgi, M. Gehde, P. Mayr, *Hybrid Joining Technology – A new method for joining thermoplastic-metal-mixed components*, PPS 2013, Nuremberg, Germany, July 2013
8. E. Brueckner, W. Georgi, M. Gehde, P. Mayr, *Relaxation behavior of metal-polymer hybrid structures mechanically joined by clinching*, ANTEC 2017, Anaheim, California, USA, May 2017, ISBN: 978-0-692-88309-9

Infrared Welding of Continuous Fibre-Reinforced Thermoplastics – Investigations on Overlapping Joints

Marios Constantinou [1, a)] and Michael Gehde [1, b)]

[1]Chemnitz University of Technology, Institute of Materials Handling, Conveying and Plastics Engineering, Professorship of Plastics Engineering, Reichenhainer Str. 70, 09126 Chemnitz, Germany

a)Corresponding author: marios.constantinou@mb.tu-chemnitz.de
b)michael.gehde@mb.tu-chemnitz.de

Abstract. Continuous fibre-reinforced thermoplastics are offered as impregnated and consolidated semi-finished products that are known as organo sheets. Parts made of organo sheets are processed in presses and formed to half shells. In order to produce larger components, hollow bodies can be produced by joining the half shells. However, current manufacturing technologies allow only cap profile shaped joints, which lead to a deflection of the fibres in the joint plane. Thus, this paper reveals the high potential of overlapping welds in organo sheets that result in a fibre orientation in load direction and lead to weld strengths close to the interlaminar shear strengths of the unwelded materials. The results show that, by using high welding pressures a matrix depletion and a change of the fibre alignment in the weld plane may occur that lead to a decrease in weld strengths. Furthermore, it is demonstrated that these two effects can be hindered by additional polymer matrix on the organo sheets which causes a more ductile fracture behaviour of the welds as well.

INTRODUCTION

Semi-finished products made of continuous fibre-reinforced thermoplastic are often referred to as organo sheets. Parts made of organo sheets are produced by thermoforming the sheets in presses which results in half shell shaped components [1, 2]. In order to stiffen these half shells, reinforcing ribs can be injection moulded onto them by using the SpriForm technology which is also known as In-Mould-Forming process [1] or reinforcing segments are joined to the organo sheet during thermoforming by using the Tailored-Blank-Technology [1]. In order to manufacture larger and more complex components, hollow body structures can be produced by joining two half shells. One possible variant for the manufacturing of hollow bodies made of organo sheets is the Twin-O-Sheet (TOS) process where the sheets are thermoformed and joined in an injection moulding tool [3]. One major advantage of hollow bodies over half shells is the higher part stiffness. This leads to the unnecessity of reinforcing ribs in hollow body parts and thus to a possible weight reduction compared to half shell shaped parts. However, the joint geometries in current hollow bodies made of organo sheets are cap profile shaped (Fig. 1 (a)). This causes the deflection of fibres in the joint plane that prevents the use of fibres across the joint plane. Since the fibre orientation in load direction is required for the transmission of stresses in fibre-reinforced plastics [1], cap profile shaped welds are weak spots in parts made of organo sheets. Overlapping welds enable the fibre orientation in direction of the load path (Fig. 1 (b)) and thus the fibre utilisation across the joint plane.

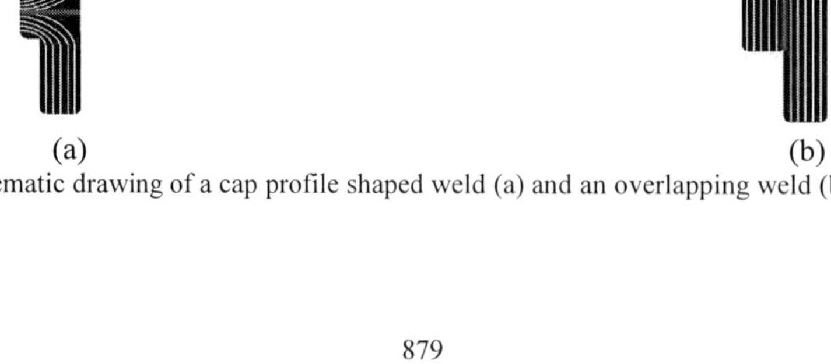

(a) (b)

FIGURE 1. Schematic drawing of a cap profile shaped weld (a) and an overlapping weld (b) in organo sheets.

Although the overlapping joining of continuous fibre-reinforced thermoplastics is possible by using the adhesive bonding as well as mechanical and thermal joining methods [4, 5], the infrared (IR) welding was chosen as joining method. The IR welding is one of the most suitable methods for the welding of organo sheets since it can be used independent from the material to be welded (no particular filler, matrix properties required), the joining areas are variable in their sizes and complexities and short heating times are achievable. [5, 6]

Hence, the objective of this paper is to show influences on the weld strengths of overlapping joints in organo sheets manufactured by IR welding in order to enable the use of fibres across the joint plane.

EXPERIMENTAL

The investigated organo sheets are shown in table 1. The materials consist of black polypropylene (PP) matrices and glass fibres and have a thickness of 1.5 mm. The organo sheets differ in the weave patterns and fibre alignments.

TABLE 1. Technical data of investigated materials.

Abbreviation	Manufacturer	Trade name	Weave pattern	Fibre alignment	Fibre content
PP1	Bond-Laminates GmbH, Germany	Tepex dynalite 104-RG600(3)/47% black	Woven fabric (twill weave 2/2)	0°/90° (50%/50%)	47 vol.%
PP2		Tepex dynalite 104-RGUD600(3)/47% black	Woven fabric (plain weave)	0°/90° (80%/20%)	

As shown in Fig. 2 (a), the overlapping areas (25 x15 mm²) are applied with IR radiation from two short-wave quartz glass radiators with a power of 1500 W each and an emitter distance of 30 mm. In the subsequent joining phase, the specimens were joined with a defined welding pressure p_w (Fig. 2 (b)). As can be seen from Fig. 2 (c), the used specimens had an area of 100 x25 mm² (length x width). Partly the specimens were modified by applying additional polymer matrix with a thickness of approx. 0.45 mm on one of the surfaces with the area of 100 x 25 mm². The modified specimens were produced by extruding the polypropylene Hostacom CR1171G1 by LyondellBasell onto one side of the specimens and consolidating them in a column press. The dimensions of the welded joints are shown in Fig. 2 (c) and were chosen based on the standard DIN EN 1465 which defines the determination of tensile lap-shear strengths of bonded assemblies.

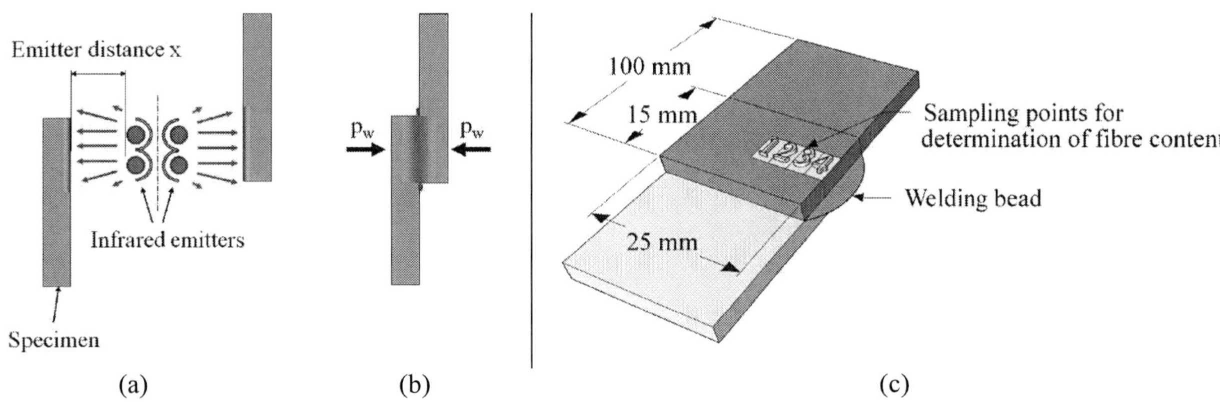

FIGURE 2. Schematic drawings of the heating phase (a) and the joining phase (b) during the infrared welding. In (c) a resulting overlapping joint with its dimensions and the sampling points for the determination of the fibre content in the weld is illustrated.

Thermogravimetric analyses (TGA) with a gas exchange from N_2 to O_2 at 550°C and a heating rate of 20 K/min were carried out on three specimens per investigated welding parameter to establish the fibre volume contents (fvc) in the overlapping areas. Samples from four different positions haven been removed from the overlapping area in order to determine differences in the fvc from the middle to the welding bead (Fig. 2 (c)). Moreover, microscopic analyses on fracture surfaces of the welds were done with a 3D Laser Scanning Microscope.

Moreover, mechanical and thermal investigations were carried out on the unwelded materials given in table 1 and on the welded joints (Fig. 2 (c)). In order to characterise the base material properties, the interlaminar shear strength (ILSS) was determined by a short bending test acc. to the standard DIN EN 2563 (Fig. 3 (a) and (b)). The ILSS is a value for the maximum shear stress in the neutral fibres parallel to the layers of the laminate at failure. The determination of the ILSS enables the evaluation of the fibre-matrix-adhesion and cannot be exceeded with welded specimens [7, 8]. The measured ILSS values were used as reference and have been compared to the tensile lap-shear strengths (TLSS) determined on the welded joints acc. to DIN EN 1465 (Fig. 3 (c)).

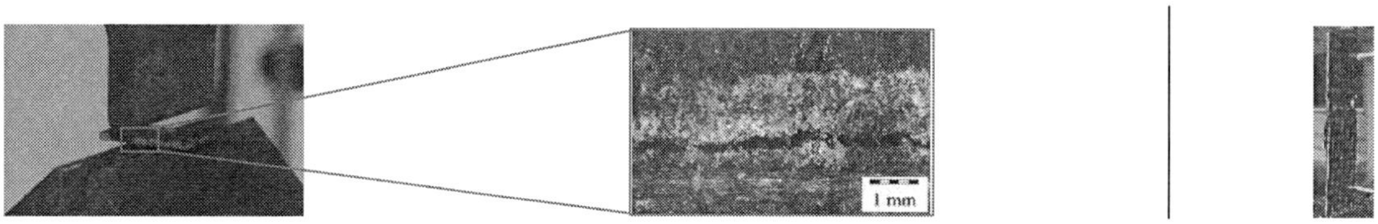

(a) (b) (c)

FIGURE 3. Test setup for the determination of the interlaminar shear strength (a) and the shear fracture of a specimen (b) as well as a tensile lap-shear test on a welded overlapping joint (c).

RESULTS AND DISCUSSION

In previous investigations, it was demonstrated that overlapping IR welds in the organo sheets considered in this paper are capable of achieving TLSS close to the ILSS of the base materials. The welding pressure p_w and the weave pattern were detected as determinants concerning the resulting weld strengths. [5]

In case of the PP1 with a twill weave and 50 % of the fibres oriented in 0° and 50 % in 90° direction (table 1), a p_w variation between 1 and 5 N/mm² leads to similar joint strengths close to the ILSS. A decrease in strength is only visible after reaching a p_w of 6 N/mm². In contrast to the PP1, the TLSS of the welded joints in PP2 (table 1) decrease at p_w higher than 1 N/mm². Due to the robust twill weave of the PP1, a mechanical interlocking occurs between the weave patterns during the joining process and the fibres weaves stay in their original state even at high p_w. In case of PP2 with a plain weave and 80 % of the fibres oriented in 0° direction, the joining partners slide into each other which leads to a matrix depletion and an increase of the fvc in the joining plane at $p_w > 1$ N/mm². [5]

Besides the described effect of matrix depletion, this paper is intended to show the influence of the fibre alignment in the weld plane on the properties of overlapping organo sheet joints manufactured by IR welding. Figure 4 shows the result of a microscopic analysis of the fracture surfaces of overlapping welded PP2 specimens. As can be seen from the fracture surface of the PP2 welded with a p_w of 1 N/mm² (Fig. 4 (a)), the fibre alignment in the weld plane stays in its original state with the majority of fibres oriented in 0° direction. Although the plain weave has a high stability against fibre displacement [9], an increase to p_w = 4 N/mm² (Fig. 4 (b)) leads to a fibre deflection and polymer melt flow in the direction of the welding bead. This is visible from the microscopic photos but also from the widening in the welding bead area compared to p_w = 1 N/mm² (Fig. 4 (a)).

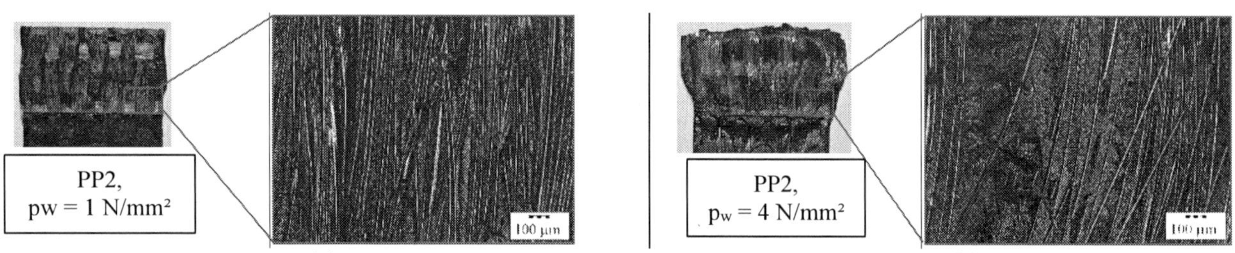

(a) (b)

FIGURE 4. Photographs of fracture surfaces resulting from tensile lap-shear tests on PP2 welds manufactured with welding pressures of 1 (a) and 4 N/mm² (b) each taken with a reflex camera (left) and a 3D Laser Scanning Confocal Microscope (right).

It can be seen from Fig. 4 (b), for PP2 joined with p_w = 4 N/mm², that the fibre orientation in load path direction is not given along the complete width of the weld, the distribution of polymer matrix between the fibres is not homogenous and partly a matrix depletion in the weld plane is visible.

The optical results described and shown in Fig. 4, were confirmed by the TGA results on the PP2 welds manufactured with p_w = 1 and 4 N/mm². The average fvc in overlapping areas welded with p_w = 4 N/mm² measured in the border areas of the weld (sampling points 3 and 4 in Fig. 2 (c)) were 7 % higher than for welds manufactured with p_w = 1 N/mm². At the same time the inner areas of the welds (sampling points 1 and 2 in in Fig. 2 (c)) produced with p_w = 4 N/mm² showed lower fvc than the welds produced with lower welding pressures. However, the average fvc of the complete overlapping areas were differing only slightly from each other (Fig. 5 (a), left).

Since the fibre orientation, fibre-matrix-adhesion and a sufficient matrix content are main influences on high strengths in fibre-reinforced plastics [1, 2], the load transfer by the matrix between the fibres is not optimal for the PP2 welded with p_w = 4 N/mm². From these effects, decreasing mechanical weld properties of PP2 at high welding pressures are following, which can be seen from the tensile shear strengths determined on PP2 joined with p_w of 1 and 4 N/mm² shown in the left part of Fig. 5 (a). The application of polymer matrix on the PP2 organo sheet specimens leads to high TLSS independent from the used welding pressures (Fig. 5 (a) right) since the matrix operates as a protection against the matrix depletion and a change of the original fibre alignment in the weld seam.

The tensile shear strengths received from tests on PP1 welds show a different behaviour compared to PP2. From the left part of Fig. 5 (b) it can be seen that the TLSS of PP1 rise with increasing welding pressures which results from the mechanical interlocking of the joining partners at high p_w for this weave pattern as explained previously. The application of matrix causes similar TLSS compared to the specimens without coating.

It can be concluded that a matrix application on organo sheets to be welded can be reasonable but is not necessary in all cases. In case of weave patterns with a low robustness against alignment changes, e.g. by pressure application, such as proven for the investigated PP2, a matrix application is to be recommended. The mechanical properties of the welds are assured even if varying joining pressures are applied along the weld area, which can follow from e.g. varying part thicknesses or complex part geometries.

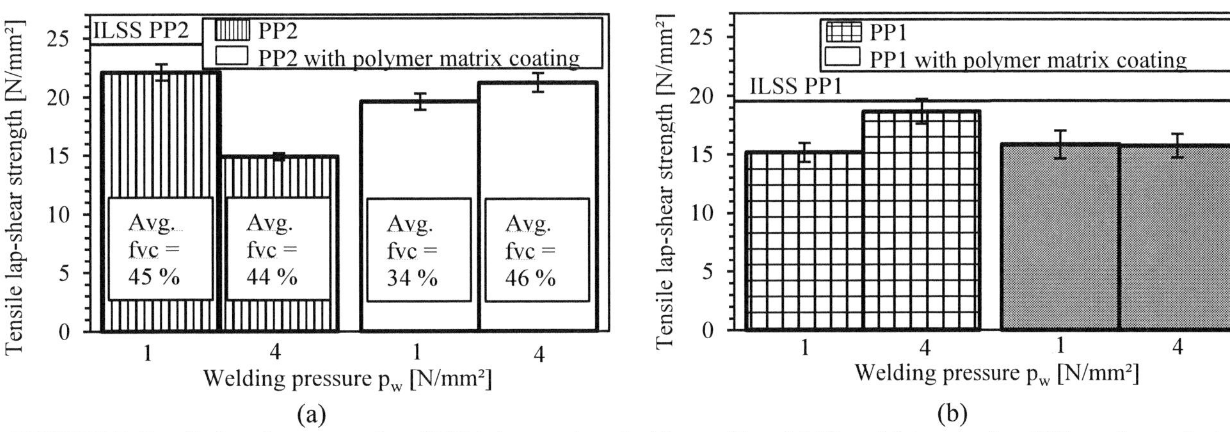

FIGURE 5. Tensile lap-shear strengths of PP2 joints produced with p_w of 1 and 4 N/mm² determined on PP2 specimens (a, left) and on PP2 specimens with a polymer matrix coating (a, right). Heating time and used emitter powers for PP2: 17 s (5 s 100 % / 5 s 80 % / 7 s 60 %). Tensile lap-shear strengths of PP1 joints produced with p_w of 1 and 4 N/mm² determined on PP1 specimens (b, left) and on PP1 specimens with a polymer matrix coating (b, right). Heating for PP1: 12 s with an emitter power of 100 %.

The described protection of the polymer matrix coating concerning the weave pattern in the specimens was proven by TGA measured on PP2 welds manufactured with specimens with polymer matrix coating and p_w of 1 and 4 N/mm². Only minor differences in the fvc between the sampling points 1 to 4 (Fig. 2 (c))were detected along the overlapping area. However, it is noticeable that the average fvc of joints welded with p_w = 1 and 4 N/mm² differed a lot from each other. Although the tensile shear strengths are almost identical, the right part of Fig. 5 (a) shows that the average fvc in the welds amounted to 34 vol.% (p_w = 1 N/mm²) and 46 vol.% (p_w = 4 N/mm²).

Another advantage of specimens with polymer matrix coating over the usual specimens is the more ductile fracture behaviour of the resulting joints. As visualised in Fig. 6, after the maximum stress has been passed and abrupt shear fracture in the entire welding area occurs, the tensile shear stress drops and the joint fails at a considerably higher strain (6 %) than the weld on PP2 without an additional polymer matrix (2%). The matrix coating on the specimens causes the fibres in the joining plane to be impregnated with additional polymer matrix. , Due to the good fibre-matrix-adhesion between additional matrix and existing fibres, the fibres are pulled out of the opposite joining part, which also leads to a stretching of the matrix in the weld. This finally results in a high ductility of the joint as well as a ductile post-fracture behaviour (Fig. 6). Due to the detection of this behaviour on welds manufactured with PP1 and PP2 with a polymer matrix coating, it can be stated that it appears independent of the weave pattern of the organo sheet.

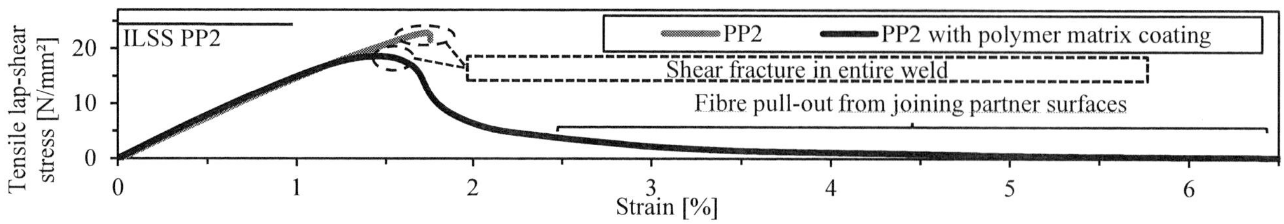

FIGURE 6. Stress-strain curves from tensile shear tests on welded PP2 specimens with and without polymer matrix coating manufactured with a welding pressure of 1 N/mm².

CONCLUSION

The present study shows that overlapping joining by infrared welding leads to weld strengths close to the ILSS of the unwelded organo sheets. The optimum welding parameters are material dependent. The main influences on the weld strengths are the weave pattern and the layer composition of the organo sheets to be welded.

If the applied joining pressure is too high, a matrix depletion and a change of the fibre alignment in the weld plane will occur and the tensile shear strengths of the joints will decrease. Organo sheets with robust weave patterns show high weld strengths even at high welding pressures whereas organo sheets with fibre alignments that are not stable against compression loads, show a decrease in strength when applying joining pressures > 1 N/mm².

A polymer matrix coating resulted in high weld strengths close to the ILSS independent of the applied joining pressures and weave patterns of the organo sheets. Furthermore, a more ductile fracture behaviour of the welds during tensile shear testing was observed. The polymer matrix coating on organo sheets can become a benefit for future parts since varying joining pressure along the welding area do not influence the weld quality.

One main finding, resulting from this paper, is that the maintaining of the fibre alignment in the weld is of high importance for high weld strengths of overlapping joints in organo sheets. A deflection of fibres, in combination with melt flow in direction of the welding bead, may cause a poor load transfer between the joining partners due to the fibres partly not oriented in loading direction and a missing wetting of the fibres with a sufficient polymer matrix content. In contrast to the grade of fibre deflection, the average fvc in the welds allow only restricted statements regarding the resulting weld strengths. High fvc in the welds are not mandatorily leading to low weld strengths.

In future works, applicable criteria for the successful welding of organo sheets shall be defined (e.g. the grade of fibre deflection or ratio of weld path to molten layer thickness). Furthermore, organo sheets will be thermoformed and hollow body structures are going to be manufactured by infrared welding.

ACKNOWLEDGMENTS

The authors are thankful for support provided by budgetary funds of the German Federal Ministry for Economic Affairs and Energy (BMWi) via the German Federation of Industrial Research Associations (AiF).

REFERENCES

1. M. Neitzel, P. Mitschang, U. Breuer, *Handbuch Verbundwerkstoffe: Werkstoffe, Verarbeitung, Anwendung - 2nd Editio*n (Carl Hanser Verlag, München, 2014).
2. H. Lengsfeld, F. Wolff-Fabris, J. Krämer, J. Lacalle, V. Altstädt, *Composite Technology – Prepregs and Monolithic Part Fabrication Technologies* (Carl Hanser Verlag, München, 2016).
3. C. Gröschel and D. Drummer, "Wall Thickness Distribution of Continuous Glass Fiber Reinforced Polyamide 6 Composite Parts Formed by Gas Pressure," *SPE ANTEC Orlando 2015*, pp. 675-681.
4. C. Ageorges, L. Ye and M. Hou, Composites: Part A 32, 839-857 (2001).
5. M. Constantinou, M. Gehde and P. Kube, Joining Plastics 69, No. 1, 48-55 (2017).
6. R. Fuhrich, „Infrarotschweißen von Kunststoffen mit thermischen Strahlungsemittern." Ph.D. thesis, Chemnitz University of Technology, 2013.
7. W. Grellmann, S. Seidler, *Kunststoffprüfung – 2nd edition* (Carl Hanser Verlag, München, 2011).
8. R. Rudolf, "Entwicklung einer neuartigen Prozess- und Anlagentechnik zum wirtschaftlichen Fügen von thermoplastischen Faser-Kunststoff-Verbunden." Ph.D. thesis, Technische Universität Kaiserslautern, 2000.
9. C. Cherif, *Textile Materials for Lightweight Constructions* (Springer-Verlag, Berlin Heidelberg, 2016).

Synthesis and Characterization of Graphene Derivatives for Application in Magnetic High-Field Induction Heating

C. Alves da Silva[1], P. Pötschke[1,b)], F. Simon[1], M. Holzschuh[1], J. Pionteck[1]. Heinrich[1,2], S. Wießner[1,3], C. Zimmerer[1,a)]

[1]Leibniz Institute of Polymer Research Dresden, Hohe Str. 6, 01069 Dresden, Germany
[2]Institut für Textilmaschinen und Textile Hochleistungswerkstofftechnik, Technische Universität Dresden, D-01062 Germany
[3]Institut für Werkstoffwissenschaft, Technische Universität Dresden, 01062 Germany

[a)]zimmerer@ipfdd.de
[b)]poe@ipfdd.de

Abstract. In the present work, different graphene derivatives were prepared and characterized envisaging their use as susceptors for magnetic high-field induction heating at polymer interfaces. For optimization of this new kind of polymer bonding method, first graphene oxide (GO) with different degrees of oxidation were synthetized from graphite through a modified Hummers method. Then, the reduction of GO was achieved using dopamine hydrochloride to obtain polydopamine-coated graphene oxide (PDA-GO). Finally, these graphene derivatives were characterized by X-Ray Photoelectron Spectroscopy (XPS) and Raman Spectroscopy.

INTRODUCTION

Magnetic high-field induction heating is a promising technique to initiate chemical reactions at polymer interfaces and to replace traditional joining and welding technologies. [1] A successful bonding between a bisphenol-A based polycarbonate (PC) and a polyvinylamine (PVAm) using silver as susceptor was already reported in the literature. Susceptors for high magnetic field induction heating must possess high electrical conductivity. [2] Therefore, graphene and graphene derivatives appear to be suitable materials for this purpose.

Graphene consists of a two-dimensional monolayer of sp2-hybridized carbon atoms arranged in a hexagonal lattice with a carbon-carbon bond length of 0.142 nm. [3] Graphene's properties have been extensively studied since it was first isolated in 2004 by Geim and Novoselov. [4] It has great potential for reinforcement in composite materials [4,5] due to its extremely high Young's modulus of 1 TPa, intrinsic strength of 130 GPa, [6] room temperature electron mobility of 250000 cm2·V-1·s-1, [4] and optical transmittance of 97.7%. [7] Other promising application areas for graphene include photonics, optoelectronics, energy generation and storage, sensors for gas detection, and biomedical area. [4] Graphene is a material with hydrophobic behavior and considerable chemical inertia. However, it is highly susceptible to physical adsorption through π-π interactions [8] and it can be chemically functionalized via non-covalent or covalent approaches, such as oxidation. [4,9] Moreover, non-functionalized graphene sheets tend to restack through van der Waals interactions to re-form graphite. [10] The functionalization of graphene often enables its dispersion to form stable suspensions and the interaction between graphene and other materials. [10,11] Graphene oxide (GO) is a graphene derivative that has been extensively employed as an alternative or as a precursor of graphene. [12] A considerable number of chemical models have been suggested to describe the GO structure, among which the Lerf and Klinowski model, Fig. 1, is the most accepted one. This model states that hydroxyl and epoxy groups are present on both sides of the basal plane while carbonyl along with carboxyl acids are distributed at the sheet edges. [13]

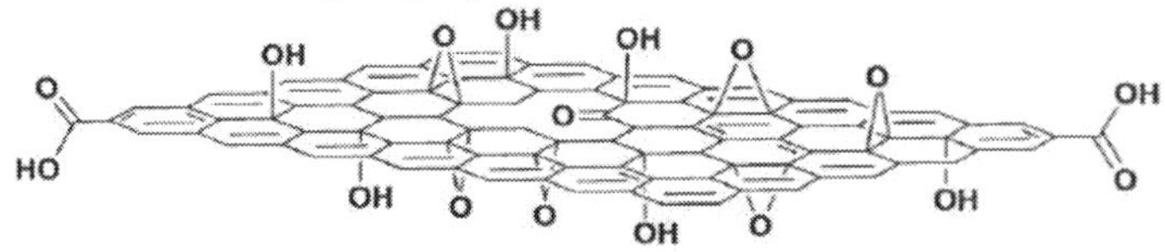

Figure 1. Chemical structure of GO according to Lerf and Klinowski [14]

The reduction of GO appears to be a viable route to produce single-layer graphene [7]. GO can be prepared through various methods, such as those ones reported by Brodie, Staudenmeier, and Hummers, either following their original protocols or introducing some variations. [14] These three methods involve the oxidation of graphite using strong oxidizing acids followed by exfoliation achieved through sonication. [15] The reduction of GO is usually accomplished by chemical or thermal routes. In the chemical approach, strong reducing agents such as hydrazine and sodium borohydrate are used. However, this process results in the formation of graphene with a defected structure and a high amount of toxic waste. Second, thermal reduction is performed through heat treatment and leads to non-perfect graphene structures as well. Other reduction methods, particularly eco-friendly methods employing nature-based reagents, hydrothermal, and electrochemical approaches, have been explored in order to prepare graphene. [16, 17] Among these, dopamine hydrochloride has been successfully used to convert GO in reduced GO (RGO). [18]

In view of the above observations, the present study aims to achieve a balance between the degree of oxidation of GO and the thickness of the polydopamine (PDA) layer in PDA-GO obtained during the reduction. By that the potential of such graphene derivatives for the use as susceptors for magnetic field induction can be evaluated afterwards.

EXPERIMENTAL SECTION

Materials

Natural graphite flakes (99%; -325 mesh) were purchased from Sigma Aldrich® (Germany). The oxidation of graphite was performed using sulfuric acid (H_2SO_4; 98 %) and potassium permanganate ($KMnO_4$; ≥99%), both from VRW® (Germany). For the purification step, hydrogen peroxide (H_2O_2; 30% w/v) purchased from Merck (Germany), hydrochloric acid (HCl; 10% v/v) from VRW® (Germany), and distilled water (DW) were used. Regarding the reduction of GO, tris(hydroxymethyl)aminomethane (Tris base; ≥99.8%), dopamine hydrochloride (DA), both from Sigma Aldrich® (Germany), and DW were employed.

Synthesis of GO

GO powders were synthetized through a modified Hummers method. Briefly, 400 ml of H_2SO_4 were added to 5 g of graphite and stirred vigorously for 2 h at room temperature (RT). Then, the suspension was placed in an ice bath and 5 g or 15 g of $KMnO_4$ was slowly added. Different GO powders were prepared, GO1 and GO3, using a mass ratio between $KMnO_4$ and graphite of 1 and 3, respectively.

The ice bath was removed and the suspension was kept stirring for 24 h at RT. After the 24 h, the suspension was placed once again in an ice bath and 300 ml of DW were slowly added, followed by 30 ml of H_2O_2, drop-by-drop, and the mixture was left to stir for 30 min. Afterwards, the suspension was poured into 80 ml polypropylene tubes and centrifuged at 10000 rpm during 15 min. The supernatant was discarded and the precipitate was washed with DW for a total of 3 cycles. The resulting precipitate was collected, 250 ml of HCl were added and the mixture was left to stir during 15 min. Then, the mixture was centrifuged at 10000 rpm during 15 min and washed with DW for a total of 5 cycles. After the washing, DW was added to the precipitate and then vacuum filtered with a nylon filter membrane (0.45 μm pore size, Whatman UK). Finally, the powders were collected, dried at 100 °C for 48 h, and milled with a mortar and pestle.

Synthesis of PDA-GO

First, GO1 and GO3 were dispersed in TRIS base solution (0.1 M) through sonication in a sonication bath for 4 h. Then, DA was added to the suspension in a mass ratio DA to GO of 2, and the mixture was left stirring for 24 h at 60 °C. After that, the suspension was poured into 80 ml polypropylene tubes and centrifuged at 10000 rpm for 20 min, the supernatant was discarded and DW was added. This procedure was repeated until a clear supernatant was obtained. Finally, the suspension was filtered using nylon membranes (0.45 μm pore size, Whatman, UK). The powders were collected and dried at 60 °C, under vacuum, for 24 h. Two different PDA-GO samples were prepared, PDA-GO1 and PDA-GO3, using GO1 and GO3 as starting materials, respectively.

Characterization of GO and PDA-GO

XPS measurements were carried out in an Axis Ultra photoelectron spectrometer (Kratos Analytical, Manchester, UK) equipped with a monochromatic Al Kα (h ν = 1486.6 eV) X-ray source of 300 W at 15 kV. Raman spectroscopy was performed on a RAMAN Imaging System alpha300R (WITec, Germany) using a laser excitation wavelength of 532 nm.

RESULTS AND DISCUSSION

X-Ray Photoelectron Spectroscopy (XPS)

The results of the XPS measurements for both GO and PDA-GO materials are presented in Fig. 2. In the spectra of GO1 and GO3, the peaks related to sp^2-hybridyzed carbon atoms arranged in aromatic rings, *Ph*, and to sp^3-hybridized carbon atoms from saturated hydrocarbons ($^{A}\underline{C}_xH_y$), *A*, are present at 284.6 eV and 285.0 eV, respectively. Regarding GO1, the C1s spectrum is constituted by three more peaks (*B*, *C*, and *D*). The O-C-O bonds led to the appearance of the component *B* (BE ≈ 286.5 eV) while the C=O bonds are responsible for the component *C* (BE ≈ 287.5 eV). [19] The component *D* (BE ≈ 288.7) might be related to the C=O bonds on O-C=O groups. [19, 20] The C1s spectrum of GO3 is constituted by two more peaks (*B* and *C*), besides *Ph* and *A*. The O-C-O bonds are responsible for the component *B* (BE ≈ 286.7 eV) [21] while the O-C=O bonds are led to the appearance of the component *C* (BE ≈ 288.7 eV). [19, 20] Shake-up peaks, *Sh*, associated to the π conjugated system of graphene were also found in both GO1 and GO3 spectra (BE > 289 eV). GO1 and GO3 have O/C ratios of 0.173 and 0.268 which means that the degree of oxidation of GO3 is higher than GO1, as expected, considering the ratios between $KMnO_4$ and graphite used during the synthesis of these materials.

Regarding the PDA-GO materials, the C1s spectra is constituted by five different peaks (*Ph*, *A*, *B*, *C*, and *D*). The peaks from the components *Ph* (BE ≈ 284.7 eV) are related to sp^2-hybridyzed carbon atoms arranged in aromatic rings, with no bonding to heteroatoms. The sp^3-hybridized carbon atoms from saturated hydrocarbons ($^{A}\underline{C}_xH_y$) led to the appearance of the component *A* (BE = 285.0 eV). Moreover, the $^{C}\underline{C}$–O bond on phenolic $^{C}\underline{C}$–OH groups gave rise to the component *C* (BE = 286.4 eV). Also, the carbon-nitrogen bonds on amines and imines are responsible for the component *B* (BE ≈ 285.55 eV). However, considering the values of the binding energies, it was not possible to distinguish primary from secondary amines and amino groups from imino groups. Nonetheless, nitrogen atoms are usually present in polydopamine as secondary amines ($^{B}\underline{C}$–NH–$^{B}\underline{C}$) and/or as imines ($^{B}\underline{C}$–N=$^{B}\underline{C}$ ↔ $^{B}\underline{C}$=N–$^{B}\underline{C}$). [22-24] The origin of the component *D* (BE ≈ 287.5 eV) is not clear. The binding energies of these components are typical of quinone or ketone groups ($^{D}\underline{C}$=O). This range of binding energies (peak of component D) is also obtained for carbonyl bound carbon atoms in amides (O=^{D}C-NH-^{B}C). Moreover, shake-up peaks, *Sh*, associated to the π conjugated system of graphene were found (BE > 288 eV). Overall, the thickness of PDA layer on GO material appears to be too high to investigate the reduction grade of covered GO.

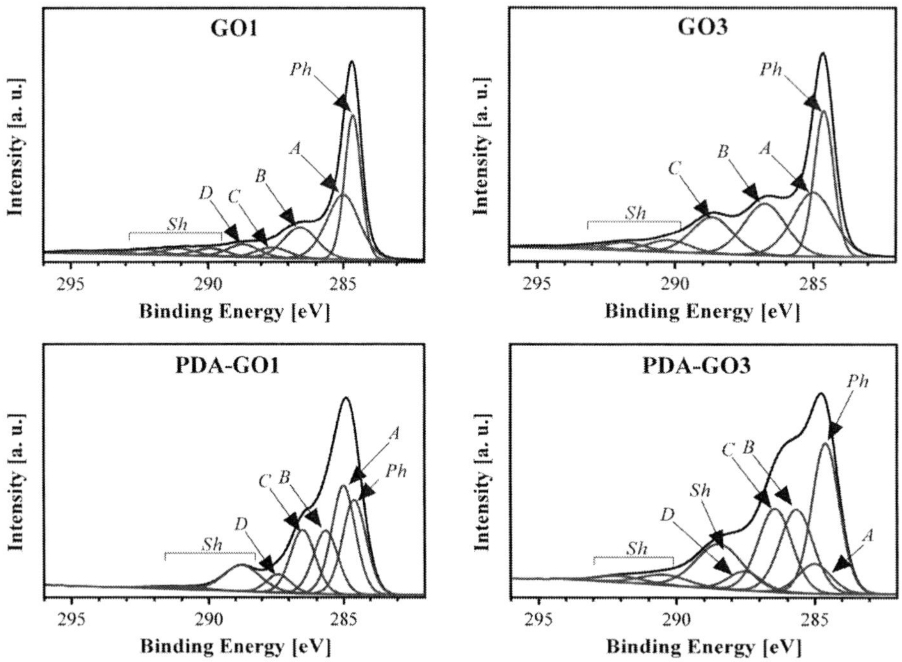

Figure 2. C1s XPS spectra and their deconvolution of GO1, GO3, PDA-GO1, and PDA-GO3.

Raman Spectroscopy

The Raman spectra, presented in Fig. 3 a) and b), show the three major bands characteristic of sp^2 carbon materials. The D band, near 1350 cm^{-1}, is related to the presence of structural defects in the hexagonal sp^2 carbon lattice of graphene and to edge effects. [25,26] The G band, at approximately 1580 cm^{-1}, is related to the in-plane vibration of the sp^2 carbon atoms. [25] The band near 2700 cm^{-1}, the 2D band, originates on a second order Raman scattering process and its shape, width, and position is related to the number of layers for n-layer graphene. [26, 27] The Raman spectra in Fig. 3 a) show an intensity increase of the D to G band ratio, I_D/I_G, when pristine graphite was chemically converted into GO1 and GO3 from 0.16 to 0.57 and 0.86, respectively. This is a consequence of an increase of the content of structural defects caused by the oxidation process. After the oxidation, the 2D band almost disappears into GO3, contrary to GO1, which is indicative of a significant loss of the sp^2 hexagonal structure of the graphene sheets.

Regarding the PDA-GO1 and PDA-GO3 (Fig. 3 b), an increase in I_D/I_G was observed after the deposition process with PDA, contrary to what is expected considering that PDA should have the ability to reduce GO. However, after analyzing the spectrum of PDA, Fig. 3 c, it can be concluded that PDA has bands in the same range like the D and G bands of graphene materials and so, the I_D/I_G values of both PDA-GO cannot reflect properly the disorders on the structure of these materials. No significant differences were found between the 2D bands of both materials in relation to their respective GO counterparts.

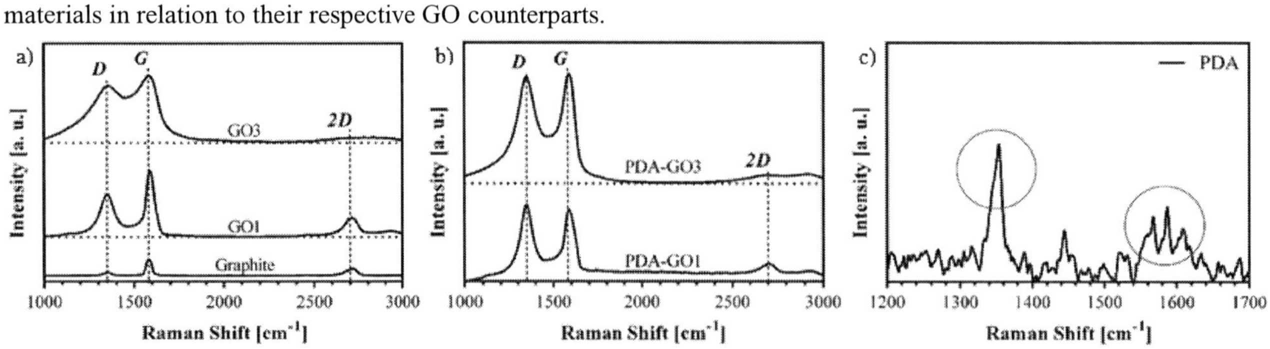

Figure 3. Raman spectra of a) graphite, GO1, and GO3; b) PDA-GO1 and PDA-GO3; c) PDA

CONCLUSIONS

In the present work, GO was synthetized with different degrees of oxidation and characterized by XPS and Raman spectroscopy. GO1 possesses a lower amount of structural defects than GO3 which can make this graphene derivative more suited for the purpose of this work, since graphene with a less defective structure usually has higher electrical conductivity and it can be reduced more easily. The polymerization of dopamine on the surface of GO was also investigated by XPS. However, the thickness of the polydopamine layer was thicker than the information depth of collected photoelectrons. No signal from covered GO was obtainable. Moreover, through the Raman spectra, it was not possible to evaluate the extent of GO reduction by DA due to the overlapping D and G bands of both graphene and PDA. In the future, different parameters of PDA deposition will be used in order to apply XPS for analyzing the reduction grade of GO beside electrical measurements. In the next step, the suitability of such modified GO structures for the function of susceptors will be evaluated.

REFERENCES

1. C. Zimmerer, G. Heinrich, F. Wolff-Fabris, E. Koch, G. Steiner, Polymer **54**, 6732-6738 (2013).
2. W. Suwanwatana, S. Yarlagadda, J. W. Gillespie, Compos. Sci. Technol. **66**, 2825-2836 (2006).
3. A. H. Castro Neto, F. Guinea, N. M. R. Peres, K. S. Novoselov, A. K. Geim, Rev. Mod. Phys. **81**, 109-145 (2009).
4. K. S. Novoselov, V. I. Fal'ko, L. Colombo, P. R. Gellert, M. G. Schwab, K. Kim, Nature **490**,192-200 (2012).
5. X. Huang, X. Qi, F. Boey, H. Zhang, Chem. Soc. Rev **41**, 666-686 (2012).
6. C. Lee, X. Wei, Q. Li, R. Carpick, J. W. Kysar, J. Hone, Phys. Status Solidi (B) **246**, 2562-2567 (2009).
7. Y. Zhu, S. Murali, W. Cai, X. Li , J. W. Suk , J. R. Potts, R. S. Ruoff, Adv. Mater. **22**, 3906-3924 (2010).
8. M. Terrones, A. R. Botello-Méndez, J. Campos-Delgado, F. López-Urías, Y. I. Vega-Cantú, F. J. Rodríguez-Macías, A. L. Elías, E. Muñoz-Sandoval, A. G. Cano-Márquez, J. C. Charlier, H. Terrones, Nano Today **5**, 351–372 (2010).
9. M. Pumera, Chem. Rec. **9**, 211-223 (2009).
10. D. Li, M. B. Müller, S. Gilje, R. B. Kaner, G. G. Wallace, Nat. Nanotechnol. **3**, 101-105 (2008).
11. N. Pu, C. Wang, Y. Liu, Y. Sung, D. Wang, M. Ger, J. Taiwan Inst. Chem. Eng. **43**, 140-146 (2012).
12. K. Hu, D. D. Kulkarni, I. Choi, V. V. Tsukruk, Progress in Polymer Science **39**, 1934-1972 (2014).
13. Y. Zhang, Y.-W- Tan, H. L. Stormer, P. Kim, Nature **438**, 201-204 (2005).
14. O. C. Compton, S. T. Nguyen, Small **6**, 711-723 (2010).
15. J. Zhao, L. Liu, F. Li, *Graphene Oxide: Physics and Applications* (Springer-Verlag, Berlin, 2015).
16. V. Singh, D. Joung, L. Zhai, S. Das, S. I. Khondaker, S. Seal, Progress in Materials Science **56**, 1178-1271 (2011).
17. W-W. Liu, S.-P. Chai, A. R. Mohamed, U. Hashim, Journal of Industrial and Engineering Chemistry **20**, 1171-1185 (2014).
18. L. Q. Xu, W. J. Yang, K.-G. Neoh, E.-T. Kang, G. D. Fu, Macromolecules **43**, 8336-8339 (2010).
19. Y.-L. Huang, H.-W. Tien, C.-C. M. Ma, S.-Y. Yang, S.-Y. Wu, H.-Y. Liu, Y.-W. Mai, J.Mater. Chem. **21**, 18236-18241 (2011).
20. S. Stankovich, D. A. Dikin, R. D. Piner, K. A. Kohlhaas, A. Kleinhammes, Y. Jia, Y. Wu, S. T. Nguyen, R. S. Ruoff, Carbon **45**, 1558-1565 (2007).
21. X. Wang, S. Huang, L.Zhu, X. Tian, S. Li,H. Tang, Carbon **69**, 101-112 (2014).
22. R. Subair, B. Tripathi, P. Formanek,; F. Simon, P. Uhlmann, M. Stamm, Polydopamine assisted gold nanoparticle decorated membranes for catalytic and environmental applications. Environmental Science: Nano, submitted in January 2015.
23. B. P. Tripathi, C. N. Dubey, S. Choudhury, F. Simon, M. Stamm, Journal of Materials Chemistry – Materials for biology and medicine B **27,** 3397-3409 (2013).
24. B. P. Tripathi, P. Das, F. Simon, M. Stamm, Molecular functionalization of polydopamine: A facile surface modification strategy to create ultralow fouling membranes. In preparation (2017).
25. L. M. Malard, M. A. Pimenta, G. Dresselhaus, M. S. Dresselhaus, Phys. Rep. **473**, 51-87 (2009).
26. A. C. Ferrari, Solid State Commun. **143**, 47-57 (2007).
27. A. C. Ferrari, J. C. Meyer, V. Scardaci, C. Casiraghi, M. Lazzeri, F. Mauri, S. Piscanec, D. Jiang, K. S. Novoselov, S. Roth, A. K. Geim, Phys. Rev. Lett. **97**, 187401 (2006).

Effects of Vibration welding process parameters on the weldability of the thermoplastics

DAS CHAPAL KUMAR* (1)

(1) FAKIR MOHAN UNIVERSITY - ODISHA - India

Joining of Plastics is mainly done at the places where the product cannot be fabricated in a single shot manufacturing process. Plastic Welding Processes can be divided into 2 groups; one involves mechanical movement and the other requires external heating. In this study the vibration welding, which involves mechanical movement has been applied to join the two thermo plastics of different character. It is very easy to weld two crystalline plastics or two semi crystalline plastics but it is very difficult to join the amorphous thermoplastics either together or crystalline or semi crystalline plastics as another partner. In this case, either in-situ compatibilizer or externally applied compatibilizer is a must. We tried to join the PMMA-PBT system and PMMA- nylon 66 systems using various welding process parameters. Externally applied compatibilizers used are PMMA-PBT= 50-50%, 70-30%, 30-70%. Similar proportions were used for PMMA- Nylon-66 system. It has been observed that the viscosity play a vital role in vibration welding process which changes the process parameters in order to get the efficient welding. The various parameters like amplitude, welding pressure, Dwell Time and melt down depth (distance) were studied. It has also been found that in case of PMMA-PBT system the more welding pressure and higher amplitude is required. Melt down depth however is little bit lower may be because of higher viscosity of PBT. The weld line was characterized by FESEM and found to be folded down in the case of PMMA- Nylon-66 system. However for PBT system, the folding is in the same plane of weld line. Better weldability have been observed when the two types of externally applied compatibilizer are applied separately on to the two substrates and then vibration welded. Two types of solution compatibilizer in the ratio of 70-30% and 30-70% is the better option than the 50-50% variety. Keywords: Vibration welding, PMMA- PBT, PMMA-Nylon66, Welding Pressure, welding Amplitude.

Biodegradable 3D printed bone anchoring bolts to enhance tendon-bone healing

Chou Ying-Chao [1], Chao Chien-Lin [2], Liu Shih-Jung [2]*

[1] Chang Gung Memorial Hospital - Tao-Yuan - Taiwan, [2] Chang Gung University - Tao-Yuan - Taiwan

This study developed a composite biodegradable polymeric model to enhance tendon graft anchoring and tendon bone healing. The in vitro and in vivo outcomes showed that the PLGA/collagen blended nanofibrous membrane functioned as a hydrophilic, stable and biocompatible scaffold. The PLA bolt was also proved to be a durable anchor for maintenance of the tendon-bone anchoring. In addition, the histological results demonstrated adequate biocompatibility of the PLA bolt on medial cortex with progressive bone ingrowths without tissue overreactions. The PLGA nanofibers within the bone tunnel also revealed the reduction of bone tunnel enlargement and enhanced the tendon-bone integration.

Property distribution and property gradient polymer products for enhanced functionality

Coates Phil **[(1)]*, *Caton-Rose Fin*** **[(1)], *Whiteside Ben*** **[(1)], *Vgenopoulos Dimitris*** **[(1)], *Nair Karthik*** **[(1)], *Thomson Brian*** **[(1)], *Hedbda Micheal*** **[(1)], *Thompson Glen*** **[(1)], *Twigg Pete*** **[(1)], *Ryabenkova Yulia*** **[(1)], *Paradkar Anant*** **[(1)]**

[(1)] University of Bradford - W Yorks - UnitedKingdom

Process structuring of polymers is of great interest in enhancing polymer product properties or functionality. We are exploring novel manufacturing routes to develop either spatial distributions of properties in regions of a product, or more continuous property gradients, using both solid phase orientation, precision moulding and 3-d printing technologies or combinations of these. One of the major application areas is in medical devices aimed at tissue fixation or tissue repair.

Optimization of collagen extraction from various cattle species hide by comparative methods

Noorzai Safiya (1)*

(1) University of Waikato - Hamilton - NewZealand

The use of regenerative medicine is on an all-time high resulting in a huge demand for renewable, biodegradable and biocompatible biomaterials. Collagen comprises 30% of the mammalian body's protein, being the main component of skin, connective tissue and cartilage. Bovine hides are a by-product of the meat industry and contain a significant amount of collagen. In this study the methods of acid-solubilisation (AS), acid-enzyme solubilisation (AES1) and a modified acid-enzyme solubilisation (AES2) were carried out to for comparative purposes to extract collagen from bull, calf, cow, face-piece and ox-hides. Acid solubilisation aids in removal of acidic proteins due to weakening of interactions between the acidic proteins and collagen fibrils, whereas acid-enzyme solubilisation maximises the extraction process by the action of proteolytic enzyme pepsin cleaving to the natural cross-linking sites of collagen (telopeptides). Purification was carried out by repetitive precipitation and dialysis. By AS extraction, collagen dry yields of 2.3-25.4% were obtained with highest yield being from calf hide. AES1 resulted in collagen dry yields of 18.2-31.3%, with bull hide resulting in the highest dry yield. Collagen dry yields ranged between 20.6-40.1% with AES2 extraction method. Hydroxyproline analysis resulted in collagen contents of 1.3-10.6% by one-stage purification with the method of AS extraction. However AS extraction with three-stage purification resulted in collagen contents of 3.8-25.7%. Collagen contents of 8.2-12.6% were obtained by AES1 extraction method with one-stage purification and 15.4-30.2% with three-stage purification. AES2 resulted in collagen contents of 23.95-35.99% with one-stage purification and 48.9-74.5% with three-stage purification which was significantly higher than found in recent literature. All the collagen samples revealed were type I with α, β, and γ chains as indicated by SDS-PAGE pattern.

Investigation of immiscible polymer blends to control drug release rate of polymeric pharmaceutical compounds

***Kelly Adrian* [(1)]*, *Enokela Ochegeinu* [(1)], *Gough Tim* [(1)], *Paradkar Anant* [(1)]**

[(1)] University of Bradford - Bradford - UnitedKingdom

The use of polymer matrices to control the release of poorly soluble drugs is of growing interest in the pharmaceutical sector. Melt processes such as twin screw extrusion are used to mix polymer(s), drugs and other excipients where the drug is dispersed, typically in an amorphous form mixed at a molecular level. Choice of polymer or polymer blend can be used to control the rate and pH dependency of drug release. Here, the use of immiscible polymer blends has been investigated in order to control release rate of a poorly soluble drug. Two pharmaceutically acceptable polymers (ethyl cellulose and shellac) were selected to provide an immiscible blend, to which a poorly soluble drug could be added. The polymers and drug were melt compounded using twin screw extrusion at different drug loadings and mixing conditions. Physical properties of the samples were characterized by scanning electron microscopy, Raman imaging, differential and scanning calorimetry whilst drug release rate was measured by dissolution. Results indicated that the drug was dissolved primarily in only one of the polymeric phases and that the drug release rate was influenced by drug loading and mixing conditions.

Injectable hydrogels for controlled release drug delivery

Shriky Bana [(1)]*, Gough Tim [(1)], Kelly Adrian [(1)], Isreb Mohammad [(1)]

[(1)] university of bradford - bradford - UnitedKingdom

This study seeks to develop hydrogel formulations for injectable controlled drug delivery from Pluronics incorporating multiple drugs, enhancing patient compliance, decreasing side effects, reducing dose and frequency. Pharmaceutically, these systems are attractive due to their unique sol-gel phase transition in the body, biocompatibility and safety being injected into the body as solutions before transforming to gel matrices at body temperature without external stimulation. Pluronic F-127 is a block copolymer (PEO-PPO-PEO) that gels at low concentrations and can be tuned using other excipients. Real-time full structural studies of F-127 systems mixed with a control drug and/or different molecular weight Pluronics under controlled temperature gradients and shear were performed. New experiments combining multiple characterisation techniques including rheology, X-ray scattering and Raman were developed to provide practical measurements of the system dynamics, essential for both application and manufacture.

Transparency and suture ability of alginate hydrogel reinforced nanofiber scaffolds

Stafiej Piotr (1)*, Küng Florian (1), Kruse Friedrich E. (1), Fuchsluger Thomas A. (1), Schubert Dirk W. (1)

(1) Friedrich-Alexander-Universität Erlangen-Nürnberg - Bavaria - Germany

The cornea is the outer most and therefore the most compromised part of the eye. The shape, the smoothness and the transparency of the cornea are responsible for 65-75 % of the eye's focusing. Corneal infection or scarring as result of injuries can lead to restricted or even to loss of vision. For corneal surface treatment the human amniotic membrane is applied traditionally [1]. Possible infections, immunologic reactions, shortcoming in donor tissue and inconsistent mechanical and optical properties are some drawbacks of this membrane. Preparation of synthetic scaffolds caught the interest of clinical research to overcome these disadvantages. Nevertheless most invented scaffolds are not able to fulfill all needed properties like mechanical strength (e.g. suture ability) and optical transparency. In this study we prepared random and aligned nanofiber scaffolds of polycaprolactone (PCL) and its blend with chitosan (CHI) by electrospinning. Sodium alginate hydrogel was incorporated into these membranes to improve the mechanical properties and bring in a possible drug eluting device. UV-Vis transmission spectroscopy was performed in the spectrum of 400 to 800 nm to quantify the transparency of the scaffolds. To visualize the transparency a method described by Schubert et. al. [1] was used. Suture retention strength measurements were performed on the different scaffolds to evaluate the resistance of the scaffolds against the pull-out of a suture [2]. The results demonstrate the semitransparent character as well as the suture ability of our scaffolds what makes them promising candidates for the use in corneal surface treatment [4]. [1] Berthold M.D.; et. al.; Cornea, 2011, 30(3), 269-272 [2] Schubert, D.W.; et. al.; Macromol. Symp., 2016, 365, 87-94 [3] Kueng, F.; et. al.; Mater Sci Eng C, 2016, 69, 941-946 [4] Schubert, D.W.; et. al.; EU-Patent No.:16185130.8-1455

Influence of the molar mass on the suture retention properties of polycaprolactone

Küng Florian [(1)]*, Stafiej Piotr [(1)], Fuchsluger Thomas A. [(1)], Kruse Friedrich E. [(1)], Schubert Dirk W. [(1)]

[(1)] Friedrich-Alexander-Universität Erlangen-Nürnberg - Bayern - Germany

Biopolymers like polycaprolactone (PCL) are common materials for soft tissue engineering. As many other biopolymers PCL degrades during its residence time in the human body. Even if the mechanism of degradation strongly is depending on the applied polymer, any degradation process causes a change in the molar mass of the polymer. Due to this change in chain-length the mechanic properties are also affected. Most soft scaffolds are attached to patients by sutures. When force is applied on a suture a high local stress is introduced in the sample. By a suture retention test [1] the resistance against this specific stress scenario is measured. Because of the degradation of the polymer during its time of application the suture retention properties of different molar masses are of importance for the mid- to long-term performance of a scaffold. PCL with a molecular number average of 80 kDa, 45 kDa, and 15 kDa has been used. In order to achieve a higher elastic modulus bimodal blends of the 80 kDa with the other above mentioned molar masses were investigated. The blends were dissolved in chloroform and cast into petri dishes. After the solvent evaporated, samples were taken from the films and suture retention tests were performed. The results of the suture retention tests show a dependency on the molar mass and on the blend composition. By blending different molar masses in a matching composition the resistance against suture pull out even can be increased. Using a tailor-made law of mixtures the effect of two component blends can be modeled. [1] Küng F., Schubert D. W., Stafiej P., Kruse F. E., Fuchsluger T. A.; Mater Sci Eng C, 2016, 69, 941-946

Failure of Silicone Breast Implants - Novel Insights on the Mechanical Properties of Implant Shells

Schubert Dirk W. [(1)*], ***Horch Raymund E.*** [(1)], ***Kaschta Jochen*** [(1)], ***Daenicke Jonas*** [(1)], ***Walter Bastian*** [(1)], ***Lämmlein Michael*** [(1)], ***Werner Siegfried*** [(1)], ***von Hanstein Harald*** [(1)]

[(1)] Friedrich-Alexander University Erlangen-Nürnberg - Bayern - Germany

This talk summarizes the knowledge on rupture of gel-filled silicone breast implants which is an open question since the first implantation in 1962. We present recent progress which was made by a mapping of the mechanical properties of the shell material together with a statistical failure analysis by means of a Weibull fit considering cyclic loading and various textures of the implant shell surface. Reference implants and intact as well as ruptured explants are tested after several years in vivo by tensile tests. Up to 110 dumbbell-shaped specimens per implant are measured allowing a detailed mapping of the mechanical properties of the silicone shell. Therefore, it is possible to illustrate the obtained results of the tensile tests by mapping the measured values of the implants in form of a contour plot revealing novel insights. The contour plots clearly reveal that the mechanical behaviour (e.g. stress at break) varies significantly over the shell. Additionally, a map is proposed which clearly shows separated clusters for the different manufacturers and product categories indicating a large variety of shell resilience. In particular we make firstly an attempt to mirror effects on implants under conditions during the use under a bulletproof vest. As a remarkable result e.g. it turns out that most implants are resilient up to a load of typically 1500 kg. References 1. Schubert, D. W.; Kaschta, J.; Horch, R.E.; Walter, B.; Denicke, J.; "On the failure of silicone breast implants – new insights by mapping of the mechanical properties of implant shells" review to be published 2. Bachelor Thesis, Werner, S. , Friedrich-Alexander University Erlangen-Nuremberg 3. Bachelor Thesis, Lämmlein, M. , Friedrich-Alexander University Erlangen-Nuremberg 4. Bachelor Thesis, Daenicke, J. , Friedrich-Alexander University Erlangen-Nuremberg 5. Master Thesis, Walter, B., Friedrich-Alexander University Erlangen-Nuremberg 6. Patent EP 15197017.05 Schubert, D. W et al

Electrospinning of spring-like fibres from elastomeric copolyesters

Sonseca Agueda (1), Kaczmarek Aleksandra (1), Michon Marcin (1), Sobolewski Peter (1), El Fray Miroslawa (1)*

(1) West Pomeranian University of Technology, Szczecin - zachodniopomorskie - Poland

Scaffolds using biodegradable polymers are a new standard in regenerative medicine. Cells can be immobilized on a support material that provides a platform for differentiation and proliferation. This approach may enable engineered cardiac constructs, capable bringing cardiomyocytes directly to the infarct region, and, thus revolutionize heart therapies. Ideally, such a construct ensures delivery of cells directly to the desired site, i.e., the damaged myocardium, and enables integration of cells with the host tissue, regenerating the myocardium, while the scaffold undergoes biodegradation [1,2]. Towards this aim, electrospinning is a technique that provides a straightforward way to produce fibrous mats with a large surface area to volume ratio and interconnected porosity. These properties are well suited to producing scaffolds mimicking the native cardiac extracellular matrix. Further, spring-shaped, coiled fibers are preferred for cardiac scaffolds as this biomimetic morphology allows extension and re-coiling in harmony with the stretching and contraction movements of the cardiac muscle. Thus, electrospun scaffolds must possess sufficient elasticity to support tissue contraction and relaxation. The electrospinning of poly(butylene succinate-butylene dilinoleate) (PBS-DLA) copolymer was successfully applied to prepare scaffolds composed of elastic, spring-like fibers. The synthesis of PBS-DLA is an enzyme-catalyzed process [3] thus providing material without any toxic residues of organometallic catalyst used in conventional synthesis. References 1. Jawad H. et al., Adv. Eng. Mater., 2010, 12(12), B664-B674 2. Tallawi M., et al. , Macromolecular Symposia, 2013, 334(1), 57-67 3. Sonsea A., McClain, A., Puskas, J. E., El Fray M, Engineering of Biomaterials, 2016, 19 (138), 31 ACKNOWLEDGMENTS The project is financially supported by National Science Centre, Poland, under the HARMONIA scheme (agreement No: UMO-2014/14/M/ST8/00610).

Obtaining Biodegradable, Elastic Electrospun Polyurethane/Graphene Oxide Grafts for Tissue Scaffolds

Ivanoska-Dacikj Aleksandra (1)*, Bogoeva-Gaceva Gordana (2), Krumme Andres (3), Tarasova Elvira (3), Plamus Tiia (3), calera Chiara (4)

(1) Research Center for Environment and Materials, Macedonian Academy of Sciences and Arts - Skopje - R. Macedonia, (2) Faculty of Technology and Metallurgy, Ss. Cyril and Methodius University - Skopje - R. Macedonia, (3) Department of Polymer Materials, Tallinn University of Technology - Tallinn - Estonia, (4) ab medica - Milan - Italy

Modern tissue engineering aims to create materials which promote, possibly in a controlled way, cell adhesion. These artificial three-dimensional structures should be capable of promoting the cell growth and gradually be replaced by new tissue. They cannot be considered as a standing substitute of a tissue, but as a temporary structure which promotes regeneration. Here, a biodegradable DegraPol® (DP), a polyester urethane, was used as a matrix which mechanical, electrical and cell adhesion properties ought to be modified by addition of graphene oxide (GO). Using solution electrospinning method four samples containing different quantities of GO (0, 0.5, 1.0, 2.0 wt%) were obtained. In order to produce fibrous scaffolds with increased porosity, solutions of DP and GO were co-spun with a water soluble polymer, polyethylene oxide (PEO), which was subsequently removed. Morphological and mechanical properties of the scaffolds were studied. Proper fiber formation was possible in the range of 8-14 wt% DP in a mixture of chloroform and ethanol in a ratio of 80:20 (wt/wt). The best homogeneity of the fibers was achieved for 12 wt% DP. The addition of GO had not changed much the morphology of the fibers. The co-electrospining of the polymers was successfully performed resulting in a mesh formation consisted of both polymers. The tensile test showed that the presence of GO in the DP matrix first increased the tensile strength, reaching the maximum value of 7.15 MPa, and then for higher GO content it started to decrease. Elongation at break (EB) had the highest value for the unfilled DP of 511 % and then it decreased with addition of GO and increase of its content. After removal of PEO from co-electrospun scaffolds, the values of the mechanical properties, TS and EB, of these porous samples significantly drop.

Drug delivery in Carboxymethylcellulose / Graphene Oxide / Hydroxyapatite composites systems

Khalifi Taghzouti Othman [1], El Mabrouk Khalil [1]*

[1] UEMF - Fès - Morocco

Carboxymethylcellulose scaffold combined with graphene oxide and hydroxyapatite (CMC/GO/HAP) were prepared to serve as a matrix for amoxicillin drug delivery during surgical treatment of musculoskeletal diseases. The nanocompistes materials were synthesized through in situ precipitation. Prepared materials loaded with different amount of amoxicillin were characterized using infrared spectroscopy (FT-IR), scanning electron microscopy (SEM), thermogravimetric analysis (TGA), energy dispersive X-ray spectroscopy (EDX), and X-ray diffraction (XRD). The release profile of amoxicillin as a model drug in buffer solution has established the efficiency of the nancomposites as a sustained delivery system. The amoxicillin-loaded nanocomposites exhibited a significant antibacterial compared to virgin HAP loaded amoxicillin .The nanocomposites developed in this study exhibited a great potential in bone tissue engineering and drug delivery application.

Development and Evaluation of Fabric modified with Polypyrrole/CuO Nanocomposite for Antimicrobial and Antistatic Applications

Bhandari Dr.Hema [1]*, Dhawan Dr.Sundeep Kumar [2]

[1] MAITREYI COLLEGE UNIVERSITY OF DELHI, New Delhi India - New Delhi - India, [2] National Physical Laboratory, CSIR, New Delhi - New Delhi - India

Present work is based on the antimicrobial and antistatic performance of conducting polymer nanocomposite impregnated fabric based on polypyrrole (PPY) and copper oxide nanoparticles (CuO NPs). Conducting polymer nanocomposites (PPY-CuO NPs) were synthesized by insitu chemical oxidative polymerization of pyrrole using sodium lauryl sulphate as a surfactant and ferric chloride as a dopant. Coating of PPY-CuO-nanocomposites on the cotton fabric was carried out during polymerization. Interaction of CuO NPs with PPY matrix was determined by the FTIR spectra, Thermo-gravimetric Analysis, X-Ray diffraction, Scanning Electron Microscopy (SEM), High Resolution Transmission Electron Microscopy (HRTEM) and conductivity measurements. The conductivity of PPY-CuO NPs coated fabric was found in the range of 10-8 to 10-3 S/cm depending on the loading concentration of CuO NPs in polymer matrix. TEM and HRTEM images showed that the PPY-CuO nanocomposites with an average diameter of 50-60 nm and were found to be nicely dispersed in the polymer matrix. The PPY-CuO NPs coated cotton fabric were tested against Escherichia coli (Gram negative) and Staphylococcus aureus (Gram positive) cultures and showed a significant antimicrobial activity; whereas its analogous uncoated and polypyrrole coated material showed almost negligible activity against the bacteria. Antistatic performance of the nanocomposite impregnated fabric was investigated by John Chubb Instrument. The static decay time of the nanocomposites coated fabric was found to be in the range 0.4 to 1.0 s on recording the decay time from 5000 V to 500 V. This indicated that the nanocomposite based on PPY-CuO nanocomposites has a great potential to be used as an effective antimicrobial and antistatic material. Moreover, the synergistic role of conducting polymer and CuO nanoparticles for antimicrobial and antistatic applications has also been studied.

Mechanical stress relaxation of medical tubing for infusion therapy: predicting product performance by material modelling

Kremser Thomas (1)*, Gläß Liesa (2), Roth Stefan (2), Schubert Dirk Wolfram (3)

(1) University of Erlangen-Nuremberg, B.Braun Melsungen AG - Melsungen - Germany, (2) B.Braun Melsungen AG - Melsungen - Germany, (3) Head of institute Polymer Materials, University of Erlangen-Nuremberg - Erlangen - Germany

Typical applications of tubing in the medical field represent flexible hoses for the infusion of fluids directly into the human blood circulation. In this application field, thermoplastic elastomers (TPEs) have found to be good alternatives to the common used plasticized polyvinyl chloride (PVC-P). There are special requirements for medical tubing like transparency, flexibility or an appropriate kinking performance to name only a few. Kinking performance of medical tubes is a crucial requirement with regard to safety and packaging. For that reason, different TPEs as well as PVC-Ps were benchmarked towards their kinking performance (short-term, long-term and recovery). In addition, mechanical stress relaxation was carried out in order to find correlation between viscoelastic behaviour and kinking performance. The goal of this approach was to figure out wether product performance, i.e. kinking performance, is predictable by using a standardized mechanical test method. Therefore, experimental data were generated by stress relaxation method iTSSR (Isothermic Temperature Scanning Stress Relaxation, Brabender GmbH & Co. KG) using tube shaped samples. The viscoelastic behaviour analyzed by iTSSR can be described by spring-dashpot systems with three maxwell elements connected in parallel. It was found, that there is correlation between viscoelastic behavior and a distinct product performance, i.e. kinking performance. Thus, material modelling using a multi-mode maxwell model for viscoelastic behavior represents an effective way of predicting kinking performance and benchmarking various materials. For this reason, further studies on additional features will be carried out.

Analysis of the spinning process and batch-to-batch consistency of aligned PCL/Collagen nanofibers for skeletal muscle

Dippold Dirk (1)*, Schubert Dirk W. (1), Beier Justus P. (2), Boccaccini Aldo R. (1)

(1) Friedrich-Alexander Universtity Erlangen-Nuremberg - Erlangen - Germany, (2) University Clinic Erlangen - Erlangen - Germany

The application of tissue engineered constructs is a promising approach in regenerative therapies for damaged skeletal muscle tissue. It aims to deliver healthy and functional cells on a mechanically supporting matrix to the damaged area of the tissue. Moreover, aligned and patterned topographies can guide cell alignment and support the formation of aligned myotubes. Therefore, the development of highly aligned scaffolds based on polycaprolactone (PCL) has been studied extensively [1]. In this study, aligned polycaprolactone/collagen (PCL/Col) biocomposite nanofibers were fabricated via electrospinning using environmentally benign acetic acid as solvent. Acetic acid was used only once with the PCL/Col material combination, resulting in very inhomogeneous fibers and bead formation [2]. Therefore the solubility and spinnability of different PCL/Col ratios and different collagen batches were investigated. The most suitable solutions were spun at various spinning parameters and collected with different collector setups in order to determine ideal processing conditions for the fabrication of highly aligned nanofibers. Light microscopy and SEM were used to investigate the fiber morphology, yield and alignment. The results demonstrate that ultrasonic treatment helps to fully dissolve PCL and that it is possible to fabricate homogenous PCL/Col fibers. Nanoscale fiber diameters can be achieved, depending on the process parameters. Additionally, various parameter fields were investigated in order to determine the most stable and suitable parameters for nanofiber fabrication [3]. In the final step, various collagen batches were analyzed according to their viscosity and spinnability, to reveal how settings have to be adapted to the challenging batch-to-batch consistency of natural polymers. [1] Doergens, A.; Materials Letters, 2015, 140, 99. [2] Yogeshwar C., V.; Journal of Applied Polymer Science, 2012, 125, 3221. [3] Dippold, D.; Materials Science and Engineering C, 2017, 72.

Understanding the Temperature Field in Fused Filament Fabrication for Enhanced Mechanical Part Performance

Natalie Rudolph [a)], Jianxing Chen [b)] and Travis Dick

[1]University of Wisconsin-Madison, Department of Mechanical Engineering,
1513 University Ave, Madison, WI 53706

[a)]Corresponding author: natalie.rudolph@wisc.edu
[b)]jchen245@wisc.edu

Abstract. The Fused Filament Fabrication process is the most widely used process for prototyping. The use of variable feedstock material, specimen geometries, pre-processing software, and printers makes comparability and reproducibility challenging, because the part properties not only change with varying print parameters, but also with changes in the slicing routine and tool path generation. The sequential part build-up causes a transient temperature field that affects the local microstructure and interfacial bonding and thus the macroscopic properties. The weld line formation between neighboring beads within layers as well as between layers has been identified as one of the biggest factors affecting mechanical properties. Using a custom python™ program for tool path generation, the cooling and re-heating effect during printing was studied using an IR thermal camera. It was shown that the temperature of the build platform has significant effect on the cooling and reheating behavior. A numerical analysis in ANSYS® Mechanical using the element death and birth effect proved that radiation should be included because of the initially high deposition temperatures, and that the presence and size of voids affect the re-heating or cooling rate during the deposition process.

INTRODUCTION

Additive manufacturing (AM) is a rapidly expanding technology with distinct advantages of creating near-zero waste during fabrication processes and having the ability to create complex and optimized component geometries. The Fused Filament Fabrication (FFF) process is the most widely used process for prototyping because of the wide variety of print materials, the low investment cost and the ease-of-use. A typical FFF process starts with the drive wheels of the printer feeding the polymer filament to a heated nozzle, where the polymer melts [1]. The solid filament acts as a piston and pushes the liquid melt through the nozzle to extrude the bead. This extruded bead of polymer adheres and solidifies to the previous layer or the build platform forming a 2D layer. Subsequent layers are stacked on top of each other to create a fully constructed 3D part.

Because polymers are insulators, large temperature gradients occur during this layer formation. As a result, the various cooling rates create residual stresses within the polymer part [2]. Warpage and delamination occurs as a result of these large internal stresses, which has a significant impact on the overall quality of the printed component. Previous studies examined the effects of varying print settings, e.g., FFF bead width, extrusion rate, and build layer thickness, and how they affect the surface quality, the phase transition, the issue of filament buckling at the nozzle entry, and the internal residual stresses caused by internal part temperature gradients [3, 4]. However, there has been little effort on measuring and simulating the deposition process and the effect of the print settings and resulting void formation between beads and layers on the cooling and reheating behavior. Therefore, this work establishes a basic model to simulate the FFF process as well as measuring the cooling and reheating effect.

EXPERIMENTAL

Modeling the FFF Process

The first step to model the deposition process involved creating two-dimensional, idealized models with known boundary conditions and an applied convection load in ANSYS Mechanical (APDL 16.2). The 2D single-bead control model represented a cross-section slice of a deposited bead, and served as a benchmark model to compare the temperature gradients of more complex 2D models in order to ensure accurate deposition simulation and model validity. Then, after determining subsequent 2D models produced similar results to the control 2D model, the third dimension (z-direction) was added along with radiation effects and a control 3D model was created. All 2D and 3D models utilized ANSYS element death and birth effects to represent the addition of polymer during the deposition process. The element death effect works by reducing the conductivity of selected areas or volumes to effectively represent the thermal resistance of the surrounding air. Then, using the element birth effect during designated load steps in the ANSYS simulations, selected areas or volumes were "reactivated" and assigned the conductivity of the deposition material [5]. Element death and birth effects have been successfully utilized to simulate other similar applications, like welding and material melting, with the same generalized solution strategy [6,7].

Overall, 8 virtual models were simulated, and analyzed, with each proceeding model adding a level of complexity to better simulate a more realistic deposition process. For simplification the boundary conditions and material properties are kept constant and are summarized in Table 1.

TABLE 1. Constant boundary conditions and material properties of Polylactic acid (PLA) used during FFF simulation.

FFF Operating Conditions	Value	PLA Material Properties	Value
Print Temperature [°]	190	Density [kg/m^3]	1250
Build Plate Temperature [°C]	60	Conductivity [W/m-K]	0.13
Ambient Temperature [°C]	25	Specific Heat Capacity [J/kg-K]	1800
Air Heat Transfer Coefficient @ 0.5 m/s air velocity [W/m^2-K]	7.5	Emissivity [-]	0.95

To initially justify the addition of radiation, the Biot number of Polylactic acid (PLA) was calculated for a 3D model just after deposition (Supplementary Documents), which was later confirmed with 3D models.

The surrounding FFF air properties can fluctuate depending on mild indoor drafts, frequency of human traffic, temperature, pressure, and other various factors. However, in a well-controlled environment these fluctuations are low, and for a simplified FFF model the conditions can be considered constant. An external flow over a flat plate (bed of a 3D printer) with air velocity of 0.5 m/s under ambient conditions produced a heat transfer coefficient of about 7.5 W/m^2-K (Supplementary Documents).

During the deposition voids and incomplete bonding may occur between bead layers [8]. Incomplete bonding is neglected for all models *except* for the 3D void models. The purpose and set-up of all models is explained in detail in the Supplementary Documents.

Temperature Measurement Using IR Camera

The online measurement of the FFF temperature field is conducted using a FLIR® E40 infrared (IR) camera. In order to accurately measure the temperature, several calibration and correction steps are needed. The true temperature of the polymer during the print is calibrated from the IR signal using an established calibration curve. This is achieved by taking temperature measurements of a pre-printed sample of the same polymer material using a thermocouple and a well-controlled hot stage. Secondly, the effect of the camera pixel size on the IR response cannot be ignored. An existing method to correct for this problem, referred to as the spatial resolution, was applied [9]. In our study, one pixel corresponds to a single layer of the print. Lastly, monitoring the temperature during the FFF process involves various levels of reflection depending on process environment and equipment. In this study, an Ultimaker® 2+ FFF printer is used, which has a glass build platform and shiny metal connecting rods. The build platform along with the reflective rods can introduce visible reflection effects during the measurements. A proven measurement strategy [10], which utilizes the emissivity and repeatability of the FFF process, was implemented to overcome the reflection problems.

The IR camera is mounted stationary in front of the printer with a tripod. A single wall rectangle structure is printed on top of a printed platform of the same material. This raises the part enough to better visualize the print and improve adhesion. In this experiment, Acrylonitrile Butadiene Styrene (ABS) filament was used, which was produced in house using SABIC® MG-94 material.

FLIR Tools+® was used for the data analysis. By applying the previously mentioned correction and measurement strategies, the frontal area of the print, can be selected to provide temperature data points by pixels. With known pixel to layer ratio of 1:1, the temperature field can be obtained as a function of time during all video frames. Then the cooling profiles and temperature distribution along each layer or vertically across the part can be obtained. Table 2 lists the most important printing conditions and measurement parameters.

TABLE 2. FFF Process Conditions, Measurement and Material Parameters

FFF Operating Conditions	Value	Measurement on ROI & Material	Value
Print Temperature [°]	245	Front Area [mm*mm]	15*20
Build Plate Temperature [°C]	100	Recording Rate [FPS]	30
Ambient Temperature [°C]	25	Distance [cm]	30
Number of Wall [-]	1	MFI [-]	11.7
Print speed [mm/min]	2000	Glass Transition Temp. Tg [°C]	110
Layer height [mm]	0.4		

RESULTS

Simulation Results: Effect of Radiation and Voids

All of the 2D models were analyzed and compared to validate the element death and birth feature for the simulation of the FFF process. The temperatures were all very similar at each node at 3 different locations. Overall, the maximum temperature difference from the control model was around 4 °C, which occurred on the top face node on the six-bead model (see Supplementary Documents). The right and left face nodes for the three-bead and six-bead models had a maximum temperature difference from the control model of approximately 2 °C. Based on the achievable measurement resolution, a 4 °C difference from the control model is acceptable enough to use this procedure for the 3D modeling.

Two 3D single-bead model simulations, one with convection only and one including additional radiation effects, were performed and the temperature data from the front face, back face, left face, right face and top face nodes were collected and compared. The 3D single-bead convection and radiation model resulted in lower overall temperatures at the same analyzed nodes; this is due to the additional considered mode of radiation energy transfer along with convection. The Front/Back/Left/Right nodes of the convection and radiation model resulted in approximately 5 °C less than the same nodes of the convection only model. The top node of the convection and radiation model resulted in a maximum difference of approximately 8 °C from the same top node on the convection only model. Therefore, further models that simulate the FFF process should include radiation effects during the cooling of beads.

Next, a 3D void-free four-bead model was compared with two 3D four-bead void models to examine how voids at the bead interfaces affect the internal temperature gradients. To accurately represent a "typical" void formed during a deposition process, an equivalent void cross-sectional area of 0.02618 mm^2 was used. Additionally, a model with a double-sized cross-section void area was created. Temperature versus time data was collected at select nodes during the three simulations. The node locations examined during the simulation were between beads at the center of their face, and at the center of the bead. Because of different geometries and void sizes, node selection remained as consistent as possible.

After selecting similar node locations for all of the 3D wall models and carrying out the simulation in ANSYS, the temperature as a function of time data was recorded and analyzed in EES (Engineering Equation Solver). The bead center temperature will be used for the analysis. Figure 1 (a) and (b) show the plotted results of the 3D wall model without voids and a comparison between typical sized void and double-sized void, respectively.

In Fig. 1 (a) the cyclic heating and cooling of the beads during the FFF process can be followed. The lower left bead center (blue) is cooling after deposition for 15 s and is then slightly reheated when the right bead (green) is deposited next to it. It is heated again when a bead is deposited on top of it (black). Each newly deposited bead has a reheating effect on the surrounding beads and the cooling again depends on that reheating effect.

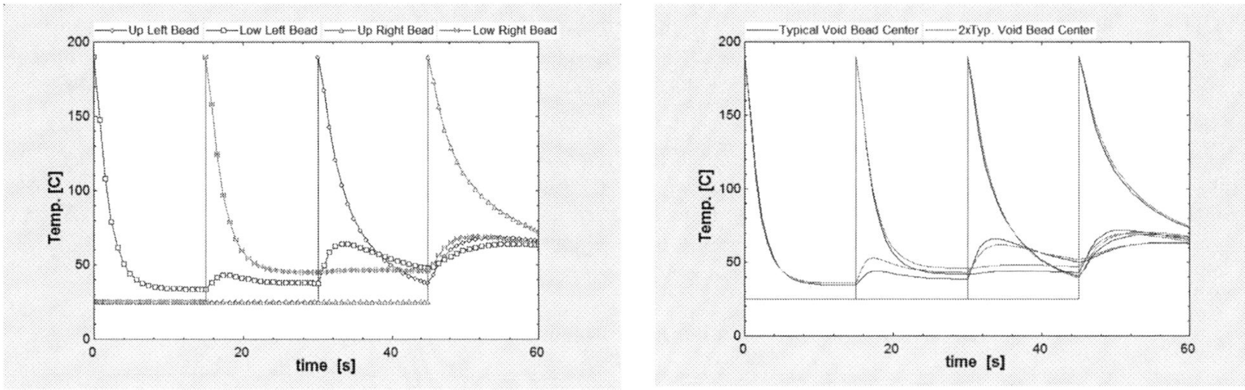

(a) (b)

FIGURE 1. Bead center temperature as a function of time plotted results for a 2 x 2 x 2 mm build (a) with no voids, (b) comparison of a typically sized void (black) to a 2 times typically sized void (red).

The same effect can be observed when voids are included in the simulation, Fig. 1 (b). However, the temperature profiles are different based on the presence and size of the void. Overall, the general trend was that the initial cooling of the bead, with or without voids, produced similar temperature profiles. However, the noticeable difference occurred in the cyclic reheating and cooling of existing beads coming into contact with newly deposited beads. Specifically, the presence of a void appeared to affect the rate at which the bead reheated or cooled (depending on location). Furthermore, larger voids enhanced the noticeable rate change of reheating and cooling, where a possible explanation is that with a larger void there is more thermally insulating air to prevent the exchange of energy between beads or to the surroundings. The no-void model and the typical void model comparison, and the typical void model and the 2 times typical void model comparisons both produced final temperature results at 60 s that were approximately 0.5 °C different (higher or lower depending on the bead center examined).

Temperature Measurement: Effect of Build Plate Temperature

Two measured temperature profiles for the single wall structure will be discussed. Firstly, it is important to analyze the reheating effect of the newly deposited material, which is much hotter than the already cooled previous layers. The number of layers re-heated over the glass transition temperature is in particular important to study since this has a huge effect on the bonding quality between layers and even across the entire part. By positioning the node at the center of each layer, at the same instant the material is being deposited at the very top, the temperature profile from top to bottom is plotted in Figure 2 (a). It can be seen that there are about 3 layers heated above glass transition temperature by the top layer. In lower layers, the temperature is gradually decreasing towards the middle of the printed structure. However, the layers closer to the build platform see a temperature increase, which approaches the set build platform temperature of 100 °C. This suggests that the bottom layers stay constant at the platform temperature for the duration of the print, which is only 10 K below the glass transition temperature. Therefore, some molecular movement is possible and the bonding between these layers should be high [11]. However, this requires further validations of the mechanical properties throughout the printed part. The temperature in the center of the part will always be the coolest and the top layers will be reheated for a couple of times above Tg, which equals only a couple of seconds in total.

In Figure 2 (b), the cooling profiles at different heights of the printed single wall are obtained right after the FFF process is finished, after the print head is moved away from the part. Four regions are chosen: the very bottom of the wall, 5 [mm], 10 [mm], and 15 [mm] above it. It can be seen that all regions show different cooling behavior. The bottom layer remains at the temperature of the build platform and cools down slowly with the cooling of the platform. From 5 [mm] to 15 [mm], an increasing cooling rate can be noticed by an increasing slope of the curves. In addition, the temperature difference over time is increasing with distance to the platform. Furthermore, after 20 [s] of cooling, the temperature at the bottom of the print is higher than at the other regions. Between the 10 [mm] region and the 15 [mm] region, a temperature difference of approximately 8 °C is shown. However, the difference between the bottom region and the rest are over 30 °C. In addition, the 5 mm and 10 mm regions cooled to a similar temperature, the higher regions cooled even more.

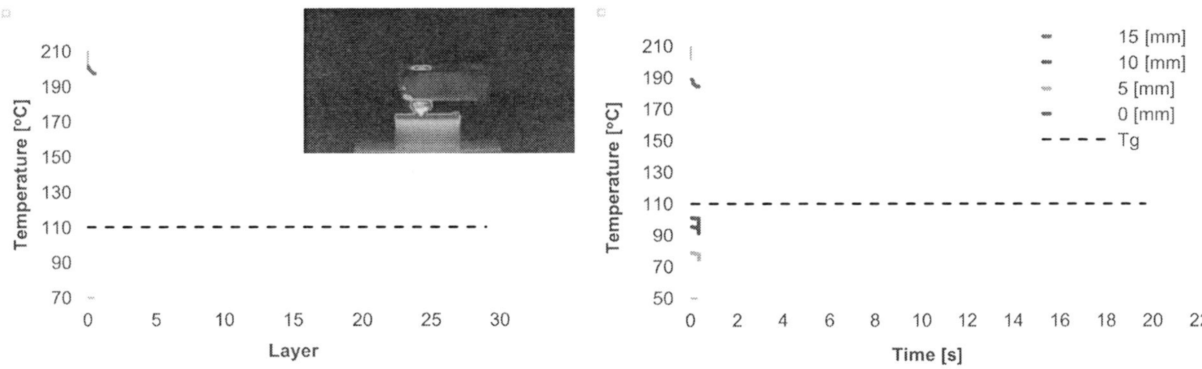

(a) (b)

FIGURE 2. Temperature distributions across the printed part (a) at the center line of the test piece with a view of the print through the IR camera, (b) at different heights of the test piece after the print is finished (cooling profiles).

CONCLUSION AND OUTLOOK

The ANSYS simulation results proved the applicability of element death and birth effects for the FFF deposition process. It further showed the importance of radiation and inclusion of voids into the model at the bead interface level. The IR measurements highlighted the different resulting cooling rates in different sections of the print, which corresponds to the density, void content, surface finish, and potentially mechanical properties of that section.

Next steps will be extended IR measurements to investigate the effects of nozzle temperature, print speed, and number of walls on the temperature filed and its correlation to mechanical properties. Furthermore, these results will be validated with continued simulations on the micro level.

REFERENCES

[1] S.-H. Ahn, M. Montero, D. Odell, S. Roundy, and P. K. Wright, "Anisotropic material properties of fused deposition modeling ABS," Rapid. Prototyp. J. **8**(4), 248–257 (2002).

[2] C. Li, Y. Wang, H. Zhan, T. Han, B. Han, and W. Zhao, "Three-dimensional finite element analysis of temperatures and stresses in wide-band laser surface melting processing," Mat. Des. **31**(7), 3366-3373 (2010).

[3] H. Bikas, P. Stavropoulos, and G. Chryssolouris, "Additive manufacturing methods and modelling approaches: a critical review," Int. J. Adv. Manu. Tech. **83**(1-4), 389-405 (2015).

[4] I. P. Ilyas, "A Study of Melt Flow Analysis of Polycarbonate (PC) in Fused Deposition Modelling Process," Int. Conf. Exhib. Sustain. Energy Adv. Mater. **14**(June), 244–252 (2011).

[5] ANSYS® Mechanical APDL (Version 16.2), ANSYS Help Viewer: Understanding Element Birth and Death (9.2)

[6] X. Shan, C. M. Davies, T. Wangsdan, N. P. O'Dowd, and K. M. Nikbin, "Thermo-mechanical modelling of a single-bead-on-plate weld using the finite element method," Int. J. Press. Vessel. Pip. **86**(1), 110–121 (2009).

[7] C. K. Lee, S. P. Chiew, and J. Jiang, "Residual stress study of welded high strength steel thin-walled plate-to-plate joints part 2: Numerical modeling," Thin-Walled Struct. **59,** 120–131 (2012).

[8] N. a Langrana, D. Qiu, E. Bossett, S. C. Danforth, M. Jafari, and A. Safari, "Virtual simulation and video microscopy for fused deposition methods," Mater. Des., **21** (2), 5–82 (2000).

[9] V. T. Marla, R. L. Shambaugh, and D. V. Papavassiliou, "Use of an Infrared Camera for Accurate Determination of the Temperature of Polymer Filaments," Ind. Eng. Chem. Res. **46**, 336-344 (2007).

[10] J. E. Seppala, and K. D. Migler, "Infrared Thermography of Welding Zones Produced by Polymer Extrusion Additive Manufacturing," Add. Manu. **12**, 71-76 (2016).

[11] Q. Sun, G. M. Rizvi, C. T. Bellehumeur, and P. Gu, "Effect of Processing Conditions on the Bonding Quality of FDM," Rapid. Prototyp. J. **14**(2), 72-80 (2008).

Supplementary Documents

Convection vs. Radiation

Biot number calculation in the Engineering Equation Solver (EES) to determine if radiation (in addition to convection) must be considered during ANSYS simulation. Calculations showed radiation should be considered for more accurate modeling.

Significance of Radiation

Objective:

The following analysis compares the resistive effects of convection and radiation during a FFF process to determine if radiation is negligible.

Resistance Network:

The resistance network for a typical FFF process can be seen below. We have already determined that conduction and convection effects will be considered during thermal modeling of FFF.

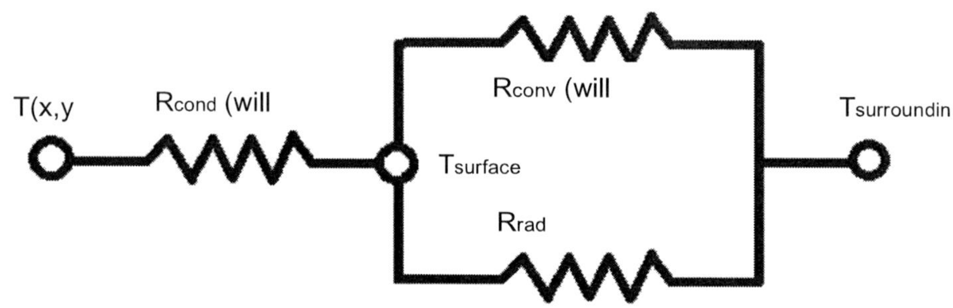

We will use the Biot number (Bi) to compare the resistance of convection to the resistance of radiation.

$$Bi = \frac{Resistance\ want\ to\ neglect}{Resistance\ that\ we\ are\ considering} = \frac{R_{radiation}}{R_{convection}} = \frac{\frac{1}{\varepsilon\ \sigma\ A_s\ (T_{surf}^2 + T_{surr}^2)(T_{surf} + T_{surr})}}{\frac{1}{h\ A_s}}$$

Where:

ε = Emissivity
σ =Stefan-Boltzmann Constant
$A_{s=}$Surface Area in Contact with Surroundings
T_{surf} = Surface Temperature of Bead
T_{surr} = Ambient Air Temperature
h = Convection Heat Transfer Coefficient

If Bi << 1, then radiation effects can be ignored.

Calculations:

EES was used to perform the calculations necessary to determine the Bi number. The nozzle rate was incorporated into the resistance equations to simulate the addition of surface area during a FFF process. Note that the calculations assumed constant surface temperature, heat transfer coefficient, emissivity, and nozzle rate.

Discussion:

Below are the resistances and maximum Biot number (during deposition) found in EES. If we are assuming constant nozzle rate the surface area should increase rather linearly, and assuming constant surface and surrounding temperature, emissivity, and natural convection heat transfer coefficient indicates that radiation should not be neglected as the maximum Biot number was 1.104 [-].

Main | **Key Variables**

(Table 1, Run 284)

bi_{num} = 1.104 [-]	*Biot Number*
R"rad = 0.1471 [K-m²/W]	*Resistance from Radiation*
R"conv = 0.1333 [K-m²/W]	*Resistance from Convection*

Figure 1: Key variables from EES calculations, which include the Biot number and the resistances from radiation and convection during a FFF process. Note that the surface area is the same for radiation and convective effects.

Main | Key Variables

Unit Settings: SI C kPa kJ mass deg

(Table 1, Run 284)

As = 0.03124 [m²]	beads = 625 [-]	bi_{num} = 1.104 [-]
η = 0.95 [-]	h_{nat} = 7.5 [W/m²-K]	k_{PLA} = 0.13 [W/m-K]
L = 0.025 [m]	$nozz_{rate}$ = 0.055 [m/s]	$process_{time}$ = 284.1 [s]
R"comb = 0.06995 [K-m²/W]	R"conv = 0.1333 [K-m²/W]	R"rad = 0.1471 [K-m²/W]
R_{comb} = 2.239 [K/W]	R_{conv} = 4.268 [K/W]	R_{rad} = 4.71 [K/W]
SBC = 5.670E-08 [W/m²-K⁴]	t = 284 [s]	tot_{length} = 15.63 [m]
T_{app} = 333.2 [K]	T_{inf} = 298.2 [K]	x = 0.001 [m]

No unit problems were detected.

Figure 2: Complete solutions window from EES.

Air heat transfer coefficient:

Below are the calculations performed in EES to determine the air heat transfer coefficient.

```
"heat transfer coefficient of air - low forced convection over flat plate (3D printed bed)"
"article had arithmetic mean of indoor velocities of 0.3 [m/s] --- 0.5 [m/s] worst case scenario"

"surrounding conditions"
T_amb = 25 [C]                              "ambient temperature"
T_amb_K = converttemp(C,K,T_amb)
P_amb = 1 [atm]                             "ambient pressure"
P_amb_kPa = P_amb*convert(atm,kPa)
L_plate = 0.25 [m]                          "length of bed - flate plate assumption"
W = 0.05 [m]                                "5cm square build part - approximate"
u_air = 0.5 [m/s]

"properties of air at ambient conditions"
rho_air=density(Air_ha,T=T_amb,P=P_amb_kPa)
mu_air=viscosity(Air_ha,T=T_amb,P=P_amb_kPa)
k_air =conductivity(Air_ha,T=T_amb,P=P_amb_kPa)

"calculated dimensionless #'s"
Re = rho_air*u_air*(L_plate/2)/mu_air
```

Pr = **prandtl**(***Air_ha***,***T***=T_amb,***P***=P_amb_kPa)

"Dimensionless - external forced flow"
***Call* external_flow_plate_nd**(Re,Pr: Nusselt1,C_f)

"calculation of h"
Nusselt1 = h1*(L_plate/2)/k_air

Main | Key Variables

Unit Settings: SI C kPa kJ mass deg

C_f = 0.02097 [-]	h1 = 7.73 [W/m^2-K]	k_{air} = 0.02625 [W/m-K]	L_{plate} = 0.25 [m]
μ_{air} = 0.00001845 [kg/m-s]	Nusselt1 = 36.81 [-]	Pr = 0.7073 [-]	P_{amb} = 1 [atm]
$P_{amb,kPa}$ = 101.3 [kPa]	Re = 4012 [-]	ρ_{air} = 1.184 [kg/m^3]	T_{amb} = 25 [C]
$T_{amb,K}$ = 298.2 [K]	u_{air} = 0.5 [m/s]	W = 0.05 [m]	

No unit problems were detected.

Calculation time = .0 sec.

A. 2D Single-Bead Control Model

The purpose of the 2D single-bead control model was to serve as a benchmark model to compare subsequent and more complex 2D models. In particular, the model was used to determine the validity and accuracy of subsequent models that utilized the element death and birth effects. The 2D single-bead control model had dimensions of 1 [mm] width by 1 [mm] height, and had square mesh elements with each side of the element having a length 0.1 [mm]. **Figure 3** shows the 2D single-bead control model in ANSYS.

Figure 3: The geometry for the 2D single-bead control model with 1 [mm] width and 1 [mm] height, and square 0.1 [mm] mesh simulation elements (shown above).

After creating the geometry and applying boundary conditions in ANSYS, the simulation process for the 2D single-bead control model was as follows:

1) Apply an initial condition to set the PLA material to equal the application temperature.
2) Start simulation, observe cooling for 15 [s], and create an animated temperature contour plot.
3) Collect temperature data at selected nodes for analysis and visually analyze animated temperature contour plot.

Then, after examining the control model to ensure results conceptually made sense, two additional beads were added along with the element death and birth effect in the subsequent model – the 2D three-bead model.

B. 2D Three-Bead Model

The main purpose of the 2D three-bead model was to implement the element death and birth effects to simulate the deposition of new beads in the x-direction (build width direction), and to compare the temperature data results to that of the single-bead control model. If the transient temperature data between the control and 2D three-bead models were similar, the element death and birth effects could be determined as a valid method to simulate the deposition process. The 2D three-bead model had dimensions of 3 [mm] width by 1 [mm] height, and had square mesh elements with each side of the element having a length 0.1 [mm]. **Figure 4** shows the 2D three-bead model in ANSYS.

Figure 4: The geometry for the 2D three-bead model with 3 [mm] width (3 beads) and 1 [mm] height, and square 0.1 [mm] mesh simulation elements.

The simulation process for the 2D three-bead control model was as follows:

1) Initially constrain temperature of all elements to that of surrounding air and deactivate all elements (element death effect); this represents an empty 3D printer deposition bed.
2) At beginning of first time step (very small time), activate elements (element birth effect) that represent the bead cross-section and constrain their temperature to application temperature; this represents the instantaneous start of the deposition process.
3) After a small time interval later the temperature constraint on the elements representing the first bead was removed, which represented the end of the deposition process.
4) Observe cooling for 15 [s].
5) After cooling observation time (15 [s]), the next bead in the x-direction started at step 2) from above. The process was repeated until all beads in the x-direction were "built".
6) Collect temperature data at selected nodes for analysis and visually analyze animated temperature contour plot.

It should be noted that a temperature initial condition command in ANSYS could not be utilized, because initial conditions can only be issued during the first time step of the simulation. Therefore, to simulate multiple bead depositions the temperature constraint method was used.

Again, after examining the 2D three-bead model to ensure results conceptually made sense, the next step in the modeling process involved adding beads in the y-direction (build height direction); this subsequent model was the 2D six-bead model.

C. 2D Six-Bead Model

The purpose of the 2D six-bead model was to add another level of complexity by again implementing the element death and birth effects, this time additionally in the y-direction (build height direction) to simulate the deposition of new beads. In particular, the six-bead temperature data was compared to the single-bead control model on both x and y faces to determine model validity. The 2D six-bead model had dimensions of 3 [mm] width by 2 [mm] height (3 by 2 beads), and had square mesh elements with each side of the element having a length 0.1 [mm]. **Figure 5** shows the 2D six-bead model in ANSYS.

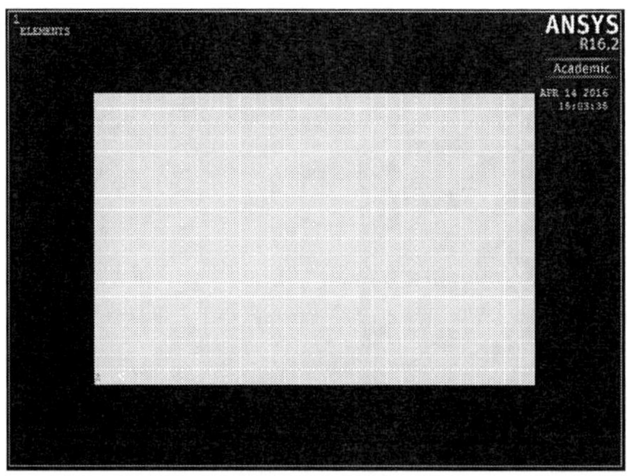

Figure 5: The geometry for the 2D six-bead model with 3 [mm] width and 2 [mm] height (3 by 2 beads), and square 0.1 [mm] mesh simulation elements (shown above).

The simulation process for the 2D six-bead control model was more complex than a single layer deposition model, and it required two command "do" loops (one nested in the other) to simulate a two-dimensional build. The process for the 2D six-bead model can be summarized as the following:

1) Initially constrain temperature of all elements to that of surrounding air and deactivate all elements (element death effect); this represents an empty 3D printer deposition bed.
2) At beginning of first time step (very small time), activate elements (element birth effect) that represent deposited bead cross-section and constrain the temperature to application temperature; this represents the instantaneous start of the deposition process.
3) After a small time interval later the temperature constraint on the elements representing the bead was removed, which represented the end of the deposition process.
4) Observe cooling for 15 [s].
5) After cooling observation time (15 [s]), the next bead in the x-direction started at step 2) from above. The process was repeated until all beads in the x-direction and current y-layer were "built".
6) After one y-layer was formed, the deposition process repeated itself (starting at step 2)) at the next layer (y+1 layer).
7) After all layers were simulated temperature data at selected nodes were collected for analysis and visually analyzed with animated temperature contour plots.

Following the simulation process of the 2D six-bead model was the 3D single-bead model, which was the next step in adding a level of complexity and simulating a more realistic deposition process in ANSYS.

D. 3D Single-Bead Model

The purpose of the 3D one-bead model was create a more realistic deposition model by adding the third z-direction dimension (build depth direction), and by adding radiation effects during the cooling process (to determine if radiation was significant). The 3D single-bead model also served as a benchmark model to compare temperature data to that of subsequent models to determine model accuracy. The 3D single-bead model had dimensions of 1 [mm] width by 1 [mm] height by 1 [mm] depth, and had square mesh elements with each side of the element having a length 0.1 [mm]. **Figure 6** shows the 3D single-bead model in ANSYS.

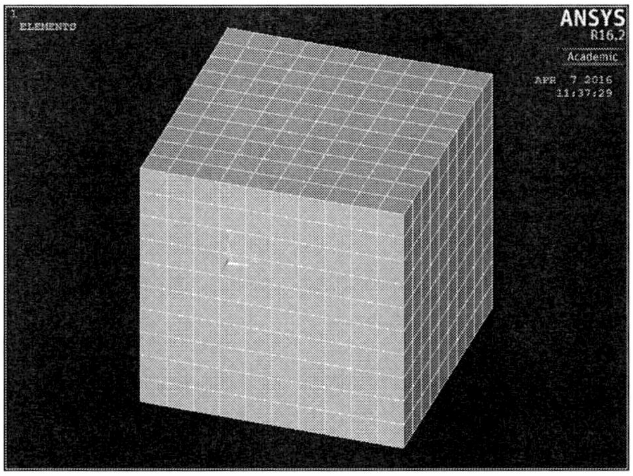

Figure 6: The geometry for the 3D single-bead model with 1 [mm] width, 1 [mm] height, 1 [mm] depth, and square 0.1 [mm] mesh simulation elements.

After creating the geometry, the simulation process was similar to the 2D six-bead model in the fact that element death and birth, temperature constraints, and nested command "do" loops were utilized to perform the simulation. However, three "do" doops – one nested inside of another, which was nested in the final loop – were used to simulate the 3D temperature gradient modeling. Next, following the 3D single-bead simulation was the 3D wall model.

E. 3D Wall Model – No Voids

The 3D wall model (no voids) simulated the deposition process of a larger geometry to better represent a real FFF process and further understand how temperature gradients form within the printed PLA component. The geometry of the build had dimensions of 2 [mm] width, 2 [mm] height, and 2 [mm] depth, which represented a cube consisting of 4 beads with 2 [mm] (z-direction) depth. Tetrahedral elements with side lengths of 0.1 [mm] were used instead of squarely meshed elements, as subsequent models with voids required tetrahedral meshes. **Figure 7** shows the 3D wall model (no voids) in ANSYS.

Figure 7: The geometry for the 3D wall model (no voids) with 2 [mm] width, 2 [mm] height, 2 [mm] depth, and square tetrahedral mesh simulation elements (0.1 [mm] element side lengths).

The simulation carried out in ANSYS was similar to the process outlined for the 2D six-bead model (with the additional steps from the 3D single-bead model), along with the addition of beads in the x an z-directions. After simulation, an animated temperature contour video was created to thermally represent the deposition process. It should be noted that this model contains zero voids between bead layers, which represents idealized bonding between beads. The next model, the 3D wall model with voids, will examine how the presence of voids affects the temperature gradients within the printed PLA component.

F. 3D Wall Model – With Voids

After establishing an idealized 3D wall model without voids, the next logical step was to create a more realistic 3D model with idealized voids. To accurately represent a "typical" void formed during a deposition process, an equivalent void cross-sectional area of 0.02618 [mm^2], or a square with 0.1618 [mm] side lengths, was used for void geometry. Additionally, a model with a double-sized cross-section void area was created. Both the "typical" sized and "large" sized void models can later be compared to temperature profile data of the void-free model with the same geometry to determine how voids and void sizes affect the internal temperature gradients. The square void geometry was rotated 45 degrees (from horizontal) to better represent actual rhombus shape voids usually form during deposition. The void dimensions used were based off a previous study that examined how voids and defects affect fused deposition technology build quality [8]. It should be noted that mapped-face meshing techniques using square elements can no longer be used with the 3D wall with voids model because of the rotated void geometry. As a result, a free meshing technique with triangular elements was used in ANSYS with the 3D void model, and caution was used when selecting the same node locations for comparison between void and no-void geometries. **Figure 8** shows a 3D wall geometry front view (four beads total with one void) with triangular free-mapped meshing, and **Figure 9** shows the same build at an oblique view for better visual representation of the component build.

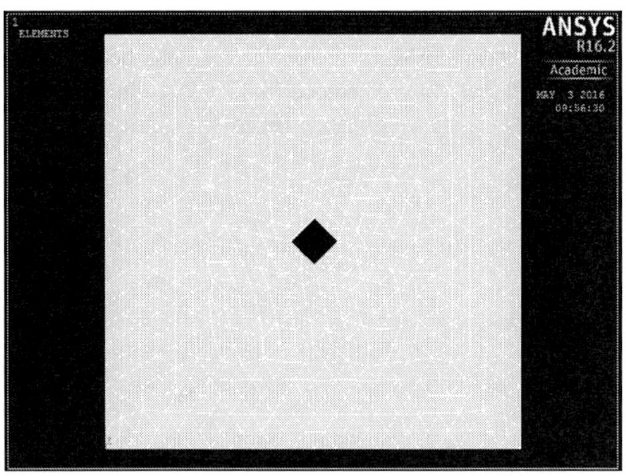

Figure 8: A 3D with void model geometry (front face view). This particular geometry represents four beads with one "typical" void formed during the deposition process.

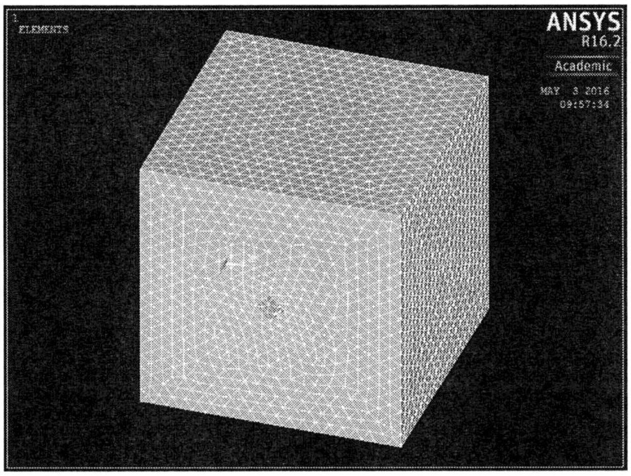

Figure 9: A 3D with void model geometry (oblique view). This particular geometry represents four beads with one "typical" void formed during the deposition process.

Finally, after creating the 3D geometry with voids and meshing with triangular elements, similar steps as the 3D wall without voids simulation were followed and the temperature profiles of select and same nodes for both 3D no-void and with void models were analyzed to determine if voids play a significant role with internal temperature gradients.

2D Model Simulation Results:

All of the 2D models, including the control model, were analyzed and compared by examining a total of three nodes on the right, top, and left side of the first bead cross-sectional area. The first node (**node 32** for all 2D models) was at the mid-way point on the right side of the bead ((x,y)=(1.0[mm], 0.5[mm])), the second node (**node 52** for all 2D models) was at the mid-way point on top of the bead ((x,y)=(0.5[mm], 1.0[mm])), and the third node (**node 71** for all 2D models) was at the mid-way point on the left side of the bead ((x,y)=(0.0[mm], 0.5[mm])). The significance of comparing these nodes for all 2D models was to determine if the element death and birth effect in ANSYS allowed for accurate deposition simulation. The temperature profiles for nodes 32, 52, and 71 for the 2D single-bead control model, 2D three-bead model, and 2D six-bead model can be seen below in **Figure 10**, **Figure 11**, **Figure 12**, respectively. Additionally, for better representation and explanation of the 2D six-bead deposition process, screenshots of the ANSYS simulation can be seen in **Figure 13**.

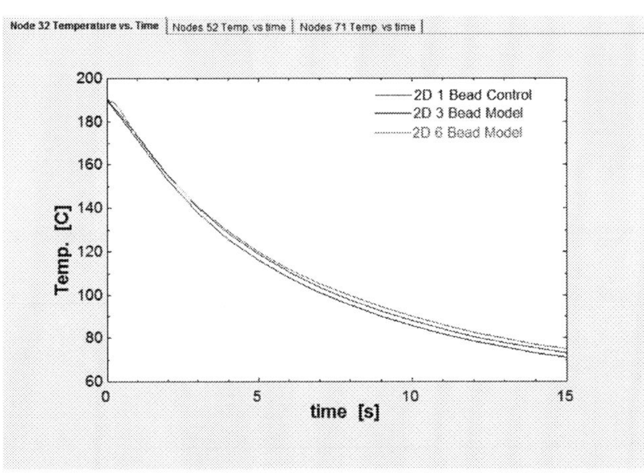

Figure 10: Right face node 32 ((x,y)=(1.0[mm], 0.5[mm])) temperature profiles over 15 [s] for all 2D models.

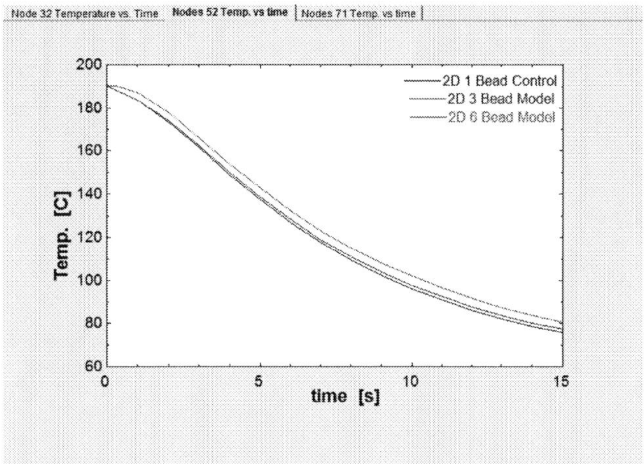

Figure 11: Top face node 52 ((x,y)=(0.5[mm], 1.0[mm])) temperature profiles over 15 [s] for all 2D models.

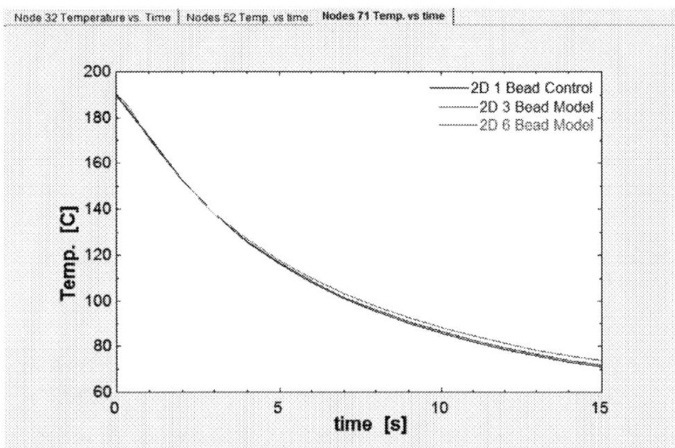

Figure 12: Left face node 71 ((x,y)=(0.0[mm], 0.5[mm])) temperature profiles over 15 [s] for all 2D models.

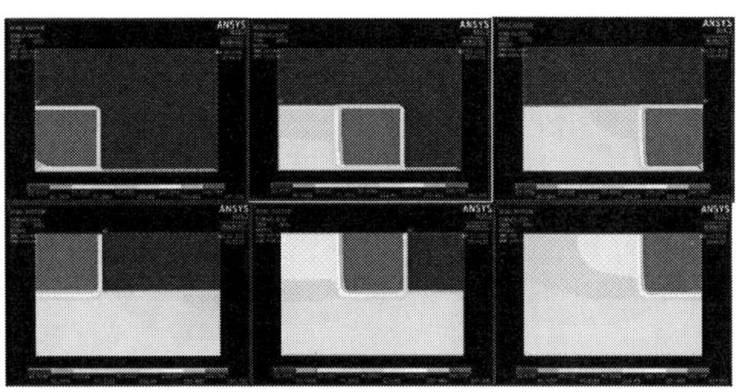

Figure 13: A series of snap shots of the beginning of the deposition simulation for a six-bead (2-row 3-column) PLA wall. The upper left snapshot is the first bead, and the lower right snapshot is the last bead. The solid dark blue represents the surrounding air.

Comparing the temperatures of nodes 32, 52, and 71 for the 2D single-bead, three-bead, and six-bead models, they all produced similar temperatures at each node. Overall, the maximum temperature difference from the control model was around 4 [C], which occurred on the top face node 52 on the six-bead model. The right and left face nodes (32 and 71) for the three-bead and six-bead models had a maximum temperature difference from the control model of approximately 2 [C]. Initially, a 4 [C] difference from the control model seems significant. However, current methods to examine the temperature gradients during a FFF build (i.e. thermal imaging) typically do not have such a precise resolution, and a maximum temperature difference of 4 [C] from the control model is acceptable enough to move on to the next step in the modeling process; 3D modeling.

3D Single Bead Model Simulation Results:

After performing the 3D single-bead simulation process in ANSYS as previously outlined, it was determined that radiation effects should *not* be neglected when modeling the FFF using PLA filament. The two 3D single-bead model simulations, one with convection only and one with convection and radiation effects, were carried out under the same operating conditions. Then, temperature data from the same nodes on each model – front face node 202, back face node 81, left face node 562, right face node 391, and top face node 481 – were collected and compared. **Figure 14** shows the cooling single-bead (with radiation effects) at four times throughout the simulation, and **Figure 15** shows the temperature data as a function of time for the two compared models.

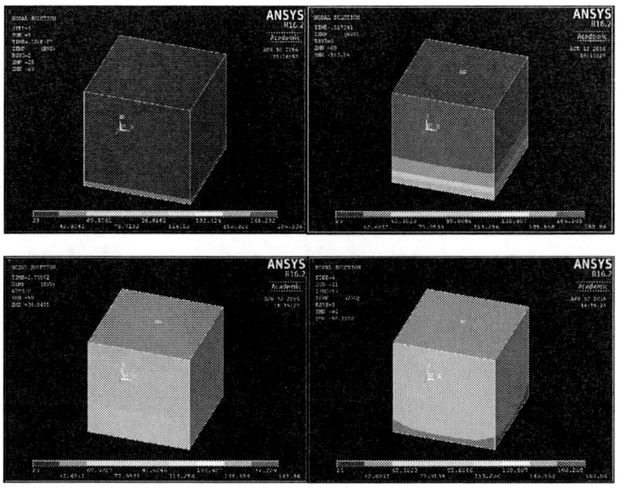

Figure 14: Temperature contour plots of the 3D single-bead with radiation effects deposition process at four times. The upper left plot is the start of the simulation (time = 0 [s] with no deposition), the upper right plot is a small time step after PLA deposition, the lower left plot represents 50% completion of the cooling observation time period, and the lower right plot represents completion of the cooling observation time period.

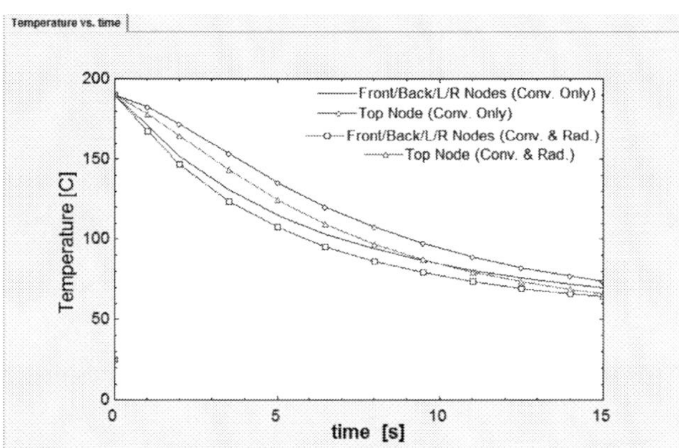

Figure 15: Temperature vs. time plot for the 3D single-bead models – with and without radiation effects. Note that the Front/Back/Left/Right face nodes for the select model produced the same temperature profile due to symmetry.

Examining **Figure 15**, the 3D single-bead convection and radiation model resulted in lower overall temperatures at the same analyzed nodes; this is due to the additional considered mode of radiation energy transfer along with convection. The Front/Back/Left/Right nodes of the convection and radiation model resulted in approximately 5 [C] less than the same nodes of the convection only model. The top node of the convection and radiation model resulted in a maximum difference of approximately 8 [C] from the same top node on the convection only model. Therefore, further models that simulate the FFF process should include radiation effects during the cooling of beads.

Production of Spherical Micron-Sized Polymer Particles for Additive Manufacturing by Liquid Phase Processes

Maximilian A. Dechet[a], Stephanie Kloos[a], Wolfgang Peukert[a] and Jochen Schmidt[a]

[a]*Friedrich-Alexander-Universität Erlangen-Nürnberg, Institute of Particle Technology, Cauerstr. 4, D-91058 Erlangen, Germany*

jochen.schmidt@fau.de

Abstract. Selective laser sintering (SLS) is an additive manufacturing process that yields excellent part qualities. This process employs micron-sized polymer particles, which are selectively fused by a laser. While there seem to be hardly any boundaries regarding design, there are quite some restrictions concerning the variety of commercially available SLS materials. At the moment, the most widely used polymeric material for SLS is polyamide 12 (PA12), with a market share of roughly 95%. In order to broaden the field of application of SLS, novel polymer powders with good handling properties are needed. In this contribution, different liquid phase processes enabling the manufacturing of spherical micron-sized polymer particles are outlined.

INTRODUCTION

The application of polymer based additive manufacturing technologies, in academia and industry, is on the rise. More and more companies use or consider using additive manufacturing. The variety of technologies and the field of application of parts made thereof is steadily increasing. With technologies like fused deposition modeling, stereolithography or binder jetting, various solid to flexible parts are readily available. However, if functional parts with excellent mechanical properties are needed, only few additive manufacturing technologies are feasible, which mostly are powder based technologies like selective laser sintering (SLS), also known as laser beam melting (LBM). While e.g. filament based technologies offer a variety of different materials including blends and composites and the possibility to easily modify and customize them, powder materials often are polyamide (PA) based with only few available commercial products and even less commercial producers [1–3]. The main routes for production of these powders are sophisticated precipitation processes optimized for polyamides [4]. Furthermore, as a more general approach to obtain fine powders, grinding processes are applied, which can be classified into dry (cryogenic) [5,6,7], and wet grinding [8,9]. To enhance the flowability of such powders, rounding of the irregular shaped particles in a heated downer reactor is feasible [10]. To readily obtain spherical micron sized particles, other than by precipitation [11], methods like melt emulsification [12] and spray agglomeration [13] have been investigated. In this contribution, we will sketch some of the aforementioned liquid phase-based manufacturing processes for production of spherical polymer microparticles.

METHODS

Particle Sizing

Particle size distributions were analyzed by static laser diffraction particle sizing using a Mastersizer 2000 / Hydro 2000S wet dispersion unit (Malvern). The product suspensions have been diluted as appropriate. Small amounts of sodium dodecyl sulphate solution (SDS, 98% (Merck)) and ethanol (96%, denatured with 1% MEK (VWR)) have been added to assure dispersion stability. Prior to measurements ultrasonication of the dispersions has been performed.

Electron microscopy

Particle size, shape and morphology analysis was performed using a Gemini Ultra 55 (Carl Zeiss). The electron microscope was operated using an acceleration voltage of 1.0 kV and the SE2 detector.

Powder flowability

The flowability of the polymer powders was investigated with a modified Zimmermann tensile strength tester [14,15]. The device allows to determine the adhesion forces between adjacent powder layers in a nearly uncompacted powder bed. The applied load to the bed was 153 Pa in all experiments. Generally, low tensile strength values indicate high flowability.

RESULTS

Wet Grinding and Rounding

Wet comminution of PBT (Ultradur, BASF) was performed in a custom-made agitator bead mill. The feed material for wet grinding of all polymers was obtained by grinding of the respective polymer granules in a rotor-impact mill using a 0.5 mm sieve ring. In the grinding experiments a feed mass concentration of 18.5 wt.% PBT was applied, the process was done at ambient temperature with ethanol as solvent. Grinding beads were 2.0 mm yttrium-stabilized zirconia. Figure 1 depicts the dependency of the product particle size on process time. The wet grinding process is superior to dry grinding, if fine products ($x_{50,3}$ < 25 µm) are desired. After grinding for around 240 minutes, the particle suspension is dried to obtain the polymer powder. To further increase the flowability of the powder, a rounding step can be applied, in order to reshape the flat, plate-like, polymer particles into spheres. The effects of this rounding treatment are shown in Figure 2.

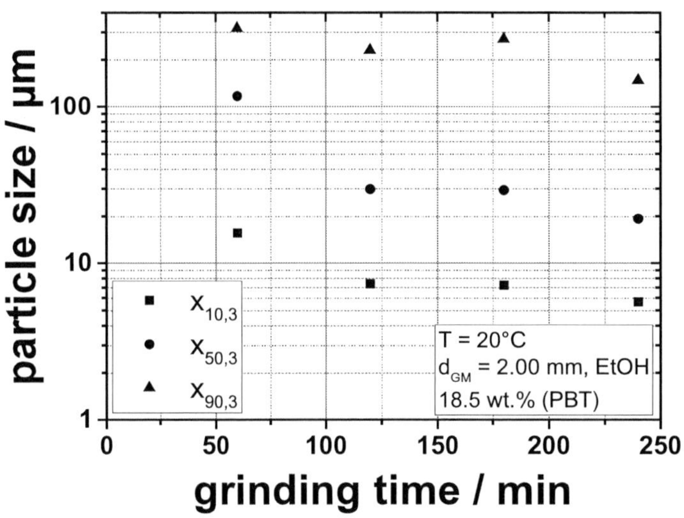

FIGURE 1. Evolution of product particle size $x_{50,3}$ over process time during wet comminution of PBT in ethanol at ambient temperature [16].

Rounding of the particles has been performed in a heated downer reactor with a length of 4.5 m and a diameter of 100 mm [17]. The comminuted particles are dispersed via a powder disperser into a nitrogen carrier gas flow of 7.5 m^3/h. In the reactor a distinct temperature profile is applied, which allows for fast melting, then rounding and subsequently cooling of the polymer particles. The obtained particles are of spherical shape and show good flowability. To further enhance the flowability, dry coating with fumed silica nanoparticles can be applied [18]: dry-coated PBT powder shows an improved flowability, respectively lower tensile strength of (2.9 +/- 0.6) Pa as compared to rounded ((10.7 +/- 1.7) Pa) or wet comminuted ((15.7 +/- 0.8) Pa) PBT particles [16].

FIGURE 2. Effect of rounding on wet comminuted PBT particles. Left: SEM image of PBT comminution product. Right: SEM image of the PBT particles after rounding [16].

Melt Emulsification

In melt emulsification a liquid phase being a solid at ambient temperature is dispersed as a melt in a continuous liquid phase at elevated temperature. In order to emulsify a semi-crystalline polymer, typically process temperatures above the melting point and high shear forces are needed. The following results, depicted in Figure 3 were obtained for emulsification of PEwax (PE EO 42, Deurex) granules in a melt emulsification pilot plant. The continuous phase was water, the emulsifier Tween 85 and the high shear was introduced by a rotor stator device operating at 10500 rpm. The produced PE particles are of spherical shape and show a smooth, slightly textured surface. Further information can be found in [12].

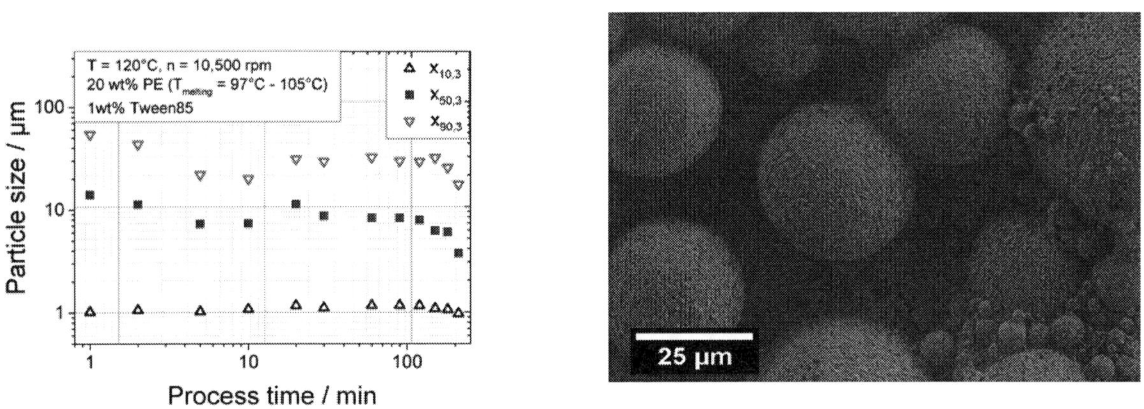

FIGURE 3. Particle size distribution and particle morphology of PE particles obtained via melt emulsification. Left: Particle size of PE particles in dependence on process time. Right: SEM images of melt emulsified spherical PE particles [12].

Precipitation

Another approach to obtain spherical micron-sized polymer particles is precipitation. While frequently triggered by addition of an anti-solvent, in the case of polymers often thermally induced precipitation is applied [19, 20]. PA12 as the most commonly used powder material for SLS typically is produced by a precipitation process [4,21]. The polymer is heated in a moderate solvent, which does not dissolve the polymer at ambient temperature. After exceeding the solubility temperature, dissolution of the polymer takes place. After the polymer is fully dissolved, the temperature is decreased, until the solvent does not act as a solvent anymore and the polymer precipitates. By proper choice of the polymer solvent system and careful temperature control, spherical particles of narrow size distribution are obtainable by precipitation and crystallization. In Figure 4 particle size distributions (volume sum Q_3) and a SEM image of PA11 particles are shown. The particles were produced by heating PA11 granules (Rilsan, Arkema) with ethanol (96%, denatured with 1% MEK (VWR)) in an autoclave. Amongst others the product particle size distribution can be tailored by the choice of polymer concentration in the system (Figure 4 left). Moreover, the addition of additives like flowing aids or pigments for an in-situ functionalization is possible. The process temperature was set to 190°C and the system was kept isothermal for 10 min under constant stirring, then the temperature was decreased to 50 °C and the product particles were collected. The spherical shape of the precipitated PA11 product particles becomes obvious from SEM imaging (Figure 4, right). The tensile strength of PA11 particles is (3.33 +/- 1.5) Pa without fumed silica, respectively (0.43 +/- 0.06) Pa for the dry-coated PA11 particles, indicating very good flowability.

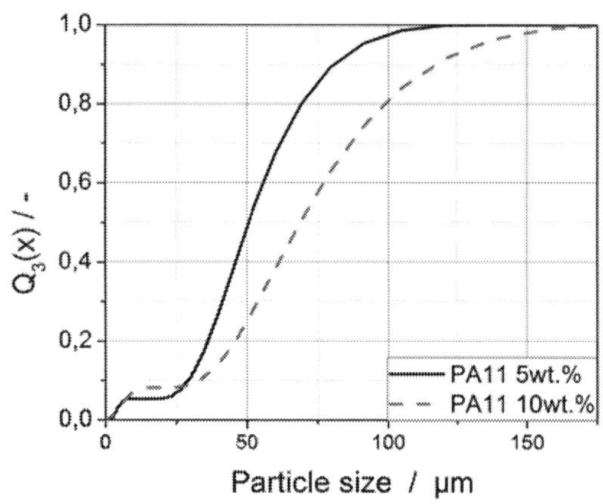

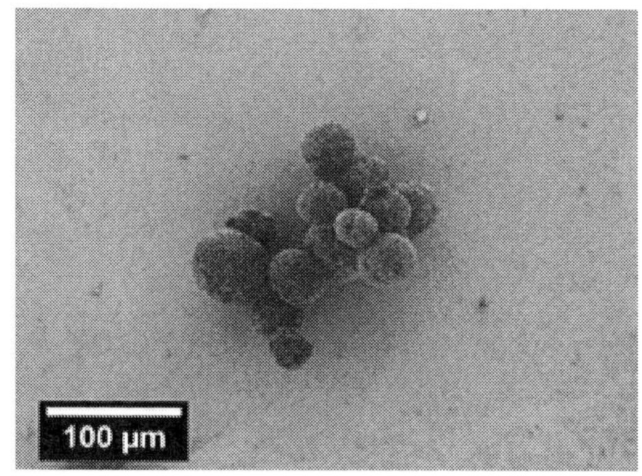

FIGURE 4. Particle morphology and particle size distributions of PA11 particles precipitated from ethanol. Left: Particle size distributions (cumulative volume, Q_3) at 5 wt.% and 10 wt.% PA11 precipitated from ethanol. Right: SEM image of precipitated PA11.

Spray Agglomeration

Spray agglomeration allows to transfer fine primary particles into spherical agglomerates of bigger size. For spray drying a feed suspension consisting of the fine polymer particles produced by wet comminution, a binder and a low boiling liquid (e.g. water or ethanol) is prepared. Upon evaporation of the sprayed liquid droplets, stable agglomerates of the formerly suspended particles are obtained. Figure 5 depicts polycarbonate spray agglomerates. Further information on the spray drying process can be found in [13].

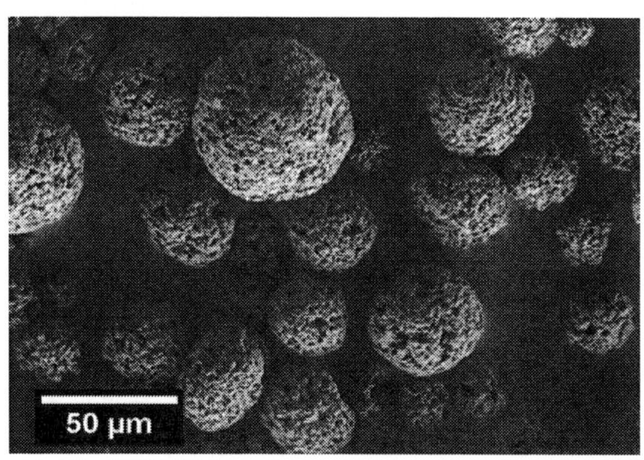

Figure 5. SEM image of spray agglomerates produced in a spray dryer from a suspension of fine polycarbonate particles [13].

CONCLUSIONS

In order to broaden the fields of application of polymer powder-based additive manufacturing technologies, new and innovative processes for production of powders of good flowability are needed. The liquid phase processes addressed in this contribution –either as a single stage process (precipitation, melt emulsification) or combined with following processes (wet comminution / rounding, wet comminution / spray agglomeration)-allow for production of spherical micron-sized polymer particles. The presented approaches are scalable and applicable to polymer materials being of interest for selective laser sintering, thus, giving the opportunity to obtain novel powders for additive manufacturing.

ACKNOWLEDGMENTS

This study has been supported by Deutsche Forschungsgemeinschaft (DFG) in the framework of the collaborative research center CRC 814 "additive manufacturing" (project A1). Financial support is gratefully acknowledged.

REFERENCES

1. T.T. Wohlers, Wohlers Report 2014 (Wohlers Associates, Fort Collins, Col., 2014).
2. I. Gibson, D. Rosen, and B. Stucker, *additive manufacturing Technologies*, 2. ed. (Springer, New York, NY, 2015).
3. M. Schmid, *Selektives Lasersintern (SLS) Mit Kunststoffen: Technologie, Prozesse Und Werkstoffe* (Hanser, 2015).
4. F.E. Baumann and N. Wilczok, DE19708946 (10 September 1998).
5. M. Liang, C. Lu, Y. Huang, and C. Zhang, J. Appl. Polym. Sci. **106**, 3895 (2007).
6. M. Wilczek, J. Bertling, and D. Hintemann, Int. J. Miner. Process. **74**, 425 (2004).
7. C. Gath and D. Drummer, in *2016 12th Int. Congr. Molded Interconnect Devices* (IEEE, 2016), pp. 1–6.
8. J. Schmidt, M. Plata, S. Tröger, and W. Peukert, Powder Technol. **228**, 84 (2012).
9. M.F.H. Wolff, S. Antonyuk, S. Heinrich, and G.A. Schneider, Particuology **17**, 92 (2014).
10. J. Schmidt, M. Sachs, C. Blümel, B. Winzer, F. Toni, K.-E. Wirth, and W. Peukert, Powder Technol. **261**, 78 (2014).
11. S.-J. Wang, J.-Y. Liu, L.-Q. Chu, H. Zou, S.-J. Zhang, and C.-J. Wu, J. Polym. Sci. Part B Polym. Phys. **55**, 320 (2017).
12. S. Fanselow, S.E. Emamjomeh, K.E. Wirth, J. Schmidt, and W. Peukert, Chem. Eng. Sci. **141**, 282 (2016).
13. S. Kloos, W. Peukert, and J. Schmidt, in *Proc. 6th Int. Conf. Addit. Technol.* (2016), pp. 147–151.
14. A. Schweiger and I. Zimmermann, Powder Technol. **101**, 1 (1998).
15. K. Meyer and I. Zimmermann, Powder Technol. **139**, 40 (2004).
16. J. Schmidt, M. Sachs, S. Fanselow, M. Zhao, S. Romeis, D. Drummer, K.E. Wirth, and W. Peukert, Chem. Eng. Sci. **156**, 1 (2016).
17. M. Sachs, M. Friedle, J. Schmidt, W. Peukert, and K.-E. Wirth, Powder Technol. **316**, 357 (2017).
18. C. Blümel, M. Sachs, T. Laumer, B. Winzer, J. Schmidt, M. Schmidt, W. Peukert, and K.-E. Wirth, Rapid Prototyp. J. **21**, 697 (2015).
19. P. van de Witte, P.J.J. Dijkstra, J.W. a. W. a van den Berg, and J. Feijen, J. Memb. Sci. **117**, 1 (1996).
20. H. Matsuyama, M. Teramoto, M. Kuwana, and Y. Kitamura, Polymer (Guildf). **41**, 8673 (2000).
21. F.-E. Baumann, M. Grebe, and S. Monsheimer, DE10334496 (24 February 2005).

Fused Deposition Modeling of Poly(vinyl alcohol) Based Filaments

Chen Gang [(1)]**_, Chen Ning_** [(1)*]**_, Wang Qi_** [(1)]

[(1)] State Key Lab of Polymer Materials Engineering, Polymer Research Institute, Sichuan University - Sichuan - China

Fused deposition modeling (FDM) is one of the most widely used 3D printing technologies. However, the commercially available filaments for FDM are ABS, PLA, etc., which are much higher price and lacking functionality. Therefore, preparation of new filaments suitable for FDM is promising but undoubtedly challenging. Poly(vinyl alcohol)(PVA) with excellent comprehensive properties is a good biodegradable material for FDM. The key is to realize the thermal processing of PVA. In this paper, based on our former research work, the hydroxyl ionic liquid (IL) was used to control the supramolecular structure of PVA. The melting point of PVA gradually decreases from 217.3 to 146.3 along with the glass transition temperature promptly decreases from 76.61 to 11.39 upon the addition of IL, revealing that IL could act as an ideal plasticizer for PVA. The Flory-Huggins interaction parameter of PVA/IL system is at 160oC, which suggested that the binary blends of PVA/IL system was fully miscible in the molten state. The O-H stretching bands shifts to a higher wavenumber about 19cm-1 together with the striking change of peak shape in amorphous region in FT-IR, which guaranteed the interaction between IL and the hydroxyl groups of PVA. This interaction also improved the thermal stability of the PVA/IL system during processing. Then a new kind of PVA-based filament with good flowability and toughness suitable for FDM is prepared via this simple and facile strategy. The obtained PVA/IL filament exhibit excellent ionic conductivity up to 2.82×10-3S/cm with 35wt% content of IL. Furthermore, the effects of the parameters such as the layer thickness, air gap, as well as filling angle on the structure and properties of PVA/IL parts fabricated by FDM were discussed.Finally, the human jaw model and the biological scaffold with complex structure and good compressive modulus were fabricated. This work was supported by the National Natural Science Foundation of China (51433006).

Processability of Perlite-Filled Polypropylene Composites in Extrusion-Based Additive Manufacturing

Schuschnigg Stephan (1)*, Spörk Martin (1), Sapkota Janak (1), Weingrill Georg (2), Fischinger Thomas (3), Arbeiter Florian (4), Holzer Clemens (1)

(1) Chair of Polymer Processing, Montanuniversitaet Leoben - Styria - Austria, (2) Chair of Mineral Processing, Montanuniversitaet Leoben - Styria - Austria, (3) Functional Surfaces and Nanostructures, Profactor GmbH - Upper Austria - Austria, (4) Chair of Polymer Testing and Material Characterization, Montanuniversitaet Leoben - Styria - Austria

Extrusion-based additive manufacturing, also known as fused filament fabrication, is a versatile and popular additive manufacturing technology, which relies on the extrusion of thermoplastic filaments to produce a three-dimensional object in a layer-by-layer technique. So far only a limited number of commercial materials are available for this technology. Especially technologically relevant semi-crystalline polymers lack studies devoted to extrusion-based additive manufacturing. This work aims at extending the material database for extrusion-based additive manufacturing to polypropylene. In order to overcome the critical shrinkage and warpage issues of polyolefins, polypropylene was filled with the spherical filler expanded-perlite, a volcanic silicate found freely in nature. The impact of filler size and content as well as the addition of an amorphous polymer was studied for specific properties that are most relevant in connection with 3D-printing. A detailed analysis on the material's shrinkage, mechanical, thermal and morphological properties was performed. Moreover, both the processability and the printability of the filaments were studied by means of industrial-scale extrusion and the printing of specific test specimens explicitly prone to warpage. In summary, a polypropylene compound containing small-sized expanded-perlite and amorphous polyolefins led to a well printable product that distinctly improved the warpage, shrinkage and mechanical properties. Thereby, our findings prove the successful application of semi-crystalline materials for extrusion-based additive manufacturing under appropriate conditions.

Developing a Patient Individualized Flexible Silicone Implant using SLS and Vacuum Die Casting

Martin Launhardt[1, a)], Nina Ebel[2, b)], Markus Kondruweit[2, c)], Michael Weyand[2, d)], Tilmann Volk[3, e)], Dietmar Drummer[1, f)]

[1]*Institute of Polymer Technology (LKT), Friedrich-Alexander-University (FAU) Erlangen-Nürnberg, Am Weichselgarten 9, 91058 Erlangen-Tennenlohe, Germany.*
[2]*Department of Cardiac Surgery, Universitätsklinikum Erlangen, Krankenhausstrasse 12, 91054 Erlangen, Germany.*
[3]*Institute of Cellular and Molecular Physiology, Friedrich-Alexander-University (FAU) Erlangen-Nürnberg, Waldstrasse 6, 91054 Erlangen, Germany.*

[a)]*Corresponding author: launhardt@lkt.uni-erlangen.de*
[b)] *nina.ebel@uk-erlangen.de*
[c)] *markus.kondruweit@uk-erlangen.de*
[d)] *michael.weyand@uk-erlangen.de*
[e)] *tilmann.volk@fau.de*
[f)] *drummer@lkt.uni-erlangen.de*

Abstract. Heart failure is a progressive and medically complex disease, which eventually requires heart transplantation. However, an internal flexible cardiac supporting device could support the patient's heart until a donor organ is available or might even improve cardiac function to an extent that transplantation becomes unnecessary. Technically, this requires patient individualized geometries, flexible structures and reinforcements such that compression forces only affect the heart but not the surrounding organs. The Selective Laser Sintering process (SLS) allows the manufacturing of complex thermoplastic parts without the need for an expensive mold. However, the available material portfolio is very narrow and yet does not offer a silicone rubber type material, which can be used for medical applications. Thus, in this approach the applicability of SLS vacuum die casts for the manufacturing of patient individualized flexible silicone implants shall be analyzed. Basic studies on the molding and demolding of silicone specimens using SLS vacuum die casts are investigated as well as the possibility for including flexible reinforcement structures. First, simplified mechanical tests show the feasibility of the manufacturing process as well as optimization potential using patient individualized geometries. First functional silicone heart supporting structures are successfully manufactured and can be used for medical in-vitro tests.

INTRODUCTION

Cardiac insufficiency is presently one of the most common diseases in Germany and the number of patients affected continues to rise annually [1]. Despite the improvement of a drug therapy, a heart transplantation represents the first choice for chronical cardiac insufficiency in the terminal stage [2,3]. This therapeutic option, however is not available for all patients because the limited number of donor hearts [4]. Thus, new routes are investigated with the goal to either bridge the waiting time for a donor heart, or even fully avoid heart transplantation. Heart supporting systems have been an alternative to cardiac transplantation since the 1980's for patients with chronical heart failure [5]. The commonly applied heart supporting systems are in direct blood contact, which comes along with the risk for blood clot formation and infections. Therefore, new systems are investigated which are placed around the heart and provide a heart compressing massage supporting the contraction and the relaxation phase of the heart mechanically from the outside, without direct blood contact [5–10]. However, these systems are not yet clinically applied and cannot be manufactured patient individualized. This can lead to an impairment of the coronary blood vessels which are located on the surface of the heart and supply the heart muscle with blood. A constriction of the vessels and the subsequent decrease of blood flow can lead to cardiac death in the worst case [11]. Therefore, a heart supporting structure should be patient individualized such that the main coronary blood vessels are not affected.

The Selective Laser Sintering Process (SLS) is an additive manufacturing technique for polymers, which provides increased freedom of design and complex parts with increased mechanical properties. Even with one raw material defined part properties can be adjusted only by changing the process parameters and the geometry. Especially, the ability to manufacture patient individualized parts, in a short time period provides a large field of applications. This ranges e.g. from the manufacturing of prosthetics, to individualized supporting tools for surgeries manufactured by SLS [12,13]. The commonly used SLS material polyamide 12 (PA12) is non-certified for medical implants and thus cannot be applied directly. However, indirectly SLS can be used to manufacture complex vacuum die casts and investment castings for the manufacturing of medical components in a second step using medically certified material such as liquid silicone [14].

Based on a previous study for the rapid manufacturing of elastomeric parts, in this paper the combination of SLS, vacuum dies casting, investment casting and eventually dip coating, a silicone based heart supporting structure shall be investigated as it can be used for first small animal tests. In perspective, supporting systems can be manufactured patient individualized and shall not influence adjacent organs. Therefore, the system shall be small and internally reinforced such that the pressure is guided only towards the heart. Using these first prototype implants, simplified mechanical tests on the deformation and the volumetric displacement are conducted.

METHODOLOGY AND PROTOTYPE MANUFACTURING

Two feasible manufacturing methods for patient individualized flexible heart supporting structures that can be used in a functional heart supporting system shall be investigated. The challenges are to realize an elastic but pressure resistant structure, which allows an efficient contraction movement on the inside and preferably has no influence on the adjacent organs. For this, the investigated manufacturing route integrates SLS, vacuum die casting, dip coating and investment casting which derive from various fields but have not yet been used in combination. Only in combination, the advantages of the additive manufacturing method and the other methods can be fully exploited and applied for the manufacturing of functional and individualized elastomeric heart supporting structures. Therefore, the two manufacturing routes are investigated. Both are intended to represent a first heart supporting structure that can be used for small animal tests later. If the basic principle can be approved, larger prototypes can be manufactured for human heart supporting systems. Cup A consists of an elastomeric membrane, which can be filled with a fluid and thus through expansion, guides a force towards the heart, Fig. 1, left. As a second component, the elastomeric membrane shall be surrounded by a stiff shell, which prevents the expansion towards adjacent organs. The second cup (cup B) completely consists of elastomeric silicone material which prevents the influence of adjacent organs and may provide optimized carrying comfort for the patient, Fig 1, right. The expansion on the outside of the elastomeric part is prevented using an inner reinforcement.

The varying design of the elastic structures requires differentiating manufacturing routes. For cup A, the outer shell as well as the joining pin between shell and membrane are manufactured by SLS. The flexible membrane shows a complex geometry and thus needs decisive manufacturing methods. Here, a lost core manufactured by SLS is used in the dip coating process to create a thin and flexible membrane after removing the core. Thus, cup A consists of three different parts, which need to be assembled. Cup B shall be manufactured by vacuum die casting using SLS die casts. These provide exact outer dimensions and the manufacturing of complex geometries. As there is hardly any SLS part certified for incorporal implants no SLS part may be exposed to the body. However, the cup needs to provide a reinforced outer shell to ensure that the compression force is guided towards the heart. Thus, an inner SLS structure shall be tested inside the elastomeric wall of cup B. The die casting method with SLS die casts however, is not suited for the manufacturing of very thin membranes. Thus, the inner membrane shall be manufactured through dip coating of a lost core, comparable to the manufacturing route of cup A.

All SLS parts are manufactured using the SLS machine Formiga P110 by EOS, GmbH, Krailling with PA12 powder, type PA2200 (EOS) with a refreshing rate of 50 weight-%. For the outer shell parts and the SLS reinforcement structure, standard parameter settings provided by the machine supplier are used. The lost cores are manufactured with adapted parameters and are object of the experimental setup. All build jobs are processed using insulation layers below the build job of 6 mm and on top 4 mm with a layer thickness of 0.1 mm. The building chamber temperature is kept at 167 °C and the exchange frame chamber temperature at 145 °C. After the SLS process all parts are cleaned from adhering powder particles using glass bead blasting, brushing and vacuum cleaning.

The elastic structure of the heart supporting system requires a flexible and durable biocompatible material, such as e. g. the silicone Med4-4220 by nusil, Silicone Technology, Carpinteria, CA. However, biocompatible and medically certified materials are very cost expensive and thus the presented investigations are conducted using a technical silicone with comparable material properties. For the present die casting and dip coating manufacturing processes a two-component silicone with a crosslinking at room temperature is selected, RTV-2 which provides low manufacturing viscosity, increased part flexibility and mechanical properties for the application as dynamic implant. Thus, for this work the RTV-2 silicone PlatSILITA 13.45H by the company PolyConForm GmbH, Düsseldorf is chosen, [14,15].

Investment casting or lost cores can be used for the forming of complex shapes with e. g. undercuts. Using SLS, usually technical parts with increased mechanical properties are manufactured. However, the process parameters as well as the geometry design can be adapted such that the lost cores provide sufficient stability for the die casting process but can easily be destroyed and removed after the forming process. Based on previously conducted investigations the lost cores for the dip coating process are designed with a thickness of 0.5 mm and a laser power of 10 W. Here, the lost core technology is combined with the dip coating process. Especially the inner membrane, which conducts the pressure onto the heart, needs to be very thin and durable. In order to manufacture parts with undercuts a lost core is required. Even though the silicone is flexible cup A would not be manufacturable without a lost core due to the torus shape of the silicone membrane. The opening in the bottom is required for the attachment to the outer SLS shell. The outer SLS shell for cup A is manufactured using SLS with a wall thickness of 1.0 mm. Eight flaps on the top of the shell are designed in order to tie the heart supporting system to the heart. The silicone membrane and the SLS shell need to be connected in the bottom such that the heart will not be pushed out of the system due to the applied pressure. For cup B this opening is not required because a silicone bridge between inner and outer membrane is designed. The elongated shape on the bottom side later acts as silicone hose and thus as connector to the controller. During the dip coating process, this can be used as handle, Fig. 2.

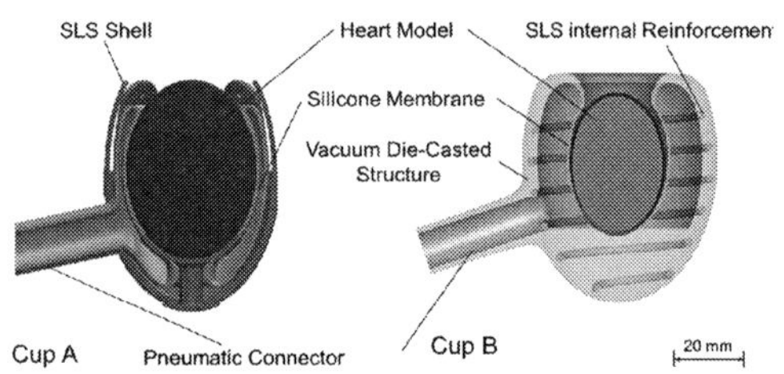

FIGURE 1. Schematic of the cup designs A, (left) and B, (right)

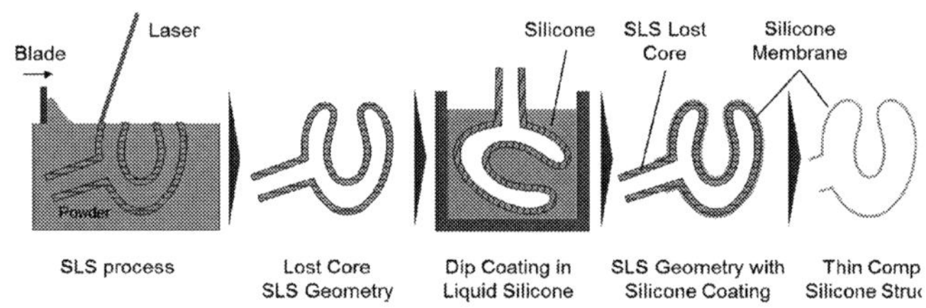

FIGURE 2. Schematic of the manufacturing of a complex silicone structure using SLS lost cores and dip coating process

The main body of cup B is built up using die casting and subsequential assembly of the main body with dip coated membrane and SLS reinforcement helix. SLS reinforcement helices can be positioned around the silicone membrane such that this is fully integrated into the silicone structure. The silicone mixture is injected into the mold via a gate using disposable syringes. One helix with a circular profile and a diameter of 2 mm and one helix with elliptical profile (1.5 x 2.5 mm) are manufactured in order to study the suitability as a supporting structure.

The basic function of the silicone cups is characterized using the controller for an Intra-Aaortic Balloon Pump System (IABP), Type Arrow Auto CAT 2 Wave by Teleflex Inc., PA, USA that provides defined and cyclic gas shifts. The volumetric displacement towards the heart is measured using a compressive dummy heart consisting of a balloon filled with water up to 10 ml height in order to simulate the counter pressure of the heart tissue. Using a leakage-testing device by Dr. Wiesner Steuerungstechnik GmbH, Remshalden a constant pneumatic pressure of 100 mbar is applied which equals approximately the heart pressure at diastole of 80 mmHg. According to the shifted water height the volumetric displacement onto the heart organ is measured. In a second test, an oval incompressible dummy heart is inserted into the cup. Using the same pressure device the volumetric displacement of the cup to the outside is characterized optically.

RESULTS AND DISCUSSION

Using the IABP controller the basic pneumatic function of the heart supporting structures is tested. During contraction without a heart inside the actor, cup A shows a three folded membrane contraction, Fig. 3, whereas cup B shows a four-folded configuration. Eventually the cups are tested using a constant pressure of 100 mbar in order to quantify the expansion to the outside as well as the contraction. Due to its stiff SLS shell cup A does not show any deformation to the outside. Cup B was tested on its displacement using optical methods, Fig. 4. It can be seen that the reinforced cups do not show any expansion in width. However, they show certain expansion in height. The cup without reinforcement structure shows most expansion in width, up to 17 %, which could influence adjacent organs.

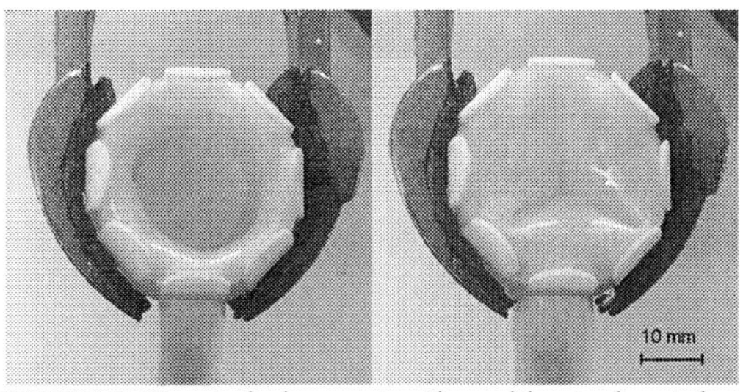

FIGURE 3. Cup A during contraction without a heart shows a three folded silicone membrane shape

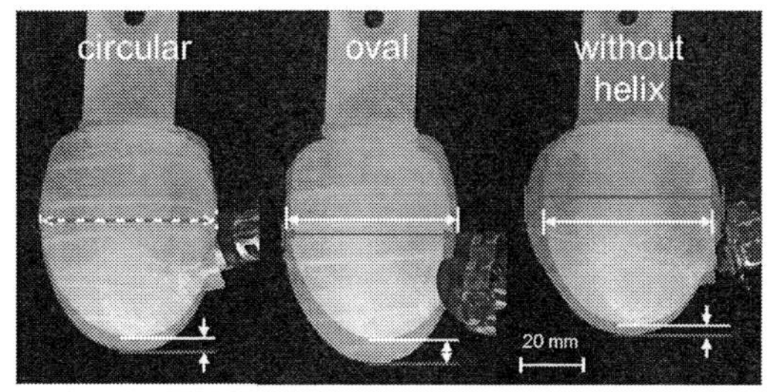

FIGURE 4. Expansion test on cup B

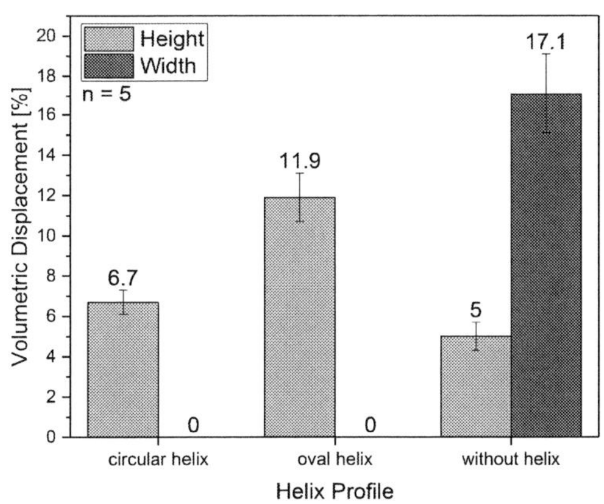

FIGURE 5. Volumetric displacement of cup B towards the outside divided into width and height

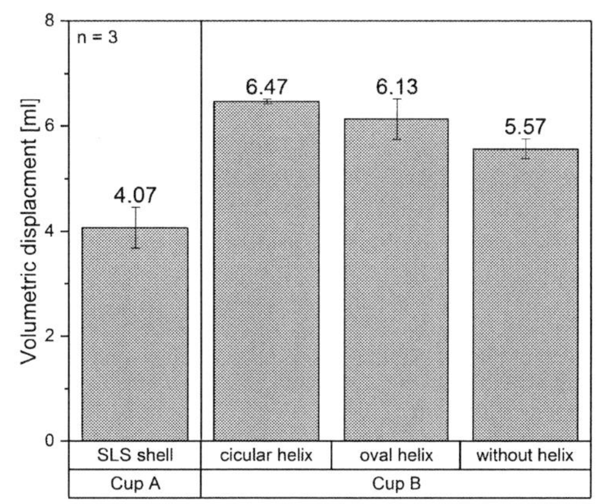

FIGURE 6. Volumetric displacement of cup A and B towards the heart

In Fig. 5 – 6 the volumetric displacement of the cup to the outside and towards the heart are depicted. It can be seen that cup B with a circular profiled helix provides the highest and cup A the lowest volumetric displacement. When comparing with the outer displacement of cup B it becomes obvious that the more energy is used for the deflection on the outside less energy remains for the deflection on the inside. Thus, the internal SLS reinforcement shows good results for the manufacturing of an energy efficient silicone structure.

CONCLUSION AND FUTURE WORK

It could be shown that a combination of SLS vacuum die casts with SLS lost core dip coating can be used for the manufacturing of complex silicone parts, as they are necessary for a mechanical heart supporting system. Furthermore, it was demonstrated that with defined SLS parameters lost cores and investment castings can be manufactured in order to realize complex geometries. Only the integration of the shown processes SLS, dip coating, vacuum die casting and the usage of lost cores allows the manufacturing of such complex parts.

It was proven that it is possible to fully integrate a SLS reinforcement structure into the silicone. Furthermore, the integration of an SLS structure provides a flexible structure and at the same time guides any expansion towards the heart. Furthermore, it was shown, that the cup B with an internal helix reinforcement shows the least volumetric displacement on the outside and thus does not influence adjacent organs. Furthermore, this cup also provides highest volumetric displacement on the inside of the cup and thus operates efficiently.

In future research the dynamic long time strength of the material and component needs to be tested. As silicone is generally certified for long time implants, cup B can be examined in first in-vitro experiments using hearts of small animals. Furthermore, using this technique allows the patient individualized shaping of the silicone structure and thus can be adapted to the individual coronary heart vessels. Simplified pressure tests may provide an inside on the impact of customized cup shapes. The overall goal is to generate a process route which allows the rapid manufacturing of patient individualized heart supporting systems.

REFERENCES

1. Statistisches Bundesamt, *Die 10 häufigsten Todesfälle durch Herz-Kreislauf-Erkrankungen*, https://www.destatis.de/, accessed July 2017.
2. S. J. Park, C. A. Milano, and A. J. Tatooles, et al., "Outcomes in advanced heart failure patients with left ventricular assist devices for destination therapy," Circulation. Heart Failure **5** (2), 241–248 (2012).
3. H. Gorler and A. Haverich, "Herztransplantation und Assist Devices," Herz **29** (7), 695–701 (2004).
4. Deutsche Stiftung Organtransplantation, *Informationen für den Organspendeprozess Koordinierung der Organspende in Deutschland*, https://www.dso.de/, accessed July 2017.
5. U. Boeken, *Mechanische Herz-Kreislauf-Unterstützung. Indikationen, Systeme, Implantationstechniken* (Springer, Berlin, 2013).
6. M. Anstadt, G. Anstadt, and S. MacDonald, et al., Method and apparatus for direct mechanical ventricular actuation with favorable conditioning and minimal heart stress, US20060211909 A1 .
7. M. Struber, A. L. Meyer, and D. Malehsa, et al., "The current status of heart transplantation and the development of "artificial heart systems"," Deutsches Ärzteblatt international **106** (28-29), 471–477 (2009).
8. M. N. Kavarana, D. N. Helman, and M. R. Williams, et al., "Circulatory support with a direct cardiac compression device," The Journal of Thoracic and Cardiovascular Surgery **122** (4), 786–787 (2001).
9. J. C. Criscione, S. Biswas, S. Hall, L. Harrison, and D. Robbins, *Device for the modulation of cardiac end diastolic volume* (US7871366 B2) (2011).
10. E. T. Roche, M. A. Horvath, and I. Wamala, et al., "Soft robotic sleeve supports heart function," Science translational medicine **9** (373) (2017).
11. E. Erdmann, *Klinische Kardiologie. Krankheiten des Herzens, des Kreislaufs und der herznahen Gefäße*, 8.th ed. (Springer, Berlin, 2011).
12. C. Leiggener, E. Messo, and A. Thor, et al., "A selective laser sintering guide for transferring a virtual plan to real time surgery in composite mandibular reconstruction with free fibula osseous flaps," International journal of oral and maxillofacial surgery **38** (2), 187–192 (2009).
13. J. Breuninger, R. Becker, and A. Wolf, et al., *Generative Fertigung mit Kunststoffen. Konzeption und Konstruktion für Selektives Lasersintern* (Springer Verlag, Berlin, Heidelberg, 2013).
14. M. Launhardt, M. Drexler, and D. Drummer, "Rapid-Manufacturing-Process for Elastomeric Components," International Journal of Recent Contributions from Engineering, Science & IT (iJES) **3** (2), 4–9 (2015).
15. Datenblatt PolyConForm GmbH, *Silikone PlatSilita 13-45 H.*

Control of Fused Deposition Modeling Melt Extrusion and Laydown

Hebda Michael James [(1)]*, Caton-Rose Philip [(2)], Whiteside Ben [(2)], Coates Philip [(2)]

[(1)] University of Bradford - Faculty of Engineering and Informatics - UnitedKingdom, [(2)] University of Bradford - West Yorkshire - UnitedKingdom

Melt extrusion and laydown of polymer within the FDM process is common practice but little is understood about laydown of polymer in relation to melt flow and die swell. This paper takes a standard 0.4 mm FDM nozzle and stepper motor controlled extruder setup and studies the laydown and extrusion techniques associated with FDM through high-speed camera work (Fig 1) and advanced data analysis techniques. Custom software has been created in order to measure and control extrusion speed, pressure and temperature with a view to linking laydown width and depth during single path extrusion to process settings. Higher rates of polymer extrusion speeds (335 mm/min) increased (Fig 2) die swell which must be considered in order to control path width during the manufacturing process preventing manufacturing issues such as negative gap width and model overfill.

Temperature Induced Ageing of PA12 Powder During Selective Laser Sintering Process

J. Benz [a)] and C. Bonten

IKT, Institut für Kunststofftechnik, University of Stuttgart, Pfaffenwaldring 32, 70569 Stuttgart, Germany
[a)]Corresponding author: johannes.benz@ikt.uni-stuttgart.de

Abstract. The selective laser sintering process has the potential to permanently overcome the boundary between model construction and functional components. The basic prerequisite for this is that components with reproducible quality can be manufactured. For this purpose, the virgin material must already be of the same quality. During the process, the laser sintered powder is heated to temperatures just below the melting temperature and held until the end of the building process. This results in thermal aging of the powder. Powder, which is not sintered, gets refreshed after the production process with new powder and can be reused. Due to the repeated refreshing process of the aged powder, a mixture is produced whose quality is difficult to define. By means of rheometry and differential scanning calorimetry (DSC), both new powder and powder, which was taken from a laser sintering process, are characterized. At the same time, the thermal aging is simulated by means of an oven heat. The rheological investigations show that the storage results in an increase of the viscosity, both in the process and in the oven. The reason is a post-condensation and thus the extension of the molecular chains. In addition, extended molecular chains lead to a faster reaching of the sol-gel transition and thus to a change from viscous to elastic material behavior. The post-condensation also leads to an increase in the melting temperature, which can be detected by means of DSC. The aging of the laser sinter powder can be reproduced by means of the oven storage and permits further investigations to ensure the quality of the virgin material.

INTRODUCTION

In order to save time and money in the development of new products, the use of additive manufacturing processes is becoming more and more frequent. At the same time, the main goal in the production of design patterns or prototypes is to recognize potential errors at an early stage and counteract them.

In addition, additive manufacturing processes offer great freedom of form. It is thus possible to produce almost any form independent of its complexity. This allows highly individualized products which are increasingly required in the market. At the same time, it makes it possible for companies to offer a wide range of products.

The process of selective laser sintering is described shortly. Plastic powder is locally melted by a laser. Any contours can be traced and thus a part can be created point by point. After one layer has cooled, the powder bed is moved downwards so that a new powder layer can be laid off. By repeating this procedure a part is produced in layers. In order to minimize the energy that the laser introduces and to avoid distortion of the part, the powder is heated up and kept just below the melting temperature of the material during the building process. Since building processes can often go on for more than one day, the non-sintered powder is kept at this temperature for a very long time. Over this period thermal aging of the powder occurs.

At the end of the laser sintering process, the complete building chamber is filled with the laser sinter powder but only a small share is sintered. The non-sintered powder is subjected to an enormous thermal stress over a long period of time which leads to an aging of the powder.

Aging is based on internal and external aging causes. The internal causes of aging are thermodynamically unstable states of the material, such as incomplete poly-condensation or crystallization. The physical or chemical action of the environment on the plastic describes the external aging. These include the supply of energy in the form of heat or radiation, chemical influences as well as mechanical stress. Aging is a process which is strongly time and temperature dependent. The influence of time is decisive for the strength of the change. The rate of change is influenced by the temperature. The dominant cause of aging in the selective laser sintering is the high temperature in the process chamber. This is an external cause of aging. The residence time of the powder in the selective laser sintering machine represents the aging period [2].

Increased temperatures near the melting point and long reaction times, as occur in the selective laser sintering, offer optimal conditions for chain extension by post-condensation. Due to the high temperature in the building chamber, the chain mobility in the solid is increased and there is the possibility that reactive end groups meet and react with each other. By means of this so-called solid-state post-condensation, the molecular weight of the powder increases, which leads to a viscosity increase and in extreme cases a gel formation. Especially in the case of SLS powder used several times, the viscosity increases so strongly that no homogenic laser melt track can form during the melting process in the laser sintering process [3].

In principle within the SLS parts an increase in the molecular weight leads in to significantly better mechanical properties. Young's modulus, tensile strength and elongation at break are increased. In addition, the layer adhesion of the SLS components can be significantly increased by the post-condensation effect. However, with the aim of saving material costs the non-sintered old powder is reused as a mixture with new powder in the subsequent building processes [3]. According to Gebhardt the recycling rate for polyamide powders is between 30 and 50 % of the used powder for the new construction process [4]. Therefore in the SLS process mixtures of the powder of several material states are constantly being worked on.

By repeatedly refreshing the aged powder, a mixture is produced, the quality of which is difficult to define. An already unknown quality of the material in the beginng makes it very difficult to predict the property profile of the subsequent part. That means that reproducibility is currently not available. Especially since the properties of new and already used powder can be very different and that is the reason why the process stability suffers.

EXPERIMENTAL

A PA12 powder from Evonik commercially available under the brand name Vestosint was used for the investigations. This was prepared by EOS and marketed under the trade name PA2200 [5].

In order to adjust the thermal aging in the process chamber of a selective laser sintering system virgin powder was stored in a vacuum oven APT.line from Binder for different periods of time. For this purpose the oven was evacuated at the beginning of storage and subsequently filled with nitrogen in order to prevent oxidation of the powder. Corresponding to the laser sintering process, the thermal storage is carried out at 170 °C. The storage times were up to 24 hours, which is clearly below those occurring in the laser sintering process. It has to be noted that the part is usually kept at this temperature in the building chamber for more than one day. A defined cooling process was not executed. As comparative material a PA12 was used, which has passed through a real laser sintering process in an unknown frequency.

Due to the different storage periods the descriptions in Table 1 are used in the following text.

TABLE 1. Description of the power state

description	power state
PA_{new}	virgin powder
PA_{1h}	Powder that has been stored in the oven for one hour.
PA_{3h}	Powder that has been stored in the oven for three hours.
PA_{8h}	Powder that has been stored in the oven for eight hours.
PA_{24h}	Powder that has been stored in the oven for 24 hours.
$PA_{process}$	Powder that has been taken from a real selective laser sintering process.

The rheological analysis of the powders are carried out in a plate-plate rheometer SR200 from Rheometrics with a plate diameter of 250 mm and a gap distance of 1 mm. The temperature of the rheometer was set to ϑ = 200 °C. The total time of the measurement was 3600 and 5400 s. In order to investigate the deformation characteristics of the SLS powder, an oscillation measurement is carried out (dynamic time-sweep). The oscillation frequency was ω = 1 rad / s and the shear stress value τ = 100 Pa.

The thermal analysis of the laser sintered powder was carried out using the method of Dynamic Differential Calorimetry (DSC). The used DSC device was a Netzsch DSC 204, Selb, Germany. With a heating rate of

10 K/min, the samples were heated from ϑ = 30 °C starting temperature to ϑ = 230 °C. Thermal effects such as a different crystallinity of the powder can be detected by the first heating run. Subsequently the PA12 was cooled to ϑ = 30 °C with a cooling rate of 10 K/min, in order to be heated again to 10 K/min to ϑ = 230 °C. The second heating cycle gives information about the material composition of the powder.

RESULTS AND DISCUSSION

By means of a rheological analysis changes in the flow properties are identified. Figure 1 shows the viscosity profiles of all the investigated powders. In all measurements the viscosity increases with increasing measuring time. Comparing the powders in Figure 2, the powders can be roughly divided into three viscosity ranges. These are shown in the diagram in the colors black, bright blue and dark blue. PA_{new} (black) has by far the lowest viscosity. The bright blue curves show the viscosity profiles of the powders after the oven storage. An increase in viscosity due to a post-condensation effect which occurs due to the thermal storage can be indicated over time. However, the viscosities vary as a function of the oven storage period, since only small amounts have been used and surface effects during storage cannot be excluded.

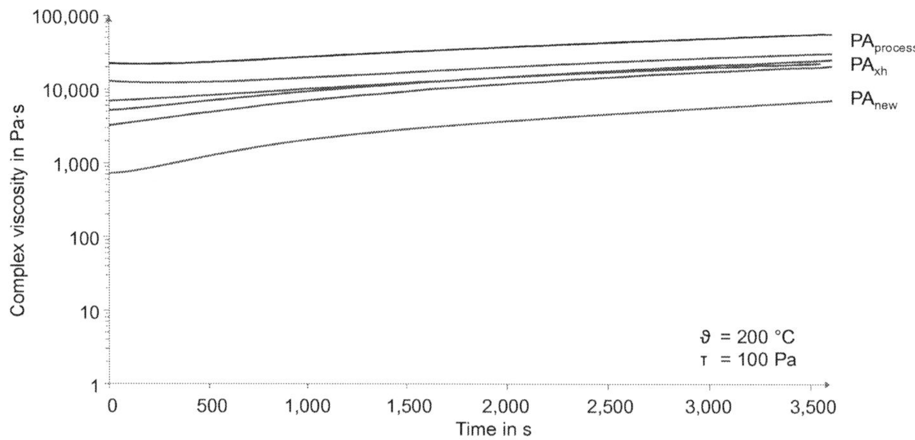

FIGURE 1. The longer the storage time the higher the viscosity

A further measurement criterion that can be determined with the oscillating rheometer is the time until the storage modulus G' exceeds the value of the loss modulus G". This characteristic point is referred to as a sol-gel transition or crossover point. When the loss modulus dominates the material has a viscous behavior. When the storage modulus G' dominates the material has an elastic behavior. The slopes of the storage and loss modulus are shown in Fig. 3 as an example for PA_{24}. In this case the crossover point occurs approximately after 1800 s.

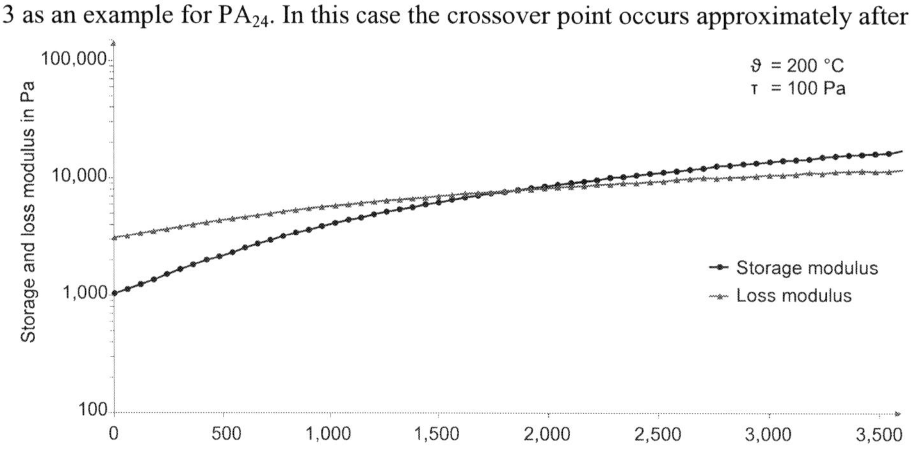

FIGURE 2. Crossover point of PA_{24}

The loss modulus initially dominates that means the material has viscous properties at 200 °C. As a result of the high temperature a chain extension in the liquid phase occurs with the advancing time caused by post-condensation effect. Due to the longer polymer chains and the greater number of entanglements the polymer movability is limited. As soon as the storage modulus G' exceeds the loss modulus G" the material shows an elastic behavior.

The diagram in figure 3 shows the time until the entry of the gel point as a function of the storage time in the oven. Since PA_{new} has the lowest viscosity of all the powders tested the time until the sol-gel transition is the longest. After one hour of oven storage the time to the crossover point already decreases strongly. The powder taken from the SLS process has the shortest duration until the crossover point is reached. Since the storage time under thermal load for this powder is unknown. The values $PA_{process}$ in the diagram are shown as a horizontal line.

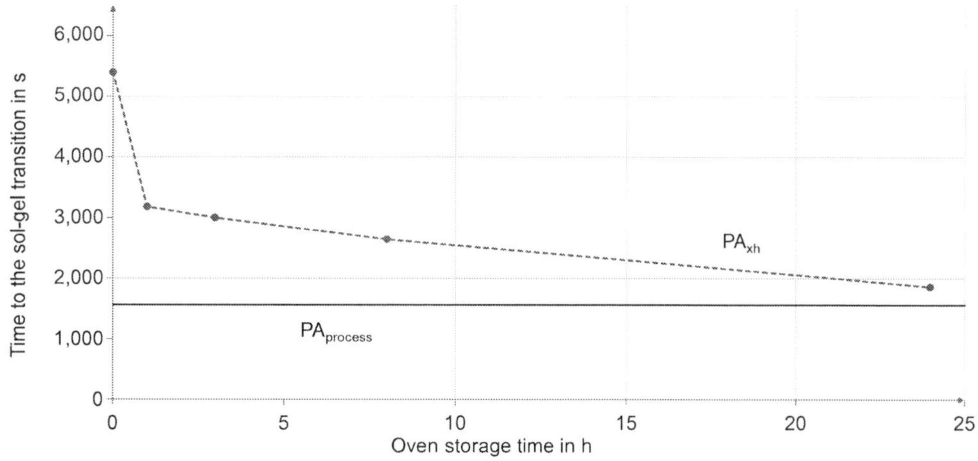

FIGURE 3. Time to sol-gel transition for all investigated powders

The decrease in time to sol-gel transition with increasing storage time in the oven can be explained by a post-condensation in the solid phase of the SLS powder. The longer the powder has already been subjected to the thermal stress the faster the rheological measurements lead to a change in flow behavior. The time to crossover point is the lowest for the powder taken from the process. This shows that the chain extension has already taken place to an extent which cannot be adjusted with a storage of 24 hours.

The results of the thermal analysis are shown in Table 1. The new powder PA_{new} shows the lowest melting temperature in the first heating process. The melting temperature increases with increasing storage time in the oven. The highest value shows powder that has been taken from the process. This increase in melting temperatures can also be explained by the post-condensation of the polymer chains. The longer the polymer chains become the more entanglements arise. In order to solve these entanglements a higher temperature is required and thus increased chain mobility is required. An increase in the crystallinity degree by recrystallization during the SLS process can be largely excluded since the enthalpy of fusion remains virtually unchanged.

TABLE 2. Thermal characteristics of the investigated powders

powder	melting point	melt enthalpy
PA_{new}	188,6 °C	102,6 J/g
PA_{1h}	189,7 °C	101,5 J/g
PA_{3h}	189,6 °C	100,6 J/g
PA_{8h}	189,9 °C	101,3 J/g
PA_{24h}	190,1 °C	100,6 J/g
$PA_{process}$	190,5 °C	99,21 J/g

The results are consistent with the results from the rheological analysis. In the case of the oscillating measurements by means of a plate-plate rheometer the time to the gel point is an indicator of the degree of the postcondensation which has already elapsed. The time to the sol-gel transition decreases in a tempering process in the oven. If the storage time is longer the time to the gel point is shorter. Even after 24 hours of oven storage it does not reach the values obtained from the powder taken from the real SLS process.

CONCLUSION

Thermal and rheological studies and a comparison of new powders and aged powders were used to analyze aging processes that occurs in the SLS process.

A rheological analysis with a plate-plate rheometer leads to the result that the thermal load of the powder results in post-condensation. However, the measurements of the complex viscosity do not provide a logical correlation between the duration of the thermal load and the degree of post-condensation. Instead, the time until the sol-gel transition is a measure for the thermal stress. Virgin powder shows an extremely long time until the sol-gel transition. The duration until the crossover point is reached decreases the longer the powder has been subjected to thermal stress.

The thermal analysis of new powder by means of DSC leads to the conclusion that the powder particles have a very high degree of crystallinity. Due to the post-condensation at a thermal load in the SLS process the melting point of the powder increases. An increase in the degree of crystallinity by a recrystallization can be excluded, since the enthalpy of fusion remains unchanged despite the thermal influences. By storing the powders in an oven flooded with nitrogen, the thermal aging of the PA during the sintering can be simulated. Post-condensation in the amorphous phase increases the melting point of the powder with increasing oven storage time as in the real SLS process. The melt enthalpy remains the same as melt enthalpy of new powder and the powder taken from the SLS process, regardless of the duration of the thermal load.

Two indicators that describe the degree of ageing to the powder are the time to the crossover point, which is measured by an oscillating measurement with a plate-plate rheometer and the melting temperature, which can be determined by a DSC analysis. Ideally, the aging process by post-condensation is reversible, so that the properties of new powder can be regenerated again by recycling the unsintered powder.

REFERENCES

1. Bonten, C. *Kunststofftechnik: Einführung und Grundlagen*. München: Hanser, 2014. ISBN 3-446-44093-7
2. Mielicki, C.: *Prozessnahes Qualitätsmanagement beim Lasersintern von Polyamid 12*. Duisburg-Essen, Universität, Fakultät für Ingenieurswissenschaften: Abteilung Maschinenbau und Verfahrenstechnik, Diss., 2014
3. Kaddar, W.: *Die generative Fertigung mittels Laser-Sintern : Scanstrategien, Einflüsse ver-schiedener Prozessparameter auf die mechanischen und optischen Eigenschaften beim LS von Thermoplasten und deren Nachbearbeitungsmöglichkeiten*. Duisburg-Essen, Universität, Fakultät für Ingenieurwissenschaften: Abteilung Maschinenbau und Verfahrenstechnik, Diss., 2010
4. Gebhardt, A.: *Generative Fertigungsverfahren : Additive Manufacturing und 3D Drucken für Prototyping – Tooling – Produktion.* 4th ed. München: Hanser, 2013. – ISBN 978-3-446-43652-7
5. Schmid, M.: *Selektives Lasersintern (SLS) mit Kunststoffen : Technologie, Prozesse und Werkstoffe*. München: Hanser, 2015. – ISBN 978-3-446-44550-5

A New Approach for Getting More Homogeneous Temperature Fields in Selective Laser Sintering of Polymers

Sandra Greiner[1, 2, a], Lydia Lanzl[1, 2], Katrin Wudy[1, 2], Dietmar Drummer[1, 2]

[1]*Institute of Polymer Technology, Friedrich-Alexander-Universität Erlangen-Nürnberg, Am Weichselgarten 9, 91058 Erlangen, Germany*
[2]*Collaborative Research Center 814-Additive Manufacturing, Friedrich-Alexander-Universität Erlangen-Nürnberg, Am Weichselgarten 9, 91058 Erlangen, Germany*

[a)] *Corresponding author: greiner@lkt.uni-erlangen.de*

Abstract. During the last 10 years, selective laser sintering of polymers (SLS), one of the most promising additive manufacturing technologies, strongly developed from rapid prototyping towards rapid manufacturing. However, main drawbacks have to be overcome, the limited range of processible materials and the lack of reproducibility of geometrical and mechanical properties in dependence on part positioning and processing parameters. The precise configuration of a homogeneous temperature field might be one of the most crucial factors for stable and reproducible processing, especially when new materials are to be tested. To date, commercially available systems work with build chamber temperature gradients ranging between 7 to 15 K, ignoring the thermal influence of exposure. Within this work, to date available systems for temperature measurements will be evaluated in terms of applicability in a powder based process. In addition, a new method for calibrating the temperature field of a SLS machine with an eight zone heating system will be introduced. Finally, the resulting temperature field is applied and verified by the powder based process.

Keywords: Selective Laser Sintering, Temperature Fields, Heating Strategy.

INTRODUCTION

Additive manufacturing techniques offer the possibility to generate parts and assemblies of high complexity in a single step. Especially selective laser sintering of polymers (SLS) is one of the most promising additive manufacturing technologies [1], meeting the requirements of flexible industrial series manufacturing of small lot numbers. However, the main drawbacks have to be overcome to conduct the technology towards rapid manufacturing. With over 90 % of material processed being polyamide (PA) based [2], the range of commercially available powder materials has to be further extended. In addition, the demand in enhanced reproducibility of geometrical and mechanical properties in dependence on part positioning and processing parameters has to be satisfied [3-6].

For achieving these two objectives, understanding and controlling the surface temperatures' influences on the fabrication result is of significant importance. The precise configuration of a homogeneous temperature field is one of the basic factors for stable and reproducible processing, especially when new materials are to be tested. Figure 1 a) shows the differential scanning calorimetry (DSC)-curve of a semi-crystalline polyamide 12 (PA 12) material (PA2200, EOS GmbH). The polymer's initiation temperatures for crystallization (T_{ic}) and melting (T_{im}) define the possible processing temperature range and the build temperature (T_b) to achieve a two-phase mixture area according to the model of quasi-isothermal laser sintering [7, 8]. Processing beyond this range will lead to process interruption, below T_{ic} by layer curling and above T_{im} by wave formation and polymer melting. However, in the process, there are many influences on the resulting surface temperature as shown in fig 1 b). Wegner [5] summarized that common SLS systems operate with surface temperature gradients ranging from 7 to 15 K, neglecting the influence of exposure. The different heat sources such as powder preheating, surface temperature heaters, cylinder or platform heaters, are implemented to guarantee stable process temperature conditioning, causing a complex thermal environment. Furthermore, different thermal transition areas lead to cooling, especially at the edge of the build chamber [9]. In [6], Goodridge highlighted the influence of inhomogeneous temperature distributions, depending on the used material system. For PA 12, significantly lower values for the Young's modulus and tensile strength were obtained when processed at colder edge regions [6].

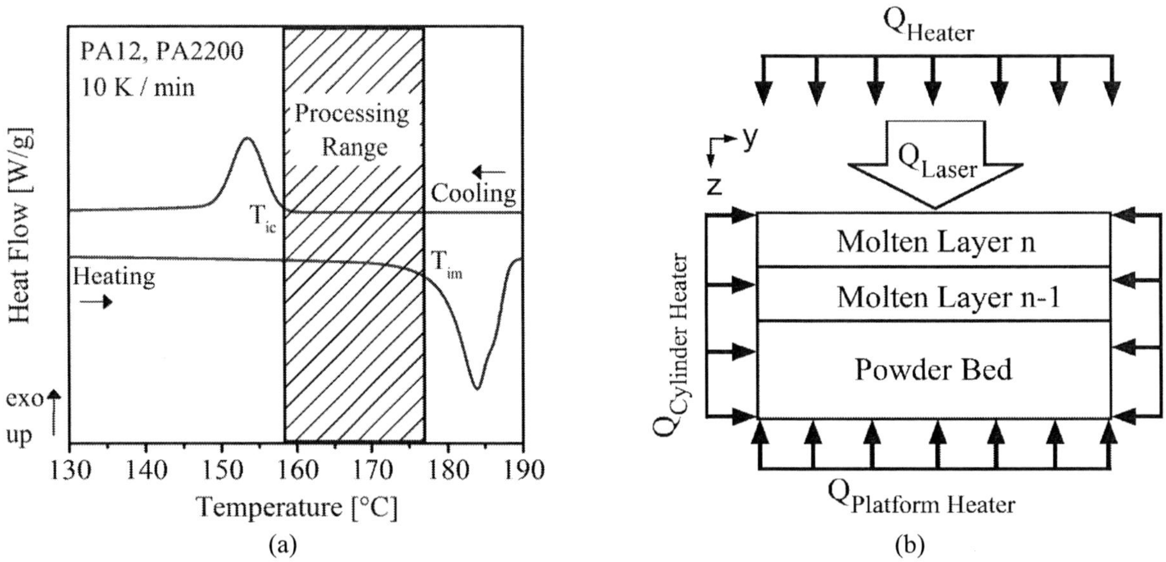

FIGURE 1. a) DSC-curve indicating the possible processing temperature range and b) heat sources influencing surface temperature distribution and temperatures in z-direction, according to [10].

Measuring the surface temperature is another challenge for the integration of new material systems into the SLS process. In SLS, commonly optical temperature sensors are applied, requiring exact knowledge of the relevant temperature-dependent emission coefficients, which change drastically when filled materials are considered or the powder begins to melt [11]. In addition, the risk of fogged-up pyrometer increases, when new materials are used. Often, this leads to rather qualitative measurement results, when working with pyrometers or thermographic cameras. However, tactile measurements, for example by using thermocouples cannot be applied to obtain information on the surface temperature, but can be implemented in deeper layers of the process, e. g. at the edges of the build chamber without exposure or after the process. In order to study and optimize the surface temperature profile before fabrication, a new method for calibrating the temperature field of a SLS system will be discussed. The resulting values can be correlated to optically measured surface temperatures by calculating thermal attenuation coefficients.

EXPERIMENTAL SETUP

Laser Sintering System

For providing constant boundary conditions for processing new materials, a special temperature stabilizable and homogenizable laser system is required. Especially, when new thermoplastic powder materials with unknown or narrow processing window are to be tested, an adjustable system is necessary for generating homogeneous temperatures across the whole build chamber. In this work, a heating system with eight independently controllable and adjustable arrays of IR-lamps is used, being able to provide surface temperature fields with temperature deviations below 3 K. Hence, the risk of process interruption can be decreased. However, because of quality losses or replacement the surface temperature field has to be characterized and optimized constantly. In this work, a measurement set-up will be introduced, which can be applied to quantify the maximum temperature deviations over the entire build chamber surface.

Measuring Plate

For setting more homogeneous surface temperatures within the SLS process, the heating system of the laser sintering system has to precisely be calibrated and the resulting temperature field characterized. For this reason, a measuring plate was fabricated. A thermally conductive steel plate with dimensions of 370 x 370 x 2 mm³ was laser cut according to the layout shown in figure 1. For thermal decoupling, circular and rounded oblong shaped air gaps of 5 mm in diameter were arranged around the measuring sections. The connection between the measuring areas and the plate was established by three thin strands sizing 300 μm. Heat transfer from the SLS system to the measuring plate was minimized as the plate was mounted on four thin bearings without further contact areas.

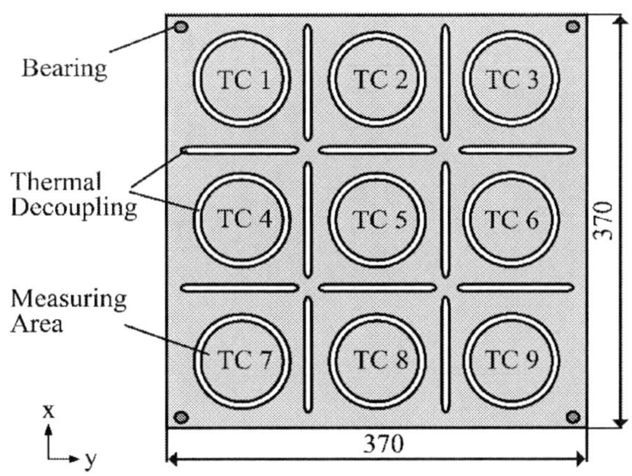

FIGURE 2. Schematic description of the applied measuring steel plate, which is mounted on four bearings and consists of nine thermally decoupled measuring areas

Nine self-adhesive thermocouples (TC) k-type (SA2F-KI-2M, OMEGA Engineering GmbH) were fixed on the underside of the measuring area centrally, measuring the build chamber surface temperature distribution. The surface of the steel plate was coated with a thin layer of carbon black lacquer to enhance absorption of the emitted IR radiation. The measuring plate was used for different experiments, initially starting from the state of lamp replacement. At first, different heating profiles were set within the laser sintering system to detect the heating and measuring dynamics of the used system. In each case, the target temperature was raised by steps of 10 K, whereas the isothermal hold times were varied between 1 min, 3 min and 5 min. Results for the minimum and the maximum hold time will be shown. In a second experiment, the control loops of the eight surface heating arrays were optimized in order to minimize the resulting temperature deviations. Due to higher process-proximity, the optimization experiments were performed with lowest hold times. Starting from room temperature, the platform was heated up until reaching the target temperature of 100 °C in the center of the plate (thermocouple TC5) as it equals the pyrometer's measuring spot ($\varepsilon = 0.98$) The maximum temperature deviation was calculated by subtracting the minimum from the maximum value obtained by the nine thermocouples. The measurement results are shown for the first six optimization steps.

RESULTS AND DISCUSSION

The results for the variation of the heating steps are shown in fig. 3. As described before, the target temperature was set according to a step function. In fig. 3 a) the temperature was raised by 10 K every minute and in fig. 3b) slower by 10 K every 5 minutes. It can be seen, that the measured temperatures are always below the target temperature. Several estimates have to be taken into account. As the thermocouples measure the temperature after heat transfer through the metal plate, dissipation and thermal diffusivity are factors influencing the sensitivity of the system. In addition, the pyrometer's emission coefficient might be too low for the "almost" black body, leading to excessive temperatures. Preliminary oven tests, keeping the temperature constant at 180 °C for three hours, showed system inherent temperature deviations of maximum 2 K.

Nevertheless, with increasing target temperatures the local deviations increase, showing a cold spot at the upper area (TC2, TC3) and a hot spot on the left side (TC4, TC7). This phenomenon is reproducible for the three different heating profiles. However, the slower the heating steps the lower the temperature deviation, which can be assigned to the remaining heat conduction processes within the SLS system and the heating plate. For the following experiments, the faster heating rate was applied being closer to the fast heating up cycles, which are performed during powder coating. In addition to that, this procedure can be seen as worst-case optimization, as the thermal conductivity of the polymer is lower, the deviations in the process will be lower, too.

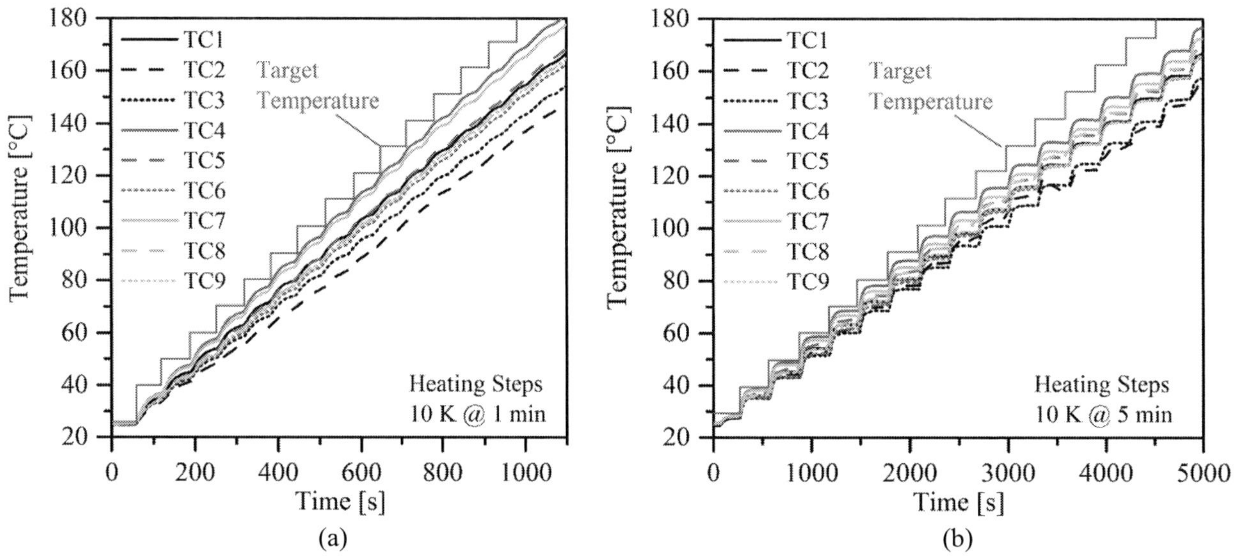

FIGURE 3. Results for the temperature measurement with the measuring plate at different heating steps a) 10 K @ 1 min and b) 10 K @ 5 min

In fig. 4 a) the results of the optimization experiments are shown, whereas in figure 4 b), the temperature distribution is visualized. Experiment number 1 is the initial state before adaption, showing that the mean temperature of the platform is below the centre temperature and a maximum temperature deviation at 100 °C values 25 K. With given temperature distribution, a processing experiment showed melting on the left side of the build chamber, which was caused by local superheating of over 10 K, whereas the cold spot in the right corner in the back did not affect the process, significantly.

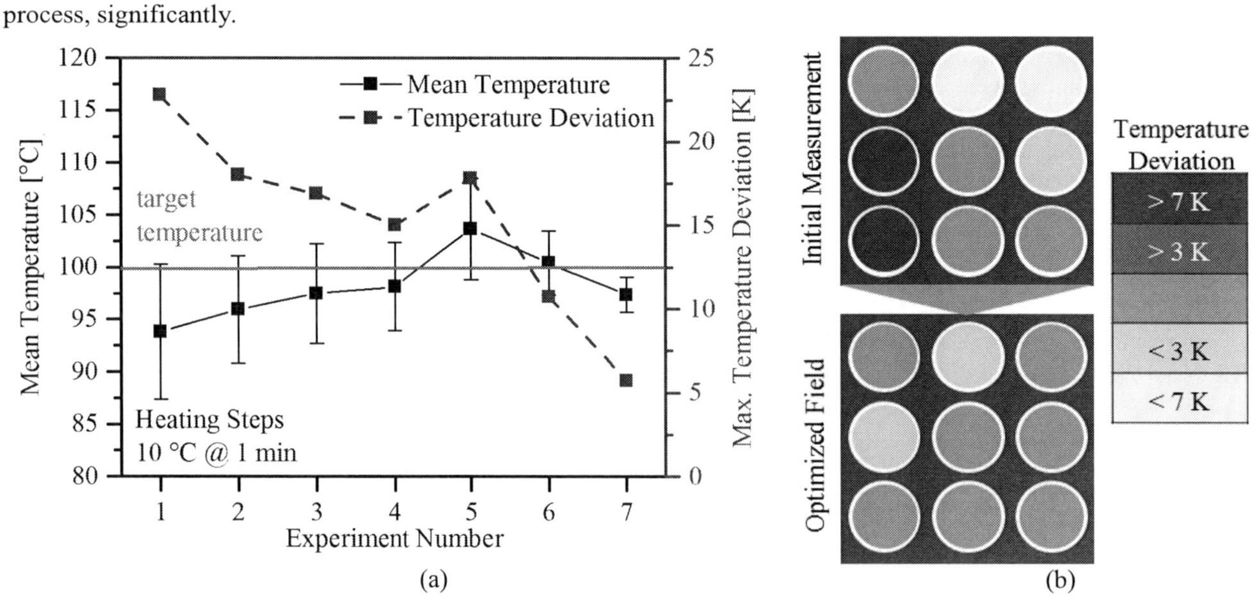

FIGURE 4. a) Results for the optimization measurement and b) spatial visualization the temperature distribution

After analysing the temperature distribution of the nine thermocouples, the parameter and control setup was adapted. It can be seen that with increasing number of experiments the mean temperature can be optimized and the maximum temperature deviation can be decreased to around 5 K, enabling the fabrication of PA 12 parts within a thermally stable SLS process. In general, by iterating these experiments further optimizations to deviations of around 3 K are possible.

CONCLUSION

Within this work, a new method for adapting the temperature field of a SLS system with an eight zone heating system was shown. As expected, higher heating rates result in higher thermal deviations from the target temperature and also across the nine measuring areas within the build chamber. The optimization results indicate that it is possible to use the shown system to measure and homogenize the surface temperature. Within six experiments, the temperature deviation can be lowered from over 20 to 5 K. Further work of the authors will focus on the correlation of the measurements to the SLS process, for example by coating the measuring plate with a defined layer of polymer powder and by additional measurements of the surface temperature with a thermographic camera system.

ACKNOWLEDGMENTS

The authors want to thank the German Research foundation (DFG) for funding the Collaborative Research Center 814 (CRC 814), sub-project B03.

REFERENCES

1. A. Gebhardt, "Generative Fertigungsverfahren: Rapid Prototyping - Rapid Tooling - Rapid Manufacturing," Hanser, München, 2007.
2. T. Wohlers, "Wohlers Report 2011-State of the industry," Annual Worldwide Progress Report, Wohlers Associates, 2011.
3. K. Wudy and D. Drummer, "Aging Behavior of Polyamide 12 Interrelation between Bulk Characteristics and Part Properties," in Proceedings of the 27th Annual International Solid Freeform Fabrication Symposium, 2016, pp. 770-781.
4. K. Wudy and D. Drummer, "Aging Behavior of Polyamide 12 during Selective Laser Melting Process - Influence on Mechanical Properties," in Proceedings of the ANTEC Conference 2015, Society of Plastics Engineers, Orlando, 2015, pp. 2356-2363.
5. A. Wegner and G. Witt, "Ursachen für eine mangelnde Reproduzierbarkeit beim Laser-Sintern von Kunststoffbauteilen," in RTejournal 10, 2013.
6. R.D. Goodridge, R.J.M. Hague, C.J. Tuck, An Empirical Study into Laser Sintering of Ultra-High Molecular Weight Polyethylene (UHMWPE) in Journal of Materials Processing Technology, 210, pp. 72-80, 2010.
7. B. Wendel, D. Rietzel, F. Kühnlein, R. Feulner, G. Hülder and E. Schmachtenberg, „Additive Processing of Polymers," in Macromolecular Materials and Engineering 293, 2008, S. 799-809.
8. E. Schmachtenberg and T. Seul, "Model of Isothermic Laser-Sintering" in Proceedings of the ANTEC Conference 2002, Society of Plastics Engineers, San Francisco, 2002, pp. 3030-3035.
9. A. Wegner and G. Witt, "Thermographie zur Temperaturmessung beim Laser-Sintern - ein Beitrag zur Qualitätssicherung" in RTejournal 8, 2011.
10. M. Drexler, „Zum Laserstrahlschmelzen von Polyamid 12 – Analyse zeitabhängiger Einflüsse", Ph.D. Thesis, Institute of Polymer Technology, University of Erlangen-Nuremberg, 2017.
11. A. Wegner, "Theorie über die Fortführung von Aufschmelzvorgängen als Grundvoraussetzung für eine robuste Prozessführung beim Lasersintern von Thermoplasten," Ph.D. Thesis, Universität Duisburg-Essen, 2015.

Polyetheretherketone hybrid nanocomposite filaments for 3D printing

Covas Jose Antonio [(1)*], Paiva Maria Conceiçao [(1)], Lima Patricia [(2)], Gonçalves Jordana [(2)], Lafont Ugo [(3)]

[(1)] University of Minho - Guimaraes - Portugal, [(2)] PIEP - Guimaraes - Portugal, [(3)] European Space Agency - Noordwijk - Netherlands

3D printing is currently recognized as a disruptive polymer processing technology. However, although recent research efforts have focused on the identification of the materials requirements for successful printing (physical, chemical, thermal and rheological properties), the commercial range of polymer filaments continues to be surprisingly limited. This is particularly noticeable when the aim is to add functionality to printed parts, such as thermal or electrical conductivity. Electrically conductive composites are attractive for several applications depending on their conductivity range. Examples are electrostatic dissipation (10-8 – 10-6 S.cm-1), electrostatic painting (10-6 – 10-4 S.cm-1), electromagnetic interference (EMI) shielding (10-3 – 10-1 S.cm-1) and lightning strike protection (>101 S.cm-1) applications1. Polymers with aromatic carbon backbone characteristically present excellent mechanical properties, high thermal resistance and may establish strong interactions with graphene-like surfaces via π-π stacking2. The present work reports the preparation of an electrically conductive thermoplastic composite by melt mixing using polyetheretherketone (PEEK), graphite nanoplates (GnP) and carbon nanotubes (CNT), its subsequent processing into conductive filaments and a few preliminary printing experiments. Characterization encompasses electrical conductivity, tensile strength and morphology. References: 1. Díez-Pascual, A. M., Naffakh, M., Marco, C., Ellis, G. & Gómez-Fatou, M. A. High-performance nanocomposites based on polyetherketones. Prog. Mater. Sci. 57, 1106–1190 (2012). 2. Yang, J. et al. Preparation and properties of poly (p-phenylene sulfide)/multiwall carbon nanotube composites obtained by melt compounding. Compos. Sci. Technol. 69, 147–153 (2009)

Flow Behaviour of Laser Sintering Powders at Elevated Temperatures

Van den Eynde Michael (1)*, Van Puyvelde Peter (1)

(1) KU Leuven - Vlaams-Brabant - Belgium

Powder flowability is a determining factor in many industrial processes. In laser sintering (LS), for example, a well-flowing powder is required to enable the deposition of smooth layers with a thickness of only 100 μm[1]. Characterising what powders meet these flow requirements is, however, not straightforward. Most industrially used characterisation techniques are indexers, which means their results cannot just be extrapolated to any flow field. This research evaluates the flowability of multiple powders intended for laser sintering. A dynamic angle of repose (dAoR) and Hausner ratio (HR) are used to provide a quick first screening. Powders passing these first tests are then subjected to a spreading experiment, which evaluates layer quality and density. The layer density is directly relevant for the sintering process, as denser powder layers yield mechanically stronger products[2]. Results show that the dAoR is able to provide a quick screening of some powders. However, powders scoring the best in the dAoR are not always the optimal powders for LS. Moreover, flowability at ambient conditions is not a sufficient criterion for LS powders, as the entire process occurs well above room temperature. Powder rheometry, in combination with DSC measurements, is therefore used to assess a maximum bed temperature at which the powder is still spreadable. The spreading experiment is also expanded to incorporate powder heating, mimicking LS even further. Finally, also humidity is taken into account. The moisture content and uptake rate are examined via thermogravimetric analysis. Results on the standard PA12 2200 show that small amounts of moisture decrease the flowability, compared to the dried powder. Moisture content should thus thoroughly be monitored in lab conditions, to provide directly relevant insights for the laser sintering process. 1 R.D. Goodride et al, Prog. in Mater. Sci. 57, 229-267 (2012) 2 T. Niino, K. Sato. Proc. SFF Symposium, Austin, 193-205 (2009)

Effects and optimization of processing parameters in fabrication of polyphenylsulfone microspheres via spray drying

Mys Nicolas [(1)]*, Verberckmoes An [(1)], Cardon Ludwig [(1)]

[(1)] University Ghent - Oost-Vlaanderen - Belgium

This article describes the processing of polyphenylsulfone (PPSU) into microspheres via the physicochemical method called spray drying. These microspheres have the potential to be used as build material for selective laser sintering provided they have spherical morphology and have a mean particle size in the range of 45-95 micrometer. The effects of the different processing parameters on these two response factors have been investigated in spray drying of PPSU by Design of Experiment. A plot is predicted in which size and morphology are optimized and combined in order to find the ideal parameter set for production. The effect of polymer concentration, solution feed rate, atomization flow rate, inlet temperature and polymer grade was optimized using a five-factor and two-level factorial design with center points. Experimental validation is performed to certify the conclusion. The particles were characterized by SEM, DSC and GPC. Particles showed no significant degradation due to processing and had overall good morphology.

Welding of 3D Printed Carbon Nanotube-Polymer Composites by Locally Induced Microwave Heating

Sweeney Charles [1], Saed Mohammad [2], Green Micah [1]*

[1] Texas A&M University - Texas - USA, [2] Texas Tech University - Texas - USA

Additive manufacturing through material extrusion (ME), often termed 3D printing, is a burgeoning method for manufacturing thermoplastic components. However, a key obstacle facing 3D-printed plastic parts in engineering applications is the weak weld between successive filament traces, which often leads to delamination and mechanical failure. This is the chief obstacle to the use of thermoplastic additive manufacturing.(1) Here we report a novel concept for welding 3D-printed thermoplastic interfaces using intense localized heating of carbon nanotubes (CNTs) by microwave irradiation. The microwave heating of the CNT-polymer composites are a function of CNT percolation, as shown through in situ infrared imaging and simulation. We apply CNT-loaded coatings to 3D printer filament; after printing, microwave irradiation is shown to improve the weld fracture strength by 275%. These remarkable results open up entirely new design spaces for additive manufacturing and also yield new insight into the coupling between dielectric properties and RF field response for nanomaterial networks.

Structure-property relationship of additive manufactured thermoplastic polymers processed with Arburg Freeformer

Kaut Franziska (1)*, Cepus Valentin (2), Grellmann Wolfgang (3)

(1) Procter&Gamble - Hessen - Germany, (2) Polymer Service GmbH,Merseburg, Hochschule Merseburg - Sachsen-Anhalt - Germany, (3) Martin-Luther-Universität Halle-Wittenberg, Zentrum für Ingenieurwissenschaften - Sachsten-Anhalt - Germany

The most additive manufacturing technologies share the similarity in production of parts layer by layer on the basis of physical models created with three-dimensional computer aided design (CAD) models. In contrast to the longer on market available additive manufacturing methods, the principle of Arburg Freeformer base on melting of conventional plastic granules with an injection molding plasticizing unit as base material. A stationary discharge unit with an integrated nozzle- valve system applies thin plastic droplets layer by layer onto a component carrier using high frequency piezo technology to open and close a nozzle-valve system at a specified cycle of 60Hz to 200Hz. Compared to fused filament fabrication (FFF), which is the most common process in additive manufacturing with material extrusion, the Arburg Freeform System (AFS) has a higher degree of freedom in material variation. In this work, structural properties are correlated with the mechanical and physical properties of additive manufactured thermoplastic polymers. Focused are investigations of interlayer adhesion of layerwised build specimen and parts.

Fatigue crack growth propagation in 3D-printed polymer structures – a comparison of different additive manufacturing technologies

***Knöchel Johannes Friedrich* (1)*, *Kropka Micheal* (1), *Neumeyer Thomas* (1), *Altstädt Volker* (2)**

(1) Neue Materialien Bayreuth GmbH - Bayern - Germany, (2) Department of Polymer Engineering - University of Bayreuth - Bayern - Germany

Additive manufacturing technologies enable the production of plastic components with complex geometries that cannot be realized by conventional mold-based methods in small production lots. The combination of design freedom with sufficient mechanical properties enables the cost-efficient production of functional prototypes. In order to elevate 3D printing from a "rapid prototyping" to a "rapid manufacturing" technique for serial products, the fatigue behavior of 3D-printed parts has to be deeply understood. Hence, the aim of this study is to analyze the fatigue properties of 3D-printed ABS parts to evaluate the lifespan in daily use in comparison to injection molded (IM) parts. For this purpose compact tension (CT) samples, manufactured by Fused Filament Fabrication (FFF) and Arburg Plastic Freeforming (APF), were analyzed under dynamic mechanical load and compared to IM samples by investigating the fatigue crack propagation (FCP) behavior. Using microscopic analysis of the fracture surface, characteristic regions of the crack path are investigated in detail in order to identify predominant crack mechanisms. It can be concluded that the crack propagation rate is quite similar for FFF and IM specimens. The stress intensity factor at which the crack growth is initiated (Kth) is on the same level as for IM samples. These findings are proven by the corresponding fracture surfaces, which reveal a homogenous material structure in both cases. APF samples show a slightly slower crack propagation speed compared to IM and FDM samples. Analysis of fracture surfaces indicates, that the lower propagation rate is induced by a high porosity of more than 5 %, which causes multiple crack branching. The evaluation of the FCP behavior in 3D-printed polymer structures reveals the potential of 3D printing with regards to its development from a "rapid prototyping" to a "rapid manufacturing" technique for technical products.

A novel process for tailored stiffness and strength in extrusion based additive manufacturing

Van De Steene Willem [1]*

[1] Ghent University, Department of Materials, Textiles and Chemical Engineering - Oost-Vlaanderen - Belgium

Extrusion based additive manufacturing (AM) techniques have gained interest and were continuously developed during the last quarter-century. However, the currently integrated processes have their limitations in different aspects. When compared to injection moulded parts, additively produced parts often suffer from limited mechanical properties, uncontrollable anisotropy, higher surface roughness, the need for support structures during production, limited process speed, etc. In this research, a novel (AM) process to reduce the aforementioned limitations is proposed. Therefore, an additive manufacturing platform with two additional degrees of freedom compared to conventional AM machines and with custom-made software was developed. This system enables Curved Layer Additive Manufacturing (CLAM). This process allows for the development of innovative build schemes, which result in a severe reduction of required support material. Another advantage of this extra flexibility is the possibility to tailor the part's strength and stiffness according to the loads it will be subjected to. This characteristic is promising since the machine is able to process both thermoplastic and continuous fibre reinforced thermoplastics. Such an engineering solution allows for weight reduction and reduced material consumption. Furthermore, the manufactured part's surface roughness can be reduced by adding a finishing layer. Finally, this technology allows automated filament placement on curved layers, non-geodesic filament winding of complex shapes, the creation of continuous fibre reinforced metamaterials, etc. This study compares and assesses build time, surface roughness, mechanical properties such as flexural strength and modulus, impact strength, etc. of samples produced with both regular extrusion based AM and with the own CLAM technique.

Determination of the Short Time Stability of Polymers

Schawe Juergen E.K. [(1)*]

[(1)] Mettler Toledo GmbH - Switzerland - Switzerland

The thermal short time stability is relevant for fast heating processing techniques like fused filament fabrication, selective laser sintering or polymer welding. Also the application of polymeric materials with temporary stress at high temperatures requires knowledge of the short term stability at high temperatures. Thermogravimetric analysis (TGA) is a standard technique to measure the thermal stability. This technique has two drawbacks: • The decomposition at high temperatures does not necessarily contain the same reaction steps as measured by TGA. • TGA is not sensitive for degradation steps which are not related to mass loss. However, such reactions can significantly influence the properties of polymeric materials. It is shown on differently stressed polymers that the crystallization behavior changes sensitively with molecular modifications due to the stresses. The modification in the crystallization behavior due to thermal stress can be studied by fast scanning calorimetry (FSC). The change of the crystallization kinetics due to thermal stress is analyzed. From the measured curve a stability parameter can be derived which characterized the thermal degradation. The stability of polyamides and polymer materials developed for selective laser sintering (SLS)is analyzed for fast heating processes. The results show a remarkable discrepancy to the TGA results and the high sensitivity of the change of the crystallization behavior for the early steps of decomposition reaction.

Comparison of the mechanical properties of 3D printed polymers and TiAL6V4 used in medical applications

Weißmann Volker (1)*

(1) Hochschule Wismar - Mecklenburg-Vorpommern - Germany

Additive manufacturing technologies such as 3D printing have enabled the production of complex structures and parts which find widespread use in engineering applications. Medical engineering with its need for materials with exceptional properties has taken on a key role. In addition to chemical resistance and biomedical tolerability (sterilization resistant, non-toxic and non-sensitizing), the suitability of these materials depends in particular on their mechanical properties. Biomedical applications such as, for example, bone substitutes often require mechanically optimized structures whose properties are determined not only by manufacturing conditions but also by the sizes and forms of the structures as well as the materials used. The priority of objectives in structure development include besides a reduction of implant stiffness, the achievement of bone-like mechanical characteristics and the realization of the substitute structure in a porous (micro, macro) structure or surface. While several medical technical problems with a requirement for defined mechanical properties have drawn upon titanium alloy-based applications, the use of polymers has just begun. The present paper compares the mechanical properties of 3D printed, load carrying structural elements (scaffolds) made of TiAl6V4 powder that are suited for use as bone substitute and scaffolds made of polymer materials (PEEK, ABS). Characterization was performed by assessing the functional relationship between the experimentally determined porosity and the modulus of elasticity. Special emphasis has been placed on their specific flexibility as the determining factor for deformation resistance. It offers an excellent possibility to establish a direct relationship between porosity and elastic modulus. Besides the material characteristics of the materials themselves, scaffolds with different porosities have been assessed and compared directly with the characteristics of the human bone (cortex, cancellous bone).

Development of novel Nanocarbon composites based on functionalised MWCNTs and Poly Lactic Acid for Fused Deposition Modeling applications

Cano Manuela (1), Parra Julián (1)*, Ramirez Maria Dolores (1), Morales Gabriel (1), Malik Sharali (2)

(1) Andaltec - Jaén - Spain, (2) Institute of Nanotechnology - Karlsruhe - Germany

Additive manufacturing is an emerging technology which could potentially initialise another industrial revolution. Important advances in the optimization of the 3D printers have been carried out in the last 10 years(1). One of the most promising techniques for rapid design and prototyping is Fused Modeling Deposition (FDM) due to the facility to extrude thermoplastic polymers in the form of filaments2. The 3D market demands a wide range of polymer materials to be printed in different applications such as automotive, aeronautics, food packaging and consumer goods1 etc. Thermoplastics reinforced with carbon nanomaterials shows an interest for the automotive or aeronautic sectors due to high mechanical, thermal and electrical properties(2). One of the main goals focuses on the optimization of the industrial processing of these polymers composites and their adaptation to the 3D specification printing. In this work we present a straightforward non-wet route to fabricate PLA/MWCNTs composites filaments by small-scale industrial extrusion process to be used in 3D printers and the studies of mechanical properties of the developed filaments. MWCNTs were first functionalised following the procedure described(3) to produce the composite with the aim to increase the nanomaterials dispersion in the polymer matrix. Pre-dispersion of MWCNTs in PLA was carried out by using a mechanical milling process. Subsequently the pre-dispersion was extruded into a filament of 1.75 mm mean diameter. Preliminary results show an enhancement of tensile load (8-10%) on filaments reinforced with non-functionalised MWCNTs and even more (16-18%) when these filaments are made of functionalised MWCNTs. [1] "A third industrial revolution," The Economist, (2012) [2] R. H. A. Haq, et al. Applied Mechanics and Materials. 8-12, 465-466 (2014). [3] S. Mali et al. Beilstein J Nanotechnol. 7, 1260–1266 (2016)

AUTHOR INDEX

AUTHOR INDEX

AUTHOR INDEX

AUTHOR INDEX

AUTHOR INDEX

AUTHOR INDEX

AUTHOR INDEX

AUTHOR INDEX

AUTHOR INDEX

AUTHOR INDEX

AUTHOR INDEX

Polymer Processing Society
810 E. 10th St.
Lawrence, KS 66044
USA

CURRAN ASSOCIATES INC.
proceedings.com

ISBN 978-1-5108-5487-1